Tensor Calculus &Physics: A General Treatise

AuthorHouse™
1663 Liberty Drive
Bloomington, IN 47403
www.authorhouse.com
Phone: 1-800-839-8640

The Modern Elements   Second Edition

Published by AuthorHouse 5/18/2015

ISBN: 978-1-4969-5274-5 (sc)
ISBN: 978-1-4969-5275-2 (e)

Print information available on the last page.

Tensor Calculus &Physics: A General Treatise

Volume II: Foundational Mathematics Continued

Table of Contents

List of Equations

List of Tables

List of Figures

List of Observations

List of Givens

Chapter 9 The Theory of Calculus and Calculus of Variations

Section 9.1 The Study of the Infinite Large and Infinitely Small

Definition 9.1.1 Infinity

A count so large that adding one to it only results in an even larger number is called infinite, which is represented by the symbolic number ***infinity*** [∞].

Definition 9.1.2 Infinitesimal

Infinitesimal is a count so small nothing remains, empty, which by definition D4.1.2 "The Number Zero" is represented by the symbolic number ***zero*** [**0**].

Definition 9.1.3 Calculus

Calculus is the study of the infinitely large and the infinitely small as in analytic geometry.

Section 9.2 Rules of Limits

A limit is the concept of bounding a quantity in order to validate that it exist. The limit is also a controllable process that bounds a measured value that approaches a fixed number, which is summarized in the following definition. [PRO70, pg 97]

Definition 9.2.1 Existences of a Limit on a Function

Given a function [f] and numbers [a] and [L], we say that f(x) tends to L as a limit, as [x] tends to [a] if for each positive number [ε] there exists a positive number [δ], such that f(x) is defined and

$| f(x) - L | < \varepsilon$ Equation A

whenever

$0 < | x - a | < \delta$ Equation B

In abbreviated notation, we write

$f(x) \rightarrow L$ Equation C

Read as the function of [x] approaches a limiting value [L].

as

$x \rightarrow a$ Equation D

Read as the value [x] approaches a limiting constant value [a].

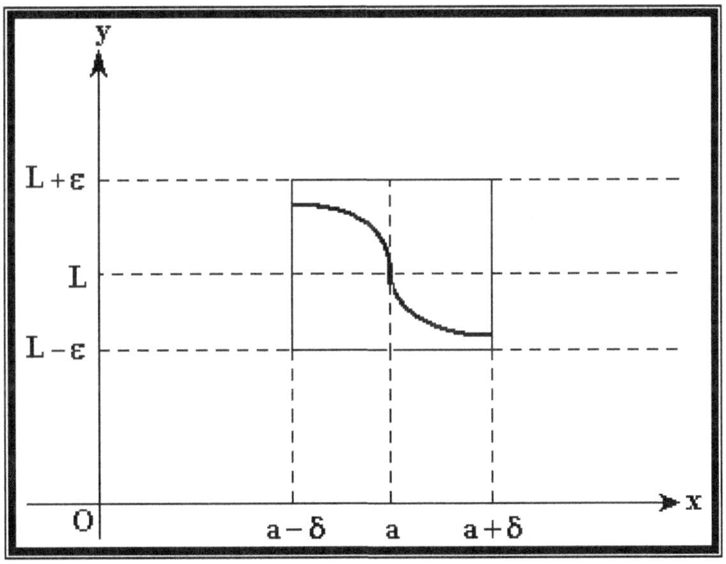

Figure 9.2.1 Bounding a Point (a, L) as a Limit

Observation 9.2.1 Decreasing Exponent Numbers

Let uses observe how negative exponents become smaller with increasing size of their power:

Exponent	Decimal Positional Equivalent Number
10^{-1}	0.1
10^{-2}	0.01
10^{-3}	0.001
10^{-4}	0.0001
10^{-5}	0.00001
10^{-6}	0.000001
10^{-7}	0.0000001
10^{-8}	0.00000001
10^{-9}	0.000000001
10^{-10}	0.0000000001
10^{-11}	0.00000000001
10^{-12}	0.000000000001

Hence, as powers are selected of ever increasing size the exponent approaches the infinitesimally small number, which is concluded by definition D9.1.2 "Infinitesimal" to be zero [0].

This can be stated generally as [n] being selected infinitely large, such that 10^{-n} decreases in size or as [n] become infinite the exponent tends to zero.

$$\text{If } n \rightarrow \infty \text{ then } 10^{-n} \rightarrow 0 \text{ for } n > 0$$

Using limit notation

$$\lim_{n \rightarrow \infty} 10^{-n} = 0$$

Converting base 10 to some other base [b] gives

$$\lim_{n \rightarrow \infty} 10^{-n} = \lim_{m \rightarrow \infty} b^{-m}$$

for

$$m = (\ln 10 / \ln b)\, n$$

since [m] and [n] are related by a direct positive constant of proportionality, then as [n] increases [m] likewise increases, hence the above relation holds for all real base numbers [b]

$$\lim_{m \rightarrow \infty} b^{-m} = \lim_{n \rightarrow \infty} 10^{-n}$$

and in the limit for [n]

$$\lim_{m \rightarrow \infty} b^{-m} = 0$$

So, for any real based exponent with a set of positive, increasing integers, then its exponent also discretely goes to zero. The limit can only be true if it is bounded and progresses in a well-behaved fashion within its boundaries, which may or may not be provable, so these issues will be side, stepped by treating this property of numbers as an axiom.

Observation 9.2.2 Increasing Exponent Numbers

Let uses observe how positive exponents become larger with increasing size of their power:

Exponent	Decimal Positional Equivalent Number
10^0	1
10^1	10
10^2	100
10^3	1000
10^4	10000
10^5	100000
10^6	1000000
10^7	10000000
10^8	100000000
10^9	1000000000
10^{10}	10000000000
10^{11}	100000000000
10^{12}	1000000000000

Hence, as powers are selected of ever increasing size the exponent approaches an infinitely large number, which is concluded by definition D9.1.1 "Infinity" to be infinite [∞].

This can be stated generally as [n] is selected infinitely large, such that 10^n increases in size or as [n] becomes infinite the exponent tends to infinity.

$$\text{If } n \rightarrow \infty \text{ then } 10^n \rightarrow \infty \text{ for } n > 0$$

Using limit notation

$$\lim_{n \rightarrow \infty} 10^n = \infty$$

Converting base 10 to some other base number [b] such that $(0 \le n, m) \in \text{IG}$ and $(0 < b) \in \text{R}$

$$\lim_{n \rightarrow \infty} 10^n = \lim_{m \rightarrow \infty} b^m$$

for

$$m = (\ln 10 / \ln b)\, n$$

since [m] and [n] are related by a directly proportional positive constant, then as [n] increases [m] likewise increases, hence the above relation holds

$$\lim_{m \rightarrow \infty} b^m = \lim_{n \rightarrow \infty} 10^n$$

and in the limit for [n]

$$\lim_{m \rightarrow \infty} b^m = \infty$$

So, for any real based exponent with a set of positive, increasing integers, then its exponent also discretely increases to infinity. Again, the limit can only be true if it is bounded and progresses in a well-behaved fashion within its boundaries, which may or may not be provable, so again these issues will be side stepped by treating this property of numbers as an axiom.

Note even the use of the ***power series ratio test*** begs the question, because it does not derive the value of these limits just proves weather its possible or not.

Observations 9.2.1 "Decreasing Exponent Numbers" and Observations 9.2.2 "Increasing Exponent Numbers" constitute a special property of integer numbers, unfortunately these limits are not proven, but intuitively extrapolated by definition. This can lead to a play on words resulting in confusing circular arguments, in fact while these arguments may seem obvious standing outside the system of numbers they might never be proven in a conclusive way. In order to avoid these dilemmas, they are simply stated here as properties of discrete positive real numbers in an axiomatic form.

Axiom 9.2.1 Limits for Discreet Exponent Numbers

	Equations	Equ	Description
Ax9.2.1.1	$\lim_{m \to \infty} b^{-m} = \lim_{n \to \infty} 10^{-n} = 0$	A	Exponents decrease infinitesimally without bounds as the negative power increases
Ax9.2.1.2	$\lim_{m \to \infty} b^{m} = \lim_{n \to \infty} 10^{n} = \infty$	A	Exponents increase infinitely without bounds as the positive power increases
Ax9.2.1.3 for	$(0 < n, m) \in IG$, $(0 < b) \in R$ and $m = (\ln 10 / \ln b)\, n$	A	Conditions of validity for all real number exponent basses
or	$n = (\ln b / \ln 10)\, m$	B	

Lema 9.2.1 Rules of Limits

	Equations	Description
Lx9.2.1.1	If $f(x) \to L_1$ as $x \to a$, and $g(x) \to L_2$ as $x \to a$. then $L_1 = L_2$.	Uniqueness of Limits
	$f(x) = g(x)$	
	$\lim_{x \to a} f(x) = \lim_{x \to a} g(x)$	
Lx9.2.1.2	If [c] is a constant and f(x) = c for all values of x, then for any number x the result will always be c.	Limit of a Constant
	$\lim_{x \to a} c = c$	
Lx9.2.1.3	If [a] is a real number and f(x) = x for all x, then x is obviously [a] in the limit.	Obvious Limit
	$\lim_{x \to a} x = a$	
Lx9.2.1.4	If f(x) and g(x) are two functions, with [a] limiting values L_1 and L_2 as [x] tends to then the limit of the sums is the sums of the limit.	Limit of the Sum
	$\lim_{x \to a} (f(x) + g(x)) = \lim_{x \to a} f(x) + \lim_{x \to a} g(x)$	
Lx9.2.1.5	If f(x) and g(x) are two functions, with [a] limiting values L_1 and L_2 as [x] tends to then the limit of the products is the products of the limit.	Limit of the Products
	$\lim_{x \to a} (f(x)\, g(x)) = \lim_{x \to a} f(x) \lim_{x \to a} g(x)$ EQ A	
	$\lim_{x \to a} (g(x)\, f(x)) = \lim_{x \to a} g(x) \lim_{x \to a} f(x)$ EQ B	

Lx9.2.1.6	If f(x) and g(x) are two functions, with [a] limiting values L_1 and L_2 as [x] tends to then the limit of the quotients is the quotients of the limit.	Limit of the Quotients
	$\lim_{x\to a} (f(x) / g(x)) = \lim_{x\to a} f(x) / \lim_{x\to a} g(x)$	
Lx9.2.1.7	If $f(g) \to L_1$ as $g \to L_2$, $g(x) \to L_2$ as $x \to a$. then	Limit of Composite Functions
	$\lim_{x\to a} f[g(x)] = L_1$	
Lx9.2.1.8	Taking the limit from of function f(x) as x tends toward $[a^-]$ or $[a^+]$ side may yield different limits. In such a case one-sided limits must be considered.	One-sided Limits
	$\lim_{x\to a-} f(x) = L^-$ and $\lim_{x\to a+} f(x) = L^+$	
	if $L^- = L^+$ then it is a two-sided limit.	
Lx9.2.1.9	If f(x) and g(x) are two functions, whose quotient has an irremovable singularity as [x] tends to [a] then the derivatives of the two functions should remove the singularity in the limit and if not their derivatives and so on.	l' Hôpital's Rule
	$\lim_{x\to a} (f(x) / g(x)) =$	
	$\lim_{x\to a} (f^I(x) / g^I(x)) = \ldots =$	
	$\lim_{x\to a} (f^m(x) / g^m(x))$	
Lx9.2.1.10	if $\lim_{x\to a} f(x) = L$, $\lim_{x\to a} g(x) = L$ and if for all [x] near the value [a] we have $f(x) \le h(x) \le g(x)$ then	Sandwiching Theorem
	$\lim_{x\to a} h(x) = L$	
Lx9.2.1.11	if $\lim_{x\to a} f(x, y) = L_x$, $\lim_{y\to b} f(x, y) = L_y$ and if for all [x] and [y] are independent of one another than the product of the limits are commutative	Dual Function Commutativety of Limits
	$\lim_{x\to a} \lim_{y\to b} f(x, y) = L_x L_y$ likewise $\lim_{y\to b} \lim_{x\to a} f(x, y) = L_y L_x$	
Lx9.2.1.12	If $f(x) \to L_1$ as $x \to a$, $g(x) \to L_2$ as $x \to a$, $L_1 > L_2$ and / or $f(x) > g(x)$ iff $R_f \cap R_g \equiv \varnothing$ (the range of functions [g] and [f]) then	Inequality of Limits
	$\lim_{x\to a} f(x) > \lim_{x\to a} g(x)$	

Theorem 9.2.2 Reciprocal Numbers Tend to Zero as Denominator Increases without Limit

1g 2g	Given	$x \in R$ $x = a \times 10^n$ as a decimal number for a = 1, 2, …, 9 such that $0 < x$
Step	Hypothesis	$\lim_{x \to \infty} 1/x = 0$ valid for x continuous or discrete ∍ $0 < x$
1	From 2g, A4.2.12 Identity Multp and T4.4.11 Equalities: Reversal of Cross Product of Proportions	$1/x = 1/(a \times 10^n)$
2	From 1 and T4.5.1 Equalities: Reciprocal Products	$1/x = (1/a) \times (1/10^n)$
3	From 2 and D4.1.18 Negative Exponential	$1/x = (1/a) \times 10^{-n}$
4	From 2g and taking the reciprocal of all possible integers [a] might be	$1/a = \{1.0, 0.5, 0.\underline{3}, 0.25, 0.2, 0.1\underline{6}, 0.\underline{142857}, 0.125, 0.\underline{1}\}$
5	From 4 and bounding the reciprocal of [a]	$0.1 \le 1/a \le 1$
6	From 5 and T4.7.5 Inequalities: Multiplication by Positive Number	$0.1 \times 10^{-n} \le 1/a \times 10^{-n} \le 1 \times 10^{-n}$
7	From 1, 6 and A4.2.3 Substitution	$0.1 \times 10^{-n} \le 1/x \le 1 \times 10^{-n}$
8	From 7 and D4.1.22 Decimal Point	$10^{-1} \times 10^{-n} \le 1/x \le 1 \times 10^{-n}$
9	From 8, A4.2.18 Summation Exp and A4.2.12 Identity Multp	$10^{-(n+1)} \le 1/x \le 10^{-n}$
10	From 9 and Lx9.2.1.10 Sandwiching Theorem	$\lim_{n \to \infty} 10^{-(n+1)} < \lim_{n \to \infty} 1/x \le \lim_{n \to \infty} 10^{-n}$
11	From 10 and Ax9.2.1.1 Exponents decrease in size as the power increases	$0 < \lim_{n \to \infty} 1/x \le 0$
12	From 11 and Lx9.2.1.10 Sandwiching Theorem	$\lim_{n \to \infty} 1/x = 0$
13	From 2g, Lx9.2.1.1 Uniqueness of Limits	$\lim_{n \to \infty} x = \lim_{n \to \infty} a \times 10^n$
14	From 13, Lx9.2.1.5 Limit of the Products and Lx9.2.1.2 Limit of a Constant	$\lim_{n \to \infty} x = a \times \lim_{n \to \infty} 10^n$
15	From 14 and Ax9.2.1.2 Exponents increase in size without bounds as the power increases	$\lim_{n \to \infty} x = a \times \infty$
16	From 15 and D4.1.19 Primitive Definition for Rational Arithmetic	$\lim_{n \to \infty} x = \infty$ hence $x \to \infty$ can replace $n \to \infty$
∴	From 2g, 12, 16 and Lx9.2.1.3 Obvious Limit	$\lim_{x \to \infty} 1/x = 0$ valid for x continuous or discrete ∍ $0 < x$

Table 9.2.1 Integer Decimal Division with Fractional Remainder

a / b	1	2	3	4	5	6	7	8	9
1	1	2	3	4	5	6	7	8	9
2	0.5	1	1.5	2	2.5	3	3.5	4	4.5
3	$0.\underline{3}$	$0.\underline{6}$	1	$1.\underline{3}$	$1.\underline{6}$	2	$2.\underline{3}$	$2.\underline{6}$	3
4	0.25	0.5	0.75	1	1.25	1.2	1.75	2	2.25
5	0.2	0.4	0.6	0.8	1	1.2	1.4	1.6	1.8
6	$1.1\underline{6}$	$0.\underline{3}$	0.5	$0.\underline{6}$	$0.8\underline{3}$	1	$1.1\underline{6}$	$1.\underline{3}$	1.5
7	0.14285	0.2857	0.42857	0.57142	0.71428	0.85714	1	1.14285	1.28571
8	0.125	0.25	0.375	0.5	0.625	0.75	0.875	1	1.125
9	$0.\underline{1}$	$0.\underline{2}$	$0.\underline{3}$	$0.\underline{4}$	$0.\underline{5}$	$0.\underline{6}$	$0.\underline{7}$	$0.\underline{8}$	1

Theorem 9.2.3 Ratio Less then One Tends to Zero Raised to a Power without Limit

1g	Given	Table 9.2.1 Integer Decimal Division
2g		$(x, y, a, b, n, m, p) \in R$
3g		$0 < x < y$
4g		$x = a \times 10^n$ as a decimal number for a = 1, 2, …, 9 such that $x \ge 0$
5g		$y = b \times 10^m$ as a decimal number for b = 1, 2, …, 9 such that $y \ge 0$
6g		$0 < p$
Step	Hypothesis	$\lim_{p \to \infty} (x / y)^p = 0$ for $(x, y) \in R$ such that $x / y < 1$
1	From 3g, 4g, 5g and A4.2.3 Substitution	$0 < a \times 10^n < b \times 10^m$
2	From 1, A4.2.17 Correspondence of Equality and Inequality and T4.4.6A Equalities: Left Cancellation by Multiplication	$0 < (a / b) \times 10^n < 10^m$
3	From 1g, 2 and A4.2.16 The Trichotomy Law of Ordered Numbers, the maximum ratio is:	$0 < (a / b) \times 10^n \le 9 \times 10^n < 10^m$
4	From 3	$9 \times 10^n < 10^m$
5	From 4 and T4.4.6B Equalities: Left Cancellation by Multiplication	$9 < 10^m / 10^n$
6	From 5 and A4.2.19 Difference Exp	$9 < 10^1 \le 10^{m-n}$
7	From 6, A4.2.16A The Trichotomy Law of Ordered Numbers and equating exponents	$0 < m - n$
8	From 7 and D4.1.7 Greater Than Inequality [>]	$n < m$
9	From 8 and T4.7.2 Inequalities: Uniqueness of Addition by a Positive Number	$1 < m - n + 1$

10	From 9, From 7, T4.7.6 Inequalities: Multiplication by Negative Number Reverses Order	$-(m - n + 1) < -1$
11	From 6, T4.7.5 Inequalities: Multiplication by Positive Number, A4.2.18 Summation Exp, A4.2.8 Inverse Add and A4.2.24 Inverse Exp	$9 \times 10^{-(m-n)} < 10^1 \times 10^{-(m-n)} \le 1$
12	From 11, A4.2.18 Summation Exp, T4.8.2 Integer Exponents: Negative One Squared and A4.2.14 Distribution	$9 \times 10^{-(m-n)} < 10^{-(m-n-1)} \le 1$
13	From 1, A4.2.12 Identity Multp, A4.2.17 Correspondence of Equality and Inequality, T4.4.6A Equalities: Left Cancellation by Multiplication, A4.2.19 Difference Exp, T4.8.2 Integer Exponents: Negative One Squared and A4.2.14 Distribution	$0 < (a / b) \times 10^{-(m-n)} < 1$
14	From 1g and 13 the maximum ratio is:	$0 < (a / b) \times 10^{-(m-n)} \le 9 \times 10^{-(m-n)}$
15	From 12 and 14 the ratio is bounded	$0 < (a / b) \times 10^{-(m-n)} \le 9 \times 10^{-(m-n)} < 10^{-(m-n-1)}$
16	From 15	$0 < (a / b) \times 10^{-(m-n)} < 10^{-(m-n-1)}$
17	From A4.2.2A Equality	$x / y = x / y$
18	From 4g, 5g, 17, A4.2.3 Substitution and A4.2.19 Difference Exp	$x / y = (a / b) \times 10^{-(m-n)}$
19	From 16, 18 and A4.2.3 Substitution	$0 < x / y < 10^{-(m-n-1)}$
20	From 19, A4.2.27 Correspondence Exp and T4.8.6 Integer Exponents: Uniqueness of Exponents	$0^p < (x / y)^p < (10^{-(m-n-1)})^p$
21	From 20, A4.8.5 Integer Exponents: Zero Raised to the Positive Power-n and A4.2.20 Commutative Exp	$0 < (x / y)^p < 10^{-p(m-n-1)}$
22	From 21 and D4.1.18 Negative Exponential	$0 < (x / y)^p < 1 / 10^{p(m-n-1)}$
23	From 22, D9.2.1 Existences of a Limit on a Function and allowing [p] to increase without bound	$\lim_{p \to \infty} (0 < (x / y)^p < 1 / 10^{p(m-n-1)})$
24	From 23, A4.2.17 Correspondence of Equality and Inequality and Lx9.2.1.1 Uniqueness of Limits	$0 < \lim_{p \to \infty} (x / y)^p < \lim_{p \to \infty} 1 / 10^{p(m-n-1)}$ for $0 < 10^{p(m-n-1)}$
25	From 6g, 10, 24, D4.1.18 Negative Exponential and T9.2.2 Reciprocal Numbers Tend to Zero as Denominator Increases without Limit	$0 < \lim_{p \to \infty} (x / y)^p < 0$
26	From 25 and Lx9.2.1.10 Sandwiching Theorem	$\lim_{p \to \infty} (x / y)^p = 0$

27	From 3g, A4.2.12 Identity Multp A4.2.17 Correspondence of Equality and Inequality and T4.4.6A Equalities: Left Cancellation by Multiplication	$x / y < 1$
∴	From 2g, 26 and 27	$\lim_{p \to \infty} (x / y)^p = 0$ for $(x, y) \in R$ such that $x / y < 1$

Theorem 9.2.4 Ratio Greater then One Tends to Infinity Raised to a Power without Limit

1g	Given	Table 9.2.1 Integer Decimal Division
2g		$(x, y, a, b, n, m, p) \in R$
3g		$0 < x < y$
4g		$x = a \times 10^n$ as a decimal number for $a = 1, 2, \ldots, 9$ such that $x \geq 0$
5g		$y = b \times 10^m$ as a decimal number for $b = 1, 2, \ldots, 9$ such that $y \geq 0$
6g		$0 < p$
Step	Hypothesis	$\lim_{p \to \infty} (y / x)^p = \infty$ for $(x, y) \in R$ such that $x / y < 1$
1	From A4.2.2A Equality	$y / x = y / x$
2	From 4g, 5g, 1, A4.2.3 Substitution and A4.2.19 Difference Exp	$y / x = (b / a) \times 10^{m-n}$
3	From 3g, A4.2.12 Identity Multp A4.2.17 Correspondence of Equality and Inequality and T4.4.6B Equalities: Left Cancellation by Multiplication	$1 < y / x$
4	From 2, 3 and A4.2.3 Substitution	$1 < (b / a) \times 10^{m-n}$
5	From 1g, 4 and A4.2.16 The Trichotomy Law of Ordered Numbers, the minimum ratio is:	$1 < 1 \times 10^{m-n} \leq (b / a) \times 10^{m-n}$
6	From 5 and D4.1.22 Decimal Point	$1 \times 10^0 < 1 \times 10^{m-n} \leq (b / a) \times 10^{m-n}$
7	From 6 and A4.2.16A The Trichotomy Law of Ordered Numbers and equating exponents, corresponding exponents	$0 < m - n$
8	From 2, 6, A4.2.12 Identity Multp and A4.2.3 Substitution	$10^{m-n} \leq y / x$
9	From 8, A4.2.27 Correspondence Exp and T4.8.6 Integer Exponents: Uniqueness of Exponents	$(10^{m-n})^p \leq (y / x)^p$
10	From 9 and A4.2.22 Product Exp	$10^{p(m-n)} \leq (y / x)^p$
11	From 10, A4.2.17 Correspondence of Equality and Inequality and Lx9.2.1.1 Uniqueness of Limits for [p] increasing without limit	$\lim_{p \to \infty} 10^{p(m-n)} \leq \lim_{p \to \infty} (y / x)^p$
12	From 6g, 7, 11 and Ax9.2.1.2	$\infty \leq \lim_{p \to \infty} (y / x)^p$
∴	From 2g, 3, 12 and taking the or condition on the inequality since the lower bound is infinite	$\lim_{p \to \infty} (x / y)^p = \infty$ for $(x, y) \in R$ such that $1 < y / x$

Theorem 9.2.5 Reciprocal Numbers Tend to Zero by l' Hôpital's Rule

Step	Hypothesis	$\lim_{x \to \infty} 1/x = 0$
1	From A4.2.2A Equality	$1/x = 1/x$
2	From 1 and Lx9.2.1.1 Uniqueness of Limits	$\lim_{x \to \infty} 1/x = \lim_{x \to \infty} 1/x$
3	From 2, Lx9.2.2 l' Hôpital's Rule, Lx9.2.3.1A Differential of a Constant and Lx9.3.3.2A Differential Identity	$\lim_{x \to \infty} 1/x = 0 / 1$
∴	From 3 and D4.1.19 Primitive Definition for Rational Arithmetic	$\lim_{x \to \infty} 1/x = 0$

Theorem 9.2.6 Dividing by Zero Blows Up to Infinity

Step	Hypothesis	$1/0 = \infty$
1	From A4.2.2A Equality	$1/(1/x) = 1/(1/x)$
2	From 1 and T4.5.3 Equalities: Compound Reciprocal Products	$1/(1/x) = x$
3	From 2 and Lx9.2.1.1 Uniqueness of Limits	$\lim_{x \to \infty} 1/(1/x) = \lim_{x \to \infty} x$
4	From 3 and Lx9.2.1.6 Limit of the Quotients	$\dfrac{\text{Lim}_{x \to \infty} 1}{\lim_{x \to \infty} 1/x} = \lim_{x \to \infty} x$
∴	From 3, Lx9.2.1.2 Limit of a Constant, T9.2.5 Reciprocal Numbers Tend to Zero by l' Hôpital's Rule and Lx9.2.1.3 Obvious Limit	$1/0 = \infty$

Observation 9.2.3 Best Approximate Limit of Sine and Cosine for a Small Angle

When does a radian angle become small enough to approximate the sine as it tends to its radian angle and the cosine to unity?

Table 9.2.2 Best Subjective Approximation of a Small Angle

Radian Angle $\delta\theta$	$\sin(\delta\theta)$	$\cos(\delta\theta)$
0.00	0.000000000	1.000000000
0.09	0.089878549	0.995952733
0.10	0.099833417	0.995004165
0.11	0.109778301	0.993956098
0.12	0.119712207	0.992808636
0.13	0.129634143	0.991561894
0.14	0.139543115	0.990215996
0.15[9.2.1]	0.149438132	0.988771078
0.16	0.159318207	0.987227283
0.17	0.169182349	0.985584767
0.18	0.179029573	0.983843693
0.19	0.188858895	0.982004235
0.20	0.198669331	0.980066578
0.21	0.208459900	0.978030915
0.22	0.218229623	0.975897449
0.23	0.227977524	0.973666395
0.24	0.237702626	0.971337975
0.25	0.247403959	0.968912422
0.26	0.257080552	0.966389978
0.27	0.266731437	0.963770896
0.28	0.276355649	0.961055438
0.29	0.285952225	0.958243876
0.30[9.2.1]	0.295520207	0.955336489
0.31	0.305058636	0.952333570
0.32	0.314566561	0.949235418
0.33	0.324043028	0.946042344
0.34	0.333487092	0.942754666
0.35	0.342897807	0.939372713
0.36	0.352274233	0.935896824
0.37	0.361615432	0.932327346
0.38	0.370920469	0.928664636
0.39	0.380188415	0.924909060
0.40	0.380418342	0.921060994

$\sin(\delta\theta) \approx \delta\theta$ by first term approximation sine series LxE.3.1.24 Equation A

$\cos(\delta\theta) \approx 1$ by first term approximation cosine series LxE.3.1.25 Equation B

[9.2.1]Note: 0.3 is accurate to ±0.01 for sine and ±0.001 cosines, but 0.3 radians is about 17° a large number in degrees, so dividing by 3 gives 5.73° about 0.15 radians a better comparative magnitude between degrees and radians, which gives a clue when to approximate the orthogonal trigonometric functions with their first term equations A and B.

Hence, as powers are selected of ever increasing size the exponent approaches the infinitesimally small number, which is concluded by definition D9.2.2 "Infinitesimal" to be zero [0].

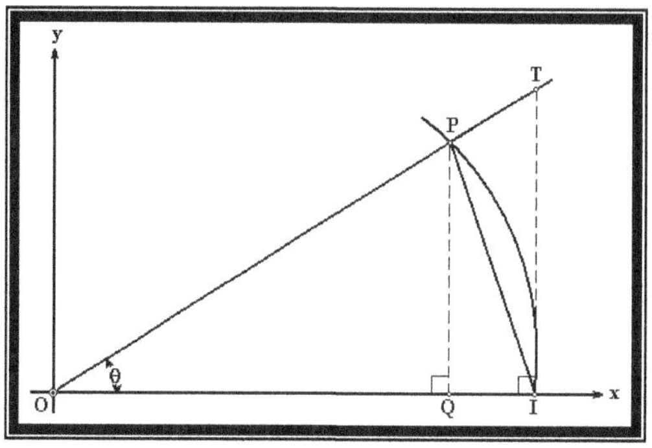

Figure 9.2.2 Sine X / X by Area

Theorem 9.2.7 Removing Singularity at Zero for Sine(X) / X by Area

1g	Given	the above geometry of Figure 9.2.2 "Sine X / X by Area"
2g		All x is in radians for
3g		$0 < x < \frac{1}{2}\pi$
4g		r = OP = OI = 1 Radius
5g		$\lvert OI\rvert = \lvert OQ\rvert + \lvert QI\rvert$
6g		$\overset{\frown}{IP}$ of a circle
Step	Hypothesis	$\lim_{\theta \to 0} \dfrac{\sin\theta}{\theta} = 1$
1	From 1g	Area ΔIOP < Area sector IOP ≤ Area ΔIOT
2	From 1g and T5.8.1 Right Triangle: Algebraic Area	Area ΔIOP = ½ $\lvert PQ\rvert\lvert OQ\rvert$ + ½ $\lvert PQ\rvert\lvert QI\rvert$ for ΔPOQ and ΔPQI
3	From 2 and A4.2.14 Distribution	Area ΔIOP = ½ $\lvert PQ\rvert(\lvert OQ\rvert + \lvert QI\rvert)$
4	From 5g, 3 and A4.2.3 Substitution	Area ΔIOP = ½ $\lvert PQ\rvert\,\lvert OI\rvert$
5	From 4g, 4, A4.2.3 Substitution and A4.2.12 Identity Multp	Area ΔIOP = ½ $\lvert PQ\rvert$
6	From 6g and T5.18.1 Finding Area of a Circle by Bounding with Regular Polygons	Area sector IOP = $\frac{1}{2}\theta r^2$
7	From 4g, 6, A4.2.3 Substitution and A4.2.12 Identity Multp	Area sector IOP = $\frac{1}{2}\theta$
8	From 1g and T5.8.1 Right Triangle: Algebraic Area	Area ΔIOT = ½ $\lvert IT\rvert\,\lvert OI\rvert$
9	From 4g, 8, A4.2.3 Substitution and A4.2.12 Identity Multp	Area ΔIOT = ½ $\lvert IT\rvert$
10	From 1g and DxE.1.1.1 sine	$\sin\theta = \lvert PQ\rvert / \lvert OP\rvert$
11	From 4g, 10, A4.2.3 Substitution and A4.2.12 Identity Multp	$\sin\theta = \lvert PQ\rvert$
12	From 1g and DxE.1.1.3 tangent	$\tan\theta = \lvert IT\rvert / \lvert OI\rvert$
13	From 12, A4.2.3 Substitution and A4.2.12 Identity Multp	$\tan\theta = \lvert IT\rvert$
14	From 1, 5, 11, 7, 9, 13 and A4.2.3 Substitution	$\frac{1}{2}\sin\theta < \frac{1}{2}\theta \le \frac{1}{2}\tan\theta$

15	From 14, T4.7.17 Inequalities: Cancellation by Multiplication for a Positive Number	$\sin\theta < \theta \le \tan\theta$
16	From 15, T4.7.31 Inequalities: Cross Multiplication with Positive Reciprocal Product	$1 < \theta / \sin\theta \le \tan\theta / \sin\theta$
17	From 16, LxE.1.15 Tangent Quotient Relations and A4.2.3 Substitution	$1 < \theta / \sin\theta \le (\sin\theta / \cos\theta) / \sin\theta$
18	From 17, A4.2.11 Associative Multp, A4.2.10 Commutative Multp, A4.2.13 Inverse Multp, and A4.2.12 Identity Multp	$1 < \theta / \sin\theta \le 1 / \cos\theta$
19	From 18, T4.7.33 Inequalities: Ternary Cross Multiplication with Positive Products and T4.7.34 Inequalities: Ternary Cross Multiplication with Negative Products	$1 \ge \sin\theta / \theta > \cos\theta$ for QI and QIV
20	From 19 and Lx9.2.1.11 Inequality of Limits	$\lim_{\theta\to 0} (1) \ge \lim_{\theta\to 0} (\sin\theta / \theta) > \lim_{\theta\to 0} (\cos\theta)$
21	From 19, Lx9.2.1.1 Uniqueness of Limits	$(1 \ge \lim_{\theta\to 0} \sin\theta / \theta > \cos 0)$
22	From 20 and DxE.1.6.1 Evaluation of Cosine	$(1 \ge \lim_{\theta\to 0} \sin\theta / \theta > 1)$
∴	From 22 and Lx9.2.1.10 Sandwiching Theorem	$\operatorname{Lim}_{\theta\to 0} \dfrac{\sin\theta}{\theta} = 1$

Theorem 9.2.8 Removing Singularity at Zero for Sine(X) / X by L' Hôpital's Rule

Step	Hypothesis	$\lim_{x \to 0} \frac{\sin x}{x} = 1$
1	From A4.2.2A Equality	$\lim_{x \to 0} \frac{\sin x}{x} = \lim_{x \to 0} \frac{\sin x}{x}$
2	From 1g and Lx9.2.1.9 l' Hôpital's Rule, Lx9.2.3.8A Differential of sine and Lx9.3.3.2 Differential Identity	$\lim_{x \to 0} \frac{\sin x}{x} = \lim_{x \to 0} \frac{\cos x}{1}$
3	From 2, Lx9.2.1.3 Obvious Limit, A5.2.18 Transitivity of Angles and DxE.1.6.1	$\lim_{x \to 0} \frac{\sin x}{x} = 1 / 1$
∴	From 3 and T4.4.9 Equalities: Any Quantity Divided by One is that Quantity	$\lim_{x \to 0} \frac{\sin x}{x} = 1$

Theorem 9.2.9 Removing Singularity at Zero for Cosine(X) / X by L' Hôpital's Rule

Step	Hypothesis	$\lim_{x \to 0} \frac{\cos x}{x} = 0$
1	From A4.2.2A Equality	$\lim_{x \to 0} \frac{\cos x}{x} = \lim_{x \to 0} \frac{\cos x}{x}$
2	From 1g and Lx9.2.1.9 l' Hôpital's Rule, Lx9.2.3.8B Differential of cosine and Lx9.3.3.2 Differential Identity	$\lim_{x \to 0} \frac{\cos x}{x} = \lim_{x \to 0} \frac{-\sin x}{1}$
3	From 2, Lx9.2.1.3 Obvious Limit, A5.2.18 Transitivity of Angles and DxE.1.6.1	$\lim_{x \to 0} \frac{\cos x}{x} = 0 / 1$
∴	From 3 and T4.4.9 Equalities: Any Quantity Divided by One is that Quantity	$\lim_{x \to 0} \frac{\cos x}{x} = 0$

Theorem 9.2.10 Removing Singularity at Zero for Tangent(X) / X

Step	Hypothesis	$\lim_{x \to 0} \frac{\tan x}{x} = 1$
1	From A4.2.2A Equality	$\lim_{x \to 0} \frac{\tan x}{x} = \lim_{x \to 0} \frac{\tan x}{x}$
2	From 1g, LxE.1.15 and Lx9.2.1.6 Limit of the Quotients	$\lim_{x \to 0} \frac{\tan x}{x} = \lim_{x \to 0} \frac{\sin x}{x} \frac{1}{\lim_{x \to 0} \cos x}$
3	From 2, T9.2.7 Removing Singularity at Zero for Sine(X) / X, Lx9.2.1.3 Obvious Limit, A5.2.18 Transitivity of Angles and DxE.1.6.1	$\lim_{x \to 0} \frac{\tan x}{x} = 1 / 1$
∴	From 3 and T4.4.9 Equalities: Any Quantity Divided by One is that Quantity	$\lim_{X \to 0} \frac{\tan x}{x} = 1$

Section 9.3 Formulas of Differentiation

Definition 9.3.1 Definition of a Derivative as a Limit

	Equations	Description
Dx9.3.1.1	$\Delta x \equiv x_2 - x_1$	Delta: Increment by Difference
Dx9.3.1.2	$f'(x) \equiv \lim_{\Delta x \to 0} \dfrac{f(x + \Delta x) - f(x)}{\Delta x}$	Definition of a Derivative as a Limit
Dx9.3.1.3	$f'(x) \equiv \lim_{\Delta x \to 0} \dfrac{\Delta f}{\Delta x}$	Definition of a Derivative Alternate Form[9.3.1]

In the three hundred and some twenty years since calculus was invented the derivative has, had plenty of time to mature, and unrepentantly, a variety of notation were devised to represent the derivative, but they all mean the same thing.

Definition 9.3.2 Definitions on Differential Notation

	Equations	Equ	Description
Dx9.3.2.1	$y = f(t)$	A	Zero Parametric Differentiation Single Variable Equation
	$\dot{y} \equiv \dot{f}(t) \equiv \dfrac{dy}{dt} \equiv \dfrac{d f(t)}{dt}$	B	First Parametric Differentiation with Newton's dot mark notation, and Leibniz fraction notation
	$\ddot{y} \equiv \ddot{f}(t) \equiv \dfrac{d^2 y}{dt^2} \equiv \dfrac{d^2 f(t)}{dt^2}$ [9.3..2]	C	Second Parametric Differentiation with mark, fraction and dot notation
	$\dddot{y} \equiv \dddot{f}(t) \equiv \dfrac{d^3 y}{dt^3} \equiv \dfrac{d^3 f(t)}{dt^3}$	D	Third Parametric Differentiation with mark, fraction and dot notation
	$y^m \equiv f^m(t) \equiv \dfrac{d^m y}{dt^m} \equiv \dfrac{d^m f(t)}{dt^m}$ [9.3.3]	E	Multiple Parametric Differentiation with mark, fraction and dot notation
	$D^m{}_t \equiv \dfrac{d^m}{dt^m} = \dfrac{\prod^m_{i=1} d}{\prod^m_{i=1} dt^i}$	F	As a general exact differential operator, for m = 1 drop the one

	$y' \equiv f'(s) \equiv \dfrac{dy}{ds} \equiv \dfrac{d\,f(s)}{ds}$	G	First Parametric Differentiation with Lagrange's prime mark notation, and Leibniz fraction notation
	$y'' \equiv f''(s) \equiv \dfrac{d^2y}{ds^2} \equiv \dfrac{d^2f(s)}{ds^2}$ [9.3..2]	H	Second Parametric Differentiation with mark, fraction and dot notation
	$y''' \equiv f'''(s) \equiv \dfrac{d^3y}{ds^3} \equiv \dfrac{d^3f(s)}{ds^3}$	I	Third Parametric Differentiation with mark, fraction and dot notation
	$y^m \equiv f^m(s) \equiv \dfrac{d^m y}{ds^m} \equiv \dfrac{d^m f(s)}{ds^m}$ [9.3.3]	J	Multiple Parametric Differentiation with mark, fraction and dot notation
	$D^m{}_s \equiv \dfrac{d^m}{ds^m} = \dfrac{\Pi^m_{i=1} d}{\Pi^m_{i=1} ds^i}$	K	General Exact Differential Operator, for m = 1 drop the one
Dx9.3.2.2	$y = f(\bullet\, x_j \bullet)$	A	Zero Partial Differentiation of Multivariable Equation
	$y,i \equiv f(\bullet\, x_j \bullet),i \equiv \dfrac{\partial y}{\partial x_i} \equiv \dfrac{\partial f}{\partial x_i}(\bullet\, x_j \bullet)$	B	Multi-variable Partial Notation
	$y^m,i \equiv f^m(\bullet\, x_j \bullet),i \equiv \dfrac{\partial^m y}{\partial x_i{}^m} \equiv \dfrac{\partial f^m}{\partial x_i{}^m}(\bullet\, x_j \bullet)$	C	
	$\partial^m,i \equiv \dfrac{\partial^m}{\partial x_i{}^m}$	D	Multiple Partial Differential Operator, for m = 1 drop the one
	$\partial^m, \bullet j_i \bullet \equiv \dfrac{\partial^m}{\coprod^m_{i=1} \partial x_{j_i}} = \dfrac{\coprod^m_{i=1} \partial}{\coprod^m_{i=1} \partial x_{j_i}}$	E	
Dx9.3.2.3	***Leibniz***'s differential fraction notation, composite denominator is a product of independent algebraic quantities, hence can follow the axiom A4.2.10 Commutative Multp and any two such quantities can be freely interchanged.	A	Leibniz's Commutative Differential Denominators, see above notation Dx9.3.2.1K and Dx9.3.2.2E

[9.3.1]Note: The small divisor of the derivative is called the *perturbation difference factor Δx on the x-axis.*

[9.3.2]Note: As long as the derivative yields a well-behaved function there is no reason why that function cannot be differentiated as well, hence the notation can neatly be extended for multiple operations and numbered as such.

[9.3.3]Note: Three and more Newtonian primes become cumbersome so there simply numbered.

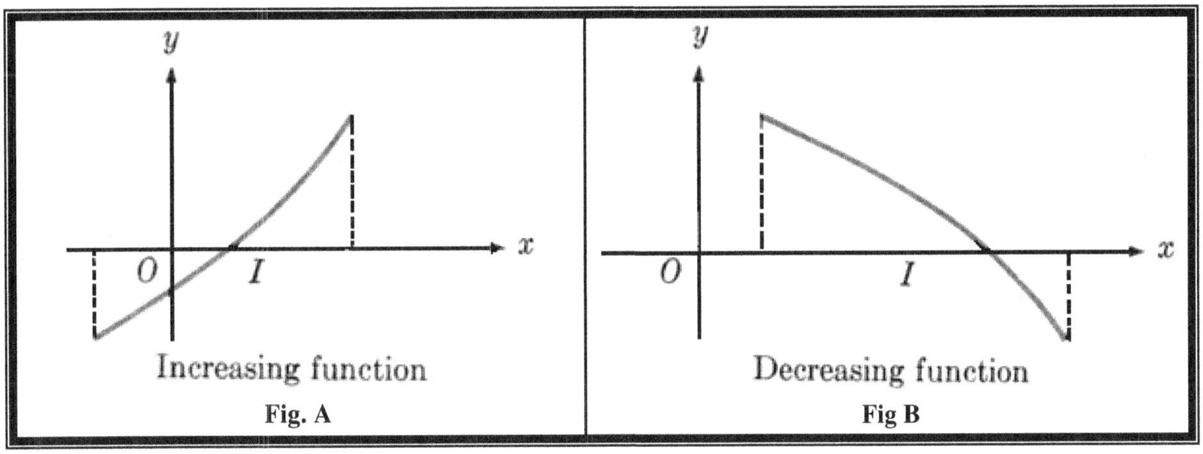

Fig. A Fig B

Figure 9.3.1 Increasing and Decreasing Functions

Definition 9.3.3 Increasing Function

A function f is said to be increasing on the interval I if $f(x_2) > f(x_1)$ whenever $x_2 > x_1$, so long as both x_1 and x_2 are in I. See Figure 9.3.1A.

Definition 9.3.4 Decreasing Function

A function f is said to be decreasing on the interval I if $f(x_2) > f(x_1)$ whenever $x_2 > x_1$, so long as both x_1 and x_2 are in I. See Figure 9.3.1B.

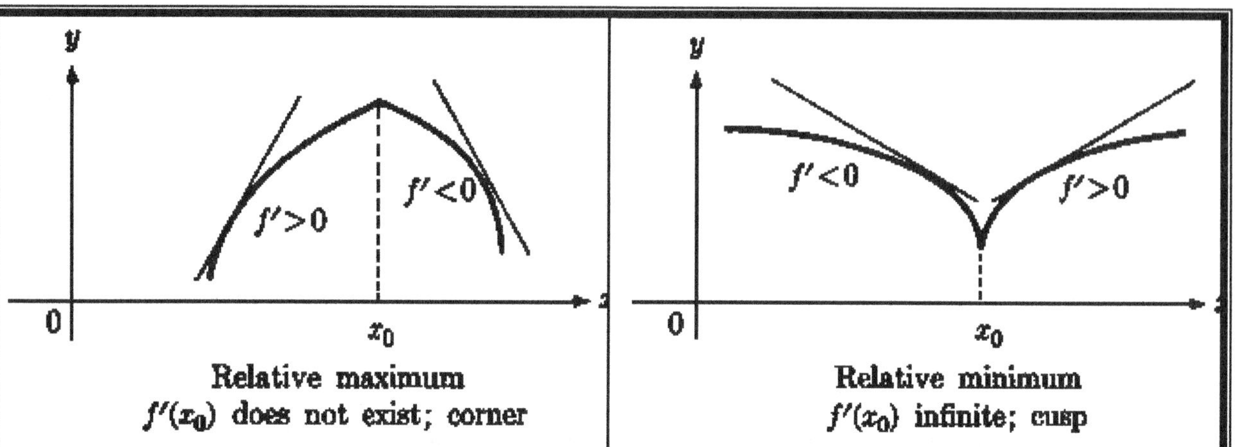

Fig. A **Fig. B**

Figure 9.3.2 Corner and Cusp of a Curve

Definition 9.3.5 Corner

A corner is a relative maximum in the interval of a curve with no derivative. See Figure 9.3.2A.

Definition 9.3.6 Cusp

A cusp is a relative minimum with a derivative having infinite slope. See Figure 9.3.2B.

Definition 9.3.7 Critical Value

A critical value of a function [f] is a value of [x] where f '(x) = 0.

Definition 9.3.8 Critical Point

A critical point of a function f is the point (x, f(x)) on the graph corresponding to the critical value [x].

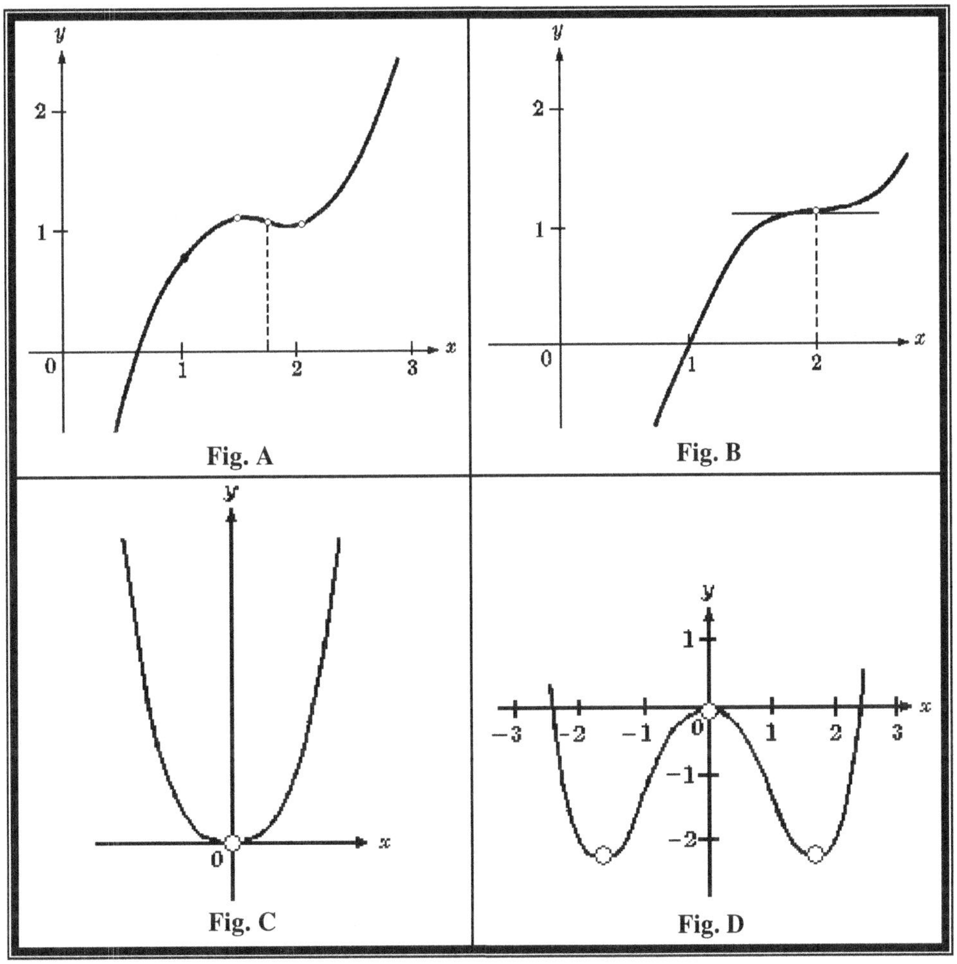

Figure 9.3.3 Inflection Points at an open dot

Definition 9.3.9 Inflection Point

An inflection point is the point at which direction of slope changes from a neutral slope of zero at a given point [x_0], where

$$f''(x_0) = 0, \qquad \text{Equation A}$$

See examples in Figure 9.3.3.

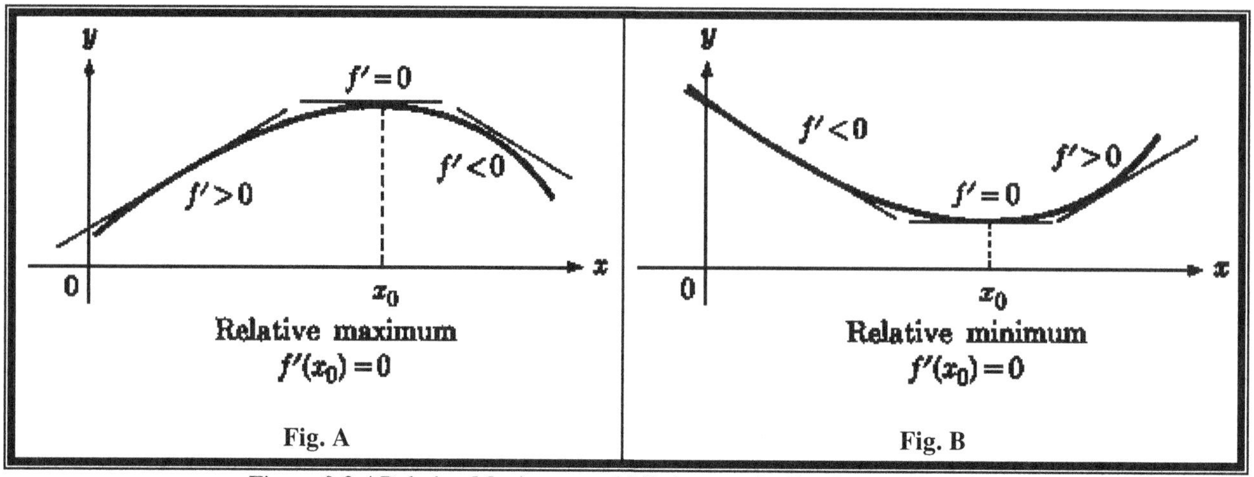

Figure 9.3.4 Relative Maximum and Minimum of a Continuous Function

Definition 9.3.10 Maximum

A ***maximum*** is a relatively high inflection point above other such points. See examples in Figure 9.3.3A.

Definition 9.3.11 Minimum

A ***minimum*** is a relatively low inflection point below other such points. See examples in Figure 9.3.3B.

Definition 9.3.12 Extrema

A maximum or minimum denotes the extremes of a function; all together they are called the ***extrema*** of a function.

Definition 9.3.13 Convex Curve

A convex curve is a curve that has a maximum peak point (x_{mx}, y_{mx}) in the interval of slop going from (+) plus to (0) zero and then (–) negative in a well defined interval about the peak from left to right.

Definition 9.3.14 Concave Curve

A concave curve is a curve that has a minimum peak point (x_{mn}, y_{mn}) in the interval of slop going from (–) negative to (0) zero and then (+) plus in a well defined interval about the peak from left to right.

Rules of differentiation and limits are simply stated here for ease of reference, but can be found as fully developed Theorems by Protter and Morrey. [PRO70]

Lema 9.3.1 Differential Theorems

	Theorem	Description
Lx9.3.1.1	$f'(x_0) \equiv 0$ f(a) = f(b) for f continuous in the interval of $a \le x \le b$	Rolle's Theorem then ∃ a differential at $a \le x_0 \le b$ that is zero
Lx9.3.1.2	$f'(x_0) \equiv \dfrac{f(b) - f(a)}{b - a}$ for [f] continuous and differentiable in the interval of $a \le x \le b$	Theorem of the Mean then ∃ a differential at $a < x_0 < b$
Lx9.3.1.3	$f'(x) > 0$ $-\infty \le x \le x_0$ $f'(x) < 0$ $x_0 \le x \le +\infty$	$f'(x_0)$ is a Relative Maximum See Fig. 9.3.1.3A
Lx9.3.1.4	$f'(x) < 0$ $-\infty \le x \le x_0$ $f'(x) > 0$ $x_0 \le x \le +\infty$	$f'(x_0)$ is a Relative Minimum See Fig. 9.3.1.3B
Lx9.3.1.5	$0 < f$ or $f < 0$ over the entire interval of x for $f \neq 0$	f positive or negative function
Lx9.3.1.6	a) If $f''(x_0) > 0$ then [f] has a relative minimum at x_0. b) If $f''(x_0) < 0$ then [f] has a relative maximum at x_0. c) If $f''(x_0) = 0$ then [f] fails the test.	f " continuous and $f'(x_0)$ is a critical value at the inflection point.
Lx9.3.1.7	$dx = dx$ EQ A $f'(x) = g'(x)$ EQ B $f'(x)\,dx = g'(x)\,dx$ EQ C $df(x) = dg(x)$ EQ D	Uniqueness Theorem of Differentiation Given functions on either sides of the equality are continuous and differentiable then they can be differentiated across the equality.

Rules of differentiation and limits are simply stated here for ease of reference, but can be found as fully developed Theorems by Protter and Morrey. [PRO70]

Lema 9.3.2 Differential Operators

	Equations	Equ	Description
Lx9.3.2.1	$\phi^1(t) \equiv \sum_n f(\bullet\, x_i \bullet),i\, dx_i$ (exact differential)	A	On Product Functions, Chain Rule
	$f(\, h(u)\,)^1 = f^1(\, h(u)\,)\, h^1(u)$	B	
	$f(\bullet\, x_i(\bullet\, u_j \bullet)\bullet),j = \sum_n f(\bullet\, x_i \bullet),i\, x_{i,\,j}$ for all j	C	
Lx9.3.2.2	$(\alpha + \beta)^m = \alpha^m + \beta^m$	A	Differential as a Linear Operator
	$(f + g),i = f,i + g,i$	B	
Lx9.3.2.3	$(\kappa\alpha)^m = \kappa\alpha^m$	A	Distribute Constant in/out of Product
	$(k\, f),i = k\, f,i$	B	
Lx9.3.2.4	$(\alpha\beta)^1 = \alpha^1\, \beta + \alpha\, \beta^1$	A	Chain Rule Across Products
	$(f\, g)_{,i} = f_{,i}\, g + f\, g_{,i}$	B	
	$(\prod_{i=1}^{n} f_i)_{,k} = \sum_{j=1}^{n-1} (\prod_{i=1}^{j-1} f_i)\, f_{j,k}\, (\prod_{i=j+1}^{n} f_i)$ where $(\prod_{i=1}^{0} f_i) = 1$ for $j = 1$ $(\prod_{i=n}^{n} f_i) = 1$ for $j = n-1$	C	C can be derived from B by parsing the product repeatedly while using B
Lx9.3.2.5	$(\alpha\beta)^m = \sum_{i=0}^{m} {}_mC_i\, \alpha^{m-i}\, \beta^i$	A	m-Differential of a Binary Product
Lx9.3.2.6	$(\alpha / \beta)^1 = (\alpha)^1 / \beta - \alpha\, (\beta)^1 / \beta^2$	A	Differential of a Binary Ratio
Lx9.3.2.7	$(f^n)^m = f^{n-m}$	A	Differential of a Differential
	$(f_{,n})_{,m} = f_{,n-m}$	B	
Lx9.3.2.8	$f^{m,n} = f^{n,m}$	A	Symmetry of Differential Order
	$f_{,i,k} = f_{,k,i}$	B	
Lx9.3.2.9	$\frac{d}{dq}\int_p^q f(x)\, dx = f(q),$ [p constant]	A	Differential of an Integral; evaluated at the upper limit
	$\frac{d}{dp}\int_p^q f(x)\, dx = -f(p)$ [q constant]	B	Differential of an Integral; evaluated at the lower limit
	$\frac{d}{da}\int_p^q f(x, a)\, dx = \int_p^q \frac{\partial}{\partial a}[f(x, a)]\, dx + f(q, a)\frac{dq}{da} - f(p, a)\frac{dp}{da}$	C	Differential of an Integral; evaluated at the upper and lower limit

Lema 9.3.3 Differential of Power and Log Functions

	Equations	Equ	Description
Lx9.3.3.1	$(\kappa)^m = 0$	A	Differential of a Constant
	$(k)_{,i} = 0$	B	
Lx9.3.3.2	$(t)^1 = 1$	A	Differential identity
	$(x_i)_{,i} = 1$	B	
Lx9.3.3.3	$(t^n)^1 = nt^{(n-1)}$	A	Differential of a power
	$(x_i^n)_{,i} = nx_i^{(n-1)}$	B	
	$(\sum^n a_i x^i)^m = \sum^n a_i x^{i-m}$	C	Differentiation of a polynomial
	$(\sum^n a_i x^i)_{,m} = \sum^n a_i x^{i-m}$	D	
	$(t^{-m})^1 = -mt^{(-m-1)}$ for $t \neq 0$ and $0 < m$	E	Differential of a reciprocal
	$(x_i^{-m})_{,i} = -mx_i^{(-m-1)}$ for $x_i \neq 0$ and $0 < m$	F	
Lx9.3.3.4	$(\ln(t))^1 = (t)^1/t$ for $t \neq 0$	A	Differential of a Natural Log base [e]
	$(\ln(x_i))_{,i} = 1/x_i$ for $x_i \neq 0$	B	
Lx9.3.3.5	$(\log_a(t))^1 = (\log_a e)\,(t)^1/t$ for $t \neq 0$	A	Differential of a Log base [a]
	$(\log_a(x_i))_{,i} = (\log_a e) 1/x_i$ for $x_i \neq 0$	B	
Lx9.3.3.6	$(e^t)^1 = e^t\,(t)^1$	A	Differential of a Natural base [e]
	$(e^{xi})_{,i} = e^{xi}$	B	
Lx9.3.3.7	$(a^t)^1 = (\log_e a) a^t\,(t)^1$	A	Differential of a base [a]
	$(a^{xi})_{,i} = (\log_e a) a^{xi}$	B	

Lema 9.3.4 Differential of Trigonometric Functions

	Equations	Equ	Description	
Lx9.3.4.1	$(\sin u)^1 = \cos u\, u^1$	A	Differential of sine	
	$(\cos u)^1 = -\sin u\, u^1$	B	Differential of cosine	
Lx9.3.4.2	$(\tan u)^1 = \sec^2 u\, u^1$	A	Differential of tangent	
	$(\cot u)^1 = -\csc^2 u\, u^1$	B	Differential of cotangent	
Lx9.3.4.3	$(\sec u)^1 = \sec u \tan u\, u^1$	A	Differential of secant	
	$(\csc u)^1 = -\csc u \cot u\, u^1$	B	Differential of cosecant	
Lx9.3.4.4	$(\mathrm{vers}\ u)^1 = \sin u\, u^1$	A	Differential of versine	
	$(\mathrm{covers}\ u)^1 = -\cos u\, u^1$	A	Differential of coversine	
Lx9.3.4.5	$(\arcsin u)^1 = [\,1 / \sqrt{}\ 1 - u^2\,]\, u^1$	A	Differential of arc sine	$-½\pi \leq \arcsin u \leq ½\pi$
	$(\arccos u)^1 = [-1 / \sqrt{}\ 1 - u^2\,]\, u^1$	B	Differential of arc cosine	$0 \leq \arccos u \leq \pi$
Lx9.3.4.6	$(\arctan u)^1 = [\,1 / 1 + u^2\,]\, u^1$	A	Differential of arc tangent	$-½\pi \leq \arctan u \leq ½\pi$
	$(\mathrm{arc}\cot u)^1 = [-1 / 1 + u^2\,]\, u^1$	B	Differential of arc cotangent	$0 \leq \mathrm{arc}\cot u \leq \pi$
Lx9.3.4.7	$(\mathrm{arc}\sec u)^1 = [\,1 / u \sqrt{}\ u^2 - 1]\, u^1$	A	Differential of arc secant	$0 \leq \mathrm{arc}\sec u \leq ½\pi$, $-\pi \leq \mathrm{arc}\sec u \leq -½\pi$
	$(\mathrm{arc}\csc u)^1 = [-1 / u \sqrt{}\ u^2 - 1]\, u^1$	B	Differential of arc cosecant	$0 \leq \mathrm{arc}\csc u \leq ½\pi$, $-\pi \leq \mathrm{arc}\csc u \leq -½\pi$
Lx9.3.4.8	$(\mathrm{arc\ vers}\ u)^1 = [\,1 / \sqrt{}\ 2u - u^2]\, u^1$	A	Differential of arc versine	$0 \leq \mathrm{arc\ vers}\ u \leq \pi$
	$(\mathrm{arc\ covers}\ u)^1 = [-1 / \sqrt{}\ 2u - u^2]\, u^1$	B	Differential of arc coversine	$-½\pi \leq \mathrm{arc\ covers}\ u \leq ½\pi$

Theorem 9.3.1 Exact Differential from First Principles and Separation of Variable

1g	Given	$\Delta Y \approx A_x \Delta X$
2g		$B_x = \lim_{\Delta X \to 0} A_x$
Step	Hypothesis	$dY = B_x\, dX$
1	From 1g	$\Delta Y \approx A_x \Delta X$
2	From 1 and Lx9.2.1.1 Uniqueness of Limits	$\lim_{\Delta X \to 0} \Delta Y / \Delta X = \lim_{\Delta X \to 0} A_t$
3	From 2, Dx9.3.1.2 Definition of a Derivative alternate form, Dx9.2.1.1B Given Equality of Parametric Eq	$dY / dX = \lim_{\Delta X \to 0} A_x$
∴	From 1g, 3 A4.2.3 Substitution and T4.4.7B Equalities: Reversal of Left Cancellation by Multiplication	$dY = B_x\, dX$

Theorem 9.3.2 Uniqueness of Differentials

1g	Given	$f(x) = g(x)$
Step	Hypothesis	$f\,'(x) = g'(x)$
1	From Dx9.3.1.2 Definition of a Derivative	$f\,'(x) = \lim_{\Delta x \to 0} \dfrac{f(x + \Delta x) - f(x)}{\Delta x}$
2	From 1g, 1 and A4.2.3 Substitution	$f\,'(x) = \lim_{\Delta x \to 0} \dfrac{g(x + \Delta x) - g(x)}{\Delta x}$
∴	From 2 and Dx9.3.1.2 Definition of a Derivative	$f\,'(x) = g'(x)$

Theorem 9.3.3 Single-variable Lemma of Differentiation

1g	Given	$F'(x)$ exists
2g		$G(\Delta x) \equiv 0$ at $h = 0$ It is called a spacer function satisfying the equality. It is also analogous to the analytic bases vector being invariant in an analytic space.
Step	Hypothesis	$F(x + \Delta x) - F(x) = [F'(x) + G(\Delta x)]\, \Delta x$
1	From Dx9.3.1.2 Definition of a Derivative, approximated without a limit.	$F'(x) \approx \dfrac{F(x + \Delta x) - F(x)}{\Delta x}$ if $\Delta x \neq 0$
2	From 1 and T4.3.3B Equalities: Right Cancellation by Addition for b = 0	$0 \approx \dfrac{F(x + \Delta x) - F(x)}{\Delta x} - F'(x)$, if $\Delta x \neq 0$
3	From 2, ∃l a function of [Δx] that satisfies the equality for $\Delta x \neq 0$	$G(\Delta x) \equiv \dfrac{F(x + \Delta x) - F(u)}{\Delta x} - F'(x)$, if $\Delta x \neq 0$

4	In order for h to be continuous at zero G(Δx) is defined as zero.	$G(\Delta x)$	$\equiv$	0	if $\Delta x = 0$
∴	From 3, T4.3.4B Equalities: Reversal of Right Cancellation by Addition, T4.4.5B Equalities: Reversal of Right Cancellation by Multiplication	$F(x + \Delta x) - F(x)$	$=$	$[F'(x) + G(\Delta x)]\, \Delta x$	

Theorem 9.3.4 Single Variable Chain Rule

1g	Given	$f(x) = g(u(x))$ for g and u differentiable.	
2g		$\Delta f = f(x + \Delta x) - f(x)$	
3g		$\Delta u = u(x + \Delta x) - u(x)$	
Step	Hypothesis	$f'(x) = g'(u)\, u'(x)$ EQ A	$df / dx = (dg / du)\, (du/dx)$ EQ B
1	From 1g, 2g and A4.2.3 Substitution of functions and corresponding arguments	$\Delta f = g(u(x + \Delta x)) - g(u(x))$	
2	From 1, A4.2.7 Identity Add and A4.2.8 Inverse Add	$\Delta f = g(\, u(x) + u(x + \Delta x) - u(x)\,) - g(u(x))$	
3	From 3g, 2 and A4.2.3 Substitution	$\Delta f = g(u + \Delta u) - g(u)$	
4	From 3, T9.3.3 Single-variable Lemma of Differentiation and A4.2.3 Substitution	$\Delta f = [g'(u) + G(\Delta u)]\, \Delta u$	
5	From 4 and T4.4.23 Equalities: Division by a Constant	$\dfrac{\Delta f}{\Delta x} = [g'(u) + G(\Delta u)]\, \dfrac{\Delta u}{\Delta x}$	
6	From 5, Lx9.2.1.1 Uniqueness of Limits and Lx9.2.1.5 Limit of the Products	$\lim_{\Delta x \to 0} \dfrac{\Delta f}{\Delta x} = \lim_{\Delta x \to 0} [g'(u) + G(\Delta u)]\, \lim_{\Delta x \to 0} \dfrac{\Delta u}{\Delta x}$	
7	From 6 and Dx9.3.1.2 Definition of a Derivative	$f'(x) = \lim_{\Delta x \to 0} [g'(u) + G(\Delta u)]\, u'(x)$	
8	From 3g, evaluating the limit and A4.2.8 Inverse Add	$\lim_{\Delta x \to 0} \Delta u = u(x) - u(x) = 0$	
9	From 7, 8 and Lx9.2.1.4 Limit of the Sum, Lx9.2.1.2 Limit of a Constant	$f'(x) = [g'(u) + G(0)\,]\, u'(x)$	
∴	From 9, T9.3.3 Single-variable Lemma of Differentiation-Step 4 and A4.2.7 Identity Add and Dx9.3.2.1B Second Parametric Differentiation	$f'(x) = g'(u)\, u'(x)$ EQ A	$df / dx = (dg / du)\, (du/dx)$ EQ B

Theorem 9.3.5 Single Variable Chain Rule Alternate[9.3.3]

1g	Given	$F(x) = f(g(x))$
2g		$F(x + \Delta x) = f(g(x + \Delta x))$
3g		$u = g(x)$
4g		$\Delta u = g(x + \Delta x) - g(x)$
Step	Hypothesis	$\dfrac{dF}{dx} = \dfrac{\partial f(g(x))}{\partial g} \dfrac{d\, g(x)}{dx}$
1	From Dx9.3.1.2 Definition of a Derivative	$\dfrac{dF}{dx} = \lim_{\Delta x \to 0} \dfrac{F(x + \Delta x) - F(x)}{\Delta x}$
2	From 1g, 2g, 1 and A4.2.3 Substitution	$\dfrac{dF}{dx} = \lim_{\Delta x \to 0} \dfrac{f(g(x + \Delta x)) - f(g(x))}{\Delta x}$
3	From 3g, 4g, A4.2.3 Substitution, T4.3.4A Equalities: Reversal of Right Cancellation by Addition and A4.2.5 Commutative Add	$u + \Delta u = g(x + \Delta x)$
4	From 3g, 2, 3 and A4.2.3 Substitution	$\dfrac{dF}{dx} = \lim_{\Delta x \to 0} \dfrac{f(u + \Delta u) - f(u)}{\Delta x}$
5	From 4, A4.2.12 Identity Multp and A4.2.13 Inverse Multp	$\dfrac{dF}{dx} = \lim_{\Delta x \to 0} \dfrac{f(u + \Delta u) - f(u)}{\Delta x} \dfrac{\Delta u}{\Delta u}$
6	From 5 A4.2.10 Commutative Multp	$\dfrac{dF}{dx} = \lim_{\Delta x \to 0} \dfrac{f(u + \Delta u) - f(u)}{\Delta u} \dfrac{\Delta u}{\Delta x}$
7	From 6, Lx9.2.1.5 Limit of the Products, 4g and A4.2.3 Substitution	$\dfrac{dF}{dx} = \lim_{\Delta x \to 0} \dfrac{f(u + \Delta u) - f(u)}{\Delta u} \lim_{\Delta x \to 0} \dfrac{g(x + \Delta x) - g(x)}{\Delta x}$
8	From 4g and $\Delta x \to 0$	$\Delta u = g(x) - g(x)$
9	From 8 and A4.2.8 Inverse Add	$\Delta u = 0$
10	From 8 and 9 as	$\Delta x \to 0 \quad \Delta u \to 0$ in the limit
11	From 7 and 10	$\dfrac{dF}{dx} = \lim_{\Delta u \to 0} \dfrac{f(u + \Delta u) - f(u)}{\Delta u} \lim_{\Delta x \to 0} \dfrac{g(x + \Delta x) - g(x)}{\Delta x}$
12	From Dx9.3.1.2 Definition of a Derivative	$\dfrac{dF}{dx} = \dfrac{\partial f(u)}{\partial u} \dfrac{d\, g(x)}{dx}$
∴	From 3g, 12 and A4.2.3 Substitution	$\dfrac{dF}{dx} = \dfrac{\partial f(g(x))}{\partial g} \dfrac{d\, g(x)}{dx}$

[9.3.3] Note: The only problem with this proof is that $k = g(x+h) - g(x)$ could be 0 for some h's arbitrarily close to 0, but not 0. A more complete proof is given by Theorem 9.3.4 "Single Variable Chain Rule".

Observation 9.3.1 About the Development of the Chain Rule

The Lx9.3.1.2 "Theorem of the Mean" is the keystone to the proof of the chain rule. Each variable starting with step zero is expanded to represent a cascade of differences representing each possible differentiated variable. This is done in such a way as to leave the difference to each corresponding argument as zero or simply the variable held constant except in the direction of differentiation. The variable of differentiation satisfies the conditions placed on the bounding values over the interval and in the direction of differentiation.

As an example:

0	Δf	$\equiv$	$f(x_1 + \Delta x_1, x_2 + \Delta x_2, x_3 + \Delta x_3, x_4 + \Delta x_4, x_5 + \Delta x_5)$	–	$f(x_1, x_2, x_3, x_4, x_5)$	Diff.
1	Δf	=	$f(x_1 + \Delta x_1, x_2, x_3, x_4, x_5)$	–	$f(x_1, x_2, x_3, x_4, x_5)$	Δx_1
2		+	$f(x_1 + \Delta x_1, x_2 + \Delta x_2, x_3, x_4, x_5)$	–	$f(x_1 + \Delta x_1, x_2, x_3, x_4, x_5)$	Δx_2
3		+	$f(x_1 + \Delta x_1, x_2 + \Delta x_2, x_3 + \Delta x_3, x_4, x_5)$	–	$f(x_1 + \Delta x_1, x_2 + \Delta x_2, x_3, x_4, x_5)$	Δx_3
4		+	$f(x_1 + \Delta x_1, x_2 + \Delta x_2, x_3 + \Delta x_3, x_4 + \Delta x_4, x_5)$	–	$f(x_1 + \Delta x_1, x_2 + \Delta x_2, x_3 + \Delta x_3, x_4, x_5)$	Δx_4
5		+	$f(x_1 + \Delta x_1, x_2 + \Delta x_2, x_3 + \Delta x_3, x_4 + \Delta x_4, x_5 + \Delta x_5)$	–	$f(x_1 + \Delta x_1, x_2 + \Delta x_2, x_3 + \Delta x_3, x_4 + \Delta x_4, x_5)$	Δx_5

In general:

$\Delta f = \sum_i f(\bullet\, x_j + [\delta_i^j + \tau_{Li}^j]\, \Delta x_j\, \bullet) - f(\bullet x_j + [\tau_{Li}^j]\, \Delta x_j\, \bullet)$ Equation A

Applying the "Theorem of the Mean"

1	Δf	=	$f_{,1}(\xi_1, x_2, x_3, x_4, x_5)$	$x_1 + \Delta x_1 - x_1$	Δx_1
2		+	$f_{,2}(x_1, \xi_2, x_3, x_4, x_5)$	$x_2 + \Delta x_2 - x_2$	Δx_2
3		+	$f_{,3}(x_1, x_2, \xi_3, x_4, x_5)$	$x_3 + \Delta x_3 - x_3$	Δx_3
4		+	$f_{,4}(x_1, x_2, x_3, \xi_4, x_5)$	$x_4 + \Delta x_4 - x_4$	Δx_4
5		+	$f_{,5}(x_1, x_2, x_3, x_4, \xi_5)$	$x_5 + \Delta x_5 - x_5$	Δx_5

In general:

$\Delta f = \sum_i f_{,i}(\bullet\, x_i + \delta_i^j[\xi_j - x_j]\, \bullet)(x_j + \Delta x_j - x_j)$ Equation B

where $x_j \le \xi_j \le x_j + \Delta x_j$ is a variable along the direction of differentiation.

Given 9.3.1 Multi-variable Chain Rule Theorems

1g		$f(\bullet x_i \bullet)$		Multi-variable function
2g		x_j	$\le$	$\xi_i \le x_i + \Delta x_i$
3g		$G_i(k: \xi_j, \bullet\Delta x_i \bullet)$	$\equiv$	$f_{,i}(\bullet\, x_i + \delta_i^j[\xi_j - x_j] \bullet) - f_{,i}(\bullet x_i \bullet)$ spacer function for j = 1, … , k arguments

Theorem 9.3.6 Multi-variable Lemma of Differentiation

1g	Given	$\Delta f \equiv f(\bullet\, x_j + \Delta x_j \bullet) - f(\bullet x_j \bullet)$
2g		$b_i = x_j + \Delta x_j$ and
3g		$a_i = x_j$
Step	Hypothesis	$\Delta f = \sum_i \{ f_{,i}(\bullet x_j \bullet) + G_i(i: \xi_j, \bullet\Delta x_j \bullet) \}\Delta x_j$
1	From G1g Multi-variable Chain Rule Theorems, 1g and O9.3.1A About the Development of the Chain Rule, expanding along the differential path using D6.1.6 Kronecker Delta and D6.1.9 Lower Tau, matrix elements	$\Delta f = \sum_i f(\bullet\, x_j + [\delta_i^j + \tau_{Li}{}^j]\, \Delta x_j \bullet) - f(\bullet x_j + [\tau_{Li}{}^j]\, \Delta x_j \bullet)$
2	From 2g, 3g, 1, O9.3.1B About the Development of the Chain Rule, Lx9.3.1.2 Theorem of the Mean, D1.7.3 Differential Notation: Partial, T4.4.5B Equalities: Reversal of Right Cancellation by Multiplication and A4.2.3 Substitution over the bound interval b_i and a_i	$\Delta f = \sum_i f_{,i}(\bullet\, x_i + \delta_i^j[\xi_j - x_j] \bullet)(\, x_j + \Delta x_j - x_j)$
3	From 2 A4.2.5 Commutative Add, A4.2.8 Inverse Add and A4.2.7 Identity Add	$\Delta f = \sum_i f(\bullet\, x_i + \delta_i^j[\xi_j - x_j] \bullet)\, \Delta x_j$
4	From 3, A4.2.7 Identity Add, A4.2.8 Inverse Add, A4.2.5 Commutative Add and A4.2.6 Associative Add	$\Delta f = \sum_i \{ f_{,i}(\bullet x_j \bullet) + [f_{,i}(\bullet\, x_i + \delta_i^j[\xi_j - x_j] \bullet) - f_{,i}(\bullet x_j \bullet)] \}\Delta x_j$
∴	From G3g Multi-variable Chain Rule Theorems, 4 and A4.2.3 Substitution	$\Delta f = \sum_i \{ f_{,i}(\bullet x_j \bullet) + G_i(i: \xi_j, \bullet\Delta x_j \bullet) \}\Delta x_j$

Theorem 9.3.7 Limits on Multi-variable Chain Rule

Step	Hypothesis	as $\Delta x_j \to 0$ then $\xi_j \to x_j$ and as $\Delta x_j \to 0$ then $G_i(k: x_j, \bullet 0 \bullet) \to 0$
1	From G2g Multi-variable Chain Rule Theorems	$x_j \le \xi_j \le x_j + \Delta x_j$
2	From 1 and $\Delta x_j \to 0$	$x_j \le \xi_j \le x_j$
3	From 2 and Lx9.2.1.10 Sandwiching Theorem	$\xi_j = x_j$ or as $\Delta x_j \to 0$ then $\xi_j \to x_j$
4	From G3g Multi-variable Chain Rule Theorems	$G_i(k: \xi_j, \bullet \Delta x_j \bullet) = f_{,i}(\bullet\, x_i + \delta_i^j[\xi_j - x_j] \bullet) - f_{,i}(\bullet x_j \bullet)$
5	From 3, 4, $\Delta x_j \to 0$ and A4.2.3 Substitution	$G_i(k: x_j, \bullet 0 \bullet) = f_{,i}(\bullet\, x_i + \delta_i^j[x_i - x_j] \bullet) - f_{,i}(\bullet x_j \bullet)$
6	From 5, A4.2.8 Inverse Add, T4.4.1 Equalities: Any Quantity Multiplied by Zero is Zero and A4.2.7 Identity Add	$G_i(k: x_j, \bullet 0 \bullet) = f_{,i}(\bullet\, x_i \bullet) - f_{,i}(\bullet\, x_j \bullet)$
7	From 6 and A4.2.8 Inverse Add	$G_i(k: x_j, \bullet 0 \bullet) = 0$ or as $\Delta x_j \to 0$ then $G_i(k: x_j, \bullet 0 \bullet) \to 0$
∴	From 3 and 7	as $\Delta x_j \to 0$ then $\xi_j \to x_j$ and as $\Delta x_j \to 0$ then $G_i(k: x_j, \bullet 0 \bullet) \to 0$

Theorem 9.3.8 Multi-variable Chain Rule

1g	Given	$x_j \equiv w_i(\bullet\, u_k \bullet)$ for $k = 1, \ldots, m$
2g		$\Delta u_k \equiv u_k - u_{0k}$ and
3g		as $\Delta x_j \to 0$ then $\Delta u_k \to 0$
Step	Hypothesis	$f_{,k} = \sum_i f_{,i}\, x_{j,k}$
1	From T9.3.6 Multi-variable Lemma of Differentiation	$\Delta f = \sum_i \{ f_{,i}(\bullet x_j \bullet) + G_i(i: \xi_j, \bullet \Delta x_j \bullet) \} \Delta x_j$
2	From 1g, 2g, 1 and T4.4.23 Equalities: Division by a Constant	$\dfrac{\Delta f}{\Delta u_k} = \sum_i \{ f_{,i}(\bullet x_j \bullet) + G_i(i: \xi_j, \bullet \Delta x_j \bullet) \} \dfrac{\Delta x_j}{\Delta u_k}$
3	From 3g, 2, Dx9.3.1.2 Definition of a Derivative, T9.3.7 Limits on Multi-variable Chain Rule and Dx9.3.2.2 Multiple Partial Differential, as $\Delta u_k \to 0$	$\dfrac{\partial f}{\partial u_k} = \sum_i \{ f_{,i}(\bullet x_j \bullet) + 0 \} \dfrac{\partial x_j}{\partial u_k}$
4	From 3 and A4.2.7 Identity Add	$\dfrac{\partial f}{\partial u_k} = \sum_i f_{,i}(\bullet x_j \bullet) \dfrac{\partial x_j}{\partial u_k}$
∴	From 4 and Dx9.3.2.2B Multi-variable Partial Notation	$f_{,k} = \sum_i f_{,i}\, x_{j,k}$

Theorem 9.3.9 Exact Differential Parametric Chain Rule

1g	Given	$x_j \equiv w_i(t)$
2g		$\Delta t \equiv t - t_0$ and
3g		as $\Delta x_j \rightarrow 0$ then $\Delta t \rightarrow 0$

Step	Hypothesis	$f' = \sum_i f_{,i}\, x_j'$
1	From T9.3.6 Multi-variable Lemma of Differentiation	$\Delta f = \sum_i \{ f_{,i}(\bullet x_j \bullet) + G_i(i: \xi_j, \bullet\Delta x_j \bullet) \}\Delta x_j$
2	From 1g, 2g, 1 and T4.4.23 Equalities: Division by a Constant	$\frac{\Delta f}{\Delta t} = \sum_i \{ f_{,i}(\bullet x_j \bullet) + G_i(i: \xi_j, \bullet\Delta x_j \bullet) \} \frac{\Delta x_j}{\Delta t}$
3	From 3g, 2, Dx9.3.1.2 Definition of a Derivative, T9.3.7 Limits on Multi-variable Chain Rule and Dx9.3.2.1B Parametric Differentiation, as $\Delta t \rightarrow 0$	$\frac{df}{dt} = \sum_i \{ f_{,i}(\bullet x_j \bullet) + 0 \} \frac{dx_j}{dt}$
4	From 3 and A4.2.7 Identity Add	$\frac{df}{dt} = \sum_i f_{,i}(\bullet x_j \bullet) \frac{dx_j}{dt}$
∴	From 4 and Dx9.3.2.1B Parametric Differentiation	$f' = \sum_i f_{,i}\, x_j'$

Section 9.4 Formulas of Integration

Definition 9.4.1 The Limit of an Integral Having a Constant Area Under the Curve

	Equations	Description
Dx9.4.1.1	$\left\lvert \sum_{k=1}^{n} f(\xi_k)\Delta_k x - [F(b) - F(a)] \right\rvert < \varepsilon$	Definition of an Integral
		with constant bounded limits [a. b]
Dx9.4.1.2	$F(b) - F(a) \equiv \lim_{\lVert\Delta\rVert \to 0} \sum_{k=1}^{n} f(\xi_k)\Delta_k x$	Definition of an Integral
		alternate form
Dx9.4.1.3	$\operatorname{Lim} \lVert\Delta\rVert \to 0 \Rightarrow \operatorname{Lim}_{\Delta x \to 0}$ as $\lim_{n \to \infty}$	Definition of Interval of Subdivisions for an Integral
		f is integrable on the interval [a, b] for $\varepsilon > 0$ there is a $\delta > \lVert\Delta\rVert > 0$
Dx9.4.1.4	$F(b) - F(a) \equiv \int_a^b f(x)dx \equiv F(x) \vert_a^b$	Definition of a Definite Integral evaluation of [f] bound [a to b].

Definition 9.4.2 The Limit of an Integral Having Variable Area Under the Curve

	Equations	Description
Dx9.4.2.1	$\left\lvert \sum_{k=1}^{n} f(\xi_k)\Delta_k x - (F[b(x)] - F[a(x)]) \right\rvert < \varepsilon$	Definition of an Integral
		with varying bounded limits [a(x), b(x)]
Dx9.4.2.2	$F[b(x)] - F[a(x)] \equiv \lim_{\lVert\Delta\rVert \to 0} \sum_{k=1}^{n} f(\xi_k)\Delta_k x$	Definition of an Integral
		alternate form
Dx9.4.2.3	$\operatorname{Lim} \lVert\Delta\rVert \to 0 \Rightarrow \operatorname{Lim}_{\Delta x \to 0}$ as $\lim_{n \to \infty}$	Definition of Interval of Subdivisions for an Integral
		f is integrable on the interval [a, b] for $\varepsilon > 0$ there is a $\delta > \lVert\Delta\rVert > 0$
Dx9.4.2.4	$F[b(x)] - F[a(x)] \equiv \int_{a(x)}^{b(x)} f(x)dx \equiv F(x) \vert_{a(x)}^{b(x)}$	Definition of a Definite Integral evaluation of [f] bound a(x) to b(x).

Definition 9.4.3 Integrand

Integrand is the function [f] being integrated.

Definition 9.4.4 Limits of Integration

In the interval of integration [a, b] were [b] is the ***upper limit*** and [a] the ***lower limit*** of integration.

Definition 9.4.5 Variable of Integration

The variable [x] is the variable over which the integration occurs. It is a dummy variable being eliminated by the limits of integration, hence may be replaced by any other variable.

Definition 9.4.6 Indefinite Integral

An indefinite integral is the ***antiderivative*** [F(x)] of the function [f(x)] with no limits being defined, in a general way they are replaced by adding an arbitrary constant [C], the ***constant of integration***, which implies one possible antiderivative after another.

$$\int f(x)\, dx = F(x) + C \qquad \text{Equation A}$$

Lema 9.4.1 Fundamental Integral Theorems		
	Theorem	**Description**
Lx9.4.1.1	If f(x) is defined and increasing (or at least nondecreasing) on the closed interval $a \le x \le b$, then the integral is there.	Integral under an increasing curve.
Lx9.4.1.2	If f(x) is continuous on [a, b], then the integral is on [a, b]	Integral under a continuous curve.
Lx9.4.1.3	If [c] is any number and [f] is on the interval [a, b], then the function cf(x) is can be integrated on [a, b] and $\int_a^b c\, f(x)dx = c \int_a^b f(x)dx$	Integral on the integrand with a constant.
Lx9.4.1.4	If f(x) and g(x) are integrable on [a, b], then f(x) + g(x) is integrable on [a, b] and $\int_a^b [f(x) + g(x)]\, dx = \int_a^b f(x)dx + \int_a^b g(x)\, dx$	Integral over the addition of integrands.
Lx9.4.1.5	If f(x) is integrable on an interval [a, b], then [f] is bounded there. $m \le f(x) \le M$	Integral integrand is bounded.
Lx9.4.1.6	If f(x) is integrable on an interval [a, b] and [m] and [M] are numbers such that $m \le f(x) \le M$ for $a \le x \le b$, then it follows that $m(b - a) \le \int_a^b f(x)\, dx \le M(b - a)$	A bounded integrand has a bounded antiderivative.
Lx9.4.1.7	If [f] and [g] are integrable on the interval [a, b] and $f(x) \le g(x)$ for [x] in [a, b] then $\int_a^b f(x)dx \le \int_a^b g(x)dx$	Trichotomy over the interval of integration.
Lx9.4.1.8	If f(x) is continuous on the interval [a, b] and [c] is any number in that interval $a \le c \le b$, then $\int_a^b f(x)dx = \int_a^c f(x)dx + \int_c^b f(x)dx$	Parsing an Integral
Lx9.4.1.9	If [f] is the integral on the interval [a, a] then $\int_a^a f(x)dx = 0$	Integrated on no interval.
Lx9.4.1.10	If [f] is the integral on the interval [a, b] then it is integral on the reverse interval [b, a] $\int_a^b f(x)dx = -\int_b^a f(x)dx$	Integrability on a reverse interval
Lx9.4.1.11	If [uv] is the integral on the interval [a, b] then $\int_a^b u\, dv = uv\vert_a^b - \int_a^b v\, du$	Integration by parts in an interval

Theorem 9.4.2 Integrals: Uniqueness of Integration

1g	Given	f(x) and g(x) are integral over the range $a \le x \le b$
2g		$f(x) \equiv g(x)$
Step	Hypothesis	$\int_a^b f(x)\,dx = \int_a^b g(x)\,dx$
1	From D9.4.1 The Limit of an Integral Having a Constant Area Under the Curve	$\Sigma_k f(x_k)\,\Delta x_k \approx F(x)$
2	From 2g, 1 and A4.2.3 Substitution	$\Sigma_k g(x_k)\,\Delta x_k \approx F(x)$
3	From 1, 2 and A4.2.2A Equality	$\Sigma_k f(x_k)\,\Delta x_k = \Sigma_k g(x_k)\,\Delta x_k$
∴	From 1g, 3, D9.4.1 The Limit of an Integral Having a Constant Area Under the Curve and Lx9.2.1.1 Uniqueness of Limits	$\int_a^b f(x)\,dx = \int_a^b g(x)\,dx$

Rules of integration are simply stated here for ease of reference, but can be found in fully developed tables by I.S. Gradshteyn and I.W. Ryzhik. [GRA65, page 54] where [+ C] is the constant of integration.

Lema 9.4.2 Basic Integrals

	Equations	Description
Lx9.4.2.1	$\int a\,dx = a\,x + C$	Integration of power functions
Lx9.4.2.2	$\int x^n\,dx = 1/(n+1)x^{n+1} + C$ for $n \ne -1$	
Lx9.4.2.3	$\int x^{-1}\,dx = \ln x + C$ for $n = -1$	
Lx9.4.2.4	$\int e^{ax}\,dx = (1/a)\,e^{ax} + C$	
Lx9.4.2.5	$\int a^{bx}\,dx = 1/(b \ln a)\,a^{bx} + C$	
Lx9.4.2.6	$\int a^x \ln a\,dx = a^x + C$ for $a > 0$	
Lx9.4.2.7	$\int \sin a\,x\,dx = -(1/a)\cos x + C$	Integration of trigonometric functions
Lx9.4.2.8	$\int \cos a\,x\,dx = (1/a)\sin a\,x + C$	
Lx9.4.2.9	$\int \tan a\,x\,dx = -(1/a)\ln \cos a\,x = (1/a)\ln \sec a\,x + C$	
Lx9.4.2.10	$\int \cot a\,x\,dx = -(1/a)\ln \csc a\,x = (1/a)\ln \sin a\,x + C$	
Lx9.4.2.11	$\int \sec a\,x\,dx = (1/a)\ln(\sec a\,x + \tan a\,x) + C$	
Lx9.4.1.12	$\int \csc a\,x\,dx = (1/a)\ln(\csc a\,x - \cot a\,x) + C$	

Lema 9.4.3 General Formulas		
	Equations	**Description**
Lx9.4.3.1	$\int a\,f\,dx = a\int f\,dx + C$	Distribution of Constant Out Of Integrand
Lx9.4.3.2	$\int (f + g)\,dx = \int f\,dx + \int g\,dx + C$	Distribution of Integrand Over Summation
Lx9.4.3.3	$\int f\,dg = fg - \int g\,df + C$	Integration by Parts not Bounded by an Interval
Lx9.4.3.4	$\int \frac{f'}{f}\,dx = \ln f + C$	Integration of Natural Log
Lx9.4.3.5	$\int \frac{f'}{2\sqrt{f}}\,dx = \sqrt{f} + C$	Integration of Root

Lema 9.4.4 Forms Containing $(\sqrt{a^2 - x^2})$		
	Equations	**Description**
Lx9.4.4.1	$\int \sqrt{(a^2 - x^2)}\,dx = ½[\,x(\sqrt{a^2 - x^2}) + a^2 \sin^{-1} x / \lvert a\rvert\,] + C$	
Lx9.4.4.2	$\int x\sqrt{(a^2 - x^2)}\,dx = -1/3[\sqrt{(a^2 - x^2)}^3 + C$	
Lx9.4.4.3	$\int x^2\sqrt{(a^2 - x^2)}\,dx = ½[\,x(\sqrt{a^2 - x^2}) + a^2 \sin^{-1} x / \lvert a\rvert\,] + C$	
Lx9.4.4.4	$\int \frac{dx}{\sqrt{(a^2 - x^2)}} = \sin^{-1}\frac{x}{\lvert a\rvert} + C$	
Lx9.4.4.5	$\int \frac{dx}{x\sqrt{(a^2 - x^2)}} = -(1/a)\ln\left(\frac{a + \sqrt{a^2 - x^2}}{x}\right)$	
Lx9.4.4.6	$\int \frac{x\,dx}{\sqrt{(a^2 - x^2)^3}} = \frac{1}{\sqrt{a^2 - x^2}} + C$	

Lema 9.4.5 Definite Integrals		
	Equations	**Description**
Lx9.4.5.1	$\int_0^\infty x^{p-1}\, e^{-qx}\, dx = q^{-p}\,\Gamma(p)$	Gamma function [GRA65, page 317]
Lx9.4.5.2	$\Gamma(n) = (n-1)!$ for $0 < n$ and integer	Factorial
Lx9.4.5.3	$\zeta(2n) = \dfrac{(2\pi)2^n}{2(2n)!} \lvert B_{2n}\rvert$	Riemann Zeta Function [GRA65, page 1074]
Lx9.4.5.4	$B_n = \begin{array}{c\|c\|c\|c\|c\|c\|c\|c} n & 0 & 1 & 2 & 3 & 4 & 5 & 6 \\ \hline nu & 1 & -1 & 1 & 0 & -1 & 0 & 1 \\ \hline de & 1 & 2 & 6 & 1 & 30 & 1 & 42 \end{array}$	Bernoulli Numbers [GRA65, page 1079]
Lx9.4.5.5	$\int_0^\infty \dfrac{x^{p-1}}{e^{qx}-1}\, dx = \dfrac{1}{q^p}\,\Gamma(p)\,\zeta(p)$	Combinations of rational functions of powers and exponentials [GRA65, page 325]
Lx9.4.5.6	$F_s(\theta, k) = \int_0^\theta \dfrac{d\varphi}{\sqrt{1 - k^2 \sin^2\varphi}}$	Elliptic integral of the first kind with the k modulus
Lx9.4.5.7	$F_c(\theta, \tau) = \int_0^\theta \dfrac{d\varphi}{\sqrt{1 + \tau^2 \cos^2\varphi}}$	Elliptic integral of the first kind with Normalized modulus
Lx9.4.5.8	$E_s(\theta, k) = \int_0^\theta \sqrt{1 - k^2 \sin^2\varphi}\; d\varphi$	Elliptic integral of the second kind with the k modulus
Lx9.4.5.9	$E_c(\theta, \tau) = \int_0^\theta \sqrt{1 + \tau^2 \cos^2\varphi}\; d\varphi$	Elliptic integral of the second kind with Normalized modulus
Lx9.4.5.10	$\tau = \dfrac{k}{\sqrt{1-k^2}}$	Normalized modulus
Lx9.4.5.11	$k' = \sqrt{1-k^2}$	Complementary modulus
Lx9.4.5.12	$k\, F_s(\theta, k) = \tau\, F_c(\theta, \tau)$	Elliptic integral of the first kind with complementary relation
Lx9.4.5.13	$\tau\, E_s(\theta, k) = k\, E_c(\theta, \tau)$	Elliptic integral of the second kind with complementary relation
Lx9.4.5.14	$F_x(2\pi n, \tau) = 4n\, F_x(½\pi, \tau)$ for x = [sine, cosine]	Elliptic integral of the first kind with quadruplicate symmetry
Lx9.4.5.15	$E_x(2\pi n, k) = 4n\, E_x(½\pi, \tau)$ for x = [sine, cosine]	Elliptic integral of the second kind quadruplicate symmetry

Section 9.5 Vector Differential Calculus

Definition 9.5.1 Vector Function

There exists a scalar variable [u] associated with a vector **A**, then **A** is called a ***vector function of*** **(u)** denoted by **A**(u). Vectors will be written in bold face so as not to annotate it with a vector arrow on top, thereby simplify its writing.

$\mathbf{A}(u) \equiv \sum_i A_i(u)\mathbf{b}_i$ Equation A

Where $A_i(u)$ is the ***component of the vector*** and $\mathbf{b}_i$ the ***bases vector*** spanning that n-space. The bases vector also defines the shape of the n-space.

Definition 9.5.2 Coordinate Vector Function of a Manifold

If a vector exists at a coordinate point $P(\bullet\ x_i\ \bullet)$, then its components are a function of that n-space, at that point and called a ***coordinate vector function***.

$\mathbf{A}(\bullet\ x_i\ \bullet) \equiv \sum_i A_i(\bullet\ x_i\ \bullet)\mathbf{b}_i$ Equation A

Definition 9.5.3 Vector Field

$\mathbf{A}(\bullet\ x_i\ \bullet)$ defines a ***vector field*** since it associates a vector with each point in that region of space.

Definition 9.5.4 Scalar Field

$\phi(\bullet\ x_i\ \bullet)$ defines a ***scalar field*** since it associates a scalar with each point of that region of space.

Definition 9.5.5 Volume Parallelepiped Product

$\mathbf{A} \bullet (\mathbf{B} \times \mathbf{C})$ ***The Volume Parallelepiped Product*** is sometimes called the scalar triple product or box product.

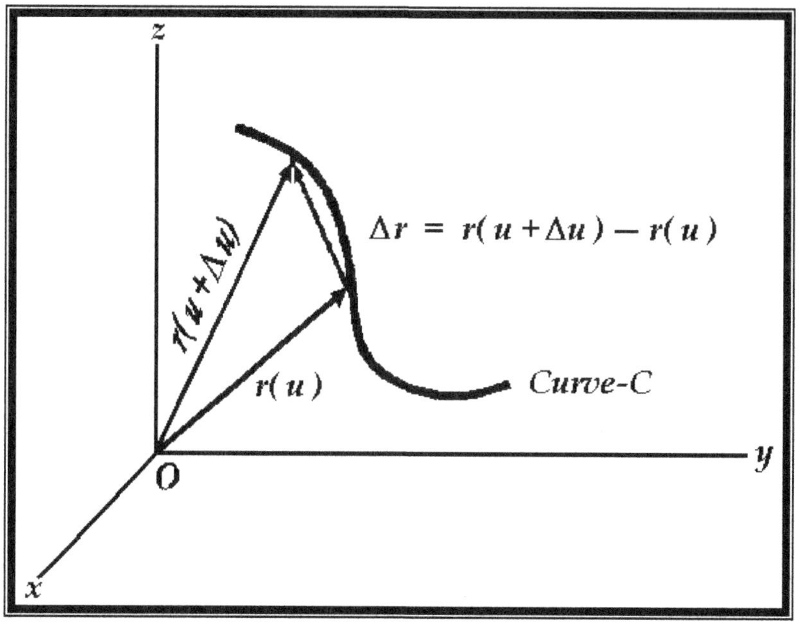

Figure 9.5.1 Approximating the Tangent Vector to the Curve

Definition 9.5.6 Unit Tangent Vector to a Curve

There exists a vector **r** = OP starting at a coordinate point of origin-O and ending on a point P(• x_i •), which traces out a curve-C of some infinitesimal arc length ds, for a parametric parameter [u] $x_j \equiv w_i(u)$.

For $\Delta\mathbf{r} = \mathbf{r}(u + \Delta u) - \mathbf{r}(u)$ as $\Delta u \to 0$ $\Delta\mathbf{r} \to \mathbf{T}$ a unit tangent vector in the direction of the tangent on the curve-C, at that point-P, and finally in the limit

$$\frac{d\mathbf{r}}{ds} \equiv \mathbf{T} \textit{ unit tangent vector}$$

Definition 9.5.7 Velocity Vector

If the parametric parameter [u] is a parameter of time [t], then the derivative of r is the velocity

$$\frac{d\mathbf{r}}{dt} \equiv \mathbf{v} \textit{ velocity}$$

Definition 9.5.8 Speed or Celerity

The magnitude of the velocity $|\mathbf{v}| = (ds / dt) = v$ is called the ***speed*** or sometimes the ***celerity***.

$$\frac{d\mathbf{r}}{dt} = \frac{d\mathbf{r}}{ds}\frac{ds}{dt} = \mathbf{T}\frac{ds}{dt} = \mathbf{T}\,v$$

Definition 9.5.9 Acceleration

The change in velocity with respect to time is called the ***acceleration***.

$$\frac{d\,\mathbf{v}}{dt} = \frac{d}{dt}\frac{d\mathbf{r}}{dt} = \frac{d^2\mathbf{r}}{dt^2} = \mathbf{a} \qquad \text{Exact Differential} \qquad \text{Equation A}$$

$$\frac{\partial\,\mathbf{v}}{\partial t} = \mathbf{a} \qquad \text{Partial Differential} \qquad \text{Equation B}$$

Axiom 9.5.1 Limit of a Continuous Vector

The vector function $\mathbf{A}(u)$ is said to be continuous at u_0 if given any positive number [ε], we can find some positive number [δ] such that

$$|\,\mathbf{A}(u) - \mathbf{A}(u_0)\,| < \varepsilon \quad \text{whenever} \quad |\,u - u_0\,| < \delta, \qquad \text{Equation A}$$

then there exists a limit for the vector $\mathbf{A}(u)$

$$\lim_{u \to u0} \mathbf{A}(u) = \mathbf{A}(u_0) \qquad \text{Equation B}$$

Axiom 9.5.2 Derivative of a Vector

If a limit for a vector exists for a given point in space, then the vector can have a derivative at that point:

$$\frac{d\,\mathbf{A}}{d\,u} \equiv \lim_{u \to u0} \frac{\mathbf{A}(\,u + \Delta u) - \mathbf{A}(u)}{\Delta u}$$

Axiom 9.5.3 Uniqueness of Differentiation of a Vector

$$\frac{d\,\mathbf{A}}{d\,u} \equiv \frac{d\,\mathbf{B}}{d\,u} \qquad \text{Exact Differential Equation A}$$

$$\frac{\partial\,\mathbf{A}}{\partial\,x} \equiv \frac{\partial\,\mathbf{B}}{\partial\,x} \qquad \text{Partial Differential Equation B}$$

Lema 9.5.1 Vector Product Identities		
	Equations	Description
Lx9.5.1.1	$\mathbf{A} \times \mathbf{B} = -\mathbf{B} \times \mathbf{A}$	Asymmetric Property of Cross Product
Lx9.5.1.2	$\mathbf{A} \times (\mathbf{B} + \mathbf{C}) = \mathbf{A} \times \mathbf{B} + \mathbf{A} \times \mathbf{C}$	Distributive Property of Cross Products
Lx9.5.1.3	$k(\mathbf{A} \times \mathbf{B}) = (k\mathbf{A}) \times \mathbf{B} =$ $\mathbf{A} \times (k\mathbf{B}) = (\mathbf{A} \times \mathbf{B})\, k$	Commutative property of Scalar to Cross Product
Lx9.5.1.4	$\mathbf{A} \times \mathbf{A} = \mathbf{0}$	Parallel Vectors of Cross Product
Lx9.5.1.5	$(\mathbf{A} \bullet \mathbf{B})\, \mathbf{C} \neq \mathbf{A}\, (\mathbf{B} \bullet \mathbf{C})$	Non Commutative of Dot Triple Property
Lx9.5.1.6	$\mathbf{A} \bullet (\mathbf{B} \times \mathbf{C}) = \mathbf{B} \bullet (\mathbf{C} \times \mathbf{A}) = \mathbf{C} \bullet (\mathbf{A} \times \mathbf{B})$	Volume of Parallelepiped
Lx9.5.1.7	$\mathbf{A} \bullet (\mathbf{B} \times \mathbf{C}) = \begin{vmatrix} A_1 & A_2 & A_3 \\ B_1 & B_2 & B_3 \\ C_1 & C_2 & C_3 \end{vmatrix}$	Volume of Parallelepiped Matrix
Lx9.5.1.8	$\mathbf{A} \bullet (\mathbf{A} \times \mathbf{C}) = 0$ EQ A $\mathbf{A} \bullet (\mathbf{B} \times \mathbf{A}) = 0$ EQ B	Parallelepiped Matrix with Two Parallel sides
Lx9.5.1.9	$\mathbf{A} \times (\mathbf{B} \times \mathbf{C}) = (\mathbf{A} \bullet \mathbf{C})\mathbf{B} - (\mathbf{A} \bullet \mathbf{B})\mathbf{C}$	Vector Triple Cross Product Right
Lx9.5.1.10	$(\mathbf{A} \times \mathbf{B}) \times \mathbf{C} = (\mathbf{A} \bullet \mathbf{C})\mathbf{B} - (\mathbf{B} \bullet \mathbf{C})\mathbf{A}$	Vector Triple Cross Product Left

Lema 9.3.2 Vector Differential Formulas		
	Equations	
Lx9.5.2.1	$\frac{d(\mathbf{A} + \mathbf{B})}{du} = \frac{d\mathbf{A}}{du} + \frac{d\mathbf{B}}{du}$	Differential Linear Vector Distribution
Lx9.5.2.2	$\frac{d(\mathbf{A} \bullet \mathbf{B})}{du} = \mathbf{A} \bullet \frac{d\mathbf{B}}{du} + \frac{d\mathbf{A}}{du} \bullet \mathbf{B}$	Differential of Dot Product
Lx9.5.2.3	$\frac{d(\mathbf{A} \times \mathbf{B})}{du} = \mathbf{A} \times \frac{d\mathbf{B}}{du} + \frac{d\mathbf{A}}{du} \times \mathbf{B}$	Differential of Cross Product
Lx9.5.2.4	$\frac{d(\phi \mathbf{A})}{du} = \mathbf{A} \frac{d\phi}{du} + \frac{d\mathbf{A}}{du} \phi$	Differential of Scalar-Vector Product
Lx9.5.2.5	$\frac{d[\mathbf{A} \bullet (\mathbf{B} \times \mathbf{C})]}{du} = \mathbf{A} \bullet \mathbf{B} \times \frac{d\mathbf{C}}{du} + \mathbf{A} \bullet \frac{d\mathbf{B}}{du} \times \mathbf{C} + \frac{d\mathbf{A}}{du} \bullet \mathbf{B} \times \mathbf{C}$	Diff. Scalar Triple Product
Lx9.5.2.6	$\frac{d[\mathbf{A} \times (\mathbf{B} \times \mathbf{C})]}{du} = \mathbf{A} \times (\mathbf{B} \times \frac{d\mathbf{C}}{du}) + \mathbf{A} \times (\frac{d\mathbf{B}}{du} \times \mathbf{C}) + \frac{d\mathbf{A}}{du} \times (\mathbf{B} \times \mathbf{C})$	Diff. Vector Triple Product
Lx9.5.2.7	$\frac{d(\mathbf{A} \bullet \mathbf{A})}{du} = 2\mathbf{A} \bullet \frac{d\mathbf{A}}{du}$ for $\mathbf{A} = \mathbf{B}$ Lx9.2.1.2	Differential of Identical Dot Product
Lx9.5.2.8	$a \frac{da}{du} = \mathbf{A} \bullet \frac{d\mathbf{A}}{du}$ for $a^2 = \mathbf{A} \bullet \mathbf{A}$	Differential of Vector Magnitude
Lx9.5.2.9	$0 = \mathbf{A} \bullet \frac{d\mathbf{A}}{du}$ for [a] constant	Differential of Constant Vector Magnitude Hence $\mathbf{A} \perp d\mathbf{A}/du$

Theorem 9.3.1 Derivative Distribution with Invariant Parallelogram Law

1g	Given	$\mathbf{A} \equiv \sum_i A_i(u)\mathbf{b}_i$	for A_i continuously differentiable
2g		$\mathbf{b}_i(u) \equiv \mathbf{b}_i(u+\Delta u)$	invariant parallelogram law, hence no curl or directional change in analytic space
Step	Hypothesis	$\dfrac{d\,\mathbf{A}}{d\,u} = \sum_i \dfrac{dA_i(u)}{du}\,\mathbf{b}_i$	
1	From A9.5.1 Derivative of a Vector	$\dfrac{d\,\mathbf{A}}{d\,u} = \lim_{u \to u0} \dfrac{\mathbf{A}(u + \Delta u) - \mathbf{A}(u)}{\Delta u}$	
2	From 1g, 2g, 1 and A7.9.1 Equivalence of Vector component by component	$\dfrac{d\,\mathbf{A}}{d\,u} = \lim_{u \to u0} \dfrac{\sum_i A_i(u + \Delta u)\mathbf{b}_i - \sum_i A_i(u)\mathbf{b}_i}{\Delta u}$	
3	From 2 and T7.10.9 Distribution of Vector over Addition of Scalars	$\dfrac{d\,\mathbf{A}}{d\,u} = \lim_{u \to u0} \sum_i \dfrac{[A_i(u + \Delta u) - A_i(u)]}{\Delta u}\,\mathbf{b}_i$	
4	From 3 and Lx9.2.1.4 Limit of the Sum	$\dfrac{d\,\mathbf{A}}{d\,u} = \sum_i \lim_{u \to u0} \dfrac{[A_i(u + \Delta u) - A_i(u)]}{\Delta u}\,\mathbf{b}_i$	
∴	From 4 and Dx9.4.1.2 Definition of a Derivative	$\dfrac{d\,\mathbf{A}}{d\,u} = \sum_i \dfrac{dA_i(u)}{du}\,\mathbf{b}_i$	

Theorem 9.5.2 Multi-variable Chain Rule for a Vector

1g 2g	Given	$\mathbf{A} \equiv \sum_i A_i(\bullet\, x_j \bullet)\mathbf{b}_i$ for A_i continuously differentiable $x_j \equiv w_i(\bullet\, u_k \bullet)$ for k = 1, …, m
Step	Hypothesis	$\mathbf{A}_{,k} = \sum_j \mathbf{A}_{,j}\, x_{j,k}$
1	From 1g, 2g and distributing Lx9.4.1.7 Uniqueness Theorem of Differentiation and T9.3.1 Derivative Distribution Across Analytic Vector Components	$\mathbf{A}_{,k} = \sum_i A_{i,k}\, \mathbf{b}_i$
2	From 1 and T9.4.8 Multi-variable Chain Rule	$\mathbf{A}_{,k} = \sum_i (\sum_j A_{i,j}\, x_{j,k})\, \mathbf{b}_i$
3	From 2, A7.8.5 Parallelogram Law: Commutative and Scalability of Collinear Vectors and summations are interchanged	$\mathbf{A}_{,k} = \sum_j (\sum_i A_i \mathbf{b}_i)_{,j}\, x_{j,k}$
∴	From 1g, 3 and A7.9.2 Substitution of Vectors	$\mathbf{A}_{,k} = \sum_j \mathbf{A}_{,j}\, x_{j,k}$

Theorem 9.5.3 Chain Rule for Exact Differential of a Vector

1g 2g	Given	$\mathbf{A} \equiv \sum_i A_i(\bullet\, x_j \bullet)\mathbf{b}_i$ for A_i continuously differentiable $x_j \equiv w_i(u)$
Step	Hypothesis	$\mathbf{A}' = \sum_j \mathbf{A}_{,j}\, x_j'$
1	From 1g, 2g and distributing Lx9.4.1.7 Uniqueness Theorem of Differentiation and T9.3.1 Derivative Distribution Across Analytic Vector Components	$\mathbf{A}' = \sum_i A_i'\mathbf{b}_i$
2	From 1 and T9.3.9 Exact Differential, Parametric, Chain Rule	$\mathbf{A}' = \sum_i (\sum_j A_{i,j}\, x_j')\, \mathbf{b}_i$
3	From 2, A7.8.5 Parallelogram Law: Commutative and Scalability of Collinear Vectors and summations are interchanged	$\mathbf{A}' = \sum_j (\sum_i A_i\, \mathbf{b}_i)_{,j}\, x_j'$
∴	From 1g, 3 and A7.9.2 Substitution of Vectors	$\mathbf{A}' = \sum_j \mathbf{A}_{,j}\, x_j'$

Theorem 9.5.4 Chain Rule for Single Variable Vector

1g	Given	$\mathbf{A}$	$\equiv$	$\sum_i A_i(w)\mathbf{b}_i$ for A_i continuously differentiable
2g		w	$\equiv$	w(u)
Step	Hypothesis	$d\mathbf{A} / du$	=	$(dw / du)\ (d\mathbf{A} / dw)$
1	From 1g, 2g, Lx9.4.1.7 Uniqueness Theorem of Differentiation and T9.3.1 Derivative Distribution Across Analytic Vector Components	$d\mathbf{A} / du$	=	$\sum_i (dA_i / du)\ \mathbf{b}_i$
2	From 1 and T9.4.4B Single Variable Chain Rule	$d\mathbf{A} / du$	=	$\sum_i (dA_i / dw)\ (dw / du)\ \mathbf{b}_i$
3	From 2 and A7.10.3 Distribution of a Scalar over Addition of Vectors	$d\mathbf{A} / du$	=	$(dw / du) \sum_i (dA_i / dw)\ \mathbf{b}_i$
4	From 3 and T9.3.1 Derivative Distribution Across Analytic Vector Components	$d\mathbf{A} / du$	=	$(dw / du)\ d(\sum_i A_i\ \mathbf{b}_i) / dw)$
∴	From 1g, 4 and A7.9.2 Substitution of Vectors	$d\mathbf{A} / du$	=	$(dw / du)\ (d\mathbf{A} / dw)$

Section 9.6 The Del Operator and Its Calculus

With the advent of differential geometry systems of vectors had to be able to transform from one coordinate system to another and maintain an invariant magnitude and direction, but in order to do that a dual system of vectors had to be established that were invertible and invariant within a specific manifold. These internal dual systems of vectors span the manifold in which they are embedded forming a special set of vectors called contra and covariant bases vectors, (see respectively in Chapter 12 definitions of 12.1.1 and 12.1.2):

Definition 9.6.1 Covariant Differential Operator, DEL [9.6.1]	
	Equations
Dx9.6.1.1A	$\nabla \equiv \sum_i \frac{\partial}{\partial x_i} \vec{b}^i$ [9.6.2] for n = 3
Dx9.6.1.1B	$\nabla \equiv \sum_i \vec{b}^i \frac{\partial}{\partial x_i}$
Dx9.6.1.2A	$\nabla \equiv \sum_i \partial_i \vec{b}^i$
Dx9.6.1.2B	$\nabla \equiv \sum_i \vec{b}^i \partial_i$
Dx9.6.1.3A	$\nabla \equiv \sum_i \nabla_i \vec{b}^i$
Dx9.6.1.3B	$\nabla \equiv \sum_i \vec{b}^i \nabla_i$

Definition 9.6.2 Covariant Laplacian Operator	
	Equations
Dx9.6.2.1A	$\nabla^2 \equiv \sum_i \sum_j \frac{\partial^2}{\partial x_i \partial x_j} g_{ij}$ for n = 3 and m = 3
Dx9.6.2.1B	$\nabla^2 \equiv \sum_i \sum_j g_{ij} \frac{\partial}{\partial x_i \partial x_j}$
Dx9.6.2.2A	$\nabla^2 \equiv \sum_i \sum_j \partial^2{}_{i,j} g_{ij}$
Dx9.6.2.2B	$\nabla^2 \equiv \sum_i \sum_j g_{ij} \partial^2{}_{i,j}$
Dx9.6.2.3A	$\nabla^2 \equiv \sum_i \sum_j \nabla^2{}_{i,j} g_{ij}$
Dx9.6.2.3B	$\nabla^2 \equiv \sum_i \sum_j g_{ij} \nabla^2{}_{i,j}$

Definition 9.6.3 Covariant Gradient Operator

	Equations
Dx9.6.3.1A	$\nabla\Phi \equiv \sum_i \frac{\partial\Phi}{\partial x_i} \vec{b}^i$
Dx9.6.3.1B	$\nabla\Phi \equiv \sum_i \vec{b}^i \frac{\partial\Phi}{\partial x_i}$
Dx9.6.3.2A	$\nabla\Phi \equiv \sum_i \Phi_{,i} \vec{b}^i$
Dx9.6.3.2B	$\nabla\Phi \equiv \sum_i \vec{b}^i \Phi_{,i}$

Definition 9.6.4 Divergent Operating on a Covariant Vector

	Equations
Given:	$\vec{A} = \sum_j A_j \vec{b}_j$ covariant vector
Dx9.6.4.1	$\nabla\bullet\vec{A} \equiv \sum_i \sum_j \frac{\partial(A_j g_{ij})}{\partial x_i}$
Dx9.6.4.2	$\nabla\bullet\vec{A} \equiv \sum_i \sum_j (A_j g_{ij})_{,i}$

Definition 9.6.5 Divergence Operating on a Contravariant Vector

	Equations
Given:	$\vec{A} = \sum_j A^j \vec{b}^j$ contravariant vector
Dx9.6.5.1	$\nabla\bullet\vec{A} \equiv \sum_i \sum_j \frac{\partial(A^j \delta_i^{\ j})}{\partial x_i} = \sum_i \frac{\partial A^i}{\partial x_i}$
Dx9.6.5.2	$\nabla\bullet\vec{A} \equiv \sum_i \sum_j (A^j \delta_i^{\ j})_{,i} = \sum_i A^i_{\ ,i}$

Definition 9.6.6 Covariant Tensor Dyadic Divergent Operator

	Equations
Given:	$\vec{A} = \sum_j A_j \vec{b}_j$ covariant vector
Dx9.6.6.1	$(\vec{A}\bullet\nabla)\vec{B} \equiv \sum_i \sum_j A_j g_{ij} \frac{\partial\vec{B}}{\partial x_i}$
Dx9.6.6.2	$(\vec{A}\bullet\nabla)\vec{B} \equiv \sum_i \sum_j A_j g_{ij} \vec{B}_{,i}$

Definition 9.6.7 Mixed Tensor Dyadic Divergent Operator	
	Equations
Given:	$\vec{A} = \sum_j A^j \vec{b}^j$ contravariant vector
Dx9.6.7.1	$(\vec{A}\bullet\nabla)\vec{B} \equiv \sum_i\sum_j A^j \delta_i^j \frac{\partial\vec{B}}{\partial x_i} = \sum_i A^i \frac{\partial\vec{B}}{\partial x_i}$
Dx9.6.7.2	$(\vec{A}\bullet\nabla)\vec{B} \equiv \sum_i\sum_j A^j \delta_i^j \vec{B}_{,i} = \sum_i A^i\vec{B}_{,i}$

Definition 9.6.8 Covariant Curl Operator[9.6.3]		
	Equations	
Dx9.6.8.1	$\nabla\times\prod_{i=3}^{n}\times\vec{A}_i \equiv \begin{vmatrix} e_{11} & e_{12} & \cdots & e_{1n} \\ \frac{1}{h_{21}}\frac{\partial}{\partial x_{21}} & \frac{1}{h_{22}}\frac{\partial}{\partial x_{22}} & \cdots & \frac{1}{h_{2n}}\frac{\partial}{\partial x_{2n}} \\ A^{31} & A^{32} & \cdots & A^{3n} \\ \vdots & & & \vdots \\ A^{n1} & A^{3n} & \cdots & A^{nn} \end{vmatrix}$	Curl of Contravariant tensors spanning a space over a Covariant gradient operator
Dx9.6.8.2	$\nabla\times\prod_{i=3}^{n}\times\vec{A}^i \equiv \begin{vmatrix} e^{11} & e^{12} & \cdots & e^{1n} \\ \frac{1}{h^{21}}\frac{\partial}{\partial x_{21}} & \frac{1}{h^{22}}\frac{\partial}{\partial x_{22}} & \cdots & \frac{1}{h^{2n}}\frac{\partial}{\partial x_{2n}} \\ A_{31} & A_{32} & \cdots & A_{3n} \\ \vdots & & & \vdots \\ A_{3n} & A_{3n} & \cdots & A_{mn} \end{vmatrix}$	Curl of Covariant tensors spanning a space over a Contravariant gradient operator

Definition 9.6.9 Covariant Quaternion Operator, D'Alembert and Poincare [9.5.4]	
	Equations
Dx9.6.9.1A	$\Box \equiv \sum_i \frac{\partial}{\partial x_i} \vec{b}_I$ for $n \geq 4$
Dx9.6.9.1B	$\Box \equiv \sum_i \vec{b}_i \frac{\partial}{\partial x_i}$
Dx9.6.9.2A	$\Box \equiv \sum_i \partial_{,i} \vec{b}_i$
Dx9.6.9.2B	$\Box \equiv \sum_i \vec{b}_i \partial_{,i}$
Dx9.6.9.3A	$\Box \equiv \sum_i \nabla_{,i} \vec{b}_i$
Dx9.6.9.3B	$\Box \equiv \sum_i \vec{b}_i \nabla_{,i}$

Definition 9.6.10 Covariant D'Alembert Quaternion Operator		
	Equations	
Dx9.6.10.1A	$\Box^2 \equiv \sum_i \sum_j \frac{\partial^2}{\partial x_i \partial x_j} g_{ij}$	for n ≥ 4 and m ≥ 4
Dx9.6.10.1B	$\Box^2 \equiv \sum_i \sum_j g_{ij} \frac{\partial}{\partial x_i \partial x_j}$	
Dx9.6.10.2A	$\Box^2 \equiv \sum_i \sum_j \partial^2{}_{,i,j}\ g_{ij}$	
Dx9.6.10.2B	$\Box^2 \equiv \sum_i \sum_j g_{ij}\ \partial^2{}_{,i,j}$	
Dx9.6.10.3A	$\Box^2 \equiv \sum_i \sum_j \nabla^2{}_{,i,j}\ g_{ij}$	
Dx9.6.10.3B	$\Box^2 \equiv \sum_i \sum_j g_{ij}\ \nabla^2{}_{,i,j}$	

Lema 9.6.1 Forms Involving ∇			
	Equations		
Lx9.6.1.1A	$\nabla(\phi) \equiv \nabla(\phi) \equiv \text{grad}(\phi)$		Uniqueness of Gradient Operator
Lx9.6.1.1B	$\nabla\bullet(\vec{A}) \equiv \nabla\bullet(\vec{A}) \equiv \text{div}(\vec{A})$		Uniqueness of Divergence Operator
Lx9.6.1.1C	$\nabla\times(\vec{A}) \equiv \nabla\times(\vec{A}) \equiv \text{curl}(\vec{A})$		Uniqueness of Curl Operator
Lx9.6.1.2A	$\nabla(\phi + \psi)$	$\equiv \nabla\phi + \nabla\psi$	Distribution of Gradient
Lx9.6.1.2B	$\text{grad}(\phi + \psi)$	$\equiv \text{grad}\ \phi + \text{grad}\ \psi$	
Lx9.6.1.3A	$\nabla\bullet(\vec{A} + \vec{B})$	$\equiv \nabla\bullet\vec{A} + \nabla\bullet\vec{B}$	Distribution of Divergence
Lx9.6.1.3B	$\text{div}(\vec{A} + \vec{B})$	$\equiv \text{div}\ \vec{A} + \text{div}\ \vec{B}$	
Lx9.6.1.4A	$\nabla\times(\vec{A} + \vec{B})$	$\equiv \nabla\times\vec{A} + \nabla\times\vec{B}$	Distribution of Curl
Lx9.6.1.4B	$\text{curl}(\vec{A} + \vec{B})$	$\equiv \text{curl}\ \vec{A} + \text{curl}\ \vec{B}$	
Lx9.6.1.5	$\nabla\bullet(\phi\vec{A})$	$\equiv (\nabla\phi)\bullet\vec{A} + \phi\nabla\bullet\vec{A}$	Divergence of Scalar Vector Product
Lx9.6.1.6	$\nabla\times(\phi\vec{A})$	$\equiv (\nabla\phi)\times\vec{A} + \phi(\nabla\times\vec{A})$	Curl of Scalar Vector Product
Lx9.6.1.7	$\nabla\bullet(\vec{A}\times\vec{B})$	$\equiv \vec{B}\bullet(\nabla\times\vec{A}) - \vec{A}\bullet(\nabla\times\vec{B})$	Divergence of Cross Product
Lx9.6.1.8	$\nabla\times(\vec{A}\times\vec{B})$	$\equiv (\vec{B}\bullet\nabla)\vec{A} - \vec{B}\ \nabla\bullet\vec{A} - (\vec{A}\bullet\nabla\ \vec{B}) + \vec{A}\ \nabla\bullet\vec{B}$	Curl of Cross Product
Lx9.6.1.9	$\nabla(\vec{A}\bullet\vec{B})$	$\equiv (\vec{B}\bullet\nabla)\vec{A} + (\vec{A}\bullet\nabla)\vec{B} + \vec{B}\times(\nabla\times\vec{A}) + \vec{A}\times(\nabla\times\vec{B})$	Gradient of Dot Product
Lx9.6.1.10	$\nabla\bullet(\nabla\phi)$	$\equiv \nabla^2\phi$ [9.6.5]	Covariant Laplacian Operator
Lx9.6.1.11	$\nabla\times(\nabla\phi)$	$\equiv \mathbf{0}$	Curl of Gradient Operator
Lx9.6.1.12	$\nabla\bullet(\nabla\times\vec{A})$	$\equiv 0$	Divergence of Curl Operator
Lx9.6.1.13	$\nabla\times(\nabla\times\vec{A})$	$\equiv \nabla(\nabla\bullet\vec{A}) - \nabla^2\vec{A}$	Curl of Curl Operator
Lx9.6.1.14	$\vec{A}\times(\nabla\times\vec{A})$	$\equiv \nabla(\vec{A}\bullet\vec{A}) - (\vec{A}\bullet\nabla)\vec{A}$	Triple Product of Curl Operator
Lx9.6.1.15	$\vec{A}\times(\nabla\times\vec{B})$	$\equiv \nabla(\vec{A}\bullet\vec{B}) - (\vec{A}\bullet\nabla)\vec{B}$	Dual-Triple Product of Curl Operator

Theorem 9.6.2 GRV: Curl of Divergence for Scalar Function

1g	Given	$\nabla\times(\pm\nabla\phi) = \mathbf{0}$ from [SPI59 pg 58 formula 10]	
2g		$\nabla\times\vec{F} = \mathbf{0}$ for $\vec{F}$ vector function	
Steps	Hypothesis	$\nabla\times\vec{F} = \mathbf{0}$ EQ A	$\vec{F} = \pm\nabla\phi$ EQ B
1	From 2g	$\nabla\times\vec{F} = \mathbf{0}$	
2	From 1g, 1 and A6.2.3A Equality of Vectors:	$\vec{F} = \pm\nabla\phi$	
∴	From 1 and 2	$\nabla\times\vec{F} = \mathbf{0}$ EQ A	$\vec{F} = \pm\nabla\phi$ EQ B

Theorem 9.6.3 GRV: Reciprocal Squared Functions Have No Curl

1g	Given	$\vec{F} = F_r\,\hat{r} + F_\theta\hat{\theta} + F_\phi\hat{\Phi}$
2g		$F_r = K\,r^{-2}$ for $K \neq f(r, \theta, \phi)$
3g		$F_\theta = 0$
4g		$F_\phi = 0$
5g		$\nabla\times\vec{F} = (r^2\sin\theta)^{-1}$ [$\{\nabla_\theta (r \sin\theta\, F_\phi) - \nabla_\phi\, r\, F_\theta)\ \}\hat{r}$ + $\{\nabla_\phi\, F_r - \nabla_r\, (r \sin\theta\, F_\phi)\ \}r\hat{\theta}$ + $\{\nabla_r\, (r\, F_\theta) - \nabla_\theta\, F_r\}\ r \sin\theta\, \hat{\Phi}]$ from [SPI59 pg 154 problem(a)] Spherical coordinates for gradient
Steps	Hypothesis	If $\vec{F} = (K\,r^{-2})\,\hat{r}$ then $\nabla\times\vec{F} = \mathbf{0}$
1	From 1g, 3g, 4g, 5g and A4.2.3 Substitution	$\nabla\times\vec{F} = (r^2\sin\theta)^{-1}$ [$\{\nabla_\theta (r \sin\theta\, 0) - \nabla_\phi\, r\, 0)\ \}\hat{r}$ + $\{\nabla_\phi\, F_r - \nabla_r\, (r \sin\theta\, 0)\ \}r\hat{\theta}$ + $\{\nabla_r\, (r\, 0) - \nabla_\theta\, F_r\}\ r \sin\theta\, \hat{\Phi}]$
2	From 1 and T4.4.1 Equalities: Any Quantity Multiplied by Zero is Zero	$\nabla\times\vec{F} = (r^2\sin\theta)^{-1}$ [$\{\nabla_\theta (0) - \nabla_\phi\, 0)\ \}\hat{r}$ + $\{\nabla_\phi\, F_r - \nabla_r\, (0)\ \}r\hat{\theta}$ + $\{\nabla_r\, (\,0) - \nabla_\theta\, F_r\}\ r \sin\theta\, \hat{\Phi}]$
3	From 2 and Lx9.2.3.1B Differential of a Constant	$\nabla\times\vec{F} = (r^2\sin\theta)^{-1}$ [$\{0 - 0)\ \}\hat{r}$ + $\{0 - 0\ \}r\hat{\theta}$ + $\{\ 0 - 0\}\ r \sin\theta\, \hat{\Phi}]$
4	From 3 and A4.2.7 Identity Add	$\nabla\times\vec{F} = (r^2\sin\theta)^{-1}$ [$0\hat{r}$ + $0r\hat{\theta}$ + $0\ r \sin\theta\, \hat{\Phi}]$
5	From 4 and T4.4.1 Equalities: Any Quantity Multiplied by Zero is Zero	$\nabla\times\vec{F} = (r^2\sin\theta)^{-1}$ [$0\hat{r}$ + $0\hat{\theta}$ + $0\hat{\Phi}]$

6	From 5, A6.2.5A Scalar-Vector Multiplication and T4.4.1 Equalities: Any Quantity Multiplied by Zero is Zero	$\nabla\times\vec{F} = 0\hat{r} + 0\hat{\theta} + 0\hat{\phi}$
∴	From 6 and D7.6.8 Zero Vector	$\nabla\times\vec{F} = \mathbf{0}$

[9.6.1]Note: The DEL operator also known as ***nabla***.

[9.6.2]Note: One of the criteria for physical gravitational field is that it operates on Newton's gravitational field potential, which can be placed in Poisson's covariant, Laplaceian equation; see [PER52, pg 148]. This idea is one of the major principles that Einstein used, which lead him to discover his covariant gravitational tensor in General Relativity; forming the basic tenet of all DEL operators in this thesis.

[9.6.3]Note: Also known as the ***cross-product*** operator as it rotates the tensor about a gradient path.

[9.6.4]Note: The D'Alembert Operator is the same as the DEL operator except it is limited to a 4-dimensional and/or greater spaces obeying all the rules of the DEL and quaternion operations and stems from electromagnetic physics were the fourth dimension is a parametric parameter $x_4 = ict$. [LOR62, pg 232]

[9.6.5]Note: Lx9.6.1.10 to Lx9.6.1.13 ϕ and $\vec{A}$ are continuously differentiable to the second partial derivative.

Section 9.7 Vector Integral Calculus

Lemma	Flux	Descriptive Name
Lema 9.7.1 Divergence and Green's Theorems		
LxVIE.9.7.1.1	$\vec{\Phi} = \oint_S F\, d\vec{S} = \int_V \nabla F\, dV$	Divergence Theorem of Gauss Scalar Multiplication
LxVIE.9.7.1.2	$\Phi = \oint_S \vec{F} \bullet d\vec{S} = \int_V \nabla \bullet \vec{F}\, dV$	Divergence Theorem of Gauss Dot Product Multiplication
LxVIE.9.7.1.3	$\vec{\Phi} = \oint_S d\vec{S} \times \vec{F} = \int_V \nabla \times \vec{F}\, dV$	Divergence Theorem of Gauss Cross Product Multiplication
LxVIE.9.7.1.4	$\Phi = \oint_S (\phi \nabla \psi) \bullet d\vec{S} = \int_V (\phi \nabla^2 \psi - \nabla \phi \bullet \nabla \psi)\, dV$	Green's First Theorem
LxVIE.9.7.1.5	$\Phi = \oint_S (\phi \nabla \psi - \psi \nabla \phi) \bullet d\vec{S} = \int_V (\phi \nabla^2 \psi - \psi \nabla^2 \phi)\, dV$	Green's Second Theorem
Lema 9.7.2 Stokes' Theorems		
LxVIE.9.7.2.6	$\oint_C F\, d\vec{r} = \int_S d\vec{S} \times \nabla F$	Stokes' Theorem Scalar Multiplication
LxVIE.9.7.2.7	$\oint_C \vec{F} \bullet d\vec{r} = \int_S (\nabla \times \vec{F}) \bullet d\vec{S}$	Stokes' Theorem Dot Product Multiplication
LxVIE.9.7.2.8	$\oint_C d\vec{r} \times \vec{F} = \int_S (d\vec{S} \times \nabla) \times \vec{F}$	Stokes' Theorem Cross Product Multiplication

Section 9.8 Calculus of Complex Variables

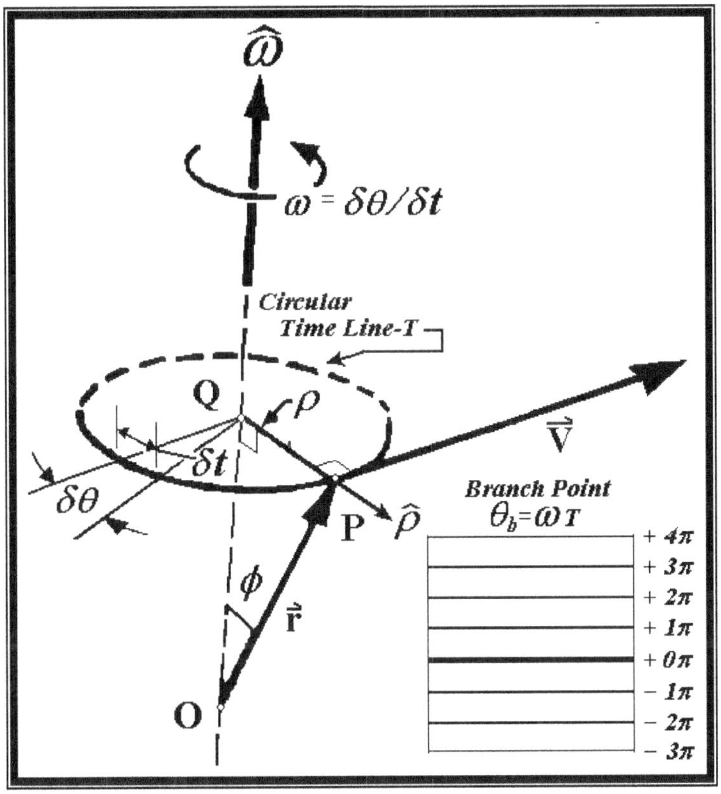

Figure 9.8.1 Branch Point θ_b

Definition 9.8.1 Branch Point

A branch point is any specific point that repeats itself a set of integer multiple times. See Fig 9.9.1.

Section 9.9 Calculus of Linear Differential Equations

Definition 9.9.1 Explicit and Implicit Functions

Given an equation of relation involving the two variables x and y f(x, y) = 0, y is explicit if it can be written independently of [x] as y = g(x). If not it is implicit. [SPI67, pg 1]

Definition 9.9.2 Linear First-Order Differential Equation

$y' + P(x)\, y = Q(x)$ [SPI67, pg 43]

Evaluated:

$$y\, e^{(\int p\, dx)} = \int Q\, e^{(\int p\, dx)}\, dx + c$$

Definition 9.9.3 Linear n^th -Order Differential Equation

$a_0(x)\, y + \sum_{i=1}^{n} a_i(x) y^i = Q(x)$ [SPI67, pg 139]

Evaluated:

$y = y_c + y_p$

y_c (complementary solution)
y_p (particular solution)

Section 9.10 The Beginning of Calculus of Variations

Definition 9.10.1 Calculus of Variations

The calculus of variations is a mathematical process that bounds various internal parameters having an infinitesimal perturbation within a system acting as a Lagrange multiplier, thereby optimizing, trajectories or object shapes.

As an example the optimal trajectory path that free falling objects take's in a gravitational or electrical field. That is the optimal path for a particle resulting in the minimal time of travel within the field.

A spatial object that is optimal would be a free floating soap bubble and the minimal shape it forms in space, without any external forces being applied to it, results in a minimal surface having an area of a perfect sphere.

Section 9.11 Ordinary Maximum and Minimum

While Kepler had observed that the increment of a function becomes vanishing small in the neighborhood of an ordinary maximum or minimum value. It was Fermat who translated this fact into a process for determining such a maximum and minimum. John Wallis and Isaac Barrow followed with their own methods, but it was Isaac Newton who came up with a formal analytic method to do this. The tool that made it possible was Newton's differential for analytically derived slope. What they all had noticed was that analytic functions at their maximum and minimum points had a neutral slope of zero as the function moves passed the extreme point. [EVE76 pg 315]

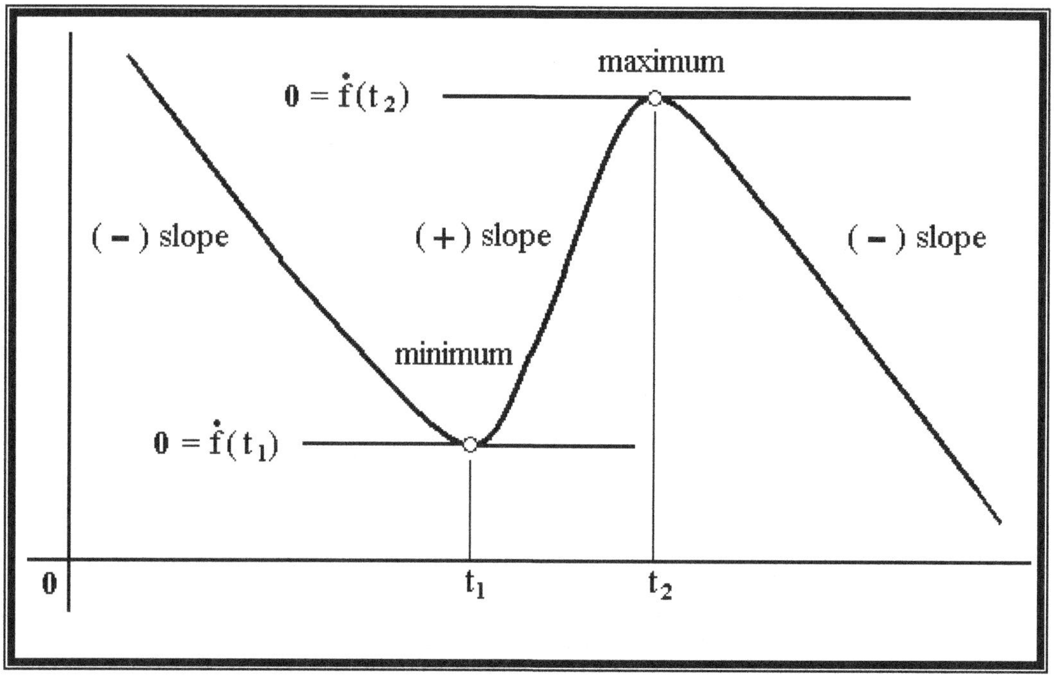

Figure 9.11.1 Zero Newtonian Slope at Maximum and Minimum Inflections

Definition 9.11.1 Inflection Point

A point on a curve is a point of inflection iff

$$f''(x_0) \equiv 0 \qquad \text{Equation A}$$

at the point $[x_0]$ where the graph is concave upward on one side and relatively convex downward on the other. This implies that between any two extrema there is a change in directed slope or an inflection.

Axiom 9.11.1 Inflection Points by Extrema

∃| a function f(x) such that it is continuous in a selected interval of $a \le x \le b$, so for any given pair of consecutive extrema there is one inflection point between them.

Theorem 9.11.1 Theorems of Extrema: Numeration of Inflection Points from Extrema

1g		$f'(x_i) = 0$ for i = 1, 2, …, N extremes
Step	Hypothesis	N % 2 = M inflection points
∴	From 1g and A9.10.1 Inflection Points by Extrema	N % 2 = M inflection points

Theorem 9.11.2 Theorems of Extrema: Numeration of Extrema from Inflection Points

1g		M = inflection points
Step	Hypothesis	2 x M = N extrema
∴	From 1g and A9.10.1 Inflection Points by Extrema	2 x M = N extrema

Section 9.12 Theorems that Bound Slope

Ordinary maximum and minimum slopes solve a lot of problems, but there are slopes that are not zero however a maximum and minimum exist between bounds. This requires a way of bounding the extreme, to find them "The Theorem of The Mean" provides such a way of achieving this.

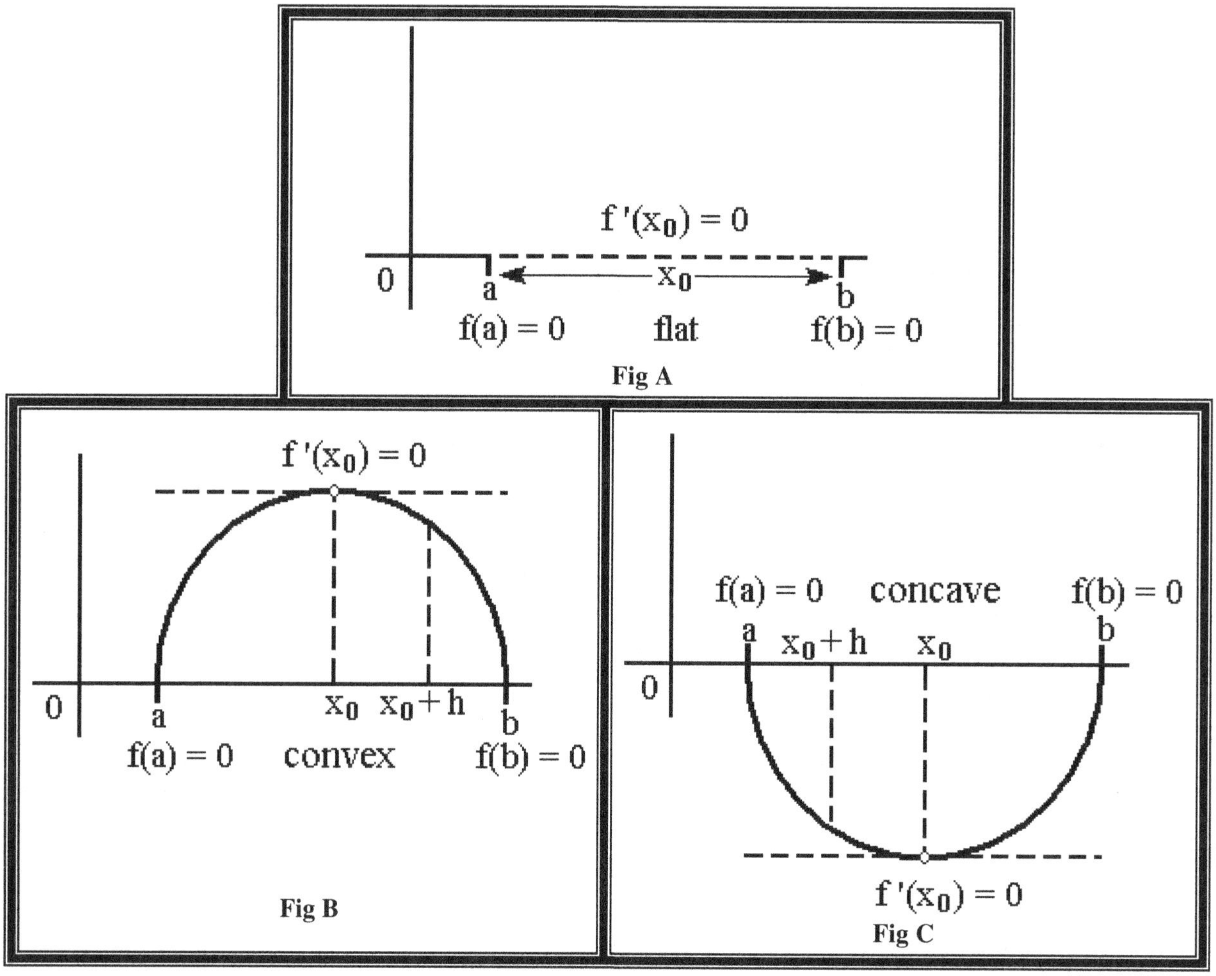

Figure 9.12.1 Rolle's Theorem for Flat, Convex and Concave Slope

Theorem 9.12.1 Theorems of Slope: Rolle's Theorem

1g	Given	From the geometry of Figures F9.12.1A "Rolle's Theorem for Flat, Convex and Concave Slope"
2g		From the geometry of Figures F9.12.1B "Rolle's Theorem for Flat, Convex and Concave Slope"
3g		From the geometry of Figures F9.12.1C "Rolle's Theorem for Flat, Convex and Concave Slope"
Step	Hypothesis	if [f] is continuous in the interval $a \le x \le b$ and $f(a) = f(b) = 0$ then $f'(x_0) = 0$ then there is at least one point that is an extrema in the interval.
1	CASE I: (trivial case) From 1g	$f(x) = 0$ in the interval $a \le x \le b$ then $f'(x_0) = 0$ for any [x] chosen to be x_0 within the interval.
		CASE II maximum; CASE III minimum
2	From 2g and 3g	$f(x_0 + h) \le f(x_0)$; $f(x_0 + h) \le f(x_0)$
3	From 2, A4.2.7 Identity Add, A4.2.5 Commutative Add and T4.3.3A Equalities: Right Cancellation by Addition,	$f(x_0 + h) - f(x_0) \le 0$; $f(x_0 + h) - f(x_0) \le 0$
4	From 3, T4.7.7 Inequalities: Multiplication by Positive Number and T4.7.8 Inequalities: Multiplication by Negative Number Reverses Order; for CASE II h > 0 and CASE II h < 0 maintaining inequality	$\dfrac{f(x_0 + h) - f(x_0)}{h} \le 0$; $\dfrac{f(x_0 + h) - f(x_0)}{h} \le 0$
5	From 4, Lx9.2.1.8 One-sided Limits; h > 0 limit moves toward the left and h < 0 toward the right.	$\lim_{h \to 0+} \dfrac{f(x_0 + h) - f(x_0)}{h} = 0$; $\lim_{h \to 0-} \dfrac{f(x_0 + h) - f(x_0)}{h} = 0$
6	From 5, Dx9.4.1.2 Definition of a Derivative as a Limit	$f'(x_0) = 0$; $f'(x_0) = 0$
∴	From 1 CASE I, and 6 CASE II and CASE III	if [f] is continuous in the interval $a \le x \le b$ and $f(a) = f(b) = 0$ then $f'(x_0) = 0$ then there is at least one point that is an extrema in the interval.

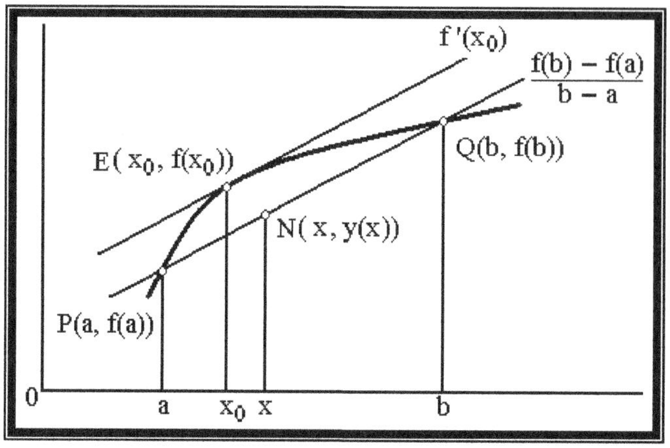

Figure 9.12.2 Theorem of the Mean

Theorem 9.12.2 Theorems of Slope: Theorem of the Mean

1g	Given	From the geometry of Figures F9.12.2 "Theorem of the Mean"
Step	Hypothesis	$f\,'(x_0) = [f(b) - f(a)] / (b - a)$
1	From 1g and T5.28.7 Three Point Straight Line	$[y(x) - f(a)] / (x - a) = [f(b) - f(a)] / (b - a)$
2	From 1 and T4.4.7B Equalities: Reversal of Left Cancellation by Multiplication	$y(x) - f(a) = \{[f(b) - f(a)] / (b - a)\}\ (x - a)$
3	From 2 and T4.3.14 Equalities: Equality by Difference Reversed	$y(x) - \{[f(b) - f(a)] / (b - a)\}\ (x - a) - f(a) = 0$
4	From 3, Lx9.3.1.7B Uniqueness Theorem of Differentiation, Lx9.3.2.2B Differential as a Linear Operator and Lx9.3.2.3B Distribute Constant in/out of Product	$y'(x) - \{[f(b) - f(a)] / (b - a)\}\ '(x - a) - 'f(a) = '0$
5	From 4, Theorem of Differentiation, Lx9.3.2.2B Differential as a Linear Operator, Lx9.3.3.1B Differential of a Constant and A4.2.7 Identity Add	$y'(x) - \{[f(b) - f(a)] / (b - a)\}\ ('x - 'a) = 0$
6	From 5, Lx9.3.3.1B Differential of a Constant and Lx9.3.3.2B Differential identity	$y'(x) - \{[f(b) - f(a)] / (b - a)\}\ (1 - 0) = 0$
7	From 6, A4.2.7 Identity Add and A4.2.12 Identity Multp	$y'(x) - [f(b) - f(a)] / (b - a) = 0$
8	From 7 and T4.3.13 Equalities: Equality by Difference	$y'(x) = [f(b) - f(a)] / (b - a)$
9	From 1, 8, T5.28.1 Constant Slope of the Three Point Straight Line and T9.11.1 Theorems of Slope: Rolle's Theorem; out of an infinite number of slopes in the interval only one will match.	$f\,'(x_0) = y'(x_0)$
∴	From 8, 9 and A4.2.3 Substitution	$f\,'(x_0) = [f(b) - f(a)] / (b - a)$

Section 9.13 Lagrange Functions with Constrained Extrema

Definition 9.13.1 Lagrange Function

The ***Lagrange function*** $L(\bullet\ x_i\ \bullet,\ \bullet\ \lambda_j\ \bullet)$ conjoins a schema function $f(\bullet\ x_i\ \bullet)$ under system constraints $\Phi_j(\bullet\ x_i\ \bullet)$ through a constant λ_j called the ***Lagrange multiplier***.

$$L(\bullet\ x_i\ \bullet,\ \bullet\ \lambda_j\ \bullet) = f(\bullet\ x_i\ \bullet) + \Sigma\ \lambda_j\ [b_j - \Phi_j(\bullet\ x_i\ \bullet)] \qquad \text{Equation A}$$

Newtonian solution of finding extrema is simple and straightforward and general without bounds, which is why it is called ***free extrema***.

Axiom 9.13.1 Free Extrema

For $F(\bullet\ x_{i0}\ \bullet)_{,j} = 0$ extrema for all i, without conditions, then there exists a ***critical point*** $P(\bullet\ x_{i0}\ \bullet)$ satisfying the free extrema.

Consider the function F(x, y, z, w) subject to the constraint condition w = f(x, y, z) a necessary condition to reduce the rank-4 F(x, y, z, w) to G(x, y, z) ≡ F[x, y, z, f(x, y, z)] of rank-3. Now when the generalizing the function G is differentiated using the chain rule, it to should be an extrema, but constrained to [w] when differentiated. [SPI63 pg 172]

$$0 = G' = F_x + F_w\ w_x + F_y + F_w\ w_y + F_z + F_w\ w_z \qquad \text{Equation A}$$

$$0 = G_x = F_x + F_w\ w_x \qquad \text{Equation B}$$
$$0 = G_y = F_y + F_w\ w_y \qquad \text{Equation C}$$
$$0 = G_z = F_z + F_w\ w_z \qquad \text{Equation D}$$

Equation A is not general for a function of n-variables, but its form lead Lagrange to consider a set of linear functions with constraints.

Given the set of constants $[b_j]$ constraining functions $[\Phi_j]$

$$b_j = \Phi_j(\bullet\ x_i\ \bullet) \qquad \text{Equation E}$$

on $f(\bullet\ x_i\ \bullet)$ and a set of corresponding conjoining constants $[\lambda_j]$ such that he could create a general function

$$L(\bullet\ x_i\ \bullet,\ \bullet\ \lambda_j\ \bullet) = f(\bullet\ x_i\ \bullet) + \Sigma\ \lambda_j\ [b_j - \Phi_j(\bullet\ x_i\ \bullet)] \qquad \text{Equation F}$$

corresponding to the differential Equation A, such that when differentiated gives a set of equations;

$$L(\bullet\ x_i\ \bullet,\ \bullet\ \lambda_j\ \bullet)_{,i} = f(\bullet\ x_i\ \bullet)_{,i} - \Sigma_j\ \lambda_j\ \Phi_j(\bullet\ x_i\ \bullet)_{,i} \qquad \text{Equation G}$$

where $G' \rightarrow L(\bullet\ x_i\ \bullet,\ \bullet\ \lambda_j\ \bullet)_{,i}$
$(F_x, F_y, F_z) \rightarrow f(\bullet\ x_i\ \bullet)_{,i}$ and
$(F_w\ w_x, F_w\ w_y, F_w\ w_z) \rightarrow \lambda_j\ \Phi_j(\bullet\ x_i\ \bullet)_{,i}$

Likewise as in axiom A9.12.1 "Free Extrema" it will also have a critical point $Q(\bullet\ x_{i0}\ \bullet)$, but resulting from the constrained conditions.

Theorem 9.13.1 Lagrange Function: Constrained Extrema

Step	Hypothesis	$\partial f(\bullet\, x_i \bullet) / \partial x_j = \sum_j \lambda_j \partial \Phi_j(\bullet\, x_i \bullet) / \partial x_j$
1	From D9.12.1 Lagrange Function and T9.3.2 Uniqueness of Differentials	$\partial L(\bullet\, x_i \bullet, \bullet\, \lambda_j \bullet) / \partial x_j = \partial\{f(\bullet\, x_i \bullet) + \sum_j \lambda_j [b_j - \Phi_j(\bullet\, x_i \bullet)]\}/ \partial x_j$
2	From 1, Lx9.3.2.2B Differential as a Linear Operator and Lx9.3.2.3B Distribute Constant in/out of Product	$\partial L(\bullet\, x_i \bullet, \bullet\, \lambda_j \bullet) / \partial x_j =$ $\partial f(\bullet\, x_i \bullet) / \partial x_j + \partial\{\sum_j \lambda_j [b_j - \Phi_j(\bullet\, x_i \bullet)]\}/ \partial x_j$
3	From 2, Lx9.3.2.2B Differential as a Linear Operator and Lx9.3.2.4B Chain Rule Across Products	$\partial L(\bullet\, x_i \bullet, \bullet\, \lambda_j \bullet) / \partial x_j =$ $\partial f(\bullet\, x_i \bullet) / \partial x_j + \sum_j \partial\lambda_j / \partial x_j [b_j - \Phi_j(\bullet\, x_i \bullet)]$ $\sum_j \lambda_j \partial [b_j - \Phi_j(\bullet\, x_i \bullet)] / \partial x_j$
4	From 3, Lx9.3.3.1B Differential of a Constant and Lx9.3.2.2B Differential as a Linear Operator	$\partial L(\bullet\, x_i \bullet, \bullet\, \lambda_j \bullet) / \partial x_j =$ $\partial f(\bullet\, x_i \bullet) / \partial x_j + \sum_j 0\, [b_j - \Phi_j(\bullet\, x_i \bullet)]$ $\sum_i \lambda_i [\partial b_i / \partial x_j - \partial\Phi_i(\bullet\, x_i \bullet) / \partial x_j]$
5	From 4, T4.4.1 Equalities: Any Quantity Multiplied by Zero is Zero, A4.2.7 Identity Add and Lx9.3.3.1B Differential of a Constant	$\partial L(\bullet\, x_i \bullet, \bullet\, \lambda_j \bullet) / \partial x_j =$ $\partial f(\bullet\, x_i \bullet) / \partial x_j + \sum_j \lambda_j [0 - \partial\Phi_j(\bullet\, x_i \bullet) / \partial x_j]$
6	From 5 and A4.2.7 Identity Add	$\partial L(\bullet\, x_i \bullet, \bullet\, \lambda_j \bullet) / \partial x_j = \partial f(\bullet\, x_i \bullet) / \partial x_j - \sum_j \lambda_j \partial \Phi_j(\bullet\, x_i \bullet) / \partial x_j$
7	From 6 and A9.12.1 Free Extrema	$0 = \partial f(\bullet\, x_i \bullet) / \partial x_j - \sum_j \lambda_j \partial \Phi_j(\bullet\, x_i \bullet) / \partial x_j$
∴	From 7, T4.3.4B Equalities: Reversal of Right Cancellation by Addition and T4.3.3B Equalities: Right Cancellation by Addition	$\partial f(\bullet\, x_i \bullet) / \partial x_j = \sum_j \lambda_j \partial \Phi_j(\bullet\, x_i \bullet) / \partial x_j$

Theorem 9.13.2 Lagrange Function: Constrained Extrema about Lagrange Multipliers

1g	Given	Now assume $i \neq j$, which implies $\lambda_i \neq f(\lambda_j)$ or $\lambda_i = c_x$ a constant, likewise hence (λ_i, λ_j) are independent of (i, j) [SPI59I pg 179-pr11]
Step	Hypothesis	$b_k = \Phi_k(\bullet\, x_i \bullet)$
1	From D9.12.1 Lagrange Function and T9.3.2 Uniqueness of Differentials	$\partial L(\bullet\, x_i \bullet, \bullet\, \lambda_j \bullet) / \partial \lambda_k =$ $\partial\{f(\bullet\, x_i \bullet) + \sum_j \lambda_j [b_j - \Phi_j(\bullet\, x_i \bullet)]\}/ \partial \lambda_k$
2	From 1, Lx9.3.2.2B Differential as a Linear Operator and Lx9.3.2.3B Distribute Constant in/out of Product	$\partial L(\bullet\, x_i \bullet, \bullet\, \lambda_j \bullet) / \partial \lambda_k =$ $\partial f(\bullet\, x_i \bullet) / \partial \lambda_k + \{\sum_j \lambda_j \partial [b_j - \Phi_j(\bullet\, x_i \bullet)]\}/ \partial \lambda_k$
3	From 2 and Lx9.3.2.2B Differential as a Linear Operator	$\partial L(\bullet\, x_i \bullet, \bullet\, \lambda_j \bullet) / \partial \lambda_k =$ $\partial f(\bullet\, x_i \bullet) / \partial \lambda_k + \sum_j \lambda_j[\partial b_j / \partial \lambda_k - \partial\Phi_j(\bullet\, x_i \bullet) / \partial \lambda_k] +$ $\sum_i \partial\lambda_i / \partial \lambda_k [b_i - \Phi_i(\bullet\, x_i \bullet)]$
4	From 1g, 3, Lx9.3.3.2B Differential identity, Lx9.3.3.1B Differential of a Constant and D6.1.6B Identity Matrix	$\partial L(\bullet\, x_i \bullet, \bullet\, \lambda_j \bullet) / \partial \lambda_k =$ $0 + \sum_j \lambda_j[0 - 0] + \sum_j \delta_j^{\ k} [b_j - \Phi_j(\bullet\, x_i \bullet)]$

5	From 4 and A4.2.7 Identity Add	$\partial L(\bullet\, x_i \bullet, \bullet\, \lambda_j \bullet) / \partial \lambda_k = \sum_i \delta_i^{\ k} [b_i - \Phi_i(\bullet\, x_i \bullet)]$
6	From 5 and D6.1.6B Identity Matrix; evaluating Kronecker Delta over the sum of j	$\partial L(\bullet\, x_i \bullet, \bullet\, \lambda_j \bullet) / \partial \lambda_k = b_k - \Phi_k(\bullet\, x_i \bullet)$
7	From 6 and A9.12.1 Free Extrema	$0 = b_k - \Phi_k(\bullet\, x_i \bullet)$
∴	From 7, T4.3.4B Equalities: Reversal of Right Cancellation by Addition and T4.3.3B Equalities: Right Cancellation by Addition	$b_k = \Phi_k(\bullet\, x_i \bullet)$

The differential interpretation of the Lagrange multiplier is given by:

Theorem 9.13.3 Lagrange Function: Multiplier as a Derivative

1g	Given	Now assume $i \neq j$, which implies $b_i \neq f(b_j)$ or $b_i = c_x$ a constant, likewise hence (b_i , b_j) are independent of (i, j) [SPI59I pg 179-pr11]
Step	Hypothesis	$\lambda_k = \partial L(\bullet\, x_i \bullet, \bullet\, \lambda_j \bullet) / \partial b_k$
1	From D9.12.1 Lagrange Function and T9.3.2 Uniqueness of Differentials	$\partial L(\bullet\, x_i \bullet, \bullet\, \lambda_j \bullet) / \partial b_k = \partial\{f(\bullet\, x_i \bullet) + \sum_j \lambda_j [b_j - \Phi_j(\bullet\, x_i \bullet)]\} / \partial b_k$
2	From 1, Lx9.3.2.2B Differential as a Linear Operator and Lx9.3.2.3B Distribute Constant in/out of Product	$\partial L(\bullet\, x_i \bullet, \bullet\, \lambda_j \bullet) / \partial b_k = \partial f(\bullet\, x_i \bullet) / \partial b_k + \sum_j \lambda_j \partial [b_k - \Phi_j(\bullet\, x_i \bullet)] / \partial b_k$
3	From 2 and Lx9.3.2.2B Differential as a Linear Operator	$\partial L(\bullet\, x_i \bullet, \bullet\, \lambda_j \bullet) / \partial b_k = \partial f(\bullet\, x_i \bullet) / \partial b_k + \sum_i \lambda_i [\partial b_i / \partial b_k - \partial \Phi_i(\bullet\, x_i \bullet) / \partial b_k]$
4	From 1g, 3, Lx9.3.3.2B Differential identity, Lx9.3.3.1B Differential of a Constant and D6.1.6B Identity Matrix	$\partial L(\bullet\, x_i \bullet, \bullet\, \lambda_j \bullet) / \partial b_k = 0 + \sum_j \delta_k^{\ j} \lambda_j [1 - 0]$
5	From 4, A4.2.7 Identity Add, and D6.1.6B Identity Matrix; evaluating Kronecker Delta over the sum of j	$\partial L(\bullet\, x_i \bullet, \bullet\, \lambda_j \bullet) / \partial b_k = \lambda_k$
∴	From 5 and A4.2.2B Equality	$\lambda_k = \partial L(\bullet\, x_i \bullet, \bullet\, \lambda_j \bullet) / \partial b_k$

Variables constructed from a Lagrange function for investigating problems on conditional extrema make it possible to obtain in a uniform way necessary optimality conditions in problems on conditional extrema.

In applied problems $L(\bullet\, x_i \bullet, \bullet\, \lambda_j \bullet)$ is often interpreted as profit or cost, and the right-hand sides, b_i, as losses of certain resources. Then the absolute value of λ_i is the ratio of the unit cost to the unit i^{th} resource. The numbers λ_i show how the maximum profit (or maximum cost) changes if the amount of the ith resource is increased by one. This interpretation of Lagrange multipliers can be extended to the case of constraints in the form of inequalities and to the case when the variables x_i are subject to the requirement of being non-negative.

In the calculus of variations one conveniently obtains by means of Lagrange multipliers necessary conditions for optimality in the problem on a conditional extrema as necessary conditions for an unconditional extrema of a certain composite functional. Lagrange multipliers in the calculus of variations are not constants, but certain functions. In the theory of optimal control and in the Pontryagin maximum principle, Lagrange multipliers have been called ***conjugate variables***.

References

[1] G.G. Hadley, "Nonlinear and dynamic programming", Addison-Wesley (1964)
[2] G.A. Bliss, "Lectures on the calculus of variations", Chicago Univ. Press (1947)
[3] A.E. Bryson, Y.-C. Ho, "Applied optimal control" , Blaisdell (1969)
[4] R.T. Rockafellar, "Convex analysis" , Princeton Univ. Press (1970)
This text originally appeared in Encyclopaedia of Mathematics - ISBN 1402006098

I.B. Vapnyarskii

Section 9.14 Calculus of Variation by Perturbation

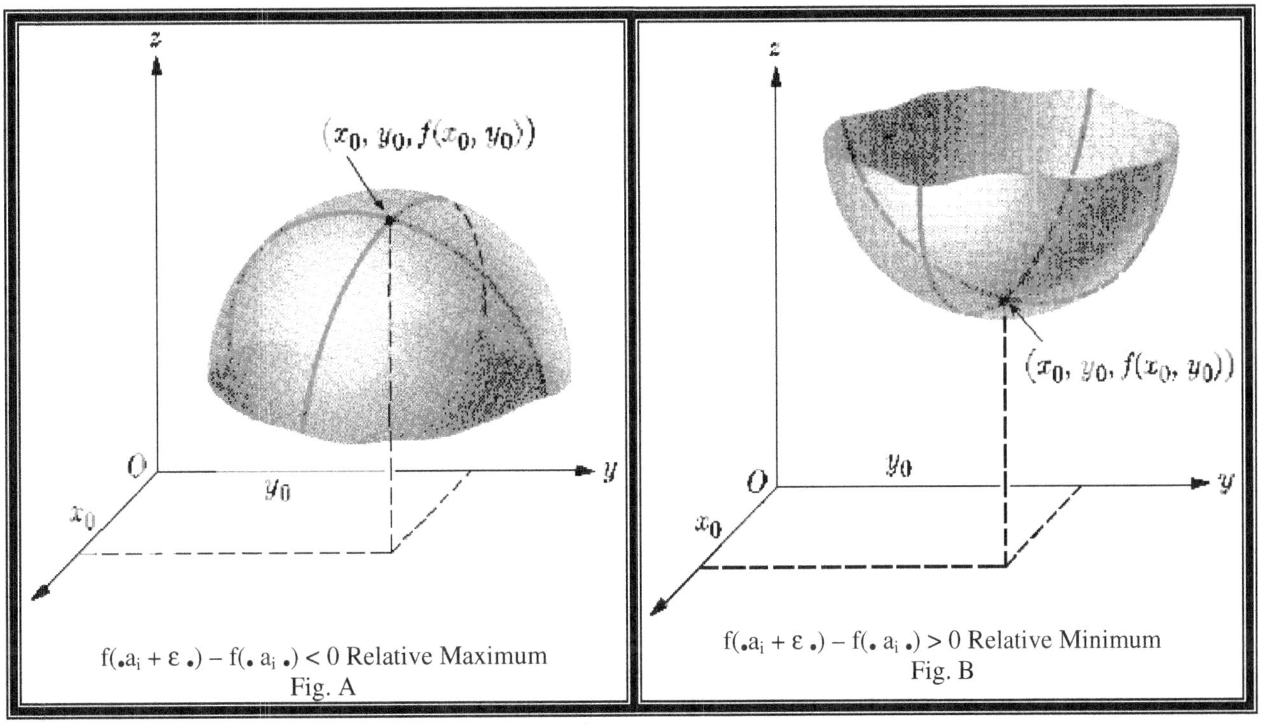

$f(\bullet a_i + \varepsilon \bullet) - f(\bullet a_i \bullet) < 0$ Relative Maximum
Fig. A

$f(\bullet a_i + \varepsilon \bullet) - f(\bullet a_i \bullet) > 0$ Relative Minimum
Fig. B

Figure 9.14.1 Relative Extrema

Theorem 9.14.1 Analytic Manifold $\mathfrak{M}^n$ Point: Talyor's Series Expanded About the Point

1g	Given	$\phi(\eta) \equiv f(\bullet\, x_i \bullet)$ at the point $A(\bullet\, a_i \bullet)$ within a radial distance δ having directional cosines μ_i for all [i].
Steps	Hypothesis	$f(\bullet\, x_i \bullet) = f(\bullet\, a_i \bullet) + (\sum_i f(\bullet\, a_i \bullet)_{,i}\, \Delta x_i) +$ $\frac{1}{2} (\sum_{\rho\in L2} f(\bullet\, a_i \bullet)_{,\rho(k)} \prod^2 \Delta x_{\lambda[i,k]}) + \frac{1}{6}\varepsilon^2$ HOT( )
1	From T11.8.6 Talyor's Series at a Point in/on a Manifold $\mathfrak{M}^n$	$f(\bullet\, x_i \bullet) = \sum_{m=0}^{3} (\sum_{\rho\in Lm} f(\bullet\, a_i \bullet)_{,\rho(k)} \prod^m \Delta x_{\lambda[i,k]}) / m!$
2	From 1g, 1, T11.8.5 Differential at a Point on or in a Manifold $\mathfrak{M}^n$, A4.2.3 Substitution and DE.3.2 HOT	$f(\bullet\, x_i \bullet) =$ $(\sum_{\rho\in L0} f(\bullet\, a_i \bullet)_{,\rho(k)} \prod^0 \Delta x_{\lambda[i,k]}) / 0! +$ $(\sum_{\rho\in L1} f(\bullet\, a_i \bullet)_{,\rho(k)} \prod^1 \Delta x_{\lambda[i,k]}) / 1! +$ $(\sum_{\rho\in L2} f(\bullet\, a_i \bullet)_{,\rho(k)} \prod^2 \Delta x_{\lambda[i,k]}) / 2! +$ HOT( )
∴	From 2, D6.5.1(A, B) Notation of Factorials; evaluate $\rho(k) \in L_k$ for k = 0, 1, 2	$f(\bullet\, x_i \bullet) = f(\bullet\, a_i \bullet) + [\sum_i f(\bullet\, a_i \bullet)_{,i}\, \Delta x_i] +$ $\frac{1}{2} [\sum_{\rho\in L2} f(\bullet\, a_i \bullet)_{,\rho(k)} \prod^2 \Delta x_{\lambda[i,k]}] + \frac{1}{6}\varepsilon^2$ HOT( )

Theorem 9.14.2 Analytic Manifold $\mathfrak{M}^n$ Point: Variation of Parameters

1g	Given	let P(• x_i •) → E(• a_i •) an extrema on the manifold
Steps	Hypothesis	$f(\bullet a_i + \varepsilon \bullet) - f(\bullet a_i \bullet) =$ $\varepsilon\sum_i f(\bullet a_i \bullet)_{,i} + ½\, \varepsilon^2 \sum_{\rho\in L2} f(\bullet a_i \bullet)_{,\rho(k)} + HOT(\varepsilon^3)$
1	From D11.8.1 Difference about a point A(• a_i •) and T9.13.1 Analytic Manifold $\mathfrak{M}^n$ Point: Talyor's Series Expanded About the Point; parameter perturbation by a small integer amount of ε[9.13.1]	$f(\bullet x_i + \varepsilon \bullet) = f(\bullet a_i \bullet) + (\sum_i f(\bullet a_i \bullet)_{,i} (x_i + \varepsilon - a_i)) +$ $½ (\sum_{\rho\in L2} f(\bullet a_i \bullet)_{,\rho(k)} \prod^2 x_{\lambda[i, k]} + \varepsilon - a_{\lambda[i, k]}) + HOT(\varepsilon^3)$
2	From 1g, 1 and A4.2.5 Commutative Add; as the point P(• x_i •) moves to the extrema of the manifold	$f(\bullet a_i + \varepsilon \bullet) = f(\bullet a_i \bullet) + (\sum_i f(\bullet a_i \bullet)_{,i} (a_i - a_i + \varepsilon)) +$ $½ (\sum_{\rho\in L2} f(\bullet a_i \bullet)_{,\rho(k)} \prod^2 a_{\lambda[i, k]} - a_{\lambda[i, k]} + \varepsilon) + HOT(\varepsilon^3)$
3	From 2, A4.2.8 Inverse Add, A4.2.7 Identity Add and A4.2.14 Distribution	$f(\bullet a_i + \varepsilon \bullet) = f(\bullet a_i \bullet) + \varepsilon\sum_i f(\bullet a_i \bullet)_{,i} +$ $½ (\sum_{\rho\in L2} f(\bullet a_i \bullet)_{,\rho(k)} \prod^2 \varepsilon) + HOT(\varepsilon^3)$
4	From 3, D4.1.17 Exponential Notation and A4.2.14 Distribution	$f(\bullet a_i + \varepsilon \bullet) = f(\bullet a_i \bullet) +$ $\varepsilon\sum_i f(\bullet a_i \bullet)_{,i} + ½\varepsilon^2 \sum_{\rho\in L2} f(\bullet a_i \bullet)_{,\rho(k)} + HOT(\varepsilon^3)$
∴	From 4 and T4.3.3A Equalities: Right Cancellation by Addition	$f(\bullet a_i + \varepsilon \bullet) - f(\bullet a_i \bullet) =$ $\varepsilon\sum_i f(\bullet a_i \bullet)_{,i} + ½\varepsilon^2 \sum_{\rho\in L2} f(\bullet a_i \bullet)_{,\rho(k)} + HOT(\varepsilon^3)$

Observation 9.14.1 Inflection Point Between Extrema

$f(\bullet a_i + \varepsilon \bullet) - f(\bullet a_i \bullet) < 0$	F9.14.1A maximum	$f''(\bullet a_i \bullet) < 0$	$f'(\bullet x_i \bullet)$ decreasing
$f(\bullet a_i + \varepsilon \bullet) - f(\bullet a_i \bullet) > 0$	F9.14.1B minimum	$f''(\bullet a_i \bullet) > 0$	$f'(\bullet x_i \bullet)$ increasing
$f(\bullet a_i + \varepsilon \bullet) - f(\bullet a_i \bullet) = 0$	no maximum and minimum	$f''(\bullet a_i \bullet) = 0$	$f'''(\bullet a_i \bullet) \neq 0$
$f(\bullet a_i + \varepsilon \bullet) = f(\bullet a_i \bullet)$	periodic or stationary for root values of $f'(\bullet x_{i0} \bullet)$ and periodic multiples of [ε]		

Theorem 9.14.3 Analytic Manifold $\mathfrak{M}^n$ Point: Variation of Scalar Gradient Yields Extrema

1g	Given	$\lim_{\varepsilon \to 0} \frac{1}{6}\varepsilon^2 \text{ HOT}(\varepsilon^3) = 0$
Steps	Hypothesis	$d\, f(\bullet\, a_i \bullet) / d\varepsilon = \sum_i f(\bullet\, a_i \bullet)_{,i}$
1	From T9.13.3 Analytic Manifold $\mathfrak{M}^n$ Point: Variation of Parameters	$f(\bullet a_i + \varepsilon \bullet) - f(\bullet\, a_i \bullet) =$ $\varepsilon \sum_i f(\bullet\, a_i \bullet)_{,i} + \frac{1}{2} \varepsilon^2 \sum_{\rho \in L2} f(\bullet\, a_i \bullet)_{,\rho(k)} + \frac{1}{6}\varepsilon^3 \text{ HOT}(\varepsilon^3)$
2	From 1, D4.1.17 Exponential Notation, A4.2.12 Identity Multp, A4.2.13 Inverse Multp and A4.2.14 Distribution	$f(\bullet\, a_i + \varepsilon \bullet - f(\bullet\, a_i \bullet) =$ $\varepsilon\, [\sum_i f(\bullet\, a_i \bullet)_{,i} + \frac{1}{2} \varepsilon \sum_{\rho \in L2} f(\bullet\, a_i \bullet)_{,\rho(k)} + \frac{1}{6}\varepsilon^2 \text{ HOT}(\varepsilon^3)]$
3	From 2 and T4.4.4A Equalities: Right Cancellation by Multiplication	$[f(\bullet\, a_i + \varepsilon \bullet - f(\bullet\, a_i \bullet)] / \varepsilon =$ $\sum_i f(\bullet\, a_i \bullet)_{,i} + \frac{1}{2} \varepsilon \sum_{\rho \in L2} f(\bullet\, a_i \bullet)_{,\rho(k)} + \frac{1}{6}\varepsilon^2 \text{ HOT}(\varepsilon^3)$
4	From 3, Lx9.2.1.1 Uniqueness of Limits and Lx9.2.1.4 Limit of the Sum; as $\varepsilon \to 0$	$\lim_{\varepsilon \to 0} [f(\bullet\, a_i + \varepsilon \bullet) - f(\bullet\, a_i \bullet)] / \varepsilon =$ $\lim_{\varepsilon \to 0} \sum_i f(\bullet\, a_i \bullet)_{,i} + \frac{1}{2} \lim_{\varepsilon \to 0} \varepsilon \sum_{\rho \in L2} f(\bullet\, a_i \bullet)_{,\rho(k)} +$ $\lim_{\varepsilon \to 0} \frac{1}{6}\varepsilon^2 \text{ HOT}(\varepsilon^3)$
5	From 1g, 4, Dx9.4.1.2 Definition of a Derivative as a Limit; variation with respect to the ith parameter; Lx9.2.1.2 Limit of a Constant and Lx9.2.1.3 Obvious Limit	$d\, f(\bullet\, a_i \bullet) / d\varepsilon = \sum_i f(\bullet\, a_i \bullet)_{,i} + \frac{1}{2}\, 0 \sum_{\rho \in L2} f(\bullet\, a_i \bullet)_{,\rho(k)} + 0$
∴	From 5, T4.4.1 Equalities: Any Quantity Multiplied by Zero is Zero and A4.2.7 Identity Add	$d\, f(\bullet\, a_i \bullet) / d\varepsilon = \sum_i f(\bullet\, a_i \bullet)_{,i}$

[9.13.1]Note: Perturbation methods lead to a smooth integer variation of parameters. [NAY73, pg 2]

Section 9.15 Weak Variations

From theorem T12.3.23 "Fundamental Quadratic: Covariant Integration over Arc Length" gives rise to a

$$\Delta s = \pm \int_{s_0}^{s} \sqrt{(\Sigma_i \Sigma_j g_{ij} x^{i'} x^{j'})}\, ds$$

integral line segment with unique interrelated properties bounding an enclosed region of space. A more general integral line function can be constructed that surrounds some internal locality described, by a multivariable function-F, in terms of arc-length, position and speed.

$$J \equiv \int_{s_0}^{s} F(s, \bullet x_i \bullet, \bullet x_i' \bullet)\, ds$$

Where [J] is called the ***cost function*** surrounding or bounding the manifold. It would follow that if the curvature path were optimized, then the internal manifold region function-F must also be optimal, hence enclosing the extrema of the system. So like the integrals of Green's theorems, which on the opposite sides of the equality relate unequal dimensions, in this case a being single to an n-dimensional system. Here things diverge from Green's theorems, because the manifold function-F is comprised of kinematic parameters, arc-length, position and velocity in order to describe interacting mechanics within the manifold. This then is the perfect tool to describe the extrema of kinematic physics for a manifold. Such a mathematical construct has unique properties in terms of extrema, which calculus of variations explores yielding solutions to problems in extrema that cannot be effectively solved in any other way.

The analysis starts by looking at the variation of the cost function as a constrained, single variation function.

Axiom 9.15.1 Cost Function

Let ∃| a manifold $\mathfrak{B}^n$ that is bounded by a line integral path-C that surrounds some internal locality described, by a multivariable function-F that is continuously differentiable at any given point within the region. It is constructed with a kinematic parameter set of [s] ***trajectory length***, [$\bullet$ x_i $\bullet$] ***positional vector*** and [$\bullet$ x_i' $\bullet$] ***velocity vector*** and [x_i'] velocity component is a measure of speed.

$$J = \int_{s_0}^{s} F(s, \bullet x_i \bullet, \bullet x_i' \bullet)\, ds \qquad \text{Equation A}$$

Theorem 9.15.1 Weak Variations: Variation About Point $P(\bullet\, a_i \bullet)$ on Manifold $\mathfrak{B}^n$

1g	Given	let $P(\bullet\, x_i \bullet) \rightarrow E(\bullet\, a_i \bullet)$ an extrema on the manifold
2g		$J_a \equiv \int_{s_0}^{s} F\, ds$
3g		$J_{,i} \equiv \int_{s_0}^{s} (F_{,ai}\, \xi_i + F_{,ai'}\, \xi_i')\, ds$
4g		$J_{,\rho(k)} \equiv \int_{s_0}^{s} (F_{,\rho(k)}\, \xi_{\rho(k)} + F_{,\rho(k)'}\, \xi_{\rho(k)}')\, ds$
5g		$hot \equiv \int_{s_0}^{s} HOT(\varepsilon^3)\, ds$
Steps	Hypothesis	$J(\varepsilon) = J_a + \varepsilon \sum_i J_{,i} + ½\, \varepsilon^2 \sum_{\rho\in L2} J_{,\rho(k)} + ⅙\varepsilon^3\, hot$
1	From A9.14.1 Cost Function and D11.8.1 Difference about a point $A(\bullet\, a_i \bullet)$; with a directed parameter that deviates by a small amount of $\varepsilon\xi_i$ [9.13.1]	$J(\varepsilon) = \int_{s_0}^{s} F(s, \bullet\, x_i + \varepsilon\xi_i \bullet, \bullet\, (x_i + \varepsilon\xi_i)' \bullet)\, ds$
2	From 1 and Lx9.3.2.2A Differential as a Linear Operator	$J(\varepsilon) = \int_{s_0}^{s} F(s, \bullet\, x_i + \varepsilon\xi_i \bullet, \bullet\, x_i' + \varepsilon\xi_i' \bullet)\, ds$
3	From 2; the integrand; T9.13.1 Analytic Manifold $\mathfrak{M}^n$ Point: Talyor's Series Expanded About the Point	$F(s, \bullet\, x_i + \varepsilon\xi_i \bullet, \bullet\, x_i' + \varepsilon\xi_i' \bullet) =$ $F(s, \bullet\, a_i \bullet, \bullet\, a_i')$ $+ [\sum_i F(s, \bullet\, a_i \bullet, \bullet\, a_i')_{,xi}\, (x_i + \varepsilon\xi_i - a_i)\,] +$ $+ [\sum_i F(s, \bullet\, a_i \bullet, \bullet\, a_i')_{,xi'}\, (x_i' + \varepsilon\xi_i' - a_i')\,] +$ $½\, [\sum_{\rho\in L2} F(s, \bullet\, a_i \bullet, \bullet\, a_i')_{,\rho(k)} \prod^2 x_{\lambda[i,k]} + \varepsilon\xi_i - a_{\lambda[i,k]}\,] +$ $½\, [\sum_{\rho\in L2} F(s, \bullet\, a_i \bullet, \bullet\, a_i')_{,\rho(k)'} \prod^2 x_{\lambda[i,k]}' + \varepsilon\xi_i' - a_{\lambda[i,k]}'\,] +$ $⅙\varepsilon^3\, HOT(\varepsilon^3)$
4	From 1g; at the point $E(\bullet\, a_i \bullet)$	$F(s, \bullet\, a_i + \varepsilon\xi_i \bullet, \bullet\, a_i' + \varepsilon\xi_i' \bullet) =$ $F(s, \bullet\, a_i \bullet, \bullet\, a_i') +$ $[\sum_i F(s, \bullet\, a_i \bullet, \bullet\, a_i')_{,ai}\, (a_i + \varepsilon\xi_i - a_i)\,] +$ $[\sum_i F(s, \bullet\, a_i \bullet, \bullet\, a_i')_{,ai'}\, (a_i' + \varepsilon\xi_i' - a_i')\,] +$ $½\, [\sum_{\rho\in L2} F(s, \bullet\, a_i \bullet, \bullet\, a_i')_{,\rho(k)} \prod^2 a_{\lambda[i,k]} + \varepsilon\xi_i - a_{\lambda[i,k]}\,] +$ $½\, [\sum_{\rho\in L2} F(s, \bullet\, a_i \bullet, \bullet\, a_i')_{,\rho(k)'} \prod^2 a_{\lambda[i,k]}' + \varepsilon\xi_i' - a_{\lambda[i,k]}'\,] +$ $⅙\varepsilon^3\, HOT(\varepsilon^3)$
5	From 4 and A4.2.5 Commutative Add	$F(s, \bullet\, a_i + \varepsilon\xi_i \bullet, \bullet\, a_i' + \varepsilon\xi_i' \bullet) =$ $F(s, \bullet\, a_i \bullet, \bullet\, a_i') +$ $+ [\sum_i F(s, \bullet\, a_i \bullet, \bullet\, a_i')_{,ai}\, (a_i - a_i + \varepsilon\xi_i)\,] +$ $+ [\sum_i F(s, \bullet\, a_i \bullet, \bullet\, a_i')_{,ai'}\, (a_i' - a_i' + \varepsilon\xi_i')\,] +$ $½\, [\sum_{\rho\in L2} F(s, \bullet\, a_i \bullet, \bullet\, a_i')_{,\rho(k)} \prod^2 a_{\lambda[i,k]} - a_{\lambda[i,k]} + \varepsilon\xi_i] +$ $½\, [\sum_{\rho\in L2} F(s, \bullet\, a_i \bullet, \bullet\, a_i')_{,\rho(k)'} \prod^2 a_{\lambda[i,k]}' - a_{\lambda[i,k]}' + \varepsilon\xi_i'] +$ $⅙\varepsilon^3\, HOT(\varepsilon^3)$
6	From 5, A4.2.8 Inverse Add, A4.2.7 Identity Add and D4.1.17 Exponential Notation	$F(s, \bullet\, a_i + \varepsilon\xi_i \bullet, \bullet\, a_i' + \varepsilon\xi_i' \bullet) =$ $F(s, \bullet\, a_i \bullet, \bullet\, a_i') +$ $+ [\sum_i F(s, \bullet\, a_i \bullet, \bullet\, a_i')_{,ai}\, \varepsilon\xi_i\,] +$ $+ [\sum_i F(s, \bullet\, a_i \bullet, \bullet\, a_i')_{,ai'}\, \varepsilon\xi_i'\,] +$ $½\, [\sum_{\rho\in L2} F(s, \bullet\, a_i \bullet, \bullet\, a_i')_{,\rho(k)}\, \varepsilon^2\, \xi_{\rho(k)}] +$ $½\, [\sum_{\rho\in L2} F(s, \bullet\, a_i \bullet, \bullet\, a_i')_{,\rho(k)'}\, \varepsilon^2\, \xi_{\rho(k)}'] +$ $⅙\varepsilon^3\, HOT(\varepsilon^3)$

7	From 6 and A4.2.14 Distribution	$F(s, \bullet\, a_i + \varepsilon\xi_i \bullet, \bullet\, a_i' + \varepsilon\xi_i' \bullet) =$ $F(s, \bullet\, a_i \bullet, \bullet\, a_i') +$ $+ \varepsilon \sum_i [F(s, \bullet\, a_i \bullet, \bullet\, a_i')_{,ai}\, \xi_i + F(s, \bullet\, a_i \bullet, \bullet\, a_i')_{,ai'}\, \xi_i'] +$ $\frac{1}{2} \varepsilon^2 \sum_{\rho\in L2} [F(s, \bullet\, a_i \bullet, \bullet\, a_i')_{,\rho(k)}\, \xi_{\rho(k)} + F(s, \bullet\, a_i \bullet, \bullet\, a_i')_{,\rho(k)'}\, \xi_{\rho(k)}'] +$ $\frac{1}{6}\varepsilon^3\, HOT(\varepsilon^3)$
8	From 2, 7 and A4.2.3 Substitution	$J(\varepsilon) = \int_{s_0}^{s} \{F(s, \bullet\, a_i \bullet, \bullet\, a_i') +$ $\varepsilon \sum_i [F(s, \bullet\, a_i \bullet, \bullet\, a_i')_{,ai}\, \xi_i + F(s, \bullet\, a_i \bullet, \bullet\, a_i')_{,ai'}\, \xi_i'] +$ $\frac{1}{2} \varepsilon^2 \sum_{\rho\in L2} [F(s, \bullet\, a_i \bullet, \bullet\, a_i')_{,\rho(k)}\, \xi_{\rho(k)} + F(s, \bullet\, a_i \bullet, \bullet\, a_i')_{,\rho(k)'}\, \xi_{\rho(k)}'] +$ $\frac{1}{6}\varepsilon^3\, HOT(\varepsilon^3) \}\, ds$
9	From 8 and Lx9.3.1.4 Integrable over the addition of integrands	$J(\varepsilon) = \int_{s_0}^{s} F\, ds +$ $\varepsilon[\sum_i \int_{s_0}^{s} (F_{,ai}\, \xi_i + F_{,ai'}\, \xi_i')]\, ds +$ $\frac{1}{2} \varepsilon^2[\sum_{\rho\in L2} \int_{s_0}^{s} (F_{,\rho(k)}\, \xi_{\rho(k)} + F_{,\rho(k)'}\, \xi_{\rho(k)}')]\, ds +$ $\frac{1}{6}\varepsilon^3 \int_{s_0}^{s} HOT(\varepsilon^3)\, ds$
∴	From 2g, 3g, 4g, 5g, 10 and A4.2.3 Substitution	$J(\varepsilon) = J_a + \varepsilon\sum_i J_{,i} + \frac{1}{2} \varepsilon^2 \sum_{\rho\in L2} J_{,\rho(k)} + \frac{1}{6}\varepsilon^3\ hot$

Theorem 9.15.2 Weak Variations: Averaged Cost Function

1g	Given	$< J(\varepsilon) > \equiv J(\varepsilon) - J_a$
Steps	Hypothesis	$< J(\varepsilon) > = \varepsilon\sum_i J_{,i} + \frac{1}{2} \varepsilon^2 \sum_{\rho\in L2} J_{,\rho(k)} + \frac{1}{6}\varepsilon^3\ hot$
1	From T9.14.1 Weak Variations: Variation About Point P(• a_i •) on Manifold $\mathfrak{B}^n$	$J(\varepsilon) = J_a + \varepsilon\sum_i J_{,i} + \frac{1}{2} \varepsilon^2 \sum_{\rho\in L2} J_{,\rho(k)} + \frac{1}{6}\varepsilon^3\ hot$
2	From 1 and T4.3.3A Equalities: Right Cancellation by Addition	$J(\varepsilon) - J_a = \varepsilon\sum_i J_{,i} + \frac{1}{2} \varepsilon^2 \sum_{\rho\in L2} J_{,\rho(k)} + \frac{1}{6}\varepsilon^3\ hot$
∴	From 1g, 2 and A4.2.3 Substitution	$< J(\varepsilon) > = \varepsilon\sum_i J_{,i} + \frac{1}{2} \varepsilon^2 \sum_{\rho\in L2} J_{,\rho(k)} + \frac{1}{6}\varepsilon^3\ hot$

Theorem 9.15.3 Weak Variations: Extrema over Cost Function

1g	Given	$0 \equiv < J(\varepsilon) >$
2g		$0 \neq \varepsilon$
Steps	Hypothesis	$0 = \sum_i J_{,i}$
1	From T9.14.2 Weak Variations: Averaged Cost Function	$< J(\varepsilon) > = \varepsilon\sum_i J_{,i} + \frac{1}{2} \varepsilon^2 \sum_{\rho\in L2} J_{,\rho(k)} + \frac{1}{6}\varepsilon^3\ hot$
2	From 1g, 1 and A4.2.3 Substitution	$0 = \varepsilon\sum_i J_{,i} + \frac{1}{2} \varepsilon^2 \sum_{\rho\in L2} J_{,\rho(k)} + \frac{1}{6}\varepsilon^3\ hot$
∴	From 2g, 2; for every [ε] term	$0 = \sum_i J_{,i}$

Theorem 9.15.4 Weak Variations: Euler's Function For Multiple Arguments

Steps	Hypothesis	$J_{,i} = F_{,ai'}\xi_i \mid^s_{s0} + \int_{s_0}^{s} \{F_{,ai} - [d(F_{,ai'}) / ds\ ]\}\ \xi_i\ ds$
1	From T9.14.1 Weak Variations: Variation About Point P(• a_i •) on Manifold $\mathfrak{B}^n$, step-3g	$J_{,i} = \int_{s_0}^{s} (F_{,ai}\ \xi_i + F_{,ai'}\xi_i')\ ds$
2	From 1, Lx9.3.1.4 Integrable over the addition of integrands	$J_{,i} = \int_{s_0}^{s} F_{,ai}\ \xi_i\ ds + \int_{s_0}^{s} F_{,ai'}\xi_i'\ ds$
3	From 2, A4.2.11 Associative Multp, Lx9.3.1.11 Integration by parts	$\int_{s_0}^{s} F_{,ai'}(\xi_i'\ ds) = \int u\ dv$
4	From 3; equating integrand factors	$u = F_{,ai'}$
5	From 4 and Lx9.3.1.7D Uniqueness Theorem of Differentiation	$du = d(F_{,ai'})$
6	From 3; equating integrand factors	$dv = \xi_i'\ ds$
7	From 6, Dx9.3.2.1G First Parametric Differentiation, A4.2.13 Inverse Multp and A4.2.12 Identity Multp	$dv = d\ \xi_i$
8	From 7 and T9.4.2 Integrals: Uniqueness of Integration	$v = \xi_i$
9	From 4, 5, 7, 8, Lx9.3.1.11 Integration by parts and A4.2.3 Substitution	$\int_{s_0}^{s} F_{,ai'}(\xi_i'\ ds) = F_{,ai'}\xi_i \mid^s_{s0} - \int_{s_0}^{s} \xi_i\ d(F_{,ai'})$
10	From 9, A4.2.12 Identity Multp, A4.2.13 Inverse Multp and Lx9.3.1.7C Uniqueness Theorem of Differentiation	$\int_{s_0}^{s} F_{,ai'}\ \xi_i'\ ds = F_{,ai'}\xi_i \mid^s_{s0} - \int_{s_0}^{s} \xi_i\ [d(F_{,ai'}) / ds\ ]\ ds$
11	From 3, 10 and Lx9.3.1.4 Integrable over the addition of integrands	$J_{,i} = F_{,ai'}\xi_i \mid^s_{s0} + \int_{s_0}^{s} \xi_i\ F_{,ai}\ ds - \xi_i\ [d(F_{,ai'}) / ds\ ]\ ds$
∴	From 11 and A4.2.14 Distribution	$J_{,i} = F_{,ai'}\xi_i \mid^s_{s0} + \int_{s_0}^{s} \{F_{,ai} - [d(F_{,ai'}) / ds\ ]\}\ \xi_i\ ds$

Theorem 9.15.5 Weak Variations: Euler's Function For Stationary End Points

1g	Given ξ_i for all [i] are arbitrary and stationary for weak variations, which satisfy the differential equation, hence it can be said that they vanish equally at the end points of the interval	$\xi_i(s) = \xi_i(s_0) = 0$
2g		$\xi_i(s) \neq 0$ for $s_0 < \sigma < s$
Steps	Hypothesis	$0 = F_{,ai} - [d(F_{,ai'}) / ds\ ]$ for all [i]
1	From T9.14.3 Weak Variations: Extrema over Cost Function	$0 = \sum_i J_{,i}$
2	From 1, T9.14.4 Weak Variations: Euler's Function For Multiple Arguments and A4.2.3 Substitution; summation over the addition	$0 = \sum_i F_{,ai'}\xi_i \mid^s_{s0} + \sum_i \int_{s_0}^{s} \{F_{,ai} - [d(F_{,ai'}) / ds\ ]\}\ \xi_i\ ds$
3	From 1 and 2; stationary terms are zero	$0 = \sum_i F_{,ai'}\xi_i(s) - F_{,ai'}\xi_i(s_0)$
4	From 1g, 3 and A4.2.3 Substitution	$0 = \sum_i F_{,ai'}0 - F_{,ai'}0$
5	From 4, T4.4.1 Equalities: Any Quantity Multiplied by Zero is Zero and A4.2.7 Identity Add	$0 = 0$

6	From 2, 5 and A4.2.7 Identity Add	$0 = \sum_i \int_{s_0}^{s} \{F_{,ai} - [d(F_{,ai'}) / ds]\} \xi_i \, ds$
7	From 6; term-by-term	$0 = \int_{s_0}^{s} \{F_{,ai} - [d(F_{,ai'}) / ds]\} \xi_i \, ds$
∴	From 2g and 7	$0 = F_{,ai} - [d(F_{,ai'}) / ds]$ for all [i]

Theorem 9.15.6 Weak Variations: Euler's Exact Differential Function

1g	Given	let $E(\bullet\, a_i \bullet) \rightarrow P(\bullet\, x_i \bullet)$ an extrema on the manifold
Steps	Hypothesis	$0 = F_{xi} - (F_{xi'xj})(dx_j / ds) - (F_{xi'xj'})(dx_j' / ds) - (F_{xi's})$
1	From T9.3.2 Uniqueness of Differentials	$d(F_{xi'}) / ds = d(F_{xi'}) / ds$
2	From 1 and T9.3.8 Multi-variable Chain Rule	$d(F_{xi'}) / ds = (F_{xi'xj})(dx_j / ds) + (F_{xi'xj'})(dx_j' / ds) + (F_{xi's})(ds / ds)$
3	From 2, Lx9.3.3.2A Differential identity and A4.2.12 Identity Multp	$d(F_{xi'}) / ds = (F_{xi'xj})(dx_j / ds) + (F_{xi'xj'})(dx_j' / ds) + (F_{xi's})$
4	From 3, T9.14.5 Weak Variations: Euler's Function For Stationary End Points	$0 = F_{ai} - d(F_{ai'}) / ds$
5	From 1g, 4; at the point $P(\bullet\, x_i \bullet)$	$0 = F_{xi} - d(F_{xi'}) / ds$
∴	From 3, 5 and A4.2.3 Substitution	$0 = F_{xi} - (F_{xi'xj})(dx_j / ds) - (F_{xi'xj'})(dx_j' / ds) - (F_{xi's})$

Tensor Calculus &Physics: A General Treatise

Section 9.16 Strong Variation Legendre Test

Tensor Calculus & Physics: A General Treatise

Table of Contents

List of Tables

List of Figures

List of Observations

Chapter 10 Tensor Algebra

Section 10.1 Tensor Algebra: Definitions

With tensors, the concept of an n-tuple is recast to a polyadic a grouping of many or n-objects. As seen from Chapter 1 a tensor is a mathematical construction that describes many attributes having a common theme. So, from optics a calcite crystal can have indices of refraction in one direction being ordinary and different indices in an off direction being extraordinary. Set theory lends itself nicely to creating just such a mathematic artifice as found in Descartes' Cartesian product of sets. Since each axial field-set can contain unique and distinct family of elements different than the other axes this is one of the requirements that constitutes a tensor. If these sets are bound to each other through a common theme than this satisfies our notation of what a tensor is. Binding these axes in a superset would satisfy this second condition.

Definition 10.1.1 A Tensor Set

$\overset{*}{A}^m \equiv \{\Pi_{j=1}{}^m\ \mathbf{A}_j\ \}$ iff the axial field sets are $\mathbf{A}_j \neq \mathbf{A}_j$ and $i \neq j$ yet the sets A_i has a property p_i along the axis and are related by a common theme.

The asterisk placed over the letter denotes a tensor, the asterisks symbolizing multiple directions for a bundle of vectors.

Definition 10.1.2 Tensor[10.1.1]

A ***tensor*** is a collection of vectors that are uniquely bound together by geometry, a common theme, and treated as a single mathematical quantity. As can be seen in Figure 1.5.2 "Three Physical examples of Tensors" the vectors comprising a tensor are packets of individual vectors, with unique individual characteristics, a set of topic attributes collectively grouped together. Formally, we say:

Let $\exists$ a set of m-vector quantities representing different attributes (**.** p_i **.**) called a ***tensor set or quiver*** (a packet or case to hold multiple quantities of the same theme) iff they are members of an order product tensor set T_m.

Definition 10.1.3 Tensor Analysis

Tensor Analysis is the study of tensors, their geometry, algebra and calculus.
From the development of definition D1.7.1L "Set Summation Notation over Long Indices", the product vectors can be broken down in the tensor summation and the terms into to ***scalar coefficient*** and ***base vector*** components:

$$\prod^m A_{i\lambda[i,k]}\mathbf{b}_{i\lambda[i,\,k]} \equiv \prod^m A_{i\lambda[i,\,k]} \prod^m \mathbf{b}_{i\lambda[i,\,k]} \text{ for } k = 1, 2, \ldots, r \text{ and } r = n^m$$

[10.1.1]Note: It is not mentioned here that a tensor is defined in terms of transformation from one coordinate system to another. In many books on tensors the authors define a tensors' existence and what they are in terms of their ability to transform, that is not what makes a tensor and such definitions are 100% wrong! The ability to transform from one coordinate system to another is a property of manifolds. Transformations are only meaningful between manifolds, not between objects; a tensor is an object not a manifold, which will be demonstrated later on in this paper. All of this came about because tensors being made up of sets of vectors, which are invariant (see axiom A7.8.7 "Parallelogram Law: Affine Transformation of Vectors"), must also be invariant under transformation. This is an attribute of the manifold space to maintain affine relationships with the vectors embedded in them; it is not inclusive of what makes a tensor, a tensor.

Definition 10.1.4 Polyadic Coefficient of a Tensor

The product $[\prod^{m} a_{i\lambda[i, k]}]$, the coefficient cluster, is called the ***polyadic coefficient*** or ***polyadic*** for short. The name for a multi-adic becomes poly-adic a grouping of many objects. [OXF71, Vol. I, pg 821 see Dyad or Dyadic]

Table 10.1.1 Tensor Polyadic Types

Name[10.1.2]	Definition
Definition 10.1.5 Dyadic	2-element set such that $(\bullet\, a_{i\lambda[i, j]} \bullet) \in \overset{*}{\mathring{A}}^{2}$ for $i = 1, 2$
Definition 10.1.6 Triadic	3-element set such that $(\bullet\, a_{i\lambda[i, j]} \bullet) \in \overset{*}{\mathring{A}}^{3}$ for $i = 1, 2, 3$
Definition 10.1.7 Quadadic	4-element set such that $(\bullet\, a_{i\lambda[i, j]} \bullet) \in \overset{*}{\mathring{A}}^{4}$ for $i = 1, 2, 3, 4$
Definition 10.1.8 Penadic	5-element set such that $(\bullet\, a_{i\lambda[i, j]} \bullet) \in \overset{*}{\mathring{A}}^{5}$ for $i = 1, 2, 3, 4, 5$
Definition 10.1.9 Hexadic	6-element set such that $(\bullet\, a_{i\lambda[i, j]} \bullet) \in \overset{*}{\mathring{A}}^{6}$ for $i = 1, 2, 3, 4, 5, 6$
Definition 10.1.10 Heptadic	7-element set such that $(\bullet\, a_{i\lambda[i, j]} \bullet) \in \overset{*}{\mathring{A}}^{7}$ for $i = 1, 2, 3, 4, 5, 6, 7$
Definition 10.1.11 Octadic	8-element set such that $(\bullet\, a_{i\lambda[i, j]} \bullet) \in \overset{*}{\mathring{A}}^{8}$ for $i = 1, 2, 3, 4, 5, 6, 7, 8$
Definition 10.1.12 Nonadic	9-element set such that $(\bullet\, a_{i\lambda[i, j]} \bullet) \in \overset{*}{\mathring{A}}^{9}$ for $i = 1, 2, 3, 4, 5, 6, 7, 8, 9$
Definition 10.1.13 Decadic	10-element set such that $(\bullet\, a_{i\lambda[i, j]} \bullet) \in \overset{*}{\mathring{A}}^{10}$ for $i = 1, 2, 3, 4, 5, 6, 7, 8, 9, 10$
Definition 10.1.14 Hendecadic	11-element set such that $(\bullet\, a_{i\lambda[i, j]} \bullet) \in \overset{*}{\mathring{A}}^{11}$ for $i = 1, 2, 3, 4, 5, 6, 7, 8, 9, 10, 11$
Definition 10.1.15 Dodecadic or duodeciadic	12-element set such that $(\bullet\, a_{i\lambda[i, j]} \bullet) \in \overset{*}{\mathring{A}}^{12}$ for $i = 1, 2, 3, 4, 5, 6, 7, 8, 9, 10, 11, 12$

[10.1.2]Note: The origins of tuple names lead with a Greek numerical prefix name.

Definition 10.1.16 Polyadic Base Tensor

$\prod^m \mathbf{b}_{i\lambda[i,\,k]}$ the product of base vectors is called the ***polyadic base tensor*** or ***base tensor*** for short.

Definition 10.1.17 Unity Tensor of Rank-0

$\overset{*}{U}_0 \equiv 1$ for m = 0

Definition 10.1.18 Zero Tensor

$\overset{*}{Z}_m = \sum_{\rho(k)\in Ln} \prod^m Z_{i\lambda[i,\,k]} \prod^m \mathbf{b}_{i\lambda[i,\,k]}$ where Equation A

$Z_{i\lambda[i,\,k]} \equiv 0$ while $m \sim\vee= 0$ Equation B

Section 10.2 Tensor Algebra: Axioms

A tensor can be expressed as defined from definition D2.19.1 "A Tensor Set". What is required however is a more formal systematic development as rendered in the following axiomatic steps.

Axiom 10.2.1 Tensor of Rank–m

Let ∃| a set of m-vector quantities $(\bullet, \mathbf{V}_{k}, \bullet)$ representing different attributes $(\bullet\ p_k\ \bullet)$ that belonging to a Tensor Set $[\overset{*}{\mathrm{V}}_m \in \mathrm{T}_m]$, having the following relationships and properties.

P1) A tensor is embedded in some type of n-dimensional space, such as a, see D7.6.2 "A Cartesian coordinate n-Space, or Riemannian n-Space".

$\mathcal{S}^n \equiv \{\prod_n \times C_i\}$ for	Product Set for n-Axes	Equation A
$C_i = \{\ x^i \mid x^i \in (\mathbf{R} \vee \mathbf{C}^2)\ \}$	Axial Set	Equation B

The tensor takes its name from the space that it is embedded in and defines the magnitude of its component vectors, such as a Cartesian or Euclidian Tensor, or a Riemannian Tensor.

P2) Associated at every point in that space are a set of ***dual base vectors***, $(\bullet\ \mathbf{b}_i \bullet)$ spanning the space as construction components to build vectors from, the dual type of vector is called ***covariant***. There is another orthogonal set of basis vectors to them $(\bullet\ \mathbf{b}^i \bullet)$ called ***contravariant***.

In this section, since covariant is orthogonal to contravariant and mirror one another, they will be considered to be the same, hence any rules for covariant are rules for contravariant and will be treated equally. However in the sections to follow on tensor calculus it will be seen this is not necessarily the case, and then they will have to be considered separately. In the interim this co- and contravariant notation will be suppressed.

$\mathrm{B}^n \equiv \{\ \mathcal{S}^n, (\bullet\ \mathbf{b}_i \bullet), (\bullet\ \mathbf{b}^i \bullet) \mid$ where for all i, $(\mathbf{b}_i \rightarrow x_i\ ) \wedge (\ \theta_{ij} \sim\vee= \pi/2$ for $i \neq j) \wedge (\ |\mathbf{b}_i| \sim\vee= 1)$ likewise
$(\mathbf{b}^i \rightarrow x_i\ ) \wedge (\ \theta^{ij} = \theta_{ij} + \pi/2) \wedge (\ |\mathbf{b}^i| \sim\vee= 1)\ \}$

P3) There exist uniquely a set, or quiver, of vectors $(\ \bullet\ \mathbf{V}_k\ \bullet\ )$ constructed from the set of basis vectors spanning the space.

$\mathbf{V}_k \equiv \sum_i \mathrm{V}_{\lambda[k,i]}\ \mathbf{b}_{\lambda[k,i]}$	spanning the space $\mathcal{S}^n$	Equation A
$\mathrm{Q}_\mathbf{b}^{\ n} \equiv \{\ \mathcal{S}^n, \mathrm{B}^n \mid \mathrm{B}^n \in \mathcal{S}^n\ \}$	quiver set of bases vectors	Equation B

P4) Associated with every vector along its shaft (axis) is a single specific attribute or property-p.

$\mathrm{Q}_{pm} \equiv \{\ p_k, \mathbf{V}_k \mid$ where for all k, $(\ p_k \rightarrow \mathbf{V}_k\ )\ \}$ — Quiver Set of Properties — Equation A

P5) A tensor is a simple mathematical construction. Following the product of sets for a set-tensor, it is simply the product of all the vectors. There is no regard to any operation on the basis's vectors comprising them, and distributed accordingly for every term.

$\overset{*}{\mathrm{V}}_m \equiv \prod_{k=1}^{m} \mathbf{V}_k$ — Tensor as a vector product — Equation A

P6) Bring all of these properties together into a single set called the ***tensor set***,

$\mathrm{T}_m \equiv \{\ \mathrm{B}^n, \mathrm{Q}_\mathbf{b}^{\ n}, \mathrm{Q}_{pm}, \overset{*}{\mathrm{V}}_m \mid \overset{*}{\mathrm{V}}_m \in \mathcal{S}^n, \mathrm{Q}_\mathbf{b}^{\ n} \in \overset{*}{\mathrm{V}}_m, \mathrm{Q}_{pm} \rightarrow \overset{*}{\mathrm{V}}_m\ \}$ — Tensor Set — Equation A

P7) $\mathcal{V}_M \equiv \{\ \mathrm{T}_{m1}, \ldots, \mathrm{T}_{mi}, \ldots, \mathrm{T}_{mM}\ \}$ — Tensor Superset — Equation A

P8) A tensor defines magnitude of a vector by Pythagorean length

$|\mathbf{V}_k| = |\ \mathbf{X}_k - \mathbf{A}_k\ | \equiv \sqrt{\sum_i (\Delta x_{ki})^2}$

Here, [m] is called the ***rank of the tensor set*** or simply the ***rank of the tensor***. Observe that it is not necessary for [m] to have any dependency on the size of n-space, since it is not a set of bases vectors; in fact it can be more or less than the space it is embedded in. Of course, for [m = 1] the quiver has only one vector in it, hence a vector is a tensor and for [m = 0] by definition is a scalar.

Also, tensor space is not necessarily a field. For it to be field, the axioms on tensors would have to follow the pattern of a Galois Group. As found in Chapter 7 on vectors those axioms do follow that form, but here on tensors their axioms do not, but the theorems do, just because tensors are based on sets of vectors its does not necessarily follow that they are configured as a group, hence they cannot be considered a field.

Axiom 10.2.2 Rank is a Countable Number

Rank-m is a number that counts the number of elements in a tensor set and as a countable number

$0 \le m$ for D4.1.1 "Positive Integers [+n]: Natural or Countable numbers"

Correspondence between Vectors and Tensors

Axiom 10.2.3 Tensor of Rank-0 is a Scalar

A tensor of rank-0 is a scalar so [$\overset{*}{T}_0 \equiv s$], such that [$s \in (R \vee C^2)$].

Axiom 10.2.4 Tensor of Rank-1 is a Vector

A tensor of rank-1 is a vector so [$\overset{*}{T}_1 \equiv \vec{V}$].

Axiom 10.2.5 Tensor as a Number has a Reciprocal Product

If $\overset{*}{A}_m / \overset{*}{B}_p$ and $p \le m$ then $\mathcal{B}_P \subseteq \mathcal{A}_M$

Axiom 10.2.6 Equality of Tensors

$\overset{*}{A}_m = \overset{*}{B}_r$ Equation A

$\overset{*}{B}_r = \overset{*}{A}_m$ Equation B

iff $\mathbf{A}_i \leftrightarrow \mathbf{B}_k$, $\mathbf{A}_i = \mathbf{B}_k$ and $p_i = q_k$ then $m = r$ and $i = k$ Equation C

Observation 10.2.1 The Importance of Directivity of the Base Tensor

Table 10.2.1 Commuting the Product of Base Vectors with Coefficients

	a_x	i	a_y	j	a_z	k
1	a_x	a_y	i	j	a_z	k
2	a_x	a_y	i	a_z	j	k
3	a_x	a_y	a_z	i	j	k

Commuting base vectors with the vector coefficients leaves an isolated base tensor (lower right hand corner of the above table), a couple of things to observe:

1) The order of the base vectors are left unaltered after a binary combinatorial series of commutative operations.
2) The base tensor provides a sense of ***direction*** or ***directivity*** for each term of the tensor.
3) The base vectors are not coupled in any way that would hamper commutative operation to occur.

If order were not maintained, then directivity would not hold and the base tensor as a whole **could not** be distributed within a tensor, and certain theorems on tensors, could not be proven. These ideas give rise to the following axiom:

Axiom 10.2.7 Base Tensor Directivity Invariant Under Binary Commutation

Base Tensor Directivity is invariant under binary commutation for independent base tensors and as such they are invariant under free binary commutation.

Axiom 10.2.8 Tensor Closure under Addition

$(\overset{*}{A}_m + \overset{*}{B}_m) \wedge (\overset{*}{B}_m + \overset{*}{A}_m) \in T_m$

Axiom 10.2.9 Tensor Closure under Subtraction

$(\overset{*}{A}_m - \overset{*}{B}_m) \wedge (\overset{*}{B}_m - \overset{*}{A}_m) \in T_m$

Axiom 10.2.10 Tensor Closure under Multiplication

$(\overset{*}{A}_m \overset{*}{B}_r) \wedge (\overset{*}{B}_r \overset{*}{A}_m) \in T_{m+r}$

Axiom 10.2.11 Tensor Closure under Reciprocal Product

$(\overset{*}{A}_m / \overset{*}{B}_r) \in T_{m-r}$ for $r \leq m$

Axiom 10.2.12 Distribution of Vectors over Vectors without a Polyadic Operator [10.2.1]

$\mathbf{A}\,(\mathbf{B} + \mathbf{C}) = \mathbf{A\,B} + \mathbf{A\,C}$ for $(\mathbf{A\,B}, \mathbf{A\,C}) \in T_2$

Axiom 10.2.13 Symmetric-Reflexive Rank of a Tensor

Rank-m = Rank-n and — Equation A
Rank-n = Rank-m — Equation B
must always balance for any tensor equation.

Axiom 10.2.14 Correspondence of Tensors and Tensor Coefficients

For the interrelation of the tensor to its coefficient $[\overset{*}{A}_m = \sum_{\rho(k)\in Lm} \prod^m A_{i\lambda[i,\,k]}\, \mathbf{b}_{i\lambda[i,\,k]}]$, then any proof on a tensor is a proof on its coefficient and conversely as long as the polyadic is independent of coefficient in the proof.

Axiom 10.2.15 Associate Scalar with Tensor Group

$\kappa\,(\overset{*}{A}_p \overset{*}{B}_q) = (\kappa \overset{*}{A}_p)\,\overset{*}{B}_q$

[10.2.1]Note: Axiom A10.2.12 "Distribution of Vectors over Vectors without a Polyadic Operator" must hold under axiom A10.2.8 "Tensor Closure under Addition".

Tensor Algebra: Tools for Constructing Tensor Theorems

There is a one-to-one correspondence between the algebra of Multiplication Sets of Chapter 2 and the algebra of tensors. As a result, proofs can be interchanged between the two disciplines as long as it is understood the differences exist and appropriate justifications made for existences.

Table 10.2.2 Correspondence Between Multiplication Sets and Tensors

	Axiom	Equation-MS	Axiom	Equation-Tensor
Product	A2.19.1A Existence of Product and Operator	$A^m \equiv \prod^m \lozenge A_i$	A10.2.1P5ATensor Quantity	$\overset{*}{V}_m \equiv \Pi_{j=1}{}^m V_j$
Set	A2.19.1B Existence of Product and Operator	$\mathcal{A}^m \equiv \{ A^m \}$	A10.2.1P7 Tensor Quantity	$\mathcal{V}_M \equiv \{\overset{*}{V}_m\}$
Rank	A2.19.1C Existence of Product and Operator	$0 < m$	A10.2.2 Rank is a Countable Number	$0 \le m$
Equality	A2.19.2 Equality	$A^m = B^m$ Equation A $B^m = A^m$ Equation B iff $A_i = B_i$ for all i	A10.2.6A Equality of Tensors	$\overset{*}{A}_m = \overset{*}{B}_m$ for $A_i = B_i$ and all i
Substitution	T2.21.1 Multiplication Sets: Substitution for Rank-m	$A^m = B^m$ $B^m = C^m$ $A^m = C^m$	T7.9.2 Substitution of Vectors	$A_i = B_i$ $B_i = C_i$ $A_i = C_i$
Commutative	A2.19.5 Commutative	$A_i \lozenge B_i = B_i \lozenge A_i$ for all i	A7.8.5 Parallelogram Law: Commutative and Scalability of Collinear Vectors	$a A = A a$
Reciprocal	A2.19.6 Set as a Number Reciprocal	A^m / B^p iff $p < m$ and $\mathcal{B}^p \subseteq \mathcal{A}^m$	A10.4.9 Product of Tensors: Reduction of Tensors by Subtraction of Rank-m–q	$\overset{*}{A}_m / \overset{*}{C}_q = \overset{*}{B}_p$ for $m - q = p \ge 0$ and $\mathcal{C}_Q \subseteq \mathcal{A}_M$
Identity	A2.19.7 Identity	$1 \lozenge A_i = A_i$ for any i	T10.3.10 Tool: One Times a Tensor is a Tensor	$1\overset{*}{A}_m = \overset{*}{A}_m$
Inverse	A2.19.8 Inverse	$A_i / A_i = 1$ iff $A_i \cong A_i$ for all i	T10.4.10 Product of Tensors: Reduction of Tensors to Identity of Rank-m–m	$\overset{*}{A}_m / \overset{*}{A}_m = 1$ for $z = 0$ and $\mathcal{A}_M = \mathcal{A}_M$

Theorem 10.2.1 Tool: Tensor Series Expansion

1g	Given	$\overset{*}{A}_m \in T_m$
Steps	Hypothesis	$\overset{*}{A}_m = \sum_{\rho(k)\in Lm} \prod^m A_{i\lambda[i,k]}\, \mathbf{b}_{i\lambda[i,k]}$ expansion from
1	From 1g and A10.2.1P5A Tensor of Rank–m	$\overset{*}{A}_m = \Pi_{i=1}{}^m\, \mathbf{A}_i$
∴	From 1 and D1.7.1L Set Summation Notation over Long Indices	$\overset{*}{A}_m = \sum_{\rho(k)\in Lm} \prod^m A_{i\lambda[i,k]}\, \mathbf{b}_{i\lambda[i,k]}$ expansion from

Theorem 10.2.2 Tool: Tensor Series Contraction

1g 2g	Given	$\overset{*}{A}_m \in T_m$ $\overset{*}{A}_m = \sum_{\rho(k)\in Lm} \prod^m A_{i\lambda[i,k]}\, \mathbf{b}_{i\lambda[i,k]}$
Steps	Hypothesis	$\overset{*}{A}_m = \overset{*}{A}^m$ contracted from $\overset{*}{A}_m = \sum_{\rho(k)\in Lm} \prod^m A_{i\lambda[i,k]}\, \mathbf{b}_{i\lambda[i,k]}$
1 2	From 2g From 1 and D1.7.1L Set Summation Notation over Long Indices	$\overset{*}{A}_m = \sum_{\rho(k)\in Lm} \prod^m A_{i\lambda[i,k]}\, \mathbf{b}_{i\lambda[i,k]}$ $\overset{*}{A}_m = \Pi_{i=1}{}^m\, A_i$
∴	From 1 and A10.2.1P5A Tensor of Rank–m	$\overset{*}{A}_m = \overset{*}{A}_m$ contracted from $\overset{*}{A}_m = \sum_{\rho(k)\in Lm} \prod^m A_{i\lambda[i,k]}\, \mathbf{b}_{i\lambda[i,k]}$

Theorem 10.2.3 Tool: Base Tensor Parsed from Polyadic Coefficient

Steps	Hypothesis	$\sum_{\rho(k)\in Lm} \prod^m A_{i\lambda[i,k]}\, \mathbf{b}_{i\lambda[i,k]} =$ $\sum_{\rho(k)\in Lm} \prod^m A_{i\lambda[i,k]} \prod^m \mathbf{b}_{i\lambda[i,k]}$
1	From A10.2.6A Equality of Tensors	$\sum_{\rho(k)\in Lm} \prod^m A_{i\lambda[i,k]}\, \mathbf{b}_{i\lambda[i,k]} = \sum_{\rho(k)\in Lm} \prod^m A_{i\lambda[i,k]}\, \mathbf{b}_{i\lambda[i,k]}$
∴	From 1, T2.21.10 Multiplicative Sets: Binary Distribution of Rank and A10.2.7 Base Tensor Directivity Invariant Under Binary Commutation	$\sum_{\rho(k)\in Lm} \prod^m A_{i\lambda[i,k]}\, \mathbf{b}_{i\lambda[i,k]} =$ $\sum_{\rho(k)\in Lm} \prod^m A_{i\lambda[i,k]} \prod^m \mathbf{b}_{i\lambda[i,k]}$

Theorem 10.2.4 Tool: Base Tensor Distributed over Polyadic Coefficient Summation

Steps	Hypothesis	$\sum_{\rho(k)\in Lm} (\prod^m A_{i\lambda[i,k]}\mathbf{b}_{\lambda[i,k]} + \prod^m B_{i\lambda[i,k]}\mathbf{b}_{\lambda[i,k]}) =$ $\sum_{\rho(k)\in Lm} (\prod^m A_{i\lambda[i,k]} + \prod^m b_{i\lambda[i,k]}) \prod^m \mathbf{b}_{\lambda[i,k]}$
1	From A10.2.6A Equality of Tensors and D1.7.1L Set Summation Notation over Long Indices	$\sum_{\rho(k)\in Lm} (\prod^m A_{i\lambda[i,k]}\mathbf{b}_{\lambda[i,k]} + \prod^m B_{i\lambda[i,k]}\mathbf{b}_{\lambda[i,k]}) =$ $\sum_{\rho(k)\in Lm} (\prod^m A_{i\lambda[i,k]}\mathbf{b}_{\lambda[i,k]} + \prod^m B_{i\lambda[i,k]}\mathbf{b}_{\lambda[i,k]})$
∴	From 1, T10.2.3 Tool: Base Tensor Parsed from Polyadic Coefficient and T7.10.9 Distribution of Vector over Addition of Scalars	$\sum_{\rho(k)\in Lm} (\prod^m A_{i\lambda[i,k]}\mathbf{b}_{\lambda[i,k]} + \prod^m B_{i\lambda[i,k]}\mathbf{b}_{\lambda[i,k]}) =$ $\sum_{\rho(k)\in Lm} (\prod^m A_{i\lambda[i,k]} + \prod^m B_{i\lambda[i,k]}) \prod^m \mathbf{b}_{\lambda[i,k]}$

Theorem 10.2.5 Tool: Tensor Substitution

1g	Given	$\overset{*}{A}_m = \overset{*}{B}_p$
2g		$\overset{*}{B}_p = \overset{*}{C}_q$
Steps	Hypothesis	$\overset{*}{A}_m = \overset{*}{C}_q$ for m = q
1	From 1g and A10.2.1P5A Tensor of Rank–m	$\overset{*}{A}_m = \Pi_{i=1}{}^m \mathbf{A}_i$
2	From 1g and A10.2.1P5A Tensor of Rank–m	$\overset{*}{B}_p = \Pi_{j=1}{}^p \mathbf{B}_j$
3	From 2g and A10.2.1P5A Tensor of Rank–m	$\overset{*}{C}_q = \Pi_{k=1}{}^q \mathbf{C}_k$
4	From 1 and A10.2.6A Equality of Tensors	$\mathbf{A}_i \leftrightarrow \mathbf{B}_j$, $\mathbf{A}_i = \mathbf{B}_j$ and $p_i = q_j$, m = p and i = j
5	From 2 and A10.2.6A Equality of Tensors	$\mathbf{B}_j \leftrightarrow \mathbf{C}_k$, $\mathbf{B}_j = \mathbf{C}_k$ and $q_j = r_k$, p = q and j = k
6	From 4, 5, A7.9.2 Substitution of Vectors and A4.2.3 Substitution	$\mathbf{A}_i \leftrightarrow \mathbf{C}_k$, $\mathbf{A}_i = \mathbf{C}_k$ and $p_i = r_k$, m = q and i = k
∴	From 6 and A10.2.1P5A Tensor of Rank–m	$\overset{*}{A}_m = \overset{*}{C}_q$ for m = q

Theorem 10.2.6 Tool: Commutative Scalar Times a Tensor

Steps	Hypothesis	$a\,\overset{*}{A}_m = \overset{*}{A}_m\,a$	for any given vector of [m]
1	From A10.2.6 Equality of Tensors	$\overset{*}{A}_m = \overset{*}{A}_m$	
2	From 1 and T7.10.1 Uniqueness of Scalar Multiplication to Vectors	$a\,\overset{*}{A}_m = a\,\overset{*}{A}_m$	
3	From 2 and A10.2.1P5A Tensor of Rank–m	$a\,\overset{*}{A}_m = a \prod^m \mathbf{A}_i$	
4	From 3 and A7.8.5 Parallelogram Law: Commutative and Scalability of Collinear Vectors	$a\,\overset{*}{A}_m = \prod^m \mathbf{A}_i\,a$	
∴	From 4 and A7.8.5 Parallelogram Law: Commutative and Scalability of Collinear Vectors	$a\,\overset{*}{A}_m = \overset{*}{A}_m\,a$	for any given vector of [m]

Theorem 10.2.7 Tool: Zero Times a Tensor is a Zero Tensor

Steps	Hypothesis	$0\,\overset{*}{A}_m = \overset{*}{Z}_m$ EQ A	$\overset{*}{A}_m\,0 = \overset{*}{Z}_m$ EQ B
1	From A10.2.6 Equality of Tensors	$\overset{*}{A}_m = \overset{*}{A}_m$	
2	From 1 andT7.10.1 Uniqueness of Scalar Multiplication to Vectors	$0\,\overset{*}{A}_m = 0\,\overset{*}{A}_m$	
3	From 2, T10.2.1 Tool: Tensor Series Expansion	$0\,\overset{*}{A}_m = 0\,\Sigma_{\rho(k)\in Lm}\prod^m A_{i\lambda[i,k]}\,\mathbf{b}_{i\lambda[i,k]}$	
4	From 3, T10.2.3 Tool: Base Tensor Parsed from Polyadic Coefficient and T7.10.3 Distribution of a Scalar over Addition of Vectors	$0\,\overset{*}{A}_m = \Sigma_{\rho(k)\in Lm}(0\prod^m A_{i\lambda[i,k]})\prod^m \mathbf{b}_{i\lambda[i,k]}$	
5	From 4 and T4.8.5B Integer Exponents: Zero Raised to the Positive Power-n	$0\,\overset{*}{A}_m = \Sigma_{\rho(k)\in Lm}(\prod^m 0\prod^m A_{i\lambda[i,k]})\prod^m \mathbf{b}_{i\lambda[i,k]}$	
6	From 5 and T2.21.10 Multiplicative Sets: Binary Distribution of Rank	$0\,\overset{*}{A}_m = \Sigma_{\rho(k)\in Lm}(\prod^m 0\,A_{i\lambda[i,k]})\prod^m \mathbf{b}_{i\lambda[i,k]}$	
7	From 5, T4.4.1 Equalities: Any Quantity Multiplied by Zero is Zero and D10.1.18B Zero Tensor	$0\,\overset{*}{A}_m = \Sigma_{\rho(k)\in Lm}\prod^m Z_{i\lambda[i,k]}\prod^m \mathbf{b}_{i\lambda[i,k]}$	
∴	From 5, T10.2.5 Tool: Tensor Substitution and T10.2.6 Tool: Commutative Scalar Times a Tensor	$0\,\overset{*}{A}_m = \overset{*}{Z}_m$ EQ A	$\overset{*}{A}_m\,0 = \overset{*}{Z}_m$ EQ B

Theorem 10.2.8 Tool: Zero Tensor Times a Tensor is a Zero Tensor

Steps	Hypothesis	$\overset{*}{Z}_m\,\overset{*}{A}_q = \overset{*}{Z}_m$
1	From A10.2.6 Equality of Tensors	$\overset{*}{Z}_m\,\overset{*}{A}_q = \overset{*}{Z}_m\,\overset{*}{A}_q$
2	From 1, T10.2.1 Tool: Tensor Series Expansion and A10.2.6 Equality of Tensors	$\overset{*}{Z}_m\,\overset{*}{A}_q = (\Sigma_{\rho(k)\in Ln}\prod^m Z_{i\lambda[i,k]}\prod^m \mathbf{b}_{\lambda[i,k]})(\Sigma_{\rho(r)\in Lq}\prod^q A_{i\lambda[i,r]}\prod^q \mathbf{b}_{i\lambda[i,r]})$
3	From 2 and A10.2.12 Distribution of Vectors over Vectors without a Polyadic Operator	$\overset{*}{Z}_m\,\overset{*}{A}_q = \Sigma_{\rho(k)\in Ln}\Sigma_{\rho(r)\in L}\prod^m Z_{i\lambda[i,k]}\prod^q A_{i\lambda[i,r]}\prod^m \mathbf{b}_{\lambda[i,k]}\prod^q \mathbf{b}_{i\lambda[i,r]}$
4	From 3 and T2.21.10 Multiplicative Sets: Binary Distribution of Rank	$\overset{*}{Z}_m\,\overset{*}{A}_q = \Sigma_{\rho(k)\in Ln}\Sigma_{\rho(r)\in L}\prod^m Z_{i\lambda[i,k]}\prod^q A_{i\lambda[i,r]}\prod^m \mathbf{b}_{\lambda[i,k]}\prod^q \mathbf{b}_{i\lambda[i,r]}$
5	From 4 and D10.1.13B Zero Tensor	$\overset{*}{Z}_m\,\overset{*}{A}_q = \Sigma_{\rho(k)\in Ln}\Sigma_{\rho(r)\in L}\prod^m 0\prod^q A_{i\lambda[i,r]}\prod^m \mathbf{b}_{\lambda[i,k]}\prod^q \mathbf{b}_{i\lambda[i,r]}$
6	From 5, T4.4.1 Equalities: Any Quantity Multiplied by Zero is Zero and D10.1.13B Zero Tensor	$\overset{*}{Z}_m\,\overset{*}{A}_q = \Sigma_{\rho(r)\in Ln}\prod^m Z_{i\lambda[i,r]}\prod^m \mathbf{b}_{i\lambda[i,r]}$
∴	From 6 and D10.1.18A Zero Tensor	$\overset{*}{Z}_m\,\overset{*}{A}_q = \overset{*}{Z}_m$

Theorem 10.2.9 Tool: Zero Tensor Plus a Zero Tensor is a Zero Tensor

1g	Given	$\overset{*}{Z}_m = \overset{*}{Z}_m + \overset{*}{Z}_m$ by assumption
Steps	Hypothesis	$\overset{*}{Z}_m = \overset{*}{Z}_m + \overset{*}{Z}_m$
1	From 1g	$\overset{*}{Z}_m = \overset{*}{Z}_m + \overset{*}{Z}_m$
2	From 1, D10.1.18A Zero Tensor and T10.2.5 Tool: Tensor Substitution	$\overset{*}{Z}_m = \sum_{\rho(r)\in Ln} \prod^m Z_{i\lambda[i,\, r]} \prod^m \mathbf{b}_{i\lambda[i,\, r]} + \sum_{\rho(r)\in Ln} \prod^m Z_{i\lambda[i,\, r]} \prod^m \mathbf{b}_{i\lambda[i,\, r]}$
3	From 1, A10.2.6A Equality of Tensors, and combining like terms	$\overset{*}{Z}_m = \sum_{\rho(r)\in Ln} (\prod^m Z_{i\lambda[i,\, r]} \prod^m \mathbf{b}_{i\lambda[i,\, r]} + \prod^m Z_{i\lambda[i,\, r]} \prod^m \mathbf{b}_{i\lambda[i,\, r]})$
4	From 2 and T10.2.4 Tool: Base Tensor Distributed over Polyadic Coefficient Summation	$\overset{*}{Z}_m = \sum_{\rho(r)\in Ln} (\prod^m Z_{i\lambda[i,\, r]} + \prod^m Z_{i\lambda[i,\, r]}) \prod^m \mathbf{b}_{i\lambda[i,\, r]}$
5	From 3 and D10.1.18B Zero Tensor	$\overset{*}{Z}_m = \sum_{\rho(r)\in Ln} [\prod^m (0) + \prod^m (0)] \prod^m \mathbf{b}_{i\lambda[i,\, r]}$
6	From 4 and T4.8.5 (A,B) Integer Exponents: Zero Raised to the Positive Power-n	$\overset{*}{Z}_m = \sum_{\rho(r)\in Ln} (0 + 0) \prod^m \mathbf{b}_{i\lambda[i,\, r]}$
7	From 5 and D4.1.19 Primitive Definition for Rational Arithmetic	$\overset{*}{Z}_m = \sum_{\rho(r)\in Ln} (0) \prod^m \mathbf{b}_{i\lambda[i,\, r]}$
8	From 6 and T4.8.5 (A,B) Integer Exponents: Zero Raised to the Positive Power-n	$\overset{*}{Z}_m = \sum_{\rho(r)\in Ln} (\prod^m 0) \prod^m \mathbf{b}_{i\lambda[i,\, r]}$
9	From 7 and D10.1.18B Zero Tensor	$\overset{*}{Z}_m = \sum_{\rho(r)\in Ln} \prod^m Z_{i\lambda[i,\, r]} \prod^m \mathbf{b}_{i\lambda[i,\, r]}$
10	From 8 and D10.1.18A Zero Tensor	$\overset{*}{Z}_m = \overset{*}{Z}_m$
∴	From 1, 10 and by identity	$\overset{*}{Z}_m = \overset{*}{Z}_m + \overset{*}{Z}_m$

Theorem 10.2.10 Tool: Identity Multiplication of a Tensor

Steps	Hypothesis	$1\overset{*}{A}_m = \overset{*}{A}_m$ EQ A	$\overset{*}{A}_m 1 = \overset{*}{A}_m$ EQ B
1	From A10.2.6A Equality of Tensors	$1\overset{*}{A}_m = 1\overset{*}{A}_m$	
2	From 1 and A10.2.1P5A Tensor Quantity	$1\overset{*}{A}_m = 1\ \Pi_{k=1}{}^m\ A_k$	
3	From 2, T4.8.1 Integer Exponents: Unity Raised to any Integer Value and A4.2.3 Substitution	$1\overset{*}{A}_m = \Pi_{k=1}{}^m\ 1\ \Pi_{k=1}{}^m\ A_k$	
4	From 3 and T2.21.10 Multiplicative Sets: Binary Distribution of Rank	$1\overset{*}{A}_m = \Pi_{k=1}{}^m\ 1\ A_k$	
5	From 4 and T7.10.5 Identity with Scalar Multiplication to Vectors	$1\overset{*}{A}_m = \Pi_{k=1}{}^m\ A_k$	
∴	From 5, A10.2.1AP5 Tensor of Rank-m and T10.2.6 Tool: Commutative Scalar Times a Tensor	$1\overset{*}{A}_m = \overset{*}{A}_m$ EQ A	$\overset{*}{A}_m 1 = \overset{*}{A}_m$ EQ B

Section 10.3 Tensor Algebra: Theorems on Algebra

Theorem 10.3.1 Product of Tensors: Parsing a Tensor into of Rank for p and m – p

1g	Given	$\overset{*}{A}_m = \overset{*}{A}_m$
Steps	Hypothesis	$\overset{*}{A}_m = \overset{*}{A}_p \overset{*}{B}_{m-p}$ for $\mathbf{B}_s = \mathbf{A}_{k-p}$ and $s = k - p$
1	From 1g and A10.2.1P5A Tensor of Rank–m	$\overset{*}{A}_m = \Pi_{i=1}{}^m \mathbf{A}_i$
2	From 1 and parse-p vector	$\overset{*}{A}_m = \Pi_{j=1}{}^p \mathbf{A}_i \Pi_{k=p+1}{}^m \mathbf{A}_k$
3	From 2, group vectors p+1 to m into new tensor group	$\overset{*}{A}_m = \Pi_{j=1}{}^p \mathbf{A}_i \Pi_{s=1}{}^{m-p} \mathbf{B}_s$ for $\mathbf{B}_s = \mathbf{A}_{k-p}$ and $s = k - p$
∴	From 3 and A10.2.1P5A Tensor of Rank–p and m – p	$\overset{*}{A}_m = \overset{*}{A}_p \overset{*}{B}_{m-p}$ for $\mathbf{B}_s = \mathbf{A}_{k-p}$ and $s = k - p$

Theorem 10.3.2 Product of Tensors: Addition of Rank for p+q

1g	Given	$\overset{*}{A}_m = \overset{*}{B}_p \overset{*}{C}_q$
Steps	Hypothesis	$\overset{*}{A}_m = \overset{*}{B}_p \overset{*}{C}_q$ for $m = p + q$
1	From 1g	$\overset{*}{A}_m = \overset{*}{B}_p \overset{*}{C}_q$
2	From 1 and A10.2.1P5A Tensor of Rank–m	$\Pi_{i=1}{}^m \mathbf{A}_i = \Pi_{j=1}{}^p \mathbf{B}_j \Pi_{k=1}{}^q \mathbf{C}_k$
3	From 2, grouping indices into index function λ(i) and κ(i) and A4.2.18 Summation Exp	$\Pi_{i=1}{}^m \mathbf{A}_i = \Pi_{i=1}{}^{p+q} \mathbf{B}_{\lambda(i)}\mathbf{C}_{\kappa(i)}$ where $\lambda(i) = \{i \text{ for } i \le p \text{ and } \varnothing \text{ for } p < i\}$, $\kappa(i) = \{\varnothing \text{ for } i \le p \text{ and } i - p \text{ for } p < i\}$ and $m = p + q$
4	From 3 and under the same product equating vector components	$\mathbf{A}_i = \mathbf{B}_{\lambda(i)}\mathbf{C}_{\kappa(i)}$
5	From 4 the product components have both properties, hence the union of the properties	$p_i = q_{\lambda(i)} \cup r_{\kappa(i)}$
∴	From 3, 4, 5 and A10.2.1P5A Tensor of Rank–m	$\overset{*}{A}_m = \overset{*}{B}_p \overset{*}{C}_q$ for $m = p + q$

Theorem 10.3.3 Product of Tensors: Products of Permutated Sets Expand to a New Set

1g	Given	$\overset{*}{A}_m = \overset{*}{B}_p \overset{*}{C}_q$
Steps	Hypothesis	$A_{i\lambda[i,k]} = B_{i\lambda[i,k]} C_{i\lambda[i,k]}$ for $L_p \cup L_q \rightarrow L_m = L_{(p+q)}$ EQ A $\sum_{\rho(k)\in L(p+q)} \prod_i^{(p+q)} A_{i\lambda[i,k]} \mathbf{d}_{i\lambda[i,k]} = \sum_{\rho(k)\in L(p+q)} \prod_i^{(p+q)} B_{i\lambda[i,k]} C_{i\lambda[i,k]} \mathbf{d}_{i\lambda[i,k]}$ EQ B
1	From 1g and A10.2.10 Tensor Closure under Multiplication	$\overset{*}{A}_m = \overset{*}{B}_p \overset{*}{C}_q$
2	From 1 and A10.2.1P5A Tensor of Rank–m	$\Pi_{i=1}{}^m \mathbf{A}_i = \Pi_{j=1}{}^p \mathbf{B}_j \Pi_{k=1}{}^q \mathbf{C}_k$
3	From 2, A10.2.1P3A Tensor of Rank–m and A7.9.2 Substitution of Vectors	$\Pi_{i=1}{}^m (\sum_\alpha A_{\lambda[j,\alpha]} \mathbf{b}_{\lambda[j,\alpha]}) = \Pi_{i=1}{}^p (\sum_\beta B_{\lambda[i,\beta]} \mathbf{b}_{\lambda[i,\beta]}) \Pi_{k=1}{}^q (\sum_\eta C_{\lambda[k,\eta]} \mathbf{b}_{\lambda[k,\eta]})$
4	From 3 and T10.2.1 Tool: Tensor Series Expansion	$\sum_{\rho(k)\in Lm} \prod_i^m A_{i\lambda[i,k]} \mathbf{b}_{i\lambda[i,k]} = (\sum_{\gamma(u)\in Lp} \prod_i^p B_{j\lambda[i,u]} \mathbf{b}_{j\lambda[i,u]})(\sum_{v(v)\in Lq} \prod_k^q C_{k\lambda[k,v]} \mathbf{b}_{k\lambda[k,v]})$
5	From 4 and A10.2.12 Distribution of Vectors over Vectors without a Polyadic Operator	$\sum_{\rho(k)\in Lm} \prod_i^m A_{i\lambda[i,k]} \mathbf{b}_{i\lambda[i,k]} = \sum_{\gamma(u)\in Lp} \sum_{v(v)\in Lq} \prod_j^p B_{j\lambda[j,u]} \mathbf{b}_{j\lambda[j,u]} \prod_k^q C_{k\lambda[k,v]} \mathbf{b}_{k\lambda[k,v]}$
6	From 5 and A7.8.5 Parallelogram Law: Commutative and Scalability of Collinear Vectors	$\sum_{\rho(k)\in Lm} \prod_i^m A_{i\lambda[i,k]} \mathbf{b}_{i\lambda[i,k]} = \sum_{\gamma(u)\in Lp} \sum_{v(v)\in Lq} \prod_j^p \prod_k^q B_{j\lambda[j,u]} C_{k\lambda[k,v]} \mathbf{b}_{j\lambda[j,u]} \mathbf{b}_{k\lambda[k,v]}$
7	From 6 and T10.4.2 Product of Tensors: Addition of Rank for p+q	$\sum_{\rho(k)\in L(p+q)} \prod_i^{(p+q)} A_{i\lambda[i,k]} \mathbf{b}_{i\lambda[i,k]} = \sum_{\gamma(u)\in Lp} \sum_{v(v)\in Lq} \prod_i^p \prod_k^q B_{j\lambda[i,u]} C_{k\lambda[k,v]} \mathbf{b}_{j\lambda[i,u]} \mathbf{b}_{k\lambda[k,v]}$
8	From 7, A10.2.10 Tensor Closure under Multiplication and reordering the summation over all terms, sets up a corresponding permutation set to the left hand side $L_{(p+q)}$.	$\sum_{\rho(k)\in L(p+q)} \prod_i^{(p+q)} A_{i\lambda[i,k]} \mathbf{b}_{i\lambda[i,k]} = \sum_{\rho(k)\in L(p+q)} \prod_i^{(p+q)} B_{i\lambda[i,k]} C_{i\lambda[i,k]} \mathbf{b}_{i\lambda[i,k]}$
9	From 8 and equating term-by-term	$A_{i\lambda[i,k]} = B_{i\lambda[i,k]} C_{i\lambda[i,k]}$
∴	From 8 and 9	$A_{i\lambda[i,k]} = B_{i\lambda[i,k]} C_{i\lambda[i,k]}$ for $L_p \cup L_q \rightarrow L_m = L_{(p+q)}$ EQ A $\sum_{\rho(k)\in L(p+q)} \prod_i^{(p+q)} A_{i\lambda[i,k]} \mathbf{b}_{i\lambda[i,k]} = \sum_{\rho(k)\in L(p+q)} \prod_i^{(p+q)} B_{i\lambda[i,k]} C_{i\lambda[i,k]} \mathbf{b}_{i\lambda[i,k]}$ EQ B

Theorem 10.3.4 Product of Tensors: Union of Tensor Supersets

Steps	Hypothesis	$\mathcal{A}_M = \mathcal{B}_P \cup \mathcal{C}_Q$ for m = p + q
1	From T10.4.2 Product of Tensors: Addition of Rank-p+q	$\overset{*}{A}_m = \overset{*}{B}_p \overset{*}{C}_q$ for m = p + q
∴	From 1 and T2.21.5 Multiplication Sets: Union of Supersets	$\mathcal{A}_M = \mathcal{B}_P \cup \mathcal{C}_Q$ for m = p + q

Theorem 10.3.5 Product of Tensors: Parsed Tensors into Supersubsets

Steps	Hypothesis	$\mathcal{B}_P \subseteq \mathcal{A}_M$	$\mathcal{C}_Q \subseteq \mathcal{A}_M$
1	From T10.4.4 Product of Tensors: Union of Tensor Sets	$\mathcal{A}_M = \mathcal{B}_P \cup \mathcal{C}_Q$ for m = p + q	
∴	From 1 and T2.21.6 Multiplication Sets: Parsed as Subsets	$\mathcal{B}_P \subseteq \mathcal{A}_M$	$\mathcal{C}_Q \subseteq \mathcal{A}_M$

Theorem 10.3.6 Product of Tensors: Uniqueness by Multiplication of Rank for p + q

1g	Given	$(\overset{*}{A}_p, \overset{*}{B}_p, \overset{*}{C}_q, \overset{*}{D}_q, \overset{*}{F}_m) \in T_m$
2g		$\overset{*}{A}_p = \overset{*}{B}_p$
3g		$\overset{*}{C}^q = \overset{*}{D}_q$
Steps	Hypothesis	$\overset{*}{A}_p\, \overset{*}{C}_q = \overset{*}{B}_p\, \overset{*}{D}_q$ for $\overset{*}{A}^p = \overset{*}{B}_p$ and $\overset{*}{C}_q = \overset{*}{D}_q$
1	From 1g and A10.2.10 Tensor Closure under Multiplication	$\overset{*}{A}_p\, \overset{*}{C}_q = \overset{*}{F}_m$
2	From 2g, 3g, 1 and T10.2.5 Tool: Tensor Substitution	$\overset{*}{B}_p\, \overset{*}{D}_q = \overset{*}{F}_m$
∴	From 2g, 3g, 1, 2 and T10.2.5 Tool: Tensor Substitution	$\overset{*}{A}_p\, \overset{*}{C}_q = \overset{*}{B}_p\, \overset{*}{D}_q$ for $\overset{*}{A}^p = \overset{*}{B}_p$ and $\overset{*}{C}_q = \overset{*}{D}_q$

Theorem 10.3.7 Product of Tensors: Commutative by Multiplication of Rank p + q → q + p

Steps	Hypothesis	$\overset{*}{A}_p\, \overset{*}{B}_q = \overset{*}{B}_q\, \overset{*}{A}_p$ p + q → q + p
1	From A10.2.6A Equality of Tensors	$\overset{*}{A}_p\, \overset{*}{B}_q = \overset{*}{A}_p\, \overset{*}{B}_q$
2	From 1 and A10.2.1P5A Tensor of Rank–m	$\overset{*}{A}_p\, \overset{*}{B}_q = \prod^p \mathbf{A}_i \prod^q \mathbf{B}_i$
3	From 2, T10.4.2 Product of Tensors: Addition of Rank-p+q	$\overset{*}{A}_p\, \overset{*}{B}_q = \prod^{p+q} \mathbf{A}_i\, \mathbf{B}_i$
4	From 3, A4.2.5 Commutative Add, A2.17.2 Commutative Product by commuting by pairs $(p+q-1)^p$ times	$\overset{*}{A}_p\, \overset{*}{B}_q = \prod^{q+p} \mathbf{B}_i\, \mathbf{A}_i$
5	From 4 and T10.4.2 Product of Tensors: Addition of Rank-p+q	$\overset{*}{A}_p\, \overset{*}{B}_q = \prod^q \mathbf{B}_i \prod^p \mathbf{A}_i$
∴	From 6 and A10.2.1P5A Tensor of Rank–m	$\overset{*}{A}_p\, \overset{*}{B}_q = \overset{*}{B}_q\, \overset{*}{A}_p$ p + q → q + p

Theorem 10.3.8 Product of Tensors: Association by Multiplication of Rank for p + q + r

1g	Given	$\overset{*}{A}_p\, (\overset{*}{B}_q\, \overset{*}{C}_r) \equiv (\overset{*}{A}_p\, \overset{*}{B}_q)\, \overset{*}{C}_r$ assume valid
Steps	Hypothesis	$\overset{*}{A}_p\, (\overset{*}{B}_q\, \overset{*}{C}_r) \equiv (\overset{*}{A}_p\, \overset{*}{B}_q)\, \overset{*}{C}_r$ for rank p + (q + r) = (p + q) + r
1	From 1g	$\overset{*}{A}_p\, (\overset{*}{B}_q\, \overset{*}{C}_r) \equiv (\overset{*}{A}_p\, \overset{*}{B}_q)\, \overset{*}{C}_r$
2	From 1 and A10.2.10 Tensor Closure under Multiplication	$\overset{*}{D}_{q+r} = \overset{*}{B}_q\, \overset{*}{C}_r$ for rank q + r
3	From 2 and A10.2.10 Tensor Closure under Multiplication	$\overset{*}{E}_{p+q} = \overset{*}{A}_p\, \overset{*}{B}_q$ for rank p + q
4	From1, 2, 3 and T10.2.5 Tool: Tensor Substitution	$\overset{*}{A}_p\, \overset{*}{D}_{q+r} = \overset{*}{E}_{p+q}\, \overset{*}{C}_r$
5	From 4 and A10.2.10 Tensor Closure under Multiplication	$\overset{*}{F}_{p+q+r} = \overset{*}{A}_p\, \overset{*}{D}_{q+r}$ for rank p + (q + r)
6	From 4 and A10.2.10 Tensor Closure under Multiplication	$\overset{*}{G}_{p+q+r} = \overset{*}{E}_{p+q}\, \overset{*}{C}_r$ for rank (p + q) + r
7	From 4, 5, 6 and T10.2.5 Tool: Tensor Substitution	$\overset{*}{F}_{p+q+r} = \overset{*}{G}_{p+q+r}$
∴	From 1 and 7	$\overset{*}{A}_p\, (\overset{*}{B}_q\, \overset{*}{C}_r) \equiv (\overset{*}{A}_p\, \overset{*}{B}_q)\, \overset{*}{C}_r$ for rank p + (q + r) = (p + q) + r

Theorem 10.3.9 Product of Tensors: Reduction of Tensors by Subtraction of Rank for m–q

Steps	Hypothesis	$\overset{*}{A}_m / \overset{*}{C}_q = \overset{*}{B}_p$ for $m - q = p$ and $\mathcal{C}_q \subseteq \mathcal{A}_m$
1	From T10.4.2 Product of Tensors: Addition of Rank-p+q and T10.4.4 Product of Tensors: Union of Tensor Supersets	$\overset{*}{A}_m = \overset{*}{B}_p \overset{*}{C}_q$ for $m = p + q$ and $\mathcal{A}_m = \mathcal{B}_p \cup \mathcal{C}_q$
∴	From 1 and T2.21.11 Multiplication Sets: Reduction by Subtraction of Rank-m–q	$\overset{*}{A}_m / \overset{*}{C}_q = \overset{*}{B}_p$ for $m - q = p \geq 0$ and $\mathcal{C}_q \subseteq \mathcal{A}_m$

Theorem 10.3.10 Product of Tensors: Reduction of Tensors by Inverse of Rank for m–m

1g	Given	$\overset{*}{B}_z = \overset{*}{U}_z$ for $A_i = U_i$ and all i assume
2g		$\overset{*}{C}_q = \overset{*}{A}_m$ for $C_i = A_i$ and all i and $q = m$ assume
Steps	Hypothesis	$\overset{*}{A}_m / \overset{*}{A}_m = 1$ for $0 = z \geq 0$ and $\mathcal{A}_m = \mathcal{A}_m$
1	From T10.4.9 Product of Tensors: Reduction of Tensors by Subtraction of Rank-m–q	$\overset{*}{A}_m / \overset{*}{C}_q = \overset{*}{B}_z$ for $m - q = z \geq 0$ and $\mathcal{C}_q \subseteq \mathcal{A}_m$
2	From 1g, 2g, 1, T10.2.5 Tool: Tensor Substitution	$\overset{*}{A}_m / \overset{*}{A}_m = \overset{*}{U}_z$ for $m - m = z \geq 0$ and $\mathcal{A}_m = \mathcal{A}_m$
3	From 2, A4.2.8 Inverse Add and A4.2.3 Substitution	$\overset{*}{A}_m / \overset{*}{A}_m = \overset{*}{U}_0$ for $0 = z \geq 0$ and $\mathcal{A}_m = \mathcal{A}_m$
4	From 3 and A10.2.3 Tensor of Rank-0 as a Scalar	$a / a = 1$
5	From 4 and A4.2.13 Inverse Multp	$1 = 1$
∴	From 1g, 2g, 3, 4, A4.2.3 Substitution and by identity	$\overset{*}{A}_m / \overset{*}{A}_m = 1$ for $0 = z \geq 0$ and $\mathcal{A}_m = \mathcal{A}_m$

Theorem 10.3.11 Addition of Tensors: Closer of Tensor Coefficients by Addition

Steps	Hypothesis	$\prod^m C_{i\lambda[i,k]} = \prod^m A_{i\lambda[i,k]} + \prod^m B_{i\lambda[i,k]}$
1	From A10.2.8 Tensor Closure under Addition	$\overset{*}{C}_m = \overset{*}{A}_m + \overset{*}{B}_m$
2	From 1 and T10.2.1 Tool: Tensor Series Expansion	$\sum_{\rho(k)\in Lm} \prod^m C_{i\lambda[i,k]}\mathbf{d}_{\lambda[i,k]} = \sum_{\rho(k)\in Lm} \prod^m A_{i\lambda[i,k]}\mathbf{d}_{\lambda[i,k]}$ $+ \sum_{\rho(k)\in Ln} \prod^m B_{i\lambda[i,k]}\mathbf{d}_{\lambda[i,k]}$
3	From 2, A7.9.3 Commutative by Vector Addition and A7.9.4 Associative by Vector Addition	$\sum_{\rho(k)\in Lm} \prod^m C_{i\lambda[i,k]}\mathbf{d}_{\lambda[i,k]} =$ $\sum_{\rho(k)\in Lm} (\prod^m A_{i\lambda[i,k]}\mathbf{d}_{\lambda[i,k]} + \prod^m B_{i\lambda[i,k]}\mathbf{d}_{\lambda[i,k]})$
4	From 3 and T10.2.4 Tool: Base Tensor Distributed over Polyadic Coefficient Summation	$\sum_{\rho(k)\in Lm} (\prod^m C_{i\lambda[i,k]}) \prod^m \mathbf{d}_{\lambda[i,k]} =$ $\sum_{\rho(k)\in Lm} (\prod^m A_{i\lambda[i,k]} + \prod^m B_{i\lambda[i,k]}) \prod^m \mathbf{d}_{\lambda[i,k]}$
∴	From 4 and equating terms	$\prod^m C_{i\lambda[i,k]} = \prod^m A_{i\lambda[i,k]} + \prod^m B_{i\lambda[i,k]}$

Theorem 10.3.12 Addition of Tensors: Addition of the ith-Uncommon Tensor Coefficient

1g 2g	Given	$\prod^m A_{i\lambda[i,k]} = (\prod^{i-1} C_{i\lambda[i,k]})\, U_{i\lambda[i,k]}\, (\prod_{i+1}{}^m C_{i\lambda[i,k]})$ $\prod^m B_{i\lambda[i,k]} = (\prod^{i-1} C_{i\lambda[i,k]})\, V_{i\lambda[i,k]}\, (\prod_{i+1}{}^m C_{i\lambda[i,k]})$
Steps	Hypothesis	$C_{i\lambda[i,k]} = U_{i\lambda[i,k]} + V_{i\lambda[i,k]}$
1	From T10.4.11 Addition of Tensors: Closer of Polyadics by Addition	$\prod^m C_{i\lambda[i,k]} = \prod^m A_{i\lambda[i,k]} + \prod^m B_{i\lambda[i,k]}$
2	From 1g, 2g, 1, A4.2.3 Substitution and expanding about the ith factor	$(\prod^{i-1} C_{i\lambda[i,k]})\, C_{i\lambda[i,k]}\, (\prod_{i+1}{}^m C_{i\lambda[i,k]}) =$ $(\prod^{i-1} C_{i\lambda[i,k]})\, U_{i\lambda[i,k]}\, (\prod_{i+1}{}^m C_{i\lambda[i,k]}) +$ $(\prod^{i-1} C_{i\lambda[i,k]})\, V_{i\lambda[i,k]}\, (\prod_{i+1}{}^m C_{i\lambda[i,k]})$
3	From 2 and A4.2.14 Distribution	$(\prod^{i-1} C_{i\lambda[i,k]})\, C_{i\lambda[i,k]}\, (\prod_{i+1}{}^m C_{i\lambda[i,k]}) =$ $(\prod^{i-1} C_{i\lambda[i,k]})\, (U_{i\lambda[i,k]} + V_{i\lambda[i,k]})\, (\prod_{i+1}{}^m C_{i\lambda[i,k]})$
∴	From 3 and T4.4.3 Equalities: Cancellation by Multiplication	$C_{i\lambda[i,k]} = U_{i\lambda[i,k]} + V_{i\lambda[i,k]}$

Theorem 10.3.13 Addition of Tensors: Closer of Tensor Coefficient by Subtraction

Steps	Hypothesis	$\prod^m C_{i\lambda[i,k]} = \prod^m A_{i\lambda[i,k]} - \prod^m B_{i\lambda[i,k]}$
1	From A10.2.9 Tensor Closure under Subtraction	$\overset{*}{C}_m = \overset{*}{A}_m - \overset{*}{B}_m$
2	From 1 and T10.2.1 Tool: Tensor Series Expansion	$\sum_{\rho(k)\in Lm} \prod^m C_{i\lambda[i,k]}\mathbf{b}_{i\lambda[i,k]} = \sum_{\rho(k)\in Lm} \prod^m A_{i\lambda[i,k]}\mathbf{b}_{i\lambda[i,k]}$ $- \sum_{\rho(k)\in Ln} \prod^m B_{i\lambda[i,k]}\mathbf{b}_{i\lambda[i,k]}$
3	From 2, A7.9.3 Commutative by Vector Addition and A7.9.4 Associative by Vector Addition	$\sum_{\rho(k)\in Lm} \prod^m C_{i\lambda[i,k]}\mathbf{b}_{i\lambda[i,k]} =$ $\sum_{\rho(k)\in Lm} (\prod^m A_{i\lambda[i,k]}\mathbf{b}_{i\lambda[i,k]} - \prod^m B_{i\lambda[i,k]}\mathbf{b}_{i\lambda[i,k]})$
4	From 3 and T10.2.4 Tool: Base Tensor Distributed over Polyadic Coefficient Summation	$\sum_{\rho(k)\in Lm} (\prod^m C_{i\lambda[i,k]}) \prod^m \mathbf{b}_{i\lambda[i,k]} =$ $\sum_{\rho(k)\in Lm} (\prod^m A_{i\lambda[i,k]} - \prod^m B_{i\lambda[i,k]}) \prod^m \mathbf{b}_{i\lambda[i,k]}$
∴	From 4 and equating terms	$\prod^m C_{i\lambda[i,k]} = \prod^m A_{i\lambda[i,k]} - \prod^m B_{i\lambda[i,k]}$

Theorem 10.3.14 Addition of Tensors: Subtraction of the ith-Uncommon Tensor Coefficient

1g 2g	Given	$\prod^m A_{i\lambda[i,k]} = (\prod^{i-1} C_{i\lambda[i,k]})\, U_{i\lambda[i,k]}\, (\prod_{i+1}{}^m C_{i\lambda[i,k]})$ $\prod^m B_{i\lambda[i,k]} = (\prod^{i-1} C_{i\lambda[i,k]})\, V_{i\lambda[i,k]}\, (\prod_{i+1}{}^m C_{i\lambda[i,k]})$
Steps	Hypothesis	$C_{i\lambda[i,k]} = U_{i\lambda[i,k]} - V_{i\lambda[i,k]}$
1	From T10.4.11 Addition of Tensors: Closer of Polyadics by Addition	$\prod^m C_{i\lambda[i,k]} = \prod^m A_{i\lambda[i,k]} - \prod^m B_{i\lambda[i,k]}$
2	From 1g, 2g, 1, A4.2.3 Substitution and expanding about the ith factor	$(\prod^{i-1} C_{i\lambda[i,k]})\, C_{i\lambda[i,k]}\, (\prod_{i+1}{}^m C_{i\lambda[i,k]}) =$ $(\prod^{i-1} C_{i\lambda[i,k]})\, U_{i\lambda[i,k]}\, (\prod_{i+1}{}^m C_{i\lambda[i,k]}) -$ $(\prod^{i-1} C_{i\lambda[i,k]})\, V_{i\lambda[i,k]}\, (\prod_{i+1}{}^m C_{i\lambda[i,k]})$
3	From 2 and A4.2.14 Distribution	$(\prod^{i-1} C_{i\lambda[i,k]})\, C_{i\lambda[i,k]}\, (\prod_{i+1}{}^m C_{i\lambda[i,k]}) =$ $(\prod^{i-1} C_{i\lambda[i,k]})\, (U_{i\lambda[i,k]} - V_{i\lambda[i,k]})\, (\prod_{i+1}{}^m C_{i\lambda[i,k]})$
∴	From 3 and T4.4.3 Equalities: Cancellation by Multiplication	$C_{i\lambda[i,k]} = U_{i\lambda[i,k]} - V_{i\lambda[i,k]}$

Theorem 10.3.15 Addition of Tensors: Uniqueness by Addition

1g	Given	$\overset{*}{A}_p = \overset{*}{C}_q$ for $A^i = B^i$ and all i
2g		$\overset{*}{B}_p = \overset{*}{D}_q$ for $C^i = D^i$ and all i
Steps	Hypothesis	$\overset{*}{A}_p + \overset{*}{B}_p = \overset{*}{C}_q + \overset{*}{D}_q$
1	From A10.2.6A Equality of Tensors	$\overset{*}{A}_p + \overset{*}{B}_p = \overset{*}{A}_p + \overset{*}{B}_p$
∴	From 1g, 2g, 1 and T10.2.5 Tool: Tensor Substitution	$\overset{*}{A}_p + \overset{*}{B}_p = \overset{*}{C}_q + \overset{*}{D}_q$

Theorem 10.3.16 Addition of Tensors: Commutative by Addition

Steps	Hypothesis	$\overset{*}{A}_m + \overset{*}{B}_m = \overset{*}{B}_m + \overset{*}{A}_m$
1	From A10.2.6A Equality of Tensors	$\overset{*}{A}_m + \overset{*}{B}_m = \overset{*}{A}_m + \overset{*}{B}_m$
2	From 1 and T10.2.1 Tool: Tensor Series Expansion	$\overset{*}{A}_m + \overset{*}{B}_m = \sum_{\rho(k)\in Lm} \prod^m A_{i\lambda[i,k]}\mathbf{b}_{i\lambda[i,k]} + \sum_{\rho(k)\in Lm} \prod^m B_{i\lambda[i,k]}\mathbf{b}_{i\lambda[i,k]}$
3	From 2, A7.9.3 Commutative by Vector Addition and A7.9.4 Associative by Vector Addition	$\overset{*}{A}_m + \overset{*}{B}_m = \sum_{\rho(k)\in Lm} (\prod^m A_{i\lambda[i,k]}\mathbf{b}_{\lambda[i,k]} + \prod^m B_{i\lambda[i,k]}\mathbf{b}_{\lambda[i,k]})$
4	From 3 and T10.2.4 Tool: Base Tensor Distributed over Polyadic Coefficient Summation	$\overset{*}{A}_m + \overset{*}{B}_m = \sum_{\rho(k)\in Lm} (\prod^m A_{i\lambda[i,k]} + \prod^m B_{i\lambda[i,k]}) \prod^m \mathbf{b}_{i\lambda[i,k]}$
5	From 4 and A4.2.5 Commutative Add	$\overset{*}{A}_m + \overset{*}{B}_m = \sum_{\rho(k)\in Lm} (\prod^m B_{i\lambda[i,k]} + \prod^m A_{i\lambda[i,k]}) \prod^m \mathbf{b}_{i\lambda[i,k]}$
6	From 5 and T10.2.4 Tool: Base Tensor Distributed over Polyadic Coefficient Summation	$\overset{*}{A}_m + \overset{*}{B}_m = \sum_{\rho(k)\in Lm} (\prod^m B_{i\lambda[i,k]}\, \mathbf{b}_{\lambda[i,k]} + \prod^m A_{i\lambda[i,k]}\, \mathbf{b}_{\lambda[i,k]})$
7	From 6, A7.9.3 Commutative by Vector Addition and A7.9.4 Associative by Vector Addition	$\overset{*}{A}_m + \overset{*}{B}_m = \sum_{\rho(k)\in Lm} \prod^m B_{i\lambda[i,k]}\, \mathbf{b}_{i\lambda[i,k]} + \sum_{\rho(k)\in Ln} \prod^m A_{i\lambda[i,k]}\, \mathbf{b}_{i\lambda[i,k]}$
∴	From 7 and T10.2.2 Tool: Tensor Series Contraction	$\overset{*}{A}_m + \overset{*}{B}_m = \overset{*}{B}_m + \overset{*}{A}_m$

Theorem 10.3.17 Addition of Tensors: Associative by Addition

Steps	Hypothesis	$\overset{*}{A}_m + (\overset{*}{B}_m + \overset{*}{C}_m) = (\overset{*}{A}_m + \overset{*}{B}_m) + \overset{*}{C}_m$
1	From A10.2.6A Equality of Tensors	$\overset{*}{A}_m + (\overset{*}{B}_m + \overset{*}{C}_m) = \overset{*}{A}_m + (\overset{*}{B}_m + \overset{*}{C}_m)$
2	From 1 and T10.2.1 Tool: Tensor Series Expansion	$\overset{*}{A}_m + (\overset{*}{B}_m + \overset{*}{C}_m) = \sum_{\rho(k)\in Ln} \prod^m A_{i\lambda[i,k]}\, \mathbf{b}_{i\lambda[i,k]}$ $+ (\sum_{\rho(k)\in Lm} \prod^m B_{i\lambda[i,k]}\, \mathbf{b}_{i\lambda[i,k]}$ $+ \sum_{\rho(k)\in Lm} \prod^m C_{i\lambda[i,k]}\, \mathbf{b}_{i\lambda[i,k]})$
3	From2 and T10.2.3 Tool: Base Tensor Parsed from Polyadic Coefficient	$\overset{*}{A}_m + (\overset{*}{B}_m + \overset{*}{C}_m) = \sum_{\rho(k)\in Ln} \prod^m A_{i\lambda[i,k]} \prod^m \mathbf{b}_{i\lambda[i,k]}$ $+ (\sum_{\rho(k)\in Lm} \prod^m B_{i\lambda[i,k]} \prod^m \mathbf{b}_{i\lambda[i,k]}$ $+ \sum_{\rho(k)\in Lm} \prod^m C_{i\lambda[i,k]} \prod^m \mathbf{b}_{i\lambda[i,k]})$
4	From 3 and T10.2.4 Tool: Base Tensor Distributed over Polyadic Coefficient Summation	$\overset{*}{A}_m + (\overset{*}{B}_m + \overset{*}{C}_m) = \sum_{\rho(k)\in Ln} [\, \prod^m A_{i\lambda[i,k]}$ $+ (\prod^m B_{i\lambda[i,k]}$ $+ \prod^m C_{i\lambda[i,k]})\,] \prod^m \mathbf{b}_{\lambda[i,k]}$
5	From 4 and A4.2.6 Associative Add[10.4.1]	$\overset{*}{A}_m + (\overset{*}{B}_m + \overset{*}{C}_m) = \sum_{\rho(k)\in Ln} [\, (\prod^m A_{i\lambda[i,k]} + \prod^m B_{i\lambda[i,k]})$ $+ \prod^m C_{i\lambda[i,k]}\,] \prod^m \mathbf{b}_{i\lambda[i,k]}$
6	From 5 and T10.2.4 Tool: Base Tensor Distributed over Polyadic Coefficient Summation	$\overset{*}{A}_m + (\overset{*}{B}_m + \overset{*}{C}_m) = (\sum_{\rho(k)\in Ln} \prod^m A_{i\lambda[i,k]} \prod^m \mathbf{b}_{i\lambda[i,k]}$ $+ \sum_{\rho(k)\in Ln} \prod^m B_{i\lambda[i,k]} \prod^m \mathbf{b}_{i\lambda[i,k]})$ $+ \sum_{\rho(k)\in Ln} \prod^m C_{i\lambda[i,k]} \prod^m \mathbf{b}_{i\lambda[i,k]}$
∴	From 6, T10.2.3 Tool: Base Tensor Parsed from Polyadic Coefficient and T01.3.2 Tool: Tensor Series Contraction	$\overset{*}{A}_m + (\overset{*}{B}_m + \overset{*}{C}_m) = (\overset{*}{A}_m + \overset{*}{B}_m) + \overset{*}{C}_m$

Theorem 10.3.18 Addition of Tensors: Identity by Addition

1g	Given	$\overset{*}{A}_m = \overset{*}{A}_m + \overset{*}{Z}_m$ assume
Steps	Hypothesis	$\overset{*}{A}_m = \overset{*}{A}_m + \overset{*}{Z}_m$
1	From 1g	$\overset{*}{A}_m = \overset{*}{A}_m + \overset{*}{Z}_m$
2	From 1 and T10.2.1 Tool: Tensor Series Expansion	$\overset{*}{A}_m = \sum_{\rho(k)\in Lm} \prod^m A_{i\lambda[i,k]}\mathbf{b}_{i\lambda[i,k]}$ $+ \sum_{\rho(k)\in Ln} \prod^m Z_{i\lambda[i,k]}\mathbf{b}_{i\lambda[i,k]}$
3	From 2, A7.9.3 Commutative by Vector Addition and A7.9.4 Associative by Vector Addition	$\overset{*}{A}_m = \sum_{\rho(k)\in Lm} (\, \prod^m A_{i\lambda[i,k]}\mathbf{b}_{i\lambda[i,k]} + \prod^m Z_{i\lambda[i,k]}\mathbf{b}_{i\lambda[i,k]}\,)$
4	From 3 and T10.2.4 Tool: Base Tensor Distributed over Polyadic Coefficient Summation	$\overset{*}{A}_m = \sum_{\rho(k)\in Lm} (\, \prod^m A_{i\lambda[i,k]} + \prod^m Z_{i\lambda[i,k]}\,) \prod^m \mathbf{b}_{i\lambda[i,k]}$
5	From 4, D10.1.18B Zero Tensor and A4.2.3 Substitution	$\overset{*}{A}_m = \sum_{\rho(k)\in Lm} (\, \prod^m A_{i\lambda[i,k]} + \prod^m 0\,) \prod^m \mathbf{b}_{i\lambda[i,k]}$
6	From 5, D4.1.19 Primitive Definition for Rational Arithmetic and A4.2.7 Identity Add	$\overset{*}{A}_m = \sum_{\rho(k)\in Lm} \prod^m A_{i\lambda[i,k]} \prod^m \mathbf{b}_{i\lambda[i,k]}$
7	From 6 and T10.2.3 Tool: Base Tensor Parsed from Polyadic Coefficient	$\overset{*}{A}_m = \sum_{\rho(k)\in Lm} \prod^m A_{i\lambda[i,k]}\, \mathbf{b}_{i\lambda[i,k]}$
8	From 7 and T10.2.2 Tool: Tensor Series Contraction	$\overset{*}{A}_m = \overset{*}{A}_m$
∴	From 1, 7 and by identity	$\overset{*}{A}_m = \overset{*}{A}_m + \overset{*}{Z}_m$

Theorem 10.3.19 Addition of Tensors: Inverse by Addition

Steps	Hypothesis	$\overset{*}{A}_m - \overset{*}{A}_m = \overset{*}{Z}_m$
1	From A10.2.6A Equality of Tensors	$\overset{*}{A}_m - \overset{*}{A}_m = \overset{*}{A}_m - \overset{*}{A}_m$
2	From 1 and T10.2.1 Tool: Tensor Series Expansion	$\overset{*}{A}_m - \overset{*}{A}_m = \sum_{\rho(k)\in Lm} \prod^m A_{i\lambda[i,\,k]}\, \mathbf{b}_{i\lambda[i,\,k]} - \sum_{\rho(k)\in Lm} \prod^m A_{i\lambda[i,\,k]}\, \mathbf{b}_{i\lambda[i,\,k]}$
3	From 2, A7.9.3 Commutative by Vector Addition and A7.9.4 Associative by Vector Addition	$\overset{*}{A}_m - \overset{*}{A}_m = \sum_{\rho(k)\in Lm} (\prod^m A_{i\lambda[i,\,k]}\, \mathbf{b}_{i\lambda[i,\,k]} - \prod^m A_{i\lambda[i,\,k]}\, \mathbf{b}_{i\lambda[i,\,k]})$
4	From 3 and T10.2.4 Tool: Base Tensor Distributed over Polyadic Coefficient Summation	$\overset{*}{A}_m - \overset{*}{A}_m = \sum_{\rho(k)\in Lm} (\prod^m A_{i\lambda[i,\,k]} - \prod^m A_{i\lambda[i,\,k]})\prod^m \mathbf{b}_{i\lambda[i,\,k]}$
5	From 4 and A4.2.8 Inverse Add	$\overset{*}{A}_m - \overset{*}{A}_m = \sum_{\rho(k)\in Lm} 0 \prod^m \mathbf{b}_{i\lambda[i,\,k]}$
6	From 5, D10.1.18B Zero Tensor and A4.2.3 Substitution	$\overset{*}{A}_m - \overset{*}{A}_m = \sum_{\rho(k)\in Lm} Z_{i\lambda[i,\,k]} \prod^m \mathbf{b}_{i\lambda[i,\,k]}$
∴	From 6 and D10.1.18A Zero Tensor	$\overset{*}{A}_m - \overset{*}{A}_m = \overset{*}{Z}_m$

Theorem 10.3.20 Addition of Tensors: Distribution of a Tensor over Addition of Tensors

Steps	Hypothesis	$\overset{*}{A}_q (\overset{*}{B}_m + \overset{*}{C}_m) = \overset{*}{A}_q \overset{*}{B}_m + \overset{*}{A}_q \overset{*}{C}_m$
1		$\overset{*}{A}_q (\overset{*}{B}_m + \overset{*}{C}_m) = \overset{*}{A}_q (\overset{*}{B}_m + \overset{*}{C}_m)$
2	From 1 and A10.2.1P5A Tensor of Rank–m	$\overset{*}{A}_q (\overset{*}{B}_m + \overset{*}{C}_m) = \Pi_{r=1}{}^q\, \mathbf{A}_r\, (\Pi_{k=1}{}^m\, \mathbf{B}_k + \Pi_{k=1}{}^m\, \mathbf{C}_k)$
3	From 2 and A10.2.12 Distribution of Vectors over Vectors without a Polyadic Operator	$\overset{*}{A}_q (\overset{*}{B}_m + \overset{*}{C}_m) = \Pi_{r=1}{}^q\, \mathbf{A}_r\, \Pi_{k=1}{}^m\, \mathbf{B}_k + \Pi_{r=1}{}^q\, \mathbf{A}_r\, \Pi_{k=1}{}^m\, \mathbf{C}_k$
∴	From 3 and A10.2.1P5A Tensor of Rank–m	$\overset{*}{A}_q (\overset{*}{B}_m + \overset{*}{C}_m) = \overset{*}{A}_q \overset{*}{B}_m + \overset{*}{A}_q \overset{*}{C}_m$

Theorem 10.3.21 Addition of Tensors: Right Cancellation by Addition

1g	Given	$\overset{*}{A}_m = \overset{*}{B}_m + \overset{*}{C}_m$	
Steps	Hypothesis	$\overset{*}{A}_m - \overset{*}{C}_m = \overset{*}{B}_m$ Equation A	$\overset{*}{B}_m = \overset{*}{A}_m - \overset{*}{C}_m$ Equation B
1	From A10.2.6A Equality of Tensors	$-\overset{*}{C}_m = -\overset{*}{C}_m$	
2	From 1, 1g and T10.4.15 Addition of Tensors: Uniqueness by Addition	$\overset{*}{A}_m - \overset{*}{C}_m = (\overset{*}{B}_m + \overset{*}{C}_m) - \overset{*}{C}_m$	
3	From 2 and T10.4.17 Addition of Tensors: Associative by Addition	$\overset{*}{A}_m - \overset{*}{C}_m = \overset{*}{B}_m + (\overset{*}{C}_m - \overset{*}{C}_m)$	
4	From 3 and 10.4.19 Addition of Tensors: Inverse by Addition	$\overset{*}{A}_m - \overset{*}{C}_m = \overset{*}{B}_m + \mathbf{0}$	
∴	From 4 and T10.4.18 Addition of Tensors: Identity by Addition	$\overset{*}{A}_m - \overset{*}{C}_m = \overset{*}{B}_m$ by A10.2.6A	$\overset{*}{B}_m = \overset{*}{A}_m - \overset{*}{C}_m$ by A10.2.6B

Theorem 10.3.22 Addition of Tensors: Reversal of Right Cancellation by Addition

1g	Given	$\overset{*}{B}_m = \overset{*}{A}_m - \overset{*}{C}_m$			
Steps	Hypothesis	$\overset{*}{B}_m + \overset{*}{C}_m = \overset{*}{A}_m$	Equation A	$\overset{*}{A}_m = \overset{*}{B}_m + \overset{*}{C}_m$	Equation B
∴	From T10.4.21 Addition of Tensors: Right Cancellation by Addition and T3.7.2 Reversibility of a Logical Argument	$\overset{*}{B}_m + \overset{*}{C}_m = \overset{*}{A}_m$	by A10.2.6A	$\overset{*}{A}_m = \overset{*}{B}_m + \overset{*}{C}_m$	by A10.2.6B

Theorem 10.3.23 Addition of Tensors: Left Cancellation by Addition

1g	Given	$\overset{*}{B}_m + \overset{*}{C}_m = \overset{*}{A}_m$			
Steps	Hypothesis	$\overset{*}{A}_m - \overset{*}{C}_m = \overset{*}{B}_m$	Equation A	$\overset{*}{B}_m = \overset{*}{A}_m - \overset{*}{C}_m$	Equation B
1	From 1g and A10.2.6B Equality	$\overset{*}{A}_m = \overset{*}{B}_m + \overset{*}{C}_m$			
∴	From 1 and A10.2.9 Tensor Closure under Subtraction	$\overset{*}{A}_m - \overset{*}{C}_m = \overset{*}{B}_m$	by A10.2.6A	$\overset{*}{B}_m = \overset{*}{A}_m - \overset{*}{C}_m$	by A10.2.6B

Theorem 10.3.24 Addition of Tensors: Reversal of Left Cancellation by Addition

1g	Given	$\overset{*}{A}_m - \overset{*}{C}_m = \overset{*}{B}_m$			
Steps	Hypothesis	$\overset{*}{A}_m = \overset{*}{B}_m + \overset{*}{C}_m$	Equation A	$\overset{*}{B}_m + \overset{*}{C}_m = \overset{*}{A}_m$	Equation B
∴	From T10.4.23 Addition of Tensors: Left Cancellation by Addition and T3.7.2 Reversibility of a Logical Argument	$\overset{*}{A}_m = \overset{*}{B}_m + \overset{*}{C}_m$	by A10.2.6A	$\overset{*}{B}_m + \overset{*}{C}_m = \overset{*}{A}_m$	by A10.2.6B

[10.4.1]Note: Associative property of the tensor coefficient is questionable. The tensor coefficient for sure is not a vector or product of vectors, hence might be considered a real number and applying the associative law of algebra A4.2.6 "Associative Add" would be valid, however it is not a simple scalar number and still is tenuously associated with the polyadic base tensor, so is something more. It maybe it requires a special tensor axiom devoted to association to remove this ambiguity?

Section 10.4 Scalars and Tensors

Theorem 10.4.1 Scalars and Tensors: Uniqueness of Scalar Multiplication to Tensors

1g	Given	$\overset{*}{A}_m = \overset{*}{B}_m$	
2g		$\kappa = \kappa$ for $\kappa \in R$	
Steps	Hypothesis	$\overset{*}{A}_m = \overset{*}{B}_m$ EQ A	$\kappa\overset{*}{A}_m = \kappa\overset{*}{B}_m$ EQ B
∴	From 1g, 2g and T7.10.1 Uniqueness of Scalar Multiplication to Vectors	$\overset{*}{A}_m = \overset{*}{B}_m$ EQ A	$\kappa\overset{*}{A}_m = \kappa\overset{*}{B}_m$ EQ B

Theorem 10.4.2 Scalars and Tensors: Distribution of Scalar across a Tensor

1g	Given	$\kappa = \kappa$ for $\kappa \in R$
Steps	Hypothesis	$\kappa\overset{*}{A}_m = \sum_{\rho(k)\in Lm} \kappa\prod^m A_{i\lambda[i,k]}\mathbf{b}_{i\lambda[i,k]}$
1	From A10.2.6A Equality of Tensors and T10.5.1 Scalars and Tensors: Uniqueness of Scalar Multiplication to a Tensors	$\kappa\overset{*}{A}_m = \kappa\overset{*}{A}_m$
2	From 1 and T10.2.1 Tool: Tensor Series Expansion	$\kappa\overset{*}{A}_m = \kappa\sum_{\rho(k)\in Lm} \prod^m A_{i\lambda[i,k]}\mathbf{b}_{i\lambda[i,k]}$
∴	From 2 and T7.10.3 Distribution of a Scalar over Addition of Vectors	$\kappa\overset{*}{A}_m = \sum_{\rho(k)\in Lm} \kappa\prod^m A_{i\lambda[i,k]}\mathbf{b}_{i\lambda[i,k]}$

Theorem 10.4.3 Scalars and Tensors: Scalar Multiplication to a Polyadic

1g	Given	$\kappa = \kappa$ for $\kappa \in R$
Steps	Hypothesis	$\kappa\overset{*}{A}_m = \sum_{\rho(k)\in Lm} (\kappa\prod^m A_{i\lambda[i,k]}) \prod^m \mathbf{b}_{i\lambda[i,k]}$
1	From 1g and T10.5.2 Scalars and Tensors: Distribution of Scalar across a Tensor	$\kappa\overset{*}{A}_m = \sum_{\rho(k)\in Lm} \kappa(\prod^m A_{i\lambda[i,k]}\mathbf{b}_{i\lambda[i,k]})$
∴	From 2, T10.2.3 Tool: Base Tensor Parsed from Polyadic Coefficient and T7.10.2 Associative Scalar Multiplication to Vectors	$\kappa\overset{*}{A}_m = \sum_{\rho(k)\in Lm} (\kappa\prod^m A_{i\lambda[i,k]}) \prod^m \mathbf{b}_{i\lambda[i,k]}$

Theorem 10.4.4 Scalars and Tensors: Scaling the ith-Tensor Coefficient

1g	Given	$\kappa = \kappa$ for $\kappa \in R$
Steps	Hypothesis	$\kappa\overset{*}{A}_m = \sum_{\rho(k)\in Lm} \prod^m (\bullet\, \kappa A_{i\lambda[i,k]} \bullet) \prod^m \mathbf{b}_{i\lambda[i,k]}$
1	From 1g and T10.5.3 Scalars and Tensors: Scalar Multiplication to a Polyadic	$\kappa\overset{*}{A}_m = \sum_{\rho(k)\in Lm} (\kappa\prod^m A_{i\lambda[i,k]}) \prod^m \mathbf{b}_{i\lambda[i,k]}$
∴	From 1, A4.2.10 Commutative Multp and A4.2.11 Associative Multp	$\kappa\overset{*}{A}_m = \sum_{\rho(k)\in Lm} \prod^m (\bullet\, \kappa A_{i\lambda[i,k]} \bullet) \prod^m \mathbf{b}_{i\lambda[i,k]}$

Theorem 10.4.5 Scalars and Tensors: Scalar multiplication of a Base Tensor

1g	Given	$\kappa = \kappa$ for $\kappa \in R$
Steps	Hypothesis	$\kappa\overset{*}{A}_m = \sum_{\rho(k)\in Lm} \prod^m A_{i\lambda[i,k]} (\kappa \prod^m \mathbf{b}_{i\lambda[i,k]})$
1	From 1g and T10.5.3 Scalars and Tensors: Scalar Multiplication to a Polyadic	$\kappa\overset{*}{A}_m = \sum_{\rho(k)\in Lm} (\kappa \prod^m A_{i\lambda[i,k]}) \prod^m \mathbf{b}_{i\lambda[i,k]}$
2	From 1 and A4.3.10 Commutative Multp	$\kappa\overset{*}{A}_m = \sum_{\rho(k)\in Lm} (\prod^m A_{i\lambda[i,k]} \kappa) \prod^m \mathbf{b}_{i\lambda[i,k]}$
∴	From 2 and T7.10.3 Distribution of a Scalar over Addition of Vectors	$\kappa\overset{*}{A}_m = \sum_{\rho(k)\in Lm} \prod^m A_{i\lambda[i,k]} (\kappa \prod^m \mathbf{b}_{i\lambda[i,k]})$

Theorem 10.4.6 Scalars and Tensors: Distribution of a Scalar over Addition of Tensors

Steps	Hypothesis	$\kappa(\overset{*}{A}_m + \overset{*}{B}_m) = \kappa\overset{*}{A}_m + \kappa\overset{*}{B}_m$
1	From A10.2.6A Equality of Tensors	$\kappa(\overset{*}{A}_m + \overset{*}{B}_m) = \kappa(\overset{*}{A}_m + \overset{*}{B}_m)$
2	From 1 and T10.2.1 Tool: Tensor Series Expansion	$\kappa(\overset{*}{A}_m + \overset{*}{B}_m) = \kappa(\sum_{\rho(k)\in Lm} \prod^m A_{i\lambda[i,k]} \mathbf{b}_{i\lambda[i,k]} + \sum_{\rho(k)\in Lm} \prod^m B_{i\lambda[i,k]} \mathbf{b}_{i\lambda[i,k]})$
3	From 2, A7.9.3 Commutative by Vector Addition and A7.9.4 Associative by Vector Addition	$\kappa(\overset{*}{A}_m + \overset{*}{B}_m) = \kappa(\sum_{\rho(k)\in Lm} (\prod^m A_{i\lambda[i,k]} \mathbf{b}_{i\lambda[i,k]} + \prod^m B_{i\lambda[i,k]} \mathbf{b}_{i\lambda[i,k]}))$
4	From 3 and T10.2.4 Tool: Base Tensor Distributed over Polyadic Coefficient Summation	$\kappa(\overset{*}{A}_m + \overset{*}{B}_m) = \kappa(\sum_{\rho(k)\in Lm} (\prod^m A_{i\lambda[i,k]} + \prod^m B_{i\lambda[i,k]})) \prod^m \mathbf{b}_{i\lambda[i,k]}$
5	From 4 and T7.10.3 Distribution of a Scalar over Addition of Vectors	$\kappa(\overset{*}{A}_m + \overset{*}{B}_m) = \sum_{\rho(k)\in Lm} \kappa(\prod^m A_{i\lambda[i,k]} + \prod^m B_{i\lambda[i,k]}) \prod^m \mathbf{b}_{i\lambda[i,k]}$
6	From 5 and A4.2.14 Distribution	$\kappa(\overset{*}{A}_m + \overset{*}{B}_m) = \sum_{\rho(k)\in Lm} (\kappa \prod^m A_{i\lambda[i,k]} + \kappa \prod^m B_{i\lambda[i,k]}) \prod^m \mathbf{b}_{i\lambda[i,k]}$
7	From 6 and T10.2.4 Tool: Base Tensor Distributed over Polyadic Coefficient Summation	$\kappa(\overset{*}{A}_m + \overset{*}{B}_m) = \sum_{\rho(k)\in Lm} (\kappa \prod^m A_{i\lambda[i,k]} \mathbf{b}_{i\lambda[i,k]} + \kappa \prod^m B_{i\lambda[i,k]} \mathbf{b}_{i\lambda[i,k]})$
8	From 7 and A7.9.4 Associative by Vector Addition	$\kappa(\overset{*}{A}_m + \overset{*}{B}_m) = \sum_{\rho(k)\in Lm} (\kappa \prod^m A_{i\lambda[i,k]} \mathbf{b}_{i\lambda[i,k]}) + \sum_{\rho(k)\in Lm} (\kappa \prod^m B_{i\lambda[i,k]} \mathbf{b}_{i\lambda[i,k]})$
9	From 8 and T10.5.6 Scalars and Tensors: Distribution of a Scalar over Addition of Tensors	$\kappa(\overset{*}{A}_m + \overset{*}{B}_m) = \kappa\sum_{\rho(k)\in Lm} (\prod^m A_{i\lambda[i,k]} \mathbf{b}_{i\lambda[i,k]}) + \kappa\sum_{\rho(k)\in Lm} (\prod^m B_{i\lambda[i,k]} \mathbf{b}_{i\lambda[i,k]})$
∴	From 9 and T10.2.2 Tool: Tensor Series Contraction	$\kappa(\overset{*}{A}_m + \overset{*}{B}_m) = \kappa\overset{*}{A}_m + \kappa\overset{*}{B}_m$

Theorem 10.4.7 Scalars and Tensors: Distribution of a Tensor over Addition of Scalars

Steps	Hypothesis	$(\kappa + \eta)\ \overset{*}{A}_m = \kappa\overset{*}{A}_m + \eta\overset{*}{A}_m$
1	From A10.2.6A Equality of Tensors	$(\kappa + \eta)\ \overset{*}{A}_m = (\kappa + \eta)\ \overset{*}{A}_m$
2	From 1 and T10.2.1 Tool: Tensor Series Expansion	$(\kappa + \eta)\ \overset{*}{A}_m = (\kappa + \eta)\ \sum_{\rho(k)\in Lm} \prod^m A_{i\lambda[i,\ k]}\ \mathbf{b}_{i\lambda[i,\ k]}$
3	From 2 and T10.5.2 Scalars and Tensors: Distribution of Scalar across a Tensor	$(\kappa + \eta)\ \overset{*}{A}_m = \sum_{\rho(k)\in Lm} (\kappa + \eta) \prod^m A_{i\lambda[i,\ k]}\ \mathbf{b}_{i\lambda[i,\ k]}$
4	From 3 and T10.2.3 Tool: Base Tensor Parsed from Polyadic Coefficient	$(\kappa + \eta)\ \overset{*}{A}_m = \sum_{\rho(k)\in Lm} (\kappa + \eta)\ (\prod^m A_{i\lambda[i,\ k]}) \prod^m \mathbf{b}_{i\lambda[i,\ k]}$
5	From 4 and A4.2.14 Distribution	$(\kappa + \eta)\ \overset{*}{A}_m = \sum_{\rho(k)\in Lm} (\kappa\prod^m A_{i\lambda[i,\ k]} + \eta\prod^m A_{i\lambda[i,\ k]})$ $\prod^m \mathbf{b}_{i\lambda[i,\ k]}$
6	From 5 and T10.2.4 Tool: Base Tensor Distributed over Polyadic Coefficient Summation	$(\kappa + \eta)\ \overset{*}{A}_m = \sum_{\rho(k)\in Lm} (\kappa\prod^m A_{i\lambda[i,\ k]}\ \mathbf{b}_{i\lambda[i,\ k]}$ $+ \eta\prod^m A_{i\lambda[i,\ k]}\ \mathbf{b}_{i\lambda[i,\ k]})$
7	From 6 and A7.9.4 Associative by Vector Addition	$(\kappa + \eta)\ \overset{*}{A}_m = \sum_{\rho(k)\in Lm} \kappa\prod^m A_{i\lambda[i,\ k]}\ \mathbf{b}_{i\lambda[i,\ k]}$ $+ \sum_{\rho(k)\in Lm} \eta\prod^m A_{i\lambda[i,\ k]}\ \mathbf{b}_{i\lambda[i,\ k]}$
8	From 7and A4.2.14 Distribution	$(\kappa + \eta)\ \overset{*}{A}_m = \kappa\sum_{\rho(k)\in Lm} \prod^m A_{i\lambda[i,\ k]}\ \mathbf{b}_{i\lambda[i,\ k]}$ $+ \eta\sum_{\rho(k)\in Lm} \prod^m A_{i\lambda[i,\ k]}\ \mathbf{b}_{i\lambda[i,\ k]}$
∴	From 8 and T10.2.2 Tool: Tensor Series Contraction	$(\kappa + \eta)\ \overset{*}{A}_m = \kappa\overset{*}{A}_m + \eta\overset{*}{A}_m$

Theorem 10.4.8 Scalars and Tensors: Scalar Commutative with Tensors

Steps	Hypothesis	$(\kappa\overset{*}{A}_p)\ \overset{*}{B}_q = \overset{*}{A}_p\ (\kappa\overset{*}{B}_q)$
1	From A10.2.6A Equality of Tensors	$(\kappa\overset{*}{A}_p)\ \overset{*}{B}_q = (\kappa\overset{*}{A}_p)\ \overset{*}{B}_q$ κ scalar
2	From 1 and A10.2.1P5A Tensor of Rank–m	$(\kappa\overset{*}{A}_p)\ \overset{*}{B}_q = (\kappa\prod^p \mathbf{A}_i)\ (\prod^q \mathbf{A}_j)$
3	From 2 and A7.8.5 Parallelogram Law: Commutative and Scalability of Collinear Vectors	$(\kappa\overset{*}{A}_p)\ \overset{*}{B}_q = (\prod^p \mathbf{A}_i)\ (\kappa\prod^q \mathbf{A}_j)$
∴	From 3 and A10.2.1P5A Tensor of Rank–m	$(\kappa\overset{*}{A}_p)\ \overset{*}{B}_q = \overset{*}{A}_p\ (\kappa\overset{*}{B}_q)$

Section 10.5 Contracting Ranks: Definitions of Dyadic Operator

Section 10.6 Contracting Ranks: Axioms of Dyadic Operator

At this point, no mention was made of Subtraction of Rank for a simple reason. There is no definition for division of a base vector by another base vector. So, in the set of theorems T8.2.9 to T8.2.18 the only thing that can be done is to increase the rank.

While exponents and tensors have a similar product definition D4.1.17 (Exponential Notation) there is no general inverse division operation like there is for exponents. It follows than that to go backwards, subtracting rank, without an inverse operator is impossible.

However there are at least two operators that can take any two vectors and reduce them to a scalar, contraction from rank-m to rank-m–2, the orthogonal vector operators, cosine and sine, D8.5.1 and D8.5.2. So, it would not be too far out to apply them to a tensor in a selective way for any two vectors within the quiver.

Axiom 10.6.1 Dyadic Linear Tensor (DLT) Operator

$x[\overset{*}{A}_m \,|i, j> \equiv |\, A_i \,||\, A_j \,|\, (\, A_i \blacklozenge A_j \,)\, \overset{*}{B}_{m-2} \,|\, i, j >$ Equation A

or simply

$x[\overset{*}{A}_m \,|i, j> \equiv |\, A_i \,||\, A_j \,|\, (\, A_i \blacklozenge A_j \,)\, \overset{*}{B}_{m-2}$ Equation B

Where, [x] is the name of an operator and operator [♦] contracts the rank over any two-vectors. The linear operator has the property of being able to distribute term-by-term over the tensor series. The bases for this is found in Theorem 8.3.7 "Dot Product: Existence of Distribution by Dot Product Across Addition", specifically:

$c[\overset{*}{A}_m \,|i, j> \equiv |\, A_i \,||\, A_j \,|\, (A_i \odot A_j)\, \overset{*}{B}_{m-2}$ for cosine Equation C

$s[\overset{*}{A}_m \,|i, j> \equiv |\, A_i \,||\, A_j \,|\, (A_i \otimes A_j)\, \overset{*}{B}_{m-2}$ for sine Equation D

Axiom 10.6.2 Uniqueness of the DLT Operator

if $\overset{*}{A}_m \equiv \overset{*}{B}_m$ then

$x[\overset{*}{A}_m \,|i, j> \equiv x[\overset{*}{B}_m \,|\, i, j >$

Section 10.7 Contracting Ranks: Theorems of Dyadic Linear Tensor Operator

Theorem 10.7.1 Subtraction of Rank: Absolute Symmetry Cosine Orthogonal Operator

Steps	Hypothesis	$c[\overset{*}{A}_m \vert i, j> = \vert A^i \Vert A^j \vert \cos(\theta_{ij}) \overset{*}{B}_{m-2}$
1	From A10.2.6A Equality of Tensors and A10.5.2 Uniqueness of the DLT Operator, with cosine operator	$c[\overset{*}{A}_m \vert i, j> = c[\overset{*}{A}_m \vert i, j>$
2	From 1 and A10.2.1P5A Tensor of Rank–m	$c[\overset{*}{A}_m \vert i, j> = c[\prod^m A^i \vert i, j>$
3	From 2 and A10.6.1C Dyadic Linear Tensor (DLT) Operator	$c[\overset{*}{A}_m \vert i, j> = \vert A_i \Vert A_j \vert (A_i \odot A_j) \overset{*}{B}_{m-2}$
∴	From 3 and D8.5.1 Orthogonal Cosine Operator	$c[\overset{*}{A}_m \vert i, j> = \vert A_i \Vert A_j \vert \cos(\theta_{ij}) \overset{*}{B}_{m-2}$

Theorem 10.7.2 Subtraction of Rank: Skewed Symmetry Sine Orthogonal Operator

Steps	Hypothesis	$s[\overset{*}{A}_m \vert i, j> = (-1)^{\theta pr} \vert A_i \Vert A_j \vert \vert \sin(\theta_{ij}) \vert \overset{*}{B}_{m-2}$
1	From A10.2.6A Equality of Tensors and A10.5.2 Uniqueness of the DLT Operator, with cosine operator	$s\ s[\overset{*}{A}_m \vert i, j> = s[\overset{*}{A}_m \vert i, j>$
2	From 1 and A10.2.1P5A Tensor of Rank–m	$s[\overset{*}{A}_m \vert i, j> = s[\prod^m A^i \vert i, j>$
3	From 2 and A10.7.1D Dyadic Linear Tensor (DLT) Operator	$s[\overset{*}{A}_m \vert i, j> = \vert A_i \Vert A_j \vert (A_i \otimes A_j) \overset{*}{B}_{m-2}$
4	From 3, T8.5.8 Orthogonal Sine Operator: Commutative Asymmetrically Skewed, A4.2.3 Substitution and A4.2.10 Commutative Multp	$s[\overset{*}{A}_m \vert i, j> = (-1)^{\theta pr} \vert A_i \Vert A_j \vert \vert \mathbf{u}_i \otimes \mathbf{u}_j \vert \overset{*}{B}_{m-2}$
∴	From 3 and D8.5.2 Orthogonal Sine Operator	$s[\overset{*}{A}_m \vert i, j> = (-1)^{\theta pr} \vert A_i \Vert A_j \vert \vert \sin(\theta_{ij}) \vert \overset{*}{B}_{m-2}$

Section 10.8 TDL: Definitions, Tensor-Determinate Linear Operator

The DLT operators are all very nice for paired contraction as long as the rank is even, but if the rank were odd, the quiver would have one base vector left over. What is needed is an operator that accounts for all possible angular relationships between base vectors no matter how many even or odd. This would give an accurate measure of the spatial distortion for a non-orthogonal tensor space.
Let's take a standard base tensor:

$$\prod^{m} \mathbf{b}_{\lambda[i]}$$

and pair each base vector by staggering them as in this case for m = n = 8, such that

$\prod_{(i<j)} \mathbf{b}_{\lambda[i]} \otimes \mathbf{b}_{\lambda[j]}$ Producing $n^2 = 64$ pairs from the combinatorial set L_m.

The diagonal being collinear as well as the lower terms, which are symmetrical and redundant are not considered in this development. This leaves the upper terms of consequence for a total of ½n(n - 1) = 28 quantities.

Observation 10.8.1: Pairing Base Vectors from a Polyadic

1	2	3	4	5	6	7	8 = m
$b^{1\lambda[1]}$	$b^{2\lambda[2]}$	$b^{3\lambda[3]}$	$b^{4\lambda[4]}$	$b^{5\lambda[5]}$	$b^{6\lambda[6]}$	$b^{7\lambda[7]}$	$b^{8\lambda[8]}$
$b^{1\lambda[1]}\,b^{1\lambda[1]}$	$b^{1\lambda[1]}\,b^{2\lambda[2]}$	$b^{1\lambda[1]}\,b^{3\lambda[3]}$	$b^{1\lambda[1]}\,b^{4\lambda[4]}$	$b^{1\lambda[1]}\,b^{5\lambda[5]}$	$b^{1\lambda[1]}\,b^{6\lambda[6]}$	$b^{1\lambda[1]}\,b^{7\lambda[7]}$	$b^{1\lambda[1]}\,b^{8\lambda[8]}$
$b^{2\lambda[2]}\,b^{1\lambda[1]}$	$b^{2\lambda[2]}\,b^{2\lambda[2]}$	$b^{2\lambda[2]}\,b^{3\lambda[3]}$	$b^{2\lambda[2]}\,b^{4\lambda[4]}$	$b^{2\lambda[2]}\,b^{5\lambda[5]}$	$b^{2\lambda[2]}\,b^{6\lambda[6]}$	$b^{2\lambda[2]}\,b^{7\lambda[7]}$	$b^{2\lambda[2]}\,b^{8\lambda[8]}$
$b^{3\lambda[3]}\,b^{1\lambda[1]}$	$b^{3\lambda[3]}\,b^{2\lambda[2]}$	$b^{3\lambda[3]}\,b^{3\lambda[3]}$	$b^{3\lambda[3]}\,b^{4\lambda[4]}$	$b^{3\lambda[3]}\,b^{5\lambda[5]}$	$b^{3\lambda[3]}\,b^{6\lambda[6]}$	$b^{3\lambda[3]}\,b^{7\lambda[7]}$	$b^{3\lambda[3]}\,b^{8\lambda[8]}$
$b^{4\lambda[4]}\,b^{1\lambda[1]}$	$b^{4\lambda[4]}\,b^{2\lambda[2]}$	$b^{4\lambda[4]}\,b^{3\lambda[3]}$	$b^{4\lambda[4]}\,b^{4\lambda[4]}$	$b^{4\lambda[4]}\,b^{5\lambda[5]}$	$b^{4\lambda[4]}\,b^{6\lambda[6]}$	$b^{4\lambda[4]}\,b^{7\lambda[7]}$	$b^{4\lambda[4]}\,b^{8\lambda[8]}$
$b^{5\lambda[5]}\,b^{1\lambda[1]}$	$b^{5\lambda[5]}\,b^{2\lambda[2]}$	$b^{5\lambda[5]}\,b^{3\lambda[3]}$	$b^{5\lambda[5]}\,b^{4\lambda[4]}$	$b^{5\lambda[5]}\,b^{5\lambda[5]}$	$b^{5\lambda[5]}\,b^{6\lambda[6]}$	$b^{5\lambda[5]}\,b^{7\lambda[7]}$	$b^{5\lambda[5]}\,b^{8\lambda[8]}$
$b^{6\lambda[6]}\,b^{1\lambda[1]}$	$b^{6\lambda[6]}\,b^{2\lambda[2]}$	$b^{6\lambda[6]}\,b^{3\lambda[3]}$	$b^{6\lambda[6]}\,b^{4\lambda[4]}$	$b^{6\lambda[6]}\,b^{5\lambda[5]}$	$b^{6\lambda[6]}\,b^{6\lambda[6]}$	$b^{6\lambda[6]}\,b^{7\lambda[7]}$	$b^{6\lambda[6]}\,b^{8\lambda[8]}$
$b^{7\lambda[7]}\,b^{1\lambda[1]}$	$b^{7\lambda[7]}\,b^{2\lambda[2]}$	$b^{7\lambda[7]}\,b^{3\lambda[3]}$	$b^{7\lambda[7]}\,b^{4\lambda[4]}$	$b^{7\lambda[7]}\,b^{5\lambda[5]}$	$b^{7\lambda[7]}\,b^{6\lambda[6]}$	$b^{7\lambda[7]}\,b^{7\lambda[7]}$	$b^{7\lambda[7]}\,b^{8\lambda[8]}$
$b^{8\lambda[8]}\,b^{1\lambda[1]}$	$b^{8\lambda[8]}\,b^{2\lambda[2]}$	$b^{8\lambda[8]}\,b^{3\lambda[3]}$	$b^{8\lambda[8]}\,b^{4\lambda[4]}$	$b^{8\lambda[8]}\,b^{5\lambda[5]}$	$b^{8\lambda[8]}\,b^{6\lambda[6]}$	$b^{8\lambda[8]}\,b^{7\lambda[7]}$	$b^{8\lambda[8]}\,b^{8\lambda[8]}$

Using the DLT sine operator in its skewed form on the above table, for all n^2 factors, with common indices eliminates any base tensor having any collinear vector pairs, since the sine is zero. In a tensor product there are n^m terms over the combinatorial set L_m, any term having a zero factor would ***cross out*** all duplicate combinations (including diagonal elements) leaving a set of permutations S_r with [½n(n–1)] quantities. With this in mind the above product can be rewritten as follows:

Definition 10.8.1 Sine Space Distortion Factor (SSDF)

$\mu_s(k) \equiv \mu_s(\rho(k)) \equiv \prod_{(i<j)} |\sin(\theta_{\lambda[i,k]\lambda[j,k]})|$ for $(\lambda[i,k], \lambda[j,k]) \in \rho(k)$ Equation A

$\mu_s = \prod_{(i<j)} |\sin(\theta_{i,j})|$ for all k, absolute value eliminates relative vector signs Equation B

Definition 10.8.2 Parity of Permutation for Base Vector (PPBV)

$\phi(k) \equiv \phi(\rho(k)) \equiv \sum_{(i<j)} \theta pr(\lambda[i,k], \lambda[j,k])$

for θpr defined in D8.5.4 "Associated Parity Between any Two Vectors (γ[p], γ[q])"

This has its origins deep in Chapter 6, D6.7.1 "Association of parity to permutated indices", D6.7.2 "Association of a parity to permutation" and D8.5.8 "Orthogonal Sine Operator: Commutative Asymmetrically Skewed".

Definition 10.8.3 Factor of Magnitude for Base Vector (FMBV)

$$f(\mathrm{k}) \equiv f(\rho(\mathrm{k})) \equiv \prod_{(i<j)} |\mathbf{b}_{\lambda[i,\,k]}|\,|\,\mathbf{b}_{\lambda[j,\,k]}\,|$$

Definition 10.8.4 Composite Factor of Distortion (CFD)

$$\mu_f(\mathrm{k}) \equiv \mu_f(\rho(\mathrm{k})) \equiv \mu_s\, f(\rho(\mathrm{k}))$$

Definition 10.8.5 Tensor Coefficient for Distortion (TCD)

Tensor Coefficient for Distortion (TCD) is defined over the summation set-L_m for the following definitions are true:

$$\overset{\times}{e}(\mathrm{k}) \equiv \langle |\textstyle\prod^m \mathbf{b}_{\lambda[i,\,k]}\,| \rangle \qquad \text{Equation A}$$

$$\overset{\times}{e} \equiv \prod_{(i<j)} \mathbf{b}_{\lambda[i,\,k]} \otimes \mathbf{b}_{\lambda[j,\,k]} \qquad \text{Equation B}$$

$$\overset{\times}{e} \equiv \mu_f \begin{cases} (-1)^{\varphi(k)} & \text{No repetitive indices} \\ 0 & \text{Repetitive indice} \end{cases} \qquad \text{Equation C}$$

$$\overset{\times}{e} \equiv \mu_f\, \mathbf{e}_{\rho(k)} \qquad \text{Equation D}$$

Repetitive indices are analogous to the inner product of collinear vectors.

Operator contracts a tensor of Rank-m to Rank-0 by operating on the base tensors.

Definition 10.8.6 Cross Coefficient for Distortion (CCD)

$$\mathbf{e}_{\rho(k)} \equiv \overset{\times}{e}(\mathrm{k}) \,/\, \mu_f$$

Section 10.9 TDL: Axioms, Tensor-Determinate Linear Operator

Axiom 10.9.1 Equality of the Tensor-Determinate Linear Operator

if $\overset{*}{\mathring{A}}_m \equiv \overset{*}{\mathring{A}}_m$ then

$<| \overset{*}{\mathring{A}}_m |> \equiv <| \overset{*}{\mathring{A}}_m |>$

Axiom 10.9.2 Tensor-Determinate Linear Operator (TDL)

$<| \overset{*}{\mathring{A}}_m |> \equiv \sum_{\rho(k) \in Lm} <| \prod^m \mathbf{b}_{\lambda[i,\, k]} |> \prod^m A_{i\lambda[i,\, k]}$

It is linear, because it can be distributed over the tensor summation term-by-term.

Section 10.10 TDL: Theorems, Tensor-Determinate Linear Operator

Theorem 10.10.1 TDL Operator Defined with CCD

Steps	Hypothesis	$<\mid \overset{*}{A}_m \mid> = \Sigma_{\rho(k)\in Lm} \overset{\times}{e}(k) \prod^m A_{i\lambda[i,k]}$
∴	From A10.2.6A Equality of Tensors, A10.10.2 Equality of the Tensor-Determinate Linear Operator, T10.2.1 Tool: Tensor Series Expansion and D10.9.5 Tensor Coefficient of Distortion (TCD)	$<\mid \overset{*}{A}_m \mid> = \Sigma_{\rho(k)\in Lm} \overset{\times}{e}(k) \prod^m A_{i\lambda[i,k]}$

Theorem 10.10.2 Distribution of TDL Operator

1g	Given	$\prod^m C_{i\lambda[i,k]} = \prod^m A_{i\lambda[i,k]} + \prod^m B_{i\lambda[i,k]}$
Steps	Hypothesis	$<\mid \overset{*}{C}_m \mid> \equiv <\mid \overset{*}{A}_m \mid> + <\mid \overset{*}{B}_m \mid>$
1	From T10.11.1 TDL Operator Defined with CCD	$<\mid \overset{*}{C}_m \mid> = \Sigma_{\rho(k)\in Lm} \overset{\times}{e}(k) \prod^m C_{i\lambda[i,k]}$
2	From 1, 1g and T4.2.3 Substitution	$<\mid \overset{*}{C}_m \mid> = \Sigma_{\rho(k)\in Lm} \overset{\times}{e}(k) (\prod^m A_{i\lambda[i,k]} + \prod^m B_{i\lambda[i,k]})$
3	From 2 and A4.2.14 Distribution	$<\mid \overset{*}{C}_m \mid> = \Sigma_{\rho(k)\in Lm} (\overset{\times}{e}(k) \prod^m A_{i\lambda[i,k]} + \overset{\times}{e}(k) \prod^m B_{i\lambda[i,k]})$
4	From 3, A4.2.5 Commutative Add and A4.2.6 Associative Add	$<\mid \overset{*}{C}_m \mid> = \Sigma_{\rho(k)\in Lm} \overset{\times}{e}(k) \prod^m A_{i\lambda[i,k]} + \Sigma_{\rho(k)\in Lm} \overset{\times}{e}(k) \prod^m B_{i\lambda[i,k]}$
∴	From 4 and T10.11.1 TDL Operator Defined with CCD	$<\mid \overset{*}{C}_m \mid> \equiv <\mid \overset{*}{A}_m \mid> + <\mid \overset{*}{B}_m \mid>$

Theorem 10.10.3 The Cross Coefficient of Distortion

Steps	Hypothesis	$\overset{\times}{e}(k) = \mu_f(k)\ (-1)^{\phi(\rho(k))}$
1	From D10.8.5 Tensor Coefficient for Distortion (TCD)	$\overset{\times}{e}(k) = \prod_{(i<j)} \mathbf{b}_{\lambda[i,k]} \otimes \mathbf{b}_{\lambda[j,k]}$
2	From 1 and T7.10.8 Vector Based on Unit Vector	$\overset{\times}{e}(k) = \prod_{(i<j)} (\mid\mathbf{b}_{\lambda[i,k]}\mid \mathbf{U}_{\lambda[i,k]}) \otimes (\mid\mathbf{b}_{\lambda[j,k]}\mid \mathbf{U}_{\lambda[j,k]})$
3	From 2, A4.2.10 Commutative Multp, T8.5.8 Orthogonal Sine Operator: Commutative Asymmetrically Skewed	$\overset{\times}{e}(k) = \prod_{(i<j)} \mid\mathbf{b}_{\lambda[i,k]}\mid \mid\mathbf{b}_{\lambda[j,k]}\mid (-1)^{\theta pr(\lambda[i,k]\lambda[j,k])} \mid \mathbf{U}_{\lambda[j,k]} \otimes \mathbf{U}_{\lambda[i,k]} \mid$
4	From 3 and D8.5.2 Orthogonal Sine Operator	$\overset{\times}{e}(k) = \prod_{(i<j)} \mid\mathbf{b}_{\lambda[i,k]}\mid \mid\mathbf{b}_{\lambda[j,k]}\mid (-1)^{\theta pr(\lambda[i,k]\lambda[j,k])} \mid \sin(\theta_{\lambda[i,k]\lambda[i,k]}) \mid$
5	From 4, A4.2.11 Associative Multp, T2.21.9 Multiplication Sets: Binary Distribution of Rank for Separation of Sets, D10.9.1B Sine **Space Distortion Factor (SSDF)**, A4.2.3 Substitution and A4.2.10 Commutative Multp	$\overset{\times}{e}(k) = \mu_s \prod_{(i<j)} \mid\mathbf{b}_{\lambda[i,k]}\mid \mid\mathbf{b}_{\lambda[j,k]}\mid \prod_{(i<j)} (-1)^{\theta pr(\lambda[i,k]\lambda[j,k])}$
6	From 5, A4.2.11 Associative Multp and T2.21.9 Multiplication Sets: Binary Distribution of Rank for Separation of Sets	$\overset{\times}{e}(k) = \mu_s \prod_{(i<j)} \mid\mathbf{b}_{\lambda[i,k]}\mid \mid\mathbf{b}_{\lambda[j,k]}\mid \prod_{(i<j)} (-1)^{\theta pr(\lambda[i,k]\lambda[j,k])}$
7	From 6 and A4.2.10 Commutative Multp	$\overset{\times}{e}(k) = \mu_s \prod_{(i<j)} \mid\mathbf{b}_{\lambda[i,k]}\mid \mid\mathbf{b}_{\lambda[j,k]}\mid \prod_{(i<j)} (-1)^{\theta pr(\lambda[i,k]\lambda[j,k])}$
8	From 7 and A4.2.18 Summation Exp	$\overset{\times}{e}(k) = \mu_s \prod_{(i<j)} \mid\mathbf{b}_{\lambda[i,k]}\mid \mid\mathbf{b}_{\lambda[i,k]}\mid (-1)^{\Sigma(i<j)\ \theta pr(\lambda[i,k]\lambda[j,k])}$
9	From 8, D10.9.1 Sine Space Distortion Factor (SSDF), D10.9.2 Parity of Permutation for Base Vector (PPBV), D10.9.3 Factor of Magnitude for Base Vector (FMBV) and T4.2.3 Substitution	$\overset{\times}{e}(k) = \mu_s f(\rho(k))\ (-1)^{\phi(\rho(k))}$
∴	From 9, D10.8.4 Composite Factor of Distortion (CFD) and T4.2.3 Substitution	$\overset{\times}{e}(k) = \mu_f(k)\ (-1)^{\phi(\rho(k))}$

Corollary 10.10.3.1 The Cross Coefficient as Parity Sign for Permuted Base Vectors

Steps	Hypothesis	$\mathbf{e}_{\rho(k)} = (-1)^{\phi(k)}$
1	From D10.8.6 Cross Coefficient of Distortion (CCD)	$\mathbf{e}_{\rho(k)} = \overset{\times}{e}(k) / \mu_f(k)$
2	From 1, T10.10.3 The Cross Coefficient of Distortion and T10.2.5 Tool: Tensor Substitution	$\mathbf{e}_{\rho(k)} = [\mu_f(k)\ (-1)^{\phi(\rho(k))}] / \mu_f(k)$
∴	From 1, T10.3.7 Product of Tensors: Commutative by Multiplication of Rank p + q → q + p, T10.3.8 Product of Tensors: Association by Multiplication of Rank for p + q + r, T10.3.10 Product of Tensors: Reduction of Tensors by Inverse of Rank for m–m and A4.2.12 Identity Multp	$\mathbf{e}_{\rho(k)} = (-1)^{\phi(k)}$

Theorem 10.10.4 Development of the Coefficient of Distortion for Sine

Steps	Hypothesis	$\mu_s(k) = \prod_{(i<j)} \vert \sin(\theta_{i,j}) \vert$
1	From D10.9.1B Sine Space Distortion Factor (SSDF)	$\mu_s(k) = \prod_{(i<j)} \vert \sin(\theta_{i,j}) \vert$ for all k
2	From 1g, 1, A5.2.18 GIV Transitivity of Angles, DxE.1.6.1 and T4.4.1 Equalities: Any Quantity Multiplied by Zero is Zero	$\mu_s(k) = \prod_{(i<j)} \vert \sin(\theta_{i,j}) \vert = 0$ $\theta_{i,j} = 0$ at least one pair is collinear
3	From 2g, 1 and A5.2.18 GIV Transitivity of Angles	$\mu_s(k) = \prod_{(i<j)} \vert \sin(\theta_{i,j}) \vert \neq 0$ $\theta_{i,j} \neq 0$ all pairs are non-collinear
∴	From 2 and 3	$\mu_s(k) = \prod_{(i<j)} \vert \sin(\theta_{i,j}) \vert$

Theorem 10.10.5 TDL Operator Summed over the Permutation Set S_r

1g 2g	Given	for $m \geq n$ $\lambda[i, k] = \lambda[j, k]$ at least one pair is collinear
Steps	Hypothesis	$<\vert \overset{*}{A}_m \vert> = \sum_{\sigma(k)\in Sr} \mu_f(k)\ \mathrm{sgn}(\sigma(k)) \prod^n A_{i\lambda[i,k]}$
1	From T10.11.1 TDL Operator Defined with CCD	$<\vert \overset{*}{A}_m \vert> = \sum_{\rho(k)\in Lm} \overset{\times}{e}(k) \prod^m A_{i\lambda[i,k]}$
2	From 1, C10.11.3.1 The Cross Coefficient as Composite Factor of Distortion and T4.2.3 Substitution	$<\vert \overset{*}{A}_m \vert> = \sum_{\rho(k)\in Lm} \mu_f(k)\ (-1)^{\phi(k)} \prod^m A_{i\lambda[i,k]}$
3	From 2g, 2, T10.11.4 Development of the Coefficient of Distortion for Sine, T10.11.5 TDL Operator Summed over the Permutation Set S_r, T4.4.1 Equalities: Any Quantity Multiplied by Zero is Zero and A4.2.7 Identity Add, dropping all zeroed terms leaves the set of permutated terms belonging to S_r and summed over all $\sigma(k)$ for the appropriate parity or sign.	$<\vert \overset{*}{A}_m \vert> = \sum_{\sigma(k)\in Sr} \mu_f(k)\ (-1)^{\sigma(k)} \prod^n A_{i\lambda[i,k]}$
∴	From 3 and D6.6.3A Association of sign to parity	$<\vert \overset{*}{A}_m \vert> = \sum_{\sigma(k)\in Sr} \mu_f(k)\ \mathrm{sgn}(\sigma(k)) \prod^n A_{i\lambda[i,k]}$

Corollary 10.10.5.1 Evaluation of $\mu_s(k)$ as Orthogonal

1g	Given	$\| \theta_{i,j} \| = 90^o$ for all orthogonal angles spanning the space
Steps	Hypothesis	$\mu_s = 1$
1	From T10.11.4 Development of the Coefficient of Distortion for Sine	$\mu_s = \prod_{(i<j)} \|\sin(\theta_{i,j})\|$ the non-collinear base
2	From 1g, 1 and A5.2.18 GIV Transitivity of Angles	$\mu_s = \prod_{(i<j)} \|\sin(90^o)\|$
3	From 2, DxE.1.6.7 and D4.1.11 Absolute Value | a |	$\mu_s = \prod_{(i<j)} 1$
∴	From 2 and A4.2.12 Identity Multp	$\mu_s = 1$

Corollary 10.10.5.2 Evaluation of $f(k)$ as Unitary

1g	Given	$\|\mathbf{b}_{\lambda[i,k]}\| = 1$ for all i
Steps	Hypothesis	$f(k) = 1$
1	From D10.9.3 Factor of Magnitude for Base Vector (FMBV)	$f(k) = \prod_{(i<j)} \|\mathbf{b}_{\lambda[i,k]}\| \|\mathbf{b}_{\lambda[j,k]}\|$
2	From 1g, 1 and T4.2.3 Substitution	$f(k) = \prod_{(i<j)} 1\ 1$
∴	From 2 and A4.2.12 Identity Multp	$f(k) = 1$

Corollary 10.10.5.3 Bounds for Sine Space Factor Distortion

1g	Given	$\| \theta_{i,j} \| = 90^o$ for all orthogonal angles spanning the space	
Steps	Hypothesis	$0 < \mu_s \le 1$	
1	From T10.11.4 Development of the Coefficient of Distortion for Sine	$\mu_s = 0$	for the collinear base, not allowed degenerate case
2	From 1g, 1 and A5.2.18 GIV Transitivity of Angles	$\mu_s = \prod_{(i<j)} \|\sin(90^o)\|$	
3	From 2, DxE.1.6.7 and D4.1.11 Absolute Value | a |	$\mu_s = \prod_{(i<j)} 1$	
∴	From 2 and A4.2.12 Identity Multp	$0 < \mu_s \le 1$	

Corollary 10.10.5.4 Absolute Value for Sine Space Distortion Factor

Steps	Hypothesis	$\mid \mu_s(k) \mid = +\mu_s(k)$ EQ A	$+\mu_s(k) = \mid \mu_s(k) \mid$ EQ B
1	From D10.8.1B Sine Space Distortion Factor (SSDF)) and D4.1.11B Absolute Value | a |; product is absolute	$0 \le \mu_s = \prod_{(i<j)} \mid \sin(\theta_{i,j}) \mid$	
∴	From 1 D4.1.11B Absolute Value | a | and A4.2.2B Equality	$\mid \mu_s(k) \mid = +\mu_s(k)$ EQ A	$+\mu_s(k) = \mid \mu_s(k) \mid$ EQ B

Corollary 10.10.5.5 Absolute Value for Composite Factor of Distortion

Steps	Hypothesis	$\mid \mu_f(k) \mid = +\mu_f(k)$ EQ A	$+\mu_f(k) = \mid \mu_f(k) \mid$ EQ B
1	From D10.8.4 Composite Factor of Distortion (CFD)	$\mu_f(k) = \mu_s f(\rho(k))$	
2	From D10.8.3 Factor of Magnitude for Base Vector (FMBV and D4.1.11B Absolute Value | a |; product is absolute	$0 \le f(k) = \prod_{(i<j)} \mid \mathbf{b}_{\lambda[i,k]} \mid \mid \mathbf{b}_{\lambda[j,k]} \mid$	
3	From D10.8.1B Sine Space Distortion Factor (SSDF)) and D4.1.11B Absolute Value | a |; product is absolute	$0 \le \mu_s = \prod_{(i<j)} \mid \sin(\theta_{i,j}) \mid$	
4	From 2, 3 and T4.7.21 Inequalities: Multiplication for Positive Numbers	$0 \le \mu_s f(k)$	
5	From 4, T4.7.21 Inequalities: Multiplication for Positive Numbers and A4.2.3 Substitution	$0 \le \mu_f(k)$	
∴	From 5 D4.1.11B Absolute Value | a | and A4.2.2B Equality	$\mid \mu_f(k) \mid = +\mu_f(k)$ EQ A	$+\mu_f(k) = \mid \mu_f(k) \mid$ EQ B

Section 10.11 TDL as Determinate: Definitions

The Tensor under the TDL operator has the same equation as defined for a determinate, except for the composite factor of distortion, CFD. Since the TDL works on the polyadic of the tensor than it should be independent of the CFD. This would imply that all of the theorems for determinate work for TDL operations as well the proof of this is in the following theorems.

Definition 10.11.1 TDL as a Determinate Operator

$\det(\overset{*}{A}_m) \equiv | \overset{*}{A}_m | \equiv < \overset{*}{A}_m > = \sum_{\sigma(k)\in Sr} \mu_f(k)\ (-1)^{\phi(k)} \prod^n A_{i\lambda[i,\, k]}$ Equation A

$| \overset{*}{A}_m | \equiv \sum_{\sigma(k)\in Sr} \mu_f(k)\ \text{sgn}(\ \text{parityn}(\sigma(k))\ ) \prod^n A_{i\lambda[i,\, k]}$ Equation B

$| \overset{*}{A}_m | \equiv \sum_{\sigma(k)\in Sr} \mu_f(k)\ \text{sgn}(\sigma,\, k) \prod^n A_{i\lambda[i,\, k]}$ Equation C

$| \overset{*}{A}_m | \equiv \sum_{\sigma(k)\in Sr} \overset{*}{e}(k) \prod^n A_{i\lambda[i,\, k]}$ Equation D

Definition 10.11.2 Permutation Identity Distortion Factor (PIDF)

$\mu_f(\varepsilon) \ = \mu_\sigma f(\varepsilon) = \mu_s f(\sigma(1))$ Equation A

for

$\mu_f \ \equiv \mu_f(\varepsilon)$ Equation B

$\mu_\sigma \ \equiv \mu_\sigma(\varepsilon)$ Equation C

Section 10.12 TDL as Determinate: Axioms

Section 10.13 TDL as Determinate: Theorems

Theorem 10.13.1 CFDs Reduce to a Identity Permutation CFD

1g	Given	$\sigma \in Sr$
Steps	Hypothesis	$\mu_f(\varepsilon) = \mu_\sigma f(\varepsilon)$
1	From 1g and D10.9.4 Composite Factor of Distortion (CFD)	$\mu_f(\sigma(k)) \equiv \mu_s f(\sigma(k))$
2	From 1, the permutation set Sr, $\mu(\sigma(k))$ and $f(\sigma(k))$ are absolute magnitudes, hence order independent, reordering.	$k = 1$
3	From 1 and 2	$\mu_f(\sigma(1)) = \mu_s f(\sigma(1))$
∴	From 3, D6.5.5 The identity permutation, D10.11.2A Permutation Identity Distortion Factor (PIDF) and A4.2.2 Equality	$\mu_f(\varepsilon) = \mu_\sigma f(\varepsilon)$

Theorem 10.13.2 TDL Relationship: And the Determinate Operator

Steps	Hypothesis	$<\mid \overset{*}{A}_m \mid> \equiv \mu_f \mid A \mid$ for [A] a matrix comprised of tensor components by row
1	From T10.10.5 TDL Operator Summed over the Permutation Set Sr	$<\mid \overset{*}{A}_m \mid> = \sum_{\sigma(k) \in Sr} \mu_f(k) \operatorname{sgn}(\sigma(k)) \prod^n A_{i\lambda[i, k]}$
2	From 1, T10.13.1 CFDs Reduce to a Identity Permutation CFD and T4.2.3 Substitution	$<\mid \overset{*}{A}_m \mid> = \sum_{\sigma(k) \in Sr} \mu_f \operatorname{sgn}(\sigma(k)) \prod^n A_{i\lambda[i, k]}$
3	From 2 and A4.2.14 Distribution	$<\mid \overset{*}{A}_m \mid> = \mu_f \sum_{\sigma(k) \in Sr} \operatorname{sgn}(\sigma(k)) \prod^n A_{i\lambda[i, k]}$
∴	From 3, D6.9.1E Determinant of A and T4.2.3 Substitution	$<\mid \overset{*}{A}_m \mid> = \mu_f \mid A \mid$ for [A] a matrix comprised of tensor components by row

Theorem 10.13.3 TDL Relationship: TDL Constant of Proportionality

Steps	Hypothesis	$\mu_f \equiv <\mid \overset{*}{A}_m \mid> / \mid A \mid$
1	From T10.13.2 TDL Relationship: And the Determinate Operator and A4.2.2B Equality	$\mu_f \mid A \mid = <\mid \overset{*}{A}_m \mid>$
∴	From 1 and T4.4.6B Equalities: Left Cancellation by Multiplication	$\mu_f = <\mid \overset{*}{A}_m \mid> / \mid A \mid$

Theorem 10.13.4 TDL Relationship: Identical Rows

1g 2g	Given	Row $a_{ji} = a_{ki}$ for i = k and all j's Yielding new determinate $\lvert B \rvert$
Steps	Hypothesis	$<\lvert \overset{*}{A}_m \rvert> = 0$ for row $a_{ji} = a_{ki}$ for i = k and all j's
1	From T10.13.2 TDL Relationship: And the Determinate Operator	$<\lvert \overset{*}{A}_m \rvert> = \mu_f \lvert A \rvert$
2	From 1 and T6.9.5A Expanding Determinant [A] of order [n] about a Row or Column	$<\lvert \overset{*}{A}_m \rvert> = \mu_f (\Sigma_{j=1}^{n} a_{ji} A_{ji})$ the ith row
3	From 1g, 2 and A7.9.2 Substitution of Vectors	$<\lvert \overset{*}{A}_m \rvert> = \mu_f (\Sigma_{j=1}^{n} a_{ki} A_{ji})$ the ith row
4	From 2g, 3, T6.9.5A Expanding Determinant [A] of order [n] about a Row or Column and A4.2.3 Substitution	$<\lvert \overset{*}{A}_m \rvert> = \mu_f \lvert B \rvert$ for row $a_{ji} = a_{ki}$ for i = k and all j's
5	From 4 and T6.9.10 A determinant with two identical rows or columns is zero	$<\lvert \overset{*}{A}_m \rvert> = \mu_f 0$
∴	From 5 and T4.4.1 Equalities: Any Quantity Multiplied by Zero is Zero	$<\lvert \overset{*}{A}_m \rvert> = 0$ for row $a_{ji} = a_{ki}$ for i = k and all j's

Theorem 10.13.5 TDL Relationship: Having Identical Columns

1g 2g	Given	Row $a_{ij} = a_{ik}$ for j = k and all i's Yielding new determinate $\lvert B \rvert$
Steps	Hypothesis	$<\lvert \overset{*}{A}_m \rvert> = 0$ for columns $a_{ij} = a_{ik}$ for j = k and all i's
1	From T10.13.2 TDL Relationship: And the Determinate Operator	$<\lvert \overset{*}{A}_m \rvert> = \mu_f \lvert A \rvert$
2	From 1 and T6.9.5B Expanding Determinant [A] of order [n] about a Row or Column	$<\lvert \overset{*}{A}_m \rvert> = \mu_f (\Sigma_{i=1}^{n} a_{ji} A_{ji})$ the jth columns
3	From 1g, 2 and A7.9.2 Substitution of Vectors	$<\lvert \overset{*}{A}_m \rvert> = \mu_f (\Sigma_{j=1}^{n} a_{ik} A_{ji})$ the jth columns
4	From 2g, 3, T6.9.5B Expanding Determinant [A] of order [n] about a Row or Column and A4.2.3 Substitution	$<\lvert \overset{*}{A}_m \rvert> = \mu_f \lvert B \rvert$ for columns $a_{jj} = a_{ik}$ for j = k and all i's
5	From 4 and T6.9.10 determinant with two identical rows or columns is zero	$<\lvert \overset{*}{A}_m \rvert> = \mu_f 0$
∴	From 5 and T4.4.1 Equalities: Any Quantity Multiplied by Zero is Zero	$<\lvert \overset{*}{A}_m \rvert> = 0$ for columns $a_{jj} = a_{ik}$ for j = k and all i's

Theorem 10.13.6 TDL Relationship: Having Zero Rows or Columns

1g 2g	Given	$a_{ij} = 0$ for any i-row $a_{ij} = 0$ for any j-column
Steps	Hypothesis	$<\mid \overset{*}{\mathring{A}}_m \mid> = 0$ for any zeroed row or column
1	From T10.13.2 TDL Relationship: And the Determinate Operator	$<\mid \overset{*}{\mathring{A}}_m \mid> = \mu_f \mid A \mid$
2	From 1g, 2g, 1 and T6.9.6 The determinant of a matrix having a row or column of zeros is zero	$<\mid \overset{*}{\mathring{A}}_m \mid> = \mu_f\, 0$
∴	From 5 and T4.4.1 Equalities: Any Quantity Multiplied by Zero is Zero	$<\mid \overset{*}{\mathring{A}}_m \mid> = 0$ for any zeroed row or column

Section 10.14 Cross Product Operator: Definitions

Definition 10.14.1 Cross Product

The cross product is a set-H of (n–1)-vectors $\mathbf{A}_\beta$ spanning Riemannian n-space and when multiplied together follows, the Hamilton and Grassmann's sense of rotation, yielding an orthogonal vector $\mathfrak{X}_h$ to the set:

$\mathfrak{X}_h \equiv \prod_{\beta=2}^{n} \times \mathbf{A}_\beta$ Equation A

$H \equiv \{\mathbf{A}_\beta, \mathfrak{X}_h \mid (\mathbf{A}_\beta, \mathfrak{X}_h) \in \mathfrak{R}^n, \mathfrak{X}_h \perp \mathbf{A}_\beta \text{ for } \beta = 2, 3, \ldots n\}$ Equation C

Section 10.15 Cross Product Operator: Axioms

Lets consider a set of tensor vectors crossing one another yielding a single resulting vector by using the T10.13.4 "TDL Relationship: Identical Rows" and T8.3.11 "Dot Product: Perpendicular Vectors" theorems. These theorems preserve the orthogonal cross product for set-H so by definition the cross product is setup in determinate form. This requires embedding the set of vectors and having them span a Riemannian n-space. In order to achieve this the first row of polyadic coefficients is set as base vectors with the appropriate sign orientation as defined in the following axiom:

Axiom 10.15.1 The Cross Product Operator in Algebraic non-Vector Form

$\prod_{\beta=2}^{n} \times \mathbf{A}_\beta \equiv \sum_{\sigma(k)\in S_r} \mathrm{sgn}(\sigma(k)) \prod_{i=1}^{n} A_{i\lambda[i,k]}$ for Equation A

$A_{1\lambda[1,k]} \equiv \mathbf{b}_{\lambda[1,k]}$ Equation B

Axiom 10.15.2 The Cross Product Cyclic Permutated Bases Vectors

The right hand, side places the base vectors into a one-to-one correspondence with the base tensor.

$\mathrm{sgn}(\sigma(k))\mathbf{b}_{\lambda[1,k]} \equiv \prod_{i=2} \times \mathbf{b}_{\lambda[i,k]}$ for $\sigma(k) \in S_r$ Equation A

Where the resulting base vector is derived from the base tensor under a special operation called the ***cross product multiplier***.

Axiom 10.15.3 Cross Product Operator: Sense of Rotation for Cyclic Permutation

Iff the Cross Product Operator is taken over the ordered permutation set S_r the sense of rotation will always be Counter-Clockwise as the sense of rotation of angle is defined through quadrants.

This is also Hamilton and Grassmann's sense of rotation or right-hand rule; see D7.6.1 "The Right-hand Rule".

Section 10.16 Cross Product Operator: Theorems

Theorem 10.16.1 Cross Product Operator: Expanded Algebraic Vector Form

1g	Given	$A_{1\lambda[1,k]} = \mathbf{b}_{\lambda[1,k]}$
Steps	Hypothesis	$\prod_{\beta=2}^{n} \times \mathbf{A}_\beta = \sum_{\sigma(k)\in Sr} sgn(\sigma(k))\, \mathbf{b}_{\lambda[1,k]} \prod_{i=2}^{n} A_{i\lambda[i,k]}$
1	From A10.15.1A The Cross Product Operator in Algebraic non-Vector Form	$\prod_{\beta=2}^{n} \times \mathbf{A}_\beta = \sum_{\sigma(k)\in Sr} sgn(\sigma(k)) \prod_{i=1}^{n} A_{i\lambda[i,k]}$
2	From 1 and expanding product by first factor	$\prod_{\beta=2}^{n} \times \mathbf{A}_\beta = \sum_{\sigma(k)\in Sr} sgn(\sigma(k))\, A_{1\lambda[1,k]} \prod_{i=2}^{n} A_{i\lambda[i,k]}$
∴	From 2, A10.15.1B The Cross Product Operator in Algebraic non-Vector Form, T10.2.5 Tool: Tensor Substitution and A10.2.14 Correspondence of Tensors and Tensor Coefficients	$\prod_{\beta=2}^{n} \times \mathbf{A}_\beta = \sum_{\sigma(k)\in Sr} sgn(\sigma(k))\, \mathbf{b}_{\lambda[1,k]} \prod_{i=2}^{n} A_{i\lambda[i,k]}$

Theorem 10.16.2 Cross Product Operator: Determinate Form

Steps	Hypothesis	$\prod_{\beta=2}^{n} \times \mathbf{A}_\beta = \begin{vmatrix} \bar{b}_1 & \bar{b}_2 & \cdots & \bar{b}_n \\ A_{12} & A_{22} & \cdots & A_{n2} \\ \vdots & \vdots & & \vdots \\ A_{1\beta} & A_{2\beta} & \cdots & A_{n\beta} \\ \vdots & \vdots & & \vdots \\ A_{1n} & A_{2n} & \cdots & A_{nn} \end{vmatrix}$
1	From T10.16.1 Cross Product Operator: Expanded Algebraic Vector Form	$\prod_{\beta=2}^{n} \times \mathbf{A}_\beta = \sum_{\sigma(k)\in Sr} sgn(\sigma(k))\, \mathbf{b}_{\lambda[1,k]} \prod_{i=2}^{n} A_{i\lambda[i,k]}$
∴	From 1 and D6.9.1C Determinant of A	$\prod_{\beta=2}^{n} \times \mathbf{A}_\beta = \begin{vmatrix} \bar{b}_1 & \bar{b}_2 & \cdots & \bar{b}_n \\ A_{12} & A_{22} & \cdots & A_{n2} \\ \vdots & \vdots & & \vdots \\ A_{1\beta} & A_{2\beta} & \cdots & A_{n\beta} \\ \vdots & \vdots & & \vdots \\ A_{1n} & A_{2n} & \cdots & A_{nn} \end{vmatrix}$

Theorem 10.16.3 Cross Product Operator: Vector Offset Index Starts from 2

1g	Given	$j = i-1$
Steps	Hypothesis	$\prod_{\gamma=1}^{n-1} \times \mathbf{A}_\gamma = \begin{vmatrix} \bar{b}_1 & \bar{b}_2 & \cdots & \bar{b}_n \\ A_{11} & A_{21} & \cdots & A_{n1} \\ \vdots & \vdots & & \vdots \\ A_{1\alpha} & A_{2\alpha} & \cdots & A_{n\alpha} \\ \vdots & \vdots & & \vdots \\ A_{1(n-1)} & A_{2(n-1)} & \cdots & A_{n(n-1)} \end{vmatrix}$
1	From T10.16.1 Cross Product Operator: Expanded Algebraic Vector Form	$\prod_{\gamma=1}^{n-1} \times \mathbf{A}_\gamma = \sum_{\sigma(k)\in S_r} \mathrm{sgn}(\sigma(k))\, \mathbf{b}_{\lambda[1,k]} \prod_{i=2}^{n} A_{i\lambda[i,k]}$
2	From 1 and new index set-$\eta(j)$	$\prod_{\gamma=1}^{n-1} \times \mathbf{A}_\gamma = \sum_{\sigma(k)\in S_r} \mathrm{sgn}(\sigma(k))\, \mathbf{b}_{\eta[1,k]} \prod_{i=1}^{n} A_{(i-1)\,\eta[i-1,k]}$
3	From 1g, 2 and A4.2.3 Substitution	$\prod_{\gamma=1}^{n-1} \times \mathbf{A}_\gamma = \sum_{\sigma(k)\in S_r} \mathrm{sgn}(\sigma(k))\, \mathbf{b}_{\eta[1,k]} \prod_{j=1}^{n-1} A_{j\,\eta[j,k]}$
∴	From 3 and D6.9.1C Determinant of A	$\prod_{\gamma=1}^{n-1} \times \mathbf{A}_\gamma = \begin{vmatrix} \bar{b}_1 & \bar{b}_2 & \cdots & \bar{b}_n \\ A_{11} & A_{21} & \cdots & A_{n1} \\ \vdots & \vdots & & \vdots \\ A_{1\alpha} & A_{2\alpha} & \cdots & A_{n\alpha} \\ \vdots & \vdots & & \vdots \\ A_{1(n-1)} & A_{2(n-1)} & \cdots & A_{n(n-1)} \end{vmatrix}$

Tensor Calculus & Physics: A General Treatise

Table of Contents

List of Equations

List of Tables

List of Figures

List of Observations

Chapter 11 The Manifold as a Set

Section 11.1 Fundamental Principles for Manifold Sets

"All the world's a stage,
And all the men and women merely players:
They have their exits and their entrances;
And one man in his time plays many parts,"

From As You Like It Act 2 Scene 7 by William Shakespeare

Manifolds are the stage upon which we create mathematical models and build the tensor systems that this treatise uses.

From the definition of a manifold as explained by the Oxford Dictionary, see Chapter 1, Section 1.9 "Modern Origins of the Manifold", a manifold is the composition of parts or components, such as pipes comprising a pipe organ. A mathematical interpretation confirms this idea as defined by Gerretsen [GER62 ,pg 61] for n-dimensional manifolds of class κ and Sokolnikoff [SOK67, pg 9] for space (or manifold) of n-dimensions.

Sokolnikoff's definition clearly specifies this, "We define a space (or manifold) of N dimensions as any set of objects that can be placed in a one–one correspondence with the totality of ordered sets of N (real or complex) numbers $x_1, x_2, \ldots, x_N$ such that $|x_i - A_i| < k_i$, $(i = 1, 2, \ldots, N)$ where $A_1, A_2, \ldots, A_N$ are constants and the $k_1, k_2, \ldots, k_N$ are real numbers." Here in this definition the manifold is comprised of coordinate points $P(x_1, x_2, \ldots, x_N)$ having a well defined range about the positional number $[x_i]$. Sokolnikoff has taken great care in doing this to set the stage for one of the most important properties that a manifold must possess in order to be able to carry out differential calculus, which requires a well defined range for a set of numbers in order for a limit to exist. This must be clearly stated for every point throughout the manifold. So here, the bound points in a manifold are analogous to the component pipes in the pipe organ, also note that the use of space, space of n-dimensions is interchangeable with the word manifold the same will be true here in this treatise.

There is also at least one other property a manifold must possess in tensor calculus, the ability to transform bilaterally between any two manifolds. Without this ability, Einstein would not have been able to prove that his physics of relativity applies not just to every point in space, but any type of space that might be found anywhere in the physical manifold, hence a general theory for the universe.

From the theory of transformations between sets, the fundamental principle of manifolds is defined in terms of the existence of those transformations comprising the manifold set, which can only be true iff their inverse transformation exits. The following table explores typical manifolds from linear equations and quadric surfaces [BEY88 pg 226] showing how the product of any two sets can be mapped. These finite mappings illustrate the basic transformation types for a manifold.

Table 11.1.1 Typical Types of Manifolds

Orthogonal Systems	Transformation	Transformation Inverse	Rendered Manifold in Set X	Rendered Manifold Graphical Drawing
Galilean Linear	$x = u$ $y = v$ $z = w$	$u = x$ $v = y$ $w = z$	$U = IX$	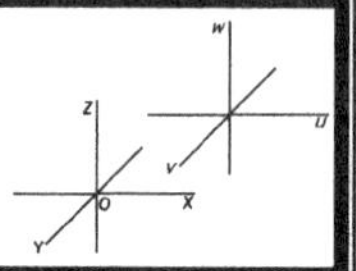
Galilean Rotational	$x = a_{xu}u + b_{xv}v + c_{xw}w$ $y = a_{yu}u + b_{yv}v + c_{yw}w$ $z = a_{zu}u + b_{zv}v + c_{zw}w$	$u = a_{ux}x + b_{uy}y + c_{uz}z$ $v = a_{vx}x + b_{vy}y + c_{vz}z$ $w = a_{wx}x + b_{wy}y + c_{wz}z$	$U = AX$	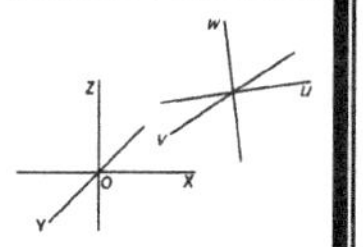
Cylindrical	$x = r\cos(\theta)$ $y = r\sin(\theta)$ $z = u$	$r = \pm\sqrt{x^2 + y^2}$ [11.1.1] $\theta = \arctan(y/x)$ $u = z$	$x^2 + y^2 = r^2$ for $z = z$	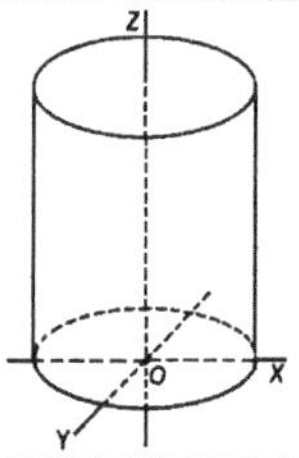
Spherical	$x = r\sin(\phi)\cos(\theta)$ $y = r\sin(\phi)\sin(\theta)$ $z = r\cos(\phi)$	$r = \pm\sqrt{x^2 + y^2 + z^2}$ [11.1.1] $\theta = \arctan[y/x]$ $\phi = \arctan[(\sqrt{x^2 + y^2})/z]$	$x^2 + y^2 + z^2 = r^2$	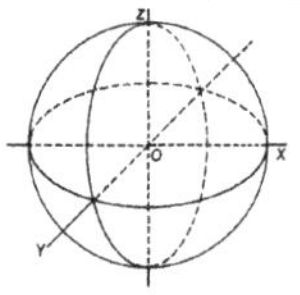
Hyperboloid of 1-Sheet	$x = r\sin(\phi)\cosh(\theta)$ $y = r\sin(\phi)\sinh(\theta)$ $z = r\cos(\phi)$	$r = \pm\sqrt{x^2 - y^2 + z^2}$ [11.1.1] $\theta = \text{arctanh}[y/x]$ $\phi = \arctan[(\sqrt{x^2 - y^2})/z]$	$x^2 - y^2 + z^2 = r^2$	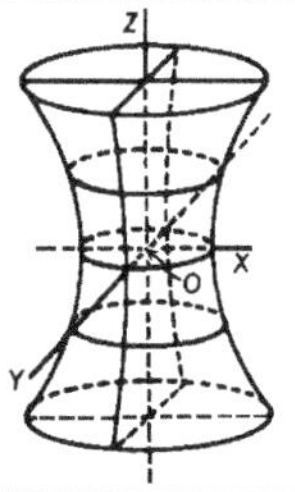
Hyperboloid of 2-Sheet	$x = r\sinh(\phi)\cos(\theta)$ $y = r\sinh(\phi)\sin(\theta)$ $z = r\cosh(\phi)$	$r = \pm\sqrt{-x^2 - y^2 + z^2}$ [11.1.1] $\theta = \arctan[y/x]$ $\phi = \text{arctanh}[(\sqrt{x^2 + y^2})/z]$	$-x^2 - y^2 + z^2 = r^2$	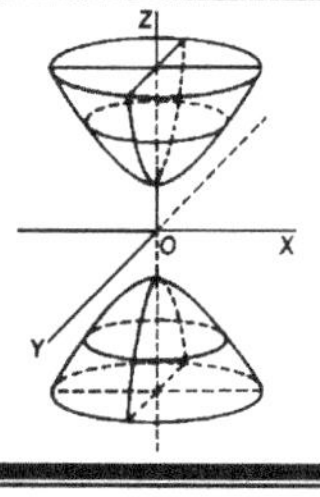

Table 11.1.1 Typical Types of Manifolds (Continued)

Orthogonal Systems	Transformation	Transformation Inverse	Rendered Manifold in Set X	Rendered Manifold Graphical Drawing
Orthogonal Hyperboloid of 1-Sheet	$x = r\sinh(\phi)\cosh(\theta)$ $y = r\sinh(\phi)\sinh(\theta)$ $z = r\cosh(\phi)$	$r = \pm\sqrt{-x^2 + y^2 + z^2}$ [11.1.1] $\theta = \text{arctanh}[y/x]$ $\phi = \text{arctanh}[(\sqrt{x^2 - y^2})/z]$	$-x^2 + y^2 + z^2 = r^2$	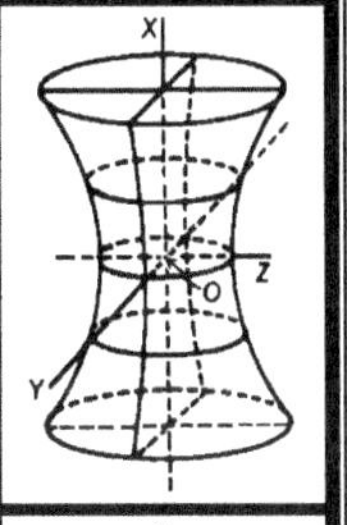
Elliptic Paraboloid	$x = r\cos(\theta)$ $y = r\sin(\theta)$ $z = r^2 / a$	$r = \pm\sqrt{x^2 + y^2}$ [11.1.1] $\theta = \arctan(y/x)$ $r = \pm\sqrt{az}$	$x^2 + y^2 = az$ or $x^2 + y^2 - az = 0$	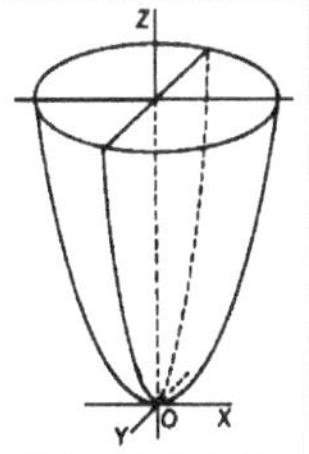
Hyperbolic Paraboloid	$x = r\cosh(\theta)$ $y = r\sinh(\theta)$ $z = r^2 / a$	$r = \pm\sqrt{x^2 - y^2}$ [11.1.1] $\theta = \text{arctanh}(y/x)$ $r = \pm\sqrt{az}$	$x^2 - y^2 = az$ or $x^2 - y^2 - az = 0$	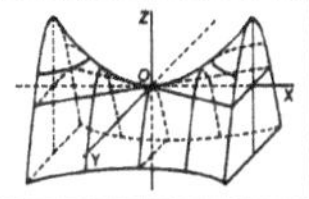
Elliptic Cone	$x = r\cos(\theta)$ $y = r\sin(\theta)$ $z = r$	$r = \pm\sqrt{x^2 + y^2}$ [11.1.1] $\theta = \arctan(y/x)$ $r = z$	$x^2 + y^2 - z^2 = 0$	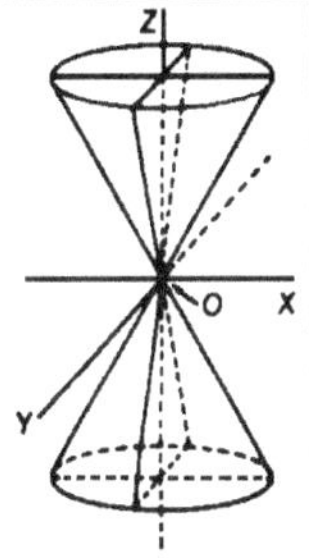

A manifold can be designed to have many properties as required by its application, however there are fundamental properties that all manifolds may have as demonstrated by the above table and observations. As manifolds are used in this treatise, it is observed that they have fundamental properties, which conform to the following basic properties.

[11.1.1]Note: In the spherical and cylindrical coordinates r takes on positive or negative depending on which side of the manifold rendering is being considered, hence its image can be found in the range of x, y or z as appropriate. It follows there exists an inverse relationship between coordinates [x, y, z] and [r, θ, ϕ] iff the predecessor is recovered by the successor under transformation.

Axiom 11.1.1 A Rendered Manifold

The rendered equation of a manifold has the form:

$R(\bullet x_i \bullet) = c$ Equation A

where [c] is independent of X, hence constant and can be separated into parametric forms comprising transformable sets X and U.

Axiom 11.1.2 Manifold functions are One–To–One and Onto

For m-functions that comprise a manifold and having, n-elements connect ***one-to-one*** (1-1) in n-ways and the mirror image of that function is an onto-function.

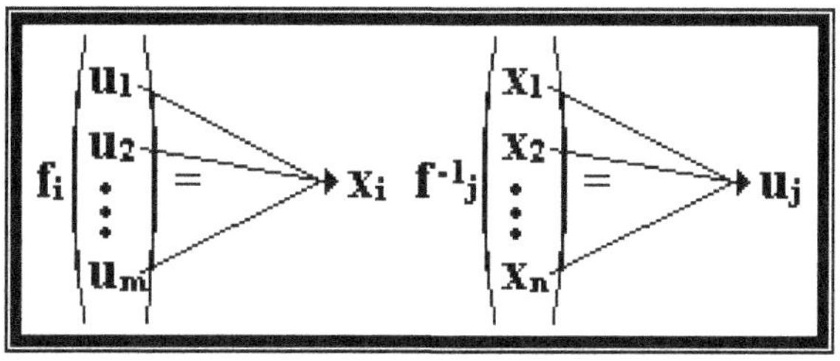

Figure 11.1.1 Connection of Manifold Functions

Axiom 11.1.3 Existence of an Inverse Function for a Manifold

Manifolds have an inverse iff their parametric functions are 1-1 and onto. This is demonstrated in the table as set U is mapped to set X and visa versa in the inverse. Note the arrowhead points to a single parameter satisfying a parametric function that is onto and the inverse is a mirror image of the original onto mapping satisfying the requirement for the function being 1-1.

Axiom 11.1.4 Parametric Functions of a Manifold

A manifold is composed of two composite sets X and U of parametric functions that are invertible, hence
$f_x(U) = X \rightarrow f_u^{-1}(X) = U$.

Axiom 11.1.5 Contraction of Orthogonal Parametric Functions

A manifold's set of parametric functions can only be rendered into an equation of set X through the binding equation $b_x(\ )$ iff they are orthogonal.

Given $b_x(U)$ and binding equation over the U coordinate set than
$r(X) = b_x(\ f_u^{-1}(X)\ )$ rendered equation of a manifold.

Axiom 11.1.6 Fundamental Property of Continuum for a Manifold

In order for a manifold to be continuously differentiable, a manifold must be a continuum of points within a region of investigation and at every point a well-defined range for every positional number $[x_i]$ within a point

Axiom 11.1.7 Fundamental Property of Transformation between Manifolds

Clearly the property of transformation from Table 11.1.1 requires that there exists a 1–1 correspondence between points of any pair of manifolds and that their be established a set parametric equations of transformation that map one coordinate system not necessarily *onto* the other and that an equivalent mirror image inverse also exist.

Table 11.1.1 Mapping of Invertible Manifold Types

Orthogonal Systems	U	Mapping; Set U to X Predecessor	X	Inverse; Set X to U Successor	U	1-1 [11.1.1]	Onto [11.1.1]
Galilean Linear	u		x		u	YES	YES
	v		y		v	YES	YES
	w		z		w	YES	YES
Galilean Rotational	u		x		u	YES	YES
	v		y		v	YES	YES
	w		z		w	YES	YES
Cylindrical	r		x		r	YES	YES
	θ		y		θ	YES	YES
	u		z		u	YES	YES
Spherical	r		x		r	YES	YES
	θ		y		θ	YES	YES
	ϕ		z		ϕ	YES	YES
Hyperboloid of 1-Sheet	r		x		r	YES	YES
	θ		y		θ	YES	YES
	ϕ		z		ϕ	YES	YES
Hyperboloid of 2-Sheet	r		x		r	YES	YES
	θ		y		θ	YES	YES
	ϕ		z		ϕ	YES	YES
Orthogonal Hyperboloid of 1 Sheet	r		x		r	YES	YES
	θ		y		θ	YES	YES
	ϕ		z		ϕ	YES	YES
Elliptic Paraboloid			x			NA	NA
	θ		y		θ	YES	NO
	r		z		r	YES	NO
Hyperbolic Paraboloid			x			NA	NA
	θ		y		θ	YES	NO
	r		z		r	YES	NO
Elliptic Cone			x			NA	NA
	θ		y		θ	YES	NO
	r		z		r	YES	NO

Section 11.2 Definitions for Manifold Sets

Definition 11.2.1 **Finite Manifold**

A ***manifold*** $\mathfrak{M}_{(\bullet m[k]\bullet)}{}^{n} \equiv \{\prod^{n} x F_{m[k]}{}^{k}\}$ sometimes called a ***space*** is a collection of a finite product fields having properties associated the axial field, such as a finite set of coordinate points. Example of associated properties would be, $F_{m[x]}{}^{x}$, $F_{m[y]}{}^{y}$, which represents axes x-y in coordinate distance or axes of ordinary and extraordinary index of refraction for optical media.

Definition 11.2.2 **Continuous (Continuum) Manifold**

A ***space filling manifold*** or ***continuum*** has an infinite number of points to fill it; this is true when a manifold $\mathfrak{M}_{(\bullet\infty\bullet)}{}^{n}$ is comprised of some or all infinite fields then the manifold set notation can be simplified by dropping the infinite number of elements in the fields to, $\mathfrak{M}^{n}$.

Definition 11.2.3 **Rendered Equation of a Manifold**

A rendered form of a manifold is that equation $R(\bullet x_i \bullet) = c$; see Table 11.1.1, which characterizes the manifold with respect to a specific coordinate space.

Definition 11.2.4 **Linear Manifold**

A linear manifold's components are linear that is the axial components can be placed into linear matrix form.

As an example, see Table 11.1.1 Orthogonal Galilean Linear U = IX and
Rotational $U = [a_{ux}x + a_{uy}y + a_{uz}z, a_{vx}x + a_{vy}y + a_{vz}z, a_{wx}x + a_{wy}y + a_{wz}z]$ system. Equation A

Definition 11.2.5 **Nonlinear Manifold**

A nonlinear manifold's components are nonlinear that is the axial components cannot be placed into linear matrix form.

As an example, see Table 11.1.1 Elliptic Paraboloid $U = [\pm\sqrt{az}, \arctan(y/x), z]$.

Definition 11.2.6 **Invariance within a Manifold**

Vector magnitude and relative affine direction are left unaltered under transformation within a manifold (see A8.2.1 and O11.5.1), such manifolds remain invariant were the product of the covariant and contravariant components are related by the identity matrix, hence orthogonal

$$I = GG^{-1}. \qquad \text{Equation A}$$

As an example, see Table 11.1.1 any Orthogonal Quadratic Surface.

Definition 11.2.7 **Metric Space**

Is any space where distance or magnitude is defined between any two points $\overline{PQ}$, such a space is called a measured or ***Metric Space***.

Section 11.3 Manifolds as Object-Oriented Sets

Borrowing Back from Computer Programming:

In this paper, the flow of logical development finds a modularization with three column frameworks and the more complex logical stratifications, which groups proof development into sets. Originally, the first computer codes at IBM borrowed the idea of a function from mathematics to modularize the computer code and simplifying it for ease of programming. Further modularization came along with Fortran that used subroutines and then the great leap to computerized sets with Object-Oriented programming, which introduced inheritance, promoting data hiding and breaking up the linear code development, thereby, minimizing errors or as they are commonly known in that field, ***bugs***. In the development of this systematic development, I have now borrowed back from those ideas and applied them to the foundation of this thesis on manifolds. Now, the kernel of this dissertation is that manifolds can inherit from one another or divest properties forming greater or lesser-structured types of manifolds. This has brought the mathematics around full circle making sets analogous to computer programming objects.

Axiom 11.3.1 Properties of a Base Manifold $\mathfrak{B}_x^n$

Let ∃| a set of properties for a space, where space is made from Axial Field Sets comprised of field elements $[x_{ik}]$

P1) $A_i \equiv \{ x_{ik} \mid (x_{ik}) \in (\mathbf{R} \vee \mathbf{C}^2) \text{ for } k = 1, 2, \ldots, m_i\}$ ***arithmetic axes***.

P2) $[\ |x_i - a_i| < k_i$ bounded for all $(i = 1, 2, \ldots, n)\ ]$
where $K(\bullet\ k_i\ \bullet)$ are real numbers, constants, that can be found to bound the inequality about the point $A(\bullet\ a_i\ \bullet)$. Absolute value in terms of complex numbers is treated as absolute magnitude.

These base properties that bound the X-positional numbers for an infinite number of points P_k in a continuous space, allows the existence of limits, and establishing differentiability with its inverse the antidifferential or the ability to integrate over that region of space.

P3) Derived units of measure associated with an axis.
$A_{xi} \equiv \{ P1, x_i, \mid P1 \vee (x_i \rightarrow [U]\ )\ \}$, for x_i a derived unit of measure, see tables TA.3.5[II] "Fundamental Physical Quantities", TA.1.1[II] "Correspondence between Rectilinear vs. Rotational Physical Equations" and TA.3.1[II] "Symbols, dimensions and Units for Physical Quantities" and for examples}

P4) $\mathcal{P}_x^n \equiv [\ \prod_{i=1}^n \times A_{ui}\ ]$ base manifold property for X-coordinate space

Such that there are a set of n-tuple points $P_k(\bullet\ x_i\ \bullet)$ for space X generated from the coordinate product of the axes

P5) $P_k(\bullet\ x_i\ \bullet) \in \mathcal{P}_x$ for $k = 1, 2, \ldots, r \wedge < r = \prod_{i=1}^n m_i >$

Axiom 11.3.2 Base Manifold (Sokolnikoff Type) $\mathfrak{B}_s^{\,n}$

Now at every point in that space, has the manifold properties 1–1 and onto assigned to it with a well-defined range for the positional numbers between any two manifolds:

$\mathcal{P}_{bx}^{\,n} \equiv$ (P1, P2, P3, P4$_x^{\,n}$, P5) list of base properties for a manifold.

$\mathfrak{B}_x^{\,n} \equiv \{\ \mathcal{P}_{bx}^{\,n}\ \}$

Where $\mathfrak{B}_x^{\,n}$ ***is called the base manifold of space X and read, as the manifold is the set of base properties*** P1, P2, P3, P4$_x^{\,n}$, and P5**.**

Axiom 11.3.3 Manifolds have the Property of Commutative Inheritance

A manifold is also a superset $\mathfrak{H}_{x,h}^{\,n}$, being able to add and remove properties $\mathcal{P}_{h}$ to the superset through the process of inheritance, thereby customizing the properties of the manifold $\mathfrak{B}_x^{\,n}$.

$\mathfrak{H}_{x,h}^{\,n} \equiv \mathfrak{B}_x^{\,n} + \{\ \mathcal{P}_h\ \}$	***inherit right***	Equation A
$\mathfrak{H}_{h,x}^{\,n} \equiv \{\ \mathcal{P}_h\ \} + \mathfrak{B}_x^{\,n}$	***inherit left***	Equation B

or

$\mathfrak{H}_{x,h}^{\,n} \equiv \{\ \mathcal{P}_{bx}^{\,n}, \mathcal{P}_h\ \}$	***inherit right***	Equation C
$\mathfrak{H}_{h,x}^{\,n} \equiv \{\ \mathcal{P}_h\ , \mathcal{P}_{bx}^{\,n}\ \}$	***inherit left***	Equation D

Read as ***assigning property $\mathcal{P}_h$ to manifold*** $\mathfrak{B}_x^{\,n}$ or $\mathcal{P}_h \rightarrow \mathfrak{B}_x^{\,n}$, or

manifold inherits modifying property $\mathfrak{B}_x^{\,n} + \{\ \mathcal{P}_h\ \}$or $\{\ \mathcal{P}_h\ \} + \mathfrak{B}_x^{\,n}$

for

$\mathfrak{B}_x^{\,n} \subset \mathfrak{H}_{x,h}^{\,n}$ Equation E

Axiom 11.3.4 Manifolds have the Property of Commutative Disinheritance

A property $\mathcal{P}_d$ in a superset manifold $\mathfrak{D}_x^{\,n} \equiv \mathfrak{M}_{d,x}^{\,n} - \mathfrak{P}_d^{\,n}$ can be disinherited from that superset through the process of disinheritance, thereby reducing manifold of a property.

Now

$\mathfrak{D}_x^{\,n} \equiv \mathfrak{M}_{d,x}^{\,n} - \{\mathcal{P}_d\ \}$	***disinherit right***	Equation A
$\mathfrak{D}_x^{\,n} \equiv -\{\mathcal{P}_d^{\,n}\ \} + \mathfrak{M}_{d,x}^{\,n}$	***disinherit left***	Equation B

therefore

$\mathfrak{D}_x^{\,n} \equiv \{\ \mathcal{P}_x^{\,n}, \mathcal{P}_b^{\,n}\}$	***disinherit***	Equation C
$\mathfrak{D}_x^{\,n} \equiv \mathfrak{B}_x^{\,n}$	***disinherit***	Equation D

Note the simplest number of properties a manifold can have is two, the base manifold, never less. There is no such thing as a NULL manifold the base manifold prohibits that.

Axiom 11.3.5 Manifold of Transformation

$\mathfrak{U}_{\chi}^{m} \cap \mathfrak{T}_{y}^{n}$ are ***manifold transformations*** iff they are 1–1 and onto, hence will have a property of transformation $\mathcal{W}_{\chi y}^{mn}$

$$\mathfrak{U}_{\chi}^{m} \cap \mathfrak{T}_{y}^{n} \equiv [\, f_j(\bullet\, x_i \bullet) = y_j \wedge f_i^{-1}(\bullet\, y_j \bullet) = x_i \wedge \text{for all } (i, j)\,] \qquad \text{Equation A}$$

or

$$\mathcal{W}_{\chi y}^{mn} \equiv \mathfrak{U}_{\chi}^{m} \cap \mathfrak{T}_{y}^{n} \qquad \text{Equation B}$$

$\mathfrak{U}_{\chi}^{m}$ maps onto $\mathfrak{T}_{y}^{n}$ maps back by inverse operation onto $\mathfrak{U}_{\chi}^{m}$

$$\mathfrak{U}_{\chi}^{m} \xrightarrow{f} \mathfrak{T}_{y}^{n} \xrightarrow{f^{-1}} \mathfrak{U}_{\chi}^{m}$$

where the ***Components of a Manifold*** are the equations, $x_i = f_i(\bullet y_i \bullet)$ and $y_i = f_i^{-1}(\bullet x_i \bullet)$ for all coordinate indices [i], allowing transformation between two manifolds.

Axiom 11.3.6 Manifold of Surfaces

Let ∃l a set of properties for a space, where a manifold $\mathfrak{H}_{u}^{n-1}$ embedded in another manifold $\mathfrak{M}_{\chi}^{n}$ is said to be a ***manifold of surface*** it can also go by the names ***hypersurface*** or sheet or ***hypersheet***. Such special surfaces have a 1–1 and onto correspondence, hence has a property of transformation $\mathcal{S}_{u\chi}^{(n-1)n}$.

$$\mathfrak{H}_{u}^{n-1} \cap \mathfrak{M}_{\chi}^{n} \equiv [\, h_j(\bullet\, x_i \bullet) = u_j \wedge f_i^{-1}(\bullet\, u_j \bullet) = x_i \wedge \text{for all } (i, j)\,] \qquad \text{Equation A}$$

or

$$\mathcal{S}_{u\chi}^{(n-1)n} \equiv \mathfrak{H}_{u}^{n-1} \cap \mathfrak{M}_{\chi}^{n} \qquad \text{property of transformation} \qquad \text{Equation A}$$

$$\mathfrak{M}_{\chi}^{n} \xrightarrow{h} \mathfrak{H}_{u}^{n-1} \xrightarrow{f^{-1}} \mathfrak{M}_{\chi}^{n}$$

Such that

$$\mathfrak{H}_{u}^{n-1} \subset \mathfrak{M}_{\chi}^{n} \qquad \text{Equation B}$$

Axiom 11.3.7 Metamorphic Manifolds

Let ∃l a set of properties for a space, where a manifold $\mathfrak{N}_{\chi}^{n-1}$ is embedded in another manifold $\mathfrak{M}_{\chi\theta}^{n}$, then it is said to be a ***metamorphic manifold*** iff there exists an n-tuple $\mathcal{D}_e^{(n-1)}(\bullet\, \theta_i \bullet)$ is called the ***eccentricity*** parameter of the manifold altering its form as it changes the orientation within $\mathfrak{N}_{\chi}^{n-1}$ relative to the space $\mathfrak{M}_{\chi\theta}^{n}$, thereby resulting in a new projected topological space $\mathcal{D}_e^{n-1}$.

$$\mathcal{D}_e^{(n-1)}(\bullet\, \theta_i \bullet) \equiv \mathfrak{A}_{\chi}^{n-1}(\bullet\, \alpha_i \bullet) \cap \mathfrak{B}_{\chi}^{n-1}(\bullet\, \beta_i \bullet) \qquad \text{Equation A}$$

or

$$\mathfrak{A}_{\chi}^{n-1}(\bullet\, \alpha_i \bullet) \xrightarrow{\theta} \mathfrak{B}_{\chi}^{n-1}(\bullet\, \beta_i \bullet) \xrightarrow{\theta^{-1}} \mathfrak{A}_{\chi}^{n-1}(\bullet\, \alpha_i \bullet) \qquad \text{Equation B}$$

$$(\, \mathfrak{A}_{\chi}^{n-1}(\bullet\, \alpha_i \bullet), \mathfrak{B}_{\chi}^{n-1}(\bullet\, \beta_i \bullet)\,) \subset \mathfrak{M}_{\chi\theta}^{n} \qquad \text{Equation C}$$

Axiom 11.3.8 Intersection of Manifolds (Worm-Hole)

Let ∃| a set of properties for a space, with the following abilities:

$\mathfrak{K}_{\chi}^{n} \equiv \mathfrak{N}_{\chi}^{n} \cap \mathfrak{M}_{\chi}^{n}$ Equation A

$\mathfrak{K}_{t}^{n} \subset \mathfrak{M}_{\chi}^{n}$ Equation B

$\mathfrak{K}_{t}^{n} \subset \mathfrak{N}_{\chi}^{n}$ Equation C

Section 11.4 Types of Manifold $\mathfrak{M}^n$ Sets

Axiom 11.4.1 Spatial Manifold $\mathfrak{S}^n$

Let ∃| a set of properties for a space, called a ***spatial manifold*** that has the following special properties: [11.4.1]

P1) A spatial manifold is comprised of a base Manifold $\mathfrak{B}_x{}^n$ having the property of a ***natural or coordinate vector base set*** $\mathbf{B}^n$ associated at every point P(• x_i •), thereby spanning the space and characterizing the shape or distortion of that space[11.4.2]. Where the point P(• x_i •) has the properties of a real **R** or complex field $\mathbf{C}^2$.

$$\mathfrak{M}^n \equiv \mathfrak{B}_x{}^n + \{ P(x_i) \wedge B^n \mid [P(x_i) \in (\mathbf{R} \vee \mathbf{C}^2)] \wedge B^n \in P(x_i) \text{ for all i} \}$$

P2) A spatial manifold has the ability to map from any given coordinate point in its space to another of like kind and back again. The spatial manifold $\mathfrak{S}^n$ ***does constitute a one-to-one correspondence and onto*** the manifold set $\mathfrak{M}^n$ and back again (see K.1.1.1 and 5).

$$\mathfrak{S}^n \xrightarrow{\ p\ } \mathfrak{M}^n \xrightarrow{\ p^{-1}\ } \mathfrak{S}^n$$

where $\boldsymbol{p}$: P(• x_i •) → M(• y_i •) and $\boldsymbol{p}^{-1}$: M(• y_i •) → P(• x_i •)

that is the function $\boldsymbol{p}$ of $\mathfrak{S}^n$ maps every point P(• x_i •) onto a corresponding point M(• y_i •) in $\mathfrak{M}^n$.
Likewise function $\boldsymbol{p}^{-1}$ of $\mathfrak{M}^n$ maps every point M(• y_i •) in reverse onto a corresponding point P(• x_i •) in $\mathfrak{S}^n$.

P3) Associated at every point in that space are a set of dual basis vectors (• $\mathbf{b}_i$•) covariant and (• $\mathbf{b}^i$•) contravariant spanning that space to construct vectors from.

$B_n \equiv \{ \mathcal{P}_x{}^n, (\bullet\, \mathbf{b}_i \bullet) \mid \text{where for all i}, (\mathbf{b}_i \rightarrow x_i) \wedge (\theta_{ij} \sim\!\vee\!= \pi/2 \text{ for } i \neq j) \wedge (|\mathbf{b}_i| \sim\!\vee\!= 1) \} +$
$+ \{ \mathcal{P}_x{}^n, (\bullet\, \mathbf{b}^i \bullet) \mid \text{where for all i}, (\mathbf{b}^i \rightarrow x_i) \wedge (\theta^{ij} \sim\!\vee\!= \pi/2 \text{ for } i \neq j) \wedge (|\mathbf{b}^i| \sim\!\vee\!= 1) \}$

[11.4.1]Note: This definition is designed to allow for creation of new types of manifolds by either modifying the base properties or adding to the set.
[11.4.2]Note: This vector base set is not to be confused with a tensor field, they may look similar but have distinctly different definitions, and the two vector sets are totally different with different properties unless otherwise specified as coinciding.

Axiom 11.4.2 Hypersurface Manifold $\mathfrak{H}^{n-1}$

P1) A hypersurface manifold (sometimes called a ***hypersheet***) is simply a manifold of one less dimension than the space it is embedded in, hence:

$$\mathfrak{H}^{n-1} \subset \mathfrak{M}^{n}.$$

The spatial manifold $\mathfrak{M}^n$ ***does not constitute a one-to-one correspondence and onto*** the manifold set $\mathfrak{H}^{n-1}$ and back again (see K.1.1.9 and 11), but <u>does support</u> an inverse transformation.

$$\mathfrak{M}^n \xrightarrow{\;p\;} \mathfrak{H}^{n-1} \xrightarrow{\;p^{-1}\;} \mathfrak{M}^n$$

where $\boldsymbol{p}$: P(• x_i •) → H(• u_h •) and $\boldsymbol{p}^{-1}$: H(• u_h •) → P(• x_i •)
for i = 1, 2, 3, … n and h = 1, 2, 3, … n-1

that is the function $\boldsymbol{p}$ of $\mathfrak{M}^n$ maps every point P(• x_i •) onto a corresponding point H(• u_h •) in $\mathfrak{H}^{n-1}$.
Likewise function $\boldsymbol{p}^{-1}$ of $\mathfrak{H}^{n-1}$ maps every point H(• u_h •) in reverse onto a corresponding point P(• x_i •) in $\mathfrak{M}^n$.

Axiom 11.4.3 Orthogonal Manifold $\mathfrak{O}^n$

Let ∃| a set of properties for a space, whose dual bases vectors spanning the space are always perpendicular to one another, hence its metrics are Kronecker Delta's.

P1) An orthogonal manifold is a spatial manifold $\mathfrak{S}^n$, and

P2) $\delta_{ij} \equiv (\mathbf{b}_i \bullet \mathbf{b}_j) / b_i b_j$ for $\{(\theta_{ij} = 0 \text{ for } i=j \ni \mathbf{b}_i \parallel \mathbf{b}_j) \vee (\theta_{ij} = ½\pi \text{ for } i\neq j \ni \mathbf{b}_i \perp \mathbf{b}_j)\}$ EQ A
$\delta^{ij} \equiv (\mathbf{b}^i \bullet \mathbf{b}^j) / b^i b^j$ for $\{(\theta^{ij} = 0 \text{ for } i=j \ni \mathbf{b}^i \parallel \mathbf{b}^j) \vee (\theta^{ij} = ½\pi \text{ for } i\neq j \ni \mathbf{b}^i \perp \mathbf{b}^j)\}$ EQ B

Axiom 11.4.4 Oblique Manifold $\mathfrak{Q}^n$

That is any manifold whose dual bases vectors spanning the space are not necessarily perpendicular to one another and none are degenerate ($\theta_{ij} \neq 0$ for any i and j).

P1) An oblique manifold is a spatial manifold $\mathfrak{S}^n$, and

P2) $\cos(\theta_{ij}) \equiv (\mathbf{b}_i \bullet \mathbf{b}_j) / b_i b_j$ for $\mathbf{b}_i \sim\vee\perp \mathbf{b}_j$ and
$\cos(\theta^{ij}) \equiv (\mathbf{b}^i \bullet \mathbf{b}^j) / b^i b^j$ for $\mathbf{b}^i \sim\vee\perp \mathbf{b}^j$

Axiom 11.4.5 Euclidian Manifold $\mathfrak{E}^n$

That is, an orthogonal manifold having distance measured, between any two points, using Pythagoras' hypotenuse of length within that space (see A9.8.1 "A Cartesian coordinate space, or n-Dimensional Space, or n-Space").

P1) $b_i \equiv b^i \equiv 1$ and $g_{ij} = \delta_{ij}$ or $g^{ij} = \delta^{ij}$ [$\mathfrak{O}^n$P2] respectively in an orthogonal-spatial manifold $\mathfrak{O}^n + \mathfrak{S}^n$ manifold with <u>unit bases vectors</u>, and

P2) $\overline{PQ} = \sqrt{\sum_n (x_{pi} - x_{qi})^2}$ (see T8.2.4) Pythagorean distance between any two points.

Axiom 11.4.6 Riemannian Manifold $\mathfrak{R}^n$

A Riemannian Manifold is a ***hypercurved manifold***, not necessarily orthogonal, which contains between any two, points P and Q, an infinitesimal measured distance given by an approximation of a Pythagoras' hypotenuse for distance.

P1)	$d\,\overline{PQ} = \sqrt{\sum_i \sum_j g_{ij} dx^i dx^j}$	Riemann's covariant fundamental spatial length.
P2)	$d\,\overline{PQ} = \sqrt{\sum_i \sum_j g^{ij} dx_i dx_j}$	Riemann's contravariant fundamental spatial length.

Section 11.5 Theorems on Finite and Infinite Manifold Sets

Theorem 11.5.1 Inequality of Finite Manifolds

1g	Given	$\mathfrak{M}_{(\bullet\,\eta i\,\bullet)}{}^{n}$ for number of r points, $r = \prod_{i=1}{}^{n} \eta_i$
2g		$\mathfrak{N}_{(\bullet\mu j\,\bullet)}{}^{n}$ for number of ρ points, $\rho = \prod_{j=1}{}^{m} \mu_j$
3g		$\rho \leq r$
Step	Hypothesis	$\mathfrak{N}_{(\bullet\mu j\,\bullet)}{}^{n} \subseteq \mathfrak{M}_{(\bullet\,\eta i\,\bullet)}{}^{n}$ for $\rho \leq r$
∴	From 1g, 2g, 3g and D2.13.1A,B Subset of Sets	$\mathfrak{N}_{(\bullet\mu j\,\bullet)}{}^{n} \subseteq \mathfrak{M}_{(\bullet\,\eta i\,\bullet)}{}^{n}$ for $\rho \leq r$

Theorem 11.5.2 Inequality of Infinite Manifolds

1g	Given	$\mathfrak{M}_{(\bullet\,\eta i\,\bullet)}{}^{n}$ for number of r points, $r = \prod_{i=1}{}^{n} \eta_i$
2g		$\mathfrak{N}_{(\bullet\mu j\,\bullet)}{}^{n}$ for number of ρ points, $\rho = \prod_{j=1}{}^{m} \mu_j$
3g		$\rho \leq r$
Step	Hypothesis	$\mathfrak{N}^{n} \subseteq \mathfrak{M}^{n}$ for $\infty_\rho \leq \infty_r$ or simply $\mathfrak{N}^{n} \subseteq \mathfrak{M}^{n}$
1	From T2.32.1 Inequality of Finite Manifolds	$\mathfrak{N}_{(\bullet\mu j\,\bullet)}{}^{n} \subseteq \mathfrak{M}_{(\bullet\,\eta i\,\bullet)}{}^{n}$ for $\rho \leq r$
2	From 1	Let $\rho \leq r$ and increase without bound maintain the inequality between them
3	From 1 and 2	$\mathfrak{N}_{(\bullet\mu j\,\bullet)}{}^{n} \subseteq \mathfrak{M}_{(\bullet,\,\eta i\,,\bullet)}{}^{n}$ for $\infty_\rho \leq \infty_r$
∴	From 3 and D2.30.2 Continuous (Continuum) Manifold	$\mathfrak{N}^{n} \subseteq \mathfrak{M}^{n}$ for $\infty_\rho \leq \infty_r$ or simply $\mathfrak{N}^{n} \subseteq \mathfrak{M}^{n}$

Section 11.6 Fundamental Principles for Parametric Manifold Sets

The concept of a parametric variable comes from the parametric set ($\bullet$ $x_i = f_i(t)$ $\bullet$ for all i) $\in P_a$ having a common variable with the same range $| t | < \infty$, such that $|t - A| < k$ for (t, k, A) $\in$ R. To understand this concept we need to look at what a directed line is which intern leads to the beginning of the analytic idea of a Vector.

Definition 11.6.1 **Directed Line**
An arrow at one-end marks ***directed line*** so that one of the two possible directions on the line is distinguished. A directed line is symbolically written as $\vec{L}$, see Figure 11.6.1A.

Definition 11.6.2 **Undirected Line**
If no arrow is placed on the line L, than it is called ***undirected***, see Figure 11.6.1B.

Definition 11.6.3 **Directed Angles**
Let there be a set of angular parameters ($\bullet$ θ_i $\bullet$ for all i), see Figure 11.6.2D, made by a directed line measured from the positive direction of the axes CCW ($\bullet$ x_i $\bullet$ for all i) on some point on that line respectively, such quantities are called ***directed angles***.

Definition 11.6.4 **Directed Numbers**
For multi-dimensional straight lines directional numbers ($\bullet$ m_i $\bullet$) are the leading coefficients of the parametric variable and are analogous to the slope of a line in 2-dimensions.

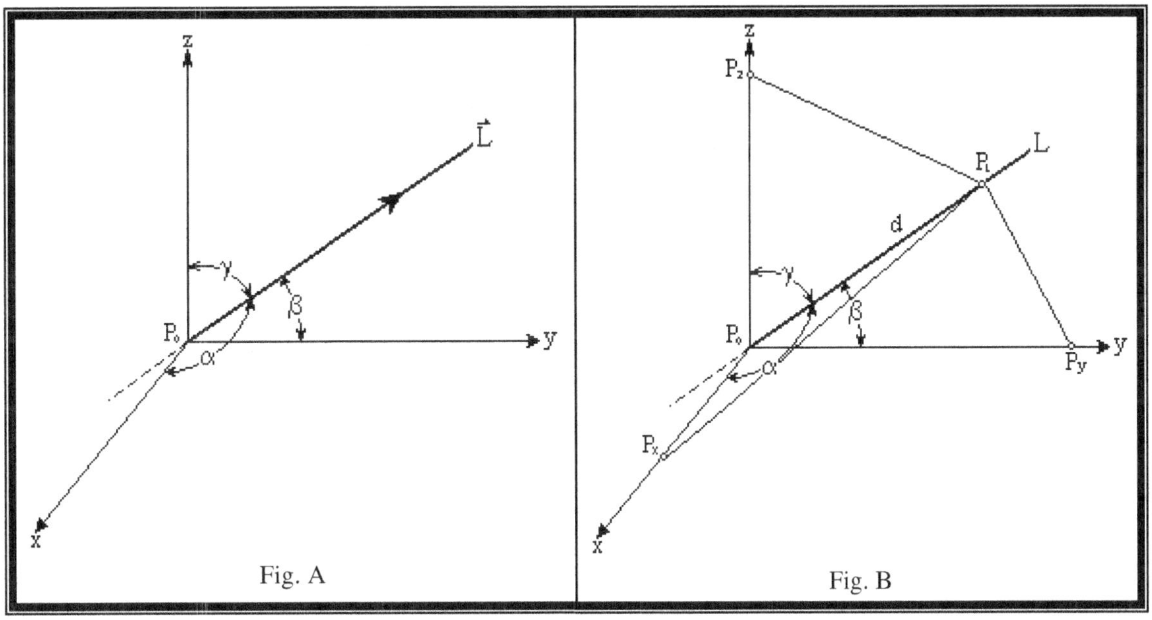

Figure 11.6.1 Directed and Undirected Lines

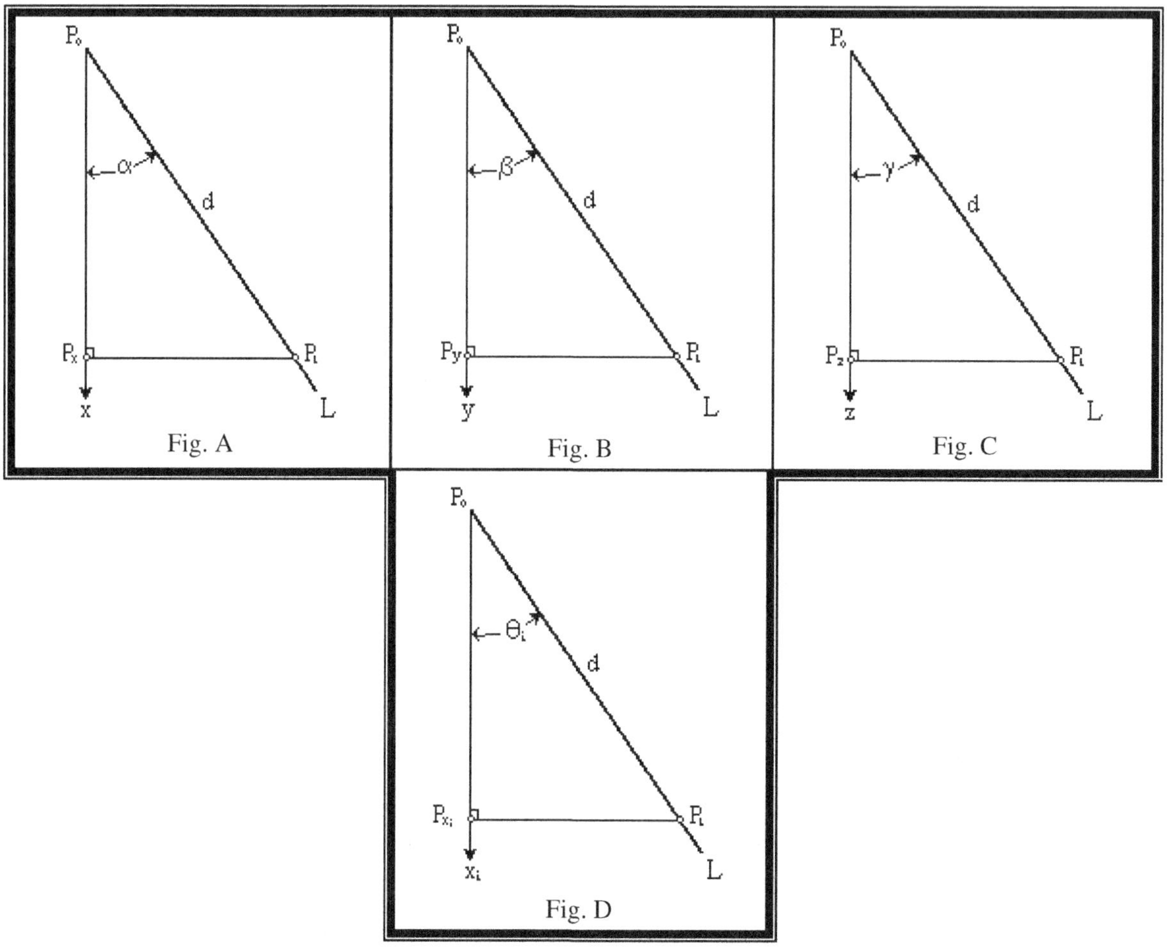

Figure 11.6.2 Directed Cosine in 3 and n-Dimensions

As seen from Figure 11.6.2A the directed angles are in threes for a 3-dimensional line oriented in that space. In Figure 11.6.2B given some point P_L on that line perpendicular lines P_LP_x, P_LP_y, and P_LP_z, relative to the system axes. Now in the planes $P_LP_0P_x$, $P_LP_0P_y$, and $P_LP_0P_z$ are right triangles by virtue of their perpendicular geometry see Figure 11.6.2A, B and C. It would be natural to extend construction of such a triangle into n-dimensions as in Figure 11.6.2D, so this would have to be a natural property true in any dimension.

Definition 11.6.5 **Directed Cosine in n-Dimensions**
From Figure 11.6.2D gives a general description of a special geometric ratio called a ***directed cosine*** expressed as

$$\mu_i \equiv \cos(\theta_i) \equiv x_i / d \qquad \text{Equation A}$$

Axiom 11.6.1 Parametric Hypercurve Manifold $\mathfrak{P}^n$

The coordinates at every point P(• x_i •) within a manifold map onto a single parametric domain defined as continuous in an interval X_i, between two points by (P2) of the manifold (• $|x_i - a_i| < k_i$ •), such that the parametric parameter-t is also continuous in the corresponding interval I_t by ($0 < |t - \alpha| < \kappa$), thereby relating all coordinates of the point to a corresponding parametric parameter. Now for that varying coordinate point there exists a set of parametric functions that can likewise be defined as P(• $x_i(t)$ •) whose inverse exists, such that

P1) $x_i \equiv f_i(t)$ and its inverse — Equation A

$t \equiv f_i^{-1}(x_i)$ were $x_i \in X_i$ for each and every [i] — Equation B.

P2) The set of points within the interval

$I_t \equiv \{a \le t \le b\}$ or abbreviated [a, b]

traces out a continuous ***hypercurve***, or ***path***, or ***trajectory*** within the manifold, $\mathfrak{P}^n \subset \mathfrak{M}^n$.

P3) Parametric magnitude between two points exists within an interval [a, b], iff a value [k] can befound, such that

$|x_i(t) - a_i(t)| < k_i < k$ for all coordinates [i].

P4) It can be concluded from this axiom that $x_i(t)$ is analytically continuous in the parametric interval I_t, as such can be differentiated a finite amount of times [r] with respect to [t].

$$\frac{d^r x_i}{dt^r}$$

P5) There are no constraints that the positional number has to be a set of linear parametric equations in fact; any set of parametric equations, $x_i \equiv f_i(t)$ for all [i] as in P1A can be allowed defining a ***hypercurve line***. Such curves can loop a finite number of times or infinitely taking on any shape.

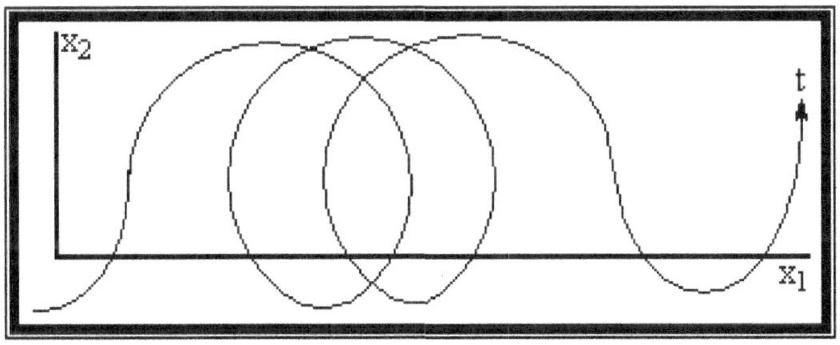

Figure 11.6.3 Finite Loops for 2-Dimensional Parametric Curve Line

P6) Given a closed curve than it has a property called ***closer***, and periodically repeats in a constant interval T.

$x_i(t + Tn) = x_i(t)$ for [n] any number of integer cycles either CW or CCW.

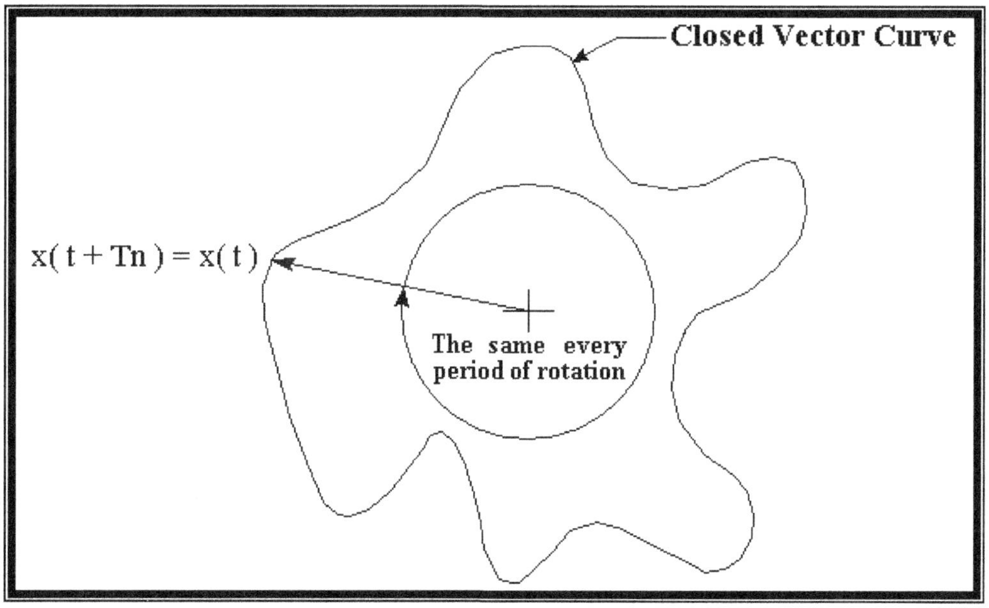

Figure 11.6.4 Periodicity On A Closed Vector Curve

P7) The correct choice of a path for a curve can remove a singularity, a cusp or corner at a ***node of intersection or tangent***, allowing the curve to be differentiated at every point within the interval.

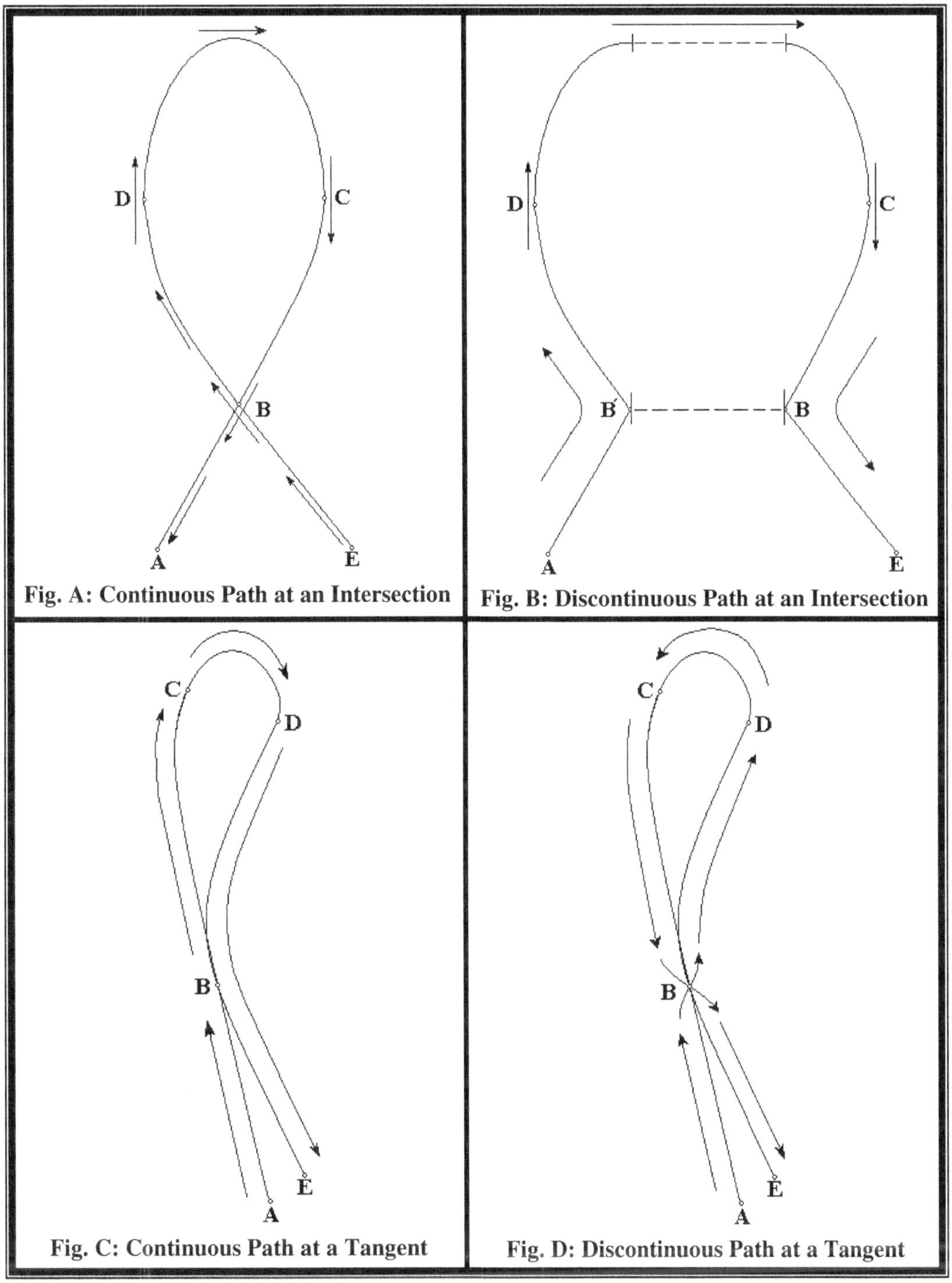

Figure 11.6.5 Existence of a Differential at the Intersection or Tangent Node

P8) Finding the function for a curve is done by parsing the path, as an example see Figure 11.6.4B Path-AB', B'D, DC, CB, BE.

Axiom 11.6.2 Straight hyperline $\mathfrak{L}^1$

Let there exist a ***straight hyperline*** $\mathfrak{L}^1$ embedded in manifold $\mathfrak{M}^n$ and let three ordered, collinear points $P(\bullet, x_i, \bullet)$, $P_1(\bullet, x_{1i}, \bullet)$, $P_0(\bullet, x_{0i}, \bullet)$ lye on that line. It follows from Theorem 11.7.4 "Two Point Parametric Equation and Its Inverse"

$$x_i = x_{i0} + m_i t \qquad \text{for all } i \qquad \text{Equation A}$$

and its inverse

$$t = (x_i - x_{i0}) / m_i \qquad \text{for all } i \qquad \text{Equation B}$$

were

$$\mathfrak{L}^1 \subset \mathfrak{M}^n \qquad \text{Equation C}$$

Observation 11.6.1 Hypercurves have an Inverse

The above axiom says that each positional number $[x_i]$ for a straight hyperline can be represented in a more general way by a parametric function in $[t]$:

$$x_i = f_i(t), \text{ for each and every } i.$$

within a range $[a_i \le x_i < b_i]$ on a hypercurve. Also there exists transformation $t = f_i^{-1}(x_i)$ analogous to the inverse as demonstrated by straight hyperline.

Axiom 11.6.3 Manifold of Curve Line

Any manifold $\mathfrak{K}_t^1$ embedded in another manifold $\mathfrak{M}_\chi^n$ is said to be a ***manifold of a curve line*** it can also go by the name ***hypercurve line***. Such spatial curve lines have a 1–1 and onto correspondence, hence have a property of transformation $\mathcal{L}_{t\chi}{}^{1\,n}$.

$$\mathfrak{K}_t^1 \cap \mathfrak{M}_\chi^n \equiv [x_i = f_i(t) \wedge t = f_i^{-1}(\bullet, x_i, \bullet) \wedge \text{for all } i\,] \qquad \text{Equation A}$$

or

$$\mathcal{L}_{t\chi}{}^{1\,n} \equiv \mathfrak{K}_t^1 \cap \mathfrak{M}_\chi^n \qquad \text{Equation B}$$

$$\mathfrak{M}_\chi^n \xrightarrow{f} \mathfrak{K}_t^1 \xrightarrow{f^{-1}} \mathfrak{M}_\chi^n$$

Such that

$$\mathfrak{K}_t^1 \subset \mathfrak{M}_\chi^n \qquad \text{Equation C}$$

Axiom 11.6.4 Manifold of a Discrete Line

Section 11.7 Theorems on Straight Hyperline Manifolds

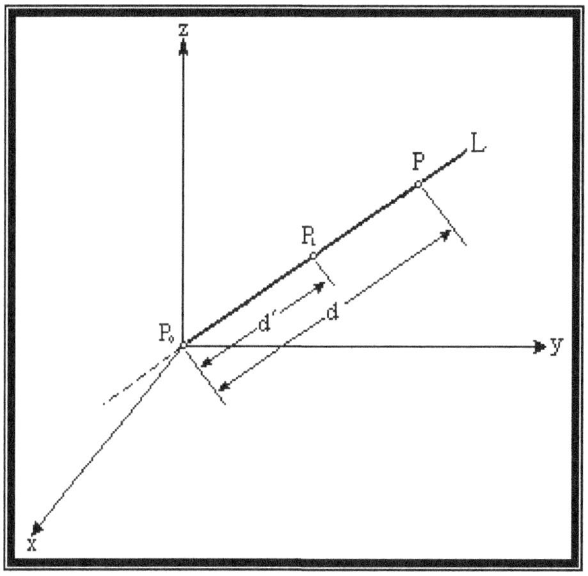

Figure 11.7.1 Three Points Lying on a Line

Theorem 11.7.1 Direction Cosines of L

1g	Given	the above geometry from figure F11.6.2D
2g		the above geometry from figure F11.7.1
3g		A set of angular parameters $(\bullet\ \theta_i \bullet$ for all i) referenced to the directed line-L
4g		$g_{\alpha xi}{}^n \equiv \{\ \Delta P_L P_0 P_{xi}$ a right triangle for $P_L(\bullet\ x_{iL} \bullet), P_0(\bullet\ x_{i0} \bullet), P_{xi}(\bullet\ x_{(i-1)0}, x_i, x_{(i+1)0} \bullet)\}$
Step	Hypothesis	$\cos(\theta_i) = (x_i - x_{i0}) / d$ for all i between points P_0P_{xi}
1	From 1g, 4g and D4.1.16C Distance or Magnitude on a Number Line	$\lvert P_0P_{xi}\rvert = x_i - x_{i0}$ for $x_{i0} < x_i$
2	From 1g, 3g and DxE.1.2.2	$\cos(\theta_i) = \lvert P_0P_{xi}\rvert / d$
3	From 1, 2 and A4.2.3 Substitution	$\cos(\theta_i) = (x_i - x_{i0}) / d$ is true for $g_{\alpha xi}{}^n$
4	From F11.6.2A, 3 and i = 1	$\cos(\theta_1) = (x_1 - x_{1(0)}) / d$ is true for $g_{\alpha xi}{}^n$
∴	From 3, 4 and A3.12.1 Equivalency Property of Constructible Geometric Components	$\cos(\theta_i) = (x_i - x_{i0}) / d$ for all i between points P_0P_{xi}

Theorem 11.7.2 Two Point Parametric Line-L

1g	Given	the above geometry from figure F11.6.2D
2g		A set of angular parameters ($\bullet$, θ_i, $\bullet$ for all i) referenced to the directed line-L
3g		$g_{\alpha\mathcal{L}}{}^{n} \equiv \{$ $P_L P_0 P$ three collinear points on a line-L $P_L(\bullet\ x_{iL}\ \bullet)$, $P_0(\bullet\ x_{i0}\ \bullet)$, $P(\bullet x_i\ \bullet)\}$
4g		$\nu_m\ t \equiv d / d'$ conversion to parametric variable with ν_m ***constant of parametric transformation*** having dimension $[T^{-1}]$ [11.7.1]
Step	Hypothesis	$\nu_m\ t = (x_i - x_{i0}) / (x_{iL} - x_{i0})$ for all i
1	From 1g, 2g and T11.7.1 Direction Cosines of L	$\cos(\theta_i) = (x_i - x_{i0}) / d$ for all i and $g_{\alpha\mathcal{L}}{}^{n}$
2	From 1g, 2g and T11.7.1 Direction Cosines of L	$\cos(\theta_i) = (x_{iL} - x_{i0}) / d'$ for all i and $g_{\alpha\mathcal{L}}{}^{n}$
3	From 1, 2 and A4.2.3 Substitution	$\cos(\theta_i) = (x_i - x_{i0}) / d = (x_{iL} - x_{i0}) / d'$
4	From 3, T4.4.8 Equalities: Cross Product of Proportions and D4.1.35 Constant of Proportional Quantities	$(x_i - x_{i0}) / (x_{iL} - x_{i0}) = d / d'$ constant of proportionality for all i
∴	From 4g, 4 and A4.2.3 Substitution	$\nu_m\ t = (x_i - x_{i0}) / (x_{iL} - x_{i0})$ for all i

Theorem 11.7.3 Two Point Parametric Equation of Slope for a Line-L

1g	Given	$m_i \equiv [\nu_m\ (x_{iL} - x_{i0})]$ with dimensions $[L][T^{-1}]$
Step	Hypothesis	$x_i = x_{i0} + m_i\ t$ for all i
1	From T11.7.2 Two Point Parametric Line-L	$(x_i - x_{i0}) / (x_{iL} - x_{i0}) = \nu_m\ t$ for all i
2	From 1 and T4.4.7B Equalities: Reversal of Left Cancellation by Multiplication	$x_i - x_{i0} = t\ [\nu_m\ (x_{iL} - x_{i0})]$
3	From 1g, 2, A4.2.3 Substitution, A4.2.10 Commutative Multp and T4.3.6A Equalities: Reversal of Left Cancellation by Addition	$x_i = m_i\ t + x_{i0}$
∴	From 3 and A4.2.5 Commutative Add	$x_i = x_{i0} + m_i\ t$ for all i

Theorem 11.7.4 Two Point Parametric Equations and their Directed Numbers

Step	Hypothesis	$x_i = x_{i0} + m_i\ t$ all i [11.7.2] EQ A	$t = (x_i - x_{i0}) / m_i$ all i EQ B[11.7.3] $m_i = (x_i - x_{i0}) / t$ all i EQ C
1	From T11.7.3 Two Point Parametric Equation of Slope for a Line-L	$x_i = x_{i0} + m_i\ t$ for all i	
2	From 1, A4.2.25 Commutative Add and T4.3.3A Equalities: Right Cancellation by Addition	$x_i - x_{i0} = m_i\ t$	
∴	From 1, 2 and T4.4.4A Equalities: Right Cancellation by Multiplication	$x_i = x_{i0} + m_i\ t$ all i EQ A	$t = (x_i - x_{i0}) / m_i$ all i EQ B $m_i = (x_i - x_{i0}) / t$ all i EQ C

Theorem 11.7.5 Directed Numbers are Slopes for Parametric Line-L

Step	Hypothesis	$x'_i = m_i$ for all i[11.7.4]
1	From T11.7.4A Two Point Parametric Equations and their Directed Numbers	$x_{1i} = x_{i0} + m_i\, t_1$ for all i and point $P_1(\bullet\, x_{1i}\, \bullet)$ at t_1 on line-L
2	From T11.7.4A Two Point Parametric Equations and their Directed Numbers	$x_{2i} = x_{i0} + m_i\, t_2$ for all i and point $P_2(\bullet\, x_{2i}\, \bullet)$ at t on line-L
3	From 1, 2 and taking the difference	$x_{2i} - x_{1i} = x_{i0} + m_i\, t_2 - (x_{i0} + m_i\, t_1)$
4	From 3, D4.1.20 Negative Coefficient, A4.2.14 Distribution and A4.2.5 Commutative Add	$x_{2i} - x_{1i} = x_{i0} - x_{i0} + m_i\, t_2 - m_i\, t_1$
5	From 4, A4.2.8 Inverse Add, A4.2.7 Identity Add and A4.2.14 Distribution	$x_{2i} - x_{1i} = m_i\, (t_2 - t_1)$
6	From 5 and DxK.3.1.1 Delta: Increment by Difference	$\Delta x_i = m_i\, \Delta t$
7	From 6 and T4.4.4A Equalities: Right Cancellation by Multiplication	$\Delta x_i / \Delta t = m_i$
8	From 7 and as P_1 is selected closer and closer toward P_2 then in the limit $\Delta t \rightarrow 0$	$\lim_{\Delta t \to 0} \dfrac{\Delta x_i}{\Delta t} = m_i$
∴	From 8 and DxK.3.1.3 Definition of a Derivative	$x'_i = m_i$ for all i

Theorem 11.7.6 Bounded Parametric Magnitude Between Two Points

1g	Given	$d^2 \equiv \Sigma^n [\, x_i(t) - a_i(t) \,]^2$
2g	for	$0 < d$
3g		$\mid X(b) - A(a) \mid \equiv \Sigma^n [\, x_i(t) - a_i(t) \,]^2$ in the interval I_t
Steps	Hypothesis	$0 < \mid X(b) - A(a) \mid / \sqrt{n} < k$ in the normalized interval δI_t[11.7.5]
1	From A11.6.1P3 Parametric Hypercurve Manifold $\mathfrak{P}^n$	$\lvert x_i(t) - a_i(t) \rvert < k$
2	From 1, A4.2.17 Correspondence of Equality and Inequality, T4.6.2 Equalities: Square of an Absolute Value and T4.8.6 Integer Exponents: Uniqueness of Exponents	$[\, x_i(t) - a_i(t) \,]^2 < k^2$
3	From 2 and summing over both sides of the inequality for all coordinate [i]	$\Sigma^n [\, x_i(t) - a_i(t) \,]^2 < \Sigma^n k^2$
4	From 3, T4.3.9 Equalities: Summation of Repeated Terms	$\Sigma^n [\, x_i(t) - a_i(t) \,]^2 < nk^2$
5	From 1g, 4 and A4.2.3 Substitution	$d^2 < nk^2$
6	From 5 and T4.10.4 Radicals: Identity Radical Raised to a Power	$d^2 < (\sqrt{n})^2 k^2$
7	From 6, A4.2.21 Distribution and A4.2.10 Commutative Multp	$d^2 < (\, k \sqrt{n})^2$
8	From 2g, 7, T4.8.6 Integer Exponents: Uniqueness of Exponents, A4.2.16 The Trichotomy Law of Ordered Numbers	$0 < d < k\sqrt{n}$
9	From 1g, 2g, T4.10.1 Radicals: Uniqueness of Radicals and T4.10.3 Radicals: Identity Power Raised to a Radical	$d = \sqrt{\Sigma^n [\, x_i(t) - a_i(t) \,]^2}$ for $0 < d$
10	From 8, 9 and A4.2.3 Substitution	$0 < \sqrt{\Sigma^n [\, x_i(t) - a_i(t) \,]^2} < k\sqrt{n}$
11	From 3g, 10 and A4.2.3 Substitution	$0 < \mid X(b) - A(a) \mid < k\sqrt{n}$
∴	From 11, A4.2.12 Identity Multp, and T4.7.31 Inequalities: Cross Multiplication with Positive Reciprocal Product	$0 < \mid X(b) - A(a) \mid / \sqrt{n} < k$ in the normalized interval δI_t

[11.7.1]Note: The dimension [T] in this case does not necessarily mean time, but any parametric dimension as specified.

[11.7.2]Note: Clearly from the inverse equations of T11.7.4B the m_i's are the directed numbers from Definition 11.6.4 "Directed Numbers".

[11.7.3]Note: From Theorem 11.7.4 "Two Point Parametric Equation and Its Inverse", equation B, gives rise to the possibility of dividing by zero, because it is permissible to have a line lying horizontal to the base axis with a slope of zero. Since this is a geometric real possibility, the singularity is removed by not dividing, but it is understood to mean the slope is horizontal.

[11.7.4]Note: It is seen that the directed numbers are nothing more than the derivative or slope for the variable $[x_i]$ at an instant of the parametric parameter [t].

[11.7.5]Note: $(\mid X(b) - A(a) \mid / \sqrt{n} < k) \rightarrow (\lvert t - \alpha \rvert < \kappa)$ within the interval δI_t so it can now be said that the magnitude is continuous within the interval [a, b], hence differentiable.

Section 11.8 Functions at a Point in or on a Manifold $\mathfrak{M}^n$

Establishing the existence of a function at a point in or on a manifold $\mathfrak{M}^n$ is done by using a Taylor's series expansion about a point. This is useful in establishing various theorems of calculus for manifolds and their properties.

Let's consider a hypersphere of radius δ defining a region in or on a manifold having directional cosines ν_i bounding a point A(• a_i •).

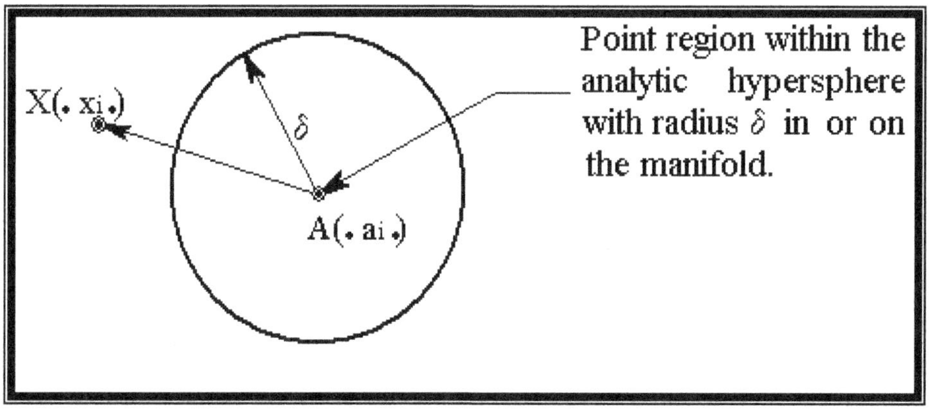

Figure 11.8.1: Point within a hypersphere of radial distance δ

Definition 11.8.1 **Difference about a point A(• a_i •)**

$\Delta x_i \equiv x_i - a_i$

Definition 11.8.2 **Parametric Directional Cosine about a point A(• a_i •)**

$\mu_i \equiv \cos(\phi_{xi})$ see T11.7.1 "Direction Cosines of L" Equation A

$\mu_i \equiv \Delta x_i / \delta$ Equation B

Theorem 11.8.1 Creating a Parametric Function from a Multi-variable Function

1g	Given	$\phi(\delta) \equiv f(\bullet\ x_i\ \bullet)$ at the point $A(\bullet\ a_i\ \bullet)$ within a radial distance δ having directional cosines μ_i for all [i].
Steps	Hypothesis	$\phi(\delta) = f(\bullet\ a_i + \mu_i\ \delta\ \bullet)$
1	From 1g and A4.2.7 Identity Add	$\phi(\delta) = f(\bullet\ 0 + x_i\ \bullet)$
2	From 1, A4.2.8 Inverse Add and A4.2.5 Commutative Add	$\phi(\delta) = f(\bullet\ a_i + x_i - a_i\ \bullet)$
3	From 2, A4.2.6 Associative Add and A4.2.12 Identity Multp	$\phi(\delta) = f(\bullet\ a_i + (x_i - a_i)\ 1\ \bullet)$
4	From 2g, 3 and A4.2.13 Inverse Multp	$\phi(\delta) = f(\bullet\ a_i + (x_i - a_i)\ \delta\ /\ \delta\ \bullet)$
5	From 4, D4.1.4B Rational Numbers, A4.2.10 Commutative Multp and A4.2.11 Associative Multp	$\phi(\delta) = f(\bullet\ a_i + (\ (x_i - a_i)\ /\ \delta\ )\ \delta\ \bullet)$
∴	From 5, D11.6.5 Directed Cosine in n-Dimensions and A4.2.3 Substitution	$\phi(\delta) = f(\bullet\ a_i + \mu_i\ \delta\ \bullet)$

Theorem 11.8.2 First Order Parametric Differentiation

1g	Given	$\phi^0(\delta) = f^0(\bullet\ a_i + \mu_i\ \delta\ \bullet)$ continuously differentiable at every point $(\bullet\ a_i\ \bullet)$ through out the manifold.
2g	Generalizing ϕ and λ_i	μ_i is a constant of the space hence independent of $[x_i]$ for all [i] and the varying parametric parameter [δ].
3g		$\phi^1\ (\delta) = f^1$
Steps	Hypothesis	$\phi^1 = \sum_n \phi^0{,i}\ \mu_i$
1	From 1g and TK.3.9 Exact Differential, Parametric, Chain Rule.	$\phi^1 = \sum_n f^0(\bullet\ a_i + \mu_i\ \delta\ \bullet){,i}\ d(a_i + \mu_i\ \delta)\ /\ d\delta$
2	From 1, 2g, LxK.3.2.2A Differentiation of a polynomial, LxK.3.3.1A Differential of a Constant, LxK.3.2.3A Distribute Across Constant Product, LxK.3.3.2A Differential identity, A4.2.7 Identity Add and A4.2.12 Identity Multp	$\phi^1 = \sum_n f^0{,i}\ \mu_i$
∴	From 1g, 2 and A4.2.3 Substitution	$\phi^1 = \sum_n \phi^0{,i}\ \mu_i$

Theorem 11.8.3 mth Order Parametric Differential: Recursion

1g	Given	$\phi^{m-1} = h(\delta)$ observe ϕ^{m-1} is just a parametric function.
Steps	Hypothesis	$\phi^m = \sum_n \phi^{m-1}{,i}\ \mu_i$
1	From 1g and T11.8.2 First Order Parametric Differentiation	$h^1 = \sum_n h^0{,i}\ \mu_i$
2	From 1g, 1 and A4.2.3 Substitution	$(\phi^{m-1})^1 = \sum_n (\phi^{m-1})^0{,i}\ \mu_i$
∴	From 2 and LxK.3.2.7A Differential of a Differential	$\phi^m = \sum_n \phi^{m-1}{,i}\ \mu_i$

Theorem 11.8.4 mth Order Parametric Differential: Continuous

1g	Given	To start the recursion set m = 1
Steps	Hypothesis	$\phi^m = \sum_{\rho\in Lm} \phi^0{}_{,\rho(k)} \prod^m \mu_{\lambda[i,k]}$
1	From 1g and T11.8.2 First Order Parametric Differentiation	$\phi^1 = {}_1\sum^n \phi^0{,}i\ \mu_i$
2	From 1, m = 2, T11.8.3 mth Order Parametric Differential: Recursion and A4.2.3 Substitution	$\phi^2 = {}_2\sum^n ({}_1\sum^n \phi^0{,}i\ \mu_i){,}j\ \mu_j$
3	From 2 and LxK.3.2.2B Differential as a Linear Operator	$\phi^2 = {}_2\sum^n \phi^0{,}i{,}j\ \mu_i\mu_j$
4	Over index k repeat steps 2 through 3 up to m	$\phi^m = {}_m\sum^n\ \phi^0{\bullet}\ i_k \bullet \prod^m \mu i_k$
5	From 4, D2.16.14 Positional Unitary Index Set and over the summation D2.16.16 Ordered Universal or Long Index Set-L_T	$\phi^m = \sum_{\rho\in Lm} \phi^0{\bullet}\ \lambda[i,k]{\bullet} \prod^m \mu_{\lambda[i,k]}$
∴	From 5 and simplifying with D2.16.15 Set of Coordinate Points; ρ-ordered indices	$\phi^m = \sum_{\rho\in Lm} \phi^0{}_{,\rho(k)} \prod^m \mu_{\lambda[i,k]}$

Theorem 11.8.5 Differential at a Point on or in a Manifold $\mathfrak{M}^n$

1g	Given	With in the region of the point A(• a_i •) lim δ → 0
Steps	Hypothesis	$\phi^m(0) = \sum_{\rho\in Lm} f(\bullet\ a_i \bullet)_{,\rho(k)} \prod^m \mu_{\lambda[i,k]}$ for $\delta = 0$
1	From T11.8.1 Creating a Parametric Function from a Multi-variable Function	$\phi(\delta) = f(\bullet\ a_i + \mu_i\ \delta \bullet)$
2	From T11.8.4 mth Order Parametric Differential: Continuous	$\phi^m(\delta) = \sum_{\rho\in Lm} f(\bullet\ a_i + \mu_i\ \delta \bullet)_{,\rho(k)} \prod^m \mu_{\lambda[i,k]}$
∴	From 1g and converging onto the point A(• a_i •)	$\phi^m(0) = \sum_{\rho\in Lm} f(\bullet\ a_i \bullet)_{,\rho(k)} \prod^m \mu_{\lambda[i,k]}$ for $\delta = 0$

Theorem 11.8.6 Talyor's Series at a Point in/on a Manifold $\mathfrak{M}^n$

1g	Given	$\phi(\eta) \equiv f(\bullet\, x_i \,\bullet)$ at the point $A(\bullet\, a_i \,\bullet)$ within a radial distance δ having directional cosines μ_i for all [i].
2g		$\eta = [\Sigma_i^n (x_i - a_i)^2]^{1/2}$
Steps	Hypothesis	$f(\bullet\, x_i \,\bullet) = \Sigma_{m=0}^{p} (\Sigma_{\rho\in Lm} f(\bullet\, a_i \,\bullet)_{,\rho(k)} \prod^m \Delta x_{\lambda[i,k]}) / m!$
1	From 2g and LxE.3.1.1 MacLaurin's Series at a point	$\phi(\eta) = \Sigma_{m=0}^{p} \phi^m(0)\, \eta^m / m!$
2	From 1g, 1, T11.8.5 Differential at a Point on or in a Manifold $\mathfrak{M}^n$ and A4.2.3 Substitution	$f(\bullet\, x_i \,\bullet) = \Sigma_{m=0}^{p} (\Sigma_{\rho\in Lm} f(\bullet\, a_i \,\bullet)_{,\rho(k)} \prod^m \mu_{\lambda[i,k]})\, \eta^m / m!$
3	From 2 and T2.21.10 Multiplicative Sets: Binary Distribution of Rank	$f(\bullet\, x_i \,\bullet) = \Sigma_{m=0}^{p} (\Sigma_{\rho\in Lm} f(\bullet\, a_i \,\bullet)_{,\rho(k)} \prod^m (\mu_{\lambda[i,k]}\, \eta) / m!$
4	From D11.8.2B Parametric Directional Cosine about a point $A(\bullet\, a_i \,\bullet)$	$\mu_{\lambda[i,k]} = \Delta x_{\lambda[i,k]} / \eta$
5	From 4, T4.4.5B Equalities: Reversal of Right Cancellation by Multiplication and DxK.3.1.1 Delta: Increment by Difference	$\mu_{\lambda[i,k]}\, \eta = \Delta x_{\lambda[i,k]}$
∴	From 3, 5 and A4.2.3 Substitution	$f(\bullet\, x_i \,\bullet) = \Sigma_{m=0}^{p} (\Sigma_{\rho\in Lm} f(\bullet\, a_i \,\bullet)_{,\rho(k)} \prod^m \Delta x_{\lambda[i,k]}) / m!$

Notice that MaClaurin's Series is for a finite number of terms that is for two reasons;

1) To look at the last term and determine if the series converges and
2) there may be only a finite number of differentials that can be taken at the point in or on the manifold. So, the series may only be able to be constructed for a finite number of terms, hence the series terminates at the finite index-p+1.

Theorem 11.8.7 Talyor's Series Combinatorial Contraction[11.3.1]

Steps	Hypothesis	$f(\bullet\, x_i \,\bullet) = f(\bullet\, a_i \,\bullet) + \Sigma_{m=1}^{p+1} (\Sigma_{s=1}^{n} \Sigma_{t=1}^{Q(m,s)}\ {}_mK_{\kappa(s,t)}$ $\Sigma_{u=1}^{S(n,s)} \prod_{v=1}^{s} \Delta x_{\gamma[s,t,u,v]}^{\eta[s,t,v]}$ $f(\bullet\, a_i \,\bullet)_{,\gamma[s,t,u,v]}) / m!$
1	From T11.8.6 Talyor's Series at a Point in/on a Manifold $\mathfrak{M}^n$	$f(\bullet\, x_i \,\bullet) = \Sigma_{m=0}^{p+1} (\Sigma_{\rho\in Lm} f(\bullet\, a_i \,\bullet), \rho(k) \prod^m \Delta x_{\lambda[i,k]}) / m!$
2	From 1 expand first term initiating index-m at 1 to match the expanded product.	$f(\bullet\, x_i \,\bullet) = f(\bullet\, a_i \,\bullet) + \Sigma_{m=1}^{p+1} (\Sigma_{\rho\in Lm} f(\bullet\, a_i \,\bullet), \rho(k)$ $\prod^m \Delta x_{\lambda[i,k]}) / m!$
∴	From 2 and D1.6.4 Set Summation Notation over Combinatorial Indices	$f(\bullet\, x_i \,\bullet) = f(\bullet\, a_i \,\bullet) + \Sigma_{m=1}^{p+1} (\Sigma_{s=1}^{n} \Sigma_{t=1}^{Q(m,s)}\ {}_mK_{\kappa(s,t)}$ $\Sigma_{u=1}^{S(n,s)} \prod_{v=1}^{s} \Delta x_{\gamma[s,t,u,v]}^{\eta[s,t,v]}$ $f(\bullet\, a_i \,\bullet)_{,\gamma[s,t,u,v]}) / m!$ for [n] the dimension of the manifold.

[11.3.1]Note: $\gamma[s, t, u, v]$ is indexed over the product for index-v, however used as a partial derivative it refers to the product of all partials index over-v. So $\prod_{v=1}^{2} f(\bullet\, a_i \,\bullet)_{,\gamma[2,t,u,v]} = f(\bullet\, a_i \,\bullet)_{,\gamma[2,t,u,1],\gamma[2,t,u,2]}$.

Theorem 11.8.8 Talyor's Series Converges at a Point in/on the Manifold $\mathfrak{M}^n$

1g	Given the last term in the Taylor's Series using LxE.3.4 Talyor's Remainder from Table E.3.1	$R_{p+1} = (\sum_{\rho \in L(p+1)} f(\bullet\ a_i\ \bullet), \rho(k) \prod^{p+1} \Delta x_{\lambda[i,k]}) / (p+1)!$
2g		$\lim_{p \to \infty} R_{p+1} \rightarrow 0$
Steps	Hypothesis	$\lim_{p \to \infty} R_{p+1} \rightarrow 0$ series converges.
∴	From 1g and 2g	is the conditions for the series to converge.

Observation 11.8.1: Evaluation of the Taylor's at the expansion point $A(\bullet\ a_i\ \bullet)$

$f(\bullet\ x_i\ \bullet) = f(\bullet\ a_i\ \bullet) + \sum_{m=1}^{p+1} (\sum_{\rho \in Lm} f(\bullet\ a_i\ \bullet), \rho(k) \prod^{m} \Delta x_{\lambda[i,k]}\ 1) / m!$ and at the point $A(\bullet\ a_i\ \bullet)$
$f(\bullet\ a_i\ \bullet) = f(\bullet\ a_i\ \bullet)$

Observation 11.8.2: δ Removable Singularity

T11.3.6 removes the radius δ defining the region, thereby eliminating it as a singularity making the function general for any point $X(\bullet\ x_i\ \bullet)$.

Section 11.9 Hyperplane Manifold ${}_p\mathfrak{H}^{n-1}$ Embedded in ${}_x\mathfrak{E}^n$

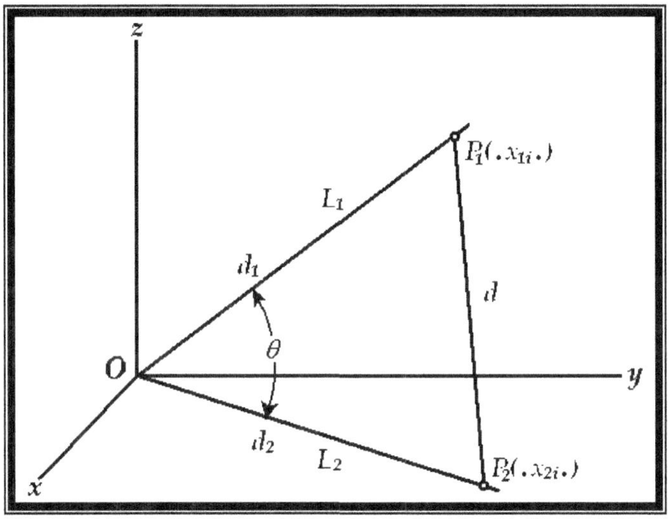

Figure 11.9.1 Angle between Two Intersecting Lines

Theorem 11.9.1 Angle between Two Intersecting Lines By Direction Cosines

1g	Given	see Figure 11.9.1
2g		$\alpha_{1i} \equiv x_{1i} / d_1$ direction cosines for L_1
3g		$\alpha_{2i} \equiv x_{2i} / d_2$ direction cosines for L_2
4g		$d_1^2 \equiv \sum_i x_{1i}^2$ distance from point-O to point-P_1
5g		$d_2^2 \equiv \sum_i x_{2i}^2$ distance from point-O to point-P_2
6g		$d^2 \equiv \sum_i (x_{2i} - x_{1i})^2$ distance between point-P_1 to point-P_2
Step	Hypothesis	$\cos\theta = \sum_i \alpha_{1i}\,\alpha_{2i}$
1	From 1g and LxE.2.5 Law of Cosine	$\cos\theta \equiv \dfrac{d_1^2 + d_2^2 - d^2}{2d_1d_2}$
2	From 4g, 5g, 6g, 1, A4.2.3 Substitution, A4.2.5 Commutative Add and A4.2.6 Associative Add	$\cos\theta = \dfrac{\sum_i x_{1i}^2 + x_{2i}^2 - (x_{2i} - x_{1i})^2}{2d_1d_2}$
3	From 2 and T4.12.2 Polynomial Quadratic: The Perfect Square by Difference	$\cos\theta = \dfrac{\sum_i x_{1i}^2 + x_{2i}^2 - (x_{2i}^2 - 2x_{1i}\,x_{2i} + x_{1i}^2)}{2d_1d_2}$
4	From 3, A4.2.14 Distribution, D4.1.20A Negative Coefficient, D4.1.19 Primitive Definition for Rational Arithmetic and A4.2.5 Commutative Add	$\cos\theta = \dfrac{\sum_i x_{1i}^2 - x_{1i}^2 + x_{2i}^2 - x_{2i}^2 + 2x_{1i}\,x_{2i}}{2d_1d_2}$
5	From 4, A4.2.8 Inverse Add, A4.2.7 Identity Add, A4.2.13 Inverse Multp and A4.2.12 Identity Multp	$\cos\theta = \dfrac{\sum_i x_{1i}\,x_{2i}}{d_1d_2}$
6	From 5, A4.2.14 Distribution and A4.2.11 Associative Multp	$\cos\theta = \sum_i \dfrac{x_{1i}}{d_1}\;\dfrac{x_{2i}}{d_2}$
∴	From 2g, 3g, 5 and A4.2.3 Substitution	$\cos\theta = \sum_i \alpha_{1i}\,\alpha_{2i}$

Corllary 11.9.1.1 Direction Cosines Sum to Zero for Line L1 and L2 Perpendicular

Step	Hypothesis			
	Hypothesis	0	=	$\sum_i \alpha_{1i} \alpha_{2i}$
1	From T11.9.1 Angle between Two Intersecting Lines By Direction Cosines	$\cos \theta$	=	$\sum_i \alpha_{1i} \alpha_{2i}$
2	For L1 and L2 Perpendicular and D5.1.7-1 Right Angle	θ	=	$\frac{1}{2}\pi$
3	From 1, 2 and A4.2.3 Substitution	$\cos (\frac{1}{2}\pi)$	=	$\sum_i \alpha_{1i} \alpha_{2i}$
∴	From 3 and DxE.1.6.7 for cosine	0	=	$\sum_i \alpha_{1i} \alpha_{2i}$

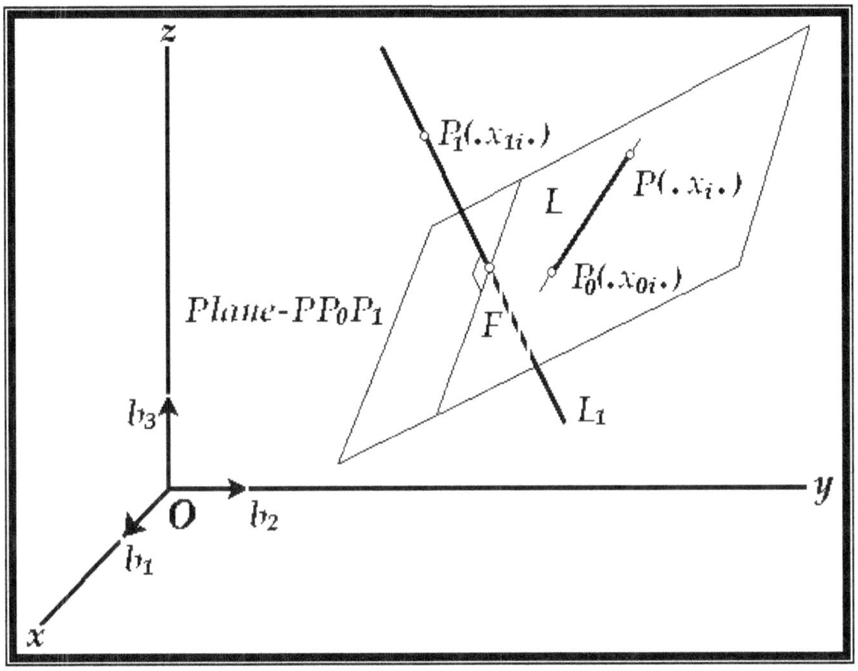

Figure 11.9.2 Three Point Plane

Theorem 11.9.2 Analytic Equation of a Three Point Plane

1g	Given	See Figure 11.9.2
Step	Hypothesis	$0 = \sum_i A_i (x_i - x_{0i})$ [11.9.1]
1	From 1g, D11.6.5A where the Three Point Plane between points P and Q on line L yields direction numbers	$\alpha_i \equiv (x_i - x_{0i}) / d$
2	From 1g, D11.6.5A where the Three Point Plane between points P_1 and F on line L_1 yields direction numbers	$\alpha_{1i} \equiv A_i / D$ directed line normal to the plane
3	From 1g and T11.9.1 Angle between Two Intersecting Lines By Direction Cosines	$\cos\theta = \sum_i \alpha_i \alpha_{1i}$
4	From T5.22.21 Line L_1 through P_0 and P is perpendicular to L in a Plane	$\theta = ½\pi$ $L \perp L_1$ defining the plane-PP_0P_1
5	From 3, 4, A4.2.3 Substitution and DxE.1.6.7 cos(90°)	$0 = \sum_i \alpha_i \alpha_{1i}$
6	From 1, 2, 5 and A4.2.3 Substitution	$0 = \sum_i \dfrac{A_i}{D} \dfrac{(x_i - x_{0i})}{d}$
∴	From 6, A4.2.14 Distribution, T4.4.1 Equalities: Any Quantity Multiplied by Zero is Zero and T4.4.13 Equalities: Cancellation by Division	$0 = \sum_i A_i (x_i - x_{0i})$

[11.9.1]Note: Theorem 11.9.2 "Analytic Equation of a Three Point Plane" solves the problem for the locus of all points-$P(\bullet\, x_i\, \bullet)$ on the plane-PP_0P_1.

Section 11.10 Propeller Spinning in a Manifold

Section 11.11 Conic and Metamorphic Manifolds

Section 11.12 Directed Coordinate Point R in Manifold ${}_x\mathfrak{C}^n$

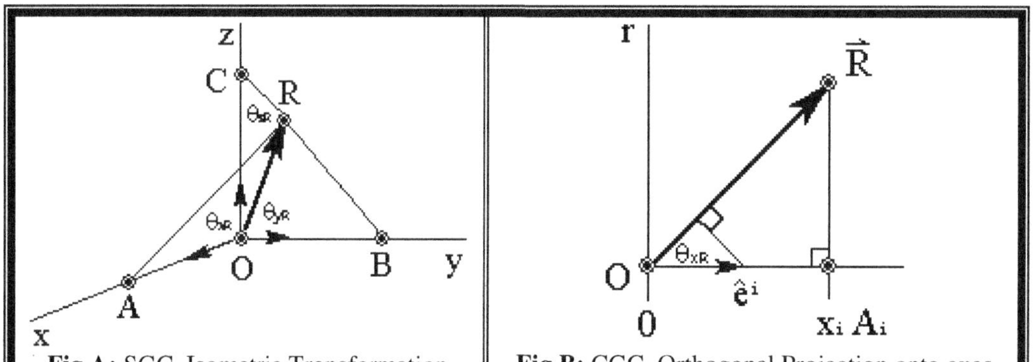

Fig A: SGC, Isometric Transformation **Fig B:** CGC, Orthogonal Projection onto axes

Figure 11.12.1: Directed Coordinate Point

Given 11.12.1 Directed Coordinate Axes between Manifolds ${}_x\mathfrak{C}^n$ and ${}_\theta\mathfrak{C}^n$

1g	Given	$i = 1, 2, \ldots, n$ Axial Index
2g		$\theta_{iR} \equiv \angle A_iOR$ ${}_x\mathfrak{C}^n$
3g		$R \equiv \overline{RO}$ common ${}_x\mathfrak{C}^n$ and ${}_\theta\mathfrak{C}^n$
4g		$x_i \equiv \overline{AO}_i$ ${}_x\mathfrak{C}^n$
5g		$\vec{R} \equiv \sum^n x_i\, \hat{e}^i$ ${}_x\mathfrak{C}^n$

Theorem 11.12.1 Directed Coordinate Point

Steps	Hypothesis	Point R(• θ$_i$ •) for all i	Point R(• x$_i$ •) for all i
1	From 3g, 5g and T9.3.6	$R^2 = \sum^n x_i^2$	
2	From 1 and A4.2.25	$R = \sqrt{\sum^n x_i^2}$	
3	From F11.4.1B		$\overline{AO}_i = \overline{RO} \cos(\angle A_iOR)$
4	From 3, 4g, 3g, 2g and A4.2.3		$x_i = R \cos(\theta_i)$
5	From 4 and TD.1.13	$\theta_i = \arccos(x_i / R)$	
∴	From 2 and by A2.12.1	Point R(• θ$_i$ •) for all i	Point R(• x$_i$ •) for all i

Quantities in a Manifold

- Skew Pythagorean Theorem
- Skew Dot Operator
- Skew Orthogonal Operator

Table of Contents

Tensor Calculus & Physics: A General Treatise

List of Tables

List of Figures

List of Observations

Chapter 12 Tensor Calculus: Metrics and their Differential Geomtry

Section 12.1 Manifold Definitions for Dual Base Vector System

With the advent of differential geometry, systems of vectors had to be able to translate from one coordinate point in a manifold, to another, and maintain an invariant magnitude and direction, but in order to do that a system of vectors has to be established that are invariant with respect to one another. These internal systems of vectors span the manifold in which they are embedded forming a spatial base set of vectors called:

Definition 12.1.1 **Covariant Vector**

Covariant vectors act as a variable vector field conjoined with the manifold, marked by lower or subscripted indices on the bases vector and upper on the coefficient. Such vectors are written as:

$\mathbf{A} \equiv \sum_i A^i \mathbf{b}_i$ Equation A

Definition 12.1.2 **Contravariant Vector**

Contravariant vectors are contrary (orthogonal) to the covariant vector field, marked by upper or superscripted indices on the bases vector and lower on the coefficient.

$\mathbf{A} \equiv \sum_i A_i \mathbf{b}^i$ Equation A

The co- and contravariant vector establishes invariance of magnitude and direction as an isometric transformation within that space. This is very important for Affined Geometry, which preserves magnitude and direction as vectors move about in the manifold.

In considering magnitude lets deliberate what is meant by the inner product for the basis vectors within the manifold system.

Definition 12.1.3 **Covariant Metric Dyadic**

$g_{ij} \equiv \mathbf{b}_i \bullet \mathbf{b}_j$	component form	Equation A
$(g_{ij}) \equiv (\mathbf{b}_i \bullet \mathbf{b}_j)$	component in matrix form	Equation B
$G \equiv (g_{ij})$	matrix form (n x n)	Equation C
$G \equiv S^{-1}\{B_c B_c^{\ T}\}$	matrix form (n x 1)(1 x n)	Equation D

for

$S\{ (\mathbf{b}_i \bullet \mathbf{b}_j) \} \equiv B_c B_c^{\ T}$	matrix form (n x 1)(1 x n)	Equation E
$(\mathbf{b}_i \bullet \mathbf{b}_j) \equiv S^{-1}\{B_c B_c^{\ T}\}$	matrix form (n x 1)(1 x n)	Equation F
$B_c \equiv [\beta_i]$	matrix form (n x 1)	Equation G

Definition 12.1.4 **Contravariant Metric Dyadic**

$g^{ij} \equiv \mathbf{b}^i \bullet \mathbf{b}^j$	component form	Equation A
$(g^{ij}) \equiv (\mathbf{b}^i \bullet \mathbf{b}^j)$	component in matrix form	Equation B
$g^{ij} \equiv g^{-1}\, G^{ij}$	where G^{ij} is the ***cofactor*** of the element g^{ij} in [g].	Equation C
$G^{-1} \equiv S^{-1}\{B^c B^{cT}\}$	matrix form (n x 1)(1 x n)	Equation D

for

$S\{ (\mathbf{b}^i \bullet \mathbf{b}^j) \} \equiv B^c B^{cT}$	matrix form (n x 1)(1 x n)	Equation E
$(\mathbf{b}^i \bullet \mathbf{b}^j) \equiv S^{-1}\{B^c B^{cT}\}$	matrix form (n x 1)(1 x n)	Equation F

Sometimes known as the ***reciprocal metric tensor***.

Observation 12.1.1 Why use column matrices for the contravariant inverse?

The column matrices alow the covariant dot products to be seperated or parsed, thereby more easly manipulated. They also revieal certain hidden propertise that otherwise could not be seen, such as permuatating columns, which in tern give rise to proofs for the existence of contravariant quantities. This is a similar to the idea of using Fourie, Lapace or Hilberet transformations, were mapping a quantity into another domain allows ease of manipulation or other otherwise simplifying complicated operations.

Definition 12.1.5 **Mixed Metric Dyadic**

$g_i^{\ j} \equiv \delta_i^{\ j} \equiv \mathbf{b}_i \bullet \mathbf{b}^j$	mix basis vectors	Equation A
$(g_i^{\ j}) \equiv (\delta_i^{\ j}) \equiv (\mathbf{b}_i \bullet \mathbf{b}^j)$	Kronecker Delta as a mixed metric	Equation B
$I \equiv S^{-1}\{B_c B^{cT}\}$	matrix form (n x 1)(1 x n)	Equation C
$g^i_{\ j} \equiv \delta^i_{\ j} \equiv \mathbf{b}^i \bullet \mathbf{b}_j$	mix basis vectors	Equation D
$(g^i_{\ j}) \equiv (\delta^i_{\ j}) \equiv (\mathbf{b}^i \bullet \mathbf{b}_j)$	Kronecker Delta as a mixed metric	Equation E
$I \equiv S^{-1}\{B^c B_c^{\ T}\}$	matrix form (n x 1)(1 x n)	Equation F

Placing g_{ij} and g^{ij} components into n x n matrices allows an orthogonal system to be constructed. This also shows the orthogonal relationship of the system with $\delta_i^{\ j} = \mathbf{b}_i \bullet \mathbf{b}^j$, hence $\mathbf{b}_i \perp \mathbf{b}^i$.

A ***dual base vector system*** has closure, because being orthogonal permits the system to come full circle with the Greek's idea of invariance for moving vectors using a parallelogram, see A7.8.7 "Parallelogram Law: Affine Transformation of Vectors" in a Euclidian flat space. This allows the system to carryout the affine geometric operation of invariance.

Definition 12.1.6 **Covariant Base Vector**

A ***covariant vector coefficient*** is

$\mathbf{b}_i \equiv b_i\, \mathbf{e}_i$ Equation A

for $[\mathbf{e}_i]$ a ***covariant unit vector*** and $[b_i]$ a ***covariant scalar factor***.

Definition 12.1.7 **Contravariant Base Vector**

A ***contravariant vector coefficient*** is

$\mathbf{b}^i \equiv b^i\, \mathbf{e}^i$ Equation A

for $[\mathbf{e}^i]$ a ***contravariant unit vector*** and $[b^i]$ a ***contravariant scalar factor***.

Definition 12.1.8 **Covariant Vector Coefficient**

A ***covariant vector coefficient*** is $[A_i]$.

Definition 12.1.9 **Contravariant Vector Coefficient**

A ***contravariant vector coefficient*** is $[A^i]$.

Definition 12.1.10 **Covariant Displacement Vector Coefficient**

A ***covariant vector coefficient*** is $[dx_i]$.

Definition 12.1.11 **Contravariant Displacement Vector Coefficient**

A ***contravariant vector coefficient*** is $[dx^i]$.

Definition 12.1.12 **Covariant Meteric Determinate**

$g = |\, g_{ij} \,|$ Equation A

Definition 12.1.13 **Space Density Factor**

$\rho_v = \prod_i b_i$ Equation A

Definition 12.1.14 **Cosine Space Distortion Factor**

The measure of how distorted space is deviating from an ideal orthogonal base vector system.

$\mu\ (c) \equiv |\cos\theta_{\alpha\beta}|$ Equation A

$\mu_c \equiv \mu\ (c)$ Equation B

Observation 12.1.2 Rule for Naming Tensor Quantities

A tensor quantity takes its name from the polyadic either co- or contra. If it is made up of a mixed polyadic, then the quantity is named as a mixed tensor.

From axiom A10.2.14 "Correspondence of Tensors and Tensor Coefficients" allows the proof writer to work on the tensor coefficient exclusively so it is often forgotten that there is a polyadic associated with it. As a result many authors often use the tensor's covariant or contravariant coefficient to name it. All of this becomes very confusing; do we name the tensor after the polyadic or the coefficient? The tensor coefficient is only one peice of the whole that comprises the tensor; it is not the whole. All of these things have lead to a lot of bad name calling in tensor calculus, but as a matter of history going back about a hundred years the names that we have, have become traditional and so in this treatises to avoid confusion are retained, but in the future the rule of thumb is go with the co- or contravariant name of the polyadic in naming the tensor. A rule of thumb is follow the basis vector idex raised or lowered?

Axiom 12.1.1 Covariant Determinate Defines the Existence of a Spatial Manifold

Let ∃ a Spatial Manifold $\mathfrak{S}^n$ such that under the dot product operator contra and covariant metrics exist, where

$g \neq 0$ Equation A

or

$|g_{ij}| \neq 0$ Equation B

Section 12.2 Manifold Tangential Plane On a Time Line

A manifold changes shape with a velocity – ($\mathbf{V} = \dot{\mathbf{r}}$) having its directional numbers change with time.

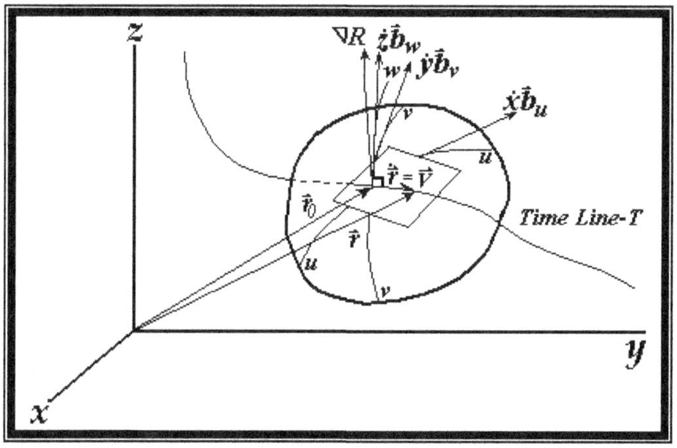

Figure 12.2.1 Manifold Tangential Plane On Time Line

Theorem 12.2.1 Manifold Rendered to Tangent Plane On Time Line

1g	Given	$x^i \equiv x^i(t)$	Function of time off of time line-T
2g		$\nabla R \equiv \sum_i R_{,i}\, \mathbf{b}^i$	Contravariant gradient vector
3g		$\mathbf{V}^\mu \equiv \sum_i \mu^j\, \mathbf{b}_j$	Covariant directional velocity vector
Step	Hypothesis	$0 = \nabla R \bullet \mathbf{V}^\mu$ built on a dual orthogonal base vector system	
1	From D11.2.3 Rendered Equation of a Manifold	$c = R(\bullet x^i \bullet)$	
2	From 1, TK.3.2 Uniqueness of Differentials and LxK.3.3.1A Differential of a Constant	$0 = R'$	
3	From 1g, 3, TK.3.9 Exact Differential, Parametric, Chain Rule and T11.9.2 Analytic Equation of a Three Point Plane	$0 = \sum_i R_{,i}\, x^{i\prime}$ the plane has two perpendicular directed lines	
4	From 3, D8.3.3C Orthogonal Vectors and D12.1.5A Mixed Metric Dyadic	$0 = \sum_i \sum_j \delta_j{}^i\, R_{,i}\, x^{j\prime}$ requires orthogonal dual base set of vectors, contravariant normal and covariant tangent to the plane.	
5	From 4 and D12.1.5B Mixed Metric Dyadic	$0 = \sum_i \sum_j (\mathbf{b}_j \bullet \mathbf{b}^i)\, R_{,i}\, x^{j\prime}$	
6	From 5 and A4.2.10 Commutative Multp	$0 = \sum_i \sum_j (R_{,i}\, \mathbf{b}^i) \bullet (x^{j\prime}\, \mathbf{b}_j)$	
7	From 6 and A7.9.4 Associative by Vector Addition	$0 = (\sum_i R_{,i}\, \mathbf{b}^i) \bullet (\sum_j x^{j\prime}\, \mathbf{b}_j)$	
8	From 7, T11.7.5 Directed Numbers are Slopes for Parametric Line-L	$0 = (\sum_i R_{,i}\, \mathbf{b}^i) \bullet (\sum_j \mu^j\, \mathbf{b}_j)$	
9	From 8, DxK.6.1.3A Covariant Differential Operator, DEL	$0 = (\sum_i \nabla R_{,i}\, \mathbf{b}^i) \bullet (\sum_j \mu^j\, \mathbf{b}_j)$	
∴	From 2g, 3g, 4, 9, A4.2.3 Substitution and D8.3.3 Orthogonal Vectors	$0 = \nabla R \bullet \mathbf{V}^\mu$	for $\delta_i{}^j = (\mathbf{b}_i \bullet \mathbf{b}^j)$ hence $\mathbf{b}_i \perp \mathbf{b}^j$.

Observation 12.2.1 Existence Dual Base Vector Set for a Manifold

Expanding the directional vector to approximate a difference [$0 = \sum_i R_{,i} (x^i - x_0^i)$] it can be clearly seen that $(x^i - x_0^i)$ comprises the directed line in the plane, which intern forces a conclusion to be made that the gradient ∇R is normal to the plane as shown by T11.9.2 "Analytic Equation of a Three Point Plane".

The physical interpretation of the equation [$0 = \nabla R \bullet \mathbf{V}^\mu$] can be seen from Figure 12.2.1, it is an infinitesimal plane at a point on the manifold. The directional vector ($\bullet\ dx^i\ \mathbf{b}_i\ \bullet$) is tangent along the infinitesimal plane and the gradient's components ($\bullet\ R_{,i}\ \mathbf{b}^i\ \bullet$) are directed normally away from the manifold plane.

This orthogonal property is brought about by two different kinds of base vectors that reside at every point within the manifold. Theorem 12.2.1 "Rendered Tangent Plane to a Manifold" clearly proves that all manifolds, having the form of the Rendered Equation, exist with an orthogonal dual set of basis vectors. As with the gradient vector the basis vectors are of the covariant type while the directed vector is contravariant in nature. Using Elwin Bruno Christoffel's notation to distinguish contra (raised) verses covariant (lowered) indices from now on in this treatises this notation is standard. Since matrices are another system of dual orthogonal basis vectors this is the reason why matrices in this paper are represented with the contra and covariant notation for rows and columns respectively.

Theorem 12.2.2 Manifold Rendered to Tangent Plane with Covariant Directional Vectors

1g	Given	$x^i \equiv x^i(\bullet\ u^k\ \bullet)$	
2g		$\nabla R \equiv \sum_i R_{,i}\ \mathbf{b}^i$	Covariant gradient vector
3g		$\mathbf{D}_k \equiv \sum_j \mu^j_k\ \mathbf{b}_j$	Contravariant directional vector
Step	Hypothesis	$0 = \nabla R \bullet \mathbf{D}_k$	for all k directed line coordinate L_k
1	From D11.2.3 Rendered Equation of a Manifold	$c = R(\bullet x^i \bullet)$	
2	From 1, TK.3.2 Uniqueness of Differentials and LxK.3.3.1A Differential of a Constant	$0 = R_{,k}$	
3	From 1g, 3, TK.3.8 Multi-variable Chain Rule and T11.9.2 Analytic Equation of a Three Point Plane	$0 = \sum_i R_{,i}\ x^i_{,k}$ the plane has two perpendicular directed lines	
4	From 3, D8.3.3C Orthogonal Vectors and D12.1.5A Mixed Metric Dyadic	$0 = \sum_i \sum_j \delta_j^i\ R_{,i}\ x^j_{,k}$ requires orthogonal dual base set of vectors, contravariant normal and covariant tangent to the plane.	
5	From 4 and D12.1.5B Mixed Metric Dyadic	$0 = \sum_i \sum_j (\mathbf{b}_j \bullet \mathbf{b}^i)\ R_{,i}\ x^j_{,k}$	
6	From 5 and A4.2.10 Commutative Multp	$0 = \sum_i \sum_j (R_{,i}\ \mathbf{b}^i) \bullet (x^j_{,k}\ \mathbf{b}_j)$ for directed lines coordinate L_k	
7	From 6 and A7.9.4 Associative by Vector Addition	$0 = (\sum_i R_{,i}\ \mathbf{b}^i) \bullet (\sum_j x^j_{,k}\ \mathbf{b}_j)$	
8	From 7, T11.7.5 Directed Numbers are Slopes for Parametric Line-L	$0 = (\sum_i R_{,i}\ \mathbf{b}^i) \bullet (\sum_j \mu^j_k\ \mathbf{b}_j)$	
9	From 8, DxK.6.1.3A Covariant Differential Operator, DEL	$0 = (\sum_i \nabla R_{,i}\ \mathbf{b}^i) \bullet (\sum_j \mu^j_k\ \mathbf{b}_j)$	
∴	From 2g, 3g, 4, 9, A4.2.3 Substitution and D8.3.3 Orthogonal Vectors	$0 = \nabla R \bullet \mathbf{D}_k$	for all k directed line coordinate L_k

Theorem 12.2.3 Mixed Bases Vector Transpose

Step	Hypothesis	$I = S^{-1}\{B_c{}^T B^c\}$
1	From D12.1.5C Mixed Metric Dyadic	$I = S^{-1}\{B_c B^{cT}\}$ for matrix form (n x n) = (n x 1)(1 x n)
2	From 1 and T6.4.1 Uniqueness of Transposition of Matrices	$I^T = S^{-1}\{ (B_c B^{cT})^T\}$
3	From 2, T6.11.2 Commute Column Vectors under Conservation of Transposition	$I^T = S^{-1}\{B^c B_c{}^T\}$ for matrix form (n x n) = (n x 1)(1 x n)
∴	From 3 and T6.4.3 Invariance of the transpose operation	$I = S^{-1}\{B^c B_c{}^T\}$ for matrix form (n x n) = (n x 1)(1 x n)

Theorem 12.2.4 Contravariant Metric Bases Vector Inverse

Step	Hypothesis	$I = G\, G^{-1}$ EQ A	$I = G^{-1}\, G$ EQ B
1	From D12.1.5C Mixed Metric Dyadic	$I = S^{-1}\{B_c B^{cT}\}$	
2	From 1, T6.3.12 Matrix Identity right of Multiplication, T12.2.3 Mixed Bases Vector Transpose and T6.3.1 Substitution of Matrices	$I = S^{-1}\{B_c B^{cT}\ B^c B_c{}^T\}$	
3	From 2, T6.11.1 Commute Column Vectors under Conservation of Order 3x and T6.3.11 Matrix Association for Multiplication	$I = S^{-1}\{B_c B^{cT}\ B^c \mathbf{B_c{}^T}\}$ $I = S^{-1}\{B_c B^{cT}\ \mathbf{B_c{}^T} B^c\}$ $I = S^{-1}\{B_c \mathbf{B_c{}^T}\ B^{cT}\ B^c\}$ $I = S^{-1}\{B_c B_c{}^T\}\ S^{-1}\{\ B^c B^{cT}\}$	$I = S^{-1}\{\mathbf{B_c} B^{cT}\ B^c B_c{}^T\}$ $I = S^{-1}\{B^{cT}\ \mathbf{B_c} B^c B_c{}^T\}$ $I = S^{-1}\{B^{cT}\ \mathbf{B^c B_c} B_c{}^T\}$ $I = S^{-1}\{\mathbf{B^c} B^{cT}\}\ S^{-1}\{B_c B_c{}^T\}$
∴	From 3, D12.1.3D Covariant Metric Dyadic, D12.1.4D Covariant Metric Dyadic and T6.3.1 Substitution of Matrices	$I = G\, G^{-1}$ EQ A	$I = G^{-1}\, G$ EQ B

Theorem 12.2.5 Contravariant Metric Reciprocal

Step	Hypothesis	$G^{-1} = g^{-1}\, G^{ij}$
1	From T6.9.18 Inverse Square Matrix for Multiplication, D6.9.1F Determinant of A and A4.2.3 Substitution	$G^{-1} = (\text{adj } G) / g$ for $g \neq 0$
2	From 1, D6.9.5 Adjoint of a Square Matrix A, D4.1.18 Negative Exponential and A4.2.3 Substitution; with **cofactor**	$G^{-1} = g^{-1}\ {}^t(G^{ji})$ for $g \neq 0$
∴	From 2 and A6.2.11 Matrix Transposition	$G^{-1} = g^{-1}\, G^{ij}$

Theorem 12.2.6 Co-Contravariant Metrics in Matrix Form

Step	Hypothesis	$G = (g_{ij})$ EQ A	$G^{-1} = (g^{ij})$ EQ B
1	From D12.1.4C Contravariant Metric Dyadic; placed in matrix form	$(g^{ij}) = (g^{-1} G^{ij})$	
2	From T12.2.5 Contravariant Metric Reciprocal	$G^{-1} = (g^{-1} G^{ij})$	
3	1, 2 and T6.3.1 Substitution of Matrices	$G^{-1} = (g^{ij})$	
∴	From 3 and D12.1.3C Covariant Metric Dyadic	$G = (g_{ij})$ EQ A	$G^{-1} = (g^{ij})$ EQ B

Theorem 12.2.7 Transpose of Mixed Kronecker Delta

Step	Hypothesis	$\delta_i^{\ j} = \delta^j_{\ i}$ EQ A	$\delta^i_{\ j} = \delta_j^{\ i}$ EQ B
1	From A4.2.2A Equality	$\delta_i^{\ j} = \delta_i^{\ j}$	$\delta^i_{\ j} = \delta^i_{\ j}$
2	From 1, D12.1.5D Mixed Metric Dyadic and A4.2.3 Substitution	$\delta_i^{\ j} = \mathbf{b}_i \bullet \mathbf{b}^j$	$\delta^i_{\ j} = \mathbf{b}^i \bullet \mathbf{b}_j$
3	From 2 and T8.3.3 Dot Product: Commutative Operation	$\delta_i^{\ j} = \mathbf{b}^j \bullet \mathbf{b}_i$	$\delta^i_{\ j} = \mathbf{b}_j \bullet \mathbf{b}^i$
∴	From 3, D12.1.5A Mixed Metric Dyadic and A4.2.3 Substitution	$\delta_i^{\ j} = \delta^j_{\ i}$ EQ A	$\delta^i_{\ j} = \delta_j^{\ i}$ EQ B

Theorem 12.2.8 Covariant and Contravariant Metric Inverse

Step	Hypothesis	$\delta_i^{\ j} = g_i^{\ j} = \sum_k g_{ik}g^{kj}$ for $\mid g_{ij} \mid \neq 0$ EQ A	$\delta^i_{\ j} = g^i_{\ j} = \sum_k g^{ik}g_{kj}$ for $\mid g_{ij} \mid \neq 0$ EQ B
1	From T6.9.19 Closure with respect to the Identity Matrix for Multiplication	$I = G * G^{-1}$	$I = G^{-1} * G$ for $\mid G \mid \neq 0$
2	From 1, T12.2.6A,B Co-Contravariant Metrics in Matrix Form and T6.3.1 Substitution of Matrices	$I = (g_{ij}) * (g^{ij})$ for $\mid g_{ij} \mid \neq 0$	$I = (g^{ij}) * (g_{ij})$ for $\mid g_{ij} \mid \neq 0$
3	From 2 and T6.2.8C Matrix Multiplication; in matrix form	$I = (\sum_k g_{ik}\, g^{kj})$	$I = (\sum_k g^{ik}\, g_{kj})$
4	From 3, D6.1.6A Identity Matrix and T12.2.7A Transpose of Mixed Kronecker Delta	$(\delta_i^{\ j}) = (\sum_k g_{ik}\, g^{kj})$	$(\delta^i_{\ j}) = (\sum_k g^{ik}\, g_{kj})$
∴	From 6, D12.1.5A,D Mixed Metric Dyadic and equating matrix elements	$\delta_i^{\ j} = g_i^{\ j} = \sum_k g_{ik}g^{kj}$ for $\mid g_{ij} \mid \neq 0$ EQ A	$\delta^i_{\ j} = g^i_{\ j} = \sum_k g^{ik}g_{kj}$ for $\mid g_{ij} \mid \neq 0$ EQ B

Theorem 12.2.9 Covariant and Contravariant Metric Inverse Alternate Proof

Step	Hypothesis	$\delta_i^{\ j} = g_i^{\ j} = \sum_k g_{ik} g^{kj}$ EQ A	$\delta^j_{\ i} = g^j_{\ i} = \sum_k g^{jk} g_{ki}$ EQ B
1	From A6.2.2 Closure Multiplication	$C_m = G\ G^{-1}$	$C^m = G^{-1}\ G$
2	From 1, T12.2.6A,B Co-Contravariant Metrics in Matrix, and T6.3.1 Substitution of Matrices	$(c_i^{\ j}) = (g_{ij})\ (g^{ij})$	$(c^i_{\ j}) = (g^{ij})\ (g_{ij})$
3	From 2 and T6.2.8C Matrix Multiplication; in matrix form	$(c_i^{\ j}) = (\sum_k g_{ik}\ g^{kj})$	$(c^i_{\ j}) = (\sum_k g^{ik}\ g_{kj})$
4	From 1, T12.2.4A,B Contravariant Metric Bases Vector Inverse and A6.2.7A Equality of Matrices	$I = C_m$	$I = C^m$
5	From 4 and D6.1.6A Identity Matrix	$(\delta_i^{\ j}) = (c_i^{\ j})$	$(\delta^i_{\ j}) = (c^i_{\ j})$
6	From 2, 5 and T6.3.1 Substitution of Matrices	$(\delta_i^{\ j}) = (\sum_k g_{ik}\ g^{kj})$	$(\delta^i_{\ j}) = (\sum_k g^{jk} g_{ki})$
∴	From 6, D12.1.5A,D Mixed Metric Dyadic and equating matrix elements	$\delta_i^{\ j} = g_i^{\ j} = \sum_k g_{ik} g^{kj}$ EQ A	$\delta^j_{\ i} = g^j_{\ i} = \sum_k g^{jk} g_{ki}$ EQ B

Theorem 12.2.10 Covariant and Contravariant Scale Factor; Reciprocal Relationship

Step	Hypothesis	$\mid \mathbf{b}^i \mid = 1 / \mid \mathbf{b}_i \mid$ for all i EQ A	$\mid \mathbf{b}_i \mid = 1 / \mid \mathbf{b}^i \mid$ for all i EQ B
1	From D12.1.5E Mixed Metric Dyadic	$\delta_i^{\ j} \equiv \mathbf{b}_i \bullet \mathbf{b}^j$	
2	From 1, D8.3.3C Orthogonal Vectors and A4.2.3 Substitution	$\delta_i^{\ j} = \mid \mathbf{b}_i \mid\mid \mathbf{b}^j \mid \delta_i^{\ j}$	
3	From 2 and Identity Matrix Kronecker Delta evaluated	$1 = \mid \mathbf{b}_i \mid\mid \mathbf{b}^i \mid$ for $i = j$	
∴	From 3 and T4.4.6 Equalities: Left Cancellation by Multiplication	$\mid \mathbf{b}^i \mid = 1 / \mid \mathbf{b}_i \mid$ for all i EQ A	$\mid \mathbf{b}_i \mid = 1 / \mid \mathbf{b}^i \mid$ for all i EQ B

Theorem 12.2.11 Covariant Metric Inner Product

Step	Hypothesis	$g_{ij} = b_i\ b_j \cos \theta_{ij}$
1	From D12.1.3A Covariant Metric Dyadic	$g_{ij} = \mathbf{b}_i \bullet \mathbf{b}_j$
∴	From 1 and D8.3.2B Fundamental Base Vector Dot Product	$g_{ij} = b_i\ b_j \cos \theta_{ij}$

Theorem 12.2.12 Contravariant Metric Inner Product

Step	Hypothesis	$g^{ij} = b^i\ b^j \cos \theta^{ij}$
1	From D12.1.4A Contravariant Metric Dyadic	$g^{ij} = \mathbf{b}^i \bullet \mathbf{b}^j$
∴	From 1 and D8.3.2B Fundamental Base Vector Dot Product	$g^{ij} = b^i\ b^j \cos \theta^{ij}$

Theorem 12.2.13 Covariant Metric Inner Product is Symmetrical

Step	Hypothesis	$g_{ji} = g_{ij}$
1	From T12.2.11 Covariant Metric Inner Product	$g_{ij} = b_i\, b_j \cos \theta_{ij}$
2	From 1, DxE.1.9.2 Periodic Reduction Formulas for Multiples of Leading 90°, A4.2.2A Equality and A4.2.10 Commutative Multp	$b_j\, b_i \cos \theta_{ji} = b_i\, b_j \cos \theta_{ij} = g_{ij}$
∴	From 1, 2 and A4.2.3 Substitution	$g_{ji} = g_{ij}$

Theorem 12.2.14 Contravariant Metric Inner Product is Symmetrical

Step	Hypothesis	$g^{ji} = g^{ij}$
1	From T12.2.12 Contravariant Metric Inner Product	$g^{ij} = b^i\, b^j \cos \theta^{ij}$
2	From 1, DxE.1.9.2 Periodic Reduction Formulas for Multiples of Leading 90°, A4.2.2A Equality and A4.2.10 Commutative Multp	$b^j\, b^i \cos \theta^{ji} = b^i\, b^j \cos \theta^{ij} = g^{ij}$
∴	From 1, 2 and A4.2.3 Substitution	$g^{ji} = g^{ij}$

Theorem 12.2.15 Covariant-Contravariant Metric in Orthogonal Curvilinear Coordinates

Step	Hypothesis	$g_{ij} = b_i b_j\, \delta_{ij}$ EQ A	$g^{ij} = b^i b^j\, \delta^{ij}$ EQ B
1	From T12.2.11 Covariant Metric Inner Product and T12.2.12 Contravariant Metric Inner Product	$g_{ij} = b_i b_j \cos(\theta_{ij})$	$g^{ij} = b^i b^j \cos(\theta^{ij})$
2	From 1 and A11.4.3P2(A, B) Orthogonal Manifold $\mathfrak{O}^n$	$g_{ij} = b_i b_j \cos(\frac{1}{2}\pi)$ for $i \neq j$	$g^{ij} = b^i b^j \cos(\frac{1}{2}\pi)$ for $i \neq j$
3	From 2 and DxE.1.6.7 for $\cos(\frac{1}{2}\pi)$	$g_{ij} = b_i b_j\ 0$ for $i \neq j$	$g^{ij} = b^i b^j\ 0$ for $i \neq j$
4	From 3 and A11.4.3P2(A, B) Orthogonal Manifold $\mathfrak{O}^n$	$g_{ij} = b_i b_j \cos(0)$ for $i=j$	$g^{ij} = b^i b^j \cos(0)$ for $i=j$
5	From 4 and DxE.1.6.1 for cos(0)	$g_{ij} = b_i b_j\ 1$ for $i=j$	$g^{ij} = b^i b^j\ 1$ for $i=j$
∴	From 3, 5 and D6.1.6 Identity Matrix defined by the Kronecker Delta	$g_{ij} = b_i b_j\, \delta_{ij}$ EQ A	$g^{ij} = b^i b^j\, \delta^{ij}$ EQ B

Theorem 12.2.16 Covariant-Contravariant Kronecker Delta Summation of like Indices

Step	Hypothesis	$n = \sum_i \delta_i^{\ i}$ EQ A	$n = \sum_i \delta^i_{\ i}$ EQ B
1	From T6.1.6B Identity Matrix and evaluate with LxE.3.1.37 Sums of Powers of 1 for the First n-Integers along the diagonal of the matrix	$n = \sum_i \delta^i_{\ i}$	
∴	From 1, T12.2.7A Transpose of Mixed Kronecker Delta, T10.2.5 Tool: Tensor Substitution and A10.2.14 Correspondence of Tensors and Tensor Coefficients	$n = \sum_i \delta_i^{\ i}$ EQ A	$n = \sum_i \delta^i_{\ i}$ EQ B

Theorem 12.2.17 Determinate of Covariant Metric Dyadic

Step	Hypothesis	$g = \rho_v{}^2 \mu_c$
1	From 1, D12.1.12 Covariant Metric Determinate, D12.1.3A Covariant Metric Dyadic, D8.3.2B Fundamental Base Vector Dot Product and A4.2.3 Substitution	$g = \mid b_i\, b_j \cos \theta_{ij} \mid$
2	From 2 and T6.9.11 Multiply a row or column of a determinant [A] by a scalar k	$g = \prod_i b_i \prod_i b_i \mid \cos \theta_{ij} \mid$
3	From 3, D12.1.13 **Space Density Factor**, D12.1.14B **Space Distortion Factor**, A4.2.3 Substitution and interchanging dummy indices i = j	$g = \rho_v\, \rho_v\, \mu_c$
∴	From 4 and D4.1.17 Exponential Notation	$g = \rho_v{}^2 \mu_c$

Theorem 12.2.18 Bounds for Cosine Space Distortion Factor

Step	Hypothesis	$0 < \mu_c \leq 1$ EQ A	$\mu_c = 1$ EQ B for orthognal system
1	From D12.1.14 **Space Distortion Factor**	$\mu_c = \mid \cos \theta_{ij} \mid$	
2	From 1 and A11.4.3P2 Orthogonal Manifold $\mathfrak{O}^n$	$\mu_c = \mid \delta_{ij} \mid$	
3	From 2 and C6.9.20.2 Determinant of a Identity Matrix	$\mu_c = 1$ for orthognal system	
4	From 1, degenerate case any two column or rows are collinear, DxE.1.6.1 Cosine Zero Degrees and T6.9.10 A determinant with two identical rows or columns is zero	$\mu_c = 0$ a degenerate case, hence not permitted.	
∴	From 3, 4 and A4.2.16 The Trichotomy Law of Ordered Numbers	$0 < \mu_c \leq 1$ EQ A	$\mu_c = 1$ EQ B for orthognal system

Observation 12.2.2 The Possibility of E–Space

Though not likely in the real world the possibility that $\mu_c < 0$, might arise, which would lead to a space turned inside out. In other words base vectors might have angles between them of $\frac{3}{2}\pi > \theta_{ij} > \frac{1}{2}\pi$, clearly another type of degenerate case, to avoid this situation the cosine distortion factor is not allowed to exist in such a region by defining it is greater than zero. But in this physical world one can never-say-never, who knows, if it should exist, words that might be used to describe such a world, like irrational and imaginary are already in use so such a space is simply dubbed as ***e–space***. E-Space is adopted from the British science fiction television series Doctor Who: "Full Circle" a saga of stories where the TARDIS, the doctor's time machine, tumbles helplessly through a "Charged Vacuum Emboitement" to emerge into a parallel antimatter world of opposite **e**-charge in time and space. [HAI83, pg 222]

Theorem 12.2.19 Covariant Metric Determinate in Orthogonal Curvilinear Coordinates

Step	Hypothesis	$\sqrt{g} = \pm\rho_v$
1	T12.2.18B Bounds for Cosine Space Distortion Factor	$\mu_c = 1$
1	From T12.2.17 Determinate of Covariant Metric Dyadic	$g = \rho_v{}^2\, \mu(c)$
2	From 1, T4.10.1 Radicals: Uniqueness of Radicals, T4.10.7 Radicals: Reciprocal Exponent by Positive Square Root and T4.10.5 Radicals: Distribution Across a Product	$\sqrt{g} = \pm\sqrt{\rho_v{}^2}\, \sqrt{\mu(c)}$
3	From 1, A4.2.3 Substitution and T4.10.3 Radicals: Identity Power Raised to a Radical	$\sqrt{g} = \pm\rho_v\, \sqrt{1}$
∴	From 3, T4.9.1 Rational Exponent: Integer of a Positive One, and A4.2.25 Reciprocal Exp and A4.2.12 Identity Multp	$\sqrt{g} = \pm\rho_v$

Theorem 12.2.20 Space Density in Euclidian Manifold $\mathfrak{E}^n$

Step	Hypothesis	$\rho_v = 1$
1	From D12.1.13 Space Density Factor	$\rho_v = \prod_i b_i$
2	From 1, A11.4.5P1 Euclidian Manifold $\mathfrak{E}^n$ and A4.2.3 Substitution	$\rho_v = \prod_i (1)$
∴	From 1 and T4.8.1 Integer Exponents: Unity Raised to any Integer Value	$\rho_v = 1$

Theorem 12.2.21 Covariant Metric Determinate in Euclidian Manifold $\mathfrak{E}^n$

Step	Hypothesis	$\sqrt{g} = \pm 1$ EQ A	$g = 1$ EQ B
1	From 12.2.19 Covariant Metric Determinate in Orthogonal Curvilinear Coordinates	$\sqrt{g} = \pm\rho_v$	$g^2 = \rho_v{}^2$
∴	From 1, T12.2.20 Space Density in Euclidian Manifold $\mathfrak{E}^n$, A4.2.3 Substitution and T4.8.1 Integer Exponents: Unity Raised to any Integer Value	$\sqrt{g} = \pm 1$ EQ A	$g^2 = 1$ EQ B

Section 12.3 Manifold Tangential Plane On a Space Curve

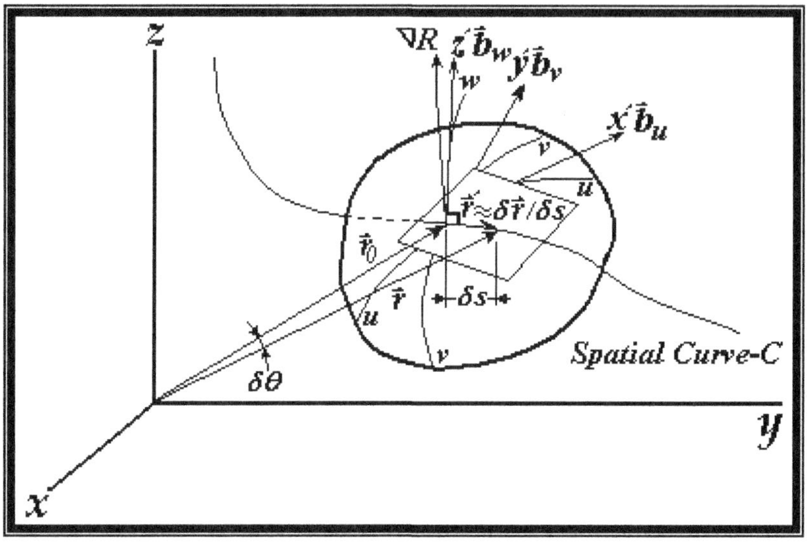

Figure 12.3.1 Manifold Tangential Plane On Space Curve

Observation 12.3.1 Riemann Covariant Fundamental Tangent and Normal Vector

From Figure 12.3.1 "Manifold Tangential Plane On Space Curve" the gradient of the manifold ∇R is clearly understood, but what of **r**'? Laying it out results in a Tangent vector in its full formula form:

$\mathbf{T} \equiv d\mathbf{r} / ds$ Equation A

$d\mathbf{r} / ds \equiv \sum_i (dx^i / ds)\, \mathbf{b}_i$ Equation B

$\mathbf{r} \perp \mathbf{T}$ Equation C

$|\, d\mathbf{r}\, | = ds$ Equation D

which is what Riemann discovered a vector that changed with respect to arc length. Here is where curvelinear coordinates starts and what lead to to the development of tensor calculus. However it is in covariant form, what Riemann found was the contravarient

$\nabla R \equiv \sum_i R_{,i}\, \mathbf{b}^i$ Equation E

A principle normal vector perpendicular to the tangent vector on the spatial curve-C

$\mathbf{N} \equiv \nabla R$ Equation F

as d**T** rotates an infinitesimal amount dθ about the priniciple normal vector on the curve-C, it is seen than that it can be rewritten as

$\mathbf{N} \equiv d\mathbf{T} / d\theta$ Equation G

Definition 12.3.1 **Contraction of Tensor Rank**

Contraction of tensor rank is a process by which a tensor's rank can be reduced to a lesser value though not less than zero. Any theorem that can reduce rank is called a ***theorm of Tenor contraction***.

Definition 12.3.2 **Positional Vector**

As seen in Figure 12.3.1 "Manifold Tangential Plane On Space Curve" the vector **r** traces the position of the trojectory path along the spatial curve-C in a manifold coordinate system, hence is called the ***positional vector***. Its infinitesmal magnitude is given by:

$| d\mathbf{r} | \equiv ds$ Equation A

Axiom 12.3.1 Covariant Tangent Vector

The change in position by the positional vector-**r** along a spatial curve-C creates a tangent covariant vector-**T**.

$\mathbf{T} \equiv d\mathbf{r} / ds$ Equation A

$d\mathbf{r} / ds \equiv \sum_i (dx^i / ds)\, \mathbf{b}_i$ Equation B

$\mathbf{r} \perp \mathbf{T}$ Equation C

Axiom 12.3.2 Contravariant Normal Vector

The change in position by the positional vector-**r** along a spatial curve-C creates a normal contravariant vector-**N** to the path-C.

$\mathbf{N} \equiv \nabla R$ Equation A

$\nabla R \equiv \sum_i R_{,i}\, \mathbf{b}^i$ Equation B

$\mathbf{N} \equiv d\mathbf{T} / d\theta$ Equation C

Theorem 12.3.1 Manifold Rendered Tangent Plane On Spatial Curve

1g	Given	$x^j \equiv x^j(s)$	Function of arc length off of curve-C
2g		$\nabla R \equiv \sum_i R_{,i} \mathbf{b}^i$	Contravariant gradient vector
3g		$\mathbf{r'} \equiv \sum_j x^{j\prime} \mathbf{b}_j$	Covariant directional vector
Step	Hypothesis	$0 = \nabla R \bullet \mathbf{r'}$ built on a dual orthogonal base vector system	
1	From D11.2.3 Rendered Equation of a Manifold	$c = R(\bullet x^i \bullet)$	
2	From 1, TK.3.2 Uniqueness of Differentials and LxK.3.3.1A Differential of a Constant	$0 = R'$	
3	From 1g, 3, TK.3.9 Exact Differential, Parametric, Chain Rule and T11.9.2 Analytic Equation of a Three Point Plane	$0 = \sum_i R_{,i} x^{i\prime}$ the plane has two perpendicular directed lines	
4	From 3, D8.3.3C Orthogonal Vectors and D12.1.5A Mixed Metric Dyadic	$0 = \sum_i \sum_j \delta_j^{\ i} R_{,i} x^{j\prime}$ requires orthogonal dual base set of vectors, contravariant normal and covariant tangent to the plane.	
5	From 4 and D12.1.5B Mixed Metric Dyadic	$0 = \sum_i \sum_j (\mathbf{b}_j \bullet \mathbf{b}^i) R_{,i} x^{j\prime}$	
6	From 5 and A4.2.10 Commutative Multp	$0 = \sum_i \sum_j (R_{,i} \mathbf{b}^i) \bullet (x^{j\prime} \mathbf{b}_j)$	
7	From 6 and A7.9.4 Associative by Vector Addition	$0 = (\sum_i R_{,i} \mathbf{b}^i) \bullet (\sum_j x^{j\prime} \mathbf{b}_j)$	
8	From 7, DxK.6.1.3A Covariant Differential Operator, DEL	$0 = (\sum_i R_{,i} \mathbf{b}^i) \bullet (\sum_j x^{j\prime} \mathbf{b}_j)$	
∴	From 2g, 3g, 4, 9, A4.2.3 Substitution and D8.3.3 Orthogonal Vectors	$0 = \nabla R \bullet \mathbf{r'}$	for $\delta_i^{\ j} = (\mathbf{b}_i \bullet \mathbf{b}^j)$ hence $\mathbf{b}_i \perp \mathbf{b}^j$.

Theorem 12.3.2 Form of a Covariant Vector

1g	Given	$\mathbf{A} \equiv d\mathbf{r} / ds$
2g		$A^i \equiv dx^i / ds$
Step	Hypothesis	$\mathbf{A} = \sum_i A^i \mathbf{b}_i$
∴	From 1g, 2g, A12.3.1B Covariant Tangent Vector, A7.9.2 Substitution of Vectors and A4.2.3 Substitution	$\mathbf{A} = \sum_i A^i \mathbf{b}_i$

Theorem 12.3.3 Contracted Covariant Bases Vector with Contravariant Vector

Step	Hypothesis	$\mathbf{b}_i \bullet \mathbf{A} = A_i$ EQ A	$A_i = \sum_j \delta_i^{\ j} A_j$ EQ B	$A_i = \sum_j A_j \delta^j_{\ i}$ EQ C
1	From D12.1.2 Contravariant Vector	$\mathbf{A} = \sum_j A_j \mathbf{b}^j$		
2	From 1, T8.3.1 Dot Product: Uniqueness Operation, by covariant bases vector and A11.4.6 Riemannian Manifold $\mathfrak{R}^n$	$\mathbf{b}_i \bullet \mathbf{A} = \mathbf{b}_j \bullet \sum_j A_j \mathbf{b}^j$		
3	From 2 and T8.3.7 Dot Product: Distribution of Dot Product Across Addition of Vectors	$\mathbf{b}_i \bullet \mathbf{A} = \sum_j (\mathbf{b}_i \bullet \mathbf{b}^j) A_j$		
4	From 3, D12.1.5A Mixed Metric Dyadic and A4.2.3 Substitution	$\mathbf{b}_i \bullet \mathbf{A} = \sum_j \delta_i^{\ j} A_j$		
5	From 4 and D6.1.6 Identity Matrix Kronecker Delta; evaluated	$\mathbf{b}_i \bullet \mathbf{A} = 1\ A_{(j=j)} + \sum_{(j\neq i)} 0\ A_j$		
∴	From 4, 5, A4.2.12 Identity Multp, T4.4.1 T Equalities: Any Quantity Multiplied by Zero is Zero, A4.2.7 Identity Add and T7.10.1B Uniqueness of Scalar Multiplication to Vectors	$\mathbf{b}_i \bullet \mathbf{A} = A_i$ EQ A	$A_i = \sum_j \delta_i^{\ j} A_j$ EQ B	$A_i = \sum_j A_j \delta_i^{\ j}$ EQ C

Theorem 12.3.4 Contracted Contravariant Bases Vector with Covariant Vector

Step	Hypothesis	$\mathbf{b}^j \bullet \mathbf{A} = A^j$ EQA	$A^j = \sum_i A^i \delta_i^{\ j}$ EQB	$A^j = \sum_i \delta^j_{\ i} A^i$ EQC
1	From D12.1.1 Contravariant Vector	$\mathbf{A} = \sum_i A^i \mathbf{b}_i$		
2	From 1, T8.3.1 Dot Product: Uniqueness Operation, by contravariant bases vector and A11.4.6 Riemannian Manifold $\mathfrak{R}^n$	$\mathbf{b}^j \bullet \mathbf{A} = \mathbf{b}^j \bullet \sum_i A^i \mathbf{b}_i$		
3	From 2 and T8.3.7 Dot Product: Distribution of Dot Product Across Addition of Vectors	$\mathbf{b}^j \bullet \mathbf{A} = \sum_i A^i (\mathbf{b}_i \bullet \mathbf{b}^j)$		
4	From 3, D12.1.5A Mixed Metric Dyadic and A4.2.3 Substitution	$\mathbf{b}^j \bullet \mathbf{A} = \sum_i A^i \delta_i^{\ j}$		
5	From 4 and D6.1.6 Identity Matrix Kronecker Delta evaluated	$\mathbf{b}^j \bullet \mathbf{A} = A^{(i=j)}\ 1 + \sum_{(i\neq j)} A^i\ 0$		
∴	From 4, 5, A4.2.12 Identity Multp, T4.4.1 Equalities: Any Quantity Multiplied by Zero is Zero and A4.2.7 Identity Add, T7.10.1B Uniqueness of Scalar Multiplication to Vectors and T12.2.7B Transpose of Mixed Kronecker Delta	$\mathbf{b}^j \bullet \mathbf{A} = A^j$ EQA	$A^j = \sum_i A^i \delta_i^{\ j}$ EQB	$A^j = \sum_i \delta^j_{\ i} A^i$ EQC

Theorem 12.3.5 Metric Transformation from Contravariant to Covariant Coefficient

1g	Given	$\mathbf{A} = \mathbf{B}$ covariance and contravariance invariant relation	
2g		$\mathbf{B} = \sum_j B_j \mathbf{b}^j$ contravarient vector	
Step	Hypothesis	$A^i = \sum_j g^{ij} B_j$ EQ A	$A^i = \sum_j B_j g^{ij}$ EQ B
1	From T12.3.2 Form of a Covariant Vector	$\mathbf{A} = \sum_i A^i \mathbf{b}_i$	
2	From 1, T8.3.1 Dot Product: Uniqueness Operation, by contravariant bases vector and A11.4.6 Riemannian Manifold $\mathfrak{R}^n$	$\mathbf{b}^j \bullet \mathbf{A} = \mathbf{b}^k \bullet \sum_j B_j \mathbf{b}^j$	
3	From 2 and T8.3.7 Dot Product: Distribution of Dot Product Across Addition of Vectors	$\mathbf{b}^j \bullet \mathbf{A} = \sum_j B_j (\mathbf{b}^j \bullet \mathbf{b}^j)$	
4	From 3, D12.1.3A Covariant Metric Dyadic and A4.2.3 Substitution	$\mathbf{b}^j \bullet \mathbf{A} = \sum_j B_j g^{ij}$	
5	From 4 and T12.3.4B Contracted Contravariant Bases Vector with Covariant Vector	$A^i = \sum_j B_j g^{ij}$	
∴	From 4, 5 and A4.2.10 Commutative Multp	$A^i = \sum_j g^{ij} B_j$ EQ A	$A^i = \sum_j B_j g^{ij}$ EQ B

Theorem 12.3.6 Transformation from Contravariant to Covariant Basis Vector

1g	Given	$\mathbf{b}_i \equiv \sum_j g_{ij} \mathbf{b}^j$ contravarient vector assume valid	
Step	Hypothesis	$\mathbf{b}_i = \sum_j g_{ij} \mathbf{b}^j$ EQ A	$\mathbf{b}_i = \sum_j \mathbf{b}^j g_{ij}$ EQ **A**
1	From 1g	$\mathbf{b}_i \equiv \sum_j g_{ij} \mathbf{b}^j$	
2	From 1, T8.3.1 Dot Product: Uniqueness Operation, by covariant bases vector and A11.4.6 Riemannian Manifold $\mathfrak{R}^n$	$\mathbf{b}^k \bullet \mathbf{b}_i = \mathbf{b}^k \bullet \sum_j g_{ij} \mathbf{b}^j$	
3	From 2, T8.3.7 Dot Product: Distribution of Dot Product Across Addition of Vectors and T8.3.3 Dot Product: Commutative Operation	$\mathbf{b}_i \bullet \mathbf{b}^k = \sum_j g_{ij} (\mathbf{b}^j \bullet \mathbf{b}^k)$	
4	From 3, D12.1.4A Contravariant Metric Dyadic, D12.1.5A Mixed Metric Dyadic and A4.2.3 Substitution	$\delta_i^{\ k} = \sum_j g_{ij} g^{jk}$	
5	From 4 and T12.2.8A Covariant and Contravariant Metric Inverse	$\delta_i^{\ k} = \delta_i^{\ k}$	
∴	From 1, 5, A7.8.5 Parallelogram Law: Commutative and Scalability of Collinear Vectors and by identity	$\mathbf{b}_i = \sum_j g_{ij} \mathbf{b}^j$ EQ A	$\mathbf{b}_i = \sum_j \mathbf{b}^j g_{ij}$ EQ **A**

Theorem 12.3.7 Form of a Contravariant Vector

1g	Given	$\mathbf{A} \equiv \nabla R$
2g		$A_i \equiv R_{,i}$
Step	Hypothesis	$\mathbf{A} = \sum_i A_i \mathbf{b}^i$
∴	From 1g, 2g, A12.3.2B Covariant Tangent Vector, A7.9.2 Substitution of Vectors and A4.2.3 Substitution	$\mathbf{A} = \sum_i A_i \mathbf{b}^i$

Theorem 12.3.8 Contracted Contravariant Bases Vector with Covariant Vector

Step	Hypothesis	$\mathbf{b}^i \bullet \mathbf{A} = A^i$ EQA	$A^i = \sum_j A^j \delta^i_j$ EQB	$A^i = \sum_j \delta_j^{\ i} A^j$ EQC
1	From D12.1.1 Covariant Vector	$\mathbf{A} = \sum_i A^j \mathbf{b}_i$		
2	From 1, T8.3.1 Dot Product: Uniqueness Operation, by contravariant bases vector and A11.4.6 Riemannian Manifold $\mathfrak{R}^n$	$\mathbf{b}^i \bullet \mathbf{A} = \mathbf{b}^i \bullet \sum_j A^j \mathbf{b}_j$		
3	From 2 and T8.3.7 Dot Product: Distribution of Dot Product Across Addition of Vectors	$\mathbf{b}^i \bullet \mathbf{A} = \sum_j A^j (\mathbf{b}^i \bullet \mathbf{b}_j)$		
4	From 3, D12.1.5D Mixed Metric Dyadic and A4.2.3 Substitution	$\mathbf{b}^i \bullet \mathbf{A} = \sum_j A^j \delta^i_j$		
5	From 4 and D6.1.6 Identity Matrix Kronecker Delta evaluated	$\mathbf{b}^i \bullet \mathbf{A} = A^{(j=i)}\ 1 + \sum_{(j\neq i)} A^j\ 0$		
∴	From 4, 5, A4.2.12 Identity Multp, T4.4.1 T Equalities: Any Quantity Multiplied by Zero is Zero, A4.2.7 Identity Add, T7.10.1B Uniqueness of Scalar Multiplication to Vectors and T12.2.7A Transpose of Mixed Kronecker Delta	$\mathbf{b}^i \bullet \mathbf{A} = A^i$ EQA	$A^i = \sum_j A^j \delta^i_j$ EQB	$A^i = \sum_j \delta_j^{\ i} A^j$ EQC

Theorem 12.3.9 Metric Transformation from Covariant to Contravariant Coefficient

1g	Given	$\mathbf{A} = \mathbf{B}$ covariance and contravariance invariant relation	
2g		$\mathbf{B} = \sum_k B^k \mathbf{b}_k$ covarient vector	
Step	Hypothesis	$A_j = \sum_j g_{ij} B^j$ EQ A	$A_j = \sum_j B^j g_{ij}$ EQ B
1	From T12.3.7 Form of a Contravariant Vector	$\mathbf{A} = \sum_j A_j \mathbf{b}^j$	
2	From 1, T8.3.1 Dot Product: Uniqueness Operation, by covariant bases vector and A11.4.6 Riemannian Manifold $\mathfrak{R}^n$	$\mathbf{b}_i \bullet \mathbf{A} = \mathbf{b}_i \bullet \sum_j B^j \mathbf{b}_j$	
3	From 2 and T8.3.7 Dot Product: Distribution of Dot Product Across Addition of Vectors	$\mathbf{b}_i \bullet \mathbf{A} = \sum_j B^j (\mathbf{b}_i \bullet \mathbf{b}_j)$	
4	From 3, D12.1.3A Covariant Metric Dyadic and A4.2.3 Substitution	$\mathbf{b}_i \bullet \mathbf{A} = \sum_j B^j g_{ij}$	
5	From 4 and T12.3.3B Contracted Covariant Bases Vector with Contravariant Vector	$A_j = \sum_j B^j g_{ij}$	
∴	From 4, 5 and A4.2.10 Commutative Multp	$A_j = \sum_j g_{ij} B^j$ EQ A	$A_j = \sum_j B^j g_{ij}$ EQ B

Theorem 12.3.10 Transformation from Covariant to Contravariant Basis Vector

1g	Given	$\mathbf{b}^i \equiv \sum_j g^{ij} \mathbf{b}_j$ contravarient vector assume valid	
Step	Hypothesis	$\mathbf{b}^i = \sum_j g^{ij} \mathbf{b}_j$ EQ A	$\mathbf{b}^i = \sum_j \mathbf{b}_j g^{ij}$ EQ B
1	From 1g	$\mathbf{b}^i \equiv \sum_i g^{ij} \mathbf{b}_i$	
2	From 1, T8.3.1 Dot Product: Uniqueness Operation, by covariant bases vector and A11.4.6 Riemannian Manifold $\mathfrak{R}^n$	$\mathbf{b}_k \bullet \mathbf{b}^i = \mathbf{b}_k \bullet \sum_j g^{ij} \mathbf{b}_j$	
3	From 2, T8.3.7 Dot Product: Distribution of Dot Product Across Addition of Vectors and T8.3.3 Dot Product: Commutative Operation	$\mathbf{b}_k \bullet \mathbf{b}^i = \sum_j g^{ij} (\mathbf{b}_j \bullet \mathbf{b}_k)$	
4	From 3, D12.1.3A covariant Metric Dyadic, D12.1.5A Mixed Metric Dyadic and A4.2.3 Substitution	$\delta_k{}^i = \sum_j g_{kj} g^{ji}$	
5	From 4 and T12.2.8A Covariant and Contravariant Metric Inverse	$\delta_k{}^i = \delta_k{}^i$	
∴	From 1, 5 and A7.8.5 Parallelogram Law: Commutative and Scalability of Collinear Vectors and by identity	$\mathbf{b}_i \equiv \sum_j g_{ij} \mathbf{b}^j$ EQ A	$\mathbf{b}_i \equiv \sum_j \mathbf{b}^j g_{ij}$ EQ B

Theorem 12.3.11 Riemann Contravariant Fundamental Tangent Vector

Step	Hypothesis	$\mathbf{dr} / ds = \sum_j (dx_j / ds)\, \mathbf{b}^j$
1	From A12.3.1B Covariant Tangent Vector	$\mathbf{dr} / ds = \sum_i (dx^i / ds)\, \mathbf{b}_i$
2	From 1, T12.3.6A Transformation from Contravariant to Covariant Basis Vector and A7.9.2 Substitution of Vectors	$\mathbf{dr} / ds = \sum_i (dx^i / ds) (\sum_j g_{ij}\, \mathbf{b}^j)$
3	From 2, T7.10.3 Distribution of a Scalar over Addition of Vectors and reorder summation	$\mathbf{dr} / ds = \sum_j [\sum_i (dx^i / ds)\, g_{ij}]\, \mathbf{b}^j$
∴	From 3, T12.3.9B Metric Transformation from Covariant to Contravariant Coefficient and A4.2.3 Substitution	$\mathbf{dr} / ds = \sum_j (dx_j / ds)\, \mathbf{b}^j$

Theorem 12.3.12 Invariance of Contra and Covariant Vectors

Step	Hypothesis		
1g	Given	$\mathbf{A} = \sum_i A_i \mathbf{b}^i$ contravarient vector	
2g		$\mathbf{B} = \sum_k A^k \mathbf{b}_k$ covarient vector	
Step	Hypothesis	$\mathbf{A} = \mathbf{B}$ EQ A	$\sum_i A_i \mathbf{b}^i = \sum_k A^k\, \mathbf{b}_k$ EQ B
1	From 1g	$\mathbf{A} = \sum_i A_i \mathbf{b}^i$	
2	From 1, T12.3.10A Transformation from Covariant to Contravariant Basis Vector and A7.9.2 Substitution of Vectors	$\mathbf{A} = \sum_i A_i (\sum_k g^{ik}\, \mathbf{b}_k)$	
3	From 2, T7.10.3 Distribution of a Scalar over Addition of Vectors and reorder summation	$\mathbf{A} = \sum_k (\sum_i A_i\, g^{ik})\, \mathbf{b}_k$	
4	From 3, T12.3.5B Metric Transformation from Contravariant to Covariant Coefficient and A4.2.3 Substitution	$\mathbf{A} = \sum_k A^k\, \mathbf{b}_k$	
∴	From 2g, 1, 4 and A7.9.1 Equivalence of Vector	$\mathbf{A} = \mathbf{B}$ EQ A	$\sum_i A_i \mathbf{b}^i = \sum_k A^k\, \mathbf{b}_k$ EQ B

Theorem 12.3.13 Dot Product of Covariant Vectors

1g	Given	$\mathbf{A} = \sum_i A^i \mathbf{b}_i$		
2g		$\mathbf{B} = \sum_i B^i \mathbf{b}_i$		
3g		$\alpha^i = A^i / A$ directional numbers for vector **A**		
4g		$\beta^j = B^j / B$ directional numbers for vector **B**		
5g		$\theta_{ab} = 0$ radians iff collinear vectors		
Step	Hypothesis	$(\sum_i A^i \mathbf{b}_i) \bullet (\sum_i B^i \mathbf{b}_i) = \sum_i \sum_j A^i B^j g_{ij}$ EQ A	$\cos(\theta_{ab}) = \sum_i \sum_j \alpha^i \beta^j g_{ij}$ EQ B	$1 = \sum_i \sum_j \alpha^i \beta^j g_{ij}$ EQ C
1	From D8.3.1 The Inner or Dot Product Operator	$\lvert\mathbf{A}\rvert \lvert\mathbf{B}\rvert \cos(\theta_{ab}) \equiv \mathbf{A} \bullet \mathbf{B}$		
2	From 1g, 2g, 1, D7.6.4 Magnitude of a Vector and T7.9.2 Substitution of Vectors	$A\, B \cos(\theta_{ab}) = (\sum_i A^i \mathbf{b}_i) \bullet (\sum_i B^i \mathbf{b}_i)$		
3	From 2 and T8.3.8 Dot Product: Distribution of Dot Product Across Another Vector	$A\, B \cos(\theta_{ab}) = \sum_i \sum_j A^i B^j (\mathbf{b}_i \bullet \mathbf{b}_j)$		
4	From 3, D7.6.4 Magnitude of a Vector, D12.1.3A Covariant Metric Dyadic and A4.2.3 Substitution	$A\, B \cos(\theta_{ab}) = \sum_i \sum_j A^i B^j g_{ij}$		
5	From 2, 4 and A4.2.3 Substitution	$(\sum_i A^i \mathbf{b}_i) \bullet (\sum_i B^i \mathbf{b}_i) = \sum_i \sum_j A^i B^j g_{ij}$		
6	From 4, T4.4.6B Equalities: Left Cancellation by Multiplication and A4.2.14 Distribution	$\cos(\theta_{ab}) = \sum_i \sum_j (A^i / A)(B^j / B)\, g_{ij}$		
7	From 3g, 4g, 6 and A4.2.3 Substitution	$\cos(\theta_{ab}) = \sum_i \sum_j \alpha^i \beta^j g_{ij}$		
8	From 5g, 7 and A4.2.3 Substitution	$\cos(0) = \sum_i \sum_j \alpha^i \beta^j g_{ij}$		
9	From 8 and DxE.1.6.1 for Cosine	$1 = \sum_i \sum_j \alpha^i \beta^j g_{ij}$		
∴	From 5, 7 and 9	$(\sum_i A^i \mathbf{b}_i) \bullet (\sum_i B^i \mathbf{b}_i) = \sum_i \sum_j A^i B^j g_{ij}$ EQ A	$\cos(\theta_{ab}) = \sum_i \sum_j \alpha^i \beta^j g_{ij}$ EQ B	$1 = \sum_i \sum_j \alpha^i \beta^j g_{ij}$ EQ C

Theorem 12.3.14 Dot Product of Contravariant Vectors

1g	Given	$\mathbf{A} = \sum_i A_i \mathbf{b}^i$		
2g		$\mathbf{B} = \sum_i B_i \mathbf{b}^i$		
3g		$\alpha_i = A_i / A$ directional numbers for vector **A**		
4g		$\beta_i = B_i / B$ directional numbers for vector **B**		
5g		$\theta_{ab} = 0$ radians iff collinear vectors		
Step	Hypothesis	$(\sum_i A_i \mathbf{b}^i) \bullet (\sum_i B_i \mathbf{b}^i) = \sum_i \sum_j A_i B_j g^{ij}$ EQ A	$\cos(\theta^{ab}) = \sum_i \sum_j \alpha_i \beta_j g^{ij}$ EQ B	$1 = \sum_i \sum_j \alpha_i \beta_j g^{ij}$ EQ C
1	From D8.3.1 The Inner or Dot Product Operator	$\lvert\mathbf{A}\rvert \lvert\mathbf{B}\rvert \cos(\theta^{ab}) \equiv \mathbf{A} \bullet \mathbf{B}$		
2	From 1g, 2g, 1, D7.6.4 Magnitude of a Vector and T7.9.2 Substitution of Vectors	$A\,B \cos(\theta^{ab}) = (\sum_i A_i \mathbf{b}^i) \bullet (\sum_i B_i \mathbf{b}^i)$		
3	From 2 and T8.3.8 Dot Product: Distribution of Dot Product Across Another Vector	$A\,B \cos(\theta^{ab}) = \sum_i \sum_j A_i B_j (\mathbf{b}^i \bullet \mathbf{b}^j)$		
4	From 3, D7.6.4 Magnitude of a Vector, D12.1.4A Contravariant Metric Dyadic and A4.2.3 Substitution	$A\,B \cos(\theta^{ab}) = \sum_i \sum_j A_i B_j g^{ij}$		
5	From 2, 4 and A4.2.3 Substitution	$(\sum_i A_i \mathbf{b}^i) \bullet (\sum_i B_i \mathbf{b}^i) = \sum_i \sum_j A_i B_j g^{ij}$		
6	From 4, T4.4.6B Equalities: Left Cancellation by Multiplication and A4.2.14 Distribution	$\cos(\theta^{ab}) = \sum_i \sum_j (A_i / A)(B_j / B) g^{ij}$		
7	From 3g, 4g, 5 and A4.2.3 Substitution	$\cos(\theta^{ab}) = \sum_i \sum_j \alpha_i \beta_j g^{ij}$		
8	From 5g, 7 and A4.2.3 Substitution	$\cos(0) = \sum_i \sum_j \alpha_i \beta_j g^{ij}$		
9	From 8 and DxE.1.6.1 for Cosine	$1 = \sum_i \sum_j \alpha_i \beta_j g^{ij}$		
∴	From 5, 7 and 9	$(\sum_i A_i \mathbf{b}^i) \bullet (\sum_i B_i \mathbf{b}^i) = \sum_i \sum_j A_i B_j g^{ij}$ EQ A	$\cos(\theta^{ab}) = \sum_i \sum_j \alpha_i \beta_j g^{ij}$ EQ B	$1 = \sum_i \sum_j \alpha_i \beta_j g^{ij}$ EQ C

Theorem 12.3.15 Dot Product of Mixed Vectors (Tensor Contraction)

1g	Given	$\mathbf{A} = \sum_i A_i \mathbf{b}^i$	
2g		$\mathbf{B} = \sum_i B^i \mathbf{b}_i$	
3g		$\alpha^i = A^i / A$	directional numbers for vector **A**
4g		$\beta_i = B_i / B$	directional numbers for vector **B**
5g		$\theta_{ab} = 0$	in radians and iff collinear vectors
Step	Hypothesis	$(\sum_i A^i \mathbf{b}_i) \bullet (\sum_i B_i \mathbf{b}^i) = \sum_i \sum_j A^i B_j \delta_i^j$ EQ A	$(\sum_i A^i \mathbf{b}_i) \bullet (\sum_i B_i \mathbf{b}^i) = \sum_i A^i B_i$ EQ B
		$\cos(\theta_{ab}) = \sum_i \alpha^i \beta_j$ EQ C	$1 = \sum_i \alpha^i \beta_j$ for $\theta_{ab} = 0$ EQ D
1	From D8.3.1 The Inner or Dot Product Operator	$\lvert\mathbf{A}\rvert \lvert\mathbf{B}\rvert \cos(\theta_{ab}) \equiv \mathbf{A} \bullet \mathbf{B}$	
2	From 1g, 2g, 1, D7.6.4 Magnitude of a Vector and T7.9.2 Substitution of Vectors	$A\,B \cos(\theta_{ab}) = (\sum_i A^i \mathbf{b}_i) \bullet (\sum_i B_i \mathbf{b}^i)$	
3	From 2 and T8.3.8 Dot Product: Distribution of Dot Product Across Another Vector	$A\,B \cos(\theta_{ab}) = \sum_i \sum_j A^i B_j (\mathbf{b}_i \bullet \mathbf{b}^j)$	
4	From 3, D7.6.4 Magnitude of a Vector, D12.1.5A Mixed Metric Dyadic and A4.2.3 Substitution	$A\,B \cos(\theta_{ab}) = \sum_i \sum_j A^i B_j \delta_i^j$	
5	From 2, 4 and A4.2.3 Substitution	$(\sum_i A^i \mathbf{b}_i) \bullet (\sum_i B_i \mathbf{b}^i) = \sum_i \sum_j A^i B_j \delta_i^j$	
6	From 5 and D6.1.6 Identity Matrix-Kronecker Delta evaluated	$A\,B \cos(\theta_{ab}) = \sum_{(i=j)} A^i B_j\, 1 + \sum_{(i\neq j)} A^i B_j\, 0$ hence co-and contravariant are collinear	
7	From 6, A4.2.12 Identity Multp, T4.4.1 Equalities: Any Quantity Multiplied by Zero is Zero and A4.2.7 Identity Add	$A\,B \cos(\theta_{ab}) = \sum_i A^i B_i$	
8	From 2, 7 and A4.2.3 Substitution	$(\sum_i A^i \mathbf{b}_i) \bullet (\sum_i B_i \mathbf{b}^i) = \sum_i A^i B_i$	
9	From 7, T4.4.24 Equalities: Division by a Constant, A4.2.13 Inverse Multp, A4.2.14 Distribution and D4.1.1A Rational Numbers	$\cos(\theta_{ab}) = \sum_i (A^i / A)(B_i / B)$	
10	From 3g, 4g, 9 and A4.2.3 Substitution	$\cos(\theta_{ab}) = \sum_i \alpha^i \beta_j$	

11	From 5g, 10, and A4.2.3 Substitution	$\cos(0) = \Sigma_i \alpha^i \beta_j$	
12	From 10 and DxE.1.6.1 for Cosine	$1 = \Sigma_i \alpha^i \beta_j$	
∴	From 4, 7, 10 and 12	$(\Sigma_i A^i \mathbf{b}_i) \bullet (\Sigma_i B_i \mathbf{b}^i) = \Sigma_i \Sigma_j A^i B_j \delta_i^j$ EQ A $\cos(\theta_{ab}) = \Sigma_i \alpha^i \beta_j$ EQ C	$(\Sigma_i A^i \mathbf{b}_i) \bullet (\Sigma_i B_i \mathbf{b}^i) = \Sigma_i A^i B_i$ EQ B $1 = \Sigma_i \alpha^i \beta_j$ for $\theta_{ab} = 0$ EQ D

This theorem is an important proposition in tensors. When two co- and contravariant indices are the same it is said, they contract from a tensor of rank-1 to a tensor of rank-0.

Theorem 12.3.16 Magnitude of Covariant Vector

Step	Hypothesis	$A^2 = \Sigma_i \Sigma_j g_{ij} A^i A^j$ EQ A	$A^2 = \Sigma_i \Sigma_j A^i A_i$ EQ B	$A = \surd\Sigma_i \Sigma_j g_{ij} A^i A^j$ EQ C
1	From T12.3.2 Form of a Covariant Vector	$\mathbf{A} = \Sigma_i A^i \mathbf{b}_i$		
2	From 1 and T8.3.4 Dot Product: Magnitude	$\mid \mathbf{A} \mid^2 = \mathbf{A} \bullet \mathbf{A}$		
3	From 1, 2 and T7.9.2 Substitution of Vectors	$\mid \mathbf{A} \mid^2 = (\Sigma_i A^i \mathbf{b}_i) \bullet (\Sigma_i A^i \mathbf{b}_i)$		
4	From 3 and T12.3.13A Dot Product of Covariant Vectors	$\mid \mathbf{A} \mid^2 = \Sigma_i \Sigma_j A^i A^j g_{ij}$		
5	From 4, D7.6.4 Magnitude of a Vector, D12.1.4A Contravariant Metric Dyadic and A4.2.3 Substitution	$A^2 = \Sigma_i \Sigma_j A^i A^j g_{ij}$		
6	From 5 and T12.3.5B Metric Transformation from Contravariant to Covariant Coefficient	$A^2 = \Sigma_i \Sigma_j A^i A_i$		
7	From 5 and T4.10.7 Radicals: Reciprocal Exponent by Positive Square Root	$A = \surd\Sigma_i \Sigma_j A^i A^j g_{ij}$	magnitude is always positive	
∴	From From 5, 6, 7, T10.3.7 Product of Tensors: Commutative by Multiplication of Rank p + q → q + p and A10.2.14 Correspondence of Tensors and Tensor Coefficients	$A^2 = \Sigma_i \Sigma_j g_{ij} A^i A^j$ EQ A	$A^2 = \Sigma_i \Sigma_j A^i A_i$ EQ B	$A = \surd\Sigma_i \Sigma_j g_{ij} A^i A^j$ EQ C

Theorem 12.3.17 Magnitude of Contravariant Vector

Step	Hypothesis	$A^2 = \Sigma_i \Sigma_j g^{ij} A_i A_j$ EQ A	$A^2 = \Sigma_i A_i A^i$ EQ B	$A = \surd\Sigma_i \Sigma_j g^{ij} A_i A_j$ EQ C
1	From T12.3.7 Form of a Contravariant Vector	$\mathbf{A} = \Sigma_i A_i \mathbf{b}^i$		
2	From 1 and T8.3.4 Dot Product: Magnitude	$\mid \mathbf{A} \mid^2 = \mathbf{A} \bullet \mathbf{A}$		
3	From 1, 2 and T7.9.2 Substitution of Vectors	$\mid \mathbf{A} \mid^2 = (\Sigma_i A_i \mathbf{b}^i) \bullet (\Sigma_i A_i \mathbf{b}^i)$		
4	From 3 and T12.3.14A Dot Product of Contravariant Vectors	$\mid \mathbf{A} \mid^2 = \Sigma_i \Sigma_j A_i B_j g^{ij}$		
5	From 4, D7.6.4 Magnitude of a Vector, D12.1.4A Contravariant Metric Dyadic and A4.2.3 Substitution	$A^2 = \Sigma_i \Sigma_j A_i A_j g^{ij}$		
6	From 5 and T12.3.5B Metric Transformation from Contravariant to Covariant Coefficient	$A^2 = \Sigma_i A_i A^i$		

7	From 5 and T4.10.7 Radicals: Reciprocal Exponent by Positive Square Root	$A = \sqrt{\sum_i \sum_j A_i A_j g^{ij}}$	magnitude is always positive	
∴	From 5, 6, 7, T10.3.7 Product of Tensors: Commutative by Multiplication of Rank p + q → q + p and A10.2.14 Correspondence of Tensors and Tensor Coefficients	$A^2 = \sum_i \sum_j g^{ij} A_i A_j$ EQ A	$A^2 = \sum_i A_i A^i$ EQ B	$A = \sqrt{\sum_i \sum_j g^{ij} A_i A_j}$ EQ C

Theorem 12.3.18 Magnitude of Normalized Vector

1g	Given	$\alpha^i = A^i / A$	directional number, covariant vector **A**		
2g		$\alpha_i = A_i / A$	directional number, contravariant vector **A**		
Step	Hypothesis	$1 = \sum_i \alpha^i \alpha_i$	EQ A	$1 = \sum_i \alpha_i \alpha^i$	EQ B
1	From 12.3.16A Magnitude of Covariant Vector	$A^2 = \sum_i \sum_j g_{ij} A^i A^j$			
2	From 12.3.17A Magnitude of Contravariant Vector	$A^2 = \sum_i \sum_j g^{ij} A_i A_j$			
3	From 1, T10.3.7 Product of Tensors: Commutative by Multiplication of Rank p + q → q + p, A10.2.14 Correspondence of Tensors and Tensor Coefficients, T12.3.9A Metric Transformation from Covariant to Contravariant Coefficient and summing over j	$A^2 = \sum_i A^i A_i$			
4	From 2, T12.3.5A Metric Transformation from Covariant to Contravariant Coefficient and summing over i and transposing dummy indices from j → i	$A^2 = \sum_i A^i A_i$			
5	From 3, 4, D4.1.17 Exponential Notation, T4.4.6B Equalities: Left Cancellation by Multiplication and A4.2.14 Distribution	$(A / A)(A / A) = \sum_i (A^i / A)(A_i / A)$			
6	From 5, A4.2.13 Inverse Multp and A4.2.12 Identity Multp	$1 = \sum_i (A^i / A)(A_i / A)$			
7	1g, 2g, 6 and A4.2.3 Substitution	$1 = \sum_i \alpha^i \alpha_i$			
∴	From 7 and A4.2.10 Commutative Multp	$1 = \sum_i \alpha^i \alpha_i$	EQ A	$1 = \sum_i \alpha_i \alpha^i$	EQ B

Theorem 12.3.19 Splitting an Orthogonal Covariant Contraction

Step	Hypothesis	$b_i A^i = A_i / b_i$
1	From T12.3.9A Metric Transformation from Covariant to Contravariant Coefficient	$A_i = \sum_j g_{ij} A^j$
2	From 1, T12.2.15A Covariant-Contravariant Metric in Orthogonal Curvilinear Coordinates and T10.2.5 Tool: Tensor Substitution	$A_j = \sum_j b_i b_j \delta_{ij} A^j$
3	From 2 and D6.1.6 Identity Matrix evaluating the Kronecker Delta; Einsteinian Summation Notation does not apply.	$A_i = b_i b_i A^i$
∴	From 3 and T4.4.4B Equalities: Right Cancellation by Multiplication	$b_i A^i = A_i / b_i$

Theorem 12.3.20 Holding Indices Constant by Using a Mixed Metric Dyadic

1g	Given	$\mathbf{b}_i = \sum_j g_i{}^j \mathbf{b}_j$ assume	$\mathbf{b}^i = \sum_j g^i{}_j \mathbf{b}^j$ assume
Step	Hypothesis	$\mathbf{b}_i = \sum_j g_i{}^j \mathbf{b}_j$ EQ A $\mathbf{b}_i = \sum_j \mathbf{b}_j g_i{}^j$ EQ B	$\mathbf{b}^i = \sum_j g^i{}_j \mathbf{b}^j$ EQ C $\mathbf{b}^i = \sum_j \mathbf{b}^j g^i{}_j$ EQ D
1	From 1g and expanding about the ith term	$\mathbf{b}_i = g_i{}^i \mathbf{b}_i + \sum_{i\neq j} g_i{}^j \mathbf{b}_j$	$\mathbf{b}^i = g^i{}_i \mathbf{b}^i + \sum_{i\neq j} g^i{}_j \mathbf{b}^j$
2	From 1, D12.1.5(A, D) Mixed Metric Dyadic, D6.1.6B Identity Matrix and evaluating the Kronecker Delta's	$\mathbf{b}_i = 1\, \mathbf{b}_i + \sum_{i\neq j} 0\, \mathbf{b}_j$	$\mathbf{b}^i = 1\, \mathbf{b}^i + \sum_{i\neq j} 0\, \mathbf{b}^j$
3	From 2, T7.10.5 Identity with Scalar Multiplication to Vectors, T7.10.4 Grassmann's Zero Vector and T7.11.12 Vector Addition: Identity; 1g proven by identity	$\mathbf{b}_i = \mathbf{b}_i$	$\mathbf{b}^i = \mathbf{b}^i$
4	From 1g and T10.3.7 Product of Tensors: Commutative by Multiplication of Rank p + q → q + p and A10.2.14 Correspondence of Tensors and Tensor Coefficients	$\mathbf{b}_i = \sum_j \mathbf{b}_j g_i{}^j$	$\mathbf{b}^i = \sum_j \mathbf{b}^j g^i{}_j$
∴	From 3 and 4	$\mathbf{b}_i = \sum_j g_i{}^j \mathbf{b}_j$ EQ A $\mathbf{b}_i = \sum_j \mathbf{b}_j g_i{}^j$ EQ B	$\mathbf{b}^i = \sum_j g^i{}_j \mathbf{b}^j$ EQ C $\mathbf{b}^i = \sum_j \mathbf{b}^j g^i{}_j$ EQ D

Theorem 12.3.21 Fundamental Quadratic: Covariant Arc Length

Step	Hypothesis	$ds^2 = \sum_i \sum_j g_{ij} dx^i dx^j$
1	From T8.3.2 Dot Product: Equality	$\mathbf{dr} / ds \bullet \mathbf{dr} / ds = \mathbf{dr} / ds \bullet \mathbf{dr} / ds$
2	From 1, A12.3.1B Covariant Tangent Vector and A7.9.2 Substitution of Vectors	$\mathbf{dr} / ds \bullet \mathbf{dr} / ds = (\sum_i (dx^i / ds) \mathbf{b}_i) \bullet (\sum_i (dx^i / ds) \mathbf{b}_i)$
3	From 2 and A7.8.5 Parallelogram Law: Commutative, Scalability of Collinear Vectors, D4.1.7 Exponential Notation and T7.10.3 Distribution of a Scalar over Addition of Vectors	$(\mathbf{dr} \bullet \mathbf{dr}) / ds^2 = [(\sum_i dx^i \mathbf{b}_i) \bullet (\sum_i dx^i \mathbf{b}_i)] / ds^2$
4	From 3 and T4.4.13 Equalities: Cancellation by Division	$\mathbf{dr} \bullet \mathbf{dr} = (\sum_i dx^i \mathbf{b}_i) \bullet (\sum_i dx^i \mathbf{b}_i)$
5	From 4 and T8.3.8 Dot Product: Distribution of Dot Product Across Another Vector	$\mathbf{dr} \bullet \mathbf{dr} = \sum_i \sum_j dx^i dx^j (\mathbf{b}_i \bullet \mathbf{b}_j)$
6	From 5, D12.1.3A Covariant Metric Dyadic and A4.2.3 Substitution	$ds^2 = \sum_i \sum_j dx^i dx^j g_{ij}$
∴	From 6 and A4.2.10 Commutative Multp	$ds^2 = \sum_i \sum_j g_{ij} dx^i dx^j$

Theorem 12.3.22 Fundamental Quadratic: Contravariant Arc Length

Step	Hypothesis	$ds^2 = \sum_i \sum_j g^{ij} dx_i dx_j$
1	From T8.3.2 Dot Product: Equality	$\mathbf{dr} / ds \bullet \mathbf{dr} / ds = \mathbf{dr} / ds \bullet \mathbf{dr} / ds$
2	From 1, T12.3.11 Riemann Contravariant Fundamental Tangent Vector and A7.9.2 Substitution of Vectors	$\mathbf{dr} / ds \bullet \mathbf{dr} / ds = (\sum_i (dx_i / ds) \mathbf{b}^i) \bullet (\sum_i (dx_i / ds) \mathbf{b}^i)$
3	From 2 and A7.8.5 Parallelogram Law: Commutative, Scalability of Collinear Vectors, D4.1.7 Exponential Notation and T7.10.3 Distribution of a Scalar over Addition of Vectors	$(\mathbf{dr} \bullet \mathbf{dr}) / ds^2 = [(\sum_i dx_i \mathbf{b}^i) \bullet (\sum_i dx_i \mathbf{b}^i)] / ds^2$
4	From 3 and T4.4.13 Equalities: Cancellation by Division	$\mathbf{dr} \bullet \mathbf{dr} = (\sum_i dx_i \mathbf{b}^i) \bullet (\sum_i dx_i \mathbf{b}^i)$
5	From 4 and T8.3.8 Dot Product: Distribution of Dot Product Across Another Vector	$\mathbf{dr} \bullet \mathbf{dr} = \sum_i \sum_j dx_i dx_j (\mathbf{b}^i \bullet \mathbf{b}^j)$
6	From 5, D12.1.4A Contravariant Metric Dyadic and A4.2.3 Substitution	$ds^2 = \sum_i \sum_j dx_i dx_j g^{ij}$
∴	From 6 and A4.2.10 Commutative Multp	$ds^2 = \sum_i \sum_j g^{ij} dx_i dx_j$

From T12.3.12 "Invariance of Contra and Covariant Vectors", T12.3.22 "Fundamental Quadratic: Contravariant Arc Length" and T12.3.21 "Fundamental Quadratic: Covariant Arc Length" it is seen that magnitude of arc length is invariant for contra and covariant vectors. The above equations are Riemann's famous equaitons for infinitesmal arc length. The following theorems demonstrate how to calculate the length of arc along path-C.

Theorem 12.3.23 Fundamental Quadratic: Covariant Integration over Arc Length

1g	Given	$x_i = w_i(s)$
Step	Hypothesis	$\Delta s = \pm \int_{s_0}^{s} \surd\, [\Sigma_i \Sigma_j\, g_{ij}\, (dx^i / ds)(dx^j / ds)]\, ds$
1	From T12.3.21 Fundamental Quadratic: Covariant Arc Length	$ds^2 = \Sigma_i \Sigma_j\, g_{ij}\, dx^i\, dx^j$
2	From 1, TK.3.4B Single Variable Chain Rule, A4.2.10 Commutative Multp, D4.1.17 Exponential Notation and A4.2.14 Distribution	$ds^2 = [\Sigma_i \Sigma_j\, g_{ij}\, (dx^i / ds)(dx^j / ds)]\, ds^2$
3	From 1, T4.10.7 Radicals: Reciprocal Exponent by Positive Square Root, T4.10.5 Radicals: Distribution Across a Product and T4.10.3 Radicals: Identity Power Raised to a Radical	$ds = \pm\surd\, [\Sigma_i \Sigma_j\, g_{ij}\, (dx^i / ds)(dx^j / ds)]\, ds$
4	From 2, TK.4.1 Integrals: Uniqueness of Integration, D4.1.20 Negative Coefficient and LxK.4.1.3 Integrable on the integrand with a constant	$\int_{s_0}^{s} ds = \pm \int_{s_0}^{s} \surd\, [\Sigma_i \Sigma_j\, g_{ij}\, (dx^i / ds)(dx^j / ds)]\, ds$
5	From 4 and LxK.4.2.1 Integration of powers for n = 0 and DxK.4.1.4 The Limit of an Integral Having a Constant Area Under the Curve	$s - s_0 = \pm \int_{s_0}^{s} \surd\, [\Sigma_i \Sigma_j\, g_{ij}\, (dx^i / ds)(dx^j / ds)]\, ds$
∴	From 5 and DxK.3.1.1 Delta: Increment by Difference	$\Delta s = \pm \int_{s_0}^{s} \surd\, [\Sigma_i \Sigma_j\, g_{ij}\, (dx^i / ds)(dx^j / ds)]\, ds$

Theorem 12.3.24 Fundamental Quadratic: Contravariant Integration over Arc Length

1g	Given	$x_i = w_i(s)$
Step	Hypothesis	$\Delta s = \pm \int_{s_0}^{s} \sqrt{[\Sigma_i \Sigma_j g^{ij} (dx_i / ds)(dx_j / ds)]}\, ds$
1	From T12.3.22 Fundamental Quadratic: Contravariant Arc Length	$ds^2 = \Sigma_i \Sigma_j g^{ij} dx_i dx_j$
2	From 1, TK.3.4B Single Variable Chain Rule, A4.2.10 Commutative Multp, D4.1.17 Exponential Notation and A4.2.14 Distribution	$ds^2 = [\Sigma_i \Sigma_j g^{ij} (dx_i / ds)(dx_j / ds)]\, ds^2$
3	From 1, T4.10.7 Radicals: Reciprocal Exponent by Positive Square Root, T4.10.5 Radicals: Distribution Across a Product and T4.10.3 Radicals: Identity Power Raised to a Radical	$ds = \pm \sqrt{[\Sigma_i \Sigma_j g^{ij} (dx_i / ds)(dx_j / ds)]}\, ds$
4	From 2, TK.4.1 Integrals: Uniqueness of Integration, D4.1.20 Negative Coefficient and LxK.4.1.3 Integrable on the integrand with a constant	$\int_{s_0}^{s} ds = \pm \int_{s_0}^{s} \sqrt{[\Sigma_i \Sigma_j g^{ij} (dx_i / ds)(dx_j / ds)]}\, ds$
5	From 4 and LxK.4.2.1 Integration of powers for n = 0 and DxK.4.1.4 The Limit of an Integral Having a Constant Area Under the Curve	$s - s_0 = \pm \int_{s_0}^{s} \sqrt{[\Sigma_i \Sigma_j g^{ij} (dx_i / ds)(dx_j / ds)]}\, ds$
∴	From 5 and DxK.3.1.1 Delta: Increment by Difference	$\Delta s = \pm \int_{s_0}^{s} \sqrt{[\Sigma_i \Sigma_j g^{ij} (dx_i / ds)(dx_j / ds)]}\, ds$

Observation 12.3.2 Open Path: [±] Direction of Tracing the Arc Path

If the path is traced in a positive way it will be plus and the opposite negative.

$$\Delta s = \begin{cases} +\Delta s_\sigma & \text{for } s > s_0 \text{ traced in a positive direction} \\ -\Delta s_\sigma & \text{for } s < s_0 \text{ traced in the opposite direction} \end{cases}$$

Observation 12.3.3 Closed Path: Integration of Periodic Arc-Length

Let the arc length be on a closed path with one complete path having traveled an arc L_c. Within a peroid, however the path traveled may only be a fraction [f], hence the arc length completed would be $\Delta s = f L_c$.

Theorem 12.3.25 Closed Path: Integration of Periodic Arc-Length

1g	Given	see Figure 11.6.4 Periodicity On A Closed Vector Curve
2g		$\Delta s \equiv f L_c$
3g		$Ⱡ \equiv L_c / 2\pi$
Step	Hypothesis	$\Delta s = n\,\theta\,Ⱡ$ where [+] is taken to be CCW and [–] CW periodic rotation and $n = 0, \pm 1, \pm 2, \pm 3, \ldots$
1	From 1g, A4.2.12 Identity Multp, A4.2.13 Inverse Multp, A4.2.10 Commutative Multp and A4.2.11 Associative Multp	$\Delta s = 2\pi f\,(L_c / 2\pi)$
2	From D5.1.36C Subtended Arc length of a Cord	$\theta \equiv 2\pi f$ in radians
3	From 1g, 2g, 3g, 2 and A4.2.3 Substitution	$\Delta s = \theta\,Ⱡ$
4	From 3; Now if the arc length loops around one time, $\Delta s = 1\,\theta Ⱡ$, if two times $\Delta s = 2\,\theta Ⱡ$,, if three times $\Delta s = 3\,\theta Ⱡ$ and n-times $\Delta s = n\,\theta Ⱡ$.	$\Delta s = n\,\theta\,Ⱡ$
∴	From 4, T12.3.24 Fundamental Quadratic: Contravariant Integration over Arc Length; integration over Arc Length and rotation convention is based on quadrature coordinate system notation.	$\Delta s = n\,\theta\,Ⱡ$ where [+] is taken to be CCW and [–] CW periodic rotation and $n = 0, \pm 1, \pm 2, \pm 3, \ldots$

Theorem 12.3.26 Closed Path: Covariant Integration over Arc Length

1g	Given	Integration of arc length is a closed path.
Step	Hypothesis	$n\,\theta = \dfrac{\pm 1}{Ⱡ} \oint_c \sqrt{[\Sigma_i \Sigma_j g_{ij}\,(dx^i / ds)(dx^j / ds)]}\,ds$
1	From T12.3.23 Fundamental Quadratic: Covariant Integration over Arc Length	$\Delta s = \pm \int_{s_0}^{s} \sqrt{[\Sigma_i \Sigma_j g_{ij}\,(dx^i / ds)(dx^j / ds)]}\,ds$
2	From 1g, 1 and T12.3.25 Closed Path: Integration of Periodic Arc-Length	$n\,\theta\,Ⱡ = \pm \oint_c \sqrt{[\Sigma_i \Sigma_j g_{ij}\,(dx^i / ds)(dx^j / ds)]}\,ds$
∴	From 2, D4.1.20A Negative Coefficient and T4.4.6B Equalities: Left Cancellation by Multiplication	$n\,\theta = \dfrac{\pm 1}{Ⱡ} \oint_c \sqrt{[\Sigma_i \Sigma_j g_{ij}\,(dx^i / ds)(dx^j / ds)]}\,ds$

Theorem 12.3.27 Closed Path: Contravariant Integration over Arc Length

1g	Given	Integration of arc length is a closed path.
Step	Hypothesis	$n\,\theta = \dfrac{\pm 1}{\text{ƀ}} \oint_c \sqrt{[\Sigma_i \Sigma_j g^{ij} (dx_i / ds)(dx_j / ds)]}\, ds$
1	From T12.3.24 Fundamental Quadratic: Contravariant Integration over Arc Length	$\Delta s = \pm \int_{s_0}^{s} \sqrt{[\Sigma_i \Sigma_j g^{ij} (dx_i / ds)(dx_j / ds)]}\, ds$
2	From 1g, 1 and T12.3.25 Closed Path: Integration of Periodic Arc-Length	$n\,\theta\,\text{ƀ} = \pm \oint_c \sqrt{[\Sigma_i \Sigma_j g^{ij} (dx_i / ds)(dx_j / ds)]}\, ds$
∴	From 2, D4.1.20A Negative Coefficient and T4.4.6B Equalities: Left Cancellation by Multiplication	$n\,\theta = \dfrac{\pm 1}{\text{ƀ}} \oint_c \sqrt{[\Sigma_i \Sigma_j g^{ij} (dx_i / ds)(dx_j / ds)]}\, ds$

Table 12.3.1 Summary of Theormes on Contraction

Theorem	Discription	Gained Rank	Reduced Rank
T12.2.16	Co-Contravariant Kronecker Delta Sum of like Indices	0	2
T12.3.13	Dot Product of Covariant Vectors	2	0
T12.3.14	Dot Product of Contravariant Vectors	2	0
T12.3.15	Dot Product of Mixed Vectors (Tensor Contraction)	2	0
T12.3.16	Magnitude of Covariant Vector	2	0
T12.3.17	Magnitude of Contravariant Vector	2	0
T12.3.18	Magnitude of Normalized Vector	2	0
T12.3.23	Fundamental Quadratic: Covariant Arc Length	2	0
T12.3.24	Fundamental Quadratic: Contravariant Arc Length	2	0

Observation 12.3.4 Rules of Tensor Contraction

1) Contraction is proformed by summing over a set of indices resulting in a constant of rank-(r–[the number of indices being summed].
2) These specific theorems the covariant, contravariant and mixed metrics are used as an operator of contraction by summing over their two indices resulting in a constant of rank-(r–2), hence in general

$a = \Sigma_i\Sigma_i A^{ij} g_{ij}$ T12.3.13A Equation A
$a = \Sigma_i\Sigma_i A_{ij} g^{ij}$ T12.3.14A Equation B
$a = \Sigma_i\Sigma_i A_i^{\,j} g_i^{\,j}$ T12.3.15A Equation C

or commutating the metrics

$a = \Sigma_i\Sigma_i g_{ij} A^{ij}$ T12.3.16A Equation D
$a = \Sigma_i\Sigma_i g^{ij} A_{ij}$ T12.3.17A Equation E
$a = \Sigma_i\Sigma_i g_i^{\,j} A_i^{\,j}$ T12.3.18A Equation F
$n = \Sigma_i g_i^{\,i}$ T12.2.16 Equation G

3) These rules are symetrical and can be used in reverse gaining the rank-(r + idx) back.

Table 12.3.2 Summary of Theormes for Raisng or Lowering Tensor Indices

Theorem	Discription	Lower	Raised
T12.3.3	Contracted Covariant Bases Vector with Contravariant Vector	low	low
T12.3.5	Metric Transformation from Contravariant to Covariant Coefficient	low	high
T12.3.10	Transformation from Covariant to Contravariant Basis Vector	low	high
T12.3.4	Contracted Contravariant Bases Vector with Covariant Vector	high	high
T12.3.6	Transformation from Contravariant to Covariant Basis Vector	high	low
T12.3.8	Contracted Contravariant Bases Vector with Covariant Vector	high	high
T12.3.9	Metric Transformation from Covariant to Contravariant Coefficient	high	low

Observation 12.3.5 Rules for Raising or Lowering Tensor Indices

1) These specific theorems the covariant, contravariant and mixed metrics are used as an operator for raising and lowering indices by summing over any one indices resulting in a lowering or raising the index, hence in general

$A_i = \sum_j A^j g_{ij}$ T12.3.9B Equation A
$A^i = \sum_j A_j g^{ij}$ T12.3.5B Equation B
$A^i = \sum_j A^j g^i{}_j$ T12.3.8B Equation C
$A_i = \sum_j A_j g_i{}^j$ T12.3.3C Equation D

or commutating the metrics

$A_i = \sum_j g_{ij} A^j$ T12.3.9A Equation E
$A^i = \sum_j g^{ij} A_j$ T12.3.5A Equation F
$A^i = \sum_j g^i{}_j A^j$ T12.3.8C Equation G
$A_i = \sum_j g_i{}^j A_j$ T12.3.3B Equation H

2) These rules are symetrical and can be used in reverse gaining a rank of two from a scalar.

Section 12.4 Curvilinear Coordinates

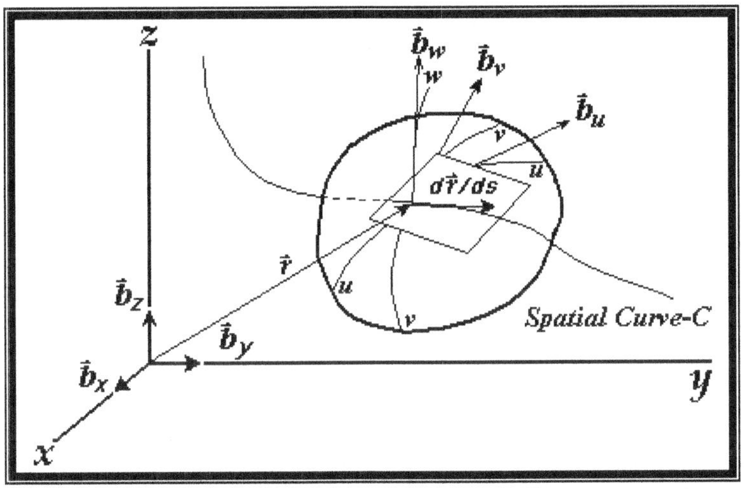

Figure 12.4.1 Cartesian to Curvilinear Coordinate on a Manifold

Theorem 12.4.1 Differential Relationship for Covariant Base Vectors

1g	Given	See Figure 12.4.1 Cartesian to Curvilinear Coordinate Manifold
Step	Hypothesis	$\mathbf{b}_i = \partial\mathbf{r} / \partial x_i$
1	From 1g	$\mathbf{r} = \sum_i x_i\, \mathbf{b}_i$
2	From A12.3.1B Covariant Tangent Vector	$d\mathbf{r} / ds = \sum_i (dx_i / ds)\, \mathbf{b}_i$
3	From 1 and TK.5.3 Chain Rule for Exact Differential of a Vector	$d\mathbf{r} / ds = \sum_i (\partial\mathbf{r} / \partial x^i)\, (dx^i / ds)$
4	From 2 and A7.8.5 Parallelogram Law: Commutative and Scalability of Collinear Vectors	$d\mathbf{r} / ds = \sum_i \mathbf{b}_i\, (dx_i / ds)$
5	From 3, 4 and A7.11.3 Vector Addition: Inverse	$\mathbf{0} = d\mathbf{r} / ds - d\mathbf{r} / ds$
6	From 3, 4 and A7.9.7 Transitivity	$\mathbf{0} = \sum_i \mathbf{b}_i\, (dx_i / ds) - \sum_i (\partial\mathbf{r} / \partial x_i)\, (dx_i / ds)$
7	From 6, D4.1.20 Negative Coefficient, T7.10.3 Distribution of a Scalar over Addition of Vectors and combining terms	$\mathbf{0} = \sum_i \mathbf{b}_i\, (dx_i / ds) - (\partial\mathbf{r} / \partial x_i)\, (dx_i / ds)$
8	From 7 and T7.10.3 Distribution of a Scalar over Addition of Vectors	$\mathbf{0} = \sum_i [\, \mathbf{b}_i - (\partial\mathbf{r} / \partial x_i)\,]\, (dx_i / ds)$
9	From 8 and (dx^i / ds) cannot be a zeroed vector so it must be assumed that the vector kernel of the summation must be zero.	$\mathbf{0} = \mathbf{b}_i - (\partial\mathbf{r} / \partial x_i)$
∴	From 9, T7.11.5A Vector Addition: Reversal of Right Cancellation by Addition, A7.9.11 Identity Vector Object Add and A7.9.6B Congruency	$\mathbf{b}_i = \partial\mathbf{r} / \partial x_i$

Theorem 12.4.2 Partial Differential Transpose of Base Vector Indices

Step	Hypothesis	
1g	Given	See Figure 12.4.1 Cartesian to Curvilinear Coordinate Manifold
Step	Hypothesis	$\partial \mathbf{b}_i / \partial x_j = \partial \mathbf{b}_j / \partial x_i$
1	From T12.4.1 Differential Relationship for Covariant Base Vectors	$\mathbf{b}_i = \partial \mathbf{r} / \partial x_i$
2	From 1 and TK.3.2 Uniqueness of Differentials	$\partial \mathbf{b}_i / \partial x_j = \partial (\partial \mathbf{r} / \partial x_i) / \partial x_j$
3	From 2 and A4.2.10 Commutative Multp	$\partial \mathbf{b}_i / \partial x_j = \partial (\partial \mathbf{r} / \partial x_j) / \partial x_i$
∴	From 3, T12.4.1 Differential Relationship for Covariant Base Vectors and A7.9.2 Substitution of Vectors	$\partial \mathbf{b}_i / \partial x_j = \partial \mathbf{b}_j / \partial x_i$

Theorem 12.4.3 Partial Differential Defines the Kronecker Delta

Step	Hypothesis	$\partial x_i / \partial x_j = \delta_i^{\ j}$ EQ A	$\partial y_i / \partial y_j = \delta^j_{\ i}$ EQ **B**
1	From T9.3.2 Uniqueness of Differentials	$\partial x_i / \partial x_j = \partial x_i / \partial x_j$	$\partial y_i / \partial y_j = \partial y_i / \partial y_j$
2	From 1 and assume i = j	$\partial x_i / \partial x_i = \partial x_i / \partial x_i$	$\partial y_i / \partial y_i = \partial y_i / \partial y_i$
3	From 2 and LxK.3.3.2B Differential identity	$\partial x_i / \partial x_j = 1$	$\partial y_i / \partial y_j = 1$
4	From 1 and now assume i ≠ j, which implies $x_i \neq f(x_j)$ or $x_i = c_x$ a constant, likewise $y_j \neq f(y_j)$ or $y_j = c_y$ a constant, likewise hence (x_i, y_j) are independent of (i, j) [SPI59I pg 179-pr11]	$\partial x_i / \partial x_j = \partial c_x / \partial x_i$	$\partial y_i / \partial y_j = \partial c_y / \partial y_i$
5	From 4 and LxK.3.3.1B Differential of a Constant	$\partial x_i / \partial x_j = 0$	$\partial y_i / \partial y_j = 0$
∴	From 3, 5 and D6.1.6B Identity Matrix	$\partial x_i / \partial x_j = \delta_i^{\ j}$ EQ A	$\partial y_i / \partial y_j = \delta^j_{\ i}$ EQ **B**

Observation 12.4.1 Caught in Between Worlds

The Kronecker Delta in theorem T12.4.3 "Partial Differential Defines the Kronecker Delta" is written as a mixed metric tensor, because there is no alternate coordinate system to transform to. So the coordinate net simplifies to an orthogonal frame work, an between worlds, in a Riemannian Manifold $\mathfrak{R}^n$, however its still a dual transpose system so covariant and contravariance still must be considered as in equation A and B.

Section 12.5 Differentation of Base Vectors

Definition 12.5.1 **Christoffel Symbol of the First Kind**

$\Gamma ij,k \equiv [ij, k] \equiv ½ (\partial g_{ik} / \partial x_j + \partial g_{jk} / \partial x_i - \partial g_{ij} / \partial x_k)$ [12.5.1]

Definition 12.5.2 **Christoffel Symbol of the Second Kind**

$\Gamma ij^{k} \equiv \left\{ \begin{matrix} k \\ ij \end{matrix} \right\} \equiv \{ij, k\} \equiv \Gamma_{ij,\alpha}\, g^{\alpha k}$ [12.5.1]

Definition 12.5.3 **Christoffel Symbol of the Third Kind**

$\Gamma^{ki}{}_{j} \equiv \left\{ \begin{matrix} ki \\ j \end{matrix} \right\} \equiv \{j, ki\} \equiv \Gamma_{\alpha j}{}^{k}\, g^{\alpha i}$ [12.5.1]

Definition 12.5.4 **Simple Differental Vector Notation**

$\mathbf{A}_{i,j} \equiv \partial (\mathbf{A}_i) / \partial x_j$	Covariant Differentiation	Equation A
$\mathbf{A}^{i,j} \equiv \partial (\mathbf{A}^i) / \partial x_j$	Contravariant Differentiation	Equation B

See DxK.3.2.2B Multi-variable Partial Notation for origins.

[12.5.1]Note: Cristoffel Symbol references are found in [SOK67, pg 79]. Also Einsteinian Summation Notation does not apply over the repeated indices the relations are valid for each component.

Untill now there has been no consideration as to what happens when a given point in a manifold is translated to another point in the space. In order to demonstrate this at that point P(•x_i•), let the second point be a translation of the first point by some small amount P + ΔP and expressed by a new point P_Δ(•x_i + Δx_i •).

As long a P stayed in the local neigborhood where the manifold was approximately linear any vector would remain invariant in magnetude and direction, however magnetude and direction are functions of the coordinate at the localized point P. Now it is being implied that such a vector no longer can be invariant because at a new point P_Δ most likely the magnetude and direction will have changed. Now as the manifold distorts magnetude and direction from point-to-point how can this be handled?

To answer this question lets consider what happenes when a vector changes direciton:

Theorem 12.5.1 Derivative of a Covariant Vector Under Change

1g	Given	$\mathbf{A} = \mathbf{F}(\bullet x_i \bullet)$
2g		Einsteinian Summation Notation does apply.
Step	Hypothesis	$\dfrac{\partial \mathbf{A}}{\partial x_i} = \dfrac{\partial A^i}{\partial x_i} \mathbf{b}_i + A^i \dfrac{\partial \mathbf{b}_i}{\partial x_i}$
1	From 2g and D12.1.1 Covariant Vector	$\mathbf{A} \equiv A^i \mathbf{b}_i$
2	From 1 and deviate from the vector by a small change in direction	$\mathbf{A} + \Delta\mathbf{A} = (A^i + \Delta A^i) (\mathbf{b}_i + \Delta\mathbf{b}_i)$
3	From 2. T7.11.1 Vector Addition: Uniqueness for Addition, T7.11.3 Vector Addition: Inverse and T7.11.2 Vector Addition: Identity	$\Delta\mathbf{A} = (A^i + \Delta A^i) (\mathbf{b}_i + \Delta\mathbf{b}_i) - (A^i \mathbf{b}_i)$
4	From 3 and T7.10.9 Distribution of Vector over Addition of Scalars	$\Delta\mathbf{A} = (A^i \mathbf{b}_i) + \Delta A^i \mathbf{b}_i + A^i\Delta\mathbf{b}_i + \Delta A^i\Delta\mathbf{b}_i - (A^i \mathbf{b}_i)$
5	From 4, T7.11.3 Vector Addition: Inverse and T7.11.2 Vector Addition: Identity	$\Delta\mathbf{A} = \Delta A^i \mathbf{b}_i + A^i\Delta\mathbf{b}_i + \Delta A^i\Delta\mathbf{b}_i$
6	From 5 and T7.10.9 Distribution of Vector over Addition of Scalars	$\Delta\mathbf{A} = \Delta A^i \mathbf{b}_i + (A^i + \Delta A^i) \Delta\mathbf{b}_i$
7	From 6 and LxK.2.1.1 Uniqueness of Limits	$\lim_{\substack{\Delta xi \to 0 \\ \Delta Ai \to 0}} \dfrac{\Delta\mathbf{A}}{\Delta x_i} = \lim_{\substack{\Delta xi \to 0 \\ \Delta Ai \to 0}} \left(\dfrac{\Delta A^i}{\Delta x_i} \mathbf{b}_i + (A^i + \Delta A^i) \dfrac{\Delta\mathbf{b}_i}{\Delta x_i} \right)$
∴	From 7, LxK.2.1.4 Limit of the Sum, LxK.2.1.3 Obvious Limit, A4.2.8 Inverse Add, A4.2.7 Identity Add and DxK.3.1.3 Definition of a Derivative alternate form	$\dfrac{\partial \mathbf{A}}{\partial x_i} = \dfrac{\partial A^i}{\partial x_i} \mathbf{b}_i + A^i \dfrac{\partial \mathbf{b}_i}{\partial x_i}$

Theorem 12.5.2 Derivative of a Contravariant Vector Under Change

1g	Given	$\mathbf{A} = \mathbf{F}(\bullet x_i \bullet)$
2g		Einsteinian Summation Notation does apply.
Step	Hypothesis	$\dfrac{\partial \mathbf{A}}{\partial x_i} = \dfrac{\partial A_i}{\partial x_i} \mathbf{b}^i + A_i \dfrac{\partial \mathbf{b}^i}{\partial x_i}$
1	From 2g and D12.1.2 Contravariant Vector	$\mathbf{A} \equiv A_i \mathbf{b}^i$
2	From 1 and deviate from the vector by a small change in direction	$\mathbf{A} + \Delta\mathbf{A} = (A_i + \Delta A_i) (\mathbf{b}^i + \Delta\mathbf{b}^i)$
3	From 2. T7.11.1 Vector Addition: Uniqueness for Addition, T7.11.3 Vector Addition: Inverse and T7.11.2 Vector Addition: Identity	$\Delta\mathbf{A} = (A_i + \Delta A_i) (\mathbf{b}^i + \Delta\mathbf{b}^i) - (A_i \mathbf{b}^i)$
4	From 3 and T7.10.9 Distribution of Vector over Addition of Scalars	$\Delta\mathbf{A} = (A_i \mathbf{b}^i) + \Delta A_i \mathbf{b}^i + A_i\Delta\mathbf{b}^i + \Delta A_i\Delta\mathbf{b}^i - (A_i \mathbf{b}^i)$
5	From 4, T7.11.3 Vector Addition: Inverse and T7.11.2 Vector Addition: Identity	$\Delta\mathbf{A} = \Delta A_i \mathbf{b}^i + A_i\Delta\mathbf{b}^i + \Delta A_i\Delta\mathbf{b}^i$
6	From 5 and T7.10.9 Distribution of Vector over Addition of Scalars	$\Delta\mathbf{A} = \Delta A_i \mathbf{b}^i + (A_i + \Delta A_i) \Delta\mathbf{b}^i$
7	From 6 and LxK.2.1.1 Uniqueness of Limits	$\lim_{\substack{\Delta xi \to 0 \\ \Delta Ai \to 0}} \dfrac{\Delta\mathbf{A}}{\Delta x_i} = \lim_{\substack{\Delta xi \to 0 \\ \Delta Ai \to 0}} \left(\dfrac{\Delta A_i}{\Delta x_i} \mathbf{b}^i + (A_i + \Delta A_i) \dfrac{\Delta\mathbf{b}^i}{\Delta x_i} \right)$
∴	From 7, LxK.2.1.4 Limit of the Sum, LxK.2.1.3 Obvious Limit, A4.2.8 Inverse Add, A4.2.7 Identity Add and DxK.3.1.3 Definition of a Derivative alternate form	$\dfrac{\partial \mathbf{A}}{\partial x_i} = \dfrac{\partial A_i}{\partial x_i} \mathbf{b}^i + A_i \dfrac{\partial \mathbf{b}^i}{\partial x_i}$

Observation 12.5.1 Differential of a Base Vectors

Theorem 12.4.1 "Derivative of a Vector Under Change" reafiremes the notoon that change of position in a manifold distorts a vector in two ways:

1) Change in the components Ai as the values of $(\bullet x_i \bullet)$ are changed.
2) Change in the base vectors spanning the space $(\bullet \mathbf{b}_i \bullet)$ as a position of the point $P(\bullet x_i \bullet)$ is altered.

While change in the amplitude of the component A^i changes, what happens in the direction of the other base vectors spannong the space? Lets write that last term in a general way to find out.

$$\frac{\partial \mathbf{b}_i}{\partial x_i} \equiv B(i, j, \beta) \mathbf{b}^\beta \qquad \text{Equation A}$$

Theorem 12.5.3 Christoffel Symbol of the First Kind

Step	Hypothesis	$[\partial (\mathbf{b}_i) / \partial x_j] \bullet \mathbf{b}_k = [ij, k]$ EQ A	$\Gamma ij,k = ½ (\partial g_{ik} / \partial x_j + \partial g_{jk} / \partial x_i - \partial g_{ij} / \partial x_k)$ EQ B
1	From D12.1.3A Covariant Metric Dyadic	$g_{ij} = \mathbf{b}_i \bullet \mathbf{b}_j$	
2	From 1 and TK.3.2 Uniqueness of Differentials	$\partial g_{ij} / \partial x_k = \partial (\mathbf{b}_i \bullet \mathbf{b}_j) / \partial x_k$	
3	From2 and TK.5.3 Chain Rule for Exact Differential of a Vector	$\partial g_{ij} / \partial x_k = [\partial (\mathbf{b}_i) / \partial x_k] \bullet \mathbf{b}_j + \mathbf{b}_i \bullet [\partial (\mathbf{b}_j) / \partial x_k]$	
4	From 3 reorder the indices in the second possible direction of change	$\partial g_{ik} / \partial x_j = [\partial (\mathbf{b}_i) / \partial x_j] \bullet \mathbf{b}_k + \mathbf{b}_i \bullet [\partial (\mathbf{b}_k) / \partial x_j]$	
5	From 3 reorder the indices in the third possible direction of change	$\partial g_{jk} / \partial x_i = [\partial (\mathbf{b}_j) / \partial x_i] \bullet \mathbf{b}_k + \mathbf{b}_j \bullet [\partial (\mathbf{b}_k) / \partial x_i]$	
6	From 4, 5 add results, subtract 3, A41.20 Negative Coefficient and A4.2.14 Distribution	$\partial g_{ik} / \partial x_j + \partial g_{jk} / \partial x_i - \partial g_{ij} / \partial x_k =$ $[\partial (\mathbf{b}_i) / \partial x_j] \bullet \mathbf{b}_k + \mathbf{b}_i \bullet [\partial (\mathbf{b}_k) / \partial x_j] +$ $[\partial (\mathbf{b}_j) / \partial x_i] \bullet \mathbf{b}_k + \mathbf{b}_j \bullet [\partial (\mathbf{b}_k) / \partial x_i] -$ $[\partial (\mathbf{b}_i) / \partial x_k] \bullet \mathbf{b}_j - \mathbf{b}_i \bullet [\partial (\mathbf{b}_j) / \partial x_k]$	
7	From 6 and T12.4.2 Partial Differential Transpose of Base Vector Indices	$\partial g_{ik} / \partial x_j + \partial g_{jk} / \partial x_i - \partial g_{ij} / \partial x_k =$ $[\partial (\mathbf{b}_i) / \partial x_j] \bullet \mathbf{b}_k + \mathbf{b}_i \bullet [\partial (\mathbf{b}_k) / \partial x_j] +$ $[\partial (\mathbf{b}_i) / \partial x_i] \bullet \mathbf{b}_k + \mathbf{b}_j \bullet [\partial (\mathbf{b}_k) / \partial x_i] -$ $[\partial (\mathbf{b}_k) / \partial x_i] \bullet \mathbf{b}_i - \mathbf{b}_i \bullet [\partial (\mathbf{b}_k) / \partial x_i]$	
8	From 7, A4.2.8 Inverse Add, A4.2.7 Identity Add and T4.3.10 Equalities: Summation of Repeated Terms by 2	$\partial g_{ik} / \partial x_j + \partial g_{jk} / \partial x_i - \partial g_{ij} / \partial x_k = 2 [\partial (\mathbf{b}_i) / \partial x_j] \bullet \mathbf{b}_k$	
9	From 8, A4.2.2B Equality and T4.4.6B Equalities: Left Cancellation by Multiplication	$[\partial (\mathbf{b}_i) / \partial x_j] \bullet \mathbf{b}_k = ½ (\partial g_{ik} / \partial x_j + \partial g_{jk} / \partial x_i - \partial g_{ij} / \partial x_k)$	
∴	From 9, D12.5.1 Christoffel Symbol of the First Kind, A4.2.3 Substitution and A4.2.2A Equality	$[\partial (\mathbf{b}_i) / \partial x_j] \bullet \mathbf{b}_k = [ij, k]$ EQ A	$\Gamma ij,k = ½ (\partial g_{ik} / \partial x_j + \partial g_{jk} / \partial x_i - \partial g_{ij} / \partial x_k)$ EQ B

Theorem 12.5.4 Covariant Derivative: Christoffel Symbol of the First Kind

1g	Given	Einsteinian Summation Notation does not apply.	
Step	Hypothesis	$\partial (\mathbf{b}_i) / \partial x_j = \Sigma_k [ij, k] \mathbf{b}^k$ EQ A	$\partial (\mathbf{b}_i) / \partial x_j = \Sigma_k \Gamma_{ij,k} \mathbf{b}^k$ EQ B
1	From T12.5.3A Christoffel Symbol of the First Kind and A4.2.12 Identity Multp	$[\partial (\mathbf{b}_i) / \partial x_j] \bullet \mathbf{b}_k = [ij, k] 1$	
2	From 1g, 1 and D6.1.6 Identity Matrix	$[\partial (\mathbf{b}_i) / \partial x_j] \bullet \mathbf{b}_k = \Sigma_k [ij, k] \delta_k^k$	
3	From 2 and D12.1.5A Mixed Metric Dyadic	$[\partial (\mathbf{b}_i) / \partial x_j] \bullet \mathbf{b}_k = (\Sigma_k [ij, k] \mathbf{b}^k) \bullet \mathbf{b}_k$	
∴	From 3 and T8.3.1 Dot Product: Uniqueness Operation, D12.5.1 Christoffel Symbol of the First Kind and A4.2.3 Substitution	$\partial (\mathbf{b}_i) / \partial x_j = \Sigma_k [ij, k] \mathbf{b}^k$ EQ A	$\partial (\mathbf{b}_i) / \partial x_j = \Sigma_k \Gamma_{ij,k} \mathbf{b}^k$ EQ B

Theorem 12.5.5 Covariant Derivative: Covariant Bases Vector

Step	Hypothesis	$\partial (\mathbf{b}_i) / \partial x_j = \Sigma_k \Gamma_{ij}{}^k \mathbf{b}_k$
1	From T12.5.4A Covariant Derivative: Christoffel Symbol of the First Kind	$\partial (\mathbf{b}_i) / \partial x_j = \Sigma_\alpha [ij, \alpha] \mathbf{b}^\alpha$
2	From 1 and T8.3.1 Dot Product: Uniqueness Operation	$\mathbf{b}^k \bullet \partial (\mathbf{b}_i) / \partial x_j = \mathbf{b}^k \bullet \Sigma_\alpha [ij, \alpha] \mathbf{b}^\alpha$
3	From 2 and T8.3.7 Dot Product: Distribution of Dot Product Across Addition of Vectors	$\mathbf{b}^k \bullet \partial (\mathbf{b}_i) / \partial x_j = \Sigma_\alpha [ij, \alpha] \mathbf{b}^k \bullet \mathbf{b}^\alpha$
4	From 3 and T8.3.3 Dot Product: Commutative Operation	$\mathbf{b}^k \bullet \partial (\mathbf{b}_i) / \partial x_j = \Sigma_\alpha [ij, \alpha] \mathbf{b}^\alpha \bullet \mathbf{b}^k$
5	From 4, D12.1.4A Contravariant Metric Dyadic and A4.2.3 Substitution	$\mathbf{b}^\alpha \bullet \partial (\mathbf{b}_i) / \partial x_j = \Sigma_\alpha [ij, \alpha] g^{\alpha k}$
6	From 5, D12.5.1 Christoffel Symbol of the First Kind and A4.2.3 Substitution	$\mathbf{b}^k \bullet \partial (\mathbf{b}_i) / \partial x_j = \Sigma_\alpha \Gamma_{ij, \alpha} g^{\alpha k}$
7	From 6, D12.5.2 Christoffel Symbol of the Second Kind and A4.2.3 Substitution	$\partial (\mathbf{b}_i) / \partial x_j \bullet \mathbf{b}^k = \Gamma_{ij}{}^k$
8	From 7, A4.2.12 Identity Multp and D6.1.6 Identity Matrix	$\partial (\mathbf{b}_i) / \partial x_j \bullet \mathbf{b}^k = \Sigma_k \Gamma_{ij}{}^k \delta_k{}^k$
9	From 8, D12.1.5A Mixed Metric Dyadic and A4.2.3 Substitution	$\partial (\mathbf{b}_i) / \partial x_j \bullet \mathbf{b}^k = \Gamma_{ij}{}^k \mathbf{b}_k \bullet \mathbf{b}^k$
∴	From 9 and T8.3.1 Dot Product: Uniqueness Operation	$\partial (\mathbf{b}_i) / \partial x_j = \Sigma_k \Gamma_{ij}{}^k \mathbf{b}_k$

Theorem 12.5.6 Contravariant Derivative: Contravariant Bases Vector

1g	Given	Einsteinian Summation Notation does not apply.
Step	Hypothesis	$\partial (\mathbf{b}^i) / \partial x_j = -\Sigma_k \Gamma_{kj}{}^i \mathbf{b}^k$
1	From 1g, D12.1.5A Mixed Metric Dyadic	$\delta_k{}^i = \mathbf{b}_k \bullet \mathbf{b}^i$
2	From 1 and TK.3.2 Uniqueness of Differentials	$\partial (\delta_k{}^i) / \partial x_j = \partial (\mathbf{b}_k \bullet \mathbf{b}^i) / \partial x_j$
3	From2, LxK.3.3.1 Differential of a Constant and TK.5.3 Chain Rule for Exact Differential of a Vector	$0 = [\partial (\mathbf{b}_k) / \partial x_j] \bullet \mathbf{b}^i + \mathbf{b}_k \bullet[\partial (\mathbf{b}^i) / \partial x_j]$
4	From 3 and T4.3.5B Equalities: Left Cancellation by Addition	$\mathbf{b}_k \bullet[\partial (\mathbf{b}^i) / \partial x_j] = -[\partial (\mathbf{b}_k) / \partial x_j] \bullet \mathbf{b}^i$
5	From 4, T12.5.4 Covariant Derivative: Christoffel Symbol of the First Kind and A4.2.3 Substitution	$\mathbf{b}_k \bullet[\partial (\mathbf{b}^i) / \partial x_j] = (-\Sigma_\alpha \Gamma_{kj,\alpha} \mathbf{b}^\alpha) \bullet \mathbf{b}^i$
6	From 5, T8.3.7 Dot Product: Distribution of Dot Product Across Addition of Vectors, D12.1.4A Contravariant Metric Dyadic and A4.2.3 Substitution	$\mathbf{b}_k \bullet[\partial (\mathbf{b}^i) / \partial x_j] = -\Sigma_\alpha \Gamma_{kj,\alpha}\, g^{\alpha i}$
7	From 6, D12.5.4B Christoffel Symbol of the First Kind and A4.2.3 Substitution	$\mathbf{b}_k \bullet[\partial (\mathbf{b}^i) / \partial x_j] = -\Gamma_{kj}{}^i$
8	From 7, A4.2.12 Identity Multp and D6.1.6 Identity Matrix	$\mathbf{b}_k \bullet[\partial (\mathbf{b}^i) / \partial x_j] = -\Sigma_\alpha \Gamma_{\alpha j}{}^i \delta_k{}^\alpha$
9	From 8, D12.1.5A Mixed Metric Dyadic and A4.2.3 Substitution	$\mathbf{b}_k \bullet[\partial (\mathbf{b}^i) / \partial x_j] = -\Sigma_\alpha \Gamma_{\alpha j}{}^i \mathbf{b}_k \bullet \mathbf{b}^\alpha$
10	From 9 and T8.3.3 Dot Product: Commutative Operation	$\mathbf{b}_k \bullet[\partial (\mathbf{b}^i) / \partial x_j] = \mathbf{b}_k \bullet [-\Sigma_\alpha \Gamma_{\alpha j}{}^i \mathbf{b}^\alpha]$
∴	From 10, T8.3.1 Dot Product: Uniqueness Operation and substituting dummy index α → k	$\partial (\mathbf{b}^i) / \partial x_j = -\Sigma_k \Gamma_{kj}{}^i \mathbf{b}^k$

Theorem 12.5.7 Contravariant Derivative: Covariant Bases Vector

Step	Hypothesis	$\partial (\mathbf{b}^i) / \partial x_j = -\Sigma_\alpha \Sigma_k \Gamma_{kj}{}^i g^{k\alpha} \mathbf{b}_\alpha$
1	From T12.5.6 Contravariant Derivative: Contravariant Bases Vector	$\partial (\mathbf{b}^i) / \partial x_j = -\Sigma_k \Gamma_{kj}{}^i \mathbf{b}^k$
2	From 1 and T8.3.1 Dot Product: Uniqueness Operation	$[\partial (\mathbf{b}^i) / \partial x_j] \bullet \mathbf{b}^\gamma = -\Sigma_k \Gamma_{kj}{}^i (\mathbf{b}^k \bullet \mathbf{b}^\alpha)$
3	From 2, D12.1.4A Contravariant Metric Dyadic and A4.2.3 Substitution	$[\partial (\mathbf{b}^i) / \partial x_j] \bullet \mathbf{b}^\gamma = -\Sigma_k \Gamma_{kj}{}^i g^{k\alpha}$
4	From 3, A4.2.12 Identity Multp and D6.1.6 Identity Matrix	$[\partial (\mathbf{b}^i) / \partial x_j] \bullet \mathbf{b}^\gamma = -\Sigma_k \Sigma_\alpha \Gamma_{kj}{}^i g^{k\alpha} \delta_\alpha{}^\gamma$
5	From 4, D12.1.5A Mixed Metric Dyadic and A4.2.3 Substitution	$[\partial (\mathbf{b}^i) / \partial x_j] \bullet \mathbf{b}^\gamma = -\Sigma_k \Sigma_\alpha \Gamma_{kj}{}^i g^{k\alpha} (\mathbf{b}_\alpha \bullet \mathbf{b}^\gamma)$
6	From 5 and T8.3.7 Dot Product: Distribution of Dot Product Across Addition of Vectors	$[\partial (\mathbf{b}^i) / \partial x_j] \bullet \mathbf{b}^\gamma = (-\Sigma_\alpha \Sigma_k \Gamma_{kj}{}^i g^{k\alpha} \mathbf{b}_\alpha) \bullet \mathbf{b}^\gamma$
∴	From 6 and T8.3.1 Dot Product: Uniqueness Operation	$\partial (\mathbf{b}^i) / \partial x_j = -\Sigma_\alpha \Sigma_k \Gamma_{kj}{}^i g^{k\alpha} \mathbf{b}_\alpha$

Theorem 12.5.8 Christoffel Symbol: Second to the First Kind

Step	Hypothesis	$\Gamma_{ij,\beta} = \Sigma_k \Gamma_{ij}{}^k g_{k\beta}$ EQ A	$\Gamma_{ij,\beta} = \Sigma_k g_{k\beta} \Gamma_{ij}{}^k$ EQ B
1	From D12.5.2 Christoffel Symbol of the Second Kind	$\Gamma_{ij}{}^k = \Sigma_\alpha \Gamma_{ij,\alpha} g^{\alpha k}$	
2	From 1, T12.3.9B Metric Transformation from Covariant to Contravariant Coefficient and A10.2.13 Symmetric-Reflexive Rank of a Tensor	$\Sigma_k \Gamma_{ij}{}^k g_{k\beta} = \Sigma_\alpha \Gamma_{ij,\alpha} \Sigma_k g^{\alpha k} g_{k\beta}$	
3	From 2, T12.2.8B Covariant and Contravariant Metric Inverse and A4.2.3 Substitution	$\Sigma_k \Gamma_{ij}{}^k g_{k\beta} = \Sigma_\alpha \Gamma_{ij,\alpha} \delta^\alpha{}_\beta$	
4	From 3 and D6.1.6 Identity Matrix-Kronecker Delta evaluated	$\Sigma_k \Gamma_{ij}{}^k g_{k\beta} = \Gamma_{ij,\beta}$	
∴	From 4, A10.2.6 Equality of Tensors and T10.4.7 Product of Tensors: Commutative by Multiplication of Rank for p + q	$\Gamma_{ij,\beta} = \Sigma_k \Gamma_{ij}{}^k g_{k\beta}$ EQ A	$\Gamma_{ij,\beta} = \Sigma_k g_{k\beta} \Gamma_{ij}{}^k$ EQ B

Theorem 12.5.9 Christoffel Symbol: First Kind Symmetry of First Two Indices

1g	Given	$\Gamma_{ij,\beta} = \Gamma_{ji,\beta}$ by assumption
Step	Hypothesis	$\Gamma_{ij,k} = \Gamma_{ji,k}$
1	From 1g	$\Gamma_{ii,k} = \Gamma_{ji,k}$
2	From 1 and T12.5.3B Christoffel Symbol of the First Kind	$\Gamma_{ij,k} =$ $\frac{1}{2} (\partial g_{jk} / \partial x_i + \partial g_{ik} / \partial x_j - \partial g_{ji} / \partial x_k)$
3	From 2, T12.2.11 Covariant Metric Inner Product is Symmetrical, T10.3.5 Tool: Tensor Substitution and T10.4.16 Addition of Tensors: Commutative by Addition	$\Gamma_{ij,k} =$ $\frac{1}{2} (\partial g_{ik} / \partial x_j + \partial g_{jk} / \partial x_i - \partial g_{ij} / \partial x_k)$
4	From 3 and T12.5.3B Christoffel Symbol of the First Kind	$\Gamma_{ij,k} = \Gamma_{ij,k}$
∴	From 4 step 1 is concluded by identity as valid, hence from step 1	$\Gamma_{ij,k} = \Gamma_{ji,k}$

Theorem 12.5.10 Christoffel Symbol: First Kind Double Bases Vector Indices

Step	Hypothesis		
		$\Gamma_{\eta\eta,k} = \partial g_{\eta k} / \partial x_\eta - \frac{1}{2} \partial g_{\eta\eta} / \partial x_k$ EQ A	$\Gamma_{\eta\eta,k} = g_{\eta k,\eta} - \frac{1}{2} g_{\eta\eta,k}$ EQ B
1	From T12.5.3B Christoffel Symbol of the First Kind	$\Gamma_{\eta\eta,k} = \frac{1}{2} (\partial g_{\eta k} / \partial x_\eta + \partial g_{\eta k} / \partial x_\eta - \partial g_{\eta\eta} / \partial x_k)$	
2	From 1 and T4.3.10 Equalities: Summation of Repeated Terms by 2	$\Gamma_{\eta\eta,k} = \frac{1}{2} (2\partial g_{\eta k} / \partial x_\eta - \partial g_{\eta\eta} / \partial x_k)$	
∴	From 2, T10.5.6 Scalars and Tensors: Distribution of a Scalar over Addition of Tensors, T4.4.18 Equalities: Product by Division, Factorization by 2 and D12.5.4A Simple Differental VectorNotation	$\Gamma_{\eta\eta,k} = \partial g_{\eta k} / \partial x_\eta - \frac{1}{2} \partial g_{\eta\eta} / \partial x_k$ EQ A	$\Gamma_{\eta\eta,k} = g_{\eta k,\eta} - \frac{1}{2} g_{\eta\eta,k}$ EQ B

Theorem 12.5.11 Christoffel Symbol: First Kind Duplicate Last Two Indices

Step	Hypothesis	$\Gamma_{i\eta,\eta} = \frac{1}{2} \partial g_{\eta\eta} / \partial x_i$ EQ A	$\Gamma_{i\eta,\eta} = \frac{1}{2} g_{\eta\eta,i}$ EQ B
1	From T12.5.3B Christoffel Symbol of the First Kind	$\Gamma_{i\eta,\eta} = \frac{1}{2} (\partial g_{i\eta} / \partial x_\eta + \partial g_{\eta\eta} / \partial x_i - \partial g_{i\eta} / \partial x_\eta)$	
2	From 1 and T4.3.10 Equalities: Summation of Repeated Terms by 2	$\Gamma_{i\eta,\eta} = \frac{1}{2} \partial g_{\eta\eta} / \partial x_i$	
∴	From 2, T10.5.6 Scalars and Tensors: Distribution of a Scalar over Addition of Tensors, T4.4.18 Equalities: Product by Division, Factorization by 2 and D12.5.4A Simple Differental VectorNotation	$\Gamma_{i\eta,\eta} = \frac{1}{2} \partial g_{\eta\eta} / \partial x_i$ EQ A	$\Gamma_{i\eta,\eta} = \frac{1}{2} g_{\eta\eta,i}$ EQ B

Theorem 12.5.12 Christoffel Symbol: First Kind Asymmetry of Last Two Indices

Step	Hypothesis	$\Gamma_{ik,j} = g_{kj,i} - \Gamma_{ij,k}$ EQ A	$\Gamma_{ij,k} + \Gamma_{ik,j} = g_{kj,i}$ EQ B
1	From T12.5.3B Christoffel Symbol of the First Kind	$\Gamma_{ik,j} = ½ (\partial g_{ij} / \partial x_k + \partial g_{kj} / \partial x_i - \partial g_{ik} / \partial x_j)$	
2	From 1, T12.2.13 Covariant Metric Inner Product is Symmetrical and T10.4.16 Addition of Tensors: Commutative by Addition	$\Gamma_{ik,j} = ½ (-\partial g_{ik} / \partial x_j + \partial g_{kj} / \partial x_i + \partial g_{ij} / \partial x_k)$	
3	From 2, D4.1.20B Negative Coefficient, D4.1.19 Primitive Definition for Rational Arithmetic and T10.5.6 Scalars and Tensors: Distribution of a Scalar over Addition of Tensors	$\Gamma_{ik,j} = -½ (\partial g_{ik} / \partial x_j - \partial g_{kj} / \partial x_i - \partial g_{ij} / \partial x_k)$	
4	From 3, T10.4.18 Addition of Tensors: Identity by Addition, T10.4.19 Addition of Tensors: Inverse by Addition, T12.2.13 Covariant Metric Inner Product is Symmetrical and T4.3.10 Equalities: Summation of Repeated Terms by 2	$\Gamma_{ik,j} = -½ (\partial g_{ik} / \partial x_j + \partial g_{kj} / \partial x_i - 2\partial g_{jk} / \partial x_i - \partial g_{ij} / \partial x_k)$	
5	From 4, T10.4.16 Addition of Tensors: Commutative by Addition, T10.4.17 Addition of Tensors: Associative by Addition and T10.5.6 Scalars and Tensors: Distribution of a Scalar over Addition of Tensors	$\Gamma_{ik,j} = -(-½)2\partial g_{kj} / \partial x_i - ½(\partial g_{ik} / \partial x_j + \partial g_{jk} / \partial x_i - \partial g_{ij} / \partial x_k)$	
6	From 5, T4.4.18 Equalities: Product by Division, Factorization by 2 and D4.1.19 Primitive Definition for Rational Arithmetic	$\Gamma_{ik,j} = \partial g_{kj} / \partial x_i - ½(\partial g_{ik} / \partial x_j + \partial g_{jk} / \partial x_i - \partial g_{ij} / \partial x_k)$	
7	From 6, T12.5.3B Christoffel Symbol of the First Kind, T10.3.5 Tool: Tensor Substitution and D12.5.4A Simple Differental VectorNotation	$\Gamma_{ik,j} = g_{kj,i} - \Gamma_{ij,k}$	
∴	From 7 and T10.4.22A Addition of Tensors: Reversal of Right Cancellation by Addition	$\Gamma_{ik,j} = g_{kj,i} - \Gamma_{ij,k}$ EQ A	$\Gamma_{ij,k} + \Gamma_{ik,j} = g_{kj,i}$ EQ B

Theorem 12.5.13 Christoffel Symbol: Of the First Kind in Orthogonal Space

1g	Given	$g_{ij} = b_i b_j \delta_{ij}$ orthogonal independent coordinates
Step	Hypothesis	$\Gamma_{ij,k} = 0$ orthogonal independent coordinates
1	From T12.5.3B Christoffel Symbol of the First Kind	$\Gamma_{ij,k} = ½ (g_{ik,j} + g_{jk,i} - g_{ij,k})$
2	From 1g, 1 and T10.3.5 Tool: Tensor Substitution	$\Gamma_{ij,k} = ½ [(b_i b_k \delta_{ik})_{,j} + (b_j b_k \delta_{jk})_{,i} - (b_i b_j \delta_{ij})_{,k}]$
3	From 2 and LxK.3.2.3B Distribute Across Constant Product	$\Gamma_{ij,k} = ½ [b_i b_k (\delta_{ik})_{,j} + b_j b_k (\delta_{jk})_{,i} - b_i b_j (\delta_{ij})_{,k}]$
4	From 3, D6.1.6B Identity Matrix, and LxK.3.3.1B Differential of a Constant differentiating the Konecker Delta's components	$\Gamma_{ij,k} = ½ [b_i b_k 0 + b_j b_k 0 - b_i b_j 0]$
∴	From 4, T4.4.1 Equalities: Any Quantity Multiplied by Zero is Zero and A4.2.7 Identity Add	$\Gamma_{ij,k} = 0$ orthogonal independent coordinates

Theorem 12.5.14 Christoffel Symbol: Second Kind Symmetry for First Two Indices

1g	Given	$\Gamma_{ij}{}^{k} = \Gamma_{ji}{}^{k}$ by assumption
Step	Hypothesis	$\Gamma_{ij}{}^{k} = \Gamma_{ji}{}^{k}$
1	From 1g	$\Gamma_{ij}{}^{k} = \Gamma_{ji}{}^{k}$
2	From 1, D12.5.2 Christoffel Symbol of the Second Kind and T10.3.5 Tool: Tensor Substitution	$\Gamma_{ij}{}^{k} = \Gamma_{ji,\alpha}\, g^{\alpha k}$
3	From 2, T12.5.9 Christoffel Symbol: First Kind Symmetry of First Two Indices and T10.3.5 Tool: Tensor Substitution	$\Gamma_{ij}{}^{k} = \Gamma_{ij,\alpha}\, g^{\alpha k}$
4	From 3, D12.5.2 Christoffel Symbol of the Second Kind and T10.3.5 Tool: Tensor Substitution	$\Gamma_{ij}{}^{k} = \Gamma_{ij}{}^{k}$
∴	From 4 step 1 is by identiy concluded as valid, hence from step 1	$\Gamma_{ij}{}^{k} = \Gamma_{ji}{}^{k}$

Theorem 12.5.15 Christoffel Symbol: Exchanging Index between Second and First Symbols

1g	Given	$\sum_b \Gamma_{ij,b}\, \Gamma_{ak}{}^{b} = \sum_b \Gamma_{ij,b}\, \Gamma_{ak}{}^{b}$	
Step	Hypothesis	$\sum_b \Gamma_{ij,b}\, \Gamma_{ak}{}^{b} = \sum_b \Gamma_{ij}{}^{b}\, \Gamma_{ak,b}$ EQ A	$\sum_b \Gamma_{ij,b}\, \Gamma_{ak}{}^{b} = \sum_b \Gamma_{ak,b}\, \Gamma_{ij}{}^{b}$ EQ B
1	From 1g	$\sum_b \Gamma_{ij,b}\, \Gamma_{ak}{}^{b} = \sum_b \Gamma_{ij,b}\, \Gamma_{ak}{}^{b}$	
2	From 1 and D12.5.2 Christoffel Symbol of the Second Kind	$\sum_b \Gamma_{ij,b}\, \Gamma_{ak}{}^{b} = \sum_b \Gamma_{ij,b} \sum_\eta \Gamma_{ak,\eta}\, g^{b\eta}$	
3	From 2 and T10.4.7 Product of Tensors: Commutative by Multiplication of Rank for p + q	$\sum_b \Gamma_{ij,b}\, \Gamma_{ak}{}^{b} = \sum_\eta \sum_b g^{b\eta} \Gamma_{ij,b}\, \Gamma_{ak,\eta}$	
4	From 3 1 and D12.5.2 Christoffel Symbol of the Second Kind	$\sum_b \Gamma_{ij,b}\, \Gamma_{ak}{}^{b} = \sum_\eta \Gamma_{ij}{}^{\eta}\, \Gamma_{ak,\eta}$	
∴	From 4 and changing dummy indices η → b	$\sum_b \Gamma_{ij,b}\, \Gamma_{ak}{}^{b} = \sum_b \Gamma_{ij}{}^{b}\, \Gamma_{ak,b}$ EQ A	$\sum_b \Gamma_{ij,b}\, \Gamma_{ak}{}^{b} = \sum_b \Gamma_{ak,b}\, \Gamma_{ij}{}^{b}$ EQ B

Theorem 12.5.16 Kronecker Delta Simple Differential

Step	Hypothesis	$\sum_\alpha g_{i\alpha,k}\, g^{\alpha j} = -\sum_\alpha g_{i\alpha}\, g^{\alpha j,k}$ EQ A	$\sum_\alpha g^{i\alpha,k}\, g_{\alpha j} = -\sum_\alpha g^{i\alpha}\, g_{\alpha j,k}$ EQ B
1	From 12.2.8(A, B) Covariant and Contravariant Metric Inverse	$\delta_i^{\ j} = g_i^{\ j} = \sum_\alpha g_{i\alpha} g^{\alpha j}$ for $\mid g_{ij} \mid \neq 0$	$\delta^i_{\ j} = g^i_{\ j} = \sum_k g^{i\alpha} g_{\alpha j}$ for $\mid g_{ij} \mid \neq 0$
2	From 1 and Lx9.3.1.7 Uniqueness Theorem of Differentiation	$(\delta_i^{\ j})_{,k} = (\sum_\alpha g_{i\alpha} g^{\alpha j})_{,k}$	$(\delta^i_{\ j})_{,k} = (\sum_\alpha g^{i\alpha} g_{\alpha j})_{,k}$
3	From 2, D6.1.6B Identity Matrix, Lx9.3.3.1B Differential of a Constant and Lx9.3.2.4 Distributed Across Products	$0 = (\sum_\alpha g_{i\alpha,k}\, g^{\alpha j}) + (\sum_\alpha g_{i\alpha}\, g^{kj,k})$	$0 = (\sum_\alpha g^{i\alpha,k}\, g_{\alpha j}) + (\sum_\alpha g^{i\alpha}\, g_{\alpha j,k})$
∴	From 3 and T10.3.21B Addition of Tensors: Right Cancellation by Addition	$\sum_\alpha g_{i\alpha,k}\, g^{\alpha j} = -\sum_\alpha g_{i\alpha}\, g^{\alpha j,k}$ EQ A	$\sum_\alpha g^{i\alpha,k}\, g_{\alpha j} = -\sum_\alpha g^{i\alpha}\, g_{\alpha j,k}$ EQ B

Theorem 12.5.17 Christoffel Symbol: Second Kind Double Bases Vector Indices

Step	Hypothesis	$\Gamma_{ii}{}^{j} = -g_{i\alpha}\, g^{\alpha j,i} - ½\, g_{ii,\alpha}\, g^{\alpha j}$ EQ A	$\Gamma_{ii}{}^{j} = (g_{i\alpha,i}\, g^{\alpha j} - ½\, g_{ii,\alpha}\, g^{\alpha j})$ EQ B
1	From D12.5.2 Christoffel Symbol of the Second Kind; with double bases vector indices	$\Gamma_{ii}{}^{j} = \Gamma_{ii,\alpha}\, g^{\alpha j}$	
2	From 1, T12.5.10B Christoffel Symbol: First Kind Double Bases Vector Indices and T10.3.5 Tool: Tensor Substitution	$\Gamma_{ii}{}^{j} = (g_{i\alpha,i} - ½\, g_{ii,\alpha})\, g^{\alpha j}$	
3	From 2 and T10.3.20 Addition of Tensors: Distribution of a Tensor over Addition of Tensors	$\Gamma_{ii}{}^{j} = (g_{i\alpha,i}\, g^{\alpha j} - ½\, g_{ii,\alpha}\, g^{\alpha j})$	
∴	From 3, T12.5.6A Kronecker Delta Simple Differential and T10.3.5 Tool: Tensor Substitution	$\Gamma_{ii}{}^{j} = -g_{i\alpha}\, g^{\alpha j,i} - ½\, g_{ii,\alpha}\, g^{\alpha j}$ EQ A	$\Gamma_{ii}{}^{j} = (g_{i\alpha,i}\, g^{\alpha j} - ½\, g_{ii,\alpha}\, g^{\alpha j})$ EQ B

Theorem 12.5.18 Christoffel Symbol: Of the Second Kind in Orthogonal Space

Step	Hypothesis	$\Gamma_{ij}{}^{k} = 0$ orthogonal independent coordinates
1	From D12.5.2 Christoffel Symbol of the Second Kind	$\Gamma_{ij}{}^{k} = \Gamma_{ij,\alpha}\, g^{\alpha k}$
∴	From 1, T12.5.17 Christoffel Symbol: Of the First Kind in Orthogonal Space and T4.4.1 Equalities: Any Quantity Multiplied by Zero is Zero	$\Gamma_{ij}{}^{k} = 0$ orthogonal independent coordinates

Table 12.5.1 Differentiation of Covariant and Contravariant Base Vectors

	$\partial\,(\mathbf{b}_i) / \partial\, x_j$	Theorem	$\partial\,(\mathbf{b}^i) / \partial\, x_j$	Theorem
$\mathbf{b}_k$	$\sum_k (\Gamma_{ij}{}^{k})\, \mathbf{b}_k$	T12.5.5	$-\sum_k (\sum_\alpha \Gamma_{\alpha j}{}^{i}\, g^{\alpha k})\, \mathbf{b}_k$	T12.5.7
$\mathbf{b}^k$	$\sum_k (\Gamma_{ij,k})\, \mathbf{b}^k$	T12.5.4B	$-\sum_k (\Gamma_{kj}{}^{i})\, \mathbf{b}^k$	T12.5.6

Section 12.6 Differentiation for a Tensor of Rank-1

Definition 12.6.1 **Differental Tensor Notation**

$\mathbf{A}^{;j} \equiv A^{i;j}\, \mathbf{b}_i$	Covariant Differentiation	Equation A
$\mathbf{A}_{;j} \equiv A_{i;j}\, \mathbf{b}^i$	Contravariant Differentiation	Equation B

Observation 12.6.1 Rules for Covariant and Contravariant Differential Notation

Rule 1: Scalar functions use a comma to symbolize partial differeentation since direction by definition has no meaning for a scalar. See theroem T12.6.1 "Differentiation of Tensor: Rank-0" for an example, $A_{;i} = A_{,i}$.

Rule 2: Therorems T12.6.2 "Differentiation of Tensor: Rank-1 Covariant", T12.6.3 "Differentiation of Tensor: Rank-1 Contravariant", T12.7.7 "Tensor Product: Covariant Chain Rule" and T12.7.8 "Tensor Product: Contravariant Chain Rule" note that when only magnitude is being differentated for a vector the simple partial differential notation is a comma, however the moment when direction is considered the difference is represented by a change to a semi-colon.

Rule 3: Also notice that it is lowered in conconjunction with the covariant index on the **bases vector** and raised with the contravariant index, this is to avoid confusion when using mixed index notation. Also it helps identify the type of dirivative, contra or covariant.

Theorem 12.6.1 Differentiation of Tensor: Rank-0

1g	Given	A(•x$_i$•) scalar and invariant over the manifold			
Step	Hypothesis	$A_{;i} = A_{,i}$	EQ A	$A^{;i} = A_{,i}$	EQ B
∴	From 1g, TK.3.2 Uniqueness of Differentials and D12.6.1 Differental TensorNotation, since a scalar by definition is not a vector the notion of covariance and contravariance has no meaning so standard differental notation is used.	$A_{;i} = A_{,i}$	EQ A	$A^{;i} = A_{,i}$	EQ B

Theorem 12.6.2 Differentiation of Tensor: Rank-1 Covariant

Step	Hypothesis	$\mathbf{A}^{;j} = \sum_i A^{i;j}\, \mathbf{b}_i$ EQ A	$\mathbf{A}^{;j} = \sum_i (A^{i,j} + \sum_k A^k \Gamma_{kj}{}^i)\, \mathbf{b}_i$ EQ B
1	From D12.1.1A Covariant Vector	$\mathbf{A} \equiv \sum_i A^i\, \mathbf{b}_i$	
2	From 1 and TK.3.2 Uniqueness of Differentials	$\partial (\mathbf{A}) / \partial x_j = \partial (\sum_i A^i\, \mathbf{b}_i) / \partial x_j$	
3	From 2, TK.5.1 Derivative Distribution Across Vector Components	$\partial (\mathbf{A}) / \partial x_j = \sum_i \partial (A^i\, \mathbf{b}_i) / \partial x_j$	
4	From 3, T12.5.1 Derivative of a Covariant Vector Under Change	$\partial (\mathbf{A}) / \partial x_j = \sum_i \partial (A^i) / \partial x_j\, \mathbf{b}_i + \sum_i A^i\, \partial (\mathbf{b}_i) / \partial x_j$	
5	From 4, T12.5.5 Covariant Derivative: Covariant Bases Vector and A4.2.3 Substitution	$\partial (\mathbf{A}) / \partial x_j = \sum_i A^{i,j}\, \mathbf{b}_i + \sum_i \sum_k A^i\, \Gamma_{ij}{}^k\, \mathbf{b}_k$	
6	From 5 over the complete summation rename the indices, k → i	$\partial (\mathbf{A}) / \partial x_j = \sum_i A^{i,j}\, \mathbf{b}_i + \sum_i \sum_k A^k\, \Gamma_{kj}{}^i\, \mathbf{b}_i$	
7	From 6, sum like terms having common base vectors	$\partial (\mathbf{A}) / \partial x_j = \sum_i A^{i,j}\, \mathbf{b}_i + \sum_i \sum_k A^k\, \Gamma_{kj}{}^i\, \mathbf{b}_i$	
8	From 7 and T7.10.9 Distribution of Vector over Addition of Scalars	$\partial (\mathbf{A}) / \partial x_j = \sum_i (A^{i,j} + \sum_k A^k\, \Gamma_{kj}{}^i)\, \mathbf{b}_i$	
∴	From 7, 8, D12.6.1A Differental Tensor Notation and A4.2.3 Substitution	$\mathbf{A}^{;j} = \sum_i A^{i;j}\, \mathbf{b}_i$ EQ A	$\mathbf{A}^{;j} = \sum_i (A^{i,j} + \sum_k A^k\, \Gamma_{kj}{}^i)\, \mathbf{b}_i$ EQ B

Theorem 12.6.3 Differentiation of Tensor: Rank-1 Contravariant

Step	Hypothesis	$\mathbf{A}_{;j} = \sum_i A_{i;j}\, \mathbf{b}^i$ EQ A	$\mathbf{A}_{;j} = \sum_i (A_{i,j} - \sum_k A_k\, \Gamma_{ij}{}^k)\, \mathbf{b}^i$ EQ B
1	From D12.1.2A Contravariant Vector	$\mathbf{A} \equiv \sum_i A_i\, \mathbf{b}^i$	
2	From 1 and TK.3.2 Uniqueness of Differentials	$\partial (\mathbf{A}) / \partial x_j = \partial (\sum_i A_i\, \mathbf{b}^i) / \partial x_j$	
3	From 2 and TK.5.1 Derivative Distribution Across Vector Components	$\partial (\mathbf{A}) / \partial x_j = \sum_i \partial (A_i\, \mathbf{b}^i) / \partial x_j$	
4	From 3, T12.5.2 Derivative of a Contravariant Vector Under Change	$\partial (\mathbf{A}) / \partial x_j = \sum_i \partial (A_i) / \partial x_j\, \mathbf{b}^i + \sum_i A_i\, \partial (\mathbf{b}^i) / \partial x_j$	
5	From 4, T12.5.6 Contravariant Derivative: Contravariant Bases Vector and A4.2.3 Substitution	$\partial (\mathbf{A}) / \partial x_j = \sum_i A_{i,j}\, \mathbf{b}^i - \sum_k \sum_i A_i\, \Gamma_{kj}{}^i\, \mathbf{b}^k$	
6	From 5 over the complete summation rename the indices, k → i	$\partial (\mathbf{A}) / \partial x_j = \sum_i A_{i,j}\, \mathbf{b}^i - \sum_i \sum_k A_k\, \Gamma_{ij}{}^k\, \mathbf{b}^i$	
7	From 7 and T7.10.9 Distribution of Vector over Addition of Scalars	$\partial (\mathbf{A}) / \partial x_j = \sum_i (A_{i,j} - \sum_k A_k\, \Gamma_{ij}{}^k)\, \mathbf{b}^i$	
∴	From 7, 8, D12.6.1A Differental Tensor Notation and A4.2.3 Substitution	$\mathbf{A}_{;j} = \sum_i A_{i;j}\, \mathbf{b}^i$ EQ A	$\mathbf{A}_{;j} = \sum_i (A_{i,j} - \sum_k A_k\, \Gamma_{ij}{}^k)\, \mathbf{b}^i$ EQ B

Theorem 12.6.4 Differentiation of Tensor: Rank-1 Uniqueness of Differentiation

Step	Hypothesis		
		$\mathbf{A}_{;j} = \mathbf{B}_{,j}$ EQ A $A^{i;j} = B^{i;j}$ EQ C	$\mathbf{A}^{;j} = \mathbf{B}^{;j}$ EQ B $A_{i;j} = B_{i;j}$ EQ D
1	From A7.9.1 Equivalence of Vector	$\mathbf{A} = \mathbf{B}$	$\mathbf{A} = \mathbf{B}$
2	From 1 and AK.5.3B Uniqueness of Differentiation of a Vector	$\mathbf{A}_{;j} = \mathbf{B}_{;j}$	$\mathbf{A}^{;j} = \mathbf{B}^{;j}$
3	From 2, T12.6.2A Differentiation of Tensor: Rank-1 Covariant and T12.6.3A Differentiation of Tensor: Rank-1 Contravariant	$\sum_i A^{i;j}\,\mathbf{b}_i = \sum_i B^{i;j}\,\mathbf{b}_i$	$\sum_i A_{i;j}\,\mathbf{b}^i = \sum_i B_{i;j}\,\mathbf{b}^i$
∴	From 2, 3 and equating by base vectors	$\mathbf{A}_{;j} = \mathbf{B}_{,j}$ EQ A $A^{i;j} = B^{i;j}$ EQ C	$\mathbf{A}^{;j} = \mathbf{B}^{;j}$ EQ B $A_{i;j} = B_{i;j}$ EQ D

Theorem 12.6.5 Differentiation of Tensor: Rank-1 Distribution of Differentiation Across Vectors

Step	Hypothesis		
		$\mathbf{A}_{;j} = \mathbf{B}_{;j} + \mathbf{C}_{;j}$ EQ A $A^{i;j} = B^{i;j} + C^{i;j}$ EQ C	$\mathbf{A}^{;j} = \mathbf{B}^{;j} + \mathbf{C}^{;j}$ EQ B $A_{i;j} = B_{i;j} + C_{i;j}$ EQ D
1	From A7.9.1 Equivalence of Vector	$\mathbf{A} = \mathbf{B} + \mathbf{C}$	$\mathbf{A} = \mathbf{B} + \mathbf{C}$
2	From 1 and AK.5.3B Uniqueness of Differentiation of a Vector	$\mathbf{A}_{;j} = (\,\mathbf{B} + \mathbf{C}\,)_{;j}$	$\mathbf{A}^{;j} = (\,\mathbf{B} + \mathbf{C}\,)^{;j}$
3	From 2 and LxK.5.1.1 Linear Distribution of Differential	$\mathbf{A}_{;j} = \mathbf{B}_{;j} + \mathbf{C}_{;j}$	$\mathbf{A}^{;j} = \mathbf{B}^{;j} + \mathbf{C}^{;j}$
4	From 3, T12.6.2A Differentiation of Tensor: Rank-1 Covariant and T12.6.3A Differentiation of Tensor: Rank-1 Contravariant	$\sum_i A^{i;j}\,\mathbf{b}_i = \sum_i B^{i;j}\,\mathbf{b}_i + C^{i;j}\,\mathbf{b}_i$	$\sum_i A_{i;j}\,\mathbf{b}^i = \sum_i B_{i;j}\,\mathbf{b}^i + C_{i;j}\,\mathbf{b}^i$
5	From 4, T7.10.9 Distribution of Vector over Addition of Scalars and T7.11. Vector Addition: Equivalency by Matching Components	$A^{i;j} = B^{i;j} + C^{i;j}$	$A_{i;j} = B_{i;j} + C_{i;j}$
∴	From 3 and 5	$\mathbf{A}_{;j} = \mathbf{B}_{;j} + \mathbf{C}_{;j}$ EQ A $A^{i;j} = B^{i;j} + C^{i;j}$ EQ C	$\mathbf{A}^{;j} = \mathbf{B}^{;j} + \mathbf{C}^{;j}$ EQ B $A_{i;j} = B_{i;j} + C_{i;j}$ EQ D

Table 12.6.1 Differentiation of Covariant and Contravariant Vectors

	Equation A	Equation B	Theorem
Covariant	$\mathbf{A}^{;j} = \sum_i A^{i;j}\,\mathbf{b}_i$	$\mathbf{A}^{;j} = \sum_i (\,A^{i,j} + \sum_k A^k\,\Gamma_{kj}{}^{i}\,)\,\mathbf{b}_i$	T12.6.2
Contravariant	$\mathbf{A}_{;j} = \sum_i A_{i;j}\,\mathbf{b}^i$	$\mathbf{A}_{;j} = \sum_i (\,A_{i,j} - \sum_k A_k\,\Gamma_{ij}{}^{k}\,)\,\mathbf{b}^i$	T12.6.3

Section 12.7 Tensor Chain Rule: Binary Product

Definition 12.7.1 **Differential Tensor Chain Rule Placeholder**

The tensor expands with differentiation about the indices of the tensor the differential index slides along the set of indices marking the position of the index being differentiated. This requires a special index operator called the ***Differential Tensor Chain Rule Placeholder***.

$$\tau[j, s, k] \equiv \begin{cases} \lambda[j,k] = s & \text{for } i = j \\ \lambda[i,k] \neq s & \text{for } i \neq j \end{cases} \qquad \text{Equation A}$$

Theorem 12.7.1 Tensor Uniqueness: Tensor Differentiation

Step	Hypothesis	$(\overset{*}{A}_p)^{;j} = (\overset{*}{B}_p)^{;j}$ EQ A	$(\overset{*}{A}^p)_{;j} = (\overset{*}{B}^p)_{;j}$ EQ B
1	From A10.2.6 Equality of Tensors	$\overset{*}{A}_m = \overset{*}{B}_m$	$\overset{*}{A}^m = \overset{*}{B}^m$
2	From 1, AK.5.3B Uniqueness of Differentiation of a Vector, D12.5.4 (A, B) Simple Differental VectorNotation and O12.6.1R3 Rules for Covariant and Contravariant Differential Notation	$(\overset{*}{A}_m)^{,j} = (\overset{*}{B}_m)^{,j}$	$(\overset{*}{A}^m)_{,j} = (\overset{*}{B}^m)_{,j}$
∴	From 2, T12.6.3A Differentiation of Tensor: Rank-1 Contravariant, T12.6.2A Differentiation of Tensor: Rank-1 Covariant and O12.6.1R2 Rules for Covariant and Contravariant Differential Notation	$(\overset{*}{A}_m)^{;j} = (\overset{*}{B}_m)^{;j}$ EQ A	$(\overset{*}{A}^m)_{;j} = (\overset{*}{B}^m)_{;j}$ EQ B

Observation 12.7.1 Tensor as a Mathaematical Quantity

A tensor is a mathaematical quantity that is a function of the manifold's coordinates, $\overset{*}{A}_m(\bullet x_i \bullet)$, A10.2.1P1B "Tensor of Rank–m".

Theorem 12.7.2 Tensor Summation: Covariant Derivative as a Linear Operator

1g	Given	$\overset{*}{C}_m(\bullet x_i \bullet) = \overset{*}{A}_m(\bullet x_i \bullet) + \overset{*}{B}_m(\bullet x_i \bullet)$
2g		$\Delta\overset{*}{C}_m \equiv \overset{*}{C}_m(\bullet x_i + \Delta x_i \bullet) - \overset{*}{C}_m(\bullet x_i \bullet)$
3g		$\Delta\overset{*}{A}_m \equiv \overset{*}{A}_m(\bullet x_i + \Delta x_i \bullet) - \overset{*}{A}_m(\bullet x_i \bullet)$
4g		$\Delta\overset{*}{B}_m \equiv \overset{*}{B}_m(\bullet x_i + \Delta x_i \bullet) - \overset{*}{B}_m(\bullet x_i \bullet)$
Step	Hypothesis	$[\overset{*}{A}_m(\bullet x_i \bullet) + \overset{*}{B}_m(\bullet x_i \bullet)]_{;i} = \overset{*}{A}_m(\bullet x_i \bullet)_{;i} + \overset{*}{B}_m(\bullet x_i \bullet)_{;i}$
1	From 1g and adding in a small perturbation difference factor Δx_i on the $[x_i]$ axis	$\overset{*}{C}_m(\bullet x_i + \Delta x_i \bullet) = \overset{*}{A}_m(\bullet x_i + \Delta x_i \bullet) + \overset{*}{B}_m(\bullet x_i + \Delta x_i \bullet)$
2	From 1g, T10.4.1 Scalars and Tensors: Uniqueness of Scalar Multiplication to a Tensors and T10.4.6 Scalars and Tensors: Distribution of a Scalar over Addition of Tensors and D4.1.20A Negative Coefficient	$-\overset{*}{C}_m(\bullet x_i \bullet) = -\overset{*}{A}_m(\bullet x_i \bullet) - \overset{*}{B}_m(\bullet x_i \bullet)$
3	From 1g, 1 and T10.3.15 Addition of Tensors: Uniqueness by Addition	$\overset{*}{C}_m(\bullet x_i + \Delta x_i \bullet) - \overset{*}{C}_m(\bullet x_i \bullet) = [\overset{*}{A}_m(\bullet x_i + \Delta x_i \bullet) - \overset{*}{A}_m(\bullet x_i \bullet)]$ $+ [\overset{*}{B}_m(\bullet x_i + \Delta x_i \bullet) - \overset{*}{B}_m(\bullet x_i \bullet)]$
4	From 2g, 3g, 4g, 3 and T10.2.5 Tool: Tensor Substitution	$\Delta\overset{*}{C}_m = \Delta\overset{*}{A}_m + \Delta\overset{*}{B}_m$
5	From 4 and T10.4.1 Scalars and Tensors: Uniqueness of Scalar Multiplication to a Tensors	$\Delta\overset{*}{C}_m / \Delta x_i = (\Delta\overset{*}{A}_m + \Delta\overset{*}{B}_m) / \Delta x_i$
6	From 5 and T10.4.6 Scalars and Tensors: Distribution of a Scalar over Addition of Tensors	$\Delta\overset{*}{C}_m / \Delta x_i = \Delta\overset{*}{A}_m / \Delta x_i + \Delta\overset{*}{B}_m / \Delta x_i$
7	From 6, LxK.2.1.1 Uniqueness of Limits and LxK.2.1.4 Limit of the Sum	$\lim_{\Delta xi \to 0} \Delta\overset{*}{C}_m / \Delta x_i = \lim_{\Delta xi \to 0} \Delta\overset{*}{A}_m / \Delta x_i + \lim_{\Delta xi \to 0} \Delta\overset{*}{B}_m / \Delta x_i$
8	From 7, DxK.3.1.3 Definition of a Derivative Alternate Form and T12.6.2A Differentiation of Tensor: Rank-1 Covariant	$\overset{*}{C}_m(\bullet x_i \bullet)_{;i} = \overset{*}{A}_m(\bullet x_i \bullet)_{;i} + \overset{*}{B}_m(\bullet x_i \bullet)_{;i}$
∴	From 1g, 8 and T10.2.5 Tool: Tensor Substitution	$[\overset{*}{A}_m(\bullet x_i \bullet) + \overset{*}{B}_m(\bullet x_i \bullet)]_{;i} = \overset{*}{A}_m(\bullet x_i \bullet)_{;i} + \overset{*}{B}_m(\bullet x_i \bullet)_{;i}$

Theorem 12.7.3 Tensor Summation: Contravariant Derivative as a Linear Operator

1g	Given	$\overset{*}{C}^m(\bullet x_i \bullet) = \overset{*}{A}^m(\bullet x_i \bullet) + \overset{*}{B}^m(\bullet x_i \bullet)$
2g		$\Delta\overset{*}{C}^m \equiv \overset{*}{C}^m(\bullet x_i + \Delta x_i \bullet) - \overset{*}{C}^m(\bullet x_i \bullet)$
3g		$\Delta\overset{*}{A}^m \equiv \overset{*}{A}^m(\bullet x_i + \Delta x_i \bullet) - \overset{*}{A}^m(\bullet x_i \bullet)$
4g		$\Delta\overset{*}{B}^m \equiv \overset{*}{B}^m(\bullet x_i + \Delta x_i \bullet) - \overset{*}{B}^m(\bullet x_i \bullet)$
Step	Hypothesis	$[\overset{*}{A}^m(\bullet x_i \bullet) + \overset{*}{B}^m(\bullet x_i \bullet)]^{;i} = \overset{*}{A}^m(\bullet x_i \bullet)^{;i} + \overset{*}{B}^m(\bullet x_i \bullet)^{;i}$
1	From 1g and adding in a small perturbation difference factor Δx_i on the $[x_i]$ axis	$\overset{*}{C}^m(\bullet x_i + \Delta x_i \bullet) = \overset{*}{A}^m(\bullet x_i + \Delta x_i \bullet) + \overset{*}{B}^m(\bullet x_i + \Delta x_i \bullet)$
2	From 1g, T10.4.1 Scalars and Tensors: Uniqueness of Scalar Multiplication to a Tensors and T10.4.6 Scalars and Tensors: Distribution of a Scalar over Addition of Tensors and D4.1.20A Negative Coefficient	$-\overset{*}{C}^m(\bullet x_i \bullet) = -\overset{*}{A}^m(\bullet x_i \bullet) - \overset{*}{B}^m(\bullet x_i \bullet)$
3	From 1g, 1 and T10.3.15 Addition of Tensors: Uniqueness by Addition	$\overset{*}{C}^m(\bullet x_i + \Delta x_i \bullet) - \overset{*}{C}^m(\bullet x_i \bullet) = [\overset{*}{A}^m(\bullet x_i + \Delta x_i \bullet) - \overset{*}{A}^m(\bullet x_i \bullet)] + [\overset{*}{B}^m(\bullet x_i + \Delta x_i \bullet) - \overset{*}{B}^m(\bullet x_i \bullet)]$
4	From 2g, 3g, 4g, 3 and T10.2.5 Tool: Tensor Substitution	$\Delta\overset{*}{C}^m = \Delta\overset{*}{A}^m + \Delta\overset{*}{B}^m$
5	From 4 and T10.4.1 Scalars and Tensors: Uniqueness of Scalar Multiplication to a Tensors	$\Delta\overset{*}{C}^m / \Delta x_i = (\Delta\overset{*}{A}^m + \Delta\overset{*}{B}^m) / \Delta x_i$
6	From 5 and T10.4.6 Scalars and Tensors: Distribution of a Scalar over Addition of Tensors	$\Delta\overset{*}{C}^m / \Delta x_i = \Delta\overset{*}{A}^m / \Delta x_i + \Delta\overset{*}{B}^m / \Delta x_i$
7	From 6, LxK.2.1.1 Uniqueness of Limits and LxK.2.1.4 Limit of the Sum	$\lim_{\Delta xi \to 0} \Delta\overset{*}{C}^m / \Delta x_i = \lim_{\Delta xi \to 0} \Delta\overset{*}{A}^m / \Delta x_i + \lim_{\Delta xi \to 0} \Delta\overset{*}{B}^m / \Delta x_i$
8	From 7, DxK.3.1.3 Definition of a Derivative Alternate Form and T12.6.3A Differentiation of Tensor: Rank-1 Contravariant	$\overset{*}{C}^m(\bullet x_i \bullet)^{;i} = \overset{*}{A}^m(\bullet x_i \bullet)^{;i} + \overset{*}{B}^m(\bullet x_i \bullet)^{;i}$
∴	From 1g, 8 and T10.2.5 Tool: Tensor Substitution	$[\overset{*}{A}^m(\bullet x_i \bullet) + \overset{*}{B}^m(\bullet x_i \bullet)]^{;i} = \overset{*}{A}^m(\bullet x_i \bullet)^{;i} + \overset{*}{B}^m(\bullet x_i \bullet)^{;i}$

Theorem 12.7.4 Tensor Product: Tensor Series Expansion Distributing Covariant Derivative

Step	Hypothesis	$(\sum_{\rho(p)\in Lm} \prod_i^m A^{i\lambda[i,p]}\, \mathbf{b}_{i\lambda[i,p]})^{;k} = \sum_{\rho(p)\in Lm} (\prod_i^m A^{i\lambda[i,p]})^{;k} \prod_i^m \mathbf{b}_{i\lambda[i,p]}$
1	From T10.2.1 Tool: Tensor Series Expansion and O12.7.1 Tensor as a Mathaematical Quantity	$(\sum_{\rho(p)\in Lm} \prod_i^m A^{i\lambda[i,p]}\, \mathbf{b}_{i\lambda[i,p]})(\bullet\, x_k \bullet) = \sum_{\rho(p)\in Lm} (\prod_i^m A^{i\lambda[i,p]})(\bullet\, x_k \bullet)\, \mathbf{b}_{i\lambda[i,p]}$
2	From 1 and deviating by a small change in delta	$(\sum_{\rho(p)\in Lm} \prod_i^m A^{i\lambda[i,p]}\, \mathbf{b}_{i\lambda[i,p]})(\bullet\, x_k + \Delta x_k \bullet) = \sum_{\rho(p)\in Lm} (\prod_i^m A^{i\lambda[i,p]})(\bullet\, x_k + \Delta x_k \bullet)\, \mathbf{b}_{i\lambda[i,p]}$
3	From 1, 2 D4.1.20 Negative Coefficient, T10.4.1 Scalars and Tensors: Uniqueness of Scalar Multiplication to a Tensors and T10.3.15 Addition of Tensors: Uniqueness by Addition	$(\sum_{\rho(p)\in Lm} \prod_i^m A^{i\lambda[i,p]}\, \mathbf{b}_{i\lambda[i,p]})(\bullet\, x_k + \Delta x_k \bullet) - (\sum_{\rho(p)\in Lm} \prod_i^m A^{i\lambda[i,p]}\, \mathbf{b}_{i\lambda[i,p]})(\bullet\, x_k \bullet) = \sum_{\rho(p)\in Lm} (\prod_i^m A^{i\lambda[i,p]})(\bullet\, x_k + \Delta x_k \bullet)\, \mathbf{b}_{i\lambda[i,p]} - \sum_{\rho(p)\in Lm} (\prod_i^m A^{i\lambda[i,p]})(\bullet\, x_k \bullet)\, \mathbf{b}_{i\lambda[i,p]}$
4	From 3, T10.4.6 Scalars and Tensors: Distribution of a Scalar over Addition of Tensors, T10.2.4 Tool: Base Tensor Distributed over Polyadic Coefficient Summation	$(\sum_{\rho(p)\in Lm} \prod_i^m A^{i\lambda[i,p]}\, \mathbf{b}_{i\lambda[i,p]})(\bullet\, x_k + \Delta x_k \bullet) - (\sum_{\rho(p)\in Lm} \prod_i^m A^{i\lambda[i,p]}\, \mathbf{b}_{i\lambda[i,p]})(\bullet\, x_k \bullet) = \sum_{\rho(p)\in Lm} [\, (\prod_i^m A^{i\lambda[i,p]})(\bullet\, x_k + \Delta x_k \bullet) - (\prod_i^m A^{i\lambda[i,p]})(\bullet\, x_k \bullet)\,] \prod_i^m \mathbf{b}_{i\lambda[i,p]}$
5	From 4, T10.4.1 Scalars and Tensors: Uniqueness of Scalar Multiplication to a Tensors and T10.4.6 Scalars and Tensors: Distribution of a Scalar over Addition of Tensors	$[(\sum_{\rho(p)\in Lm} \prod_i^m A^{i\lambda[i,p]}\, \mathbf{b}_{i\lambda[i,p]})(\bullet\, x_k + \Delta x_k \bullet) - (\sum_{\rho(p)\in Lm} \prod_i^m A^{i\lambda[i,p]}\, \mathbf{b}_{i\lambda[i,p]})(\bullet\, x_k \bullet)] / \Delta x_k = \sum_{\rho(p)\in Lm} \{\, [\, (\prod_i^m A^{i\lambda[i,p]})(\bullet\, x_k + \Delta x_k \bullet) - (\prod_i^m A^{i\lambda[i,p]})(\bullet\, x_k \bullet)\,] / \Delta x_k \,\} \prod_i^m \mathbf{b}_{i\lambda[i,p]}$
∴	From 5, LxK.2.1.1 Uniqueness of Limits, LxK.2.1.4 Limit of the Sum, DxK.3.1.3 Definition of a Derivative Alternate Form and T12.7.1A Tensor Product: Uniqueness of Tensor Differentiation	$(\sum_{\rho(p)\in Lm} \prod_i^m A^{i\lambda[i,p]}\, \mathbf{b}_{i\lambda[i,p]})^{;k} = \sum_{\rho(p)\in Lm} (\prod_i^m A^{i\lambda[i,p]})^{;k} \prod_i^m \mathbf{b}_{i\lambda[i,p]}$

Theorem 12.7.5 Tensor Product: Tensor Series Expansion Distributing Contravariant Derivative

Step	Hypothesis	$(\sum_{\rho(p)\in Lm} \prod_i^m A_{i\lambda[i,p]} \mathbf{b}^{i\lambda[i,p]})_{;k} = \sum_{\rho(p)\in Lm} (\prod_i^m A_{i\lambda[i,p]})_{;k} \prod_i^m \mathbf{b}^{i\lambda[i,p]}$
1	From T10.2.1 Tool: Tensor Series Expansion and O12.7.1 Tensor as a Mathaematical Quantity	$(\sum_{\rho(p)\in Lm} \prod_i^m A_{i\lambda[i,p]} \mathbf{b}^{i\lambda[i,p]})(\bullet\, x_k \bullet) = \sum_{\rho(p)\in Lm} (\prod_i^m A_{i\lambda[i,p]})(\bullet\, x_k \bullet)\, \mathbf{b}^{i\lambda[i,p]}$
2	From 1 and deviating by a small change in delta	$(\sum_{\rho(p)\in Lm} \prod_i^m A_{i\lambda[i,p]} \mathbf{b}^{i\lambda[i,p]})(\bullet\, x_k + \Delta x_k \bullet) = \sum_{\rho(p)\in Lm} (\prod_i^m A_{i\lambda[i,p]})(\bullet\, x_k + \Delta x_k \bullet)\, \mathbf{b}^{i\lambda[i,p]}$
3	From 1, 2 D4.1.20 Negative Coefficient, T10.4.1 Scalars and Tensors: Uniqueness of Scalar Multiplication to a Tensors and T10.3.15 Addition of Tensors: Uniqueness by Addition	$(\sum_{\rho(p)\in Lm} \prod_i^m A_{i\lambda[i,p]} \mathbf{b}^{i\lambda[i,p]})(\bullet\, x_k + \Delta x_k \bullet) - (\sum_{\rho(p)\in Lm} \prod_i^m A_{i\lambda[i,p]} \mathbf{b}^{i\lambda[i,p]})(\bullet\, x_k \bullet) = \sum_{\rho(p)\in Lm} (\prod_i^m A_{i\lambda[i,p]})(\bullet\, x_k + \Delta x_k \bullet)\, \mathbf{b}^{i\lambda[i,p]} - \sum_{\rho(p)\in Lm} (\prod_i^m A_{i\lambda[i,p]})(\bullet\, x_k \bullet)\, \mathbf{b}^{i\lambda[i,p]}$
4	From 3, T10.4.6 Scalars and Tensors: Distribution of a Scalar over Addition of Tensors, T10.2.4 Tool: Base Tensor Distributed over Polyadic Coefficient Summation	$(\sum_{\rho(p)\in Lm} \prod_i^m A_{i\lambda[i,p]} \mathbf{b}^{i\lambda[i,p]})(\bullet\, x_k + \Delta x_k \bullet) - (\sum_{\rho(p)\in Lm} \prod_i^m A_{i\lambda[i,p]} \mathbf{b}^{i\lambda[i,p]})(\bullet\, x_k \bullet) = \sum_{\rho(p)\in Lm} [\, (\prod_i^m A_{i\lambda[i,p]})(\bullet\, x_k + \Delta x_k \bullet) - (\prod_i^m A_{i\lambda[i,p]})(\bullet\, x_k \bullet)\,]\, \prod_i^m \mathbf{b}^{i\lambda[i,p]}$
5	From 4, T10.4.1 Scalars and Tensors: Uniqueness of Scalar Multiplication to a Tensors and T10.4.6 Scalars and Tensors: Distribution of a Scalar over Addition of Tensors	$[(\sum_{\rho(p)\in Lm} \prod_i^m A_{i\lambda[i,p]} \mathbf{b}^{i\lambda[i,p]})(\bullet\, x_k + \Delta x_k \bullet) - (\sum_{\rho(p)\in Lm} \prod_i^m A_{i\lambda[i,p]} \mathbf{b}^{i\lambda[i,p]})(\bullet\, x_k \bullet)] / \Delta x_k = \sum_{\rho(p)\in Lm} \{\, [\, (\prod_i^m A_{i\lambda[i,p]})(\bullet\, x_k + \Delta x_k \bullet) - (\prod_i^m A_{i\lambda[i,p]})(\bullet\, x_k \bullet)\,] / \Delta x_k \,\}\, \prod_i^m \mathbf{b}^{i\lambda[i,p]}$
∴	From 5, LxK.2.1.1 Uniqueness of Limits, LxK.2.1.4 Limit of the Sum, DxK.3.1.3 Definition of a Derivative Alternate Form and T12.7.1B Tensor Product: Uniqueness of Tensor Differentiation	$(\sum_{\rho(p)\in Lm} \prod_i^m A_{i\lambda[i,p]} \mathbf{b}^{i\lambda[i,p]})_{;k} = \sum_{\rho(p)\in Lm} (\prod_i^m A_{i\lambda[i,p]})_{;k} \prod_i^m \mathbf{b}^{i\lambda[i,p]}$

Theorem 12.7.6 Tensor Product: Chain Rule for Binary Covariant Product

1g	Given	$\overset{*}{C}_p = \overset{*}{A}_q(\bullet x_i \bullet)\, \overset{*}{B}_r(\bullet x_i \bullet)$ rank $p = q + r$
2g		$\Delta\overset{*}{C}_p \equiv \overset{*}{C}_p(\bullet x_i + \Delta x_i \bullet) - \overset{*}{C}_p(\bullet x_i \bullet)$
3g		$\Delta\overset{*}{A}_q \equiv \overset{*}{A}_q(\bullet x_i + \Delta x_i \bullet) - \overset{*}{A}_q(\bullet x_i \bullet)$
4g		$\Delta\overset{*}{B}_r \equiv \overset{*}{B}_r(\bullet x_i + \Delta x_i \bullet) - \overset{*}{B}_r(\bullet x_i \bullet)$
Step	Hypothesis	$(\overset{*}{A}_q\, \overset{*}{B}_r)_{;i} = \overset{*}{A}_{q;i}\, \overset{*}{B}_r + \overset{*}{B}_{r;i}\, \overset{*}{A}_q$ for rank $p = q + r$
1	From 2g	$\Delta\overset{*}{C}_p = \overset{*}{C}_p(\bullet x_i + \Delta x_i \bullet) - \overset{*}{C}_p(\bullet x_i \bullet)$
2	From 1g, 1 and T10.2.5 Tool: Tensor Substitution	$\Delta\overset{*}{C}_p = \overset{*}{A}_q(\bullet x_i + \Delta x_i \bullet)\, \overset{*}{B}_r(\bullet x_i + \Delta x_i \bullet) - \overset{*}{A}_q(\bullet x_i \bullet)\, \overset{*}{B}_r(\bullet x_i \bullet)$
3	From 2, T10.3.18 Addition of Tensors: Identity by Addition, T10.3.19 Addition of Tensors: Inverse by Addition and T10.3.16 Addition of Tensors: Commutative by Addition	$\Delta\overset{*}{C}_p = [\overset{*}{A}_q(\bullet x_i \bullet) + \overset{*}{A}_q(\bullet x_i + \Delta x_i \bullet) - \overset{*}{A}_q(\bullet x_i \bullet)]$ $[\overset{*}{B}_r(\bullet x_i \bullet) + \overset{*}{B}_r(\bullet x_i + \Delta x_i \bullet) - \overset{*}{B}_r(\bullet x_i \bullet)]$ $- \overset{*}{A}_q(\bullet x_i \bullet)\, \overset{*}{B}_r(\bullet x_i \bullet)$
4	From 3g, 4g, 3 and T10.2.5 Tool: Tensor Substitution	$\Delta\overset{*}{C}_p = [\overset{*}{A}_q + \Delta\overset{*}{A}_q]\, [\overset{*}{B}_r + \Delta\overset{*}{B}_r] - \overset{*}{A}_q\, \overset{*}{B}_r$
5	From 4 and T10.3.20 Addition of Tensors: Distribution of a Tensor over Addition of Tensors	$\Delta\overset{*}{C}_p = [\overset{*}{A}_q\overset{*}{B}_r + \Delta\overset{*}{A}_q\overset{*}{B}_r + \Delta\overset{*}{B}_r\overset{*}{A}_q + \Delta\overset{*}{B}_r\Delta\overset{*}{A}_q] - \overset{*}{A}_q\, \overset{*}{B}_r$
6	From 5, T10.3.16 Addition of Tensors: Commutative by Addition, T10.3.19 Addition of Tensors: Inverse by Addition and T10.3.18 Addition of Tensors: Identity by Addition	$\Delta\overset{*}{C}_p = \overset{*}{A}_q\overset{*}{B}_r + \Delta\overset{*}{A}_q\overset{*}{B}_r + \Delta\overset{*}{B}_r\overset{*}{A}_q + \Delta\overset{*}{B}_r\Delta\overset{*}{A}_q$
7	From 6, T10.4.1 Scalars and Tensors: Uniqueness of Scalar Multiplication to a Tensors, T10.4.6 Scalars and Tensors: Distribution of a Scalar over Addition of Tensors and A10.2.15 Associate Scalar with Tensor Group	$\Delta\overset{*}{C}_p / \Delta x_i = (\Delta\overset{*}{A}_q / \Delta x_i)\, \overset{*}{B}_r + (\Delta\overset{*}{B}_r / \Delta x_i)\, \overset{*}{A}_q + (\Delta\overset{*}{B}_r / \Delta x_i)\, \Delta\overset{*}{A}_q$

8	From 7, DxK.3.1.3 Definition of a Derivative Alternate Form and DxK.3.2.2B Multi-variable Partial Notation	$\lim_{\Delta xi \to 0} \dfrac{\Delta \overset{*}{C}_p}{\Delta x_i} = \dfrac{\partial \overset{*}{C}_p}{\partial x_i} = \overset{*}{C}_{p;i}$
9	From 7, DxK.3.1.3 Definition of a Derivative Alternate Form and DxK.3.2.2B Multi-variable Partial Notation	$\lim_{\Delta xi \to 0} \dfrac{\Delta \overset{*}{A}_q}{\Delta x_i} = \dfrac{\partial \overset{*}{A}_q}{\partial x_i} = \overset{*}{A}_{q;i}$
10	From 7, DxK.3.1.3 Definition of a Derivative Alternate Form and DxK.3.2.2B Multi-variable Partial Notation	$\lim_{\Delta xi \to 0} \dfrac{\Delta \overset{*}{B}_r}{\Delta x_i} = \dfrac{\partial \overset{*}{B}_r}{\partial x_i} = \overset{*}{B}_{r;i}$
11	From 3g, 7 and T10.3.19 Addition of Tensors: Inverse by Addition	$\lim_{\Delta xi \to 0} \Delta \overset{*}{A}_q = \mathbf{0}$
12	From 7, 8, 9, 10, 11 and T10.2.5 Tool: Tensor Substitution	$\overset{*}{C}_{p;i} = \overset{*}{A}_{q;i} \overset{*}{B}_r + \overset{*}{B}_{r;i} \overset{*}{A}_q + \overset{*}{B}_{r;i} \mathbf{0}$
∴	From 1g, 12, T10.2.8 Tool: Zero Tensor Times a Tensor is a Zero Tensor, T10.3.18 Addition of Tensors: Identity by Addition and T10.2.5 Tool: Tensor Substitution	$(\overset{*}{A}_q \overset{*}{B}_r)_{;i} = \overset{*}{A}_{q;i} \overset{*}{B}_r + \overset{*}{B}_{r;i} \overset{*}{A}_q$ for rank $p = q + r$

Theorem 12.7.7 Tensor Product: Chain Rule for Contravariant Binary Product

1g	Given	$\overset{*}{C}^p = \overset{*}{A}^q(\bullet x_i \bullet)\, \overset{*}{B}^r(\bullet x_i \bullet)$ rank $p = q + r$
2g		$\Delta\overset{*}{C}^p \equiv \overset{*}{C}^p(\bullet x_i + \Delta x_i \bullet) - \overset{*}{C}^p(\bullet x_i \bullet)$
3g		$\Delta\overset{*}{A}^q \equiv \overset{*}{A}^q(\bullet x_i + \Delta x_i \bullet) - \overset{*}{A}^q(\bullet x_i \bullet)$
4g		$\Delta\overset{*}{B}^r \equiv \overset{*}{B}^r(\bullet x_i + \Delta x_i \bullet) - \overset{*}{B}^r(\bullet x_i \bullet)$
Step	Hypothesis	$(\overset{*}{A}^q \overset{*}{B}^r)^{;i} = \overset{*}{A}^{q;i} \overset{*}{B}^r + \overset{*}{B}^{r;i} \overset{*}{A}^q$ for rank $p = q + r$
1	From 2g	$\Delta\overset{*}{C}^p = \overset{*}{C}^p(\bullet x_i + \Delta x_i \bullet) - \overset{*}{C}_p(\bullet x_i \bullet)$
2	From 1g, 1 and T10.2.5 Tool: Tensor Substitution	$\Delta\overset{*}{C}^p = \overset{*}{A}^q(\bullet x_i + \Delta x_i \bullet)\, \overset{*}{B}^r(\bullet x_i + \Delta x_i \bullet) - \overset{*}{A}^q(\bullet x_i \bullet)\, \overset{*}{B}^r(\bullet x_i \bullet)$
3	From 2, T10.3.18 Addition of Tensors: Identity by Addition, T10.3.19 Addition of Tensors: Inverse by Addition and T10.3.16 Addition of Tensors: Commutative by Addition	$\Delta\overset{*}{C}^p = [\overset{*}{A}^q(\bullet x_i \bullet) + \overset{*}{A}^q(\bullet x_i + \Delta x_i \bullet) - \overset{*}{A}^q(\bullet x_i \bullet)]$ $[\overset{*}{B}^r(\bullet x_i \bullet) + \overset{*}{B}^r(\bullet x_i + \Delta x_i \bullet) - \overset{*}{B}^r(\bullet x_i \bullet)]$ $- \overset{*}{A}^q(\bullet x_i \bullet)\, \overset{*}{B}^r(\bullet x_i \bullet)$
4	From 3g, 4g, 3 and T10.2.5 Tool: Tensor Substitution	$\Delta\overset{*}{C}^p = [\overset{*}{A}^q + \Delta\overset{*}{A}^q]\, [\overset{*}{B}^r + \Delta\overset{*}{B}^r] - \overset{*}{A}^q \overset{*}{B}^r$
5	From 4 and T10.3.20 Addition of Tensors: Distribution of a Tensor over Addition of Tensors	$\Delta\overset{*}{C}^p = [\overset{*}{A}^q \overset{*}{B}^r + \Delta\overset{*}{A}^q \overset{*}{B}^r + \Delta\overset{*}{B}^r \overset{*}{A}^q + \Delta\overset{*}{B}^r \Delta\overset{*}{A}^q] - \overset{*}{A}^q \overset{*}{B}^r$
6	From 5, T10.3.16 Addition of Tensors: Commutative by Addition, T10.3.19 Addition of Tensors: Inverse by Addition and T10.3.18 Addition of Tensors: Identity by Addition	$\Delta\overset{*}{C}^p = \overset{*}{A}^q \overset{*}{B}^r + \Delta\overset{*}{A}^q \overset{*}{B}^r + \Delta\overset{*}{B}^r \overset{*}{A}^q + \Delta\overset{*}{B}^r \Delta\overset{*}{A}^q$

7	From 6, T10.4.1 Scalars and Tensors: Uniqueness of Scalar Multiplication to a Tensors, T10.4.6 Scalars and Tensors: Distribution of a Scalar over Addition of Tensors and A10.2.15 Associate Scalar with Tensor Group	$\Delta\overset{*}{C}{}^{p} / \Delta x_i = (\Delta\overset{*}{A}{}^{q} / \Delta x_i)\, \overset{*}{B}{}^{r} + (\Delta\overset{*}{B}{}^{r} / \Delta x_i)\, \overset{*}{A}{}^{q} + (\Delta\overset{*}{B}{}^{r} / \Delta x_i)\, \overset{*}{A}{}^{q}$
8	From 7, DxK.3.1.3 Definition of a Derivative Alternate Form and DxK.3.2.2B Multi-variable Partial Notation	$\lim_{\Delta xi \to 0} \dfrac{\Delta\overset{*}{C}{}^{p}}{\Delta x_i} = \dfrac{\partial\overset{*}{C}{}^{p}}{\partial x_i} = \overset{*}{C}{}^{p;i}$
9	From 7, DxK.3.1.3 Definition of a Derivative Alternate Form and DxK.3.2.2B Multi-variable Partial Notation	$\lim_{\Delta xi \to 0} \dfrac{\Delta\overset{*}{A}{}^{q}}{\Delta x_i} = \dfrac{\partial\overset{*}{A}{}^{q}}{\partial x_i} = \overset{*}{A}{}^{q;i}$
10	From 7, DxK.3.1.3 Definition of a Derivative Alternate Form and DxK.3.2.2B Multi-variable Partial Notation	$\lim_{\Delta xi \to 0} \dfrac{\Delta\overset{*}{B}{}^{r}}{\Delta x_i} = \dfrac{\partial\overset{*}{B}{}^{r}}{\partial x_i} = \overset{*}{B}{}^{r;i}$
11	From 3g, 7 and T10.3.19 Addition of Tensors: Inverse by Addition	$\lim_{\Delta xi \to 0} \Delta\overset{*}{A}{}^{q} = \mathbf{0}$
12	From 7, 8, 9, 10, 11 and T10.2.5 Tool: Tensor Substitution	$\overset{*}{C}{}^{p;i} = \overset{*}{A}{}^{q;i}\, \overset{*}{B}{}^{r} + \overset{*}{B}{}^{r;i}\, \overset{*}{A}{}^{q} + \overset{*}{B}{}^{r;i}\, \mathbf{0}$
∴	From 1g, 12, T10.2.8 Tool: Zero Tensor Times a Tensor is a Zero Tensor, T10.3.18 Addition of Tensors: Identity by Addition and T10.2.5 Tool: Tensor Substitution	$(\overset{*}{A}{}^{q}\, \overset{*}{B}{}^{r})^{;i} = \overset{*}{A}{}^{q;i}\, \overset{*}{B}{}^{r} + \overset{*}{B}{}^{r;i}\, \overset{*}{A}{}^{q}$ for rank p = q + r

Theorem 12.7.8 Tensor Product: Chain Rule for Mixed Binary Product

1g	Given	$\overset{*}{C}_q{}^r = \overset{*}{A}_q(\,\bullet x_i\bullet\,)\,\overset{*}{B}^r(\,\bullet x_i\bullet\,)$
2g		$\Delta\overset{*}{C}_q{}^r \equiv \overset{*}{C}_q{}^r(\,\bullet x_i + \Delta x_i\,\bullet\,) - \overset{*}{C}_q{}^r(\,\bullet x_i\bullet\,)$
3g		$\Delta\overset{*}{A}_q \equiv \overset{*}{A}_q(\,\bullet x_i + \Delta x_i\,\bullet\,) - \overset{*}{A}_q(\,\bullet x_i\bullet\,)$
4g		$\Delta\overset{*}{B}^r \equiv \overset{*}{B}^r(\,\bullet x_i + \Delta x_i\,\bullet\,) - \overset{*}{B}^r(\,\bullet x_i\bullet\,)$
Step	Hypothesis	$(\overset{*}{A}_q\,\overset{*}{B}^r)_{;i} = \overset{*}{A}_{q;i}\,\overset{*}{B}^r + \overset{*}{B}^{r;i}\,\overset{*}{A}_q$
1	From 2g	$\Delta\overset{*}{C}_q{}^r = \overset{*}{C}_q{}^r(\,\bullet x_i + \Delta x_i\,\bullet\,) - \overset{*}{C}_q{}^r(\,\bullet x_i\bullet\,)$
2	From 1g, 1 and T10.2.5 Tool: Tensor Substitution	$\Delta\overset{*}{C}_q{}^r = \overset{*}{A}_q(\,\bullet x_i + \Delta x_i\,\bullet\,)\,\overset{*}{B}^r(\,\bullet x_i + \Delta x_i\,\bullet\,) - \overset{*}{A}_q(\,\bullet x_i\bullet\,)\,\overset{*}{B}^r(\,\bullet x_i\bullet\,)$
3	From 2, T10.3.18 Addition of Tensors: Identity by Addition, T10.3.19 Addition of Tensors: Inverse by Addition and T10.3.16 Addition of Tensors: Commutative by Addition	$\Delta\overset{*}{C}_q{}^r = [\overset{*}{A}_q(\,\bullet x_i\bullet\,) + \overset{*}{A}_q(\,\bullet x_i + \Delta x_i\,\bullet\,) - \overset{*}{A}_q(\,\bullet x_i\bullet\,)\,]$ $[\overset{*}{B}^r(\,\bullet x_i\bullet\,) + \overset{*}{B}^r(\,\bullet x_i + \Delta x_i\,\bullet\,) - \overset{*}{B}^r(\,\bullet x_i\bullet\,)]$ $- \overset{*}{A}_q(\,\bullet x_i\bullet\,)\,\overset{*}{B}^r(\,\bullet x_i\bullet\,)$
4	From 3g, 4g, 3 and T10.2.5 Tool: Tensor Substitution	$\Delta\overset{*}{C}_q{}^r = [\overset{*}{A}_q + \Delta\overset{*}{A}_q\,]\,[\overset{*}{B}^r + \Delta\overset{*}{B}^r\,] - \overset{*}{A}_q\,\overset{*}{B}^r$
5	From 4 and T10.3.20 Addition of Tensors: Distribution of a Tensor over Addition of Tensors	$\Delta\overset{*}{C}_q{}^r = [\overset{*}{A}_q\overset{*}{B}^r + \Delta\overset{*}{A}_q\overset{*}{B}^r + \Delta\overset{*}{B}^r\overset{*}{A}_q + \Delta\overset{*}{B}^r\Delta\overset{*}{A}_q\,] - \overset{*}{A}_q\,\overset{*}{B}^r$
6	From 5, T10.3.16 Addition of Tensors: Commutative by Addition, T10.3.19 Addition of Tensors: Inverse by Addition and T10.3.18 Addition of Tensors: Identity by Addition	$\Delta\overset{*}{C}_q{}^r = \overset{*}{A}_q\overset{*}{B}^r + \Delta\overset{*}{A}_q\overset{*}{B}^r + \Delta\overset{*}{B}^r\overset{*}{A}_q + \Delta\overset{*}{B}^r\Delta\overset{*}{A}_q$
7	From 6, T10.4.1 Scalars and Tensors: Uniqueness of Scalar Multiplication to a Tensors, T10.4.6 Scalars and Tensors: Distribution of a Scalar over Addition of Tensors and A10.2.15 Associate Scalar with Tensor Group	$\Delta\overset{*}{C}_q{}^r / \Delta x_i = (\Delta\overset{*}{A}_q / \Delta x_i\,)\,\overset{*}{B}^r + (\Delta\overset{*}{B}^r / \Delta x_i\,)\,\overset{*}{A}_q + (\Delta\overset{*}{B}^r / \Delta x_i\,)\,\Delta\overset{*}{A}_q$
8	From 7, DxK.3.1.3 Definition of a Derivative Alternate Form and DxK.3.2.2B Multi-variable Partial Notation	$\lim_{\Delta xi \to 0} \dfrac{\Delta\overset{*}{C}_q{}^r}{\Delta x_i} = \dfrac{\partial\,\overset{*}{C}_q{}^r}{\partial x_i} = \Delta\overset{*}{C}_q{}^r{}_{;i}$
9	From 7, DxK.3.1.3 Definition of a Derivative Alternate Form and DxK.3.2.2B Multi-variable Partial Notation	$\lim_{\Delta xi \to 0} \dfrac{\Delta\overset{*}{A}_q}{\Delta x_i} = \dfrac{\partial\overset{*}{A}_q}{\partial x_i} = \overset{*}{A}_{q;i}$
10	From 7, DxK.3.1.3 Definition of a Derivative Alternate Form and DxK.3.2.2B Multi-variable Partial Notation	$\lim_{\Delta xi \to 0} \dfrac{\Delta\overset{*}{B}^r}{\Delta x_i} = \dfrac{\partial\overset{*}{B}^r}{\partial x_i} = \overset{*}{B}^{r;i}$
11	From 3g, 7 and T10.3.19 Addition of Tensors: Inverse by Addition	$\lim_{\Delta xi \to 0} \Delta\overset{*}{A}_q = \mathbf{0}$

12	From 7, 8, 9, 10, 11 and T10.2.5 Tool: Tensor Substitution	$(\overset{*}{\mathring{C}}_q{}^r)_{;i} = \overset{*}{\mathring{A}}_{q;i}\, \overset{*}{\mathring{B}}^r + \overset{*}{\mathring{B}}^{r;i}\, \overset{*}{\mathring{A}}_q + \overset{*}{\mathring{B}}^{r;i}\, \mathbf{0}$
∴	From 1g, 12, T10.2.8 Tool: Zero Tensor Times a Tensor is a Zero Tensor, T10.3.18 Addition of Tensors: Identity by Addition and T10.2.5 Tool: Tensor Substitution	$(\overset{*}{\mathring{A}}_q\, \overset{*}{\mathring{B}}^r)_{;i} = \overset{*}{\mathring{A}}_{q;i}\, \overset{*}{\mathring{B}}^r + \overset{*}{\mathring{B}}^{r;i}\, \overset{*}{\mathring{A}}_q$

Theorem 12.7.9 Tensor Product: Covariant Chain Rule

Step	Hypothesis	$\overset{*}{\mathring{A}}_m{}^{;k} = \mathbf{A}_1{}^{;k}\,(\prod_{i=2}{}^m \mathbf{A}_i) + \sum_{j=2}{}^{m-1} (\prod_{i=1}{}^{j-1} \mathbf{A}_i)\, \mathbf{A}_j{}^{;k}\,(\prod_{i=j+1}{}^m \mathbf{A}_i) + (\prod_{i=1}{}^{m-1} \mathbf{A}_i)\, \mathbf{A}_m{}^{;k}$
1	From A10.2.1P5A Tensor of Rank–m and expand first factor and T12.7.1A Tensor Product: Uniqueness of Tensor Differentiation	$\overset{*}{\mathring{A}}_m{}^{;k} = (\mathbf{A}_1 \prod_{i=2}{}^m \mathbf{A}_i)^{;k}$
2	From 1 and T12.7.6 Tensor Product: Chain Rule for Binary Covariant Product	$\overset{*}{\mathring{A}}_m{}^{;k} = \mathbf{A}_1{}^{;k}\,(\prod_{i=2}{}^m \mathbf{A}_i) + \mathbf{A}_1\,(\prod_{i=2}{}^m \mathbf{A}_i)^{;k}$
3	From 2 and last term expand first factor second product	$\overset{*}{\mathring{A}}_m{}^{;k} = \mathbf{A}_1{}^{;k}\,(\prod_{i=2}{}^m \mathbf{A}_i) + \mathbf{A}_1\,(\mathbf{A}_2 \prod_{i=3}{}^m \mathbf{A}_i)^{;k}$
4	From 3 and T12.7.6 Tensor Product: Chain Rule for Binary Covariant Product	$\overset{*}{\mathring{A}}_m{}^{;k} = \mathbf{A}_1{}^{;k}\,(\prod_{i=2}{}^m \mathbf{A}_i) + \mathbf{A}_1\,[\,\mathbf{A}_2{}^{;k}\,(\prod_{i=3}{}^m \mathbf{A}_i) + \mathbf{A}_2\,(\prod_{i=3}{}^m \mathbf{A}_i)^{;k}\,]$
5	From 4 and T10.3.20 Addition of Tensors: Distribution of a Tensor over Addition of Tensors	$\overset{*}{\mathring{A}}_m{}^{;k} = \mathbf{A}_1{}^{;k}\,(\prod_{i=2}{}^m \mathbf{A}_i) + \mathbf{A}_1\mathbf{A}_2{}^{;k}\,(\prod_{i=3}{}^m \mathbf{A}_i) + \mathbf{A}_1\mathbf{A}_2\,(\prod_{i=3}{}^m \mathbf{A}_i)^{;k}$
6	From 2 and last term expand first factor second product	$\overset{*}{\mathring{A}}_m{}^{;k} = \mathbf{A}_1{}^{;k}\,(\prod_{i=2}{}^m \mathbf{A}_i) + \mathbf{A}_1\mathbf{A}_2{}^{;k}\,(\prod_{i=3}{}^m \mathbf{A}_i) + \mathbf{A}_1\mathbf{A}_2\,(\mathbf{A}_3 \prod_{i=4}{}^m \mathbf{A}_i)^{;k}$
7	From 6 and T12.7.6 Tensor Product: Chain Rule for Binary Covariant Product	$\overset{*}{\mathring{A}}_m{}^{;k} = \mathbf{A}_1{}^{;k}\,(\prod_{i=2}{}^m \mathbf{A}_i) + \mathbf{A}_1\mathbf{A}_2{}^{;k}\,(\prod_{i=3}{}^m \mathbf{A}_i) + \mathbf{A}_1\mathbf{A}_2\,[\mathbf{A}_3{}^{;k}\,(\prod_{i=4}{}^m \mathbf{A}_i) + \mathbf{A}_3\,(\prod_{i=4}{}^m \mathbf{A}_i)^{;k}\,]$
8	From 7 and T10.3.20 Addition of Tensors: Distribution of a Tensor over Addition of Tensors	$\overset{*}{\mathring{A}}_m{}^{;k} = \mathbf{A}_1{}^{;k}\,(\prod_{i=2}{}^m \mathbf{A}_i) + \mathbf{A}_1\mathbf{A}_2{}^{;k}\,(\prod_{i=3}{}^m \mathbf{A}_i) + \mathbf{A}_1\mathbf{A}_2\mathbf{A}_3{}^{;k}\,(\prod_{i=4}{}^m \mathbf{A}_i) + \mathbf{A}_1\mathbf{A}_2\mathbf{A}_3\,(\prod_{i=4}{}^m \mathbf{A}_i)^{;k}$
9	From 8 and summation is generalized. Term-1, term-2 to term-(m–1) and term-m by correspondence of counting terms	$\overset{*}{\mathring{A}}_m{}^{;k} = \mathbf{A}_1{}^{;k}\,(\prod_{i=2}{}^m \mathbf{A}_i) + \sum_{j=2}{}^{m-1} (\prod_{i=1}{}^{j-1} \mathbf{A}_i)\, \mathbf{A}_j{}^{;k}\,(\prod_{i=j+1}{}^m \mathbf{A}_i) + (\prod_{i=1}{}^{j-1} \mathbf{A}_i)\,(\prod_{i=j}{}^m \mathbf{A}_i)^{;k}$
10	From 9 and j = m for the last term, terminates the expansion for the last vector factor	$\overset{*}{\mathring{A}}_m{}^{;k} = \mathbf{A}_1{}^{;k}\,(\prod_{i=2}{}^m \mathbf{A}_i) + \sum_{j=2}{}^{m-1} (\prod_{i=1}{}^{j-1} \mathbf{A}_i)\, \mathbf{A}_j{}^{;k}\,(\prod_{i=j+1}{}^m \mathbf{A}_i) + (\prod_{i=1}{}^{m-1} \mathbf{A}_i)\, \mathbf{A}_m{}^{;k}$
∴	From 10 and T12.6.2A Differentiation of Tensor: Rank-1 Covariant	$\overset{*}{\mathring{A}}_m{}^{;k} = \mathbf{A}_1{}^{;k}\,(\prod_{i=2}{}^m \mathbf{A}_i) + \sum_{j=2}{}^{m-1} (\prod_{i=1}{}^{j-1} \mathbf{A}_i)\, \mathbf{A}_j{}^{;k}\,(\prod_{i=j+1}{}^m \mathbf{A}_i) + (\prod_{i=1}{}^{m-1} \mathbf{A}_i)\, \mathbf{A}_m{}^{;k}$

Theorem 12.7.10 Tensor Product: Contravariant Chain Rule

Step	Hypothesis	
		$\overset{*}{A}{}^{m}{}_{;k} = \mathbf{A}^{1}{}_{;k}\,(\prod_{i=2}^{m}\mathbf{A}^{i}) + \sum_{j=2}^{m-1}(\prod_{i=1}^{j-1}\mathbf{A}^{i})\,\mathbf{A}^{j}{}_{;k}\,(\prod_{i=j+1}^{m}\mathbf{A}^{i}) + (\prod_{i=1}^{m-1}\mathbf{A}^{i})\,\mathbf{A}^{m}{}_{;k}$
1	From A10.2.1P5A Tensor of Rank–m and expand first factor and T12.7.1B Tensor Product: Uniqueness of Tensor Differentiation	$\overset{*}{A}{}^{m}{}_{;k} = (\mathbf{A}^{1}\prod_{i=2}^{m}\mathbf{A}^{i})_{,k}$
2	From 1 and T12.7.7 Tensor Product: Chain Rule for Contravariant Binary Product	$\overset{*}{A}{}^{m}{}_{;k} = \mathbf{A}^{1}{}_{;k}\,(\prod_{i=2}^{m}\mathbf{A}^{i}) + \mathbf{A}^{1}\,(\prod_{i=2}^{m}\mathbf{A}^{i})_{;k}$
3	From 2 and last term expand first factor second product	$\overset{*}{A}{}^{m}{}_{;k} = \mathbf{A}^{1}{}_{;k}\,(\prod_{i=2}^{m}\mathbf{A}^{i}) + \mathbf{A}^{1}\,(\mathbf{A}^{2}\prod_{i=3}^{m}\mathbf{A}^{i})_{,k}$
4	From 3 and T12.7.7 Tensor Product: Chain Rule for Contravariant Binary Product	$\overset{*}{A}{}^{m}{}_{;k} = \mathbf{A}^{1}{}_{;k}\,(\prod_{i=2}^{m}\mathbf{A}^{i}) + \mathbf{A}^{1}\,[\,\mathbf{A}^{2}{}_{;k}\,(\prod_{i=3}^{m}\mathbf{A}^{i}) + \mathbf{A}^{2}\,(\prod_{i=3}^{m}\mathbf{A}^{i})_{,k}\,]$
5	From 4 and T10.3.20 Addition of Tensors: Distribution of a Tensor over Addition of Tensors	$\overset{*}{A}{}^{m}{}_{;k} = \mathbf{A}^{1}{}_{;k}\,(\prod_{i=2}^{m}\mathbf{A}^{i}) + \mathbf{A}^{1}\mathbf{A}^{2}{}_{;k}\,(\prod_{i=3}^{m}\mathbf{A}^{i}) + \mathbf{A}^{1}\mathbf{A}^{2}\,(\prod_{i=3}^{m}\mathbf{A}^{i})_{;k}$
6	From 2 and last term expand first factor second product	$\overset{*}{A}{}^{m}{}_{;k} = \mathbf{A}^{1}{}_{;k}\,(\prod_{i=2}^{m}\mathbf{A}^{i}) + \mathbf{A}^{1}\mathbf{A}^{2}{}_{;k}\,(\prod_{i=3}^{m}\mathbf{A}^{i}) + \mathbf{A}^{1}\mathbf{A}^{2}\,(\mathbf{A}^{3}\prod_{i=4}^{m}\mathbf{A}^{i})_{,k}$
7	From 6 and T12.7.7 Tensor Product: Chain Rule for Contravariant Binary Product	$\overset{*}{A}{}^{m}{}_{;k} = \mathbf{A}^{1}{}_{;k}\,(\prod_{i=2}^{m}\mathbf{A}^{i}) + \mathbf{A}^{1}\mathbf{A}^{2}{}_{;k}\,(\prod_{i=3}^{m}\mathbf{A}^{i}) + \mathbf{A}^{1}\mathbf{A}^{2}\,[\mathbf{A}^{3}{}_{;k}\,(\prod_{i=4}^{m}\mathbf{A}^{i}) + \mathbf{A}^{3}\,(\prod_{i=4}^{m}\mathbf{A}^{i})_{,k}\,]$
8	From 7 and T10.3.20 Addition of Tensors: Distribution of a Tensor over Addition of Tensors	$\overset{*}{A}{}^{m}{}_{;k} = \mathbf{A}^{1}{}_{;k}\,(\prod_{i=2}^{m}\mathbf{A}^{i}) + \mathbf{A}^{1}\mathbf{A}^{2}{}_{;k}\,(\prod_{i=3}^{m}\mathbf{A}^{i}) + \mathbf{A}^{1}\mathbf{A}^{2}\mathbf{A}^{3}{}_{;k}\,(\prod_{i=4}^{m}\mathbf{A}^{i}) + \mathbf{A}^{1}\mathbf{A}^{2}\mathbf{A}^{3}\,(\prod_{i=4}^{m}\mathbf{A}^{i})_{,k}$
9	From 8 and summation is generalized. Term-1, term-2 to term-(m–1) and term-m by correspondence of counting terms	$\overset{*}{A}{}^{m}{}_{;k} = \mathbf{A}^{1}{}_{;k}\,(\prod_{i=2}^{m}\mathbf{A}^{i}) + \sum_{j=2}^{m-1}(\prod_{i=1}^{j-1}\mathbf{A}^{i})\,\mathbf{A}^{j}{}_{;k}\,(\prod_{i=j+1}^{m}\mathbf{A}^{i}) + (\prod_{i=1}^{j-1}\mathbf{A}^{i})\,(\prod_{i=j}^{m}\mathbf{A}^{i})_{,k}$
10	From 9 and j = m for the last term, terminates the expansion for the last vector factor	$\overset{*}{A}{}^{m}{}_{;k} = \mathbf{A}^{1}{}_{;k}\,(\prod_{i=2}^{m}\mathbf{A}^{i}) + \sum_{j=2}^{m-1}(\prod_{i=1}^{j-1}\mathbf{A}^{i})\,\mathbf{A}^{j}{}_{;k}\,(\prod_{i=j+1}^{m}\mathbf{A}^{i}) + (\prod_{i=1}^{m-1}\mathbf{A}^{i})\,\mathbf{A}^{m}{}_{;k}$
∴	From 10 and T12.6.3A Differentiation of Tensor: Rank-1 Contravariant	$\overset{*}{A}{}^{m}{}_{;k} = \mathbf{A}^{1}{}_{;k}\,(\prod_{i=2}^{m}\mathbf{A}^{i}) + \sum_{j=2}^{m-1}(\prod_{i=1}^{j-1}\mathbf{A}^{i})\,\mathbf{A}^{j}{}_{;k}\,(\prod_{i=j+1}^{m}\mathbf{A}^{i}) + (\prod_{i=1}^{m-1}\mathbf{A}^{i})\,\mathbf{A}^{m}{}_{;k}$

Theorem 12.7.11 Tensor Product: Covariant Chain Rule for Tensor Coefficient

1g	Given	$A^{\bullet\lambda[i,k]\bullet} \equiv \prod_{i=1}^{m} A^{i\lambda[i,k]}$
2g		$A^{\bullet\tau[i,j,s]\bullet} \equiv (\prod_{i=1}^{j-1} A^{i\lambda[i,k]})\, A^{s\lambda[j,k]}\, (\prod_{i=j+1}^{m} A^{i\lambda[i,k]})$
Step	Hypothesis	$(A^{\bullet\lambda[i,k]\bullet})^{;r} = (A^{\bullet\lambda[i,k]\bullet})^{,r} + \sum_s \sum_{j=1}^{m} A^{\bullet\tau[j,s,k]\bullet}\, \Gamma_{sr}^{\lambda[j,k]}$
1	From T12.7.1B Tensor Uniqueness: Tensor Differentiation and T10.2.1 Tensor Uniqueness: Tensor Differentiation	$\overset{*}{A}_m{}^{;r} = (\sum_{\rho(k)\in Lm} \prod_i^m A^{i\lambda[i,k]}\, \mathbf{b}_{i\lambda[i,k]})^{;r}$
2	From 1 and T10.2.3 Tool: Base Tensor Parsed from Polyadic Coefficient	$\overset{*}{A}_m{}^{;r} = (\sum_{\rho(k)\in Lm} \prod_i^m A^{i\lambda[i,k]} \prod_i^m \mathbf{b}_{i\lambda[i,k]})^{;r}$
3	From 2 and T12.7.5 Tensor Product: Tensor Series Expansion Distributing Covariant Derivative	$\overset{*}{A}_m{}^{;r} = \sum_{\rho(k)\in Lm} (\prod_i^m A^{i\lambda[i,k]})^{;r} \prod_i^m \mathbf{b}_{i\lambda[i,k]}$
4	From 3 and T12.7.10 Tensor Product: Contravariant Chain Rule	$\overset{*}{A}_m{}^{;r} = \sum_{\rho(k)\in Lm}$ $\sum_{j=1}^m (\prod_{i=1}^{j-1} A^{i\lambda[i,k]}\, \mathbf{b}_{i\lambda[i,k]})\, (A^{j\lambda[j,k]}\, \mathbf{b}_{j\lambda[j,k]})^{;r}\, (\prod_{i=j+1}^m A^{i\lambda[i,k]}\, \mathbf{b}_{i\lambda[i,k]})$
5	From 4, T12.6.2 Differentiation of Tensor: Rank-1 Covariant and T10.2.5 Tool: Tensor Substitution	$\overset{*}{A}_m{}^{;r} = \sum_{\rho(k)\in Lm}$ $\sum_{j=1}^m (\prod_{i=1}^{j-1} A^{i\lambda[i,k]}\, \mathbf{b}_{i\lambda[i,k]})$ $(A^{j\lambda[j,k],r} + \sum_s A^{s\lambda[j,k]}\, \Gamma_{sr}^{\lambda[j,k]})\, \mathbf{b}_{j\lambda[j,k]}$ $(\prod_{i=j+1}^m A^{i\lambda[i,k]}\, \mathbf{b}_{i\lambda[i,k]})$
6	From 7, T10.2.4 Tool: Base Tensor Parsed from Polyadic Coefficient, D4.1.20 Negative Coefficient and T10.4.6 Scalars and Tensors: Distribution of a Scalar over Addition of Tensors	$\overset{*}{A}_m{}^{;r} = \sum_{\rho(k)\in Lm}$ $\sum_{j=1}^m$ $(\prod_{i=1}^{j-1} A^{i\lambda[i,k]}\, \mathbf{b}_{i\lambda[i,k]})\, (A^{j\lambda[j,k],r})\, \mathbf{b}_{j\lambda[j,k]}\, (\prod_{i=j+1}^m A^{i\lambda[i,k]}\, \mathbf{b}_{i\lambda[i,k]})$ $+\sum_s \sum_{j=1}^m$ $(\prod_{i=1}^{j-1} A^{i\lambda[i,k]}\, \mathbf{b}_{i\lambda[i,k]})\, (A^{s\lambda[j,k]}\, \Gamma_{sr}^{\lambda[j,k]})\, \mathbf{b}_{j\lambda[j,k]}$ $(\prod_{i=j+1}^m A^{i\lambda[i,k]}\, \mathbf{b}_{i\lambda[i,k]})$
7	From 6, T10.4.8 Scalars and Tensors: Scalar Commutative with Tensors, A10.2.15 Associate Scalar with Tensor Group and D10.1.16 Polyadic Base Tensor	$\overset{*}{A}_m{}^{;r} = \sum_{\rho(k)\in Lm}$ $\sum_{j=1}^m$ $(\prod_{i=1}^{j-1} A^{i\lambda[i,k]})\, (A^{j\lambda[j,k],r})\, (\prod_{i=j+1}^m A^{i\lambda[i,k]})\, (\prod_{i=1}^m \mathbf{b}_{i\lambda[i,k]})$ $+\sum_s \sum_{j=1}^m$ $(\prod_{i=1}^{j-1} A^{i\lambda[i,k]})\, (A^{s\lambda[j,k]}\, \Gamma_{sr}^{\lambda[j,k]})\, (\prod_{i=j+1}^m A^{i\lambda[i,k]})\, (\prod_{i=1}^m \mathbf{b}_{i\lambda[i,k]})$
8	From 7 and T10.2.4 Tool: Base Tensor Distributed over Polyadic Coefficient Summation	$\overset{*}{A}_m{}^{;r} = \sum_{\rho(k)\in Lm}$ $[\sum_{j=1}^m$ $(\prod_{i=1}^{j-1} A^{i\lambda[i,k]})\, (A^{j\lambda[j,k],r})\, (\prod_{i=j=1}^m A^{i\lambda[i,k]})$ $+\sum_s \sum_{j=1}^m$ $(\prod_{i=1}^{j-1} A^{i\lambda[i,k]})\, (A^{s\lambda[j,k]}\, \Gamma_{sr}^{\lambda[j,k]})\, (\prod_{i=j=1}^m A^{i\lambda[i,k]})]\, (\prod_{i=1}^m \mathbf{b}_{i\lambda[i,k]})$
9	From 8 and LxK.3.2.4C Distributed Across Products	$\overset{*}{A}_m{}^{;r} = \sum_{\rho(k)\in Lm}$ $[(\prod_{i=1}^m A^{i\lambda[i,k]})^{,r}$ $+\sum_s \sum_{j=1}^m$ $(\prod_{i=1}^{j-1} A^{i\lambda[i,k]})\, (A^{s\lambda[j,k]}\, \Gamma_{sr}^{\lambda[j,k]})\, (\prod_{i=j+1}^m A^{i\lambda[i,k]})]\, (\prod_{i=1}^m \mathbf{b}_{i\lambda[i,k]})$
10	From 3, 9 and comparing like polyadics term-by-term and equating coefficients on the right and left hand side of the equality	$(\prod_i^m A^{i\lambda[i,k]})^{;r} = (\prod_{i=1}^m A^{i\lambda[i,k]})^{,r}$ $+ \sum_s \sum_{j=1}^m (\prod_{i=1}^{j-1} A^{i\lambda[i,k]})\, (A^{s\lambda[j,k]}\, \Gamma_{sr}^{\lambda[j,k]})\, (\prod_{i=j+1}^m A^{i\lambda[i,k]})$

11	From 1g, 10, T10.2.5 Tool: Tensor Substitution, T10.3.20 Addition of Tensors: Distribution of a Tensor over Addition of Tensors and A10.2.14 Correspondence of Tensors and Tensor Coefficients	$(A^{\bullet\lambda[i,k]\bullet})^{;r} =$ $(A^{\bullet\lambda[i,k]\bullet})^{,r}$ $+ \sum_s \sum_{j=1}^{m} [(\prod_{i=1}^{j-1} A^{i\lambda[i,k]}) A^{s\lambda[j,k]} (\prod_{i=j+1}^{m} A^{i\lambda[i,k]})] \Gamma_{sr}^{\lambda[j,k]}$
∴	From 2g, 11, T10.2.5 Tool: Tensor Substitution	$(A^{\bullet\lambda[i,k]\bullet})^{;r} = (A^{\bullet\lambda[i,k]\bullet})^{,r} + \sum_s \sum_{j=1}^{m} A^{\bullet\tau[j,s,k]\bullet} \Gamma_{sr}^{\lambda[j,k]}$

Theorem 12.7.12 Tensor Product: Contravariant Chain Rule for Tensor Coefficient

1g	Given	$A_{\bullet\lambda[i,\,k]\bullet} \equiv \prod_{i=1}^{m} A_{i\lambda[i,\,k]}$
2g		$A_{\bullet\tau[j,\,s,\,k]\bullet} \equiv (\prod_{i=1}^{j-1} A_{i\lambda[i,\,k]})\, A_{s\lambda[j,\,k]}\, (\prod_{i=j+1}^{m} A_{i\lambda[i,\,k]})$
Step	Hypothesis	$(A_{\bullet\lambda[i,\,k]\bullet})_{;r} = (A_{\bullet\lambda[i,\,k]\bullet})_{,r} - \sum_s \sum_{j=1}^{m} A_{\bullet\tau[j,\,s,\,k]\bullet}\, \Gamma_{\lambda[j,\,k]r}{}^{s}$
1	From T12.7.1B Tensor Uniqueness: Tensor Differentiation and T10.2.1 Tensor Uniqueness: Tensor Differentiation	$\overset{*}{A}{}^{m}{}_{;r} = (\sum_{\rho(k)\in Lm} \prod_{i}^{m} A_{i\lambda[i,\,k]}\, \mathbf{b}^{i\lambda[i,\,k]})_{;r}$
2	From 1 and T10.2.3 Tool: Base Tensor Parsed from Polyadic Coefficient	$\overset{*}{A}{}^{m}{}_{;r} = (\sum_{\rho(k)\in Lm} \prod_{i}^{m} A_{i\lambda[i,\,k]} \prod_{i}^{m} \mathbf{b}^{i\lambda[i,\,k]})_{;r}$
3	From 2 and T12.7.4 Tensor Product: Tensor Series Expansion Distributing Covariant Derivative	$\overset{*}{A}{}^{m}{}_{;r} = \sum_{\rho(k)\in Lm} (\prod_{i}^{m} A_{i\lambda[i,\,k]})_{;r} \prod_{i}^{m} \mathbf{b}^{i\lambda[i,\,k]}$
4	From 3 and T12.7.10 Tensor Product: Contravariant Chain Rule	$\overset{*}{A}{}^{m}{}_{;r} = \sum_{\rho(k)\in Lm}$ $\sum_{j=1}^{m} (\prod_{i=1}^{j-1} A_{i\lambda[i,\,k]}\, \mathbf{b}^{i\lambda[i,\,k]})\, (A_{j\lambda[j,\,k]}\, \mathbf{b}^{j\lambda[j,\,k]})_{;r}\, (\prod_{i=j+1}^{m} A_{i\lambda[i,\,k]}\, \mathbf{b}^{i\lambda[i,\,k]})$
5	From 4, T12.6.3 Differentiation of Tensor: Rank-1 Contravariant and T10.2.5 Tool: Tensor Substitution	$\overset{*}{A}{}^{m}{}_{;r} = \sum_{\rho(k)\in Lm}$ $\sum_{j=1}^{m} (\prod_{i=1}^{j-1} A_{i\lambda[i,\,k]}\, \mathbf{b}^{i\lambda[i,\,k]})$ $(A_{j\lambda[j,\,k],r} - \sum_s A_{s\lambda[j,\,k]}\, \Gamma_{\lambda[j,\,k]r}{}^{s})\, \mathbf{b}^{j\lambda[j,\,k]}$ $(\prod_{i=j+1}^{m} A_{i\lambda[i,\,k]}\, \mathbf{b}^{i\lambda[i,\,k]})$
6	From 7, T10.2.4 Tool: Base Tensor Parsed from Polyadic Coefficient, D4.1.20 Negative Coefficient and T10.4.6 Scalars and Tensors: Distribution of a Scalar over Addition of Tensors	$\overset{*}{A}{}^{m}{}_{;r} = \sum_{\rho(k)\in Lm}$ $\sum_{j=1}^{m}$ $(\prod_{i=1}^{j-1} A_{i\lambda[i,\,k]}\, \mathbf{b}^{i\lambda[i,\,k]})\, (A_{j\lambda[j,\,k],r})\, \mathbf{b}^{j\lambda[j,\,k]}\, (\prod_{i=j+1}^{m} A_{i\lambda[i,\,k]}\, \mathbf{b}^{i\lambda[i,\,k]})$ $-\sum_s \sum_{j=1}^{m}$ $(\prod_{i=1}^{j-1} A_{i\lambda[i,\,k]}\, \mathbf{b}^{i\lambda[i,\,k]})\, (A_{s\lambda[j,\,k]}\, \Gamma_{\lambda[j,\,k]r}{}^{s})\, \mathbf{b}^{j\lambda[j,\,k]}$ $(\prod_{i=j+1}^{m} A_{i\lambda[i,\,k]}\, \mathbf{b}^{i\lambda[i,\,k]})$
7	From 6, T10.4.8 Scalars and Tensors: Scalar Commutative with Tensors, A10.2.15 Associate Scalar with Tensor Group and D10.1.16 Polyadic Base Tensor	$\overset{*}{A}{}^{m}{}_{;r} = \sum_{\rho(k)\in Lm}$ $\sum_{j=1}^{m}$ $(\prod_{i=1}^{j-1} A_{i\lambda[i,\,k]})\, (A_{j\lambda[j,\,k],r})\, (\prod_{i=j+1}^{m} A_{i\lambda[i,\,k]})\, (\prod_{i=1}^{m} \mathbf{b}^{i\lambda[i,\,k]})$ $-\sum_s \sum_{j=1}^{m}$ $(\prod_{i=1}^{j-1} A_{i\lambda[i,\,k]})\, (A_{s\lambda[j,\,k]}\, \Gamma_{\lambda[j,\,k]r}{}^{s})\, (\prod_{i=j+1}^{m} A_{i\lambda[i,\,k]})\, (\prod_{i=1}^{m} \mathbf{b}^{i\lambda[i,\,k]})$
8	From 7 and T10.2.4 Tool: Base Tensor Distributed over Polyadic Coefficient Summation	$\overset{*}{A}{}^{m}{}_{;r} = \sum_{\rho(k)\in Lm}$ $[\sum_{j=1}^{m}$ $(\prod_{i=1}^{j-1} A_{i\lambda[i,\,k]})\, (A_{j\lambda[j,\,k],r})\, (\prod_{i=j+1}^{m} A_{i\lambda[i,\,k]})$ $-\sum_s \sum_{j=1}^{m}$ $(\prod_{i=1}^{j-1} A_{i\lambda[i,\,k]})\, (A_{s\lambda[j,\,k]}\, \Gamma_{\lambda[j,\,k]r}{}^{s})\, (\prod_{i=j+1}^{m} A_{i\lambda[i,\,k]})\,]\, (\prod_{i=1}^{m} \mathbf{b}^{i\lambda[i,\,k]})$
9	From 8 and LxK.3.2.4C Distributed Across Products	$\overset{*}{A}{}^{m}{}_{;r} = \sum_{\rho(k)\in Lm}$ $[(\prod_{i=1}^{m} A_{i\lambda[i,\,k]})_{,r}$ $-\sum_s \sum_{j=1}^{m}$ $(\prod_{i=1}^{j-1} A_{i\lambda[i,\,k]})\, (A_{s\lambda[j,\,k]}\, \Gamma_{\lambda[j,\,k]r}{}^{s})\, (\prod_{i=j+1}^{m} A_{i\lambda[i,\,k]})\,]\, (\prod_{i=1}^{m} \mathbf{b}^{i\lambda[i,\,k]})$

10	From 3, 9 and comparing like polyadics term-by-term and equating coefficients on the right and left hand side of the equality	$(\prod_i^m A_{i\lambda[i,k]})_{;r} = (\prod_{i=1}^m A_{i\lambda[i,k]})_{,r} - \sum_s \sum_{j=1}^m (\prod_{i=1}^{j-1} A_{i\lambda[i,k]})(A_{s\lambda[j,k]}\Gamma_{\lambda[j,k]r}{}^s)(\prod_{i=j+1}^m A_{i\lambda[i,k]})$
11	From 1g, 10, T10.2.5 Tool: Tensor Substitution, T10.3.20 Addition of Tensors: Distribution of a Tensor over Addition of Tensors and A10.2.14 Correspondence of Tensors and Tensor Coefficients	$(A_{\bullet\lambda[i,k]\bullet})_{;r} = (A_{\bullet\lambda[i,k]\bullet})_{,r} - \sum_s \sum_{j=1}^m [(\prod_{i=1}^{j-1} A_{i\lambda[i,k]}) A_{s\lambda[j,k]} (\prod_{i=j+1}^m A_{i\lambda[i,k]})] \Gamma_{\lambda[j,k]r}{}^s$
∴	From 2g, 11, T10.2.5 Tool: Tensor Substitution	$(A_{\bullet\lambda[i,k]\bullet})_{;r} = (A_{\bullet\lambda[i,k]\bullet})_{,r} - \sum_s \sum_{j=1}^m A_{\bullet\tau[j,s,k]\bullet}\Gamma_{\lambda[j,k]r}{}^s$

Theorem 12.7.13 Tensor Product: Differential of Covariant Tensor Sets

Step	Hypothesis	$(\prod_{i=1}^m \overset{*}{A}_{q[i]})^{;k} = \overset{*}{A}_{q[1]}{}^{;k}(\prod_{i=2}^m \overset{*}{A}_{q[i]}) + \sum_{j=2}^{m-1}\prod_{i=1}^{j-1}\overset{*}{A}_{q[i]}\overset{*}{A}_{q[j]}{}^{;k}(\prod_{i=j+1}^m \overset{*}{A}_{q[i]}) + \prod_{i=1}^{m-1}\overset{*}{A}_{q[i]}\overset{*}{A}_{q[m]}{}^{;k}$
1	From A10.2.6A Equality of Tensors and T12.7.1A Tensor Uniqueness: Tensor Differentiation	$(\prod_{i=1}^m \overset{*}{A}_{q[i]})^{;k} = (\prod_{i=1}^m \overset{*}{A}_{q[i]})^{;k}$
2	From 1 and expanding the tensor product set	$(\prod_{i=1}^m \overset{*}{A}_{q[i]})^{;k} = (\overset{*}{A}_{q[1]}\prod_{i=2}^m \overset{*}{A}_{q[i]})^{;k}$
3	From 2 and T12.7.6 Tensor Product: Chain Rule for Binary Covariant Product	$(\prod_{i=1}^m \overset{*}{A}_{q[i]})^{;k} = \overset{*}{A}_{q[1]}{}^{;k}(\prod_{i=2}^m \overset{*}{A}_{q[i]}) + \overset{*}{A}_{q[1]}(\prod_{i=2}^m \overset{*}{A}_{q[i]})^{;k}$
4	From 3 and expanding the tensor product set	$(\prod_{i=1}^m \overset{*}{A}_{q[i]})^{;k} = \overset{*}{A}_{q[1]}{}^{;k}(\prod_{i=2}^m \overset{*}{A}_{q[i]}) + \overset{*}{A}_{q[1]}(\overset{*}{A}_{q[2]}\prod_{i=3}^m \overset{*}{A}_{q[i]})^{;k}$
5	From 4, T12.7.6 Tensor Product: Chain Rule for Binary Covariant Product, T10.3.20 Addition of Tensors: Distribution of a Tensor over Addition of Tensors and A10.2.14 Correspondence of Tensors and Tensor Coefficients	$(\prod_{i=1}^m \overset{*}{A}_{q[i]})^{;k} = \overset{*}{A}_{q[1]}{}^{;k}(\prod_{i=2}^m \overset{*}{A}_{q[i]}) + \overset{*}{A}_{q[1]}\overset{*}{A}_{q[2]}{}^{;k}(\prod_{i=3}^m \overset{*}{A}_{q[i]}) + \overset{*}{A}_{q[1]}\overset{*}{A}_{q[2]}(\prod_{i=3}^m \overset{*}{A}_{q[i]})^{;k}$
6	From 5 and expanding the tensor product set	$(\prod_{i=1}^m \overset{*}{A}_{q[i]})^{;k} = \overset{*}{A}_{q[1];k}(\prod_{i=2}^m \overset{*}{A}_{q[i]}) + \overset{*}{A}_{q[1]}\overset{*}{A}_{q[2];k}(\prod_{i=3}^m \overset{*}{A}_{q[i]}) + \overset{*}{A}_{q[1]}\overset{*}{A}_{q[2]}(\overset{*}{A}_{q[3]}\prod_{i=4}^m \overset{*}{A}_{q[i]})^{;k}$
7	From 6, T12.7.6 Tensor Product: Chain Rule for Binary Covariant Product, T10.3.20 Addition of Tensors: Distribution of a Tensor over Addition of Tensors and A10.2.14 Correspondence of Tensors and Tensor Coefficients	$(\prod_{i=1}^m \overset{*}{A}_{q[i]})^{;k} = \overset{*}{A}_{q[1]}{}^{;k}(\prod_{i=2}^m \overset{*}{A}_{q[i]}) + \overset{*}{A}_{q[1]}\overset{*}{A}_{q[2]}{}^{;k}(\prod_{i=3}^m \overset{*}{A}_{q[i]}) + + \overset{*}{A}_{q[1]}\overset{*}{A}_{q[2]}\overset{*}{A}_{q[3]}{}^{;k}(\prod_{i=3}^m \overset{*}{A}_{q[i]}) + \overset{*}{A}_{q[1]}\overset{*}{A}_{q[2]}\overset{*}{A}_{q[3]}(\prod_{i=4}^m \overset{*}{A}_{q[i]})^{;k}$
8	From 7 and expanding the tensor product set	$(\prod_{i=1}^m \overset{*}{A}_{q[i]})^{;k} = \overset{*}{A}_{q[1]}{}^{;k}(\prod_{i=2}^m \overset{*}{A}_{q[i]}) + \overset{*}{A}_{q[1]}\overset{*}{A}_{q[2]}{}^{;k}(\prod_{i=3}^m \overset{*}{A}_{q[i]}) + \overset{*}{A}_{q[1]}\overset{*}{A}_{q[2]}\overset{*}{A}_{q[3]}{}^{;k}(\prod_{i=3}^m \overset{*}{A}_{q[i]}) + \overset{*}{A}_{q[1]}\overset{*}{A}_{q[2]}\overset{*}{A}_{q[3]}(\overset{*}{A}_{q[4]}\prod_{i=5}^m \overset{*}{A}_{q[i]})^{;k}$

9	From 8, T12.7.6 Tensor Product: Chain Rule for Binary Covariant Product, T10.3.20 Addition of Tensors: Distribution of a Tensor over Addition of Tensors and A10.2.14 Correspondence of Tensors and Tensor Coefficients	$(\prod_{i=1}^{m} \overset{*}{A}_{q[i]})^{;k} = \overset{*}{A}_{q[1]}{}^{;k} (\prod_{i=2}^{m} \overset{*}{A}_{q[i]}) + \overset{*}{A}_{q[1]} \overset{*}{A}_{q[2]}{}^{;k} (\prod_{i=3}^{m} \overset{*}{A}_{q[i]}) +$ $\overset{*}{A}_{q[1]} \overset{*}{A}_{q[2]} \overset{*}{A}_{q[3]}{}^{;k} (\prod_{i=3}^{m} \overset{*}{A}_{q[i]}) +$ $\overset{*}{A}_{q[1]} \overset{*}{A}_{q[2]} \overset{*}{A}_{q[3]} \overset{*}{A}_{q[4]}{}^{;k} (\prod_{i=5}^{m} \overset{*}{A}_{q[i]}) +$ $\overset{*}{A}_{q[1]} \overset{*}{A}_{q[2]} \overset{*}{A}_{q[3]} \overset{*}{A}_{q[4]} (\prod_{i=5}^{m} \overset{*}{A}_{q[i]})^{;k}$
10	From 9; generalize index-j between summation of j=2 to j= m–1, first and last terms are fixed to remove singularities	$(\prod_{i=1}^{m} \overset{*}{A}_{q[i]})^{;k} = \overset{*}{A}_{q[1]}{}^{;k} (\prod_{i=2}^{m} \overset{*}{A}_{q[i]}) +$ $\sum_{j=2}^{m-1} \prod_{i=1}^{j-1} \overset{*}{A}_{q[i]} \overset{*}{A}_{q[j]}{}^{;k} (\prod_{i=j+1}^{m} \overset{*}{A}_{q[i]}) +$ $\prod_{i=1}^{m-2} \overset{*}{A}_{q[i]} \overset{*}{A}_{q[m-1]} (\prod_{i=j+1}^{m} \overset{*}{A}_{q[i]})^{;k}$
11	From 10, A4.2.3 Substitution and A4.2.8 Inverse Add; for last summation term j = m – 1	$(\prod_{i=1}^{m} \overset{*}{A}_{q[i]})^{;k} = \overset{*}{A}_{q[1]}{}^{;k} (\prod_{i=2}^{m} \overset{*}{A}_{q[i]}) +$ $\sum_{j=2}^{m-1} \prod_{i=1}^{j-1} \overset{*}{A}_{q[i]} \overset{*}{A}_{q[j]}{}^{;k} (\prod_{i=j+1}^{m} \overset{*}{A}_{q[i]}) +$ $\prod_{i=1}^{m-1} \overset{*}{A}_{q[i]} (\prod_{i=m}^{m} \overset{*}{A}_{q[i]})^{;k}$
12	From 11 and evaluate product of last term: Now the first and last terms match with 3, 5, 7 and 9 bounding the center summation. So as m-increases it will always be properly bounded and maintain the correct form.	$(\prod_{i=1}^{m} \overset{*}{A}_{q[i]})^{;k} = \overset{*}{A}_{q[1]}{}^{;k} (\prod_{i=2}^{m} \overset{*}{A}_{q[i]}) +$ $\sum_{j=2}^{m-1} \prod_{i=1}^{j-1} \overset{*}{A}_{q[i]} \overset{*}{A}_{q[j]}{}^{;k} (\prod_{i=j+1}^{m} \overset{*}{A}_{q[i]}) + \prod_{i=1}^{m-1} \overset{*}{A}_{q[i]} \overset{*}{A}_{q[m]}{}^{;k}$
13	From 12; increase [m] by one	$(\prod_{i=1}^{m} \overset{*}{A}_{q[i]})^{;k} = \overset{*}{A}_{q[1]}{}^{;k} (\prod_{i=2}^{m+1} \overset{*}{A}_{q[i]}) +$ $\sum_{j=2}^{m} \prod_{i=1}^{j-1} \overset{*}{A}_{q[i]} \overset{*}{A}_{q[j]}{}^{;k} (\prod_{i=j+1}^{m+1} \overset{*}{A}_{q[i]}) + \prod_{i=1}^{m} \overset{*}{A}_{q[i]} \overset{*}{A}_{q[m+1]}{}^{;k}$
∴	From 12 and 13: The end terms are left unaltered in form; hence the center summation is properly bounded.	$(\prod_{i=1}^{m} \overset{*}{A}_{q[i]})^{;k} = \overset{*}{A}_{q[1]}{}^{;k} (\prod_{i=2}^{m} \overset{*}{A}_{q[i]}) +$ $\sum_{j=2}^{m-1} \prod_{i=1}^{j-1} \overset{*}{A}_{q[i]} \overset{*}{A}_{q[j]}{}^{;k} (\prod_{i=j+1}^{m} \overset{*}{A}_{q[i]}) + \prod_{i=1}^{m-1} \overset{*}{A}_{q[i]} \overset{*}{A}_{q[m]}{}^{;k}$

Theorem 12.7.14 Tensor Product: Differential of Contravariant Tensor Sets

Step	Hypothesis	
Step	Hypothesis	$(\prod_{i=1}^{m} \overset{*}{A}^{q[i]})_{;k} = \overset{*}{A}^{q[1]}{}_{;k} (\prod_{i=2}^{m} \overset{*}{A}^{q[i]}) + \sum_{j=2}^{m-1} \prod_{i=1}^{j-1} \overset{*}{A}^{q[i]} \overset{*}{A}^{q[j]}{}_{;k} (\prod_{i=j+1}^{m} \overset{*}{A}^{q[i]}) + \prod_{i=1}^{m-1} \overset{*}{A}^{q[i]} \overset{*}{A}^{q[m]}{}_{;k}$
1	From A10.2.6A Equality of Tensors and T12.7.1B Tensor Uniqueness: Tensor Differentiation	$(\prod_{i=1}^{m} \overset{*}{A}^{q[i]})_{;k} = (\prod_{i=1}^{m} \overset{*}{A}^{q[i]})_{;k}$
2	From 1 and expanding the tensor product set	$(\prod_{i=1}^{m} \overset{*}{A}^{q[i]})_{;k} = (\overset{*}{A}^{q[1]} \prod_{i=2}^{m} \overset{*}{A}^{q[i]})_{;k}$
3	From 2 and T12.7.6 Tensor Product: Chain Rule for Binary Covariant Product	$(\prod_{i=1}^{m} \overset{*}{A}^{q[i]})_{;k} = \overset{*}{A}^{q[1]}{}_{;k} (\prod_{i=2}^{m} \overset{*}{A}^{q[i]}) + \overset{*}{A}^{q[1]} (\prod_{i=2}^{m} \overset{*}{A}^{q[i]})_{;k}$
4	From 3 and expanding the tensor product set	$(\prod_{i=1}^{m} \overset{*}{A}^{q[i]})_{;k} = \overset{*}{A}^{q[1]}{}_{;k} (\prod_{i=2}^{m} \overset{*}{A}^{q[i]}) + \overset{*}{A}^{q[1]} (\overset{*}{A}^{q[2]} \prod_{i=3}^{m} \overset{*}{A}^{q[i]})_{;k}$
5	From 4, T12.7.6 Tensor Product: Chain Rule for Binary Covariant Product, T10.3.20 Addition of Tensors: Distribution of a Tensor over Addition of Tensors and A10.2.14 Correspondence of Tensors and Tensor Coefficients	$(\prod_{i=1}^{m} \overset{*}{A}^{q[i]})_{;k} = \overset{*}{A}^{q[1]}{}_{;k} (\prod_{i=2}^{m} \overset{*}{A}^{q[i]}) + \overset{*}{A}^{q[1]} \overset{*}{A}^{q[2]}{}_{;k} (\prod_{i=3}^{m} \overset{*}{A}^{q[i]}) + \overset{*}{A}^{q[1]} \overset{*}{A}^{q[2]} (\prod_{i=3}^{m} \overset{*}{A}^{q[i]})_{;k}$
6	From 5 and expanding the tensor product set	$(\prod_{i=1}^{m} \overset{*}{A}^{q[i]})_{;k} = \overset{*}{A}^{q[1]}{}_{;k} (\prod_{i=2}^{m} \overset{*}{A}^{q[i]}) + \overset{*}{A}^{q[1]} \overset{*}{A}^{q[2]}{}_{;k} (\prod_{i=3}^{m} \overset{*}{A}^{q[i]}) + \overset{*}{A}^{q[1]} \overset{*}{A}^{q[2]} (\overset{*}{A}^{q[3]} \prod_{i=4}^{m} \overset{*}{A}^{q[i]})_{;k}$
7	From 6, T12.7.6 Tensor Product: Chain Rule for Binary Covariant Product, T10.3.20 Addition of Tensors: Distribution of a Tensor over Addition of Tensors and A10.2.14 Correspondence of Tensors and Tensor Coefficients	$(\prod_{i=1}^{m} \overset{*}{A}^{q[i]})_{;k} = \overset{*}{A}^{q[1]}{}_{;k} (\prod_{i=2}^{m} \overset{*}{A}^{q[i]}) + \overset{*}{A}^{q[1]} \overset{*}{A}^{q[2]}{}_{;k} (\prod_{i=3}^{m} \overset{*}{A}^{q[i]}) + + \overset{*}{A}^{q[1]} \overset{*}{A}^{q[2]} \overset{*}{A}^{q[3]}{}_{;k} (\prod_{i=3}^{m} \overset{*}{A}^{q[i]}) + \overset{*}{A}^{q[1]} \overset{*}{A}^{q[2]} \overset{*}{A}^{q[3]} (\prod_{i=4}^{m} \overset{*}{A}^{q[i]})_{;k}$
8	From 7 and expanding the tensor product set	$(\prod_{i=1}^{m} \overset{*}{A}^{q[i]})_{;k} = \overset{*}{A}^{q[1]}{}_{;k} (\prod_{i=2}^{m} \overset{*}{A}^{q[i]}) + \overset{*}{A}^{q[1]} \overset{*}{A}^{q[2]}{}_{;k} (\prod_{i=3}^{m} \overset{*}{A}^{q[i]}) + \overset{*}{A}^{q[1]} \overset{*}{A}^{q[2]} \overset{*}{A}^{q[3]}{}_{;k} (\prod_{i=3}^{m} \overset{*}{A}^{q[i]}) + \overset{*}{A}^{q[1]} \overset{*}{A}^{q[2]} \overset{*}{A}^{q[3]} (\overset{*}{A}_{q[4]} \prod_{i=5}^{m} \overset{*}{A}^{q[i]})_{;k}$
9	From 8, T12.7.6 Tensor Product: Chain Rule for Binary Covariant Product, T10.3.20 Addition of Tensors: Distribution of a Tensor over Addition of Tensors and A10.2.14 Correspondence of Tensors and Tensor Coefficients	$(\prod_{i=1}^{m} \overset{*}{A}^{q[i]})_{;k} = \overset{*}{A}^{q[1]}{}_{;k} (\prod_{i=2}^{m} \overset{*}{A}^{q[i]}) + \overset{*}{A}^{q[1]} \overset{*}{A}^{q[2]}{}_{;k} (\prod_{i=3}^{m} \overset{*}{A}^{q[i]}) + \overset{*}{A}^{q[1]} \overset{*}{A}^{q[2]} \overset{*}{A}^{q[3]}{}_{;k} (\prod_{i=3}^{m} \overset{*}{A}^{q[i]}) + \overset{*}{A}^{q[1]} \overset{*}{A}^{q[2]} \overset{*}{A}^{q[3]} \overset{*}{A}^{q[4]}{}_{;k} (\prod_{i=5}^{m} \overset{*}{A}^{q[i]}) + \overset{*}{A}^{q[1]} \overset{*}{A}^{q[2]} \overset{*}{A}^{q[3]} \overset{*}{A}^{q[4]} (\prod_{i=5}^{m} \overset{*}{A}^{q[i]})_{;k}$
10	From 9; generalize index-j between summation of j=2 to j= m–1, first and last terms are fixed to remove singularities	$(\prod_{i=1}^{m} \overset{*}{A}^{q[i]})_{;k} = \overset{*}{A}^{q[1]}{}_{;k} (\prod_{i=2}^{m} \overset{*}{A}^{q[i]}) + \sum_{j=2}^{m-1} \prod_{i=1}^{j-1} \overset{*}{A}^{q[i]} \overset{*}{A}^{q[j]}{}_{;k} (\prod_{i=j+1}^{m} \overset{*}{A}^{q[i]}) + \prod_{i=1}^{m-2} \overset{*}{A}^{q[i]} \overset{*}{A}^{q[m-1]} (\prod_{i=j+1}^{m} \overset{*}{A}^{q[i]})_{;k}$
11	From 10, A4.2.3 Substitution and A4.2.8 Inverse Add; for last summation term j = m – 1	$(\prod_{i=1}^{m} \overset{*}{A}^{q[i]})_{;k} = \overset{*}{A}^{q[1]}{}_{;k} (\prod_{i=2}^{m} \overset{*}{A}^{q[i]}) + \sum_{j=2}^{m-1} \prod_{i=1}^{j-1} \overset{*}{A}^{q[i]} \overset{*}{A}^{q[j]}{}_{;k} (\prod_{i=j+1}^{m} \overset{*}{A}^{q[i]}) + \prod_{i=1}^{m-1} \overset{*}{A}^{q[i]} (\prod_{i=m}^{m} \overset{*}{A}^{q[i]})_{;k}$

12	From 11 and evaluate product of last term: Now the first and last terms match with 3, 5, 7 and 9 bounding the center summation. So as m-increases it will always be properly bounded and maintain the correct form.	$(\prod_{i=1}^{m} \overset{*}{A}^{q[i]})_{;k} = \overset{*}{A}^{q[1]}{}_{;k} (\prod_{i=2}^{m} \overset{*}{A}^{q[i]}) + \sum_{j=2}^{m-1} \prod_{i=1}^{j-1} \overset{*}{A}^{q[i]} \overset{*}{A}^{q[j]}{}_{;k} (\prod_{i=j+1}^{m} \overset{*}{A}^{q[i]}) + \prod_{i=1}^{m-1} \overset{*}{A}^{q[i]} \overset{*}{A}^{q[m]}{}_{;k}$
13	From 12; increase [m] by one	$(\prod_{i=1}^{m+1} \overset{*}{A}^{q[i]})_{;k} = \overset{*}{A}^{q[1]}{}_{;k} (\prod_{i=2}^{m+1} \overset{*}{A}^{q[i]}) + \sum_{j=2}^{m} \prod_{i=1}^{j-1} \overset{*}{A}^{q[i]} \overset{*}{A}^{q[j]}{}_{;k} (\prod_{i=j+1}^{m+1} \overset{*}{A}^{q[i]}) + \prod_{i=1}^{m} \overset{*}{A}^{q[i]} \overset{*}{A}^{q[m+1]}{}_{;k}$
∴	From 12 and 13: The end terms are left unaltered in form; hence the center summation is properly bounded.	$(\prod_{i=1}^{m} \overset{*}{A}^{q[i]})_{;k} = \overset{*}{A}^{q[1]}{}_{;k} (\prod_{i=2}^{m} \overset{*}{A}^{q[i]}) + \sum_{j=2}^{m-1} \prod_{i=1}^{j-1} \overset{*}{A}^{q[i]} \overset{*}{A}^{q[j]}{}_{;k} (\prod_{i=j+1}^{m} \overset{*}{A}^{q[i]}) + \prod_{i=1}^{m-1} \overset{*}{A}^{q[i]} \overset{*}{A}^{q[m]}{}_{;k}$

Section 12.8 Differentiation of Metric Tensors

The bases vectors of a manifold span the space thereby defining its shape like our skeleton shapes us. The measurement of the shape or deviation from an orthogonal structure is given by the space distortion factor. If a manifold should change form then a change in the space distortion factor would be observed. It would also mean that a particular manifold would be some new or other space, since each a manifold is of a unique and different shape it must be treated as another manifold.

If a manifold changes with time, again the argument could be made that a different manifold is being considered. The same argument could also be made for the parametric parameter of arc length, so as a general rule it can be stated manifolds do not change with any parametric parameter. If a manifold should change then it must be stated so and treated as another manifold.

The covariant metric tensor of the manifold is unique to the manifold as well; this can be seen from its definition being comprised of two bases vectors.

The contravariant tensor is defined [$g^{ij} = g^{-1} G^{ij}$], see D12.1.4D "Contravariant Metric Dyadic", the cofactor [G^{ij}], is comprised of covariant tensors, being the inverse matrix for the covariant tensor. It follows if the covariant tensors are constant with respect to the shape of the manifold, then the contravariant tensor must also be, hence [$\Delta G^{ij} / \Delta g_{ij} = 0$].

Axiom 12.8.1 Constancy of Distorting a Manifold

Let ∃| a manifold comprised of bases vectors shaping that region space it would follow then that the bases vectors would neither wax nor wan in shaping the manifold.

P1 Manifolds do not change shape with any variation in a parametric parameter, such as time or trajectory path.

P2 If a variation should take place, then the manifold must be treated as a separate and different space.

P3 If the manifold has constancy of shape, then the following conditions must hold:

$\Delta\mu_c \equiv 0$	Cosine Space Distortion Factor	Equation A
$\Delta\mu_s \equiv 0$	Sine Space Distortion Factor	Equation B
$\delta G^{ij} / \delta g_{ij} \equiv 0$	No change in contravariant with respect to covariant	Equation C

Notice that equation C is a conditional

Theorem 12.8.1 Differential of the Contravariant Cofactor to the Metric Dyadic

Step	Hypothesis	$\partial (G^{ik}) / \partial g_{ij} = 0$
1	From A12.8.1CP3 Constancy of Distorting a Manifold and LxK.2.1.1 Uniqueness of Limits	$\lim_{\delta gij \to 0} (\delta G^{ik} / \delta g_{ij}) = \lim_{\delta gij \to 0} (0)$
∴	From 1, DxK.3.1.3 Definition of a Derivative Alternate Form, DxK.3.2.2D Multiple Partial Differential Operator and LxK.3.3.1 Differential of a Constant	$\partial (G^{ik}) / \partial g_{ij} = 0$

Theorem 12.8.2 Differentiation of the Metric Tensor of the First Kind

Step	Hypothesis	$\partial g_{ij}/\partial x_k = \Gamma_{ik,j} + \Gamma_{jk,i}$ EQ A	$(g_{ij})_{,k} = \Gamma_{ik,j} + \Gamma_{jk,i}$ EQ B
1	From T12.5.3B Christoffel Symbol of the First Kind	$\Gamma_{ij,k} = ½ (\partial g_{ik} / \partial x_k + \partial g_{jk} / \partial x_i - \partial g_{ij} / \partial x_k)$	
2	From 1 and permutation of indices	$\Gamma_{ik,j} = ½ (\partial g_{ij} / \partial x_k + \partial g_{jk} / \partial x_i - \partial g_{ik} / \partial x_j)$	
3	From 1 and permutation of indices	$\Gamma_{jk,i} = ½ (\partial g_{ij} / \partial x_k + \partial g_{ik} / \partial x_j - \partial g_{jk} / \partial x_i)$	
4	From 2, 3, T4.3.10 Equalities: Summation of Repeated Terms by 2 and adding equalities, A4.2.5 Commutative Add and A4.2.6 Associative Add	$\Gamma_{ik,j} + \Gamma_{jk,i} = ½ [2\partial g_{ij} / \partial x_k + (\partial g_{jk} / \partial x_i - \partial g_{jk} / \partial x_i) + (\partial g_{ik} / \partial x_j - \partial g_{ik} / \partial x_j)]$	
5	From 4, A4.2.8 Inverse Add and A4.2.7 Identity Add	$\Gamma_{ik,j} + \Gamma_{jk,i} = ½ (2\partial g_{ij} / \partial x_k)$	
∴	From 5, T4.4.7B Equalities: Product by Division, Factorization by 2, A4.2.2B Equality and D12.5.4 Simple Differental VectorNotation	$\partial g_{ij}/\partial x_k = \Gamma_{ik,j} + \Gamma_{jk,i}$ EQ A	$(g_{ij})_{,k} = \Gamma_{ik,j} + \Gamma_{jk,i}$ EQ B

Theorem 12.8.3 Differentiation of the Covariant Metric Tensor (Ricci's Theorem-A)

Step	Hypothesis	$(g_{ij})_{,k} = g_{j\alpha}\,\Gamma_{ik}{}^{\alpha} + g_{i\alpha}\,\Gamma_{jk}{}^{\alpha}$ EQ A	$(g_{ij})_{;k} = \mathbf{0}$ EQ B
1	From T12.8.2B Differentiation of the Metric Tensor of the First Kind	$(g_{ij})_{,k} = \Gamma_{ik,j} + \Gamma_{jk,i}$	
2	From 1g, 1, A4.2.12 Identity Multp and D6.1.6 Identity Matrix the Kronecker Delta	$(g_{ij})_{,k} = \delta_j^{\ j}\,\Gamma_{ik,j} + \delta_i^{\ i}\,\Gamma_{jk,i}$	
3	From 2, T12.2.8A Covariant and Contravariant Metric Inverse, T10.3.5 Tool: Tensor Substitution and A4.2.11 Associative Multp	$(g_{ij})_{,k} = g_{j\alpha}\,(g^{\alpha j}\Gamma_{ik,j}) + g_{i\alpha}\,(g^{\alpha i}\,\Gamma_{jk,i})$	
4	From 3, D12.5.3 Christoffel Symbol of the Second Kind and T10.3.5 Tool: Tensor Substitution	$(g_{ij})_{,k} = g_{j\alpha}\,\Gamma_{ik}{}^{\alpha} + g_{i\alpha}\,\Gamma_{jk}{}^{\alpha}$	
5	From 4, T10.4.18 Addition of Tensors: Identity by Addition and T10.4.21A Addition of Tensors: Right Cancellation by Addition	$(g_{ij})_{,k} - g_{j\alpha}\,\Gamma_{ik}{}^{\alpha} - g_{i\alpha}\,\Gamma_{jk}{}^{\alpha} = \mathbf{0}$	
6	From 5 and T12.6.3A Differentiation of Tensor: Rank-1 Contravariant	$(g_{ij})_{;k} = \mathbf{0}$	
∴	From 4 and 6	$(g_{ij})_{,k} = g_{j\alpha}\,\Gamma_{ik}{}^{\alpha} + g_{i\alpha}\,\Gamma_{jk}{}^{\alpha}$ EQ A	$(g_{ij})_{;k} = \mathbf{0}$ EQ B

Theorem 12.8.4 Simple Differential of the Contravariant Metric Tensor

Step	Hypothesis	$\partial g^{ij} / \partial x_k = -g^{i\alpha} \Gamma_{\alpha k}{}^{j} - g^{\alpha j} \Gamma_{\alpha k}{}^{i}$ EQ A	$(g^{ij})_{,k} = -g^{i\alpha} \Gamma_{\alpha k}{}^{j} - g^{\alpha j} \Gamma_{\alpha k}{}^{i}$ EQ B
1	From D12.1.5A Mixed Metric Dyadic	$\delta_i^{\ j} = g_{i\alpha} g^{\alpha j}$	
2	From 1 and TK.3.2 Uniqueness of Differentials	$\partial\delta_i^{\ j} / \partial x_k = \partial g_{i\alpha} g^{\alpha j} / \partial x_k$	
3	From 2, LxK.3.3.1B Differential of a Constant and LxK.3.2.4B Distributed Across Multiplication	$0 = g_{i\alpha} \partial g^{\alpha j} / \partial x_k + g^{\alpha j} \partial g_{i\alpha} / \partial x_k$	
4	From 3, T4.3.3B Equalities: Right Cancellation by Addition and A4.2.7 Identity Add	$g_{i\alpha} \partial g^{\alpha j} / \partial x_k = -g^{\alpha j} \partial g_{i\alpha} / \partial x_k$	
5	From 4, A4.2.2 Equality, D4.1.20 Negative Coefficient and T4.4.2 Equalities: Uniqueness of Multiplication	$g^{\beta i} g_{i\alpha} \partial g^{\alpha j} / \partial x_k = -g^{\beta i} g^{\alpha j} \partial g_{i\alpha} / \partial x_k$	
6	From 5, T12.2.8B Covariant and Contravariant Metric Inverse and T10.3.5 Tool: Tensor Substitution	$\delta^{\beta}{}_{\alpha} \partial g^{\alpha j} / \partial x_k = -g^{\alpha j} g^{\beta i} \partial g_{i\alpha} / \partial x_k$	
7	From 6, D6.1.6B Identity Matrix, and summing over $\alpha \rightarrow \beta$	$\partial g^{\beta j} / \partial x_k = -g^{\alpha j} g^{\beta i} \partial g_{i\alpha} / \partial x_k$	
8	From 8, T12.8.2 Differentiation of the Metric Tensor of the First Kind and T10.3.5 Tool: Tensor Substitution	$\partial g^{\beta j} / \partial x_k = -g^{\alpha j} g^{\beta i} (\Gamma_{ik,\alpha} + \Gamma_{\alpha k,i})$	
9	From 8, A4.2.14 Distribution, A4.2.10 Commutative Multp, T12.2.14 Contravariant Metric Inner Product is Symmetrical and T10.3.5 Tool: Tensor Substitution	$\partial g^{\beta j} / \partial x_k = -(g^{\beta i} \Gamma_{ik,\alpha} g^{\alpha j} + g^{\alpha j} \Gamma_{\alpha k,i} g^{i\beta})$	
10	From 9, D12.5.2 Christoffel Symbol of the Second Kind and T10.3.5 Tool: Tensor Substitution	$\partial g^{\beta j} / \partial x_k = -(g^{\beta i} \Gamma_{ik}{}^{j} + g^{\alpha j} \Gamma_{\alpha k}{}^{\beta})$	
11	From 10, replace dummy indices with $\beta = i$, $i = \alpha$ and $\alpha = \beta$	$\partial g^{ij} / \partial x_k = -(g^{i\alpha} \Gamma_{\alpha k}{}^{j} + g^{\alpha j} \Gamma_{\alpha k}{}^{i})$	
∴	From 11, D4.1.20 Negative Coefficient and A4.2.14 Distribution and D12.5.4B Simple Differental VectorNotation	$\partial g^{ij} / \partial x_k = -g^{i\alpha} \Gamma_{\alpha k}{}^{j} - g^{\alpha j} \Gamma_{\alpha k}{}^{i}$ EQ A	$(g^{ij})_{,k} = -g^{i\alpha} \Gamma_{\alpha k}{}^{j} - g^{\alpha j} \Gamma_{\alpha k}{}^{i}$ EQ B

Theorem 12.8.5 Differentiation of the Contravariant Metric Tensor (Ricci's Theorem-B)

Step	Hypothesis	$(g^{ij})_{,k} = -\Gamma_{\alpha k}{}^{j}\, g^{\alpha i} - \Gamma_{\beta k}{}^{i}\, g^{\beta j}$ EQ A	$(g^{ij})^{;k} = \mathbf{0}$ EQ B
1	From T12.8.4 Simple Differential of the Contravariant Metric Tensor and D12.5.4B Simple Differental VectorNotation	$(g^{ij})_{,k} = -g^{i\alpha}\, \Gamma_{\alpha k}{}^{j} - g^{\beta j}\, \Gamma_{\beta k}{}^{i}$	
2	From 1, A4.2.10 Commutative Multp and T12.2.14 Contravariant Metric Inner Product is Symmetrical	$(g^{ij})_{,k} = -\Gamma_{\alpha k}{}^{j}\, g^{\alpha i} - \Gamma_{\beta k}{}^{i}\, g^{\beta j}$	
3	From 2, D12.5.3 Christoffel Symbol of the Third Kind and T10.3.5 Tool: Tensor Substitution	$(g^{ij})_{,k} = -\Gamma^{ji}{}_{k} - \Gamma^{ij}{}_{k}$	
4	From 3, T10.4.18 Addition of Tensors: Identity by Addition and T10.4.22A Addition of Tensors: Reversal of Right Cancellation by Addition	$(g^{ij})_{,k} + \Gamma_{\alpha k}{}^{j}\, g^{\alpha i} + \Gamma_{\beta k}{}^{i}\, g^{\beta j} = \mathbf{0}$	
5	From 5 and T12.6.2A Differentiation of Tensor: Rank-1 Covariant	$(g^{ij})^{;k} = \mathbf{0}$	
∴	From 2 and 5	$(g^{ij})_{,k} = -\Gamma_{\alpha k}{}^{j}\, g^{\alpha i} - \Gamma_{\beta k}{}^{i}\, g^{\beta j}$ EQ A	$(g^{ij})^{;k} = \mathbf{0}$ EQ B

Observation 12.8.1 Covariant and Contravariant Derivatives of the Fundamental Tensor

From T12.8.3B "Differentiation of the Covariant Metric Tensor", $(g_{ij})_{;k} = \mathbf{0}$ and T12.8.5B "Differentiation of the Contravariant Metric Tensor", $(g^{ij})^{;k} = \mathbf{0}$ implying the fundamental tensors create a tensor field that is parallel so that direction is constant.

The following corollaries can be proven similarly with the following theorems T12.7.6 "Tensor Product: Chain Rule for Binary Covariant Product", T12.7.7 "Tensor Product: Chain Rule for Contravariant Binary Product", T12.7.8 "Tensor Product: Chain Rule for Mixed Binary Product", T12.8.3B "Differentiation of the Covariant Metric Tensor (Ricci's Theorem-A)" and T12.8.5B "Differentiation of the Contravariant Metric Tensor (Ricci's Theorem-B)", as an example see the corollary C12.8.5.7 "Differentiation of Covariant Metric and Mixed Tensor Rank-3" bellow:

Table 12.8.1 Distribution of Fundamental Tensors Under Tenor Differentiation

Corllary 12.8.5.1 Differentiation of Covariant Metric with Covariant Coefficient	$(g_{\alpha j} A_{\alpha})_{;r} = g_{\alpha j} A_{\alpha;r}$
Corllary 12.8.5.2 Differentiation of Covariant Metric with Contravariant Coefficient	$(g_{\alpha j} A^{\alpha})^{;r} = g_{\alpha j} A^{\alpha;r}$
Corllary 12.8.5.3 Differentiation of Covariant Metric with Mixed Coefficient	$(g_{\alpha k} A_k{}^{\alpha})^{;r} = g_{\alpha k} A_k{}^{\alpha;r}$
Corllary 12.8.5.4 Differentiation of Contravariant Metric with Covariant Coefficient	$(g^{\alpha j} A_{\alpha})_{;r} = g^{\alpha j} A_{\alpha;r}$
Corllary 12.8.5.5 Differentiation of Contravariant Metric with Contravariant Coefficient	$(g^{\alpha j} A^{\alpha})^{;r} = g^{\alpha j} A^{\alpha;r}$
Corllary 12.8.5.6 Differentiation of Contravariant Metric with Mixed Coefficient	$(g^{\alpha k} A_k{}^{\alpha})^{;r} = g^{\alpha k} A_k{}^{\alpha;r}$

Observation 12.8.2 Factoring the Fundamental Tensor from a Differential

As an immediate corollary of Ricci's theorem we note that the fundamental tensors may be taken outside the sign of covariant differentiation, and hence the operations of lowering and raising indices are permutable with covariant differentiation.

Theorem 12.8.6 Derivative Covariant Determinate to the Covariant Metric

1g	Given	Einsteinian Summation Notation does not apply.
Step	Hypothesis	$\partial g / \partial g_{ij} = G^{ij}$
1	From T6.9.5A Expanding Determinant [A] of order [n] about a Row or Column	$g = \Sigma_k g_{ik} G^{ik}$ expanding over determinate rows
2	From 1 and TK.3.2 Uniqueness of Differentials	$\partial g / \partial g_{ij} = \partial (\Sigma_k g_{ik} G^{ik}) / \partial g_{ij}$
3	From 2 and LxK.3.2.2B Differential as a Linear Operator	$\partial g / \partial g_{ij} = \Sigma_k \partial (g_{ik} G^{ik}) / \partial g_{ij}$
4	From 2, K.3.8 Multi-variable Chain Rule	$\partial g / \partial g_{ij} = \Sigma_k [\partial (g_{ik}) / \partial g_{ij}] G^{ik} + g^{ik} [\partial (G^{ik}) / \partial g_{ij}]$
5	From 4, T12.8.1 Differential of the Contravariant Cofactor to the Metric Dyadic	$\partial (G^{ik}) / \partial g_{ij} = 0$
6	From 5 and LxK.3.3.2 Differential identity	$\partial (g_{ik}) / \partial g_{ij} = 1$ for $k = j$
7	From 6 and LxK.3.3.1 Differential of a Constant	$\partial (g_{ik}) / \partial g_{ij} = 0$ for $g^{ik} \neq f(g^{ik})$ independent, constant or $k \neq j$
8	From 6, 7 and D6.1.6 Identity Matrix the Kronecker Delta	$\partial (g_{ik}) / \partial g_{ij} = \delta_k^{\ j}$
9	From 4, 5, 8, T10.3.5 Tool: Tensor Substitution and A4.2.3 Substitution	$\partial g / \partial g_{ij} = \Sigma_k [\delta_k^{\ j}] G^{ik} + g^{ik} [0]$
∴	From 9, D6.1.6 Identity Matrix is one, T4.4.1 Equalities: Any Quantity Multiplied by Zero is Zero, A4.2.12 Identity Multp and A4.2.7 Identity Add	$\partial g / \partial g_{ij} = G^{ij}$

Theorem 12.8.7 Derivative of the Natural Log of the Covariant Determinate

1g	Given	Einsteinian Summation Notation does apply for [α].	
Step	Hypothesis	$\partial \ln \sqrt{|g_e|} / \partial x_j = \Sigma_\alpha \Gamma_{\alpha i}{}^{\alpha}$ EQ A	$\partial \ln \sqrt{|g_e|} / \partial x_j = \Sigma_\alpha \Gamma_{i\alpha}{}^{\alpha}$ EQ B
1	From 12.1.4D Contravariant Metric Dyadic and T4.4.5A Equalities: Reversal of Right Cancellation by Multiplication	$g\, g^{ij} = G^{ij}$	
2	From A4.2.2A Equality and TK.3.2 Uniqueness of Differentials	$\partial g / \partial x_j = \partial g / \partial x_j$	
3	From 2 and TK.3.4B Single Variable Chain Rule	$\partial g / \partial x_j = \Sigma_\alpha \Sigma_\beta (\partial g / \partial g_{\alpha\beta}) (\partial g_{\alpha\beta} / \partial x_j)$	
4	From 1, 3, T12.8.6 Derivative of the Covariant Determinate by Covariant Metric and T10.3.5 Tool: Tensor Substitution	$\partial g / \partial x_j = \Sigma_\alpha \Sigma_\beta\, g\, g^{\alpha\beta} (\partial g_{\alpha\beta} / \partial x_j)$	
5	From 4, T12.8.3A Differentiation of the Covariant Metric Tensor (Ricci's Theorem-A) and T10.3.5 Tool: Tensor Substitution	$\partial g / \partial x_j = g\, \Sigma_\alpha \Sigma_\beta\, g^{\alpha\beta} (\Sigma_x g_{\beta x} \Gamma_{\alpha i}{}^{x} + \Sigma_y g_{\alpha y} \Gamma_{\beta i}{}^{y})$	
6	From 5 and T12.2.13 Covariant Metric Inner Product is Symmetrical	$\partial g / \partial x_j = g\, \Sigma_\alpha \Sigma_\beta\, g^{\alpha\beta} (\Sigma_x g_{\beta x} \Gamma_{\alpha i}{}^{x} + \Sigma_y g_{y\alpha} \Gamma_{\beta i}{}^{y})$	
7	From 6 and A4.2.14 Distribution	$\partial g / \partial x_j = g\, (\Sigma_\alpha \Sigma_x \Sigma_\beta\, g^{\alpha\beta} g_{\beta x} \Gamma_{\alpha i}{}^{x} + \Sigma_\beta \Sigma_y \Sigma_\alpha\, g^{\beta\alpha} g_{\alpha y} \Gamma_{\beta i}{}^{y})$	
8	From 7, T12.2.8B Covariant and Contravariant Metric Inverse and T10.3.5 Tool: Tensor Substitution	$\partial g / \partial x_j = g\, \Sigma_\alpha (\Sigma_x \delta^{\alpha}{}_{x} \Gamma_{\alpha i}{}^{x}) + \Sigma_\beta (\Sigma_y \delta^{\beta}{}_{y} \Gamma_{\beta i}{}^{y})$	
9	From 8 and D6.1.6 Identity Matrix then x = α and y = β	$\partial g / \partial x_j = g\, (\Sigma_\alpha \Gamma_{\alpha i}{}^{\alpha} + \Sigma_\beta \Gamma_{\beta i}{}^{\beta})$	
10	From 9 and substitute dummy indices $\alpha = \beta$	$\partial g / \partial x_j = g\, (\Sigma_\alpha \Gamma_{\alpha i}{}^{\alpha} + \Sigma_\alpha \Gamma_{\alpha i}{}^{\alpha})$	
11	From 10 and T4.3.10 Equalities: Summation of Repeated Terms by 2	$\partial g / \partial x_j = 2\, g\, \Sigma_\alpha \Gamma_{\alpha i}{}^{\alpha}$	
12	From 11 and T4.4.4A Equalities: Right Cancellation by Multiplication	$\frac{1}{2} (1/g)\, \partial g / \partial x_j = \Sigma_\alpha \Gamma_{\alpha i}{}^{\alpha}$	
13	From 12 and LxK.3.3.4B Differential a Natural Log base [e]	$\frac{1}{2}\, \partial \ln (g) / \partial x_j = \Sigma_\alpha \Gamma_{\alpha i}{}^{\alpha}$	
14	From 13 and LxK.3.2.3B Distribute Across Constant Product	$\partial\, \frac{1}{2} \ln (g) / \partial x_j = \Sigma_\alpha \Gamma_{\alpha i}{}^{\alpha}$	
∴	From 1g, 14, T4.11.4 Real Exponents: Exponent of Logarithm, T4.2.25 Reciprocal Exp and T12.5.13 Christoffel Symbol: Second Kind Symmetry for First Two Indices	$\partial \ln \sqrt{|g_e|} / \partial x_j = \Sigma_\alpha \Gamma_{\alpha i}{}^{\alpha}$ EQ A argument of the log function cannot be negative, hence the positive sign is taken	$\partial \ln \sqrt{|g_e|} / \partial x_j = \Sigma_\alpha \Gamma_{i\alpha}{}^{\alpha}$ EQ B argument of the log function cannot be negative, hence the positive sign value is taken

Theorem 12.8.8 Christoffel Symbol: Second Kind Difference between Duplicate Last Indices

Step	Hypothesis	$\Gamma_{ik}{}^{k} = ½\,(g_{i\alpha,k} - g_{ik,\alpha})\,g^{\alpha k}$ EQ A	$\Gamma_{ik}{}^{k} = ½\,(g_{i\alpha,k}\,g^{\alpha k} - g_{ik,\alpha}\,g^{\alpha k})$ EQ B
1	From D12.5.2 Christoffel Symbol of the Second Kind	$\Gamma_{ik}{}^{k} = \Gamma_{ik,\alpha}\,g^{\alpha k}$	
2	From 1, D12.5.1 Christoffel Symbol of the First Kind and T10.2.5 Tool: Tensor Substitution	$\Gamma_{ik}{}^{k} = ½\,(g_{i\alpha,k} + g_{k\alpha,i} - g_{ik,\alpha})\,g^{\alpha k}$	
3	From 2 and T10.3.20 Addition of Tensors: Distribution of a Tensor over Addition of Tensors	$\Gamma_{ik}{}^{k} = ½\,(g_{i\alpha,k}\,g^{\alpha k} + g_{k\alpha,i}\,g^{\alpha k} - g_{ik,\alpha}\,g^{\alpha k})$	
4	From 3 and C12.8.5.4 Differentiation of Contravariant Metric with Covariant Coefficient	$\Gamma_{ik}{}^{k} = ½\,(g_{i\alpha,k}\,g^{\alpha k} + (g_{k\alpha}\,g^{\alpha k})_{,i} - g_{ik,\alpha}\,g^{\alpha k})$	
5	From 4 and T12.2.8A Covariant and Contravariant Metric Inverse	$\Gamma_{ik}{}^{k} = ½\,(g_{i\alpha,k}\,g^{\alpha k} + (\delta_{k}{}^{k})_{,i} - g_{ik,\alpha}\,g^{\alpha k})$	
6	From 5, T12.2.16 Covariant-Contravariant Kronecker Delta Summation of like Indices and A4.2.3 Substitution	$\Gamma_{ik}{}^{k} = ½\,(g_{i\alpha,k}\,g^{\alpha k} + (n)_{,i} - g_{ik,\alpha}\,g^{\alpha k})$	
7	From 6, Lx9.3.3.1B Differential of a Constant and T10.3.18 Addition of Tensors: Identity by Addition	$\Gamma_{ik}{}^{k} = ½\,(g_{i\alpha,k}\,g^{\alpha k} - g_{ik,\alpha}\,g^{\alpha k})$	
8	From 7 and T10.3.20 Addition of Tensors: Distribution of a Tensor over Addition of Tensors	$\Gamma_{ik}{}^{k} = ½\,(g_{i\alpha,k} - g_{ik,\alpha})\,g^{\alpha k}$	
∴	From 7 and 8	$\Gamma_{ik}{}^{k} = ½\,(g_{i\alpha,k} - g_{ik,\alpha})\,g^{\alpha k}$ EQ A	$\Gamma_{ik}{}^{k} = ½\,(g_{i\alpha,k}\,g^{\alpha k} - g_{ik,\alpha}\,g^{\alpha k})$ EQ B

Theorem 12.8.9 Christoffel Symbol: Second Kind Duplicate Last Indices

Step	Hypothesis	$\Gamma_{ib}{}^{b} = ½ \Sigma_\alpha g_{b\alpha,i} g^{\alpha b}$
1	From T12.5.3 Christoffel Symbol of the First Kind and D12.5.2 Christoffel Symbol of the Second Kind	$\Gamma_{ij}{}^{k} = ½ \Sigma_\alpha (g_{i\alpha,j} + g_{j\alpha,i} - g_{ij,\alpha}) g^{\alpha k}$
2	From 1, T10.3.16 Addition of Tensors: Commutative by Addition and setting j = k = b while Einsteinian Summation Notation	$\Sigma_b \Gamma_{ib}{}^{b} = ½ \Sigma_b \Sigma_\alpha (g_{b\alpha,i} + g_{i\alpha,b} - g_{ib,\alpha}) g^{\alpha b}$
3	From 2, T10.3.20 Addition of Tensors: Distribution of a Tensor over Addition of Tensors	$\Sigma_b \Gamma_{ib}{}^{b} = ½ (\Sigma_\alpha \Sigma_b g_{b\alpha,i} g^{\alpha b} + \Sigma_\alpha \Sigma_b g_{i\alpha,b} g^{\alpha b} - \Sigma_\alpha \Sigma_b g_{ib,\alpha} g^{\alpha b})$
4	From 3 and summing over all tensor components all elements are present, hence the equality holds	$\Sigma_\alpha \Sigma_b g_{i\alpha,b} g^{\alpha b} = \Sigma_\alpha \Sigma_b g_{ib,\alpha} g^{\alpha b}$
5	From 3, 4 and T10.3.5 Tool: Tensor Substitution	$\Sigma_b \Gamma_{ib}{}^{b} = ½ (\Sigma_\alpha \Sigma_b g_{b\alpha,i} g^{\alpha b} + \Sigma_\alpha \Sigma_b g_{i\alpha,b} g^{\alpha b} - \Sigma_\alpha \Sigma_b g_{i\alpha,b} g^{\alpha b})$
6	From 5 and T10.3.20 Addition of Tensors: Distribution of a Tensor over Addition of Tensors	$\Sigma_b \Gamma_{ib}{}^{b} = ½ \Sigma_\alpha \Sigma_b g_{b\alpha,i} g^{\alpha b} + \Sigma_\alpha \Sigma_b (g_{i\alpha,b} - g_{i\alpha,b})g^{\alpha b}$
7	From 6 and T10.3.19 Addition of Tensors: Inverse by Addition	$\Sigma_b \Gamma_{ib}{}^{b} = ½ \Sigma_\alpha \Sigma_b g_{b\alpha,i} g^{\alpha b} + \Sigma_\alpha \Sigma_b (\mathbf{0})g^{\alpha b}$
8	From 7 and T10.2.7 Tool: Zero Times a Tensor is a Zero Tensor and T10.3.18 Addition of Tensors: Identity by Addition	$\Sigma_b \Gamma_{ib}{}^{b} = \Sigma_b (½ \Sigma_\alpha g_{b\alpha,i} g^{\alpha b})$
∴	From 7 and equating coefficients term-by-term	$\Gamma_{ib}{}^{b} = ½ \Sigma_\alpha g_{b\alpha,i} g^{\alpha b}$

Section 12.9 Generalization of the Kronecker Delta

The Kronecker Delta as seen from definition 12.1.5A "Mixed Metric Dyadic" is actually the product of two bases vectors forming a tensor of rank-2. Even though it evaluates to two real numbers one and zero, it still is a tensor and as such has all of the properties of any tensor of that rank, including differentiation.

Theorem 12.9.1 Differentiated Kronecker Delta and Fundamental Tensors

Step	Hypothesis	$\mathbf{0} = (g_{i\alpha})_{;k}\, g^{\alpha j} + g_{i\alpha}\, (g^{\alpha j})_{;k}$ EQ A	$(\delta_i^{\ j})_{;k} = \mathbf{0}$ EQ B
1	From T12.2.8A Covariant and Contravariant Metric Inverse	$\delta_i^{\ j} = g_{i\alpha}\, g^{\alpha j}$	
2	From 1, TK.3.2 Uniqueness of Differentials, D12.5.4A Simple Differental VectorNotation and K.5.2 Multi-variable Chain Rule for a Vector	$(\delta_i^{\ j})_{;k} = (g_{i\alpha})_{;k}\, g^{\alpha j} + g_{i\alpha}\, (g^{\alpha j})_{;k}$	
3	From 2, T12.8.3A Differentiation of the Covariant Metric Tensor (Ricci's Theorem-A), T12.8.4 Simple Differential of the Contravariant Metric Tensor and T10.3.5 Tool: Tensor Substitution	$(\delta_i^{\ j})_{;k} = (g_{\alpha\beta}\, \Gamma_{ik}^{\ \ \beta} + g_{i\beta}\, \Gamma_{\alpha k}^{\ \ \beta})g^{\alpha j} + g_{i\alpha}\, (-g^{\alpha\beta}\, \Gamma_{\beta k}^{\ \ j} - g^{\beta j}\, \Gamma_{\beta k}^{\ \ \alpha})$	
4	From 3, D4.1.21A Negative Coefficient, T10.4.20 Addition of Tensors: Distribution of a Tensor over Addition of Tensors	$(\delta_i^{\ j})_{;k} = g_{\alpha\beta}\, g^{\alpha j}\, \Gamma_{ik}^{\ \ \beta} + g_{i\beta}\, g^{\alpha j}\, \Gamma_{\alpha k}^{\ \ \beta} - g_{i\alpha}\, g^{\alpha\beta}\, \Gamma_{\beta k}^{\ \ j} - g_{i\alpha}\, g^{\beta j}\, \Gamma_{\beta k}^{\ \ \alpha}$	
5	From 4 and T12.2.13 Covariant Metric Inner Product is Symmetrical	$(\delta_i^{\ j})_{;k} = g_{\beta\alpha}\, g^{\alpha j}\, \Gamma_{ik}^{\ \ \beta} + g_{i\beta}\, g^{\alpha j}\, \Gamma_{\alpha k}^{\ \ \beta} - g_{i\alpha}\, g^{\alpha\beta}\, \Gamma_{\beta k}^{\ \ j} - g_{i\alpha}\, g^{\beta j}\, \Gamma_{\beta k}^{\ \ \alpha}$	
6	From 5, T12.2.8A Covariant and Contravariant Metric Inverse and T10.3.5 Tool: Tensor Substitution	$(\delta_i^{\ j})_{;k} = \delta_\beta^{\ j}\, \Gamma_{ik}^{\ \ \beta} + g_{i\beta}\, g^{\alpha j}\, \Gamma_{\alpha k}^{\ \ \beta} - \delta_i^{\ \beta}\, \Gamma_{\beta k}^{\ \ j} - g_{i\alpha}\, g^{\beta j}\, \Gamma_{\beta k}^{\ \ \alpha}$	
7	From 6 and D6.1.6 Identity Matrix-Kronecker Delta; evaluated	$(\delta_i^{\ j})_{;k} = \Gamma_{ik}^{\ \ j} + g_{i\beta}\, g^{\alpha j}\, \Gamma_{\alpha k}^{\ \ \beta} - \Gamma_{ik}^{\ \ j} - g_{i\alpha}\, g^{\beta j}\, \Gamma_{\beta k}^{\ \ \alpha}$	
8	From 7, T10.4.16 Addition of Tensors: Commutative by Addition and T10.4.17 Addition of Tensors: Associative by Addition	$(\delta_i^{\ j})_{;k} = (\Gamma_{ik}^{\ \ j} - \Gamma_{ik}^{\ \ j}) + (g_{i\beta}\, g^{\alpha j}\, \Gamma_{\alpha k}^{\ \ \beta} - g_{i\beta}\, g^{\alpha j}\, \Gamma_{\alpha k}^{\ \ \beta})$	
9	From 8, T10.4.19 Addition of Tensors: Inverse by Addition and T10.4.18 Addition of Tensors: Identity by Addition	$(\delta_i^{\ j})_{;k} = (g_{i\beta}\, g^{\alpha j}\, \Gamma_{\alpha k}^{\ \ \beta}) - (g_{i\beta}\, g^{\alpha j}\, \Gamma_{\alpha k}^{\ \ \beta})$	
10	From 9 and T10.4.19 Addition of Tensors: Inverse by Addition	$(\delta_i^{\ j})_{;k} = \mathbf{0}$	
11	From 2, 10 and T10.3.5 Tool: Tensor Substitution	$\mathbf{0} = (g_{i\alpha})_{;k}\, g^{\alpha j} + g_{i\alpha}\, (g^{\alpha j})_{;k}$	
∴	From 10 and 11	$\mathbf{0} = (g_{i\alpha})_{;k}\, g^{\alpha j} + g_{i\alpha}\, (g^{\alpha j})_{;k}$ EQ A	$(\delta_i^{\ j})_{;k} = \mathbf{0}$ EQ B

Theorem 12.9.2 Differentiation of Kronecker Delta and Covariant Coefficient

1g	Given	$\delta^{\alpha}{}_{j}\, A_{\alpha} = \delta^{\alpha}{}_{j}\, A_{\alpha}$
Step	Hypothesis	$(\delta^{\alpha}{}_{j}\, A_{\alpha})_{;r} = \delta^{\alpha}{}_{j}\, A_{\alpha;r}$
1	From 1g, T12.7.1(A, B) Tensor Product: Uniqueness of Tensor Differentiation	$(\delta^{\alpha}{}_{j}\, A_{\alpha})_{;r} = (\delta^{\alpha}{}_{j}\, A_{\alpha})_{;r}$
2	From 1 and T12.7.8 Tensor Product: Chain Rule for Mixed Binary Product	$(\delta^{\alpha}{}_{j}\, A_{\alpha})_{;r} = (\delta^{\alpha}{}_{j})_{;r}\, A_{\alpha} + \delta^{\alpha}{}_{j}\, A_{\alpha;r}$
3	From 2, T12.9.1B Differentiated Kronecker Delta and Fundamental Tensors and T10.3.5 Tool: Tensor Substitution	$(\delta^{\alpha}{}_{j}\, A_{\alpha})_{;r} = (\mathbf{0})\, A_{\alpha} + \delta^{\alpha}{}_{j}\, A_{\alpha;r}$
∴	From 3, T10.3.7 Tool: Zero Times a Tensor is a Zero Tensor and T10.4.18 Addition of Tensors: Identity by Addition	$(\delta^{\alpha}{}_{j}\, A_{\alpha})_{;r} = \delta^{\alpha}{}_{j}\, A_{\alpha;r}$

Theorem 12.9.3 Differentiation of Kronecker Delta and Contravariant Tensor

1g	Given	$\delta_{\alpha}^{\ j}\, A^{\alpha} = \delta_{\alpha}^{\ j}\, A^{\alpha}$
Step	Hypothesis	$(\delta_{\alpha}^{\ j}\, A^{\alpha})^{;r} = \delta_{\alpha}^{\ j}\, A^{\alpha;r}$
1	From 1g, T12.7.1(A, B) Tensor Product: Uniqueness of Tensor Differentiation	$(\delta_{\alpha}^{\ j}\, A^{\alpha})^{;r} = (\delta_{\alpha}^{\ j}\, A^{\alpha})^{;r}$
2	From 1 and T12.7.8 Tensor Product: Chain Rule for Mixed Binary Product	$(\delta_{\alpha}^{\ j}\, A^{\alpha})^{;r} = (\delta_{\alpha}^{\ j})^{;r}\, A^{\alpha} + \delta_{\alpha}^{\ j}\, A^{\alpha;r}$
3	From 2, T12.9.1B Differentiated Kronecker Delta and Fundamental Tensors and T10.3.5 Tool: Tensor Substitution	$(\delta_{\alpha}^{\ j}\, A^{\alpha})^{;r} = (\mathbf{0})\, A_{jk}{}^{\alpha} + \delta_{\alpha}^{\ j}\, A^{\alpha;r}$
∴	From 3, T10.3.7 Tool: Zero Times a Tensor is a Zero Tensor and T10.4.18 Addition of Tensors: Identity by Addition	$(\delta_{\alpha}^{\ j}\, A^{\alpha})^{;r} = \delta_{\alpha}^{\ j}\, A^{\alpha;r}$

Theorem 12.9.4 Christoffel Symbol: Second Kind Diagonal Indices

1g	Given	$g_{i\alpha,j} = g_{j\alpha,i} = g_{ij,\alpha} = 0$ for $i \neq j$ iff an orthogonal system
2g		$g_{b\alpha,b} \sim\neq g_{bb,\alpha} \sim\neq 0$ for $i = j = b$ on the diagonal
Step	Hypothesis	$\Gamma_{bb}{}^{k} = -½ \sum_\alpha g_{bb,\alpha}\, g^{\alpha k}$
1	From T12.5.3 Christoffel Symbol of the First Kind and D12.5.2 Christoffel Symbol of the Second Kind	$\Gamma_{ij}{}^{k} = ½ \sum_\alpha (g_{i\alpha,j} + g_{j\alpha,i} - g_{ij,\alpha})\, g^{\alpha k}$
2	From 2g, 1 and substitution of indices	$\Gamma_{bb}{}^{k} = ½ \sum_\alpha (g_{b\alpha,b} + g_{b\alpha,b} - g_{bb,\alpha})\, g^{\alpha k}$ on the diagonal
3	From 1, T10.3.7 Product of Tensors: Commutative by Multiplication of Rank p + q → q + p, T10.2.10 Tool: Identity Multiplication of a Tensor, and D4.1.19 Primitive Definition for Rational Arithmetic	$\Gamma_{bb}{}^{k} = \sum_\alpha ½\, g^{\alpha k} (2\, g_{b\alpha,b} - g_{bb,\alpha})$
4	From 2, T10.4.8 Scalars and Tensors: Scalar Commutative with Tensors, T4.4.16B Equalities: Product by Division, Common Factor, T10.2.10 Tool: Identity Multiplication of a Tensor, T10.3.20 Addition of Tensors: Distribution of a Tensor over Addition of Tensors, C12.8.5.4 Differentiation of Contravariant Metric and Covariant Coefficient and T10.3.7 Product of Tensors: Commutative by Multiplication of Rank p + q → q + p	$\Gamma_{bb}{}^{k} = (\sum_\alpha g_{b\alpha}\, g^{\alpha k})_{,b} - ½ \sum_\alpha g^{\alpha k}\, g_{bb,\alpha}$
5	From 4 and T12.2.8A Covariant and Contravariant Metric Inverse	$\Gamma_{bb}{}^{k} = (\delta_b{}^{k})_{,b} - ½ \sum_\alpha g^{\alpha k}\, g_{bb,\alpha}$
∴	From 5, T12.9.1B Differentiated Kronecker Delta and Fundamental Tensors and T10.3.18 Addition of Tensors: Identity by Addition and T10.3.7 Product of Tensors: Commutative by Multiplication of Rank p + q → q + p	$\Gamma_{bb}{}^{k} = -½ \sum_\alpha g_{bb,\alpha}\, g^{\alpha k}$

Section 12.10 Cross Product Operator: Theorems

The following theorems use standard covariant-contravariant notation, see section 12.1.

Theorem 12.10.1 Cross Product: Product of (n-1)-Covariant Vectors Spanning an n-Space

1g 2g	Given	$A_1\ (\bullet\ \mathbf{b}_i \bullet)$ $A_{1j} = \mathbf{b}_j$ for i = 1 first row	
Steps	Hypothesis	$\prod_{\beta=2}{}^{n} \times \mathbf{A}_\beta = \sum_j A_{1j}\ \mathbf{b}_j$ EQ A	$\langle| \overset{*}{\mathrm{A}}_m |\rangle \equiv \mu_f \prod_{\beta=2}{}^{n} \times \mathbf{A}_\beta$ EQ B
1	From T10.14.2 Relationship between TDL and the Determinate Operator	$\langle| \overset{*}{\mathrm{A}}_m |\rangle = \mu_f \mid A \mid$	
2	From 1 and T6.9.5 Expanding Determinant [A] of order [n] about a Row or Column	$\langle| \overset{*}{\mathrm{A}}_m |\rangle = \mu_f \sum_j \mathbf{b}_j A_{1j}$ cofactor about row 1	
3	From 1g, 2g, 2 and T7.9.2 Substitution of Vectors	$\langle| \overset{*}{\mathrm{A}}_m |\rangle = \mu_f \sum_j \mathbf{b}_j A_{1j}$ cofactor about row 1	
4	From 3 and T6.9.5 Expanding Determinant [A] of order [n] about a Row or Column	$\langle| \overset{*}{\mathrm{A}}_m |\rangle \equiv \mu_f \prod_{\beta=2}{}^{n} \times \mathbf{A}_\beta$ determinate of base vectors spanning a n-space	
5	From 3 and 4	$\mu_f \prod_{\beta=2}{}^{n} \times \mathbf{A}_\beta = \mu_f \sum_j \mathbf{b}_j A_{1j}$	
∴	From 4, 5, T4.4.3 Equalities: Cancellation by Multiplication and A7.8.5 Parallelogram Law: Commutative and Scalability of Collinear Vectors	$\prod_{\beta=2}{}^{n} \times \mathbf{A}_\beta = \sum_j A_{1j}\ \mathbf{b}_j$ EQ A	$\langle| \overset{*}{\mathrm{A}}_m |\rangle \equiv \mu_f \prod_{\beta=2}{}^{n} \times \mathbf{A}_\beta$ EQ B

Theorem 12.10.2 Cross Product: Product of (n-1)-Contravariant Vectors Spanning an n-Space

1g 2g	Given	$A^1\ (\bullet\ \mathbf{b}^j \bullet)$ $A^{1j} = \mathbf{b}^j$ for i = 1 first row	
Steps	Hypothesis	$\prod_{\beta=2}{}^{n} \times \mathbf{A}^\beta = \sum_j A^{1j}\ \mathbf{b}^j$ EQ A	$\langle| \overset{*}{\mathrm{A}}_m |\rangle \equiv \mu_f \prod_{\beta=2}{}^{n} \times \mathbf{A}^\beta$ EQ B
1	From T10.14.2 Relationship between TDL and the Determinate Operator	$\langle| \overset{*}{\mathrm{A}}_m |\rangle = \mu_f \mid A \mid$	
2	From 1 and T6.9.5 Expanding Determinant [A] of order [n] about a Row or Column	$\langle| \overset{*}{\mathrm{A}}_m |\rangle = \mu_f \sum_j \mathbf{b}^j A^{1j}$ cofactor about row 1	
3	From 1g, 2g, 2 and T7.9.2 Substitution of Vectors	$\langle| \overset{*}{\mathrm{A}}_m |\rangle = \mu_f \sum_j \mathbf{b}^j A^{1j}$ cofactor about row 1	
4	From 3 and T6.9.5 Expanding Determinant [A] of order [n] about a Row or Column	$\langle| \overset{*}{\mathrm{A}}_m |\rangle \equiv \mu_f \prod_{\beta=2}{}^{n} \times \mathbf{A}^\beta$ determinate of base vectors spanning a n-space	
5	From 3 and 4	$\mu_f \prod_{\beta=2}{}^{n} \times \mathbf{A}^\beta = \mu_f \sum_j \mathbf{b}^j A^{1j}$	
∴	From 4, 5, T4.4.3 Equalities: Cancellation by Multiplication and A7.8.5 Parallelogram Law: Commutative and Scalability of Collinear Vectors	$\prod_{\beta=2}{}^{n} \times \mathbf{A}^\beta = \sum_j A^{1j}\ \mathbf{b}^j$ EQ A	$\langle| \overset{*}{\mathrm{A}}_m |\rangle \equiv \mu_f \prod_{\beta=2}{}^{n} \times \mathbf{A}^\beta$ EQ B

This is probably the most crucial theorem in tensor algebra it shows two things:

1) A tensor can be embedded inside a tensor, and
2) it takes (n-1) vectors to span an n-space to multiply them and gain a resulting vector in return.

Theorem 12.10.3 Cross Product: Covariant Expansion

1g	Given	$A^{1j} = \mathbf{b}_j$ for i = 1 first row
Steps	Hypothesis	$\prod_{\beta=2}^{n} \times \mathbf{A}_\beta = \sum_{\sigma(k)\in Sn} \prod_{i=2}^{n} A^{i\lambda[i,k]} \times \mathbf{b}_{\lambda[i,k]}$
1	From T10.11.1B Cross Product: Multiplication of (n-1)-Vectors Spanning an n-Space	$<\mid \overset{*}{\mathring{A}}_m \mid> = \mu_f \prod_{\beta=2}^{n} \times \mathbf{A}_\beta$
2	From 1, D6.9.1E Determinant of A and factoring the first factor	$<\mid \overset{*}{\mathring{A}}_m \mid> = \mu_f \sum_{\sigma(k)\in Sn} \mathrm{sgn}(\sigma(k))\, \mathbf{b}_{\lambda[1,k]} \prod_{i=2}^{n} A^{i\lambda[i,k]}$
3	From 2, A10.15.1 The Cross Product Operator and T10.3.5 Tool: Tensor Substitution	$<\mid \overset{*}{\mathring{A}}_m \mid> = \mu_f \sum_{\sigma(k)\in Sn} (\prod_{i=2}^{n} \times \mathbf{b}_{\lambda[i,k]})\, (\prod_{i=2}^{n} A^{i\lambda[i,k]})$
4	From 3 and T10.4.7 Product of Tensors: Commutative by Multiplication of Rank for p + q	$<\mid \overset{*}{\mathring{A}}_m \mid> = \mu_f \sum_{\sigma(k)\in Sn} (\prod_{i=2}^{n} A^{i\lambda[i,k]})\, (\prod_{i=2}^{n} \times \mathbf{b}_{\lambda[i,k]})$
5	From 4 and T10.3.4 Tool: Base Tensor Distributed over Polyadic Coefficient Summation	$<\mid \overset{*}{\mathring{A}}_m \mid> = \mu_f \sum_{\sigma(k)\in Sn} \prod_{i=2}^{n} A^{i\lambda[i,k]} \times \mathbf{b}_{\lambda[i,k]}$
6	From 1 and 5	$\mu_f \prod_{\beta=2}^{n} \times \mathbf{A}_\beta = \mu_f \sum_{\sigma(k)\in Sn} \prod_{i=2}^{n} A^{i\lambda[i,k]} \times \mathbf{b}_{\lambda[i,k]}$
∴	From 6 and T4.4.3 Equalities: Cancellation by Multiplication	$\prod_{\beta=2}^{n} \times \mathbf{A}_\beta = \sum_{\sigma(k)\in Sn} \prod_{i=2}^{n} A^{i\lambda[i,k]} \times \mathbf{b}_{\lambda[i,k]}$

Theorem 12.10.4 Cross Product: Contravariant Expansion

1g	Given	$A_{1j} = \mathbf{b}^j$ for i = 1 first row
Steps	Hypothesis	$\prod_{\beta=2}^{n} \times \mathbf{A}^\beta = \sum_{\sigma(k)\in Sn} \prod_{i=2}^{n} A_{i\lambda[i,k]} \times \mathbf{b}^{\lambda[i,k]}$
1	From T10.11.1B Cross Product: Multiplication of (n-1)-Vectors Spanning an n-Space	$<\mid \overset{*}{\mathring{A}}_m \mid> = \mu_f \prod_{\beta=2}^{n} \times \mathbf{A}^\beta$
2	From 1, D6.9.1E Determinant of A and factoring the first factor	$<\mid \overset{*}{\mathring{A}}_m \mid> = \mu_f \sum_{\sigma(k)\in Sn} \mathrm{sgn}(\sigma(k))\, \mathbf{b}^{\lambda[1,k]} \prod_{i=2}^{n} A_{i\lambda[i,k]}$
3	From 2, A10.15.1 The Cross Product Operator and T10.3.5 Tool: Tensor Substitution	$<\mid \overset{*}{\mathring{A}}_m \mid> = \mu_f \sum_{\sigma(k)\in Sn} (\prod_{i=2}^{n} \times \mathbf{b}^{\lambda[i,k]})\, (\prod_{i=2}^{n} A_{i\lambda[i,k]})$
4	From 3 and T10.4.7 Product of Tensors: Commutative by Multiplication of Rank for p + q	$<\mid \overset{*}{\mathring{A}}_m \mid> = \mu_f \sum_{\sigma(k)\in Sn} (\prod_{i=2}^{n} A_{i\lambda[i,k]})\, (\prod_{i=2}^{n} \times \mathbf{b}^{\lambda[i,k]})$
5	From 4 and T10.3.4 Tool: Base Tensor Distributed over Polyadic Coefficient Summation	$<\mid \overset{*}{\mathring{A}}_m \mid> = \mu_f \sum_{\sigma(k)\in Sn} \prod_{i=2}^{n} A_{i\lambda[i,k]} \times \mathbf{b}^{\lambda[i,k]}$
6	From 1 and 5	$\mu_f \prod_{\beta=2}^{n} \times \mathbf{A}^\beta = \mu_f \sum_{\sigma(k)\in Sn} \prod_{i=2}^{n} A_{i\lambda[i,k]} \times \mathbf{b}^{\lambda[i,k]}$
∴	From 6 and T4.4.3 Equalities: Cancellation by Multiplication	$\prod_{\beta=2}^{n} \times \mathbf{A}^\beta = \sum_{\sigma(k)\in Sn} \prod_{i=2}^{n} A_{i\lambda[i,k]} \times \mathbf{b}^{\lambda[i,k]}$

Theorem 12.10.5 Cross Product: Generalization of CCW Cyclic Covariant Permutation

Steps	Hypothesis	$\mathrm{sgn}(\sigma(k))\mathbf{b}_{\lambda[1,k]} = \prod_{i=2} \times \mathbf{b}_{\lambda[i,k]}$ cycling through $\sigma(k)$ yields cyclic permutation of the polyadic.
∴	From A10.15.2 Cross Product Operator: Sense of Rotation for Cyclic Permutation	$\mathrm{sgn}(\sigma(k))\mathbf{b}_{\lambda[1,k]} = \prod_{i=2} \times \mathbf{b}_{\lambda[i,k]}$ cycling through $\sigma(k)$.

Theorem 12.10.6 Cross Product: Generalization of CCW Cyclic Contravariant Permutation

Steps	Hypothesis	$\text{sgn}(\sigma(k))\mathbf{b}^{\lambda[1,\,k]} = \prod_{i=2} \times \mathbf{b}^{\lambda[i,\,k]}$ cycling through $\sigma(k)$ yields cyclic permutation of the polyadic.
∴	From A10.15.2 Cross Product Operator: Sense of Rotation for Cyclic Permutation	$\text{sgn}(\sigma(k))\mathbf{b}^{\lambda[1,\,k]} = \prod_{i=2} \times \mathbf{b}^{\lambda[i,\,k]}$ cycling through $\sigma(k)$.

So the Cross Product of bases vectors A10.15.1 "The Cross Product Operator" is the generalization of Hamilton's and Grassmann's Counter Clockwise Cyclic permutation of bases vectors for n-space.

Theorem 12.10.7 Cross Product: Covariant; Dotted with Contravariant Quiver Vector

1g	Given	$\prod_{\beta=2}^{n} \times \mathbf{A}_\beta = (\mathbf{b}_k \mid A^{2k} \mid A^{3k} \mid \ldots \mid A^{nk})$ by k-columns
2g		$\mathbf{C}^1 = \sum_i c_{1i}\, \mathbf{b}^i$ out of the quiver $\overset{*}{\mathbf{A}}_m$ 1-row
3g		$\mid A \mid = (c_{1k} \mid A^{2k} \mid A^{3k} \mid \ldots \mid A^{nk})$ by k-column
Steps	Hypothesis	$\prod_{\beta=2}^{n} \times \mathbf{A}_\beta \bullet \mathbf{C}^1 = \mid A \mid$ contraction about cofactors
1	From T12.101A Cross Product: Product of (n-1)-Covariant Vectors Spanning an n-Space	$\prod_{\beta=2}^{n} \times \mathbf{A}_\beta = \sum_j A^{1j}\mathbf{b}_j$
2	From 1, T12.3.6A Transformation from Contravariant to Covariant Basis Vector and T10.2.5 Tool: Tensor Substitution	$\prod_{\beta=2}^{n} \times \mathbf{A}_\beta = \sum_j A^{1j} \sum_k g_{jk}\mathbf{b}^k$ expansion about cofactors
3	From 2g, 1 and T8.3.1 Dot Product: Uniqueness Operation	$\prod_{\beta=2}^{n} \times \mathbf{A}_\beta \bullet \mathbf{C}^1 = (\sum_j A^{1j} \sum_k g_{jk}\mathbf{b}^k) \bullet \mathbf{C}^1$
4	From 2 and T8.3.8 Dot Product: Distribution of Dot Product Across Another Vector	$\prod_{\beta=2}^{n} \times \mathbf{A}_\beta \bullet \mathbf{C}^1 = (\sum_j \sum_k A^{1j}\, g_{jk}\mathbf{b}^k) \bullet \mathbf{C}^1$
5	From 2g, 3, and A7.9.2 Substitution of Vectors	$\prod_{\beta=2}^{n} \times \mathbf{A}_\beta \bullet \mathbf{C}^1 = (\sum_j \sum_k A^{1j}\, g_{jk}\mathbf{b}^k) \bullet (\sum_i c_{1i}\, \mathbf{b}^i)$
6	From 5 and T12.3.14 Dot Product of Contravariant Vectors	$\prod_{\beta=2}^{n} \times \mathbf{A}_\beta \bullet \mathbf{C}^1 = \sum_j A^{1j} [\sum_i c_{1i} \sum_k g_{jk}\, g^{ki})]$
7	From 6, T12.2.8A Covariant and Contravariant Metric Inverse and T10.2.5 Tool: Tensor Substitution	$\prod_{\beta=2}^{n} \times \mathbf{A}_\beta \bullet \mathbf{C}^1 = \sum_j A^{1j} (\sum_i c_{1i}\, \delta_j^{\,i})$
8	From 7, T12.3.3B Contracted Covariant Bases Vector with Contravariant Vector and T10.3.5 Tool: Tensor Substitution	$\prod_{\beta=2}^{n} \times \mathbf{A}_\beta \bullet \mathbf{C}^1 = \sum_j A^{1j}\, c_{1j}$
∴	From 9, 3g and T6.9.5A Expanding Determinant [A] of order [n] about a Row or Column	$\prod_{\beta=2}^{n} \times \mathbf{A}_\beta \bullet \mathbf{C}^1 = \mid A \mid$ contraction about cofactors

Theorem 12.10.8 Cross Product: Contravariant; Dotted with Covariant Quiver Vector

1g	Given	$\prod_{\beta=2}^{n} \times \mathbf{A}^{\beta} = (\mathbf{b}^{k} \mid A_{2k} \mid A_{3k} \mid \ldots \mid A_{nk})$ by k-columns
2g		$\mathbf{C}_1 = \sum_i c^{1i}\, \mathbf{b}_i$ out of the quiver $\overset{*}{A}{}^{m}$ 1-row
3g		$\mid A \mid = (c^{1k} \mid A_{2k} \mid A_{3k} \mid \ldots \mid A_{nk})$ k-by k-column
Steps	Hypothesis	$\prod_{\beta=2}^{n} \times \mathbf{A}^{\beta} \bullet \mathbf{C}_1 = \mid A \mid$ contraction about cofactors
1	From T12.102A Cross Product: Product of (n-1)-Contravariant Vectors Spanning an n-Space	$\prod_{\beta=2}^{n} \times \mathbf{A}^{\beta} = \sum_j A_{1j}\mathbf{b}^{j}$
2	From 1, T12.3.10A Transformation from Covariant to Contravariant Basis Vector and T10.2.5 Tool: Tensor Substitution	$\prod_{\beta=2}^{n} \times \mathbf{A}^{\beta} = \sum_j A_{1j} \sum_k g^{jk}\mathbf{b}_k$ expansion about cofactors
3	From 2g, 1 and T8.3.1 Dot Product: Uniqueness Operation	$\prod_{\beta=2}^{n} \times \mathbf{A}^{\beta} \bullet \mathbf{C}_1 = (\sum_j A_{1j} \sum_k g^{jk}\mathbf{b}_k) \bullet \mathbf{C}_1$
4	From 2 and T8.3.7 Dot Product: Distribution of Dot Product Across Addition of Vectors	$\prod_{\beta=2}^{n} \times \mathbf{A}^{\beta} \bullet \mathbf{C}_1 = (\sum_j \sum_k A_{1j}\, g^{jk}\mathbf{b}_k) \bullet \mathbf{C}_1$
5	From 2g, 3, and A7.9.2 Substitution of Vectors	$\prod_{\beta=2}^{n} \times \mathbf{A}^{\beta} \bullet \mathbf{C}_1 = (\sum_j \sum_k A_{1j}\, g^{jk}\mathbf{b}_k) \bullet (\sum_i c^{1i}\, \mathbf{b}_i)$
6	From 5 and T12.3.13A Dot Product of Covariant Vectors	$\prod_{\beta=2}^{n} \times \mathbf{A}^{\beta} \bullet \mathbf{C}_1 = \sum_j A_{1j} [\sum_i c^{1i} \sum_k g^{jk}\, g_{ki})]$
7	From 6, T12.2.8B Covariant and Contravariant Metric Inverse and T10.2.5 Tool: Tensor Substitution	$\prod_{\beta=2}^{n} \times \mathbf{A}^{\beta} \bullet \mathbf{C}_1 = \sum_j A_{1j} (\sum_i c^{1i}\, \delta^{j}{}_{i})$
8	From 7, T12.3.4C Contracted Contravariant Bases Vector with Covariant Vector and T10.3.5 Tool: Tensor Substitution	$\prod_{\beta=2}^{n} \times \mathbf{A}^{\beta} \bullet \mathbf{C}_1 = \sum_j A_{1j}\, c^{1j}$
∴	From 9, 3g and T6.9.5A Expanding Determinant [A] of order [n] about a Row or Column	$\prod_{\beta=2}^{n} \times \mathbf{A}^{\beta} \bullet \mathbf{C}_1 = \mid A \mid$ contraction about cofactors

Theorem 12.10.9 Cross Product: Contravariant; Dotted with Covariant Row Vector

1g	Given	$\mathbf{A}_i = \sum_k A^{ik}\, \mathbf{b}_k$ out of the quiver $\overset{*}{A}{}^{m}$ row 1
2g		$\mid A \mid = (A_{ik} \mid A_{2k} \mid \ldots \mid A_{ik} \mid \ldots \mid A_{nk})$ by i-row
Steps	Hypothesis	$\prod_{\beta=2}^{n} \times \mathbf{A}^{\beta} \bullet \mathbf{A}_i = 0$
1	From 1g and 12.10.8 Cross Product: Contravariant; Dotted with Covariant Quiver Vector	$\prod_{\beta=2}^{n} \times \mathbf{A}^{\beta} \bullet \mathbf{A}_i = \mid A \mid$ row-1 replaced by $\mathbf{A}^{i}$
∴	From 2g, 1, T6.9.5 Expanding Determinant [A] of order [n] about a Row or Column and T6.9.10 A determinant with two identical rows or columns is zero	$\prod_{\beta=2}^{n} \times \mathbf{A}^{\beta} \bullet \mathbf{A}_i = 0$

Theorem 12.10.10 Cross Product: Covariant; Dotted with Contravariant Row Vector

1g	Given	$\mathbf{A}^i = \sum_k A_{ik}\, \mathbf{b}^k$ out of the quiver $\overset{*}{A}{}^m$ row 1
2g		$\mid A \mid = (A^{ik}\mid A^{2k}\mid \ldots \mid A^{ik}\mid \ldots \mid A^{nk})$ by i-row
Steps	Hypothesis	$\prod_{\beta=2}{}^n \times \mathbf{A}_\beta \bullet \mathbf{A}^i = 0$
1	From 1g and 12.10.7 Cross Product: Covariant; Dotted with Contravariant Quiver Vector	$\prod_{\beta=2}{}^n \times \mathbf{A}_\beta \bullet \mathbf{A}^i = \mid A \mid$ row-1 replaced by $\mathbf{A}_i$
∴	From 2g, 1, T6.9.5 Expanding Determinant [A] of order [n] about a Row or Column and T6.9.10 A determinant with two identical rows or columns is zero	$\prod_{\beta=2}{}^n \times \mathbf{A}_\beta \bullet \mathbf{A}^i = 0$

This theorem is important as well, because it shows that any row vector from the tensor quiver is orthogonal to all other row vectors. These two theorems lay the foundation for analytic geometry in n-space. This is how a hyper-plane is constructed with the resultant vector being the orientation vector for the plane.

Section 12.11 e-Tensor System

In order to understand the idea of a cross product and how to analytically represent it a new normalized operator needs to be considered the e-Tensor and the system it resides in. From D10.11.1A "TDL as a Determinate Operator" it springs from the determinate operator. It is constant bound with respect to the manifold coordinates. Being a coefficient of the determinate tensor it has Euclidian like properties with in its own space.

From axiom A7.8.6 "Parallelogram Law: Multiplying a Negative Number Times a Vector" vectors have a sense of direction, but a vector is also a tensor of rank-1, so it follows tensors have a sense of direction as well. The net direction is the permutated product of each sign as demonstrated in corollary C10.10.3.1 "The Cross Coefficient as Parity Sign for Permuted Base Vectors" and sign of parity D6.6.3B "Association of Sign to Parity".

$$\mathrm{sgn}(\sigma(k)) = (-1)^{\phi(k)}$$ Equation A

$$\mathbf{e}_{\rho(k)} = (-1)^{\phi(k)}$$ Equation B

It can be symbolically rewritten as a coefficient by using the "Parity of Permutation for Base Vector" with aggregate dot indices.

$$\mathbf{e}_{i_1 i_2 \ldots i_j \ldots i_m(k)} \rightarrow (-1)^{\phi(k)}$$ Equation C

$$\mathbf{e}_{\bullet i_j \bullet(k)} \rightarrow (-1)^{\phi(k)}$$ Equation D

or using box dot and long index notation

$$\mathbf{e}_{\bullet\lambda[i,\,k]\bullet} \rightarrow (-1)^{\phi(k)}$$ Equation E

Unlike a conventional vector, which can change direction in space, and results in the emergence of the Christoffel Symbol, the space with the e-tensor embeded in it has only a directed sense like on a number line, having its own unique spacial geometry and independent of the space the tensor it is associated with very like a Euclidian space. Like a Euclidian space direction is constant much as it is in a geodesic space where vectors move parallel to one another. In fact the e-Tensor System can be thought of as a type of geodesic space and the tensor density field has the e-tensor component $[\mathbf{e}_{\bullet\lambda[i,\,k]\bullet}]$ at every point independent of the real geometry coordinates of the space.

However as an operator on the polyadic of a tensor it must also account for the case where two of the quiver vectors are collinear and when the indices are equivalent result in a zero value. The following definition can now be stated.

Definition 12.11.1 **Covariant and Contravariant e-Skew Tensor Components**[12.12.1]

Covariant and Contravariant e-Skew Tensor Components is defined over the summation set-L_m the following definition of Tensor Coefficient for Distortion is true:

$$\mathbf{e}_{\bullet\lambda[i,\,k]\bullet} \equiv \begin{cases} (-1)^{\varphi(k)} & \text{for} \quad \lambda[i,k] \neq \lambda[j,k] \\ 0 & \text{for} \quad \lambda[i,k] = \lambda[j,k] \end{cases}$$ Equation A

for e-Skew Tensor covariant relative paired indices

and

$$\mathbf{e}^{\bullet\lambda[i,\,k]\bullet} \equiv \begin{cases} (-1)^{\varphi(k)} & \text{for} \quad \lambda[i,k] \neq \lambda[j,k] \\ 0 & \text{for} \quad \lambda[i,k] = \lambda[j,k] \end{cases}$$ Equation B

for e-Skew Tensor contravariant relative paired indices

Axiom 12.11.1 e-Tensor of Rank-m

The ***e-Tensor System Space*** is a Euclidian affine type space where the tensor density field has the e-tensor component $[\mathbf{e}_{\bullet\lambda[i,\,k]\bullet}]$ and $[\mathbf{e}^{\bullet\lambda[i,\,k]\bullet}]$ defined at every point. The tensor is of rank-m and as an operator has only a directed sense so its direction in space is constant implying the covariant determinate $[g_e]$ as independent of the manifold spatial coordinate system, $P(\bullet x_{i}\bullet)$.

[12.12.1]Note: Since the e-Tensor Operator is the operation on the tensor's polyadic it would follow then that if the polyadic were covariant so to would be the e-Tensor. This one-to-one correspondence of polyadic and the operator is just the opposite of the relationship of the polyadic to the tensor coefficient. The user of the e-Tensor should be aware of this to avoid confusion in symbolism and nomenclature.

Theorem 12.11.1 e-Tensor System: Skew Symmetry by Transposing any Two Indices

1g	Given	$\phi(k) = \text{parityn}(\kappa)$ is (even / odd)
2g		$\text{parityn}(\eta)$ is (odd / odd)
3g		$\text{parityn}(\sigma)$ is (even / odd)
Steps	Hypothesis	$\mathbf{e}_{\rho(k)b} = -\mathbf{e}_{\rho(k)a}$
1	From 1g and E12.12B	$\mathbf{e}_{\rho(k)a} = (-1)^{\text{parityn}(\kappa) \text{ is (even / odd)}}$
2	From 1 and A6.12.3 Transposition of Any Two Indices Gives an Opposite Odd or Even Parity	$\mathbf{e}_{\rho(k)b} = (-1)^{\text{parityn}(\tau) \text{ is (odd / even)}}$
3	From 2, A6.7.2 Association of Parity with the cyclic Composition Operator and A4.2.3 Substitution	$\mathbf{e}_{\rho(k)b} = (-1)^{[\text{parityn}(\eta) + \text{parityn}(\sigma)] \text{ is (odd + even / odd + odd)}}$
4	From 3 and A4.2.18 Summation Exp	$\mathbf{e}_{\rho(k)b} = (-1)^{\text{parityn}(\eta)}(-1)^{\text{parityn}(\sigma)}$
5	From 1g, 3g and T6.6.1 Uniqueness of the parity operator by correspondence of even/odd	$\text{parity}(\sigma) = \text{parity}(\kappa)$
6	From 1g, 2g, 5, T4.8.9 Integer Exponents: Uniqueness of Unequal Even Exponents Raised to a Negative One and T4.8.10 Integer Exponents: Uniqueness of Unequal Odd Exponents Raised to a Negative One	$(-1)^{\text{parityn}(\sigma)} = (-1)^{\text{parityn}(\kappa)}$
7	From 4, 6 and A4.2.3 Substitution	$\mathbf{e}_{\rho(k)b} = (-1)^{\text{parityn}(\eta)}(-1)^{\text{parityn}(\kappa)}$
8	From 2g, 7 and T4.8.4 Integer Exponents: Negative One Raised to an Odd Number	$\mathbf{e}_{\rho(k)b} = (-1)\,(-1)^{\text{parityn}(\kappa)}$
∴	From 1, 8, A4.2.3 Substitution and D4.1.20A Negative Coefficient	$\mathbf{e}_{\rho(k)b} = -\mathbf{e}_{\rho(k)a}$

Theorem 12.11.2 e-Tensor System: Skew Symmetry Covariant e-Tensor Operator

Steps	Hypothesis	$\mathbf{e}_{\bullet\lambda[j,\,k]\,\bullet\bullet\,\lambda[i,\,k]\bullet} = -\mathbf{e}_{\bullet\lambda[i,\,k]\,\bullet\bullet\lambda[j,\,k]\bullet}$
1	From T12.12.1 e-Tensor System: Skew Symmetry by Transposing any Two Indices	$\mathbf{e}_{\rho(k)b} = -\mathbf{e}_{\rho(k)a}$ by Transpose of any Two Indices
∴	From 1 and D12.12.1A Covariant and Contravariant e-Tensor Components	$\mathbf{e}_{\bullet\lambda[j,\,k]\,\bullet\bullet\,\lambda[i,\,k]\bullet} = -\mathbf{e}_{\bullet\lambda[i,\,k]\,\bullet\bullet\lambda[j,\,k]\bullet}$

Theorem 12.11.3 e-Tensor System: Skew Symmetry Contravariant e-Tensor Operator

Steps	Hypothesis	$\mathbf{e}^{\bullet\lambda[j,\,k]\,\bullet\bullet\,\lambda[i,\,k]\bullet} = -\mathbf{e}^{\bullet\lambda[i,\,k]\,\bullet\bullet\lambda[j,\,k]\bullet}$
1	From T12.12.1 e-Tensor System: Skew Symmetry by Transpose of any Two Indices	$\mathbf{e}_{\rho(k)b} = -\mathbf{e}_{\rho(k)a}$
∴	From 1 and D12.12.1B Covariant and Contravariant e-Tensor Components	$\mathbf{e}^{\bullet\lambda[j,\,k]\,\bullet\bullet\,\lambda[i,\,k]\bullet} = -\mathbf{e}^{\bullet\lambda[i,\,k]\,\bullet\bullet\lambda[j,\,k]\bullet}$

Theorem 12.11.4 e-Tensor System: Simple Differential of Covariant e-Tensor Operator

1g	Given	$\phi(k) = \text{parityn}(\kappa)$ where parity can only be even or odd
Steps	Hypothesis	$(\mathbf{e}_{\bullet\lambda[i,\,k]\bullet})_{,r} = \mathbf{0}$
1	From A10.2.6 Equality of Tensors and T12.7.1B Tensor Product: Uniqueness of Tensor Differentiation	$(\mathbf{e}_{\bullet\lambda[i,\,k]\bullet})_{,r} = (\mathbf{e}_{\bullet\lambda[i,\,k]\bullet})_{,r}$
2	From 1, D12.12.1A Covariant and Contravariant e-Skew Tensor Components and T10.2.5 Tool: Tensor Substitution	$(\mathbf{e}_{\bullet\lambda[i,\,k]\bullet})_{,r} = \begin{cases} (-1)^{\varphi(k)} & \text{for } \lambda[i,k] \neq \lambda[j,k] \\ 0 & \text{for } \lambda[i,k] = \lambda[j,k] \end{cases}_{,r}$
3	From 1g, 2, evaluating the parity of permutation for base vectors as an even value	$(\mathbf{e}_{\bullet\lambda[i,\,k]\bullet})_{,r} = \begin{cases} +1 & \text{for } \lambda[i,k] \neq \lambda[j,k] \\ 0 & \text{for } \lambda[i,k] = \lambda[j,k] \end{cases}_{,r}$ where for $\varphi(k)$ even
4	From 1g, 3, evaluating the parity of permutation for base vectors as an odd value	$(\mathbf{e}_{\bullet\lambda[i,\,k]\bullet})_{,r} = \begin{cases} -1 & \text{for } \lambda[i,k] \neq \lambda[j,k] \\ 0 & \text{for } \lambda[i,k] = \lambda[j,k] \end{cases}_{,r}$ where for $\varphi(k)$ odd
5	From 3, 4 and combining steps then differentiating components	$(\mathbf{e}_{\bullet\lambda[i,\,k]\bullet})_{,r} = \begin{cases} (\pm 1)_{,r} & \text{for } \lambda[i,k] \neq \lambda[j,k] \\ (0)_{,r} & \text{for } \lambda[i,k] = \lambda[j,k] \end{cases}$
6	From 3 and LxK.3.3.1B Differential of a Constant	$(\mathbf{e}_{\bullet\lambda[i,\,k]\bullet})_{,r} = \begin{cases} 0 & \text{for } \lambda[i,k] \neq \lambda[j,k] \\ 0 & \text{for } \lambda[i,k] = \lambda[j,k] \end{cases}$
∴	From 6 and superimposing cases	$(\mathbf{e}_{\bullet\lambda[i,\,k]\bullet})_{,r} = \mathbf{0}$

Theorem 12.11.5 e-Tensor System: Simple Differential of Contravariant e-Tensor Operator

1g	Given	$\phi(k) = \text{parityn}(\kappa)$ where parity can only be even or odd
Steps	Hypothesis	$(\mathbf{e}^{\bullet\lambda[i,\,k]\bullet})^{,r} = \mathbf{0}$
1	From A10.2.6 Equality of Tensors and T12.7.1B Tensor Product: Uniqueness of Tensor Differentiation	$(\mathbf{e}^{\bullet\lambda[i,\,k]\bullet})^{,r} = (\mathbf{e}^{\bullet\lambda[i,\,k]\bullet})^{,r}$
2	From 1, D12.12.1B Covariant and Contravariant e-Skew Tensor Components and T10.2.5 Tool: Tensor Substitution	$(\mathbf{e}^{\bullet\lambda[i,\,k]\bullet})^{,r} = \begin{cases} (-1)^{\varphi(k)} & \text{for } \lambda[i,k] \neq \lambda[j,k] \\ 0 & \text{for } \lambda[i,k] = \lambda[j,k] \end{cases}^{,r}$
3	From 1g, 2, evaluating the parity of permutation for base vectors as an even value	$(\mathbf{e}^{\bullet\lambda[i,\,k]\bullet})^{,r} = \begin{cases} +1 & \text{for } \lambda[i,k] \neq \lambda[j,k] \\ 0 & \text{for } \lambda[i,k] = \lambda[j,k] \end{cases}^{,r}$ where $\varphi(k)$ even
4	From 1g, 3, evaluating the parity of permutation for base vectors as an odd value	$(\mathbf{e}^{\bullet\lambda[i,\,k]\bullet})^{,r} = \begin{cases} -1 & \text{for } \lambda[i,k] \neq \lambda[j,k] \\ 0 & \text{for } \lambda[i,k] = \lambda[j,k] \end{cases}^{,r}$ where $\varphi(k)$ odd
5	From 3, 4 and combining steps then differentiating components	$(\mathbf{e}^{\bullet\lambda[i,\,k]\bullet})^{,r} = \begin{cases} (\pm 1)^{,r} & \text{for } \lambda[i,k] \neq \lambda[j,k] \\ (0)^{,r} & \text{for } \lambda[i,k] = \lambda[j,k] \end{cases}$
6	From 3 and LxK.3.3.1B Differential of a Constant	$(\mathbf{e}^{\bullet\lambda[i,\,k]\bullet})^{,r} = \begin{cases} 0 & \text{for } \lambda[i,k] \neq \lambda[j,k] \\ 0 & \text{for } \lambda[i,k] = \lambda[j,k] \end{cases}$
∴	From 6 and superimposing cases	$(\mathbf{e}^{\bullet\lambda[i,\,k]\bullet})^{,r} = \mathbf{0}$

Theorem 12.11.6 e-Tensor System: Covariant e-Tensor as a Mixed Kronecker Delta

1g	Given	$\phi(k) = \text{parityn}(\kappa)$ where parity can only be even or odd	
Steps	Hypothesis	$\lvert \mathbf{e}_{\bullet\tau[i,j,s]\bullet} \rvert = \delta_s^{\lambda[j,k]}$ EQ A	$\delta_s^{\lambda[j,k]} = \begin{cases} 1 & \text{for } \lambda[j,k]=s \text{ such that } i=j \\ 0 & \text{for } \lambda[i,k]\neq s \text{ such that } i\neq j \end{cases}$ EQ B
1	From A10.2.6 Equality of Tensors and T12.7.1B Tensor Product: Uniqueness of Tensor Differentiation	$\mathbf{e}_{\bullet\tau[i,j,s]\bullet} = \mathbf{e}_{\bullet\tau[i,j,s]\bullet}$	
2	From 1, D12.12.1A Covariant and Contravariant e-Skew Tensor Components and T10.2.5 Tool: Tensor Substitution	$\mathbf{e}_{\bullet\tau[i,j,s]\bullet} = \begin{cases} (-1)^{\varphi(k)} & \text{for } \lambda[j,k]=s \text{ such that } i=j \\ 0 & \text{for } \lambda[i,k]\neq s \text{ such that } i\neq j \end{cases}$	
3	From 2 and T4.6.1 Equalities: Uniqueness of Absolute Value	$\lvert \mathbf{e}_{\bullet\tau[i,j,s]\bullet} \rvert = \left\lvert \begin{cases} (-1)^{\varphi(k)} & \text{for } \lambda[j,k]=s \text{ such that } i=j \\ 0 & \text{for } \lambda[i,k]\neq s \text{ such that } i\neq j \end{cases} \right\rvert$	
4	From 3 and taking the absolute value of all components of the e-tensor	$\lvert \mathbf{e}_{\bullet\tau[i,j,s]\bullet} \rvert = \begin{cases} \lvert(-1)^{\varphi(k)}\rvert & \text{for } \lambda[j,k]=s \text{ such that } i=j \\ \lvert 0 \rvert & \text{for } \lambda[i,k]\neq s \text{ such that } i\neq j \end{cases}$	
5	From 1g and 4 the permutation number φ could be said to be odd or even, hence when raised to a negative one result in a plus or minus one	$\lvert \mathbf{e}_{\bullet\tau[i,j,s]\bullet} \rvert = \begin{cases} \lvert(\pm 1)\rvert & \text{for } \lambda[j,k]=s \text{ such that } i=j \text{ and } \varphi \text{ odd or even} \\ \lvert 0 \rvert & \text{for } \lambda[i,k]\neq s \text{ such that } i\neq j \end{cases}$	
6	From 3 and D4.1.11 Absolute Value $\lvert a \rvert$	$\lvert \mathbf{e}_{\bullet\tau[i,j,s]\bullet} \rvert = \begin{cases} 1 & \text{for } \lambda[j,k]=s \text{ such that } i=j \\ 0 & \text{for } \lambda[i,k]\neq s \text{ such that } i\neq j \end{cases}$	
∴	From 6 and D6.1.6B Identity Matrix the definition of the Kronecker Delta	$\lvert \mathbf{e}_{\bullet\tau[i,j,s]\bullet} \rvert = \delta_s^{\lambda[j,k]}$ EQ A	$\delta_s^{\lambda[j,k]} = \begin{cases} 1 & \text{for } \lambda[j,k]=s \text{ such that } i=j \\ 0 & \text{for } \lambda[i,k]\neq s \text{ such that } i\neq j \end{cases}$ EQ B

Theorem 12.11.7 e-Tensor System: Contravariant e-Tensor as a Mixed Kronecker Delta

1g	Given	$\phi(k) = \text{parityn}(\kappa)$ where parity can only be even or odd	
Steps	Hypothesis	$\lvert \mathbf{e}^{\bullet\tau[i,j,s]\bullet} \rvert = \delta_{\lambda[j,k]}^{\ \ s}$ EQ A	$\delta_{\lambda[j,k]}^{\ \ s} = \begin{cases} 1 & \text{for } \lambda[j,k]=s \text{ such that } i=j \\ 0 & \text{for } \lambda[i,k]\neq s \text{ such that } i\neq j \end{cases}$ EQ B
1	From A10.2.6 Equality of Tensors and T12.7.1B Tensor Product: Uniqueness of Tensor Differentiation	$\mathbf{e}^{\bullet\tau[i,j,s]\bullet} = \mathbf{e}^{\bullet\tau[i,j,s]\bullet}$	
2	From 1, D12.12.1A Covariant and Contravariant e-Skew Tensor Components and T10.2.5 Tool: Tensor Substitution	$\mathbf{e}^{\bullet\tau[i,j,s]\bullet} = \begin{cases} (-1)^{\varphi(k)} & \text{for } \lambda[j,k]=s \text{ such that } i=j \\ 0 & \text{for } \lambda[i,k]\neq s \text{ such that } i\neq j \end{cases}$	
3	From 2 and T4.6.1 Equalities: Uniqueness of Absolute Value	$\lvert \mathbf{e}^{\bullet\tau[i,j,s]\bullet} \rvert = \left\lvert \begin{cases} (-1)^{\varphi(k)} & \text{for } \lambda[j,k]=s \text{ such that } i=j \\ 0 & \text{for } \lambda[i,k]\neq s \text{ such that } i\neq j \end{cases} \right\rvert$	
4	From 3 and taking the absolute value of all components of the e-tensor	$\lvert \mathbf{e}^{\bullet\tau[i,j,s]\bullet} \rvert = \begin{cases} \lvert(-1)^{\varphi(k)}\rvert & \text{for } \lambda[j,k]=s \text{ such that } i=j \\ \lvert 0 \rvert & \text{for } \lambda[i,k]\neq s \text{ such that } i\neq j \end{cases}$	
5	From 1g and 4 the permutation number φ could be said to be odd or even, hence when raised to a negative one result in a plus or minus one	$\lvert \mathbf{e}^{\bullet\tau[i,j,s]\bullet} \rvert = \begin{cases} \lvert(\pm 1)\rvert & \text{for } \lambda[j,k]=s \text{ such that } i=j \text{ and } \varphi \text{ odd or even} \\ \lvert 0 \rvert & \text{for } \lambda[i,k]\neq s \text{ such that } i\neq j \end{cases}$	
6	From 3 and D4.1.11 Absolute Value $\lvert a \rvert$	$\lvert \mathbf{e}^{\bullet\tau[i,j,s]\bullet} \rvert = \begin{cases} 1 & \text{for } \lambda[j,k]=s \text{ such that } i=j \\ 0 & \text{for } \lambda[i,k]\neq s \text{ such that } i\neq j \end{cases}$	
∴	From 6 and D6.1.6B Identity Matrix the definition of the Kronecker Delta	$\lvert \mathbf{e}^{\bullet\tau[i,j,s]\bullet} \rvert = \delta_s^{\lambda[j,k]}$ EQ A	$\delta_s^{\lambda[j,k]} = \begin{cases} 1 & \text{for } \lambda[j,k]=s \text{ such that } i=j \\ 0 & \text{for } \lambda[i,k]\neq s \text{ such that } i\neq j \end{cases}$ EQ B

Theorem 12.11.8 e-Tensor System: Differential of Covariant e-Tensor Operator

1g	Given	$(A_{\bullet\lambda[i, k]\bullet})_{;r} = (A_{\bullet\lambda[i, k]\bullet})_{,r} - \sum_s \sum_{j=1}^{m} A_{\bullet\tau[j, s, k]\bullet}\, \Gamma_{\lambda[j, k]r}{}^{s}$
Steps	Hypothesis	$(\mathbf{e}_{\bullet\lambda[i, k]\bullet})_{;r} = \mathbf{0}$
1	From A10.2.6 Equality of Tensors and T12.7.1B Tensor Product: Uniqueness of Tensor Differentiation	$(\mathbf{e}_{\bullet\lambda[i, k]\bullet})_{;r} = (\mathbf{e}_{\bullet\lambda[i, k]\bullet})_{;r}$
2	From 1 and T12.7.12 Tensor Product: Contravariant Chain Rule for Tensor Coefficient	$(\mathbf{e}_{\bullet\lambda[i, k]\bullet})_{;r} = (\mathbf{e}_{\bullet\lambda[i, k]\bullet})_{,r} - \sum_s \sum_{j=1}^{m} \mathbf{e}_{\bullet\tau[j, s, k]\bullet}\, \Gamma_{\lambda[j, k]r}{}^{s}$
3	From 2, T12.12.4 e-Tensor System: Simple Differential of Contravariant e-Tensor Coefficient, T10.3.18 Addition of Tensors: Identity by Addition and T10.3.18 Addition of Tensors: Identity by Addition	$(\mathbf{e}_{\bullet\lambda[i, k]\bullet})_{;r} = -\sum_s \sum_{j=1}^{m} \mathbf{e}_{\bullet\tau[j, s, k]\bullet}\, \Gamma_{\lambda[j, k]r}{}^{s}$
4	From 1 and A4.1.11B Absolute Value |a|	$0 \leq \| (\mathbf{e}_{\bullet\lambda[i, k]\bullet})_{;r} \|$
5	From 3, 4 and T10.2.5 Tool: Tensor Substitution	$0 \leq \| (\mathbf{e}_{\bullet\lambda[i, k]\bullet})_{;r} \| = \| -\sum_s \sum_{j=1}^{m} \mathbf{e}_{\bullet\tau[j, s, k]\bullet}\, \Gamma_{\lambda[j, k]r}{}^{s} \|$
6	From 5, D4.1.20 Negative Coefficient and T4.6.3 Equalities: Absolute Product is the Absolute of the Products	$0 \leq \| (\mathbf{e}_{\bullet\lambda[i, k]\bullet})_{;r} \| = \| -1 \| \|\sum_s \sum_{j=1}^{m} \mathbf{e}_{\bullet\tau[j, s, k]\bullet}\, \Gamma_{\lambda[j, k]r}{}^{s} \|$
7	From 6, A4.1.11A Absolute Value |a| and T4.7.31 Inequalities: Absolute Summation is Less than the Sum of all Absolute Values	$0 \leq \| (\mathbf{e}_{\bullet\lambda[i, k]\bullet})_{;r} \| \leq \sum_s \sum_{j=1}^{m} \| \mathbf{e}_{\bullet\tau[j, s, k]\bullet}\, \Gamma_{\lambda[j, k]r}{}^{s} \|$
8	From 7 and T4.6.3 Equalities: Absolute Product is the Absolute of the Products	$0 \leq \| (\mathbf{e}_{\bullet\lambda[i, k]\bullet})_{;r} \| \leq \sum_s \sum_{j=1}^{m} \| \mathbf{e}_{\bullet\tau[j, s, k]\bullet} \| \| \Gamma_{\lambda[j, k]r}{}^{s} \|$
9	From 8, T12.12.6A e-Tensor System: Covariant e-Tensor as a Mixed Kronecker Delta, and T10.2.5 Tool: Tensor Substitution	$0 \leq \| (\mathbf{e}_{\bullet\lambda[i, k]\bullet})_{;r} \| \leq \sum_{j=1}^{m} (\sum_s \delta_s{}^{\lambda[j, k]} \| \Gamma_{\lambda[j, k]r}{}^{s} \|)$
10	From 9 and T4.7.31 Inequalities: Absolute Summation is Less than the Sum of all Absolute Values	$0 \leq \| (\mathbf{e}_{\bullet\lambda[i, k]\bullet})_{;r} \| \leq \| \sum_{j=1}^{m} (\sum_s \delta_s{}^{\lambda[j, k]}\, \Gamma_{\lambda[j, k]r}{}^{s}) \|$
11	From 10, T12.12.6B e-Tensor System: Covariant e-Tensor as a Mixed Kronecker Delta and evaluating over the summation of [s]	$0 \leq \| (\mathbf{e}_{\bullet\lambda[i, k]\bullet})_{;r} \| \leq \| \sum_{j=1}^{m} \Gamma_{\lambda[j, k]r}{}^{\lambda[j, k]} \|$
12	From 11, T12.8.7A Derivative of the Natural Log of the Covariant Determinate,	$0 \leq \| (\mathbf{e}_{\bullet\lambda[i, k]\bullet})_{;r} \| \leq \| (\ln \sqrt{\| g_e \|})_{,r} \|$
13	From 12, A12.12.1 e-Tensor of Rank-m and LxK.3.3.1B Differential of a Constant	$0 \leq \| (\mathbf{e}_{\bullet\lambda[i, k]\bullet})_{;r} \| \leq \| 0 \|$ for $[g_e \neq f(_\bullet x_{i\bullet})]$
14	From 13 and D4.1.11 Absolute Value | a |	$0 \leq \| (\mathbf{e}_{\bullet\lambda[i, k]\bullet})_{;r} \| \leq 0$
∴	From 14 and LxK.2.1.10 Sandwiching Theorem	$(\mathbf{e}_{\bullet\lambda[i, k]\bullet})_{;r} = \mathbf{0}$

Theorem 12.11.9 e-Tensor System: Differential of Contravariant e-Tensor Operator

1g	Given	$(A^{\bullet\lambda[i,\,k]\bullet})^{;r} = (A^{\bullet\lambda[i,\,k]\bullet})^{,r} + \sum_s \sum_{j=1}^{m} A^{\bullet\tau[j,\,s,\,k]\bullet}\, \Gamma_{sr}^{\lambda[j,\,k]}$
Steps	Hypothesis	$(\mathbf{e}^{\bullet\lambda[i,\,k]\bullet})^{;r} = \mathbf{0}$
1	From A10.2.6 Equality of Tensors and T12.7.1B Tensor Product: Uniqueness of Tensor Differentiation	$(\mathbf{e}^{\bullet\lambda[i,\,k]\bullet})^{;r} = (\mathbf{e}^{\bullet\lambda[i,\,k]\bullet})^{;r}$
2	From 1 and T12.7.12 Tensor Product: Contravariant Chain Rule for Tensor Coefficient	$(\mathbf{e}^{\bullet\lambda[i,\,k]\bullet})^{;r} = (\mathbf{e}^{\bullet\lambda[i,\,k]\bullet})^{,r} + \sum_s \sum_{j=1}^{m} \mathbf{e}^{\bullet\tau[j,\,s,\,k]\bullet}\, \Gamma_{sr}^{\lambda[j,\,k]}$
3	From 2, T12.12.5 e-Tensor System: Simple Differential of Contravariant e-Tensor Coefficient, T10.3.18 Addition of Tensors: Identity by Addition and T10.3.18 Addition of Tensors: Identity by Addition	$(\mathbf{e}^{\bullet\lambda[i,\,k]\bullet})^{;r} = \sum_s \sum_{j=1}^{m} \mathbf{e}^{\bullet\tau[j,\,s,\,k]\bullet}\, \Gamma_{sr}^{\lambda[j,\,k]}$
4	From 1 and A4.1.11B Absolute Value |a|	$0 \le \lvert(\mathbf{e}^{\bullet\lambda[i,\,k]\bullet})^{;r}\rvert$
5	From 3, 4 and T10.2.5 Tool: Tensor Substitution	$0 \le \lvert(\mathbf{e}^{\bullet\lambda[i,\,k]\bullet})^{;r}\rvert = \lvert \sum_s \sum_{j=1}^{m} \mathbf{e}^{\bullet\tau[j,\,s,\,k]\bullet}\, \Gamma_{sr}^{\lambda[j,\,k]} \rvert$
6	From 5, A4.1.11A Absolute Value |a| and T4.7.31 Inequalities: Absolute Summation is Less than the Sum of all Absolute Values	$0 \le \lvert(\mathbf{e}^{\bullet\lambda[i,\,k]\bullet})^{;r}\rvert \le \sum_s \sum_{j=1}^{m} \lvert \mathbf{e}_{\bullet\tau[j,\,s,\,k]\bullet}\, \Gamma_{sr}^{\lambda[j,\,k]} \rvert$
7	From 6 and T4.6.3 Equalities: Absolute Product is the Absolute of the Products	$0 \le \lvert(\mathbf{e}^{\bullet\lambda[i,\,k]\bullet})^{;r}\rvert \le \sum_s \sum_{j=1}^{m} \lvert \mathbf{e}_{\bullet\tau[j,\,s,\,k]\bullet} \rvert\, \lvert \Gamma_{sr}^{\lambda[j,\,k]} \rvert$
8	From 7, T12.12.7A e-Tensor System: Contravariant e-Tensor as a Mixed Kronecker Delta, and T10.2.5 Tool: Tensor Substitution	$0 \le \lvert(\mathbf{e}^{\bullet\lambda[i,\,k]\bullet})^{;r}\rvert \le \sum_{j=1}^{m} (\sum_s \delta_{\lambda[j,\,k]}^{\ s}\, \lvert \Gamma_{sr}^{\lambda[j,\,k]} \rvert)$
9	From 8 and T4.7.31 Inequalities: Absolute Summation is Less than the Sum of all Absolute Values	$0 \le \lvert(\mathbf{e}^{\bullet\lambda[i,\,k]\bullet})^{;r}\rvert \le \lvert \sum_{j=1}^{m} (\sum_s \delta_{\lambda[j,\,k]}^{\ s}\, \Gamma_{sr}^{\lambda[j,\,k]}) \rvert$
10	From 9, T12.12.5B e-Tensor System: Contravariant e-Tensor as a Mixed Kronecker Delta and evaluating over the summation of [s]	$0 \le \lvert(\mathbf{e}^{\bullet\lambda[i,\,k]\bullet})^{;r}\rvert \le \lvert \sum_{j=1}^{m} \Gamma_{\lambda[j,\,k]r}^{\lambda[j,\,k]} \rvert$
11	From 11, T12.8.7A Derivative of the Natural Log of the Covariant Determinate,	$0 \le \lvert(\mathbf{e}^{\bullet\lambda[i,\,k]\bullet})^{;r}\rvert \le \lvert (\ln \sqrt{\lvert g_e \rvert})_{,r} \rvert$
12	From 12, A12.12.1 e-Tensor of Rank-m and LxK.3.3.1B Differential of a Constant	$0 \le \lvert(\mathbf{e}^{\bullet\lambda[i,\,k]\bullet})^{;r}\rvert \le \lvert 0 \rvert$ for $[g_e \neq f(\bullet x_i \bullet)]$
13	From 13 and D4.1.11 Absolute Value | a |	$0 \le \lvert(\mathbf{e}^{\bullet\lambda[i,\,k]\bullet})^{;r}\rvert \le 0$
∴	From 13 and LxK.2.1.10 Sandwiching Theorem	$(\mathbf{e}^{\bullet\lambda[i,\,k]\bullet})^{;r} = \mathbf{0}$

Observation 12.11.1 e-Tensor Proof of Existence

Theorems T12.12.8 "e-Tensor System: Differential of Covariant e-Tensor Operator" and T12.12.9 "e-Tensor System: Differential of Contravariant e-Tensor Operator" reaffirms the existence of the e-Tensor as a true tensor, since it vanishes everywhere in its tensor density field.

The question arises how would the proof end if the e-Tensor density field were a true Euclidian Space $\mathfrak{E}^n$?

Theorem 12.11.10 e-Tensor System: Covariant-Contravariant e-Tensor in a Euclidian Space

1g	Given	e-tensor density field is embedded in Euclidian Spaces $\mathfrak{E}^n$	
Steps	Hypothesis	$(\mathbf{e}_{\bullet\lambda[i,\,k]\bullet})_{;r} = \mathbf{0}$	$(\mathbf{e}^{\bullet\lambda[i,\,k]\bullet})^{;r} = \mathbf{0}$
1	From T12.12.8S12 e-Tensor System: Differential of Covariant e-Tensor Operator and T12.12.9S11 e-Tensor System: Differential of Contravariant e-Tensor Operator	$0 \le \vert (\mathbf{e}_{\bullet\lambda[i,\,k]\bullet})_{;r} \vert \le \vert (\ln \sqrt{\vert g_e \vert})_{,r} \vert$	$0 \le \vert (\mathbf{e}^{\bullet\lambda[i,\,k]\bullet})^{;r} \vert \le \vert (\ln \sqrt{\vert g_e \vert})_{,r} \vert$
2	From 1g, 1 and T12.2.20 Covariant Metric Determinate in Euclidian Manifold $\mathfrak{E}^n$	$0 \le \vert (\mathbf{e}_{\bullet\lambda[i,\,k]\bullet})_{;r} \vert \le \vert [\ln (1)]_{,r} \vert$	$0 \le \vert (\mathbf{e}^{\bullet\lambda[i,\,k]\bullet})^{;r} \vert \le \vert [\ln (1)]_{,r} \vert$
3	From 2 and T4.11.2 Real Exponents: The Logarithm of Unity	$0 \le \vert (\mathbf{e}_{\bullet\lambda[i,\,k]\bullet})_{;r} \vert \le \vert 0 \vert$	$0 \le \vert (\mathbf{e}^{\bullet\lambda[i,\,k]\bullet})^{;r} \vert \le \vert 0 \vert$
4	From 3 and D4.1.11 Absolute Value $\vert a \vert$	$0 \le \vert (\mathbf{e}_{\bullet\lambda[i,\,k]\bullet})_{;r} \vert \le 0$	$0 \le \vert (\mathbf{e}^{\bullet\lambda[i,\,k]\bullet})^{;r} \vert \le 0$
∴	From 4 and LxK.2.1.10 Sandwiching Theorem	$(\mathbf{e}_{\bullet\lambda[i,\,k]\bullet})_{;r} = \mathbf{0}$	$(\mathbf{e}^{\bullet\lambda[i,\,k]\bullet})^{;r} = \mathbf{0}$

Observation 12.11.2 e-Tensor Operator Space Independent

So from theorem T12.12.9 "e-Tensor System: Differential of Contravariant e-Tensor Operator" and T12.12.9 "e-Tensor System: Differential of Contravariant e-Tensor Operator" weather the space is coordinate independent of by theorem T12.12.10 "e-Tensor System: Covariant-Contravariant e-Tensor in a Euclidian Space" a Euclidian Space $\mathfrak{E}^n$ the result is the same it vanishes everywhere in the e-Tensor density field making the proof general. This is exactly what one would want for a leading skew-asymmetric tensor operator and the conditions imposed by the space.

Theorem 12.11.11 e-Tensor System: The Product of the Kronecker Delta and Sign of Parity

Steps	Hypothesis	$\mathbf{e}_{\bullet\lambda[i,k]\bullet} = \delta_{\lambda[i,k]}{}^{\lambda[j,k]}\,\mathrm{sgn}(k)$ EQ A	$\mathbf{e}^{\bullet\lambda[i,k]\bullet} = \delta^{\lambda[i,k]}{}_{\lambda[j,k]}\,\mathrm{sgn}(k)$ EQ B
1	D12.11.1(A, B) Covariant and Contravariant e-Skew Tensor Components	$\mathbf{e}_{\bullet\lambda[i,k]\bullet} = \begin{cases}(-1)^{\varphi(k)} & \text{for } \lambda[i,k] \neq \lambda[j,k] \\ 0 & \text{for } \lambda[i,k] = \lambda[j,k]\end{cases}$	$\mathbf{e}^{\bullet\lambda[i,k]\bullet} = \begin{cases}(-1)^{\varphi(k)} & \text{for } \lambda[i,k] \neq \lambda[j,k] \\ 0 & \text{for } \lambda[i,k] = \lambda[j,k]\end{cases}$
2	From 1, A4.2.12 Identity Multp and T4.4.1 Equalities: Any Quantity Multiplied by Zero is Zero	$\mathbf{e}_{\bullet\lambda[i,k]\bullet} = \begin{cases}1\,(-1)^{\varphi(k)} & \text{for } \lambda[i,k] \neq \lambda[j,k] \\ 0\,(-1)^{\varphi(k)} & \text{for } \lambda[i,k] = \lambda[j,k]\end{cases}$	$\mathbf{e}^{\bullet\lambda[i,k]\bullet} = \begin{cases}1\,(-1)^{\varphi(k)} & \text{for } \lambda[i,k] \neq \lambda[j,k] \\ 0\,(-1)^{\varphi(k)} & \text{for } \lambda[i,k] = \lambda[j,k]\end{cases}$
3	From 2; factor common factor	$\mathbf{e}_{\bullet\lambda[i,k]\bullet} = \begin{Bmatrix}1 & \text{for } \lambda[i,k] \neq \lambda[j,k] \\ 0 & \text{for } \lambda[i,k] = \lambda[j,k]\end{Bmatrix} (-1)^{\varphi(k)}$	$\mathbf{e}^{\bullet\lambda[i,k]\bullet} = \begin{Bmatrix}1 & \text{for } \lambda[i,k] \neq \lambda[j,k] \\ 0 & \text{for } \lambda[i,k] = \lambda[j,k]\end{Bmatrix} (-1)^{\varphi(k)}$
∴	From 3, D6.1.6B Identity Matrix, D6.6.3A Association of Sign to Parity and A4.2.3 Substitution	$\mathbf{e}_{\bullet\lambda[i,k]\bullet} = \delta_{\lambda[i,k]}{}^{\lambda[j,k]}\,\mathrm{sgn}(k)$ EQ A	$\mathbf{e}^{\bullet\lambda[i,k]\bullet} = \delta^{\lambda[i,k]}{}_{\lambda[j,k]}\,\mathrm{sgn}(k)$ EQ B

Now we can come full-circle combining e-tensor operator with the cross product operator as used in the classical sense coming up with a more general interpretation of the covariant and contravariant cross product.

Theorem 12.11.12 Cross Product: Covariant Unified with the e-Tensor Operator

Steps	Hypothesis	$\prod_{\beta=2}{}^{n} \times \mathbf{A}_\beta = \sum_{\rho(k)\in Lr} \mathbf{e}_{\bullet\lambda[i,k]\bullet}\, \mathbf{b}_{\lambda[1,k]} \prod_{i=2}{}^{n} A^{i\lambda[i,k]}$
1	From T10.16.1 Cross Product Operator: Expanded Algebraic Vector Form	$\prod_{\beta=2}{}^{n} \times \mathbf{A}_\beta = \sum_{\sigma(k)\in Sr} \mathrm{sgn}(\sigma(k))\, \mathbf{b}_{\lambda[1,k]} \prod_{i=2}{}^{n} A^{i\lambda[i,k]}$
2	From 1 and transforming from permutated index set-S_r to long index set-L_r by adding back zeroed terms	$\prod_{\beta=2}{}^{n} \times \mathbf{A}_\beta = \sum_{\rho(k)\in Lr} \begin{cases}\mathrm{sgn}(\sigma(k)) & \text{for } \lambda[i,k] \neq \lambda[j,k] \\ 0 & \text{for } \lambda[i,k] = \lambda[j,k]\end{cases} \mathbf{b}_{\lambda[1,k]} \prod_{i=2}{}^{n} A^{i\lambda[i,k]}$
3	From 2, D6.6.3A Association of Sign to Parity, T10.2.5 Tool: Tensor Substitution and A10.2.14 Correspondence of Tensors and Tensor Coefficients	$\prod_{\beta=2}{}^{n} \times \mathbf{A}_\beta = \sum_{\rho(k)\in Lr} \begin{cases}(-1)^{\varphi(k)} & \text{for } \lambda[i,k] \neq \lambda[j,k] \\ 0 & \text{for } \lambda[i,k] = \lambda[j,k]\end{cases} \mathbf{b}_{\lambda[1,k]} \prod_{i=2}{}^{n} A^{i\lambda[i,k]}$
∴	From 3, D12.12.1A Covariant and Contravariant e-Skew Tensor Components T10.2.5 Tool: Tensor Substitution and A10.2.14 Correspondence of Tensors and Tensor Coefficients	$\prod_{\beta=2}{}^{n} \times \mathbf{A}_\beta = \sum_{\rho(k)\in Lr} \mathbf{e}_{\bullet\lambda[i,k]\bullet}\, \mathbf{b}_{\lambda[1,k]} \prod_{i=2}{}^{n} A^{i\lambda[i,k]}$

Theorem 12.11.13 Cross Product: Contravariant Unified with the e-Tensor Operator

Steps	Hypothesis	$\prod_{\beta=2}{}^{n} \times \mathbf{A}^{\beta} = \sum_{\rho(k)\in Lr} \mathbf{e}^{\bullet\lambda[i,k]\bullet}\, \mathbf{b}^{\lambda[1,k]} \prod_{i=2}{}^{n} A_{i\lambda[i,k]}$
1	From T10.16.1 Cross Product Operator: Expanded Algebraic Vector Form	$\prod_{\beta=2}{}^{n} \times \mathbf{A}^{\beta} = \sum_{\sigma(k)\in Sr} \text{sgn}(\sigma(k))\, \mathbf{b}^{\lambda[1,k]} \prod_{i=2}{}^{n} A_{i\lambda[i,k]}$
2	From 1 and transforming from permutated index set-S_r to long index set-L_r by adding back zeroed terms	$\prod_{\beta=2}{}^{n} \times \mathbf{A}^{\beta} = \sum_{\rho(k)\in Lr} \begin{cases} \text{sgn}(\sigma(k)) & \text{for } \lambda[i,k] \neq \lambda[j,k] \\ 0 & \text{for } \lambda[i,k] = \lambda[j,k] \end{cases} \mathbf{b}^{\lambda[1,k]} \prod_{i=2}{}^{n} A_{i\lambda[i,k]}$
3	From 2, D6.6.3A Association of Sign to Parity, T10.2.5 Tool: Tensor Substitution and A10.2.14 Correspondence of Tensors and Tensor Coefficients	$\prod_{\beta=2}{}^{n} \times \mathbf{A}^{\beta} = \sum_{\rho(k)\in Lr} \begin{cases} (-1)^{\varphi(\mathbf{k})} & \text{for } \lambda[i,k] \neq \lambda[j,k] \\ 0 & \text{for } \lambda[i,k] = \lambda[j,k] \end{cases} \mathbf{b}^{\lambda[1,k]} \prod_{i=2}{}^{n} A_{i\lambda[i,k]}$
∴	From 3, D12.12.1A Covariant and Contravariant e-Skew Tensor Components T10.2.5 Tool: Tensor Substitution and A10.2.14 Correspondence of Tensors and Tensor Coefficients	$\prod_{\beta=2}{}^{n} \times \mathbf{A}^{\beta} = \sum_{\rho(k)\in Lr} \mathbf{e}^{\bullet\lambda[i,k]\bullet}\, \mathbf{b}^{\lambda[1,k]} \prod_{i=2}{}^{n} A_{i\lambda[i,k]}$

Theorem 12.11.14 Cross Product: Covariant Uniqueness

1g 2g	Given	$\mathbf{A}_i = \mathbf{B}_i$ for all i $A^{i\lambda[i,k]} = B^{i\lambda[i,k]}$
Steps	Hypothesis	$\prod_{\beta=2}{}^{n} \times \mathbf{A}_{\beta} = \prod_{\beta=2}{}^{n} \times \mathbf{B}_{\beta}$
1	From T12.11.12 Cross Product: Covariant Unified with the e-Tensor Operator	$\prod_{\beta=2}{}^{n} \times \mathbf{A}_{\beta} = \sum_{\rho(k)\in Lr} \mathbf{e}_{\bullet\lambda[i,k]\bullet}\, \mathbf{b}_{\lambda[1,k]} \prod_{i=2}{}^{n} A^{i\lambda[i,k]}$
2	From 1g, 2g, 1 and A7.9.2 Substitution of Vectors	$\prod_{\beta=2}{}^{n} \times \mathbf{A}_{\beta} = \sum_{\rho(k)\in Lm} \mathbf{e}_{\bullet\lambda[i,k]\bullet}\, \mathbf{b}_{i\lambda[i,k]} \prod_{i=2}{}^{m} B^{i\lambda[i,k]}$
∴	From 2 and T12.11.12 Cross Product: Covariant Unified with the e-Tensor Operator	$\prod_{\beta=2}{}^{n} \times \mathbf{A}_{\beta} = \prod_{\beta=2}{}^{n} \times \mathbf{B}_{\beta}$

Theorem 12.11.15 Cross Product: Contravariant Uniqueness

1g 2g	Given	$\mathbf{A}^i = \mathbf{B}^i$ for all i $A_{i\lambda[i,k]} = B_{i\lambda[i,k]}$
Steps	Hypothesis	$\prod_{\beta=2}{}^{n} \times \mathbf{A}^{\beta} = \prod_{\beta=2}{}^{n} \times \mathbf{B}^{\beta}$
1	From T12.11.12 Cross Product: Covariant Unified with the e-Tensor Operator	$\prod_{\beta=2}{}^{n} \times \mathbf{A}^{\beta} = \sum_{\rho(k)\in Lr} \mathbf{e}^{\bullet\lambda[i,k]\bullet}\, \mathbf{b}^{\lambda[1,k]} \prod_{i=2}{}^{n} A_{i\lambda[i,k]}$
2	From 1g, 2g, 1 and A7.9.2 Substitution of Vectors	$\prod_{\beta=2}{}^{n} \times \mathbf{A}^{\beta} = \sum_{\rho(k)\in Lm} \mathbf{e}^{\bullet\lambda[i,k]\bullet}\, \mathbf{b}^{i\lambda[i,k]} \prod_{i=2}{}^{m} B_{i\lambda[i,k]}$
∴	From 2 and T12.11.12 Cross Product: Covariant Unified with the e-Tensor Operator	$\prod_{\beta=2}{}^{n} \times \mathbf{A}^{\beta} = \prod_{\beta=2}{}^{n} \times \mathbf{B}^{\beta}$

Section 12.12 Serret-Frenet Formulas

In this section a set of three remarkable formulas, generally known as Frenet's formulas characterize all essential geometric properties of space curves. With the Serret-Frenet Formulas allows a bridge of theoretical mathematics to the physical world. This allows an explanation or interpretation of the meaning of the first and second differential along the spatial curve-C.

Definition 12.12.1 **Curvature Vector**

A vector **r**(s) that traces a point along a spatial curve-C is called a ***curvature vector***.

Definition 12.12.2 **Unit Tangent Vector**

The ***unit tangent vector*** **T**(s) is a vector of unit length at a specific point along a spatial curve-C.

Definition 12.12.3 **Unit Principle Normal Vector**

A ***unit normal vector*** **N**(s) is a unit vector at a point to a unit tangent vector along a spatial curve-C.

Definition 12.12.4 **Unit Binormal Vector**

A ***unit binormal vector*** forms a perpendicular ***trihedral*** set with a tangent and normal unit vector at a point along a spatial curve-C, given by:

$\mathbf{B} \equiv \mathbf{T} \times \mathbf{N}$ Equation A

Definition 12.12.5 **Osculating Plane**

The ***osculating plane*** is a plane formed from the Unit Tangent and Normal vectors [**T**, **N**] were the binormal vector represents the perpendicular normal vector to the plan. The osculating plane acquires its name as it oscillates along the spatial curve tracing out a spatial path.

Definition 12.12.6 **Radius of Curvature**

The radius of curvature [ρ] arises from an infinitesimal segment along a spatial curve that exactly matches the curvature of another infinitesimal segment from a comparable circle at a point-P along the spatial path. The radius of the circle lays parallel along the unit normal vector and center of the circle terminates at the opposite end of the radius from the point-P on the spatial curve-C.

Definition 12.12.7 **Curvature to a Spatial Curve-C**

The measure of curvature [κ] is the reciprocal of the radius of curvature at any given point along the spatial curve-C.

$\kappa \equiv 1 / \rho$	Measure of Curvature	Equation A
$\boldsymbol{\kappa} \equiv \kappa \mathbf{N}$	Curvature Vector	Equation B
$R \equiv \lvert \rho \rvert$	Radius of Curvature	Equation C

Definition 12.12.8 **Radial or Centripetal Acceleration**

Centripetal acceleration [a_ρ] is an acceleration brought on by some aspect of a rigid body in rotation.

$a_\rho \equiv v^2 / \rho$	Radial Acceleration	Equation A
$a_\rho \equiv \kappa v^2$		Equation B

Definition 12.12.9 **Torque and Torsion**

Torque arises as a type of counter force that opposes a change in the momentum of a rotating rigid body.

$\boldsymbol{\tau} \equiv \boldsymbol{\rho} \times d\mathbf{P} / dt$	Torque[HAL66II, pg 261]	Equation A
$\tau \equiv \lvert \boldsymbol{\tau} \rvert$	Torsion	Equation B

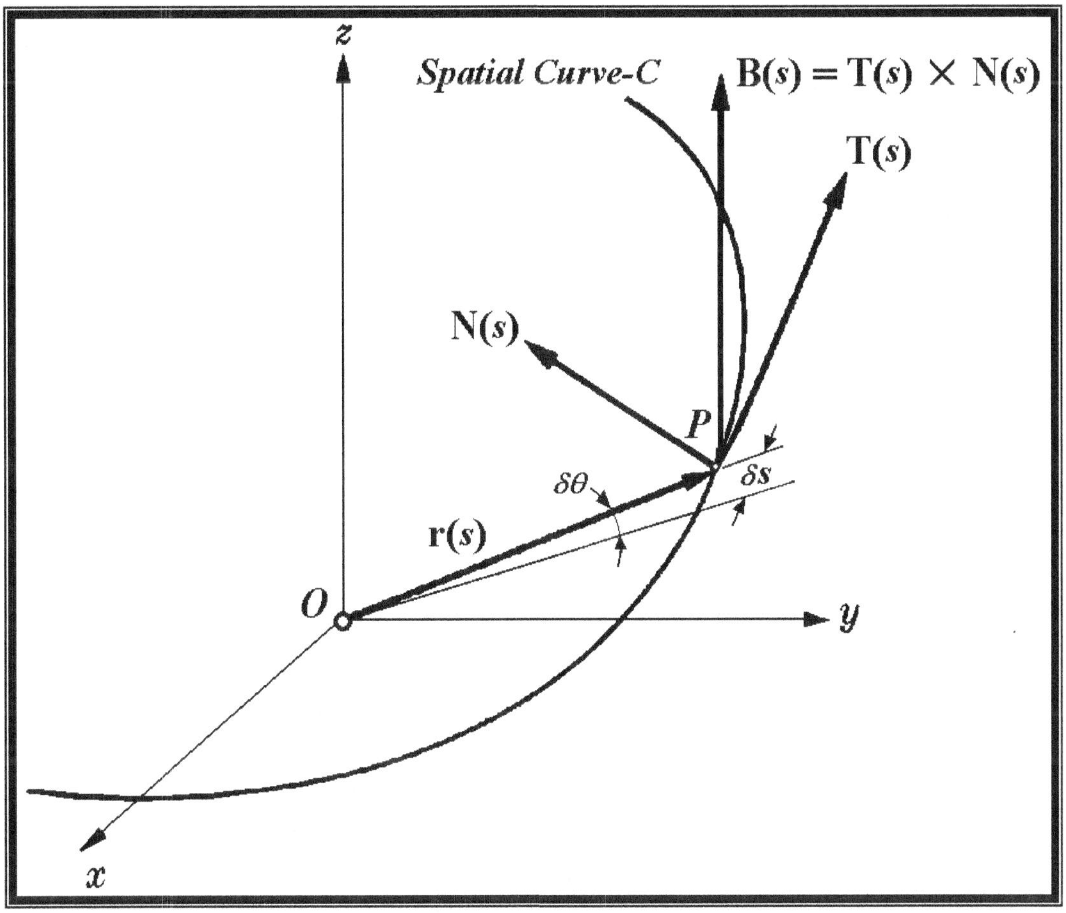

Figure 12.12.1 Moving Trihedral Unit Vectors

Theorem 12.12.1 On Space Curve-C: Unit Vector

1g	Given	$q^2 \equiv d\mathbf{r} / ds \bullet d\mathbf{r} / ds$
Step	Hypothesis	$1 = d\mathbf{r} / ds \bullet d\mathbf{r} / ds$
1	From 1g	$q^2 = d\mathbf{r} / ds \bullet d\mathbf{r} / ds$
2	From 1 and A7.8.5 Parallelogram Law: Commutative and Scalability of Collinear Vectors	$q^2 = (d\mathbf{r} \bullet d\mathbf{r}) / ds\, ds$
3	From 2 and D4.1.7 Exponential Notation	$q^2 = (d\mathbf{r} \bullet d\mathbf{r}) / ds^2$
4	From 3 and T8.3.4 Dot Product: Magnitude	$q^2 = \lvert d\mathbf{r} \rvert^2 / ds^2$
5	From 4 and O12.3.1D Properties of Riemann Arc Length	$q^2 = ds^2 / ds^2$
6	From 5 and A4.2.13 Inverse Multp	$q^2 = 1$
∴	From 1 and 6	$1 = d\mathbf{r} / ds \bullet d\mathbf{r} / ds$

Theorem 12.12.2 On Space Curve-C: Dot Product of Unit Tangent Vector

Step	Hypothesis	$1 = \mathbf{T} \bullet \mathbf{T}$
1	From T12.12.1 On Space Curve-C: Unit Vector and T8.4.5 Dot Product: Unit Vector	$1 = d\mathbf{r} / ds \bullet d\mathbf{r} / ds$
∴	From 1, T12.10.1 On Space Curve-C: Unit Vector and A7.9.2 Substitution of Vectors	$1 = \mathbf{T} \bullet \mathbf{T}$

Theorem 12.12.3 On Space Curve-C: Unit Tangent Vector

Step	Hypothesis	$\mathbf{T} = d\mathbf{r} / ds$ unit tangent vector
1	From 12.12.2 On Space Curve-C: Dot Product of Unit Tangent Vector	$1 = \mathbf{T} \bullet \mathbf{T}$
2	From 1 and T8.3.4 Dot Product: Magnitude	$1 = \mid \mathbf{T} \mid^2$
3	From 2, T4.10.1 Radicals: Uniqueness of Radicals, T4.10.3 Radicals: Identity Power Raised to a Radical and T4.9.4 Rational Exponent: Square Root of a Positive One	$1 = \mid \mathbf{T} \mid$ taking positive sign for a positive magnitude
∴	From 3 and O12.3.1B Properties of Riemann Arc Length	$\mathbf{T} = d\mathbf{r} / ds$ unit tangent vector

Theorem 12.12.4 On Space Curve-C: Velocity Vector

Step	Hypothesis	$\mathbf{V} = v\,\mathbf{T}$ EQ A	$\mathbf{V} = (ds / dt)\,\mathbf{T}$ EQ B
1	From DK.5.6 Velocity Vector	$\mathbf{V} = d\mathbf{r} / dt$	
2	From 1 and TK.5.3 Chain Rule for Exact Differential of a Vector	$\mathbf{V} = (ds / dt)\,(d\mathbf{r} / ds)$	
∴	From 2, DK.5.7 Speed or Celerity, O12.3.1B Properties of Riemann Arc Length, A4.2.3 Substitution and A7.9.2 Substitution of Vectors	$\mathbf{V} = v\,\mathbf{T}$ EQ A	$\mathbf{V} = (ds / dt)\,\mathbf{T}$ EQ B

Theorem 12.12.5 On Space Curve-C: Approximation of Circles

1g	Given	See Figure 12.3.1 Manifold Tangential Plane On Space Curve	
Step	Hypothesis	$ds = \rho\, d\theta$ EQ A	$1 / \rho = d\theta / ds$ EQ B
1	From D5.1.36B Subtended Arc length of a Cord	$s = \rho\, \theta$	
2	From 1g, 1, TK.3.2 Uniqueness of Differentials, TK.3.4B Single Variable Chain Rule, LxK.3.3.1A Differential of a Constant and holding ρ constant	$ds = \rho\, d\theta$ infinitesimal approximation of circles to every point on the curvature of arc with a radii ρ	
3	From 2 and T4.4.12 Equalities: Reversal of Cross Product of Proportions	$1 / \rho = d\theta / ds$ ρ the radius to the curve at that point	
∴	From 2 and 3	$ds = \rho\, d\theta$ EQ A	$1 / \rho = d\theta / ds$ EQ B

Theorem 12.12.6 On Space Curve-C: First Order Derivative of Tangent Vector

Step	Hypothesis	$d\mathbf{T} / ds = \kappa\, \mathbf{N}$ EQ A	$0 = \mathbf{T} \bullet \mathbf{N}$ EQ B
1	From T12.12.2 On Space Curve-C: Dot Product of Unit Tangent Vector	$1 = \mathbf{T} \bullet \mathbf{T}$	
2	From 1 and TK.3.2 Uniqueness of Differentials	$d\,(1) / ds = d\,(\mathbf{T} \bullet \mathbf{T}) / ds$	
3	From2, LxK.3.3.1 Differential of a Constant and TK.5.3 Chain Rule for Exact Differential of a Vector	$0 = \mathbf{T} \bullet d\mathbf{T} / ds + d\mathbf{T} / ds \bullet \mathbf{T}$	
4	From 3 and T4.3.10 Equalities: Summation of Repeated Terms by 2	$0 = 2\, \mathbf{T} \bullet d\mathbf{T} / ds$	
5	From 4, T4.4.1 Equalities: Any Quantity Multiplied by Zero is Zero and T4.4.3 Equalities: Cancellation by Multiplication	$0 = \mathbf{T} \bullet d\mathbf{T} / ds$	
6	From 5 and TK.5.3 Chain Rule for Exact Differential of a Vector	$d\mathbf{T} / ds = (d\theta / ds)(d\mathbf{T} / d\theta)$	
7	From 7, O12.3.1E Riemann Covariant Fundamental Tangent Vector, T12.10.4B On Space Curve-C: Approximation of Circles, A7.9.2 Substitution of Vectors and A4.2.3 Substitution	$d\mathbf{T} / ds = (1 / \rho)\, \mathbf{N}$ where $\mathbf{N}$ the primary normal to $\mathbf{T}$	
8	From 8, D12.12.7A Curvature to a Spatial Curve-C and A4.2.3 Substitution	$d\mathbf{T} / ds = \kappa\, \mathbf{N}$ hence $\mathbf{N} \parallel d\mathbf{T} / ds$	
9	From 5, 8 and A7.9.2 Substitution of Vectors	$0 = \mathbf{T} \bullet \kappa\mathbf{N}$	
10	From 9, T8.3.10 Dot Product: Right Scalar-Vector Association, T4.4.6A Equalities: Left Cancellation by Multiplication, T4.4.10 Equalities: Zero Divided by a Non-Zero Number and D8.3.3B Orthogonal Vectors	$0 = \mathbf{T} \bullet \mathbf{N}$ hence $\mathbf{T} \perp \mathbf{N}$	
∴	From 9 and 10	$d\mathbf{T} / ds = \kappa\, \mathbf{N}$ EQ A	$0 = \mathbf{T} \bullet \mathbf{N}$ EQ B

Theorem 12.12.7 On Space Curve-C: Dot Product of Unit Normal Vector

Step	Hypothesis	$1 = \mathbf{N} \bullet \mathbf{N}$
1	From T12.12.6A On Space Curve-C: First Order Derivative of Tangent Vector and T8.3.1 Dot Product: Uniqueness Operation	$d\mathbf{T} / ds \bullet d\mathbf{T} / ds = \kappa\, \mathbf{N} \bullet \kappa\, \mathbf{N}$
2	From 1, T12.12.6S8 On Space Curve-C: First Order Derivative of Tangent Vector, T4.4.6A Equalities: Left Cancellation by Multiplication, T8.3.9B Dot Product: Left Scalar-Vector Association and T8.3.10B Dot Product: Right Scalar-Vector Association	$[(d\mathbf{T} / ds) / \kappa] \bullet [(d\mathbf{T} / ds) / \kappa] = \mathbf{N} \bullet \mathbf{N}$
∴	From 2 and curvature κ normalizes the rate of the tangent vector	$1 = \mathbf{N} \bullet \mathbf{N}$

Theorem 12.12.8 On Space Curve-C: Acceleration Vector

Step	Hypothesis	$\mathbf{A} = a\, \mathbf{T} + \kappa v^2\, \mathbf{N}$
1	From DK5.8 Acceleration	$\mathbf{A} = d\, \mathbf{V} / dt$
2	From 1, T12.12.4A On Space Curve-C: Velocity Vector and A7.9.2 Substitution of Vectors	$\mathbf{A} = d\, (\, v\, \mathbf{T}\,) / dt$
3	From 2 and LxK.5.1.4 Differential of a Scalar and Vector Function	$\mathbf{A} = (d\, v / dt)\, \mathbf{T} + v\, (d\, \mathbf{T} / dt)$
4	From 3 and TK.5.4 Chain Rule for Single Variable Vector	$\mathbf{A} = a\, \mathbf{T} + v\, (d\, s / dt)\, (d\, \mathbf{T} / ds)$
5	From 4, DK.5.7 Speed or Celerity, A4.2.3 Substitution and D4.1.17 Exponential Notation	$\mathbf{A} = a\, \mathbf{T} + v^2\, (d\, \mathbf{T} / ds)$
6	From 5, T12.12.6 A On Space Curve-C: On Space Curve-C: First Order Derivative of Tangent Vector and A7.9.2 Substitution of Vectors	$\mathbf{A} = a\, \mathbf{T} + v^2\, (\kappa\, \mathbf{N})$
∴	From 6 and A4.2.10 Commutative Multp	$\mathbf{A} = a\, \mathbf{T} + \kappa v^2\, \mathbf{N}$

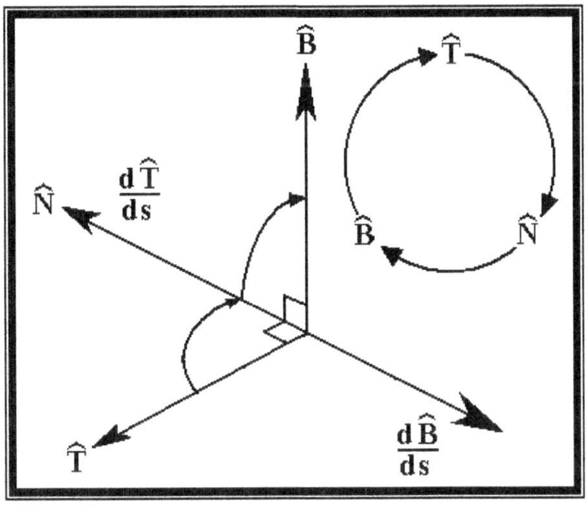

Figure 12.12.2 Binormal Rate Vector in a Trihedral Ordinate System

Theorem 12.12.9 On Space Curve-C: Binormal Unit Vector

Step	Hypothesis	B ≡ (1/τ) (d **N** / ds + κ **T**) the Binormal Unit Vector
1	From T12.12.6B On Space Curve-C: First Order Derivative of Tangent Vector and A4.2.2B Equality	0 = **T** • **N**
2	From 1 and LxK.5.2.2 Differential of Dot Product	0 = d **T** / ds • **N** + **T** • d **N** / ds
3	From 2, T4.3.3B Equalities: Right Cancellation by Addition and A4.2.7 Identity Add	**T** • d **N** / ds = –d **T** / ds • **N**
4	From 3, T12.12.6A On Space Curve-C: First Order Derivative of Tangent Vector and A4.2.3 Substitution	**T** • d **N** / ds = –κ **N** • **N**
5	From 4, T12.12.7 On Space Curve-C: Dot Product of Unit Normal Vector and A4.2.3 Substitution	**T** • d **N** / ds = –κ 1
6	From 5, T12.12.2 On Space Curve-C: Dot Product of Unit Tangent Vector	**T** • d **N** / ds = –κ **T** • **T**
7	From 6, A4.2.7 Identity Add and T4.3.4B Equalities: Reversal of Right Cancellation by Addition	0 = **T** • d **N** / ds + κ **T** • **T**
8	From 7, T8.3.7 Dot Product: Distribution of Dot Product Across Addition of Vectors	0 = **T** • (d **N** / ds + κ **T**) hence **T** ⊥ (d **N** / ds + κ **T**)
∴	From 8 and let a new unit vector be defined by normalizing it with the proportional torsion factor τ	B ≡ (1/τ) (d **N** / ds + κ **T**) the Binormal Unit Vector

Theorem 12.12.10 On Space Curve-C: Dot Product of Unit Binormal Vector

Step	Hypothesis	$1 = \mathbf{B} \bullet \mathbf{B}$
∴	From T12.12.9 On Space Curve-C: Binormal Unit Vector and T8.3.1 Dot Product: Uniqueness Operation	$1 = \mathbf{B} \bullet \mathbf{B}$

Theorem 12.12.11 On Space Curve-C: Rate Normal Vector

Step	Hypothesis	$d\,\mathbf{N} / ds = \tau\mathbf{B} - \kappa\,\mathbf{T}$
1	From T12.12.9 On Space Curve-C: Binormal Unit Vector	$B = (1/\tau)\,(d\,\mathbf{N} / ds + \kappa\,\mathbf{T})$
2	From 1 and T4.4.5A Equalities: Reversal of Right Cancellation by Multiplication	$\tau B = d\,\mathbf{N} / ds + \kappa\,\mathbf{T}$
∴	From 2 and T4.3.5B Equalities: Left Cancellation by Addition	$d\,\mathbf{N} / ds = \tau\mathbf{B} - \kappa\,\mathbf{T}$

Theorem 12.12.12 On Space Curve-C: Rate of Binormal Vector Rate of Normal Vector

Step	Hypothesis	$d\,\mathbf{B} / ds = \mathbf{T} \times (\,d\,\mathbf{N} / ds\,)$
1	From D12.10.4A Unit Binormal Vector and LxK.3.1.7B Uniqueness Theorem of Differentiation	$d\,\mathbf{B} / ds = d\,(\,\mathbf{T} \times \mathbf{N}\,) / ds$
2	From 1 and LxK.5.1.3 Differential of Cross Product	$d\,\mathbf{B} / ds = (\,d\,\mathbf{T} / ds\,) \times \mathbf{N} + \mathbf{T} \times (\,d\,\mathbf{N} / ds\,)$
3	From 2, T12.12.6 On Space Curve-C: First Order Derivative of Tangent Vector and A7.9.2 Substitution of Vectors	$d\,\mathbf{B} / ds = \kappa\,(\mathbf{N} \times \mathbf{N}) + \mathbf{T} \times (\,d\,\mathbf{N} / ds\,)$
∴	From 3, LxK.5.1.4 Parallel Vectors of Cross Product and A7.9.11 Identity Vector Object Add	$d\,\mathbf{B} / ds = \mathbf{T} \times (\,d\,\mathbf{N} / ds\,)$

Theorem 12.12.13 On Space Curve-C: Dot Product Tangent to Binormal Rate Vector

1g 2g	Given	$d\,\mathbf{B} / ds = d\,\mathbf{B} / ds$ $\mathbf{T} = \mathbf{T}$
Step	Hypothesis	$\mathbf{T} \bullet d\,\mathbf{B} / ds = 0$ for $\mathbf{T} \perp d\,\mathbf{B} / ds$
1	From 1g, 2g and T8.3.1 Dot Product: Uniqueness Operation	$\mathbf{T} \bullet d\,\mathbf{B} / ds = \mathbf{T} \bullet d\,\mathbf{B} / ds$
2	From 1, T12.12.12 On Space Curve-C: Rate of Binormal Vector Drives Rate Normal Vector and A7.9.2 Substitution of Vectors	$\mathbf{T} \bullet d\,\mathbf{B} / ds = \mathbf{T} \bullet [\mathbf{T} \times (\,d\,\mathbf{N} / ds\,)]$
∴	From 2 and LxK.5.1.8A Parallelepiped Matrix with Two Parallel sides and D8.3.3B Orthogonal Vectors	$\mathbf{T} \bullet d\,\mathbf{B} / ds = 0$ for $\mathbf{T} \perp d\,\mathbf{B} / ds$

Theorem 12.12.14 On Space Curve-C: Dot Product Binormal to Rate of Binormal Vector

Step	Hypothesis	$\mathbf{B} \bullet d\,\mathbf{B} / ds = 0$ for $\mathbf{B} \perp d\,\mathbf{B} / ds$
1	From D12.12.5 Unit Binormal Vector and T8.4.5 Dot Product: Unit Vector	$\mathbf{B} \bullet \mathbf{B} = 1$
∴	From 1, LxK.5.2.9 Differential of Constant Vector Magnitude	$\mathbf{B} \bullet d\,\mathbf{B} / ds = 0$ for $\mathbf{B} \perp d\,\mathbf{B} / ds$

Observation 12.12.1 Proportionality of Binormal Rate to Normal Vector

Theorem T12.12.13 "On Space Curve-C: Dot Product Tangent to Binormal Rate Vector" and T12.12.14 "On Space Curve-C: Dot Product of Binormal with Rate of Binormal Vector" $\mathbf{T} \perp d\,\mathbf{B} / ds$ and $\mathbf{B} \perp d\,\mathbf{B} / ds$ forcing us to conclude that the Binormal rate vector is parallel to the Normal vector in the Triad ordinate vector system, see Figure 12.12.2 "Binormal Rate Vector in a Trihedral Ordinate System", hence it can be concluded that

$$d\,\mathbf{B} / ds \propto \mathbf{N}. \qquad \text{Equation A}$$

Theorem 12.12.15 On Space Curve-C: Rate of Binormal Vector

1g	Given	F12.12.2 Binormal Rate Vector in a Trihedral Ordinate System
Step	Hypothesis	$d\,\mathbf{B} / ds = -\tau\,\mathbf{N}$
1	From D12.12.5 Unit Binormal Vector and TK.3.2 Uniqueness of Differentials	$d\,\mathbf{B} / ds = d\,(\mathbf{T} \times \mathbf{N}) / ds$
2	From 1 and LxK.5.2.3 Differential of Cross Product	$d\,\mathbf{B} / ds = (d\,\mathbf{T}/ ds) \times \mathbf{N} + \mathbf{T} \times (d\,\mathbf{N} / ds)$
3	From 2, T12.12.6A On Space Curve-C: First Order Derivative of Tangent Vector, T12.12.11 On Space Curve-C: Rate Normal Vector and A7.9.2 Substitution of Vectors	$d\,\mathbf{B} / ds = \kappa\,\mathbf{N} \times \mathbf{N} + \mathbf{T} \times (\tau\,\mathbf{B} - \kappa\,\mathbf{T})$
4	From 3, LxK.5.1.2 Distributive Property of Cross Products and LxK.5.1.3 Commutative property of Scalar to Cross Product	$d\,\mathbf{B} / ds = \kappa\,\mathbf{N} \times \mathbf{N} + \tau\mathbf{T} \times \mathbf{B} - \kappa\,\mathbf{T} \times \mathbf{T}$
5	From 4, LxK.5.1.4 Parallel Vectors of Cross Product, T7.10.3 Distribution of a Scalar over Addition of Vectors and T7.10.4 Grassmann's Zero Vector	$d\,\mathbf{B} / ds = \mathbf{0} + \tau\mathbf{T} \times \mathbf{B} - \mathbf{0}$
6	From 5 and A7.9.11 Identity Vector Object Add	$d\,\mathbf{B} / ds = \tau\mathbf{T} \times \mathbf{B}$
∴	From 1g, 6, Cyclic permutation of Trihedral Ordinate System and A7.9.2 Substitution of Vectors	$d\,\mathbf{B} / ds = -\tau\,\mathbf{N}$

Theorem 12.12.16 On Space Curve-C: Perpendicular Tangent to Binormal Vector

Step	Hypothesis	$\mathbf{T} \bullet \mathbf{B} = 0$
1	From T8.3.2 Dot Product: Equality	$\mathbf{T} \bullet \mathbf{B} = \mathbf{T} \bullet \mathbf{B}$
2	From 1, D12.10.4A Unit Binormal Vector and A7.9.2 Substitution of Vectors	$\mathbf{T} \bullet \mathbf{B} = \mathbf{T} \times (\mathbf{T} \times \mathbf{N})$
∴	From 2, LxK.5.1.8A Parallelepiped Matrix with Two Parallel sides	$\mathbf{T} \bullet \mathbf{B} = 0$

Theorem 12.12.17 On Space Curve-C: Perpendicular Normal to Binormal Vector

Step	Hypothesis	$\mathbf{N} \bullet \mathbf{B} = 0$
1	From T8.3.2 Dot Product: Equality	$\mathbf{N} \bullet \mathbf{B} = \mathbf{N} \bullet \mathbf{B}$
2	From 1, D12.10.4A Unit Binormal Vector and A7.9.2 Substitution of Vectors	$\mathbf{N} \bullet \mathbf{B} = \mathbf{N} \times (\mathbf{T} \times \mathbf{N})$
∴	From 2, LxK.5.1.8B Parallelepiped Matrix with Two Parallel sides	$\mathbf{N} \bullet \mathbf{B} = 0$

Theorem 12.12.18 On Space Curve-C: Perpendicular Tangent to Binormal Vector

Step	Hypothesis	$\mathbf{T} \bullet d(\mathbf{B} / ds) = 0$
1	From T12.10.10 On Space Curve-C: Perpendicular Tangent to Binormal Vector	$\mathbf{T} \bullet \mathbf{B} = 0$
2	From 1 and LxK.3.1.7B Uniqueness Theorem of Differentiation	$d(\mathbf{T} \bullet \mathbf{B}) / ds = d(0) / ds$
3	From 2, LxK.3.3.1A Differential of a Constant and LxK.5.2.2 Differential of Dot Product	$d(\mathbf{T} / ds) \bullet \mathbf{B} + \mathbf{T} \bullet d(\mathbf{B} / ds) = 0$
4	From 3, T12.10.5A On Space Curve-C: First Order Derivative of Tangent Vector and A7.9.2 Substitution of Vectors	$\kappa \mathbf{N} \bullet \mathbf{B} + \mathbf{T} \bullet d(\mathbf{B} / ds) = 0$
5	From 4, T12.10.11 On Space Curve-C: Perpendicular Normal to Binormal Vector and T4.4.1 Equalities: Any Quantity Multiplied by Zero is Zero	$0 + \mathbf{T} \bullet d(\mathbf{B} / ds) = 0$
∴	From 5 and A4.2.7 Identity Add	$\mathbf{T} \bullet d(\mathbf{B} / ds) = 0$

Theorem 12.12.19 On Space Curve-C: Second Order Differential of Curvature Vector

Step	Hypothesis	$d^2 \mathbf{r} / ds^2 = \kappa \mathbf{N}$
1	LxK.3.1.7B Uniqueness Theorem of Differentiation and DxK.3.2.1H Second Parametric Differentiation with mark, fraction and dot notation	$d^2 \mathbf{r} / ds^2 = d^2 \mathbf{r} / ds^2$
2	From 1 and DxK.3.2.1K Second Parametric Differentiation with mark, fraction and dot notation	$d^2 \mathbf{r} / ds^2 = d (d \mathbf{r} / ds) / ds$
3	From 1, T12.12.3 On Space Curve-C: Unit Tangent Vector and A7.9.2 Substitution of Vectors	$d^2 \mathbf{r} / ds^2 = d \mathbf{T} / ds$
∴	From 2, T12.12.6A On Space Curve-C: First Order Derivative of Tangent Vector and A7.9.2 Substitution of Vectors	$d^2 \mathbf{r} / ds^2 = \kappa \mathbf{N}$

Theorem 12.12.20 On Space Curve-C: Third Order Differential of Curvature Vector

Step	Hypothesis	$d^3 \mathbf{r} / ds^3 = \kappa\tau \mathbf{B} + \kappa' \mathbf{N} - \kappa^2 \mathbf{T}$
1	LxK.3.1.7B Uniqueness Theorem of Differentiation and DxK.3.2.1I Second Parametric Differentiation with mark, fraction and dot notation	$d^3 \mathbf{r} / ds^3 = d^3 \mathbf{r} / ds^3$
2	From 1 and DxK.3.2.1K General Exact Differential Operator	$d^3 \mathbf{r} / ds^3 = d (d^2 \mathbf{r} / ds^2) / ds$
3	From 2, T12.12.19 On Space Curve-C: Second Order Differential of Curvature Vector and A7.9.2 Substitution of Vectors	$d^3 \mathbf{r} / ds^3 = d (\mathbf{T}) / ds$
4	From 3, T12.12.6A On Space Curve-C: First Order Derivative of Tangent Vector and A7.9.2 Substitution of Vectors	$d^3 \mathbf{r} / ds^3 = d (\kappa \mathbf{N}) / ds$
5	From 4 and LxK.5.2.4 Differential of Scalar-Vector Product	$d^3 \mathbf{r} / ds^3 = (d \kappa / ds) \mathbf{N} + \kappa (d \mathbf{N} / ds)$
6	From 5, DxK.3.2.1G First Parametric Differentiation, T12.12.11 On Space Curve-C: Rate Normal Vector and A7.9.2 Substitution of Vectors	$d^3 \mathbf{r} / ds^3 = \kappa' \mathbf{N} + \kappa (\tau\mathbf{B} - \kappa \mathbf{T})$
∴	From 6, T7.10.3 Distribution of a Scalar over Addition of Vectors, D4.1.17 Exponential Notation and A7.9.9 Commutative Vector Object Add	$d^3 \mathbf{r} / ds^3 = \kappa\tau \mathbf{B} + \kappa' \mathbf{N} - \kappa^2 \mathbf{T}$

Table 12.12.1 Serret-Frenet Orthogonal Trihedral Formulas[12.12.1]

Theorem 12.12.6B On Space Curve-C: First Order Derivative of Tangent Vector	$\mathbf{T} \bullet \mathbf{N} = 0$
Theorem 12.12.21 On Space Curve-C: Perpendicular Tangent Vector	$\mathbf{T} \bullet d\mathbf{T} /ds = 0$
Theorem 12.12.22 On Space Curve-C: Perpendicular Normal Vector	$\mathbf{N} \bullet d\mathbf{N} /ds = 0$
Theorem 12.12.14 On Space Curve-C: Dot Product Binormal to Rate of Binormal Vector	$\mathbf{B} \bullet d\mathbf{B} /ds = 0$

Table 12.12.2 First, Second and Third Order Derivatives for Curvature Vector

Theorem 12.12.3 On Space Curve-C: Unit Tangent Vector	$d\mathbf{r} /ds = \mathbf{T}$
Theorem 12.12.19 On Space Curve-C: Second Order Differential of Curvature Vector	$d^2 \mathbf{r} / ds^2 = \kappa \mathbf{N}$
Theorem 12.12.20 Third Order Differential of Curvature Vector	$d^3 \mathbf{r} / ds^3 = \kappa\tau \mathbf{B} + \kappa' \mathbf{N} - \kappa^2 \mathbf{T}$

Table 12.12.3 In summary: Serret-Frenet Perpendicular Relationships

By Theorem	Perpendicular Relationships
Serret-Frenet Perpendicular Vectors	
O12.3.1B "Riemann Covariant Fundamental Tangent and Normal Vector"	$\mathbf{r} \perp \mathbf{T}$
T12.12.6 "On Space Curve-C: First Order Derivative of Tangent Vector"	$\mathbf{T} \perp \mathbf{N}$
T12.12.16 "On Space Curve-C: Perpendicular Tangent to Binormal Vector"	$\mathbf{T} \perp \mathbf{B}$
T12.12.17 "On Space Curve-C: Perpendicular Normal to Binormal Vector"	$\mathbf{N} \perp \mathbf{B}$
Serret-Frenet Trihedral Formulas	
O12.3.1B "Riemann Covariant Fundamental Tangent and Normal Vector"	$\mathbf{r}' \perp \mathbf{r}$
T12.12.16 "On Space Curve-C: Perpendicular Tangent Vector"	$\mathbf{T}' \perp \mathbf{T}$
T12.13.18 "On Space Curve-C: Perpendicular Normal Vector"	$\mathbf{N}' \perp \mathbf{N}$
T12.12.13 "On Space Curve-C: Dot Product Tangent to Binormal Rate Vector"	$\mathbf{B}' \perp \mathbf{T}$
T12.12.14 "On Space Curve-C: Dot Product of Binormal with Rate of Binormal Vector"	$\mathbf{B}' \perp \mathbf{B}$

[12.12.1]Note: The cornerstone to prove observations O12.3.1B and Theorems 12.12.16 to 12.12.14 is LxK.5.2.9 "Differential of Constant Vector Magnitude".

Section 12.13 Flat Plane Curvature

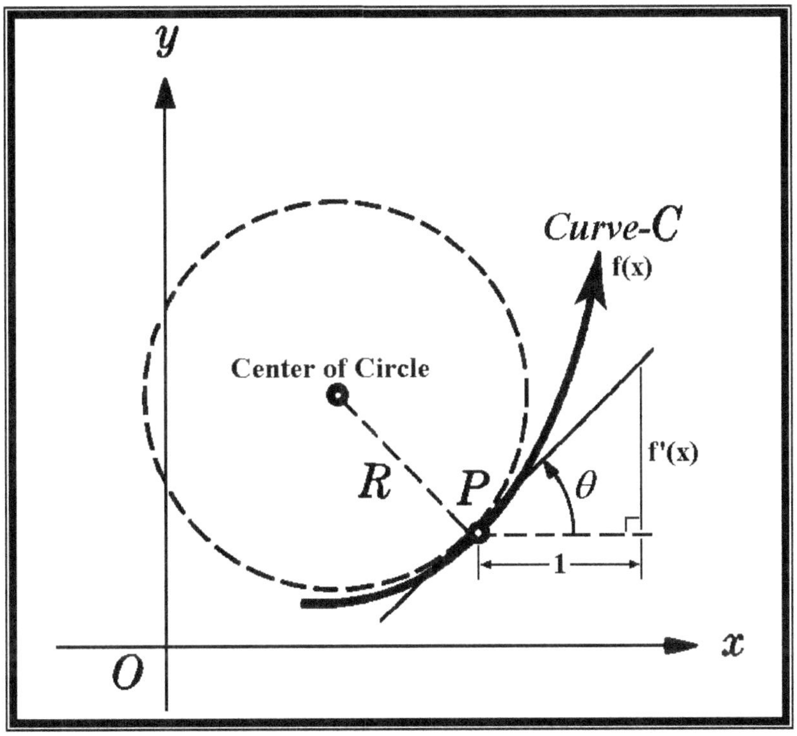

Figure 12.13.1 Curvature of Arc in a Flat Euclidian Plane

The questions now arise by how much does a curve-C deviate from a straight line? More important how can the deviation be measured? The answers are solved by Euclid with a series of theorems dealing with a circle and a line tangent to the circle at a point on the circumference, theorems T5.13.1 "Radius is Perpendicular to Tangent", T5.13.2 "Radius is Perpendicular Line Implies a Tangent" and T5.13.4 "Radii-Tangent Angles". Now if the curve-C passes through the point on the circumference as shown in figure F12.13.1 "Curvature of Arc in a Flat Euclidian Plane" in just the right way so that geometrically, circle, tangent line and the curve share the tangent point then the radius of the circle can proved away to typify the deviation of the curve from the tangent line.

Also what if the tangent line of the circle were the tangent of the curve then, a radius can be associated with the curve at that point. So a curve can now be thought of as being made up of an infinite number of circles, each circle with its own unique radius, giving away of measuring the curvature at any given point along the curve.

Another way of rationalizing having circles associated at every point along the curve is if each circle were chosen in just such away, so that a very small piece, arc, segment or portion is selected. As the arc becomes smaller it appears to approximate a straight line, thereby representing the tangent line of the curve.

Using a radius to measure curvature is a good idea, but what of the most common problem of when part of the curve is a straight line? The radius of a straight line is infinite in length. To compensate for this singularity the reciprocal of the radius is taken so that the measure is now zero, this then is the definition of curvature and how it is measured.

Theorem 12.13.1 Curvature On Curve-C: Calculating Curvature in a Flat Euclidian Plane

1g	Given	F12.14.1 Curvature of Arc in a Flat Euclidian Plane
2g		Euclidian Space $\mathfrak{E}^n$
Step	Hypothesis	$\kappa = f'' / \sqrt{[1 + (f')^2]^3}$
1	From T12.12.5B On Space Curve-C: Approximation of Circles, D12.12.7A Curvature to a Spatial Curve-C and A4.2.3 Substitution	$\kappa = d\theta / ds$
2	From 1 and TK.3.4B Single Variable Chain Rule	$\kappa = (d\theta / dx)(dx / ds)$
3	From T12.3.22 Fundamental Quadratic: Contravariant Arc Length	$ds^2 = \Sigma_i \Sigma_j g^{ij} dx_i dx_j$
4	From 2g, 3, A11.4.5P1 Euclidian Manifold $\mathfrak{E}^n$ and T10.2.5 Tool: Tensor Substitution	$ds^2 = \Sigma_i \Sigma_j \delta^{ij} dx_i dx_j$
5	From 4, D6.1.6 Identity Matrix and evaluating, A4.2.3 Identity Exp, A4.2.18 Summation Exp and D4.1.19 Primitive Definition for Rational Arithmetic	$ds^2 = \Sigma_i dx_i^2$
6	From 1g, 2g and 5	$ds = \sqrt{dx^2 + df^2}$ at point-P in Euclidian Space $\mathfrak{E}^2$
7	From 6, A4.2.12 Identity Multp, A4.2.13 Inverse Multp, D4.1.4(A, B) Rational Numbers, A4.2.10 Commutative Multp and A4.2.11 Associative Multp	$ds = \sqrt{[dx^2 + (df^2 / dx^2) dx^2]}$
8	From 7, A4.2.12 Identity Multp A4.2.14 Distribution, T4.10.3 Radicals: Identity Power Raised to a Radical and T4.8.7 Integer Exponents: Distribution Across a Rational Number	$ds = [\sqrt{1 + (df / dx)^2}]\, dx$
9	From 8, A4.2.12 Identity Multp and T4.4.12 Equalities: Reversal of Cross Product of Proportions	$dx / ds = 1 / [\sqrt{1 + (f')^2}]$
10	From 1g and DxE.1.2.3 Analytic Tangent of Angle θ	$\tan\theta = f' / 1$
11	From 10 and TE.1.13 Quick Lookup: Inverse Trigonometric Functions, for tangent	$\theta = \arctan f'$
12	From 11, LxK.3.4.6 Derivative of arc tangent	$d\theta / dx = \{1 / [1 + (f')^2]\}(df' / dx)$
13	From 12 and DxK.3.2.1H Second Parametric Differentiation with mark	$d\theta / dx = \{1 / [1 + (f')^2]\} f''$
14	From 13, A4.2.10 Commutative Multp and A4.2.12 Identity Multp	$d\theta / dx = f'' / [1 + (f')^2]$

15	From 2, 9, 14 and A4.2.3 Substitution	$d\,\theta / ds = \{f'' / [\,1 + (f')^2\,]\}\{\,1 / [\sqrt{1 + (f')^2}\,]\}$
16	From 15, T4.5.4 Equalities: Product of Two Rational Fractions and A4.2.12 Identity Multp	$d\,\theta / ds = f'' / \{\,[\,1 + (f')^2\,]\,[\sqrt{1 + (f')^2}\,]\}$
17	From 16 and T4.10.11A Radicals: Addition by Product and Radical Number	$d\,\theta / ds = f'' / \sqrt{[1 + (f')^2\,]^3}$
∴	From 1, 17 and A4.2.3 Substitution	$\kappa = f'' / \sqrt{[1 + (f')^2\,]^3}$

Theorem 12.13.2 Curvature On Curve-C: Second Derivative of the Curve Function

1g	Given	$f' = dy / dx$
2g		$x = x(t)$ a function of a parametric timeline
3g		$y = y(t)$ a function of a parametric timeline
Step	Hypothesis	$f'' = (1/\dot{x})^3\,[\dot{x}\,\ddot{y} - \dot{y}\,\ddot{x}\,]$
1	From 1g, LxK.3.1.7B Uniqueness Theorem of Differentiation and DxK.3.2.1H Second Parametric Differentiation with mark, fraction and dot notation	$f'' = d(\,dy/dx\,) / dx$
2	From 2g, 3g, 1 and TK.3.4B Single Variable Chain Rule	$f'' = d[(\,dy/dt\,)(\,dt/dx\,)] / dt\,(dt/dx)$
3	From 2, DxK.3.2.1B First Parametric Differentiation and A4.2.10 Commutative Multp	$f'' = (1/\dot{x})\,[d\,(\dot{y}/\dot{x}) / dt]$
4	From 3, LxK.3.2.6 Differential of a Binary Ratio and DxK.3.2.1C Second Parametric Differentiation with mark, fraction and dot notation	$f'' = (1/\dot{x})\,[\,(1/\dot{x})\ddot{y} - \dot{y}\ddot{x}/\dot{x}^2]$
5	From 4, A4.2.12 Identity Multp, A4.2.13 Inverse Multp, T4.5.4 Equalities: Product of Two Rational Fractions, D4.1.17 Exponential Notation and A4.2.10 Commutative Multp	$f'' = (1/\dot{x})\,[\dot{x}\,\ddot{y}/\dot{x}^2 - \ddot{x}/\dot{x}^2]$
∴	From 5, A4.2.14 Distribution T4.5.4 Equalities: Product of Two Rational Fractions and D4.1.17 Exponential Notation	$f'' = (1/\dot{x})^3\,[\dot{x}\,\ddot{y} - \dot{y}\,\ddot{x}\,]$

Theorem 12.13.3 Curvature On Curve-C: First Derivative of the Arc Length Function

1g	Given	$f' = dy / dx$
2g		$x = x(t)$ a function of a parametric timeline
3g		$y = y(t)$ a function of a parametric timeline
Step	Hypothesis	$\sqrt{(1 + f'^2)^3} = (1 / \dot{x})^3 \sqrt{(\dot{x}^2 + \dot{y}^2)^3}$
1	From A4.2.2A Equality	$\sqrt{(1 + f'^2)^3} = \sqrt{(1 + f'^2)^3}$
2	From 1g, 1 and A4.2.3 Substitution	$\sqrt{(1 + f'^2)^3} = \sqrt{(1 + (dy/dx)^2)^3}$
3	From 2g, 3g, 2 and TK.3.4B Single Variable Chain Rule	$\sqrt{(1 + f'^2)^3} = \sqrt{(1 + (dy/dt\ dt/dx)^2)^3}$
4	From 3 and DxK.3.2.1B First Parametric Differentiation	$\sqrt{(1 + f'^2)^3} = \sqrt{(1 + (\dot{y} / \dot{x})^2)^3}$
5	From 4 and A4.2.13 Inverse Multp	$\sqrt{(1 + f'^2)^3} = \sqrt{((\dot{x} / \dot{x})^2 + (\dot{y} / \dot{x})^2)^3}$
6	From 5 and T4.8.7 Integer Exponents: Distribution Across a Rational Number	$\sqrt{(1 + f'^2)^3} = \sqrt{((\dot{x}^2 / \dot{x}^2) + (\dot{y}^2 / \dot{x}^2))^3}$
7	From 6 and A4.2.14 Distribution T4.5.4	$\sqrt{(1 + f'^2)^3} = \sqrt{[(\dot{x}^2 + \dot{y}^2)(1 / \dot{x}^2)]^3}$
8	From 7 and A4.2.21 Distribution Exp	$\sqrt{(1 + f'^2)^3} = \sqrt{(\dot{x}^2 + \dot{y}^2)^3 (1 / \dot{x}^2)^3}$
9	From 8, T4.8.7 Integer Exponents: Distribution Across a Rational Number and A4.2.20 Commutative Exp	$\sqrt{(1 + f'^2)^3} = \sqrt{(\dot{x}^2 + \dot{y}^2)^3 (1 / \dot{x}^3)^2}$
∴	From 9, T4.10.3 Radicals: Identity Power Raised to a Radical and A4.2.10 Commutative Multp	$\sqrt{(1 + f'^2)^3} = (1 / \dot{x})^3 \sqrt{(\dot{x}^2 + \dot{y}^2)^3}$

Theorem 12.13.4 Curvature On Curve-C: Calculating Implicit Curvature

Step	Hypothesis	$\kappa = (\dot{x}\ddot{y} - \ddot{x}\dot{y}) / \sqrt{(\dot{x}^2 + \dot{y}^2)^3}$
1	T12.14.1 Curvature On Curve-C: Calculating Curvature in a Flat Euclidian Plane	$\kappa = f'' / \sqrt{[1 + (f')^2]^3}$
2	From 1, T12.14.2 Curvature On Curve-C: Second Derivative of the Curve Function, T12.14.3 Curvature On Curve-C: First Derivative of the Arc Length Function and A4.2.3 Substitution	$\kappa = (1 / \dot{x})^3 [\dot{x}\ddot{y} - \dot{y}\ddot{x}] / (1 / \dot{x})^3 \sqrt{(\dot{x}^2 + \dot{y}^2)^3}$
∴	From 2, A4.2.13 Inverse Multp and A4.2.12 Identity Multp	$\kappa = (\dot{x}\ddot{y} - \ddot{x}\dot{y}) / \sqrt{(\dot{x}^2 + \dot{y}^2)^3}$

Section 12.14 Torsion of an Osculating Plane

Theorem 12.14.1 Torsion: Calculating Torsion in Intrinsic Form

1g	Given	$\tau\kappa^2 \equiv \mathbf{r}' \bullet (\mathbf{r}'' \times \mathbf{r}''')$ assume
2g		Figure 12.12.2 Binormal Rate Vector in a Trihedral Ordinate System
Step	Hypothesis	$\tau = \mathbf{r}' \bullet (\mathbf{r}'' \times \mathbf{r}''') / \kappa^2$
1	From 1g	$\tau\kappa^2 = \mathbf{r}' \bullet (\mathbf{r}'' \times \mathbf{r}''')$
2	From 1, T12.12.3 On Space Curve-C: Unit Tangent Vector, T12.12.19 On Space Curve-C: Second Order Differential of Curvature Vector, T12.12.20 Third Order Differential of Curvature Vector and A7.9.2 Substitution of Vectors	$\tau\kappa^2 = \mathbf{T} \bullet [\kappa\, \mathbf{N} \times (\kappa\tau\, \mathbf{B} + \kappa'\, \mathbf{N} - \kappa^2\, \mathbf{T})]$
3	From 2 and LxK.5.1.2 Distributive Property of Cross Products	$\tau\kappa^2 = \kappa\mathbf{T} \bullet (\kappa\tau\mathbf{N} \times \mathbf{B} + \kappa'\, \mathbf{N} \times \mathbf{N} - \kappa^2\, \mathbf{N} \times \mathbf{T})$
4	From 2g, 3, LxK.5.1.4 Parallel Vectors of Cross Product and cyclic permutation	$\tau\kappa^2 = \kappa\mathbf{T} \bullet [\kappa\tau\mathbf{T} + \mathbf{0} - \kappa^2(-\,\mathbf{B})]$
5	From 4, A7.9.11 Identity Vector Object Add and T4.8.2B Integer Exponents: Negative One Squared	$\tau\kappa^2 = \kappa\mathbf{T} \bullet (\kappa\tau\mathbf{T} + \kappa^2\, \mathbf{B})$
6	From 5 and T8.3.7 Dot Product: Distribution of Dot Product Across Addition of Vectors	$\tau\kappa^2 = \kappa\, (\kappa\tau\mathbf{T} \bullet \mathbf{T} + \kappa^2\, \mathbf{T} \bullet \mathbf{B})$
7	From 6, T12.12.2 On Space Curve-C: Dot Product of Unit Tangent Vector and T12.12.16 On Space Curve-C: Perpendicular Tangent to Binormal Vector	$\tau\kappa^2 = \kappa\, (\kappa\tau\, 1 + 0\,)$
8	From 7, A4.2.12 Identity Multp, A4.2.7 Identity Add and D4.1.17 Exponential Notation	$\tau\kappa^2 = \tau\kappa^2$
∴	From 1, 8, T4.4.6B Equalities: Left Cancellation by Multiplication and by identity	$\tau = \mathbf{r}' \bullet (\mathbf{r}'' \times \mathbf{r}''') / \kappa^2$

Theorem 12.14.2 Torsion: Calculating Torsion in Matrix Form

1g	Given	r = (x, y, z) continuously differentiable in Euclidian Space $\mathfrak{E}^3$
Step	Hypothesis	$\tau = \dfrac{\begin{vmatrix} x' & y' & z' \\ x'' & y'' & z'' \\ x''' & y''' & z''' \end{vmatrix}}{x''^2 + y''^2 + z''^2}$
1	From 1g, 2g, AK.5.3A Uniqueness of Differentiation of a Vector and TK.5.1 Derivative Distribution Across Vector Components	r' = (x', y', z')
2	From 1, AK.5.3A Uniqueness of Differentiation of a Vector and TK.5.1 Derivative Distribution Across Vector Components	r'' = (x'', y'', z'')
3	From 2, AK.5.3A Uniqueness of Differentiation of a Vector and TK.5.1 Derivative Distribution Across Vector Components	r''' = (x''', y''', z''')
4	From 1g, AK.5.3A Uniqueness of Differentiation of a Vector and TK.5.1 Derivative Distribution Across Vector Components, T12.12.3 On Space Curve-C: Unit Tangent Vector and A7.9.1 Equivalence of Vector	**T** = (x', y', z')
5	From 4, AK.5.3A Uniqueness of Differentiation of a Vector and TK.5.1 Derivative Distribution Across Vector Components	d**T**/ds = (x'', y'', z'')
6	From 5, T12.12.6A On Space Curve-C: First Order Derivative of Tangent Vector and A7.9.1 Equivalence of Vector	κ **N** = (x'', y'', z'')
7	From 6 and T8.3.1 Dot Product: Uniqueness Operation	κ **N** • κ **N** = (x'', y'', z'') • (x'', y'', z'')
8	From 7, T8.3.10A Dot Product: Right Scalar-Vector Association, D4.1.17 Exponential Notation, T12.12.2 On Space Curve-C: Dot Product of Unit Tangent Vector, A4.2.12 Identity Multp and T8.4.4 Dot Product: Magnitude of a Vector	$\kappa^2 = x''^2 + y''^2 + z''^2$

9	From 8 and T12.14.1 Torsion: Calculating Torsion	$\tau = \mathbf{r}' \bullet (\mathbf{r}'' \times \mathbf{r}''') / \kappa^2$
∴	From 1, 2, 3, 8, 9, LxK.5.1.7 Volume of Parallelepiped Matrix and A4.2.3 Substitution	$\tau = \dfrac{\begin{vmatrix} x' & y' & z' \\ x'' & y'' & z'' \\ x''' & y''' & z''' \end{vmatrix}}{x''^2 + y''^2 + z''^2}$

Section 12.15 Trihedral System

Definition 12.15.1 **Complimentary, Conjugate, Normal Rate Vector**

$d\mathbf{N}^* / ds \equiv \tau\, \mathbf{T} + \kappa\, \mathbf{B}$ Equation A

Definition 12.15.2 **Spin Coefficient and Its Conjugate**

The internal coefficient made from the spinning systems parameters are called the ***spin coefficient***

$s \equiv \tau + \kappa$ Equation A

and the ***conjugate spin coefficient***

$s \equiv \tau - \kappa$ Equation B

Theorem 12.15.1 Trihedral System: Serret-Frenet Tangent Rate Vector

1g	Given	d**T** / ds = **N*'** x **T** assume
2g		**N*'** = τ **T** + κ **B**
3g		Figure 12.12.2 "Binormal Rate Vector in a Trihedral Ordinate System"
Step	Hypothesis	d**T** / ds = **N*'** x **T**
1	From 1g	d**T** / ds = **N*'** x **T**
2	From 2g, 1, T12.12.6A On Space Curve-C: First Order Derivative of Tangent Vector and A7.9.2 Substitution of Vectors	κ **N** = (τ **T** + κ **B**) x **T**
3	From 2 and LxK.5.1.2 Distributive Property of Cross Products	κ **N** = τ **T** x **T** + κ **B** x **T**
4	From 3, LxK.5.1.4 Parallel Vectors of Cross Product, Cyclic permutation of Trihedral Ordinate System and A7.9.2 Substitution of Vectors	κ **N** = **0** + κ **N**
5	From 4 and A7.9.11 Identity Vector Object Add	κ **N** = κ **N**
∴	From 1, 5 and by identity	d**T** / ds = **N*'** x **T**

Theorem 12.15.2 Trihedral System: Serret-Frenet Normal Rate Vector

1g	Given	d**N** / ds = **N*'** x **N** assume
2g		**N*'** = τ **T** + κ **B**
3g		Figure 12.12.2 "Binormal Rate Vector in a Trihedral Ordinate System"
Step	Hypothesis	d **N** / ds = **N*'** x **N**
1	From 1g	d **N** / ds = **N*'** x **N**
2	From 2g, 1, T12.12.11 On Space Curve-C: Rate Normal Vector and A7.9.2 Substitution of Vectors	τ**B** – κ **T** = (τ **T** + κ **B**) x **N**
3	From 2 and LxK.5.1.2 Distributive Property of Cross Products	τ**B** – κ **T** = τ **T** x **N** + κ **B** x **N**
4	From 1g, 3, Cyclic permutation of Trihedral Ordinate System and A7.9.2 Substitution of Vectors	τ**B** – κ **T** = τ **B** – κ **T**
∴	From 1, 4 and by identity	d **N** / ds = **N*'** x **N**

Theorem 12.15.3 Trihedral System: Serret-Frenet Binormal Rate Vector

1g	Given	$d\mathbf{B} / ds = \mathbf{N}^{*\prime} \times \mathbf{B}$ assume
2g		$\mathbf{N}^{*\prime} = \tau\, \mathbf{T} + \kappa\, \mathbf{B}$
3g		Figure 12.12.2 "Binormal Rate Vector in a Trihedral Ordinate System"
Step	Hypothesis	$d\,\mathbf{B} / ds = \mathbf{N}^{*\prime} \times \mathbf{B}$
1	From 1g	$d\,\mathbf{B} / ds = \mathbf{N}^{*\prime} \times \mathbf{B}$
2	From 2g, 1, T12.12.15 On Space Curve-C: Rate of Binormal Vector and A7.9.2 Substitution of Vectors	$-\tau\, \mathbf{N} = (\tau\, \mathbf{T} + \kappa\, \mathbf{B}) \times \mathbf{B}$
3	From 2 and LxK.5.1.2 Distributive Property of Cross Products	$-\tau\, \mathbf{N} = \tau\, \mathbf{T} \times \mathbf{B} + \kappa\, \mathbf{B} \times \mathbf{B}$
4	From 1g, 3, LxK.5.1.4 Parallel Vectors of Cross Product, Cyclic permutation of Trihedral Ordinate System and A7.9.2 Substitution of Vectors	$-\tau\, \mathbf{N} = -\tau\, \mathbf{N} + \mathbf{0}$
5	From 4 and T7.9.11 Identity Vector Object Add	$-\tau\, \mathbf{N} = -\tau\, \mathbf{N}$
∴	From 1, 5 and by identity	$d\,\mathbf{B} / ds = \mathbf{N}^{*\prime} \times \mathbf{B}$

Table 12.15.1 Summary of Trihedral Serret-Frenet Rotation Rate Vectors

Rotation Rate Vector	Theorems
$d\,\mathbf{T} / ds = \mathbf{N}^{*\prime} \times \mathbf{T}$	T12.15.2 Trihedral System: Serret-Frenet Tangent Rate Vector
$d\,\mathbf{N} / ds = \mathbf{N}^{*\prime} \times \mathbf{N}$	T12.15.3 Trihedral System: Serret-Frenet Normal Rate Vector
$d\,\mathbf{B} / ds = \mathbf{N}^{*\prime} \times \mathbf{B}$	T12.15.4 Trihedral System: Serret-Frenet Binormal Rate Vector

Observation 12.15.1 Complimentary and Conjugate Normal Quantities

The equation from theorem T12.12.11 "On Space Curve-C: Rate Normal Vector"

$$d\,\mathbf{N} / ds = \tau\, \mathbf{B} - \kappa\, \mathbf{T} \qquad \text{Equation A}$$

looks a great deal like ***complimentary, conjugate, normal rate vector***

$$d\mathbf{N}^{*} / ds \equiv \tau\, \mathbf{T} + \kappa\, \mathbf{B} \qquad \text{Equation B}$$

Two mathematical quantities that are similar are said to be complements of one another, however if they are opposite in form by a negative sign then they are said to be conjugates to one another, hence the name for the derivative transforming quantity is ***conjugate normal rate vector***. The superscripted asterisk symbol is taken from the conjugate operator used with complex numbers.

Observation 12.15.2 Embedded Trihedral, Serret-Frenet, Rate Vectors in a Spin System

The trihedral, Serret-Frenet, Rate Vectors are unified in a similar way by a spinning trihedral system by replacing the unit vectors in O12.15.1B "Complimentary and Conjugate Normal Quantities" with a set of spin unit vectors, so it follows that

$$\mathbf{N}^{*\prime} = \tau\, \mathbf{v} + \kappa\, \boldsymbol{\omega} \qquad \text{Equation A}$$

Where N*' is still the conjugate normal rate vector, but abbreviated with Lagrange's prime mark notation for derivatives.

Since the normal rate equations are complimentary conjugate numbers games can be played with them by adding and subtracting and investigating how they behave under different vector operations.

Theorem 12.15.4 Trihedral System: Addition of a Normal and its Conjugate

Step	Hypothesis	$\mathbf{N}' + \mathbf{N}^{*\prime} = (\tau - \kappa)\, \mathbf{T} + (\tau + \kappa)\, \mathbf{B}$
1	From T7.11.1 Vector Addition: Uniqueness for Addition	$\mathbf{N}' + \mathbf{N}^{*\prime} = \mathbf{N}' + \mathbf{N}^{*\prime}$
2	From 1, O12.15.1A Complimentary and Conjugate Normal Quantities, O12.15.1B Complimentary and Conjugate Normal Quantities and A7.9.2 Substitution of Vectors	$\mathbf{N}' + \mathbf{N}^{*\prime} = \tau\, \mathbf{B} - \kappa\, \mathbf{T} + \tau\, \mathbf{T} + \kappa\, \mathbf{B}$
3	From 2, A7.9.3 Commutative by Vector Addition and by like terms	$\mathbf{N}' + \mathbf{N}^{*\prime} = \tau\, \mathbf{T} - \kappa\, \mathbf{T} + \kappa\, \mathbf{B} + \tau\, \mathbf{B}$
∴	From 3 and T7.10.9 Distribution of Vector over Addition of Scalars	$\mathbf{N}' + \mathbf{N}^{*\prime} = (\tau - \kappa)\, \mathbf{T} + (\tau + \kappa)\, \mathbf{B}$

Theorem 12.15.5 Trihedral System: Subtraction of a Normal and its Conjugate

Step	Hypothesis	$\mathbf{N}' - \mathbf{N}^{*\prime} = -(\tau + \kappa)\, \mathbf{T} + (\tau - \kappa)\, \mathbf{B}$
1	From T7.11.1 Vector Addition: Uniqueness for Addition	$\mathbf{N}' - \mathbf{N}^{*\prime} = \mathbf{N}' - \mathbf{N}^{*\prime}$
2	From 1, O12.15.1A Complimentary and Conjugate Normal Quantities, O12.15.1B Complimentary and Conjugate Normal Quantities and A7.9.2 Substitution of Vectors	$\mathbf{N}' - \mathbf{N}^{*\prime} = \tau\, \mathbf{B} - \kappa\, \mathbf{T} - (\, \tau\, \mathbf{T} + \kappa\, \mathbf{B}\,)$
3	From 2, A7.8.6 Parallelogram Law: Multiplying a Negative Number Times a Vector and T7.10.3 Distribution of a Scalar over Addition of Vectors	$\mathbf{N}' - \mathbf{N}^{*\prime} = \tau\, \mathbf{B} - \kappa\, \mathbf{T} - \tau\, \mathbf{T} - \kappa\, \mathbf{B}$
	From 2, A7.9.3 Commutative by Vector Addition and by like terms	$\mathbf{N}' - \mathbf{N}^{*\prime} = -\, \tau\, \mathbf{T} - \kappa\, \mathbf{T} + \tau\, \mathbf{B} - \kappa\, \mathbf{B}$
∴	From 3, A7.8.6 Parallelogram Law: Multiplying a Negative Number Times a Vector and T7.10.9 Distribution of Vector over Addition of Scalars	$\mathbf{N}' - \mathbf{N}^{*\prime} = -(\tau + \kappa)\, \mathbf{T} + (\tau - \kappa)\, \mathbf{B}$

Theorem 12.15.6 Trihedral System: Cross Multiplication Complimentary Normal Rates

Step	Hypothesis	$(\mathbf{N'} + \mathbf{N^{*'}}) \times (\mathbf{N'} - \mathbf{N^{*'}}) = -[\,(\tau + \kappa)^2 + (\tau - \kappa)^2\,]\,\mathbf{N}$
1	From T12.15.5 Trihedral Ordinate System: Addition of a Normal and its Conjugate, T12.15.6 Trihedral Ordinate System: Subtraction of a Normal and its Conjugate and T12.11.15 Cross Product: Contravariant Uniqueness	$(\mathbf{N'} + \mathbf{N^{*'}}) \times (\mathbf{N'} - \mathbf{N^{*'}}) = [(\tau - \kappa)\,\mathbf{T} + (\tau + \kappa)\,\mathbf{B}] \times [-(\tau + \kappa)\,\mathbf{T} + (\tau - \kappa)\,\mathbf{B}]$
2	From 1 and LxK.5.1.2 Distributive Property of Cross Products	$(\mathbf{N'} + \mathbf{N^{*'}}) \times (\mathbf{N'} - \mathbf{N^{*'}}) = [(\tau - \kappa)\,\mathbf{T} \times [-(\tau + \kappa)\,\mathbf{T}] + (\tau + \kappa)\,\mathbf{B} \times [-(\tau + \kappa)\,\mathbf{T}] + [(\tau - \kappa)\,\mathbf{T} \times (\tau - \kappa)\,\mathbf{B}] + (\tau + \kappa)\,\mathbf{B} \times (\tau - \kappa)\,\mathbf{B}$
3	From 2 and T7.10.2 Associative Scalar Multiplication to Vectors	$(\mathbf{N'} + \mathbf{N^{*'}}) \times (\mathbf{N'} - \mathbf{N^{*'}}) = -(\tau + \kappa)\,(\tau - \kappa)\,\mathbf{T} \times \mathbf{T} - (\tau + \kappa)(\tau + \kappa)\,\mathbf{B} \times \mathbf{T} + + (\tau + \kappa)(\tau - \kappa)\,\mathbf{B} \times \mathbf{B} + (\tau - \kappa)(\tau - \kappa)\,\mathbf{T} \times \mathbf{B}$
4	From 3 and LxK.5.1.1 Asymmetric Property of Cross Product	$(\mathbf{N'} + \mathbf{N^{*'}}) \times (\mathbf{N'} - \mathbf{N^{*'}}) = -(\tau + \kappa)\,(\tau - \kappa)\,\mathbf{T} \times \mathbf{T} - (\tau + \kappa)(\tau + \kappa)\,\mathbf{B} \times \mathbf{T} + + (\tau + \kappa)(\tau - \kappa)\,\mathbf{B} \times \mathbf{B} - (\tau - \kappa)(\tau - \kappa)\,\mathbf{B} \times \mathbf{T}$
5	From 4, LxK.5.1.4 Parallel Vectors of Cross Product and Cyclic permutation of Trihedral Ordinate	$(\mathbf{N'} + \mathbf{N^{*'}}) \times (\mathbf{N'} - \mathbf{N^{*'}}) = -(\tau + \kappa)\,(\tau - \kappa)\,\mathbf{0} - (\tau + \kappa)(\tau + \kappa)\,\mathbf{N} + + (\tau + \kappa)(\tau - \kappa)\,\mathbf{0} - (\tau - \kappa)(\tau - \kappa)\,\mathbf{N}$
6	From 5 and A7.9.11 Identity Vector Object Add	$(\mathbf{N'} + \mathbf{N^{*'}}) \times (\mathbf{N'} - \mathbf{N^{*'}}) = -\,(\tau + \kappa)(\tau + \kappa)\,\mathbf{N} - (\tau - \kappa)(\tau - \kappa)\,\mathbf{N}$
7	From 6 and D4.1.17 Exponential Notation	$(\mathbf{N'} + \mathbf{N^{*'}}) \times (\mathbf{N'} - \mathbf{N^{*'}}) = -\,(\tau + \kappa)^2\,\mathbf{N} - (\tau - \kappa)^2\,\mathbf{N}$
∴	From 7, A7.8.6 Parallelogram Law: Multiplying a Negative Number Times a Vector, A4.2.14 Distribution and T7.10.9 Distribution of Vector over Addition of Scalars	$(\mathbf{N'} + \mathbf{N^{*'}}) \times (\mathbf{N'} - \mathbf{N^{*'}}) = -[\,(\tau + \kappa)^2 + (\tau - \kappa)^2\,]\,\mathbf{N}$

Theorem 12.15.7 Trihedral System: Dot Product of Complimentary Normal Rates

Step	Hypothesis	$(\mathbf{N'} + \mathbf{N^{*}}') \bullet (\mathbf{N'} - \mathbf{N^{*}}') = 0$ for $(\mathbf{N'} + \mathbf{N^{*}}') \perp (\mathbf{N'} - \mathbf{N^{*}}')$
1	From T12.15.5 Trihedral Ordinate System: Addition of a Normal and its Conjugate, T12.15.6 Trihedral Ordinate System: Subtraction of a Normal and its Conjugate and T8.3.1 Dot Product: Uniqueness Operation	$(\mathbf{N'} + \mathbf{N^{*}}') \bullet (\mathbf{N'} - \mathbf{N^{*}}') = [(\tau - \kappa)\, \mathbf{T} + (\tau + \kappa)\, \mathbf{B}] \bullet [-(\tau + \kappa)\, \mathbf{T} + (\tau - \kappa)\, \mathbf{B}]$
2	From 1 and T8.3.8 Dot Product: Distribution of Dot Product Across Another Vector	$(\mathbf{N'} + \mathbf{N^{*}}') \bullet (\mathbf{N'} - \mathbf{N^{*}}') = [(\tau - \kappa)\, \mathbf{T} \bullet [-(\tau + \kappa)\, \mathbf{T}] + (\tau + \kappa)\, \mathbf{B} \bullet [-(\tau + \kappa)\, \mathbf{T}] + [(\tau - \kappa)\, \mathbf{T} \bullet (\tau - \kappa)\, \mathbf{B}] + (\tau + \kappa)\, \mathbf{B} \bullet (\tau - \kappa)\, \mathbf{B}$
3	From 2 and T8.3.10A Dot Product: Right Scalar-Vector Association	$(\mathbf{N'} + \mathbf{N^{*}}') \text{ x } (\mathbf{N'} - \mathbf{N^{*}}') = -(\tau + \kappa)\, (\tau - \kappa)\, \mathbf{T} \bullet \mathbf{T} - (\tau + \kappa)(\tau + \kappa)\, \mathbf{B} \bullet \mathbf{T} + + (\tau + \kappa)(\tau - \kappa)\, \mathbf{B} \bullet \mathbf{B} + (\tau - \kappa)(\tau - \kappa)\, \mathbf{T} \bullet \mathbf{B}$
4	From 3 and T8.3.3 Dot Product: Commutative Operation	$(\mathbf{N'} + \mathbf{N^{*}}') \bullet (\mathbf{N'} - \mathbf{N^{*}}') = -(\tau + \kappa)\, (\tau - \kappa)\, \mathbf{T} \bullet \mathbf{T} - (\tau + \kappa)(\tau + \kappa)\, \mathbf{B} \bullet \mathbf{T} + + (\tau + \kappa)(\tau - \kappa)\, \mathbf{B} \bullet \mathbf{B} - (\tau - \kappa)(\tau - \kappa)\, \mathbf{B} \bullet \mathbf{T}$
5	From 4, T12.12.2 On Space Curve-C: Dot Product of Unit Tangent Vector, T12.12.10 On Space Curve-C: Dot Product of Unit Binormal Vector and T12.12.16 On Space Curve-C: Perpendicular Tangent to Binormal Vector	$(\mathbf{N'} + \mathbf{N^{*}}') \bullet (\mathbf{N'} - \mathbf{N^{*}}') = -(\tau + \kappa)\, (\tau - \kappa)\, 1 - (\tau + \kappa)(\tau + \kappa)\, 0 + + (\tau + \kappa)(\tau - \kappa)\, 1 - (\tau - \kappa)(\tau - \kappa)\, 0$
6	From 5, A4.2.12 Identity Multp, T4.4.1 Equalities: Any Quantity Multiplied by Zero is Zero and A4.2.7 Identity Add	$(\mathbf{N'} + \mathbf{N^{*}}') \bullet (\mathbf{N'} - \mathbf{N^{*}}') = -(\tau + \kappa)\, (\tau - \kappa) + (\tau + \kappa)(\tau - \kappa)$
7	From 6 and A4.2.8 Inverse Add	$(\mathbf{N'} + \mathbf{N^{*}}') \bullet (\mathbf{N'} - \mathbf{N^{*}}') = 0$
∴	From 7 and T8.3.11 Dot Product: Perpendicular Vectors	$(\mathbf{N'} + \mathbf{N^{*}}') \bullet (\mathbf{N'} - \mathbf{N^{*}}') = 0$ for $(\mathbf{N'} + \mathbf{N^{*}}') \perp (\mathbf{N'} - \mathbf{N^{*}}')$

Section 12.16 Relative Particle Dynamics

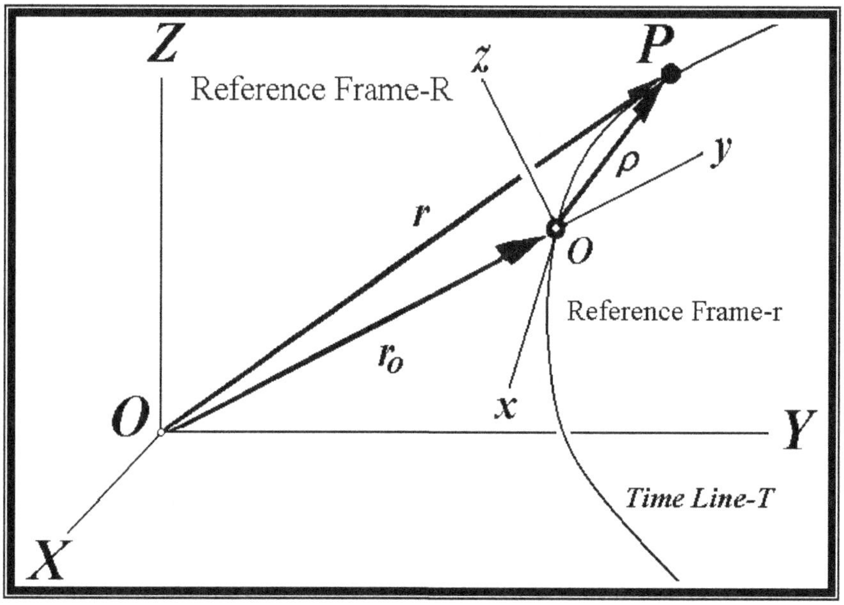

Figure 12.16.1 Kinematic System of a Relative Particle-P in Motion

Definition 12.16.1 **Kinematic Equation of Position**

The ***kinematic equation of position*** is an equation that is represent by a particle-P on a Time Line-T, which intern is embedded in a frame-r. Within that frame ***relative positional vector*** [**ρ**] points to particle-P, all of which is embedded in the frame-R and pointed to by vector [**r**]. There is another relative position vector, from frame-R to frame-r, [$\mathbf{r}_0$]. The reference frame-r is free to move by translation and/or rotation within frame-R, while particle-P being common to both reference frames is invariant.

Definition 12.16.2 **Kinematic Angular Speed**

Change in circular angle with respect to time is the ***kinematic angular speed*** of an object in rotation about a point.

$\omega \rightarrow [A][T^{-1}]$	dimension of angular speed	Equation A
$\omega \equiv 2\pi f$	where f is the periodic frequency of spin	Equation B
$f \equiv 1 / T$	where T is the period of rotation	Equation C

Given 12.16.1 Kinematics: Equation of Position

1g	Given	$\mathbf{r} \equiv X(t)\mathbf{b}^x + Y(t)\mathbf{b}^y + Z(t)\mathbf{b}^z$
2g		$\mathbf{P} \equiv x(t)\mathbf{b}^x + y(t)\mathbf{b}^y + z(t)\mathbf{b}^z$
3g		$\mathbf{r}_0 \equiv (X - x)\mathbf{b}^x + (Y - y)\mathbf{b}^y + (Z - z)\mathbf{b}^z$ Relative Origin
4g		$\mathbf{r} \equiv \mathbf{r}_0 + \mathbf{P}$ Curvature Vector
5g		$\mathbf{P} \equiv \rho\, \boldsymbol{\rho}$ Relative Positional Vector
6g		$\boldsymbol{\Omega} \equiv \omega\, \boldsymbol{\omega}$ Kinematic Angular Speed Vector
7g		$\mathbf{V} \equiv v\, \mathbf{v}$ Tangential or Trajectory Velocity
8g		$\boldsymbol{\Omega}_b \equiv \omega\, \mathbf{B}$ Serret-Frenet Angular Vector
9g		$d\boldsymbol{\Omega} / dt \equiv d\omega / dt\, \boldsymbol{\omega}$ Kinematic Angular Acceleration Vector

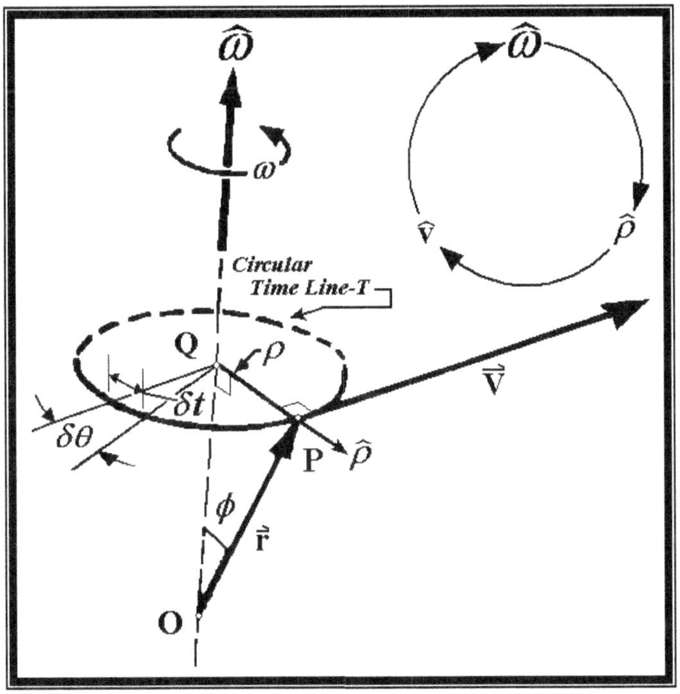

Figure 12.16.2 Spin System

Theorem 12.16.1 Kinematics Spin System: Angular Speed

1g	Given	Figure 12.16.2 "Spin System"
Step	Hypothesis	$d\theta / dt = \omega$
1	From 1g and D12.16.11A Kinematic Angular Speed	$\delta\theta \approx \omega\delta t$
2	From 1, T4.4.4A Equalities: Right Cancellation by Multiplication	$\delta\theta / \delta t \approx \omega$
3	From 2 and LxK.2.1.1 Uniqueness of Limits	$\lim_{\delta t \to 0} \delta\theta/\delta t \approx \lim_{\delta t \to 0} \omega$
∴	From 3, DxK.3.1.3 Definition of a Derivative Alternate Form and LxK.2.1.2 Limit of a Constant	$d\theta / dt = \omega$

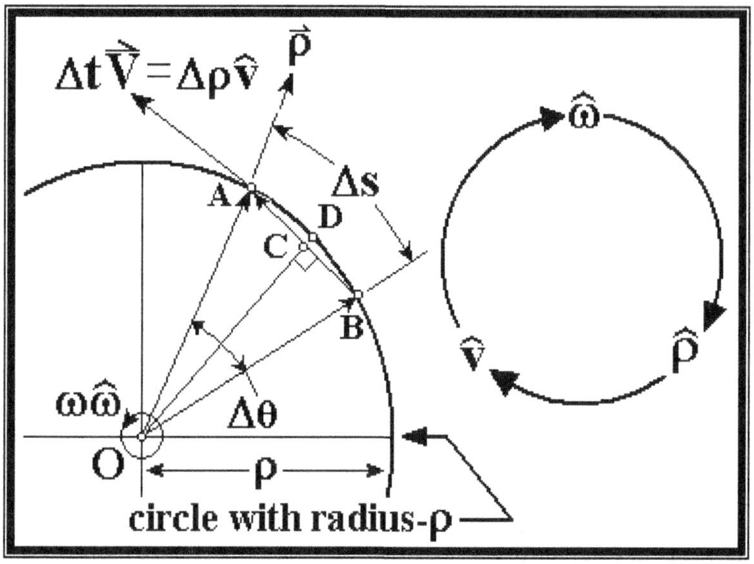

Figure 12.16.3 Circular Angular Velocity

Theorem 12.16.2 Kinematics: Rate of Change for Rotational Radius

1g	Given	Figure 12.16.1 "Circular Angular Velocity"
Step	Hypothesis	$d\,\rho / dt = \omega\rho$
1	From 1g	ΔAOB inscribed in a circle with radius-ρ
2	From 3 and by inscribed circle	$AO = BO = \rho$
3	From 4 and D5.1.17 Triangle; Isosceles	ΔAOB and isosceles triangle
4	From 1g	$\Delta\rho = AB$
5	From 1, 4T5.9.8 Isosceles Triangle: Trigonometry; Base Length	$\Delta\rho = 2\rho \sin \frac{1}{2}\Delta\theta$
6	From 5 and LxE.3.22 sine Trigonometric series	$\sin \frac{1}{2}\Delta\theta = \frac{1}{2}\Delta\theta - (\frac{1}{2}\Delta\theta)3 / 3! + (\frac{1}{2}\Delta\theta)4 / 4! - HOT$
7	From 6 and small angles	$\frac{1}{2}\Delta\theta << 0.0495$ radians or 3° and $HOT \approx 0$
8	From 6 and 7	$\sin \frac{1}{2}\Delta\theta \approx \frac{1}{2}\Delta\theta$
9	From 5, 8 and A4.2.3 Substitution	$\Delta\rho \approx 2\rho\ \frac{1}{2}\Delta\theta$ angular argument very small
10	From 9 and T4.4.18A Equalities: Product by Division, Factorization by 2	$\Delta\rho \approx \rho\Delta\theta$
11	From 10 and T4.4.4A Equalities: Right Cancellation by Multiplication	$\Delta\rho / \Delta\theta \approx \rho$
12	From 11 and LxK.2.1.1 Uniqueness of Limits	$\lim_{\Delta\theta \to 0} \Delta\rho / \Delta\theta \approx \lim_{\Delta\theta \to 0} \rho$
13	From 12, DxK.3.1.3 Definition of a Derivative Alternate Form and LxK.2.1.2 Limit of a Constant	$d\,\rho / d\,\theta = \rho$
14	From 13 and TK.3.4B Single Variable Chain Rule	$(d\,\rho / dt)(dt / d\,\theta) = \rho$

15	From 14, T12.16.1 Kinematics Spin System: Angular Speed and T4.4.14 Equalities: Formal Cross Product	$(d\rho / dt)(1 / \omega) = \rho$
∴	From 15 and T4.4.7B Equalities: Reversal of Left Cancellation by Multiplication	$d\rho / dt = \omega\rho$

Theorem 12.16.3 Kinematics: Rotation of Positional Vector about a Point

1g	Given	Figure 12.16.3 "Circular Angular Velocity"
Step	Hypothesis	$d\mathbf{P} / dt = \mathbf{\Omega} \times \mathbf{P}$
1	From 1g	$\mathbf{AO} = \mathbf{AB} + \mathbf{BO}$
2	From 1 and T7.11.4B Vector Addition: Right Cancellation by Addition	$\mathbf{AB} = \mathbf{AO} - \mathbf{BO}$
3	From 1g, 2, A7.9.1 Equivalence of Vector	$\mathbf{AB} = \Delta\rho\ \mathbf{v}$
4	From 1g and 3	$\Delta t\ \mathbf{V} = \Delta\rho\ \mathbf{v}$
5	From 4 and T4.4.6B Equalities: Left Cancellation by Multiplication	$\mathbf{V} = \Delta\rho / \Delta t\ \mathbf{v}$
6	From 5 and LxK.2.1.1 Uniqueness of Limits	$\lim_{\Delta t \to 0} \mathbf{V} = (\lim_{\Delta t \to 0} \Delta\rho / \Delta t)\ \mathbf{v}$
7	From 6, DxK.3.1.3 Definition of a Derivative Alternate Form and LxK.2.1.2 Limit of a Constant	$\mathbf{V} = d\rho / dt\ \mathbf{v}$
8	From 7, T12.16.2 Kinematics Spin System: Instantaneous Change In Rotational Radius and A4.2.3 Substitution	$d\rho / dt\ \mathbf{v} = \omega\rho\ \mathbf{v}$
9	From 1g, 8, Cyclic permutation of Trihedral Ordinate System and A7.9.2 Substitution of Vectors	$d\rho / dt\ \mathbf{v} = \omega\rho\ \boldsymbol{\omega} \times \boldsymbol{\rho}$
10	From 9 and 10, T12.4.1 Differential Relationship for Covariant Bases Vectors [12.16.1]	$d\rho\boldsymbol{\rho} / dt = \omega\ \boldsymbol{\omega} \times \rho\ \boldsymbol{\rho}$
∴	From G12.16.1 Kinematics: Equation of Position, 5g, 6g, 7g, 10 and A7.9.2 Substitution of Vectors	$d\mathbf{P} / dt = \mathbf{\Omega} \times \mathbf{P}$

[12.16.1]Note: The velocity vector exists after the differentiation of the vector; prior to being in the direction of the curvature vector.

Theorem 12.16.4 Kinematics: Frenet Rotation Unit Tangent Vector about Point

Step	Hypothesis	$d\mathbf{T} / dt = (\omega\mathbf{B}) \times \mathbf{T}$ EQ A	$d\mathbf{T} / dt = \mathbf{\Omega}_b \times \mathbf{T}$ EQ B
1	From T12.12.6A On Space Curve-C: First Order Derivative of Tangent Vector	$d\mathbf{T} / ds = \kappa\, \mathbf{N}$	
2	From 1, D12.12.7A Curvature to a Spatial Curve-C and A4.2.3 Substitution	$d\mathbf{T} / ds = (1 / \rho)\, \mathbf{N}$	
3	From 2, T4.4.5A Equalities: Reversal of Right Cancellation by Multiplication	$\rho\, d\mathbf{T} / ds = \mathbf{N}$	
4	From 3 and TK.3.4B Single Variable Chain Rule	$\rho\, (d\theta / ds)\, d\mathbf{T} / d\theta = \mathbf{N}$	
5	From 4, T12.12.5A On Space Curve-C: Approximation of Circles and A4.2.3 Substitution	$\rho\, (1 / \rho)\, d\mathbf{T} / d\theta = \mathbf{N}$	
6	From 5, T4.4.15A Equalities: Product by Division, A4.2.13 Inverse Multp and A4.2.12 Identity Multp	$d\mathbf{T} / d\theta = \mathbf{N}$	
7	From 6, Cyclic permutation of Trihedral Ordinate System and A7.9.2 Substitution of Vectors	$d\mathbf{T} / d\theta = \mathbf{B} \times \mathbf{T}$	
8	From 7 and TK.3.4B Single Variable Chain Rule	$dt / d\theta\, d\mathbf{T} / dt = \mathbf{B} \times \mathbf{T}$	
9	From 8, T4.5.9 Equalities: Ratio and Reciprocal of a Ratio, T12.16.1 Kinematics Spin System: Angular Speed and A4.2.3 Substitution	$(1 / \omega)\, d\mathbf{T} / dt = \mathbf{B} \times \mathbf{T}$	
∴	From 9, T4.4.5B Equalities: Reversal of Right Cancellation by Multiplication, LxK.5.1.3 Commutative property of Scalar to Cross Product, G12.16.1 Kinematics: Equation of Position, 8g and A4.2.3 Substitution	$d\mathbf{T} / dt = (\omega\mathbf{B}) \times \mathbf{T}$ EQ A	$d\mathbf{T} / dt = \mathbf{\Omega}_b \times \mathbf{T}$ EQ B

Theorem 12.16.5 Kinematics: Velocity Positional Vector in Kinetic Coordinates

1g	Given	$d\,\mathbf{P}_r / dt \equiv d\,x(t) / dt\,\mathbf{b}^x + d\,y(t) / dt\,\mathbf{b}^y + d\,z(t) / dt\,\mathbf{b}^z$
Step	Hypothesis	$d\,\mathbf{P} / dt = d\,\mathbf{P}_r / dt + \mathbf{\Omega} \times \mathbf{P}$
1	From G12.16.1 Kinematics: Equation of Position, 2g and LxK.3.1.7B Uniqueness Theorem of Differentiation	$d\,\mathbf{P} / dt = d\,(x(t)\mathbf{b}^x + y(t)\mathbf{b}^y + z(t)\mathbf{b}^z) / dt$
2	From 1, LxK.5.2.1 Differential Linear Vector Distribution and TK.3.4B Single Variable Chain Rule	$d\,\mathbf{P} / dt =$ $d\,(x(t)\mathbf{b}^x) / dt + x(t)\,d(\mathbf{b}^x) /dt +$ $d\,(y(t)\mathbf{b}^y) / dt + y(t)\,d(\mathbf{b}^y) /dt +$ $d\,(z(t)\mathbf{b}^z) / dt + z(t)\,d(\mathbf{b}^x) /dt$
7	From 1g, 2, A7.9.2 Substitution of Vectors, A7.9.3 Commutative by Vector Addition, T12.16.3 Kinematics: Rotation of Positional Vector about a Point and A7.9.2 Substitution of Vectors	$d\,\mathbf{P} / dt = d\,\mathbf{P}_r / dt +$ $x(t)\,\mathbf{\Omega} \times (\mathbf{b}^x) + y(t)\,\mathbf{\Omega} \times (\mathbf{b}^y) + z(t)\,\mathbf{\Omega} \times (\mathbf{b}^x)$
8	From 7 and LxK.5.1.2 Distributive Property of Cross Products	$d\,\mathbf{P} / dt = d\,\mathbf{P}_r / dt +$ $\mathbf{\Omega} \times [\,x(t)\,\mathbf{b}^x + y(t)\,\mathbf{b}^y + z(t)\,\mathbf{b}^z\,]$
∴	From 8, G12.16.1 Kinematics: Equation of Position, 2g and A7.9.2 Substitution of Vectors	$d\,\mathbf{P} / dt = d\,\mathbf{P}_r / dt + \mathbf{\Omega} \times \mathbf{P}$

Theorem 12.16.6 Kinematics: Velocity Vector

1g	Given	$d\,\mathbf{P}_r / dt \equiv d(x(t) / dt\,\mathbf{b}^x + d(y(t) / dt\,\mathbf{b}^y + d(z(t) / dt\,\mathbf{b}^z$
Step	Hypothesis	$d\,\mathbf{r} / dt = d\,\mathbf{r}_0 / dt + d\,\mathbf{P}_r / dt + \mathbf{\Omega} \times \mathbf{P}$
1	From G12.16.1 Kinematics: Equation of Position, 4g	$\mathbf{r} = \mathbf{r}_0 + \mathbf{P}$
2	From 1, LxK.3.1.7B Uniqueness Theorem of Differentiation and LxK.5.2.1 Differential Linear Vector Distribution	$d\,\mathbf{r} / dt = d\,\mathbf{r}_0 / dt + d\,\mathbf{P} / dt$
∴	From 2, T12.16.5 Kinematics: Velocity Positional Vector Kinetic Coordinates and A7.9.2 Substitution of Vectors	$d\,\mathbf{r} / dt = d\,\mathbf{r}_0 / dt + d\,\mathbf{P}_r / dt + \mathbf{\Omega} \times \mathbf{P}$

Theorem 12.16.7 Kinematics: Acceleration Vector

1g 2g	Given	$d\, \mathbf{P}_r / dt \equiv d\, x(t) / dt\, \mathbf{b}^x + d\, y(t) / dt\, \mathbf{b}^y + d\, z(t) / dt\, \mathbf{b}^z$ $d^2\, \mathbf{P}_r / dt^2 \equiv d^2\, x(t) / dt^2\, \mathbf{b}^x + d^2\, y(t) / dt^2\, \mathbf{b}^y + d^2\, z(t) / dt^2\, \mathbf{b}^z$
Step	Hypothesis	$d^2\, \mathbf{r} / dt^2 = d^2\, \mathbf{r}_0 / dt^2 + d^2\, \mathbf{\rho}_r / dt^2 +$ $d\, (\, \mathbf{\Omega}\,) / dt \times \mathbf{P} + \mathbf{\Omega} \times \mathbf{\Omega} \times \mathbf{P} + 2\, \mathbf{\Omega} \times d\, (\, \mathbf{P}_r\,) / dt$
1	From T12.16.6 Kinematics: Velocity Vector	$d\, \mathbf{r} / dt = d\, \mathbf{r}_0 / dt + d\, \mathbf{P}_r / dt + \mathbf{\Omega} \times \mathbf{P}$
2	From 1, LxK.3.1.7B Uniqueness Theorem of Differentiation and LxK.5.2.1 Differential Linear Vector Distribution	$d^2\, \mathbf{r} / dt^2 = d^2\, \mathbf{r}_0 / dt^2 + d\, (d\, \mathbf{P}_r / dt) / dt + d\, (\, \mathbf{\Omega} \times \mathbf{P}\,) / dt$
3	From 1g, 2 and A7.9.2 Substitution of Vectors	$d^2\, \mathbf{r} / dt^2 = d^2\, \mathbf{r}_0 / dt^2 +$ $d\, (d\, x(t) / dt\, \mathbf{b}^x + d\, y(t) / dt\, \mathbf{b}^y + d\, z(t) / dt\, \mathbf{b}^z) / dt +$ $d\, (\, \mathbf{\Omega} \times \mathbf{P}\,) / dt$
4	From 2g, 3, LxK.5.2.1 Differential Linear Vector Distribution, TK.3.4B Single Variable Chain Rule, A7.9.3 Commutative by Vector Addition and A7.9.2 Substitution of Vectors	$d^2\, \mathbf{r} / dt^2 = d^2\, \mathbf{r}_0 / dt^2 + d^2\, \mathbf{P}_r / dt^2 +$ $d(x(t) / dt\, d(\, \mathbf{b}^x\,) / dt +$ $d(y(t) / dt\, d(\, \mathbf{b}^y\,) / dt +$ $d(z(t) / dt\, d(\, \mathbf{b}^z\,) / dt +$ $d\, (\, \mathbf{\Omega} \times \mathbf{P}\,) / dt$
5	From 4, T12.16.3 Kinematics: Rotation of Positional Vector about a Point and A7.9.2 Substitution of Vectors	$d^2\, \mathbf{r} / dt^2 = d^2\, \mathbf{r}_0 / dt^2 + d^2\, \mathbf{P}_r / dt^2 +$ $d(x(t) / dt\, \mathbf{\Omega} \times (\, \mathbf{b}^x\,) +$ $d(y(t) / dt\, \mathbf{\Omega} \times (\, \mathbf{b}^y\,) +$ $d(z(t) / dt\, \mathbf{\Omega} \times (\, \mathbf{b}^z\,) +$ $d\, (\, \mathbf{\Omega} \times \mathbf{P}\,) / dt$
6	From 5 and LxK.5.1.2 Distributive Property of Cross Products	$d^2\, \mathbf{r} / dt^2 = d^2\, \mathbf{r}_0 / dt^2 + d^2\, \mathbf{P}_r / dt^2 +$ $\mathbf{\Omega} \times [d(x(t) / dt\, \mathbf{b}^x + d(y(t) / dt\, \mathbf{b}^y + d(z(t) / dt\, \mathbf{b}^z\,] +$ $d\, (\, \mathbf{\Omega} \times \mathbf{P}\,) / dt$
7	From 1g, 6 and A7.9.2 Substitution of Vectors	$d^2\, \mathbf{r} / dt^2 = d^2\, \mathbf{r}_0 / dt^2 + d^2\, \mathbf{P}_r / dt^2 + \mathbf{\Omega} \times d\, \mathbf{P}_r / dt +$ $d\, (\, \mathbf{\Omega} \times \mathbf{P}\,) / dt$
8	From 7 and LxK.5.2.3 Differential of Cross Product	$d^2\, \mathbf{r} / dt^2 = d^2\, \mathbf{r}_0 / dt^2 + d^2\, \mathbf{P}_r / dt^2 + \mathbf{\Omega} \times d\, \mathbf{P}_r / dt +$ $d\, (\, \mathbf{\Omega}\,) / dt \times \mathbf{P} + \mathbf{\Omega} \times d\, (\, \mathbf{P}\,) / dt$
9	From 8, T12.16.5 Kinematics: Velocity Positional Vector Kinetic Coordinates and A7.9.2 Substitution of Vectors	$d^2\, \mathbf{r} / dt^2 = d^2\, \mathbf{r}_0 / dt^2 + d^2\, \mathbf{P}_r / dt^2 + \mathbf{\Omega} \times d\, \mathbf{P}_r / dt +$ $d\, (\, \mathbf{\Omega}\,) / dt \times \mathbf{P} + \mathbf{\Omega} \times (d\, \mathbf{P}_r / dt + \mathbf{\Omega} \times \mathbf{P}\,)$
10	From 9 and LxK.5.1.2 Distributive Property of Cross Products	$d^2\, \mathbf{r} / dt^2 = d^2\, \mathbf{r}_0 / dt^2 + d^2\, \mathbf{P}_r / dt^2 + \mathbf{\Omega} \times d\, \mathbf{P}_r / dt +$ $d\, (\, \mathbf{\Omega}\,) / dt \times \mathbf{P} + \mathbf{\Omega} \times d\, (\, \mathbf{P}_r\,) / dt + \mathbf{\Omega} \times \mathbf{\Omega} \times \mathbf{P}$
∴	From 10, A7.9.3 Commutative by Vector Addition, T7.10.5 Identity with Scalar Multiplication to Vectors, T7.10.9 Distribution of Vector over Addition of Scalars and D4.1.19 Primitive Definition for Rational Arithmetic	$d^2\, \mathbf{r} / dt^2 = d^2\, \mathbf{r}_0 / dt^2 + d^2\, \mathbf{P}_r / dt^2 +$ $d\, (\, \mathbf{\Omega}\,) / dt \times \mathbf{P} + \mathbf{\Omega} \times \mathbf{\Omega} \times \mathbf{P} + 2\, \mathbf{\Omega} \times d\, (\, \mathbf{P}_r\,) / dt$

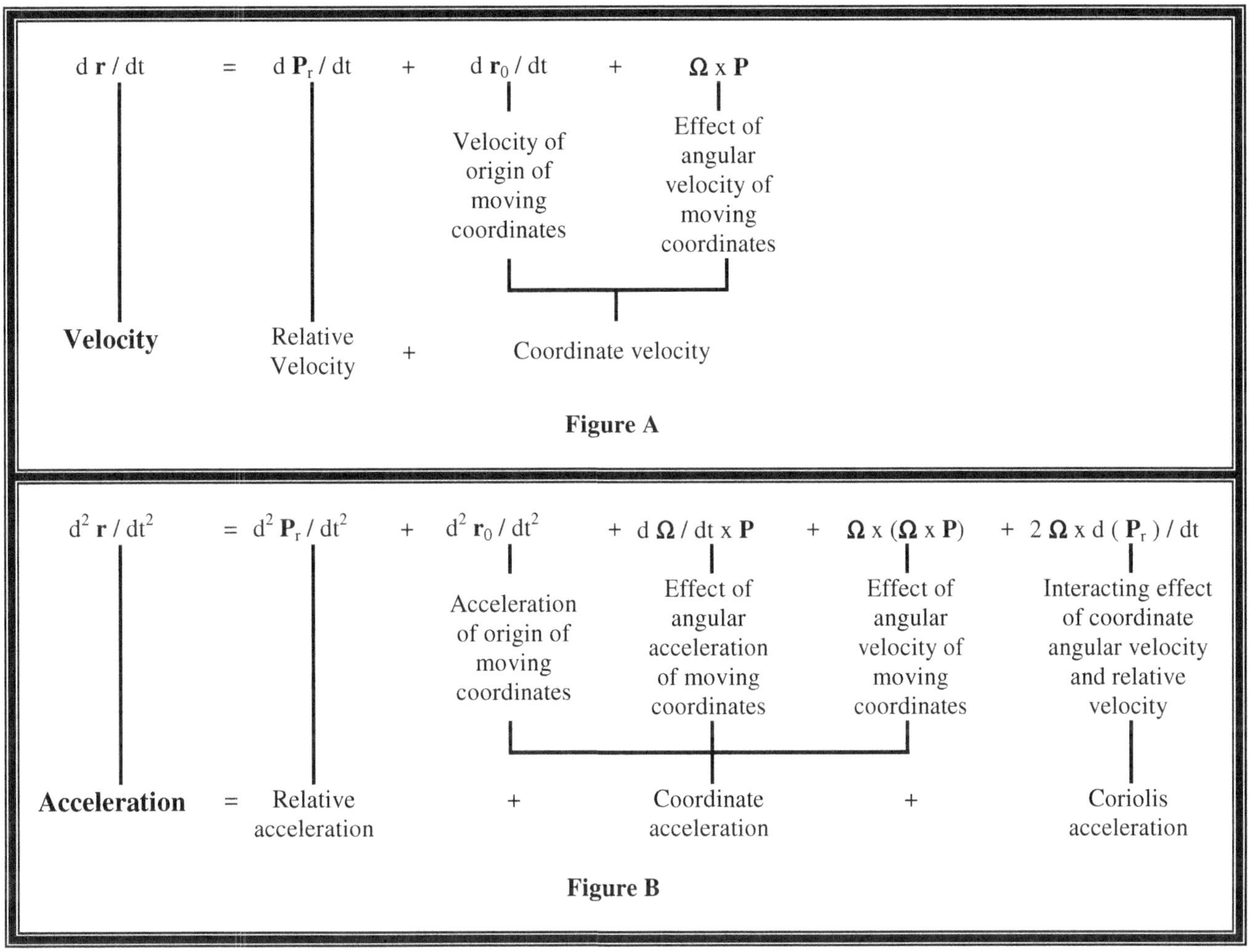

Figure 12.16.4 Kinematics: Components of Velocity and Acceleration Vectors

Observation 12.16.1 Certain Differentials Seem To Have Common Relationships

Rate	=	Translation	+	Rotation	Description	Location
$A_{;j}$	=	$\sum_i (A^{i,j}$	+	$\sum_k A^k \Gamma_{kj}{}^{i}) \mathbf{b}_i$	Differentiation of Covariant Vectors	T12.6.2
$A^{;j}$	=	$\sum_i (A_{i,j}$	−	$\sum_k A_k \Gamma_{ij}{}^{k}) \mathbf{b}^i$	Differentiation of Contravariant Vectors	T12.6.3
$d\,\mathbf{N^*} / ds$	=	$\tau\,\mathbf{v}$	+	$\kappa\,\boldsymbol{\omega}$	Trihedral, Serret-Frenet, Rate Vectors Spin System	O12.15.2
$d\,\mathbf{P} / dt$	=	$d\,\mathbf{P}_r / dt$	+	$\boldsymbol{\Omega} \times \mathbf{P}$	Kinematics: Velocity Positional Vector in Kinetic Coordinates	T12.16.5

These relationships while they were derived totally independent of one another seem to have a common relationship in terms of rate, translation and rotation. If true this can give a physical explanation of the terms and show how they are bound to one another again a possible further verification of Langlands' Hypothesis.

The development of these topics is to demonstrate completeness and relationships, but dynamics of kinematics is an extremely broad topic having many facets, which go beyond what this treatise is prepared to cover, see "Engineering Mechanics Dynamics" [HUA67I] for a more complete development.

Table of Contents

List of Tables

List of Figures

List of Observations

Chapter 13 Tensor Transformations, Tensor Like Quntities and Manifolds

Alice: "Now, if you'll only attend, Kitty, and not talk so much, I'll tell you all my ideas about Looking-Glass House. First, there's the room you can see through the glass --- that's just the same as our drawing–room, only the things go the other way. I can see all of it when I get upon a chair --- all but the bit just behind the fireplace. Oh! I do so wish I could see that bit! I want so much to know whether they've a fire in the winter: you never can tell, you know, unless our fire smokes, and then smoke comes up in that room too --- but that may be only pretence, just to make it look as if they had a fire. Well then, the books are something like our books, only the words go the wrong way: I know that, because I've held up one of our books to the glass, and then they hold up one in the other room."

"How would you like to live in Looking-Glass House, Kitty? I wonder if they'd give you milk in there? Perhaps looking-glass milk isn't good to drink --- but oh, Kitty, now we come to the passage. ..."

From Lewis Carroll's immortal story
"Through the Looking-Glass and What Alice Found There"
page 5, 1832

Section 13.1 Tensor Transformations

Clearly the Alice that climbs through the mirror is the same Alice; she does not change, but remains invariant as a literary frame of reference through out the story. Yet the world she enters is another coordinate system with a negative axis going into the mirror and reversing the horizontal axis. So as she passes across the boundary of the mirror she moves into an asymmetrical world, clearly if Kitty came with her and drinks the looking-glass milk a number of possibilities might occur:

One, on a molecular level the DNA would be a helix winding in the opposite direction making it incompatible with the world she came from, in which case, at best, it would have no nutritional value and Kitty would starve, at worst if there is chemical interaction it might poison Kitty.

Two, on a nuclear level the charge of every piece of nuecular matter might be flipped in the opposite direction and the moment Kitty's tongue made contact with the looking-glass milk, the comparable anti-mass would annihilate that matter in a violent nuclear explosion leaving a very big hole in the mirror room. Who would be left to clean up that mess!!! ?

So Kitty being poisoned is the least of Kitty's problems. In this analogy Alice is akin to a vector in a manifold existing in the same space, but two different coordinate systems yet she remains invariant, however she has serious compatibility issues.

From our study of manifolds in Chapter 11 it can be seen that they have a special property of representing coordinate systems. One of the most prominent properties is their ability to transform between different coordinates, just as Alice moves across the boundary of the mirror. This allows the use of those different systems in ways appropriate to solve problems. A problem in one coordinate system may not be soluble, but doable in another. Also if it can be solved in that particular coordinate system it can be demonstrated it is a viable proof for any other system, hence confirmation for all. Einstein found this very handy, because it allowed him a universal method of proof that applied to all points in the universe in any shape or form, which intern allowed a general substantiation of physical principles, since they are rendered mathematically in relativite way.

Section 13.2 Invariance Under Transformation

Another important property of transformation between manifolds is invariance. It can be shown certain mathematical quantities and operations remain the same for all coordinate systems in a manifold. For example lets say there is a regulation baseball with a density of about 0.5 ounces per cubic inch in one coordinate system it will still have the same density in another system, likewise for a vector of 5 feet length and points due North-by-West in one coordinate a system then under manifold transformation it will still be measured 5 feet and point due North-by-West in another. This is important if you need to know that measurement, so an experiment done in one corner of your universe will still give the same results if repeated in another corner of the universe. This is a very important feature in order to have consistency for mathematics and physics. Certainly an important universal ability for Einsteinian physics, for special and general relativity, to work correctly.

Axiom 13.2.1 Riemannian Manifold $\mathfrak{R}^n$: Congruency of Points Under Transformation

Let ∃| a point P($\bullet x_i \bullet$) in manifold $\mathfrak{R}^n$ corresponds to the same point Q($\bullet y_i \bullet$) in an alternate coordinate system, , then point Q is the inverse of point P and vice a versa:

P1) A bounded point is continuously differentiable within the neighborhood of that point

$|x_i - a_i| < k_i$ bounded for all (i = 1, 2, …, n) Equation A

where K($\bullet$ k_i $\bullet$) are real numbers and constant, that can be found to bound the inequality about the point A($\bullet$ a_i $\bullet$). The absolute value in terms of complex numbers is treated as absolute magnitude.

These base properties that bound the X-positional numbers for an infinite number of points P_k in a continuous space, allows the existence of limits, and establishes differentiability with its inverse the antidifferential or the ability to integrate over that region of space.

P2) $\mathfrak{U}_{\mathcal{X}}^m$ maps onto $\mathfrak{T}_y^n$ maps back by inverse operation onto $\mathfrak{U}_{\mathcal{X}}^m$

$$\mathfrak{U}_{\mathcal{X}}^m \xrightarrow{f} \mathfrak{T}_y^n \xrightarrow{f^{-1}} \mathfrak{U}_{\mathcal{X}}^m$$

where the ***Components of a Manifold*** are the equations, $x_i = f_i(\bullet y_i \bullet)$ and $y_i = f_i^{-1}(\bullet x_i \bullet)$ for all coordinate indices [i], allowing the existence of transformation and its inverse between any two coordinates in a manifold.

Axiom 13.2.2 Riemannian Manifold $\mathfrak{R}^n$: Invariance of a Common Vector in Two Coordinate Systems

Let ∃| a vector [**A**] lying in coordinate system-x and also residing equally in coordinate system-y, then

$${}_y\mathbf{A} \equiv {}_x\mathbf{A} \qquad \text{Equation A}$$

$$\textstyle\sum_j {}_yA^j\, {}_y\mathbf{b}_j = \sum_i {}_xA^i\, {}_x\mathbf{b}_i \qquad \text{covariant} \qquad \text{Equation B}$$

$$\textstyle\sum_j {}_yA_j\, {}_y\mathbf{b}^j = \sum_i {}_xA_i\, {}_x\mathbf{b}^i \qquad \text{contravariant} \qquad \text{Equation C}$$

such vectors are said to be ***invariant*** in magnitude and direction in either coordinate system.

Axiom 13.2.3 Riemannian Manifold $\mathfrak{R}^n$: Invariance of a Tensor in Two Coordinate Systems

Let ∃| in a Riemannian Manifold $\mathfrak{R}^n$ a set of common vectors grouped as a tensor ${}_x\overset{*}{\mathbf{A}}$ in coordinate system-x and also residing equally in coordinate system-y yeilding tensor ${}_y\overset{*}{\mathbf{A}}$ result in a dual covariant and contravariant system,

$$ {}_x\overset{*}{\mathbf{A}}_p \equiv {}_y\overset{*}{\mathbf{A}}_p \qquad \text{Equation A}$$

$$ \sum_{\rho(k)\in Lp} \prod^p {}_xA_i^{\lambda[i,k]}\,{}_x\mathbf{b}_{i\lambda[i,k]} = \sum_{\rho(k)\in Lp} \prod^p {}_yA^{i\lambda[i,k]}\,{}_y\mathbf{b}_{i\lambda[i,k]} \qquad \text{covariant} \qquad \text{Equation B}$$

$$ {}_x\overset{*}{\mathbf{A}}^p \equiv {}_y\overset{*}{\mathbf{A}}^p \qquad \text{Equation C}$$

$$ \sum_{\rho(k)\in Lp} \prod^p {}_xA_{i\lambda[i,k]}\,{}_x\mathbf{b}^{i\lambda[i,k]} = \sum_{\rho(k)\in Lp} \prod^p {}_yA_{i\lambda[i,k]}\,{}_y\mathbf{b}^{i\lambda[i,k]} \qquad \text{contravariant} \qquad \text{Equation D}$$

such tensors are said to be ***invariant*** in magnitude and direction in either coordinate system.
From observation O13.4.2 “Testing for Tensors”:

Axiom 13.2.4 Testing for a Tensor

A tensor like quantity is a tensor if under transformation, between coordinate systems, no additional nonlinear term arises.

Axiom 13.2.5 Testing for a non-Tensor

A tensor like quantity is *not* a tensor, if under transformation, between coordinate systems, an additional nonlinear term arises.

Given 13.2.1 Dual Coordinate System in Manifold $\mathfrak{R}^n$: Their Inverse Relation

1g	Given	$x_i = x_i(\bullet\, y_i \bullet)$ for i = 1, 2, …, n
2g	and its inverse	$y_j = y_j(\bullet\, x_i \bullet)$ for j = 1, 2, …, n

Theorem 13.2.1 Transformation Derivatives: Exact Differential

Step	Hypothesis	$dx_i = \sum_j x_{i,j}\, dy_j$ EQ A	$dy_j = \sum_i y_{j,i}\, dx_i$ EQ B
1	From G13.2.1;1g and G13.2.1;2g	$x_i = x_i(\bullet\, y_i \bullet)$	$y_j = y_j(\bullet\, x_i \bullet)$
∴	From 1 and TK.3.9 Exact Differential, Parametric, Chain Rule	$dx_i = \sum_j x_{i,j}\, dy_j$ EQ A	$dy_j = \sum_i y_{j,i}\, dx_i$ EQ B

Theorem 13.2.2 Transformation Derivatives: As Covariant Orthogonal System

Step	Hypothesis	$\delta_i^{\,j} = \sum_k x_{i,k}\, y_{k,j}$
1	From T13.2.1A Invariance Under Transformation: Exact Differential	$dx_i = \sum_k x_{i,k}\, dy_k$
2	From 1, T13.2.1B Invariance Under Transformation: Exact Differential, T10.2.5 Tool: Tensor Substitution and A10.2.14 Correspondence of Tensors and Tensor Coefficients	$dx_i = \sum_k x_{i,k}\, (\sum_j y_{k,j}\, dx_j)$
3	From 2, T10.3.20 Addition of Tensors: Distribution of a Tensor over Addition of Tensors and A10.2.14 Correspondence of Tensors and Tensor Coefficients	$dx_i = \sum_k (\sum_j x_{i,k}\, y_{k,j}\, dx_j)$
4	From 3, T10.3.20 Addition of Tensors: Distribution of a Tensor over Addition of Tensors and A10.2.14 Correspondence of Tensors and Tensor Coefficients; exchange summation k → j	$dx_i = \sum_j (\sum_k x_{i,k}\, y_{k,j})\, dx_j$
5	From 4, A4.2.10 Commutative Multp, A4.2.13 Inverse Multp, and A4.2.12 Identity Multp; evaluated over the summation-k	$dx_i = \sum_j x_{i,j}\, dx_j$
6	From 5 and T12.4.3A Partial Differential Defines the Kronecker Delta	$dx_i = \sum_j \delta_i^{\,j}\, dx_j$
7	From 6 and D6.1.6B Identity; evaluated over the summation-j	$dx_i = dx_i$
∴	From 4, 6; equating term-by-term over the differential [dx_j] and by identity	$\delta_i^{\,j} = \sum_k x_{i,k}\, y_{k,j}$

Theorem 13.2.3 Transformation Derivatives: As Transpose Covariant Orthogonal System

Step	Hypothesis	$\delta^j_i = \Sigma_k y_{j,k} x_{k,i}$
1	From T13.2.1B Invariance Under Transformation: Exact Differential	$dy_j = \Sigma_k y_{j,k} dx_k$
2	From 1, T13.2.1A Invariance Under Transformation: Exact Differential, T10.2.5 Tool: Tensor Substitution and A10.2.14 Correspondence of Tensors and Tensor Coefficients	$dy_j = \Sigma_k y_{j,k} (\Sigma_i x_{k,i} dy_i)$
3	From 2, T10.3.20 Addition of Tensors: Distribution of a Tensor over Addition of Tensors and A10.2.14 Correspondence of Tensors and Tensor Coefficients	$dy_j = \Sigma_k (\Sigma_i y_{j,k} x_{k,i} dy_i)$
4	From 3, T10.3.20 Addition of Tensors: Distribution of a Tensor over Addition of Tensors, A10.2.14 Correspondence of Tensors and Tensor Coefficients and reorder summation while converting dummy indices k → i	$dy_j = \Sigma_i (\Sigma_k y_{j,k} x_{k,i}) dy_i$
5	From 4, A4.2.10 Commutative Multp, A4.2.13 Inverse Multp, and A4.2.12 Identity Multp; evaluated over the summation-k	$dy_j = \Sigma_i y_{j,i} dy_i$
6	From 5 and T12.4.3B Partial Differential Defines the Kronecker Delta	$dy_j = \Sigma_i \delta^j_i dy_i$
7	From 6 and D6.1.6B Identity; evaluated over the summation-i	$dy_j = dy_j$
∴	From 4, 6; equating term-by-term over the differential [dy_i] and by identity	$\delta^j_i = \Sigma_k y_{j,k} x_{k,i}$

Theorem 13.2.4 Transformation Derivatives: Transpose of Covariant Kronecker Delta Indices

Step	Hypothesis	$\sum_k x_{i,k}\, y_{k,j} = \sum_k y_{j,k}\, x_{k,i}$
1	From A10.2.6A Equality of Tensors and A10.2.14 Correspondence of Tensors and Tensor Coefficients	$\delta_i^{\,j} = \delta_i^{\,j}$
2	From 1 and T12.2.7A Transpose of Mixed Kronecker Delta	$\delta_i^{\,j} = \delta^j{}_i$
3	From T13.2.2 Transformation Derivatives: As Covariant Orthogonal System	$\delta_i^{\,j} = \sum_k x_{i,k}\, y_{k,j}$
4	From T13.2.3 Transformation Derivatives: As Transpose Covariant Orthogonal System	$\delta^j{}_i = \sum_k \sum_k y_{j,k}\, x_{k,i}$
∴	From 2, 3, 4, T10.2.5 Tool: Tensor Substitution and A10.2.14 Correspondence of Tensors and Tensor Coefficients	$\sum_k x_{i,k}\, y_{k,j} = \sum_k y_{j,k}\, x_{k,i}$

Theorem 13.2.5 Invariance Under Transformation: Scalar Function

1g 2g	Given	${}_y\Phi(\bullet\ y_i\ \bullet)$ scalar function ${}_x\Phi(\bullet\ x_i\ \bullet) \equiv {}_y\Phi(\bullet\ y_j(\bullet\ x_i\ \bullet)\ \bullet)$ some scalar function ${}_x\Phi$ allowing for internal manipulation resulting in an invariant magnitude under transformation in terms of x_i-only
Step	Hypothesis	${}_y\Phi(\bullet\ y_j\ \bullet) = {}_x\Phi(\bullet\ x_i\ \bullet)$
1	From 1g and A4.2.2A Equality	${}_y\Phi(\bullet\ y_i\ \bullet) = {}_y\Phi(\bullet\ y_i\ \bullet)$
2	From 1, G13.2.1;2g and A4.2.3 Substitution	${}_y\Phi(\bullet\ y_j\ \bullet) = {}_y\Phi(\bullet\ y_j(\bullet\ x_i\ \bullet)\ \bullet)$
∴	From 2g, 2 and A4.2.3 Substitution	${}_y\Phi(\bullet\ y_j\ \bullet) = {}_x\Phi(\bullet\ x_i\ \bullet)$

Theorem 13.2.6 Invariance Under Transformation: Covariant Gradient Function

Step	Hypothesis	${}_y\Phi_{,j} = {}_x\Phi_{,i}\ x_{i,j}$ EQ A	${}_x\Phi_{,i} = {}_y\Phi_{,j}\ y_{j,i}$ EQ B
1	From T13.2.1 Invariance Under Transformation: Scalar Function and A4.2.2B Equality	${}_y\Phi = {}_x\Phi$	${}_x\Phi = {}_y\Phi$
2	From 1 and TK.3.2 Uniqueness of Differentials	${}_y\Phi_{,j} = {}_x\Phi_{,j}$	${}_x\Phi_{,i} = {}_y\Phi_{,i}$
∴	From 2 and TK.3.8 Multi-variable Chain Rule	${}_y\Phi_{,j} = {}_x\Phi_{,i}\ x_{i,j}$ EQ A	${}_x\Phi_{,i} = {}_y\Phi_{,j}\ y_{j,i}$ EQ B

Observation 13.2.1 Introduction of the Gradient Operator Spanning Riemannian space, $\mathfrak{R}^n$

Since ${}_y\Phi(\bullet\ y_j\ \bullet)_{,j}$ and ${}_x\Phi(\bullet\ x_i\ \bullet)_{,i}$ the indices appears in the denominator and thus acts like a subscript which indicates a covariant character. These tensor coefficients represent the coefficients of a gradient operator, which between transformations of the scalar function linearises them as they span the Riemannian space.

Theorem 13.2.7 Invariance Under Transformation: x-y Parsing Covariant Vectors

Step	Hypothesis	${}_yA^j = \Sigma_k\, y_{j,k}\, {}_xA^k$ EQ A	${}_y\mathbf{b}_j = \Sigma_k\, x_{k,j}\, {}_x\mathbf{b}_k$ EQ B
1	From A13.2.2B Riemannian Manifold $\mathfrak{R}^n$: Invariance of a Common Vector in Two Coordinate Systems	$\Sigma_j\, {}_yA^j\, {}_y\mathbf{b}_j = \Sigma_j\, {}_xA^j\, {}_x\mathbf{b}_j$	
2	From 1, T12.3.8B Contracted Contravariant Bases Vector with Covariant Vector and A7.9.7 Transitivity of Vector Objects	$\Sigma_j\, {}_yA^j\, {}_y\mathbf{b}_j = \Sigma_j\, (\Sigma_i\, {}_xA^i\, \delta^j_i)\, {}_x\mathbf{b}_j)$	
3	From 2, T13.2.3 Transformation Derivatives: As Transpose Covariant Orthogonal System, T10.2.5 Tool: Tensor Substitution and A10.2.14 Correspondence of Tensors and Tensor Coefficients	$\Sigma_j\, {}_yA^j\, {}_y\mathbf{b}_j = \Sigma_j\, [\Sigma_i\, {}_xA^i\, (\Sigma_u\, y_{u,i}\, x_{j,u})]\, {}_x\mathbf{b}_j$	
4	From 3, T10.3.20 Addition of Tensors: Distribution of a Tensor over Addition of Tensors and A10.2.14 Correspondence of Tensors and Tensor Coefficients; grouping by index factors sums over [i] and [j]	$\Sigma_j\, {}_yA^j\, {}_y\mathbf{b}_j = \Sigma_u\, (\Sigma_i\, {}_xA^i\, y_{u,i})(\Sigma_j\, x_{j,u}\, {}_x\mathbf{b}_j)$	
5	From 4, substituting dummy indices, i → k, j → k and u → j	$\Sigma_j\, {}_yA^j\, {}_y\mathbf{b}_j = \Sigma_j\, (\Sigma_k\, {}_xA^k\, y_{j,k})(\Sigma_k\, x_{k,j}\, {}_x\mathbf{b}_k)$	
∴	From 5 and equating coefficients and basis vectors term-by-term across equality	${}_yA^j = \Sigma_k\, y_{j,k}\, {}_xA^k$ EQ A	${}_y\mathbf{b}_j = \Sigma_k\, x_{k,j}\, {}_x\mathbf{b}_k$ EQ B

Theorem 13.2.8 Invariance Under Transformation: x-y Parsing Contravariant Vectors

Step	Hypothesis	${}_yA_j = \Sigma_k\, x_{k,j}\, {}_xA_k$ EQ A	${}_y\mathbf{b}^j = \Sigma_k\, y_{j,k}\, {}_x\mathbf{b}^k$ EQ B
1	From A13.2.2C Riemannian Manifold $\mathfrak{R}^n$: Invariance of a Common Vector in Two Coordinate Systems and A7.9.6B Reflexiveness of Vector Object	$\Sigma_j\, {}_yA_j\, {}_y\mathbf{b}^j = \Sigma_j\, {}_xA_j\, {}_x\mathbf{b}^j$	
2	From 1, T12.3.3B Contracted Covariant Bases Vector with Contravariant Vector and A7.9.7 Transitivity of Vector Objects	$\Sigma_j\, {}_yA_j\, {}_y\mathbf{b}^j = \Sigma_j\, (\Sigma_i\, {}_xA_i\, \delta_j^{\ i}\,)\, {}_x\mathbf{b}^j$	
3	From 2, T13.2.2 Transformation Derivatives: As Covariant Orthogonal System, T10.2.5 Tool: Tensor Substitution and A10.2.14 Correspondence of Tensors and Tensor Coefficients	$\Sigma_j\, {}_yA_j\, {}_y\mathbf{b}^j = \Sigma_j\, [\Sigma_i\, {}_xA_i\, (\Sigma_u\, x_{i,u}\, y_{u,j})]\, {}_x\mathbf{b}^j$	
4	From 3, T10.3.20 Addition of Tensors: Distribution of a Tensor over Addition of Tensors, T10.3.7 Product of Tensors: Commutative by Multiplication of Rank p + q → q + p and A10.2.14 Correspondence of Tensors and Tensor Coefficients; grouping by index factors sums over [i] and [j]	$\Sigma_j\, {}_yA_j\, {}_y\mathbf{b}^j = \Sigma_u\, (\Sigma_i\, x_{i,u}\, {}_xA_i)(\Sigma_j\, y_{u,j}\, {}_x\mathbf{b}^j)$	
5	From 4, substituting dummy indices, i →k, j → k and u → j	$\Sigma_j\, {}_yA_j\, {}_y\mathbf{b}^j = \Sigma_j\, (\Sigma_k\, x_{k,j}\, {}_xA_k)(\Sigma_k\, y_{j,k}\, {}_x\mathbf{b}^k)$	
∴	From 5 and equating coefficients and basis vectors term-by-term across equality	${}_yA_j = \Sigma_k\, x_{k,j}\, {}_xA_k$ EQ A	${}_y\mathbf{b}^j = \Sigma_k\, y_{j,k}\, {}_x\mathbf{b}^k$ EQ B

Theorem 13.2.9 Invariance Under Transformation: y-x Parsing Covariant Vectors

Step	Hypothesis	${}_xA^i = \sum_j x_{i,j}\, {}_yA^j$ EQ A	${}_x\mathbf{b}_i = \sum_j y_{j,i}\, {}_y\mathbf{b}_j$ EQ B
1	From 13.2.7(A, B) Invariance Under Transformation: x-y Parsing Covariant Vectors	${}_yA^j = \sum_k y_{j,k}\, {}_xA^k$	${}_y\mathbf{b}_j = \sum_k x_{k,j}\, {}_x\mathbf{b}_k$
2	From 1, T10.3.6 Product of Tensors: Uniqueness by Multiplication of Rank for p + q and A10.2.14 Correspondence of Tensors and Tensor Coefficients; summing over index [j]	$\sum_j x_{u,j}\, {}_yA^j =$ $\sum_j x_{u,j} \sum_k y_{j,k}\, {}_xA^k$	$\sum_j y_{j,u}\, {}_y\mathbf{b}_j =$ $\sum_j y_{j,u} \sum_k x_{k,j}\, {}_x\mathbf{b}_k$
3	From 2, T10.3.20 Addition of Tensors: Distribution of a Tensor over Addition of Tensors, T10.3.8 Product of Tensors: Association by Multiplication of Rank for p + q + r and A10.2.14 Correspondence of Tensors and Tensor Coefficients; exchanging summation j → k	$\sum_j x_{u,j}\, {}_yA^j =$ $\sum_k (\sum_j x_{u,j}\, y_{j,k})\, {}_xA^k$	$\sum_j y_{j,u}\, {}_y\mathbf{b}_j =$ $\sum_k (\sum_j x_{k,j}\, y_{j,u})\, {}_x\mathbf{b}_k$
4	From 3, T13.2.2 Transformation Derivatives: As Covariant Orthogonal System, T10.2.5 Tool: Tensor Substitution and A10.2.14 Correspondence of Tensors and Tensor Coefficients	$\sum_j x_{u,j}\, {}_yA^j = \sum_k \delta_k^{\ u}\, {}_xA^k$	$\sum_j y_{j,u}\, {}_y\mathbf{b}_j = \sum_k \delta_k^{\ u}\, {}_x\mathbf{b}_k$
5	From 4, T12.3.4B Contracted Contravariant Bases Vector with Covariant Vector and T12.3.20A Holding Indices Constant by Using a Mixed Metric Dyadic	$\sum_j x_{u,j}\, {}_yA^j = {}_xA^u$	$\sum_j y_{j,u}\, {}_y\mathbf{b}_j = {}_x\mathbf{b}_u$
∴	From 5, A10.2.6B Equality of Tensors and A10.2.14 Correspondence of Tensors and Tensor Coefficients; substituting dummy indices and u → i	${}_xA^i = \sum_j x_{i,j}\, {}_yA^j$ EQ A	${}_x\mathbf{b}_i = \sum_j y_{j,i}\, {}_y\mathbf{b}_j$ EQ B

Theorem 13.2.10 Invariance Under Transformation: y-x Parsing Contravariant Vectors

Step	Hypothesis	${}_xA_i = \sum_j y_{j,i}\ {}_yA_j$ EQ A	${}_x\mathbf{b}^i = \sum_j x_{i,j}\ {}_y\mathbf{b}^j$ EQ B
1	From 13.2.8(A, B) Invariance Under Transformation: x-y Parsing Contravariant Vectors	${}_yA_j = \sum_k x_{k,j}\ {}_xA_k$	${}_y\mathbf{b}^j = \sum_k y_{j,k}\ {}_x\mathbf{b}^k$
2	From 1, T10.3.6 Product of Tensors: Uniqueness by Multiplication of Rank for p + q and A10.2.14 Correspondence of Tensors and Tensor Coefficients; summing over index [j]	$\sum_j y_{j,u}\ {}_yA_j =$ $\sum_j y_{j,u} \sum_k x_{k,j}\ {}_xA_k$	$\sum_j x_{u,j}\ {}_y\mathbf{b}^j =$ $\sum_j x_{u,j} \sum_k y_{j,k}\ {}_x\mathbf{b}^k$
3	From 2, T10.3.20 Addition of Tensors: Distribution of a Tensor over Addition of Tensors, T10.3.8 Product of Tensors: Association by Multiplication of Rank for p + q + r, T10.3.7 Product of Tensors: Commutative by Multiplication of Rank p + q → q + p and A10.2.14 Correspondence of Tensors and Tensor Coefficients; exchanging summation j → k	$\sum_j y_{j,u}\ {}_yA_j =$ $\sum_k (\sum_j x_{k,j}\ y_{j,u})\ {}_xA_k$	$\sum_j x_{u,j}\ {}_y\mathbf{b}^j =$ $\sum_k (\sum_j x_{u,j}\ y_{j,k})\ {}_x\mathbf{b}^k$
4	From 3, T13.2.2 Transformation Derivatives: As Covariant Orthogonal System, T10.2.5 Tool: Tensor Substitution and A10.2.14 Correspondence of Tensors and Tensor Coefficients	$\sum_j y_{j,u}\ {}_yA_j = \sum_k \delta_u{}^k\ {}_xA_k$	$\sum_j x_{u,j}\ {}_y\mathbf{b}^j = \sum_k \delta_k{}^u\ {}_x\mathbf{b}^k$
5	From 4, T12.3.3B Contracted Covariant Bases Vector with Contravariant Vector and T12.3.20C Holding Indices Constant by Using a Mixed Metric Dyadic	$\sum_j y_{j,u}\ {}_yA_j = {}_xA_u$	$\sum_j x_{u,j}\ {}_y\mathbf{b}^j = {}_x\mathbf{b}^u$
∴	From 5, A10.2.6B Equality of Tensors and A10.2.14 Correspondence of Tensors and Tensor Coefficients; substituting dummy indices and u → i	${}_xA_i = \sum_j y_{j,i}\ {}_yA_j$ EQ A	${}_x\mathbf{b}^i = \sum_j x_{i,j}\ {}_y\mathbf{b}^j$ EQ B

Theorem 13.2.11 Invariance Under Transformation: x-y Parsing Covariant Tensor of Rank-p

Step	Hypothesis		
		${}_{y}A^{r\lambda[r, k]} =$ $\sum_{\tau(k)\in Ls} \prod_{q=1}^{p} y_{r\lambda[r, k],\nu[q, k]}\ {}_{x}A^{r\nu[q, k]}$ EQ A	${}_{y}\mathbf{b}_{r\lambda[r, k]} =$ $\sum_{\tau(k)\in Ls} \prod_{q=1}^{p} x_{r\sigma[q, k],\lambda[r, k]}\ {}_{x}\mathbf{b}_{r\sigma[q, k]}$ EQ B
1	From A10.2.6A Equality of Tensors	${}_{y}\overset{*}{\mathbf{A}}_{p} \equiv {}_{x}\overset{*}{\mathbf{A}}_{p}$	
2	From A10.2.1P5 Tensor of Rank–p	$\prod_{r=1}^{p} {}_{y}\mathbf{A}_{r} = \prod_{r=1}^{p} {}_{x}\mathbf{A}_{r}$	
3	From 1, T13.2.7S5 Invariance Under Transformation: x-y Parsing Covariant Vectors and A7.9.7 Transitivity of Vector Objects	$\prod_{r=1}^{p} (\sum_{j}\ {}_{y}A^{rj}\ {}_{y}\mathbf{b}_{rj}) =$ $\prod_{r=1}^{p} [\sum_{j} (\sum_{u} y_{rj,u}\ {}_{x}A^{ru})(\sum_{v} x_{rv,j}\ {}_{x}\mathbf{b}_{rv})]$	
4	From 3 and T10.2.1 Tool: Tensor Series Expansion	$\sum_{\rho(k)\in Lr} \prod_{r=1}^{p} ({}_{y}A^{r\lambda[r, k]}\ {}_{y}\mathbf{b}_{r\lambda[r, k]}) =$ $\sum_{\rho(k)\in Lr} \prod_{r=1}^{p} [(\sum_{u} y_{r\lambda[r, k],u}\ {}_{x}A^{ru})(\sum_{v} x_{rv,\lambda[r, k]}\ {}_{x}\mathbf{b}_{rv})]$	
5	From 4 and T10.2.1 Tool: Tensor Series Expansion	$\sum_{\rho(k)\in Lr} \prod_{r=1}^{p} ({}_{y}A^{r\lambda[r, k]}\ {}_{y}\mathbf{b}_{r\lambda[r, k]}) =$ $\sum_{\rho(k)\in Lr} [$ $\prod_{r=1}^{p} (\sum_{\tau(k)\in Ls} \prod_{q=1}^{p} y_{r\lambda[r, k],\nu[q, k]}\ {}_{x}A^{r\nu[q, k]})$ $\prod_{r=1}^{p} (\sum_{\tau(k)\in Ls} \prod_{q=1}^{p} x_{r\sigma[q, k],\lambda[r, k]}\ {}_{x}\mathbf{b}_{r\sigma[q, k]})]$	
∴	From 5 and equating coefficients and basis vectors term-by-term across equality	${}_{y}A^{r\lambda[r, k]} =$ $\sum_{\tau(k)\in Ls} \prod_{q=1}^{p} y_{r\lambda[r, k],\nu[q, k]}\ {}_{x}A^{r\nu[q, k]}$ EQ A	${}_{y}\mathbf{b}_{r\lambda[r, k]} =$ $\sum_{\tau(k)\in Ls} \prod_{q=1}^{p} x_{r\sigma[q, k],\lambda[r, k]}\ {}_{x}\mathbf{b}_{r\sigma[q, k]}$ EQ B

Theorem 13.2.12 Invariance Under Transformation: x-y Parsing Contravariant Tensor of Rank-p

Step	Hypothesis		
		${}_{y}A_{r\lambda[r, k]} =$ $\sum_{\tau(k)\in Ls} \prod_{q=1}^{p} x_{\nu[q, k],\lambda[r, k]}\ {}_{x}A_{\nu[q, k]}$ EQ A	${}_{y}\mathbf{b}^{r\lambda[r, k]} =$ $\sum_{\tau(k)\in Ls} \prod_{q=1}^{p} y_{\lambda[r, k],\sigma[q, k]}\ {}_{x}\mathbf{b}^{\sigma[q, k]}$ EQ B
1	From A10.2.6A Equality of Tensors	${}_{y}\overset{*}{\mathbf{A}}{}^{p} \equiv {}_{x}\overset{*}{\mathbf{A}}{}^{p}$	
2	From A10.2.1P5 Tensor of Rank–p	$\prod_{r=1}^{p} {}_{y}\mathbf{A}^{r} = \prod_{r=1}^{p} {}_{x}\mathbf{A}^{r}$	
3	From 1, T13.2.8S5 Invariance Under Transformation: x-y Parsing Contravariant Vectors and A7.9.7 Transitivity of Vector Objects	$\prod_{r=1}^{p} (\sum_{j}\ {}_{y}A_{rj}\ {}_{y}\mathbf{b}^{rj}) =$ $\prod_{r=1}^{p} [\sum_{j} (\sum_{u} x_{u,j}\ {}_{x}A_{u})(\sum_{v} y_{j,v}\ {}_{x}\mathbf{b}^{v})]$	
4	From 3 and T10.2.1 Tool: Tensor Series Expansion	$\sum_{\rho(k)\in Lr} \prod_{r=1}^{p} ({}_{y}A_{r\lambda[r, k]}\ {}_{y}\mathbf{b}^{r\lambda[r, k]}) =$ $\sum_{\rho(k)\in Lr} \prod_{r=1}^{p} [(\sum_{u} x_{u,\lambda[r, k]}\ {}_{x}A_{u})(\sum_{v} y_{\lambda[r, k],v}\ {}_{x}\mathbf{b}^{v})]$	
5	From 4 and T10.2.1 Tool: Tensor Series Expansion	$\sum_{\rho(k)\in Lr} \prod_{r=1}^{p} ({}_{y}A_{r\lambda[r, k]}\ {}_{y}\mathbf{b}^{r\lambda[r, k]}) =$ $\sum_{\rho(k)\in Lr} [$ $\prod_{r=1}^{p} (\sum_{\tau(k)\in Ls} \prod_{q=1}^{p} x_{\nu[q, k],\lambda[r, k]}\ {}_{x}A_{\nu[q, k]})$ $\prod_{r=1}^{p} (\sum_{\tau(k)\in Ls} \prod_{q=1}^{p} y_{\lambda[r, k],\sigma[q, k]}\ {}_{x}\mathbf{b}^{\sigma[q, k]})]$	
∴	From 5 and equating coefficients and basis vectors term-by-term across equality	${}_{y}A_{r\lambda[r, k]} =$ $\sum_{\tau(k)\in Ls} \prod_{q=1}^{p} x_{\nu[q, k],\lambda[r, k]}\ {}_{x}A_{\nu[q, k]}$ EQ A	${}_{y}\mathbf{b}^{r\lambda[r, k]} =$ $\sum_{\tau(k)\in Ls} \prod_{q=1}^{p} y_{\lambda[r, k],\sigma[q, k]}\ {}_{x}\mathbf{b}^{\sigma[q, k]}$ EQ B

Theorem 13.2.13 Invariance Under Transformation: x-y Simplified Covariant Tensors of Rank-p

1g	Given	${}_yA^{v[k]} \equiv \Pi_{r=1}{}^p\ {}_yA^{r;\lambda[r,\,k]}$ eliminating redundant indices	${}_y\mathbf{b}_{\eta[k]} \equiv \Pi_{r=1}{}^p\ {}_y\mathbf{b}_{r\lambda[r,\,k]}$ eliminating redundant indices
Step	Hypothesis	${}_yA^{v[k]} = \Sigma_{\tau(k)\in Ls}\ \Pi_{q=1}{}^p\ y_{r\lambda[r,\,k],v[q,\,k]}\ {}_xA^{v[k]}$ EQ A	${}_y\mathbf{b}_{\eta[k]} = \Sigma_{\tau(k)\in Ls}\ \Pi_{q=1}{}^p\ x_{r\sigma[q,\,k],\lambda[r,\,k]}\ {}_x\mathbf{b}_{\eta[k]}$ EQ B
1	From T13.2.11(A, B) Invariance Under Transformation: x-y Parsing Covariant Tensor of Rank-p	${}_yA^{r\lambda[r,\,k]} = \Sigma_{\tau(k)\in Ls}\ \Pi_{q=1}{}^p\ y_{r\lambda[r,\,k],v[q,\,k]}\ {}_xA^{rv[q,\,k]}$	${}_y\mathbf{b}_{r\lambda[r,\,k]} = \Sigma_{\tau(k)\in Ls}\ \Pi_{q=1}{}^p\ x_{r\sigma[q,\,k],\lambda[r,\,k]}\ {}_x\mathbf{b}_{r\sigma[q,\,k]}$
2	From 1, T2.21.10 Multiplicative Sets: Binary Distribution of Rank and T10.2.3 Tool: Base Tensor Parsed from Polyadic Coefficient	$\Pi_{r=1}{}^p\ {}_yA^{r\lambda[r,\,k]} = \Sigma_{\tau(k)\in Ls}\ \Pi_{q=1}{}^p\ {}_xA^{rv[q,\,k]}\ \Pi_{q=1}{}^p\ y_{rv[q,\,k],\lambda[r,\,k]}$ $\Pi_{r=1}{}^p\ {}_yA^{r\lambda[r,\,k]} = \Sigma_{\tau(k)\in Ls}\ \Pi_{q=1}{}^p\ y_{r\lambda[r,\,k],v[q,\,k]}\ \Pi_{q=1}{}^p\ {}_xA^{rv[q,\,k]}$	$\Pi_{r=1}{}^p\ {}_y\mathbf{b}_{r\lambda[r,\,k]} = \Sigma_{\tau(k)\in Ls}\ \Pi_{q=1}{}^p\ x_{r\lambda[r,\,k],\sigma[q,\,k]}\ \Pi_{q=1}{}^p\ {}_x\mathbf{b}_{r\sigma[q,\,k]}$ $\Pi_{r=1}{}^p\ {}_y\mathbf{b}_{r\lambda[r,\,k]} = \Sigma_{\tau(k)\in Ls}\ \Pi_{q=1}{}^p\ x_{r\sigma[q,\,k],\lambda[r,\,k]}\ \Pi_{q=1}{}^p\ {}_x\mathbf{b}_{r\sigma[q,\,k]}$
∴	From 1g, 1, T10.2.5 Tool: Tensor Substitution and A10.2.14 Correspondence of Tensors and Tensor Coefficients	${}_yA^{v[k]} = \Sigma_{\tau(k)\in Ls}\ \Pi_{q=1}{}^p\ y_{r\lambda[r,\,k],v[q,\,k]}\ {}_xA^{v[k]}$ EQ A	${}_y\mathbf{b}_{\eta[k]} = \Sigma_{\tau(k)\in Ls}\ \Pi_{q=1}{}^p\ x_{r\sigma[q,\,k],\lambda[r,\,k]}\ {}_x\mathbf{b}_{\eta[k]}$ EQ B

Theorem 13.2.14 Invariance Under Transformation: x-y Simplified Contravariant Tensors of Rank-p

1g	Given	${}_yA_{v[k]} \equiv \Pi_{i=1}{}^p\ {}_yA_{i;\lambda[i,\,k]}$ eliminating redundant indices	${}_y\mathbf{b}^{\eta[k]} \equiv \Pi_{r=1}{}^p\ {}_y\mathbf{b}^{r\eta[r,\,k]}$ eliminating redundant indices
Step	Hypothesis	${}_yA_{v[k]} = \Sigma_{\tau(k)\in Ls}\ \Pi_{q=1}{}^p\ x_{v[q,\,k],\lambda[r,\,k]}\ {}_xA_{v[k]}$ EQ A	${}_y\mathbf{b}^{\eta[k]} = \Sigma_{\tau(k)\in Ls}\ \Pi_{q=1}{}^p\ y_{\lambda[r,\,k],\sigma[q,\,k]}\ {}_x\mathbf{b}^{\eta[k]}$ EQ B
1	From T13.2.12(A, B) Invariance Under Transformation: x-y Parsing Contravariant Tensor of Rank-p	$\Pi_{r=1}{}^p\ {}_yA_{r\lambda[r,\,k]} = \Sigma_{\tau(k)\in Ls}\ \Pi_{q=1}{}^p\ x_{v[q,\,k],\lambda[r,\,k]}\ {}_xA_{v[q,\,k]}$	$\Pi_{r=1}{}^p\ {}_y\mathbf{b}^{r\lambda[r,\,k]} = \Sigma_{\tau(k)\in Ls}\ \Pi_{q=1}{}^p\ y_{\lambda[r,\,k],\sigma[q,\,k]}\ {}_x\mathbf{b}^{\sigma[q,\,k]}$
2	From 1, T2.21.10 Multiplicative Sets: Binary Distribution of Rank and T10.2.3 Tool: Base Tensor Parsed from Polyadic Coefficient	$\Pi_{r=1}{}^p\ {}_yA_{r\lambda[r,\,k]} = \Sigma_{\tau(k)\in Ls}\ \Pi_{q=1}{}^p\ x_{v[q,\,k],\lambda[r,\,k]}\ \Pi_{q=1}{}^p\ {}_xA_{v[q,\,k]}$	$\Pi_{r=1}{}^p\ {}_y\mathbf{b}^{r\lambda[r,\,k]} = \Sigma_{\tau(k)\in Ls}\ \Pi_{q=1}{}^p\ y_{\lambda[r,\,k],\sigma[q,\,k]}\ \Pi_{q=1}{}^p\ {}_x\mathbf{b}^{\sigma[q,\,k]}$
∴	From 1g, 1, T10.2.5 Tool: Tensor Substitution and A10.2.14 Correspondence of Tensors and Tensor Coefficients	${}_yA_{v[k]} = \Sigma_{\tau(k)\in Ls}\ \Pi_{q=1}{}^p\ x_{v[q,\,k],\lambda[r,\,k]}\ {}_xA_{v[k]}$ EQ A	${}_y\mathbf{b}^{\eta[k]} = \Sigma_{\tau(k)\in Ls}\ \Pi_{q=1}{}^p\ y_{\lambda[r,\,k],\sigma[q,\,k]}\ {}_x\mathbf{b}^{\eta[k]}$ EQ B

Observation 13.2.2 In Classical Form

In theorems T13.2.131g "Invariance Under Transformation: x-y Covariant Tensor of Rank-p" and T13.2.141g "Invariance Under Transformation: x-y Contravariant Tensor of Rank-p" the following tensor coefficients are represented as follows:

$${}_{x}A^{\nu[k]} \equiv \Pi_{i=1}{}^{p}\, {}_{x}A^{i;\lambda[i,\, k]} \qquad \text{Equation A}$$

$${}_{x}A_{\nu[k]} \equiv \Pi_{i=1}{}^{p}\, {}_{x}A_{i;\lambda[i,\, k]} \qquad \text{Equation B}$$

where the box dot notation was originally setup to replace the more clumsy sub-subscripted index D1.6.6F "Aggregate Index Notation" as familiarly found in many tensor text books:

$${}_{x}A^{i_1 i_2 \ldots i_j \ldots i_p} \equiv \Pi_{i=1}{}^{p}\, {}_{x}A^{i;i_j} \qquad \text{Equation A}$$

$${}_{x}A_{i_1 i_2 \ldots i_j \ldots i_p} \equiv \Pi_{i=1}{}^{p}\, {}_{x}A_{i;i_j} \qquad \text{Equation B}$$

Table 13.2.1 Invariance Under Transformation: Summary of Parsing Vectors

Eq	Coefficient EQ A	Basis Vector EQ B	Type	Direction	Theorem
1	${}_{y}A^{j} = \Sigma_k\, y_{j,k}\, {}_{x}A^{k}$	${}_{y}\mathbf{b}_j = \Sigma_k\, x_{k,j}\, {}_{x}\mathbf{b}_k$	Covariant	x→y	T13.2.7[13.2.1]
2	${}_{y}A_{j} = \Sigma_k\, x_{k,j}\, {}_{x}A_{k}$	${}_{y}\mathbf{b}^j = \Sigma_k\, y_{j,k}\, {}_{x}\mathbf{b}^k$	Contravariant	x→y	T13.2.8[13.2.1]
3	${}_{x}A^{i} = \Sigma_j\, x_{i,j}\, {}_{y}A^{j}$	${}_{x}\mathbf{b}_i = \Sigma_j\, y_{j,i}\, {}_{y}\mathbf{b}_j$	Covariant	y→x	T13.2.9[13.2.2]
4	${}_{x}A_{i} = \Sigma_j\, y_{j,i}\, {}_{y}A_{j}$	${}_{x}\mathbf{b}^i = \Sigma_j\, x_{i,j}\, {}_{y}\mathbf{b}^j$	Contravariant	y→x	T13.2.10

[13.2.1]Note: Equation 1 and 2, see [SPI59 pg 177] and [SOK67 pg 60].
[13.2.2]Note: Equation 3 see [SPI59 pg 179].

Section 13.3 Transformation of a Metric Tensor

Theorem 13.3.1 Invariance Under Transformation: x-y Metric Tensors

Step	Hypothesis	${}_y g_{ij} = {}_x g_{rs}\, x_{r,i}\, x_{s,j}$ EQ A	${}_y g^{ij} = {}_x g^{rs}\, y_{i,r}\, y_{j,s}$ EQ B
1	From A10.2.6A Equality of Tensors	${}_y g_{ij} = {}_y g_{ij}$	${}_y g^{ij} = {}_y g^{ij}$
2	From 1 and D12.1.3A Covariant Metric Dyadic and D12.1.4A Contravariant Metric Dyadic	${}_y g_{ij} = {}_y \mathbf{b}_i \bullet {}_y \mathbf{b}_j$	${}_y g^{ij} = {}_y \mathbf{b}^i \bullet {}_y \mathbf{b}^j$
3	From 2, T13.2.7B Invariance Under Transformation: x-y Parsing Covariant Vectors, T13.2.8B x-y Invariance Under Transformation: Parsing Contravariant Vectors and A7.9.7 Transitivity of Vector Objects; changing dummy indices k → r and k → s	${}_y g_{ij} = (x_{r,i}\, {}_x \mathbf{b}_r) \bullet (x_{s,j}\, {}_x \mathbf{b}_s)$	${}_y g^{ij} = (y_{i,r}\, {}_x \mathbf{b}^r) \bullet (y_{j,s}\, {}_x \mathbf{b}^s)$
4	From 3, 4, T8.3.10B Dot Product: Right Scalar-Vector Association and T8.3.8 Dot Product: Distribution of Dot Product Across Another Vector	${}_y g_{ij} = x_{r,i}\, x_{s,j}\, ({}_x \mathbf{b}_r \bullet {}_x \mathbf{b}_s)$	${}_y g^{ij} = y_{i,r}\, y_{j,s}\, ({}_x \mathbf{b}^r \bullet {}_x \mathbf{b}^s)$
∴	From 5, D12.1.3A Covariant Metric Dyadic, D12.1.4A Contravariant Metric Dyadic, T10.2.5 Tool: Tensor Substitution, T10.3.7 Product of Tensors: Commutative by Multiplication of Rank p + q → q + p and A10.2.14 Correspondence of Tensors and Tensor Coefficients	${}_y g_{ij} = {}_x g_{rs}\, x_{r,i}\, x_{s,j}$ EQ A	${}_y g^{ij} = {}_x g^{rs}\, y_{i,r}\, y_{j,s}$ EQ B

Theorem 13.3.2 Invariance Under Transformation: y-x Metric Tensors

Step	Hypothesis	${}_{x}g_{ij} = {}_{y}g_{rs}\, y_{r,i}\, y_{s,j}$ EQ A	${}_{x}g^{ij} = {}_{y}g^{rs}\, x_{i,r}\, x_{j,s}$ EQ B
1	From A10.2.6A Equality of Tensors	${}_{x}g_{ij} = {}_{x}g_{ij}$	${}_{x}g^{ij} = {}_{x}g^{ij}$
2	From 1 and D12.1.3A Covariant Metric Dyadic and D12.1.4A Contravariant Metric Dyadic	${}_{x}g_{ij} = {}_{x}\mathbf{b}_i \bullet {}_{x}\mathbf{b}_j$	${}_{x}g^{ij} = {}_{x}\mathbf{b}^i \bullet {}_{x}\mathbf{b}^j$
3	From 2, T13.2.9B Invariance Under Transformation: y-x Parsing Covariant Vectors, T13.2.10B Invariance Under Transformation: y-x Parsing Contravariant Vectors and A7.9.7 Transitivity of Vector Objects; changing dummy indices k → r and k → s	${}_{x}g_{ij} = (y_{r,i}\, {}_{y}\mathbf{b}_r) \bullet (y_{s,j}\, {}_{y}\mathbf{b}_s)$	${}_{x}g^{ij} = (x_{i,r}\, {}_{y}\mathbf{b}^r) \bullet (x_{j,s}\, {}_{y}\mathbf{b}^s)$
4	From 3, 4, T3.8.10B Dot Product: Right Scalar-Vector Association and T8.3.8 Dot Product: Distribution of Dot Product Across Another Vector	${}_{x}g_{ij} = y_{r,i}\, y_{s,j}\, ({}_{y}\mathbf{b}_r \bullet {}_{y}\mathbf{b}_s)$	${}_{x}g^{ij} = x_{i,r}\, x_{j,s}\, ({}_{y}\mathbf{b}^r \bullet {}_{y}\mathbf{b}^s)$
∴	From 5, D12.1.3A Covariant Metric Dyadic, D12.1.4A Contravariant Metric Dyadic, T10.2.5 Tool: Tensor Substitution, T10.3.7 Product of Tensors: Commutative by Multiplication of Rank p + q → q + p and A10.2.14 Correspondence of Tensors and Tensor Coefficients	${}_{x}g_{ij} = {}_{y}g_{rs}\, y_{r,i}\, y_{s,j}$ EQ A	${}_{x}g^{ij} = {}_{y}g^{rs}\, x_{i,r}\, x_{j,s}$ EQ B

Section 13.4 Transformation of the Christoffel Symbol

Theorem 13.4.1 Christoffel Symbol-First Kind Transformation

1g	Given	$i \leftrightarrow u$, $j \leftrightarrow v$ and $k \leftrightarrow w$ corresponding indices across transformation	
Step	Hypothesis	${}_y\Gamma_{ij,k} = x_{u,i}\, x_{v,j}\, x_{w,k}\, {}_x\Gamma_{uv,w} + x_{v,k}\, x_{u,i,j}\, {}_xg_{uv}$ EQ A	${}_x\Gamma_{uv,w} = y_{k,w}\, y_{i,u}\, y_{j,v}\, {}_y\Gamma_{ij,k} + y_{j,w}\, y_{i,u,v}\, {}_yg_{ij}$ EQ B
1	From 1g, T13.2.7B Invariance Under Transformation: x-y Parsing Covariant Vectors and T13.2.9B Invariance Under Transformation: y-x Parsing Covariant Vectors; exchanging $k \rightarrow u$ and $j \rightarrow k$	${}_y\mathbf{b}_i = x_{u,i}\, {}_x\mathbf{b}_u$ and ${}_y\mathbf{b}_k = x_{w,k}\, {}_x\mathbf{b}_w$	${}_x\mathbf{b}_u = y_{i,u}\, {}_y\mathbf{b}_i$ and ${}_x\mathbf{b}_w = y_{k,w}\, {}_y\mathbf{b}_k$
2	From 1g, 12.5.3A Christoffel Symbol of the First Kind, D12.5.1 Christoffel Symbol of the First Kind, T10.2.5 Tool: Tensor Substitution and A10.2.14 Correspondence of Tensors and Tensor Coefficients	${}_y\Gamma_{ij,k} = {}_y\mathbf{b}_k \bullet {}_y\mathbf{b}_{i,j}$	${}_x\Gamma_{uv,w} = {}_x\mathbf{b}_w \bullet {}_x\mathbf{b}_{u,v}$
3	From 1, 2 and A7.9.7 Transitivity of Vector Objects	${}_y\Gamma_{ij,k} = (x_{w,k}\, {}_x\mathbf{b}_w) \bullet (x_{u,i}\, {}_x\mathbf{b}_u)_{,j}$	${}_x\Gamma_{uv,w} = (y_{k,w}\, {}_y\mathbf{b}_k) \bullet (y_{i,u}\, {}_y\mathbf{b}_i)_{,v}$
4	From 3 and Lx9.1.2.4 Differential of Scalar-Vector Product	${}_y\Gamma_{ij,k} = (x_{w,k}\, {}_x\mathbf{b}_w) \bullet (x_{u,i}\, {}_x\mathbf{b}_{u,j} + x_{u,i,j}\, {}_x\mathbf{b}_u)$	${}_x\Gamma_{uv,w} = (y_{k,w}\, {}_y\mathbf{b}_k) \bullet (y_{i,u}\, {}_y\mathbf{b}_{i,v} + y_{i,u,v}\, {}_y\mathbf{b}_i)$
5	From 4, T3.8.10B Dot Product: Right Scalar-Vector Association and T8.3.8 Dot Product: Distribution of Dot Product Across Another Vector	${}_y\Gamma_{ij,k} = x_{w,k}\, x_{u,i}\, ({}_x\mathbf{b}_w \bullet {}_x\mathbf{b}_{u,j}) + x_{w,k}\, x_{u,i,j}\, ({}_x\mathbf{b}_w \bullet {}_x\mathbf{b}_u)$	${}_x\Gamma_{uv,w} = y_{k,w}\, y_{i,u}\, ({}_y\mathbf{b}_k \bullet {}_y\mathbf{b}_{i,v}) + y_{k,w}\, y_{i,u,v}\, ({}_y\mathbf{b}_k \bullet {}_y\mathbf{b}_i)$
6	From 5, T9.5.1 Multi-variable Chain Rule for a Vector and T8.3.10A Dot Product: Right Scalar-Vector Association and T8.3.3 Dot Product: Commutative Operation	${}_y\Gamma_{ij,k} = x_{w,k}\, x_{u,i}\, x_{v,j}\, ({}_x\mathbf{b}_w \bullet {}_x\mathbf{b}_{u,v}) + x_{w,k}\, x_{u,i,j}\, ({}_x\mathbf{b}_u \bullet {}_x\mathbf{b}_w)$	${}_x\Gamma_{uv,w} = y_{k,w}\, y_{i,u}\, y_{j,v}\, ({}_y\mathbf{b}_k \bullet {}_y\mathbf{b}_{i,j}) + y_{k,w}\, y_{i,u,v}\, ({}_y\mathbf{b}_i \bullet {}_y\mathbf{b}_k)$
7	From 1g, 6, D12.5.3A Christoffel Symbol of the First Kind, D12.5.1 Christoffel Symbol of the First Kind, D12.1.3A Covariant Metric Dyadic, T10.3.7 Product of Tensors: Commutative by Multiplication of Rank $p + q \rightarrow q + p$, T10.2.5 Tool: Tensor Substitution and A10.2.14 Correspondence of Tensors and Tensor Coefficients	${}_y\Gamma_{ij,k} = x_{u,i}\, x_{v,j}\, x_{w,k}\, {}_x\Gamma_{uv,w} + x_{w,k}\, x_{u,i,j}\, {}_xg_{uw}$	${}_x\Gamma_{uv,w} = y_{k,w}\, y_{i,u}\, y_{j,v}\, {}_y\Gamma_{ij,k} + y_{k,w}\, y_{i,u,v}\, {}_yg_{ik}$
∴	From 7 and last term changing dummy index $w \rightarrow v$ and $k \rightarrow j$	${}_y\Gamma_{ij,k} = x_{u,i}\, x_{v,j}\, x_{w,k}\, {}_x\Gamma_{uv,w} + x_{v,k}\, x_{u,i,j}\, {}_xg_{uv}$ EQ A	${}_x\Gamma_{uv,w} = y_{k,w}\, y_{i,u}\, y_{j,v}\, {}_y\Gamma_{ij,k} + y_{j,w}\, y_{i,u,v}\, {}_yg_{ij}$ EQ B

Theorem 13.4.2 Christoffel Symbol-Second Kind Transformation

Step	Hypothesis	${}_y\Gamma_{ij}{}^k = x_{u,i}\, x_{v,j}\, y_{k,s}\, {}_x\Gamma_{uv}{}^s + y_{k,s}\, x_{s,i,j}$ EQ A	${}_x\Gamma_{uv}{}^w = y_{i,u}\, y_{j,v}\, x_{w,s}\, {}_y\Gamma_{ij}{}^s + x_{w,s}\, y_{s,u,v}$ EQ B
1	From D12.5.2 Christoffel Symbol of the Second Kind	${}_y\Gamma_{ij}{}^k = {}_y\Gamma_{ij,\alpha}\, {}_yg^{\alpha k}$	${}_x\Gamma_{uv}{}^w = {}_x\Gamma_{uv,\beta}\, {}_xg^{\beta w}$
2	From 1, T13.4.1(A, B) Christoffel Symbol-First Kind Transformation, T13.3.1B Invariance Under Transformation: x-y Metric Tensors, T10.2.5 Tool: Tensor Substitution and A10.2.14 Correspondence of Tensors and Tensor Coefficients	${}_y\Gamma_{ij}{}^k =$ $(x_{u,i}\, x_{v,j}\, x_{w,\alpha}\, {}_x\Gamma_{uv,w} + x_{u,i,j}\, x_{v,\alpha}\, {}_xg_{uv})$ $({}_xg^{rs}\, y_{\alpha,r}\, y_{k,s})$	${}_x\Gamma_{uv}{}^w =$ $(y_{k,\beta}\, y_{i,u}\, y_{j,v}\, {}_y\Gamma_{ij,k} + y_{i,u,v}\, y_{j,\beta}\, {}_yg_{ij})$ $({}_yg^{rs}\, x_{\beta,r}\, x_{w,s})$
3	From 2, T10.3.20 Addition of Tensors: Distribution of a Tensor over Addition of Tensors and A10.2.14 Correspondence of Tensors and Tensor Coefficients	${}_y\Gamma_{ij}{}^k =$ $x_{u,i}\, x_{v,j}\, x_{w,\alpha}\, {}_x\Gamma_{uv,w}({}_xg^{rs}\, y_{\alpha,r}\, y_{k,s}) +$ $x_{u,i,j}\, x_{v,\alpha}\, {}_xg_{uv}\, ({}_xg^{rs}\, y_{\alpha,r}\, y_{k,s})$	${}_x\Gamma_{uv}{}^w =$ $y_{k,\beta}\, y_{i,u}\, y_{j,v}\, {}_y\Gamma_{ij,k}\, ({}_yg^{rs}\, x_{\beta,r}\, x_{w,s}) +$ $y_{i,u,v}\, y_{j,\beta}\, {}_yg_{ij}\, ({}_yg^{rs}\, x_{\beta,r}\, x_{w,s})$
4	From 3, T10.3.7 Product of Tensors: Commutative by Multiplication of Rank p + q → q + p, T10.3.7 Product of Tensors: Commutative by Multiplication of Rank p + q → q + p, T10.3.8 Product of Tensors: Association by Multiplication of Rank for p + q + r and A10.2.14 Correspondence of Tensors and Tensor Coefficients	${}_y\Gamma_{ij}{}^k =$ $x_{u,i}\, x_{v,j}\, y_{k,s}\, (x_{w,\alpha}\, y_{\alpha,r})\, ({}_x\Gamma_{uv,w}\, {}_xg^{rs}) +$ $x_{u,i,j}\, (x_{v,\alpha}\, y_{\alpha,r})\, y_{k,s}\, ({}_xg_{uv}\, {}_xg^{rs})$	${}_x\Gamma_{uv}{}^w =$ $y_{i,u}\, y_{j,v}\, x_{w,s}\, (y_{k,\beta}\, x_{\beta,r})\, {}_y\Gamma_{ij,k}\, {}_yg^{rs} +$ $y_{i,u,v}\, (y_{j,\beta}\, x_{\beta,r})\, x_{w,s}\, ({}_yg_{ij}\, {}_yg^{rs})$
5	From 4, T13.2.2 Transformation Derivatives: As Covariant Orthogonal System, T10.2.5 Tool: Tensor Substitution and A10.2.14 Correspondence of Tensors and Tensor Coefficients	${}_y\Gamma_{ij}{}^k =$ $x_{u,i}\, x_{v,j}\, y_{k,s}\, \delta_r{}^w\, ({}_x\Gamma_{uv,w}\, {}_xg^{rs}) +$ $x_{u,i,j}\, \delta_r{}^v\, y_{k,s}\, ({}_xg_{uv}\, {}_xg^{rs})$	${}_x\Gamma_{uv}{}^w =$ $y_{i,u}\, y_{j,v}\, x_{w,s}\, \delta_r{}^k\, ({}_y\Gamma_{ij,k}\, {}_yg^{rs}) +$ $y_{i,u,v}\, \delta_r{}^j\, x_{w,s}\, ({}_yg_{ij}\, {}_yg^{rs})$
6	From 5, T10.3.7 Product of Tensors: Commutative by Multiplication of Rank p + q → q + p, T10.3.8 Product of Tensors: Association by Multiplication of Rank for p + q + r and A10.2.14 Correspondence of Tensors and Tensor Coefficients	${}_y\Gamma_{ij}{}^k =$ $x_{u,i}\, x_{v,j}\, y_{k,s}\, (\delta^w{}_r\, {}_x\Gamma_{uv,w})\, {}_xg^{rs} +$ $y_{k,s}\, (\delta^v{}_r\, {}_xg_{uv}\, {}_xg^{rs})\, x_{u,i,j}$	${}_x\Gamma_{uv}{}^w =$ $y_{i,u}\, y_{j,v}\, x_{w,s}\, (\delta^k{}_r\, {}_y\Gamma_{ij,k})\, {}_yg^{rs} +$ $x_{w,s}\, (\delta^j{}_r\, {}_yg_{ij}\, {}_yg^{rs})\, y_{i,u,v}$
7	From 6, D6.1.6B Identity Matrix; evaluate Kronecker Delta; T10.3.8 Product of Tensors: Association by Multiplication of Rank for p + q + r and A10.2.14 Correspondence of Tensors and Tensor Coefficients	${}_y\Gamma_{ij}{}^k =$ $x_{u,i}\, x_{v,j}\, y_{k,s}\, ({}_x\Gamma_{uv,r}\, {}_xg^{rs}) +$ $y_{k,s}\, ({}_xg_{ur}\, {}_xg^{rs})\, x_{u,i,j}$	${}_x\Gamma_{uv}{}^w =$ $y_{i,u}\, y_{j,v}\, x_{w,s}\, ({}_y\Gamma_{ij,r}\, {}_yg^{rs}) +$ $x_{w,s}\, ({}_yg_{ir}\, {}_yg^{rs})\, y_{i,u,v}$

8	From 7, T12.2.8A Covariant and Contravariant Metric Inverse, T10.2.5 Tool: Tensor Substitution and A10.2.14 Correspondence of Tensors and Tensor Coefficients	${}_y\Gamma_{ij}{}^k =$ $x_{u,i}\, x_{v,j}\, y_{k,s}\, {}_x\Gamma_{uv,r}\, {}_xg^{rs} + y_{k,s}\, \delta_s{}^u\, x_{u,i,j}$	${}_x\Gamma_{uv}{}^w =$ $y_{i,u}\, y_{j,v}\, x_{w,s}\, {}_y\Gamma_{ij,r}\, {}_yg^{rs} + x_{w,s}\, \delta_s{}^i\, y_{i,u,v}$
9	From 8 and D6.1.6B Identity Matrix; evaluate Kronecker Delta	${}_y\Gamma_{ij}{}^k =$ $x_{u,i}\, x_{v,j}\, y_{k,s}\, {}_x\Gamma_{uv,r}\, {}_xg^{rs} + y_{k,s}\, x_{s,i,j}$	${}_x\Gamma_{uv}{}^w =$ $y_{i,u}\, y_{j,v}\, x_{w,s}\, {}_y\Gamma_{ij,r}\, {}_yg^{rs} + x_{w,s}\, y_{s,u,v}$
∴	From 9, D12.5.2 Christoffel Symbol of the Second Kind, T10.2.5 Tool: Tensor Substitution and A10.2.14 Correspondence of Tensors and Tensor Coefficients	${}_y\Gamma_{ij}{}^k = x_{u,i}\, x_{v,j}\, y_{k,s}\, {}_x\Gamma_{uv}{}^s + y_{k,s}\, x_{s,i,j}$ EQ A	${}_x\Gamma_{uv}{}^w = y_{i,u}\, y_{j,v}\, x_{w,s}\, {}_y\Gamma_{ij}{}^s + x_{w,s}\, y_{s,u,v}$ EQ B

Theorem 13.4.3 Christoffel Symbol-Third Kind Transformation

Step	Hypothesis	${}_y\Gamma^{ki}{}_j = y_{i,w}\, x_{v,j}\, y_{k,w}\, {}_x\Gamma_{\gamma v}{}^w\, {}_xg^{\gamma u} +$ $y_{i,w}\, (y_{\alpha,r}\, x_{w,\alpha,j})\, y_{k,w}\, {}_xg^{ru}$ EQ A	${}_x\Gamma^{wu}{}_v = x_{u,i}\, y_{j,v}\, x_{w,k}\, {}_y\Gamma_{\eta j}{}^k\, {}_yg^{\eta i} +$ $x_{u,i}\, (x_{\beta,s}\, y_{k,\beta,v})\, x_{w,k}\, {}_yg^{si}$ EQ B
1	From D12.5.3 Christoffel Symbol of the Third Kind	${}_y\Gamma^{ki}{}_j = {}_y\Gamma_{\alpha j}{}^k\, {}_yg^{\alpha i}$	${}_x\Gamma^{wu}{}_v = {}_x\Gamma_{\beta v}{}^w\, {}_xg^{\beta u}$
2	From 1, T13.4.2(A, B) Christoffel Symbol-Second Kind Transformation, T13.3.1B Invariance Under Transformation: x-y Metric Tensors, T13.3.2B Invariance Under Transformation: y-x Metric Tensors, T10.2.5 Tool: Tensor Substitution and A10.2.14 Correspondence of Tensors and Tensor Coefficients	${}_y\Gamma^{ki}{}_j =$ $(x_{\gamma,\alpha}\, x_{v,j}\, y_{k,w}\, {}_x\Gamma_{\gamma v}{}^w + x_{w,\alpha,j}\, y_{k,w})$ $({}_xg^{ru}\, y_{\alpha,r}\, y_{i,u})$	${}_x\Gamma^{wu}{}_v =$ $(y_{\eta,\beta}\, y_{j,v}\, x_{w,k}\, {}_y\Gamma_{\eta j}{}^k + y_{k,\beta,v}\, x_{w,k})$ $({}_yg^{si}\, x_{u,i}\, x_{\beta,s})$
3	From 2, T10.3.20 Addition of Tensors: Distribution of a Tensor over Addition of Tensors and A10.2.14 Correspondence of Tensors and Tensor Coefficients	${}_y\Gamma^{ki}{}_j =$ $x_{\gamma,\alpha}\, x_{v,j}\, y_{k,w}\, {}_x\Gamma_{\gamma v}{}^w\, ({}_xg^{ru}\, y_{\alpha,r}\, y_{i,u}) +$ $x_{w,\alpha,j}\, y_{k,w}\, ({}_xg^{ru}\, y_{\alpha,r}\, y_{i,u})$	${}_x\Gamma^{wu}{}_v =$ $y_{\eta,\beta}\, y_{j,v}\, x_{w,k}\, {}_y\Gamma_{\eta j}{}^k\, ({}_yg^{si}\, x_{u,i}\, x_{\beta,s}) +$ $y_{k,\beta,v}\, x_{w,k}\, ({}_yg^{si}\, x_{u,i}\, x_{\beta,s})$
4	From 3, T10.3.7 Product of Tensors: Commutative by Multiplication of Rank p + q → q + p, T10.3.7 Product of Tensors: Commutative by Multiplication of Rank p + q → q + p, T10.3.8 Product of Tensors: Association by Multiplication of Rank for p + q + r and A10.2.14 Correspondence of Tensors and Tensor Coefficients	${}_y\Gamma^{ki}{}_j =$ $x_{v,j}\, y_{i,u}\, y_{k,w}\, ({}_x\Gamma_{\gamma v}{}^w\, {}_xg^{ru})\, (x_{\gamma,\alpha}\, y_{\alpha,r}) +$ $y_{i,u}\, y_{k,w}\, y_{\alpha,r}\, x_{w,\alpha,j}\, {}_xg^{ru}$	${}_x\Gamma^{wu}{}_v =$ $y_{j,v}\, x_{u,i}\, x_{w,k}\, ({}_y\Gamma_{\eta j}{}^k\, {}_yg^{si})\, (y_{\eta,\beta}\, x_{\beta,s}) +$ $x_{\beta,s}\, x_{u,i}\, x_{w,k}\, y_{k,\beta,v}\, {}_yg^{si}$

5	From 4, T13.2.2 Transformation Derivatives: As Covariant Orthogonal System, T13.2.3 Transformation Derivatives: As Transpose Covariant Orthogonal System, T10.2.5 Tool: Tensor Substitution and A10.2.14 Correspondence of Tensors and Tensor Coefficients	${}_{y}\Gamma^{ki}{}_{j} =$ $x_{v,j}\, y_{i,w}\, y_{k,w}\, {}_{x}\Gamma_{\gamma v}{}^{w}\, ({}_{x}g^{ru}\, \delta_{r}{}^{\gamma}) +$ $y_{\alpha,r}\, y_{i,w}\, y_{k,w}\, x_{w,\alpha,j}\, {}_{x}g^{ru}$	${}_{x}\Gamma^{wu}{}_{v} =$ $y_{j,v}\, x_{u,i}\, x_{w,k}\, {}_{y}\Gamma_{\eta j}{}^{k}\, ({}_{y}g^{si}\, \delta^{\eta}{}_{s}) +$ $x_{\beta,s}\, x_{u,i}\, x_{w,k}\, y_{k,\beta,v}\, {}_{y}g^{si}$
6	From 5, T10.3.7 Product of Tensors: Commutative by Multiplication of Rank p + q → q + p, T10.3.8 Product of Tensors: Association by Multiplication of Rank for p + q + r and A10.2.14 Correspondence of Tensors and Tensor Coefficients	${}_{y}\Gamma^{ki}{}_{j} =$ $x_{v,j}\, y_{i,w}\, y_{k,w}\, {}_{x}\Gamma_{\gamma v}{}^{w}\, {}_{x}g^{\gamma u} +$ $y_{\alpha,r}\, y_{i,w}\, y_{k,w}\, x_{w,\alpha,j}\, {}_{x}g^{ru}$	${}_{x}\Gamma^{wu}{}_{v} =$ $y_{j,v}\, x_{u,i}\, x_{w,k}\, {}_{y}\Gamma_{\eta j}{}^{k}\, {}_{y}g^{i\eta} +$ $x_{\beta,s}\, x_{u,i}\, x_{w,k}\, y_{k,\beta,v}\, {}_{y}g^{si}$
7	From 6, D6.1.6B Identity Matrix; evaluate Kronecker Delta; T12.2.14 Contravariant Metric Inner Product is Symmetrical, T10.3.8 Product of Tensors: Association by Multiplication of Rank for p + q + r and A10.2.14 Correspondence of Tensors and Tensor Coefficients	${}_{y}\Gamma^{ki}{}_{j} =$ $y_{i,w}\, x_{v,j}\, y_{k,w}\, {}_{x}\Gamma_{\gamma v}{}^{w}\, {}_{x}g^{\gamma u} +$ $y_{i,w}\, (y_{\alpha,r}\, x_{w,\alpha,j})\, y_{k,w}\, {}_{x}g^{ru}$	${}_{x}\Gamma^{wu}{}_{v} =$ $x_{u,i}\, y_{j,v}\, x_{w,k}\, {}_{y}\Gamma_{\eta j}{}^{k}\, {}_{y}g^{\eta i} +$ $x_{u,i}\, (x_{\beta,s}\, y_{k,\beta,v})\, x_{w,k}\, {}_{y}g^{si}$
∴	From 9, D12.5.3 Christoffel Symbol of the Third Kind, T10.2.5 Tool: Tensor Substitution and A10.2.14 Correspondence of Tensors and Tensor Coefficients	${}_{y}\Gamma^{ki}{}_{j} = y_{i,w}\, x_{v,j}\, y_{k,w}\, {}_{x}\Gamma_{\gamma v}{}^{w}\, {}_{x}g^{\gamma u} + y_{i,w}\, (y_{\alpha,r}\, x_{w,\alpha,j})\, y_{k,w}\, {}_{x}g^{ru}$ EQ A	${}_{x}\Gamma^{wu}{}_{v} = x_{u,i}\, y_{j,v}\, x_{w,k}\, {}_{y}\Gamma_{\eta j}{}^{k}\, {}_{y}g^{\eta i} + x_{u,i}\, (x_{\beta,s}\, y_{k,\beta,v})\, x_{w,k}\, {}_{y}g^{si}$ EQ B

Theorem 13.4.4 Tirtionary Differental Transformation

1g	Given	$y_{\alpha,r} = y_{\beta,r}$	$x_{\beta,i} = x_{\alpha,i}$
Step	Hypothesis	$y_{\alpha,r}\, x_{w,\alpha,j} = -x_{w,\alpha}\, x_{\alpha,\beta}\, y_{\beta,r,j}$ EQ A	$x_{\beta,i}\, y_{k,\beta,v} = -y_{k,\beta}\, y_{\beta,\alpha}\, x_{\alpha,i,v}$ EQ B
1	From, T13.2.2 Transformation Derivatives: As Covariant Orthogonal System and T13.2.3 Transformation Derivatives: As Transpose Covariant Orthogonal System	$\delta_{w}{}^{r} = x_{w,\alpha}\, y_{\alpha,r}$	$\delta^{s}{}_{k} = y_{k,\beta}\, x_{\beta,s}$
2	From 1 and TK.3.2 Uniqueness of Differentials; O12.6.1R2 Rules for Covariant and Contravariant Differential Notation	$\delta_{w}{}^{r,j} = (x_{w,\alpha}\, y_{\alpha,r})_{,j}$	$\delta^{s}{}_{k,v} = (y_{k,\beta}\, x_{\beta,s})_{,v}$
3	From 2 and T12.7.6 Tensor Product: Chain Rule for Binary Covariant Product; for p = 0	$\delta_{w}{}^{r,j} = x_{w,\alpha,j}\, y_{\alpha,r} + x_{w,\alpha}\, y_{\alpha,r,j}$	$\delta^{s}{}_{k,v} = y_{k,\beta,v}\, x_{\beta,s} + y_{k,\beta}\, x_{\beta,s,v}$
4	From 3 and T12.9.1B Differentiated Kronecker Delta and Fundamental Tensors	$0 = x_{w,\alpha,j}\, y_{\alpha,r} + x_{w,\alpha}\, y_{\alpha,r,j}$	$0 = y_{k,\beta,v}\, x_{\beta,s} + y_{k,\beta}\, x_{\beta,s,v}$

5	From 1g, 4 and T13.2.6(A, B) Invariance Under Transformation: Covariant Gradient Function	$y_{\alpha,r,j} = x_{\alpha,\beta}\, y_{\beta,r,j}$	$x_{\beta,s,v} = y_{\beta,\alpha}\, x_{\alpha,s,v}$
6	From 4, 5, T10.2.5 Tool: Tensor Substitution and A10.2.14 Correspondence of Tensors and Tensor Coefficients	$0 = x_{w,\alpha,j}\, y_{\alpha,r} + x_{w,\alpha}\, x_{\alpha,\beta}\, y_{\beta,r,j}$	$0 = y_{k,\beta,v}\, x_{\beta,s} + y_{k,\beta}\, y_{\beta,\alpha}\, x_{\alpha,s,v}$
7	From 6, T10.3.21B Addition of Tensors: Right Cancellation by Addition, T10.3.18 Addition of Tensors: Identity by Addition and A10.2.14 Correspondence of Tensors and Tensor Coefficients	$x_{w,\alpha,j}\, y_{\alpha,r} = -x_{w,\alpha}\, x_{\alpha,\beta}\, y_{\beta,r,j}$	$y_{k,\beta,v}\, x_{\beta,s} = -y_{k,\beta}\, y_{\beta,\alpha}\, x_{\alpha,s,v}$
∴	T10.3.7 Product of Tensors: Commutative by Multiplication of Rank p + q → q + p and A10.2.14 Correspondence of Tensors and Tensor Coefficients	$y_{\alpha,r}\, x_{w,\alpha,j} = -x_{w,\alpha}\, x_{\alpha,\beta}\, y_{\beta,r,j}$ EQ A	$x_{\beta,s}\, y_{k,\beta,v} = -y_{k,\beta}\, y_{\beta,\alpha}\, x_{\alpha,s,v}$ EQ B

Theorem 13.4.5 Christoffel Symbol-Third Kind Skewed Transformation

Step	Hypothesis	${}_y\Gamma^{ki}{}_j = y_{i,w}\, x_{v,j}\, y_{k,w}\, {}_x\Gamma_{\gamma v}{}^{w}\, {}_xg^{\gamma u} - x_{i,\beta}\, y_{k,w}\, y_{\beta,r,j}\, {}_xg^{ru}$ EQ A	${}_x\Gamma^{wu}{}_v = x_{u,i}\, y_{j,v}\, x_{w,k}\, {}_y\Gamma_{\eta j}{}^{k}\, {}_yg^{\eta i} - x_{u,i}\, y_{w,\alpha}\, x_{\alpha,s,v}\, {}_yg^{si}$ EQ B
1	From 13.4.4 Christoffel Symbol-Third Kind Transformation	${}_y\Gamma^{ki}{}_j = y_{i,w}\, x_{v,j}\, y_{k,w}\, {}_x\Gamma_{\gamma v}{}^{w}\, {}_xg^{\gamma u} + y_{i,w}\, (y_{\alpha,r}\, x_{w,\alpha,j})\, y_{k,w}\, {}_xg^{ru}$	${}_x\Gamma^{wu}{}_v = x_{u,i}\, y_{j,v}\, x_{w,k}\, {}_y\Gamma_{\eta j}{}^{k}\, {}_yg^{\eta i} + x_{u,i}\, (x_{\beta,s}\, y_{k,\beta,v})\, x_{w,k}\, {}_yg^{si}$
2	From 1, T13.4.4 Tirtionary Differental Transformation, T10.2.5 Tool: Tensor Substitution and A10.2.14 Correspondence of Tensors and Tensor Coefficients	${}_y\Gamma^{ki}{}_j = y_{i,w}\, x_{v,j}\, y_{k,w}\, {}_x\Gamma_{\gamma v}{}^{w}\, {}_xg^{\gamma u} + y_{i,w}\, (-x_{w,\alpha}\, x_{\alpha,\beta}\, y_{\beta,r,j})\, y_{k,w}\, {}_xg^{ru}$	${}_x\Gamma^{wu}{}_v = x_{u,i}\, y_{j,v}\, x_{w,k}\, {}_y\Gamma_{\eta j}{}^{k}\, {}_yg^{\eta i} + x_{u,i}\, (-y_{k,\beta}\, y_{\beta,\alpha}\, x_{\alpha,s,v})\, x_{w,k}\, {}_yg^{si}$
3	From 2, D4.1.20A Negative Coefficient, T10.4.8 Scalars and Tensors: Scalar Commutative with Tensors, T10.3.7 Product of Tensors: Commutative by Multiplication of Rank p + q → q + p, T10.3.8 Product of Tensors: Association by Multiplication of Rank for p + q + r and A10.2.14 Correspondence of Tensors and Tensor Coefficients	${}_y\Gamma^{ki}{}_j = y_{i,w}\, x_{v,j}\, y_{k,w}\, {}_x\Gamma_{\gamma v}{}^{w}\, {}_xg^{\gamma u} - (y_{i,w}\, x_{w,\alpha})\, x_{\alpha,\beta}\, y_{\beta,r,j}\, y_{k,w}\, {}_xg^{ru}$	${}_x\Gamma^{wu}{}_v = x_{u,i}\, y_{j,v}\, x_{w,k}\, {}_y\Gamma_{\eta j}{}^{k}\, {}_yg^{\eta i} - x_{u,i}\, y_{\beta,\alpha}\, x_{\alpha,s,v}\, (x_{w,k}\, y_{k,\beta})\, {}_yg^{si}$
4	From 3, T13.2.3 Transformation Derivatives: As Transpose Covariant Orthogonal System, T10.2.5 Tool: Tensor Substitution and A10.2.14 Correspondence of Tensors and Tensor Coefficients	${}_y\Gamma^{ki}{}_j = y_{i,w}\, x_{v,j}\, y_{k,w}\, {}_x\Gamma_{\gamma v}{}^{w}\, {}_xg^{\gamma u} - \delta^{i}{}_{\alpha}\, x_{\alpha,\beta}\, y_{\beta,r,j}\, y_{k,w}\, {}_xg^{ru}$	${}_x\Gamma^{wu}{}_v = x_{u,i}\, y_{j,v}\, x_{w,k}\, {}_y\Gamma_{\eta j}{}^{k}\, {}_yg^{\eta i} - x_{u,i}\, y_{\beta,\alpha}\, x_{\alpha,s,v}\, \delta_{w}{}^{\beta}\, {}_yg^{si}$

5	From 4, T10.3.7 Product of Tensors: Commutative by Multiplication of Rank p + q → q + p and A10.2.14 Correspondence of Tensors and Tensor Coefficients	${}_y\Gamma^{ki}{}_j = y_{i,w}\, x_{v,j}\, y_{k,w}\, {}_x\Gamma_{\gamma v}{}^w\, {}_xg^{\gamma u} - \delta^i{}_\alpha\, x_{\alpha,\beta}\, y_{k,w}\, y_{\beta,r,j}\, {}_xg^{ru}$	${}_x\Gamma^{wu}{}_v = x_{u,i}\, y_{j,v}\, x_{w,k}\, {}_y\Gamma_{\eta j}{}^k\, {}_yg^{\eta i} - x_{u,i}\, \delta_w{}^\beta\, y_{\beta,\alpha}\, x_{\alpha,s,v}\, {}_yg^{si}$
∴	From 5 and T12.3.3B Contracted Covariant Bases Vector with Contravariant Vector	${}_y\Gamma^{ki}{}_j = y_{i,w}\, x_{v,j}\, y_{k,w}\, {}_x\Gamma_{\gamma v}{}^w\, {}_xg^{\gamma u} - x_{i,\beta}\, y_{k,w}\, y_{\beta,r,j}\, {}_xg^{ru}$ EQ A	${}_x\Gamma^{wu}{}_v = x_{u,i}\, y_{j,v}\, x_{w,k}\, {}_y\Gamma_{\eta j}{}^k\, {}_yg^{\eta i} - x_{u,i}\, y_{w,\alpha}\, x_{\alpha,s,v}\, {}_yg^{si}$ EQ B

Theorem 13.4.6 Addition of a Tensor and Nonlinear Term

1g	Given	$\overset{*}{\mathbf{A}}$ and $\overset{*}{\mathbf{B}}$ are assumed to be tensors
2g	also	$\overset{*}{\mathbf{B}} \equiv \mathbf{T}\overset{*}{\mathbf{A}} \pm \mathbf{N}$ where $\mathbf{T}$ is the transformation coefficient from one coordinate system to another
3g	and	$\mathbf{N}$ being a nonlinear term
Step	Hypothesis	if $\overset{*}{\mathbf{B}} \neq \mathbf{T}\overset{*}{\mathbf{A}} \pm \mathbf{N}$ then $\overset{*}{\mathbf{A}}$ and $\overset{*}{\mathbf{B}}$ are not tensors
1	From 1g and 2g	$\overset{*}{\mathbf{B}} = \mathbf{T}\overset{*}{\mathbf{A}} \pm \mathbf{N}$
2	From 3g, 1, A10.2.8 Tensor Closure under Addition, A10.2.9 Tensor Closure under Subtraction; but since **N** arises as a nonlinear term then closer is violated hence the left hand side cannot be a tensor to the right hand side	$\overset{*}{\mathbf{B}} \neq \mathbf{T}\overset{*}{\mathbf{A}} \pm \mathbf{N}$
∴	From 1g, 2 and by identity; the assumption $\overset{*}{\mathbf{A}}$ and $\overset{*}{\mathbf{B}}$ are tensors cannot hold, hence cannot be tensors	if $\overset{*}{\mathbf{B}} \neq \mathbf{T}\overset{*}{\mathbf{A}} \pm \mathbf{N}$ then $\overset{*}{\mathbf{A}}$ and $\overset{*}{\mathbf{B}}$ are not tensors

Observation 13.4.1 A Christoffel Symbol is not a Tensor

From theorem T13.4.6 "Addition of a Tensor and Nonlinear Term" and transformation equations in T13.4.1 "Christoffel Symbol-First Kind Transformation", T13.4.2 "Christoffel Symbol-Second Kind Transformation" and T14.4.5 "Christoffel Symbol-Third Kind Skewed Transformation" end with nonlinear term left over under transformation, hence the Christoffel symbol cannot transform as a tensor quantity, because the right hand doesn't match the left hand side as the same quantity.

Observation 13.4.2 Testing for Tensors

All of this implies a new condition to determine that a tensor like quantity is a tensor; that no additional nonlinear term can arise under transformation from one coordinate system to another.

Theorem 13.4.7 Riemannian Manifold $\mathfrak{R}^n$: Solving for Tirtionary Transformation Derivatives

Step	Hypothesis	$x_{w,i,j} = x_{w,\alpha}\ {}_y\Gamma_{ij}{}^{\alpha} - x_{u,i}\ x_{v,j}\ {}_x\Gamma_{uv}{}^{w}$ EQ A	$y_{k,u,v} = y_{k,\beta}\ {}_x\Gamma_{uv}{}^{\beta} - y_{i,u}\ y_{j,v}\ {}_y\Gamma_{ij}{}^{k}$ EQ B
1	From T13.4.2(A, B) Christoffel Symbol-Second Kind Transformation	${}_y\Gamma_{ij}{}^{k} = x_{u,i}\ x_{v,j}\ y_{k,s}\ {}_x\Gamma_{uv}{}^{s} + y_{k,s}\ x_{s,i,j}$	${}_x\Gamma_{uv}{}^{w} = y_{i,u}\ y_{j,v}\ x_{w,s}\ {}_y\Gamma_{ij}{}^{s} + x_{w,s}\ y_{s,u,v}$
2	From 1, T10.3.6 Product of Tensors: Uniqueness by Multiplication of Rank for p + q and A10.2.14 Correspondence of Tensors and Tensor Coefficients; k → α, w → β and summing over index [α] and [β] respectively	$x_{w,\alpha}\ {}_y\Gamma_{ij}{}^{\alpha} =$ $x_{w,\alpha}\ (x_{u,i}\ x_{v,j}\ y_{\alpha,s}\ {}_x\Gamma_{uv}{}^{s} + y_{\alpha,s}\ x_{s,i,j})$	$y_{k,\beta}\ {}_x\Gamma_{uv}{}^{\beta} =$ $y_{k,\beta}\ (y_{i,u}\ y_{j,v}\ x_{\beta,s}\ {}_y\Gamma_{ij}{}^{s} + x_{\beta,s}\ y_{s,u,v})$
3	From 2, T10.3.20 Addition of Tensors: Distribution of a Tensor over Addition of Tensors, T10.3.8 Product of Tensors: Association by Multiplication of Rank for p + q + r and A10.2.14 Correspondence of Tensors and Tensor Coefficients	$x_{w,\alpha}\ {}_y\Gamma_{ij}{}^{\alpha} =$ $x_{u,i}\ x_{v,j}\ (x_{w,\alpha}\ y_{\alpha,s})\ {}_x\Gamma_{uv}{}^{s} +$ $(x_{w,\alpha}\ y_{\alpha,s})\ x_{s,i,j}$	$y_{k,\beta}\ {}_x\Gamma_{uv}{}^{\beta} =$ $y_{i,u}\ y_{j,v}\ (y_{k,\beta}\ x_{\beta,s})\ {}_y\Gamma_{ij}{}^{s} +$ $(y_{k,\beta}\ x_{\beta,s})\ y_{s,u,v}$
4	From 3, T13.2.2 Transformation Derivatives: As Covariant Orthogonal System, T13.2.3 Transformation Derivatives: As Transpose Covariant Orthogonal System, T10.2.5 Tool: Tensor Substitution and A10.2.14 Correspondence of Tensors and Tensor Coefficients	$x_{w,\alpha}\ {}_y\Gamma_{ij}{}^{\alpha} =$ $x_{u,i}\ x_{v,j}\ \delta_w{}^{s}\ {}_x\Gamma_{uv}{}^{s} + \delta_w{}^{s}\ x_{s,i,j}$	$y_{k,\beta}\ {}_x\Gamma_{uv}{}^{\beta} =$ $y_{i,u}\ y_{j,v}\ \delta^k{}_s\ {}_y\Gamma_{ij}{}^{s} + \delta^k{}_s\ y_{s,u,v}$
5	From 4 and T12.3.3B Contracted Covariant Bases Vector with Contravariant Vector	$x_{w,\alpha}\ {}_y\Gamma_{ij}{}^{\alpha} = x_{u,i}\ x_{v,j}\ {}_x\Gamma_{uv}{}^{w} + x_{w,i,j}$	$y_{k,\beta}\ {}_x\Gamma_{uv}{}^{\beta} = y_{i,u}\ y_{j,v}\ {}_y\Gamma_{ij}{}^{k} + y_{k,u,v}$
∴	From 5, T10.3.16 Addition of Tensors: Commutative by Addition, T10.3.20B Addition of Tensors: Right Cancellation by Addition and A10.2.14 Correspondence of Tensors and Tensor Coefficients	$x_{w,i,j} = x_{w,\alpha}\ {}_y\Gamma_{ij}{}^{\alpha} - x_{u,i}\ x_{v,j}\ {}_x\Gamma_{uv}{}^{w}$ EQ A	$y_{k,u,v} = y_{k,\beta}\ {}_x\Gamma_{uv}{}^{\beta} - y_{i,u}\ y_{j,v}\ {}_y\Gamma_{ij}{}^{k}$ EQ B

Section 13.5 Transformation of the e-Tensor

Early on in this chapter it was mentioned that a number of quantities are invarient, one was the idea of certain types of operators must also be invarient in order to have consistency through out the scope of a transformation. One such operation is the Orthogonal Sine Operator as seen in chapter 8 and used in chapter 12.

Theorem 13.5.1 e-Tensor Transformation: Covariant Equivalency

Step	Hypothesis	$\sum_{\rho(k)\in Lp}\ {}_x\mu_s\ {}_x\mathbf{e}_{\bullet\lambda[i,k]}\bullet \prod^p {}_xA^{i\lambda[i,k]} = \sum_{\rho(k)\in Lp}\ {}_y\mu_s\ {}_y\mathbf{e}_{\bullet\lambda[i,k]}\bullet \prod^p {}_yA^{i\lambda[i,k]}$
1	From A13.2.3A Riemannian Manifold $\mathfrak{R}^n$: Invariance of a Tensor in Two Coordinate Systems	${}_x\overset{*}{\mathbf{A}}_p = {}_y\overset{*}{\mathbf{A}}_p$
2	From 1, T10.9.1 Equality of the Tensor-Determinate Linear Operator, T8.5.6 Orthogonal Sine Operator: Unitary Sine Operator, T8.5.8 Orthogonal Sine Operator: Commutative Asymmetrically Skewed	$<\vert{}_x\overset{*}{\mathbf{A}}_p\vert> = <\vert{}_y\overset{*}{\mathbf{A}}_p\vert>$
3	From 2 and T10.9.2 Tensor-Determinate Linear Operator (TDL)	$\sum_{\rho(k)\in Lp} <\vert \prod^p {}_x\mathbf{b}_{\lambda[i,k]} \vert> \prod^p {}_xA^{i\lambda[i,k]} =$ $\sum_{\rho(k)\in Lp} <\vert \prod^p {}_y\mathbf{b}_{\lambda[i,k]} \vert> \prod^p {}_yA^{i\lambda[i,k]}$
4	From 3, D10.8.5D Tensor Coefficient for Distortion (TCD) and A4.2.3 Substitution	$\sum_{\rho(k)\in Lp}\ {}_x\mu_f\ {}_x\mathbf{e}_{\rho(k)} \prod^p {}_xA^{i\lambda[i,k]} =$ $\sum_{\rho(k)\in Lp}\ {}_y\mu_f\ {}_y\mathbf{e}_{\rho(k)} \prod^p {}_yA^{i\lambda[i,k]}$
∴	From 4, D12.11.1A Covariant and Contravariant e-Skew Tensor Components, T10.2.5 Tool: Tensor Substitution and A10.2.14 Correspondence of Tensors and Tensor Coefficients	$\sum_{\rho(k)\in Lp}\ {}_x\mu_f\ {}_x\mathbf{e}_{\bullet\lambda[i,k]}\bullet \prod^p {}_xA^{i\lambda[i,k]} = \sum_{\rho(k)\in Lp}\ {}_y\mu_f\ {}_y\mathbf{e}_{\bullet\lambda[i,k]}\bullet \prod^p {}_yA^{i\lambda[i,k]}$

Theorem 13.5.2 e-Tensor Transformation: Contravariant Equivalency

Step	Hypothesis	$\sum_{\rho(k)\in Lp}\ {}_x\mu_f\ {}_x\mathbf{e}^{\bullet\lambda[i,k]\bullet}\ \prod^p\ {}_xA_{i\lambda[i,k]} = \sum_{\rho(k)\in Lp}\ {}_y\mu_f\ {}_y\mathbf{e}^{\bullet\lambda[i,k]\bullet}\ \prod^p\ {}_yA_{i\lambda[i,k]}$
1	From A13.2.3C Riemannian Manifold $\mathfrak{R}^n$: Invariance of a Tensor in Two Coordinate Systems	${}_x\overset{*}{\mathbf{A}}{}^p = {}_y\overset{*}{\mathbf{A}}{}^p$
2	From 1, T10.9.1 Equality of the Tensor-Determinate Linear Operator, T8.5.6 Orthogonal Sine Operator: Unitary Sine Operator, T8.5.8 Orthogonal Sine Operator: Commutative Asymmetrically Skewed	$<\mid_x\overset{*}{\mathbf{A}}{}^p\mid> = <\mid_y\overset{*}{\mathbf{A}}{}^p\mid>$
3	From 2 and T10.9.2 Tensor-Determinate Linear Operator (TDL)	$\sum_{\rho(k)\in Lp} <\mid \prod^p\ {}_x\mathbf{b}^{\lambda[i,k]} \mid> \prod^p\ {}_xA_{i\lambda[i,k]} =$ $\sum_{\rho(k)\in Lp} <\mid \prod^p\ {}_y\mathbf{b}^{\lambda[i,k]} \mid> \prod^p\ {}_yA_{i\lambda[i,k]}$
4	From 3, D10.8.5D Tensor Coefficient for Distortion (TCD) and A4.2.3 Substitution	$\sum_{\rho(k)\in Lp}\ {}_x\mu_f\ {}_x\mathbf{e}^{\rho(k)}\ \prod^p\ {}_xA_{i\lambda[i,k]} =$ $\sum_{\rho(k)\in Lp}\ {}_y\mu_f\ {}_y\mathbf{e}^{\rho(k)}\ \prod^p\ {}_yA_{i\lambda[i,k]}$
∴	From 4, D12.11.1B Covariant and Contravariant e-Skew Tensor Components, T10.2.5 Tool: Tensor Substitution and A10.2.14 Correspondence of Tensors and Tensor Coefficients	$\sum_{\rho(k)\in Lp}\ {}_x\mu_f\ {}_x\mathbf{e}^{\bullet\lambda[i,k]\bullet}\ \prod^p\ {}_xA_{i\lambda[i,k]} = \sum_{\rho(k)\in Lp}\ {}_y\mu_f\ {}_y\mathbf{e}^{\bullet\lambda[i,k]\bullet}\ \prod^p\ {}_yA_{i\lambda[i,k]}$

Observation 13.5.1 Absoulate Magnetude of the Sine Space Distortion Factor

The *sine space distortion factor* is an absolute magnitude, hence has no permutable orientations and as such is independent of the permuatation long set L_p and its index-k.

Theorem 13.5.3 e-Tensor Transformation: Covariant

Step	Hypothesis	$_y\mu_f\ {}_y\mathbf{e}_{\bullet\lambda[i,k]\bullet} = {}_x\mu_f \sum_{\rho(k)\in Lp} \prod^p x_{j,\eta[j,k]}\ {}_x\mathbf{e}_{\bullet\eta[i,k]\bullet}$ EQ A	$_x\mu_f\ {}_x\mathbf{e}_{\bullet\lambda[i,k]\bullet} = {}_y\mu_f \sum_{\rho(k)\in Lp} \prod^p y_{k,\eta[j,k]}\ {}_y\mathbf{e}_{\bullet\eta[i,k]\bullet}$ EQ B
1	From D10.8.5A Tensor Coefficient for Distortion (TCD), D10.8.5D Tensor Coefficient for Distortion (TCD), A4.2.3 Substitution, D12.11.1A Covariant and Contravariant e-Skew Tensor Components, T10.2.5 Tool: Tensor Substitution and A10.2.14 Correspondence of Tensors and Tensor Coefficients	$_y\mu_f\ {}_y\mathbf{e}_{\bullet\lambda[i,k]\bullet} = <\mid \prod^p {}_y\ \mathbf{b}_{\lambda[i,k]} \mid>$	$_x\mu_f\ {}_x\mathbf{e}_{\bullet\lambda[i,k]\bullet} = <\mid \prod^p {}_y\ \mathbf{b}_{\lambda[i,k]} \mid>$
2	From 1, T13.2.7B Invariance Under Transformation: x-y Parsing Covariant Vectors, T13.2.9B Invariance Under Transformation: y-x Parsing Covariant Vectors and A7.9.7 Transitivity of Vector Objects	$_y\mu_f\ {}_y\mathbf{e}_{\bullet\lambda[i,k]\bullet} = <\mid \prod^p \sum_k x_{k,j}\ {}_x\mathbf{b}_k \mid>$	$_x\mu_f\ {}_x\mathbf{e}_{\bullet\lambda[i,k]\bullet} = <\mid \prod^p \sum_j y_{j,i}\ {}_y\mathbf{b}_j \mid>$
3	From 2, T10.2.1 Tool: Tensor Series Expansion and A10.2.14 Correspondence of Tensors and Tensor Coefficients	$_y\mu_f\ {}_y\mathbf{e}_{\bullet\lambda[i,k]\bullet} = <\mid \sum_{\rho(k)\in Lp} \prod^p x_{j,\eta[j,k]}\ {}_x\mathbf{b}_{j\eta[j,k]} \mid>$	$_x\mu_f\ {}_x\mathbf{e}_{\bullet\lambda[i,k]\bullet} = <\mid \sum_{\rho(k)\in Lp} \prod^p y_{k,\eta[j,k]}\ {}_y\mathbf{b}_{j\eta[j,k]} \mid>$
4	From 3 and T10.9.2 Tensor-Determinate Linear Operator (TDL)	$_y\mu_f\ {}_y\mathbf{e}_{\bullet\lambda[i,k]\bullet} = \sum_{\rho(k)\in Lp} \prod^p x_{j,\eta[j,k]} <\mid {}_x\mathbf{b}_{j\eta[j,k]} \mid>$	$_y\mu_f\ {}_y\mathbf{e}_{\bullet\lambda[i,k]\bullet} = \sum_{\rho(k)\in Lp} \prod^p y_{k,\eta[j,k]} <\mid {}_y\mathbf{b}_{j\eta[j,k]} \mid>$
5	From 4, D10.8.5A Tensor Coefficient for Distortion (TCD), D10.8.5D Tensor Coefficient for Distortion (TCD), A4.2.3 Substitution, D12.11.1A Covariant and Contravariant e-Skew Tensor Components, T10.2.5 Tool: Tensor Substitution and A10.2.14 Correspondence of Tensors and Tensor Coefficients	$_y\mu_f\ {}_y\mathbf{e}_{\bullet\lambda[i,k]\bullet} = \sum_{\rho(k)\in Lp} \prod^p x_{j,\eta[j,k]}\ {}_x\mu_f\ {}_x\mathbf{e}_{\bullet\eta[i,k]\bullet}$	$_x\mu_f\ {}_x\mathbf{e}_{\bullet\lambda[i,k]\bullet} = \sum_{\rho(k)\in Lp} \prod^p y_{k,\eta[j,k]}\ {}_y\mu_f\ {}_y\mathbf{e}_{\bullet\eta[i,k]\bullet}$
∴	From 5, D10.8.1B Sine Space Distortion Factor (SSDF), T10.4.6 Scalars and Tensors: Distribution of a Scalar over Addition of Tensors and O13.5.1 Absoulate Magnetude of the Sine Space Distortion Factor; hence independent of [k]	$_y\mu_f\ {}_y\mathbf{e}_{\bullet\lambda[i,k]\bullet} = {}_x\mu_f \sum_{\rho(k)\in Lp} \prod^p x_{j,\eta[j,k]}\ {}_x\mathbf{e}_{\bullet\eta[i,k]\bullet}$ EQ A	$_x\mu_f\ {}_x\mathbf{e}_{\bullet\lambda[i,k]\bullet} = {}_y\mu_f \sum_{\rho(k)\in Lp} \prod^p y_{k,\eta[j,k]}\ {}_y\mathbf{e}_{\bullet\eta[i,k]\bullet}$ EQ B

Theorem 13.5.4 e-Tensor Transformation: Contravariant

Step	Hypothesis	${}_y\mu_f\ {}_y\mathbf{e}^{\bullet\lambda[i,k]\bullet} =$ ${}_x\mu_f \sum_{\rho(k)\in Lp} \prod^p x_{j,\eta[j,k]}\ {}_x\mathbf{e}^{\bullet\eta[i,k]\bullet}$ EQ A	${}_x\mu_f\ {}_x\mathbf{e}^{\bullet\lambda[i,k]\bullet} =$ ${}_y\mu_f \sum_{\rho(k)\in Lp} \prod^p y_{k,\eta[j,k]}\ {}_y\mathbf{e}^{\bullet\eta[i,k]\bullet}$ EQ B
1	From D10.8.5A Tensor Coefficient for Distortion (TCD), D10.8.5D Tensor Coefficient for Distortion (TCD), A4.2.3 Substitution, D12.11.1A Covariant and Contravariant e-Skew Tensor Components, T10.2.5 Tool: Tensor Substitution and A10.2.14 Correspondence of Tensors and Tensor Coefficients	${}_y\mu_f\ {}_y\mathbf{e}^{\bullet\lambda[i,k]\bullet} = <\mid \prod^p\ {}_y\mathbf{b}^{\lambda[i,k]} \mid>$	${}_x\mu_f\ {}_x\mathbf{e}^{\bullet\lambda[i,k]\bullet} = <\mid \prod^p\ {}_y\mathbf{b}^{\lambda[i,k]} \mid>$
2	From 1, T13.2.8B Invariance Under Transformation: x-y Parsing Contravariant Vectors, T13.2.10B Invariance Under Transformation: y-x Parsing Contravariant Vectors and A7.9.7 Transitivity of Vector Objects	${}_y\mu_f\ {}_y\mathbf{e}^{\bullet\lambda[i,k]\bullet} = <\mid \prod^p \sum_k x_{k,j}\ {}_x\mathbf{b}^k \mid>$	${}_x\mu_f\ {}_x\mathbf{e}^{\bullet\lambda[i,k]\bullet} = <\mid \prod^p \sum_k y_{j,k}\ {}_y\mathbf{b}^k \mid>$
3	From 2, T10.2.1 Tool: Tensor Series Expansion and A10.2.14 Correspondence of Tensors and Tensor Coefficients	${}_y\mu_f\ {}_y\mathbf{e}^{\bullet\lambda[i,k]\bullet} =$ $<\mid \sum_{\rho(k)\in Lp} \prod^p x_{j,\eta[j,k]}\ {}_x\mathbf{b}^{j\eta[j,k]} \mid>$	${}_x\mu_f\ {}_x\mathbf{e}^{\bullet\lambda[i,k]\bullet} =$ $<\mid \sum_{\rho(k)\in Lp} \prod^p y_{k,\eta[j,k]}\ {}_y\mathbf{b}^{j\eta[j,k]} \mid>$
4	From 3 and T10.9.2 Tensor-Determinate Linear Operator (TDL)	${}_y\mu_f\ {}_y\mathbf{e}^{\bullet\lambda[i,k]\bullet} =$ $\sum_{\rho(k)\in Lp} \prod^p x_{j,\eta[j,k]} <\mid {}_x\mathbf{b}^{j\eta[j,k]} \mid>$	${}_x\mu_f\ {}_x\mathbf{e}^{\bullet\lambda[i,k]\bullet} =$ $\sum_{\rho(k)\in Lp} \prod^p y_{k,\eta[j,k]} <\mid {}_y\mathbf{b}^{j\eta[j,k]} \mid>$
5	From 4, D10.8.5A Tensor Coefficient for Distortion (TCD), D10.8.5D Tensor Coefficient for Distortion (TCD), A4.2.3 Substitution, D12.11.1A Covariant and Contravariant e-Skew Tensor Components, T10.2.5 Tool: Tensor Substitution and A10.2.14 Correspondence of Tensors and Tensor Coefficients	${}_y\mu_f\ {}_y\mathbf{e}^{\bullet\lambda[i,k]\bullet} =$ $\sum_{\rho(k)\in Lp} \prod^p x_{j,\eta[j,k]}\ {}_x\mu_f\ {}_x\mathbf{e}^{\bullet\eta[i,k]\bullet}$	${}_x\mu_f\ {}_x\mathbf{e}^{\bullet\lambda[i,k]\bullet} =$ $\sum_{\rho(k)\in Lp} \prod^p y_{k,\eta[j,k]}\ {}_y\mu_f\ {}_y\mathbf{e}^{\bullet\eta[i,k]\bullet}$
∴	From 5, D10.8.1B Sine Space Distortion Factor (SSDF), T10.4.6 Scalars and Tensors: Distribution of a Scalar over Addition of Tensors and O13.5.1 Absoulate Magnetude of the Sine Space Distortion Factor; hence independent of [k]	${}_y\mu_f\ {}_y\mathbf{e}^{\bullet\lambda[i,k]\bullet} =$ ${}_x\mu_f \sum_{\rho(k)\in Lp} \prod^p x_{j,\eta[j,k]}\ {}_x\mathbf{e}^{\bullet\eta[i,k]\bullet}$ EQ A	${}_x\mu_f\ {}_x\mathbf{e}^{\bullet\lambda[i,k]\bullet} =$ ${}_y\mu_f \sum_{\rho(k)\in Lp} \prod^p y_{k,\eta[j,k]}\ {}_y\mathbf{e}^{\bullet\eta[i,k]\bullet}$ EQ B

Theorem 13.5.5 e-Tensor Transformation: Validity Covariant e-Tensor as a Tensor

Step	Hypothesis	x-y covariant e-tensors are true tensors EQ A	y-x covariant e-tensors are true tensors EQ B
1	From T13.5.3(A, B) e-Tensor Transformation: Covariant	${}_y\mu_f\, {}_y\mathbf{e}_{\bullet\lambda[i,k]\bullet} = {}_x\mu_f \sum_{\rho(k)\in Lp} \prod^p x_{j,\eta[j,k]}\, {}_x\mathbf{e}_{\bullet\eta[i,k]\bullet}$	${}_x\mu_f\, {}_x\mathbf{e}_{\bullet\lambda[i,k]\bullet} = {}_y\mu_f \sum_{\rho(k)\in Lp} \prod^p y_{k,\eta[j,k]}\, {}_y\mathbf{e}_{\bullet\eta[i,k]\bullet}$
∴	From 1, no non-linear term arises; and T13.4.6 Addition of a Tensor and Nonlinear Term, hence	x-y covariant e-tensors are true tensors EQ A	y-x covariant e-tensors are true tensors EQ B

Theorem 13.5.6 e-Tensor Transformation: Validity Contravariant e-Tensor as a Tensor

Step	Hypothesis	x-y contravariant e-tensors are true tensors EQ A	y-x contravariant e-tensors are true tensors EQ B
1	From T13.5.4(A, B) e-Tensor Transformation: Contravariant	${}_y\mu_f\, {}_y\mathbf{e}^{\bullet\lambda[i,k]\bullet} = {}_x\mu_f \sum_{\rho(k)\in Lp} \prod^p x_{j,\eta[j,k]}\, {}_x\mathbf{e}^{\bullet\eta[i,k]\bullet}$	${}_x\mu_f\, {}_x\mathbf{e}^{\bullet\lambda[i,k]\bullet} = {}_y\mu_f \sum_{\rho(k)\in Lp} \prod^p y_{k,\eta[j,k]}\, {}_y\mathbf{e}^{\bullet\eta[i,k]\bullet}$
∴	From 1; no non-linear term arises; and T13.4.6 Addition of a Tensor and Nonlinear Term, hence	x-y contravariant e-tensors are true tensors EQ A	y-x contravariant e-tensors are true tensors EQ B

Theorem 13.5.7 e-Tensor Transformation: Absolute Magnitude

Step	Hypothesis	${}_y\mu_f = {}_x\mu_f \mid {}_{yx}J \mid$ EQ A	${}_x\mu_f = {}_y\mu_f \mid {}_{xy}J \mid$ EQ B
1	From T13.5.3(A, B) e-Tensor Transformation: Covariant and	${}_y\mu_f\, {}_y\mathbf{e}_{\bullet\lambda[i,k]\bullet} = {}_x\mu_f \sum_{\rho(k)\in Lp} \prod^p x_{j,\eta[i,k]}\, {}_x\mathbf{e}_{\bullet\eta[i,k]\bullet}$	${}_x\mu_f\, {}_x\mathbf{e}_{\bullet\lambda[i,k]\bullet} = {}_y\mu_f \sum_{\rho(k)\in Lp} \prod^p y_{k,\eta[i,k]}\, {}_y\mathbf{e}_{\bullet\eta[i,k]\bullet}$
2	From 1 and T4.6.1 Equalities: Uniqueness of Absolute Value	$\mid {}_y\mu_f\, {}_y\mathbf{e}_{\bullet\lambda[i,k]\bullet} \mid = \mid {}_x\mu_f \sum_{\rho(k)\in Lp} \prod^p x_{j,\eta[i,k]}\, {}_x\mathbf{e}_{\bullet\eta[i,k]\bullet} \mid$	$\mid {}_x\mu_f\, {}_x\mathbf{e}_{\bullet\lambda[i,k]\bullet} \mid = \mid {}_y\mu_f \sum_{\rho(k)\in Lp} \prod^p y_{k,\eta[i,k]}\, {}_y\mathbf{e}_{\bullet\eta[i,k]\bullet} \mid$
3	From 2 and T4.6.3 Equalities: Absolute Product is the Absolute of the Products	$\mid {}_y\mu_f \mid \mid {}_y\mathbf{e}_{\bullet\lambda[i,k]\bullet} \mid = \mid {}_x\mu_f \mid \mid \sum_{\rho(k)\in Lp} \prod^p x_{j,\eta[j,k]}\, {}_x\mathbf{e}_{\bullet\eta[i,k]\bullet} \mid$	$\mid {}_x\mu_f \mid \mid {}_x\mathbf{e}_{\bullet\lambda[i,k]\bullet} \mid = \mid {}_y\mu_f \mid \mid \sum_{\rho(k)\in Lp} \prod^p y_{k,\eta[j,k]}\, {}_y\mathbf{e}_{\bullet\eta[i,k]\bullet} \mid$
4	From 3, C10.10.5.5B Absolute Value for Composite Factor of Distortion, D4.1.4F Rational Numbers, A4.2.3 Substitution and T4.7.31 Inequalities: Absolute Summation is Less than the Sum of all Absolute Values	${}_y\mu_f \mid {}_y\mathbf{e}_{\bullet\lambda[i,k]\bullet} \mid \le {}_x\mu_f \sum_{\rho(k)\in Lp} \mid \prod^p x_{j,\eta[j,k]}\, {}_x\mathbf{e}_{\bullet\eta[i,k]\bullet} \mid$	${}_x\mu_f \mid {}_x\mathbf{e}_{\bullet\lambda[i,k]\bullet} \mid \le {}_y\mu_f \mid \sum_{\rho(k)\in Lp} \mid \prod^p y_{k,\eta[j,k]}\, {}_y\mathbf{e}_{\bullet\eta[i,k]\bullet} \mid$
5	From 4 and T4.6.3 Equalities: Absolute Product is the Absolute of the Products	${}_y\mu_f \mid {}_y\mathbf{e}_{\bullet\lambda[i,k]\bullet} \mid \le {}_x\mu_f \sum_{\rho(k)\in Lp} \mid \prod^p x_{j,\eta[j,k]} \mid \mid {}_x\mathbf{e}_{\bullet\eta[i,k]\bullet} \mid$	${}_x\mu_f \mid {}_x\mathbf{e}_{\bullet\lambda[i,k]\bullet} \mid \le {}_y\mu_f \mid \sum_{\rho(k)\in Lp} \mid \prod^p y_{k,\eta[j,k]} \mid \mid {}_y\mathbf{e}_{\bullet\eta[i,k]\bullet} \mid$
6	From 5, T12.11.11A e-Tensor System: The Product of the Kronecker Delta and Sign of Parity, A4.2.3 Substitution and T4.6.3 Equalities: Absolute Product is the Absolute of the Products	${}_y\mu_f \mid {}_y\delta_{\eta[i,k]}{}^{\eta[j,k]} \mid \mid {}_y\mathrm{sgn}(k) \mid \le {}_x\mu_f \sum_{\rho(k)\in Lp} \mid \prod^p x_{j,\eta[j,k]} \mid \mid {}_x\delta_{\eta[i,k]}{}^{\eta[j,k]} \mid \mid {}_x\mathrm{sgn}(k) \mid$	${}_x\mu_f \mid {}_x\delta_{\eta[i,k]}{}^{\eta[j,k]} \mid \mid {}_x\mathrm{sgn}(k) \mid \le {}_y\mu_f \sum_{\rho(k)\in Lp} \mid \prod^p y_{k,\eta[j,k]} \mid \mid {}_y\delta_{\eta[i,k]}{}^{\eta[j,k]} \mid \mid {}_y\mathrm{sgn}(k) \mid$

7	From 6; evaluating collinear terms leaves only combinatoral terms to be summed over	${}_y\mu_f\ 1 \mid (-1)^{\varphi(k)} \mid \leq$ ${}_x\mu_f \sum_{\sigma(k)\in Sr} \mid\prod^r x_{j,\eta[j,k]} \mid$ $1 \mid {}_x\mathrm{sgn}(k) \mid$ for $\eta[i, k] = \eta[j, k]$	${}_x\mu_f\ 1 \mid (-1)^{\varphi(k)} \mid \leq$ ${}_y\mu_f \sum_{\sigma(k)\in Sr} \mid\prod^r y_{k,\eta[j,k]} \mid$ $1 \mid {}_x\mathrm{sgn}(k) \mid$ for $\eta[i, k] = \eta[j, k]$
8	From 7, D4.1.11B Absolute Value $\mid a \mid$ and A4.2.12 Identity Multp	${}_y\mu_f \leq$ ${}_x\mu_f \sum_{\sigma(k)\in Sr} \mid \prod^r x_{j,\eta[j,k]} \mid \mid {}_x\mathrm{sgn}(k) \mid$	${}_x\mu_f\ {}_x\mu_s \leq$ ${}_y\mu_f \sum_{\sigma(k)\in Sr} \mid \prod^r y_{k,\eta[j,k]} \mid \mid {}_y\mathrm{sgn}(k) \mid$
9	From 8 and T4.6.3 Equalities: Absolute Product is the Absolute of the Products	${}_y\mu_f \leq$ ${}_x\mu_f \sum_{\sigma(k)\in Sr} \mid {}_x\mathrm{sgn}(k) \prod^r x_{j,\eta[j,k]} \mid$	${}_x\mu_f \leq$ ${}_y\mu_f \sum_{\sigma(k)\in Sr} \mid {}_y\mathrm{sgn}(k) \prod^r y_{j,\eta[j,k]} \mid$
10	From 9 and T4.7.31 Inequalities: Absolute Summation is Less than the Sum of all Absolute Values	${}_y\mu_f =$ ${}_x\mu_f \mid \sum_{\sigma(k)\in Sr}\ {}_x\mathrm{sgn}(k) \prod^r x_{j,\eta[j,k]} \mid$	${}_x\mu_f =$ ${}_y\mu_f \mid \sum_{\sigma(k)\in Sr}\ {}_y\mathrm{sgn}(k) \prod^r y_{j,\eta[j,k]} \mid$
11	From 10 and T6.9.1(B, E) Determinant of A	${}_y\mu_f = {}_x\mu_f \parallel x_{i,j} \parallel$	${}_x\mu_f = {}_y\mu_f \parallel y_{i,j} \parallel$
∴	From 10, D13.7.2(A, B) Jacobian Determinate in a Riemannian Manifold $\mathfrak{R}^n$, A4.2.3 Substitution	${}_y\mu_f = {}_x\mu_f \mid {}_{yx}J \mid$ EQ A	${}_x\mu_f = {}_y\mu_f \mid {}_{xy}J \mid$ EQ B

Section 13.6 Transformation of Hyperarea by a Jacobian Determinate

Definition 13.6.1 **Genus of a Hypersurface $\mathfrak{H}_g^{n-1}$**
Genus of a hypesurface is the largest number of non-intersecting simple closed clurves that can be drawn on the surface without separating it into two or more parts as with a figure eight. One closed curve having a genus of zero [$\mathfrak{H}_0^{n-1}$] would be a sphere, but a torus is not a simply closed surface, hence has a genus of one [$\mathfrak{H}_1^{n-1}$]. [ADL66 pg 375]

Definition 13.6.2 **Hypersurface for Manifold Hyperarea ${}_c\mathfrak{A}^{n-1}$**
A coordinate hyperarea [${}_c\mathfrak{A}^{n-1}$] is a continuously closed, deformable, manifold surface with (no holes ${}_c\mathfrak{A}^{n-1}$) in the skin bounding a Riemannian hypervolume [$\mathfrak{R}^n$], also with no anomalies or sigularities. As such that it is always one dimension less than the volume it bounds.

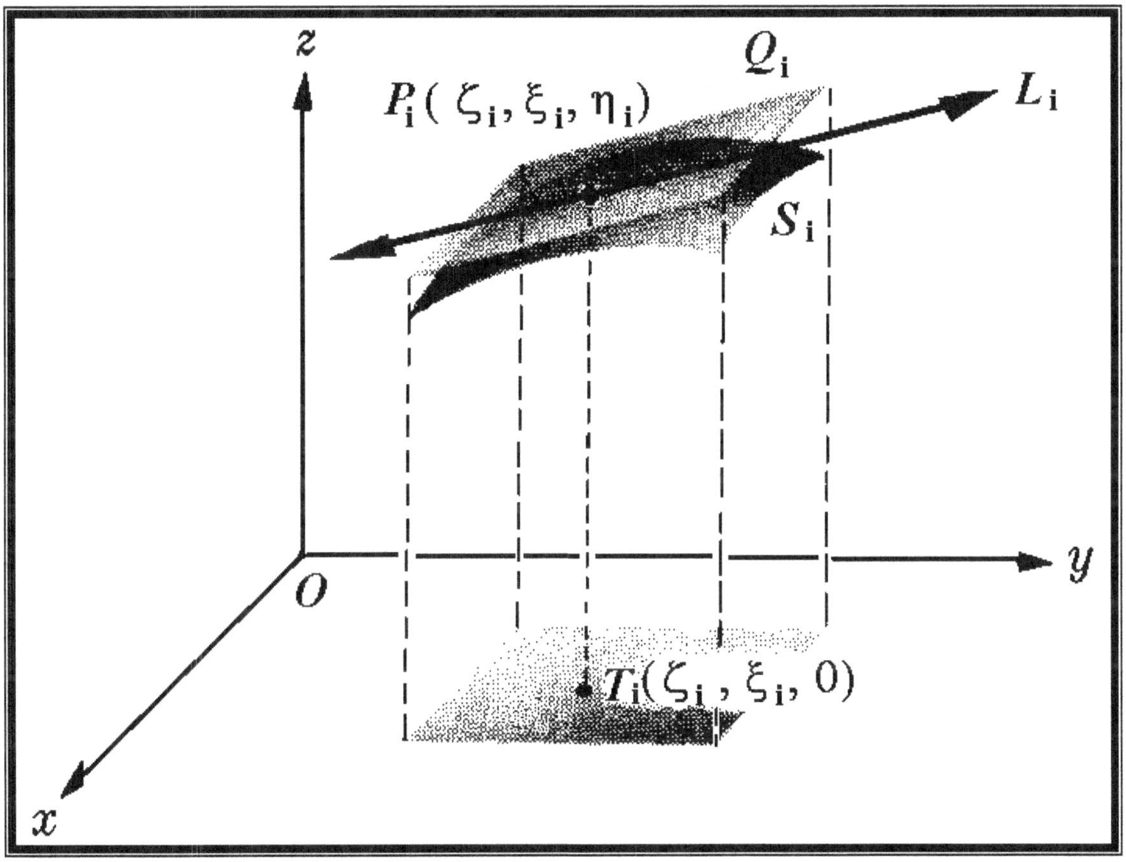

Figure 13.6.1 Approximate Finite Element of Area(T_i) to Hypersurface(S_i)

At every point P_i on a hypersurface let it be approximated by a finite plane of area. In beded in the plane is a line L_i at the point P_i that spans that plane and defines its orinatation.

Definition 13.6.3 **Differential Hyperarea Element in a Riemannian Manifold Area $\mathfrak{A}^{n-1}$**

$dA_x \equiv \prod_i{}^{n-1} dx_i$	Hyperplane element in manifold ${}_x\mathfrak{A}^{n-1}$	Equation A
$dA_y \equiv \prod_i{}^{n-1} dy_i$	Hyperplane element in manifold ${}_y\mathfrak{A}^{n-1}$	Equation B

Definition 13.6.4 **Jacobian Determinate in a Riemannian Manifold Area $\mathfrak{A}^{n-1}$**

${}_{xy}J \equiv \mid x_{i,j} \mid$	determinate Jacobian element in manifold ${}_x\mathfrak{A}^{n-1}$	Equation A
${}_{yx}J \equiv \mid y_{j,i} \mid$	determinate Jacobian element in manifold ${}_y\mathfrak{A}^{n-1}$	Equation B

Definition 13.6.5 **Jacobian Determinate as a Partial Fractional Derivative**

${}_{xy}J \equiv \partial(\bullet x_i \bullet) / \partial(\bullet y_i \bullet)$	Jacobian Fractional Derivative in ${}_x\mathfrak{A}^{n-1}$	Equation A
${}_{yx}J \equiv \partial(\bullet y_i \bullet) / \partial(\bullet x_i \bullet)$	Jacobian Fractional Derivative in ${}_y\mathfrak{A}^{n-1}$	Equation B

Theorem 13.6.1 Transformation from ${}_x\mathfrak{A}^{n-1}$ to ${}_y\mathfrak{A}^{n-1}$: Differential Jacobian Hyperarea Element

Step	Hypothesis	$dA_x = {}_{xy}J\, dA_y$ [13.6.1] EQ A	$dA_y = {}_{yx}J\, dA_x$ [13.6.1] EQ B
1	From T13.2.1(A, B) Transformation Derivatives: Exact Differential	$dx_i = \sum_j x_{i,j}\, dy_j$	$dy_j = \sum_i y_{j,i}\, dx_i$
2	From 1, D6.1.5C Diagonal Matrix, D6.1.6A Identity Matrix and A6.2.10(C, B) Matrix Multiplication; now placing in matrix format such that $[dx_i]$ and $[dy_i]$ are row - column vectors respectively	$(I\,[dx_i]) = (x_{i,j})\,(I\,[dy_i\,])$	$(I\,[dy_i]) = (y_{j,i})\,(I\,[dx_i\,])$
3	From 2 and T6.9.1 Uniqueness of a Determinate	$\lvert I\,[dx_i]\rvert = \lvert(x_{i,j})\,(I\,[dy_i\,])\rvert$	$\lvert I\,[dy_i]\rvert = \lvert(y_{j,i})\,(I\,[dx_i\,])\rvert$
4	From 3 and T6.9.27 The product of two matrices can be factored as individual determinants	$\lvert I\,[dx_i]\rvert = \lvert x_{i,j}\rvert\,\lvert I\,[dy_i\,]\rvert$	$\lvert I\,[dy_i]\rvert = \lvert y_{j,i}\rvert\,\lvert I\,[dx_i\,]\rvert$
5	From 4 and C6.9.20.1 Determinate of a Diagonal Matrix	$\Pi^{n-1}\, dx_i = \lvert x_{i,j}\rvert\, \Pi^{n-1}\, dy_i$	$\Pi^{n-1}\, dy_i = \lvert y_{j,i}\rvert\, \Pi^{n-1}\, dx_i$
∴	From 5, D13.6.3(A, B) Differential Hyperarea Element in a Riemannian Manifold $\mathfrak{A}^{n-1}$, D13.6.4(A, B) Jacobian Determinate in a Riemannian Manifold $\mathfrak{A}^{n-1}$ and A4.2.3 Substitution	$dA_x = {}_{xy}J\, dA_y$ EQ A	$dA_y = {}_{yx}J\, dA_x$ EQ B

[13.6.1]Note: Hyperarea transformation is given by a constant of proporitionality for a Jacoban (n–1)x(n–1) determinate for a weight of (w = 1), [SPI63 pg 182].

Theorem 13.6.2 Transformation from $_x\mathfrak{A}^{n-1}$ to $_y\mathfrak{A}^{n-1}$: Tangent Plane to a Jacobian Hyperarea

1g	Given: for i≠j at the point P_i	$z_i = f_i(.x_j.)$	$w_i = h(.y_j.)$
2g	angle from $\delta A(T_i)$ onto $\delta A(Q_i)$	$\delta A(Q_i) = \delta A(T_i)\cos(\varphi_i)$	$\delta B(Q_i) = \delta B(T_i)\cos(\theta_i)$
3g	altitude numbers from $\delta A(T_i)$	$(0, \ldots, 1 \ldots, 0)$	$(0, \ldots, 1 \ldots, 0)$
4g	altitude numbers projected onto $\delta A(Q_i)$ 828	$(1, f_1, \ldots, f_{(n-1)})$ [PRO70 pg 570] [SPI63 pg 199]	$(1, h_1, \ldots, h_{(n-1)})$
Steps	Hypothesis	$\delta A(T_i) = \delta A(Q_i)\sqrt{(1 + \Sigma_j^{(n-1)} f_{ij}^2)}$	$\delta B(T_i) = \delta B(Q_i)\sqrt{(1 + \Sigma_j^{(n-1)} h_{ij}^2)}$
1	From 3g, 4g, T8.6.1 Angle between Hyperplanes Φ_{12}, and A4.2.3 Substitution, T4.4.1 Equalities: Any Quantity Multiplied by Zero is Zero, and A4.2.7 Identity Add	$\cos(\varphi_i) = 1 / (\sqrt{1})(\sqrt{1 + \Sigma_j^{(n-1)} f_{ij}^2})$	$\cos(\theta_i) = 1 / (\sqrt{1})(\sqrt{1 + \Sigma_j^{(n-1)} h_{ij}^2})$
2	From 1 and T4.9.1 Rational Exponent: Integer of a Positive One: n = 2	$\cos(\varphi_i) = 1 / \sqrt{(1 + \Sigma_j^{(n-1)} f_{ij}^2)}$	$\cos(\theta_i) = 1 / \sqrt{(1 + \Sigma_j^{(n-1)} h_{ij}^2)}$
3	From 2 and T4.4.14 Equalities: Formal Cross Product	$1 / \cos(\varphi_i) = \sqrt{(1 + \Sigma_j^{(n-1)} f_{ij}^2)}$	$1 / \cos(\theta_i) = \sqrt{(1 + \Sigma_j^{(n-1)} h_{ij}^2)}$
4	From 3, DxE.1.1.5 Secant of A, LxE.3.1.5 Reciprocal Relation of Cosine and A4.2.3 Substitution	$\sec(\varphi_i) = \sqrt{(1 + \Sigma_j^{(n-1)} f_{ij}^2)}$	$\sec(\theta_i) = \sqrt{(1 + \Sigma_j^{(n-1)} h_{ij}^2)}$
5	From 2g, T4.4.4B Equalities: Right Cancellation by Multiplication, DxE.1.1.5 Secant of A, LxE.3.1.5 Reciprocal Relation of Cosine and A4.2.3 Substitution	$\delta A_x(T_i) \approx \delta A_x(Q_i)\sec(\varphi_i)$	$\delta B_y(T_i) \approx \delta B_y(Q_i)\sec(\theta_i)$
∴	From 4, 5, A4.2.3 Substitution	$\delta A_x(T_i) \approx \delta A_x(Q_i)\sqrt{(1 + \Sigma_j^{(n-1)} f_{ij}^2)}$	$\delta B_y(T_i) \approx \delta B_y(Q_i)\sqrt{(1 + \Sigma_j^{(n-1)} h_{ij}^2)}$

Theorem 13.6.3 Transformation from $_x\mathfrak{A}^{n-1}$ to $_y\mathfrak{A}^{n-1}$: Tangent Plane Shading a Jacobian Hyperarea

Step	Hypothesis	$dA_x(T_i) \approx \sqrt{(1 + \Sigma_j^{(n-1)} f_{ij}^2)}\ {}_{xy}J\ dA_y$ EQ A	$dB_y(T_i) \approx \sqrt{(1 + \Sigma_j^{(n-1)} h_{ij}^2)}\ {}_{yx}J\ dB_x$ EQ B
1	From T13.6.2 Transformation from $_x\mathfrak{A}^{n-1}$ to $_y\mathfrak{A}^{n-1}$: Tangent Plane to a Jacobian Hyperarea	$\delta A_x(T_i) \approx \delta A_x(Q_i)\sqrt{(1 + \Sigma_j^{(n-1)} f_{ij}^2)}$	$\delta B_y(T_i) \approx \delta B_y(Q_i)\sqrt{(1 + \Sigma_j^{(n-1)} h_{ij}^2)}$
∴	From 1, A4.2.10 Commutative Multp, T13.6.1(A,B) Transformation from $_x\mathfrak{A}^{n-1}$ to $_y\mathfrak{A}^{n-1}$: Differential Jacobian Hyperarea Element and A4.2.3 Substitution	$dA_x(T_i) \approx \sqrt{(1 + \Sigma_j^{(n-1)} f_{ij}^2)}\ {}_{xy}J\ dA_y$ EQ A	$dB_y(T_i) \approx \sqrt{(1 + \Sigma_j^{(n-1)} h_{ij}^2)}\ {}_{yx}J\ dB_x$ EQ B

Theorem 13.6.4 Transformation from $_x\mathfrak{A}^{n-1}$ to $_y\mathfrak{A}^{n-1}$: Surfacing a Jacobian Hyperarea

Step	Hypothesis		
		$A(S)_x = \int_S \sqrt{(1 + \Sigma_j^{(n-1)} f_j^{\ 2})_{xy}} J\, dA_y$ EQ A	$A(S)_y = \int_S \sqrt{(1 + \Sigma_j^{(n-1)} f_j^{\ 2})_{yx}} J\, dA_x$ EQ B
1	From T13.6.3 Transformation from $_x\mathfrak{A}^{n-1}$ to $_y\mathfrak{A}^{n-1}$: Tangent Plane Shading a Hyperarea, summing over all surface area elements gives an approximate area bounding the region.	$\Sigma_i\, dA_x(T_i) \approx \Sigma_i \sqrt{(1 + \Sigma_j^{(n-1)} f_{ij}^{\ 2})_{xy}} J\, dA_y$	$\Sigma_i\, dB_y(T_i) \approx \Sigma_i \sqrt{(1 + \Sigma_j^{(n-1)} h_{ij}^{\ 2})_{yx}} J\, dB_x$
2	From 1, the exact surface is found in the limit by reducing the area elements to zero for all elements	$A(S)_x = \lim_{\|\Delta\| \to 0} \Sigma_i\, dA_x(T_i)$	$A(S)_y = \lim_{\|\Delta\| \to 0} \Sigma_i\, dB_y(T_i)$
3	From 2 and Dx9.4.1.4 The Limit of an Integral Having a Constant Area Under the Curve	$A(S)_x = \int_S dA_x$	$A(S)_y = \int_S dA_y$
∴	From 5, 8 and A4.2.3 Substitution	$A(S)_x = \int_S \sqrt{(1 + \Sigma_j^{(n-1)} f_j^{\ 2})_{xy}} J\, dA_y$ EQ A	$A(S)_y = \int_S \sqrt{(1 + \Sigma_j^{(n-1)} f_j^{\ 2})_{yx}} J\, dA_x$ EQ B

Section 13.7 Transformation of Hypervolume by a Jacobian Determinate

Definition 13.7.1 **Differential Hypervolume Element in a Riemannian Manifold $\mathfrak{R}^n$**

$dV_x \equiv \prod_i dx_i$	Hypervolume element in manifold $\mathfrak{X}^n$	Equation A
$dV_y \equiv \prod_i dy_i$	Hypervolume element in manifold $\mathfrak{Y}^n$	Equation B

Definition 13.7.2 **Jacobian Determinate in a Riemannian Manifold $\mathfrak{R}^n$**

${}_{xy}J \equiv \mid x_{i,j} \mid$	determinate Jacobian element in manifold $\mathfrak{X}^n$	Equation A
${}_{yx}J \equiv \mid y_{j,i} \mid$	determinate Jacobian element in manifold $\mathfrak{Y}^n$	Equation B

Definition 13.7.3 **Jacobian Determinate as a Partial Fractional Derivative**

${}_{xy}J \equiv \partial(\bullet x_i \bullet) / \partial(\bullet y_i \bullet)$	Jacobian Fractional Derivative in $\mathfrak{X}^n$	Equation A
${}_{yx}J \equiv \partial(\bullet y_i \bullet) / \partial(\bullet x_i \bullet)$	Jacobian Fractional Derivative in $\mathfrak{Y}^n$	Equation B

Theorem 13.7.1 Riemannian XFMR ${}_x\mathfrak{R}^n$ to ${}_y\mathfrak{R}^n$: Differential Hypervolume Element

Step	Hypothesis	$dV_x = {}_{xy}J\ dV_y$ EQ A	$dV_y = {}_{yx}J\ dV_x$ EQ B
1	From T13.2.1(A, B) Transformation Derivatives: Exact Differential	$dx_i = \sum_j x_{i,j}\ dy_j$	$dy_j = \sum_i y_{j,i}\ dx_i$
2	From 1, D6.1.5C Diagonal Matrix, D6.1.6A Identity Matrix and A6.2.10(C, B) Matrix Multiplication; now placing in matrix format such that $[dx_i]$ and $[dy_i]$ are row - column vectors respectively	$(I\ [dx_i]) = (x_{i,j})\ (I\ [dy_i\ ])$	$(I\ [dy_i]) = (y_{j,i})\ (I\ [dx_i\ ])$
3	From 2 and T6.9.1 Uniqueness of a Determinate	$\lvert I\ [dx_i]\rvert = \lvert(x_{i,j})\ (I\ [dy_i\ ])\rvert$	$\lvert I\ [dy_i]\rvert = \lvert(y_{j,i})\ (I\ [dx_i\ ])\rvert$
4	From 3 and T6.9.27 The product of two matrices can be factored as individual determinants	$\lvert I\ [dx_i]\rvert = \lvert x_{i,j}\rvert\ \lvert I\ [dy_i\ ]\rvert$	$\lvert I\ [dy_i]\rvert = \lvert y_{j,i}\rvert\ \lvert I\ [dx_i\ ]\rvert$
5	From 4 and C6.9.20.1 Determinate of a Diagonal Matrix	$\Pi^n\ dx_i = \lvert x_{i,j}\rvert\ \Pi^n\ dy_i$	$\Pi^n\ dy_i = \lvert y_{j,i}\rvert\ \Pi^n\ dx_i$
∴	From 5, D13.7.1(A, B) Differential Hypervolume Element in a Riemannian Manifold $\mathfrak{R}^n$, D13.7.2(A, B) Jacobian Determinate in a Riemannian Manifold $\mathfrak{R}^n$ and A4.2.3 Substitution	$dV_x = {}_{xy}J\ dV_y$ EQ A	$dV_y = {}_{yx}J\ dV_x$ EQ B

Theorem 13.7.2 Riemannian XFMR $_x\mathfrak{R}^n$ to $_y\mathfrak{R}^n$: Jacobian Hypervolumn Element Relationships

Step	Hypothesis		
		$_{xy}J = dV_x / dV_y$ EQ A $_{xy}J^{-1} = dV_y / dV_x$ EQ B $_{xy}J^{-1} = \partial(\bullet y_i \bullet) / \partial(\bullet x_i \bullet)$ EQ C $_{xy}J^{-1} = \mid y_{j,i} \mid$ EQ D	$_{yx}J = dV_y / dV_x$ EQ E $_{yx}J^{-1} = dV_x / dV_y$ EQ F $_{yx}J^{-1} = \partial(\bullet x_i \bullet) / \partial(\bullet y_i \bullet)$ EQ G $_{yx}J^{-1} = \mid x_{i,j} \mid$ EQ H
1	From T13.7.1(A, B) Riemannian XFMR $_x\mathfrak{R}^n$ to $_y\mathfrak{R}^n$: Differential Hypervolume Element	$dV_x = {}_{xy}J\ dV_y$	$dV_y = {}_{yx}J\ dV_x$
2	From 1 and T4.4.4B Equalities: Right Cancellation by Multiplication	$_{xy}J = dV_x / dV_y$	$_{yx}J = dV_y / dV_x$
3	From 2, D13.7.2(A, B) Jacobian Determinate in a Riemannian Manifold $\mathfrak{R}^n$ and A4.2.3 Substitution	$dV_x / dV_y = \mid x_{i,j} \mid$	$dV_y / dV_x = \mid y_{j,i} \mid$
4	From 2, A4.2.12 Identity Multp and T4.4.12 Equalities: Reversal of Cross Product of Proportions	$1 / (dV_x / dV_y) = 1 / {}_{xy}J$	$1 / (dV_y / dV_x) = 1 / {}_{yx}J$
5	From 4, T4.5.9 Equalities: Ratio and Reciprocal of a Ratio and D4.1.18 Negative Exponential	$dV_y / dV_x = {}_{xy}J^{-1}$	$dV_x / dV_y = {}_{yx}J^{-1}$
6	From 2, 5, D13.7.3(A, B) Jacobian Determinate as a Partial Fractional Derivative and A4.2.3 Substitution	$_{xy}J^{-1} = \partial(\bullet y_i \bullet) / \partial(\bullet x_i \bullet)$	$_{yx}J^{-1} = \partial(\bullet x_i \bullet) / \partial(\bullet y_i \bullet$
7	From 3, 5 and A4.2.3 Substitution	$_{xy}J^{-1} = \mid y_{j,i} \mid$	$_{yx}J^{-1} = \mid x_{i,j} \mid$
∴	From 2 From 5 and A4.2.2B Equality From 6 From 7	$_{xy}J = dV_x / dV_y$ EQ A $_{xy}J^{-1} = dV_y / dV_x$ EQ B $_{xy}J^{-1} = \partial(\bullet y_i \bullet) / \partial(\bullet x_i \bullet)$ EQ C $_{xy}J^{-1} = \mid y_{j,i} \mid$ EQ D	$_{yx}J = dV_y / dV_x$ EQ E $_{yx}J^{-1} = dV_x / dV_y$ EQ F $_{yx}J^{-1} = \partial(\bullet x_i \bullet) / \partial(\bullet y_i \bullet)$ EQ G $_{yx}J^{-1} = \mid x_{i,j} \mid$ EQ H

Theorem 13.7.3 Riemannian XFMR $_x\mathfrak{R}^n$ to $_y\mathfrak{R}^n$: Jacobian Chain Rule

1g	Given	$dV_1 \equiv dV_x$	
2g		$dV_{n+1} \equiv dV_y$	
3g		$dV_k \equiv dV_{k+1}$ for all k	
Step	Hypothesis	$_{xy}J = \prod_{k=1}^{m} (dV_{x[k]} / dV_{y[k]})$ EQ A $_{xy}J = \prod_{k=1}^{m} {}_{xy[k]}J$ EQ B $_{xy}J = \prod_{k=1}^{m} \partial(\bullet y_k \bullet) / \partial(\bullet x_k \bullet)$ EQ C $_{xy}J = \prod_{k=1}^{n} \mid y_{i[k],j[k]} \mid$ EQ D	$_{yx}J = \prod_{k=1}^{m} (dV_{y[k]} / dV_{x[k]})$ EQ E $_{yx}J = \prod_{k=1}^{m} {}_{yx[k]}J$ EQ F $_{yx}J = \prod_{k=1}^{m} \partial(\bullet x_k \bullet) / \partial(\bullet y_k \bullet)$ EQ G $_{yx}J = \prod_{k=1}^{n} \mid x_{i[k],j[k]} \mid$ EQ H
1	From T13.7.2A Riemannian XFMR $\mathfrak{R}_x^n$ to $\mathfrak{R}_y^n$: Jacobian Hypervolumn Element Relationships	$_{xy}J = dV_x / dV_y$	$_{yx}J = dV_y / dV_x$
2	From 1g, 2g, 3g, 1 and T4.5.6 Equalities: Contraction of a Rational Chain	$dV_x / dV_y = \prod_{k=1}^{m} (dV_{x[k]} / dV_{y[k]})$	$dV_y / dV_x = \prod_{k=1}^{m} (dV_{y[k]} / dV_{x[k]})$
3	From 1, 2, T13.7.2(A, E) Riemannian XFMR $\mathfrak{R}_x^n$ to $\mathfrak{R}_y^n$: Jacobian Hypervolumn Element Relationships and A4.2.3 Substitution	$_{xy}J = \prod_{k=1}^{m} {}_{xy[k]}J$	$_{yx}J = \prod_{k=1}^{m} {}_{yx[k]}J$
4	From 1, 2, T13.7.2(C, G) Riemannian XFMR $\mathfrak{R}_x^n$ to $\mathfrak{R}_y^n$: Jacobian Hypervolumn Element Relationships and A4.2.3 Substitution	$_{xy}J = \prod_{k=1}^{m} \partial(\bullet y_k \bullet) / \partial(\bullet x_k \bullet)$	$_{yx}J = \prod_{k=1}^{m} \partial(\bullet x_k \bullet) / \partial(\bullet y_k \bullet)$
5	From 1, 2, T13.7.2(D, H) Riemannian XFMR $\mathfrak{R}_x^n$ to $\mathfrak{R}_y^n$: Jacobian Hypervolumn Element Relationships and A4.2.3 Substitution	$_{xy}J = \prod_{k=1}^{n} \mid y_{i[k],j[k]} \mid$	$_{yx}J = \prod_{k=1}^{n} \mid x_{i[k],j[k]} \mid$
∴	From 1, 2 and A4.2.3 Substitution	$_{xy}J = \prod_{k=1}^{m} (dV_{x[k]} / dV_{y[k]})$ EQ A	$_{yx}J = \prod_{k=1}^{m} (dV_{y[k]} / dV_{x[k]})$ EQ E
	From 3	$_{xy}J = \prod_{k=1}^{m} {}_{xy[k]}J$ EQ B	$_{yx}J = \prod_{k=1}^{m} {}_{yx[k]}J$ EQ F
	From 4	$_{xy}J = \prod_{k=1}^{m} \partial(\bullet y_k \bullet) / \partial(\bullet x_k \bullet)$ EQ C	$_{yx}J = \prod_{k=1}^{m} \partial(\bullet x_k \bullet) / \partial(\bullet y_k \bullet)$ EQ G
	From 5	$_{xy}J = \prod_{k=1}^{n} \mid y_{i[k],j[k]} \mid$ EQ D	$_{yx}J = \prod_{k=1}^{n} \mid x_{i[k],j[k]} \mid$ EQ H

Section 13.8 Transformation of Weighted Tensors

A relatiave scalar of weight one is called a ***scalar density***. The reason for this terminology may be seen from the expression for the total mass distributed from its density over the entire regional volumn,

$\rho(x_1, x_2, x_3) \equiv dM / dV,$ Equation A

the coordinates being rectangular cartesian. The mass contained in a volumn V, embedded in a manifold $\mathfrak{X}^n$, is given by the triple integral

$M \equiv \iiint {}_x\rho(x_1, x_2, x_3)\, d\, x_1\, d\, x_2\, d\, x_3$

from T13.7.1S5A "Riemannian XFMR ${}_x\mathfrak{R}^n$ to ${}_y\mathfrak{R}^n$: Differential Hypervolume Element"

$M \equiv \iiint {}_x\rho[x_1(y_1), x_2(y_2), x_3(y_2)]\, |\, x_{i,j}\, |\, d\, y_1\, d\, y_2\, d\, y_3$

from T13.2.5 "Invariance Under Transformation: Scalar Function"

$M \equiv \iiint {}_y\rho(y_1, y_2, y_3)\, |\, x_{i,j}\, |\, d\, y_1\, d\, y_2\, d\, y_3$

It is clear that the density of distribution when referred from the X to the Y-coordinates is

${}_x\rho(x_1, x_2, x_3) = {}_y\rho(y_1, y_2, y_3)\, |\, x_{i,j}\, |$ Equation A

This idea of a weighted scalar transformation can be generalized by the following definitions.

Definition 13.8.1 **Relative Scalar of Weight Zero**
A *relatiave scalar of weight zero* is of the following form;

${}_xf \equiv |\, x_{i,j}\, |^0\, {}_yf = {}_yf$ Equation A

Some times known as an ***absolute scalar***.

Definition 13.8.2 **Relative Scalar of Weight One**
A *relatiave scalar of weight one* is of the following form;

${}_xf \equiv |\, x_{i,j}\, |^1\, {}_yf = |\, x_{i,j}\, |\, {}_yf$ Equation A

Definition 13.8.3 **Relative Scalar of Weight W**
A *relatiave scalar of weight W* is of the following form;

${}_xf \equiv |\, x_{i,j}\, |^W\, {}_yf$ Equation A

where W is a rational number.

Observation 13.8.1 Continuity Rules and Definitions of Weighted Tensors

A scalar is the simplest form of a tensor so it is not unexpected that these rules and definitions of scalar transformations can be extended to covariant, contravariant and mixed tensor transformations.

Axiom 13.8.1 Relative Tensor of Weight W

∃| a set of relatiave tensor compoents that are weighted down by Jacobian transformation of weight W, such that

Covariant

$$ {}_{y}A^{r\lambda[r,\,k]} \equiv \sum_{\tau(k)\in Ls} |\, x_{i,j} \,|^{w} \prod_{q=1}^{p} y_{r\lambda[r,\,k],\nu[q,\,k]}\ {}_{x}A^{r\nu[q,\,k]} \qquad \text{Equation A} $$

$$ {}_{y}\mathbf{b}_{r\lambda[r,\,k]} \equiv \sum_{\tau(k)\in Ls} |\, x_{i,j} \,|^{w} \prod_{q=1}^{p} x_{r\sigma[q,\,k],\lambda[r,\,k]}\ {}_{x}\mathbf{b}_{r\sigma[q,\,k]} \qquad \text{Equation B} $$

Contravariant

$$ {}_{y}A_{r\lambda[r,\,k]} \equiv \sum_{\tau(k)\in Ls} |\, x_{i,j} \,|^{w} \prod_{q=1}^{p} x_{\nu[q,\,k],\lambda[r,\,k]}\ {}_{x}A_{\nu[q,\,k]} \qquad \text{Equation C} $$

$$ {}_{y}\mathbf{b}^{r\lambda[r,\,k]} \equiv \sum_{\tau(k)\in Ls} |\, x_{i,j} \,|^{w} \prod_{q=1}^{p} y_{\lambda[r,\,k],\sigma[q,\,k]}\ {}_{x}\mathbf{b}^{\sigma[q,\,k]} \qquad \text{Equation D} $$

Axiom 13.8.2 Closure for Summation of Relative Tensor of Weight W

The resulting summation of two relative tensors of the same weight must result in a relative tensor of the same weight.

Axiom 13.8.3 Closure for the Product of Two Relative Tensors of Different Weights W and Q

The resulting product of two relative tensors of weight-W and weight-Q must result in a relative tensor of the sum of the two weights (W + Q).

Axiom 13.8.4 Closure Under Contraction of a Relative Tensor Results Likewise in Weight W

The resulting contraction of a relative tensor of weight W results in a relative tensor of the same weight W.

Theorem 13.8.1 Weighted Tensor Transformation: Relative Scalar of Weight W

1g	Given	${}_{y}f \equiv {}_{y}f\,|\, y_{ij} \,|^{W}$	${}_{x}f \equiv {}_{y}f\,|\, y_{ij} \,|^{W}$
Step	Hypothesis	${}_{x}f\,|\, x_{ij} \,|^{W} = {}_{y}f\,|\, y_{ij} \,|^{W}$ EQ A	${}_{y}f\,|\, y_{ij} \,|^{W} = {}_{x}f\,|\, x_{ij} \,|^{W}$ EQ B
1	From D13.8.3 Relative Scalar of Weight W	${}_{z}f = {}_{x}f\,|\, x_{ij} \,|^{W}$	${}_{t}f = {}_{y}f\,|\, y_{ij} \,|^{W}$
2	From 1 and T9.3.8 Multi-variable Chain Rule	${}_{x}f\,|\, x_{ij} \,|^{W} = {}_{x}f\,|\, x_{ik}\, y_{kj} |^{W}$	${}_{y}f\,|\, y_{ij} \,|^{W} = {}_{y}f\,|\, y_{ik}\, x_{kj} |^{W}$
3	From 2 and A6.2.10(C, B) Matrix Multiplication	${}_{x}f\,|\, x_{ij} \,|^{W} = {}_{x}f\,|\, (x_{ij})(y_{ij}) |^{W}$	${}_{y}f\,|\, y_{ij} \,|^{W} = {}_{y}f\,|\, (y_{ij})(x_{ij}) |^{W}$
4	From 3 and T6.9.27 The product of two matrices can be factored as individual determinants	${}_{x}f\,|\, x_{ij} \,|^{W} = {}_{x}f\,(|\, x_{ij} \,|\,|\, y_{ij} \,|)^{W}$	${}_{y}f\,|\, y_{ij} \,|^{W} = {}_{y}f\,(|\, y_{ij} \,|\,|\, x_{ij} \,|)^{W}$
5	From 4 and A4.2.21 Distribution Exp	${}_{x}f\,|\, x_{ij} \,|^{W} = {}_{x}f\,|\, x_{ij} \,|^{W}\,|\, y_{ij} \,|^{W}$	${}_{y}f\,|\, y_{ij} \,|^{W} = {}_{y}f\,|\, y_{ij} \,|^{W}\,|\, x_{ij} \,|^{W}$
∴	From 1g, 5 and A4.2.3 Substitution	${}_{x}f\,|\, x_{ij} \,|^{W} = {}_{y}f\,|\, y_{ij} \,|^{W}$ EQ A	${}_{y}f\,|\, y_{ij} \,|^{W} = {}_{x}f\,|\, x_{ij} \,|^{W}$ EQ B

Theorem 13.8.2 Weighted Tensor Transformation: Jocobian as Covariant Meteric Determinates

Step	Hypothesis		
		${}_{y}g = {}_{x}g\ {}_{xy}J^2$ EQ A ${}_{xy}J = \pm\sqrt{{}_{y}g / {}_{x}g}$ EQ B	${}_{x}g = {}_{y}g\ {}_{yx}J^2$ EQ C ${}_{yx}J = \pm\sqrt{{}_{x}g / {}_{y}g}$ EQ D
1	From T13.3.1A Invariance Under Transformation: x-y Metric Tensors and T13.1.2A Invariance Under Transformation: y-x Metric Tensors	${}_{y}g_{ij} = {}_{x}g_{rs}\ x_{r,i}\ x_{s,j}$	${}_{x}g_{ij} = {}_{y}g_{rs}\ y_{r,i}\ y_{s,j}$
2	From 1 and A6.2.10B Matrix Multiplication	$({}_{y}g_{ij}) = ({}_{x}g_{rs})\ (x_{r,i})\ (x_{s,j})$	$({}_{x}g_{ij}) = ({}_{y}g_{rs})\ (y_{r,i})\ (y_{s,j})$
3	From 2 and T6.9.1 Uniqueness of a Determinate	$\mid {}_{y}g_{ij} \mid = \mid ({}_{x}g_{rs})\ (x_{r,i})\ (x_{s,j}) \mid$	$\mid ({}_{x}g_{ij}) \mid = \mid ({}_{y}g_{rs})\ (y_{r,i})\ (y_{s,j}) \mid$
4	From 3 and T6.9.27 The product of two matrices can be factored as individual determinants	$\mid {}_{y}g_{ij} \mid = \mid {}_{x}g_{rs} \mid \mid x_{r,i} \mid \mid x_{s,j} \mid$	$\mid {}_{x}g_{ij} \mid = \mid {}_{y}g_{rs} \mid \mid y_{r,i} \mid \mid y_{s,j} \mid$
5	From 4, D12.1.12 Covariant Meteric Determinate, D13.7.2(A, B) Jacobian Determinate in a Riemannian Manifold $\mathfrak{R}^n$ and A4.2.3 Substitution	${}_{y}g = {}_{x}g\ {}_{xy}J\ {}_{xy}J$	${}_{x}g = {}_{y}g\ {}_{yx}J\ {}_{yx}J$
6	From 5 and D4.1.17 Exponential Notation	${}_{y}g = {}_{x}g\ {}_{xy}J^2$	${}_{x}g = {}_{y}g\ {}_{yx}J^2$
7	From 6 and T4.4.4B Equalities: Right Cancellation by Multiplication	${}_{xy}J^2 = {}_{y}g / {}_{x}g$	${}_{yx}J^2 = {}_{x}g / {}_{y}g$
8	From 7, T4.10.7 Radicals: Reciprocal Exponent by Positive Square Root	${}_{xy}J = \pm\sqrt{{}_{y}g / {}_{x}g}$	${}_{yx}J = \pm\sqrt{{}_{x}g / {}_{y}g}$
∴	From 6 From 8	${}_{y}g = {}_{x}g\ {}_{xy}J^2$ EQ A ${}_{xy}J = \pm\sqrt{{}_{y}g / {}_{x}g}$ EQ B	${}_{x}g = {}_{y}g\ {}_{yx}J^2$ EQ C ${}_{yx}J = \pm\sqrt{{}_{x}g / {}_{y}g}$ EQ D

Section 13.9 Transformation of Space Factors

Theorem 13.9.1 Space Factor Transformations: All Ternary Relations

Step	Hypothesis		
		${}_x\mu_c\,{}_x\rho_v^{\,2}\,{}_{xy}J^2 = {}_y\mu_c\,{}_y\rho_v^{\,2}$ EQ A ${}_x\mu_c\,({}_x\mu_f\,{}_x\rho_v)^2 = {}_y\mu_c\,({}_y\mu_f\,{}_y\rho_v)^2$ EQ B	${}_y\mu_c\,{}_y\rho_v^{\,2}\,{}_{yx}J^2 = {}_x\mu_c\,{}_x\rho_v^{\,2}$ EQ C ${}_y\mu_c\,({}_y\mu_f\,{}_y\rho_v)^2 = {}_x\mu_c\,({}_x\mu_f\,{}_x\rho_v)^2$ EQ D
1	From T13.8.2(A, C) Weighted Tensor Transformation: Jocobian as Covariant Meteric Determinates and T4.4.5A Equalities: Reversal of Right Cancellation by Multiplication	${}_xg\,{}_{xy}J^2 = {}_yg$	${}_yg\,{}_{yx}J^2 = {}_xg$
2	From 1, T12.2.17 Determinate of Covariant Metric Dyadic and A4.2.3 Substitution	${}_x\mu_c\,{}_x\rho_v^{\,2}\,{}_{xy}J^2 = {}_y\mu_c\,{}_y\rho_v^{\,2}$	${}_y\mu_c\,{}_y\rho_v^{\,2}\,{}_{yx}J^2 = {}_x\mu_c\,{}_x\rho_v^{\,2}$
3	From 2 and A4.2.21 Distribution Exp	${}_x\mu_c\,({}_x\rho_v\,\lvert{}_{xy}J\rvert)^2 = {}_y\mu_c\,{}_y\rho_v^{\,2}$	${}_y\mu_c\,({}_y\rho_v\,\lvert{}_{yx}J\rvert)^2 = {}_x\mu_c\,{}_x\rho_v^{\,2}$
4	From 3, T4.4.2 Equalities: Uniqueness of Multiplication and A4.2.10 Commutative Multp	${}_x\mu_c\,{}_y\mu_f^{\,2}\,({}_x\rho_v\,\lvert{}_{xy}J\rvert)^2 = {}_y\mu_c\,{}_y\mu_f^{\,2}\,{}_y\rho_v^{\,2}$	${}_y\mu_c\,{}_x\mu_f^{\,2}\,({}_y\rho_v\,\lvert{}_{yx}J\rvert)^2 = {}_x\mu_c\,{}_x\mu_f^{\,2}\,{}_x\rho_v^{\,2}$
5	From 4 and A4.2.21 Distribution Exp	${}_x\mu_c\,({}_y\mu_f\,\lvert{}_{xy}J\rvert\,{}_x\rho_v)^2 = {}_y\mu_c\,({}_y\mu_f\,{}_y\rho_v)^2$	${}_y\mu_c\,({}_x\mu_f\,\lvert{}_{yx}J\rvert\,{}_y\rho_v)^2 = {}_x\mu_c\,({}_x\mu_f\,{}_x\rho_v)^2$
∴	From 2	${}_x\mu_c\,{}_x\rho_v^{\,2}\,{}_{xy}J^2 = {}_y\mu_c\,{}_y\rho_v^{\,2}$ EQ A	${}_y\mu_c\,{}_y\rho_v^{\,2}\,{}_{yx}J^2 = {}_x\mu_c\,{}_x\rho_v^{\,2}$ EQ C
	From 5, T13.5.7(A, B) and A4.2.3 Substitution	${}_x\mu_c\,({}_x\mu_f\,{}_x\rho_v)^2 = {}_y\mu_c\,({}_y\mu_f\,{}_y\rho_v)^2$ EQ B	${}_y\mu_c\,({}_y\mu_f\,{}_y\rho_v)^2 = {}_x\mu_c\,({}_x\mu_f\,{}_x\rho_v)^2$ EQ D

Section 13.10 Geodesic Coordinates and Geodesic Space $\mathfrak{G}^n$

If the affinity $\Gamma_{ij}{}^k$ is symmetric in its subscripts, it is always possible to find a coordinate frame in which all components of the affinity vanish at any chosen point.

With a careful choice of a coordinate frame in which the chosen point P is the origin and hence has coordinates $y_i = 0$ for all i. A transformation can be done into a special space having coordinates $(\bullet\, y_i \bullet)$:

$$x^k = y^k + a_{ij}{}^k y^i y^j \qquad \text{assume} \qquad \text{Equation A}$$

where the $[a_{ij}{}^k]$ are constants whose values will be chosen later and it shall be assumed, without loss of generality, that $[a_{ij}{}^k]$ is symmetric with respect to i and j. Differentiating equation-A with respect to $[y_i]$ it is seen that:

Theorem 13.10.1 Geodesic Space $\mathfrak{G}^n$: Christoffel Symbol Vanishes at Every Point within

1g	Given	$x_i = x_i(\bullet\, y_i \bullet)$ and its inverse
2g		$y_i = y_i(\bullet\, x_i \bullet)$
3g		$x_k = y_k + a_{ij}{}^k\, y_i\, y_j$ assume
Step	Hypothesis	${}_g\Gamma_{ij}{}^k = 0$
1	From 1g, 2g, 3g, LxK.3.1.7B Uniqueness Theorem of Differentiation and LxK.3.2.2B Differential as a Linear Operator; differentiating with respect to y_i	$x_{k,i} = \partial y_k / \partial y_i + a_{ij}{}^k (\partial y_i / \partial y_i)\, y_j$
2	From 1, T12.4.3B Partial Differential Defines the Kronecker Delta; the second term simplifies with; A4.2.13 Inverse Multp and A4.2.12 Identity Multp	$x_{k,i} = \delta^k{}_i + a_{ij}{}^k\, y_j$
3	From 2, LxK.3.1.7B Uniqueness Theorem of Differentiation, LxK.3.2.2B Differential as a Linear Operator, T12.9.1B Differentiated Kronecker Delta and Fundamental Tensors and LxK.3.2.3B Distribute Across Constant Product; differentiating with respect to y_j	$x_{k,i,j} = 0 + a_{ij}{}^k (\partial y_j / \partial y_j)$
4	From 3, A4.2.7 Identity Add A4.2.13 Inverse Multp and A4.2.12 Identity Multp	$x_{k,i,j} = a_{ij}{}^k$
5	From 2, A4.2.12 Identity Multp and A4.2.3 Substitution; at point P $[y_j = 0]$	$x_{k,i} = \delta^k{}_i + a_{ij}{}^k\, 0$
6	From 5, T4.4.1 Equalities: Any Quantity Multiplied by Zero is Zero and A4.2.7 Identity Add	$x_{k,i} = \delta^k{}_i$
7	From 1g, 2g and T13.2.2 Transformation Derivatives: As Covariant Orthogonal System	$x_{i,k}\, y_{k,j} = \delta^i{}_j$
8	From 6, 7 and A4.2.3 Substitution	$\delta^i{}_k\, y_{k,i} = \delta^i{}_i$
9	From 8 and T12.3.8C Contracted Contravariant Bases Vector with Covariant Vector	$y_{i,j} = \delta^i{}_j$

10	From T13.4.2B Christoffel Symbol-Second Kind Transformation; transforming to a geodesic space	${}_g\Gamma_{ij}^{\;k} = y_{k,r}\, x_{s,t}\, x_{t,j}\; {}_y\Gamma_{st}^{\;r} + y_{k,r}\, x_{s,i,j}$
11	From 4, 6, 9, 10, T10.2.5 Tool: Tensor Substitution and A10.2.14 Correspondence of Tensors and Tensor Coefficients	${}_g\Gamma_{ij}^{\;k} = \delta^k_{\;r}\, \delta^s_{\;i}\, \delta^t_{\;j}\; {}_y\Gamma_{st}^{\;r} + \delta^k_{\;r}\, a_{ij}^{\;r}$
12	From 11, T12.3.8C Contracted Contravariant Bases Vector with Covariant Vector and T12.3.3B Contracted Covariant Bases Vector with Contravariant Vector	${}_g\Gamma_{ij}^{\;k} = {}_y\Gamma_{ij}^{\;k} + a_{ij}^{\;k}$
13	From 12; Since the affinity is symmetric, it is now possible to choose the $[a_{ij}^{\;k}]$ to satisfy the equality	$a_{ij}^{\;k} \equiv -{}_y\Gamma_{ij}^{\;k}$
14	From 12, 13, T10.2.5 Tool: Tensor Substitution and A10.2.14 Correspondence of Tensors and Tensor Coefficients	${}_g\Gamma_{ij}^{\;k} = {}_y\Gamma_{ij}^{\;k} - {}_y\Gamma_{ij}^{\;k}$
∴	From 14, T10.3.19 Addition of Tensors: Inverse by Addition and A10.2.14 Correspondence of Tensors and Tensor Coefficients	${}_g\Gamma_{ij}^{\;k} = 0$

Observation 13.10.1 Existance of Geodesic Space $\mathfrak{G}^n$

The above theroem proves the existance of a geodesic space, because it shows ${}_g\Gamma_{ij}^{\;k} = 0$ vanishing at every point within the space as required.

The coordinate point P(• y^i •) is said to be ***geodesic*** at that point. If such coordinates are employed, it is clear that covariant and partial derivatives are identical at P. This enables a simplification for many arguments leading to tensor equations. However, if such equations can, by this means, be proved valid in the geodesic frame, they are necessarily valid in all frames. [LAW62 pg 109]

Section 13.11 Transformation by Translation; $_x\mathfrak{C}^n$ to $_y\mathfrak{C}^n$

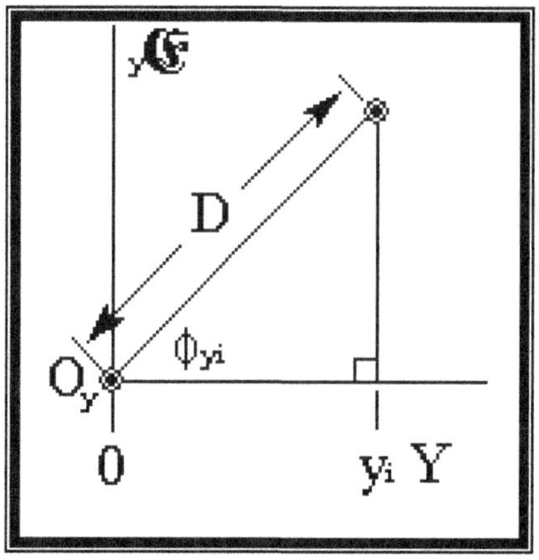

Figure 13.11.1: CGC, Directional Cosine between any given Axes

Definition 13.11.1 **Directional Cosine between any given Axes**
Directional Cosine is the angle between the hypotenuse any given n-dimensional axis directing the line segment:

$$\cos(\phi_{yi}) \equiv y_i / D \qquad \text{Equation A}$$

Theorem 13.11.1 Directed Cosines: A Hypersphere of Unit Radius

1g	Given	Let $_x\mathfrak{C}$ to $_y\mathfrak{C}$ related as in F13.11.1.
Steps	Hypothesis	$1 = \sum_n \cos(\phi_{yi})^2$
1	From F13.11.1 and T8.2.4 Pythagorean Theorem for Vector Magnitude D in n-Dimensions	$D = \sqrt{\sum_n y_i^2}$
2	From 1 and T4.8.6 Integer Exponents: Uniqueness of Exponents	$D^2 = (\sqrt{\sum_n y_i^2}\,)^2$
3	From 2 and T4.10.4	$D^2 = \sum_n y_i^2$
4	From 3 and A4.2.2 Equality	$1/ D^2 = 1/ D^2$
5	From 3, 4 and T4.4.2 Equalities: Uniqueness of Multiplication	$(1/ D^2)\, D^2 = (1/ D^2)\sum_n y_i^2$
6	From 5, A4.2.12 Identity Multp and A4.2.14 Distribution	$1 = \sum_n y_i^2 / D^2$
7	From 6, T4.8.7 Integer Exponents: Distribution Across a Rational Number	$1 = \sum_n (y_i / D)^2$
∴	From 7, D13.11.1 Directional Cosine between any given Axes and A4.2.3 Substitution	$1 = \sum_n \cos(\phi_{yi})^2$

Observation 13.11.1: On Unit Hypersphere

Notice that for T13.11.1, if the directed cosines were allowed to vary that the directed unitary radius would trace out a Hypersphere.

The concept of mapping between manifolds is key to tensor analysis and to the broader topic of topology. It is not by accident that definitions 11.1 to 5 are defined in the terminology of functional mapping. This intern sets the stage for carrying out transformations between manifolds.

The simplest of all transformations is translation in a Euclidian Manifold with these types of transformations it is easer to handle the n-dimensional components in a matrix.

Let manifolds [11.2.1] ${}_{x}\mathfrak{C}$ to ${}_{y}\mathfrak{C}$ be related to one another by the following metric distances from their origins $D(\bullet\ a_i\ \bullet)$.
Further let ${}_{x}\mathfrak{C}$ have coordinates $X(\bullet\ x_i\ \bullet)$ and ${}_{y}\mathfrak{C}$ coordinates $Y(\bullet\ y_i\ \bullet)$ than Y is related to X as follows:

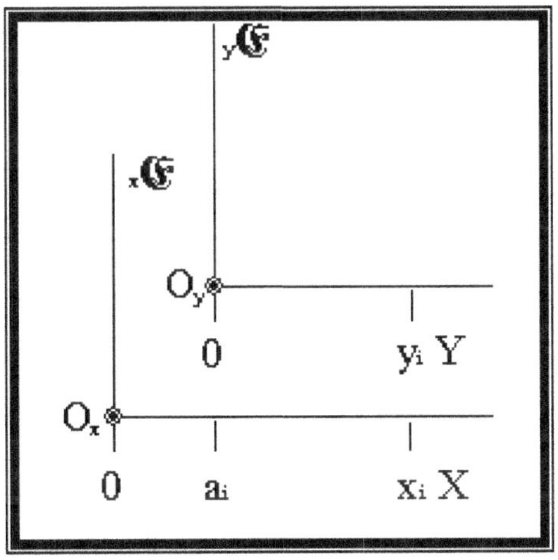

Figure 13.11.2: CGC, Translation between any given Axes

Theorem 13.11.2 Translation between any given Axes

1g	Given	Let ${}_{x}\mathfrak{C}$ and ${}_{y}\mathfrak{C}$ be related as in F13.11.2.
Steps	Hypothesis	$y_i = x_i - a_i$
1	From F13.11.2	$O_yY = y_i$
2	From F13.11.2	$O_xX = x_i$
3	From F13.11.2	$O_xO_y = a_i$
4	From F13.11.2	$O_yY = O_xX - O_xO_y$
∴	From 1, 2, 3, 4 and A4.2.3 Substitution; for all i.	$y_i = x_i - a_i$ for all axes[13.11.1]

[13.11.1]Note: For matrix form see Appendix LxE.5.1.1 First degree equaiton.

In physics, such a coordinate transformation is called a Galilean Transformation for the coordinate system is absolute not elastic like between frames of references as Einstein conceived in his measured system.

Theorem 13.11.3 Translation of Pythagorean Distance

1g	Given	Let ${}_x\mathfrak{C}$ and ${}_y\mathfrak{C}$ be related as in F13.11.2
Steps	Hypothesis	$D = \sqrt{\sum_n (x_i - a_i)^2}$
1	From 1g and T13.11.3	$y_i = x_i - a_i$
∴	From 1g, 1, T8.2.4 Pythagorean Theorem for Vector Magnitude D in n-Dimensions and A4.2.3 Substitution	$D = \sqrt{\sum_n (x_i - a_i)^2}$

Observation 13.11.2 Invariance of Magnitude for Distance

Notice that the magnitude of distance remains unchanged under translation from ${}_x\mathfrak{C}$ to ${}_y\mathfrak{C}$.

Theorem 13.11.4 Translation of the Pythagorean Distance to Directed Cosine

1g	Given	Let ${}_x\mathfrak{C}$ and ${}_y\mathfrak{C}$ be related as in F13.11.2
Steps	Hypothesis	$\cos(\phi_{xi}) = (x_i - a_i) / D$
∴	From D13.11.1, T13.11.2 Directional Cosine between any given Axes and A4.2.3 Substitution	$\cos(\phi_{xi}) = (x_i - a_i) / D$

Observation 13.11.3 Translation of Directed Cosine

Note that any two points on the directed line can be given by points $P_x(\bullet x_i \bullet)$ and $P_a(\bullet a_i \bullet)$. So as a parametric equation in n-space, it can be rewritten as:

Theorem 13.11.5 Parametric Equation of a Line in n-Space

1g 2g 3g	Given	$P_x(\bullet x_i \bullet)$ and $P_a(\bullet a_i \bullet)$ on a directed line $b_i = \cos(\phi_{xi})$ t = D as continuously differentiable
Steps	Hypothesis	$t = (x_i - a_i) / b_i$ for all i
1	From T13.11.4 Translation of the Pythagorean Distance to Directed Cosine	$\cos(\phi_{xi}) = (x_i - a_i) / D$
2	From 1g, 2g, 3g, 1 and A4.2.3 Substitution	$b_i = (x_i - a_i) / t$
3	From 2 and T4.4.5A Equalities: Reversal of Right Cancellation by Multiplication	$t\, b_i = (x_i - a_i)$
∴	From 2 and T4.4.4B Equalities: Right Cancellation by Multiplication	$t = (x_i - a_i) / b_i$ for all i

Observation 13.11.4 Dividing by Zero on a Parameteic Line

Hence any point on a line can be found. Also notice that for any situation where [t] is indeterminate:

t = 0/0 that it maybe interpreted as:	Equation A
$0 = x_i - a_i$ and	Equation B
$b_i = 0$	Equation C

Section 13.12 Directed Rotation between Manifolds $_x\mathfrak{C}^n$ and $_y\mathfrak{C}^n$

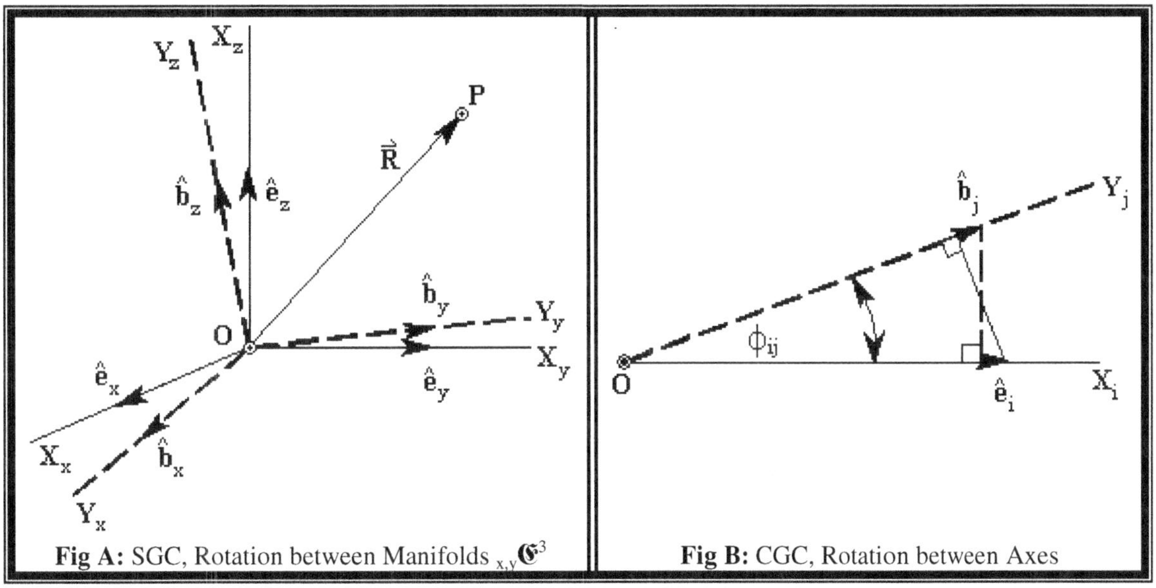

Fig A: SGC, Rotation between Manifolds $_{x,y}\mathfrak{C}^3$ **Fig B:** CGC, Rotation between Axes

Figure 13.12.1: Directed Rotation between Axes

Given 13.12.1 Directed Rotation between Manifolds $_x\mathfrak{C}^n$ and $_u\mathfrak{C}^n$

1g	Given, from F13.12.1	$(i, j) = 1, 2, \dots, n$ Axial Index
2g		$\phi_{ij} \equiv \angle X_i O Y_j$ $_x\mathfrak{C}^n$ and $_y\mathfrak{C}^n$ for $i \neq j$
3g		$\phi_{ji} \equiv \angle Y_j O X_i$ $_y\mathfrak{C}^n$ and $_x\mathfrak{C}^n$ for $i \neq j$
4g		$R \equiv \overline{RO}$ common $_x\mathfrak{C}^n$ and $_y\mathfrak{C}^n$
5g		$\vec{R} \equiv \sum^n x_i\, \hat{e}_i$ $_x\mathfrak{C}^n$
6g		$\vec{R} \equiv \sum^n y_j\, \hat{b}_j$ $_y\mathfrak{C}^n$
7g		$x_i \equiv \overline{XO}_i$ $_x\mathfrak{C}^n$
8g		$y_j \equiv \overline{YO}_j$ $_y\mathfrak{C}^n$
9g		$X \equiv (x_i)$ for column matrix (1 x n)
10g		$Y \equiv (y_i)$ for column matrix (1 x n)
11g		$C \equiv (\cos(\phi_{ij}))$ matrix of directional cosines
12g		$C_i^j \equiv \mathrm{sgn}(i + j)\, \lvert M_i^j \rvert$ cofactor of directional cosines
13g		$\nu_c \equiv \lvert \cos(\phi_{ij}) \rvert$ determinate of directional cosines

Theorem 13.12.1 Directed Rotation between Manifolds $_x\mathfrak{G}^n$ and $_y\mathfrak{G}^n$

Steps	Hypothesis	$y_j = \sum^n x_i \cos(\phi_{ji})$ Rotation between $_x\mathfrak{G}^n$ and $_y\mathfrak{G}^n$
1	From F13.12.1A and A8.3.1 Dot Product: Uniqueness Operation	$\hat{b}_k \bullet \vec{R} = \hat{b}_k \bullet \vec{R}$
2	From 1, G6g, G5g and A7.9.2 Substitution of Vectors	$\hat{b}_k \bullet \sum^n y_j \hat{b}_j = \hat{b}_k \bullet \sum^n x_i \hat{e}_i$
3	From 2 and A8.3.7 Dot Product: Distribution of Dot Product Across Addition of Vectors	$\sum^n y_j \hat{b}_k \bullet \hat{b}_j = \sum^n x_i \hat{b}_k \bullet \hat{e}_i$
4	From 3, D8.3.3 Orthogonal BasesVectors, D8.3.2 Fundamental Base Vector Dot Product and A4.2.3 Substitution	$\sum^n y_j \delta_{kj} = \sum^n x_i \cos(\angle Y_k O X_i)$
∴	From 4, G3g, A4.2.3 Substitution, and by k = j by summing over j	$y_j = \sum^n x_i \cos(\phi_{ji})$ Rotation between $_x\mathfrak{G}^n$ and $_y\mathfrak{G}^n$

Theorem 13.12.2 Directed Rotation between Manifolds $_y\mathfrak{C}^n$ and $_x\mathfrak{C}^n$

Steps	Hypothesis	$x_j = \Sigma^n y_j \cos(\phi_{ij})$ rotation between $_y\mathfrak{C}^n$ and $_x\mathfrak{C}^n$
1	From F13.12.1A and A8.3.1 Dot Product: Uniqueness Operation	$\hat{e}_k \bullet \vec{R} = \hat{e}_k \bullet \vec{R}$
2	From 1, G5g, G6g and A7.9.2 Substitution of Vectors	$\hat{e}_k \bullet \Sigma^n x_j\, \hat{e}_i = \hat{e}_k \bullet \Sigma^n y_j\, \hat{b}_j$
3	From 2 and T8.3.7 Dot Product: Distribution of Dot Product Across Addition of Vectors	$\Sigma^n x_j\, \hat{e}_k \bullet \hat{e}_i = \Sigma^n u_j\, \hat{e}_k \bullet \hat{b}_j$
4	From G2g, 3, D8.3.6 Orthonormal Unit Base Vectors, D8.3.2 Fundamental Base Vector Dot Product and A4.2.3 Substitution	$\Sigma_j^n x_i\, \delta_{ki} = \Sigma_j^n y_j \cos(\angle X_k O Y_j)$
∴	From 4, G2g, A4.2.3 Substitution, and D6.16 Identity Matrix evaluating the Kronecker Delta k = i and summing over j	$x_i = \Sigma_j^n y_j \cos(\phi_{ij})$ rotation between $_y\mathfrak{C}^n$ and $_x\mathfrak{C}^n$

Observation 13.12.1: Of Invariance

A property that bridges a manifold and leaves it unaltered is a quantity shared by the manifold leaving the maniforld unchanged. This is one of the more important properties in the mathematics of manifolds since tensor structures are based on this property as they are transformed from one manifold to another. As seen in the above example the two coordinate systems share $\overline{RO}$, hence the line segment will be measured the same in both systems this, is invariance (see Section 13.2 "Invariance Under Transformation").

Theorem 13.12.3 Directed Rotation Determinate of Transpose of Directional Cosine Matrix

Steps	Hypothesis	$\lvert {}^tC \rvert = \lvert C \rvert = \nu_c$
∴	From G11g, G13g T6.9.1 Uniqueness of a Determinate, T6.9.2 Transpose of a Determinate leaves it unaltered and A4.2.3 Substitution	$\lvert {}^tC \rvert = \lvert C \rvert = \nu_c$

Theorem 13.12.4 Directed Rotation Right Cosine Identity

Steps	Hypothesis	$\nu_c\, \delta_{ij} = \Sigma_k C_{ki} \cos(\phi_{kj})$
1	From G11g and T6.9.19 Closure with respect to the Identity Matrix for Multiplication	$I\ = C^{-1}\, C$
2	From G11g, 1, T6.9.18 Inverse Square Matrix for Multiplication and A4.2.3 Substitution	$I\ = \lvert \cos(\phi_{ij}) \rvert^{-1}\, (C_{ji})\, (\cos(\phi_{ij}))$
3	From G13g, 2 and T6.3.1 Substitution of Matrices	$I\ = \nu_c^{-1}\, (C_{ji})\, (\cos(\phi_{ij}))$
4	From 3 and A6.2.10C Matrix Multiplication	$I\ = \nu_c^{-1}\, (\Sigma_k C_{ki} \cos(\phi_{kj}))$
5	From 4 and equating elements across the equality	$\delta_{ij} = \nu_c^{-1}\ \Sigma_k C_{ki} \cos(\phi_{kj})$
∴	From 5 and T4.4.7A Equalities: Reversal of Left Cancellation by Multiplication	$\nu_c\, \delta_{ij} = \Sigma_k C_{ki} \cos(\phi_{kj})$

Theorem 13.12.5 Directed Rotation Left Cosine Identity

Steps	Hypothesis	$\nu_c\ \delta_{ij} = \sum_k C_{ki} \cos(\phi_{kj})$
1	From G11g, T6.9.19 Closure with respect to the Identity Matrix for Multiplication and T6.4.1 Uniqueness of Transposition of Matrices, T6.4.5 Distribution of the transpose operation for products	$I\ = {}^tC^{-1}\ {}^tC$
2	From G11g, 1, T6.9.18 Inverse Square Matrix for Multiplication, T6.3.1 Substitution of Matrices and A6.2.11 Matrix Transposition	$I\ = \mid \cos(\phi_{ji}) \mid^{-1} (C_{ij}) (\ \cos(\phi_{ji})\ )$
3	From 2, G13g and A4.2.3 Substitution	$I\ = \nu_c^{-1} (C_{ij}) (\ \cos(\phi_{ji})\ )$
4	From 3 and A6.2.10C Matrix Multiplication	$I\ = \nu_c^{-1} (\sum_k C_{ik} \cos(\phi_{jk})\ )$
5	From 4 and equating elements across the equality	$\delta_{ij} = \nu_c^{-1} \sum_k C_{ik} \cos(\phi_{jk})$
∴	From 5 and T4.4.7A Equalities: Reversal of Left Cancellation by Multiplication	$\nu_c\ \delta_{ij} = \sum_k C_{ik} \cos(\phi_{jk})$

Theorem 13.12.6 Directed Rotation Symmetry of Directional Cosine for i < j

Steps	Hypothesis	$\cos(\phi_{ij}) = \cos(\phi_{ji})$
1	From G2g, G3g and F13.12.1B	$\phi_{ij} = \angle X_iOY_j = -\angle Y_jOX_i = -\phi_{ji}$ for $i < j$
2	From 1 and AE.1.2 Uniqueness of all Trigonometric Functions Operating on Equal Angles	$\cos(\phi_{ij}) = \cos(\phi_{ij})$
3	From 2 and AE.1.3 Change in Direction of Angular Measurement Changes Sign; implied by a transpose of index	$\cos(\phi_{ij}) = \cos(-\phi_{ji})$
∴	From 3 and DxE.1.7.1 Quick Lookup: Periodic Reduction Formulas for Sign for Cosine	$\cos(\phi_{ij}) = \cos(\phi_{ji})$ for $i < j$

Theorem 13.12.7 Transpose of Directional Cosine Matrix

Steps	Hypothesis	${}^tC = [\ \cos(\phi_{ij}) \setminus \cos(\phi_{ii}) \setminus \cos(\phi_{ij})\ ]$ for $i < j$
1	From G11g	${}^tC = [\ \cos(\phi_{ji}) \setminus \cos(\phi_{ii}) \setminus \cos(\phi_{ij})\ ]$
∴	From 1 and A6.2.11 Matrix Transposition	${}^tC = [\ \cos(\phi_{ij}) \setminus \cos(\phi_{ii}) \setminus \cos(\phi_{ij})\ ]$ for $i < j$

Theorem 13.12.8 Invarence of Transpose of Directional Cosine Matrix

Steps	Hypothesis	$C = {}^tC$
1	From G11g, expanded	$C = [\ \cos(\phi_{ji}) \setminus \cos(\phi_{ii}) \setminus \cos(\phi_{ij})\ ]$
2	From 1 and AE.1.3Change in Direction of Angular Measurement Changes Sign	$C = [\ \cos(-\phi_{ij}) \setminus \cos(-\phi_{ii}) \setminus \cos(-\phi_{ji})\ ]$
3	From 2 and DxE.1.7.1 Quick Lookup: Periodic Reduction Formulas for Sign for Cosine	$C = [\ \cos(\phi_{ij}) \setminus \cos(\phi_{ii}) \setminus \cos(\phi_{ji})\ ]$
∴	From 3 and A6.2.11 Matrix Transposition	$C = {}^tC$

Theorem 13.12.9 Directed Rotation Symmetry of Directional Cosine Cofactors

Steps	Hypothesis	$C_{ij} = C_{ji}$ cofactors are symmetrical
1	From T6.4.6 Transpose of the Identity Matrix	$I = \overset{t}{I}$
2	From 1, T6.9.19 Closure with respect to the Identity Matrix for Multiplication and T6.3.1 Substitution of Matrices	$C^{-1}\, C = {}^tC^{-1}\, C$
∴	From 2 and equating matrix elements	$C_{ij} = C_{ji}$

Observation 13.12.2: Degrees of Freedom for Rotation

The coordinates of a manifold moves as a unit always maintaining their angle and magnetude of the bases vectors as invariant. So in two dimensions there can only be one degree of freedom of one manifold relative to another coincident manifold and that is represented by a characteristic angle of rotation. Likewise in three dimensions there can only be two degrees of freedom for rotation represented by two characteristic angles. For four dimensions three and so on. So the numbers of independent characteristic angles of rotation are given by (n – 1) for n-space.

Theorem 13.12.10 Directed Rotation between Manifolds $_x\mathfrak{C}^2$ and $_y\mathfrak{C}^2$

Steps	Hypothesis	$y_1 = +x_1 \cos(\theta) + x_2 \sin(\theta)$ $y_2 = -x_1 \sin(\theta) + x_2 \cos(\theta)$
1	From T13.12.1 Directed Rotation between Manifolds $_x\mathfrak{C}^n$ and $_y\mathfrak{C}^n$ and expand in 2-dimensions	$y_1 = x_1 \cos(\phi_{11}) + x_2 \cos(\phi_{12})$ $y_2 = x_1 \cos(\phi_{21}) + x_2 \cos(\phi_{22})$
2	From 1 and i=j characteristic angle of rotation between manifolds	$\phi_{11} = \phi_{22} = \theta$
3	From $_x\mathfrak{C}^2$ orthogonal lagging rotation	$\phi_{12} = \theta - 90^\circ$ lag space between basis vectors
4	From $_x\mathfrak{C}^2$ orthogonal leading rotation	$\phi_{21} = \theta + 90^\circ$ lead space between basis vectors
5	From 1, 2, 3, 4 and A4.2.3 Substitution	$y_1 = x_1 \cos(\theta \qquad) + x_2 \cos(\theta - 90^\circ)$ $y_2 = x_1 \cos(\theta + 90^\circ) + x_2 \cos(\theta)$
∴	From 5, DxE.1.7.3 and DxE.1.7.2 Quick Lookup: Periodic Reduction Formulas for Sign	$y_1 = +x_1 \cos(\theta) + x_2 \sin(\theta)$ $y_2 = -x_1 \sin(\theta) + x_2 \cos(\theta)$

Section 13.13 Planar Coordinate Point between Manifolds ${}_x\mathfrak{C}^n$ and ${}_\varphi\mathfrak{C}^n$

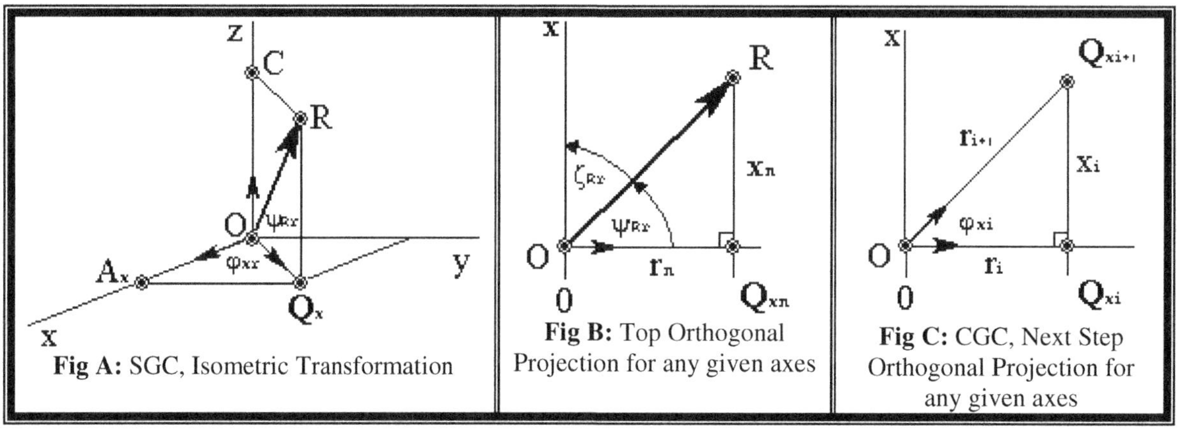

Fig A: SGC, Isometric Transformation

Fig B: Top Orthogonal Projection for any given axes

Fig C: CGC, Next Step Orthogonal Projection for any given axes

Figure 13.13.1: Planar Coordinates

Theorem 13.13.1 Planar Coordinate Point by Constructible Geometrical Components

1g	Given	$R = \overline{OR}$	
2g		$x_n = \overline{QR}_{xn}$	
3g		$r_n = \overline{OQ}_{xn}$	
4g		$r_i = \overline{OQ}_{xi}$	
5g		$x_i = \overline{QQ}_i$	
Steps	Hypothesis	$R^2 = x_n^{\ 2} + r_n^{\ 2}$ EQA	$r_{i+1}^{\ \ 2} = r_i^{\ 2} + x_i^{\ 2}$ EQB
1	From F13.13.1B and D.1.7A Pythagorean Theorem: Algebraic Version	$\overline{OR}^{\,2} = \overline{QR}_{xn}^{\ 2} + \overline{OQ}_{xn}^{\ 2}$	
2	From F13.13.1C and D.1.7A Pythagorean Theorem: Algebraic Version	$\overline{OQ}_{xi+1}^{\ 2} = \overline{OQ}_{xi}^{\ 2} + \overline{QQ}_i^{\ 2}$	
3	From 1, 1g, 2g, 3g and A4.2.3 Substitution	$R^2 = x_n^{\ 2} + r_n^{\ 2}$	
4	From 2, 3g, 4g, 5g and A4.2.3 Substitution	$r_{i+1}^{\ \ 2} = r_i^{\ 2} + x_i^{\ 2}$	
∴	From 3 and 4	$R^2 = x_n^{\ 2} + r_n^{\ 2}$ EQA	$r_{i+1}^{\ \ 2} = r_i^{\ 2} + x_i^{\ 2}$ EQB

Theorem 13.13.2 Planar Coordinate Point

Steps	Hypothesis	$R = \pm\sqrt{\sum^n x_i^2}$
1	From T13.13.1B Planar Coordinate Point by Constructible Geometrical Components and collecting S4 indices	$n = i+1$
2	From 2 and T4.3.3A Equalities: Right Cancellation by Addition	$i = n - 1$
3	From 2, T13.13.1B Planar Coordinate Point by Constructible Geometrical Components	$r_{i+1}^2 = r_i^2 + x_i^2$
4	From 2, 3, 4 and A4.2.3 Substitution	$r_i^2 = r_{i-1}^2 + x_{i-1}^2$
5	From 4 and contracting summation	$R^2 = \sum^n x_i^2$
∴	From 5, T4.10.1 Radicals: Uniqueness of Radicals, T4.10.3 Radicals: Identity Power Raised to a Radical and T4.19.7 Radicals: Reciprocal Exponent by Positive Square Root	$R = \pm\sqrt{\sum^n x_i^2}$

While the above proof has the same result of the n-dimensional Pythagorean Theorem T8.2.4 "Pythagorean Theorem for Vector Magnitude D in n-Dimensions" it needed be restated, but this time A3.12.1 "Equivalency Property of Constructible Geometric Components" is applied since the geometry readily lends itself to this technique.

Section 13.14 Planar Coordinate Axes between Manifolds $_x\mathfrak{C}^n$ and $_u\mathfrak{C}^n$

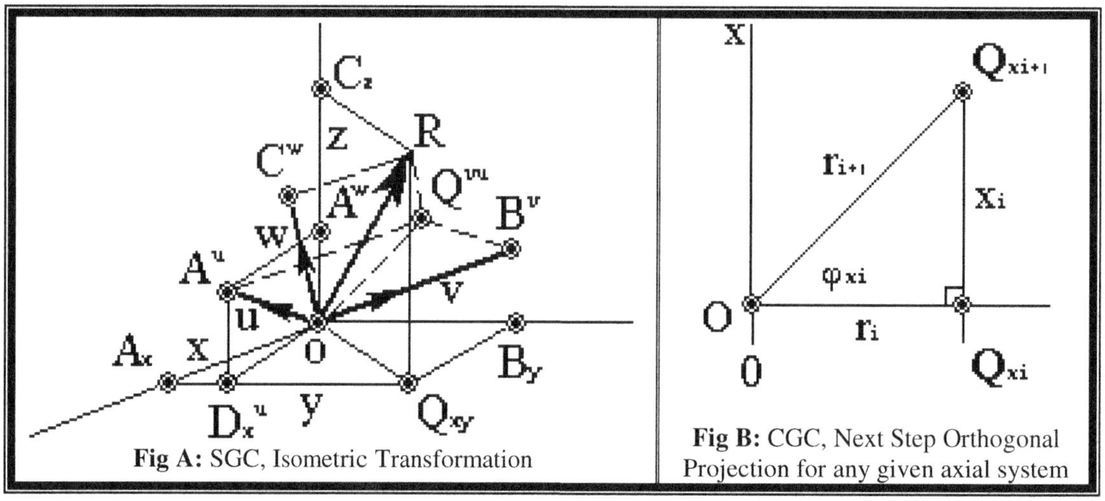

Fig A: SGC, Isometric Transformation

Fig B: CGC, Next Step Orthogonal Projection for any given axial system

Figure 13.14.1: Planar Coincident Axes

Now let the planar point coincide with an axis on another coordinate system, hence for any axial geometric component the problem can be generalized to n-dimensions. Notice unlike directional coordinate systems the cross relationships to other coordinates have been decoupled. Now each axis is its own system requiring one less angle of the dimension of the space to describe position.

Given 13.14.1 Planar Coordinate Axes between Manifolds $_x\mathfrak{C}^n$ and $_u\mathfrak{C}^n$

1g	Given	$i = 1, 2, \ldots, n-1$	Axial Index
2g		$\zeta_x \equiv \angle C_zOR$	$_x\mathfrak{C}^n$
3g		$\psi_x \equiv \angle Q_{xy}OR \equiv \varphi_{xn}$	$_x\mathfrak{C}^n$
4g		$\varphi_{xi} \equiv \angle Q_iOQ_{xi+1}$	$_x\mathfrak{C}^n$
5g		$\zeta^u \equiv \angle C_wOR$	$_u\mathfrak{C}^n$
6g		$\psi^u \equiv \angle Q_{uv}OR \equiv \varphi^{un}$	$_u\mathfrak{C}^n$
7g		$\varphi^{ui+1} \equiv \angle Q_iOQ_{ui+1}$	$_u\mathfrak{C}^n$
8g		$\zeta_x^{\ ui} \equiv \angle A_uOA_{wi}$	$_x\mathfrak{C}^n$ and $_u\mathfrak{C}^n$
9g		$\psi_x^{\ ui} \equiv \angle A_uOD_{xui} \equiv \phi_{x\ un}$	$_x\mathfrak{C}^n$ and $_u\mathfrak{C}^n$
10g		$\phi_x^{\ ui+1} \equiv \angle Q_{x\ ui}OQ_{x\ ui+1}$	$_x\mathfrak{C}^n$ and $_u\mathfrak{C}^n$
11g		$R \equiv \overline{RO} \equiv r^n \equiv r_n$ common	$_x\mathfrak{C}^n$ and $_u\mathfrak{C}^n$
12g		$r^i = \overline{QO}_{ui}$	$_u\mathfrak{C}^n$
13g		$r_i = \overline{QO}_{xi}$	$_x\mathfrak{C}^n$
14g		$\rho^i = \overline{DO}_{x\ ui}$	$_x\mathfrak{C}^n$ and $_u\mathfrak{C}^n$
15g		$x_i \equiv \overline{QQ}_{xi}$	$_x\mathfrak{C}^n$
16g		$u^i \equiv \overline{QQ}_{ui} \equiv \overline{AO}_{ui}$	$_u\mathfrak{C}^n$
17g		$a^i \equiv \overline{AD}_{x\ ui}$	$_x\mathfrak{C}^n$ and $_u\mathfrak{C}^n$
18g		$x_1 \equiv r_1 \equiv x;\ x_2 \equiv y;\ x_3 \equiv z$	$_x\mathfrak{C}^n$
19g		$u_1 \equiv r_1 \equiv u;\ x_2 \equiv v;\ x_3 \equiv w$	$_u\mathfrak{C}^n$
20g		$x_1 \equiv \rho_1;\ a_1 \equiv a;\ u_n \equiv \rho_n$	$_x\mathfrak{C}^n$ and $_u\mathfrak{C}^n$
21g		$1 \equiv \prod_{k=n+1}^{n} \cos^2(\alpha_k)$	

Theorem 13.14.1 Planar Coordinate Axes Trigonometric Relationship for $_x\mathfrak{G}^n$

Steps	Hypothesis	$x_1 = R \prod_{i=2}^{n} \cos(\varphi_{xi})$ | $x_i = R \sin(\varphi_{xi}) \prod_{k=i+1}^{n} \cos(\varphi_{xk})$
1	From F13.14.1A, G3g and DxE.1.2.2 Cosine	$\overline{QO}_{xi} = \overline{RO} \cos(\angle Q_{xy}OR)$
2	From 1, G3g, G11g, G13g and A4.2.3 Substitution	$r_{i+1} = R \cos(\psi_x)$
3	From F13.14.1B G4g and DxE.1.2.2 Cosine	$\overline{QO}_{xi} = \overline{QO}_{xi+1} \cos(\angle Q_iOQ_{xi+1})$
4	From 3, G4g, G13g and A4.2.3 Substitution	$r_i = r_{i+1} \cos(\varphi_{xi})$
5	From F13.14.1B, G3g and DxE.1.2.1 Sine	$\overline{QQ}_{xi} = \overline{QO}_{xi+1} \sin(\angle Q_iOQ_{xi+1})$
6	From 6, G4g, G13g, G15g and A4.2.3 Substitution	$x_i = r_{i+1} \sin(\varphi_{xi})$
7	From F13.14.1B, G4g and DxE.1.2.2 Cosine	$\overline{QO}_{x1} = \overline{QO}_{x2} \cos(\angle Q_1OQ_{x2})$
8	From 7, G4g, G15g, G18g and A4.2.3 Substitution	$x = r_2 \cos(\varphi_{x2})$
9	From F13.14.1B, G4g and DxE.1.2.1 Sine	$\overline{QQ}_{x1} = \overline{QO}_{x2} \sin(\angle Q_1OQ_{x2})$
10	From 9, G4g, G15g, G18g and A4.2.3 Substitution	$y = r_2 \sin(\varphi_{x2})$
11	From 2, (4 and A4.2.3 Substitution) repeat, 18g and A4.2.3 Substitution	$x_1 = R \cos(\varphi_{x2}) \cos(\varphi_{x3}) \ldots \cos(\varphi_{xn})$
12	From 2, (4 and A4.2.3 Substitution) repeat, 6 and A4.2.3 Substitution	$x_i = R \sin(\varphi_{xi}) \cos(\varphi_{xi+1}) \ldots \cos(\varphi_{xn})$ for $1 < i$
∴	From 11, 12 and contracting	$x_1 = R \prod_{i=2}^{n} \cos(\varphi_{xi})$ | $x_i = R \sin(\varphi_{xi}) \prod_{k=i+1}^{n} \cos(\varphi_{xk})$

Theorem 13.14.2 Planar Coordinate Axes Trigonometric Relationship for $_u\mathfrak{G}^n$

Steps	Hypothesis	$u_1 = R \prod_{i=2}^{n} \cos(\varphi_{ui})$ | $u_i = R \sin(\varphi_{ui}) \prod_{k=i+1}^{n} \cos(\varphi_{uk})$
1	From F13.14.1A, G3g and DxE.1.2.2 Cosine	$\overline{QO}_{ui} = \overline{RO} \cos(\angle Q_{uv}OR)$
2	From 1, G6g, G11g, G12g and A4.2.3 Substitution	$r_{i+1} = R \cos(\psi_u)$
3	From F13.14.1B, G10g and DxE.1.2.2 Cosine	$\overline{QO}_{ui} = \overline{QO}_{ui+1} \cos(\angle Q_iOQ_{ui+1})$
4	From 3, G7g, G12g and A4.2.3 Substitution	$r_i = r_{i+1} \cos(\varphi_{ui})$
5	From F13.14.1B, G7g and DxE.1.2.1 Sine	$\overline{QQ}_{ui} = \overline{QO}_{xi+1} \sin(\angle Q_iOQ_{ui+1})$
6	From 6, G7g, G12g, G16g and A4.2.3 Substitution	$u_i = r_{i+1} \sin(\varphi_{ui})$
7	From F13.14.1B, G10g and DxE.1.2.2 Cosine	$\overline{QO}_{u1} = \overline{QO}_{u2} \cos(\angle Q_1OQ_{u2})$
8	From 7, G7g, G16g, G19g and A4.2.3 Substitution	$u = r_2 \cos(\varphi_{u2})$

9	From F11.7.1B, G7g and DxE.1.2.1 Sine	$\overline{QQ}_{u1} = \overline{QO}_{u2} \sin(\angle Q_1OQ_{u2})$
10	From 9, G7g, G16g, G19g and A4.2.3 Substitution	$v = r_2 \sin(\varphi_{u2})$
11	From 2, (4 and A4.2.3 Substitution) repeat, G19g and A4.2.3 Substitution	$u_1 = R \cos(\varphi_{u2}) \cos(\varphi_{u2}) \ldots \cos(\varphi_{un})$
12	From 2, (4 and A4.2.3 Substitution) repeat, 6 and A4.2.3 Substitution	$u_i = R \sin(\varphi_{ui}) \cos(\varphi_{ui+1}) \ldots \cos(\varphi_{un})$ for $1 < i$
∴	From 11, 12 and contracting	$u_1 = R \prod_{i=2}^{n} \cos(\varphi_{ui}) \;\vert\; u_i = R \sin(\varphi_{ui}) \prod_{k=i+1}^{n} \cos(\varphi_{uk})$

Theorem 13.14.3 Planar Coordinate Axes Trigonometric Relationship between ${}_x\mathfrak{G}^n$ and ${}_u\mathfrak{G}^n$

Steps	Hypothesis	$x_1 = u_1 \prod_{i=2}^{n} \cos(\phi_{x\,ui}) \;\vert\; a_i = u_i \sin(\phi_{x\,ui}) \prod_{k=i+1}^{n} \cos(\phi_{x\,uk})$
1	From F13.14.1A, G3g and DxE.1.2.2 Cosine	$\overline{DO}_{x\,ui} = \overline{AO}_{ui} \cos(\angle A_uOD_{x\,ui})$
2	From 1, G9g, G14g, G16g and A4.2.3 Substitution	$\rho_{i+1} = u_i \cos(\psi_{x\,ui})$
3	From F13.14.1B , G10g and DxE.1.2.2 Cosine	$\overline{DO}_{x\,ui} = \overline{DO}_{x\,ui+1} \cos(\angle Q_{x\,ui}OQ_{x\,ui+1})$
4	From 3, G10g, G14g and A4.2.3 Substitution	$\rho_i = \rho_{i+1} \cos(\phi_{x\,ui})$
5	From F13.14.1B, G7g and DxE.1.2.1 Sine	$\overline{AD}_{x\,ui} = \overline{DO}_{x\,ui+1} \sin(\angle Q_{x\,ui}OQ_{x\,ui+1})$
6	From 6, G10g, G14g, G17g and A4.2.3 Substitution	$a_i = \rho_{i+1} \sin(\varphi_{ui})$
7	From F13.14.1B, G10g and DxE.1.2.2 Cosine	$\overline{DO}_{x\,u1} = \overline{DO}_{x\,u2} \cos(\angle Q_{x\,u1}OQ_{x\,u2})$
8	From 7, G10g, G14g, G20g and A4.2.3 Substitution	$x_1 = \rho_2 \cos(\phi_{x\,u1})$
9	From F13.14.1B, G10g and DxE.1.2.1 Sine	$\overline{AD}_{x\,u1} = \overline{DO}_{x\,u2} \sin(\angle Q_{x\,u1}OQ_{x\,u2})$
10	From 9, G10g, G14g, G17g, G20g and A4.2.3 Substitution	$a_1 = \rho_2 \sin(\phi_{x\,u1})$
11	From 2, (4 and A4.2.3 Substitution) repeat, G20g and A4.2.3 Substitution	$x_1 = u_1 \cos(\phi_{x\,u2}) \cos(\phi_{x\,u3}) \ldots \cos(\phi_{x\,un})$
12	From 2, (4 and A4.2.3 Substitution) repeat, 6 and A4.2.3 Substitution	$a_i = u_i \sin(\phi_{x\,ui}) \cos(\phi_{x\,ui+1}) \ldots \cos(\phi_{x\,un})$ for $1 < i$
∴	From 11, 12 and contracting	$x_1 = u_1 \prod_{i=2}^{n} \cos(\phi_{x\,ui}) \;\vert\; a_i = u_i \sin(\phi_{x\,ui}) \prod_{k=i+1}^{n} \cos(\phi_{x\,uk})$

Clearly $[a_i]$ has no real meaning in terms of coordinate position so equation T13.14.3B " Planar Coordinate Axes Trigonometric Relationship between ${}_x\mathfrak{E}^n$ and ${}_u\mathfrak{E}^n$", however equation T13.14.3A "Planar Coordinate Axes Trigonometric Relationship between ${}_x\mathfrak{E}^n$ and ${}_u\mathfrak{E}^n$" provides an exact equation for coordinate $[x_1]$. The coordinate however is independent of any other and there is no reason why it could not represent any other coordinate, hence

$$x_k = u_k \prod_{i=2}^{n} \cos(\phi_{xk\ ui}) \qquad \text{for } k = 1, 2, \ldots, n$$

Equation 13.14.1: Coordinate Transformation between ${}_x\mathfrak{E}^n$ and ${}_u\mathfrak{E}^n$

Theorem 13.14.4 Planar Coordinate First Axes: Unified Angle Relationship

Steps	Hypothesis	$\prod_{i=2}^{n} \cos(\phi_{x1\ ui}) = \prod_{i=2}^{n} \cos(\varphi_{xi}) \sec(\varphi_{ui})$
1	From T13.14.1A Planar Coordinate Axes Trigonometric Relationship for ${}_x\mathfrak{E}^n$	$x_1 = R \prod_{i=2}^{n} \cos(\varphi_{xi})$
2	From A4.2.2A Equality	$\prod_{i=2}^{n} \cos(\varphi_{ui}) = \prod_{i=2}^{n} \cos(\varphi_{ui})$
3	From 1, 2 and T4.4.2 Equalities: Uniqueness of Multiplication	$x_1 \prod_{i=2}^{n} \cos(\varphi_{ui}) = R \prod_{i=2}^{n} \cos(\varphi_{xi}) \prod_{i=2}^{n} \cos(\varphi_{ui})$
4	From 3 and A4.2.10 Commutative Multp	$x_1 \prod_{i=2}^{n} \cos(\varphi_{ui}) = R \prod_{i=2}^{n} \cos(\varphi_{ui}) \prod_{i=2}^{n} \cos(\varphi_{xi})$
5	From 4, T13.14.2A and A4.2.3 Substitution	$x_1 \prod_{i=2}^{n} \cos(\varphi_{ui}) = u^1 \prod_{i=2}^{n} \cos(\varphi_{xi})$
6	From 5, T4.4.6B Equalities: Left Cancellation by Multiplication and LxE.3.1.5 Secant Reciprocal of Cosine	$x_1 = u^1 \prod_{i=2}^{n} \cos(\varphi_{xi}) \prod_{i=2}^{n} \sec(\varphi_{ui})$
7	From 6 and A4.2.11 Associative Multp; contracting the product	$x_1 = u^1 \prod_{i=2}^{n} \cos(\varphi_{xi}) \sec(\varphi_{ui})$
8	From 7, T13.14.3A "Planar Coordinate Axes Trigonometric Relationship between ${}_x\mathfrak{E}^n$ and ${}_u\mathfrak{E}^n$"and A4.2.2A Equality	$u^1 \prod_{i=2}^{n} \cos(\phi_{x1\ ui}) = u^1 \prod_{i=2}^{n} \cos(\varphi_{xi}) \sec(\varphi_{ui})$
∴	From 8 and T4.4.3 Equalities: Cancellation by Multiplication	$\prod_{i=2}^{n} \cos(\phi_{x1\ ui}) = \prod_{i=2}^{n} \cos(\varphi_{xi}) \sec(\varphi_{ui})$

Theorem 13.14.5 Planar Coordinate Axes: Confirmation of the Pathagonian Metric

Steps	Hypothesis	$\sum^n x_i^2 = R^2$		
1	From T13.14.1A Planar Coordinate Axes Trigonometric Relationship for ${}_x\mathfrak{E}^n$, T4.8.6 Integer Exponents: Uniqueness of Exponents, A4.2.21 Distribution Exp, T13.14.1B Planar Coordinate Axes Trigonometric Relationship for ${}_x\mathfrak{E}^n$, T4.8.6 Integer Exponents: Uniqueness of Exponents, A4.2.21 Distribution Exp and sum	$x_1^2 + x_2^2$	=	$R^2 \prod_{k=2}^{n} \cos^2(\varphi_{xi}) +$ $R^2 \sin^2(\varphi_{x2}) \prod_{k=3}^{n} \cos^2(\varphi_{xk})$
2	From 1 and A4.2.14 Distribution	$x_1^2 + x_2^2$	=	$R^2 \prod_{k=3}^{n} \cos^2(\varphi_{xk})$ $(\cos^2(\varphi_{x2}) + \sin^2(\varphi_{x2}))$
3	From 2, LxE.3.1.19A Pythagorean Relations and A4.2.12 Identity Multp	$x_1^2 + x_2^2$	=	$R^2 \prod_{k=3}^{n} \cos^2(\varphi_{xk})$
4	From 3, T 13.14.1A Planar Coordinate Axes Trigonometric Relationship for ${}_x\mathfrak{E}^n$, T4.8.6 Integer Exponents: Uniqueness of Exponents, A4.2.21 Distribution Exp, T13.14.1B Planar Coordinate Axes Trigonometric Relationship for ${}_x\mathfrak{E}^n$, T4.8.6 Integer Exponents: Uniqueness of Exponents, A4.2.21 Distribution Exp and sum	$\sum^2 x_i^2 + x_3^2$	=	$R^2 \prod_{k=3}^{n} \cos^2(\varphi_{xk}) +$ $R^2 \sin(\varphi_{x3}) \prod_{k=4}^{n} \cos^2(\varphi_{xk})$
5	From 4 and A4.2.14 Distribution	$\sum^2 x_i^2 + x_3^2$	=	$R^2 \prod_{k=4}^{n} \cos^2(\varphi_{xk})$ $(\cos^2(\varphi_{x3}) + \sin^2(\varphi_{x3}))$
6	From 5, LxE.3.1.19A Pythagorean Relations and A4.2.12 Identity Multp	$\sum^2 x_i^2 + x_3^2$	=	$R^2 \prod_{k=3}^{n} \cos^2(\varphi_{xk})$
7	From 2, 6 and extrapolating by repeated steps 4 to 6, n-1 times at k=n	$\sum^{n-1} x_i^2$	=	$R^2 \cos^2(\varphi_{xn})$
8	From 7, T 13.14.1A Planar Coordinate Axes Trigonometric Relationship for ${}_x\mathfrak{E}^n$, T4.8.6 Integer Exponents: Uniqueness of Exponents, A4.2.21 Distribution Exp, T13.14.1B Planar Coordinate Axes Trigonometric Relationship for ${}_x\mathfrak{E}^n$, T4.8.6 Integer Exponents: Uniqueness of Exponents, A4.2.21 Distribution Exp and sum	$\sum^{n-1} x_i^2 + x_n^2$	=	$R^2 \prod_{k=n}^{n} \cos^2(\varphi_{xk}) +$ $R^2 \sin^2(\varphi_{xn}) \prod_{k=n+1}^{n} \cos^2(\varphi_{xk})$
9	From 8, evaluate k=n	$\sum^{n-1} x_i^2 + x_n^2$	=	$R^2 \cos^2(\varphi_{xn}) +$ $R^2 \sin^2(\varphi_{xn})$
10	From 9 and A4.2.14 Distribution	$\sum^{n-1} x_i^2 + x_n^2$	=	$R^2 (\cos^2(\varphi_{xn}) + \sin^2(\varphi_{xn}))$
11	From 10, LxE.3.1.19A Pythagorean Relations and A4.2.12 Identity Multp	$\sum^{n-1} x_i^2 + x_n^2$	=	R^2
∴	From 11 and evaluate index	$\sum^n x_i^2 = R^2$		

Section 13.15 Axial Coordinate Point between Manifolds ${}_{x}\mathfrak{C}^{n}$ and ${}_{u}\mathfrak{C}^{n}$

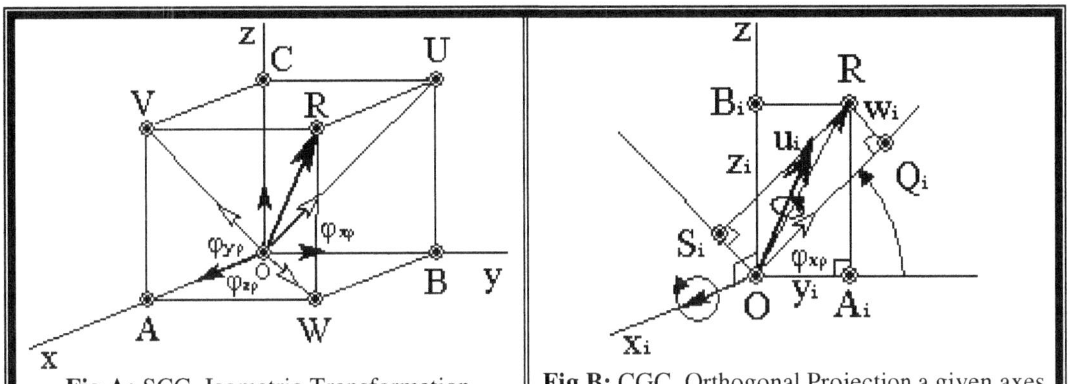

Fig A: SCG, Isometric Transformation

Fig B: CGC, Orthogonal Projection a given axes

Figure 13.15.1: Axial Coordinates

Theorem 13.15.1 Axial Coordinate Point

1g	Given	$y_i = \overline{OA_i}$	
2g		$z_i = \overline{OB_i}$	
3g		$u_i = \overline{OQ_i}$	
4g		$w_i = \overline{OS_i}$	
5g		$r = \overline{OR}$	
6g		$\alpha = \angle ROQ_i$	
Steps	Hypothesis	${}_{x}\mathfrak{C}^{n} \rightarrow {}_{u}\mathfrak{C}^{n}$	${}_{u}\mathfrak{C}^{n} \rightarrow {}_{x}\mathfrak{C}^{n}$
1	From F13.5.1A and B	$\Phi = \sum^{n} \varphi_{xi}\, e^{xi}$	$R = \sum^{n} y_i\, e^{xi} = \sum^{n} w_i\, e^{wi}$
2	From 1g, 3g and F13.15.1B	$u_i = r \cos(\alpha)$	$y_i = r \cos(\alpha + \varphi_{xi})$
3	From 2g, 4g and F13.15.1B	$w_i = r \sin(\alpha)$	$z_i = r \sin(\alpha + \varphi_{xi})$
4	From 2 and LxD.1.24		$y_i = r \cos(\alpha)\sin(\varphi_{xi})$ $- r \sin(\alpha)\sin(\varphi_{xi})$
5	From 3 and LxD.1.22		$z_i = r \sin(\alpha) \cos(\varphi_{xi})$ $+ r \cos(\alpha) \sin(\varphi_{xi})$
6	From 2, 3, 4 and A3.1.3		$y_i = u_i \cos(\varphi_{xi})$ $- w_i \sin(\varphi_{xi})$
7	From 2, 3, 5 and A3.1.3		$z_i = u_i \sin(\varphi_{xi})$ $+ w_i \cos(\varphi_{xi})$
8	From 6, 7 and T5.9.18	$u_i = y_i \cos(\varphi_{xi})$ $+ z_i \sin(\varphi_{xi})$	
9	From 6, 7 and T5.9.18	$w_i = - y_i \sin(\varphi_{xi})$ $+ z_i \cos(\varphi_{xi})$	
∴	From 9	Point $R(\bullet\, u_i, w_i\, \bullet)$	Point $R(\bullet\, y_i, z_i\, \bullet)$

Section 13.16 Oblique Coordinate Point between Manifolds $_{x}\mathfrak{C}^{n}$ and $_{\varphi}\mathfrak{S}^{n}$

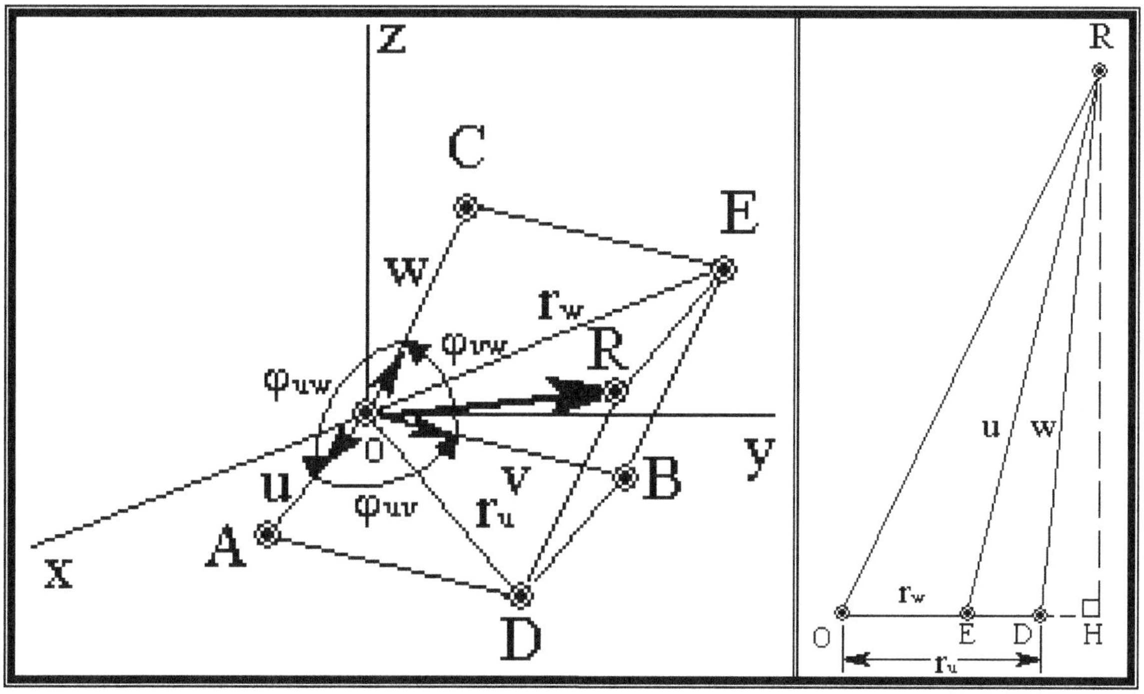

Figure 13.16.1: Skew Coordinates

1g	Given	$\varphi_{uv} \equiv \angle AOB = \angle ADB$	for	OADB parallelogram
2g		$\varphi_{uw} \equiv \angle AOC$		
3g		$\varphi_{vw} \equiv \angle COB = \angle BEC$	for	OBEC parallelogram
4g		$\varphi_{ua} \equiv \angle AOD = \angle ODB$	for	OADB parallelogram
5g		$\varphi_{ub} \equiv \angle DOB = \angle ADO$	for	OADB parallelogram
6g		$\varphi_{wa} \equiv \angle COE = \angle OEB$	for	OBEC parallelogram
7g		$\varphi_{wb} \equiv \angle EOB = \angle OEC$	for	OBEC parallelogram
8g		$u \equiv \overline{OA} = \overline{BD}$	for	OADB parallelogram
9g		$v \equiv \overline{OB} = \overline{AD} = \overline{CE}$	for	OADB and OBEC
10g		$w \equiv \overline{OC} = \overline{BE} = \overline{DR}$	for	OBEC parallelogram
11g		$r_u \equiv \overline{OD}$		
12g		$r_w \equiv \overline{OE}$		
13g		$R \equiv \overline{OR}$		

Theorem 13.16.1 Skewed Coordinate Point

Steps	Hypothesis	
1	From F11.9.1	$\varphi_{uv} \equiv \angle AOD + \angle DOB$
2	From 1, 4g, 5g and A3.1.3	$\varphi_{uv} = \varphi_{ua} + \varphi_{ub}$
3	From F11.9.1	$\varphi_{vw} \equiv \angle COE + \angle EOB$
4	From 3, 6g, 7g and A3.1.3	$\varphi_{vw} = \varphi_{wa} + \varphi_{wb}$
5	Form F11.9.1	$180^o \equiv \angle AOD + \angle OAD + \angle ADO$
6	From 5, 4g, 5g and A3.1.3	$180^o = \varphi_{ua} + \angle OAD + \varphi_{ub}$
7	From 5, A3.2.5, 2 and A3.1.3	$180^o = \varphi_{uv} + \angle OAD$
8	From 7 and T3.3.5B	$\angle OAD = 180^o - \varphi_{uv}$
9	From F11.9.1 and LxD.2.2	$\overline{OD}^2 = \overline{OA}^2 + \overline{AD}^2 - 2\ \overline{OA}\ \overline{AD} \cos(\angle OAD)$
10	From 8, 9, 8g, 9g, 11g and A3.1.3	$r_u^2 = u^2 + v^2 - 2\ uv \cos(180^o - \varphi_{uv})$
11	From 10 and xTD.1.7	$r_u^2 = u^2 + v^2 + 2\ uv \cos(\varphi_{uv})$
12		$\overline{OR}^2 = \overline{OD}^2 + \overline{DR}^2 - 2\ \overline{OD}\ \overline{DR} \cos(\angle ODR)$
		$R^2 = r_u^2 + w^2 - 2\ r_u w \cos(\angle ODR)$
		$R^2 = u^2 + v^2 + 2\ uv \cos(\varphi_{uv}) + w^2 - 2\ r_u w \cos(\angle ODR)$
		$R^2 = u^2 + v^2 + w^2 + 2\ uv \cos(\varphi_{uv}) - 2\ r_u w \cos(\angle ODR)$
		$R^2 = u^2 + v^2 + w^2 + 2\ uv \cos(\varphi_{uv}) + 2\ w$
		$(\ -r_u \cos(\angle ODR)\ )$
		$v \cos(\varphi_{vw})$
		$r_w^2 = v^2 + w^2 + 2\ vw \cos(\varphi_{vw})$
		$r_u^2 + r_w^2 = u^2 + v^2 + 2\ uv \cos(\varphi_{uv}) + v^2 + w^2 + 2\ vw$
		$\cos(\varphi_{vw})$
		$r_u^2 + r_w^2 - v^2 = u^2 + v^2 + w^2 + 2\ uv \cos(\varphi_{uv}) + 2\ vw$
		$\cos(\varphi_{vw})$
		$R^2 = r_u^2 + r_w^2 - v^2$
		$\angle RDH = 180^o - \angle ODR$
		$\angle REH = 180^o - \angle OER$
		$w \cos(\angle RDH) = -w \cos(\angle ODR)$
		$u \cos(\angle REH) = -u \cos(\angle OER)$
$\therefore$		

Section 13.17 Matrix Transformations by Translation $T_a : {}_x\mathfrak{E} \rightarrow {}_y\mathfrak{E}$

Verb: Move, Return, Move Back

Two Euclidian coordinate systems are related to each other by translation iff;

In one dimension.

$$y_1 = x_1 - a_1$$

In two dimensions.

$$y_1 = x_1 - a_1$$
$$y_2 = x_2 - a_2$$

In three dimensions.

$$y_1 = x_1 - a_1$$
$$y_2 = x_2 - a_2$$
$$y_3 = x_3 - a_3$$

and so on for n-dimensions.

$y_i = x_i - a_i$ for i = 1, 2, 3, …, n [PLA86, pg 81]

Equation 13.17.1 Translation Base

where $X(\bullet\ x_i\ \bullet)$, $Y(\bullet\ y_i\ \bullet)$, $A(\bullet\ a_i\ \bullet) \in$ E Euclidian n-space, $X(\bullet\ x_i\ \bullet) \in {}_x\mathfrak{E}$ and $Y(\bullet\ y_i\ \bullet) \in {}_y\mathfrak{E}$ Euclidian manifolds. The point A is the joining constant relationship coupling ${}_x\mathfrak{E}$ to ${}_y\mathfrak{E}$.

From a matrix point of view E9.1.1 is not very practical. With its particular index form it could legitimately be treated as three column vectors;

Y = X – A for (1 x n) matrices.

Equation 13.17.2 Translation Matrix

In order to manipulate column vectors and take advantage of inverse matrices, and such, this is not very satisfying. In the form of ***Projective Coordinates*** (Pj) however this can be made into something a little more practical. Lets look at a two-dimension system. Let

$$A = \begin{pmatrix} 1 & 0 & -a_1 \\ 0 & 1 & -a_2 \\ 0 & 0 & 1 \end{pmatrix} \text{ a (3 x 3) matrix}$$

This is nice because it is invertible.

The relationship can also be written;

$$Y = \begin{pmatrix} y_1 \\ y_2 \\ 1 \end{pmatrix} = \begin{pmatrix} x_1 - a_1 \\ x_2 - a_2 \\ 1 \end{pmatrix} = \begin{pmatrix} 1 & 0 & -a_1 \\ 0 & 1 & -a_2 \\ 0 & 0 & 1 \end{pmatrix} \begin{pmatrix} x_1 \\ x_2 \\ 1 \end{pmatrix} = A\,X$$

This can be natural extrapolated to n-dimensions with (n x n) and all of the nice properties of matrix multiplication and inverting matrices work out.

In general [A] will be renamed to [T] for translation matrix and is symbolically written in shorthand notation as follows for n-dimensions;

$$T = \left(\begin{array}{c|c} I & -a \\ \hline 0 & 1 \end{array}\right) \text{ for ([n+1] x [n+1]) matrix [PLA86, pg 81, 300]}$$

Equation 13.17.3 Translation Projective Coordinates

Where [I] is the identity matrix [a] is the classical column matrix of translation constants, [13.22.1]is just the element one and [0] is a row vector of zeros finishing filling out the matrix.

Any column vectors being translated will have the general form of;

$$Y = \begin{pmatrix} y \\ 1 \end{pmatrix} \text{ for (1 x [n+1]) in } {}_y\mathfrak{C} \text{ and [PLA86, pg 86]}$$

Equation 13.17.4 Translation Y-Projective Coordinates

$$X = \begin{pmatrix} x \\ 1 \end{pmatrix} \text{ for (1 x [n+1]) in } {}_x\mathfrak{C}$$

Equation 13.17.5 Translation X-Projective Coordinates

where the lower case x and y are actually column matrices of n-coordinate variables for their respective manifold.

These kind of matrices have a very import role in computers, specifically in three-dimensions for computer graphics and is why for 3-space they go by the name of ***Graphical Transformation Matrices***. Coupled with scaling (magnification) and rotation they go by another name ***Composite Transformation Matrices*** as various visual objects are operated upon to change shape, move and rotate graphical computer images. [PLA86, pg 80]

Every matrix has its inverse as a ([n+1] x [n+1]) square matrix, however blindly the inverse where taken of the translation matrix it would be a mathematical mess. Where the only thing that needs to be recognized is that the inverse of a translation is simply moving back to the beginning of the translation's origin.

$$T^{-1} = \left(\begin{array}{c|c} I & +a \\ \hline 0 & 1 \end{array}\right) \text{ for ([n+1] x [n+1]) matrix [PLA86, pg 86]}$$

Equation 13.17.6 Translation Inverse Projective Coordinates

Section 13.18 Matrix Transformations by Axial Rotation $R_\varphi : {}_x\mathfrak{C} \rightarrow {}_y\mathfrak{C}$

Verb: Orientate

For any Axial Coordinate system only a single plane is considered rotating through an angle φ about an axis which gives rise to the following matrix. For derivation see Section 11.5 and [PLA86, pg 88].

$y_i = u_i \cos(\varphi_{xi}) - w_i \sin(\varphi_{xi})$ by T11.5.1
$z_i = u_i \sin(\varphi_{xi}) + w_i \cos(\varphi_{xi})$

Let i = 1, 2, 3, 4, 5 and rotation takes place in the (3, 4) plane rotating about 5 by right hand rotation as well as permutation.

$y_1 = x_1\, 1 + x_2\, 0 + x_3\, 0 + x_4\, 0 + x_5\, 0$
$y_2 = x_1\, 0 + x_2\, 1 + x_3\, 0 + x_4\, 0 + x_5\, 0$
$y_3 = x_1\, 0 + x_2\, 0 + x_3 \cos(\varphi_5) - x_4 \sin(\varphi_5) + x_5\, 0$
$y_4 = x_1\, 0 + x_2\, 0 + x_3 \sin(\varphi_5) + x_4 \cos(\varphi_5) + x_5\, 0$
$y_5 = x_1\, 0 + x_2\, 0 + x_3\, 0 + x_4\, 0 + x_5\, 1$

$y_1 = x^1 r_1{}^1 + x^2 r_1{}^2 + x^3 r_1{}^3 + x^4 r_1{}^4 + x^5 r_1{}^5$ for $i \neq 3 \neq 4$
$y_2 = x^1 r_2{}^1 + x^2 r_2{}^2 + x^3 r_2{}^3 + x^4 r_2{}^4 + x^5 r_2{}^5$ for $i \neq 3 \neq 4$
$y_3 = x^1 r_3{}^1 + x^2 r_3{}^2 + x^3 r_3{}^3 + x^4 r_3{}^4 + x^5 r_3{}^5$ for i = j = 3 and for i = 3, j = 4
$y_4 = x^1 r_4{}^1 + x^2 r_4{}^2 + x^3 r_4{}^3 + x^4 r_4{}^4 + x^5 r_4{}^5$ for i = j = 4 and for i = 4, j = 3
$y_5 = x^1 r_5{}^1 + x^2 r_5{}^2 + x^3 r_5{}^3 + x^4 r_5{}^4 + x^5 r_5{}^5$ for $i \neq 3 \neq 4$

Or in general:

$$r_i^j = \begin{cases} \delta_i{}^j & \text{for } i \neq p \neq q \\ +\cos(\varphi_{pxq}) & \text{for } i = j = p \\ -\sin(\varphi_{pxq}) & \text{for } i = p \text{ and } j = q \\ +\sin(\varphi_{pxq}) & \text{for } i = q \text{ and } j = p \\ +\cos(\varphi_{pxq}) & \text{for } i = j = q \end{cases}$$

Equation 13.18.1 Rotation Matrix

$$R_p{}^q = \left(\begin{array}{c|c} r_i{}^j & 0 \\ \hline 0 & 1 \end{array}\right) \text{ for ([n+1] x [n+1]) matrix [PLA86, pg 89]}$$

Equation 13.18.2 Rotation Projective Coordinates

Now for the inverse matrix rotation the process is repeated.

$u_i = \quad y_i \cos(\varphi_{xi}) + z_i \sin(\varphi_{xi})$ by T11.5.1
$w_i = -\, y_i \sin(\varphi_{xi}) + z_i \cos(\varphi_{xi})$

Let i = 1, 2, 3, 4, 5 and rotation takes place in the (3, 4) plane rotating about 5 by right hand rotation by permutation.

$x_1 = y_1\, 1 + y_2\, 0 + y_3\, 0 + y_4\, 0 + y_5\, 0$
$x_2 = y_1\, 0 + y_2\, 1 + y_3\, 0 + y_4\, 0 + y_5\, 0$
$x_3 = y_1\, 0 + y_2\, 0 + y_3 \cos(\varphi_5) + y_4 \sin(\varphi_5) + y_5\, 0$
$x_4 = y_1\, 0 + y_2\, 0 - y_3 \sin(\varphi_5) + y_4 \cos(\varphi_5) + y_5\, 0$
$x_5 = y_1\, 0 + y_2\, 0 + y_3\, 0 + y_4\, 0 + y_5\, 1$

$x_1 = y^1 \rho_1^{\ 1} + y^2 \rho_1^{\ 2} + y^3 \rho_1^{\ 3} + y^4 \rho_1^{\ 4} + y^5 \rho_1^{\ 5}$ for $i \neq 3 \neq 4$
$x_2 = y^1 \rho_2^{\ 1} + y^2 \rho_2^{\ 2} + y^3 \rho_2^{\ 3} + y^4 \rho_2^{\ 4} + y^5 \rho_2^{\ 5}$ for $i \neq 3 \neq 4$
$x_3 = y^1 \rho_3^{\ 1} + y^2 \rho_3^{\ 2} + y^3 \rho_3^{\ 3} + y^4 \rho_3^{\ 4} + y^5 \rho_3^{\ 5}$ for $i = j = 3$ and for $i = 3, j = 4$
$x_4 = y^1 \rho_4^{\ 1} + y^2 \rho_4^{\ 2} + y^3 \rho_4^{\ 3} + y^4 \rho_4^{\ 4} + y^5 \rho_4^{\ 5}$ for $i = j = 4$ and for $i = 4, j = 3$
$x_5 = y^1 \rho_5^{\ 1} + y^2 \rho_5^{\ 2} + y^3 \rho_5^{\ 3} + y^4 \rho_5^{\ 4} + y^5 \rho_5^{\ 5}$ for $i \neq 3 \neq 4$

Or in general:

$$\rho_i^{\ j} = \begin{cases} \delta_i^{\ j} & \text{for } i \neq p \neq q \\ +\cos(\varphi_{pxq}) & \text{for } i = j = p \\ +\sin(\varphi_{pxq}) & \text{for } i = p \text{ and } j = q \\ -\sin(\varphi_{pxq}) & \text{for } i = q \text{ and } j = p \\ +\cos(\varphi_{pxq}) & \text{for } i = j = q \end{cases}$$

Equation 13.18.3 Rotation Inverse Matrix

$$R^{-1}{}_p{}^q = \left(\begin{array}{c|c} \rho_i^{\ j} & 0 \\ \hline 0 & 1 \end{array} \right) \text{ for ([n+1] x [n+1]) matrix}$$

Equation 13.18.4 Rotation Inverse Projective Coordinates

Notice that for the rotation matrix is and its inverse are setup in projective coordinates to be compatible with axial translation.

Hence, every axes contribute to an aspect of rotation such that the net rotation can be described by the angle vector.

$$\bar{\Phi} = \Sigma^n \varphi_i\, \hat{e}^i$$

Equation 13.18.5 Rotation Angle Vector

Section 13.19 Matrix Transformations by Scaling $S_s : {}_x\mathfrak{C} \rightarrow {}_y\mathfrak{C}$

Verb: Magnify, Shape, Demagnify, Distort

Scaling is the process of scaling changes the dimensions of an object. The scale factor [s] determines whether the scaling is magnification, s > 1 or a reduction s < 1.

Scaling takes place with respect to the origin remaining fixed:

$y_i = s_i \, x_i$ for all i

In matrix form this is simply the diagonal S scaling matrix I*s, where.

$$s = \begin{pmatrix} s^1 \\ s^2 \\ \vdots \\ s^i \\ \vdots \\ s^n \end{pmatrix} \text{ for } (1 \times n) \text{ column vector}$$

Equation 13.19.1 Scaling Vector

$$S = \left(\begin{array}{c|c} \text{Diag}(s) & 0 \\ \hline 0 & 1 \end{array}\right) \text{ for } ([n+1] \times [n+1]) \text{ matrix [PLA86, pg 116]}$$

Equation 13.19.2 Scaling Projective Coordinates

The interesting thing is that the inverse matrix creates the inverse magnification of the image.

$$S^{-1} = \left(\begin{array}{c|c} \text{Diag}(s^{-1}) & 0 \\ \hline 0 & 1 \end{array}\right) \text{ for } (1 \times [n+1]) \text{ column vector}$$

Equation 13.19.3 Scaling Inverse Projective Coordinates

$$s^{-1} = \begin{pmatrix} 1/s^1 \\ 1/s^2 \\ \vdots \\ 1/s^i \\ \vdots \\ 1/s^n \end{pmatrix} \text{ for } (1 \times n) \text{ column vector}$$

Equation 13.19.4 Scaling Inverse Vector

as it follows from the inverse of a diagonal matrix C6.9.22.1 "Diagonal Inverse Matrix".

Section 13.20 Coordinate Object in a System $_x\mathfrak{C}$

Definition of a coordinate object.

Let a point in an object be represented by a coordinate column vector in n-space:

$$\pi_i = \begin{pmatrix} p_i^{\ 1} \\ p_i^{\ 2} \\ \vdots \\ p_i^{\ j} \\ \vdots \\ p_i^{\ n} \end{pmatrix} \text{ for } (1 \times n) \text{ column vector}$$

Equation 13.20.1 Object Vector

For backward compatibility with translation π_i is setup in projective coordinates to be compatible with axial translation as follows:

$$p_i = \begin{pmatrix} \pi_i \\ \hline 1 \end{pmatrix} \text{ for } (1 \times [n+1]) \text{ column vector}$$

Equation 13.20.2 Object Vector Projective Coordinates

Let the object be the collection of all points m:

$O_{bj} = (p_1\ p_2 \ldots p_i \ldots p_m)$ for an $(m \times [n+1])$ matrix

Equation 13.20.3 Object Projective Coordinates

Let an object have a coordinate position in a frame at the origin of a coordinate system $_x\mathfrak{C}$ about which all coordinate points are measured from. **Condition 13.21.1**

Section 13.21 Operations on Objects $O_p : {}_x\mathfrak{S} \rightarrow {}_y\mathfrak{S}$

An object can be generally operated on in two ways:

- By placing the object.
- By specifically operating on the object.

The first way, placing the object, can be described in terms of objects as images placed in coordinate systems as pictures. In this guise, the phrases arise to create an ***instance*** of an image and the term follows ***Instance Transformation***. Also, actions describing the transformations are used such as:

Each of these cases are characterized by different matrix concatenated products.

Placing an object is just a straight sequence of matrix transformation products.

$O_{bj}(y) = \prod_k{}^t O_p{}^k O_{bj}(x)$ for an (m x [n+1]) matrix [PLA86, pg 86]

Equation 13.21.1 Operation Instance

The matrix product represents the action taken as verbs. Typical example would be; move: orientate: shape an object.

Here the actions operate on the object and then move it to a new position in the picture.

Specific operations on an object.

$O_{bj}(y) = T^{-1} (\prod_k{}^t O_p{}^k) T O_{bj}(x)$ for an (m x [n+1]) matrix [PLA86, pg 86, 89]

Equation 13.21.2 Operation Specific

The matrix product represents the action taken as verbs.

Here the actions place the object back within its frame of origin, operate on it and then return it leaving it modified but back in the picture in its original position.

As an example take the action; move back: orientate: magnify: return:

$X_T = T O_{bj}(Old)$
$X_R = R X_T$
$X_M = S X_R$
$O_{bj}(New) = T^{-1} X_M$

Upon substitution:

$O_{bj}(New) = T^{-1} (S R) T O_{bj}(Old)$

Section 13.22 Eigenvectors and Eigenvalues [CHE70, pg 35]

Definition 13.22.1 **Column or Row Vector of Eigenvalues**
For the eigenvalues $(\cdot\cdot\, \lambda_i \cdot\cdot\,) \in C^2$

$$\mathbf{l} \equiv \begin{pmatrix} \lambda^1 \\ \lambda^2 \\ \vdots \\ \lambda^j \\ \vdots \\ \lambda^n \end{pmatrix} \text{ or } (\lambda_1\ \lambda_2 \ldots \lambda_i \ldots \lambda_n) \quad \text{for } (1 \times n) \text{ column vector or } (n \times 1) \text{ row vector.}$$

Definition 13.22.2 **Eigenmatrix**

$\Lambda \equiv \mathrm{Diag}(\mathbf{l})$.

Definition 13.22.3 **Eigenvector**

For the eigenvalues $\lambda \in C^2$ of $A \in (C^2)^n$,
$\exists$ a nonzero vector $\mathbf{x}$, $\ni \lambda\, \mathbf{x} = A\, \mathbf{x}$ than $\mathbf{x}$ is called an ***Eigenvector***.

Definition 13.22.4 **Eigenvalue Matrix Generator Function**

$\Delta(\lambda) \equiv I\lambda - A$	Equation A
$\Delta(\lambda) \equiv (\delta_i^{\,j}\, \lambda - a_i^{\,j})$	Equation B

Definition 13.22.5 **Characteristic Eigenpolynomial**

$0 = \lambda^n + \alpha_1 \lambda^{n-1} + \ldots + \alpha_{n-1} \lambda^1 + \alpha_n$	Equation A
$0 = \prod^n (\lambda - \lambda^j)$	Equation B

Definition 13.22.6 **Minimal Polynomial**

A polynomial of finite terms.

Definition 13.22.7 **Jordan Canonical Matrix**

Do the characteristic and minimal Eigenvalue Matrix Generator Functions, type upper triangular, characterize a block diagonal matrix: [LIP68, pg226]

$| \Delta(\lambda) | = \prod^r (\lambda - \lambda^j)^{n(j)}$ for $n(j) = n_j$ Equation A
$| m(\lambda) | = \prod^r (\lambda - \lambda^j)^{m(j)}$ for $m(j) = m_j$ Equation B
$J = (\delta_i^j\, J_i^j)$ for Equation C
$J_i^j = (\lambda^j\, \delta_r^s + \delta_r^{s-1})$ Equation D

where J_i^j is the Jordan Block Matrix for s = 0 cycles to s = n and $\lambda^j \in R$.
Block J_i^j has the following properties:

1) There is at least one J_i^j of order m_i; all other J_i^j are of order $\leq m_i$
2) The sum of the orders of the J_i^j is n_i.
3) The number of J_i^j equals the geometric multiplicity of λ^j.
4) The number of J_i^j of each possible order is uniquely determined by the transformation matrix [T].

From T6.9.33 "Form for the Product of n Diagonal Matrices" diagonal matrices have certain advantage in terms of matrix multiplication not only in many multiplications of by itself, but with any other diagonal matrix. It would be nice if given any non-diagonal square matrix [A] that there exists a transformation matrix [T] to convert [A] to a diagonal matrix [Λ], let this be a property of our new system.

Axiom 13.22.1 Transformation from none to Diagonal Matrix

$\exists$ a matrix Λ, $\ni \Lambda \equiv T^{-1}\, A\, T$ the Eigenmatrix iff Eigen-transformation matrix [T] exists for [A].

Theorem 13.22.1 System of n-Eigenequations

Steps	Hypothesis	$T \Lambda = A T$
1	From A6.2.7A Equality of Matrices	$T = T$
2	From 1, A9.6.1 Transformation from none to Diagonal Matrix and T6.3.4 Uniqueness of Matrix Multiplication	$T \Lambda = T (T^{-1} A T)$
3	From 2 and T6.3.11 Matrix Association for Multiplication	$T \Lambda = (T T^{-1}) A T$
4	From 3 and T6.9.19 Closure with respect to the Identity Matrix for Multiplication	$T \Lambda = I A T$
∴	From 4 and T6.3.13 Matrix Identity left of Multiplication	$T \Lambda = A T$

Theorem 13.22.2 Matrix System by n-Eigenequations

Steps	Hypothesis	$A = T^{-1} \Lambda T$
1	From T9.6.1 System of n-Eigenequations	$T \Lambda = A T$
2	From 1 and A6.2.7A Equality of Matrices	$A T = T \Lambda$
3	From A6.2.7A Equality of Matrices	$T^{-1} = T^{-1}$
4	From 2, 3 and T6.3.4 Uniqueness of Matrix Multiplication	$A T T^{-1} = T \Lambda T^{-1}$
5	From 4, T6.3.11 Matrix Association for Multiplication and T6.9.19 Closure with respect to the Identity Matrix for Multiplication	$A I = T \Lambda T^{-1}$
∴	From 5 and T6.3.12 Matrix Identity right of Multiplication	$A = T \Lambda T^{-1}$

Theorem 13.22.3 Nontrivial Solution for System of n-Eigenequations

Steps	Hypothesis	$\delta_k^{\ j} \lambda^j - a_i^{\ k} = 0$ for each term
1	From T9.6.1 System of n-Eigenequations and D6.1.2 Rectangular Matrix	$(t_i^{\ j})(\delta_i^{\ j} \lambda^j) = (a_i^{\ j})(t_i^{\ j})$
2	From 1 and A6.2.10 Matrix Multiplication	$(\sum_k t_i^{\ k} \delta_k^{\ j} \lambda^j) = (\sum_k a_i^{\ k} t_k^{\ j})$
3	From A6.2.7A Equality of Matrices	$(\sum_k - a_i^{\ k} t_k^{\ j}) = (\sum_k - a_i^{\ k} t_k^{\ j})$
4	From 2, 3 and T6.3.2 Uniqueness of Matrix Addition	$(\sum t_i^{\ k} \delta_k^{\ j} \lambda^j) + (\sum - a_i^{\ k} t_k^{\ j}) = (\sum a_i^{\ k} t_k^{\ j}) + (\sum - a_i^{\ k} t_k^{\ j})$
5	From 4 and A6.2.8 Matrix Addition	$(\sum_k t_i^{\ k} \delta_k^{\ j} \lambda^j - a_i^{\ k} t_k^{\ j}) = (\sum_k a_i^{\ k} t_k^{\ j} - a_i^{\ k} t_k^{\ j})$
6	From 5 and A4.2.10 Inverse Add	$(\sum_k t_i^{\ k} \delta_k^{\ j} \lambda^j - t_k^{\ j} a_i^{\ k}) = (\sum_k 0)$
7	From 6 and the sum of zeros	$(\sum_k t_i^{\ k} \delta_k^{\ j} \lambda^j - t_k^{\ j} a_i^{\ k}) = (0)$
8	From 7 and equating elements	$\sum_k t_i^{\ k} \delta_k^{\ j} \lambda^j - t_k^{\ j} a_i^{\ k} = 0$
9	From 8 and D6.1.6 Identity Matrix	$t_i^{\ j} \lambda^j - \sum_k t_k^{\ j} a_i^{\ k} = 0$
10	From 9 and expanding all terms	$t_1^{\ j}\, 0\, \lambda^j - t_1^{\ j} a_i^{\ 1} + \ldots + t_i^{\ j}\, 1\, \lambda^j - t_i^{\ j} a_i^{\ 1} + \ldots +$ $t_n^{\ j}\, 0\, \lambda^j - t_n^{\ j} a_i^{\ n} = 0$
11	From 10 and D6.1.6 Identity Matrix, summing over k	$\sum_k t_k^{\ j} \delta_k^{\ j} \lambda^j - t_k^{\ j} a_i^{\ k} = 0$
12	From 11 and A4.2.14 Distribution	$\sum_k t_k^{\ j} (\delta_k^{\ j} \lambda^j - a_i^{\ k}) = 0$
∴	From 12 valid for nontrivial solution	$\delta_k^{\ j} \lambda^j - a_i^{\ k} = 0$ for each term

If this system of linear homogeneous equations is to have a nontrivial solution for the t_k^j, then λ^j must be a root of the determent equation, hence [λ] becomes a general variable so that the resulting polynomial can be solved yielding all the [λ^j] roots.

$$| \Delta(\lambda) | = 0$$

Equation 13.22.1 Existence of Eigenvalues

This n^{th} order algebraic equation in [λ] has n-roots, which are known as the ***characteristic*** or ***eigenvalues*** of the matrix A. If these n-roots are distinct we can readily show that the system of equations E9.6.1 "Existence of Eigenvalues" yields a set of [n] linearly independent eigenvectors [t_k^j] for k = 1, 2, … n, and hence a nonsingular matrix T, as required by A9.6.1 "Transformation from none to Diagonal Matrix" to exist. If the roots are not distinct, it may not be possible to determine the desired matrix T.

Clearly if the first vector column of T, t^1 is multiplied and so on the equality must be satisfied:

$$\lambda^1 t^1 = A t^1$$
$$\lambda^2 t^2 = A t^2$$
$$\vdots$$
$$\lambda^n t^n = A t^n$$

and all together

$$(\lambda^1 t^1 | \lambda^2 t^2 | \ldots | \lambda^n t^n) = (A t^1 | A t^2 | \ldots | A t^n)$$

$$T \, \mathrm{Diag}(\mathbf{I}) = A T \text{ from T9.6.1 System of n-Eigenequations}$$

This leads to the notation there must exist vector or a set of linear independent vectors [x] that are satisfied by the eigen-equality. Such vectors are called eigenvectors.

$$\lambda x = A x \qquad \text{(in general)}$$

Equation 13.22.2 Eigenvector Relationship

Theorem 13.22.4 Raising the Eigenmatrix System to an Integer Exponential Power

1g	Given	for 0 < m an integer
Steps	Hypothesis	$A^m = T \Lambda^m T^{-1}$
1	From T9.6.2 Matrix System by n-Eigenequations and T6.3.22 Matrix Uniqueness of Exponents	$A^m = (T \Lambda T^{-1})^m$
2	From 1 and T6.3.21 Matrix Exponential Notation	$A^m = \prod^m *(T \Lambda T^{-1})$
3	From 2 and expanding	$A^m = (T \Lambda T^{-1}) (T \Lambda T^{-1}) (T \Lambda T^{-1}) \ldots (T \Lambda T^{-1}) (T \Lambda T^{-1})$ m times
4	From 3 and T6.3.11 Matrix Association for Multiplication	$A^m = T \Lambda (T^{-1} T) \Lambda (T^{-1} T) \Lambda (T^{-1} T) \ldots (T^{-1} T) \Lambda (T^{-1} T) \Lambda T^{-1}$ m times
5	From 4 and T6.9.19 Closure with respect to the Identity Matrix for Multiplication	$A^m = T \Lambda I \Lambda I \Lambda I \ldots I \Lambda I \Lambda T^{-1}$ m times
6	From 5 and T6.3.12 Matrix Identity right of Multiplication	$A^m = T \Lambda \Lambda \Lambda \ldots \Lambda \Lambda T^{-1}$ m times
7	From 6 and T6.3.11 Matrix Association for Multiplication	$A^m = T (\prod^m *\Lambda) T^{-1}$
∴	From 7 and T6.3.21 Matrix Exponential Notation	$A^m = T \Lambda^m T^{-1}$

Theorem 13.22.5 Raising the Eigen Transformation System to an Integer Exponential Power

1g	Given	for $0 < m$ an integer
Steps	Hypothesis	$\Lambda^m = T^{-1} A^m T$
1	From A6.2.7A Equality of Matrices	$T = T$
2	From T9.6.4 Raising the Eigenmatrix System to an Integer Exponential Power	$A^m = T \Lambda^m T^{-1}$
3	From 1, 2 and T6.3.4 Uniqueness of Matrix Multiplication	$A^m T = T \Lambda^m T^{-1} T$
4	From 3 and T6.3.11 Matrix Association for Multiplication	$A^m T = T \Lambda^m (T^{-1} T)$
5	From 4 and T6.9.19 Closure with respect to the Identity Matrix for Multiplication	$A^m T = T \Lambda^m I$
6	From 5 and T6.3.12 Matrix Identity right of Multiplication	$A^m T = T \Lambda^m$
7	From A6.2.7A Equality of Matrices	$T^{-1} = T^{-1}$
8	From 6, 7 and T6.3.4 Uniqueness of Matrix Multiplication	$T^{-1} A^m T = T^{-1} T \Lambda^m$
9	From 8 and T6.3.11 Matrix Association for Multiplication	$T^{-1} A^m T = (T^{-1} T) \Lambda^m$
10	From 9 and T6.9.19 Closure with respect to the Identity Matrix for Multiplication	$T^{-1} A^m T = I \Lambda^m$
∴	From 10, T6.3.12 Matrix Identity right of Multiplication and A6.2.7B Equality of Matrices	$\Lambda^m = T^{-1} A^m T$

Theorem 13.22.6 Raising Eigen System to the Zero Power

1g 2g	Given	for $0 = m$ an integer $I = A^0 = T \Lambda^0 T^{-1}$	assume $I = \Lambda^0 = T^{-1} A^0 T$
Steps	Hypothesis	$I = A^0 = T \Lambda^0 T^{-1}$	$I = \Lambda^0 = T^{-1} A^0 T$
1	From 1g, T9.6.4 Raising the Eigenmatrix System to an Integer Exponential Power, T9.6.5 Raising the Eigen Transformation System to an Integer Exponential Power and A4.2.3 Substitution	$A^0 = T \Lambda^0 T^{-1}$	$\Lambda^0 = T^{-1} A^0 T$
2	From 1 and C6.9.19.1 Zero Exponential for a Square Matrix	$I = T I T^{-1}$	$I = T^{-1} I T$
3	From 2, T6.3.12 Matrix Identity right of Multiplication and T6.3.13 Matrix Identity left of Multiplication	$I = T T^{-1}$	$I = T^{-1} T$
4	From 3 and T6.9.19 Closure with respect to the Identity Matrix for Multiplication, by identity	$I = I$	$I = I$
∴	From 2g, 4 and by identity	$I = A^0 = T \Lambda^0 T^{-1}$	$I = \Lambda^0 = T^{-1} A^0 T$

Theorem 13.22.7 Eigen System Transform of the Zero Matrix

1g	Given	$0 = T\,0\,T^{-1} = T^{-1}\,0\,T$ assume
Steps	Hypothesis	$0 = T\,0\,T^{-1} = T^{-1}\,0\,T$
1	From 1g	$0 = T\,0\,T^{-1} = T^{-1}\,0\,T$
2	From 1 and T6.3.15 Matrix Zero Left of Multiplication	$0 = T\,0 = T^{-1}\,0$
3	From 1 and T6.3.14 Matrix Zero Right of Multiplication	$0 = 0 = 0$
∴	From 1g, 3 and by identity	$0 = T\,0\,T^{-1} = T^{-1}\,0\,T$

Theorem 13.22.8 Characteristic Eigenmatrix as a Triangular Matrix

1g	Given	$\Delta(\lambda) \equiv (\,\delta_i^{\,j}\,\lambda - a_i^{\,j}\,)$ Characteristic Eigenmatrix
2g		$a_i^{\,j}$ is either a T_U or T_L triangular matrix
Steps	Hypothesis	$\mid \Delta(\lambda) \mid = \prod^n (\lambda - a_j^{\,j})$
1	From 1g and T6.9.1 Uniqueness of a Determinate	$\mid \Delta(\lambda) \mid = \mid \delta_i^{\,j}\,\lambda - a_i^{\,j} \mid$
∴	From 1, 2g and T6.9.20 Determinant of a Triangular Matrix	$\mid \Delta(\lambda) \mid = \prod^n (\lambda - a_j^{\,j})$

Observation 13.22.1 Diagonal Elements as Eigenvalues

It follows from T9.6.8 "Characteristic Eigenmatrix as a Triangular Matrix", the diagonal elements of A are the eigenvalues of A.

$\lambda^j = a_j^{\,j}$
Equation 13.22.3 Solution to the Eigenvalues for a Triangular Matrix.

Theorem 13.22.9 Cayley-Hamilton Characteristic Eigenvalue Matrix Generator Function

Steps	Hypothesis	$0 = \prod^n (\lambda - \lambda^j)$
1	From D9.6.4 Eigenvalue Matrix Generator Function, T9.6.4 Raising the Eigenmatrix System to an Integer Exponential Power and T6.3.1 Substitution of Matrices	$\Delta(\lambda) = I\lambda - T\,\Lambda\,T^{-1}$
2	From 1 and T6.9.19 Closure with respect to the Identity Matrix for Multiplication	$\Delta(\lambda) = TT^{-1}\lambda - T\,\Lambda\,T^{-1}$
3	From 2 and T6.3.12 Matrix Identity right of Multiplication	$\Delta(\lambda) = T\,I\,T^{-1}\lambda - T\,\Lambda\,T^{-1}$
4	From 3 and A6.2.9 Matrix Multiplication by a Scalar	$\Delta(\lambda) = T\,I\lambda\,T^{-1} - T\,\Lambda^m\,T^{-1}$
5	From 4, T6.3.17 Left Distribution for Matrices and T6.3.18 Right Distribution for Matrices	$\Delta(\lambda) = T\,(\,I\lambda - \Lambda\,)\,T^{-1}$
6	From 5 and T6.9.1 Uniqueness of a Determinate	$\lvert\Delta(\lambda)\rvert = \lvert\,T\,(\,I\lambda - \Lambda\,)\,T^{-1}\rvert$
7	From 6, T6.9.27 The product of two matrices can be factored as individual determinants, 1g, and E9.6.1 Existence of Eigenvalues	$0 = \lvert T\rvert\,\lvert I\lambda - \Lambda\rvert\,\lvert T^{-1}\rvert$
8	From 7, A4.2.10 Commutative Multp and A4.2.11 Associative Multp	$0 = (\,\lvert T\rvert\,\lvert T^{-1}\rvert\,)\,\lvert I\lambda - \Lambda\rvert$
9	From 8 and C6.9.27.1 Reciprocal of the Inverse Matrix	$0 = (\,\lvert T\rvert^1\,\lvert T\rvert^{-1}\,)\,\lvert I\lambda - \Lambda\rvert$
10	From 9 and A4.2.19 Difference Exp	$0 = \lvert T\rvert^0\,\lvert I\lambda - \Lambda\rvert$
11	From 10 and A4.2.24 Inverse Exp	$0 = 1\,\lvert I\lambda - \Lambda\rvert$
12	From 11 and A4.2.12 Identity Multp	$0 = \lvert I\lambda - \Lambda\rvert$
∴	From 12 and C6.9.20.1 Determinate of a Diagonal Matrix	$0 = \prod^n (\lambda - \lambda^j)$

Theorem 13.22.10 Cayley-Hamilton Characteristic Eigenpolynomial

1g	Given	$0 = A^n + \alpha_1 A^{n-1} + \ldots + \alpha_{n-1} A^1 + \alpha_n I$
2g	Eigenpolynomial	$0 = \lambda^n + \alpha_1 \lambda^{n-1} + \ldots + \alpha_{n-1} \lambda^1 + \alpha_n$
3g		and matrix [A] convertible to an eigenmatrix
Steps	Hypothesis	$0 = A^n + \alpha_1 A^{n-1} + \ldots + \alpha_{n-1} A^1 + \alpha_n I$ Since [A] is convertible to an Eigenmatrix it satisfies the Eigenpolynomial for every element in [A].
1	From 1g, 3g, T9.6.7 Eigen System Transform of the Zero Matrix, T9.6.2 and T6.3.1 Substitution of Matrices	$T\,0\,T^{-1} = T\,\Lambda^n\,T^{-1} + \alpha_1\,T\,\Lambda^{n-1}\,T^{-1} + \ldots$ $+ \alpha_{n-1}\,T\,\Lambda^1\,T^{-1} + \alpha_n\,T\,I\,T^{-1}$

2	From 1, T6.3.17 Left Distribution for Matrices and T6.3.18 Right Distribution for Matrices	$T\,0\,T^{-1} = T\,(\,\Lambda^{n} + \alpha_1\,\Lambda^{n-1} + \ldots$ $+ \alpha_{n-1}\,\Lambda^{1} + \alpha_n\,I\,)\,T^{-1}$
3	From 2 and A6.2.7A Equality of Matrices	$0 = \Lambda^{n} + \alpha_1\,\Lambda^{n-1} + \ldots + \alpha_{n-1}\,\Lambda^{1} + \alpha_n\,I$
∴	From 3, T6.9.33 Form for the Product of n Diagonal Matrices, A6.2.9 Matrix Multiplication by a Scalar and A6.2.8 Matrix Addition	$0 = \delta_i^{\,j}\,[(\,\lambda^{j}\,)^{n} + \alpha_1\,(\,\lambda^{j}\,)^{n-1} + \ldots + \alpha_{n-1}\,(\,\lambda^{j}\,)^{1} + \alpha_n]$

The Caley-Hamilton theorem is the watershed in matrix algebra, because the implication is so profound. Here a non-matrix operation of a linear polynomial is guaranteed to work on every element in any matrix with the only condition that it be convertible to an eigenmatrix.

This idea can be extrapolated to functions of linear infinite series where the linear finite polynomial is only a limited case. In fact, one could imagine any function represented by a linear infinite series working as long as the matrix it is operating on is convertible to an eigenmatrix. So, the idea of function matrix operators comes into existence.

Theorem 13.22.11 Analytic Eigenfunction Operators

1g	Given	$g(x) = \sum_m c_m\, x^m$ a linear infinite series
2g	let	$A \rightarrow \Lambda$ to an eigenmatrix
3g		$G(A) = \sum_m c_m\, A^m$ a matrix linear infinite series
4g		$G(\Lambda) = (\,\delta_i^{\,j}\, g(\lambda^j)\,)$
Steps	Hypothesis	$G(A) = T\, G(\Lambda)\, T^{-1}$
1	From 2g, 3g, T9.6.4 Raising the Eigenmatrix System to an Integer Exponential Power and T6.3.1 Substitution of Matrices	$G(A) = \sum_m c_m\,(\,T\,\Lambda^m\,T^{-1}\,)$
2	From 1, T6.3.17 Left Distribution for Matrices and T6.3.18 Right Distribution for Matrices	$G(A) = T\,(\,\sum_m c_m\,\Lambda^m\,)\,T^{-1}$
3	From 2, T6.9.33 Form for the Product of n Diagonal Matrices, A6.2.9 Matrix Multiplication by a Scalar and A6.2.8 Matrix Addition	$G(A) = T\,(\delta_i^{\,j} \sum_m c_m\,(\lambda^j)^m\,)\,T^{-1}$
4	From 3 and 1g	$G(A) = T\,(\delta_i^{\,j}\, g(\lambda^j)\,)\,T^{-1}$
∴	From 4 and T6.3.1 Substitution of Matrices	$G(A) = T\, G(\Lambda)\, T^{-1}$

Observation 13.22.2 On use of Eigenmatrices

It is not a necessary requirement that eigenmatrices by used in order to operate on every element within a matrix series, however for the ease of people, and even computers, it is simpler to calculate a diagonal matrix raised to a power than an independent square matrix.

Observation 13.22.3 Convergences of Matrix Function Series

If the series g(x) converges for values of $x \le r$, the radius of convergence for the series, than $g(\lambda^j)$ is also convergent for eigenvalues iff $\lambda^j \le r$ for all j and a necessary requirement for a matrix series to converge. If this is not rigorously followed than some series elements in the matrix will converge, and others not, than function elements behave like landmines blowing up randomly.

Observation 13.22.4 Minimal Matrix Functions

Note that if $c_m = 0$ for all $k < m$ than the series to the right of g(x) is a minimal polynomial, hence the function and its function matrix will be minimal.

Observation 13.22.5 Differentiability of a Matrix Function
If the function g(x) is continuously differentiable with in a domain $x_0 \le x \le x_1$ and the series is convergent within that domain than

$$g^k(x) = \sum_m c_m\, (m!/(m-k)!)\, x^{m-k}$$

Equation 13.22.4 Differential of a Matrix Function

Theorem 13.22.12 Computing Coefficients for a Minimal Polynomial given the Eigenvalues

1g	Given		$g(\lambda) = \sum_k^n c_k \lambda^k$ a minimal polynomial
2g	let		$0 = \prod^m (\lambda - \lambda^j)(\lambda - \lambda^r)^p$ for n = m + p eigenvalues
3g			$G(A) = \sum_k^n c_k A^k$ a matrix minimal polynomial
4g			$G = (g(\lambda^j))$ for a column vector (1 x n)
5g			$C = (c_j)$ for a column vector (1 x n)
6g			$\Gamma = (\gamma_i^j)$ for (n x n)
Steps	Hypothesis		$C = \Gamma^{-1} G$ for G(A)
1		1	$g(\lambda^1) = \sum_k^n c_k (\lambda^1)^k$
2		2	$g(\lambda^2) = \sum_k^n c_k (\lambda^2)^k$
3		⋮	⋮
4		m	$g(\lambda^m) = \sum_k^n c_k (\lambda^m)^k$
5		1	$g(\lambda^r) = \sum_k^n c_k (\lambda^r)^k$
6		2	$g^1(\lambda^r) = \sum_k^n c_k\, k\, (\lambda^r)^{k-1}$
7		⋮	⋮
8		p	$g^p(\lambda^r) = \sum_k^n c_k (k!/(k-p)!)(\lambda^r)^{k-p}$
9	From 1 to 9 provides the matrix equation		$G = \Gamma C$ for 4g, 5g, and 6g
10	From 9 construction of Γ is independent		$\mid\Gamma\mid \ne 0$
11	From 10 and A6.2.7A Equality of Matrices		$\Gamma^{-1} = \Gamma^{-1}$
12	From 11, 9 and T6.3.4 Uniqueness of Matrix Multiplication		$\Gamma^{-1} G = \Gamma^{-1} \Gamma C$
13	From 12 and T6.9.19 Closure with respect to the Identity Matrix for Multiplication		$\Gamma^{-1} G = I C$
14	From 13 and T6.3.13 Matrix Identity left of Multiplication		$\Gamma^{-1} G = C$
∴	From 14, 3g and A6.2.7B Equality of Matrices		$C = \Gamma^{-1} G$ for G(A)

This method telescopes the matrix function from a possible infinite number of terms to a finite number. Also note this method does not require the transformation matrices [T].

Up until now diagonal matrices have been used because of their simplifying properties but there are other matrices that exhibit similar proprieties and are advantageous to use. One such matrix is the Jordan Conical matrix. So far only the Maclaurin type series of no transnational constant has been used, however a version of the Taylor series can also be considered.

Consider a Jordan Canonical Matrix of [J] with the following characteristic polynomial: $| \Delta(\lambda) | = (\lambda - \lambda^1)^n$. With a minimal characteristic polynomial of: $g(\lambda) = \sum_{k=0}^{n-1} c_k (\lambda - \lambda^1)^k$. In T9.6.12 "Computing Coefficients for a Minimal Polynomial given the Eigenvalues" the coefficients where unknown and had to be solved for, in the following proof the coefficients are Taylor coefficients and can easily be calculated. Let $c_k = g^k(\lambda^1) / k!$.

Theorem 13.22.13 Raising the Nilpotent Operator, Jordan Conical Form to a pth Power

1g	Given	$N = (\delta_i^{\,m})$ Nilpotent Operator shifts column by one
2g	for	$m = j - 1$ such that m cycles through 1, 2, … ,n
Steps	Hypothesis	$N^k = (\delta_i^{\,j-k})$
1	From 1g and T6.3.21 Matrix Exponential Notation	$N^2 = (\delta_i^{\,m})\,(\delta_i^{\,m})$
2	From 1, 1g and A6.2.10C Matrix Multiplication	$N^2 = (\sum_k \delta_i^{\,k-1}\,\delta_k^{\,m-1})$ [13.22.1]
3	From 2 and result by A6.2.10C Matrix Multiplication	$N^2 = (\delta_i^{\,m-1})$
4	From 3, 2g and A4.2.3 Substitution	$m - 1 = (j - 1) - 1$
5	From 4 and D4.1.19 Primitive Definition for Rational Arithmetic	$m \equiv j - 2$
6	From 5, 3 and T6.3.21 Matrix Exponential Notation	$N^3 = (\delta_i^{\,m})\,(\delta_i^{\,m})$
7	From 6, 1g and A6.2.10C Matrix Multiplication	$N^3 = (\sum_k \delta_i^{\,k-1}\,\delta_k^{\,m-1})$
8	From 7 and result by A6.2.10C Matrix Multiplication	$N^3 = (\delta_i^{\,m-1})$
9	From 8, 5 and A4.2.3 Substitution	$m - 1 = (j - 2) - 1$
10	From 9 and D4.1.19 Primitive Definition for Rational Arithmetic	$m \equiv j - 3$
11	From 10 by definition let	$m \equiv j - r$
12	From 11 by definition let	$N^{r+1} \equiv (\delta_i^{\,m})\,(\delta_i^{\,m})$
13	From 12, 1g and A6.2.10C Matrix Multiplication	$N^{r+1} = (\sum_k \delta_i^{\,k-1}\,\delta_k^{\,m-1})$
14	From 13 and result by A6.2.10C Matrix Multiplication	$N^{r+1} = (\delta_i^{\,m-1})$
15	From 14, 11 and A4.2.3 Substitution	$m - 1 = (j - r) - 1$
16	From 15 and D4.1.19 Primitive Definition for Rational Arithmetic	$m \equiv j - (r + 1)$
17	From 11 to 16 and A3.11.1 Induction Property of Integers	$N^p = (\delta_i^{\,(j-p+1)-1})$
∴	From 18, D4.1.19 Primitive Definition for Rational Arithmetic, by stratification [ל$_2$]	$N^p = (\delta_i^{\,j-p})$

[13.22.1] Note: Shift in column is accounted for in the summation and product of matrix elements.

Theorem 13.22.14 Jordan-Taylor Analytic Matrix Eigenfunction

1g	Given	$\mid \Delta(\lambda) \mid = (\lambda - \lambda^1)^n$ characteristic polynomial
2g		$g(\lambda) = \Sigma_{k=0}{}^{n-1} (g^k(\lambda^1) / k!) (\lambda - \lambda^1)^k$
3g		$G(J) = \Sigma_{k=0}{}^{n-1} (g^k(\lambda^1) / k!) (J - \lambda^1 I)^k$
Steps	Hypothesis	$G(J) = (\Sigma_{k=0}{}^{n-1} (g^k(\lambda^1) / k!) \delta_i{}^{j-k})$
1	From 3g and expanding the series	$G(J) = g(\lambda^1) I + (g^1(\lambda^1) / 1!) (J - \lambda^1 I)^1$ $+ (g^2(\lambda^1) / 2!) (J - \lambda^1 I)^2 + (g^3(\lambda^1) / 3!) (J - \lambda^1 I)^3$ $+ (g^4(\lambda^1) / 4!) (J - \lambda^1 I)^4 + (g^5(\lambda^1) / 5!) (J - \lambda^1 I)^5$ $\vdots$ $+ (g^{n-1}(\lambda^1) / (n-1)!) (J - \lambda^1 I)^{n-1}$
2	From 1 and D9.6.7D Jordan Canonical Matrix	$J - \lambda^1 I = (\lambda^1 \delta_i{}^j + \delta_i{}^{j-1}) - (\lambda^1 \delta_i{}^j)$
3	From 2 and A6.2.8 Matrix Addition	$J - \lambda^1 I = (\lambda^1 \delta_i{}^j + \delta_i{}^{j-1} - \lambda^1 \delta_i{}^j)$
4	From 3 and A4.2.7 Commutative Add	$J - \lambda^1 I = (\lambda^1 \delta_i{}^j - \lambda^1 \delta_i{}^j + \delta_i{}^{j-1})$
5	From 4, A4.2.8 Associative Add and A4.2.8 Inverse Add	$J - \lambda^1 I = (\delta_i{}^{j-1})$
6	From 5, 1, D6.1.6 Identity Matrix and T6.3.1 Substitution of Matrices	$G(J) = g(\lambda^1) (\delta_i{}^j) + (g^1(\lambda^1) / 1!) (\delta_i{}^{j-1})^1$ $+ (g^2(\lambda^1) / 2!) (\delta_i{}^{j-1})^2 + (g^3(\lambda^1) / 3!) (\delta_i{}^{j-1})^3$ $+ (g^4(\lambda^1) / 4!) (\delta_i{}^{j-1})^4 + (g^5(\lambda^1) / 5!) (\delta_i{}^{j-1})^5$ $\vdots$ $+ (g^{n-1}(\lambda^1) / (n-1)!) (\delta_i{}^{j-1})^{n-1}$
7	From 6 and T9.6.13 Raising the Nilpotent Operator, Jordan Conical Form to a pth Power	$G(J) = g(\lambda^1) (\delta_i{}^j) + (g^1(\lambda^1) / 1!) (\delta_i{}^{j-1})$ $+ (g^2(\lambda^1) / 2!) (\delta_i{}^{j-2}) + (g^3(\lambda^1) / 3!) (\delta_i{}^{j-3})$ $+ (g^4(\lambda^1) / 4!) (\delta_i{}^{j-4}) + (g^5(\lambda^1) / 5!) (\delta_i{}^{j-5})$ $\vdots$ $+ (g^{n-1}(\lambda^1) / (n-1)!) (\delta_i{}^{j-n+1})$
8	From 7, A6.2.9 Matrix Multiplication by a Scalar and A6.2.8 Matrix Addition	$G(J) = ($ $g(\lambda^1) \delta_i{}^j + (g^1(\lambda^1) / 1!) \delta_i{}^{j-1} + (g^2(\lambda^1) / 2!) \delta_i{}^{j-2} +$ $(g^3(\lambda^1) / 3!) \delta_i{}^{j-3} + \ldots + (g^{n-1}(\lambda^1) / (n-1)!) (\delta_i{}^{j-n+1})$
∴	From 8 contracting expansion	$G(J) = (\Sigma_{k=0}{}^{n-1} (g^k(\lambda^1) / k!) \delta_i{}^{j-k})$

Observation 13.22.6 Jordan-Taylor Analytic Matrix is a T_U

T9.6.14 constructs an upper triangular matrix.

Observation 13.22.7 Jordan-Taylor Analytic Matrix Limits

The Jordan-Taylor Analytic Matrix is based on a minimal polynomial yet a true Taylor Series is infinite, however this would give rise to a matrix that would be infinite in size. So the minimal polynomial is only valid where a finite amount of terms is yields a good approximation or converges within the finite number of terms given or upper terms are zero.

Theorem 13.22.15 Time-Invariant Dynamic State Matrix Solution for Eigenvalues

1g	Given [CHE70, pg 69, 151]	
2g		
3g		
Steps	Hypothesis	
1		
2		
3		
4		
5		
6		
7		
8		
∴		

Section 13.23 Closed Form Solution for Scalar Analytic Functions

Tensor Calculus & Physics: A General Treatise

Table of Contents

List of Tables

List of Figures

List of Observations

Chapter 14 Tensor Calculus: Intrinsic Tensors

Section 14.1 Intrinsic First and Second Order Differential

Definition 14.1.1 Intrinsic Differentiation

Intrinsic differentiation is an exact differential with respect to a parametric parameter. In the following examples it is taken to be some form of time in a physical system or geometric arc length in a geometry structure.

Definition 14.1.2 Mixed Riemann-Christoffel tensor

The ***Mixed Riemann-Christoffel tensor*** is given by the expression:

$R^{\alpha}{}_{jki} \equiv (\nabla_j \Gamma_{ik}{}^{\alpha} - \nabla_k \Gamma_{ij}{}^{\alpha}) + (\Gamma_{\beta j}{}^{\alpha} \Gamma_{ik}{}^{\beta} - \Gamma_{\beta k}{}^{\alpha} \Gamma_{ij}{}^{\beta})$

Einsteinian Summation Notation implied, remember Σ_β beta is not a free index.

This tensor also goes by other names the ***Riemann-Christoffel tensor of the second kind*** and because of its direct relationship to the Serret-Frenet curvature, sees T14.1.5B "Equivalent Components of Arc Length and Intrinsic Covariant Vector" and T14.1.11 "Mixed Riemann-Christoffel Tensor as Contravariant Curvature", it is also called the ***Riemannian Curvature Tensor***.

Definition 14.1.3 Riemann-Christoffel Tensor Coefficient of the First Kind

The ***Mixed Riemann to Christoffel Tensor*** **Coefficient** ***of the Second Kind*** transforms to the ***First Kind*** is given by the expression:

$R_{\gamma ijk} \equiv g_{\gamma\alpha} R^{\alpha}{}_{ijk}$ Equation A

Einsteinian Summation Notation implied.

Definition 14.1.4 Symmetric Tensor

A tensor is said to be ***symmetric*** iff $A_{ij} = A_{ji}$ or $A^{ij} = A^{ji}$ any two indices can be transposed and the tensor is left unaltered.

Definition 14.1.5 Asymmetric Tensor

A tensor is said to be ***asymmetric*** if $A_{ij} \neq A_{ji}$ or $A^{ij} \neq A^{ji}$ any two indices are transposed and the tensor is altered.

Definition 14.1.6 Skewed Tensor

A tensor is said to be ***skewed*** iff $A_{ij} = -A_{ji}$ or $A^{ij} = -A^{ji}$ any two indices are transposed resulting in a negative skew of the tensor. Skewed tensors are asymmetric.

Given 14.1.1 For Arc Length verses Intrinsic Covariant Vector

1g	Given	see Figure 12.4.1 Cartesian to Curvilinear Coordinate on a Manifold
2g		$\mathbf{r} \equiv \mathbf{A} = (\bullet x_i \bullet)$
3g		$x_i \equiv x_i(t)$ Function of time for all i

Theorem 14.1.1 First Order Differential Arc Length verses Intrinsic Covariant Vector A

Step	Hypothesis	$d (\mathbf{r}) / dt = [d (s) / dt]\ \mathbf{e}_T$
∴	From G14.1.1;1g, G14.1.1;2g, TK.3.2 Uniqueness of Differentials and T12.12.3B On Space Curve-C: Velocity Vector	$d (\mathbf{r}) / dt = [d (s) / dt]\ \mathbf{e}_T$

Theorem 14.1.2 First Order Differential Arc Length verses Intrinsic Covariant Vector B

Step	Hypothesis	$d (\mathbf{A}) / dt = \sum_i [\partial (\mathbf{A}) / \partial x_i] (dx_i / dt)$
∴	From G14.1.1;1g, G14.1.1;2g, TK.3.2 Uniqueness of Differentials and TK.5.3 Chain Rule for Exact Differential of a Vector	$d (\mathbf{A}) / dt = \sum_i [\partial (\mathbf{A}) / \partial x_i] (dx_i / dt)$

Theorem 14.1.3 Second Order Differential Arc Length verses Intrinsic Covariant Vector A

Step	Hypothesis	$d^2 (\mathbf{r}) / dt^2 = a\ \mathbf{e}^T + \kappa v^2\ \mathbf{e}^N$
∴	From T14.1.1 First Order Differential Arc Length verses Intrinsic Covariant Vector A, TK.3.2 Uniqueness of Differentials and T12.12.6A Space Curve-C: Acceleration Vector	$d^2 (\mathbf{r}) / dt^2 = a\ \mathbf{e}^T + \kappa v^2\ \mathbf{e}^N$

Theorem 14.1.4 Second Order Differential Arc Length verses Intrinsic Covariant Vector B

Step	Hypothesis	$d^2 (\mathbf{A}) / dt^2 = \sum_i (d^2x_i / dt^2) [\partial (\mathbf{A}) / \partial x_i] + \sum_i\sum_j (dx_i / dt) (dx_j /dt) [\partial^2 (\mathbf{A}) / \partial x_i \partial x_j]$
1	From T14.1.2 First Order Differential Arc Length verses Intrinsic Covariant Vector B, TK.3.2 Uniqueness of Differentials and TK.5.1 Derivative Distribution Across Vector Components	$d^2 (\mathbf{A}) / dt^2 = \sum_i d \{[\partial (\mathbf{A}) / \partial x_i] (dx_i / dt)\} /dt$
2	From 1 and LxK.5.1.4 Differentiation of Scalar and Vector	$d^2 (\mathbf{A}) / dt^2 = \sum_i (d^2x_i / dt^2) [\partial (\mathbf{A}) / \partial x_i] + (dx_i / dt)d [\partial (\mathbf{A}) / \partial x_i] /dt$
3	From 2 and TK.5.3 Chain Rule for Exact Differential of a Vector	$d^2 (\mathbf{A}) / dt^2 = \sum_i (d^2x_i / dt^2) [\partial (\mathbf{A}) / \partial x_i] + \sum_i (dx_i / dt) \sum_j \partial [\partial (\mathbf{A}) / \partial x_i] / \partial x_j (dx_j /dt)$
∴	From 3, A4.2.10 Commutative Multp, DxK.3.2.2E Multiple Partial Differential Operator and grouping terms	$d^2 (\mathbf{A}) / dt^2 = \sum_i (d^2x_i / dt^2) [\partial (\mathbf{A}) / \partial x_i] + \sum_i\sum_j (dx_i / dt) (dx_j /dt) [\partial^2 (\mathbf{A}) / \partial x_i \partial x_j]$

Theorem 14.1.5 Equivalent Components of Arc Length and Intrinsic Covariant Vector

1g	Given	$v_i \equiv d\, x_i / dt$	
2g		$v_j \equiv d\, x_j / dt$	
3g		$a_i \equiv d^2x_i / dt^2$	
Step	Hypothesis	$a\, \mathbf{e}^T = \sum_i a_i [\, \partial (\mathbf{A}) / \partial x_i \,]$ EQ A acceleration	$\kappa v^2\, \mathbf{e}^N = \sum_i \sum_j v_i\, v_j [\, \partial^2 (\mathbf{A}) / \partial x_i\, \partial x_j]$ EQ B curvature
1	From G14.1.1;2g	$d^2 (\mathbf{A}) / dt^2{}_{\text{arc-length}} = d^2 (\mathbf{A}) / dt^2{}_{\text{Intrinsic}}$	
2	From 1, T14.1.3 Second Order Differential Arc Length verses Intrinsic Covariant Vector A and T14.1.4 Second Order Differential Arc Length verses Intrinsic Covariant Vector B	$a\, \mathbf{e}^T + \kappa v^2\, \mathbf{e}^N = \sum_i (d^2x_i / dt^2) [\, \partial (\mathbf{A}) / \partial x_i \,] + \sum_i \sum_j (dx_i / dt) (dx_j /dt) [\, \partial^2 (\mathbf{A}) / \partial x_i\, \partial x_j]$	
∴	From 1g, 2g, 3g, 2 and A4.2.3 Substitution; acceleration and curvature are mutally exclusive beginning orthogonal, so equating terms by dimension and units can be seperated.	$a\, \mathbf{e}^T = \sum_i a_i [\, \partial (\mathbf{A}) / \partial x_i \,]$ EQ A acceleration	$\kappa v^2\, \mathbf{e}^N = \sum_i \sum_j v_i\, v_j [\, \partial^2 (\mathbf{A}) / \partial x_i\, \partial x_j]$ EQ B curvature

Theorem 14.1.6 Directivity of Curvature Along a Manifold Path

1g	Given	$\mathrm{v}_i \equiv v_i / v$ velocity directional numbers along i-path
2g		$\mathrm{v}_j \equiv v_j / v$ velocity directional numbers along j-path
Step	Hypothesis	$\kappa\, \mathbf{e}^N = \sum_i \sum_j \mathrm{v}_i\, \mathrm{v}_j [\, \partial^2 (\mathbf{A}) / \partial x_i\, \partial x_j]$
1	From T14.1.5B Equivalent Components of Arc Length and Intrinsic Covariant Vectors	$\kappa v^2\, \mathbf{e}^N = \sum_i \sum_j v_i\, v_j [\, \partial^2 (\mathbf{A}) / \partial x_i\, \partial x_j]$
2	From 1, T4.4.6B Equalities: Left Cancellation by Multiplication, A4.2.14 Distribution, D4.1.17 Exponential Notation, A4.2.10 Commutative Multp and A4.2. Associative Multp	$\kappa\, \mathbf{e}^N = \sum_i \sum_j (v_i / v) (v_j / v) [\, \partial^2 (\mathbf{A}) / \partial x_i\, \partial x_j]$
∴	From 1g, 2g, A4.2.3 Substitution	$\kappa\, \mathbf{e}^N = \sum_i \sum_j \mathrm{v}_i\, \mathrm{v}_j [\, \partial^2 (\mathbf{A}) / \partial x_i\, \partial x_j]$

Observation 14.1.1 About Intrinsic Curvature

From the above theorem T14.1.6 "Directivity of Curvature Along a Manifold Path" the curvature is built from the intrinsic covariant vector in the [i] and [j] gradient direction, along the coordinates on the manifold. In fact with [v_i] and [v_j] the curvature is directed by these velocity directional numbers and is normal at all points on the manifold.

Now the differentials can be taken in either direction, because they are symmetrically commutative as an operator. If this is true, then a transpose of indices or change in the gradient direction should not alter the curvature over the manifold? If false however then subtracting the curvature equation with respect to its transpose indices would reveal them to be non-commutative asymmetrical or skewed.

Theorem 14.1.7 Changing Direction Changes Curvature

1g	Given	$\Delta\kappa \equiv \kappa_\alpha - \kappa_\beta$ for $\kappa_\alpha > \kappa_\beta$ and $\mathbf{A}_{,i,j} > \mathbf{A}_{,j,i}$
2g		$\kappa_\alpha\, \mathbf{e}^N = \sum_i\sum_j v_i\, v_j\, \mathbf{A}_{,i,j}$
3g		$\kappa_\beta\, \mathbf{e}^N = \sum_i\sum_j v_j\, v_i\, \mathbf{A}_{,j,i}$
Step	Hypothesis	$\Delta\kappa\, \mathbf{e}^N = \sum_i\sum_j v_i\, v_j\, (\mathbf{A}_{,i,j} - \mathbf{A}_{,j,i})$ for $\kappa_\alpha > \kappa_\beta$ and $\mathbf{A}_{,i,j} > \mathbf{A}_{,j,i}$
1	From 2g, 3g and T7.11.8 Vector Addition: Uniqueness of Subtraction	$\kappa_\alpha\, \mathbf{e}^N - \kappa_\beta\, \mathbf{e}^N = \sum_i\sum_j v_i\, v_j\, \mathbf{A}_{,i,j} - \sum_i\sum_j v_j\, v_i\, \mathbf{A}_{,j,i}$
2	From 1, T7.10.9 Distribution of Vector over Addition of Scalars, A4.2.10 Commutative Multp and T7.10.3 Distribution of a Scalar over Addition of Vectors	$(\kappa_\alpha - \kappa_\beta)\, \mathbf{e}^N = \sum_i\sum_j v_i\, v_j\, (\mathbf{A}_{,i,j} - \mathbf{A}_{,j,i})$
∴	From 1g, 2 and A4.2.3 Substitution	$\Delta\kappa\, \mathbf{e}^N = \sum_i\sum_j v_i\, v_j\, (\mathbf{A}_{,i,j} - \mathbf{A}_{,j,i})$ for $\kappa_\alpha > \kappa_\beta$ and $\mathbf{A}_{,i,j} > \mathbf{A}_{,j,i}$

Observation 14.1.2 A Matter of Curvature

There are two possible major cases:
Case I: Change in curvature is zero

$\Delta\kappa = 0$ Equation A

Case II: Change in curvature is not zero

$\Delta\kappa \neq 0$ Equation B

Theorem 14.1.8 Zero Change of Curvature over a Manifold

1g	Given	Einsteinian Summation Notation does not apply.
Step	Hypothesis	$\mathbf{A}_{,i,j} = \mathbf{A}_{,j,i}$ for $\Delta\kappa = 0$
1	From O14.1.2A A Matter of Curvature, T14.1.7 Changing Direction Changes Curvature and A4.2.3 Substitution	$0\, \mathbf{e}^N = \sum_i\sum_j v_i\, v_j\, (\mathbf{A}_{,i,j} - \mathbf{A}_{,j,i})$
2	From 1, since the directional numbers are not necessarily zero, then term-by-term the differences by the covariant vectors must be zero.	$\mathbf{0} = \mathbf{A}_{,i,j} - \mathbf{A}_{,j,i}$ for $\Delta\kappa = 0$
∴	From 2 and T7.11.5A Vector Addition: Reversal of Right Cancellation by Addition	$\mathbf{A}_{,i,j} = \mathbf{A}_{,j,i}$ for $\Delta\kappa = 0$

Observation 14.1.3 Zero Change of Curvature and Euclidian Space

From the above theorem T14.1.8 "Zero Change of Curvature over a Manifold" the intrinsic covariant vectors are symmetrical, and the symmetrical gradients over the manifold are commutative and curvature is constant and does not change.

For any two directions that have a zero change in curvature then they constitute a flat surface. If all paired directions have the same curvature, being zero then the space is Euclidian.

Theorem 14.1.9 Evaluating the Second Order Intrinsic Contravariant Vector

1g	Given	Einsteinian Summation Notation does not apply.
Step	Hypothesis	$\mathbf{A}_{,i,j} = \sum_i A_{k,i,j}\, \mathbf{b}^k - \sum_k \sum_\alpha A_{\alpha,j}\, \Gamma_{ki}{}^\alpha\, \mathbf{b}^k - \sum_k \sum_\alpha A_\alpha\, \nabla_j \Gamma_{ki}{}^\alpha\, \mathbf{b}^k$ $- \sum_k \sum_\alpha A_{\alpha,i}\, \Gamma_{kj}{}^\alpha\, \mathbf{b}^k + \sum_k \sum_\beta \sum_\alpha A_\alpha\, \Gamma_{\beta i}{}^\alpha \Gamma_{kj}{}^\beta\, \mathbf{b}^k$
1	From D12.1.2 Contravariant Vector	$\mathbf{A} = A_k \mathbf{b}^k$
2	From 1, TK.3.2 Uniqueness of Differentials and T12.6.3B Differentiation of Tensor: Rank-1 Contravariant	$\mathbf{A}_{,i} = \sum_k (A_{k,i} - \sum_\alpha A_\alpha\, \Gamma_{ki}{}^\alpha)\, \mathbf{b}^k$
3	From 2 and TK.3.2 Uniqueness of Differentials	$\mathbf{A}_{,i,j} = \partial\, [\sum_k (A_{k,i} - \sum_\alpha A_\alpha\, \Gamma_{ki}{}^\alpha)\, \mathbf{b}^k] / \partial x_j$
4	From 3, TK.5.1 Derivative Distribution Across Vector Components and T12.5.2 Derivative of a Contravariant Vector Under Change	$\mathbf{A}_{,i,j} = \sum_k \partial (A_{k,i} - \sum_\alpha A_\alpha\, \Gamma_{ki}{}^\alpha) / \partial x_j\, \mathbf{b}^k$ $+ \sum_k (A_{k,i} - \sum_\alpha A_\alpha\, \Gamma_{ki}{}^\alpha)\, \partial(\mathbf{b}^k) / \partial x_j$
5	From 4, LxK.3.2.2B Differential as a Linear Operator, LxK.3.2.4B Distributed Across Products and T12.5.6 Contravariant Derivative: Contravariant Bases Vector	$\mathbf{A}_{,i,j} = \sum_k A_{k,i,j}\, \mathbf{b}^k - \sum_k \sum_\alpha (A_{\alpha,j}\, \Gamma_{ki}{}^\alpha + A_\alpha\, \nabla_j \Gamma_{ki}{}^\alpha)\, \mathbf{b}^k$ $+ \sum_k (A_{k,i} - \sum_\alpha A_\alpha\, \Gamma_{ki}{}^\alpha)\, \sum_\beta (-\Gamma_{\beta j}{}^k\, \mathbf{b}^\beta)$
6	From 5, T7.10.9 Distribution of Vector over Addition of Scalars, D4.1.20 Negative Coefficient and D4.1.19 Primitive Definition for Rational Arithmetic	$\mathbf{A}_{,i,j} = \sum_k A_{k,i,j}\, \mathbf{b}^k - \sum_k \sum_\alpha (A_{\alpha,j}\, \Gamma_{ki}{}^\alpha + A_\alpha\, \nabla_j \Gamma_{ki}{}^\alpha)\, \mathbf{b}^k$ $+ \sum_k (-\sum_\beta A_{k,i}\, \Gamma_{\beta j}{}^k\, \mathbf{b}^\beta + \sum_\alpha \sum_\beta A_\alpha\, \Gamma_{ki}{}^\alpha \Gamma_{\beta j}{}^k\, \mathbf{b}^\beta)$
7	From 6 and summing term-by-term reorders summation	$\mathbf{A}_{,i,j} = \sum_k A_{k,i,j}\, \mathbf{b}^k - \sum_k \sum_\alpha A_{\alpha,j}\, \Gamma_{ki}{}^\alpha\, \mathbf{b}^k - \sum_k \sum_\alpha A_\alpha\, \nabla_j \Gamma_{ki}{}^\alpha\, \mathbf{b}^k$ $- \sum_k \sum_\beta A_{k,i}\, \Gamma_{\beta j}{}^k\, \mathbf{b}^\beta + \sum_k \sum_\alpha \sum_\beta A_\alpha\, \Gamma_{ki}{}^\alpha \Gamma_{\beta j}{}^k\, \mathbf{b}^\beta$
8	From 7 and substituting dummy indices last two terms k → r and β → k	$\mathbf{A}_{,i,j} = \sum_k A_{k,i,j}\, \mathbf{b}^k - \sum_k \sum_\alpha A_{\alpha,j}\, \Gamma_{ki}{}^\alpha\, \mathbf{b}^k - \sum_k \sum_\alpha A_\alpha\, \nabla_j \Gamma_{ki}{}^\alpha\, \mathbf{b}^k$ $- \sum_r \sum_i A_{r,i}\, \Gamma_{kj}{}^r\, \mathbf{b}^k + \sum_r \sum_\alpha \sum_k A_\alpha\, \Gamma_{ri}{}^\alpha \Gamma_{kj}{}^r\, \mathbf{b}^k$
9	From 8 and substituting dummy indices last term α → β and r → α	$\mathbf{A}_{,i,j} = \sum_k A_{k,i,j}\, \mathbf{b}^k - \sum_k \sum_\alpha A_{\alpha,j}\, \Gamma_{ki}{}^\alpha\, \mathbf{b}^k - \sum_k \sum_\alpha A_\alpha\, \nabla_j \Gamma_{ki}{}^\alpha\, \mathbf{b}^k$ $- \sum_\alpha \sum_i A_{\alpha,i}\, \Gamma_{kj}{}^\alpha\, \mathbf{b}^k + \sum_\alpha \sum_\beta \sum_k A_\beta\, \Gamma_{\alpha i}{}^\beta \Gamma_{kj}{}^\alpha\, \mathbf{b}^k$
∴	From 9 and substituting dummy indices last term α → r, β → α and r → β	$\mathbf{A}_{,i,j} = \sum_i A_{k,i,j}\, \mathbf{b}^k - \sum_k \sum_\alpha A_{\alpha,j}\, \Gamma_{ki}{}^\alpha\, \mathbf{b}^k - \sum_k \sum_\alpha A_\alpha\, \nabla_j \Gamma_{ki}{}^\alpha\, \mathbf{b}^k$ $- \sum_k \sum_\alpha A_{\alpha,i}\, \Gamma_{kj}{}^\alpha\, \mathbf{b}^k + \sum_k \sum_\beta \sum_\alpha A_\alpha\, \Gamma_{\beta i}{}^\alpha \Gamma_{kj}{}^\beta\, \mathbf{b}^k$

Theorem 14.1.10 Vector Relation to a Contravariant Riemann-Christoffel Tensor

1g	Given	Einsteinian Summation Notation does apply.
Step	Hypothesis	$\mathbf{A},i,j - \mathbf{A},j,i = \sum_k \sum_\alpha A_\alpha R^\alpha{}_{ijk}\, \mathbf{b}^k$
1	From T14.1.9 Evaluating the Second Order Intrinsic Contravariant Vector	$\mathbf{A},i,j = \sum_k A_{k,i,j}\, \mathbf{b}^k - \sum_k \sum_\alpha A_{\alpha,j}\, \Gamma_{ki}{}^\alpha\, \mathbf{b}^k - \sum_k \sum_\alpha A_\alpha\, \nabla_j \Gamma_{ki}{}^\alpha\, \mathbf{b}^k - \sum_k \sum_\alpha A_{\alpha,i}\, \Gamma_{kj}{}^\alpha\, \mathbf{b}^k + \sum_k \sum_\beta \sum_\alpha A_\alpha\, \Gamma_{\beta i}{}^\alpha \Gamma_{kj}{}^\beta\, \mathbf{b}^k$
2	From 1 and transpose derivative indices (i, j) → (j, i)	$\mathbf{A},j,i = \sum_k A_{k,j,i}\, \mathbf{b}^k - \sum_k \sum_\alpha A_{\alpha,i}\, \Gamma_{kj}{}^\alpha\, \mathbf{b}^k - \sum_k \sum_\alpha A_\alpha\, \nabla_i \Gamma_{kj}{}^\alpha\, \mathbf{b}^k - \sum_k \sum_\alpha A_{\alpha,j}\, \Gamma_{ki}{}^\alpha\, \mathbf{b}^k + \sum_k \sum_\beta \sum_\alpha A_\alpha\, \Gamma_{\beta j}{}^\alpha \Gamma_{ki}{}^\beta\, \mathbf{b}^k$
3	From 1, 2, T7.11.8 Vector Addition: Uniqueness of Subtraction, D4.1.20 Negative Coefficient, A7.10.3 Distribution of a Scalar over Addition of Vectors, D4.1.19 Primitive Definition for Rational Arithmetic, T7.11.3 Vector Addition: Inverse and T7.11.2 Vector Addition: Identity	$\mathbf{A},i,j - \mathbf{A},j,i = \sum_k \sum_\alpha A_\alpha\, (\nabla_i \Gamma_{kj}{}^\alpha - \nabla_j \Gamma_{ki}{}^\alpha)\, \mathbf{b}^k + \sum_k \sum_\alpha \sum_\beta A_\alpha\, (\Gamma_{\beta i}{}^\alpha \Gamma_{kj}{}^\beta - \Gamma_{\beta j}{}^\alpha \Gamma_{ki}{}^\beta)\, \mathbf{b}^k$
4	From 3 and pairing term-by-term factors out $\sum_\beta$, A4.2.5 Commutative Add and A4.2.6 Associative Add	$\mathbf{A},i,j - \mathbf{A},j,i = \sum_k \sum_\alpha A_\alpha\, (\nabla_i \Gamma_{kj}{}^\alpha - \nabla_j \Gamma_{ki}{}^\alpha)\, \mathbf{b}^k + \sum_k \sum_\alpha A_\alpha\, [\sum_\beta (\Gamma_{\beta i}{}^\alpha \Gamma_{kj}{}^\beta - \Gamma_{\beta j}{}^\alpha \Gamma_{ki}{}^\beta)]\, \mathbf{b}^k$
5	From 4 and T7.10.3 Distribution of a Scalar over Addition of Vectors	$\mathbf{A},i,j - \mathbf{A},j,i = \sum_k \sum_\alpha A_\alpha\, [\, (\nabla_i \Gamma_{kj}{}^\alpha - \nabla_j \Gamma_{ki}{}^\alpha) + \sum_\beta (\Gamma_{\beta i}{}^\alpha \Gamma_{kj}{}^\beta - \Gamma_{\beta j}{}^\alpha \Gamma_{ki}{}^\beta)]\, \mathbf{b}^k$
∴	From 5, D14.1.2 Mixed Riemann-Christoffel tensor, T10.2.5 Tool: Tensor Substitution and A10.2.14 Correspondence of Tensors and Tensor Coefficients	$\mathbf{A},i,j - \mathbf{A},j,i = \sum_k \sum_\alpha A_\alpha R^\alpha{}_{ijk}\, \mathbf{b}^k$

Observation 14.1.4 The Covariant Riemann-Christoffel Tensor

From theorem T14.1.10 "Vector Relation to a Contravariant Riemann-Christoffel Tensor" the internal quantity $[(\nabla_i \Gamma_{kj}{}^\alpha - \nabla_j \Gamma_{ki}{}^\alpha) + \sum_\beta (\Gamma_{\beta i}{}^\alpha \Gamma_{kj}{}^\beta - \Gamma_{\beta j}{}^\alpha \Gamma_{ki}{}^\beta)]$ is a covariant tensor coefficient with four free indices. Summing over $[\beta]$ contracts the index making it inoperative, so as not to be used as a free index. The new tensor will be a mixed coefficient tensor of rank-4 in the form $R^\alpha{}_{ijk}$. This new coefficient tensor is independent of the mixed tensor coefficient and the contravariant bases vector, just like the Christoffel tensor comprising it and acting as a unique operator over the Riemannian space. For these reasons it takes on the name, "Mixed Riemann-Christoffel Tensor" or the "Riemann-Christoffel Tensor of the Second Kind", see D14.1.2.

As an operator the Mixed Riemann-Christoffel Tensor allows the vector coefficients to be factored out independently leaving behind $[R^\alpha{}_{ijk}]$ a quantity that represents a pure rendering of the geometry of the space that it is in.

Theorem 14.1.11 Mixed Riemann-Christoffel Tensor as Contravariant Curvature

1g	Given	Einsteinian Summation Notation does not apply.
Step	Hypothesis	$\Delta\kappa\, \mathbf{e}^N = \sum_i \sum_j \sum_k \sum_\alpha v_i\, v_j\, A_\alpha\, R^\alpha{}_{ijk}\, \mathbf{b}^k$
1	From T14.1.10 Vector Relation to a Contravariant Riemann-Christoffel Tensor	$\mathbf{A}_{,i,j} - \mathbf{A}_{,j,i} = \sum_k \sum_\alpha A_\alpha\, R^\alpha{}_{ijk}\, \mathbf{b}^k$
2	From 1, 14.1.7 Changing Directions Changes Curvature and A7.9.7 Transitivity	$\Delta\kappa\, \mathbf{e}^N = \sum_i \sum_j v_i\, v_j \sum_k \sum_\alpha A_\alpha\, R^\alpha{}_{ijk}\, \mathbf{b}^k$
∴	From 2 and T7.10.3 Distribution of a Scalar over Addition of Vectors	$\Delta\kappa\, \mathbf{e}^N = \sum_i \sum_j \sum_k \sum_\alpha v_i\, v_j\, A_\alpha\, R^\alpha{}_{ijk}\, \mathbf{b}^k$

Observation 14.1.5 When Curvature is Not Zero

From O14.1.2 "A Matter of Curvature" Case II was explored in theorem T14.1.8 "Zero Change of Curvature over a Manifold" and the intrinsic differential vectors were found to be symmetrical, but now with the mixed Riemann-Christoffel tensor the second Case II can now be explored. For $\Delta\kappa \neq 0$ then from T14.1.11 "Mixed Riemann-Christoffel Tensor as Contravariant Curvature" the equation cannot vanish everywhere in space and $[R^\alpha{}_{ijk}]$ is seen to be asymmetrical or a skewed asymmetric tensor, thereby giving rise to positive or negative curvature.

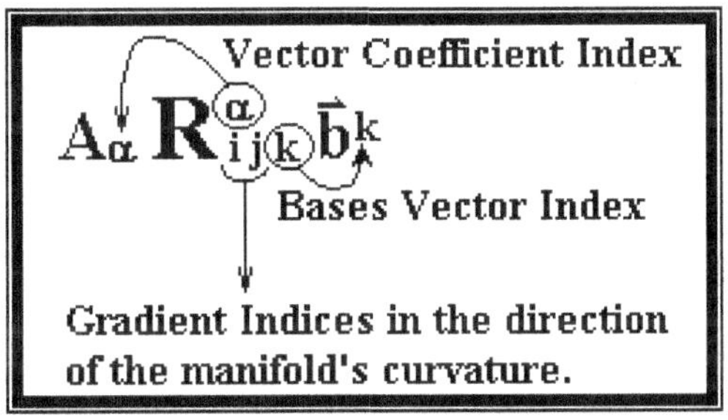

Figure 14.1.1 The Symbol Indices of the Mixed Riemann-Christoffel Tensor

The reason for Figure 14.1.1 is that many authors writing about tensor calculus immediately forget what the indices represent, having derived the Riemann-Christoffel Tensor coefficient, and blindly plow on using it out of context, and sometimes come to some very erroneous conclusions. It is anchored quiet solidly in the geometry of the manifold, therefore is very important to remember what it means in the interrelated conditions of its use.

Table 14.1.1 Correspondence of Curvature over a Manifold in Riemannian Space

O14.1.2 Expanded			**Case I**	**Case II**	**Case III**
A	**Change in Curvature**	**Rank-0**	$\Delta\kappa < 0$	$\Delta\kappa = 0$	$\Delta\kappa > 0$
B	**Change in Curvature Vector**	**Rank-1**	$\Delta\kappa\, \mathbf{e}^n < \mathbf{0}$	$\Delta\kappa\, \mathbf{e}^n = \mathbf{0}$	$\Delta\kappa\, \mathbf{e}^n > \mathbf{0}$
C	**Change in Intrinsic Vector**	**Rank-1**	$\mathbf{A}_{,i,j} - \mathbf{A}_{,j,i} < \mathbf{0}$	$\mathbf{A}_{,i,j} - \mathbf{A}_{,j,i} = \mathbf{0}$	$\mathbf{A}_{,i,j} - \mathbf{A}_{,j,i} > \mathbf{0}$
D	**Mixed Riemann-Christoffel tensor**	**Rank-4**	$R^\alpha{}_{ijk} < 0$	$R^\alpha{}_{ijk} = 0$	$R^\alpha{}_{ijk} > 0$

Observation 14.1.6 Riemann-Christoffel Curvature

It is seen from table T14.1.1 "Correspondence of Curvature over a Manifold in Riemannian Space" by comparing rows A and D that the Riemann-Christoffel Tensor coefficient has a one-to-one corresponds with changes in curvature. So it models directly spatial curvature in the surrounding space, from which it acquires its name ***Riemann-Christoffel curvature tensor***.

Theorem 14.1.12 Mixed Riemann-Christoffel Tensor in Determinate Form

1g	Given	Einsteinian Summation Notation does not apply.
Step	Hypothesis	$R^{\alpha}{}_{ijk} = \begin{vmatrix} \nabla_j & \nabla_k \\ \Gamma_{ij}{}^{\alpha} & \Gamma_{ik}{}^{\alpha} \end{vmatrix} + \Sigma_\beta \begin{vmatrix} \Gamma_{ik}{}^{\beta} & \Gamma_{ij}{}^{\beta} \\ \Gamma_{\beta k}{}^{\alpha} & \Gamma_{\beta j}{}^{\alpha} \end{vmatrix}$
1	From D14.1.2A Mixed Riemann-Christoffel tensor	$R^{\alpha}{}_{ijk} = (\nabla_j \Gamma_{ik}{}^{\alpha} - \nabla_k \Gamma_{ij}{}^{\alpha}) + \Sigma_\beta (\Gamma_{\beta j}{}^{\alpha} \Gamma_{ik}{}^{\beta} - \Gamma_{\beta k}{}^{\alpha} \Gamma_{ij}{}^{\beta})$
∴	From 1 and T6.10.2 Determinate of a 2x2 Matrix	$R^{\alpha}{}_{ijk} = \begin{vmatrix} \nabla_j & \nabla_k \\ \Gamma_{ij}{}^{\alpha} & \Gamma_{ik}{}^{\alpha} \end{vmatrix} + \Sigma_\beta \begin{vmatrix} \Gamma_{ik}{}^{\beta} & \Gamma_{ij}{}^{\beta} \\ \Gamma_{\beta k}{}^{\alpha} & \Gamma_{\beta j}{}^{\alpha} \end{vmatrix}$

Theorem 14.1.13 Alternate Proof Evaluation for the Second Order Intrinsic Contravariant Vector

1g	Given	Einsteinian Summation Notation does not apply.
Step	Hypothesis	$\mathbf{A}{,}j{,}k = \Sigma_i [\, A_{i,j,k} - \Sigma_\alpha A_{\alpha,k} \Gamma_{ij}{}^{\alpha} + A_\alpha \nabla_{,k} \Gamma_{ij}{}^{\alpha}$ $- \Sigma_\alpha A_{i,\alpha} \Gamma_{jk}{}^{\alpha} + \Sigma_\alpha \Sigma_\beta A_\beta \Gamma_{i\alpha}{}^{\beta} \Gamma_{jk}{}^{\alpha}$ $- \Sigma_\alpha A_{\alpha,j} \Gamma_{ik}{}^{\alpha} + \Sigma_\alpha \Sigma_\beta A_\beta \Gamma_{\alpha j}{}^{\beta} \Gamma_{ik}{}^{\alpha} \,]\, \mathbf{b}^i$
1	From D12.1.2 Contravariant Vector	$\mathbf{A} = A_i \mathbf{b}^i$
2	From 1, TK.3.2 Uniqueness of Differentials and T12.6.3B Differentiation of Tensor: Rank-1 Contravariant	$\mathbf{A}{,}j = \Sigma_i A_{i;j} \mathbf{b}^i = \Sigma_i (\, A_{i,j} - \Sigma_\alpha A_{i\alpha} \Gamma_{ij}{}^{\alpha} \,)\, \mathbf{b}^i$
3	From 2 and TK.3.2 Uniqueness of Differentials	$\mathbf{A}{,}j{,}k = \partial (A_{i;j} \mathbf{b}^i) / \partial x_k$
4	From 3, T12.7.12 Tensor Product: Contravariant Chain Rule for Tensor Coefficient	$\mathbf{A}{,}j{,}k = \Sigma_i [\, (A_{i;j})_{,k} - \Sigma_\alpha (A_{i;\alpha}) \Gamma_{ik}{}^{\alpha} - \Sigma_\alpha (A_{\alpha;j}) \Gamma_{ik}{}^{\alpha} \,]\, \mathbf{b}^i$
5	From 4, T12.6.3B Differentiation of Tensor: Rank-1 Contravariant and T10.3.5 Tool: Tensor Substitution	$\mathbf{A}{,}j{,}k = \Sigma_i [\, (A_{i,j} - \Sigma_\alpha A_\alpha \Gamma_{ij}{}^{\alpha})_{,k}$ $- \Sigma_\alpha (A_{i,\alpha} - \Sigma_\beta A_\beta \Gamma_{i\alpha}{}^{\beta}) \Gamma_{jk}{}^{\alpha}$ $- \Sigma_\alpha (A_{\alpha,j} - \Sigma_\beta A_\beta \Gamma_{\alpha j}{}^{\beta}) \Gamma_{ik}{}^{\alpha} \,]\, \mathbf{b}^i$
∴	From 5, D4.1.20A Negative Coefficient, T10.5.2 Scalars and Tensors: Distribution of Scalar across a Tensor and D4.1.19 Primitive Definition for Rational Arithmetic	$\mathbf{A}{,}j{,}k = \Sigma_i [\, A_{i,j,k} - \Sigma_\alpha A_{\alpha,k} \Gamma_{ij}{}^{\alpha} + \Sigma_\alpha A_\alpha \nabla_k \Gamma_{ij}{}^{\alpha}$ $- \Sigma_\alpha A_{i,\alpha} \Gamma_{jk}{}^{\alpha} + \Sigma_\alpha \Sigma_\beta A_\beta \Gamma_{i\alpha}{}^{\beta} \Gamma_{jk}{}^{\alpha}$ $- \Sigma_\alpha A_{\alpha,j} \Gamma_{ik}{}^{\alpha} + \Sigma_\alpha \Sigma_\beta A_\beta \Gamma_{\alpha j}{}^{\beta} \Gamma_{ik}{}^{\alpha} \,]\, \mathbf{b}^i$

Theorem 14.1.14 Alternate Proof Contravariant Riemann-Christoffel Tensor

1g	Given	Einsteinian Summation Notation does not apply.
Step	Hypothesis	$\mathbf{A}_{,j,k} - \mathbf{A}_{,k,j} = \Sigma_i \Sigma_\alpha A_\alpha [(\nabla_j \Gamma_{ik}{}^{\alpha} - \nabla_k \Gamma_{ij}{}^{\alpha})$ $+ \Sigma_\beta (\Gamma_{\beta j}{}^{\alpha} \Gamma_{ik}{}^{\beta} - \Gamma_{\beta k}{}^{\alpha} \Gamma_{ij}{}^{\beta})] \mathbf{b}^i$
1	From T14.1.9 Evaluating the Second Order Intrinsic Contravariant Vector	$\mathbf{A}_{,j,k} = \Sigma_i [A_{i,j,k} - \Sigma_\alpha A_{\alpha,k} \Gamma_{ij}{}^{\alpha} + \Sigma_\alpha A_\alpha \nabla_k \Gamma_{ij}{}^{\alpha}$ $- \Sigma_\alpha A_{i,\alpha} \Gamma_{jk}{}^{\alpha} + \Sigma_\alpha \Sigma_\beta A_\beta \Gamma_{i\alpha}{}^{\beta} \Gamma_{jk}{}^{\alpha}$ $- \Sigma_\alpha A_{\alpha,j} \Gamma_{ik}{}^{\alpha} + \Sigma_\alpha \Sigma_\beta A_\beta \Gamma_{\alpha j}{}^{\beta} \Gamma_{ik}{}^{\alpha}] \mathbf{b}^i$
2	From 1, transpose derivative indices (j, k) → (k, j) and compare different terms between step 1 and 2	$\mathbf{A}_{,k,j} = \Sigma_i [A_{i,k,j} - \Sigma_\alpha A_{\alpha,j} \Gamma_{ik}{}^{\alpha} + \Sigma_\alpha A_\alpha \nabla_j \Gamma_{ik}{}^{\alpha}$ $- \Sigma_\alpha A_{i,\alpha} \Gamma_{kj}{}^{\alpha} + \Sigma_\alpha \Sigma_\beta A_\beta \Gamma_{i\alpha}{}^{\beta} \Gamma_{kj}{}^{\alpha}$ $- \Sigma_\alpha A_{\alpha,k} \Gamma_{ij}{}^{\alpha} + \Sigma_\alpha \Sigma_\beta A_\beta \Gamma_{\alpha k}{}^{\beta} \Gamma_{ij}{}^{\alpha}] \mathbf{b}^i$
3	From 1, 2, T7.11.8 Vector Addition: Uniqueness of Subtraction, T10.4.16 Addition of Tensors: Commutative by Addition and match differences	$\mathbf{A}_{,j,k} - \mathbf{A}_{,k,j} = \Sigma_i [A_{i,j,k} - A_{i,k,j}$ $+ \Sigma_\alpha A_\alpha \nabla_j \Gamma_{ik}{}^{\alpha} - \Sigma_\alpha A_\alpha \nabla_k \Gamma_{ij}{}^{\alpha}$ $- \Sigma_\alpha A_{\alpha,k} \Gamma_{ij}{}^{\alpha} + \Sigma_\alpha A_{\alpha,k} \Gamma_{ij}{}^{\alpha}$ $- \Sigma_\alpha A_{\alpha,j} \Gamma_{ik}{}^{\alpha} + \Sigma_\alpha A_{\alpha,j} \Gamma_{ik}{}^{\alpha}$ $- \Sigma_\alpha A_{i,\alpha} \Gamma_{jk}{}^{\alpha} + \Sigma_\alpha A_{i,\alpha} \Gamma_{kj}{}^{\alpha}$ $+ \Sigma_\alpha \Sigma_\beta A_\beta \Gamma_{i\alpha}{}^{\beta} \Gamma_{jk}{}^{\alpha} - \Sigma_\alpha \Sigma_\beta A_\beta \Gamma_{i\alpha}{}^{\beta} \Gamma_{kj}{}^{\alpha}$ $+ \Sigma_\alpha \Sigma_\beta A_\beta \Gamma_{\alpha j}{}^{\beta} \Gamma_{ik}{}^{\alpha} - \Sigma_\alpha \Sigma_\beta A_\beta \Gamma_{\alpha k}{}^{\beta} \Gamma_{ij}{}^{\alpha}] \mathbf{b}^i$
4	From 3, DxK.3.2.2E Multiple Partial Differential Operator and first terms commute denominators	$\mathbf{A}_{,j,k} - \mathbf{A}_{,k,j} = \Sigma_i [A_{i,j,k} - A_{i,j,k}$ $+ \Sigma_\alpha A_\alpha \nabla_j \Gamma_{ik}{}^{\alpha} - \Sigma_\alpha A_\alpha \nabla_k \Gamma_{ij}{}^{\alpha}$ $- \Sigma_\alpha A_{\alpha,k} \Gamma_{ij}{}^{\alpha} + \Sigma_\alpha A_{\alpha,k} \Gamma_{ij}{}^{\alpha}$ $- \Sigma_\alpha A_{\alpha,j} \Gamma_{ik}{}^{\alpha} + \Sigma_\alpha A_{\alpha,j} \Gamma_{ik}{}^{\alpha}$ $- \Sigma_\alpha A_{i,\alpha} \Gamma_{jk}{}^{\alpha} + \Sigma_\alpha A_{i,\alpha} \Gamma_{kj}{}^{\alpha}$ $+ \Sigma_\alpha \Sigma_\beta A_\beta \Gamma_{i\alpha}{}^{\beta} \Gamma_{jk}{}^{\alpha} - \Sigma_\alpha \Sigma_\beta A_\beta \Gamma_{i\alpha}{}^{\beta} \Gamma_{kj}{}^{\alpha}$ $+ \Sigma_\alpha \Sigma_\beta A_\beta \Gamma_{\alpha j}{}^{\beta} \Gamma_{ik}{}^{\alpha} - \Sigma_\alpha \Sigma_\beta A_\beta \Gamma_{\alpha k}{}^{\beta} \Gamma_{ij}{}^{\alpha}] \mathbf{b}^i$
5	From 4, T10.4.20 Addition of Tensors: Distribution of a Tensor over Addition of Tensors and T10.4.19 Addition of Tensors: Inverse by Addition	$\mathbf{A}_{,j,k} - \mathbf{A}_{,k,j} = \Sigma_i \{ \mathbf{0}$ $+ \Sigma_\alpha A_\alpha (\nabla_j \Gamma_{ik}{}^{\alpha} - \nabla_k \Gamma_{ij}{}^{\alpha})$ $+ \mathbf{0}$ $+ \mathbf{0}$ $+ \Sigma_\alpha A_{i,\alpha} (-\Gamma_{jk}{}^{\alpha} + \Gamma_{kj}{}^{\alpha})$ $+ \Sigma_\alpha \Sigma_\beta A_\beta (\Gamma_{i\alpha}{}^{\beta} \Gamma_{jk}{}^{\alpha} - \Gamma_{i\alpha}{}^{\beta} \Gamma_{kj}{}^{\alpha} + \Gamma_{\alpha j}{}^{\beta} \Gamma_{ik}{}^{\alpha} - \Gamma_{\alpha k}{}^{\beta} \Gamma_{ij}{}^{\alpha}) \} \mathbf{b}^i$
6	From 5, T10.4.18 Addition of Tensors: Identity by Addition and T10.4.19 Addition of Tensors: Inverse by Addition	$\mathbf{A}_{,j,k} - \mathbf{A}_{,k,j} = \Sigma_i \{ \Sigma_\alpha A_\alpha (\nabla_j \Gamma_{ik}{}^{\alpha} - \nabla_k \Gamma_{ij}{}^{\alpha})$ $- \Sigma_\alpha A_{i,\alpha}\, 0$ $+ \Sigma_\alpha \Sigma_\beta A_\beta [\Gamma_{i\alpha}{}^{\beta}\, 0 + \Gamma_{\alpha j}{}^{\beta} \Gamma_{ik}{}^{\alpha} - \Gamma_{\alpha k}{}^{\beta} \Gamma_{ij}{}^{\alpha}] \} \mathbf{b}^i$
7	From 6 and substitute dummy indices on last terms $\alpha \to r$, $\beta \to \alpha$ and $r \to \beta$	$\mathbf{A}_{,j,k} - \mathbf{A}_{,k,j} = \Sigma_i [\Sigma_\alpha A_\alpha (\nabla_j \Gamma_{ik}{}^{\alpha} - \nabla_k \Gamma_{ij}{}^{\alpha})$ $+ \Sigma_\beta \Sigma_\alpha A_\alpha (\Gamma_{\beta j}{}^{\alpha} \Gamma_{ik}{}^{\beta} - \Gamma_{\beta k}{}^{\alpha} \Gamma_{ij}{}^{\beta})] \mathbf{b}^i$
∴	From 7 and group terms, distribute out summation	$\mathbf{A}_{,j,k} - \mathbf{A}_{,k,j} = \Sigma_i \Sigma_\alpha A_\alpha [(\nabla_j \Gamma_{ik}{}^{\alpha} - \nabla_k \Gamma_{ij}{}^{\alpha})$ $+ \Sigma_\beta (\Gamma_{\beta j}{}^{\alpha} \Gamma_{ik}{}^{\beta} - \Gamma_{\beta k}{}^{\alpha} \Gamma_{ij}{}^{\beta})] \mathbf{b}^i$

Observation 14.1.7 Why an Alternate Proof for Mixed Riemann-Christoffel Tensor Coefficient

The reason for doing this alternate proof is that it verifies, T14.1.9 "Evaluating the Second Order Intrinsic Contravariant Vector" and T14.1.10 "Vector Relation to a Contravariant Riemann-Christoffel Tensor", to T14.1.14 "Alternate Proof Evaluation for the Second Order Intrinsic Contravariant Vector", confirming the validity of the contravariant Riemann-Christoffel tensor by two totally independent and different proofs. It also demonstrates different construction techniques L14.2.1 "Product and Differential of Contravariant Metric" differentiating the bases vector and T14.1.14 differentiating the tensor coefficient.

Theorem 14.1.15 Riemann-Christoffel Tensor Second Kind; Transpose last two indices

1g	Given	Einsteinian Summation Notation does not apply.
2g		$R^{\alpha}{}_{ijk} = -R^{\alpha}{}_{ikj}$ by assumption
Step	Hypothesis	$R^{\alpha}{}_{ijk} = -R^{\alpha}{}_{ikj}$
1	From 2g	$R^{\alpha}{}_{ijk} = -R^{\alpha}{}_{ikj}$
2	From 1, D14.1.2A Mixed Riemann-Christoffel tensor and T10.3.5 Tool: Tensor Substitution	$R^{\alpha}{}_{ijk} = -[\ (\nabla_k\Gamma_{ij}{}^{\alpha} - \nabla_j\Gamma_{ik}{}^{\alpha}) + \Sigma_\beta\ (\Gamma_{\beta k}{}^{\alpha}\Gamma_{ij}{}^{\beta} - \Gamma_{\beta j}{}^{\alpha}\Gamma_{ik}{}^{\beta})\ ]$
3	From 2, D4.1.20A Negative Coefficient and T10.5.6 Scalars and Tensors: Distribution of a Scalar over Addition of Tensors	$R^{\alpha}{}_{ijk} = -[\ -(\nabla_j\Gamma_{ik}{}^{\alpha} - \nabla_k\Gamma_{ij}{}^{\alpha}) - \Sigma_\beta\ (\Gamma_{\beta j}{}^{\alpha}\Gamma_{ik}{}^{\beta} - \Gamma_{\beta k}{}^{\alpha}\Gamma_{ij}{}^{\beta})\ ]$
4	From 3 and A4.2.14 Distribution	$R^{\alpha}{}_{ijk} = -(-1)[\ (\nabla_j\Gamma_{ik}{}^{\alpha} - \nabla_k\Gamma_{ij}{}^{\alpha}) + \Sigma_\beta\ (\Gamma_{\beta j}{}^{\alpha}\Gamma_{ik}{}^{\beta} - \Gamma_{\beta k}{}^{\alpha}\Gamma_{ij}{}^{\beta})\ ]$
5	From 4 and D4.1.19 Primitive Definition for Rational Arithmetic	$R^{\alpha}{}_{ijk} = R^{\alpha}{}_{ijk}$
∴	From 1, 5 and by identity	$R^{\alpha}{}_{ijk} = -\ R^{\alpha}{}_{ikj}$

Theorem 14.1.16 Riemann-Christoffel Tensor Second Kind; Duplicate last two indices

1g	Given	$j = k = \eta$ by assumption
Step	Hypothesis	$R^{\alpha}{}_{i\eta\eta} = \mathbf{0}$
1	From A10.2.6A Equality of Tensors	$R^{\alpha}{}_{ijk} = R^{\alpha}{}_{ikj}$
2	From 2g and A4.2.3 Substitution	$R^{\alpha}{}_{i\eta\eta} = R^{\alpha}{}_{i\eta\eta}$
3	From 2 and D14.1.2 Mixed Riemann-Christoffel tensor	$R^{\alpha}{}_{i\eta\eta} = (\nabla_\eta\Gamma_{i\eta}{}^{\alpha} - \nabla_\eta\Gamma_{i\eta}{}^{\alpha}) + (\Gamma_{\beta\eta}{}^{\alpha}\Gamma_{i\eta}{}^{\beta} - \Gamma_{\beta\eta}{}^{\alpha}\Gamma_{i\eta}{}^{\beta})$
4	From 3 and T10.4.19 Addition of Tensors: Inverse by Addition	$R^{\alpha}{}_{i\eta\eta} = \mathbf{0} + \mathbf{0}$
∴	From 4 and T10.3.9 Tool: Zero Tensor Plus a Zero Tensor	$R^{\alpha}{}_{i\eta\eta} = \mathbf{0}$

Theorem 14.1.17 Riemann-Christoffel Tensor First Kind to Second Kind

Step	Hypothesis	$R^{\alpha}{}_{ijk} = g^{\beta\alpha}R_{\beta ijk}$
1	From D14.1.3 Riemann-Christoffel Tensor Coefficient of the First Kind and A10.2.6B Equality of Tensors	$g_{\beta\gamma}R^{\gamma}{}_{ijk} = R_{\beta ijk}$
2	From 1 and T12.3.5A Metric Transformation from Contravariant to Covariant Coefficient	$g^{\beta\alpha}g_{\beta\gamma}R^{\gamma}{}_{ijk} = g^{\beta\alpha}R_{\beta ijk}$
3	From 2 and T12.2.8B Covariant and Contravariant Metric Inverse	$\delta^{\alpha}{}_{\gamma}R^{\gamma}{}_{ijk} = g^{\beta\alpha}R_{\beta ijk}$
∴	From 3 and T12.3.8B Contracted Contravariant Bases Vector with Covariant Vector	$R^{\alpha}{}_{ijk} = g^{\beta\alpha}R_{\beta ijk}$

Theorem 14.1.18 Riemann-Christoffel Mixed Tensor and Principle Curvatures

1g	Given	$\kappa_k \equiv \sum_i \sum_j \sum_\alpha \nu_i \nu_j A_\alpha R^\alpha{}_{ijk}$ Principle Curvatures of Riemann-Christoffel Mixed Tensor along the manifolds gradient paths
Step	Hypothesis	$\Delta\kappa\, \mathbf{e}^N = \sum_k \kappa_k\, \mathbf{b}^k$
1	T14.1.11 Mixed Riemann-Christoffel Tensor as Contravariant Curvature	$\Delta\kappa\, \mathbf{e}^N = \sum_i \sum_j \sum_k \sum_\alpha \nu_i \nu_j A_\alpha R^\alpha{}_{ijk}\, \mathbf{b}^k$
2	From 1 and reordering summation by bases vectors	$\Delta\kappa\, \mathbf{e}^N = \sum_k (\sum_i \sum_j \sum_\alpha \nu_i \nu_j A_\alpha R^\alpha{}_{ijk})\, \mathbf{b}^k$
∴	From 1g, 1 and T10.2.5 Tool: Tensor Substitution	$\Delta\kappa\, \mathbf{e}^N = \sum_k \kappa_k\, \mathbf{b}^k$

Section 14.2 Riemann-Christoffel Tensor Theorems of the First and Second Kind

Lema 14.2.1 Product and Differential of Contravariant Metric

Step	Hypothesis	$\Sigma_\alpha g_{\gamma\alpha}\, \nabla_j(g^{\alpha\beta}) = -\, \Gamma_{\gamma j}{}^{\beta} - g^{\tau\beta}\, \Gamma_{\tau j,\gamma}$ EQ A	$\Sigma_\alpha g_{\gamma\alpha}\, \nabla_k(g^{\alpha\beta}) = -\, \Gamma_{\gamma k}{}^{\beta} - g^{\tau\beta}\, \Gamma_{\tau k,\gamma}$ EQ B
1	From T12.8.5B Simple Differential of the Contravariant Metric Tensor	$\nabla_p(g^{ab}) = -g^{a\tau}\, \Gamma_{\tau p}{}^{b} - g^{\tau b}\, \Gamma_{\tau p}{}^{a}$	
2	From 1, T12.3.9 Metric Transformation from Covariant to Contravariant Coefficient, T10.4.6 Product of Tensors: Uniqueness by Multiplication of Rank for p + q and T10.4.20 Addition of Tensors: Distribution of a Tensor over Addition of Tensors	$\Sigma_a g_{\gamma a}\, \nabla_p(g^{ab}) = -\, \Sigma_a g_{\gamma a}\, g^{a\tau}\, \Gamma_{\tau p}{}^{b} - \Sigma_\alpha g_{\gamma a}\, g^{\tau b}\, \Gamma_{\tau p}{}^{a}$	
3	From 2, T12.2.8A Covariant and Contravariant Metric Inverse and T10.3.5 Tool: Tensor Substitution	$\Sigma_a g_{\gamma a}\, \nabla_p(g^{ab}) = -\, \delta_\gamma{}^{\tau}\, \Gamma_{\tau p}{}^{b} - g_{\gamma a}\, g^{\tau b}\, \Gamma_{\tau p}{}^{a}$	
4	From 3 and D6.1.6 Identity Matrix evaluate τ → γ	$\Sigma_a g_{\gamma\alpha}\, \nabla_p(g^{ab}) = -\, \Gamma_{\gamma p}{}^{b} - g_{\gamma a}\, g^{\tau b}\, \Gamma_{\tau p}{}^{a}$	
5	From 4 and D12.5.2 Christoffel Symbol of the Second Kind	$\Sigma_a g_{\gamma a}\, \nabla_p(g^{ab}) = -\, \Gamma_{\gamma}{}^{b} - g_{\gamma a}\, g^{\tau b}\, g^{\eta a}\, \Gamma_{\tau p,\eta}$	
6	From 5 and T10.4.7 Product of Tensors: Commutative by Multiplication of Rank for p + q	$\Sigma_a g_{\gamma a}\, \nabla_p(g^{ab}) = -\, \Gamma_{\gamma p}{}^{b} - g^{\tau b}\, g_{\gamma a}\, g^{\eta a}\, \Gamma_{\tau p,\eta}$	
7	From 6, T12.2.8A Covariant and Contravariant Metric Inverse and T10.3.5 Tool: Tensor Substitution	$\Sigma_a g_{\gamma a}\, \nabla_p(g^{ab}) = -\, \Gamma_{\gamma p}{}^{b} - g^{\tau b}\, \delta_\gamma{}^{\eta}\, \Gamma_{\tau p,\eta}$	
8	From 7 and D6.1.6 Identity Matrix evaluate γ → η	$\Sigma_a g_{\gamma a}\, \nabla_p(g^{ab}) = -\, \Gamma_{\gamma p}{}^{b} - g^{\tau b}\, \Gamma_{\tau p,\gamma}$	
∴	From 8, p = j and p = k	$\Sigma_\alpha g_{\gamma\alpha}\, \nabla_j(g^{\alpha\beta}) = -\, \Gamma_{\gamma j}{}^{\beta} - g^{\tau\beta}\, \Gamma_{\tau j,\gamma}$ EQ A	$\Sigma_\alpha g_{\gamma\alpha}\, \nabla_k(g^{\alpha\beta}) = -\, \Gamma_{\gamma k}{}^{\beta} - g^{\tau\beta}\, \Gamma_{\tau k,\gamma}$ EQ B

Theorem 14.2.2 Riemann-Christoffel Tensor First Kind

1g	Given	$\nabla_k(g^{ij}) = -g^{i\alpha}\, \Gamma_{\alpha k}{}^{j} - g^{\alpha j}\, \Gamma_{\alpha k}{}^{i}$
Step	Hypothesis	$R_{\gamma ijk} = (\nabla_j\, \Gamma_{ik,\gamma} - \nabla_k\, \Gamma_{ij,\gamma}) + \Sigma_\beta\, (\Gamma_{\gamma k,\beta}\, \Gamma_{ij}{}^{\beta} - \Gamma_{\gamma j,\beta}\, \Gamma_{ik}{}^{\beta})$
1	From D14.1.3A Riemann-Christoffel Tensor Coefficient of the First Kind	$R_{\gamma ijk} = \Sigma_\alpha\, g_{\gamma\alpha} R^{\alpha}{}_{ijk}$
2	From 1, D14.1.2A Mixed Riemann-Christoffel tensor and T10.3.5 Tool: Tensor Substitution	$R_{\gamma ijk} = \Sigma_\alpha\, g_{\gamma\alpha}\, [(\nabla_j \Gamma_{ik}{}^{\alpha} - \nabla_k \Gamma_{ij}{}^{\alpha}) + \Sigma_\beta\, (\Gamma_{\beta j}{}^{\alpha} \Gamma_{ik}{}^{\beta} - \Gamma_{\beta k}{}^{\alpha} \Gamma_{ij}{}^{\beta})]$
3	From 2 and T10.4.20 Addition of Tensors: Distribution of a Tensor over Addition of Tensors	$R_{\gamma ijk} = \Sigma_\alpha\, g_{\gamma\alpha}\, (\nabla_j \Gamma_{ik}{}^{\alpha} - \nabla_k \Gamma_{ij}{}^{\alpha}) + \Sigma_\beta\, \Sigma_\alpha g_{\gamma\alpha}\, (\Gamma_{\beta j}{}^{\alpha} \Gamma_{ik}{}^{\beta} - \Gamma_{\beta k}{}^{\alpha} \Gamma_{ij}{}^{\beta})$
4	From 3 and D12.5.2 Christoffel Symbol of the Second Kind	$R_{\gamma ijk} = \Sigma_\alpha\, g_{\gamma\alpha}\, (\nabla_j\, g^{\alpha\beta}\, \Gamma_{ik,\beta} - \nabla_k\, g^{\alpha\beta}\, \Gamma_{ij,\beta})$ $+ \Sigma_\beta\, \Sigma_\alpha g_{\gamma\alpha}\, (g^{\alpha\tau}\, \Gamma_{\beta j,\tau} \Gamma_{ik}{}^{\beta} - g^{\alpha\tau}\, \Gamma_{\beta k,\tau}\, \Gamma_{ij}{}^{\beta})$
5	From 4 and T10.4.20 Addition of Tensors: Distribution of a Tensor over Addition of Tensors	$R_{\gamma ijk} = \Sigma_\alpha\, (g_{\gamma\alpha}\, \nabla_j\, g^{\alpha\beta}\, \Gamma_{ik,\beta} - g_{\gamma\alpha}\, \nabla_k\, g^{\alpha\beta}\, \Gamma_{ij,\beta})$ $+ \Sigma_\beta\, \Sigma_\alpha\, (g_{\gamma\alpha}\, g^{\alpha\tau}\, \Gamma_{\beta j,\tau} \Gamma_{ik}{}^{\beta} - g_{\gamma\alpha}\, g^{\alpha\tau}\, \Gamma_{\beta k,\tau}\, \Gamma_{ij}{}^{\beta})$
6	From 5, T12.2.8A Covariant and Contravariant Metric Inverse and T10.3.5 Tool: Tensor Substitution	$R_{\gamma ijk} = \Sigma_\alpha\, (g_{\gamma\alpha}\, \nabla_j\, g^{\alpha\beta}\, \Gamma_{ik,\beta} - g_{\gamma\alpha}\, \nabla_k\, g^{\alpha\beta}\, \Gamma_{ij,\beta})$ $+ \Sigma_\beta\, (\delta_\gamma{}^{\tau}\, \Gamma_{\beta j,\tau} \Gamma_{ik}{}^{\beta} - \delta_\gamma{}^{\tau}\, \Gamma_{\beta k,\tau}\, \Gamma_{ij}{}^{\beta})$
7	From 6 and D6.1.6 Identity Matrix evaluate $\gamma \rightarrow \tau$	$R_{\gamma ijk} = \Sigma_\alpha\, (g_{\gamma\alpha}\, \nabla_j\, g^{\alpha\beta}\, \Gamma_{ik,\beta} - g_{\gamma\alpha}\, \nabla_k\, g^{\alpha\beta}\, \Gamma_{ij,\beta})$ $+ \Sigma_\beta\, (\Gamma_{\beta j,\gamma}\, \Gamma_{ik}{}^{\beta} - \Gamma_{\beta k,\gamma}\, \Gamma_{ij}{}^{\beta})$
8	From 7, T12.7.8 Tensor Product: Chain Rule for Mixed Binary Product and A10.2.14 Correspondence of Tensors and Tensor Coefficients	$R_{\gamma ijk} = \Sigma_\alpha\, g_{\gamma\alpha}\, (g^{\alpha\beta}\, \nabla_j\, \Gamma_{ik,\beta} + \Gamma_{ik,\beta}\, \nabla_j\, g^{\alpha\beta}) -$ $g_{\gamma\alpha}\, (g^{\alpha\beta}\, \nabla_k\, \Gamma_{ij,\beta} + \Gamma_{ij,\beta}\, \nabla_j\, g^{\alpha\beta})$ $+ \Sigma_\beta\, (\Gamma_{\beta j,\gamma}\, \Gamma_{ik}{}^{\beta} - \Gamma_{\beta k,\gamma}\, \Gamma_{ij}{}^{\beta})$
9	From 8 and T10.3.20 Addition of Tensors: Distribution of a Tensor over Addition of Tensors	$R_{\gamma ijk} = [\Sigma_\alpha\, (g_{\gamma\alpha}\, g^{\alpha\beta}\, \nabla_j\, \Gamma_{ik,\beta} + \Gamma_{ik,\beta}\, g_{\gamma\alpha}\, \nabla_j\, g^{\alpha\beta}) - \Sigma_\alpha\, (g_{\gamma\alpha}\, g^{\alpha\beta}\, \nabla_k\, \Gamma_{ij,\beta}$ $+ \Gamma_{ij,\beta}\, g_{\gamma\alpha}\, \nabla_j\, g^{\alpha\beta})\,] + \Sigma_\beta\, (\Gamma_{\beta j,\gamma}\, \Gamma_{ik}{}^{\beta} - \Gamma_{\beta k,\gamma}\, \Gamma_{ij}{}^{\beta})$
10	From 9, T12.2.8A Covariant and Contravariant Metric Inverse and T10.3.5 Tool: Tensor Substitution	$R_{\gamma ijk} = [(\delta_\gamma{}^{\beta} \nabla_j\, \Gamma_{ik,\beta} + \Sigma_\alpha \Gamma_{ik,\beta}\, g_{\gamma\alpha}\, \nabla_j\, g^{\alpha\beta})$ $- \Sigma_\beta\, (\delta_\gamma{}^{\beta}\, \nabla_k\, \Gamma_{ij,\beta} + \Sigma_\alpha \Gamma_{ij,\beta}\, g_{\gamma\alpha}\, \nabla_k\, g^{\alpha\beta})\,] + \Sigma_\beta\, (\Gamma_{\beta j,\gamma}\, \Gamma_{ik}{}^{\beta} - \Gamma_{\beta k,\gamma}\, \Gamma_{ij}{}^{\beta})$
11	From 10 and D6.1.6 Identity Matrix evaluate $\gamma \rightarrow \beta$	$R_{\gamma ijk} = (\nabla_j\, \Gamma_{ik,\gamma} - \nabla_k\, \Gamma_{ij,\gamma})$ $+ \Sigma_\alpha \Sigma_\beta\, (\Gamma_{ik,\beta}\, g_{\gamma\alpha}\, \nabla_j\, g^{\alpha\beta} - \Gamma_{ij,\beta}\, g_{\gamma\alpha}\, \nabla_k\, g^{\alpha\beta})$ $+ \Sigma_\beta\, (\Gamma_{\beta j,\gamma}\, \Gamma_{ik}{}^{\beta} - \Gamma_{\beta k,\gamma}\, \Gamma_{ij}{}^{\beta})$
12	From 11, T10.3.16 Addition of Tensors: Commutative by Addition and T10.3.17 Addition of Tensors: Associative by Addition	$R_{\gamma ijk} = (\nabla_j\, \Gamma_{ik,\gamma} - \nabla_k\, \Gamma_{ij,\gamma}) + \Sigma_\beta\, (\Gamma_{\beta j,\gamma}\, \Gamma_{ik}{}^{\beta} - \Gamma_{\beta k,\gamma}\, \Gamma_{ij}{}^{\beta}) +$ $\Sigma_\alpha \Sigma_\beta\, (\Gamma_{ik,\beta}\, g_{\gamma\alpha}\, \nabla_j\, g^{\alpha\beta} - \Gamma_{ij,\beta}\, g_{\gamma\alpha}\, \nabla_k\, g^{\alpha\beta})$

13	From 12, L14.1.1(A,B) and T10.3.5 Tool: Tensor Substitution, D4.1.20A Negative Coefficient, T10.4.6 Scalars and Tensors: Distribution of a Scalar over Addition of Tensors and D4.1.19 Primitive Definition for Rational Arithmetic	$R_{\gamma ijk} = (\nabla_j \Gamma_{ik,\gamma} - \nabla_k \Gamma_{ij,\gamma}) + \Sigma_\beta (\Gamma_{\beta j,\gamma} \Gamma_{ik}{}^\beta - \Gamma_{\beta k,\gamma} \Gamma_{ij}{}^\beta) + \Sigma_\beta [\Gamma_{ik,\beta} (- \Gamma_{\gamma j}{}^\beta - g^{\tau\beta} \Gamma_{\tau j,\gamma}) + \Gamma_{ij,\beta} (\Gamma_{\gamma k}{}^\beta + g^{\tau\beta} \Gamma_{\tau k,\gamma})]$
14	From 13 and T10.3.20 Addition of Tensors: Distribution of a Tensor over Addition of Tensors	$R_{\gamma ijk} = (\nabla_j \Gamma_{ik,\gamma} - \nabla_k \Gamma_{ij,\gamma}) + [\Sigma_\beta \Gamma_{\beta j,\gamma} \Gamma_{ik}{}^\beta - \Sigma_\beta \Gamma_{\beta k,\gamma} \Gamma_{ij}{}^\beta - \Sigma_\beta \Gamma_{ik,\beta} \Gamma_{\gamma j}{}^\beta - \Sigma_\beta g^{\tau\beta} \Gamma_{ik,\beta} \Gamma_{\tau j,\gamma} + \Sigma_\beta \Gamma_{ij,\beta} \Gamma_{\gamma k}{}^\beta + \Sigma_\beta g^{\tau\beta} \Gamma_{ij,\beta} \Gamma_{\tau k,\gamma}]$
15	From 14 and changing dummy indices $\tau \rightarrow \beta$	$R_{\gamma ijk} = (\nabla_j \Gamma_{ik,\gamma} - \nabla_k \Gamma_{ij,\gamma}) + \Sigma_\beta [\Gamma_{\beta j,\gamma} \Gamma_{ik}{}^\beta - \Gamma_{\beta k,\gamma} \Gamma_{ij}{}^\beta - \Gamma_{ik,\beta} \Gamma_{\gamma j}{}^\beta - \Gamma_{ik}{}^\beta \Gamma_{\beta j,\gamma} + \Gamma_{ij,\beta} \Gamma_{\gamma k}{}^\beta + \Gamma_{ij}{}^\beta \Gamma_{\beta k,\gamma}]$
16	From 15, T10.3.16 Addition of Tensors: Commutative by Addition and subtracting like terms	$R_{\gamma ijk} = (\nabla_j \Gamma_{ik,\gamma} - \nabla_k \Gamma_{ij,\gamma}) + \Sigma_\beta [\Gamma_{\beta j,\gamma} \Gamma_{ik}{}^\beta - \Gamma_{\beta k,\gamma} \Gamma_{ij}{}^\beta - \Gamma_{ik,\beta} \Gamma_{\gamma j}{}^\beta - \Gamma_{ik}{}^\beta \Gamma_{\beta j,\gamma} + \Gamma_{ij,\beta} \Gamma_{\gamma k}{}^\beta + \Gamma_{ij}{}^\beta \Gamma_{\beta k,\gamma}]$
17	From 16, T10.3.19 Addition of Tensors: Inverse by Addition and T10.3.18 Addition of Tensors: Identity by Addition	$R_{\gamma ijk} = (\nabla_j \Gamma_{ik,\gamma} - \nabla_k \Gamma_{ij,\gamma}) + \Sigma_\beta (-\Gamma_{ik,\beta} \Gamma_{\gamma j}{}^\beta + \Gamma_{ij,\beta} \Gamma_{\gamma k}{}^\beta)$
18	From 17, T10.3.16 Addition of Tensors: Commutative by Addition T12.5.14 Christoffel Symbol: Exchanging Index between Second and First Symbols	$R_{\gamma ijk} = (\nabla_j \Gamma_{ik,\gamma} - \nabla_k \Gamma_{ij,\gamma}) + \Sigma_\eta (\Gamma_{\gamma k,\eta} \Gamma_{ij}{}^\eta - \Gamma_{\gamma j,\eta} \Gamma_{ik}{}^\eta)$
∴	From 18 and changing dummy indices $\eta \rightarrow \beta$	$R_{\gamma ijk} = (\nabla_j \Gamma_{ik,\gamma} - \nabla_k \Gamma_{ij,\gamma}) + \Sigma_\beta (\Gamma_{\gamma k,\beta} \Gamma_{ij}{}^\beta - \Gamma_{\gamma j,\beta} \Gamma_{ik}{}^\beta)$

Observation 14.2.1 Beta is not a free index for the Riemann-Christoffel Tensor

Einsteinian Summation Notation hides the $[\Sigma_\beta]$, but it is always there. So beta is not a free index, and to the user of these tensors in evaluating them they must always be aware of its hidden existence.

Theorem 14.2.3 Riemann-Christoffel Tensor First Kind in Metric Form

Step	Hypothesis	$R_{\gamma ijk} = ½(g_{k\gamma,i,j} - g_{ik,\gamma,j} + g_{ij,\gamma,k} - g_{j\gamma,i,k}) + \Sigma_\beta (\Gamma_{\gamma k,\beta} \Gamma_{ij}{}^\beta - \Gamma_{\gamma j,\beta} \Gamma_{ik}{}^\beta)$
1	From T14.2.2 Riemann-Christoffel Tensor First Kind	$R_{\gamma ijk} = (\nabla_j \Gamma_{ik,\gamma} - \nabla_k \Gamma_{ij,\gamma}) + \Sigma_\beta (\Gamma_{\gamma k,\beta} \Gamma_{ij}{}^\beta - \Gamma_{\gamma j,\beta} \Gamma_{ik}{}^\beta)$
2	From 1 and T12.5.3B Christoffel Symbol of the First Kind	$\nabla_j \Gamma_{ik,\gamma} = \nabla_j ½ (\partial g_{i\gamma} / \partial x_k + \partial g_{k\gamma} / \partial x_i - \partial g_{ik} / \partial x_\gamma)$
3	From 1 and T12.5.3B Christoffel Symbol of the First Kind	$\nabla_k \Gamma_{ij,\gamma} = \nabla_k ½ (\partial g_{i\gamma} / \partial x_j + \partial g_{j\gamma} / \partial x_i - \partial g_{ij} / \partial x_\gamma)$
4	From 2, LxK.3.2.3B Distribute Across Constant Product and LxK.3.2.2B Differential as a Linear Operator	$\nabla_j \Gamma_{ik,\gamma} = ½(g_{i\gamma,k,j} + g_{k\gamma,i,j} - g_{ik,\gamma,j})$
5	From 3, LxK.3.2.3B Distribute Across Constant Product and LxK.3.2.2B Differential as a Linear Operator	$\nabla_k \Gamma_{ij,\gamma} = ½(g_{i\gamma,j,k} + g_{j\gamma,i,k} - g_{ij,\gamma,k})$
6	From 4, 5, D4.1.20A Negative Coefficient and T10.3.15 Addition of Tensors: Uniqueness by Addition	$\nabla_j \Gamma_{ik,\gamma} - \nabla_k \Gamma_{ij,\gamma} = ½(g_{i\gamma,k,j} + g_{k\gamma,i,j} - g_{ik,\gamma,j}) - ½(g_{i\gamma,j,k} + g_{j\gamma,i,k} - g_{ij,\gamma,k})$
7	From 6 and T10.4.6 Scalars and Tensors: Distribution of a Scalar over Addition of Tensors, factor ½ distribute the negative coefficient	$\nabla_j \Gamma_{ik,\gamma} - \nabla_k \Gamma_{ij,\gamma} = ½(g_{i\gamma,k,j} + g_{k\gamma,i,j} - g_{ik,\gamma,j} - g_{i\gamma,j,k} - g_{j\gamma,i,k} + g_{ij,\gamma,k})$
8	From 7, LxK.3.2.8B Symmetry of Differential Order and take the difference of like terms	$\nabla_j \Gamma_{ik,\gamma} - \nabla_k \Gamma_{ij,\gamma} = ½(g_{i\gamma,j,k} - g_{i\gamma,j,k} + g_{k\gamma,i,j} - g_{ik,\gamma,j} - g_{j\gamma,i,k} + g_{ij,\gamma,k})$
9	From 8, T10.3.19 Addition of Tensors: Inverse by Addition and T10.3.18 Addition of Tensors: Identity by Addition	$\nabla_j \Gamma_{ik,\gamma} - \nabla_k \Gamma_{ij,\gamma} = ½(g_{k\gamma,i,j} - g_{ik,\gamma,j} - g_{j\gamma,i,k} + g_{ij,\gamma,k})$
∴	From 1, 9, T10.3.5 Tool: Tensor Substitution and T10.3.16 Addition of Tensors: Commutative by Addition	$R_{\gamma ijk} = ½(g_{k\gamma,i,j} - g_{ik,\gamma,j} + g_{ij,\gamma,k} - g_{j\gamma,i,k}) + \Sigma_\beta (\Gamma_{\gamma k,\beta} \Gamma_{ij}{}^\beta - \Gamma_{\gamma j,\beta} \Gamma_{ik}{}^\beta)$

Observation 14.2.2 Dimension of Riemann-Christoffel Tensor First Kind to Second Kind

From theorem T14.2.3 "Riemann-Christoffel Tensor First Kind in Metric Form" and the product of the metric yields a Riemann-Christoffel tensor of the second kind, T14.1.17 "Riemann-Christoffel Tensor First Kind to Second Kind" the contravariant metric cancels the covariant metric. This leaves first four terms dimensionally with just the two partial differentials having the dimensions of $[L^{-2}]$.

Theorem 14.2.4 Riemann-Christoffel Tensor First Kind; Transpose First Two Indices

Step	Hypothesis	$R_{i\gamma jk} = -R_{\gamma ijk}$
1	From T14.2.3 Riemann-Christoffel Tensor First Kind in Metric Form and indices [γijk]	$R_{\gamma ijk} = ½(g_{k\gamma,i,j} - g_{ik,\gamma,j} + g_{ij,\gamma,k} - g_{j\gamma,i,k}) + \Sigma_\beta (\Gamma_{\gamma k,\beta} \Gamma_{ij}{}^\beta - \Gamma_{\gamma j,\beta} \Gamma_{ik}{}^\beta)$
2	From 1 and transpose first two indices γ → i	$R_{i\gamma jk} = ½(g_{ki,\gamma,j} - g_{\gamma k,i,j} + g_{\gamma j,i,k} - g_{ji,\gamma,k}) + \Sigma_\beta (\Gamma_{ik,\beta} \Gamma_{\gamma j}{}^\beta - \Gamma_{ij,\beta} \Gamma_{\gamma k}{}^\beta)$
3	From 2, T10.3.16 Addition of Tensors: Commutative by Addition, match terms to the original Riemann-Christoffel Tensor and T12.5.14 Christoffel Symbol: Exchanging Index between Second and First Symbols	$R_{i\gamma jk} = ½(-g_{\gamma k,i,j} + g_{ki,\gamma,j} - g_{ji,\gamma,k} + g_{\gamma j,i,k}) + \Sigma_\beta (\Gamma_{ik}{}^\beta \Gamma_{\gamma j,\beta} - \Gamma_{ij}{}^\beta \Gamma_{\gamma k,\beta})$
4	From 3, D4.1.20 Negative Coefficient, T4.8.2B Integer Exponents: Negative One Squared, T10.4.6 Scalars and Tensors: Distribution of a Scalar over Addition of Tensors and T12.2.13 Covariant Metric Inner Product is Symmetrical	$R_{i\gamma jk} = -½(g_{k\gamma,i,j} - g_{ik,\gamma,j} + g_{ji,\gamma,k} - g_{j\gamma,i,k}) - \Sigma_\beta (-\Gamma_{ik}{}^\beta \Gamma_{\gamma j,\beta} + \Gamma_{ij}{}^\beta \Gamma_{\gamma k,\beta})$
5	From 4 and T10.3.16 Addition of Tensors: Commutative by Addition	$R_{i\gamma jk} = -½(g_{k\gamma,i,j} - g_{ik,\gamma,j} + g_{ji,\gamma,k} - g_{j\gamma,i,k}) - \Sigma_\beta (\Gamma_{ij}{}^\beta \Gamma_{\gamma k,\beta} - \Gamma_{ik}{}^\beta \Gamma_{\gamma j,\beta})$
6	From D4.1.20 Negative Coefficient and T10.4.6 Scalars and Tensors: Distribution of a Scalar over Addition of Tensors	$R_{i\gamma jk} = -[½(g_{k\gamma,i,j} - g_{ik,\gamma,j} + g_{ji,\gamma,k} - g_{j\gamma,i,k}) + \Sigma_\beta (\Gamma_{\gamma k,\beta} \Gamma_{ij}{}^\beta - \Gamma_{\gamma j,\beta} \Gamma_{ik}{}^\beta)]$
∴	From 1, 6 T10.3.5 Tool: Tensor Substitution	$R_{i\gamma jk} = -R_{\gamma ijk}$

Theorem 14.2.5 Riemann-Christoffel Tensor First Kind; Transpose Last Two Indices

Step	Hypothesis	$R_{\gamma ikj} = -R_{\gamma ijk}$
1	From T14.2.3 Riemann-Christoffel Tensor First Kind in Metric Form and indices [γijk]	$R_{\gamma ijk} = ½(g_{k\gamma,i,j} - g_{ik,\gamma,j} + g_{ij,\gamma,k} - g_{j\gamma,i,k}) + \Sigma_\beta (\Gamma_{\gamma k,\beta} \Gamma_{ij}{}^\beta - \Gamma_{\gamma j,\beta} \Gamma_{ik}{}^\beta)$
2	From 1, and transpose last two indices j → k	$R_{\gamma ikj} = ½(- g_{k\gamma,i,j} + g_{ik,\gamma,j} - g_{ij,\gamma,k} + g_{j\gamma,i,k}) + \Sigma_\beta (- \Gamma_{\gamma k,\beta} \Gamma_{ij}{}^\beta + \Gamma_{\gamma j,\beta} \Gamma_{ik}{}^\beta)$
3	From 2, D4.1.20 Negative Coefficient, T4.8.2B Integer Exponents: Negative One Squared and T10.4.6 Scalars and Tensors: Distribution of a Scalar over Addition of Tensors	$R_{\gamma ikj} = -[½(g_{k\gamma,i,j} - g_{ik,\gamma,j} + g_{ij,\gamma,k} - g_{j\gamma,i,k}) + \Sigma_\beta (\Gamma_{\gamma k,\beta} \Gamma_{ij}{}^\beta - \Gamma_{\gamma j,\beta} \Gamma_{ik}{}^\beta)]$
∴	From 1, 6 T10.3.5 Tool: Tensor Substitution	$R_{\gamma ikj} = -R_{\gamma ijk}$

Theorem 14.2.6 Riemann-Christoffel Tensor First Kind; Transpose Last Two Indices

Step	Hypothesis	$R_{jk\gamma i} = R_{\gamma ijk}$
1	From T14.2.3 Riemann-Christoffel Tensor First Kind in Metric Form and indices [γijk]	$R_{\gamma ijk} = ½(g_{k\gamma,i,j} - g_{ik,\gamma,j} + g_{ij,\gamma,k} - g_{j\gamma,i,k}) + \sum_\beta (\Gamma_{\gamma k,\beta}\, \Gamma_{ij}{}^\beta - \Gamma_{\gamma j,\beta}\, \Gamma_{ik}{}^\beta)$
2	From 1 and changing dummy indices γ → p, i → r, j → s, k → v	$R_{prsv} = ½(g_{vp,r,s} - g_{rv,p,s} + g_{rs,p,v} - g_{sp,r,v}) + \sum_\beta (\Gamma_{pv,\beta}\, \Gamma_{rs}{}^\beta - \Gamma_{ps,\beta}\, \Gamma_{rv}{}^\beta)$
3	From 2 and changing dummy indices p → i, r → k, s → γ, v → i	$R_{jk\gamma i} = ½(g_{ij,k,\gamma} - g_{ki,j,\gamma} + g_{k\gamma,j,i} - g_{\gamma j,k,i}) + \sum_\beta (\Gamma_{ji,\beta}\, \Gamma_{k\gamma}{}^\beta - \Gamma_{j\gamma,\beta}\, \Gamma_{ki}{}^\beta)$
4	From 3, T12.5.14 Christoffel Symbol: Exchanging Index between Second and First Symbols, LxK.3.2.8B Symmetry of Differential Order and T10.3.7 Product of Tensors: Commutative by Multiplication of Rank p + q → q + p	$R_{jk\gamma i} = ½(g_{k\gamma,j,i} - g_{ik,\gamma,j} + g_{ij,\gamma,k} - g_{j\gamma,i,k}) + \sum_\beta (\Gamma_{k\gamma,\beta}\, \Gamma_{ji}{}^\beta - \Gamma_{\gamma j,\beta}\, \Gamma_{ik}{}^\beta)$
∴	From 1, 4 and T10.3.5 Tool: Tensor Substitution	$R_{jk\gamma i} = R_{\gamma ijk}$

Theorem 14.2.7 Riemann-Christoffel Tensor First Kind; 3-Index Cyclic Identity

Step	Hypothesis	$R_{\gamma ijk} + R_{\gamma jki} + R_{\gamma kij} = \mathbf{0}$
1	From T14.2.3 Riemann-Christoffel Tensor First Kind in Metric Form	$R_{\gamma ijk} = ½\,(g_{k\gamma,i,j} - g_{ik,\gamma,j} - g_{j\gamma,i,k} + g_{ij,\gamma,k}) + g^{\alpha\beta}\,\lvert\Delta(ikj,\gamma\text{: } ijk,\gamma)\rvert$
2	From 1, T14.2.3 Riemann-Christoffel Tensor First Kind in Metric Form, cycle to end index ijk → jki	$R_{\gamma jki} = ½\,(g_{i\gamma,j,k} - g_{ji,\gamma,k} - g_{k\gamma,j,i} + g_{jk,\gamma,i}) + g^{\alpha\beta}\,\lvert\Delta(jik,\gamma\text{: } jki,\gamma)\rvert$
3	From 2, T14.2.3 Riemann-Christoffel Tensor First Kind in Metric Form, cycle to end index jki → kij	$R_{\gamma kij} = ½\,(g_{j\gamma,k,i} - g_{kj,\gamma,i} - g_{i\gamma,k,j} + g_{ki,\gamma,j}) + g^{\alpha\beta}\,\lvert\Delta(kji,\gamma\text{: } kij,\gamma)\rvert$
4	From 1, 2, 3, sum first four terms and T10.5.6 Scalars and Tensors: Distribution of a Scalar over Addition of Tensors	$½\,[\,(g_{k\gamma,i,j} - g_{ik,\gamma,j} - g_{j\gamma,i,k} + g_{ij,\gamma,k}) + (g_{i\gamma,j,k} - g_{ji,\gamma,k} - g_{k\gamma,j,i} + g_{jk,\gamma,i}) + (g_{j\gamma,k,i} - g_{kj,\gamma,i} - g_{i\gamma,k,j} + g_{ki,\gamma,j})\,]$
5	From 4 and T10.4.7 Product of Tensors: Commutative by Multiplication of Rank for p + q and compare like terms	$½\,[\,(g_{k\gamma,i,j} - g_{k\gamma,j,i} + g_{ij,\gamma,k} - g_{ji,\gamma,k}) + (g_{i\gamma,j,k} - g_{i\gamma,k,j} + g_{jk,\gamma,i} - g_{kj,\gamma,i}) + (g_{j\gamma,k,i} - g_{j\gamma,i,k} + g_{ki,\gamma,j} - g_{ik,\gamma,j})\,]$
6	From 5, T10.4.19 Addition of Tensors: Inverse by Addition and T10.3.9 Tool: Zero Tensor Plus a Zero Tensor	$½\,[\,\mathbf{0}\,]$
7	From 1 and evaluate last term	$\lvert\Delta(ikj,\gamma\text{: } ijk,\gamma)\rvert = (\Gamma_{\beta j,\gamma}\,\Gamma_{ik,\alpha} - \Gamma_{\beta k,\gamma}\,\Gamma_{ij,\alpha})$
8	From 2 and evaluate last term	$\lvert\Delta(jik,\gamma\text{: } jki,\gamma)\rvert = (\Gamma_{\beta k,\gamma}\,\Gamma_{ji,\alpha} - \Gamma_{\beta i,\gamma}\,\Gamma_{jk,\alpha})$
9	From 3 and evaluate last term	$\lvert\Delta(kji,\gamma\text{: } kij,\gamma)\rvert = (\Gamma_{\beta i,\gamma}\,\Gamma_{kj,\alpha} - \Gamma_{\beta j,\gamma}\,\Gamma_{ki,\alpha})$
10	From 8, T12.5.10 Christoffel Symbol: Symmetry for the Second Kind (ji) → (ij) and T10.3.5 Tool: Tensor Substitution	$\lvert\Delta(jik,\gamma\text{: } jki,\gamma)\rvert = (\Gamma_{\beta k,\gamma}\,\Gamma_{ij,\alpha} - \Gamma_{\beta i,\gamma}\,\Gamma_{jk,\alpha})$
11	From 9, T12.5.10 Christoffel Symbol: Symmetry for the Second Kind (ki) → (ik) and T10.3.5 Tool: Tensor Substitution	$\lvert\Delta(kji,\gamma\text{: } kij,\gamma)\rvert = (\Gamma_{\beta i,\gamma}\,\Gamma_{jk,\alpha} - \Gamma_{\beta j,\gamma}\,\Gamma_{ik,\alpha})$
12	From 7, 10, 11 and sum terms	$\lvert\Delta(ikj,\gamma\text{: } ijk,\gamma)\rvert + \lvert\Delta(jik,\gamma\text{: } jki,\gamma)\rvert + \lvert\Delta(kji,\gamma\text{: } kij,\gamma)\rvert = (\Gamma_{\beta j,\gamma}\,\Gamma_{ik,\alpha} - \Gamma_{\beta k,\gamma}\,\Gamma_{ij,\alpha}) + (\Gamma_{\beta k,\gamma}\,\Gamma_{ij,\alpha} - \Gamma_{\beta i,\gamma}\,\Gamma_{jk,\alpha}) + (\Gamma_{\beta i,\gamma}\,\Gamma_{jk,\alpha} - \Gamma_{\beta j,\gamma}\,\Gamma_{ik,\alpha})$
13	From 12 and T10.4.7 Product of Tensors: Commutative by Multiplication of Rank for p + q and compare like terms	$\lvert\Delta(ikj,\gamma\text{: } ijk,\gamma)\rvert + \lvert\Delta(jik,\gamma\text{: } jki,\gamma)\rvert + \lvert\Delta(kji,\gamma\text{: } kij,\gamma)\rvert = (\Gamma_{\beta j,\gamma}\,\Gamma_{ik,\alpha} - \Gamma_{\beta j,\gamma}\,\Gamma_{ik,\alpha}) + (\Gamma_{\beta k,\gamma}\,\Gamma_{ij,\alpha} - \Gamma_{\beta k,\gamma}\,\Gamma_{ij,\alpha}) + (\Gamma_{\beta i,\gamma}\,\Gamma_{jk,\alpha} - \Gamma_{\beta i,\gamma}\,\Gamma_{jk,\alpha})$

14	From 13, T10.4.19 Addition of Tensors: Inverse by Addition and T10.3.9 Tool: Zero Tensor Plus a Zero Tensor	$\|\Delta(ikj,\gamma\text{: } ijk,\gamma)\| + \|\Delta(jik,\gamma\text{: } jki,\gamma)\| + \|\Delta(kji,\gamma\text{: } kij,\gamma)\| = \mathbf{0}$
∴	From 1, 2, 3, summing all terms, 6, 14 evaluates results	$R_{\gamma ijk} + R_{\gamma jki} + R_{\gamma kij} = \mathbf{0}$

Theorem 14.2.8 Riemann-Christoffel Tensor First Kind; Transpose Curvature Indices

Step	Hypothesis	$R_{\gamma jik} = R_{\gamma ijk} + R_{\gamma kij}$
1	From T14.2.7 Riemann-Christoffel Tensor First Kind; Cyclic Symmetry Identity	$R_{\gamma ijk} + R_{\gamma jki} + R_{\gamma kij} = \mathbf{0}$
2	From 1 and T14.1.21 Riemann-Christoffel Tensor First Kind; Transpose Last Two Indices	$R_{\gamma ijk} - R_{\gamma jik} + R_{\gamma kij} = \mathbf{0}$
∴	From 2 and T10.3.24A Addition of Tensors: Reversal of Left Cancellation by Addition and T10.3.18 Addition of Tensors: Identity by Addition	$R_{\gamma jik} = R_{\gamma ijk} + R_{\gamma kij}$

Observation 14.2.3 Riemann-Christoffel Tensor First Kind; 4-Index Cyclic Identity

Though not developed here, the same principles can be used to prove a similar identity by cycling all four indices;

$$R_{\gamma ijk} + R_{ji\gamma k} + R_{jk\gamma i} + R_{\gamma kji} = \mathbf{0} \qquad \textbf{Equation A}$$

also for reference alternate forms of the Riemann-Christoffel Tensor by different authors:

$R_{ijkt} = ½ (g_{it,j,k} - g_{jt,i,k} - g_{ik,j,t} + g_{jk,i,t}) + g^{ab} (\Gamma_{jk,b} \Gamma_{it,a} - \Gamma_{jt,b} \Gamma_{ik,a})$ Sokolnikoff pg 90

$R_{ijkt} = ½ (g_{ti,j,k} - g_{jt,i,k} - g_{ki,j,t} + g_{jk,i,t}) + g_{sr} (\Gamma_{it}{}^{r} \Gamma_{jk}{}^{s} - \Gamma_{ik}{}^{r} \Gamma_{jt}{}^{s})$ Lawden pg 117

$R_{mnab} = ½ (g_{na,m,b} - g_{nb,m,a} - g_{ma,n,b} + g_{mb,n,a}) + g_{rs} (\Gamma_{ma}{}^{r} \Gamma_{nb}{}^{s} - \Gamma_{mb}{}^{r} \Gamma_{na}{}^{s})$ Fock pg 154

Theorem 14.2.9 Riemann-Christoffel Tensor First Kind in Determinate Form

1g	Given	Einsteinian Summation Notation does not apply.
Step	Hypothesis	$R_{\gamma ijk} = \begin{vmatrix} \nabla_j & \nabla_k \\ \Gamma_{ij,\gamma} & \Gamma_{ik,\gamma} \end{vmatrix} + \Sigma_\beta \begin{vmatrix} \Gamma_{\gamma k,\beta} & \Gamma_{\gamma i,\beta} \\ \Gamma_{ik}{}^{\beta} & \Gamma_{ij}{}^{\beta} \end{vmatrix}$
1	From T14.2.2 Riemann-Christoffel Tensor First Kind	$R_{\gamma ijk} = (\nabla_j \Gamma_{ik,\gamma} - \nabla_k \Gamma_{ij,\gamma}) + \Sigma_\beta (\Gamma_{\gamma k,\beta} \Gamma_{ij}{}^{\beta} - \Gamma_{\gamma j,\beta} \Gamma_{ik}{}^{\beta})$
∴	From 1 and T6.10.2 Determinate of a 2x2 Matrix	$R_{\gamma ijk} = \begin{vmatrix} \nabla_j & \nabla_k \\ \Gamma_{ij,\gamma} & \Gamma_{ik,\gamma} \end{vmatrix} + \Sigma_\beta \begin{vmatrix} \Gamma_{\gamma k,\beta} & \Gamma_{\gamma i,\beta} \\ \Gamma_{ik}{}^{\beta} & \Gamma_{ij}{}^{\beta} \end{vmatrix}$

Theorem 14.2.10 Riemann-Christoffel Tensor First Kind; with Two Distinct Indices

Step	Hypothesis	$R_{\gamma i\gamma i} = g_{i\gamma,\gamma,i} - ½ (g_{ii,\gamma,\gamma} + g_{\gamma\gamma,i,i}) + \Sigma_\beta (\Gamma_{\gamma i,\beta} \Gamma_{i\gamma}{}^{\beta} - \Gamma_{\gamma\gamma,\beta} \Gamma_{ii}{}^{\beta})$
1	From T14.2.3 Riemann-Christoffel Tensor First Kind in Metric Form	$R_{\gamma ijk} = ½ (g_{k\gamma,i,j} - g_{ik,\gamma,j} + g_{ij,\gamma,k} - g_{j\gamma,i,k}) + \Sigma_\beta (\Gamma_{\gamma k,\beta} \Gamma_{ij}{}^{\beta} - \Gamma_{\gamma j,\beta} \Gamma_{ik}{}^{\beta})$
2	From 1 and changing dummy indices γ → p, i → r, j → s, k → v	$R_{prsv} = ½ (g_{vp,r,s} - g_{rv,p,s} + g_{rs,p,v} - g_{sp,r,v}) + \Sigma_\beta (\Gamma_{pv,\beta} \Gamma_{rs}{}^{\beta} - \Gamma_{ps,\beta} \Gamma_{rv}{}^{\beta})$
3	From 2 and changing dummy indices p →γ, r → i, s → γ, v → i	$R_{\gamma i\gamma i} = ½ (g_{i\gamma,i,\gamma} - g_{ii,\gamma,\gamma} + g_{i\gamma,\gamma,i} - g_{\gamma\gamma,i,i}) + \Sigma_\beta (\Gamma_{\gamma i,\beta} \Gamma_{i\gamma}{}^{\beta} - \Gamma_{\gamma\gamma,\beta} \Gamma_{ii}{}^{\beta})$
4	From 3, T10.3.16 Addition of Tensors: Commutative by Addition and LxK.3.2.8B Symmetry of Differential Order	$R_{\gamma i\gamma i} = ½ (g_{i\gamma,i,\gamma} + g_{i\gamma,i,\gamma} - g_{ii,\gamma,\gamma} - g_{\gamma\gamma,i,i}) + \Sigma_\beta (\Gamma_{\gamma i,\beta} \Gamma_{i\gamma}{}^{\beta} - \Gamma_{\gamma\gamma,\beta} \Gamma_{ii}{}^{\beta})$
5	From 4, T10.4.7 Scalars and Tensors: Distribution of a Tensor over Addition of Scalars, D4.1.19 Primitive Definition for Rational Arithmetic, D4.1.20 Negative Coefficient and T10.4.6 Scalars and Tensors: Distribution of a Scalar over Addition of Tensors	$R_{\gamma i\gamma i} = ½ [2g_{i\gamma,i,\gamma} - (g_{ii,\gamma,\gamma} + g_{\gamma\gamma,i,i})] + \Sigma_\beta (\Gamma_{\gamma i,\beta} \Gamma_{i\gamma}{}^{\beta} - \Gamma_{\gamma\gamma,\beta} \Gamma_{ii}{}^{\beta})$
∴	From 5, Tensors: Distribution of a Scalar over Addition of Tensors and D4.1.19 Primitive Definition for Rational Arithmetic	$R_{\gamma i\gamma i} = g_{i\gamma,\gamma,i} - ½ (g_{ii,\gamma,\gamma} + g_{\gamma\gamma,i,i}) + \Sigma_\beta (\Gamma_{\gamma i,\beta} \Gamma_{i\gamma}{}^{\beta} - \Gamma_{\gamma\gamma,\beta} \Gamma_{ii}{}^{\beta})$

From theorems T14.2.2 "Riemann-Christoffel Tensor First Kind; T14.2.4 Transpose First Two Indices", T14.2.5 "Riemann-Christoffel Tensor First Kind; Transpose Last Two Indices", T14.2.7 "Riemann-Christoffel Tensor First Kind; Cyclic Symmetry Identity", T14.2.8 "Riemann-Christoffel Tensor First Kind; Transpose Curvature Indices" it follows from these identities that distinct, none vanishing components of $R_{\gamma ijk}$, form three types, of none independent components, that are linked by the cyclic symmetry conditions, and of course their repeating quantities. Therefore, the number of independent tensor components having four distinct suffices, are twice that of quartet of numbers while three distinct suffices, are three times of quartet of numbers:

Table 14.2.1 Calculating the Number of Riemann-Christoffel Tensor Components

Count	Tensor Components	Distinct Indices	Repeating Tensor Components	Combinatorial Components	Calculated Number of Components
1	$R_{\gamma i\gamma i}$	two	1	$1 \times {}_nC_2$	$½\, n(n-1)$
2	$R_{\gamma i\gamma k}$	three	3	$3 \times {}_nC_3$	$½\, n(n-1)(n-2)$
3	$R_{\gamma ijk}$	four	2	$2 \times {}_nC_4$	$1/12\, n(n-1)(n-2)(n-3)$

Theorem 14.2.11 Riemann-Christoffel Tensor: Counting Total Number of Components

Step	Hypothesis	$N = {}^{1}/_{12}\, n^2(n^2 - 1)$
1	From Tb14.1.2 Calculating the Number of Riemann-Christoffel Tensor Components, total count 1,2 and 3	$N = {}^{1}/_{2}\, n(n - 1) + {}^{1}/_{2}\, n(n - 1)(n - 2) + {}^{1}/_{12}\, n(n - 1)(n - 2)(n - 3)$
2	From 1 and A4.2.14 Distribution	$N = {}^{1}/_{2}\, n(n - 1)[1 + (n - 2) + {}^{1}/_{6}(n - 2)(n - 3)]$
3	From 2, A4.2.5 Commutative Add and D4.1.19 Primitive Definition for Rational Arithmetic	$N = {}^{1}/_{2}\, n(n - 1)[(n - 1) + {}^{1}/_{6}(n - 2)(n - 3)]$
4	From 3, A4.2.13 Inverse Multp, A4.2.14 Distribution and D4.1.19 Primitive Definition for Rational Arithmetic	$N = {}^{1}/_{12}\, n(n - 1)[6(n - 1) + (n - 2)(n - 3)]$
5	From 4, A4.2.14 Distribution, D4.1.19 Primitive Definition for Rational Arithmetic, T4.12.4 Polynomial Quadratic: Product of N-Quantities and D4.1.17 Exponential Notation	$N = {}^{1}/_{12}\, n(n - 1)(6n - 6 + n^2 - 5n + 6)$
6	From 5 and A4.2.5 Commutative Add, by like terms	$N = {}^{1}/_{12}\, n(n - 1)(n^2 + 6n - 5n + 6 - 6)$
7	From 6, A4.2.14 Distribution, D4.1.19 Primitive Definition for Rational Arithmetic, A4.2.8 Inverse Add and A4.2.7 Identity Add	$N = {}^{1}/_{12}\, n(n - 1)(n^2 + n)$
8	From 7, D4.1.17 Exponential Notation, A4.2.12 Identity Multp and A4.2.14 Distribution	$N = {}^{1}/_{12}\, n(n - 1)\, n\, (n + 1)$
9	From 8, A4.2.10 Commutative Multp and D4.1.17 Exponential Notation	$N = {}^{1}/_{12}\, n^2(n - 1)(n + 1)$
∴	From 9 and T4.12.3 Polynomial Quadratic: Difference of Two Squares	$N = {}^{1}/_{12}\, n^2(n^2 - 1)$

By theorems T14.1.17 "Riemann-Christoffel Tensor First Kind to Second Kind" and T12.5.13 "Christoffel Symbol: Second Kind Symmetry for First Two Indices" the following theorems can be converted from covariant tensor form into contravariant.

Table 14.2.2 Riemann-Christoffel Tensors of the Second Kind

Theorem 14.2.12 Riemann-Christoffel Tensor Second Kind in Metric Form	$R^{\alpha}{}_{ijk} = ½(g_{k\tau,i,j} - g_{ik,\tau,j} + g_{ij,\tau,k} - g_{j\tau,i,k})\, g^{\tau\alpha} + \sum_{\beta} \sum_{r} g^{\tau\alpha}\, g_{r\beta}\, (\Gamma_{\tau k}{}^{r}\, \Gamma_{ij}{}^{\beta} - \Gamma_{\tau j}{}^{r}\, \Gamma_{ik}{}^{\beta})$
T14.1.15 Riemann-Christoffel Tensor Second Kind; Transpose last two indices	$R^{\alpha}{}_{ikj} = -R^{\alpha}{}_{ijk}$
Theorem 14.2.13 Riemann-Christoffel Tensor Second Kind; Transpose Second Two Indices	$R_{i}{}^{\alpha}{}_{jk} = -R^{\alpha}{}_{ijk}$
Theorem 14.2.14 Riemann-Christoffel Tensor Second Kind; Transpose Last Two Indices	$R_{jk}{}^{\alpha}{}_{i} = R^{\alpha}{}_{ijk}$
Theorem 14.2.15 Riemann-Christoffel Tensor Second Kind; Cyclic Symmetry Identity	$R^{\alpha}{}_{ijk} + R^{\alpha}{}_{jki} + R^{\alpha}{}_{kij} = \mathbf{0}$
Theorem 14.2.16 Riemann-Christoffel Tensor Second Kind; Transpose Curvature Indices	$R^{\alpha}{}_{jik} = R^{\alpha}{}_{ijk} + R^{\alpha}{}_{kij}$
Theorem 14.2.17 Riemann-Christoffel Tensor Second Kind; with Two Distinct Indices	$R^{\alpha}{}_{i\alpha i} = ½(g_{i\tau,i,\alpha} - g_{ii,\tau,\alpha} + g_{i\alpha,\tau,i} - g_{\alpha\tau,i,i})\, g^{\alpha\tau} + \sum_{\beta} \sum_{r} g^{\tau\alpha}\, g_{r\beta}\, (\Gamma_{\tau i}{}^{r}\, \Gamma_{i\alpha}{}^{\beta} - \Gamma_{\tau\alpha}{}^{r}\, \Gamma_{ii}{}^{\beta})$

Section 14.3 Bianchi's Identity

Definition 14.3.1 Geodesic Space $\mathfrak{G}^n$

A ***geodesic space*** is an affine space where a particle falling through it will take an optimum trajectory through the manifold, such that

$R_{\gamma ijk} \equiv (\nabla_j \Gamma_{ik,\gamma} - \nabla_k \Gamma_{ij,\gamma})$ Equation A

$\mathbf{0} \equiv \Sigma_\beta (\Gamma_{\gamma k,\beta} \Gamma_{ij}{}^\beta - \Gamma_{\gamma j,\beta} \Gamma_{ik}{}^\beta)$ Equation B

identically vanishing everywhere, since the trajectory traces out a parallel vector field, hence no change in direction forcing the Christoffel Symbols to vanish to zero as well. So it would follow that

$\mathbf{A}_{;k} \equiv \mathbf{A}_{,k}$ Equation B

Applying Langlands' Program this space can provide a region where Bianchi's Identity can be proven, under the basic principle of tensor calculus, so any proof in such a space is a proof in all spaces.

Theorem 14.3.1 Bianchi Identity Covariant

Step	Hypothesis	$R_{\gamma ijk;a} + R_{\gamma ika;j} + R_{\gamma iaj;k} = \mathbf{0}$
1	From T14.2.2 Riemann-Christoffel Tensor First Kind	$R_{\gamma ijk} = (\nabla_j \Gamma_{ik,\gamma} - \nabla_k \Gamma_{ij,\gamma}) + \Sigma_\beta (\Gamma_{\gamma k,\beta} \Gamma_{ij}{}^\beta - \Gamma_{\gamma j,\beta} \Gamma_{ik}{}^\beta)$
2	From 1 and D14.3.1A Geodesic Space, in a geodesic space, $\mathfrak{R}^n \xrightarrow{f} \mathfrak{G}^n$	$R_{\gamma ijk} = (\nabla_j \Gamma_{ik,\gamma} - \nabla_k \Gamma_{ij,\gamma})$
3	From 2, T12.7.1B Tensor Product: Uniqueness of Tensor Differentiation and D14.3.1B Geodesic Space, in a geodesic space	$R_{\gamma ijk;a} = (\nabla_j \Gamma_{ik,\gamma} - \nabla_k \Gamma_{ij,\gamma})_{,a}$
4	From 3 and LxK.5.1.1 Linear Distribution of Differential	$R_{\gamma i(jk;a)} = (\nabla^2{}_{j,a} \Gamma_{ik,\gamma} - \nabla^2{}_{k,a} \Gamma_{ij,\gamma})$
5	From 4 cyclically permuting the indices a → j, k → a and j → k	$R_{\gamma i(ka;j)} = (\nabla^2{}_{k,j} \Gamma_{ia,\gamma} - \nabla^2{}_{a,j} \Gamma_{ik,\gamma})$
6	From 5 cyclically permuting the indices j → k, a → j and k → a	$R_{\gamma i(aj;k)} = (\nabla^2{}_{a,k} \Gamma_{ij,\gamma} - \nabla^2{}_{j,k} \Gamma_{ia,\gamma})$
7	From 4, 5, 6 and sum all terms	$R_{\gamma i(jk;a)} + R_{\gamma i(ka;j)} + R_{\gamma i(aj;k)} =$ $\nabla^2{}_{j,a} \Gamma_{ik,\gamma} - \nabla^2{}_{k,a} \Gamma_{ij,\gamma} + \nabla^2{}_{k,j} \Gamma_{ia,\gamma} - \nabla^2{}_{a,j} \Gamma_{ik,\gamma} + \nabla^2{}_{a,k} \Gamma_{ij,\gamma} - \nabla^2{}_{j,k} \Gamma_{ia,\gamma}$
8	From 7 and T10.3.16 Addition of Tensors: Commutative by Addition	$R_{\gamma i(jk;a)} + R_{\gamma i(ka;j)} + R_{\gamma i(aj;k)} =$ $\nabla^2{}_{j,a} \Gamma_{ik,\gamma} - \nabla^2{}_{a,j} \Gamma_{ik,\gamma} + \nabla^2{}_{k,j} \Gamma_{ia,\gamma} - \nabla^2{}_{j,k} \Gamma_{ia,\gamma} + \nabla^2{}_{a,k} \Gamma_{ij,\gamma} - \nabla^2{}_{k,a} \Gamma_{ij,\gamma}$

9	From 8 and LxK.3.2.8B Symmetry of Differential Order	$R_{\gamma i(jk;a)} + R_{\gamma i(ka;j)} + R_{\gamma i(aj;k)} =$ $\nabla^2{}_{a,j}\,\Gamma_{ik,\gamma} - \nabla^2{}_{a,j}\,\Gamma_{ik,\gamma} + \nabla^2{}_{j,k}\,\Gamma_{ia,\gamma} - \nabla^2{}_{j,k}\,\Gamma_{ia,\gamma} + \nabla^2{}_{k,a}\,\Gamma_{ij,\gamma} - \nabla^2{}_{k,a}\,\Gamma_{ij,\gamma}$
∴	From 9, T10.3.19 Addition of Tensors: Inverse by Addition and T10.3.18 Addition of Tensors: Identity by Addition and $\mathfrak{G}^n \xrightarrow{f^{-1}} \mathfrak{R}^n$	$R_{\gamma ijk;a} + R_{\gamma ika;j} + R_{\gamma iaj;k} = \mathbf{0}$

Theorem 14.3.2 Bianchi Identity Contravariant

Step	Hypothesis	$R^{\alpha}{}_{ijk;a} + R^{\alpha}{}_{ika;j} + R^{\alpha}{}_{iaj;k} = \mathbf{0}$
1	From T14.3.1 Bianchi Identity: Covariant	$R_{\gamma ijk;a} + R_{\gamma ika;j} + R_{\gamma iaj;k} = \mathbf{0}$
2	From 1, T10.3.6 Product of Tensors: Uniqueness by Multiplication of Rank for p + q and T10.3.20 Addition of Tensors: Distribution of a Tensor over Addition of Tensors	$g^{\alpha\gamma}R_{\gamma ijk;a} + g^{\alpha\gamma}R_{\gamma ika;j} + g^{\alpha\gamma}R_{\gamma iaj;k} = g^{\alpha\gamma}\,(\mathbf{0})$
∴	From 2, T10.2.8 Tool: Zero Tensor Times a Tensor is a Zero Tensor and T14.1.17 Riemann-Christoffel Tensor First Kind to Second Kind	$R^{\alpha}{}_{ijk;a} + R^{\alpha}{}_{ika;j} + R^{\alpha}{}_{iaj;k} = \mathbf{0}$

Section 14.4 Riemann-Christoffel Tensors of Rank-2

Definition 14.4.1 Ricci's Tensor

$R_{ij} \equiv R^{\alpha}{}_{ij\alpha}$ Rank-2 Equation A

Definition 14.4.2 Riemannian or Einstein Spatial Curvature

$R \equiv g^{ij} R_{ij}$ Rank-0 Equation A

Theorem 14.4.1 Riemann-Christoffel Tensor Rank-2: Ricci Tensor

Step	Hypothesis	
		$R_{ij} = (\nabla_j [\nabla_i (\sqrt{g}) / \sqrt{g}\,] - \nabla_\alpha \Gamma_{ij}{}^{\alpha}) + \Sigma_\beta [\Gamma_{\beta j}{}^{\alpha} \Gamma_{i\alpha}{}^{\beta} - \Gamma_{ij}{}^{\beta} \nabla_\beta \ln (\sqrt{g})]$ EQ A
		$R_{ij} = (\nabla_j \Gamma_{i\alpha}{}^{\alpha} - \nabla_\alpha \Gamma_{ij}{}^{\alpha}) + \Sigma_\beta (\Gamma_{\beta j}{}^{\alpha} \Gamma_{i\alpha}{}^{\beta} - \Gamma_{\beta\alpha}{}^{\alpha} \Gamma_{ij}{}^{\beta})$ EQ B
1	From D14.1.2A Mixed Riemann-Christoffel tensor	$R^{\alpha}{}_{ijk} = (\nabla_j \Gamma_{ik}{}^{\alpha} - \nabla_k \Gamma_{ij}{}^{\alpha}) + \Sigma_\beta (\Gamma_{\beta j}{}^{\alpha} \Gamma_{ik}{}^{\beta} - \Gamma_{\beta k}{}^{\alpha} \Gamma_{ij}{}^{\beta})$
2	From 1 and setting $k = \alpha$	$R^{\alpha}{}_{ij\alpha} = (\nabla_j \Gamma_{i\alpha}{}^{\alpha} - \nabla_\alpha \Gamma_{ij}{}^{\alpha}) + \Sigma_\beta (\Gamma_{\beta j}{}^{\alpha} \Gamma_{i\alpha}{}^{\beta} - \Gamma_{\beta\alpha}{}^{\alpha} \Gamma_{ij}{}^{\beta})$
3	From 2, D14.4.1A Ricci's Tensor and T10.3.5 Tool: Tensor Substitution	$R_{ij} = (\nabla_j \Gamma_{i\alpha}{}^{\alpha} - \nabla_\alpha \Gamma_{ij}{}^{\alpha}) + \Sigma_\beta (\Gamma_{\beta j}{}^{\alpha} \Gamma_{i\alpha}{}^{\beta} - \Gamma_{\beta\alpha}{}^{\alpha} \Gamma_{ij}{}^{\beta})$
4	From 3, T12.8.7B Derivative of the Natural Log of the Covariant Determinate and T10.3.5 Tool: Tensor Substitution	$R_{ij} = (\nabla_j [\nabla_i \ln (\sqrt{g})] - \nabla_\alpha \Gamma_{ij}{}^{\alpha}) + \Sigma_\beta [\Gamma_{\beta j}{}^{\alpha} \Gamma_{i\alpha}{}^{\beta} - \nabla_\beta \ln (\sqrt{g})\, \Gamma_{ij}{}^{\beta}]$
5	From 4 and LxK.3.3.4B Differential a Natural Log base [e]	$R_{ij} = (\nabla_j [\nabla_i (\sqrt{g}) / \sqrt{g}\,] - \nabla_\alpha \Gamma_{ij}{}^{\alpha}) + \Sigma_\beta [\Gamma_{\beta j}{}^{\alpha} \Gamma_{i\alpha}{}^{\beta} - \nabla_\beta \ln (\sqrt{g})\, \Gamma_{ij}{}^{\beta}]$
∴	From 3, 4, DxK.3.2.2E Multiple Partial Differential Operator and T10.3.7 Product of Tensors: Commutative by Multiplication of Rank p + q → q + p	$R_{ij} = (\nabla_j [\nabla_i (\sqrt{g})] / \sqrt{g} - \nabla_\alpha \Gamma_{ij}{}^{\alpha}) + \Sigma_\beta [\Gamma_{\beta j}{}^{\alpha} \Gamma_{i\alpha}{}^{\beta} - \Gamma_{ij}{}^{\beta} \nabla_\beta \ln (\sqrt{g})]$ EQ A
		$R_{ij} = (\nabla_j \Gamma_{i\alpha}{}^{\alpha} - \nabla_\alpha \Gamma_{ij}{}^{\alpha}) + \Sigma_\beta (\Gamma_{\beta j}{}^{\alpha} \Gamma_{i\alpha}{}^{\beta} - \Gamma_{\beta\alpha}{}^{\alpha} \Gamma_{ij}{}^{\beta})$ EQ B

Theorem 14.4.2 Riemann-Christoffel Tensor Rank-2: Ricci Tensor in Determinate Form

Step	Hypothesis	
		$R_{ij} = \begin{vmatrix} \nabla_j & \nabla_\alpha \\ \Gamma_{ij}{}^{\alpha} & \Gamma_{i\alpha}{}^{\alpha} \end{vmatrix} + \Sigma_\beta \begin{vmatrix} \Gamma_{\beta j}{}^{\alpha} & \Gamma_{\beta\alpha}{}^{\alpha} \\ \Gamma_{ij}{}^{\beta} & \Gamma_{i\alpha}{}^{\beta} \end{vmatrix}$
1	From T14.4.1B Riemann-Christoffel Tensor Rank-2: Ricci Tensor	$R_{ij} = (\nabla_j \Gamma_{i\alpha}{}^{\alpha} - \nabla_\alpha \Gamma_{ij}{}^{\alpha}) + \Sigma_\beta (\Gamma_{\beta j}{}^{\alpha} \Gamma_{i\alpha}{}^{\beta} - \Gamma_{\beta\alpha}{}^{\alpha} \Gamma_{ij}{}^{\beta})$
∴	From 1 and T6.10.2 Determinate of a 2x2 Matrix	$R_{ij} = \begin{vmatrix} \nabla_j & \nabla_\alpha \\ \Gamma_{ij}{}^{\alpha} & \Gamma_{i\alpha}{}^{\alpha} \end{vmatrix} + \Sigma_\beta \begin{vmatrix} \Gamma_{\beta j}{}^{\alpha} & \Gamma_{\beta\alpha}{}^{\alpha} \\ \Gamma_{ij}{}^{\beta} & \Gamma_{i\alpha}{}^{\beta} \end{vmatrix}$

Theorem 14.4.3 Riemann-Christoffel Tensor Rank-2: Bianchi Identity

Step	Hypothesis	$g^{ij}\, R_{ij;a} - g^{ij}\, R_{ia;j} - g^{\gamma k}\, R_{\gamma a;k} = \mathbf{0}$
1	From T14.3.1 Bianchi Identity Covariant	$R_{\gamma ijk;a} + R_{\gamma ika;j} + R_{\gamma iaj;k} = \mathbf{0}$
2	From 1 and T10.3.6 Product of Tensors: Uniqueness by Multiplication of Rank for p + q	$g^{\gamma k}\, g^{ij}\, (R_{\gamma ijk;a} + R_{\gamma ika;j} + R_{\gamma iaj;k}) = g^{\gamma k}\, g^{ij}\, (\mathbf{0})$
3	From 2, T10.3.20 Addition of Tensors: Distribution of a Tensor over Addition of Tensors	$g^{\gamma k}\, g^{ij}\, R_{\gamma ijk;a} + g^{\gamma k}\, g^{ij}\, R_{\gamma ika;j} + g^{\gamma k}\, g^{ij}\, R_{\gamma iaj;k} = \mathbf{0}$
4	From 3, T10.2.8 Tool: Zero Tensor Times a Tensor is a Zero Tensor, T10.3.7 Product of Tensors: Commutative by Multiplication of Rank p + q → q + p and T10.3.8 Product of Tensors: Association by Multiplication of Rank for p + q + r	$g^{ij}\, (g^{\gamma k}\, R_{\gamma ijk;a}) + g^{ij}\, (g^{\gamma k}\, R_{\gamma ika;j}) + g^{\gamma k}\, (g^{ij}\, R_{\gamma iaj;k}) = \mathbf{0}$
5	From 4, T14.2.5 Riemann-Christoffel Tensor First Kind; Transpose Last Two Indices, D4.1.20A Negative Coefficient and T10.4.8 Scalars and Tensors: Scalar Commutative with Tensors	$g^{ij}\, (g^{\gamma k}\, R_{\gamma ijk;a}) - g^{ij}\, (g^{\gamma k}\, R_{\gamma iak;j}) - g^{\gamma k}\, (g^{ij}\, R_{\gamma ija;k}) = \mathbf{0}$
6	From 5 and T14.1.17 Riemann-Christoffel Tensor First Kind to Second Kind	$g^{ij}\, (R_{\gamma ij}{}^{\gamma}{}_{;a}) - g^{ij}\, (R_{\gamma ia}{}^{\gamma}{}_{;j}) - g^{\gamma k}\, (R_{\gamma i}{}^{i}{}_{a;k}) = \mathbf{0}$
∴	From 6 and D14.1.1A Ricci Tensor	$g^{ij}\, R_{ij;a} - g^{ij}\, R_{ia;j} - g^{\gamma k}\, R_{\gamma a;k} = \mathbf{0}$

Theorem 14.4.4 Riemann-Christoffel Tensor Rank-2: Contraction of the Bianchi Identity

Step	Hypothesis	$(R_a{}^k - ½\, \delta^k{}_a R\,)_{;k} = \mathbf{0}$
1	From T14.4.3 Riemann-Christoffel Tensor Rank-2: Bianchi Identity	$g^{ij}\, R_{ij;a} - g^{ij}\, R_{ia;j} - g^{\gamma k}\, R_{\gamma a;k} = \mathbf{0}$
2	From 1 and C12.8.5.4 Differentiation of Contravariant Metric and Covariant Coefficient	$(g^{ij}\, R_{ij}\,)_{;a} - (g^{ij}\, R_{ia}\,)_{;j} - (g^{\gamma k}\, R_{\gamma a}\,)_{;k} = \mathbf{0}$
3	From 2, D14.4.2 Riemannian or Einstein Spatial Curvature and T12.3.5A Metric Transformation from Contravariant to Covariant Coefficient	$R_{;a} - R_a{}^j{}_{;j} - R_a{}^k{}_{;k} = \mathbf{0}$
4	From 3 and changing dummy indices j → k	$R_{;a} - R_a{}^k{}_{;k} - R_a{}^k{}_{;k} = \mathbf{0}$
5	From 4, D4.1.20 Negative Coefficient, T10.4.7 Scalars and Tensors: Distribution of a Tensor over Addition of Scalars and D4.1.19 Primitive Definition for Rational Arithmetic	$R_{;a} - 2R_a{}^k{}_{;k} = \mathbf{0}$
6	From 5, T4.4.2 Equalities: Uniqueness of Multiplication	$-½\, (R_{;a} - 2R_a{}^k{}_{;k}\,) = -½\, \mathbf{0}$
7	From 6, T4.4.1 Equalities: Any Quantity Multiplied by Zero is Zero, A4.1.14 Distribution, T4.4.18 Equalities: Product by Division, Factorization by 2 and T4.8.2B Integer Exponents: Negative One Squared	$R_a{}^k{}_{;k} - ½\, R_{;a} = \mathbf{0}$
8	From 7, T12.3.3C Contracted Covariant Bases Vector with Contravariant Vector, T10.3.7 Product of Tensors: Commutative by Multiplication of Rank p + q → q + p and A10.2.14 Correspondence of Tensors and Tensor Coefficients	$R_a{}^k{}_{;k} - ½\, \delta^k{}_a R_{;k} = \mathbf{0}$
9	From 8 and T12.9.1B Differentiated Kronecker Delta and Fundamental Tensors	$R_a{}^k{}_{;k} - ½\, (\delta^k{}_a R)_{;k} = \mathbf{0}$
10	From 9, replacing dummy indices k →b	$R_a{}^b{}_{;b} - ½\, (\delta^b{}_a R)_{;b} = \mathbf{0}$
∴	From 10 and T12.7.2 Tensor Summation: Covariant Derivative as a Linear Operator	$(R_a{}^b - ½\, \delta^b{}_a R\,)_{;b} = \mathbf{0}$

Theorem 14.4.5 Riemann-Christoffel Tensor Rank-2: Einstein's Mixed Tensor

Step	Hypothesis	$T_a{}^b\,\xi \equiv R_a{}^b - ½\,\delta^b{}_a R$
1	From T14.4.4 Riemann-Christoffel Tensor Rank-2: Contraction of the Bianchi Identity	$(R_a{}^b - ½\,\delta^b{}_a R\,)_{;b} = \mathbf{0}$
2	From 1 and A10.2.9 Tensor Closure under Subtraction	$(T_a{}^b)_{;b} = \mathbf{0}$ [14.4.1]
∴	From 1, 2 and A10.2.6A Equality of Tensors	$T_a{}^b\,\xi \equiv R_a{}^b - ½\,\delta^b{}_a R$ for ξ Einstein's constant of proportionality

Theorem 14.4.6 Riemann-Christoffel Tensor Rank-2: Einstein's Covariant Tensor

Step	Hypothesis	$T_{ab}\,\xi = R_{ab} - ½\,g_{ab}R$
1	From T14.4.5 Riemann-Christoffel Tensor Rank-2: Einstein's Mixed Tensor	$T_a{}^b\,\xi = R_a{}^b - ½\,\delta^b{}_a R$
2	From 1, T12.3.5A Metric Transformation from Contravariant to Covariant Coefficient and T12.2.8B Covariant and Contravariant Metric Inverse	$g^{kb}\,T_{ak}\,\xi = g^{kb}R_{ak} - ½\,g^{kb}g_{ak}R$
3	From 2 and T10.3.20 Addition of Tensors: Distribution of a Tensor over Addition of Tensors	$g^{kb}\,(T_{ak}\,\xi) = g^{kb}\,(R_{ak} - ½\,g_{ak}R)$
4	From 3 and T10.3.6 Product of Tensors: Uniqueness by Multiplication of Rank for p + q, equating term-by-term and	$T_{ak}\,\xi = R_{ak} - ½\,g_{ak}R$
∴	From 4 and replacing dummy indices k →b	$T_{ab}\,\xi = R_{ab} - ½\,g_{ab}R$

[14.4.1]Note: The differential of the Einstein curvature tensor $[T_a{}^k]$ is zero, hence has constant curvature over a Riemannian space.

Theorem 14.4.7 Riemann-Christoffel Tensor Rank-0: Contraction of Einstein's Covariant Tensor

Step	Hypothesis	$T\ \xi = -(\frac{1}{2} n - 1)\ R$
1	From T14.4.6 Riemann-Christoffel Tensor Rank-2: Einstein's Covariant Tensor	$T_{ab}\ \xi = R_{ab} - \frac{1}{2}\ g_{ab}R$
2	From 1, T12.3.5A Metric Transformation from Contravariant to Covariant Coefficient and T12.2.8B Covariant and Contravariant Metric Inverse	$g^{ar}T_{rb}\ \xi = g^{ar}R_{rb} - \frac{1}{2}\ g^{ar}g_{rb}R$
3	From 2, D14.4.2 Riemannian or Einstein Spatial Curvature and T12.2.8B Covariant and Contravariant Metric Inverse	$T\ \xi = R - \frac{1}{2}\ (\sum_a \delta^a{}_b)\ R$
4	From 3, D6.1.6 Identity Matrix and LxE.3.29 Sums of Powers of Zero for the First n-Integers	$T\ \xi = R - \frac{1}{2}\ nR$
5	From 4 and A4.2.14 Distribution	$T\ \xi = (1 - \frac{1}{2}\ n)\ R$
∴	From 5, T4.8.2B Integer Exponents: Negative One Squared, D4.1.20A Negative Coefficient, A4.2.14 Distribution and A4.2.5 Commutative Add	$T\ \xi = -(\frac{1}{2}\ n - 1)\ R$

Theorem 14.4.8 Riemann-Christoffel Tensor Rank-2: Einstein's Contravariant Tensor

Step	Hypothesis	$T^{ab}\ \xi = R^{ab} - \frac{1}{2}\ g^{ab}R$
1	From T14.4.5 Riemann-Christoffel Tensor Rank-2: Einstein's Mixed Tensor	$T_a{}^k\ \xi = R_a{}^k - \frac{1}{2}\ \delta^k{}_aR$
2	From 1, T12.3.9A Metric Transformation from Covariant to Contravariant Coefficient, T12.2.8B Covariant and Contravariant Metric Inverse and T10.3.7 Product of Tensors: Commutative by Multiplication of Rank p + q → q + p	$g_{ra}T^{kr}\ \xi = g_{ra}R^{kr} - \frac{1}{2}\ g_{ra}g^{kr}R$
3	From 2 and T10.3.20 Addition of Tensors: Distribution of a Tensor over Addition of Tensors	$g_{ra}\ (T^{kr}\ \xi) = g_{ra}\ (R^{kr} - \frac{1}{2}\ g^{kr}R)$

4	From 3 and T10.3.6 Product of Tensors: Uniqueness by Multiplication of Rank for p + q	$T^{kr}\ \xi = R^{kr} - ½\ g^{kr}R$
∴	From 4 and replacing dummy indices (k →a, r →b)	$T^{ab}\ \xi = R^{ab} - ½\ g^{ab}R$

Theorem 14.4.9 Riemann-Christoffel Tensor First Kind: Contracted to Spatial Curvature

Step	Hypothesis	$R = (g^{ij}\ g^{\gamma k})\ R_{\gamma ijk}$
1	From D14.4.2 Riemannian or Einstein Spatial Curvature; contracts from rank-2 to rank-0	$R \equiv g^{ij}\ R_{ij}$
2	From 1and T12.3.14A Dot Product of Contravariant Vectors; contracts from rank-4 to rank-2	$R_{ij} = g^{\gamma k}\ R_{\gamma ijk}$
3	From 1, 2, T10.2.5 Tool: Tensor Substitution and A10.2.14 Correspondence of Tensors and Tensor Coefficients	$R = g^{ij}\ (g^{\gamma k}\ R_{\gamma ijk})$
∴	From 3, T10.3.8 Product of Tensors: Association by Multiplication of Rank for p + q + r and A10.2.14 Correspondence of Tensors and Tensor Coefficients	$R = (g^{ij}\ g^{\gamma k})\ R_{\gamma ijk}$

Section 14.5 Riemann-Christoffel Tensor in Geodesic Space $\mathfrak{G}^n$

Theorem 14.5.1 Geodesic Space $\mathfrak{G}^n$: Riemann-Christoffel Tensor Coefficient of the First Kind

Step	Hypothesis	$R_{\gamma ijk} = ½(g_{k\gamma,i,j} - g_{ik,\gamma,j} + g_{ij,\gamma,k} - g_{j\gamma,i,k})$
1	From T14.2.3 Riemann-Christoffel Tensor First Kind in Metric Form	$R_{\gamma ijk} = ½(g_{k\gamma,i,j} - g_{ik,\gamma,j} + g_{ij,\gamma,k} - g_{j\gamma,i,k}) + \sum_\beta (\Gamma_{\gamma k,\beta}\, \Gamma_{ij}{}^{\beta} - \Gamma_{\gamma j,\beta}\, \Gamma_{ik}{}^{\beta})$
∴	From 1 and D14.3.1B Geodesic Space $\mathfrak{G}^n$	$R_{\gamma ijk} = ½(g_{k\gamma,i,j} - g_{ik,\gamma,j} + g_{ij,\gamma,k} - g_{j\gamma,i,k})$

Theorem 14.5.2 Geodesic Space $\mathfrak{G}^n$: Laplacian Contravariant Metric

Step	Hypothesis	$R = -\sum_i \nabla^2 g^{ii} - 2\sum_{i<j} g^{ij}{}_{,i,j}$
1	From T14.5.1 Geodesic Space $\mathfrak{G}^n$: Riemann-Christoffel Tensor Coefficient of the First Kind	$R_{\gamma ijk} = ½(g_{k\gamma,i,j} - g_{ik,\gamma,j} + g_{ij,\gamma,k} - g_{j\gamma,i,k})$
2	From 1, T14.4.9 Riemann-Christoffel Tensor First Kind: Contracted to Spatial Curvature, T10.2.5 Tool: Tensor Substitution and T10.2.6 Tool: Commutative Scalar Times a Tensor	$R = ½ (g^{ij} g^{\gamma k}) (g_{k\gamma,i,j} - g_{ik,\gamma,j} + g_{ij,\gamma,k} - g_{j\gamma,i,k})$
3	From 2 and T10.3.20 Addition of Tensors: Distribution of a Tensor over Addition of Tensors	$R = ½ [(g^{ij} g^{\gamma k}) g_{k\gamma,i,j} - (g^{ij} g^{\gamma k}) g_{ik,\gamma,j} +$ $(g^{ij} g^{\gamma k}) g_{ij,\gamma,k} - (g^{ij} g^{\gamma k}) g_{j\gamma,i,k}]$
4	From 3 and C12.8.5.4 Differentiation of Contravariant Metric and Covariant Coefficient	$R = ½ [g^{ij} (g^{\gamma k} g_{k\gamma})_{,i,j} - g^{\gamma k} (g^{ij} g_{ik})_{,\gamma,j} +$ $g^{\gamma k} (g^{ij} g_{ij})_{,\gamma,k} - g^{\gamma k} (g^{ij} g_{j\gamma})_{,i,k}]$
5	From 4, T12.2.14 Contravariant Metric Inner Product is Symmetrical and T12.2.8B Covariant and Contravariant Metric Inverse	$R = ½ [g^{ij} (\delta^{\gamma}{}_{\gamma})_{,i,j} - g^{\gamma k} (\delta^{j}{}_{k})_{,\gamma,j} +$ $g^{\gamma k} (\delta^{i}{}_{i})_{,\gamma,k} - g^{\gamma k} (\delta^{i}{}_{\gamma})_{,i,k}]$
6	From 7, T12.2.16B Covariant-Contravariant Kronecker Delta Summation of like Indices and C12.8.5.4 Differentiation of Contravariant Metric and Covariant Coefficient	$R = ½ [g^{ij} (n)_{,i,j} - (g^{\gamma k} \delta^{j}{}_{k})_{,\gamma,j} + g^{\gamma k} (n)_{,\gamma,k} - (g^{\gamma k} \delta^{i}{}_{\gamma})_{,i,k}]$

7	From 6, T12.3.8B Contracted Contravariant Bases Vector with Covariant Vector, LxK.3.3.1B Differential of a Constant and A4.2.7 Identity Add	$R = ½ (-g^{\gamma k}{}_{,\gamma,k} - g^{\gamma k}{}_{,\gamma,k})$
8	From 7 and T4.3.10 Equalities: Summation of Repeated Terms by 2	$R = ½\, 2 (-g^{\gamma k}{}_{,\gamma,k})$
9	From 8 and A4.2.13 Inverse Multp	$R = -g^{\gamma k}{}_{,\gamma,k}$
10	From 9 and T4.12.5 Polynomial Quadratic: Perfect Square of N-Quantities	$R = -\sum_i g^{ii}{}_{,i,i} - 2\sum_{i<j} g^{ij}{}_{,i,j}$
∴	From 10 and DxK.6.2.1A Covariant Laplacian Operator	$R = -\sum_i \nabla^2 g^{ii} - 2\sum_{i<j} g^{ij}{}_{,i,j}$

Theorem 14.5.3 Geodesic Space $\mathfrak{G}^n$: Laplacian Contravariant Metric in an Orthogonal Space

1g	Given	$0 = g^{ij}{}_{,i,j}$ in an orthogonal space
Step	Hypothesis	$R = -\sum_i \nabla^2 g^{ii}$
1	From T14.5.2 Geodesic Space $\mathfrak{G}^n$: Laplacian Contravariant Metric	$R = -\sum_i \nabla^2 g^{ii} - 2\sum_{i<j} g^{ij}{}_{,i,j}$
∴	From 1g, 1, T10.2.5 Tool: Tensor Substitution and A4.2.13 Inverse Multp	$R = -\sum_i \nabla^2 g^{ii}$

Tensor Calculus & Physics: A General Treatise

Table of Contents

List of Tables

List of Figures

List of Observations

Chapter 15 Tensor Calculus: Del Tensor Operators

Section 15.1 Gradient Tensor Operators

Observation 15.1.1 Dimensionality of the Gradient Operator

From the partial differential of the gradient it has the axial dimensions of $[L^{-1}]$ in the appropriate units. The bases vector carries units of suitable dimensions, units that are carried by the scaling factor of the base vector's magnitude. However this must be qualified, if the denominator of the derivative is carrying unit dimensions, then the scaling factor is unit less, on the other hand if the derivative does not carry the units for the dimension, then the scaling factor includes them. This means that the gradient and the vectors it operates on have a special form in order to generally work; this form is not a natural, logical extension.

As an example if the gradient axis has no units, lets say in radians and then the scaling factor must carry the radial component of unit length. Conversely if the gradient axes has units of radial length, then the scaling factor has no units or in units of radians. In short the net product of the scaling factor and axial denominator must have units of length one way or another. This rationale defines the construction of the gradient in terms of the covariant-contravariant interrelationship.

By definition and tradition, the gradient coefficient is covariant implying that the scaling factor must also be covariant, from theorem T12.2.10 "Covariant and Contravariant Scale Factor; Reciprocal Relationship", this leads back to the bases vector being contravariant and likewise the gradient operator.

Axiom 15.1.1 Gradient Unit Length

The net product of the scaling factor and axial denominator must have units of length one way or another,

$[h^i][x_i^{-1}] \rightarrow [\varnothing]$	Normalized Gradient	Dimensional Equation A
$[h^i] \rightarrow [L]$	Scale Factor	Dimensional Equation B
$[x_i] \rightarrow [L]$	Gradient Axis Variable	Dimensional Equation C

Axiom 15.1.2 Gradient Covariant-Contravariant Interrelationship

The scaling factor takes its vector duality from the bases vector

$[h_i] \rightarrow [b_i]$	Covariant	Equation A
$[h^i] \rightarrow [b^i]$	Contravariant	Equation B
$[x_i]$	Covariant	Equation C

In order to be consistent with the literature and maintain neutrality with the partial of the function being operated on the axis variable is always taken to be covariant.

Observation 15.1.2 Gradient Scales as an n-Dimensional Operator

The gradient operators either in covariant or contravariant forms are linear and in general dimensional structure to be naturally extended to n-dimensions. This means any linear operator comprised of a general gradient operator will natural size from n-dimensions to m-dimensions either for a greater or smaller [n, m] in scale. As an example integral are nonlinear being specifically defined or fixed dimensionally, hence do not scale

Axiom 15.1.3 Dimensionally Scaling Gradient Operators

Gradient linear operators, either in covariant or contravariant forms, are dimensionally general in structure so naturally size from [n] to [m]-dimensions.

Lemma	Flux	Descriptive Name
LxVIE.6.1	$\vec{\Phi} = \oint_S \Phi\, d\vec{S} = \int_V \nabla\Phi\, dV$	Divergence Theorem of Gauss Scalar Multiplication
Dx9.6.3.2A	$\nabla\Phi \equiv \sum_i \Phi_{,i}\, \mathbf{b}^i$ [9.6.2]	Covariant Differential Operator, DEL [9.6.1]

Theorem 15.1.1 Covariant Gradient: in Orthogonal Curvilinear Coordinates

Step	Hypothesis	$\nabla\Phi = \sum_i [(1/h_i)\, \Phi_{,i}]\, \mathbf{e}^i$ EQ A	$\nabla\Phi = \sum_i (h^i\, \Phi_{,i})\, \mathbf{e}^i$ EQ B
1	From Dx9.6.3.2A Covariant Differential Operator, DEL	$\nabla\Phi = \sum_i \Phi_{,i}\, \mathbf{b}^i$	
2	From 1, D12.1.7A Contravariant Base Vector and T10.2.6 Tool: Commutative Scalar Times a Tensor	$\nabla\Phi = \sum_i h^i\, \Phi_{,i}\, \mathbf{e}^i$	
∴	From 2, T12.2.10 Covariant and Contravariant Scale Factor; Reciprocal Relationship and A4.2.3 Substitution	$\nabla\Phi = \sum_i [(1/h_i)\, \Phi_{,i}]\, \mathbf{e}^i$ EQ A	$\nabla\Phi = \sum_i (h^i\, \Phi_{,i})\, \mathbf{e}^i$ EQ B

Theorem 15.1.2 Covariant Gradient: in Orthogonal Curvilinear Coordinates

Step	Hypothesis	$\nabla\Phi = \sum_i (1/h_i)\, \Phi_{,i}\, \mathbf{e}^i$
1	From Dx9.6.3.2A Covariant Differential Operator, DEL	$\nabla\Phi = \sum_i \Phi_{,i}\, \mathbf{b}^i$
2	From 1 and D12.1.6A Covariant Base Vector	$\nabla\Phi = \sum_i \Phi_{,i}\, h^i\, \mathbf{e}^i$
∴	From 2, T12.2.10 Covariant and Contravariant Scale Factor; Reciprocal Relationship, A4.2.3 Substitution and T10.2.6 Tool: Commutative Scalar Times a Tensor	$\nabla\Phi = \sum_i (1/h_i)\, \Phi_{,i}\, \mathbf{e}^i$

Observation 15.1.3 Anti-Contravariant Gradient

The simple differential $[\Phi_{,i}]$ can only be a covariant coefficient by A15.1.2 "Gradient Covariant-Contravariant Interrelationship", hence a contravariant coefficient proof would be exclude. It follows then that theorem T15.1.2 "Contravariant Gradient: in Orthogonal Curvilinear Coordinates" is barred.

Section 15.2 Divergent Tensor Operator

Lemma	Flux	Descriptive Name
LxVIE.6.2	$\Phi = \oint_S \vec{F} \bullet d\vec{S} = \int_V \nabla \bullet \vec{F}\, dV$	Divergence Theorem of Gauss Dot Product Multiplication
Dx9.6.5.2	$\nabla \bullet \mathbf{F} \equiv \sum_i F^i{}_{;i}$	Divergence Operating on a Contravariant Vector

Theorem 15.2.1 Covariant Divergent Operator: Definition of Divergence

1g 2g	Given	$\mathbf{F} = F^i\, \mathbf{b}_i$ in unitary system space Einsteinian Summation Notation does apply.
Step	Hypothesis	$\nabla \bullet \mathbf{F} = F^{i;i}$
1	From 1g and T15.1.1B Covariant Gradient: in Orthogonal Curvilinear Coordinates	$\nabla \bullet \mathbf{F} = (\nabla_i\, \mathbf{b}^i) \bullet (F^j\, \mathbf{b}_j)$
2	From 1 and T8.3.7 Dot Product: Distribution of Dot Product Across Addition of Vectors	$\nabla \bullet \mathbf{F} = \nabla_i\, (\mathbf{b}^i \bullet F^j\, \mathbf{b}_j\,)$
3	From 2, T12.6.2A Differentiation of Tensor: Rank-1 Covariant and T8.3.10A Dot Product: Right Scalar-Vector Association	$\nabla \bullet \mathbf{F} = (F^{j;i})\, (\mathbf{b}^i \bullet \mathbf{b}_j)$
4	From 3 and D12.1.5D Mixed Metric Dyadic	$\nabla \bullet \mathbf{F} = (F^{j;i})\, \delta^i{}_j$
∴	From 4, D6.1.6 Identity Matrix and evaluating the Kronecker Delta over summation	$\nabla \bullet \mathbf{F} = F^{i;i}$

Theorem 15.2.2 Covariant Divergent Operator: On Covariant Vector

Step	Hypothesis	$\nabla \bullet \mathbf{F} = [\sum_i (\sqrt{g}\, F^i)_{,i}\,] / \sqrt{g}$
1	From T15.2.1 Covariant Divergent Operator: Definition of Divergence	$\nabla \bullet \mathbf{F} = \sum_i F^{i;i}$
2	From 1 and T12.6.2B Differentiation of Tensor: Rank-1 Covariant	$\nabla \bullet \mathbf{F} = \sum_i F^i_{,i} + \sum_i \sum_j \Gamma_{ji}{}^i\, F^j$
3	From 2 and interchanging order of summation i ←→ j	$\nabla \bullet \mathbf{F} = \sum_i F^i_{,i} + \sum_j (\sum_i \Gamma_{ji}{}^i)\, F^j$
4	From 3, T12.8.7B Derivative of the Natural Log of the Covariant Determinate and T10.2.5 Tool: Tensor Substitution	$\nabla \bullet \mathbf{F} = \sum_i F^i_{,i} + \sum_j (\ln \sqrt{g})_{,j}\, F^j$
5	From 4 and changing dummy indices j → i	$\nabla \bullet \mathbf{F} = \sum_i F^i_{,i} + \sum_i (\ln \sqrt{g})_{,i}\, F^i$
6	From 5 and LxK.3.3.4B Differential a Natural Log base [e]	$\nabla \bullet \mathbf{F} = \sum_i F^i_{,i} + (\sqrt{g})_{,j} / \sqrt{g}\, F^j$
7	From 6, A4.2.13 Inverse Multp and T10.2.6 Tool: Commutative Scalar Times a Tensor	$\nabla \bullet \mathbf{F} = \sum_i [\sqrt{g}\, F^i_{,i}] / \sqrt{g} + \sum_j [\, (\sqrt{g})_{,j}\,] / \sqrt{g}\, F^j$
8	From 7 and T10.4.6 Scalars and Tensors: Distribution of a Scalar over Addition of Tensors	$\nabla \bullet \mathbf{F} = [\sum_i \sqrt{g}\, F^i_{,i} + (\sqrt{g})_{,i}\, F^i\,] / \sqrt{g}$
∴	From 8 and T12.7.8 Tensor Product: Chain Rule for Mixed Binary Product	$\nabla \bullet \mathbf{F} = [\sum_i (\sqrt{g}\, F^i)_{,i}\,] / \sqrt{g}$

Theorem 15.2.3 Covariant Divergent Operator: Cross product Proof Orthogonal Coordinates

1g	Given [SPI59 pg 149]	$\nabla u_1 = \mathbf{e}_1 / h_1$
2g		$\nabla u_2 = \mathbf{e}_2 / h_2$
3g		$\nabla u_3 = \mathbf{e}_3 / h_3$
Step	Hypothesis	$\nabla \bullet (A^1 \mathbf{e}_1) = \nabla_1[(\rho_v / h_1) A^1] / \rho_v$
1	From T12.4.1 Differential Relationship for Covariant Bases Vectors	$\nabla u_1 = \mathbf{e}_1 / h_1$
2	From 1 and A4.4.5B Equalities: Reversal of Right Cancellation by Multiplication	$\mathbf{e}_1 = h_1 \nabla u_1$
3	From 2g, 3g, 2, F7.6.1B+cycle Grassmann's Product of Permuted Unit Vectors and A7.9.2 Substitution of Vectors	$\mathbf{e}_1 = h_2 \nabla u_2 \text{ x } h_3 \nabla u_3$
4	From 3 and A7.8.5 Parallelogram Law: Commutative and Scalability of Collinear Vectors	$\mathbf{e}_1 = h_2\, h_3 \nabla u_2 \text{ x } \nabla u_3$
5	From 4, T7.10.1 Uniqueness of Scalar Multiplication to Vectors and LxK.6.1.1B Uniqueness of Divergence Operator	$\nabla \bullet (A^1 \mathbf{e}_1) = \nabla \bullet (A^1\, h_2\, h_3 \nabla u_2 \text{ x } \nabla u_3)$
6	LxK.6.1.5 Divergence of Scalar Vector Product	$\nabla \bullet (A^1 \mathbf{e}_1) = \nabla (A^1\, h_2\, h_3) \bullet (\nabla u_2 \text{ x } \nabla u_3)$ $+ A^1\, h_2\, h_3\, \nabla \bullet (\nabla u_2 \text{ x } \nabla u_3)$
7	From 6, LxK.6.1.7 Divergence of Cross Product, LxK.6.1.11 Curl of Gradient Operator and LxK.6.1.12 Divergence of Curl Operator	$\nabla \bullet (A^1 \mathbf{e}_1) = \nabla (A^1\, h_2\, h_3) \bullet (\nabla u_2 \text{ x } \nabla u_3) + A_1\, h_2\, h_3\, 0$
8	From 7, T4.4.1 Equalities: Any Quantity Multiplied by Zero is Zero and A4.2.7. Identity Add	$\nabla \bullet (A^1 \mathbf{e}_1) = \nabla (A^1\, h_2\, h_3) \bullet (\mathbf{e}_1 / h_2\, h_3)$
9	From 8 and T15.1.1 Covariant Gradient: in Orthogonal Curvilinear Coordinates	$\nabla \bullet (A^1 \mathbf{e}_1) = [\nabla_1 (A^1\, h_2\, h_3)\, \mathbf{e}^1 / h_1 + \nabla_2 (A^1\, h_2\, h_3)\, \mathbf{e}^2 / h_2$ $+ \nabla_3 (A^1\, h_2\, h_3)\, \mathbf{e}^3 / h_3] \bullet (\mathbf{e}_1 / h_2\, h_3)$
10	From 9 and T8.3.7 Dot Product: Distribution of Dot Product Across Addition of Vectors	$\nabla \bullet (A^1 \mathbf{e}_1) =$ $[\nabla_1 (A^1\, h_2\, h_3)\, (\mathbf{e}^1 \bullet \mathbf{e}_1) / h_1 h_2 h_3$ $+ \nabla_2 (A^1\, h_2\, h_3)\, (\mathbf{e}^2 \bullet \mathbf{e}_1) / h_2 h_2 h_3$ $+ \nabla_3 (A^1\, h_2\, h_3)\, (\mathbf{e}^3 \bullet \mathbf{e}_1) / h_2 h_3 h_3]$
11	From 10, D12.1.5D Mixed Metric Dyadic, A4.2.12 Identity Multp, T4.4.1 Equalities: Any Quantity Multiplied by Zero is Zero and A4.2.7. Identity Add	$\nabla \bullet (A^1 \mathbf{e}_1) = \nabla_1 (A^1\, h_2\, h_3) / h_1\, h_2\, h_3$

12	From 11 and A4.2.13 Inverse Multp	$\nabla \bullet (A^1\ \mathbf{e}_1) = \nabla_1(A^1\ h_1\ h_2\ h_3 / h_1) / h_1\ h_2\ h_3$
∴	From 12, D12.1.13 Space Density Factor and A4.2.3 Substitution	$\nabla \bullet (A^1\ \mathbf{e}_1) = \nabla_1[(\rho_v / h_1)\ A^1] / \rho_v$

From theorem T15.2.3 "Covariant Divergent Operator: Cross product Proof Orthogonal Coordinates" the derivative kernel goes from [$\sqrt{g}$] → [ρ_v / h_i] with a scale factor in the denominator in an orthogonal system.

Theorem 15.2.4 Covariant Divergent Operator: Orthogonal Divergent Operator

Step	Hypothesis	$\nabla \bullet \mathbf{F} = (\rho_v / h_i\ F_i)_{,i} / \rho_v$ Orthogonal System
1	From T15.2.2 Covariant Divergent Operator: On Covariant Vector	$\nabla \bullet \mathbf{F} = (\sqrt{g}\ F^i)_{,i} / \sqrt{g}$
∴	From 1, T12.2.19 Covariant Metric Determinate in Orthogonal Curvilinear Coordinates, A4.2.3 Substitution, T12.3.19 Splitting an Orthogonal Covariant Contraction and T10.2.5 Tool: Tensor Substitution	$\nabla \bullet \mathbf{F} = (\rho_v / h_i\ F_i)_{,i} / \rho_v$ Orthogonal System

Yet theorem T15.2.4 "Covariant Divergent Operator: Orthogonal Divergent Operator" works by relying on a contrived crutch of [$F_i = h_i F^i$] theorem T12.3.19 "Splitting an Orthogonal Covariant Contraction" shows the true relationship [$F_i = h_i^2 F^i$]. So theorem T15.2.4 is false. Lets try the following set of theorems.

Theorem 15.2.5 Covariant Divergent Operator: Definition of Divergence in a Unitary System

1g	Given	$\mathbf{F} = F^i\ \mathbf{e}_i$ in unitary system
Step	Hypothesis	$\nabla \bullet \mathbf{F} = h^i\ F^{i;i}$
1	From 1g, T15.1.1B Covariant Gradient: in Orthogonal Curvilinear Coordinates and T8.3.7 Dot Product: Distribution of Dot Product Across Addition of Vectors	$\nabla \bullet \mathbf{F} = \nabla_i\ \mathbf{b}^i \bullet (F^j\ \mathbf{e}_j)$
2	From 1, T12.6.2A Differentiation of Tensor: Rank-1 Covariant and T8.3.10A Dot Product: Right Scalar-Vector Association D12.1.7 Contravariant Base Vector	$\nabla \bullet \mathbf{F} = h^i\ F^{j;i}\ (\mathbf{e}^i \bullet \mathbf{e}_j)$
3	From 2 and D12.1.5D Mixed Metric Dyadic	$\nabla \bullet \mathbf{F} = h^i\ F^{j;i}\ \delta^i_j$
∴	From 3, T12.3.4C Contracted Contravariant Bases Vector with Covariant Vector and T12.6.2A Differentiation of Tensor: Rank-1 Covariant	$\nabla \bullet \mathbf{F} = h^i\ F^{i;i}$

Theorem 15.2.6 Covariant Divergent Operator: Scale Factor Distribution

Step	Hypothesis	$\nabla \bullet \mathbf{F} = \sum_i (h^i F^{i;i})$ EQ A	$\sum_i h^i (F^{i;i}) = \sum_i (h^i F^i)_{;i}$ EQ B
1	From T15.2.2 Covariant Divergent Operator: Proof by Definition and T10.2.6 Tool: Commutative Scalar Times a Tensor	$\nabla \bullet \mathbf{F} = \sum_i F^{i;i} h^i$	
2	From 1 and T12.3.3B Contracted Covariant Bases Vector with Contravariant Vector	$\nabla \bullet \mathbf{F} = \sum_i \sum_j F^{i;j} \delta_i^{\ j} h^i$	
3	From 2 and D12.1.5A Mixed Metric Dyadic	$\nabla \bullet \mathbf{F} = \sum_i \sum_j F^{i;j} (\mathbf{b}_i \bullet \mathbf{b}^j) h^i$	
4	From 3, T8.3.3 Dot Product: Commutative Operation and T8.3.7 Dot Product: Distribution of Dot Product Across Addition of Vectors	$\nabla \bullet \mathbf{F} = \sum_i (\sum_j F^{i;j} \mathbf{b}^j) \bullet \mathbf{b}_i h^i$	
5	From 4, D12.1.7A Contravariant Base Vector and T10.2.6 Tool: Commutative Scalar Times a Tensor	$\nabla \bullet \mathbf{F} = \sum_i [\sum_j (h^j F^i \mathbf{e}^j)_{;j}] \bullet \mathbf{b}_i h^i$	
6	From 5 and T12.6.2A Differentiation of Tensor: Rank-1 Covariant	$\nabla \bullet \mathbf{F} = \sum_i \sum_j (h^j F^i)_{;j} \mathbf{e}^j \bullet \mathbf{b}_i h^i$	
7	From 6 and D12.1.6A Covariant Base Vector and	$\nabla \bullet \mathbf{F} = \sum_i \sum_j (h^j F^i)_{;j} (\mathbf{e}^j \bullet \mathbf{e}_i) h_i h^i$	
8	From 7 and D12.1.5D Mixed Metric Dyadic	$\nabla \bullet \mathbf{F} = \sum_i \sum_j (h^j F^i)_{;j} \delta^j_{\ i}$	
9	From 8, D6.1.6 Identity Matrix and evaluating the Kronecker Delta over summation	$\nabla \bullet \mathbf{F} = \sum_i [(h^j F^i)_{;(i=j)} 1 + \Sigma_{(i<j)} (h^j F^i)_{;j} 0 + \Sigma_{(i>j)} (h^j F^i)_{;j} 0]$	
∴	From 1, 9, A4.2.12 Identity Multp, T4.4.1 T Equalities: Any Quantity Multiplied by Zero is Zero and A4.2.7 Identity Add and A4.2.3 Substitution	$\nabla \bullet \mathbf{F} = \sum_i (h^i F^{i;i})$ EQ A	$\sum_i h^i (F^{i;i}) = \sum_i (h^i F^i)_{;i}$ EQ B

Theorem 15.2.7 Covariant Divergent Operator: Orthogonal Divergent Operator

Step	Hypothesis	$\nabla \bullet \mathbf{F} = (\rho_v\, h_i\, F^i)_{,i} / \rho_v$
1	From T15.2.2A,B Covariant Divergent Operator: Orthogonal System Scaling Factor Distribution and T10.2.5 Tool: Tensor Substitution	$\nabla \bullet \mathbf{F} = \sum_i (h^i F^i)_{;i}$
2	From 1 and T12.6.2B Differentiation of Tensor: Rank-1 Covariant	$\nabla \bullet \mathbf{F} = \sum_i (h^i F^i)_{,i} + \sum_i \sum_j \Gamma_{ji}{}^{i} (h^j F^j)$
3	From 2 and interchanging order of summation	$\nabla \bullet \mathbf{F} = \sum_i [(h^i F^i)_{,i} + \sum_j (\sum_i \Gamma_{ji}{}^{i}) (h^j F^j)]$
4	From 3, T12.8.7B Derivative of the Natural Log of the Covariant Determinate and T10.2.5 Tool: Tensor Substitution	$\nabla \bullet \mathbf{F} = \sum_i (h^i F^i)_{,i} + \sum_j \ln(\sqrt{g})_{,j} (h^j F^j)$
5	From 4 and changing dummy indices j → i	$\nabla \bullet \mathbf{F} = \sum_i (h^i F^i)_{,i} + \sum_i \ln(\sqrt{g})_{,i} (h^i F^i)$
6	From 5 and LxK.3.3.4B Differential a Natural Log base [e], A4.2.13 Inverse Multp and T10.2.6 Tool: Commutative Scalar Times a Tensor	$\nabla \bullet \mathbf{F} = \sqrt{g}\, (h^i F^i)_{,i} / \sqrt{g} + (\sqrt{g})_{,i} (h^i F^i) / \sqrt{g}$
7	From 6 and T10.4.6 Scalars and Tensors: Distribution of a Scalar over Addition of Tensors	$\nabla \bullet \mathbf{F} = [\sqrt{g}\, (h^i F^i)_{,i} + (\sqrt{g})_{,i} (h^i F^i)] / \sqrt{g}$
8	From 7 and T12.7.8 Tensor Product: Chain Rule for Mixed Binary Product, T12.2.10 Covariant and Contravariant Scale Factor; Reciprocal Relationship and A4.2.3 Substitution	$\nabla \bullet \mathbf{F} = (\sqrt{g}\, h_i F^i)_{,i} / \sqrt{g}$
∴	From 8, T12.2.19 Covariant Metric Determinate in Orthogonal Curvilinear Coordinates and A4.2.3 Substitution	$\nabla \bullet \mathbf{F} = (\rho_v\, h_i\, F^i)_{,i} / \rho_v$

Observation 15.2.1 Conformation of Divergence Operator with Contravariant Components

The introduction of the ***physical component*** $[F_i \equiv h_i F^i]$, is contrived and forces the vector to be covariant, which from theorem T15.2.7 "Covariant Divergent Operator: Orthogonal Divergent Operator" is seen to be contravariant, hence not true. To make it covariant which is OK, would give rise to a new metric factor $[g_{ij}]$ to account for the change. Also T15.2.3 "Covariant Divergent Operator: Cross product Proof Orthogonal Coordinates" confirms the physical component to be actually contravariant.

Observation 15.2.2 Raising the Gradient Index to Contravariant Coefficient Position

As seen in theorem T15.2.1 "Covariant Divergent Operator: Definition of Divergence" the divergence of the gradient component is covariant it follows that the notation for the coefficient index is raised likewise for consistency. This introduces the idea that regardless whether the vector operated on is, covariant or contravariant, that under the Covariant Divergent Operator the coefficient index is always raised for continuity.

Section 15.3 Laplacian Tensor Operator

Lemma	Flux	Descriptive Name
LxVIE.6.4	$\Phi = \oint_S (\phi\nabla\psi)\bullet d\vec{S} = \int_V (\phi\nabla^2\psi - \nabla\phi\bullet\nabla\psi)\, dV$	Green's First Theorem

or

Lemma	Flux	Descriptive Name
LxVIE.6.5	$\Phi = \oint_S (\phi\nabla\psi - \psi\nabla\phi)\bullet d\vec{S} = \int_V (\phi\nabla^2\psi - \psi\nabla^2\phi)\, dV$	Green's Second Theorem

Theorem 15.3.1 Laplacian Operator: Definition of Laplacian

1g	Given	$\mathbf{F} = \nabla\,\Phi$
Step	Hypothesis	$\nabla^2\,\Phi = (\sqrt{g}\; g^{ij}\,\Phi_{,j})_{,i} / \sqrt{g}$
1	From T15.2.1 Covariant Divergent Operator: On Covariant Vector	$\nabla \bullet \mathbf{F} = (\sqrt{g}\; F^i)_{,i} / \sqrt{g}$
2	From 1g, T12.3.2 Form of a Covariant Vector, DxK.6.1.3A Covariant Differential Operator, DEL and T12.3.12 Invariance of Contra and Covariant Vectors	$F^i\,\mathbf{b}_i = \Phi_{,j}\,\mathbf{b}^j$
3	From 2, T10.3.6 Product of Tensors: Uniqueness by Multiplication of Rank for p + q and T10.3.7 Product of Tensors: Commutative by Multiplication of Rank p + q → q + p	$F^i\,g^{ij}\,\mathbf{b}_i = g^{ij}\,\Phi_{,j}\,\mathbf{b}^j$
4	From 3 and T12.3.10A Transformation from Covariant to Contravariant Basis Vector	$F^i\,\mathbf{b}^j = g^{ij}\,\Phi_{,j}\,\mathbf{b}^j$
5	From 4 and equating unit vector with coefficient	$F^i = g^{ij}\,\Phi_{,j}$
6	From 1, 5 and T10.2.5 Tool: Tensor Substitution	$\nabla \bullet \nabla\,\Phi = (\sqrt{g}\;[g^{ij}\,\Phi_{,j}])_{,i} / \sqrt{g}$
∴	From 6, T10.4.8 Scalars and Tensors: Scalar Commutative with Tensors and LxK.6.1.10 Laplacian Operator	$\nabla^2\,\Phi = (\sqrt{g}\; g^{ij}\,\Phi_{,j})_{,i} / \sqrt{g}$

Theorem 15.3.2 Laplacian Operator: Orthogonal Laplacian Operator

Step	Hypothesis	$\nabla^2 \Phi = [(\rho_v / h_i h_j)\Phi_{,j}]_{,i} / \rho_v$
1	From T15.3.1 Laplacian Operator: Definition of Laplacian	$\nabla^2 \Phi = [(\sqrt{g}\, g^{ij})\Phi_{,j}]_{,i} / \sqrt{g}$
2	From 1, T12.2.19 Covariant Metric Determinate in Orthogonal Curvilinear Coordinates, T12.2.15B Covariant-Contravariant Metric in Orthogonal Curvilinear Coordinates and A4.2.3 Substitution	$\nabla^2 \Phi = [(\rho_v h^i h^j)\Phi_{,j}]_{,i} / \rho_v$
∴	From 2, T12.2.10 Covariant and Contravariant Scale Factor; Reciprocal Relationship and A4.2.3 Substitution	$\nabla^2 \Phi = [(\rho_v / h_i h_j)\Phi_{,j}]_{,i} / \rho_v$

Section 15.4 Cross Product Tensor Operator

Lemma	Flux	Descriptive Name
LxVIE.6.3	$\vec{\Phi} = \oint_S d\vec{S}\times\vec{F} = \int_V \nabla\times\vec{F}\, dV$	Divergence Theorem of Gauss Cross Product Multiplication

or

Lemma	Flux	Descriptive Name
LxVIE.6.7	$\oint_C \vec{F}\bullet d\vec{r} = \int_S (\nabla\times\vec{F})\bullet d\vec{S}$	Stokes' Theorem Dot Product Multiplication

Definition 15.4.1 Contra-Covariant Rules of an n-Dimensional Curl Operator
Just as the Divergence and Laplacian are constructed from the Gradient operator so to is the Curl constructed from the cross product and the appropriate gradient. However it is made up also of a set of vectors to fill out the rest of the [nxn] array called the ***curl quiver*** or simply quiver for short. The bases vectors associated with the quiver are in the first row called the ***bases vector row***. It follows that if the bases vector row is covariant the vectors comprising the quiver are contravariant coefficients and if the bases vector row is contravariant the quiver is covariant. Mixed vectors in the quiver are not allowed because they would not conform to the dual vector system of the bases vector row.

Definition 15.4.2 Covariant Curvilinear Curl Operator over a Quiver [K.6.3]

	Equations	
Dx9.6.8.1	$\nabla\times \prod_{i=3}^{n}\times\mathbf{F}^i \equiv \begin{vmatrix} e_{11} & e_{12} & \cdots & e_{1n} \\ \dfrac{1}{h_{21}}\dfrac{\partial}{\partial x_{21}} & \dfrac{1}{h_{22}}\dfrac{\partial}{\partial x_{22}} & \cdots & \dfrac{1}{h_{2n}}\dfrac{\partial}{\partial x_{2n}} \\ F^{31} & F^{32} & \cdots & F^{3n} \\ \vdots & & & \vdots \\ F^{n1} & F^{3n} & \cdots & F^{nn} \end{vmatrix}$	Curl of Contravariant tensors spanning a space over a Covariant gradient operator

Definition 15.4.3 Contravariant Curvilinear Curl Operator over a Quiver

	Equations	
Dx9.6.8.2	$\nabla\times \prod_{i=3}^{n}\times\mathbf{F}_i \equiv \begin{vmatrix} \mathbf{e}^{11} & \mathbf{e}^{12} & \cdots & \mathbf{e}^{1n} \\ \dfrac{1}{h^{21}}\dfrac{\partial}{\partial x_{21}} & \dfrac{1}{h^{22}}\dfrac{\partial}{\partial x_{22}} & \cdots & \dfrac{1}{h^{2n}}\dfrac{\partial}{\partial x_{2n}} \\ F_{31} & F_{32} & \cdots & F_{3n} \\ \vdots & & & \vdots \\ F_{n1} & F_{n2} & \cdots & F_{nn} \end{vmatrix}$	Curl of covariant tensors spanning a space over the gradient operator

Theorem 15.4.1 Curl Operator: Curvilinear Covariant Cross Product

Step	Hypothesis			
Step	Hypothesis	$\nabla\times\prod_{i=3}^{n}\times\mathbf{F}^{i}$	=	$(1/\rho_v)\begin{vmatrix} \mathbf{b}_{11} & \mathbf{b}_{12} & \cdots & \mathbf{b}_{1n} \\ \frac{\partial}{\partial x_{21}} & \frac{\partial}{\partial x_{22}} & \cdots & \frac{\partial}{\partial x_{2n}} \\ h_{21}F^{31} & h_{22}F^{32} & \cdots & h_{2n}F^{3n} \\ \vdots & & & \vdots \\ h_{21}F^{n1} & h_{22}F^{3n} & \cdots & h_{2n}F^{nn} \end{vmatrix}$
1	From D15.4.2 Covariant Curvilinear Curl Operator over Quiver	$\nabla\times\prod_{i=3}^{n}\times\mathbf{F}^{i}$	=	$\begin{vmatrix} \mathbf{e}_{11} & \mathbf{e}_{12} & \cdots & \mathbf{e}_{1n} \\ \frac{1}{h_{21}}\frac{\partial}{\partial x_{21}} & \frac{1}{h_{22}}\frac{\partial}{\partial x_{22}} & \cdots & \frac{1}{h_{2n}}\frac{\partial}{\partial x_{2n}} \\ F^{31} & F^{32} & \cdots & F^{3n} \\ \vdots & & & \vdots \\ F^{n1} & F^{3n} & \cdots & F^{nn} \end{vmatrix}$
2	From 1, A4.2.12 Identity Multp and A4.2.13 Inverse Multp for all elements in the determinate	$\nabla\times\prod_{i=3}^{n}\times\mathbf{F}^{i}$	=	$\begin{vmatrix} \frac{h_{21}}{h_{21}}e_{11} & \frac{h_{22}}{h_{22}}e_{12} & \cdots & \frac{h_{2n}}{h_{2n}}e_{1n} \\ \frac{1}{h_{21}}\frac{\partial}{\partial x_{21}} & \frac{1}{h_{22}}\frac{\partial}{\partial x_{22}} & \cdots & \frac{1}{h_{2n}}\frac{\partial}{\partial x_{2n}} \\ \frac{h_{21}}{h_{21}}F^{31} & \frac{h_{22}}{h_{22}}F^{32} & \cdots & \frac{h_{2n}}{h_{2n}}F^{3n} \\ \vdots & & & \vdots \\ \frac{h_{21}}{h_{21}}F^{n1} & \frac{h_{22}}{h_{22}}F^{3n} & \cdots & \frac{h_{2n}}{h_{2n}}F^{nn} \end{vmatrix}$
3	From 2 and T6.9.11 Multiply a row or column of a determinant [A] by a scalar k for all denominators	$\nabla\times\prod_{i=3}^{n}\times\mathbf{F}^{i}$	=	$\prod_i(1/h_{2i})\begin{vmatrix} h_{21}\mathbf{e}_{11} & h_{22}\mathbf{e}_{12} & \cdots & h_{2n}\mathbf{e}_{1n} \\ \frac{\partial}{\partial x_{21}} & \frac{\partial}{\partial x_{22}} & \cdots & \frac{\partial}{\partial x_{2n}} \\ h_{21}F^{31} & h_{22}F^{32} & \cdots & h_{2n}F^{3n} \\ \vdots & & & \vdots \\ h_{21}F^{n1} & h_{22}F^{3n} & \cdots & h_{2n}F^{nn} \end{vmatrix}$
4	From 3 and T4.8.1 Integer Exponents: Unity Raised to any Integer Value, the product of the numerator and denominator	$\nabla\times\prod_{i=3}^{n}\times\mathbf{F}^{i}$	=	$(1/\prod_i h_{2i})\begin{vmatrix} h_{21}\mathbf{e}_{11} & h_{22}\mathbf{e}_{12} & \cdots & h_{2n}\mathbf{e}_{1n} \\ \frac{\partial}{\partial x_{21}} & \frac{\partial}{\partial x_{22}} & \cdots & \frac{\partial}{\partial x_{2n}} \\ h_{21}F^{31} & h_{22}F^{32} & \cdots & h_{2n}F^{3n} \\ \vdots & & & \vdots \\ h_{21}F^{n1} & h_{22}F^{3n} & \cdots & h_{2n}F^{nn} \end{vmatrix}$
5	From 4, D12.1.13A Space Density Factor and A4.2.3 Substitution	$\nabla\times\prod_{i=3}^{n}\times\mathbf{F}^{i}$	=	$(1/\rho_v)\begin{vmatrix} h_{21}\mathbf{e}_{11} & h_{22}\mathbf{e}_{12} & \cdots & h_{2n}\mathbf{e}_{1n} \\ \frac{\partial}{\partial x_{21}} & \frac{\partial}{\partial x_{22}} & \cdots & \frac{\partial}{\partial x_{2n}} \\ h_{21}F^{31} & h_{22}F^{32} & \cdots & h_{2n}F^{3n} \\ \vdots & & & \vdots \\ h_{21}F^{n1} & h_{22}F^{3n} & \cdots & h_{2n}F^{nn} \end{vmatrix}$
∴	From 5, D12.1.6 Covariant Base Vector and T10.2.5 Tool: Tensor Substitution	$\nabla\times\prod_{i=3}^{n}\times\mathbf{F}^{i}$	=	$(1/\rho_v)\begin{vmatrix} \mathbf{b}_{11} & \mathbf{b}_{12} & \cdots & \mathbf{b}_{1n} \\ \frac{\partial}{\partial x_{21}} & \frac{\partial}{\partial x_{22}} & \cdots & \frac{\partial}{\partial x_{2n}} \\ h_{21}F^{31} & h_{22}F^{32} & \cdots & h_{2n}F^{3n} \\ \vdots & & & \vdots \\ h_{21}F^{n1} & h_{22}F^{3n} & \cdots & h_{2n}F^{nn} \end{vmatrix}$

Theorem 15.4.2 Curl Operator: Curvilinear Contravariant Cross Product

Step	Hypothesis			
Step	Hypothesis	$\nabla\times\prod_{i=3}^{n}\times\mathbf{F}_i$	$=$	$\rho_v\begin{vmatrix}\mathbf{b}^{11} & \mathbf{b}^{12} & \cdots & \mathbf{b}^{1n}\\ \frac{\partial}{\partial x_{21}} & \frac{\partial}{\partial x_{22}} & \cdots & \frac{\partial}{\partial x_{2n}}\\ h^{21}F_{31} & h^{22}F_{32} & \cdots & h^{2n}F_{3n}\\ \vdots & & & \vdots\\ h^{21}F_{n1} & h^{22}F^{n2} & \cdots & h^{2n}F^{nn}\end{vmatrix}$
1	From D15.4.3 Contravariant Curvilinear Curl Operator over Quiver	$\nabla\times\prod_{i=3}^{n}\times\mathbf{F}_i$	$=$	$\begin{vmatrix}\mathbf{e}^{11} & \mathbf{e}^{12} & \cdots & \mathbf{e}^{1n}\\ \frac{1}{h^{21}}\frac{\partial}{\partial x_{21}} & \frac{1}{h^{22}}\frac{\partial}{\partial x_{22}} & \cdots & \frac{1}{h^{2n}}\frac{\partial}{\partial x_{2n}}\\ F_{31} & F_{32} & \cdots & F_{3n}\\ \vdots & & & \vdots\\ F_{n1} & F_{n2} & \cdots & F_{nn}\end{vmatrix}$
2	From 1, A4.2.12 Identity Multp and A4.2.13 Inverse Multp for all elements in the determinate	$\nabla\times\prod_{i=3}^{n}\times\mathbf{F}_i$	$=$	$\begin{vmatrix}\frac{h^{21}}{h^{21}}\mathbf{e}^{11} & \frac{h^{22}}{h^{22}}\mathbf{e}^{12} & \cdots & \frac{h^{2n}}{h^{2n}}\mathbf{e}^{1n}\\ \frac{1}{h^{21}}\frac{\partial}{\partial x_{21}} & \frac{1}{h^{22}}\frac{\partial}{\partial x_{22}} & \cdots & \frac{1}{h^{2n}}\frac{\partial}{\partial x_{2n}}\\ \frac{h^{21}}{h^{21}}F_{31} & \frac{h^{22}}{h^{22}}F_{32} & \cdots & \frac{h^{2n}}{h^{2n}}F_{3n}\\ \vdots & & & \vdots\\ \frac{h^{21}}{h^{21}}F_{n1} & \frac{h^{22}}{h^{22}}F_{n2} & \cdots & \frac{h^{2n}}{h^{2n}}F_{nn}\end{vmatrix}$
3	From 2 and T6.9.11 Multiply a row or column of a determinant [A] by a scalar k for all denominators	$\nabla\times\prod_{i=3}^{n}\times\mathbf{F}_i$	$=$	$\prod_i(1/h^{2i})\begin{vmatrix}h^{21}\mathbf{e}^{11} & h^{22}\mathbf{e}^{12} & \cdots & h^{2n}\mathbf{e}^{1n}\\ \frac{\partial}{\partial x_{21}} & \frac{\partial}{\partial x_{22}} & \cdots & \frac{\partial}{\partial x_{2n}}\\ h^{21}F_{31} & h^{22}F_{32} & \cdots & h^{2n}F_{3n}\\ \vdots & & & \vdots\\ h^{21}F_{n1} & h^{22}F_{n2} & \cdots & h^{2n}F_{nn}\end{vmatrix}$
4	From 3, T12.2.10A Covariant and Contravariant Scale Factor; Reciprocal Relationship, A4.2.3 Substitution, T4.5.7 Equalities: Product and Reciprocal of a Product	$\nabla\times\prod_{i=3}^{n}\times\mathbf{F}_i$	$=$	$\prod_i h_{2i}\begin{vmatrix}h^{21}\mathbf{e}^{11} & h^{22}\mathbf{e}^{12} & \cdots & h^{2n}\mathbf{e}^{1n}\\ \frac{\partial}{\partial x_{21}} & \frac{\partial}{\partial x_{22}} & \cdots & \frac{\partial}{\partial x_{2n}}\\ h^{21}F_{31} & h^{22}F_{32} & \cdots & h^{2n}F_{3n}\\ \vdots & & & \vdots\\ h^{21}F_{n1} & h^{22}F_{n2} & \cdots & h^{2n}F_{nn}\end{vmatrix}$
5	From 4, D12.1.13A Space Density Factor and A4.2.3 Substitution	$\nabla\times\prod_{i=3}^{n}\times\mathbf{F}_i$	$=$	$\rho_v\begin{vmatrix}h^{21}\mathbf{e}^{11} & h^{22}\mathbf{e}^{12} & \cdots & h^{2n}\mathbf{e}^{1n}\\ \frac{\partial}{\partial x_{21}} & \frac{\partial}{\partial x_{22}} & \cdots & \frac{\partial}{\partial x_{2n}}\\ h^{21}F_{31} & h^{22}F_{32} & \cdots & h^{2n}F_{3n}\\ \vdots & & & \vdots\\ h^{21}F_{n1} & h^{22}F_{n2} & \cdots & h^{2n}F_{nn}\end{vmatrix}$
∴	From 5, D12.1.7 Contravariant Base Vector and T10.2.5 Tool: Tensor Substitution	$\nabla\times\prod_{i=3}^{n}\times\mathbf{F}_i$	$=$	$\rho_v\begin{vmatrix}\mathbf{b}^{11} & \mathbf{b}^{12} & \cdots & \mathbf{b}^{1n}\\ \frac{\partial}{\partial x_{21}} & \frac{\partial}{\partial x_{22}} & \cdots & \frac{\partial}{\partial x_{2n}}\\ h^{21}F_{31} & h^{22}F_{32} & \cdots & h^{2n}F_{3n}\\ \vdots & & & \vdots\\ h^{21}F_{n1} & h^{22}F_{n2} & \cdots & h^{2n}F_{nn}\end{vmatrix}$

Theorem 15.4.3 Curl Operator: Riemannian $\mathfrak{R}^n$ Covariant Cross Product

1g 2g	Given	Orthogonal Space Curvilinear Space
Step	Hypothesis	$\nabla\times \prod_{i=3}^{n}\times\mathbf{F}^i = (1/\sqrt{g}) \begin{vmatrix} \mathbf{b}_{11} & \mathbf{b}_{12} & \cdots & \mathbf{b}_{1n} \\ \frac{\partial}{\partial x_{21}} & \frac{\partial}{\partial x_{22}} & \cdots & \frac{\partial}{\partial x_{2n}} \\ h_{21}F^{31} & h_{22}F^{32} & \cdots & h_{2n}F^{3n} \\ \vdots & & & \vdots \\ h_{21}F^{n1} & h_{22}F^{3n} & \cdots & h_{2n}F^{nn} \end{vmatrix}$
1	From 1g and T15.4.1 Curl Operator: Curvilinear Covariant Cross Product	$\nabla\times \prod_{i=3}^{n}\times\mathbf{F}^i = (1/\rho_v) \begin{vmatrix} \mathbf{b}_{11} & \mathbf{b}_{12} & \cdots & \mathbf{b}_{1n} \\ \frac{\partial}{\partial x_{21}} & \frac{\partial}{\partial x_{22}} & \cdots & \frac{\partial}{\partial x_{2n}} \\ h_{21}F^{31} & h_{22}F^{32} & \cdots & h_{2n}F^{3n} \\ \vdots & & & \vdots \\ h_{21}F^{n1} & h_{22}F^{3n} & \cdots & h_{2n}F^{nn} \end{vmatrix}$
∴	From 2g, 4, T12.2.19 Covariant Metric Determinate in Orthogonal Curvilinear Coordinates and A4.2.3 Substitution	$\nabla\times \prod_{i=3}^{n}\times\mathbf{F}^i = (1/\sqrt{g}) \begin{vmatrix} \mathbf{b}_{11} & \mathbf{b}_{12} & \cdots & \mathbf{b}_{1n} \\ \frac{\partial}{\partial x_{21}} & \frac{\partial}{\partial x_{22}} & \cdots & \frac{\partial}{\partial x_{2n}} \\ h_{21}F^{31} & h_{22}F^{32} & \cdots & h_{2n}F^{3n} \\ \vdots & & & \vdots \\ h_{21}F^{n1} & h_{22}F^{3n} & \cdots & h_{2n}F^{nn} \end{vmatrix}$

Theorem 15.4.4 Curl Operator: Riemannian $\mathfrak{R}^n$ Contravariant Cross Product

1g 2g	Given	Orthogonal Space Curvilinear Space
Step	Hypothesis	$\nabla\times \prod_{i=3}^{n}\times\mathbf{F}^i = \sqrt{g} \begin{vmatrix} \mathbf{b}^{11} & \mathbf{b}^{12} & \cdots & \mathbf{b}^{1n} \\ \frac{\partial}{\partial x_{21}} & \frac{\partial}{\partial x_{22}} & \cdots & \frac{\partial}{\partial x_{2n}} \\ h^{21}F_{31} & h^{22}F_{32} & \cdots & h^{2n}F_{3n} \\ \vdots & & & \vdots \\ h^{21}F_{n1} & h^{22}F_{n2} & \cdots & h^{2n}F_{nn} \end{vmatrix}$
1	From 1g and T15.4.2 Curl Operator: Curvilinear Contravariant Cross Product	$\nabla\times \prod_{i=3}^{n}\times\mathbf{F}^i = \rho_v \begin{vmatrix} \mathbf{b}^{11} & \mathbf{b}^{12} & \cdots & \mathbf{b}^{1n} \\ \frac{\partial}{\partial x_{21}} & \frac{\partial}{\partial x_{22}} & \cdots & \frac{\partial}{\partial x_{2n}} \\ h^{21}F_{31} & h^{22}F_{32} & \cdots & h^{2n}F_{3n} \\ \vdots & & & \vdots \\ h^{21}F_{n1} & h^{22}F_{n2} & \cdots & h^{2n}F_{nn} \end{vmatrix}$
∴	From 2g, 4 From 4, T12.2.19 Covariant Metric Determinate in Orthogonal Curvilinear Coordinates and A4.2.3 Substitution	$\nabla\times \prod_{i=3}^{n}\times\mathbf{F}^i = \sqrt{g} \begin{vmatrix} \mathbf{b}^{11} & \mathbf{b}^{12} & \cdots & \mathbf{b}^{1n} \\ \frac{\partial}{\partial x_{21}} & \frac{\partial}{\partial x_{22}} & \cdots & \frac{\partial}{\partial x_{2n}} \\ h^{21}F_{31} & h^{22}F_{32} & \cdots & h^{2n}F_{3n} \\ \vdots & & & \vdots \\ h^{21}F_{n1} & h^{22}F_{n2} & \cdots & h^{2n}F_{nn} \end{vmatrix}$

Theorem 15.4.5 Curl Operator: Evaluating the Determinate for Covariant Cross Product

1g	Given	$A^{2\lambda[2,k]} \equiv \nabla_{2\lambda[2,k]}$
2g		$A^{i\lambda[i,k]} \equiv h_{2\lambda[i,k]} F^{i\lambda[i,k]}$ for $3 \le j \le n$
Step	Hypothesis	$\nabla\times \prod_{i=3}{}^{n}\times \mathbf{F}^{i} = (1/\sqrt{g}) \sum_{\rho(k)\in Ln} \mathbf{e}_{\bullet\lambda[i,k]\bullet}\, \mathbf{b}_{\lambda[1,k]}\, \nabla_{2\lambda[2,k]} \prod_{i=3}{}^{n} (\, h_{2\lambda[i,k]} F^{i\lambda[i,k]}\,)$
1	From T12.11.12 Cross Product: Covariant Unified with the e-Tensor Operator	$\nabla\times \prod_{i=3}{}^{n}\times \mathbf{F}^{i} = \sum_{\rho(k)\in Ln} \mathbf{e}_{\bullet\lambda[i,k]\bullet}\, \mathbf{b}_{\lambda[1,k]} \prod_{i=2}{}^{n} A^{i\lambda[i,k]}$
∴	From 2g, 3g, 1, T15.4.3 Curl Operator: Riemannian $\mathfrak{R}^n$ Covariant Cross Product and T10.2.5 Tool: Tensor Substitution	$\nabla\times \prod_{i=3}{}^{n}\times \mathbf{F}^{i} = (1/\sqrt{g}) \sum_{\rho(k)\in Ln} \mathbf{e}_{\bullet\lambda[i,k]\bullet}\, \mathbf{b}_{\lambda[1,k]}\, \nabla_{2\lambda[2,k]} \prod_{i=3}{}^{n} (\, h_{2\lambda[i,k]} F^{i\lambda[i,k]}\,)$

Theorem 15.4.6 Curl Operator: Differential of the Covariant Cross Product Determinate

1g	Given	$\nabla\times\prod_{i=3}{}^{n}\times \mathbf{A}_{j} \equiv [\sum_{\rho(k)\in Ln} \mathbf{e}_{\bullet\lambda[i,k]\bullet}\, \mathbf{b}_{\lambda[1,k]} (\prod_{i=3}{}^{j-1} A^{i\lambda[i,k]}) \nabla_{\lambda[2,k]} (\, h_{2\lambda[j,k]} F^{j\lambda[j,k]}\,) (\prod_{i=j+1}{}^{n} A^{i\lambda[i,k]})]$
Step	Hypothesis	$\nabla\times \prod_{i=3}{}^{n}\times \mathbf{F}_{i} = (1/\sqrt{g}) \sum_{j}{}^{n-3} \nabla\times\prod_{i=3}{}^{n}\times \mathbf{A}^{j}$
1	From T15.4.5 Curl Operator: Evaluating the Determinate for Covariant Cross Product	$\nabla\times \prod_{i=3}{}^{n}\times \mathbf{F}_{i} = (1/\sqrt{g}) \sum_{\rho(k)\in Ln} \mathbf{e}_{\bullet\lambda[i,k]\bullet}\, \mathbf{b}_{\lambda[1,k]}\, \nabla_{2\lambda[2,k]} \prod_{i=3}{}^{n} (\, h_{2\lambda[i,k]} F^{i\lambda[i,k]}\,)$
2	From 1 and LxK.3.2.4C Distributed Across Products	$\nabla\times \prod_{i=3}{}^{n}\times \mathbf{F}_{i} = (1/\sqrt{g}) \sum_{\rho(k)\in Ln} \mathbf{e}_{\bullet\lambda[i,k]\bullet}\, \mathbf{b}_{\lambda[1,k]} [\sum_{j}{}^{n-3} (\prod_{i=3}{}^{j-1} A^{i\lambda[i,k]}) \nabla_{\lambda[2,k]} (\, h_{2\lambda[j,k]} F^{j\lambda[j,k]}\,) (\prod_{i=j+1}{}^{n} A^{i\lambda[i,k]})]$
3	From 2, T10.3.20 Addition of Tensors: Distribution of a Tensor over Addition of Tensors and reordering summation	$\nabla\times \prod_{i=3}{}^{n}\times \mathbf{F}_{i} = (1/\sqrt{g}) \sum_{j}{}^{n-3} [\sum_{\rho(k)\in Ln} \mathbf{e}_{\bullet\lambda[i,k]\bullet}\, \mathbf{b}_{\lambda[1,k]} (\prod_{i=3}{}^{j-1} A^{i\lambda[i,k]}) \nabla_{\lambda[2,k]} (\, h_{2\lambda[j,k]} F^{j\lambda[j,k]}\,) (\prod_{i=j+1}{}^{n} A^{i\lambda[i,k]})]$
∴	From 1g, 3 and T10.2.5 Tool: Tensor Substitution	$\nabla\times \prod_{i=3}{}^{n}\times \mathbf{F}_{i} = (1/\sqrt{g}) \sum_{j}{}^{n-3} \nabla\times\prod_{i=3}{}^{n}\times \mathbf{A}_{j}$

Theorem 15.4.7 Curl Operator: Evaluating the Determinate for Contravariant Cross Product

1g	Given	$A_{2\lambda[2,k]} \equiv \nabla_{2\lambda[2,k]}$
2g		$A_{i\lambda[i,k]} \equiv h^{2\lambda[i,k]} F_{i\lambda[i,k]}$ for $3 \le j \le n$
Step	Hypothesis	$\nabla\times \prod_{i=3}{}^{n}\times \mathbf{F}_{i} = \sqrt{g} \sum_{\rho(k)\in Ln} \mathbf{e}^{\bullet\lambda[i,k]\bullet}\, \mathbf{b}^{i\lambda[i,k]}\, \nabla_{2\lambda[2,k]} \prod_{i=3}{}^{n} (\, h^{2\lambda[i,k]} F_{i\lambda[i,k]}\,)$
1	From T12.12.12 Cross Product: Contravariant Unified with the e-Tensor Operator	$\nabla\times \prod_{i=3}{}^{n}\times \mathbf{F}_{i} = \sum_{\rho(k)\in Ln} \mathbf{e}^{\bullet\lambda[i,k]\bullet}\, \mathbf{b}^{i\lambda[i,k]} \prod_{i=2}{}^{n} A_{i\lambda[i,k]}$
∴	From 2g, 3g, 1, T15.4.4 Curl Operator: Riemannian $\mathfrak{R}^n$ Contravariant Cross Product and T10.2.5 Tool: Tensor Substitution	$\nabla\times \prod_{i=3}{}^{n}\times \mathbf{F}_{i} = \sqrt{g} \sum_{\rho(k)\in Ln} \mathbf{e}^{\bullet\lambda[i,k]\bullet}\, \mathbf{b}^{i\lambda[i,k]}\, \nabla_{2\lambda[2,k]} \prod_{i=3}{}^{n} (\, h^{2\lambda[i,k]} F_{i\lambda[i,k]}\,)$

Theorem 15.4.8 Curl Operator: Differential of the Contravariant Cross Product Determinate

1g	Given	$\nabla\times\prod_{i=3}^{n}\times\mathbf{A}^{j} \equiv [\sum_{\rho(k)\in Ln} \mathbf{e}^{\bullet\lambda[i,k]\bullet}\ \mathbf{b}^{i\lambda[i,k]}$ $(\prod_{i=3}^{j-1} A_{i\lambda[i,k]})\nabla_{\lambda[2,k]}\ (h^{2\lambda[j,k]}\ F_{j\lambda[j,k]})\ (\prod_{i=j+1}^{n} A_{i\lambda[i,k]})]$
Step	Hypothesis	$\nabla\times \prod_{i=3}^{n}\times\mathbf{F}^{i} = \sqrt{g}\ \sum_{j}^{n-3}\ \nabla\times\prod_{i=3}^{n}\times\mathbf{A}^{j}$
1	From T15.4.7 Curl Operator: Evaluating the Determinate for Contravariant Cross Product	$\nabla\times \prod_{i=3}^{n}\times\mathbf{F}^{i} = \sqrt{g}$ $\sum_{\rho(k)\in Ln} \mathbf{e}^{\bullet\lambda[i,k]\bullet}\ \mathbf{b}^{i\lambda[i,k]}\ \nabla_{2\lambda[2,k]} \prod_{i=3}^{n} (\ h^{2\lambda[i,k]}\ F_{i\lambda[i,k]}\ )$
2	From 1 and Lx9.3.2.4C Distributed Across Products	$\nabla\times \prod_{i=3}^{n}\times\mathbf{F}^{i} = \sqrt{g}\ \sum_{\rho(k)\in Ln} \mathbf{e}^{\bullet\lambda[i,k]\bullet}\ \mathbf{b}^{i\lambda[i,k]}$ $[\sum_{j}^{n-3} (\prod_{i=3}^{j-1} A_{i\lambda[i,k]})\nabla_{\lambda[2,k]}\ (\ h^{2\lambda[j,k]}\ F_{j\lambda[j,k]}\ )$ $(\prod_{i=j+1}^{n} A_{i\lambda[i,k]})]$
3	From 2, T10.3.20 Addition of Tensors: Distribution of a Tensor over Addition of Tensors and reordering summation	$\nabla\times \prod_{i=3}^{n}\times\mathbf{F}^{i} = \sqrt{g}\ \sum_{j}^{n-3}$ $[\sum_{\rho(k)\in Ln} \mathbf{e}^{\bullet\lambda[i,k]\bullet}\ \mathbf{b}^{i\lambda[i,k]}$ $(\prod_{i=3}^{j-1} A_{i\lambda[i,k]})\nabla_{\lambda[2,k]}\ (h^{2\lambda[j,k]}\ F_{j\lambda[j,k]})\ (\prod_{i=j+1}^{n} A_{i\lambda[i,k]})]$
∴	From 1g, 3 and T10.2.5 Tool: Tensor Substitution	$\nabla\times \prod_{i=3}^{n}\times\mathbf{F}^{i} = \sqrt{g}\ \sum_{j}^{n-3}\ \nabla\times\prod_{i=3}^{n}\times\mathbf{A}^{j}$

Observation 15.4.1 In the Case of a Curl for a Contravariant and Covariant Derivatives

Because of the way the curl operator is defined the differential is in front the row of bases vectors, which means they are never differentiated. So differentiation is simple, see definition D12.5.4 "Simple Differental Vector Notation". If another ordering of rows were taken a more sophisticated development would be required, changing the outcome for therorems T15.4.6 "Curl Operator: Differential of the Covariant Cross Product Determinate" and T15.4.8 "Curl Operator: Differential of the Contravariant Cross Product Determinate".

Tensor Calculus & Physics: A General Treatise

Table of Contents

List of Tables

List of Figures

List of Observations

List of Givens

Chapter 16 Tensor Calculus: Differential Intrinsic Geometry

Section 16.1 Intrinsic Geometric Definitions

It will be shown that certain properties can be phrased independently of the space in which the surface is immersed and that they are concerned solely with the structure of the differential quadratic form for the element of arc of a curve drawn on the surface; see A17.1.8P2 "Riemannian Hypersheet $\mathfrak{H}^{n-1}$". All such properties of surfaces are termed the ***intrinsic*** properties, and the geometry based on the study of this differential quadratic form is called the ***intrinsic geometry*** of the surface.

Definition 16.1.1 Intrinsic Newtonian Notation

On a time line the classical intrinsic notation that ***Newton*** devised for the first differential notation, the dot above the time variable-x(t), however it is often interchangeable with ***Leibniz***'s notation, the differential fraction.

$\dot{x} \equiv dx / dt$ Equation A[16.1.1]

$\ddot{x} \equiv d^2x / dt^2$ Equation B[16.1.1]

for repeated operations see DK.3.2B-F " Definitions on Differential Notation"

Definition 16.1.2 Intrinsic Lagrangian Notation

Alternatively for a spatial curve ***Lagrange***'s notation, the prime mark is used.

$x' \equiv dx / ds$ Equation A[16.1.1]

$x'' \equiv d^2x / ds^2$ Equation B[16.1.1]

for repeated operations see DK.3.2G-K " Definitions on Differential Notation"

Definition 16.1.3 Intrinsic Index Notation

As it is used here in this treatises given a set of indices {i, j, k} the intrinsic index is appended in the set with a colon, {•j•: t, s}, hence $x_{\bullet j \bullet :t}$ or $x_{\bullet j \bullet :s}$ or $x_{\bullet j \bullet :t,s}$, being the last index or indices in the delineated sequence.

Definition 16.1.4 Intrinsic Differential Notation

Intrinsic differential notation is patterned after the D12.5.4 "Simple Differental Vector Notation".

$\mathbf{A}_{:s} \equiv d\,\mathbf{A} / ds$ Equation A

Definition 16.1.5 Intrinsic Covariant Differential Notation

Intrinsic covariant differential notation is patterned after the TK.3.9 "Exact Differential, Parametric, Chain Rule".

$\mathbf{A}_{;s} \equiv \sum_i \mathbf{A}_{;j}\, x^{j:s}$ Equation A[16.1.2] [16.1.3]

Definition 16.1.6 Intrinsic Contravariant Differential Notation

Intrinsic contravariant differential notation is patterned after the TK.5.3 "Chain Rule for Exact Differential of a Vector".

$\mathbf{A}^{;s} \equiv \sum_j \mathbf{A}^{;j}\, x_{j:s}$ Equation A[16.1.2] [16.1.3]

[16.1.1]Note: The intrinsic Newtonian and Lagrange notation, beyond about two and three differentials, become clumsy, unwieldy and in more modern writings is why they have been abandon using higher order notation.

[16.1.2]Note: The Einsteinian summation notation does not apply to the jth index since it is a specific differential independently selected.

[16.1.3]Note: That the intrinsic differential notation $[x^{j:s}, x_{j:s}]$ for covariant and contravariant are coefficients so must follow the raising and lowering of indices of the corresponding tensor coefficient.

Definition 16.1.7 Intrinsic Differential Pseudo Vectors

There are quantities that are used just like a vector with all the properties of a vector, but there not vectors. Such quantities are found with intrinsic differentials. They acquire their vector properties, because they are associated with a set of basis vectors as follows:

$\lambda^t \mathbf{c}_t = dq^t / ds\ \mathbf{c}_t$ Equation A

and

$\lambda_t \mathbf{c}^t = dq_t / ds\ \mathbf{c}^t$ Equation B

However if the basis vectors are dropped the lambda coefficients are left, but as long as they are associated with the basis vectors they appear to have all the properties of a vector. The result is

$\lambda^t \equiv q^{t:s} \equiv dq^t / ds$ is an intrinsic covariant pseudo vector Equation C

and

$\lambda_t \equiv q_{t:s} \equiv dq_t / ds$ is an intrinsic contravariant pseudo vector Equation D

The intrinsic covariant and contravariant pseudo vectors follow the rules of raising and lowering indices by covariant and contravariant convention for coefficients, which is what they really are.

Section 16.2 Serret-Frenet, Particular, Intrinsic Differential Solutions

Theorem 16.2.1 Particular Solution: Intrinsic Covariant Differential

Step	Hypothesis	$\mathbf{A}_{:s} = \sum_i (\sum_j A^{i,j} x^{j:s} + \sum_j \sum_k A^k \Gamma_{kj}{}^i x^{j:s}) \mathbf{b}_i$
1	From TK.5.3 Chain Rule for Exact Differential of a Vector, D16.1.4 Intrinsic Exact Differential Notation, D16.1.5 Intrinsic Covariant Differential Notation and A7.9.2 Substitution of Vectors	$\mathbf{A}_{:s} = \sum_j \mathbf{A}_{;j} x^{j:s}$
2	From 1, T12.3.2 Form of a Covariant Vector and A7.9.2 Substitution of Vectors	$\mathbf{A}_{:s} = \sum_j (\sum_i A^{i;j} \mathbf{b}_i) x^{j:s}$
3	From 2, T7.10.3 Distribution of a Scalar over Addition of Vectors and reorder summation	$\mathbf{A}_{:s} = \sum_i (\sum_j A^{i;j} x^{j:s}) \mathbf{b}_i$
4	From 3, T12.6.2B Differentiation of Tensor: Rank-1 Covariant and A7.9.2 Substitution of Vectors	$\mathbf{A}_{:s} = \sum_i [(\sum_j (A^{i,j} + \sum_k A^k \Gamma_{kj}{}^i) x^{j:s}] \mathbf{b}_i$
∴	From 4, T10.3.20 Addition of Tensors: Distribution of a Tensor over Addition of Tensors and A10.2.14 Correspondence of Tensors and Tensor Coefficients	$\mathbf{A}_{:s} = \sum_i (\sum_j A^{i,j} x^{j:s} + \sum_j \sum_k A^k \Gamma_{kj}{}^i x^{j:s}) \mathbf{b}_i$

Theorem 16.2.2 Particular Solution: Intrinsic Contravariant Differential

Step	Hypothesis	$\mathbf{A}^{:s} = \sum_i (\sum_j A_{i,j}\, x_{j:s} - \sum_j \sum_k A_k\, \Gamma_{ij}{}^k\, x_{j:s})\, \mathbf{b}^i$
1	From TK.5.3 Chain Rule for Exact Differential of a Vector, D16.1.4 Intrinsic Exact Differential Notation, D16.1.6 Intrinsic Contravariant Differential Notation and A7.9.2 Substitution of Vectors	$\mathbf{A}^{:s} = \sum_j \mathbf{A}^{:j}\, x_{j:s}$
2	From 1, T12.3.7 Form of a Contravariant Vector and A7.9.2 Substitution of Vectors	$\mathbf{A}^{:s} = \sum_j (\sum_i A_{i;j}\, \mathbf{b}^i)\, x_{j:s}$
3	From 2, T7.10.3 Distribution of a Scalar over Addition of Vectors and reorder summation	$\mathbf{A}^{:s} = \sum_i (\sum_j A_{i;j}\, x_{j:s})\, \mathbf{b}^i$
4	From 3, T12.6.3B Differentiation of Tensor: Rank-1 Contravariant and A7.9.2 Substitution of Vectors	$\mathbf{A}^{:s} = \sum_i [(\sum_j (A_{i,j} - \sum_k A_k\, \Gamma_{ij}{}^k)\, x_{j:s}]\, \mathbf{b}^i$
∴	From 4, T10.3.20 Addition of Tensors: Distribution of a Tensor over Addition of Tensors and A10.2.14 Correspondence of Tensors and Tensor Coefficients	$\mathbf{A}^{:s} = \sum_i (\sum_j A_{i,j}\, x_{j:s} - \sum_j \sum_k A_k\, \Gamma_{ij}{}^k\, x_{j:s})\, \mathbf{b}^i$

Theorem 16.2.3 Particular Solution: Intrinsic Serret-Frenet Tangent Differential

Step	Hypothesis	$\kappa \mathbf{N} = \sum_i (\sum_j T^{i,j}\, x^{j:s} + \sum_j \sum_k T^k\, \Gamma_{kj}{}^i\, x^{j:s})\, \mathbf{b}_i$
1	From AK.5.3A Uniqueness of Differentiation of a Vector	$\mathbf{T}_{:s} = \mathbf{T}_{:s}$
2	From1, D16.1.4 Intrinsic Exact Differential Notation, D16.1.2A Intrinsic Lagrangian Notation and T12.12.6A On Space Curve-C: First Order Derivative of Tangent Vector	$\mathbf{T}' = \mathbf{T}_{:s}$
∴	From 2, T16.2.1 Particular Solution: Intrinsic Covariant Differential, T12.12.6A On Space Curve-C: First Order Derivative of Tangent Vector and A7.9.2 Substitution of Vectors	$\kappa \mathbf{N} = \sum_i (\sum_j T^{i,j}\, x^{j:s} + \sum_j \sum_k T^k\, \Gamma_{kj}{}^i\, x^{j:s})\, \mathbf{b}_i$

Theorem 16.2.4 Particular Solution: Intrinsic Serret-Frenet Second Order Differential

Step	Hypothesis	$\kappa \mathbf{N} = \sum_i (\sum_j x^{i:s:s} + \sum_j \sum_k \Gamma_{kj}{}^i x^{k:s} x^{j:s}) \mathbf{b}_i$
1	From T12.3.2S2g Form of a Covariant Vector and D16.1.4 Intrinsic Exact Differential Notation	$T^i \mathbf{T}_i = x^{i:s} \mathbf{b}_i$
2	From 1 and equating base vectors	$T^i = x^{i:s}$
3	From 1, T16.2.1 Particular Solution: Intrinsic Covariant Differential, O12.3.1B Riemann Covariant Fundamental Tangent and Normal Vector and A7.9.2 Substitution of Vectors	$\kappa \mathbf{N} = \sum_i (\sum_j A^{i,j} x^{j:s} + \sum_j \sum_k A^k \Gamma_{kj}{}^i x^{j:s}) \mathbf{b}_i$
4	From 2, 3, T10.2.5 Tool: Tensor Substitution and A10.2.14 Correspondence of Tensors and Tensor Coefficients	$\kappa \mathbf{N} = \sum_i [\sum_j (x^{i:s})^{,j} x^{j:s} + \sum_j \sum_k x^{k:s} \Gamma_{kj}{}^i x^{j:s}] \mathbf{b}_i$
5	From 3 and T9.5.2 Chain Rule for Exact Differential of a Vector	$\kappa \mathbf{N} = \sum_i [(x^{i:s})^{:s} + \sum_j \sum_k x^{k:s} \Gamma_{kj}{}^i x^{j:s}] \mathbf{b}_i$
6	From 4, Dx9.3.2.1H Second Parametric Differentiation with mark, fraction and dot notation, D16.1.5 Intrinsic Covariant Differential Notation and D16.1.3 Intrinsic Index Notation	$\kappa \mathbf{N} = \sum_i (x^{i:s:s} + \sum_j \sum_k x^{k:s} \Gamma_{kj}{}^i x^{j:s}) \mathbf{b}_i$
∴	From 5, T10.3.7 Product of Tensors: Commutative by Multiplication of Rank p + q → q + p and A10.2.14 Correspondence of Tensors and Tensor Coefficients	$\kappa \mathbf{N} = \sum_i (\sum_j x^{i:s:s} + \sum_j \sum_k \Gamma_{kj}{}^i x^{k:s} x^{j:s}) \mathbf{b}_i$

Theorem 16.2.5 Particular Solution: Intrinsic Serret-Frenet Normal Differential

Step	Hypothesis			
		$\tau\mathbf{B} - \kappa\,\mathbf{T}$	=	$\sum_i (\sum_j N^{i,j}\, x^{j:s} + \sum_j \sum_k N^k\, \Gamma_{kj}{}^i\, x^{j:s})\, \mathbf{b}_i$
1	From AK.5.3A Uniqueness of Differentiation of a Vector	$\mathbf{N}_{:s}$	=	$\mathbf{N}_{:s}$
2	From1, D16.1.4 Intrinsic Exact Differential Notation, D16.1.2A Intrinsic Lagrangian Notation and T12.12.11 On Space Curve-C: Rate Normal Vector	$\mathbf{N}'$	=	$\mathbf{N}_{:s}$
∴	From 2, T16.2.1 Particular Solution: Intrinsic Covariant Differential, T12.12.11 On Space Curve-C: Rate Normal Vector and A7.9.2 Substitution of Vectors	$\tau\mathbf{B} - \kappa\,\mathbf{T}$	=	$\sum_i (\sum_j N^{i,j}\, x^{j:s} + \sum_j \sum_k N^k\, \Gamma_{kj}{}^i\, x^{j:s})\, \mathbf{b}_i$

Theorem 16.2.6 Particular Solution: Intrinsic Serret-Frenet Binormal Differential

Step	Hypothesis			
		$-\tau\,\mathbf{N}$	=	$\sum_i (\sum_j B^{i,j}\, x^{j:s} + \sum_j \sum_k B^k\, \Gamma_{kj}{}^i\, x^{j:s})\, \mathbf{b}_i$
1	From AK.5.3A Uniqueness of Differentiation of a Vector	$\mathbf{B}_{:s}$	=	$\mathbf{B}_{:s}$
2	From1, D16.1.4 Intrinsic Exact Differential Notation, D16.1.2A Intrinsic Lagrangian Notation and T12.12.15 On Space Curve-C: Rate of Binormal Vector	$\mathbf{B}'$	=	$\mathbf{B}_{:s}$
∴	From 2, T16.2.1 Particular Solution: Intrinsic Covariant Differential, T12.12.15 On Space Curve-C: Rate of Binormal Vector and A7.9.2 Substitution of Vectors	$-\tau\,\mathbf{N}$	=	$\sum_i (\sum_j B^{i,j}\, x^{j:s} + \sum_j \sum_k B^k\, \Gamma_{kj}{}^i\, x^{j:s})\, \mathbf{b}_i$

Section 16.3 Affine, Complementary, Intrinsic Differential Solutions

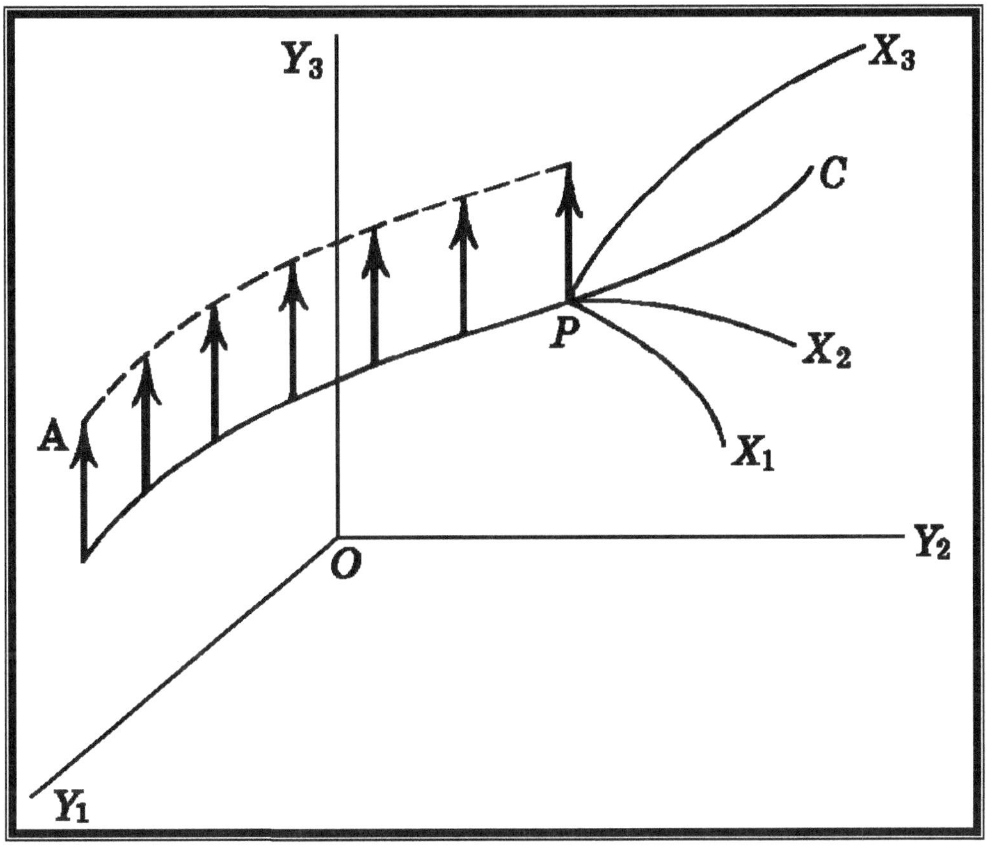

Figure 16.3.1 Parallel Vector Field

Given vector **A** an intrinsic vector as defined in definition D14.1.1 "Intrinsic Differentiation" and the given G14.1.1 "For Arc Length verses Intrinsic Covariant Vector" can be restated as follows:

Given 16.3.1 For Arc Length verses Intrinsic Parallel Vector Field

1g	Given	see Figure 16.1.1	Cartesian to Curvilinear Coordinate on a Manifold
2g		$\mathbf{A} \equiv (\bullet\, x_i \bullet)$	
3g		$x_i \equiv x_i(t)$	for all [i] tracing out the curve-C
4g		$t_a \leq t \leq t_b$	

If a field of vectors is constructed parallel at every point along the curve-C, then vectors **A's** do not change along the curve and it can be written as $d\mathbf{A}/dt = \mathbf{0}$ and the following will hold

Theorem 16.3.1 Affine Field: Intrinsic Covariant Differential along a Spatial Curve

1g	Given	$\mathbf{0} \equiv d\mathbf{A}/ds$
Step	Hypothesis	$0 = A^{i:s} + \sum_\alpha \Gamma_{\alpha k}{}^{i} A^\alpha x^{k:s}$
1	From 1g	$\mathbf{0} = d\mathbf{A}/ds$
2	From 1, T12.3.2 Form of a Covariant Vector and A7.9.2 Substitution of Vectors	$\mathbf{0} = d(A^i \mathbf{b}_i)/ds$
3	From 2 and TK.5.3 Chain Rule for Exact Differential of a Vector	$\mathbf{0} = \sum_k [\partial(A^i \mathbf{b}_i)/ \partial x_k](dx^k/ds)$
4	From 3, D12.6.1B Differental Tensor Notation and D16.1.1 Intrinsic Index Notation	$\mathbf{0} = \sum_k (A^i \mathbf{b}_i)^{;k}\, x^{k:s}$
5	From 4, T12.6.2B Differentiation of Tensor: Rank-1 Covariant and T10.3.5 Tool: Tensor Substitution	$\mathbf{0} = \sum_k (A^{i,k} + \sum_\alpha \Gamma_{\alpha k}{}^{i} A^\alpha)\, x^{k:s}\, \mathbf{b}_i$
6	From 5 and T10.3.20 Addition of Tensors: Distribution of a Tensor over Addition of Tensors	$\mathbf{0} = (\sum_k A^{i,k} x^{k:s} + \sum_k \sum_\alpha \Gamma_{\alpha k}{}^{i} A^\alpha x^{k:s})\, \mathbf{b}_i$
7	From 6 and TK.3.9 Exact Differential, Parametric, Chain Rule	$\mathbf{0} = (A^{i:s} + \sum_k \sum_\alpha \Gamma_{\alpha k}{}^{i} A^\alpha x^{k:s})\, \mathbf{b}_i$
∴	From 7 and D10.1.18B Zero Tensor	$0 = A^{i:s} + \sum_k \sum_\alpha \Gamma_{\alpha k}{}^{i} A^\alpha x^{k:s}$

Theorem 16.3.2 Affine Field: Intrinsic Contravariant Differential along a Time-Line

1g	Given	$\mathbf{0} \equiv d\mathbf{A}/dt$
Step	Hypothesis	$0 = A_{i:t} - \sum_k \sum_\alpha \Gamma_{ik}{}^{\alpha} A_\alpha x_{k:t}$
1	From 1g	$\mathbf{0} = d\mathbf{A}/dt$
2	From 1, T12.3.7 Form of a Contravariant Vector and A7.9.2 Substitution of Vectors	$\mathbf{0} = d(A_i \mathbf{b}^i)/dt$
3	From 2 and TK.5.3 Chain Rule for Exact Differential of a Vector	$\mathbf{0} = \sum_k [\partial(A_i \mathbf{b}^i)/ \partial x_k](dx_k/dt)$
4	From 3, D12.6.1B Differental Tensor Notation and D16.1.1 Intrinsic Index Notation	$\mathbf{0} = \sum_k (A_i \mathbf{b}^i)_{;k}\, x_{k:t}$
5	From 4, T12.6.3B Differentiation of Tensor: Rank-1 Contravariant and T10.3.5 Tool: Tensor Substitution	$\mathbf{0} = \sum_k (A_{i,k} - \sum_\alpha \Gamma_{ik}{}^{\alpha} A_\alpha)\, x_{k:t} \mathbf{b}^i$
6	From 5 and T10.3.20 Addition of Tensors: Distribution of a Tensor over Addition of Tensors	$\mathbf{0} = (\sum_k A_{i,k}\, x_{k:t} - \sum_k \sum_\alpha \Gamma_{ik}{}^{\alpha} A_\alpha x_{k:t}) \mathbf{b}^i$

7	From 6 and TK.3.9 Exact Differential, Parametric, Chain Rule	$\mathbf{0} = (A_{i:t} - \sum_k \sum_\alpha \Gamma_{ik}{}^\alpha A_\alpha x_{k:t})\mathbf{b}^i$
∴	From 7 and D10.1.18B Zero Tensor	$0 = A_{i:t} - \sum_k \sum_\alpha \Gamma_{ik}{}^\alpha A_\alpha x_{k:t}$

Theorem 16.3.3 Affine Field: Affine Spatial Curve

1g	Given	A_i is a Riemann Covariant Tangent Vector on a Spatial Curve-C
Step	Hypothesis	$0 = x^{i:s:s} + \sum_k \sum_\alpha \Gamma_{\alpha k}{}^i x^{\alpha:s} x^{k:s}$
1	From T16.3.1 Affine Field: Intrinsic Covariant Differential along a Spatial Curve	$0 = A^{i:s} + \sum_k \sum_\alpha \Gamma_{\alpha k}{}^i A^\alpha x^{k:s}$
2	From 1g and O12.3.1A Riemann Covariant Fundamental Tangent Vector	$A^i \equiv x^{i:s}$
3	From 1, 2, T10.2.5 Tool: Tensor Substitution and A10.2.14 Correspondence of Tensors and Tensor Coefficients	$0 = \sum_j (x^{i:s})_{,j}\, x^{j:s} + \sum_j \sum_k x^{k:s}\, \Gamma_{kj}{}^i\, x^{j:s}$
4	From 3 and T9.5.2 Chain Rule for Exact Differential of a Vector	$0 = x^{i:s:s} + \sum_j \sum_k x^{k:s}\, \Gamma_{kj}{}^i\, x^{j:s}$
∴	From 4, T10.3.7 Product of Tensors: Commutative by Multiplication of Rank p + q → q + p and A10.2.14 Correspondence of Tensors and Tensor Coefficients	$0 = x^{i:s:s} + \sum_k \sum_\alpha \Gamma_{\alpha k}{}^i x^{\alpha:s} x^{k:s}$

Theorem 16.3.4 Affine Field: Affine Temporal Time-Line

1g	Given	A_i is a Riemann Contravariant Tangent Vector on Time Line-T
Step	Hypothesis	$0 = x_{i:t:t} - \sum_j \sum_k \Gamma_{ij}{}^k\, x_{k:t}\, x_{j:t}$
1	From T16.3.2 Affine Field: Intrinsic Contravariant Differential along a Time-Line	$0 = A_{i:t} - \sum_k \sum_\alpha \Gamma_{ik}{}^\alpha\, A_\alpha x_{k:t}$
2	From 1g and F12.2. Manifold Tangential Plane On Time Line	$A_i \equiv x_{i:t}$
3	From 1, 2, T10.2.5 Tool: Tensor Substitution and A10.2.14 Correspondence of Tensors and Tensor Coefficients	$0 = (x_{i:t})_{:t} - \sum_j \sum_k \Gamma_{ij}{}^k\, (x_{k:t})\, x_{j:t}$
4	From 3, Dx9.3.2.1H Second Parametric Differentiation with mark, fraction and dot notation, D16.1.5 Intrinsic Covariant Differential Notation and D16.1.3 Intrinsic Index Notation	$0 = x_{i:t:t} - \sum_j \sum_k \Gamma_{ij}{}^k\, x_{k:t}\, x_{j:t}$
∴	From 4, T10.3.7 Product of Tensors: Commutative by Multiplication of Rank p + q → q + p and A10.2.14 Correspondence of Tensors and Tensor Coefficients	$0 = x_{i:t:t} - \sum_j \sum_k \Gamma_{ij}{}^k\, x_{k:t}\, x_{j:t}$

Theorem 16.3.5 Affine Field: Affine Spatial Straight Line

1g	Given	Orthogonal, distinct, independent coordinate system
2g		$x_{i0} \equiv b_{0i} - m\, s_0$
Step	Hypothesis	$x_i = m\, s + x_{i0}$
1	From T16.3.3 Affine Field: Spatial Curve and DxK.3.2.1H Second Parametric Differentiation with mark, fraction and dot notation	$0 = d^2x_i / ds^2 + \sum_k \sum_\alpha \Gamma_{\alpha k}{}^i\, x^{\alpha:s}\, x^{k:s}$
2	From 1 and T12.5.18 Christoffel Symbol: Of the Second Kind in Orthogonal Space	$0 = d^2x_i / ds^2 + \sum_k \sum_\alpha 0\, x^{\alpha:s}\, x^{k:s}$
3	From 2, T4.4.1 Equalities: Any Quantity Multiplied by Zero is Zero and A4.2.7 Identity Add	$0 = d^2x_i / ds^2$
4	From 3, A4.2.2B Equality, DxK.3.2.1K exact differential operator and LxK.3.1.7C Uniqueness Theorem of Differentiation	$[d\,(\,dx_i / ds\,) / ds\,]\, ds = 0\, ds$
5	From 4, A4.2.11 Associative Multp and D4.1.4(B, A)	$d\,(\,dx_i / ds\,)\,[ds / ds\,] = 0\, ds$
6	From 5, A4.2.13 Inverse Multp and A4.2.12 Identity Multp	$d\,(\,dx_i / ds\,) = 0\, ds$
7	From 6 and TK.4.2 Integrals: Uniqueness of Integration	$\int d\,(\,dx_i / ds\,) = \int 0\, ds$
8	From 7 and DK.4.6 Indefinite Integral	$dx_i / ds = m$
9	From 8 and LxK.3.1.7C Uniqueness Theorem of Differentiation	$(dx_i / ds)\, ds = m\, ds$
10	From 9, A4.2.11 Associative Multp, D4.1.4(B, A), A4.2.13 Inverse Multp and A4.2.12 Identity Multp	$dx_i = m\, ds$
11	From 10 and TK.4.2 Integrals: Uniqueness of Integration	$\int_{b_0}^{x} dx_i = \int_{s_0}^{s} m\, ds$
12	From 11, LxK.4.3.1 Distribution of Constant Out Of Integrand and DxK.4.1.4 Definition of a Definite Integral	$x_i - b_{0i} = m\,(s - s_0)$
13	From 12 and T4.3.6A Equalities: Reversal of Left Cancellation by Addition	$x_i = m\,(s - s_0) + b_{0i}$
14	From 13, A4.2.14 Distribution and A4.2.5 Commutative Add	$x_i = m\, s + b_{0i} - m\, s_0$
∴	From 2g, 14 and A4.2.3 Substitution	$x_i = m\, s + x_{i0}$

Tensor Calculus & Physics: A General Treatise

Table of Contents

List of Tables

List of Figures

List of Observations

Chapter 17 Tensor Calculus: Hypersurfaces and Hypersheets

Section 17.1 Riemannian Hypersurface $\mathfrak{H}^{n-1}$

From A11.1.1A "A Rendered Manifold"

$R(\bullet x_i \bullet) = c$ Equation A

creates a many layered surface for varying values of [c], much as the many layers of an onion, with infinitesimally thin stratum.

Definition 17.1.1 Riemannian Hypersheet $\mathfrak{H}^{n-1}$

A rendered manifold embedded in a Riemannian space $\mathfrak{R}^n$ with each value of [c] represents a surface of many dimensions called a ***Riemannian Hypersheet*** $\mathfrak{H}^{n-1}$ as described in A11.4.2 "Hypersurface Manifold $\mathfrak{H}^{n-1}$".

Definition 17.1.2 Covariant Metric Hypersheet Dyadic

$a_{\alpha\beta} \equiv \mathbf{a}_\alpha \bullet \mathbf{a}_\beta$	component form	Equation A
$(a_{\alpha\beta}) \equiv (\mathbf{a}_\alpha \bullet \mathbf{a}_\beta)$	component in matrix form	Equation B
$A \equiv (a_{\alpha\beta})$	matrix form (n x n)	Equation C
$A \equiv S^{-1}\{A_c A_c^T\}$	matrix form (n x 1)(1 x n)	Equation D

for

$S\{ (\mathbf{a}_\alpha \bullet \mathbf{a}_\beta) \} \equiv A_c A_c^T$	matrix form (n x 1)(1 x n)	Equation E
$(\mathbf{a}_\alpha \bullet \mathbf{a}_\beta) \equiv S^{-1}\{A_c A_c^T\}$	matrix form (n x 1)(1 x n)	Equation F
$A_c \equiv [\varepsilon_\alpha]$	matrix form (n x 1)	Equation G

Let ∃| a coordinate point $P(\bullet\ x_i\ \bullet)$ of a Riemannian Manifold $\mathfrak{R}^n$ and let it correspond to a parametric coordinate point $Q(\bullet\ u_\alpha\ \bullet)$ embedded in that space forming a metric, submanifold, called a Riemannian Hypersheet $\mathfrak{H}^{n-1}$.

Definition 17.1.3 Contravariant Metric Hypersheet Dyadic

$a^{\alpha\beta} \equiv \mathbf{a}^\alpha \bullet \mathbf{a}^\beta$	component form	Equation A
$(a^{\alpha\beta}) \equiv (\mathbf{a}^\alpha \bullet \mathbf{a}^\beta)$	component in matrix form	Equation B
$a^{\alpha\beta} \equiv a^{-1} A^{\alpha\beta}$	where $A^{\alpha\beta}$ is the ***cofactor*** of the element $a^{\alpha\beta}$ in [a].	Equation C
$A^{-1} \equiv S^{-1}\{A^c A^{cT}\}$	matrix form (n x 1)(1 x n)	Equation D

for

$S\{ (\mathbf{a}^\alpha \bullet \mathbf{a}^\beta) \} \equiv A^c A^{cT}$	matrix form (n x 1)(1 x n)	Equation E
$(\mathbf{a}^\alpha \bullet \mathbf{a}^\beta) \equiv S^{-1}\{A^c A^{cT}\}$	matrix form (n x 1)(1 x n)	Equation F

Sometimes known as the ***reciprocal metric tensor***.

Definition 17.1.4 Mixed Metric Hypersheet Dyadic

$a_\alpha{}^\beta \equiv \delta_\alpha{}^\beta \equiv \mathbf{a}_\alpha \bullet \mathbf{a}^\beta$	mix basis vectors	Equation A
$(a_\alpha{}^\beta) \equiv (\delta_\alpha{}^\beta) \equiv (\mathbf{a}_\alpha \bullet \mathbf{a}^\beta)$	Kronecker Delta as a mixed metric	Equation B
$I \equiv S^{-1}\{A_c A^{cT}\}$	matrix form (n x 1)(1 x n)	Equation C
$a^\alpha{}_\beta \equiv \delta^\alpha{}_\beta \equiv \mathbf{a}^\alpha \bullet \mathbf{a}_\beta$	mix basis vectors	Equation D
$(a^\alpha{}_\beta) \equiv (\delta^\alpha{}_\beta) \equiv (\mathbf{a}^\alpha \bullet \mathbf{a}_\beta)$	Kronecker Delta as a mixed metric	Equation E
$I \equiv S^{-1}\{A^c A_c{}^T\}$	matrix form (n x 1)(1 x n)	Equation F

Placing $a_{\alpha\beta}$ and $a^{\alpha\beta}$ components into n x n matrices allows an orthogonal system to be constructed. This also shows the orthogonal relationship of the system with $\delta_\alpha{}^\beta = \mathbf{a}_\alpha \bullet \mathbf{a}^\beta$, hence $\mathbf{a}_\alpha \perp \mathbf{a}^\beta$.

Definition 17.1.5 Covariant Metric Determinate for Hypersheet $\mathfrak{H}^{n-1}$

$a = | a_{\alpha\beta} |$ Covariant Determinate Equation A

Definition 17.1.6 Surface Density Factor

$\rho_u = \prod_{\alpha=1}^{(n-1)} a_\alpha$ Surface Density Factor Equation A

Definition 17.1.7 Cosine Surface Distortion Factor

The measure of how distorted the surface is deviating from an ideal orthogonal base vector system.

$\nu(c) \equiv | \cos \theta_{\alpha\beta} |$ Equation A

$\nu_c \equiv \nu(c)$ Equation B

Definition 17.1.8 Riemannian Manifold $\mathfrak{R}^n$ and Hypersheet $\mathfrak{H}^{n-1}$ Transfer Functions

A rendered function iff it can be separated into [n] ***Riemannian functions*** $x_i(\bullet\ u_\alpha\ \bullet)$ and [n–1] ***Hypersheet functions*** $u_\alpha(\bullet\ x_i\ \bullet)$ forming a set of transfer functions between manifold and hypersheet.

Definition 17.1.9 Riemannian Manifold $\mathfrak{R}^n$ Mixed Tensor of Rank-2 on a Hypersheet $\mathfrak{H}^{n-1}$

The transformation derivative in a Riemannian Manifold $\mathfrak{R}^n$ behave like a tensor of rank-2 having an inverse operation with the transformation derivative on a Hypersheet $\mathfrak{H}^{n-1}$:

$x^i{}_\alpha \equiv x^{i;\alpha}$ Equation A

Definition 17.1.10 Hypersheet $\mathfrak{H}^{n-1}$ Mixed Tensor of Rank-2 in a Riemannian Manifold $\mathfrak{R}^n$

The transformation derivative on a Hypersheet $\mathfrak{H}^{n-1}$ behaves like a tensor of rank-2 having an inverse operation with the transformation derivative in a Riemannian Manifold $\mathfrak{R}^n$:

$u^\alpha{}_i \equiv u^{\alpha;i}$ Equation A

Axiom 17.1.1 Properties of a Base Hypersheet $\mathfrak{H}_u^{n-1}$

Let $\exists$ a set of properties for a space, where space is made from Axial Field Sets comprised of field elements $[u_{\alpha\kappa}]$

P1) $A_\alpha \equiv \{\, u_{\alpha\kappa} \mid (u_{\alpha\kappa}) \in (\mathbf{R} \vee \mathbf{C}^2) \text{ for } \kappa = 1, 2, \ldots, \mu_\alpha \}$ *arithmetic axes.*

P2) $[\, \mid u_\alpha - a_\alpha \mid\, < \kappa_\alpha \text{ bounded for all } (\alpha = 1, 2, \ldots, n-1)\,]$

where $K(\bullet \kappa_\alpha \bullet)$ are real numbers, constants, that can be found to bound the inequality about the point $A(\bullet\, a_\alpha\, \bullet)$. Absolute value in terms of complex numbers is treated as absolute magnitude.

These base properties that bound the U-positional numbers for an infinite number of points P_κ in a continuous space, allows the existence of limits, and establishing differentiability with its inverse the antidifferential or the ability to integrate over that region of space.

P3) Derived units of measure associated with an axis.

$A_{u\alpha} \equiv \{\, P1, u_\alpha, \mid P1 \vee (u_\alpha \rightarrow [U]\,)\,\}$, for u_α a derived unit of measure, tables TA.3.5[II] "Fundamental Physical Quantities", TA.1.1[II] "Correspondence between Rectilinear vs. Rotational Physical Equations" and TA.3.1[II] "Symbols, dimensions and Units for Physical Quantities" and for examples$\}$ for examples$\}$

P4) $\mathcal{P}_u^{n-1} \equiv [\, \prod_{\alpha=1}^{n-1} \times A_{u\alpha}\,]$ base hypersheet property for U-coordinate space

Such that there are a set of n-tuple points $P_\kappa(\bullet\, u_\alpha\, \bullet)$ for space U generated from the coordinate product of the axes

P5) $P_\kappa(\bullet\, u_\alpha\, \bullet) \in \mathcal{P}_u$ for $\kappa = 1, 2, \ldots, r \,\wedge\, < r = \prod_{\alpha=1}^{n-1} \mu_\alpha >$

Axiom 17.1.2 Base Hypersheet (Sokolnikoff Type) $\mathfrak{B}_u^{n-1}$

Now at every point in that space, has the hypersheet properties 1–1 and onto assigned to it with a well-defined range for the positional numbers between any two hypersheets:

$\mathcal{P}_{bu}^{n-1} \equiv (\, P1, P2, P3, P4_u^{n-1}, P5\,)$ list of base properties for a hypersheet.

$\mathfrak{B}_u^{n-1} \equiv \{\, \mathcal{P}_{bu}^{n-1}\,\}$

While $\mathfrak{B}_u^{n-1}$ *is called the base hypersheet of space U and read, as the hypersheet is the set of base hypersheet properties* P1, P2, P3, P4_u^{n-1}, *and* P5.

Axiom 17.1.3 Manifold to Hypersheet 1-1 Correspondence and Onto Image Functions

The elements connecting *one-to-one* (1-1) in n-ways, such that they are onto-functions, return mirror images of those functions by an inverse operation are given in the following form:

$x_i = x_i(\bullet\, u_\alpha\, \bullet)$ Equation A

$u_\alpha = u_\alpha(\bullet\, x_i\, \bullet)$ Equation B

Orthogonal Systems	U	Mapping, Set U to X Predecessor	X	Inverse, Set X to U Successor	U	1-1 [11.1.1]	Onto [11.1.1]
Elliptic Paraboloid			x			NA	NA
	θ		y		θ	YES	NO
	r		z		r	YES	NO

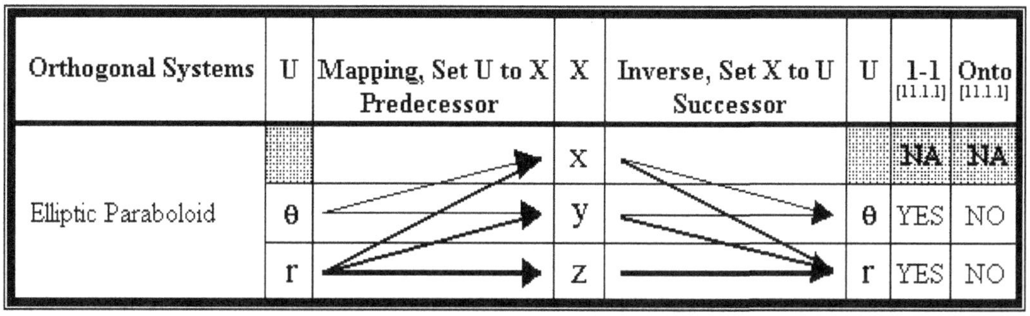

Figure 17.1.1 Invertible Elliptic Paraboloid Mapping Diagram

Table 17.1.1 Image of Elliptic Paraboloid Mapping Diagram

Step	Transformation Predecessor	1-1	Function Onto	Transformation Successor	1-1	Function Onto
1	$f_x(\theta, r) = x$	1-1	~Onto	$f_\theta(x, y) = \theta$	1-1	~Onto
2	$f_Y(\theta, r) = y$	1-1	~Onto	$f_r(x, y) = r$	1-1	~Onto
3	$g_z(r) = z$	1-1	Onto	$g_r(z) = r$	1-1	Onto

Axiom 17.1.4 Existence of Inverse Functions between a Manifold and Hypersheet

Manifolds to hypersheets are onto parametric functions, such that, $f(X) \supset U$, then there exists an inverse set of parametric functions, such that, $f^{-1}(U) \supset X$. This is demonstrated in the above table as an example set U is mapped to set X and visa versa in the inverse. Note the arrowhead points to a single parameter satisfying a parametric function that is onto, and the inverse is a mirror image of the original satisfying the requirement for an inverse set of functions.

Axiom 17.1.5 Mapping Parametric Functions between a Manifold and a Hypersheet

A manifold is composed of two composite sets X and U of parametric functions that are invertible, hence

$f(X) \rightarrow U$ Equation A

$f^{-1}(U) \rightarrow X$ Equation B

Axiom 17.1.6 Images of Parametric Functions between a Manifold and a Hypersheet

A manifold is composed of two composite sets of parametric parameters $(\bullet\, x_i \,\bullet)$ and $(\bullet\, u_\alpha \,\bullet)$ at point-Q on the manifolds, hence

$f_\alpha(\bullet\, x_i \,\bullet) = u_\alpha$ Equation A

$f_i^{-1}(\bullet\, u_\alpha \,\bullet) = x_i$ Equation B

Axiom 17.1.7 Existence of an Inverse Transformation

$\exists|$ a set of Kronecker Delta relationships that transform from an enveloping Riemannian Manifold $\mathfrak{R}^n$ to an embedded Hypersheet $\mathfrak{H}^{n-1}$ and inversely back again. Kronecker delta's can only exist if their reciprocal relationships exist, it would follow, then that if they exist that there relative inverse transformation functions, $x_i(\bullet\, u_\alpha \,\bullet)$ and $u_\alpha(\bullet\, x_i \,\bullet)$ which comprise them, must exist as well.

Axiom 17.1.8 Fundamental Property of Continuum for a Hypersheet

It follows that since the enveloping manifold is continuous that the hypersurface embedded in it also is continuous and as such continuously differentiable at every point within the region of investigation. This includes at every point it must be a well-defined range over the axis of the positional numbers $[\bullet\, x_i \,\bullet]$ as pointed to by the positional vector.

$\mathbf{r} \equiv \sum_i x_i\, \mathbf{b}_i$ Equation A

Axiom 17.1.9 Fundamental Property of a Unique Manifold

Iff the render function can be separated into its parametric parameter transfer functions then the inverse transformation exists.

Axiom 17.1.10 Covariant Determinate Defines the Existence of a Hypersheet $\mathfrak{H}^{n-1}$

Let ∃ Hypersheet $\mathfrak{H}^{n-1}$ such that under the dot product operator contra and covariant metrics exist, where

$a \neq 0$ Equation A

or

$| a_{\alpha\beta} | \neq 0$ Equation B

Axiom 17.1.11 Covariant Metric Displacement on a Riemannian Hypersheet $\mathfrak{H}^{n-1}$

A Riemannian Hypersheet is a ***hypercurved manifold***, not necessarily orthogonal, which contains between any two, points P and Q, an infinitesimal measured distance given by an approximation of a Pythagoras' hypotenuse for distance.

P1) $a_{\alpha\beta} \equiv \mathbf{a}_\alpha \bullet \mathbf{a}_\beta$
(n–1)-Riemannian covariant base vectors,

P2) $d\,\overline{PQ} = \sqrt{\sum_\alpha \sum_\beta a_{\alpha\beta}\, du^\alpha du^\beta}$
(n–1)x(n–1) Riemann's covariant fundamental spatial length.

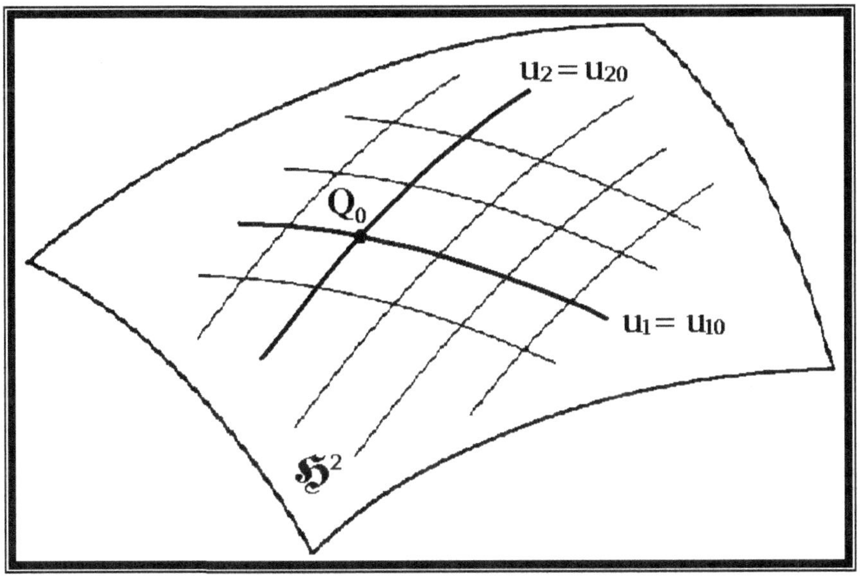

Figure 17.1.2 Hypersheet $\mathfrak{H}^2$; Constant Coordinates at Point-Q_0

From A17.1.1P2 "Properties of a Base Hypersheet $\mathfrak{H}_u{}^{n-1}$" These points are continuously differentiable within the neighborhood of the point-Q_0, so by the chain rule

Theorem 17.1.1 Manifold-Hypersheet Transformation: Exact Differential

Step	Hypothesis	$dx_i = \sum_\alpha x_{i,\alpha}\, du_\alpha$ for nx(n−1) EQ A	$du_\alpha = \sum_i u_{\alpha,i}\, dx_i$ for (n−1)xn EQ B
1	From A17.1.3(A, B) Manifold functions are One–To–One and Onto Hypersheets	$x_i = x_i(\bullet\, u_\alpha\, \bullet)$	$u_\alpha = u_\alpha(\bullet\, x_i\, \bullet)$
∴	From 1 and TK.3.9 Exact Differential, Parametric, Chain Rule	$dx_i = \sum_\alpha x_{i,\alpha}\, du_\alpha$ for nx(n−1) EQ A	$du_\alpha = \sum_i u_{\alpha,i}\, dx_i$ for (n−1)xn EQ B

Observation 17.1.1 Infinitesimal Linear Transformation at a Point

$(x_{i,\alpha}, u_{\alpha,i})$ are the transformational slopes, which provide an infinitesimal linear transformation at a Point-Q_0 between the manifold and the hypersurface.

Observation 17.1.2 Existence of Inverse Transformation Functions

From Table 17.1.1S3 "Interpretation of Elliptic Paraboloid Mapping Diagram" is invertible or has an inverse function for transformation as an, onto function by D2.22.7 "Onto Functions" and A2.23.6 "Inverse Transformation". On the other hand transformations in step 1 and 2 for predecessor and successor functions are not onto and there is no apparent way of guaranteeing that it is invertible, yet by inspection of the render function shows the inverse does exist for this specific case, also note that these functions are nonlinear.

It follows that if these transformations are shown to satisfy a Kronecker delta relationship, however the Kronecker delta can only exist if their reciprocal relationships exist, and then the inverse transformation functions must exist as well.

Theorem 17.1.2 Manifold-Hypersheet Transformation: Existence of Hypersheet $\mathfrak{H}^{n-1}$ Inverse

Step	Hypothesis	$\delta_{\alpha\beta} = \sum_i u_{\alpha,i}\, x_{i,\beta}$ for (n–1)x(n–1)
1	From T17.1.1B Manifold-Hypersheet Transformation: Exact Differential	$du_\alpha = \sum_i u_{\alpha,i}\, dx_i$
2	From 1, T17.1.1A Manifold-Hypersheet Transformation: Exact Differential, T10.2.5 Tool: Tensor Substitution and A10.2.14 Correspondence of Tensors and Tensor Coefficients	$du_\alpha = \sum_i u_{\alpha,i}\, (\sum_\alpha x_{i,\alpha}\, du_\alpha)$
3	From 2, T10.3.20 Addition of Tensors: Distribution of a Tensor over Addition of Tensors and A10.2.14 Correspondence of Tensors and Tensor Coefficients	$du_\alpha = \sum_i (\sum_\alpha u_{\alpha,i}\, x_{i,\alpha}\, du_\alpha)$
4	From 3, T10.3.20 Addition of Tensors: Distribution of a Tensor over Addition of Tensors, A10.2.14 Correspondence of Tensors and Tensor Coefficients and reorder summation	$du_\alpha = \sum_\alpha(\sum_i u_{\alpha,i}\, x_{i,\alpha})\, du_\alpha$
5	From 4, T6.3.4 Uniqueness of Matrix Multiplication; leaving $[du_\alpha]$ (n–1)x1 and D6.1.6A Identity Matrix-I	$I\,[du_\alpha] = [\sum_i u_{\alpha,i}\, x_{i,\beta}][\,du_\alpha]$
6	From 5 and equating row-column matrices	$I = [\sum_i u_{\alpha,i}\, x_{i,\beta}]$
7	From 6 and D6.1.6A Identity Matrix $[\delta_{\alpha\beta}]$	$[\delta_{\alpha\beta}] = [\sum_i u_{\alpha,i}\, x_{i,\beta}]$ for (n–1)x(n–1)
∴	From 7 and equating matrix elements	$\delta_{\alpha\beta} = \sum_i u_{\alpha,i}\, x_{i,\beta}$ for (n–1)x(n–1)

Theorem 17.1.3 Manifold-Hypersheet Transformation: Existence of Manifold $\mathfrak{R}^n$ Inverse

Step	Hypothesis	$\delta_{ij} = \Sigma_\alpha x_{i,\alpha} u_{\alpha,j}$ for nxn
1	From T17.1.1A Manifold-Hypersheet Transformation: Exact Differential	$dx_i = \Sigma_\alpha x_{i,\alpha} du_\alpha$
2	From 1, T17.1.1B Manifold-Hypersheet Transformation: Exact Differential, T10.2.5 Tool: Tensor Substitution and A10.2.14 Correspondence of Tensors and Tensor Coefficients	$dx_i = \Sigma_\alpha x_{i,\alpha} (\Sigma_i u_{\alpha,i} dx_i)$
3	From 2, T10.3.20 Addition of Tensors: Distribution of a Tensor over Addition of Tensors and A10.2.14 Correspondence of Tensors and Tensor Coefficients	$dx_i = \Sigma_\alpha (\Sigma_i x_{i,\alpha} u_{\alpha,i} dx_i)$
4	From 3, T10.3.20 Addition of Tensors: Distribution of a Tensor over Addition of Tensors, A10.2.14 Correspondence of Tensors and Tensor Coefficients and reorder summation	$dx_i = \Sigma_i (\Sigma_\alpha x_{i,\alpha} u_{\alpha,i}) dx_i$
5	From 4, T6.3.4 Uniqueness of Matrix Multiplication; leaving $[du_\alpha]$ (n–1)x1 and D6.1.6A Identity Matrix-I	$I [dx_i] = [\Sigma_\alpha x_{i,\alpha} u_{\alpha,j}][dx_i]$
6	From 5 and equating row-column matrices	$I = [\Sigma_\alpha x_{i,\alpha} u_{\alpha,j}]$
7	From 6 and D6.1.6A Identity Matrix $[\delta_{\alpha\beta}]$	$[\delta_{ij}] = [\Sigma_\alpha x_{i,\alpha} u_{\alpha,j}]$ for nxn
∴	From 7 and equating matrix elements	$\delta_{ij} = \Sigma_\alpha x_{i,\alpha} u_{\alpha,j}$ for nxn

Observation 17.1.3 Existence of the Manifold-Hypersheet relative Inverse Functions

From theorems T17.1.2 "Manifold-Hypersheet Transformation: Existence of Hypersheet $\mathfrak{H}^{n-1}$ Identity" and T17.1.3 "Manifold-Hypersheet Transformation: Existence of Manifold $\mathfrak{R}^n$ Identity" that the Kronecker Delta relationships can be found and do exists so from axiom A17.1.7 "Existence of an Inverse Transformation" the relative inverse transformation functions, $x_i(\bullet\ u_\alpha \bullet)$ and $u_\alpha(\bullet\ x_i \bullet)$ must exist as well.

Observation 17.1.4 Riemannian Manifold $\mathfrak{R}^n$ Transfer Functions as a Mixed Tensor

Since the transfer functions for a hypersurface $[x^i_{\ \alpha}]$ have an inverse relationship they can be thought of as Mixed Tensor of Rank-2 along the hyperplane.

Observation 17.1.5 Hypersheet $\mathfrak{H}^{n-1}$Transfer Functions as a Mixed Tensor

Since the transfer functions for a hypersurface $[u^\alpha_{\ i}]$ have an inverse relationship they can be thought of as Mixed Tensor of Rank-2 along the hyperplane.

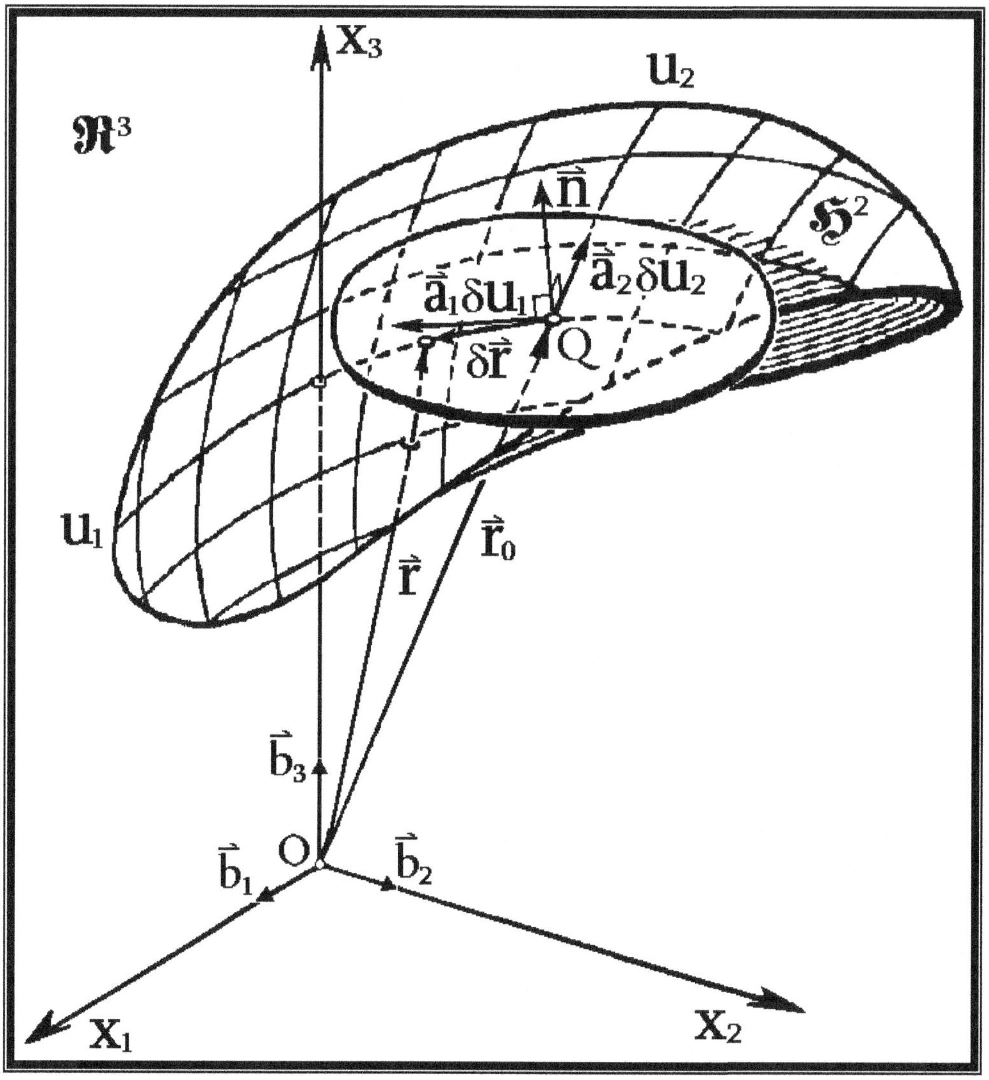

Figure 17.1.3 Embedded Hypersheet $\mathfrak{H}^2$ in Riemannian Manifold $\mathfrak{R}^3$

Theorem 17.1.4 Contravariant Bases Vectors for a Hypersheet

1g	Given	See Figure 17.1.3 Embedded Hypersheet $\mathfrak{H}^2$ in Riemannian Manifold $\mathfrak{R}^3$
Step	Hypothesis	$\mathbf{a}_\alpha = \partial\mathbf{r} / \partial u_\alpha$ for all α
1	From 1g	$\mathbf{a}_\alpha \, \delta u_\alpha \approx \delta\mathbf{r}$
2	From 1 and T4.4.4B Equalities: Right Cancellation by Multiplication	$\mathbf{a}_\alpha \approx \delta\mathbf{r} / \delta u_\alpha$
3	From 1 and LxK.2.1.3 Obvious Limit	$\mathbf{a}_\alpha = \lim_{\delta u\alpha \to 0} \delta\mathbf{r} / \delta u_\alpha$
∴	From 2 and DxK.3.1.3 Definition of a Derivative Alternate Form	$\mathbf{a}_\alpha = \partial\mathbf{r} / \partial u_\alpha$ for all α

Theorem 17.1.5 Manifold-Hypersheet Transformation: Hypersheet Contravariant Bases Vectors

1g	Given	See Figure 17.1.3 Embedded Hypersheet $\mathfrak{H}^2$ in Riemannian Manifold $\mathfrak{R}^3$	
Step	Hypothesis	$\mathbf{a}_\alpha = \sum_i x^{i;\alpha}\, \mathbf{b}_i$ for all α EQ A	$\mathbf{a}_\alpha = \sum_i x^i{}_\alpha\, \mathbf{b}_i$ for all α EQ B
1	From 1g and A17.1.8 Fundamental Property of Continuum for a Hypersheet	$\mathbf{r} = \sum_i x_i\, \mathbf{b}_i$	
2	From 1, T17.1.4 Contravariant Bases Vectors for a Hypersheet and A7.9.2 Substitution of Vectors	$\mathbf{a}_\alpha = \partial(\sum_i x_i\, \mathbf{b}_i) / \partial u_\alpha$	
3	From 2 and LxK.5.2.1 Differential Linear Vector Distribution	$\mathbf{a}_\alpha = \sum_i \partial(x_i\, \mathbf{b}_i) / \partial u_\alpha$	
∴	From 3, D17.1.9 Riemannian Manifold $\mathfrak{R}^n$ Mixed Tensor of Rank-2 on a Hypersheet $\mathfrak{H}^{n-1}$ and T12.6.2A Differentiation of Tensor: Rank-1 Covariant	$\mathbf{a}_\alpha = \sum_i x^{i;\alpha}\, \mathbf{b}_i$ for all α EQ A	$\mathbf{a}_\alpha = \sum_i x^i{}_\alpha\, \mathbf{b}_i$ for all α EQ B

Theorem 17.1.6 Manifold-Hypersheet Transformation: Angle Between Contravariant Bases Vectors

1g	Given	$\mathbf{b}^i = \mathbf{b}^i$	
Step	Hypothesis	$\cos(\theta^i{}_\alpha) = (\mathbf{b}^i \bullet \mathbf{a}_\alpha) / b^i a_\alpha$ EQ A	$\cos(\theta^i{}_\alpha) = x^{i;\alpha} / b^i a_\alpha$ EQ B
1	From 1g, T17.1.5A Manifold-Hypersheet Transformation: Hypersheet Contravariant Bases Vectors and T8.3.1 Dot Product: Uniqueness Operation	$\mathbf{b}^i \bullet \mathbf{a}_\alpha = \mathbf{b}^i \bullet \sum_j x^{j;\alpha}\, \mathbf{b}_j$	
2	From 1 and T8.3.7 Dot Product: Distribution of Dot Product Across Addition of Vectors	$\mathbf{b}^i \bullet \mathbf{a}_\alpha = \sum_j x^{j;\alpha}\, (\mathbf{b}^i \bullet \mathbf{b}_j)$	
3	From 2, D12.1.5D Mixed Metric Dyadic and A4.2.3 Substitution	$\mathbf{b}^i \bullet \mathbf{a}_\alpha = \sum_j x^{i;\alpha}\, \delta^i_j$	
4	From 3, D6.1.6B Identity Matrix and evaluate the Kronecker Delta	$\mathbf{b}^i \bullet \mathbf{a}_\alpha = x^{i;\alpha}$	
5	From 4, D8.3.1 The Inner or Dot Product Operator and A4.2.3 Substitution	$\mathbf{b}^i \bullet \mathbf{a}_\alpha = b^i a^\alpha \cos(\theta^{i\alpha})$	
∴	From 4, 5, T4.4.4B Equalities: Right Cancellation by Multiplication and A4.2.3 Substitution	$\cos(\theta^i{}_\alpha) = (\mathbf{b}^i \bullet \mathbf{a}_\alpha) / b^i a_\alpha$ EQ A	$\cos(\theta^i{}_\alpha) = x^{i;\alpha} / b^i a_\alpha$ EQ B

Theorem 17.1.7 Manifold-Hypersheet Transformation: Angle Between Mixed Bases Vectors

1g	Given	$\mathbf{b}_i = \mathbf{b}_i$	
Step	Hypothesis	$\cos(\theta_{i\alpha}) = (\mathbf{b}_i \bullet \mathbf{a}_\alpha) / b_i a_\alpha$ EQ A	$\cos(\theta_{i\alpha}) = (\sum_j x^{j;\alpha} g_{ij}) / b_i a_\alpha$ EQ B
1	From 1g, T17.1.5A Manifold-Hypersheet Transformation: Hypersheet Contravariant Bases Vectors and T8.3.1 Dot Product: Uniqueness Operation	$\mathbf{b}_i \bullet \mathbf{a}_\alpha = \mathbf{b}_i \bullet \sum_j x^{j;\alpha} \mathbf{b}_j$	
2	From 1 and T8.3.7 Dot Product: Distribution of Dot Product Across Addition of Vectors	$\mathbf{b}_i \bullet \mathbf{a}_\alpha = \sum_j x^{j;\alpha} (\mathbf{b}_i \bullet \mathbf{b}_j)$	
3	From 2, D12.1.3A Covariant Metric Dyadic and A4.2.3 Substitution	$\mathbf{b}_i \bullet \mathbf{a}_\alpha = \sum_j x^{j;\alpha} g_{ij}$	
4	From 3, D8.3.1 The Inner or Dot Product Operator and A4.2.3 Substitution	$\mathbf{b}_i \bullet \mathbf{a}_\alpha = b_i a_\alpha \cos(\theta_{i\alpha})$	
∴	From 3, 4, T4.4.4B Equalities: Right Cancellation by Multiplication and A4.2.3 Substitution	$\cos(\theta_{i\alpha}) = (\mathbf{b}_i \bullet \mathbf{a}_\alpha) / b_i a_\alpha$ EQ A	$\cos(\theta_{i\alpha}) = (\sum_j x^{j;\alpha} g_{ij}) / b_i a_\alpha$ EQ B

Theorem 17.1.8 Manifold-Hypersheet Transformation: h-g Metrics Without Change in Direction

Step	Hypothesis	$a_{\alpha\beta} = \sum_i \sum_j x_{i,\alpha} x_{j,\beta} g_{ij}$ for (n–1)x(n–1)
1	From O17.1.1A Manifold-Hypersheet Invariance of Arc and A4.2.2 Equality	$\sum_\alpha \sum_\beta a_{\alpha\beta} du_\alpha du_\beta = \sum_i \sum_j g_{ij} dx_i dx_j$
2	From 1, T17.1.1B Manifold-Hypersheet Transformation: Exact Differential and A4.2.3 Substitution	$\sum_\alpha \sum_\beta a_{\alpha\beta} du_\alpha du_\beta = \sum_i \sum_j g_{ij} (\sum_\alpha x_{i,\alpha} du_\alpha) (\sum_\beta x_{j,\beta} du_\beta)$
3	From 2 T10.3.7 Product of Tensors: Commutative by Multiplication of Rank p + q → q + p, A10.2.14 Correspondence of Tensors and Tensor Coefficients,, compare term-by-term and reorder summation	$\sum_\alpha \sum_\beta a_{\alpha\beta} du_\alpha du_\beta = \sum_\alpha\sum_\beta [\sum_i \sum_j g_{ij} x_{i,\alpha} x_{j,\beta}] du_\alpha du_\beta$
∴	From 3 and equate kernel	$a_{\alpha\beta} = \sum_i \sum_j x_{i,\alpha} x_{j,\beta} g_{ij}$ for (n–1)x(n–1)

Theorem 17.1.9 Manifold-Hypersheet Transformation: h-g Metrics With Change in Direction

Step	Hypothesis	$a_{\alpha\beta} = \sum_i \sum_j x^{i;\alpha} x^{j;\beta} g_{ij}$ for (n−1)x(n−1) EQ A	$a_{\alpha\beta} = \sum_i \sum_j x^i{}_\alpha x^j{}_\beta g_{ij}$ for (n−1)x(n−1) EQ B
1	From T8.3.2 Dot Product: Equality	$\mathbf{a}_\alpha \bullet \mathbf{a}_\beta = \mathbf{a}_\alpha \bullet \mathbf{a}_\beta$	
2	From 1, T17.1.5 Manifold-Hypersheet Transformation: Hypersheet Contravariant Bases Vectors and A7.9.2 Substitution of Vectors	$\mathbf{a}_\alpha \bullet \mathbf{a}_\beta = (\sum_i x^{i;\alpha} \mathbf{b}_i) \bullet (\sum_j x^{j;\beta} \mathbf{b}_j)$	
∴	From 2 and T12.3.13A Dot Product of Covariant Vectors	$a_{\alpha\beta} = \sum_i \sum_j x^{i;\alpha} x^{j;\beta} g_{ij}$ for (n−1)x(n−1) EQ A	$a_{\alpha\beta} = \sum_i \sum_j x^i{}_\alpha x^j{}_\beta g_{ij}$ for (n−1)x(n−1) EQ B

Theorem 17.1.10 Manifold-Hypersheet Transformation: Angle Between Hypersheet Bases Vectors

1g	Given	$\mathbf{a}_\alpha = \mathbf{a}_\alpha$	
Step	Hypothesis	$\cos(\varphi_{\alpha\beta}) = (\mathbf{a}_\alpha \bullet \mathbf{a}_\beta) / a_\alpha a_\beta$ EQ A	$\cos(\varphi_{\alpha\beta}) = a_{\alpha\beta} / a_\alpha a_\beta$ EQ B
1	From 1g, T17.1.5 Manifold-Hypersheet Transformation: Hypersheet Contravariant Bases Vectors and T8.3.1 Dot Product: Uniqueness Operation	$\mathbf{a}_\alpha \bullet \mathbf{a}_\beta = \mathbf{a}_\alpha \bullet \sum_j x^{j;\beta} \mathbf{b}_j$	
2	From 1 and T8.3.7 Dot Product: Distribution of Dot Product Across Addition of Vectors	$\mathbf{a}_\alpha \bullet \mathbf{a}_\beta = \sum_i x^{i;\alpha} (\mathbf{b}_i \bullet \mathbf{a}^\alpha)$	
3	From 2, T17.1.5S3 Manifold-Hypersheet Transformation: Angle Between Mixed Bases Vectors and A4.2.3 Substitution	$\mathbf{a}_\alpha \bullet \mathbf{a}_\beta = \sum_i x^{i;\alpha} \sum_j x^{j;\alpha} g_{ij}$	
4	From 3 and regroup terms in summations	$\mathbf{a}_\alpha \bullet \mathbf{a}_\beta = \sum_i \sum_j x^{i;\alpha} x^{j;\alpha} g_{ij}$	
5	From 4, T17.1.7 Manifold-Hypersheet Transformation: h-g Metrics With Change in Direction and A4.2.3 Substitution	$\mathbf{a}_\alpha \bullet \mathbf{a}_\beta = a_{\alpha\beta}$	
6	From 5, D8.3.1 The Inner or Dot Product Operator and A4.2.3 Substitution	$\mathbf{a}_\alpha \bullet \mathbf{a}_\beta = a_\alpha a_\beta \cos(\varphi_{\alpha\beta})$	
∴	From 5, 6, T4.4.4B Equalities: Right Cancellation by Multiplication and A4.2.3 Substitution	$\cos(\varphi_{\alpha\beta}) = (\mathbf{a}_\alpha \bullet \mathbf{a}_\beta) / a_\alpha a_\beta$ EQ A	$\cos(\varphi_{\alpha\beta}) = a_{\alpha\beta} / a_\alpha a_\beta$ EQ B

Section 17.2 Manifold-Hypersheet Invariance of Arc

Definition 17.2.1 First Fundamental Quadratic form of the Surface

The ***first fundamental quadratic form of the surface*** is given by the arc equation tracing a curve-C on the surface,

$ds^2 \equiv d\mathbf{r} \bullet d\mathbf{r}$ Equation A

Observation 17.2.1 Invariance in Position between $\mathfrak{R}^n$ and $\mathfrak{H}^{(n-1)}$

The inverse of working between the Riemannian Manifold $\mathfrak{R}^n$ and the embedded Hypersheet $\mathfrak{H}^{(n-1)}$ requires the positional vector [**r**] to be invariant between the two systems, hence it must be defined in the following way:

Definition 17.2.2 Invariance of the Positional Vector between $\mathfrak{R}^n$ and $\mathfrak{H}^{(n-1)}$

$\rho \equiv \sum_\alpha u_\alpha \mathbf{a}_\alpha$ Equation A

$\rho \equiv \mathbf{r}$ Equation B

Axiom 17.2.1 Manifold-Hypersheet Invariance of Arc

Let ∃| a length of arc of the curve-C [ds] traced out on a hypersurface $\mathfrak{H}^{n-1}$ and note that it is the same arc enveloped by the manifold $\mathfrak{R}^n$ of the hypersheet; hence one and the same, and invariant in either space.

$\sum_\alpha \sum_\beta a_{\alpha\beta}\, du_\alpha\, du_\beta \equiv \sum_i \sum_j g_{ij}\, dx_i\, dx_j$ Equation A

Theorem 17.2.1 Manifold-Hypersheet Transformation: Hypersheet Contravariant Bases Vectors

Step	Hypothesis	$\mathbf{b}_i = \Sigma_\alpha u^{\alpha;i}\, \mathbf{a}_\alpha$ for all i EQ A	$\mathbf{b}_i = \Sigma_\alpha u^{\alpha}{}_i\, \mathbf{a}_\alpha$ for all i EQ B
1	From A17.1.8A Fundamental Property of Continuum for a Hypersheet	$\mathbf{r} = \Sigma_i x_i\, \mathbf{b}_i$	
2	From D17.2.2A Invariance of the Positional Vector between $\mathfrak{R}^n$ and $\mathfrak{H}^{(n-1)}$	$\rho = \Sigma_\alpha u_\alpha\, \mathbf{a}_\alpha$	
3	From 2 and T12.7.1B Tensor Uniqueness: Tensor Differentiation	$(\rho)_{;i} = (\Sigma_\alpha u_\alpha\, \mathbf{a}_\alpha)_{;i}$	
4	From 3 and T12.7.2 Tensor Summation: Covariant Derivative as a Linear Operator	$(\rho)_{;i} = \Sigma_\alpha u^{\alpha;i}\, \mathbf{a}_\alpha$	
5	From 4, D17.2.2B Invariance of the Positional Vector between $\mathfrak{R}^n$ and $\mathfrak{H}^{(n-1)}$ and A7.9.2 Substitution of Vectors	$(\rho)_{;i} = (\mathbf{r})_{;i}$	
6	From 1, 5 and A7.9.2 Substitution of Vectors	$(\rho)_{;i} = (\Sigma_j x_j\, \mathbf{b}_j)_{;i}$	
7	From 6 and T12.7.2 Tensor Summation: Covariant Derivative as a Linear Operator	$(\rho)_{;i} = (\Sigma_j x_{j;i}\, \mathbf{b}_j)$	
8	From 7, T12.6.2B Differentiation of Tensor: Rank-1 Covariant, Differential with respect to it's self has no change in direction, hence zero.	$(\rho)_{;i} = (\Sigma_j x_{j,i}\, \mathbf{b}_j)$	
9	From 8 and differential with respect to it's self is independent and can be represented by a Kronecker Delta	$(\rho)_{;i} = (\Sigma_j \delta_{ji}\, \mathbf{b}_j)$	
10	From 9 and summing over the Kronecker Delta	$(\rho)_{;i} = \mathbf{b}_i$	
11	From 4, 10 and A7.9.1 Equivalence of Vector	$\mathbf{b}_i = \Sigma_\alpha u^{\alpha;i}\, \mathbf{a}_\alpha$	
∴	From 11 and D17.1.10 Hypersheet $\mathfrak{H}^{n-1}$ Mixed Tensor of Rank-2 in a Riemannian Manifold $\mathfrak{R}^n$	$\mathbf{b}_i = \Sigma_\alpha u^{\alpha;i}\, \mathbf{a}_\alpha$ for all i EQ A	$\mathbf{b}_i = \Sigma_\alpha u^{\alpha}{}_i\, \mathbf{a}_\alpha$ for all i EQ B

Theorem 17.2.2 Manifold-Hypersheet Transformation: g-h Metrics Without Change in Direction

Step	Hypothesis	$g_{ij} = \Sigma_\alpha \Sigma_\beta u_{\alpha,i} u_{\beta,j} a_{\alpha\beta}$ for nxn
1	From A17.2.1A Manifold-Hypersheet Invariance of Arc and A4.2.2 Equality	$\Sigma_i \Sigma_j g_{ij} dx_i dx_j = \Sigma_\alpha \Sigma_\beta a_{\alpha\beta} du_\alpha du_\beta$
2	From 1, T17.1.1B Manifold-Hypersheet Transformation: Exact Differential and A4.2.3 Substitution	$\Sigma_i \Sigma_j g_{ij} dx_i dx_j = \Sigma_\alpha \Sigma_\beta a_{\alpha\beta} (\Sigma_i u_{\alpha,i} dx_i) (\Sigma_i u_{\beta,j} dx_j)$
3	From 2, T10.3.7 Product of Tensors: Commutative by Multiplication of Rank p + q → q + p, A10.2.14 Correspondence of Tensors and Tensor Coefficients, compare term-by-term and reorder summation	$\Sigma_i \Sigma_j g_{ij} dx_i dx_j = \Sigma_i \Sigma_i [\Sigma_\alpha \Sigma_\beta u_{\alpha,i} u_{\beta,j} a_{\alpha\beta}] dx_i dx_j$
∴	From 3 and equate coordinate paths	$g_{ij} = \Sigma_\alpha \Sigma_\beta u_{\alpha,i} u_{\beta,j} a_{\alpha\beta}$ for nxn

Theorem 17.2.3 Manifold-Hypersheet Transformation: h-g Metrics Without Change in Direction

Step	Hypothesis	$a_{\alpha\beta} = \Sigma_i \Sigma_j x_{i,\alpha} x_{j,\beta} g_{ij}$ for (n–1)x(n–1)
1	From A17.2.1A Manifold-Hypersheet Invariance of Arc and A4.2.2 Equality	$\Sigma_\alpha \Sigma_\beta a_{\alpha\beta} du_\alpha du_\beta = \Sigma_i \Sigma_j g_{ij} dx_i dx_j$
2	From 1, T17.1.1A Manifold-Hypersheet Transformation: Exact Differential and A4.2.3 Substitution	$\Sigma_\alpha \Sigma_\beta a_{\alpha\beta} du_\alpha du_\beta = \Sigma_i \Sigma_j g_{ij} (\Sigma_\alpha x_{i,\alpha} du_\alpha) (\Sigma_\beta x_{j,\beta} du_\beta)$
3	From 2, T10.3.7 Product of Tensors: Commutative by Multiplication of Rank p + q → q + p, A10.2.14 Correspondence of Tensors and Tensor Coefficients, compare term-by-term and reorder summation	$\Sigma_\alpha \Sigma_\beta a_{\alpha\beta} du_\alpha du_\beta = \Sigma_\alpha \Sigma_\beta (\Sigma_i \Sigma_j x_{i,\alpha} x_{j,\beta} g_{ij}) du_\alpha du_\beta$
∴	From 3 and equate coordinate paths	$a_{\alpha\beta} = \Sigma_i \Sigma_j x_{i,\alpha} x_{j,\beta} g_{ij}$ for (n–1)x(n–1)

Theorem 17.2.4 Manifold-Hypersheet Transformation: h-g Metrics With Change in Direction

Step	Hypothesis	$g_{ij} = \Sigma_\alpha \Sigma_\beta u^{\alpha;i} u^{\beta;j} a_{\alpha\beta}$ for nxn EQ A	$g_{ij} = \Sigma_\alpha \Sigma_\beta u^\alpha{}_i u^\beta{}_j a_{\alpha\beta}$ for nxn EQ B
1	From T8.3.2 Dot Product: Equality	$\mathbf{b}_i \bullet \mathbf{b}_j = \mathbf{b}_i \bullet \mathbf{b}_j$	
2	From 1, T17.2.1 Manifold-Hypersheet Transformation: Hypersheet Contravariant Bases Vectors and A7.9.2 Substitution of Vectors	$\mathbf{b}_i \bullet \mathbf{b}_j = (\Sigma_\alpha u^{\alpha;i} \mathbf{a}_\alpha) \bullet (\Sigma_\beta u^{\beta;j} \mathbf{a}_\beta)$	
∴	From 2, T12.1.3A Covariant Metric Dyadic, T8.3.8 Dot Product: Distribution of Dot Product Across Another Vector, D17.1.10 Hypersheet $\mathfrak{H}^{n-1}$ Mixed Tensor of Rank-2 in a Riemannian Manifold $\mathfrak{R}^n$ and A17.5.1 Theorem Construction on Manifolds of Different Dimension	$g_{ij} = \Sigma_\alpha \Sigma_\beta u^{\alpha;i} u^{\beta;j} a_{\alpha\beta}$ for nxn EQ A	$g_{ij} = \Sigma_\alpha \Sigma_\beta u^\alpha{}_i u^\beta{}_j a_{\alpha\beta}$ for nxn EQ B

Theorem 17.2.5 Manifold-Hypersheet Transformation: g-h Metrics With Change in Direction

Step	Hypothesis	$a_{\alpha\beta} = \Sigma_i \Sigma_j x^{i;\alpha} x^{j;\beta} g_{ij}$ for (n–1)x(n–1) EQ A	$a_{\alpha\beta} = \Sigma_i \Sigma_j x^i{}_\alpha x^j{}_\beta g_{ij}$ for (n–1)x(n–1) EQ B
1	From T8.3.2 Dot Product: Equality	$\mathbf{a}_\alpha \bullet \mathbf{a}_\beta = \mathbf{a}_\alpha \bullet \mathbf{a}_\beta$	
2	From 1, T17.1.5A Manifold-Hypersheet Transformation: Hypersheet Contravariant Bases Vectors and A7.9.2 Substitution of Vectors	$\mathbf{a}_\alpha \bullet \mathbf{a}_\beta = (\Sigma_i x^{i;\alpha} \mathbf{b}_i) \bullet (\Sigma_j x^{j;\beta} \mathbf{b}_j)$	
∴	From 2, T17.5.1A Dot Product of Covariant Vectors, T12.3.13A Dot Product of Covariant Vectors, D17.1.9 Riemannian Manifold $\mathfrak{R}^n$ Mixed Tensor of Rank-2 on a Hypersheet $\mathfrak{H}^{n-1}$ and A17.5.1 Theorem Construction on Manifolds of Different Dimension	$a_{\alpha\beta} = \Sigma_i \Sigma_j x^{i;\alpha} x^{j;\beta} g_{ij}$ for (n–1)x(n–1) EQ A	$a_{\alpha\beta} = \Sigma_i \Sigma_j x^i{}_\alpha x^j{}_\beta g_{ij}$ for (n–1)x(n–1) EQ B

Observation 17.2.2 Manifold-Hypersheet Transformation Obeys the Rules of Transformation

Transforming from Riemannian coordinates to the hypersurface obeys the rules of coordinate transformation. The kernels of the metric differential transformations behave as asymmetric Jacobean matrices promoting transformations by specific rules.

Section 17.3 The Normal Vector to a Hypersurface

Traditionally surfaces are treated in 3-dimensions and the normal to the surface is setup by taking the cross product of any two vectors at a common point on the surface, than goes on to prove the resulting normal vector is perpendicular to the two surface vectors. But what if the surface is embedded in an n-dimensional space? Fortunately the cross product can be extended into a Riemannian manifold $\mathfrak{R}^n$ so a natural development of a normal vector can be constructed.

From theorem T6.9.10 "A determinant with two identical rows or columns is zero", is used as the bases for defining a normal vector. The normal vector than would be made up from the set of contravariant bases vectors [$\mathbf{a}^\alpha$ for all α] at a point-Q on the surface, now any vector selected from that set, when dotted with it, would guarantee normality because a duplicate row is now part of the determinate, driving it to zero. This forces a conclusion to be made that the selected vector is perpendicular. So by using a determinate to define a normal vector guarantees the construction of a vector that is always normal to any and all surface vectors built from that set.

Definition 17.3.1 The Minor of a Normal Vector

$$|N_{1i}| \equiv \begin{vmatrix} x^{1;1} & x^{2;1} & \cdots & x^{(i-1);1} & x^{(i+1);1} & \cdots & x^{n;1} \\ \vdots & \vdots & & \vdots & \vdots & & \vdots \\ x^{1;\alpha} & x^{2;\alpha} & \cdots & x^{(i-1);\alpha} & x^{(i+1);\alpha} & \cdots & x^{n;\alpha} \\ \vdots & \vdots & & \vdots & \vdots & & \vdots \\ x^{1;(n-1)} & x^{2;(n-1)} & \cdots & x^{(i-1);(n-1)} & x^{(i+1);(n-1)} & \cdots & x^{n;(n-1)} \end{vmatrix}$$

Axiom 17.3.1 Manifold-Hypersheet Invariance of a Normal Vector

Let $\exists|$ a Normal Vector to the hypersurface $\mathfrak{H}^{n-1}$ at a point-Q. Note that it is the same vector embedded in the enveloping manifold $\mathfrak{R}^n$ of the hypersheet; hence one and the same, and invariant in either space.

$\mathbf{N} \equiv \sum_\alpha n^\alpha \mathbf{a}_\alpha$ in $\mathfrak{H}^{n-1}$ space Equation A

$\mathbf{N} \equiv \prod_{\alpha=1}^{n-1} \times \mathbf{a}_\alpha$ in $\mathfrak{R}^n$ space Equation B

Axiom 17.3.2 Normal Vector is Orthogonal to Curvature Coordinates

The $N \perp \mathbf{S}$ for $\mathbf{S}$ lying on the surface of the hypersheet, hence

$0 \equiv N \bullet \mathbf{S}$ Equation A

Axiom 17.3.3 Hypersurface $\mathfrak{H}^{n-1}$: Existence of Dual Covariant Tensor Field

Let ∃ an ordered product set H^{n-1} rendered from a hypersurface $\mathfrak{H}^{n-1}$, such that it is comprised of two orthogonal components a ***tangential tensor*** $x^i{}_\alpha$ and its ***normal vector*** n^i.

$$H^{n-1} \equiv \mathbf{T} \times \mathbf{NR} \equiv \{ (\overset{*}{B}_m, \overset{*}{C}_m) \mid \overset{*}{B}_m \in \mathbf{T}, \overset{*}{C}_m \in \mathbf{NR}, \text{ yields } \overset{*}{B}_m x^i{}_\alpha + \overset{*}{C}_m n_i \}$$

where $[\overset{*}{B}_m x^i{}_\alpha + \overset{*}{C}_m n_i]$ is called a ***dual Hypersurface tensor***.

P1) By its very definition as being an ordered set:

$\mathbf{T} \times \mathbf{NR} \neq \mathbf{NR} \times \mathbf{T}$, hence non-commutative. Equation A

Any point P in the dual tensor field can be described as a set of ordered pairs, or if thought of as a vector, then the compound dual tensor is used.

P2) The dual tensor components being orthogonal are mutually exclusive one from the other

$0 = \sum_i n_i x^i{}_\alpha$ Equation B

P3) $\overset{*}{A}_{m+1} \equiv \overset{*}{B}_m x^i{}_\alpha + \overset{*}{C}_m n^i$ Covariant Equation C

P4) $\overset{*}{A}^{m+1} \equiv \overset{*}{B}^m x^i{}_\alpha + \overset{*}{C}^m n^i$ Contravariant Equation D

Axiom 17.3.4 Hypersurface $\mathfrak{H}^{n-1}$: Closure Under Addition for Dual Covariant Tensors

$$\{[(\overset{*}{A}_m x^i{}_\alpha + \overset{*}{B}_m n^i) + (\overset{*}{C}_m x^i{}_\alpha + \overset{*}{D}_m n^i)] \wedge [(\overset{*}{C}_m x^i{}_\alpha + \overset{*}{D}_m n^i) + (\overset{*}{A}_m x^i{}_\alpha + \overset{*}{B}_m n^i)]\} \wedge$$
$$\{[(\overset{*}{A}_m x^i{}_\alpha + \overset{*}{B}_m n^i) - (\overset{*}{C}_m x^i{}_\alpha + \overset{*}{D}_m n^i)] \wedge [(\overset{*}{C}_m x^i{}_\alpha + \overset{*}{D}_m n^i) - (\overset{*}{A}_m x^i{}_\alpha + \overset{*}{B}_m n^i)]\} \in H^{n-1}$$

Axiom 17.3.5 Dual Covariant Tensors: Equality

Iff $\overset{*}{A}_m x^i{}_\alpha + \overset{*}{B}_m n^i \equiv \overset{*}{C}_m x^i{}_\alpha + \overset{*}{D}_m n^i$ then the following must be true; $\overset{*}{A}_m = \overset{*}{C}_m$ and $\overset{*}{B}_m = \overset{*}{D}_m$, and the dual covariant tensors are said to be equal.

Axiom 17.3.6 Dual Covariant Tensors: Equality Converse

Iff $\overset{*}{A}_m = \overset{*}{C}_m$ and $\overset{*}{B}_m = \overset{*}{D}_m$, then the following must be true; $\overset{*}{A}_m x^i{}_\alpha + \overset{*}{B}_m n^i \equiv \overset{*}{C}_m x^i{}_\alpha + \overset{*}{D}_m n^i$ and the dual covariant tensors are said to be equal.

Axiom 17.3.7 Dual Contravariant Tensors: Correspond to Covariant Axioms and Theorems

Axioms A17.3.3 to A17.3.6 and theorems T17.3.9 to 17 hold true for contravariant tensors T17.3.18 to T17.3.26 as well, because they are comprised of the same set of orthogonal basis vectors, hence follow the same rules of operation.

Theorem 17.3.1 Hypersurface Normal Vector: Normal Covariant Quiver Vector

1g 2g	Given	$\mathbf{a}_\alpha = \sum_i x^{i;\alpha}\mathbf{b}_i$	a quiver vector
		$\mathbf{N} \equiv \begin{vmatrix} \vec{b}^1 & \vec{b}^2 & \cdots & \vec{b}^n \\ x^{1;1} & x^{2;1} & \cdots & x^{n;1} \\ \vdots & \vdots & & \vdots \\ x^{1;\alpha} & x^{2;\alpha} & \cdots & x^{n;\alpha} \\ \vdots & \vdots & & \vdots \\ x^{1;(n-1)} & x^{2;(n-1)} & \cdots & x^{n;(n-1)} \end{vmatrix}$	assume by definition
Step	Hypothesis	$\mathbf{a}_\alpha \bullet \mathbf{N} = 0$ EQ A	$\mathbf{N} = \begin{vmatrix} \vec{b}^1 & \vec{b}^2 & \cdots & \vec{b}^n \\ x^{1;1} & x^{2;1} & \cdots & x^{n;1} \\ \vdots & \vdots & & \vdots \\ x^{1;\alpha} & x^{2;\alpha} & \cdots & x^{n;\alpha} \\ \vdots & \vdots & & \vdots \\ x^{1;(n-1)} & x^{2;(n-1)} & \cdots & x^{n;(n-1)} \end{vmatrix}$ EQ B
1	From T8.3.1 Dot Product: Uniqueness Operation	$\mathbf{a}_\alpha \bullet \mathbf{N} = \mathbf{N}_\alpha \bullet \mathbf{N}$	
2	From 1g, 2g and A7.9.2 Substitution of Vectors	$\mathbf{a}_\alpha \bullet \mathbf{N} = (\sum_i x^{i;\alpha}\mathbf{b}_i) \bullet \begin{vmatrix} \vec{b}^1 & \vec{b}^2 & \cdots & \vec{b}^n \\ x^{1;1} & x^{2;1} & \cdots & x^{n;1} \\ \vdots & \vdots & & \vdots \\ x^{1;\alpha} & x^{2;\alpha} & \cdots & x^{n;\alpha} \\ \vdots & \vdots & & \vdots \\ x^{1;(n-1)} & x^{2;(n-1)} & \cdots & x^{n;(n-1)} \end{vmatrix}$	
3	From 2 and T6.9.5 Expanding Determinant [A] of order [n] about a Row or Column	$\mathbf{a}_\alpha \bullet \mathbf{N} = (\sum_i x^{i;\alpha}\mathbf{b}_i) \bullet (\sum_j (-1)^{j+1} \lvert N_{1j}\rvert (\mathbf{b}^j)_{1j})$	
4	From 3 and T12.3.15A Dot Product of Mixed Vectors (Tensor Contraction)	$\mathbf{a}_\alpha \bullet \mathbf{N} = \sum_i \sum_j x^{i;\alpha} [(-1)^{j+1} \lvert N_{1j}\rvert \delta_i^{\ j}]$	
5	From 4 and D6.1.6 Identity Matrix Kronecker Delta evaluated over summation of j	$\mathbf{a}_\alpha \bullet \mathbf{N} = \sum_i x^{i;\alpha} [(-1)^{i+1} \lvert N_{1i}\rvert]$	
6	From 5 and T6.9.5 Expanding Determinant [A] of order [n] about a Row or Column	$\mathbf{a}_\alpha \bullet \mathbf{N} = \begin{vmatrix} x^{1;\alpha} & x^{2;\alpha} & \cdots & x^{n;\alpha} \\ x^{1;1} & x^{2;1} & \cdots & x^{n;1} \\ \vdots & \vdots & & \vdots \\ x^{1;\alpha} & x^{2;\alpha} & \cdots & x^{n;\alpha} \\ \vdots & \vdots & & \vdots \\ x^{1;(n-1)} & x^{2;(n-1)} & \cdots & x^{n;(n-1)} \end{vmatrix}$	
∴	From 2g, 6 and T6.9.10 A determinant with two identical rows or columns is zero	$\mathbf{a}_\alpha \bullet \mathbf{N} = 0$ EQ A	$\mathbf{N} = \begin{vmatrix} \vec{b}^1 & \vec{b}^2 & \cdots & \vec{b}^n \\ x^{1;1} & x^{2;1} & \cdots & x^{n;1} \\ \vdots & \vdots & & \vdots \\ x^{1;\alpha} & x^{2;\alpha} & \cdots & x^{n;\alpha} \\ \vdots & \vdots & & \vdots \\ x^{1;(n-1)} & x^{2;(n-1)} & \cdots & x^{n;(n-1)} \end{vmatrix}$ EQ B

Theorem 17.3.2 Hypersurface Normal Vector: Normal Vector Offset Index Form

Step	Hypothesis	
		$\prod_{\alpha=1}^{n-1} \times \mathbf{a}_\alpha = \begin{vmatrix} \vec{b}^{1} & \vec{b}^{2} & \cdots & \vec{b}^{n} \\ x^{1;1} & x^{2;1} & \cdots & x^{n;1} \\ \vdots & \vdots & & \vdots \\ x^{1;\alpha} & x^{2;\alpha} & \cdots & x^{n;\alpha} \\ \vdots & \vdots & & \vdots \\ x^{1;(n-1)} & x^{2;(n-1)} & \cdots & x^{n;(n-1)} \end{vmatrix}$
1	A7.9.1 Equivalence of Vector	$\mathbf{N} = \mathbf{N}$
∴	From 1, A17.3.1B Manifold-Hypersheet Invariance of a Normal Vector, T17.3.1 Normal Vector to a Surface: Normal Covariant Quiver Vector and A7.9.2 Substitution of Vectors	$\prod_{\alpha=1}^{n-1} \times \mathbf{a}_\alpha = \begin{vmatrix} \vec{b}^{1} & \vec{b}^{2} & \cdots & \vec{b}^{n} \\ x^{1;1} & x^{2;1} & \cdots & x^{n;1} \\ \vdots & \vdots & & \vdots \\ x^{1;\alpha} & x^{2;\alpha} & \cdots & x^{n;\alpha} \\ \vdots & \vdots & & \vdots \\ x^{1;(n-1)} & x^{2;(n-1)} & \cdots & x^{n;(n-1)} \end{vmatrix}$

Theorem 17.3.3 Hypersurface Normal Vector: Invariance of the Normal Vector

Step	Hypothesis	$\sum_\alpha n^\alpha \mathbf{a}_\alpha \equiv \prod_{\alpha=1}^{n-1} \times \mathbf{a}_\alpha$
∴	From A17.3.1(A, B) Manifold-Hypersheet Invariance of a Normal Vector and A7.9.1 Equivalence of Vector	$\sum_\alpha n^\alpha \mathbf{a}_\alpha = \prod_{\alpha=1}^{n-1} \times \mathbf{a}_\alpha$

Theorem 17.3.4 Hypersurface Normal Vector: With Cross product of Covariant Bases Vectors

Step	Hypothesis	$\sum_i (\sum_\alpha n^\alpha x^{i;\alpha})\, \mathbf{b}_i = \prod_{\alpha=1}^{n-1} \times (\sum_i x^{i;\alpha}\, \mathbf{b}_i)$
1	From T17.1.5 Manifold-Hypersheet Transformation: Hypersheet Contravariant Bases Vectors and A7.9.2 Substitution of Vectors	$\sum_\alpha n^\alpha (\sum_i x^{i;\alpha}\, \mathbf{b}_i) = \prod_{\alpha=1}^{n-1} \times (\sum_i x^{i;\alpha}\, \mathbf{b}_i)$
∴	From 1, T7.10.3 Distribution of a Scalar over Addition of Vectors and reorder summation	$\sum_i (\sum_\alpha n^\alpha x^{i;\alpha})\, \mathbf{b}_i = \prod_{\alpha=1}^{n-1} \times (\sum_i x^{i;\alpha}\, \mathbf{b}_i)$

Theorem 17.3.5 Hypersurface Normal Vector: With Normal Contravariant Metric Coefficient

1g	Given Normal Contravariant Metric Coefficient	$\lvert N_{gk}\rvert \equiv \begin{vmatrix} g^{1k} & g^{2k} & \cdots & g^{nk} \\ x^{1;1} & x^{2;1} & \cdots & x^{n;1} \\ \vdots & \vdots & & \vdots \\ x^{1;\alpha} & x^{2;\alpha} & \cdots & x^{n;\alpha} \\ \vdots & \vdots & & \vdots \\ x^{1;(n-1)} & x^{2;(n-1)} & \cdots & x^{n;(n-1)} \end{vmatrix}$
Step	Hypothesis	$\prod_{\alpha=1}^{n-1} \times (\sum_i x^{i;\alpha}\, \mathbf{b}_i) = \sum_j \lvert N_{gj}\rvert\, \mathbf{b}_j$
1	From A7.9.1 Equivalence of Vector	$\prod_{\alpha=1}^{n-1} \times (\sum_i x^{i;\alpha}\, \mathbf{b}_i) = \prod_{\alpha=1}^{n-1} \times (\sum_i x^{i;\alpha}\, \mathbf{b}_i)$
2	From1, T17.3.2 Normal Vector to a Surface: Normal Vector Offset Index Form	$\prod_{\alpha=1}^{n-1} \times (\sum_i x^{i;\alpha}\, \mathbf{b}_i) = \begin{vmatrix} \bar{b}^{1} & \bar{b}^{2} & \cdots & \bar{b}^{n} \\ x^{1;1} & x^{2;1} & \cdots & x^{n;1} \\ \vdots & \vdots & & \vdots \\ x^{1;\alpha} & x^{2;\alpha} & \cdots & x^{n;\alpha} \\ \vdots & \vdots & & \vdots \\ x^{1;(n-1)} & x^{2;(n-1)} & \cdots & x^{n;(n-1)} \end{vmatrix}$
3	From 2 and T6.9.5 Expanding Determinant [A] of order [n] about a Row or Column	$\prod_{\alpha=1}^{n-1} \times (\sum_i x^{i;\alpha}\, \mathbf{b}_i) = \sum_i [(-1)^{i+1} \lvert N_{1i}\rvert]\, (\mathbf{b}^i)_{1i}$
4	From 3, T12.3.10A Transformation from Covariant to Contravariant Basis Vector and A7.9.2 Substitution of Vectors	$\prod_{\alpha=1}^{n-1} \times (\sum_i x^{i;\alpha}\, \mathbf{b}_i) = \sum_i [(-1)^{i+1} \lvert N_{1i}\rvert]\, (\sum_j g^{ij}\, \mathbf{b}_j)_{1i}$
5	From 4, T10.3.20 Addition of Tensors: Distribution of a Tensor over Addition of Tensors, A10.2.14 Correspondence of Tensors and Tensor Coefficients and reordering summation	$\prod_{\alpha=1}^{n-1} \times (\sum_i x^{i;\alpha}\, \mathbf{b}_i) = \sum_j \{ \sum_i (g^{ij})_{1i}\, [(-1)^{i+1} \lvert N_{1i}\rvert] \}\, \mathbf{b}_j$
6	From 5 and T6.9.5 Expanding Determinant [A] of order [n] about a Row or Column; about row-1	$\prod_{\alpha=1}^{n-1} \times (\sum_i x^{i;\alpha}\, \mathbf{b}_i) = \sum_j \begin{vmatrix} g^{1j} & g^{2j} & \cdots & g^{nj} \\ x^{1;1} & x^{2;1} & \cdots & x^{n;1} \\ \vdots & \vdots & & \vdots \\ x^{1;\alpha} & x^{2;\alpha} & \cdots & x^{n;\alpha} \\ \vdots & \vdots & & \vdots \\ x^{1;(n-1)} & x^{2;(n-1)} & \cdots & x^{n;(n-1)} \end{vmatrix} \mathbf{b}_j$
∴	From 1g, 6 and A4.2.3 Substitution	$\prod_{\alpha=1}^{n-1} \times (\sum_i x^{i;\alpha}\, \mathbf{b}_i) = \sum_j \lvert N_{gj}\rvert\, \mathbf{b}_j$

Theorem 17.3.6 Hypersurface Normal Vector: Normal Vector Coefficient

Step	Hypothesis	$n^{\alpha} = \sum_j [(-1)^{j+\alpha} \vert(x^{j;\beta})_{1(i\neq j),(\alpha\neq\beta)}\vert\,]\, g^{ij}$
1	From T17.3.4 Surface Normal Vector: Normal Vector with Hypersheet Covariant Bases Vector	$\sum_i (\sum_\alpha n^{\alpha} x^{i;\alpha})\, \mathbf{b}_i = \prod_{\alpha=1}^{n-1} \times (\sum_i x^{i;\alpha}\, \mathbf{b}_i)$
2	From 1, T17.3.4 Surface Normal Vector: With Normal Contravariant Metric Coefficient and A7.9.2 Substitution of Vectors	$\sum_i (\sum_\alpha n^{\alpha} x^{i;\alpha})\, \mathbf{b}_i = \sum_i \vert N_{gi}\vert\, \mathbf{b}_i$
3	From 2 and equating base vectors over the summation	$\sum_\alpha n^{\alpha} x^{i;\alpha} = \vert N_{gi}\vert$
4	From 3 and T6.9.5 Expanding Determinant [A] of order [n] about a Row or Column; about row-1	$\sum_\alpha n^{\alpha} x^{i;\alpha} = \sum_j (g^{ij})_{1j}\, [(-1)^{j+1} \vert N_{1j}\vert\,]$
5	From 4 and evaluating the normal minor eliminating column-i when i = j	$\sum_\alpha n^{\alpha} x^{i;\alpha} = \sum_j (g^{ij})_{1j}\, [(-1)^{j+1} \vert(x^{j;\alpha})_{1(i\neq j)}\vert\,]$
6	From 5 and T6.9.5 Expanding Determinant [A] of order [n] about a Row or Column; about column-1, eliminating row-α when α = β	$\sum_\alpha n^{\alpha} x^{i;\alpha} = \sum_j (g^{ij})_{1j}\, [(-1)^{j+1} \sum_\alpha (-1)^{1+\alpha} x^{i;\alpha} \vert(x^{j;\beta})_{1(i\neq j),(\alpha\neq\beta)}\vert\,]$
7	From 6, T7.10.3 Distribution of a Scalar over Addition of Vectors and reorder summation	$\sum_\alpha n^{\alpha} x^{i;\alpha} = \sum_\alpha\sum_j (g^{ij})_{1j}\, [(-1)^{j+\alpha+2} \vert(x^{j;\beta})_{1(i\neq j),(\alpha\neq\beta)}\vert\,]\, x^{i;\alpha}$
∴	From 7 and equating contravariant manifold transformation components	$n^{\alpha} = \sum_j [(-1)^{j+\alpha} \vert(x^{j;\beta})_{1(i\neq j),(\alpha\neq\beta)}\vert\,]\, g^{ij}$

Theorem 17.3.7 Hypersurface Normal Vector: Orthogonal System $\mathfrak{R}^n$ Transformation and Normal

1g	Given	$n^i \equiv \sum_\alpha \eta^\alpha x^i{}_\alpha$	Normalized-normal vector		
2g		$\eta^\alpha \equiv n^\alpha / N$	Intermediate normalized-normal vector		
Step	Hypothesis	$\mathbf{n} = \sum_i n^i \mathbf{b}_i$	EQ A	$\sum_i \sum_j g_{ij}\, n^j x^i{}_\alpha = 0$	EQ C
		$\mathbf{n} = \sum_i n_i \mathbf{b}^i$	EQ B	$\sum_i n_i x^i{}_\alpha = 0$	EQ D
1	From A17.3.1A Manifold-Hypersheet Invariance of a Normal Vector	$\mathbf{N} = \sum_\alpha n^\alpha \mathbf{a}_\alpha$			
2	From 1, T17.1.5B Manifold-Hypersheet Transformation: Hypersheet Contravariant Bases Vectors and A7.9.2 Substitution of Vectors; regrouping by pairing base vector term-by-term	$\mathbf{N} = \sum_i (\sum_\alpha n^\alpha x^i{}_\alpha)\, \mathbf{b}_i$			
3	From 1, D7.6.4 Magnitude of a Vector and A7.9.2 Substitution of Vectors	$N\mathbf{n} = \sum_i (\sum_\alpha n^\alpha x^i{}_\alpha)\, \mathbf{b}_i$			
4	From 3, T4.4.6B Equalities: Left Cancellation by Multiplication, T7.10.3 Distribution of a Scalar over Addition of Vectors and T7.10.2 Associative Scalar Multiplication to Vectors	$\mathbf{n} = \sum_i [\sum_\alpha (n^\alpha / N)\, x^i{}_\alpha]\, \mathbf{b}_i$			
5	From 2g, 4 and A4.2.3 Substitution	$\mathbf{n} = \sum_i (\sum_\alpha \eta^\alpha x^i{}_\alpha)\, \mathbf{b}_i$			
6	From 5, T10.2.5 Tool: Tensor Substitution, A10.2.14 Correspondence of Tensors and Tensor Coefficients and T12.3.12B Invariance of Contra and Covariant Vectors	$\mathbf{n} = \sum_i n^i \mathbf{b}_i$ Normalized vector		$\mathbf{n} = \sum_i n_i \mathbf{b}^i$	
7	From T17.3.1A Surface Normal Vector: Normal Covariant Quiver Vector, D7.6.4C Magnitude of a Vector and A7.9.2 Substitution of Vectors	$\mathbf{a}_\alpha \bullet N\, \mathbf{n} = 0$			
8	From 7, T8.3.10A Dot Product: Right Scalar-Vector Association	$N\, (\mathbf{a}_\alpha \bullet \mathbf{n}) = 0$			
9	From 8 and D7.6.4B Magnitude of a Vector since N is not zero by definition then the none trivial solution is:	$\mathbf{a}_\alpha \bullet \mathbf{n} = 0$			
10	From 6, T17.3.1A Surface Normal Vector: Normal Covariant Quiver Vector, T17.1.5B Manifold-Hypersheet Transformation: Hypersheet Contravariant Bases Vectors and A7.9.2 Substitution of Vectors	$(\sum_i x^i{}_\alpha \mathbf{b}_i) \bullet (\sum_j n^j \mathbf{b}_j) = 0$			

11	From 10, T12.3.13A Dot Product of Covariant Vectors	$\sum_i \sum_j x^i_\alpha n^j g_{ij} = 0$			
12	From 11, T10.3.7 Product of Tensors: Commutative by Multiplication of Rank p + q → q + p and A10.2.14 Correspondence of Tensors and Tensor Coefficients	$\sum_i \sum_j g_{ij} n^j x^i_\alpha = 0$			
13	From 12 and T12.3.9A Metric Transformation from Covariant to Contravariant Coefficient; lowering indices	$\sum_i n_i x^i_\alpha = 0$			
∴	From 6, 12 and 13	$\mathbf{n} = \sum_i n^i \mathbf{b}_i$ $\mathbf{n} = \sum_i n_i \mathbf{b}^i$	EQ A EQ B	$\sum_i \sum_j g_{ij} n^j x^i_\alpha = 0$ $\sum_i n_i x^i_\alpha = 0$	EQ C EQ D

Theorem 17.3.8 Hypersurface Normal Vector: Intermediate Normalized-Normal Vector

1g	Given	$n^j \equiv \sum_\alpha \eta^\alpha x^j_\alpha$	
Step	Hypothesis	$\sum_\alpha \eta^\alpha x_\alpha = 0$	
1	From 6, T17.3.7D Surface Normal Vector: $\mathfrak{R}^n$ Transformation and Normal as Orthogonal System	$\sum_i n_i x^i_\alpha = 0$	
2	From 1, T12.3.9A Metric Transformation from Covariant to Contravariant Coefficient and A7.9.2 Substitution of Vectors	$\sum_i \sum_j g_{ij} n^j x^i_\alpha = 0$	
3	From 1g, 2 and A7.9.2 Substitution of Vectors	$\sum_i \sum_j g_{ij} (\sum_\alpha \eta^\alpha x^j_\alpha) x^i_\alpha = 0$	
4	T10.3.7 Product of Tensors: Commutative by Multiplication of Rank p + q → q + p and A10.2.14 Correspondence of Tensors and Tensor Coefficients; reordering summations	$\sum_\alpha \eta^\alpha (\sum_i \sum_j g_{ij} x^i_\alpha x^j_\alpha) = 0$	
5	From 4, T12.3.16A Magnitude of Covariant Vector and A7.9.2 Substitution of Vectors	$\sum_\alpha \eta^\alpha x_\alpha^2 = 0$	
6	From 5 and D4.1.17 Exponential Notation	$\sum_\alpha (\eta^\alpha x_\alpha) x_\alpha = 0$	for x_α non-trivially zero
∴	From 6 and for $x_\alpha \neq 0$ then the non-tivial solution is	$\sum_\alpha \eta^\alpha x_\alpha = 0$	

Theorem 17.3.9 Dual Covariant Tensors: Mutually Exclusive Tangential Tensor Components

1g	Given iff	$\overset{*}{A}_m x^i_\alpha + \overset{*}{B}_m n^i = \overset{*}{C}_m x^i_\alpha + \overset{*}{D}_m n^i$
Step	Hypothesis	$\overset{*}{A}_m = \overset{*}{C}_m$ must also follow
1	From T17.1.5A Manifold-Hypersheet Transformation: Hypersheet Contravariant Bases Vectors	$\mathbf{a}_\alpha = \sum_i x^i_\alpha \mathbf{b}_i$ standard hypersurface vector
2	From 1g, 1 and adding the vector component by equating bases vector-by-bases vector	$\sum_i \overset{*}{A}_m x^i_\alpha \mathbf{b}_i + \sum_i \overset{*}{B}_m n^i \mathbf{b}_i = \sum_i \overset{*}{C}_m x^i_\alpha \mathbf{b}_i + \sum_i \overset{*}{D}_m n^i \mathbf{b}_i$
3	From 1, 2 and T8.3.1 Dot Product: Uniqueness Operation	$(\sum_i \overset{*}{A}_m x^i_\alpha \mathbf{b}_i + \sum_i \overset{*}{B}_m n^i \mathbf{b}_i) \bullet (\sum_j x^j_\alpha \mathbf{b}_j) =$ $(\sum_i \overset{*}{C}_m x^i_\alpha \mathbf{b}_i + \sum_i \overset{*}{D}_m n^i \mathbf{b}_i) \bullet (\sum_j x^j_\alpha \mathbf{b}_j)$
4	From 3, T12.3.13A Dot Product of Covariant Vectors and T12.3.9A Metric Transformation from Covariant to Contravariant Coefficient	$\overset{*}{A}_m \sum_i x^i_\alpha x_{i\alpha} + \overset{*}{B}_m \sum_i n^i x_{i\alpha} = \overset{*}{C}_m \sum_i x^i_\alpha x_{i\alpha} + \overset{*}{D}_m \sum_i n^i x_{i\alpha}$
5	From 4, T17.3.7D Hypersurface Normal Vector: Orthogonal System $\mathfrak{R}^n$ Transformation and Normal and T12.3.16B Magnitude of Covariant Vector	$\overset{*}{A}_m a_\alpha^2 + \overset{*}{B}_m 0 = \overset{*}{C}_m a_\alpha^2 + \overset{*}{D}_m 0$
6	From 5, T10.2.7 Tool: Zero Times a Tensor is a Zero Tensor and T10.3.18 Addition of Tensors: Identity by Addition	$\overset{*}{A}_m a_\alpha^2 = \overset{*}{C}_m a_\alpha^2$
∴	From 6 and T10.4.1 Scalars and Tensors: Uniqueness of Scalar Multiplication to a Tensors	$\overset{*}{A}_m = \overset{*}{C}_m$

Theorem 17.3.10 Dual Covariant Tensors: Mutually Exclusive Normalized Tensor Components

1g	Given iff	$\overset{*}{A}_m\, x^i{}_\alpha + \overset{*}{B}_m\, n^i = \overset{*}{C}_m\, x^i{}_\alpha + \overset{*}{D}_m\, n^i$
Step	Hypothesis	$\overset{*}{B}_m = \overset{*}{D}_m$ then must also follow
1	From T17.3.7A Hypersurface Normal Vector: Orthogonal System $\mathfrak{R}^n$ Transformation and Normal	$\mathbf{n} = \sum_i n^i\, \mathbf{b}_i$
2	From 1g, 2 and adding the vector component by equating bases vector-by-bases vector	$\sum_i \overset{*}{A}_m\, x^i{}_\alpha\, \mathbf{b}_i + \sum_i \overset{*}{B}_m\, n^i\, \mathbf{b}_i = \sum_i \overset{*}{C}_m\, x^i{}_\alpha\, \mathbf{b}_i + \sum_i \overset{*}{D}_m\, n^i\, \mathbf{b}_i$
3	From 1, 2 and T8.3.1 Dot Product: Uniqueness Operation	$(\sum_i \overset{*}{A}_m\, x^i{}_\alpha\, \mathbf{b}_i + \sum_i \overset{*}{B}_m\, n^i\, \mathbf{b}_i) \bullet (\sum_j n^j\, \mathbf{b}_j) =$ $(\sum_i \overset{*}{C}_m\, x^i{}_\alpha\, \mathbf{b}_i + \sum_i \overset{*}{D}_m\, n^i\, \mathbf{b}_i) \bullet (\sum_j n^j\, \mathbf{b}_j)$
4	From 3, T12.3.13A Dot Product of Covariant Vectors and T12.3.9A Metric Transformation from Covariant to Contravariant Coefficient	$\overset{*}{A}_m \sum_i x^i{}_\alpha\, n_i + \overset{*}{B}_m \sum_i n^i\, n_i = \overset{*}{C}_m \sum_i x^i{}_\alpha\, n_i + \overset{*}{D}_m \sum_i n^i\, n_i$
5	From 4, T17.3.7D Hypersurface Normal Vector: Orthogonal System $\mathfrak{R}^n$ Transformation and Normal and T12.3.18A Magnitude of Normalized Vector	$\overset{*}{A}_m\, 0 + \overset{*}{B}_m\, 1 = \overset{*}{C}_m\, 0 + \overset{*}{D}_m\, 1$
∴	From 5, T10.2.7 Tool: Zero Times a Tensor is a Zero Tensor, T10.3.18 Addition of Tensors: Identity by Addition, T10.2.10 Tool: Identity Multiplication of a Tensor and A10.2.14 Correspondence of Tensors and Tensor Coefficients	$\overset{*}{B}_m = \overset{*}{D}_m$

Theorem 17.3.11 Dual Covariant Tensors: Symmetric Reflexiveness

1g	Given	$\overset{*}{A}_m x^i_{\alpha} + \overset{*}{B}_m n^i = \overset{*}{C}_m x^i_{\alpha} + \overset{*}{D}_m n^i$	
Step	Hypothesis	$\overset{*}{A}_m x^i_{\alpha} + \overset{*}{B}_m n^i = \overset{*}{C}_m x^i_{\alpha} + \overset{*}{D}_m n^i$ EQ A	$\overset{*}{C}_m x^i_{\alpha} + \overset{*}{D}_m n^i = \overset{*}{A}_m x^i_{\alpha} + \overset{*}{B}_m n^i$ EQ B
		Step A	Step B
1	From 1g, T17.3.9 Dual Covariant Tensors: Mutually Exclusive Normalized Tensor Components and T17.3.10 Dual Covariant Tensors: Mutually Exclusive Tangential Tensor Components	$\overset{*}{A}_m = \overset{*}{C}_m$	$\overset{*}{B}_m = \overset{*}{D}_m$
2	From 1g and A17.3.5 Dual Covariant Tensors: Equality	$\overset{*}{A}_m x^i_{\alpha} + \overset{*}{B}_m n^i = \overset{*}{A}_m x^i_{\alpha} + \overset{*}{B}_m n^i$	
∴	From 1SA, 1SB, T10.2.5 Tool: Tensor Substitution; left and right hand side step by substitution.	$\overset{*}{A}_m x^i_{\alpha} + \overset{*}{B}_m n^i = \overset{*}{C}_m x^i_{\alpha} + \overset{*}{D}_m n^i$ EQ A	$\overset{*}{C}_m x^i_{\alpha} + \overset{*}{D}_m n^i = \overset{*}{A}_m x^i_{\alpha} + \overset{*}{B}_m n^i$ EQ B

Theorem 17.3.12 Dual Covariant Tensors: Substitution

1g	Given	$\overset{*}{A}_m x^i_\alpha + \overset{*}{B}_m n^i = \overset{*}{C}_m x^i_\alpha + \overset{*}{D}_m n^i$	
2g		$\overset{*}{C}_m x^i_\alpha + \overset{*}{D}_m n^i = \overset{*}{E}_m x^i_\alpha + \overset{*}{F}_m n^i$	
Step	Hypothesis	$\overset{*}{A}_m x^i_\alpha + \overset{*}{B}_m n^i = \overset{*}{E}_m x^i_\alpha + \overset{*}{F}_m n^i$	
		Step A	Step B
1	1g, T17.3.9 Dual Covariant Tensors: Mutually Exclusive Normalized Tensor Components and T17.3.10 Dual Covariant Tensors: Mutually Exclusive Tangential Tensor Components	$\overset{*}{A}_m = \overset{*}{C}_m$	$\overset{*}{B}_m = \overset{*}{D}_m$
2	2g, T17.3.9 Dual Covariant Tensors: Mutually Exclusive Normalized Tensor Components and T17.3.10 Dual Covariant Tensors: Mutually Exclusive Tangential Tensor Components	$\overset{*}{C}_m = \overset{*}{E}_m$	$\overset{*}{D}_m = \overset{*}{F}_m$
3	From 1SA, 1SB, 2SA, 2SB, T10.2.5 Tool: Tensor Substitution; left and right hand side step by substitution.	$\overset{*}{A}_m = \overset{*}{E}_m$	$\overset{*}{B}_m = \overset{*}{F}_m$
4	From 1g and A17.3.5 Dual Covariant Tensors: Equality	$\overset{*}{A}_m x^i_\alpha + \overset{*}{B}_m n^i = \overset{*}{A}_m x^i_\alpha + \overset{*}{B}_m n^i$	
∴	From 3SA, 3SB, T10.2.5 Tool: Tensor Substitution; right hand side step by substitution.	$\overset{*}{A}_m x^i_\alpha + \overset{*}{B}_m n^i = \overset{*}{E}_m x^i_\alpha + \overset{*}{F}_m n^i$	

Theorem 17.3.13 Dual Covariant Tensors: Commutative Addition

Step	Hypothesis	$(\overset{*}{A}_m x^i_\alpha + \overset{*}{B}_m n^i) + (\overset{*}{C}_m x^i_\alpha + \overset{*}{D}_m n^i) =$ $(\overset{*}{C}_m x^i_\alpha + \overset{*}{D}_m n^i) + (\overset{*}{A}_m x^i_\alpha + \overset{*}{B}_m n^i)$
1	From 1g and A17.3.5 Dual Covariant Tensors: Equality	$(\overset{*}{A}_m x^i_\alpha + \overset{*}{B}_m n^i) + (\overset{*}{C}_m x^i_\alpha + \overset{*}{D}_m n^i) =$ $(\overset{*}{A}_m x^i_\alpha + \overset{*}{B}_m n^i) + (\overset{*}{C}_m x^i_\alpha + \overset{*}{D}_m n^i)$
∴	From 1, A17.3.3P3 Hypersurface $\mathfrak{H}^{n-1}$: Existance of Dual Covariant Tensor Field and T10.3.16 Addition of Tensors: Commutative by Addition	$(\overset{*}{A}_m x^i_\alpha + \overset{*}{B}_m n^i) + (\overset{*}{C}_m x^i_\alpha + \overset{*}{D}_m n^i) =$ $(\overset{*}{C}_m x^i_\alpha + \overset{*}{D}_m n^i) + (\overset{*}{A}_m x^i_\alpha + \overset{*}{B}_m n^i)$

Theorem 17.3.14 Dual Covariant Tensors: Associative Addition

1g	Given	$[(\overset{*}{A}_m x^i_\alpha + \overset{*}{B}_m n^i) + (\overset{*}{C}_m x^i_\alpha + \overset{*}{D}_m n^i)] + (\overset{*}{E}_m x^i_\alpha + \overset{*}{F}_m n^i) =$ $(\overset{*}{A}_m x^i_\alpha + \overset{*}{B}_m n^i) + [(\overset{*}{C}_m x^i_\alpha + \overset{*}{D}_m n^i) + (\overset{*}{E}_m x^i_\alpha + \overset{*}{F}_m n^i)]$ assume
Step	Hypothesis	$[(\overset{*}{A}_m x^i_\alpha + \overset{*}{B}_m n^i) + (\overset{*}{C}_m x^i_\alpha + \overset{*}{D}_m n^i)] + (\overset{*}{E}_m x^i_\alpha + \overset{*}{F}_m n^i) =$ $(\overset{*}{A}_m x^i_\alpha + \overset{*}{B}_m n^i) + [(\overset{*}{C}_m x^i_\alpha + \overset{*}{D}_m n^i) + (\overset{*}{E}_m x^i_\alpha + \overset{*}{F}_m n^i)]$
1	From 1g, T10.3.16 Addition of Tensors: Commutative by Addition and T10.3.20 Addition of Tensors: Distribution of a Tensor over Addition of Tensors	$[(\overset{*}{A}_m + \overset{*}{C}_m)\, x^i_\alpha + (\overset{*}{B}_m + \overset{*}{D}_m)\, n^i] + (\overset{*}{E}_m x^i_\alpha + \overset{*}{F}_m n^i) =$ $(\overset{*}{A}_m x^i_\alpha + \overset{*}{B}_m n^i) + [(\overset{*}{C}_m + \overset{*}{E}_m)\, x^i_\alpha + (\overset{*}{D}_m + \overset{*}{F}_m)\, n^i]$
2	From 1, T10.3.16 Addition of Tensors: Commutative by Addition and T10.3.20 Addition of Tensors: Distribution of a Tensor over Addition of Tensors	$[(\overset{*}{A}_m + \overset{*}{C}_m) + \overset{*}{E}_m\,]\, x^i_\alpha + [(\overset{*}{B}_m + \overset{*}{D}_m) + \overset{*}{F}_m]\, n^i =$ $[\overset{*}{A}_m + (\overset{*}{C}_m + \overset{*}{E}_m)\,]\, x^i_\alpha + [\overset{*}{B}_m + (\overset{*}{D}_m + \overset{*}{F}_m)\,]\, n^i$
3	From 2 and T10.3.17 Addition of Tensors: Associative by Addition	$[(\overset{*}{A}_m + \overset{*}{C}_m) + \overset{*}{E}_m\,]\, x^i_\alpha + [\overset{*}{A}_m + (\overset{*}{C}_m + \overset{*}{E}_m)\,]\, n^i =$ $[(\overset{*}{A}_m + \overset{*}{C}_m) + \overset{*}{E}_m\,]\, x^i_\alpha + [\overset{*}{A}_m + (\overset{*}{C}_m + \overset{*}{E}_m \}\, n^i$
∴	From 1g, 3 and by identity	$[(\overset{*}{A}_m x^i_\alpha + \overset{*}{B}_m n^i) + (\overset{*}{C}_m x^i_\alpha + \overset{*}{D}_m n^i)] + (\overset{*}{E}_m x^i_\alpha + \overset{*}{F}_m n^i) =$ $(\overset{*}{A}_m x^i_\alpha + \overset{*}{B}_m n^i) + [(\overset{*}{C}_m x^i_\alpha + \overset{*}{D}_m n^i) + (\overset{*}{E}_m x^i_\alpha + \overset{*}{F}_m n^i)]$

Theorem 17.3.15 Dual Covariant Tensors: Identity Addition

Step	Hypothesis	$(\overset{*}{A}_m x^i_\alpha + \overset{*}{B}_m n^i\,) + (\overset{*}{Z}_m x^i_\alpha + \overset{*}{Z}_m n^i) = (\overset{*}{A}_m x^i_\alpha + \overset{*}{B}_m n^i)$
1	From A17.3.5 Dual Covariant Tensors: Equality	$(\overset{*}{A}_m x^i_\alpha + \overset{*}{B}_m n^i) = (\overset{*}{A}_m x^i_\alpha + \overset{*}{B}_m n^i)$
2	From 1, T10.3.18 Addition of Tensors: Identity by Addition	$(\overset{*}{A}_m + \overset{*}{Z}_m)\, x^i_\alpha + (\overset{*}{B}_m + \overset{*}{Z}_m)\, n^i = (\overset{*}{A}_m x^i_\alpha + \overset{*}{B}_m n^i)$
3	From 2 and T10.3.20 Addition of Tensors: Distribution of a Tensor over Addition of Tensors	$\overset{*}{A}_m x^i_\alpha + \overset{*}{Z}_m x^i_\alpha + \overset{*}{B}_m n^i + \overset{*}{Z}_m n^i = (\overset{*}{A}_m x^i_\alpha + \overset{*}{B}_m n^i)$
4	From 3 and T10.3.16 Addition of Tensors: Commutative by Addition	$\overset{*}{A}_m x^i_\alpha + \overset{*}{B}_m n^i + \overset{*}{Z}_m x^i_\alpha + \overset{*}{Z}_m n^i = (\overset{*}{A}_m x^i_\alpha + \overset{*}{B}_m n^i)$
∴	From 4 and A17.3.5 Dual Covariant Tensors: Equality	$(\overset{*}{A}_m x^i_\alpha + \overset{*}{B}_m n^i\,) + (\overset{*}{Z}_m x^i_\alpha + \overset{*}{Z}_m n^i) = (\overset{*}{A}_m x^i_\alpha + \overset{*}{B}_m n^i)$

Theorem 17.3.16 Dual Covariant Tensors: Inverse Addition

1g	Given	$(\overset{*}{A}_m x^i_\alpha + \overset{*}{B}_m n^i) - (\overset{*}{A}_m x^i_\alpha + \overset{*}{B}_m n^i) = (\overset{*}{Z}_m x^i_\alpha + \overset{*}{Z}_m n^i)$ assume
Step	Hypothesis	$(\overset{*}{A}_m x^i_\alpha + \overset{*}{B}_m n^i) - (\overset{*}{A}_m x^i_\alpha + \overset{*}{B}_m n^i) = (\overset{*}{Z}_m x^i_\alpha + \overset{*}{Z}_m n^i)$
1	From 1g, D4.1.20A Negative Coefficient and T10.4.6 Scalars and Tensors: Distribution of a Scalar over Addition of Tensors	$\overset{*}{A}_m x^i_\alpha + \overset{*}{B}_m n^i - \overset{*}{A}_m x^i_\alpha - \overset{*}{B}_m n^i = (\overset{*}{Z}_m x^i_\alpha + \overset{*}{Z}_m n^i)$
2	From 1, and T10.3.16 Addition of Tensors: Commutative by Addition	$\overset{*}{A}_m x^i_\alpha - \overset{*}{A}_m x^i_\alpha + \overset{*}{B}_m n^i - \overset{*}{B}_m n^i = (\overset{*}{Z}_m x^i_\alpha + \overset{*}{Z}_m n^i)$
3	From 2 and T10.3.20 Addition of Tensors: Distribution of a Tensor over Addition of Tensors	$(\overset{*}{A}_m - \overset{*}{A}_m) x^i_\alpha + (\overset{*}{B}_m - \overset{*}{B}_m) n^i = (\overset{*}{Z}_m x^i_\alpha + \overset{*}{Z}_m n^i)$
4	From 3 and T10.3.19 Addition of Tensors: Inverse by Addition	$(\overset{*}{Z}_m x^i_\alpha + \overset{*}{Z}_m n^i) = (\overset{*}{Z}_m x^i_\alpha + \overset{*}{Z}_m n^i)$
∴	From 1g, 4 and by identity	$(\overset{*}{A}_m x^i_\alpha + \overset{*}{B}_m n^i) - (\overset{*}{A}_m x^i_\alpha + \overset{*}{B}_m n^i) = (\overset{*}{Z}_m x^i_\alpha + \overset{*}{Z}_m n^i)$

Theorem 17.3.17 Dual Covariant Tensors: None Closure Under Dot Multiplication

Step	Hypothesis	$(\overset{*}{A}_m x^i_\alpha + \overset{*}{B}_m n^i) \bullet (\overset{*}{C}_m x^i_\alpha + \overset{*}{D}_m n^i) = \overset{*}{A}_m \overset{*}{C}_m x_\alpha^2 + \overset{*}{B}_m \overset{*}{D}_m$
1	From A17.3.5 Dual Covariant Tensors: Equality	$(\overset{*}{A}_m x^i_\alpha + \overset{*}{B}_m n^i) \bullet (\overset{*}{C}_m x^i_\alpha + \overset{*}{D}_m n^i) =$ $(\overset{*}{A}_m x^i_\alpha + \overset{*}{B}_m n^i) \bullet (\overset{*}{C}_m x^j_\alpha + \overset{*}{D}_m n^j)$
2	From 1 and adding the vector component by equating bases vector-by-bases vector	$(\overset{*}{A}_m x^i_\alpha + \overset{*}{B}_m n^i) \bullet (\overset{*}{C}_m x^i_\alpha + \overset{*}{D}_m n^i) =$ $(\overset{*}{A}_m x^i_\alpha \mathbf{b}_i + \overset{*}{B}_m n^i \mathbf{b}_i) \bullet (\overset{*}{C}_m x^j_\alpha \mathbf{b}_j + \overset{*}{D}_m n^j \mathbf{b}_j)$
3	From 2, A17.3.3P3 Hypersurface $\mathfrak{H}^{n-1}$: Existance of Dual Covariant Tensor Field and T10.3.20 Addition of Tensors: Distribution of a Tensor over Addition of Tensors	$(\overset{*}{A}_m x^i_\alpha + \overset{*}{B}_m n^i) \bullet (\overset{*}{C}_m x^i_\alpha + \overset{*}{D}_m n^i) =$ $[(\overset{*}{A}_m x^i_\alpha \mathbf{b}_i + \overset{*}{B}_m n^i \mathbf{b}_i) \bullet \overset{*}{C}_m x^j_\alpha \mathbf{b}_j + (\overset{*}{A}_m x^i_\alpha \mathbf{b}_i + \overset{*}{B}_m n^i \mathbf{b}_i) \bullet \overset{*}{D}_m n^j \mathbf{b}_j]$
4	From 3, A17.3.3P3 Hypersurface $\mathfrak{H}^{n-1}$: Existance of Dual Covariant Tensor Field and T10.3.20 Addition of Tensors: Distribution of a Tensor over Addition of Tensors	$(\overset{*}{A}_m x^i_\alpha + \overset{*}{B}_m n^i) \bullet (\overset{*}{C}_m x^i_\alpha + \overset{*}{D}_m n^i) =$ $\overset{*}{A}_m \overset{*}{C}_m x^i_\alpha x^j_\alpha (\mathbf{b}_i \bullet \mathbf{b}_j) + \overset{*}{B}_m \overset{*}{C}_m n^i x^j_\alpha (\mathbf{b}_i \bullet \mathbf{b}_j) +$ $\overset{*}{A}_m \overset{*}{D}_m x^i_\alpha n^j (\mathbf{b}_i \bullet \mathbf{b}_j) + \overset{*}{B}_m \overset{*}{D}_m n^i n^j (\mathbf{b}_i \bullet \mathbf{b}_j)$
5	From 3 and T10.3.7 Product of Tensors: Commutative by Multiplication of Rank p + q → q + p	$(\overset{*}{A}_m x^i_\alpha + \overset{*}{B}_m n^i) \bullet (\overset{*}{C}_m x^i_\alpha + \overset{*}{D}_m n^i) =$ $\overset{*}{A}_m \overset{*}{C}_m x^i_\alpha x^j_\alpha g_{ij} + \overset{*}{B}_m \overset{*}{C}_m n^i x^j_\alpha g_{ij} +$ $\overset{*}{A}_m \overset{*}{D}_m x^i_\alpha n^j g_{ij} + \overset{*}{B}_m \overset{*}{D}_m n^i n^j g_{ij}$
6	From 5 and T12.3.9B Metric Transformation from Covariant to Contravariant Coefficient	$(\overset{*}{A}_m x^i_\alpha + \overset{*}{B}_m n^i) \bullet (\overset{*}{C}_m x^i_\alpha + \overset{*}{D}_m n^i) =$ $\overset{*}{A}_m \overset{*}{C}_m x^i_\alpha x_{i\alpha} + \overset{*}{B}_m \overset{*}{C}_m n^i x_{i\alpha} + \overset{*}{A}_m \overset{*}{D}_m x^i_\alpha n_i + \overset{*}{B}_m \overset{*}{D}_m n^i n_i$

7	From 6, T17.3.7D Hypersurface Normal Vector: Orthogonal System $\mathfrak{R}^n$ Transformation and Normal and T12.3.16B Magnitude of Covariant Vector	$(\overset{*}{A}_m x^i_\alpha + \overset{*}{B}_m n^i) \bullet (\overset{*}{C}_m x^i_\alpha + \overset{*}{D}_m n^i) =$ $\overset{*}{A}_m \overset{*}{C}_m x_\alpha^2 + \overset{*}{B}_m \overset{*}{C}_m 0 + \overset{*}{A}_m \overset{*}{D}_m 0 + \overset{*}{B}_m \overset{*}{D}_m 1$
∴	From 7, T10.2.7 Tool: Zero Times a Tensor is a Zero Tensor and T10.3.18 Addition of Tensors: Identity by Addition	$(\overset{*}{A}_m x^i_\alpha + \overset{*}{B}_m n^i)(\overset{*}{C}_m x^i_\alpha + \overset{*}{D}_m n^i) = \overset{*}{A}_m \overset{*}{C}_m x_\alpha^2 + \overset{*}{B}_m \overset{*}{D}_m$ Product of Dual Covariant Tensor → Common Tensor

Theorem 17.3.18 Dual Covariant Tensors: Zero Tensor Rank-2

Step	Hypothesis	$\overset{*}{Z}_2 = \mathbf{0}\, x^i_\alpha + \mathbf{0}\, n^i$
1	From A10.2.4 Tensor of Rank-1 is a Vector and D10.1.18(A, B) Zero Tensor	$\mathbf{0} = 0\, \overset{*}{A}_1$
2	From 1 and T10.2.7A Tool: Zero Times a Tensor is a Zero Tensor	$0\, \overset{*}{A}_1 = \overset{*}{Z}_1$
3	From A17.3.3P3 Hypersurface $\mathfrak{H}^{n-1}$: Existance of Dual Covariant Tensor Field	$\overset{*}{Z}_2 = \overset{*}{Z}_1\, x^i_\alpha + \overset{*}{Z}_1\, n^i$
4	From 2, 3 and T10.2.5 Tool: Tensor Substitution	$\overset{*}{Z}_2 = 0\, \overset{*}{A}_1\, x^i_\alpha + 0\, \overset{*}{A}_1\, n^i$
∴	From 1, 4 and T10.2.5 Tool: Tensor Substitution	$\overset{*}{Z}_2 = \mathbf{0}\, x^i_\alpha + \mathbf{0}\, n^i$

Observation 17.3.1 None Closure for Product of Dual Covariant Tensor

From theroem T17.3.17 "Dual Covariant Tensors: None Closure Under Multiplication" there can be no distribution for a dual covariant tensor with another, because it collapses to a common tensor and cannot be reversed. However a scalar when multiplied to it leaves it unaltered so a distribution could exist for a scalar.

It also occurs to me that if a new operator for dual covariant tensors could be developed, along the lines of a Serret-Frenet operation with a binormal vector, then the resulting Galois Group would create a bi-operational set of Galois groups. This would provide the missing symmetry the axiomatic set needs preserving distribution.

Table 17.3.1 Dual Contravariant Tensors on a Hypersurface $\mathfrak{H}^{n-1}$

Theorem	Source	Equation
Theorem 17.3.19 Dual Contravariant Tensors: Mutually Exclusive Tangential Tensor Components	T17.3.10	$\overset{*}{A}{}^m = \overset{*}{C}{}^m$ for $\overset{*}{A}{}^m x^i{}_\alpha + \overset{*}{B}{}^m n^i = \overset{*}{C}{}^m x^i{}_\alpha + \overset{*}{D}{}^m n^i$
Theorem 17.3.20 Dual Contravariant Tensors: Mutually Exclusive Normalized Tensor Components	T17.3.9	$\overset{*}{B}{}^m = \overset{*}{D}{}^m$ for $\overset{*}{A}{}^m x^i{}_\alpha + \overset{*}{B}{}^m n^i = \overset{*}{C}{}^m x^i{}_\alpha + \overset{*}{D}{}^m n^i$
Theorem 17.3.21 Dual Contravariant Tensors: Symmetric Reflexiveness	T17.3.11	$\overset{*}{A}{}^m x^i{}_\alpha + \overset{*}{B}{}^m n^i =$ $\overset{*}{C}{}^m x^i{}_\alpha + \overset{*}{D}{}^m n^i$ EQ A $\overset{*}{C}{}^m x^i{}_\alpha + \overset{*}{D}{}^m n^i =$ $\overset{*}{A}{}^m x^i{}_\alpha + \overset{*}{B}{}^m n^i$ EQ B
Theorem 17.3.22 Dual Contravariant Tensors: Substitution	T17.3.12	$\overset{*}{A}{}^m x^i{}_\alpha + \overset{*}{B}{}^m n^i = \overset{*}{C}{}^m x^i{}_\alpha + \overset{*}{D}{}^m n^i$ $\overset{*}{C}{}^m x^i{}_\alpha + \overset{*}{D}{}^m n^i = \overset{*}{E}{}^m x^i{}_\alpha + \overset{*}{F}{}^m n^i$ $\overset{*}{A}{}^m x^i{}_\alpha + \overset{*}{B}{}^m n^i = \overset{*}{E}{}^m x^i{}_\alpha + \overset{*}{F}{}^m n^i$
Theorem 17.3.23 Dual Contravariant Tensors: Commutative Addition	T17.3.13	$(\overset{*}{A}{}^m x^i{}_\alpha + \overset{*}{B}{}^m n^i) + (\overset{*}{C}{}^m x^i{}_\alpha + \overset{*}{D}{}^m n^i) =$ $(\overset{*}{A}{}^m x^i{}_\alpha + \overset{*}{B}{}^m n^i) + (\overset{*}{C}{}^m x^i{}_\alpha + \overset{*}{D}{}^m n^i)$
Theorem 17.3.24 Dual Contravariant Tensors: Associative Addition	T17.3.14	$[(\overset{*}{A}{}^m x^i{}_\alpha + \overset{*}{B}{}^m n^i) + (\overset{*}{C}{}^m x^i{}_\alpha + \overset{*}{D}{}^m n^i)] +$ $(\overset{*}{E}{}^m x^i{}_\alpha + \overset{*}{F}{}^m n^i) =$ $(\overset{*}{A}{}^m x^i{}_\alpha + \overset{*}{B}{}^m n^i) + [(\overset{*}{C}{}^m x^i{}_\alpha + \overset{*}{D}{}^m n^i) +$ $(\overset{*}{E}{}^m x^i{}_\alpha + \overset{*}{F}{}^m n^i)]$
Theorem 17.3.25 Dual Contravariant Tensors: Identity Addition	T17.3.15	$(\overset{*}{A}{}^m x^i{}_\alpha + \overset{*}{B}{}^m n^i) + (\overset{*}{Z}{}^m x^i{}_\alpha + \overset{*}{Z}{}^m n^i) =$ $(\overset{*}{A}{}^m x^i{}_\alpha + \overset{*}{B}{}^m n^i)$
Theorem 17.3.26 Dual Contravariant Tensors: Inverse Addition	T17.3.16	$(\overset{*}{A}{}^m x^i{}_\alpha + \overset{*}{B}{}^m n^i) - (\overset{*}{A}{}^m x^i{}_\alpha + \overset{*}{B}{}^m n^i) =$ $(\overset{*}{Z}{}^m x^i{}_\alpha + \overset{*}{Z}{}^m n^i)$
Theorem 17.3.27 Dual Contravariant Tensors: None Closure Under Dot Multiplication	T17.3.17	$(\overset{*}{A}{}^m x^i{}_\alpha + \overset{*}{B}{}^m n^i) \bullet (\overset{*}{C}{}^m x^i{}_\alpha + \overset{*}{D}{}^m n^i) =$ $\overset{*}{A}{}^m \overset{*}{C}{}^m x_\alpha{}^2 + \overset{*}{B}{}^m \overset{*}{D}{}^m$
Theorem 17.3.28 Dual Contravariant Tensors: Zero Tensor Rank-2	T17.3.18	$\overset{*}{Z}{}^2 = \mathbf{0}\, x^i{}_\alpha + \mathbf{0}\, n^i$

Section 17.4 First Fundamental Quadratic Form of the Hypersurface $\mathfrak{H}^{n-1}$

pg 140 [a]

Observation 17.4.1 Condition for the Existence of Hypersheet Contravariant Base Vectors

From theorem T17.4.3 "Hypersheet: Contravariant Reciprocal Cofactor" the reciprocal dyadic matrix exists iff a ≠ 0 so it follows that since theorem T17.4.4 "Hypersheet: Covariant and Contravariant Dyadics in Matrix Form" relates the contravariant metric tensor directly to distance in that surface manifold, then the contravariant metric tensor must exist, and likewise the contravariant base vectors satisfying a dual vector definition D17.1.3A "Contravariant Metric Hypersheet Dyadic".

$a^{\alpha\beta} \equiv \mathbf{a}^{\alpha} \bullet \mathbf{a}^{\beta}$ Equation A

From axiom A17.1.11P1 "Riemannian Metric on a Hypersheet $\mathfrak{H}^{n-1}$" on a metric hypersheet it would follow, then that the differential distance would be

$d\sigma^2 \equiv \Sigma_\alpha \Sigma_\beta a^{\alpha\beta} du_\alpha du_\beta$ Equation B

Observation 17.4.2 A Positional and Normal Vector Undergoing Change

As the positional [**r**] and normal [**N**] vectors undergoing change in position on the hypersurface at point-Q(• $u_\alpha + \delta u_\alpha$ •) they expand about that position of displacement.

Axiom 17.4.1 Contravariant Metric Displacement on a Riemannian Hypersheet $\mathfrak{H}^{n-1}$

There exists uniquely a set of contravariant base vectors completing a closed dual base vector system for a hypersheet iff a ≠ 0;

P1) $a^{\alpha\beta} \equiv \mathbf{a}^{\alpha} \bullet \mathbf{a}^{\beta}$

and as such has a unique metric distance,

P2) $d\sigma^2 \equiv \Sigma_\alpha \Sigma_\beta a^{\alpha\beta} du_\alpha du_\beta$

Theorem 17.4.1 Hypersheet: Mixed Bases Vector Transpose

Step	Hypothesis	$I = S^{-1}\{A_c{}^T A^c\}$
1	From D17.1.4C Mixed Metric Hypersheet Dyadic	$I = S^{-1}\{A_c A^{cT}\}$ for matrix form (n–1)x(n–1) = ((n–1)x1)(1x(n–1))
2	From 1 and T6.4.1 Uniqueness of Transposition of Matrices	$I^T = S^{-1}\{ (A_c A^{cT})^T\}$
3	From 2, T6.11.2 Commute Column Vectors under Conservation of Transposition	$I^T = S^{-1}\{A^c A_c{}^T\}$ for matrix form (n–1)x(n–1) = ((n–1)x1)(1x(n–1))
∴	From 3 and T6.4.3 Invariance of the transpose operation	$I = S^{-1}\{A^c A_c{}^T\}$ for matrix form ((n–1) x (n–1)) = ((n–1) x 1)(1 x (n–1))

Theorem 17.4.2 Hypersheet: Contravariant Metric Bases Vector Inverse

Step	Hypothesis	$I = A A^{-1}$ EQ A	$I = A^{-1} A$ EQ B
1	From D17.1.4C Mixed Metric Dyadic	$I = S^{-1}\{A_c A^{cT}\}$	
2	From 1, T6.3.12 Matrix Identity right of Multiplication, T12.2.3 Mixed Bases Vector Transpose and T6.3.1 Substitution of Matrices	$I = S^{-1}\{A_c A^{cT} A^c A_c{}^T\}$	
3	From 2, T6.11.1 Commute Column Vectors under Conservation of Order 3x and T6.3.11 Matrix Association for Multiplication	$I = S^{-1}\{A_c A^{cT} A^c A_c{}^T\}$ $I = S^{-1}\{A_c A^{cT} A_c{}^T A^c\}$ $I = S^{-1}\{A_c A_c{}^T A^{cT} A^c\}$ $I = S^{-1}\{A_c A_c{}^T\} S^{-1}\{ A^c A^{cT}\}$	$I = S^{-1}\{A_c A^{cT} A^c A_c{}^T\}$ $I = S^{-1}\{A^{cT} A_c A^c A_c{}^T\}$ $I = S^{-1}\{A^{cT} A^c A_c A_c{}^T\}$ $I = S^{-1}\{A^c A^{cT}\} S^{-1}\{A_c A_c{}^T\}$
∴	From 3, D17.1.2D Covariant Metric Hypersheet Dyadic, D12.1.3D Contravariant Metric Hypersheet Dyadic and T6.3.1 Substitution of Matrices	$I = A A^{-1}$ EQ A	$I = A^{-1} A$ EQ B

Theorem 17.4.3 Hypersheet: Contravariant Reciprocal Cofactor

Step	Hypothesis	$A^{-1} = (a^{-1} A^{\alpha\beta})$
1	From T6.9.18 Inverse Square Matrix for Multiplication, D6.9.1F Determinant of A and A4.2.3 Substitution	$A^{-1} = (\text{adj } A) / a$ for $a \neq 0$
2	From 1, D6.9.5 Adjoint of a Square Matrix A, D4.1.18 Negative Exponential and A4.2.3 Substitution; with cofactor	$A^{-1} = a^{-1}\ {}^t(A^{\beta\alpha})$ for $a \neq 0$
∴	From 2, A6.2.9 Matrix Multiplication by a Scalar and A6.2.11 Matrix Transposition	$A^{-1} = (a^{-1} A^{\alpha\beta})$

Theorem 17.4.4 Hypersheet: Covariant and Contravariant Dyadics in Matrix Form

Step	Hypothesis	$A^{-1} = (a^{-1}\, A^{\alpha\beta})$ EQ A	$A^{-1} = (a^{\alpha\beta})$ EQ B
1	From D17.1.3C Contravariant Metric Hypersheet Dyadic; placed in matrix form	$(a^{\alpha\beta}) = (a^{-1}\, A^{\alpha\beta})$	
2	From T17.4.3 Hypersheet: Contravariant Reciprocal Cofactor	$A^{-1} = (a^{-1}\, A^{\alpha\beta})$	
3	1, 2 and T6.3.1 Substitution of Matrices	$A^{-1} = (a^{\alpha\beta})$	
∴	From 2 and 3	$A^{-1} = (a^{-1}\, A^{\alpha\beta})$ EQ A	$A^{-1} = (a^{\alpha\beta})$ EQ B

Theorem 17.4.5 Hypersheet: Equivalency of Contra-Covariant Kronecker Delta

Step	Hypothesis	$\delta^{\alpha}{}_{\beta} = \delta_{\beta}{}^{\alpha}$
1	From A4.2.2A Equality	$\delta^{\alpha}{}_{\beta} = \delta^{\alpha}{}_{\beta}$
2	From 1, D17.1.4D Mixed Metric Hypersheet Dyadic and A4.2.3 Substitution	$\delta^{\alpha}{}_{\beta} = \mathbf{a}^{\alpha} \bullet \mathbf{a}_{\beta}$
3	From 2 and T8.3.3 Dot Product: Commutative Operation	$\delta^{\alpha}{}_{\beta} = \mathbf{a}_{\beta} \bullet \mathbf{a}^{\alpha}$
∴	From 3, D17.1.4A Mixed Metric Hypersheet Dyadic and A4.2.3 Substitution	$\delta^{\alpha}{}_{\beta} = \delta_{\beta}{}^{\alpha}$

Theorem 17.4.6 Hypersheet: Covariant and Contravariant Metric Inverse

Step	Hypothesis	$\delta_{\alpha}{}^{\beta} = a_{\alpha}{}^{\beta} = \sum_k a_{\alpha\kappa}\, a^{\kappa\beta}$ EQ A for $\mid a_{\alpha\beta} \mid \neq 0$	$\delta^{\alpha}{}_{\beta} = a^{\alpha}{}_{\beta} = \sum_k a^{\alpha\kappa}\, a_{\kappa\beta}$ EQ B for $\mid a_{\alpha\beta} \mid \neq 0$
1	From T6.9.19 Closure with respect to the Identity Matrix for Multiplication	$I = A * A^{-1}$	$I = A^{-1} * A$ for $\mid A \mid \neq 0$
2	From 1, T17.4.4A,B Hypersheet: Covariant and Contravariant Dyadics in Matrix Form and T6.3.1 Substitution of Matrices	$I = (a_{\alpha\beta}) * (a^{\alpha\beta})$ for $\mid a_{\alpha\beta} \mid \neq 0$	$I = (a^{\alpha\beta}) * (a_{\alpha\beta})$ for $\mid a_{\alpha\beta} \mid \neq 0$
3	From 2 and T6.2.8C Matrix Multiplication; in matrix form	$I = (\sum_\kappa a_{\alpha\kappa}\, a^{\kappa\beta})$	$I = (\sum_k a^{\alpha\kappa}\, a_{\kappa\beta})$
4	From 3, D6.1.6A Identity Matrix and T12.2.7A Transpose of Mixed Kronecker Delta	$(\delta_{\alpha}{}^{\beta}) = (\sum_\kappa a_{\alpha\kappa}\, a^{\kappa\beta})$	$(\delta^{\alpha}{}_{\beta}) = (\sum_k a^{\alpha\kappa}\, a_{\kappa\beta})$
∴	From 6, D17.1.4A,D Mixed Metric Hypersheet Dyadic and equating matrix elements	$\delta_{\alpha}{}^{\beta} = a_{\alpha}{}^{\beta} = \sum_k a_{\alpha\kappa}\, a^{\kappa\beta}$ EQ A for $\mid a_{\alpha\beta} \mid \neq 0$	$\delta^{\alpha}{}_{\beta} = a^{\alpha}{}_{\beta} = \sum_k a^{\alpha\kappa}\, a_{\kappa\beta}$ EQ B for $\mid a_{\alpha\beta} \mid \neq 0$

Theorem 17.4.7 Hypersheet: Covariant and Contravariant Metric Inverse Alternate Proof

Step	Hypothesis	$\delta_\alpha{}^\beta = a_\alpha{}^\beta = \Sigma_\kappa a_{\alpha\kappa} a^{\kappa\beta}$ EQ A	$\delta^\alpha{}_\beta = a^\alpha{}_\beta = \Sigma_\kappa a^{\alpha\kappa} a_{\kappa\beta}$ EQ B
1	From A6.2.2 Closure Multiplication	$Q_m = A A^{-1}$	$Q^m = A^{-1} A$
2	From 1, T17.4.4A,B Hypersheet: Covariant and Contravariant Dyadics in Matrix Form and T6.3.1 Substitution of Matrices	$(q_\alpha{}^\beta) = (a_{\alpha\beta}) (a^{\alpha\beta})$	$(q^\alpha{}_\beta) = (a^{\alpha\beta}) (a_{\alpha\beta})$
3	From 2 and T6.2.8C Matrix Multiplication; in matrix form	$(q_\alpha{}^\beta) = (\Sigma_\kappa a_{\alpha\kappa} a^{\kappa\beta})$	$(q^\alpha{}_\beta) = (\Sigma_\kappa a^{\alpha\kappa} a_{\kappa\beta})$
4	From 1, T17.2.4A,B Hypersheet: Contravariant Metric Bases Vector Inverse and A6.2.7A Equality of Matrices	$I = Q_m$	$I = Q^m$
5	From 4 and D6.1.6A Identity Matrix	$(\delta_\alpha{}^\beta) = (q_i{}^j)$	$(\delta^\alpha{}_\beta) = (q^\alpha{}_\beta)$
6	From 2, 5 and T6.3.1 Substitution of Matrices	$(\delta_\alpha{}^\beta) = (\Sigma_\kappa a_{\alpha\kappa} a^{\kappa\beta})$	$(\delta^\alpha{}_\beta) = (\Sigma_\kappa a^{\alpha\kappa} a_{\kappa\beta})$
∴	From 6, D17.1.4A,D Mixed Metric Dyadic and equating matrix elements	$\delta_\alpha{}^\beta = a_\alpha{}^\beta = \Sigma_\kappa a_{\alpha\kappa} a^{\kappa\beta}$ EQ A	$\delta^\alpha{}_\beta = a^\alpha{}_\beta = \Sigma_\kappa a^{\alpha\kappa} a_{\kappa\beta}$ EQ B

Theorem 17.4.8 Hypersheet: Covariant and Contravariant Scale Factor; Reciprocal Relationship

Step	Hypothesis	$\lvert \mathbf{a}^\alpha \rvert = 1 / \lvert \mathbf{a}_\alpha \rvert$ for all α EQ A	$\lvert \mathbf{a}_\alpha \rvert = 1 / \lvert \mathbf{a}^\alpha \rvert$ for all α EQ B
1	From D17.1.4B Mixed Metric Hypersheet Dyadic	$\delta_\alpha{}^\beta \equiv \mathbf{a}_\alpha \bullet \mathbf{a}^\beta$	
2	From 1, D8.3.3C Orthogonal Vectors and A4.2.3 Substitution	$\delta_\alpha{}^\beta = \lvert \mathbf{a}_\alpha \rvert \lvert \mathbf{a}^\beta \rvert \delta_\alpha{}^\beta$	
3	From 2 and Identity Matrix Kronecker Delta evaluated	$1 = \lvert \mathbf{a}_\alpha \rvert \lvert \mathbf{a}^\beta \rvert$ for $\alpha = \beta$	
∴	From 3 and T4.4.6 Equalities: Left Cancellation by Multiplication	$\lvert \mathbf{a}^\alpha \rvert = 1 / \lvert \mathbf{a}_\alpha \rvert$ for all α EQ A	$\lvert \mathbf{a}_\alpha \rvert = 1 / \lvert \mathbf{a}^\alpha \rvert$ for all α EQ B

Theorem 17.4.9 Hypersheet: Covariant Metric Inner Product

Step	Hypothesis	$a_{\alpha\beta} = a_\alpha a_\beta \cos \theta_{\alpha\beta}$
1	From D17.1.2A Covariant Metric Hypersheet Dyadic	$a_{\alpha\beta} = \mathbf{a}_\alpha \bullet \mathbf{a}_\beta$
∴	From 1, D8.3.2B Fundamental Base Vector Dot Product and A4.2.3 Substitution	$a_{\alpha\beta} = a_\alpha a_\beta \cos \theta_{\alpha\beta}$

Theorem 17.4.10 Hypersheet: Contravariant Metric Inner Product

Step	Hypothesis	$a^{\alpha\beta} = a^{\alpha} a^{\beta} \cos \theta^{\alpha\beta}$
1	From D7.1.3A Contravariant Metric Hypersheet Dyadic	$a^{\alpha\beta} = \mathbf{a}^{\alpha} \bullet \mathbf{a}^{\beta}$
∴	From 1, D8.3.2B Fundamental Base Vector Dot Product and A4.2.3 Substitution	$a^{\alpha\beta} = a^{\alpha} a^{\beta} \cos \theta^{\alpha\beta}$

Theorem 17.4.11 Hypersheet: Covariant Metric Inner Product is Symmetrical

Step	Hypothesis	$a_{\beta\alpha} = a_{\alpha\beta}$
1	From T17.4.8 Hypersheet: Covariant Metric Inner Product	$a_{\alpha\beta} = a_{\alpha} a_{\beta} \cos \theta_{\alpha\beta}$
2	From 1, DxE.1.9.2 Periodic Reduction Formulas for Multiples of Leading 90°, A4.2.2A Equality, A4.2.10 Commutative Multp and A4.2.2A Equality	$a_{\beta} a_{\alpha} \cos \theta_{\beta\alpha} = a_{\alpha} a_{\beta} \cos \theta_{\alpha\beta} = a_{\alpha\beta}$
∴	From 1, 2 and A4.2.3 Substitution	$a_{\beta\alpha} = a_{\alpha\beta}$

Theorem 17.4.12 Hypersheet: Contravariant Metric Inner Product is Symmetrical

Step	Hypothesis	$a^{\beta\alpha} = a^{\alpha\beta}$
1	From T17.4.9 Hypersheet: Contravariant Metric Inner Product	$a^{\alpha\beta} = b^{\alpha} b^{\beta} \cos \theta^{\alpha\beta}$
2	From 1, DxE.1.9.2 Periodic Reduction Formulas for Multiples of Leading 90°, A4.2.2A Equality, A4.2.10 Commutative Multp and A4.2.2A Equality	$b^{\beta} b^{\alpha} \cos \theta^{\beta\alpha} = b^{\alpha} b^{\beta} \cos \theta^{\alpha\beta} = a^{\alpha\beta}$
∴	From 1, 2 and A4.2.3 Substitution	$a^{\beta\alpha} = a^{\alpha\beta}$

Theorem 17.4.13 Hypersheet: Covariant-Contravariant Metric Curvilinear Coordinates

Step	Hypothesis	$a_{\alpha\beta} = a_{\alpha} a_{\beta}\, \delta_{\alpha\beta}$ EQ A	$a^{\alpha\beta} = a^{\alpha} a^{\beta}\, \delta^{\alpha\beta}$ EQ B
1	From T17.4.8 Hypersheet: Covariant Metric Inner Product and T17.4.9 Hypersheet: Contravariant Metric Inner Product	$a_{\alpha\beta} = a_{\alpha} a_{\beta} \cos(\theta_{\alpha\beta})$	$a^{\alpha\beta} = a^{\alpha} a^{\beta} \cos(\theta^{\alpha\beta})$
2	From 1 and A11.4.3P2(A, B) Orthogonal Manifold $\mathfrak{O}^n$	$a_{\alpha\beta} = a_{\alpha} a_{\beta} \cos(\frac{1}{2}\pi)$ for $\alpha \neq \beta$	$a^{\alpha\beta} = a^{\alpha} a^{\beta} \cos(\frac{1}{2}\pi)$ for $\alpha \neq \beta$
3	From 2 and DxE.1.6.7 for $\cos(\frac{1}{2}\pi)$	$a_{\alpha\beta} = a_{\alpha} a_{\beta}$ 0 for $\alpha \neq \beta$	$a^{\alpha\beta} = a^{\alpha} a^{\beta}$ 0 for $\alpha \neq \beta$
4	From 3 and A11.4.3P2(A, B) Orthogonal Manifold $\mathfrak{O}^n$	$a_{\alpha\beta} = a_{\alpha} a_{\beta} \cos(0)$ for $\alpha = \beta$	$a^{\alpha\beta} = a^{\alpha} a^{\beta} \cos(0)$ for $\alpha = \beta$
5	From 4 and DxE.1.6.1 for cos(0)	$a_{\alpha\beta} = a_{\alpha} a_{\beta}$ 1 for $\alpha = \beta$	$a^{\alpha\beta} = a^{\alpha} a^{\beta}$ 1 for $\alpha = \beta$
∴	From 3, 5 and D6.1.6 Identity Matrix defined by the Kronecker Delta	$a_{\alpha\beta} = a_{\alpha} a_{\beta}\, \delta_{\alpha\beta}$ EQ A	$a^{\alpha\beta} = a^{\alpha} a^{\beta}\, \delta^{\alpha\beta}$ EQ B

Theorem 17.4.14 Hypersheet: Covariant-Contravariant Kronecker Delta Summation of like Indices

Step	Hypothesis	$n = \sum_\alpha \delta_\alpha{}^\alpha$ EQ A	$n = \sum_\alpha \delta^\alpha{}_\alpha$ EQ B
1	From T6.1.6B Identity Matrix and evaluate with LxE.3.1.37 Sums of Powers of 1 for the First n-Integers along the diagonal of the matrix	$n = \sum_\alpha \delta^\alpha{}_\alpha$	
∴	From 1, T17.4.5 Hypersheet: Equivalency of Contra-Covariant Kronecker Delta, T10.2.5 Tool: Tensor Substitution and A10.2.14 Correspondence of Tensors and Tensor Coefficients	$n = \sum_\alpha \delta_\alpha{}^\alpha$ EQ A	$n = \sum_\alpha \delta^\alpha{}_\alpha$ EQ B

Theorem 17.4.15 Hypersheet: Determinate of Covariant Metric Dyadic

Step	Hypothesis	$a = \rho_u{}^2 \nu_c$
1	From D17.1.5 Covariant Metric Determinate for Hypersheet $\mathfrak{H}^{n-1}$, D17.1.2A Covariant Metric Dyadic, D8.3.2B Fundamental Base Vector Dot Product and A4.2.3 Substitution	$a = \vert\, a_\alpha\, a_\beta \cos \theta_{\alpha\beta} \,\vert$
2	From 1 and T6.9.11 Multiply a row or column of a determinant [A] by a scalar k	$a = \prod_\alpha a_\alpha \prod_\beta a_\beta \,\vert \cos \theta_{\alpha\beta} \,\vert$
3	From 2, D17.1.6 Surface Density Factor, D17.1.7B Cosine Surface Distortion Factor, A4.2.3 Substitution and interchanging dummy indices i = j	$a = \rho_u\, \rho_u\, \nu_c$
∴	From 4 and D4.1.17 Exponential Notation	$a = \rho_u{}^2 \nu_c$

Theorem 17.4.16 Hypersheet: Bounds for Cosine Hypersurface Distortion Factor

Step	Hypothesis	$0 < \nu_c \leq 1$ EQ A	$\nu_c = 1$ EQ B for orthogonal system
1	From D17.1.7 Cosine Surface Distortion Factor	$\nu_c = \vert \cos \theta_{\alpha\beta} \vert$	
2	From 1 and A11.4.3P2 Orthogonal Manifold $\mathfrak{O}^n$	$\nu_c = \vert \delta_{\alpha\beta} \vert$	
3	From 2 and C6.9.20.2 Determinant of a Identity Matrix	$\nu_c = 1$ for orthogonal system	
4	From 1, degenerate case any two column or rows are collinear, DxE.1.6.1 Cosine Zero Degrees and T6.9.10 A determinant with two identical rows or columns is zero	$\nu_c = 0$ a degenerate case, hence not permitted.	
∴	From 3, 4 and A4.2.16 The Trichotomy Law of Ordered Numbers	$0 < \nu_c \leq 1$ EQ A	$\nu_c = 1$ EQ B for orthogonal system

Theorem 17.4.17 Hypersheet: Covariant Metric Determinate Curvilinear Coordinates

Step	Hypothesis	$\sqrt{a} = \pm\rho_u$
1	T17.4.16B Hypersheet: Bounds for Cosine Hypersurface Distortion Factor	$\nu_c = 1$
1	From T17.4.15 Hypersheet: Determinate of Covariant Metric Dyadic	$a = \rho_u{}^2 \nu_c$
2	From 1, T4.10.1 Radicals: Uniqueness of Radicals, T4.10.7 Radicals: Reciprocal Exponent by Positive Square Root and T4.10.5 Radicals: Distribution Across a Product	$\sqrt{a} = \pm\sqrt{\rho_u{}^2}\sqrt{\nu_c}$
3	From 1, A4.2.3 Substitution and T4.10.3 Radicals: Identity Power Raised to a Radical	$\sqrt{a} = \pm\rho_u \sqrt{1}$
∴	From 3, T4.9.1 Rational Exponent: Integer of a Positive One, and A4.2.25 Reciprocal Exp and A4.2.12 Identity Multp	$\sqrt{a} = \pm\rho_u$

Theorem 17.4.18 Hypersheet: Density in Euclidian Hypersurface $\mathfrak{H}^{n-1}$

Step	Hypothesis	$\rho_u = 1$
1	From D17.1.6 Surface Density Factor	$\rho_u = \prod_{\alpha=1}^{(n-1)} a_\alpha$
2	From 1, A11.4.5P1 Euclidian Manifold $\mathfrak{E}^n$ and A4.2.3 Substitution	$\rho_u = \prod_\alpha (1)$
∴	From 1 and T4.8.1 Integer Exponents: Unity Raised to any Integer Value	$\rho_u = 1$

Theorem 17.4.19 Hypersheet: Covariant Metric Determinate in Euclidian Hypersurface $\mathfrak{H}^{n-1}$

Step	Hypothesis	$\sqrt{a} = \pm 1$ EQ A	$a = 1$ EQ B
1	From T17.4.17 Hypersheet: Covariant Metric Determinate Curvilinear Coordinates	$\sqrt{a} = \pm\rho_u$	$a^2 = \rho_u{}^2$
∴	From 1, T17.4.18 Hypersheet: Density in Euclidian Hypersurface $\mathfrak{H}^{n-1}$, A4.2.3 Substitution and T4.8.1 Integer Exponents: Unity Raised to any Integer Value	$\sqrt{a} = \pm 1$ EQ A	$a^2 = 1$ EQ B

Theorem 17.4.20 Positional and Normal Vector Undergoing Change in Position

Step	Hypothesis	$d\mathbf{r} / ds = \sum_\alpha \mathbf{r}_{,\alpha}\, du_\alpha / ds$ EQ A	$d\mathbf{N} / ds = \sum_\alpha \mathbf{N}_{,\alpha}\, du_\alpha / ds$ EQ B
1	From O17.4.2 A Positional and Normal Vector Exactly Undergoing Change and TK.5.3 "Chain Rule for Exact Differential of a Vector"	$d\mathbf{r} / ds = \sum_\alpha \mathbf{r}_{,\alpha}\, du_\alpha / ds$	
2	From O17.4.2 A Positional and Normal Vector Exactly Undergoing Change and TK.5.3 "Chain Rule for Exact Differential of a Vector"	$d\mathbf{N} / ds = \sum_\alpha \mathbf{N}_{,\alpha}\, du_\alpha / ds$	
∴	From 1 and 2	$d\mathbf{r} / ds = \sum_\alpha \mathbf{r}_{,\alpha}\, du_\alpha / ds$ EQ A	$d\mathbf{N} / ds = \sum_\alpha \mathbf{N}_{,\alpha}\, du_\alpha / ds$ EQ B

Theorem 17.4.21 First Fundamental Quadratic Form of the Hypersurface $\mathfrak{H}^{n-1}$

Step	Hypothesis	$d\mathbf{r} \bullet d\mathbf{r} = \sum_\alpha\sum_\beta a_{\alpha\beta}\, du_\alpha\, du_\beta$ EQ A	$ds^2 = \sum_\alpha\sum_\beta a_{\alpha\beta}\, du_\alpha\, du_\beta$ EQ B
1	From T8.3.2 Dot Product: Equality	$d\mathbf{r} \bullet d\mathbf{r} = d\mathbf{r} \bullet d\mathbf{r}$	
2	From 1, T17.4.20A Positional and Normal Vector Undergoing Change in Position and A7.9.2 Substitution of Vectors	$d\mathbf{r} \bullet d\mathbf{r} = (\sum_\alpha \mathbf{r}_{,\alpha}\, du_\alpha) \bullet (\sum_\beta \mathbf{r}_{,\beta}\, du_\beta)$	
3	From 2 T17.1.4 Contravariant Bases Vectors for a Hypersheet and A7.9.2 Substitution of Vectors	$d\mathbf{r} \bullet d\mathbf{r} = (\sum_\alpha \mathbf{a}_\alpha\, du_\alpha) \bullet (\sum_\beta \mathbf{a}_\beta\, du_\beta)$	
∴	From 3, T12.3.13A Dot Product of Covariant Vectors and A17.5.1 Theorem Construction on Manifolds of Different Dimension	$d\mathbf{r} \bullet d\mathbf{r} = \sum_\alpha\sum_\beta a_{\alpha\beta}\, du_\alpha\, du_\beta$ EQ A	$ds^2 = \sum_\alpha\sum_\beta a_{\alpha\beta}\, du_\alpha\, du_\beta$ EQ B

Theorem 17.4.22 Hypersheet Fundamental Quadratic: Covariant Integration over Arc Length

1g	Given	$u_\alpha = w_\alpha(s)$
Step	Hypothesis	$\Delta\sigma = \pm \int_{s_0}^{s} \sqrt{[\Sigma_\alpha \Sigma_\beta a_{\alpha\beta} (du^\alpha / ds)(du^\beta / ds)]}\, ds$
1	From T17.4.20B First Fundamental Quadratic Form of the Hypersurface $\mathfrak{H}^{n-1}$	$d\sigma^2 = \Sigma_\alpha \Sigma_\beta a_{\alpha\beta} du^\alpha du^\beta$
2	From 1, TK.3.4B Single Variable Chain Rule, A4.2.10 Commutative Multp, D4.1.17 Exponential Notation and A4.2.14 Distribution	$d\sigma^2 = [\Sigma_\alpha \Sigma_\beta a_{\alpha\beta} (du^\alpha / ds)(du^\beta / ds)]\, ds^2$
3	From 1, T4.10.7 Radicals: Reciprocal Exponent by Positive Square Root, T4.10.5 Radicals: Distribution Across a Product and T4.10.3 Radicals: Identity Power Raised to a Radical	$d\sigma = \pm\sqrt{[\Sigma_\alpha \Sigma_\beta a_{\alpha\beta} (du^\alpha / ds)(du^\beta / ds)]}\, ds$
4	From 2, TK.4.1 Integrals: Uniqueness of Integration, D4.1.20 Negative Coefficient and LxK.4.1.3 Integrable on the integrand with a constant	$\int_{s_0}^{s} d\sigma = \pm \int_{s_0}^{s} \sqrt{[\Sigma_\alpha \Sigma_\beta a_{\alpha\beta} (du^\alpha / ds)(du^\beta / ds)]}\, ds$
5	From 4 and LxK.4.2.1 Integration of powers for n = 0 and DxK.4.1.4 The Limit of an Integral Having a Constant Area Under the Curve	$s - s_0 = \pm \int_{s_0}^{s} \sqrt{[\Sigma_\alpha \Sigma_\beta a_{\alpha\beta} (du^\alpha / ds)(du^\beta / ds)]}\, ds$
∴	From 5 and DxK.3.1.1 Delta: Increment by Difference	$\Delta s = \pm \int_{s_0}^{s} \sqrt{[\Sigma_\alpha \Sigma_\beta a_{\alpha\beta} (du^\alpha / ds)(du^\beta / ds)]}\, ds$

Theorem 17.4.23 Hypersheet Fundamental Quadratic: Contravariant Integration over Arc Length

1g	Given	$u_\alpha = w_\alpha(s)$
Step	Hypothesis	$\Delta s = \pm \int_{s_0}^{s} \surd\, [\Sigma_\alpha \Sigma_\beta\, a^{\alpha\beta} (du_\alpha / ds)(du_\beta / ds)]\, ds$
1	From A17.4.1B Contravariant Base Vectors and Displacement	$d\sigma^2 \equiv \Sigma_\alpha \Sigma_\beta\, a^{\alpha\beta}\, du_\alpha\, du_\beta$
2	From 1, TK.3.4B Single Variable Chain Rule, A4.2.10 Commutative Multp, D4.1.17 Exponential Notation and A4.2.14 Distribution	$d\sigma^2 = [\Sigma_\alpha \Sigma_\beta\, a^{\alpha\beta} (du_\alpha / ds)(du_\beta / ds)]\, ds^2$
3	From 1, T4.10.7 Radicals: Reciprocal Exponent by Positive Square Root, T4.10.5 Radicals: Distribution Across a Product and T4.10.3 Radicals: Identity Power Raised to a Radical	$d\sigma = \pm \surd\, [\Sigma_\alpha \Sigma_\beta\, a^{\alpha\beta} (du_\alpha / ds)(du_\beta / ds)]\, ds$
4	From 2, TK.4.1 Integrals: Uniqueness of Integration, D4.1.20 Negative Coefficient and LxK.4.1.3 Integrable on the integrand with a constant	$\int_{s_0}^{s} d\sigma = \pm \int_{s_0}^{s} \surd\, [\Sigma_\alpha \Sigma_\beta\, a^{\alpha\beta} (du_\alpha / ds)(du_\beta / ds)]\, ds$
5	From 4 and LxK.4.2.1 Integration of powers for n = 0 and DxK.4.1.4 The Limit of an Integral Having a Constant Area Under the Curve	$s - s_0 = \pm \int_{s_0}^{s} \surd\, [\Sigma_\alpha \Sigma_\beta\, a^{\alpha\beta} (du_\alpha / ds)(du_\beta / ds)]\, ds$
∴	From 5 and DxK.3.1.1 Delta: Increment by Difference	$\Delta s = \pm \int_{s_0}^{s} \surd\, [\Sigma_\alpha \Sigma_\beta\, a^{\alpha\beta} (du_\alpha / ds)(du_\beta / ds)]\, ds$

Observation 17.4.3 Open Path: [±] Direction of Tracing the Arc Path on a Hypersheet

If the path is traced in a positive way it will be plus and the opposite negative.

$$\Delta s = \begin{cases} +\Delta s_\sigma & \text{for } s > s_0 \text{ traced in a positive direction} \\ -\Delta s_\sigma & \text{for } s < s_0 \text{ traced in the opposite direction} \end{cases}$$

Observation 17.4.4 Closed Path: Integration of Periodic Arc-Length on a Hypersheet

Let the arc length be on a closed path with one complete path having traveled an arc L_c. Within a peroid, however the path traveled may only be a fraction [f], hence the arc length completed would be $\Delta s = f\, L_c$.

Theorem 17.4.24 Hypersheet Closed Path: Integration over Periodic Arc-Length

1g	Given	see Figure 11.6.4 Periodicity On A Closed Vector Curve	
2g		$\Delta s \equiv f\, L_c$	
3g		$Ł \equiv L_c / 2\pi$	
Step	Hypothesis	$\Delta s = n\, \theta\, Ł$	where [+] is taken to be CCW and [–] CW periodic rotation and n = 0, ±1, ±2, ±3, …
1	From 1g, A4.2.12 Identity Multp, A4.2.13 Inverse Multp, A4.2.10 Commutative Multp and A4.2.11 Associative Multp	$\Delta s = 2\pi f\, (L_c / 2\pi)$	
2	From D5.1.36C Subtended Arc length of a Cord	$\theta \equiv 2\pi f$ in radians	
3	From 1g, 2g, 3g, 2 and A4.2.3 Substitution	$\Delta s = \theta\, Ł$	
4	From 3; Now if the arc length loops around one time, $\Delta s = 1\, \theta Ł$, if two times $\Delta s = 2\, \theta Ł$,, if three times $\Delta s = 3\, \theta Ł$ and n-times $\Delta s = n\, \theta Ł$.	$\Delta s = n\, \theta\, Ł$	
∴	From 4, T17.4.20 Hypersheet Fundamental Quadratic: Contravariant Integration over Arc Length; Integration over Arc Length and rotation convention is based on quadrature coordinate system notation.	$\Delta s = n\, \theta\, Ł$	where [+] is taken to be CCW and [–] CW periodic rotation and n = 0, ±1, ±2, ±3, …

Theorem 17.4.25 Hypersheet Closed Path: Covariant Integration over Arc Length

1g	Given	Integration of arc length is a closed path.
Step	Hypothesis	$n\, \theta = \dfrac{\pm 1}{Ł} \oint_c \sqrt{[\Sigma_\alpha \Sigma_\beta\, a_{\alpha\beta}\, (du^\alpha / ds)(du^\beta / ds)]}\, ds$
1	From T17.4.22 Hypersheet Fundamental Quadratic: Covariant Integration over Arc Length	$\Delta s = \pm \int_{s_0}^{s} \sqrt{[\Sigma_\alpha \Sigma_\beta\, a_{\alpha\beta}\, (du^\alpha / ds)(du^\beta / ds)]}\, ds$
2	From 1g, 1 and Hypersheet: Closed Path Integration over Periodic Arc-Length	$n\, \theta\, Ł = \pm \oint_c \sqrt{[\Sigma_\alpha \Sigma_\beta\, a_{\alpha\beta}\, (du^\alpha / ds)(du^\beta / ds)]}\, ds$
∴	From 2, D4.1.20A Negative Coefficient and T4.4.6B Equalities: Left Cancellation by Multiplication	$n\, \theta = \dfrac{\pm 1}{Ł} \oint_c \sqrt{[\Sigma_\alpha \Sigma_\beta\, a_{\alpha\beta}\, (du^\alpha / ds)(du^\beta / ds)]}\, ds$

Theorem 17.4.26 Hypersheet Closed Path: Contravariant Integration over Arc Length

1g	Given	Integration of arc length is a closed path.
Step	Hypothesis	$n\,\theta = \dfrac{\pm 1}{\text{Ł}} \oint_c \sqrt{[\Sigma_\alpha \Sigma_\beta\, a^{\alpha\beta}\, (du_\alpha / ds)(du_\beta / ds)]}\, ds$
1	From T17.4.20 Hypersheet Fundamental Quadratic: Contravariant Integration over Arc Length	$\Delta s = \pm \int_{s_0}^{s} \sqrt{[\Sigma_\alpha \Sigma_\beta\, a^{\alpha\beta}\, (du_\alpha / ds)(du_\beta / ds)]}\, ds$
2	From 1g, 1 and T17.4.21 Hypersheet Closed Path: Integration over Periodic Arc-Length	$n\,\theta\,\text{Ł} = \pm \oint_c \sqrt{[\Sigma_\alpha \Sigma_\beta\, a^{\alpha\beta}\, (du_\alpha / ds)(du_\beta / ds)]}\, ds$
∴	From 2, D4.1.20A Negative Coefficient and T4.4.6B Equalities: Left Cancellation by Multiplication	$n\,\theta = \dfrac{\pm 1}{\text{Ł}} \oint_c \sqrt{[\Sigma_\alpha \Sigma_\beta\, a^{\alpha\beta}\, (du_\alpha / ds)(du_\beta / ds)]}\, ds$

Section 17.5 Differentiation of Vector and Metric Tensors on a Hypersurface $\mathfrak{H}^{n-1}$

Definition 17.5.1 Hypersurface Christoffel Symbol of the First Kind

$$\Gamma_{\alpha\beta,\kappa} \equiv [\alpha\beta,\kappa] \equiv \tfrac{1}{2} (\partial a_{\alpha\kappa} / \partial u_\beta + \partial a_{\beta\kappa} / \partial u_\alpha - \partial a_{\alpha\beta} / \partial u_\kappa)$$

Definition 17.5.2 Hypersurface Christoffel Symbol of the Second Kind

$$\Gamma_{\alpha\beta}{}^{\kappa} \equiv \left\{ \begin{matrix} \kappa \\ \alpha\beta \end{matrix} \right\} \equiv \{\alpha\beta,\kappa\} \equiv \Gamma_{\alpha\beta,w}\, a^{w\kappa}$$

Definition 17.5.3 Hypersurface Christoffel Symbol of the Third Kind

$$\Gamma^{\kappa\alpha}{}_{\beta} \equiv \left\{ \begin{matrix} \kappa\kappa \\ \beta \end{matrix} \right\} \equiv \{\beta,\kappa\alpha\} \equiv \Gamma_{w\beta}{}^{\kappa}\, a^{w\alpha}$$

Observation 17.5.1 Correspondence Riemann Space $\mathfrak{R}^n$ to Hypersurface $\mathfrak{H}^{n-1}$ Theorems

A space, is a space, is a space regardless of its dimension, so it would follow that as long as going from one space to another maintains the same manifold properties the construction of a proof should remain the same independent of dimension. So theorems T12.3.13 to T16.3.3 correspond one-to-one to T17.5.1 to T17.5.64. However if a proof does not remain inside the manifold and crosses over dimensional boundaries of different dimension that argument would not remain true.

Axiom 17.5.1 Theorem Construction on Manifolds of Different Dimension

Construction of a theorem within boundaries of a manifold will remain the same while crossing from one manifold to another independent of their dimensions.

Table 17.5.1 Dot Product and Magnitude of Vectors on a Hypersurface $\mathfrak{H}^{n-1}$

Theorem	Source	Equation	
Theorem 17.5.1 Dot Product of Covariant Vectors	T12.3.13	$(\Sigma_\alpha A^\alpha \mathbf{a}_\alpha) \bullet (\Sigma_\alpha B^\alpha \mathbf{a}_\alpha) = \Sigma_\alpha \Sigma_\beta A^\alpha B^\beta a_{\alpha\beta}$ EQ A	$\cos(\theta_{ab}) = \Sigma_\alpha \Sigma_\beta n^\alpha m^\beta a_{\alpha\beta}$ EQ B
Theorem 17.5.2 Dot Product of Contravariant Vectors	T12.3.14	$(\Sigma_\alpha A_\alpha \mathbf{a}^\alpha) \bullet (\Sigma_\alpha B_\alpha \mathbf{a}^\alpha) = \Sigma_\alpha \Sigma_\beta A_\alpha B_j a^{\alpha\beta}$ EQ A	$\cos(\theta^{ab}) = \Sigma_\alpha \Sigma_\beta n_\alpha m_\beta a^{\alpha\beta}$ EQ B
Theorem 17.5.3 Dot Product of Mixed Vectors (Tensor Contraction)	T12.3.15	$(\Sigma_\alpha A^\alpha \mathbf{a}_\alpha) \bullet (\Sigma_\alpha B_\alpha \mathbf{a}^\alpha) = \Sigma_\alpha \Sigma_\beta A^\alpha B_j \delta_\alpha{}^\beta$ EQ A	$(\Sigma_\alpha A^\alpha \mathbf{a}_\alpha) \bullet (\Sigma_\alpha B_\alpha \mathbf{a}^\alpha) = \Sigma_\alpha A^\alpha B_\alpha$ EQ B
Theorem 17.5.4 Magnitude of Covariant Vector	T12.3.16	$A^2 = \Sigma_\alpha \Sigma_\beta a_{\alpha\beta} A^\alpha A^j$ EQ A	$A = \sqrt{\Sigma_\alpha \Sigma_\beta a_{\alpha\beta} A^\alpha A^\beta}$ EQ B
Theorem 17.5.5 Magnitude of Contravariant Vector	T12.3.17	$A^2 = \Sigma_\alpha \Sigma_\beta a^{\alpha\beta} A_\alpha A_\beta$ EQ A	$A = \sqrt{\Sigma_\alpha \Sigma_\beta a^{\alpha\beta} A_\alpha A_\beta}$ EQ B
Theorem 17.5.6 Magnitude of Normalized Vector	T12.3.18	$1 = \Sigma_\alpha n^\alpha n_\alpha$ EQ A	$1 = \Sigma_\alpha n_\alpha n^\alpha$ EQ B

Table 17.5.2 Differential Theorems of Bases Vectors on a Hypersurface $\mathfrak{H}^{n-1}$

Theorem	Source	Equation	
Theorem 17.5.7 Derivative of a Covariant Vector Under Change	T12.5.1	$\mathbf{A}_{;\alpha} = \sum_\kappa A^{\kappa,\alpha}\, \mathbf{a}_\kappa + A^\kappa\, \mathbf{a}_{\kappa,\alpha}$	
Theorem 17.5.8 Derivative of a Contravariant Vector Under Change	T12.5.2	$\mathbf{A}^{;\alpha} = \sum_\kappa A_{\kappa,\alpha}\, \mathbf{a}^\kappa + A_\kappa\, \mathbf{a}^{\kappa,\alpha}$	
Theorem 17.5.9 Christoffel Symbol of the First Kind	T12.5.3	$\Gamma_{\alpha\beta,\kappa} =$ $\frac{1}{2}(\partial a_{\alpha\kappa} / \partial u_\beta + \partial a_{\beta\kappa} / \partial u_\alpha - \partial a_{\alpha\beta} / \partial u_\kappa)$	
Theorem 17.5.10 Covariant Derivative: Christoffel Symbol of the First Kind	T12.5.4	$\partial(\mathbf{a}_\alpha) / \partial u_\beta = \sum_\kappa \Gamma_{\alpha\beta,\kappa}\, \mathbf{a}^\kappa$	
Theorem 17.5.11 Covariant Derivative: Covariant Bases Vector	T12.5.5	$\partial(\mathbf{a}_\alpha) / \partial u_\beta = \sum_\kappa \Gamma_{\alpha\beta}{}^\kappa\, \mathbf{a}_\kappa$	
Theorem 17.5.12 Contravariant Derivative: Contravariant Bases Vector	T12.5.6	$\partial(\mathbf{a}^\alpha) / \partial u_\beta = -\sum_\kappa \Gamma_{\kappa\beta}{}^\alpha\, \mathbf{a}^\kappa$	
Theorem 17.5.13 Contravariant Derivative: Covariant Bases Vector	T12.5.7	$\partial(\mathbf{a}^\alpha) / \partial u_\beta = -\sum_w \sum_\kappa \Gamma_{\kappa\beta}{}^\alpha\, g^{\kappa w}\, \mathbf{b}_w$	
Theorem 17.5.14 Christoffel Symbol: Second to the First Kind	T12.5.8	$\Gamma_{\alpha\beta,\kappa} = \sum_w \Gamma_{\alpha\beta}{}^w\, a_{w\kappa}$ EQ A	$\Gamma_{\alpha\beta,\kappa} = \sum_w a_{w\kappa} \Gamma_{\alpha\beta}{}^w$ EQ B
Theorem 17.5.15 Christoffel Symbol: First Kind Symmetry of First Two Indices	T12.5.9	$\Gamma_{\alpha\beta,\kappa} = \Gamma_{\beta\alpha,\kappa}$	
Theorem 17.5.16 Christoffel Symbol: First Kind Duplicate First Indices	T12.5.10	$\Gamma_{\eta\eta,\kappa} = a_{\eta\kappa,\eta} - \frac{1}{2} a_{\eta\eta,\kappa}$	
Theorem 17.5.17 Christoffel Symbol: First Kind Duplicate Last Two Indices	T12.5.11	$\Gamma_{\alpha\eta,\eta} = \frac{1}{2} a_{\eta\eta,\alpha}$	
Theorem 17.5.18 Christoffel Symbol: First Kind Asymmetry of Last Two Indices	T12.5.12	$\Gamma_{\alpha\kappa,\beta} = a_{\kappa\beta,\alpha} - \Gamma_{\alpha\beta,\kappa}$ EQ A	$\Gamma_{\alpha\beta,\kappa} + \Gamma_{\alpha\kappa,\beta} = a_{\kappa\beta,\alpha}$ EQ B
Theorem 17.5.19 Christoffel Symbol: Second Kind Symmetry for First Two Indices	T12.5.13	$\Gamma_{\alpha\beta}{}^\kappa = \Gamma_{\beta\alpha}{}^\kappa$	
Theorem 17.5.20 Christoffel Symbol: Exchanging Index between Second and First Symbols	T12.5.14	$\sum_w \Gamma_{\alpha\beta,w}\, \Gamma_{u\kappa}{}^w =$ $\sum_w \Gamma_{\alpha\beta}{}^w\, \Gamma_{u\kappa,w}$ EQ A	$\sum_w \Gamma_{\alpha\beta,w}\, \Gamma_{u\kappa}{}^w =$ $\sum_w \Gamma_{u\kappa,w}\, \Gamma_{\alpha\beta}{}^w$ EQ B

Theorem 17.5.21 Christoffel Symbol: Double Indices of the Second Kind	T12.5.15	$\Gamma_{\alpha\eta}{}^{\eta} = ½ \sum_{\kappa} g_{\eta\kappa,\alpha}\, g^{\kappa\eta}$
Theorem 17.5.22 Christoffel Symbol: Alternate Double Indices of the Second Kind	T12.5.16	$\Gamma_{\alpha\eta}{}^{\eta} = ½ \sum_{w} g_{\eta w,\alpha}\, g^{w\eta}$
Theorem 17.5.23 Christoffel Symbol: Of the First Kind in Orthogonal Space	T12.5.17	$\Gamma_{\alpha\beta,\kappa} = 0$ orthogonal independent coordinates
Theorem 17.5.24 Christoffel Symbol: Of the Second Kind in Orthogonal Space	T12.5.18	$\Gamma_{\alpha\beta}{}^{\kappa} = 0$ orthogonal independent coordinates

Table 17.5.3 Differentiation for a Tensor of Rank-1 on a Hypersurface $\mathfrak{H}^{n-1}$

Theorem	Source	Equation	
Theorem 17.5.25 Differentiation of Tensor: Rank-0	T12.6.1	$A_{;\alpha} = A_{,\alpha}$ EQ A	$A^{;\alpha} = A_{,\alpha}$ EQ B
Theorem 17.5.26 Differentiation of Tensor: Rank-1 Covariant	T12.6.2	$\mathbf{A}_{;\beta} = \sum_{\alpha} A^{\alpha;\beta}\, \mathbf{a}_{\alpha}$ EQ A	$\mathbf{A}_{;\beta} =$ $\sum_{\alpha} (A^{\alpha,\beta} + \sum_{\kappa} A^{\kappa}\, \Gamma_{\kappa\beta}{}^{\alpha})\, \mathbf{a}_{\alpha}$ EQ B
Theorem 17.5.27 Differentiation of Tensor: Rank-1 Contravariant	T12.6.3	$\mathbf{A}^{;\beta} = \sum_{\alpha} A_{\alpha;\beta}\, \mathbf{a}^{\alpha}$ EQ A	$\mathbf{A}^{;\beta} =$ $\sum_{\alpha} (A_{\alpha,\beta} - \sum_{\kappa} A_{\kappa}\, \Gamma_{\alpha\beta}{}^{\kappa})\, \mathbf{a}^{\alpha}$ EQ B
Theorem 17.5.28 Differentiation of Tensor: Rank-1 Uniqueness of Differentiation	T12.6.4	$\sum_{\alpha} A^{\alpha;\beta}\, \mathbf{a}_{\alpha} =$ $\sum_{\alpha} B^{\alpha;\beta}\, \mathbf{a}_{\alpha}$ EQ A	$\sum_{\alpha} A_{\alpha;\beta}\, \mathbf{a}^{\alpha} =$ $\sum_{\alpha} B_{\alpha;\beta}\, \mathbf{a}^{\alpha}$ EQ B
Theorem 17.5.29 Differentiation of Tensor: Rank-1 Distribution of Differentiation Across Vectors	T12.6.5	$\sum_{\alpha} A^{\alpha;\beta}\, \mathbf{a}_{\alpha} =$ $\sum_{i} B^{\alpha;\beta}\, \mathbf{a}_{\alpha}$ EQ A	$\sum_{\alpha} A_{\alpha;\beta}\, \mathbf{a}^{\alpha} =$ $\sum_{\alpha} B_{\alpha;\beta}\, \mathbf{a}^{\alpha}$ EQ B

Table 17.5.4 Differentiation of Metric Tensors on a Hypersurface $\mathfrak{H}^{n-1}$

Theorem	Source	Equation	
Theorem 17.5.30 Differential of the Contravariant Cofactor to the Metric Dyadic	T12.8.1	$\partial (A^{\alpha\kappa}) / \partial a_{\alpha\beta} = 0$	
Theorem 17.5.31 Differentiation of the Metric Tensor of the First Kind	T12.8.2	$(a_{\alpha\beta})_{,\kappa} = \Gamma_{\alpha k,\beta} + \Gamma_{\beta\kappa,\alpha}$	
Theorem 17.5.32 Differentiation of the Covariant Metric Tensor (Ricci's Theorem-A)	T12.8.3	$(a_{\alpha\beta})_{,\kappa} = a_{\beta w}\, \Gamma_{\alpha\kappa}{}^{w} + a_{\alpha w}\, \Gamma_{\beta\kappa}{}^{w}$ EQ A	$(a_{\alpha\beta})_{;\kappa} = \mathbf{0}$ EQ B
Theorem 17.5.33 Simple Differential of the Contravariant Metric Tensor	T12.8.4	$(a^{\alpha\beta})_{,\kappa} = -a^{\alpha w}\, \Gamma_{w\kappa}{}^{\beta} - a^{w\beta}\, \Gamma_{w\kappa}{}^{\alpha}$	
Theorem 17.5.34 Differentiation of the Contravariant Metric Tensor (Ricci's Theorem-B)	T12.8.5	$(a^{\alpha\beta})_{,\kappa} = -\Gamma_{w\kappa}{}^{\beta}\, a^{w\alpha} - \Gamma_{w\kappa}{}^{\alpha}\, a^{w\beta}$ EQ A	$(a^{\alpha\beta})^{;\kappa} = \mathbf{0}$ EQ B
Theorem 17.5.35 Derivative Covariant Determinate to the Covariant Metric	T12.8.6	$\partial\, a / \partial\, a_{\alpha\beta} = A^{\alpha\beta}$	
Theorem 17.5.36 Derivative of the Natural Log of the Covariant Determinate	T12.8.7	$\partial \ln \lvert\sqrt{a_e}\rvert / \partial u_\alpha = \Gamma_{w\alpha}{}^{w}$ EQ A	$\partial \ln \lvert\sqrt{a_e}\rvert / \partial u_\alpha = \Gamma_{\alpha w}{}^{w}$ EQ B

Table 17.5.5 Contraction by Metric on a Hypersurface $\mathfrak{H}^{n-1}$

Theorem	Source	Equation	
Theorem 17.5.37 Contracted Covariant Bases Vector with Contravariant Vector	T12.3.3	$A_\alpha = \Sigma_\beta\, A_\beta\, \delta_\alpha{}^\beta$ EQ A	$A_\alpha = \Sigma_\beta\, A_\beta\, \delta^\beta{}_\alpha$ EQ B
Theorem 17.5.38 Contracted Contravariant Bases Vector with Covariant Vector	T12.3.8	$A^\alpha = \Sigma_\beta\, A^\beta\, \delta^\alpha{}_\beta$ EQ A	$A^\alpha = \Sigma_\beta\, \delta_\beta{}^\alpha\, A^\beta$ EQ B
Theorem 17.5.39 Contracted Contravariant Bases Vector with Covariant Vector	T12.3.4	$A^\beta = \Sigma_\alpha\, A^\alpha\, \delta_\alpha{}^\beta$ EQ A	$A^\beta = \Sigma_\alpha\, \delta^\beta{}_\alpha\, A^\alpha$ EQ B
Theorem 17.5.40 Metric Transformation from Covariant to Contravariant Coefficient	T12.3.9	$A_\alpha = \Sigma_\beta\, a_{\alpha\beta}\, B^\beta$ EQ A	$A_\alpha = \Sigma_\beta\, B^\beta\, a_{\alpha\beta}$ EQ B
Theorem 17.5.41 Metric Transformation from Contravariant to Covariant Coefficient	T12.3.5	$A^\alpha = \Sigma_\beta\, a^{\alpha\beta}\, B_\beta$ EQ A	$A^\alpha = \Sigma_\beta\, B_\beta\, a^{\alpha\beta}$ EQ B
Theorem 17.5.42 Transformation from Contravariant to Covariant Basis Vector	T12.3.6	$\mathbf{a}_\alpha = \Sigma_\beta\, a_{\alpha\beta}\, \mathbf{a}^\beta$ EQ A	$\mathbf{a}_\alpha = \Sigma_\beta\, \mathbf{a}^\beta\, a_{\alpha\beta}$ EQ A
Theorem 17.5.43 Transformation from Covariant to Contravariant Basis Vector	T12.3.10	$\mathbf{a}^\alpha = \Sigma_\beta\, a^{\alpha\beta}\, \mathbf{a}_\beta$ EQ A	$\mathbf{a}^\alpha = \Sigma_\beta\, \mathbf{a}_\beta\, a^{\alpha\beta}$ EQ B

Table 17.5.6 Generalization of the Kronecker Delta on a Hypersurface $\mathfrak{H}^{n-1}$

Theorem	Source	Equation	
Theorem 17.5.44 Differentiated Kronecker Delta and Fundamental Tensors	T12.9.1	$\mathbf{0} = (a_{\alpha w})_{;\kappa}\, a^{w\beta} + a_{\alpha w}\, (a^{w\beta})_{;\kappa}$ EQ A	$(\delta_\alpha{}^\beta)_{;\kappa} = \mathbf{0}$ EQ B
Theorem 17.5.45 Differentiation of Kronecker Delta and Covariant Coefficient	T12.9.2	$(\delta^w{}_\alpha\, A_w)_{;r} = \delta^w{}_\alpha\, A_{w;r}$	
Theorem 17.5.46 Differentiation of Kronecker Delta and Contravariant Tensor	T12.9.3	$(\delta_\alpha{}^\beta\, A^\alpha)^{;r} = \delta_\alpha{}^\beta\, A^{\alpha;r}$	

Table 17.5.7 Fundamental Tensor Distribution Under Tenor Differentiation on a Hypersurface $\mathfrak{H}^{n-1}$

Theorem	Source	Equation
Theorem 17.5.47 Differentiation of Covariant Metric and Covariant Coefficient	C12.8.5.1	$(a_{w\alpha}\, A_w)_{;r} = a_{w\alpha}\, A_{w;r}$
Theorem 17.5.48 Differentiation of Covariant Metric and Contravariant Coefficient	C12.8.5.2	$(a_{w\alpha}\, A^w)^{;r} = a_{w\alpha}\, A^{w;r}$
Theorem 17.5.49 Differentiation of Covariant Metric and Mixed Coefficient	C12.8.5.3	$(a_{w\kappa}\, A_\kappa{}^w)^{;r} = a_{w\kappa}\, A_\kappa{}^{w;r}$
Theorem 17.5.50 Differentiation of Contravariant Metric and Covariant Coefficient	C12.8.5.4	$(a^{w\alpha}\, A_w)_{;r} = a^{w\alpha}\, A_{w;r}$
Theorem 17.5.51 Differentiation of Contravariant Metric and Contravariant Coefficient	C12.8.5.5	$(a^{w\alpha}\, A^w)^{;r} = a^{w\alpha}\, A^{w;r}$
Theorem 17.5.52 Differentiation of Contravariant Metric and Mixed Coefficient	C12.8.5.6	$(a^{w\kappa}\, A_\kappa{}^w)^{;r} = a^{w\kappa}\, A_\kappa{}^{w;r}$

Table 17.5.8 Affine Vector Fields on a Hypersurface $\mathfrak{H}^{n-1}$

Theorem	Source	Equation
Theorem 17.5.53 Affine Field: Intrinsic Covariant Differential along a Spatial Curve	T16.3.1	$0 = A^{\alpha:s} + \sum_w \Gamma_{w\kappa}{}^\alpha\, A^w u^{\kappa:s}$
Theorem 17.5.54 Affine Field: Intrinsic Contravariant Differential along a Time-Line	T16.3.2	$0 = A_{\alpha:t} - \sum_\kappa \sum_w \Gamma_{\alpha\kappa}{}^w\, A_w u_{\kappa:t}$
Theorem 17.5.55 Affine Field: Affine Spatial Curve	T16.3.3	$0 = u^{\alpha:s,s} + \sum_\kappa \sum_w \Gamma_{w\kappa}{}^\alpha\, u^{w:s}\, u^{\kappa:s}$

Table 17.5.9 Riemann-Christoffel Field Tensors on a Hypersurface $\mathfrak{H}^{n-1}$

Theorem	Source	Equation
Theorem 17.5.56 Vector Relation to a Contravariant Riemann-Christoffel Tensor	T14.1.10	$\mathbf{A}_{,\alpha,\beta} - \mathbf{A}_{,\beta,\alpha} = \sum_\gamma \sum_p A_p R^p{}_{\alpha\beta\gamma} \mathbf{b}^\gamma$
Theorem 17.5.57 Riemann-Christoffel Tensor First Kind; 3-Index Cyclic Identity	T14.2.7	$R_{p\alpha\beta\gamma} + R_{p\beta\gamma\alpha} + R_{p\gamma\alpha\beta} = \mathbf{0}$
Theorem 17.5.58 Bianchi Identity Covariant	T14.3.1	$R_{p\alpha\beta\gamma;\eta} + R_{p\alpha\gamma\eta;\beta} + R_{p\alpha\eta\beta;\gamma} = \mathbf{0}$
Theorem 17.5.59 Bianchi Identity Contravariant	T14.3.2	$R^p{}_{\alpha\beta\gamma;\eta} + R^p{}_{\alpha\gamma\eta;\beta} + R^p{}_{\alpha\eta\beta;\gamma} = \mathbf{0}$
Theorem 17.5.60 Riemann-Christoffel Tensor Rank-2: Contraction of the Bianchi Identity	T14.4.4	$(R_\eta{}^\gamma - ½\, \delta^\gamma{}_\eta R)_{;\gamma} = \mathbf{0}$
Theorem 17.5.61 Riemann-Christoffel Tensor Rank-2: Einstein's Mixed Tensor	T14.4.5	$T_\eta{}^\gamma\, \xi \equiv R_\eta{}^\gamma - ½\, \delta^\gamma{}_\eta R$
Theorem 17.5.62 Riemann-Christoffel Tensor Rank-2: Einstein's Covariant Tensor	T14.4.6	$T_{\rho\eta}\, \xi = R_{\rho\eta} - ½\, a_{\rho\eta} R$
Theorem 17.5.63 Riemann-Christoffel Tensor Rank-0: Contraction of Einstein's Covariant Tensor	T14.4.7	$T\, \xi = -(½\, n - 1)\, R$
Theorem 17.5.64 Riemann-Christoffel Tensor Rank-2: Einstein's Contravariant Tensor	T14.4.8	$T_{\rho\eta}\, \xi = R_{\rho\eta} - ½\, a_{\rho\eta} R$
Theorem 17.5.65 Riemann-Christoffel Tensor Second Kind; Transpose last two indices	T14.1.15	$R^a{}_{\alpha\beta\gamma} = -R^a{}_{\alpha\gamma\beta}$
Theorem 17.5.66 Riemann-Christoffel Tensor Second Kind; Duplicate last two indices	T14.1.16	$R^a{}_{\alpha\eta\eta} = \mathbf{0}$
Theorem 17.5.67 Riemann-Christoffel Tensor First Kind to Second Kind	T14.1.17	$R^a{}_{\alpha\beta\gamma} = a^{ab} R_{b\alpha\beta\gamma}$
Theorem 17.5.68 Riemann-Christoffel Tensor First Kind; Transpose First Two Indices	T14.2.4	$R_{\alpha g\beta\gamma} = -R_{g\alpha\beta\gamma}$
Theorem 17.5.69 Riemann-Christoffel Tensor First Kind; Transpose Last Two Indices	T14.2.5	$R_{g\alpha\gamma\beta} = -R_{g\alpha\beta\gamma}$
Theorem 17.5.70 Riemann-Christoffel Tensor First Kind; Transpose Last Two Indices	T14.2.6	$R_{\beta\gamma g\alpha} = R_{g\alpha\beta\gamma}$
Theorem 17.5.71 Riemann-Christoffel Tensor First Kind: Contracted to Spatial Curvature	T14.4.9	$R = a^{\alpha\beta}\, a^{g\gamma}\, R_{g\alpha\beta\gamma}$

Table 17.5.10 Field Tensors on a Hypersurface $\mathfrak{H}^{n-1}$

Theorem	Source	Equation
Theorem 17.5.72 Tensor Product: Differential of Covariant Tensor Sets	T12.7.13	$(\prod_{\alpha=1}^{\eta} \overset{*}{\mathrm{A}}_{\tau[\alpha]})^{;\kappa} = \overset{*}{\mathrm{A}}_{\tau[1]}{}^{;\kappa} (\prod_{\alpha=2}^{\eta} \overset{*}{\mathrm{A}}_{\tau[\alpha]}) + \sum_{\beta=2}^{\eta-1} \prod_{\alpha=1}^{\beta-1} \overset{*}{\mathrm{A}}_{\tau[\alpha]} \overset{*}{\mathrm{A}}_{\sigma[\beta]}{}^{;\kappa} (\prod_{\alpha=\beta+1}^{\eta} \overset{*}{\mathrm{A}}_{\tau[\alpha]}) + \prod_{\alpha=1}^{\eta-1} \overset{*}{\mathrm{A}}_{\tau[\alpha]} \overset{*}{\mathrm{A}}_{\sigma[\eta]}{}^{;\kappa}$
Theorem 17.5.73 Tensor Product: Differential of Contravariant Tensor Sets	T12.7.14	$(\prod_{\alpha=1}^{\eta} \overset{*}{\mathrm{A}}{}^{\tau[\alpha]})_{;\kappa} = \overset{*}{\mathrm{A}}{}^{\tau[1]}{}_{;\kappa} (\prod_{\alpha=2}^{\eta} \overset{*}{\mathrm{A}}{}^{\tau[\alpha]}) + \sum_{\beta=2}^{\eta-1} \prod_{\alpha=1}^{\beta-1} \overset{*}{\mathrm{A}}{}^{\tau[\alpha]} \overset{*}{\mathrm{A}}{}^{\sigma[\beta]}{}_{;\kappa} (\prod_{\alpha=\beta+1}^{\eta} \overset{*}{\mathrm{A}}{}^{\tau[\alpha]}) + \prod_{\alpha=1}^{\eta-1} \overset{*}{\mathrm{A}}{}^{\tau[\alpha]} \overset{*}{\mathrm{A}}{}^{\sigma[\eta]}{}_{;\kappa}$

Section 17.6 Tensor Derivatives on a Hypersheet $\mathfrak{H}^{n-1}$

pg 177 [a] with Formed by Christoffel

Theorem 17.6.1 Hypersheet Derivative: Mixed Tensor of Rank-2

1g	Given	$\Phi(t) = T_{\alpha}{}^{i} A_i B^{\alpha}$
2g		$0 = \delta A_i / \delta t = dA_i / dt - \Gamma_{ij}{}^{k} A_k \, dx^j / dt$ Parallel field over path C as a surface curve.
3g		$0 = \delta B^{\beta} / \delta t = dB^{\beta} / dt + \Gamma_{\eta\gamma}{}^{\beta} B^{\eta} du^{\gamma} / dt$ Parallel field over path C as a surface curve.
Step	Hypothesis	$T^{t}{}_{\alpha;\gamma} \equiv (T^{t}{}_{\alpha})_{,\gamma} + (T^{i}{}_{\alpha}) \Gamma_{ij}{}^{t} x^{j}{}_{\gamma} - (T^{t}{}_{\alpha}) \Gamma_{\eta\gamma}{}^{\beta}$
1	From 1g and LxK.3.1.7B Uniqueness Theorem of Differentiation	$d\,\Phi(t) / dt = (T^{i}{}_{\alpha} A_i B^{\alpha}) / dt$
2	From 1 and LxK.3.2.4C Distributed Across Products	$d\,\Phi(t) / dt = d(T^{i}{}_{\alpha}) / dt\, (A_i)(B^{\alpha}) +$ $(T^{i}{}_{\alpha}) d(A_i) / dt\, (B^{\alpha}) +$ $(T^{i}{}_{\alpha})(A_i) d(B^{\alpha}) / dt$
3	From 2g, T4.3.4B Equalities: Reversal of Right Cancellation by Addition and A4.2.7 Identity Add	$d\,A_i / dt = \Gamma_{ij}{}^{k} A_k \, dx^j / dt$
4	From 3g, T4.3.5B Equalities: Left Cancellation by Addition and A4.2.7 Identity Add	$d\,B^{\beta} / dt = -\Gamma_{\eta\gamma}{}^{\beta} B^{\eta} du^{\gamma} / dt$
5	From 2, 3, 4, T10.2.5 Tool: Tensor Substitution and A10.2.14 Correspondence of Tensors and Tensor Coefficients	$d\,\Phi(t) / dt = d(T^{i}{}_{\alpha}) / dt\, (A_i)(B^{\alpha}) +$ $(T^{i}{}_{\alpha}) \Gamma_{ij}{}^{k} A_k \, dx^j / dt\, (B^{\alpha}) -$ $(T^{i}{}_{\alpha})(A_i) \Gamma_{\eta\gamma}{}^{\beta} B^{\eta} du^{\gamma} / dt$
6	From 5, LxK.3.2.1B exact differential, T10.3.7 Product of Tensors: Commutative by Multiplication of Rank p + q → q + p, A10.2.14 Correspondence of Tensors and Tensor Coefficients and transpose t → i, t → k dummy indices	$d\,\Phi(t) / dt = \partial(T^{t}{}_{\alpha}) / \partial u^{\gamma} (du^{\gamma} / dt)(A_t)(B^{\alpha}) +$ $(T^{i}{}_{\alpha}) \Gamma_{ij}{}^{t} (\partial x^j / \partial u^{\gamma})(du^{\gamma} / dt)(A_t)(B^{\alpha}) -$ $(T^{t}{}_{\alpha}) \Gamma_{\eta\gamma}{}^{\beta} (A_t)(B^{\eta})(du^{\gamma} / dt)$
7	From 6, T10.3.20 Addition of Tensors: Distribution of a Tensor over Addition of Tensors and A10.2.14 Correspondence of Tensors and Tensor Coefficients	$d\,\Phi(t) / dt = [\partial(T^{t}{}_{\alpha}) / \partial u^{\gamma} +$ $(T^{i}{}_{\alpha}) \Gamma_{ij}{}^{t} x^{j}{}_{,\gamma} -$ $(T^{t}{}_{\alpha}) \Gamma_{\eta\gamma}{}^{\beta}] (A_t)(B^{\eta})(du^{\gamma} / dt)$
∴	From 7, LxK.3.2.1B exact differential and equating coefficients	$T^{t}{}_{\alpha;\gamma} \equiv (T^{t}{}_{\alpha})_{,\gamma} + (T^{i}{}_{\alpha}) \Gamma_{ij}{}^{t} x^{j}{}_{\gamma} - (T^{t}{}_{\alpha}) \Gamma_{\eta\gamma}{}^{\beta}$

Observation 17.6.1 Hypersheet Derivative: General Mixed Tensor of Rank-3

Using principles from theorem T17.6.1 "Hypersheet Derivative: Mixed Tensor of Rank-2" other tensors of higher rank can be solved such as:

$$T^{t}{}_{\alpha\beta;\gamma} \equiv (T^{t}{}_{\alpha\beta})_{,\gamma} + (T^{i}{}_{\alpha\beta})\, \Gamma_{ij}{}^{t}\, x^{j}{}_{\gamma} - (T^{i}{}_{\eta\beta})\, \Gamma_{\alpha\gamma}{}^{\eta} - (T^{i}{}_{\alpha\eta})\, \Gamma_{\beta\gamma}{}^{\eta} \qquad \text{Equation A}$$

Section 17.7 Second Fundamental Quadratic Form of the Hypersurface $\mathfrak{H}^{n-1}$

pg 176, 180 [b] Formed by Christoffel

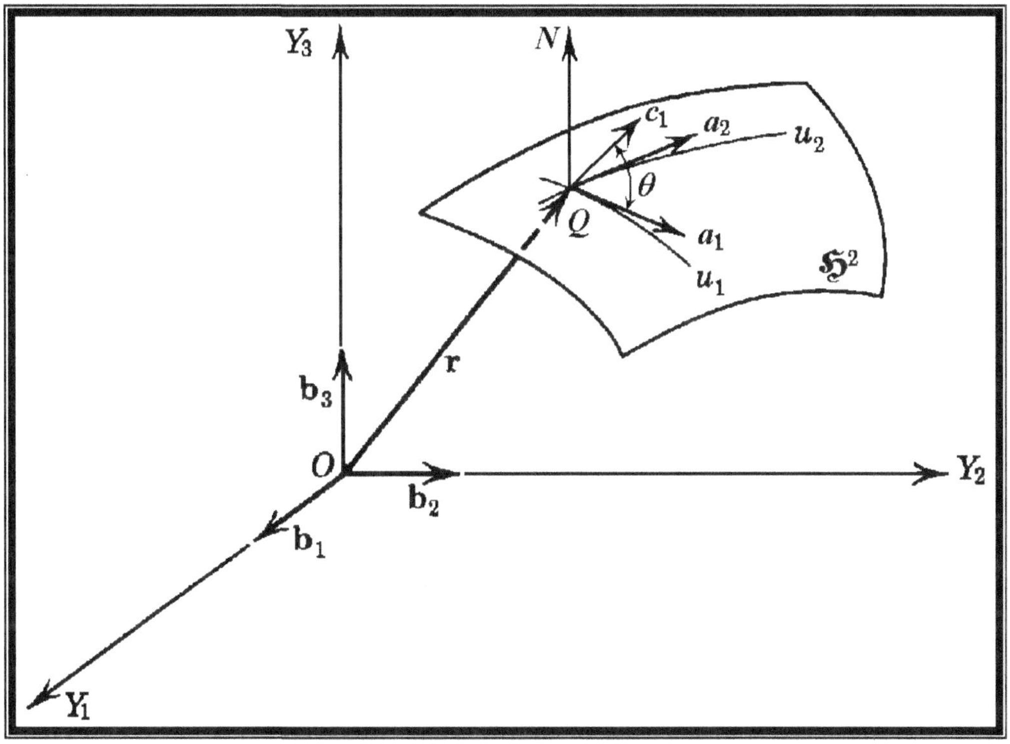

Figure 17.7.1 Second Fundamental Quadratic Bases Vector vs Normal

Definition 17.7.1 Second Fundamental Quadratic Dyadic of Covariant Metric Form

$$c_{\alpha\beta} \equiv \mathbf{a}_{\alpha,\beta} \bullet \mathbf{N} \qquad \text{Equation A}$$

$$c_{\alpha\beta} \equiv \mathbf{c}_{\alpha} \bullet \mathbf{c}_{\beta} \qquad \text{Equation B}$$

Definition 17.7.2 Arc Length of a Second Fundamental Quadratic Covariant Metric

$$d\beta^2 \equiv -d\mathbf{r} \bullet d\mathbf{N} \qquad \text{Equation A}$$

Theorem 17.7.1 Second Fundamental Covariant Metric: Dyadic Existence

Step	Hypothesis	$c_{\alpha\beta} = -\frac{1}{2}\,[(\mathbf{r},_{\alpha} \bullet \mathbf{N}_{,\beta}) + (\mathbf{r},_{\beta} \bullet \mathbf{N}_{,\alpha})]$ EQ A	$c_{\alpha\beta} = \mathbf{a}_{\alpha,\beta} \bullet \mathbf{N}$ EQ B
1	From T17.3.1A Surface Normal Vector: Normal Covariant Quiver Vector	$\mathbf{a}_{\alpha} \bullet \mathbf{N} = 0$	
2	From 1 and LxK.3.1.7C Uniqueness Theorem of Differentiation	$(\mathbf{a}_{\alpha} \bullet \mathbf{N})_{,\beta} = (\ 0\ )_{,\beta}$	
3	From 2, LxK.5.2.2 Differential of Dot Product and LxK.3.3.1B Differential of a Constant	$(\mathbf{a}_{\alpha,\beta} \bullet \mathbf{N}) + (\mathbf{a}_{\alpha} \bullet \mathbf{N}_{,\beta}) = 0$	
4	From 3 and permutating indices	$(\mathbf{a}_{\beta,\alpha} \bullet \mathbf{N}) + (\mathbf{a}_{\beta} \bullet \mathbf{N}_{,\alpha}) = 0$	
5	From 3 and 4; adding term-by-term	$(\mathbf{a}_{\alpha,\beta} \bullet \mathbf{N}) + (\mathbf{a}_{\beta,\alpha} \bullet \mathbf{N}) + (\mathbf{a}_{\alpha} \bullet \mathbf{N}_{,\beta}) + (\mathbf{a}_{\beta} \bullet \mathbf{N}_{,\alpha}) = 0$	
6	From 5 and DxK.3.2.2E Multiple Partial Differential Operator; as a Leibniz's notational fraction the denominator product can be transposed.	$(\mathbf{a}_{\alpha,\beta} \bullet \mathbf{N}) + (\mathbf{a}_{\alpha,\beta} \bullet \mathbf{N}) + (\mathbf{a}_{\alpha} \bullet \mathbf{N}_{,\beta}) + (\mathbf{a}_{\beta} \bullet \mathbf{N}_{,\alpha}) = 0$	
7	From 6 and T4.3.10 Equalities: Summation of Repeated Terms by 2	$2(\mathbf{a}_{\alpha,\beta} \bullet \mathbf{N}) + (\mathbf{a}_{\alpha} \bullet \mathbf{N}_{,\beta}) + (\mathbf{a}_{\beta} \bullet \mathbf{N}_{,\alpha}) = 0$	
8	From 7 and T4.3.5B Equalities: Left Cancellation by Addition and A4.2.7 Identity Add	$2(\mathbf{a}_{\alpha,\beta} \bullet \mathbf{N}) = -\ [(\mathbf{a}_{\alpha} \bullet \mathbf{N}_{,\beta}) + (\mathbf{a}_{\beta} \bullet \mathbf{N}_{,\alpha})]$	
9	From 8, T4.6.6B Equalities: Left Cancellation by Multiplication and D4.1.20 Negative Coefficient	$\mathbf{a}_{\alpha,\beta} \bullet \mathbf{N} = -\frac{1}{2}\,[(\mathbf{a}_{\alpha} \bullet \mathbf{N}_{,\beta}) + (\mathbf{a}_{\beta} \bullet \mathbf{N}_{,\alpha})]$	
10	From 9, D17.7.1 Second Fundamental Quadratic Dyadic of Covariant Metric Form and A4.2.3 Substitution	$c_{\alpha\beta} = -\frac{1}{2}\,[(\mathbf{a}_{\alpha} \bullet \mathbf{N}_{,\beta}) + (\mathbf{a}_{\beta} \bullet \mathbf{N}_{,\alpha})]$	
∴	From 9, 10, T17.1.4 Contravariant Bases Vectors for a Hypersheet and A7.9.2 Substitution of Vectors	$c_{\alpha\beta} = -\frac{1}{2}\,[(\mathbf{r},_{\alpha} \bullet \mathbf{N}_{,\beta}) + (\mathbf{r},_{\beta} \bullet \mathbf{N}_{,\alpha})]$ EQ A	$c_{\alpha\beta} = \mathbf{a}_{\alpha,\beta} \bullet \mathbf{N}$ EQ B

Observation 17.7.1 Existence of a New Type of Hypersurface Metric Dyadic

It now follows that $c_{\alpha\beta} = \mathbf{a}_{\alpha,\beta} \bullet \mathbf{N}$ must be a new type of metric tensor uniquely associate with the Hypersurface $\mathfrak{H}^{n-1}$. Unlike the first fundamental quadratic, the simple intrinsic vector constitutes a mixing of different types of intrinsic vectors. This gives rise to a fundamental quadratic of a second kind, hence it's name the second fundamental quadratic form on a hypersurface.

Theorem 17.7.2 Second Fundamental Quadratic Covariant Metric: Measured Distance

Step	Hypothesis	$-d\mathbf{r} \bullet d\mathbf{N} = c_{\alpha\beta}\, du_\alpha\, du_\beta$ EQ A	$d\beta^2 = c_{\alpha\beta}\, du_\alpha\, du_\beta$ EQ B
1	From T8.3.2 Dot Product: Equality	$(d\mathbf{r} / ds) \bullet (d\mathbf{N} / ds) = (d\mathbf{r} / ds) \bullet (d\mathbf{N} / ds)$	
2	From 1, T17.7.1A Positional and Normal Vector Exactly Undergoing Change, T17.7.1B Positional and Normal Vector Exactly Undergoing Change and A7.9.2 Substitution of Vectors	$(d\mathbf{r} / ds) \bullet (d\mathbf{N} / ds) = (\Sigma_\alpha\, \mathbf{r}_{,\alpha}\, du_\alpha / ds) \bullet (\Sigma_\alpha\, \mathbf{N}_{,\alpha}\, du_\alpha / ds)$	
3	From 2 and T8.3.8 Dot Product: Distribution of Dot Product Across Another Vector	$(d\mathbf{r} / ds) \bullet (d\mathbf{N} / ds) = \Sigma_\alpha\, \Sigma_\beta\, (\mathbf{r}_{,\alpha} \bullet \mathbf{N}_{,\beta})\, (du_\alpha / ds)\, (du_\beta / ds)$	
4	From 3 and permutating indices	$(d\mathbf{r} / ds) \bullet (d\mathbf{N} / ds) = \Sigma_\beta\, \Sigma_\alpha\, (\, \mathbf{r}_{,\beta} \bullet \mathbf{N}_{,\alpha})\, (du_\alpha / ds)\, (du_\beta / ds)$	
5	From 3 and 4; adding term-by-term	$(d\mathbf{r} / ds) \bullet (d\mathbf{N} / ds) + (d\mathbf{r} / ds) \bullet (d\mathbf{N} / ds) =$ $\Sigma_\alpha\, \Sigma_\beta\, \mathbf{r}_{,\alpha} \bullet \mathbf{N}_{,\beta}\, (du_\alpha / ds)\, (du_\beta / ds) +$ $\Sigma_\beta\, \Sigma_\alpha\, \mathbf{r}_{,\beta} \bullet \mathbf{N}_{,\alpha}\, (du_\alpha / ds)\, (du_\beta / ds)$	
6	From 5, T4.3.10 Equalities: Summation of Repeated Terms by 2; summing over α and β term-by-term and match differential elements.	$2\, (d\mathbf{r} / ds) \bullet (d\mathbf{N} / ds) =$ $\Sigma_\alpha\, \Sigma_\beta\, (\mathbf{r}_{,\alpha} \bullet \mathbf{N}_{,\beta})\, (du_\alpha / ds)\, (du_\beta / ds) +$ $(\mathbf{r}_{,\beta} \bullet \mathbf{N}_{,\alpha}\,)\, (du_\alpha / ds)\, (du_\beta / ds)$	
7	From 6, A4.2.12 Identity Multp, T4.8.2B Integer Exponents: Negative One Squared and T10.3.20 Addition of Tensors: Distribution of a Tensor over Addition of Tensors	$2\, (d\mathbf{r} / ds) \bullet (d\mathbf{N} / ds) =$ $(-1)\, (-1)\Sigma_\alpha\, \Sigma_\beta\, (\, \mathbf{r}_{,\alpha} \bullet \mathbf{N}_{,\beta} + \mathbf{r}_{,\beta} \bullet \mathbf{N}_{,\alpha}\,)\, (du_\alpha / ds)\, (du_\beta / ds)$	
8	From 7, T4.4.6B Equalities: Left Cancellation by Multiplication, D4.1.10A Negative Coefficient and A4.2.11 Associative Multp	$(d\mathbf{r} / ds) \bullet (d\mathbf{N} / ds) =$ $(-1)\, [-½\, \Sigma_\alpha\, \Sigma_\beta\, (\, \mathbf{r}_{,\alpha} \bullet \mathbf{N}_{,\beta} + \mathbf{r}_{,\beta} \bullet \mathbf{N}_{,\alpha}\,)]\, (du_\alpha / ds)\, (du_\beta / ds)$	
9	From 8, T17.7.1 Second Fundamental Covariant Metric: Dyadic Existence and T10.2.5 Tool: Tensor Substitution	$(d\mathbf{r} / ds) \bullet (d\mathbf{N} / ds) = -c_{\alpha\beta}\, (du_\alpha / ds)\, (du_\beta / ds)$	
10	From 9, DK.3.2H Second Parametric Differentiation and D16.1.4 Intrinsic Differential Notation; with ***Leibniz***'s notation, the differential fraction	$(d\mathbf{r} \bullet d\mathbf{N}) / ds^2 = (-c_{\alpha\beta}\, du_\alpha\, du_\beta\,) / ds^2$	

11	From 10, D16.1.4 Intrinsic Differential Notation and equating numerators; with ***Leibniz***'s notation, the differential fraction	$\mathbf{dr} \bullet \mathbf{dN} = -c_{\alpha\beta}\, du_\alpha\, du_\beta$	
12	From 11, T4.4.23 Equalities: Multiplication of a Constant, D4.1.20 Negative Coefficient, T4.8.2B Integer Exponents: Negative One Squared and A4.2.12 Identity Multp	$-\mathbf{dr} \bullet \mathbf{dN} = c_{\alpha\beta}\, du_\alpha\, du_\beta$	
∴	From 11, 12, D17.7.2 Arc Length of a Second Fundamental Quadratic Covariant Metric and A4.2.3 Substitution	$-\mathbf{dr} \bullet \mathbf{dN} = c_{\alpha\beta}\, du_\alpha\, du_\beta$ EQ A	$d\beta^2 = c_{\alpha\beta}\, du_\alpha\, du_\beta$ EQ B

Theorem 17.7.3 Second Fundamental Covariant Metric: Angle to Normal

Step	Hypothesis	$c_{\alpha\beta} = N\, \Sigma_\kappa\, \Gamma_{\alpha\beta,\kappa}\, a^\kappa \cos\theta^{\kappa n}$
1	From T17.7.1B Second Fundamental Quadratic Dyadic of Covariant Metric Form	$c_{\alpha\beta} = \mathbf{a}_{\alpha,\beta} \bullet \mathbf{N}$
2	From 1, T12.5.4B Covariant Derivative: Christoffel Symbol of the First Kind; selected, because T17.7.1 Second Fundamental Covariant Metric: Dyadic Existence, proves its existence; hence the contravariant dot product cannot identical vanish everywhere over the summation-κ	$c_{\alpha\beta} = (\Sigma_\kappa\, \Gamma_{\alpha\beta,\kappa}\, \mathbf{a}^\kappa) \bullet \mathbf{N}$
3	From 2 and T8.3.7 Dot Product: Distribution of Dot Product Across Addition of Vectors	$c_{\alpha\beta} = \Sigma_\kappa\, \Gamma_{\alpha\beta,\kappa}\, (\mathbf{a}^\kappa \bullet \mathbf{N})$
4	From 3 and D8.3.1 The Inner or Dot Product Operator	$c_{\alpha\beta} = \Sigma_\kappa\, \Gamma_{\alpha\beta,\kappa}\, (a^\kappa N \cos\theta^{\kappa n})$
∴	From 4 and A4.2.14 Distribution	$c_{\alpha\beta} = N\, \Sigma_\kappa\, \Gamma_{\alpha\beta,\kappa}\, a^\kappa \cos\theta^{\kappa n}$

Theorem 17.7.4 Second Fundamental Covariant Metric: with Contravariant Christoffel Symbol

Step	Hypothesis	$c_{\alpha\beta} = \sum_\kappa \Gamma_{\alpha\beta}{}^\kappa N_\kappa$
1	From T17.7.1B Second Fundamental Quadratic Dyadic of Covariant Metric Form	$c_{\alpha\beta} = \mathbf{a}_{\alpha,\beta} \bullet \mathbf{N}$
2	From 1, T12.5.4B Covariant Derivative: Christoffel Symbol of the First Kind; selected, because T17.7.1 Second Fundamental Covariant Metric: Dyadic Existence, proves its existence	$c_{\alpha\beta} = (\sum_\kappa \Gamma_{\alpha\beta}{}^\kappa \mathbf{a}_\kappa) \bullet \mathbf{N}$
3	From 2 and T8.3.7 Dot Product: Distribution of Dot Product Across Addition of Vectors	$c_{\alpha\beta} = \sum_\kappa \Gamma_{\alpha\beta}{}^\kappa (\mathbf{a}_\kappa \bullet \mathbf{N})$
4	From 3, A17.3.1A Manifold-Hypersheet Invariance of a Normal Vector and A7.9.2 Substitution of Vectors	$c_{\alpha\beta} = \sum_\kappa\sum_\nu \Gamma_{\alpha\beta}{}^\kappa (\mathbf{a}_\kappa \bullet N^\nu \mathbf{a}_\nu)$
5	From 4 and T8.3.10B Dot Product: Right Scalar-Vector Association	$c_{\alpha\beta} = \sum_\kappa\sum_\nu \Gamma_{\alpha\beta}{}^\kappa N^\nu (\mathbf{a}_\kappa \bullet \mathbf{a}_\nu)$
6	From 5, A17.1.11P1 Covariant Metric Displacement on a Riemannian Hypersheet $\mathfrak{H}^{n-1}$; comparing term-by-term and reordering summation-ν	$c_{\alpha\beta} = \sum_\kappa \Gamma_{\alpha\beta}{}^\kappa (\sum_\nu a_{\kappa\nu} N^\nu)$
∴	From 4 and T17.5.40A Metric Transformation from Covariant to Contravariant Coefficient	$c_{\alpha\beta} = \sum_\kappa \Gamma_{\alpha\beta}{}^\kappa N_\kappa$

Theorem 17.7.5 Riemannian Mixed Tensor of Rank-3: Relationship to Positional Vector

Step	Hypothesis	$\mathbf{r}_{,\alpha;\beta} = \sum_i x^i{}_{\alpha;\beta}\, \mathbf{b}_i$ EQ A	$\mathbf{a}_{\alpha;\beta} = \mathbf{r}_{;\alpha;\beta}$ EQ B
1	From T17.1.5B Manifold-Hypersheet Transformation: Hypersheet Contravariant Bases Vectors	$\mathbf{a}_\alpha = \sum_i x^i{}_\alpha\, \mathbf{b}_i$	
2	From 1 and T12.6.4B Differentiation of Tensor: Rank-1 Uniqueness of Differentiation	$(\mathbf{a}_\alpha)_{;\beta} = (\sum_i x^i{}_\alpha\, \mathbf{b}_i)_{;\beta}$	
3	From 2, T12.6.3A Differentiation of Tensor: Rank-1 Contravariant and T12.7.2 Tensor Summation: Covariant Derivative as a Linear Operator	$\mathbf{a}_{\alpha;\beta} = \sum_i x^i{}_{\alpha;\beta}\, \mathbf{b}_i$	
∴	From 3, T17.1.4 Contravariant Bases Vectors for a Hypersheet and A7.9.2 Substitution of Vectors	$\mathbf{r}_{;\alpha;\beta} = \sum_i x^i{}_{\alpha;\beta}\, \mathbf{b}_i$ EQ A	$\mathbf{a}_{\alpha;\beta} = \mathbf{r}_{;\alpha;\beta}$ EQ B

Theorem 17.7.6 Second Fundamental Covariant Metric: Relationship to Positional Vector

Step	Hypothesis	$c_{\alpha\beta} = (\mathbf{r}_{;\beta;\alpha}) \bullet \mathbf{N}$
1	From D17.7.1A Second Fundamental Quadratic Dyadic of Covariant Metric Form	$c_{\alpha\beta} = \mathbf{a}_{\alpha;\beta} \bullet \mathbf{N}$
∴	From 1, T17.7.5B Riemannian Mixed Tensor of Rank-3: to Positional Vector Relationship and A7.9.2 Substitution of Vectors	$c_{\alpha\beta} = (\mathbf{r}_{;\beta;\alpha}) \bullet \mathbf{N}$

Theorem 17.7.7 Second Fundamental Covariant Metric: Transpose of Indices

Step	Hypothesis	$c_{\alpha\beta} = c_{\beta\alpha}$
1	From T17.7.6 Second Fundamental Covariant Metric: Relationship to Positional Vector	$c_{\alpha\beta} = (\mathbf{r}_{;\alpha;\beta}) \bullet \mathbf{N}$
2	From 1 and transpose Leibniz fractional denominator	$c_{\alpha\beta} = (\mathbf{r}_{;\beta;\alpha}) \bullet \mathbf{N}$
∴	From 2 and D17.7.1A Second Fundamental Quadratic Dyadic of Covariant Metric Form	$c_{\alpha\beta} = c_{\beta\alpha}$

Theorem 17.7.8 Second Fundamental Covariant Metric: in Unit Normal Form

1g	Given	$\mathbf{n} = \sum_i n_i \mathbf{b}^i$
Step	Hypothesis	$c_{\alpha\beta} = \sum_i x^i_{\alpha;\beta}\, n_i$
1	From T17.7.6 Second Fundamental Covariant Metric: Relationship to Positional Vector	$c_{\alpha\beta} = (\mathbf{r}_{;\alpha;\beta}) \bullet \mathbf{n}$
2	From 1g, 1, T17.7.5A Riemannian Mixed Tensor of Rank-3: Relationship to Positional Vector and A7.9.2 Substitution of Vectors	$c_{\alpha\beta} = (\sum_i x^i_{\alpha;\beta}\, \mathbf{b}_i) \bullet (\sum_i n_i \mathbf{b}^i)$
3	From 3, T17.5.3A Dot Product of Mixed Vectors (Tensor Contraction)	$c_{\alpha\beta} = \sum_i \sum_j x^i_{\alpha;\beta}\, n_j\, \delta_i^j$
∴	From 3 and D6.1.6B Identity Matrix; evaluating the Kronecker Delta over the summation-(i, j)	$c_{\alpha\beta} = \sum_i x^i_{\alpha;\beta}\, n_i$

Theorem 17.7.9 Riemannian Mixed Tensor of Rank-3: Transpose of Indices

Step	Hypothesis	$x^i_{\alpha;\beta} = x^i_{\beta;\alpha}$
1	From A10.2.6 Equality of Tensors	$x^i_{\alpha} = x^i_{\alpha}$
2	From 1 and T12.6.4B Differentiation of Tensor: Rank-1 Uniqueness of Differentiation	$x^i_{\alpha;\beta} = x^i_{\alpha;\beta}$
3	From 2 and T17.6.1 Hypersheet Derivative: Mixed Tensor of Rank-2	$x^i_{\alpha;\beta} = x^i_{,\alpha,\beta} + \Sigma_k\, x^j_{\alpha}\, x^k_{\beta}\, \Gamma_{jk}{}^i - \Sigma_\eta\, x^i_{\eta}\, \Gamma_{\alpha\beta}{}^\eta$
4	From 3 and transpose $\alpha \rightarrow \beta$ dummy indices	$x^i_{\beta;\alpha} = x^i_{,\beta,\alpha} + \Sigma_k\, x^j_{\beta}\, x^k_{\alpha}\, \Gamma_{jk}{}^i - \Sigma_\eta\, x^i_{\eta}\, \Gamma_{\beta\alpha}{}^\eta$
5	From 4, T10.3.7 Product of Tensors: Commutative by Multiplication of Rank p + q → q + p, A10.2.14 Correspondence of Tensors and Tensor Coefficients, T12.5.13 Christoffel Symbol: Second Kind Symmetry for First Two Indices, T17.5.19 Christoffel Symbol: Second Kind Symmetry for First Two Indices and and transpose Leibniz fractional denominator	$x^i_{\beta;\alpha} = x^i_{,\alpha,\beta} + \Sigma_k\, x^k_{\alpha}\, x^j_{\beta}\, \Gamma_{kj}{}^i - \Sigma_\eta\, x^i_{\eta}\, \Gamma_{\alpha\beta}{}^\eta$
6	From 5 and transpose j → k dummy indices	$x^i_{\beta;\alpha} = x^i_{,\alpha,\beta} + \Sigma_k\, x^j_{\alpha}\, x^k_{\beta}\, \Gamma_{kj}{}^i - \Sigma_\eta\, x^i_{\eta}\, \Gamma_{\alpha\beta}{}^\eta$
∴	From 3, 6, T10.2.5 Tool: Tensor Substitution and A10.2.14 Correspondence of Tensors and Tensor Coefficients	$x^i_{\alpha;\beta} = x^i_{\beta;\alpha}$

Theorem 17.7.10 Riemannian Mixed Tensor of Rank-2: Differential g-h Transformation

Step	Hypothesis	$\mathbf{0} = \Sigma_i \Sigma_j x^i{}_{\alpha;\gamma} x^i{}_\beta g_{ij} + x^i{}_\alpha x^i{}_{\beta;\gamma} g_{ij}$
1	From T17.1.9B Manifold-Hypersheet Transformation: h-g Metrics With Change in Direction	$a_{\alpha\beta} = \Sigma_i \Sigma_j x^i{}_\alpha x^j{}_\beta g_{ij}$
2	From 1 and T12.7.1B Tensor Uniqueness: Tensor Differentiation	$(a_{\alpha\beta});_\gamma = (\Sigma_i \Sigma_j x^i{}_\alpha x^j{}_\beta g_{ij});_\gamma$
3	From 2 and T12.7.2 Tensor Summation: Covariant Derivative as a Linear Operator	$(a_{\alpha\beta});_\gamma = \Sigma_i \Sigma_j [(x^i{}_\alpha)(x^j{}_\beta g_{ij})];_\gamma$
4	From 3 and T12.7.6 Tensor Product: Chain Rule for Binary Covariant Product	$(a_{\alpha\beta});_\gamma = \Sigma_i \Sigma_j x^i{}_{\alpha;\gamma} x^j{}_\beta g_{ij} + x^i{}_\alpha (x^j{}_\beta g_{ij})_{;\gamma}$
5	From 4 and T12.7.6 Tensor Product: Chain Rule for Binary Covariant Product	$(a_{\alpha\beta});_\gamma = \Sigma_i \Sigma_j x^i{}_{\alpha;\gamma} x^j{}_\beta g_{ij} + x^i{}_\alpha x^j{}_{\beta;\gamma} g_{ij} + x^i{}_\alpha x^j{}_\beta g_{ij;\gamma}$
6	From 5, T17.5.32 Differentiation of the Covariant Metric Tensor (Ricci's Theorem-A) and T12.8.3 Differentiation of the Covariant Metric Tensor (Ricci's Theorem-A)	$\mathbf{0} = \Sigma_i \Sigma_j x^i{}_{\alpha;\gamma} x^j{}_\beta g_{ij} + x^i{}_\alpha x^j{}_{\beta;\gamma} g_{ij} + x^i{}_\alpha x^j{}_\beta \mathbf{0}$
7	From 6, T10.2.8 Tool: Zero Tensor Times a Tensor is a Zero Tensor and A10.2.14 Correspondence of Tensors and Tensor Coefficients	$\mathbf{0} = \Sigma_i \Sigma_j x^i{}_{\alpha;\gamma} x^j{}_\beta g_{ij} + x^i{}_\alpha x^j{}_{\beta;\gamma} g_{ij} + \mathbf{0}$
∴	From 7, T10.3.18 Addition of Tensors: Identity by Addition and A10.2.14 Correspondence of Tensors and Tensor Coefficients	$\mathbf{0} = \Sigma_i \Sigma_j x^i{}_{\alpha;\gamma} x^j{}_\beta g_{ij} + x^i{}_\alpha x^j{}_{\beta;\gamma} g_{ij}$

Theorem 17.7.11 Riemannian Mixed Tensor of Rank-3: Directed Along the Covariant Normal

Step	Hypothesis	$\mathbf{N}_\gamma \parallel (x^i_{\alpha;\beta}\, \mathbf{b}_i)$ EQ A	$\mathbf{0} = \sum_i \sum_j x^i_{\alpha;\beta}\, x^j_\gamma\, g_{ij}$ EQ B
1	From T17.7.8 Riemannian Mixed Tensor of Rank-2: Differential g-h Transformation	$\mathbf{0} = \sum_i \sum_j x^i_{\alpha;\gamma}\, x^j_\beta\, g_{ij} + x^i_\alpha\, x^j_{\beta;\gamma}\, g_{ij}$	
2	From 1 and permutating $(\alpha, \gamma) \rightarrow (\beta, \alpha)$	$\mathbf{0} = \sum_i \sum_j x^i_{\beta;\alpha}\, x^j_\gamma\, g_{ij} + x^i_\beta\, x^j_{\gamma;\alpha}\, g_{ij}$	
3	From 1 and permutating $(\alpha, \gamma) \rightarrow (\gamma, \beta)$	$\mathbf{0} = \sum_i \sum_j x^i_{\gamma;\beta}\, x^j_\alpha\, g_{ij} + x^i_\gamma\, x^j_{\alpha;\beta}\, g_{ij}$	
4	From 1, 2, 3, add 2, 3 and subtract 1	$\mathbf{0} = \sum_i \sum_j x^i_{\beta;\alpha}\, x^j_\gamma\, g_{ij} + x^i_\beta\, x^j_{\gamma;\alpha}\, g_{ij} + x^i_{\gamma;\beta}\, x^j_\alpha\, g_{ij} + x^i_\gamma\, x^j_{\alpha;\beta}\, g_{ij} - x^i_{\alpha;\gamma}\, x^j_\beta\, g_{ij} - x^i_\alpha\, x^j_{\beta;\gamma}\, g_{ij}$	
5	From 4, T10.3.19 Addition of Tensors: Inverse by Addition, T10.3.18 Addition of Tensors: Identity by Addition and A10.2.14 Correspondence of Tensors and Tensor Coefficients	$\mathbf{0} = \sum_i \sum_j x^i_{\beta;\alpha}\, x^j_\gamma\, g_{ij} + x^i_\gamma\, x^j_{\alpha;\beta}\, g_{ij}$	
6	From 5, T17.7.7 Riemannian Mixed Tensor of Rank-2: Transpose of Indices, T10.3.7 Product of Tensors: Commutative by Multiplication of Rank p + q → q + p, T10.2.5 Tool: Tensor Substitution and A10.2.14 Correspondence of Tensors and Tensor Coefficients	$\mathbf{0} = \sum_i \sum_j x^i_{\alpha;\beta}\, x^j_\gamma\, g_{ij} + x^j_{\alpha;\beta}\, x^i_\gamma\, g_{ij}$	
7	From 6, T10.2.10 Tool: Tool: Identity Multiplication of a Tensor, T10.4.7 Scalars and Tensors: Distribution of a Tensor over Addition of Scalars and D4.1.10 Primitive Definition for Rational Arithmetic	$\mathbf{0} = \sum_i \sum_j 2\, x^i_{\alpha;\beta}\, x^j_\gamma\, g_{ij}$	
8	From 7, T10.4.6 Scalars and Tensors: Distribution of a Scalar over Addition of Tensors, T10.2.7B Tool: Zero Times a Tensor is a Zero Tensor and equating tensors on both sides of the equality.	$\mathbf{0} = \sum_i \sum_j x^i_{\alpha;\beta}\, x^j_\gamma\, g_{ij}$	
9	From 8 and T17.5.1A Dot Product of Covariant Vectors	$\mathbf{0} = (\sum_i x^i_{\alpha;\beta}\, \mathbf{b}_i) \bullet (\sum_j x^j_\gamma\, \mathbf{b}_j)$	

10	From 9, T17.1.5B Manifold-Hypersheet Transformation: Hypersheet Contravariant Bases Vectors and A7.9.2 Substitution of Vectors	$\mathbf{0} = (\sum_i x^i_{\alpha;\beta}\, \mathbf{b}_i) \bullet \mathbf{a}_\gamma$	
∴	From 8, 10, T8.3.11 Dot Product: Perpendicular Vectors, hence it follows that $(x^i_{\alpha;\beta}\, \mathbf{b}_i)$ is a space vector normal to the hypersurface and directed along the covariant normal vector-**N**.	$\mathbf{N}_\gamma \parallel (x^i_{\alpha;\beta}\, \mathbf{b}_i)$ EQ A	$\mathbf{0} = \sum_i \sum_j x^i_{\alpha;\beta}\, x^j_\gamma\, g_{ij}$ EQ B

Theorem 17.7.12 Second Fundamental Covariant Metric: in Inverse Unit Normal Form

1g	Given	$\mathbf{n} = \sum_i n_i\, \mathbf{b}^i$
Step	Hypothesis	$c_{\alpha\beta}\, n^i = x^i_{\alpha;\beta}$
1	From T17.7.11 Riemannian Mixed Tensor of Rank-3: Directed Along the covariant unit Normal	$\sum_i Q_{\alpha\beta}\, n^i\, \mathbf{b}_i \equiv \sum_i x^i_{\alpha;\beta}\, \mathbf{b}_i$
2	From 1 and T8.3.1 Dot Product: Uniqueness Operation	$(\sum_i Q_{\alpha\beta}\, n^i\, \mathbf{b}_i) \bullet \mathbf{n} = (\sum_i x^i_{\alpha;\beta}\, \mathbf{b}_i) \bullet \mathbf{n}$
3	From 1g, 2 and A7.9.2 Substitution of Vectors	$(\sum_i Q_{\alpha\beta}\, n^i\, \mathbf{b}_i) \bullet (\sum_i n_i\, \mathbf{b}^i) = (\sum_i x^i_{\alpha;\beta}\, \mathbf{b}_i) \bullet (\sum_i n_i\, \mathbf{b}^i)$
4	From 3 and T17.5.3A Dot Product of Mixed Vectors (Tensor Contraction)	$\sum_i \sum_j Q_{\alpha\beta}\, n^i\, n_j\, \delta_i^{\ j} = \sum_i \sum_j x^i_{\alpha;\beta}\, n_j\, \delta_i^{\ j}$
5	From 4 and D6.1.6B Identity Matrix; evaluating the Kronecker Delta over the summation-(i, j)	$Q_{\alpha\beta} \sum_i n^i\, n_i = \sum_i x^i_{\alpha;\beta}\, n_i$
6	From 5 and T17.5.6A Magnitude of Normalized Vector	$Q_{\alpha\beta}\, 1 = \sum_i x^i_{\alpha;\beta}\, n_i$
7	From 7, T10.2.10 Tool: Tool: Identity Multiplication of a Tensor, T17.7.8 Second Fundamental Covariant Metric: in Unit Normal Form, T10.2.5 Tool: Tensor Substitution and A10.2.14 Correspondence of Tensors and Tensor Coefficients	$Q_{\alpha\beta} = c_{\alpha\beta}$
8	From 1, 8, T10.2.5 Tool: Tensor Substitution and A10.2.14 Correspondence of Tensors and Tensor Coefficients	$\sum_i c_{\alpha\beta}\, n^i\, \mathbf{b}_i \equiv \sum_i x^i_{\alpha;\beta}\, \mathbf{b}_i$
∴	From 8 and equating bases vector-by-bases vector	$c_{\alpha\beta}\, n^i = x^i_{\alpha;\beta}$

Section 17.8 Formulas of Weingarten and Equations of Gauss and Codazzi

pg 184 [a][b] with Riemann

Definition 17.8.1 Weingarten Tensor on a Hypersheet $\mathfrak{H}^{n-1}$

First and second fundamental domains can and do cross over into one another's regions with a mixed metric tensor called the ***Mixed Metric for Crossed First-Second Fundamental Metrics*** or simply the ***Weingarten Mixed Metric Tensor***.

$$w_\alpha{}^\gamma \equiv \sum_\eta c_{\alpha\eta}\, a^{\eta\gamma} \qquad \text{Equation A}$$

Definition 17.8.2 Covariant Determinate of Gauss

$$c \equiv |c_{\alpha\beta}| \qquad \text{Equation A}$$

$$c \equiv \begin{vmatrix} c_{\alpha\beta} & c_{\alpha\gamma} \\ c_{\sigma\beta} & c_{\sigma\gamma} \end{vmatrix} \qquad \text{Equation B}$$

$$c = c_{\alpha\beta}\, c_{\sigma\gamma} - c_{\alpha\gamma}\, c_{\sigma\beta} \qquad \text{Equation C}$$

Theorem 17.8.1 Weingarten Formulas: Auxiliary-1 Differential Normal

Step	Hypothesis	$(n^i{}_\alpha\, \mathbf{b}_i) \perp (\sum_j n^j\, \mathbf{b}_j)$ EQ A	$\mathbf{0} = \sum_i n^i{}_\alpha\, n_i$ EQ B
1	From T12.3.18A Magnitude of Normalized Vector	$1 = \sum_i n^i\, n_i$	
2	From 1, T12.3.9A Metric Transformation from Covariant to Contravariant Coefficient, T10.3.7 Product of Tensors: Commutative by Multiplication of Rank p + q → q + p and A10.2.14 Correspondence of Tensors and Tensor Coefficients	$1 = \sum_i \sum_j g_{ij}\, n^i\, n^j$	
3	From 2 and T12.7.1B Tensor Uniqueness: Tensor Differentiation	$(1)_{;\alpha} = (\sum_i \sum_j g_{ij}\, n^i\, n^j)_{;\alpha}$	
4	From 3, LxK.3.3.1B Differential of a Constant and T12.7.2 Tensor Summation: Covariant Derivative as a Linear Operator	$\mathbf{0} = \sum_i \sum_j (g_{ij}\, n^i\, n^j)_{;\alpha}$	
5	From 4, T12.7.6 Tensor Product: Chain Rule for Binary Covariant Product and again expanding the products through the chain rule	$\mathbf{0} = \sum_i \sum_j (g_{ij})_{;\alpha}\, n^i\, n^j + g_{ij}\, (n^i)_{;\alpha}\, n^j + g_{ij}\, n^i\, (n^j)_{;\alpha}$	

6	From 5, D17.1.9 Riemannian Manifold $\mathfrak{R}^n$ Mixed Tensor of Rank-2 on a Hypersheet $\mathfrak{H}^{n-1}$and T12.8.3 Differentiation of the Covariant Metric Tensor (Ricci's Theorem-A)	$\mathbf{0} = \sum_i \sum_j \mathbf{0}\ n^i\ n^j + g_{ij}\ n^i{}_\alpha n^j + g_{ij}\ n^i\ n^j{}_\alpha$	
7	From 6, T10.2.8 Tool: Zero Tensor Times a Tensor is a Zero Tensor, T10.3.18 Addition of Tensors: Identity by Addition and A10.2.14 Correspondence of Tensors and Tensor Coefficients	$\mathbf{0} = \sum_i \sum_j g_{ij}\ n^i{}_\alpha n^j + g_{ij}\ n^i\ n^j{}_\alpha$	
8	From 7, T10.2.10 Tool: Identity Multiplication of a Tensor, T10.4.7 Scalars and Tensors: Distribution of a Tensor over Addition of Scalars and D4.1.10 Primitive Definition for Rational Arithmetic	$\mathbf{0} = \sum_i \sum_j 2\ g_{ij}\ n^i{}_\alpha n^j$	
9	From 8, T10.4.6 Scalars and Tensors: Distribution of a Scalar over Addition of Tensors, T10.2.7B Tool: Zero Times a Tensor is a Zero Tensor and equating tensors on both sides of the equality.	$\mathbf{0} = \sum_i \sum_j g_{ij}\ n^i{}_\alpha n^j$	
10	From 9, T10.3.7 Product of Tensors: Commutative by Multiplication of Rank p + q → q + p and A10.2.14 Correspondence of Tensors and Tensor Coefficients	$\mathbf{0} = \sum_i n^i{}_\alpha \sum_j g_{ij}\ n^j$	
11	From 10, T12.3.9A Metric Transformation from Covariant to Contravariant Coefficient	$\mathbf{0} = \sum_i n^i{}_\alpha\ n_i$	
12	From 11 and T17.5.1A Dot Product of Covariant Vectors	$\mathbf{0} = (\sum_i n^i{}_\alpha\ \mathbf{b}_i) \bullet (\sum_j n^j\ \mathbf{b}_j)$	
∴	From 12, T8.3.11 Dot Product: Perpendicular Vectors; hence the differential normal is perpendicular to the normal and must lie in the tangent plane of the hypersheet.	$(n^i{}_\alpha\ \mathbf{b}_i) \perp (\sum_j n^j\ \mathbf{b}_j)$ EQ A	$\mathbf{0} = \sum_i n^i{}_\alpha\ n_i$ EQ B

Theorem 17.8.2 Weingarten Formulas: Auxiliary-1 Differential Normal in Linear Form

1g	Given	Assume there exist a transformation metric tensor $\xi^{\beta}{}_{\alpha}$
Step	Hypothesis	$n^{i}{}_{\alpha} = \sum_{\gamma} \xi_{\alpha}{}^{\gamma} x^{i}{}_{\gamma}$
1	From 1g and T17.8.1 Weingarten Formulas: Auxiliary-1 Differential Normal; so the auxiliary differential normal can be written in linear form	$\sum_{i} n^{i}{}_{\alpha} \mathbf{b}_{i} \equiv \sum_{i} \sum_{\gamma} \xi_{\alpha}{}^{\gamma} x^{i}{}_{\gamma} \mathbf{b}_{i}$
∴	From 1 and equating bases vector-by-bases vector	$n^{i}{}_{\alpha} = \sum_{\gamma} \xi_{\alpha}{}^{\gamma} x^{i}{}_{\gamma}$

Theorem 17.8.3 Weingarten Formulas: Weingarten Symmetrical Tensor on a Hypersheet $\mathfrak{H}^{n-1}$

Step	Hypothesis	$w_{\alpha}{}^{\gamma} = w^{\gamma}{}_{\alpha}$
1	From D17.8.1 Weingarten Tensor on a Hypersheet $\mathfrak{H}^{n-1}$	$w_{\alpha}{}^{\gamma} = \sum_{\eta} c_{\alpha\eta}\, a^{\eta\gamma}$
2	From 1 T10.3.7 Product of Tensors: Commutative by Multiplication of Rank p + q → q + p and A10.2.14 Correspondence of Tensors and Tensor Coefficients	$w_{\alpha}{}^{\gamma} = \sum_{\eta} a^{\gamma\eta}\, c_{\eta\alpha}$
∴	From 2 and D17.8.1 Weingarten Tensor on a Hypersheet $\mathfrak{H}^{n-1}$	$w_{\alpha}{}^{\gamma} = w^{\gamma}{}_{\alpha}$

Theorem 17.8.4 Weingarten Formulas: Auxiliary-2 Differential Normal to Curvature Coordinates

Step	Hypothesis	$0 = \sum_{i} \sum_{j} g_{ij}\, x^{i}{}_{\alpha;\beta}\, n^{j} + g_{ij}\, x^{i}{}_{\alpha}\, n^{j}{}_{;\beta}$
1	From T17.3.7 Surface Normal Vector: $\mathfrak{R}^{n}$ Transformation and Normal as Orthogonal System and T12.7.1B Tensor Uniqueness: Tensor Differentiation	$(0)_{;\beta} = (\sum_{i} \sum_{j} g_{ij}\, x^{i}{}_{\alpha}\, n^{j})_{;\beta}$
2	From 1, LxK.3.3.1B Differential of a Constant and T12.7.2 Tensor Summation: Covariant Derivative as a Linear Operator	$0 = \sum_{i} \sum_{j} (g_{ij}\, x^{i}{}_{\alpha}\, n^{j})_{;\beta}$
3	From 2, T12.7.6 Tensor Product: Chain Rule for Binary Covariant Product and again expanding the products through the chain rule	$0 = \sum_{i} \sum_{j} (g_{ij})_{;\beta}\, x^{i}{}_{\alpha}\, n^{j} + g_{ij}\, (x^{i}{}_{\alpha})_{;\beta}\, n^{j} + g_{ij}\, x^{i}{}_{\alpha}\, (n^{j})_{;\beta}$

4	From 3, D17.1.9 Riemannian Manifold $\mathfrak{R}^n$ Mixed Tensor of Rank-2 on a Hypersheet $\mathfrak{H}^{n-1}$ and T12.8.3 Differentiation of the Covariant Metric Tensor (Ricci's Theorem-A)	$0 = \Sigma_i \Sigma_j 0\, x^i_\alpha n^j + g_{ij} x^i_{\alpha;\beta} n^j + g_{ij} x^i_\alpha n^{j;\beta}$
∴	From 4, T10.2.8 Tool: Zero Tensor Times a Tensor is a Zero Tensor, T10.3.18 Addition of Tensors: Identity by Addition, A10.2.14 Correspondence of Tensors and Tensor Coefficients, D17.1.9 Riemannian Manifold $\mathfrak{R}^n$ Mixed Tensor of Rank-2 on a Hypersheet $\mathfrak{H}^{n-1}$ and A7.9.2 Substitution of Vectors	$0 = \Sigma_i \Sigma_j g_{ij} x^i_{\alpha;\beta} n^j + g_{ij} x^i_\alpha n^j_\beta$

Theorem 17.8.5 Weingarten Formulas: Unit Normal Differential Solution (Weingarten Formula)

Step	Hypothesis	$n^i_\alpha = -\Sigma_\gamma w_\alpha{}^\gamma x^i_\gamma$
1	From 17.7.17 Weingarten Formulas: Auxiliary-2 Differential Normal to Curvature Coordinates	$0 = \Sigma_i \Sigma_j g_{ij} x^i_{\alpha;\beta} n^j + g_{ij} x^i_\alpha n^j_\beta$
2	From 1, T17.7.12 Second Fundamental Covariant Metric: in Inverse Unit Normal Form, T17.8.2 Weingarten Formulas: Auxiliary-1 Differential Normal in Linear Form and A7.9.2 Substitution of Vectors	$0 = \Sigma_i \Sigma_j [g_{ij} c_{\alpha\beta} n^i n^j + g_{ij} x^i_\alpha (\Sigma_\gamma \xi_\beta{}^\gamma x^j_\gamma)]$
3	From 2, T10.3.7 Product of Tensors: Commutative by Multiplication of Rank p + q → q + p, A10.2.14 Correspondence of Tensors and Tensor Coefficients and Distributing summation (i, j)	$0 = \Sigma_i \Sigma_j c_{\alpha\beta} n^i g_{ij} n^j + \Sigma_\gamma \Sigma_i \Sigma_j g_{ij} x^i_\alpha x^j_\gamma \xi_\beta{}^\gamma$
4	From 3 and T12.3.9A Metric Transformation from Covariant to Contravariant Coefficient	$0 = c_{\alpha\beta} \Sigma_i n^i n_i + \Sigma_\gamma \Sigma_i \Sigma_j g_{ij} x^i_\alpha x^j_\gamma \xi_\beta{}^\gamma$
5	From 4 and T12.3.18 Magnitude of Normalized Vector	$0 = c_{\alpha\beta}\, 1 + \Sigma_\gamma \Sigma_i \Sigma_j g_{ij} x^i_\alpha x^j_\gamma \xi_\beta{}^\gamma$

6	From 5, T10.2.10 Tool: Tool: Identity Multiplication of a Tensor, T10.3.21 Addition of Tensors: Right Cancellation by Addition, T10.3.18 Addition of Tensors: Identity by Addition, T10.3.8 Product of Tensors: Association by Multiplication of Rank for p + q + r and A10.2.14 Correspondence of Tensors and Tensor Coefficients	$c_{\alpha\beta} = -\sum_\gamma (\sum_i \sum_j g_{ij}\, x^i{}_\alpha\, x^j{}_\gamma)\, \xi_\beta{}^\gamma$
7	From 6, T17.1.9B Manifold-Hypersheet Transformation: h-g Metrics With Change in Direction, T10.2.5 Tool: Tensor Substitution and A10.2.14 Correspondence of Tensors and Tensor Coefficients	$c_{\alpha\beta} = -\sum_\tau a_{\alpha\tau}\, \xi_\beta{}^\tau$
8	From 9, T10.3.6 Product of Tensors: Uniqueness by Multiplication of Rank for p + q, T10.3.8 Product of Tensors: Association by Multiplication of Rank for p + q + r and A10.2.14 Correspondence of Tensors and Tensor Coefficients; while summing over index-η	$\sum_\eta a^{\eta\gamma}\, c_{\eta\beta} = -(\sum_\tau \sum_\eta a^{\eta\gamma}\, a_{\eta\tau})\, \xi_\beta{}^\tau$
9	From 8 and T17.4.6B Hypersheet: Covariant and Contravariant Metric Inverse	$\sum_\eta a^{\eta\gamma}\, c_{\eta\beta} = -(\sum_\tau \delta^\gamma{}_\tau)\, \xi_\beta{}^\tau$
10	From 9 and T17.5.39B Contracted Contravariant Bases Vector with Covariant Vector	$\sum_\eta a^{\eta\gamma}\, c_{\eta\beta} = -\xi_\beta{}^\gamma$
11	From 10, T10.4.1 Scalars and Tensors: Uniqueness of Scalar Multiplication to a Tensors, A10.2.14 Correspondence of Tensors and Tensor Coefficients, D4.1.20A Negative Coefficient and T4.8.2B Integer Exponents: Negative One Squared	$\xi_\beta{}^\gamma = -\sum_\eta a^{\eta\gamma}\, c_{\eta\beta}$

12	From 11, T17.8.2 Weingarten Formulas: Auxiliary-1 Differential Normal in Linear Form, T10.2.5 Tool: Tensor Substitution and A10.2.14 Correspondence of Tensors and Tensor Coefficients; changing β → α dummy indices	$n^i{}_\alpha = \Sigma_\gamma (-\Sigma_\eta a^{\eta\gamma} c_{\eta\alpha})\, x^i{}_\gamma$
13	From 12 T10.3.7 Product of Tensors: Commutative by Multiplication of Rank p + q → q + p, A10.2.14 Correspondence of Tensors and Tensor Coefficients and T17.4.11 Hypersheet: Covariant Metric Inner Product is Symmetrical	$n^i{}_\alpha = -\Sigma_\gamma (\Sigma_\eta c_{\alpha\eta} a^{\eta\gamma})\, x^i{}_\gamma$
∴	From 13, D17.8.1 Weingarten Tensor on a Hypersheet $\mathfrak{H}^{(n-1)}$, T10.2.5 Tool: Tensor Substitution and A10.2.14 Correspondence of Tensors and Tensor Coefficients	$n^i{}_\alpha = -\Sigma_\gamma w_\alpha{}^\gamma x^i{}_\gamma$

Theorem 17.8.6 Weingarten Formulas: Distribution of Fundamental Tensor Under Differentiation

Step	Hypothesis	$w_{\alpha}{}^{\gamma}{}_{;\beta} = \sum_\eta c_{\alpha\beta;\eta}\, a^{\eta\gamma}$ EQ A	$(\sum_\eta c_{\alpha\eta}\, a^{\eta\gamma})_{;\beta} = \sum_\eta c_{\alpha\beta;\eta}\, a^{\eta\gamma}$ EQ B
1	From T12.7.1B Tensor Uniqueness: Tensor Differentiation	$w_{\alpha}{}^{\gamma}{}_{;\beta} = w_{\alpha}{}^{\gamma}{}_{;\beta}$	
2	From 1, D17.8.1 Weingarten Tensor on a Hypersheet $\mathfrak{H}^{(n-1)}$, T10.2.5 Tool: Tensor Substitution and A10.2.14 Correspondence of Tensors and Tensor Coefficients	$w_{\alpha}{}^{\gamma}{}_{;\beta} = (\sum_\eta c_{\alpha\eta}\, a^{\eta\gamma})_{;\beta}$	
3	From 2 and T12.7.6 Tensor Product: Chain Rule for Binary Covariant Product	$w_{\alpha}{}^{\gamma}{}_{;\beta} = \sum_\eta c_{\alpha\eta;\beta}\, a^{\eta\gamma} + \sum_\eta c_{\alpha\eta}\, a^{\eta\gamma}{}_{;\beta}$	
4	From 3 and T17.8.11B Codazzi-Gauss Equations: Equations of Codazzi	$w_{\alpha}{}^{\gamma}{}_{;\beta} = \sum_\eta c_{\alpha\beta;\eta}\, a^{\eta\gamma} + \sum_\eta c_{\alpha\eta}\, a^{\eta\gamma}{}_{;\beta}$	
5	From 3 and T17.5.32B Differentiation of the Covariant Metric Tensor (Ricci's Theorem-A)	$w_{\alpha}{}^{\gamma}{}_{;\beta} = \sum_\eta c_{\alpha\beta;\eta}\, a^{\eta\gamma} + \sum_\eta c_{\alpha\eta}\, \mathbf{0}$	
∴	From 2, 5, T10.2.8 Tool: Zero Tensor Times a Tensor is a Zero Tensor, T10.3.18 Addition of Tensors: Identity by Addition and A10.2.14 Correspondence of Tensors and Tensor Coefficients	$w_{\alpha}{}^{\gamma}{}_{;\beta} = \sum_\eta c_{\alpha\beta;\eta}\, a^{\eta\gamma}$ EQ A	$(\sum_\eta c_{\alpha\eta}\, a^{\eta\gamma})_{;\beta} = \sum_\eta c_{\alpha\beta;\eta}\, a^{\eta\gamma}$ EQ B

Theorem 17.8.7 Riemann-Christoffel Tensor on a Hypersheet: Vector Relation

1g	Given	$A_\nu = x^i{}_\nu$
Step	Hypothesis	$x^i{}_{\alpha;\beta;\gamma} - x^i{}_{\alpha;\gamma;\beta} = \sum_p x^i{}_\nu\, R^\nu{}_{\alpha\beta\gamma}$
∴	From 1g, T17.5.56 Vector Relation to a Contravariant Riemann-Christoffel Tensor, T10.2.5 Tool: Tensor Substitution and A10.2.14 Correspondence of Tensors and Tensor Coefficients	$x^i{}_{\alpha;\beta;\gamma} - x^i{}_{\alpha;\gamma;\beta} = \sum_\nu x^i{}_\nu\, R^\nu{}_{\alpha\beta\gamma}$

Theorem 17.8.8 Codazzi Equations: Tensor of Rank-4 on a Hypersheet

Step	Hypothesis	$x^i_{\alpha;\beta;\gamma} = -(\Sigma_\tau \Sigma_\eta c_{\alpha\beta} a^{\eta\tau} c_{\eta\gamma} x^i_\tau) + c_{\alpha\beta;\gamma} n^i$
1	From T17.7.12 Second Fundamental Covariant Metric: in Inverse Unit Normal Form	$x^i_{\alpha;\beta} = c_{\alpha\beta} n^i$
2	From 1 and T12.7.1B Tensor Uniqueness: Tensor Differentiation	$(x^i_{\alpha;\beta})_{;\gamma} = (c_{\alpha\beta} n^i)_{;\gamma}$
3	From 2 and T12.7.6 Tensor Product: Chain Rule for Binary Covariant Product and D17.1.9 Riemannian Manifold $\mathfrak{R}^n$ Mixed Tensor of Rank-2 on a Hypersheet $\mathfrak{H}^{n-1}$	$x^i_{\alpha;\beta;\gamma} = c_{\alpha\beta;\gamma} n^i + c_{\alpha\beta} n^i_\gamma$
4	From 3, T17.8.5 Weingarten Formulas: Unit Normal Differential Solution (Weingarten Formula), T10.2.5 Tool: Tensor Substitution and A10.2.14 Correspondence of Tensors and Tensor Coefficients; changing α → γ dummy indices	$x^i_{\alpha;\beta;\gamma} = + c_{\alpha\beta} (-\Sigma_\tau \Sigma_\eta a^{\eta\tau} c_{\eta\gamma} x^i_\tau) + c_{\alpha\beta;\gamma} n^i$
∴	From 4 T10.3.20 Addition of Tensors: Distribution of a Tensor over Addition of Tensors and A10.2.14 Correspondence of Tensors and Tensor Coefficients	$x^i_{\alpha;\beta;\gamma} = -(\Sigma_\tau \Sigma_\eta c_{\alpha\beta} a^{\eta\tau} c_{\eta\gamma} x^i_\tau) + c_{\alpha\beta;\gamma} n^i$

Theorem 17.8.9 Codazzi-Gauss Equations: Riemann-Christoffel Tensor on a Hypersheet

Step	Hypothesis	$\sum_\nu R^\nu{}_{\alpha\beta\gamma}\, x^i{}_\nu =$ $\sum_\nu [\sum_\eta a^{\eta\nu} (c_{\alpha\gamma}\, c_{\eta\beta} - c_{\alpha\beta}\, c_{\eta\gamma})]\, x^i{}_\nu + (c_{\alpha\beta;\gamma} - c_{\alpha\gamma;\beta})\, n^i$
1	From 17.8.8 Codazzi Equations: Tensor of Rank-4 on a Hypersheet	$x^i{}_{\alpha;\beta;\gamma} = -(\sum_\tau \sum_\eta c_{\alpha\beta}\, a^{\eta\tau}\, c_{\eta\gamma}\, x^i{}_\tau) + c_{\alpha\beta;\gamma}\, n^i$
2	From 1 and transposing indices $\beta \rightarrow \gamma$	$x^i{}_{\alpha;\gamma;\beta} = -(\sum_\tau \sum_\eta c_{\alpha\gamma}\, a^{\eta\tau}\, c_{\eta\beta}\, x^i{}_\tau) + c_{\alpha\gamma;\beta}\, n^i$
3	From 1, 2 and taking the difference of the equations	$x^i{}_{\alpha;\beta;\gamma} - x^i{}_{\alpha;\gamma;\beta} = -(\sum_\tau \sum_\eta c_{\alpha\beta}\, a^{\eta\tau}\, c_{\eta\gamma}\, x^i{}_\tau) + c_{\alpha\beta;\gamma}\, n^i$ $+ (\sum_\tau \sum_\eta c_{\alpha\gamma}\, a^{\eta\tau}\, c_{\eta\beta}\, x^i{}_\tau) - c_{\alpha\gamma;\beta}\, n^i$
4	From 3, T10.3.20 Addition of Tensors: Distribution of a Tensor over Addition of Tensors and A10.2.14 Correspondence of Tensors and Tensor Coefficients; matching corresponding terms	$x^i{}_{\alpha;\beta;\gamma} - x^i{}_{\alpha;\gamma;\beta} =$ $\sum_\tau \sum_\eta a^{\eta\tau} (c_{\alpha\gamma}\, c_{\eta\beta} - c_{\alpha\beta}\, c_{\eta\gamma})\, x^i{}_\tau + (c_{\alpha\beta;\gamma} - c_{\alpha\gamma;\beta})\, n^i$
∴	From 4, T17.8.7 Riemann-Christoffel Tensor on a Hypersheet: Vector Relation and changing $\tau \rightarrow \nu$ dummy indices	$\sum_\nu R^\nu{}_{\alpha\beta\gamma}\, x^i{}_\nu =$ $\sum_\nu [\sum_\eta a^{\eta\nu} (c_{\alpha\gamma}\, c_{\eta\beta} - c_{\alpha\beta}\, c_{\eta\gamma})]\, x^i{}_\nu + (c_{\alpha\beta;\gamma} - c_{\alpha\gamma;\beta})\, n^i$

Theorem 17.8.10 Codazzi-Gauss Equations: Equations of Codazzi

1g	Given	$\mathbf{n} = \sum_i n_i \mathbf{b}^i$	Normalized vector
Step	Hypothesis	$0 = c_{\alpha\beta;\gamma} - c_{\alpha\gamma;\beta}$ EQ A	$c_{\alpha\beta;\gamma} = c_{\alpha\gamma;\beta}$ EQ B
1	From 17.8.8 Codazzi-Gauss Equations: Riemann-Christoffel Tensor on a Hypersheet	$\sum_v R^v{}_{\alpha\beta\gamma}\, x^i{}_v =$ $+ [\sum_v \sum_\eta a^{\eta v} (c_{\alpha\gamma}\, c_{\eta\beta} - c_{\alpha\beta}\, c_{\eta\gamma})\, x^i{}_v] + (c_{\alpha\beta;\gamma} - c_{\alpha\gamma;\beta})\, n^i$	
2	From 1 and T17.3.15 Dual Covariant Tensors: Identity Addition	$\sum_i \sum_v R^v{}_{\alpha\beta\gamma}\, x^i{}_v + 0\, n^i =$ $\sum_i \sum_v [\sum_\eta a^{\eta v} (c_{\alpha\gamma}\, c_{\eta\beta} - c_{\alpha\beta}\, c_{\eta\gamma})\, x^i{}_v] + \sum_i (c_{\alpha\beta;\gamma} - c_{\alpha\gamma;\beta})\, n^i$	
3	From 1g, 2 and T17.3.9 Dual Covariant Tensors: Mutually Exclusive Normalized Tensor Components	$0 = c_{\alpha\beta;\gamma} - c_{\alpha\gamma;\beta}$	
4	From 3 and T10.3.22B Addition of Tensors: Reversal of Right Cancellation by Addition	$c_{\alpha\beta;\gamma} = c_{\alpha\gamma;\beta}$	
∴	From 3 and 4	$0 = c_{\alpha\beta;\gamma} - c_{\alpha\gamma;\beta}$ EQ A	$c_{\alpha\beta;\gamma} = c_{\alpha\gamma;\beta}$ EQ B

Observation 17.8.1 Loss of Ricci's Theorem with Second Fundamental Covariant Metric

Since the second fundamental covariant metric is not made out of a dual dyadic pairs, then it does not go to zero like the other metrics when differentiate, therefore Ricci's theorem cannot hold. However it is replaced by theorem T17.8.10B "Codazzi-Gauss Equations: Equations of Codazzi", Codazzi's equation, thereby maintaining distribution of the metric under differentiation, see theorem T17.8.6B "Weingarten Formulas: Distribution of Fundamental Tensor Under Differentiation".

Theorem 17.8.11 Codazzi-Gauss Equations: Equation of Gauss

1g	Given	$\mathbf{a}_\alpha = \sum_i x^i{}_\alpha\, \mathbf{b}_i$	Standard hypersurface vector
Step	Hypothesis	$R_{\sigma\alpha\beta\gamma} = c_{\alpha\gamma}\, c_{\sigma\beta} - c_{\alpha\beta}\, c_{\sigma\gamma}$ EQ A	$R_{\sigma\alpha\beta\gamma} = -\begin{vmatrix} c_{\alpha\beta} & c_{\alpha\gamma} \\ c_{\sigma\beta} & c_{\sigma\gamma} \end{vmatrix}$ EQ B
1	From 17.8.9 Codazzi-Gauss Equations: Riemann-Christoffel Tensor on a Hypersheet	$\sum_v R^v{}_{\alpha\beta\gamma}\, x^i{}_v =$ $[\sum_v \sum_\eta a^{\eta v} (c_{\alpha\gamma}\, c_{\eta\beta} - c_{\alpha\beta}\, c_{\eta\gamma})\, x^i{}_v] + (c_{\alpha\beta;\gamma} - c_{\alpha\gamma;\beta})\, n^i$	
2	From 1 and T17.3.15 Dual Covariant Tensors: Identity Addition	$\sum_i \sum_v R^v{}_{\alpha\beta\gamma}\, x^i{}_v + 0\, n^i =$ $\sum_i \sum_v [\sum_\eta a^{\eta v} (c_{\alpha\gamma}\, c_{\eta\beta} - c_{\alpha\beta}\, c_{\eta\gamma})\, x^i{}_v] + \sum_i (c_{\alpha\beta;\gamma} - c_{\alpha\gamma;\beta})\, n^i$	
3	From 1g, 2 and T17.3.10 Dual Covariant Tensors: Mutually Exclusive Tangential Tensor Components	$\sum_v R^v{}_{\alpha\beta\gamma} = \sum_v \sum_\eta a^{\eta v} (c_{\alpha\gamma}\, c_{\eta\beta} - c_{\alpha\beta}\, c_{\eta\gamma})$	

4	From 3g, 3, T10.3.6 Product of Tensors: Uniqueness by Multiplication of Rank for p + q and A10.2.14 Correspondence of Tensors and Tensor Coefficients	$\sum_\kappa a_{\sigma\kappa} R^\kappa{}_{\alpha\beta\gamma} = \sum_\eta \sum_\kappa a_{\sigma\kappa} a^{\kappa\eta} (c_{\alpha\gamma} c_{\eta\beta} - c_{\alpha\beta} c_{\eta\gamma})$
5	From 4, T17.5.67 Riemann-Christoffel Tensor First Kind to Second Kind, T17.4.6A Hypersheet: Covariant and Contravariant Metric Inverse, T10.2.5 Tool: Tensor Substitution, A10.2.14 Correspondence of Tensors and Tensor Coefficients	$R_{\sigma\alpha\beta\gamma} = \sum_\eta \delta_\sigma{}^\eta (c_{\alpha\gamma} c_{\eta\beta} - c_{\alpha\beta} c_{\eta\gamma})$
6	From 5, T10.3.20 Addition of Tensors: Distribution of a Tensor over Addition of Tensors and A10.2.14 Correspondence of Tensors and Tensor Coefficients	$R_{\sigma\alpha\beta\gamma} = (c_{\alpha\gamma} \sum_\eta \delta_\sigma{}^\eta c_{\eta\beta} - c_{\alpha\beta} \sum_\eta \delta_\sigma{}^\eta c_{\eta\gamma})$
7	From 6 and T17.5.37A Contracted Covariant Bases Vector with Contravariant Vector	$R_{\sigma\alpha\beta\gamma} = c_{\alpha\gamma} c_{\sigma\beta} - c_{\alpha\beta} c_{\sigma\gamma}$
8	From 7, T6.10.2 Determinate of a 2x2 Matrix and T6.9.9 Interchanging of any two rows or columns changes the sign in a determinant	$R_{\sigma\alpha\beta\gamma} = -\begin{vmatrix} c_{\alpha\beta} & c_{\alpha\gamma} \\ c_{\sigma\beta} & c_{\sigma\gamma} \end{vmatrix}$
∴	From 7 and 8	$R_{\sigma\alpha\beta\gamma} = c_{\alpha\gamma} c_{\sigma\beta} - c_{\alpha\beta} c_{\sigma\gamma}$ EQ A
		$R_{\sigma\alpha\beta\gamma} = -\begin{vmatrix} c_{\alpha\beta} & c_{\alpha\gamma} \\ c_{\sigma\beta} & c_{\sigma\gamma} \end{vmatrix}$ EQ B

Section 17.9 Gaussian Total Curvatures of a Surface in 2-Dimensions

pg 186 [a][b]

Definition 17.9.1 Covariant Matrix and Gaussian Curvature

There is a unique invariant constant [K] in 2-dimensional space called a ***Gaussian*** or ***total curvature based on a*** (n–1)x(n–1) covariant matrix embedded on a Hypersheet $\mathfrak{H}^{(n-1)}$.

$A_{\mathfrak{H}} = (a_{\alpha\beta})$	(n–1)x(n–1) covariant matrix	Equation A
$a = \lvert a_{\alpha\beta} \rvert$	Covariant Metric Determinate for Hypersheet $\mathfrak{H}^{(n-1)}$	Equation B
and		
$a\,K \equiv R_{1212}$ [17.9.1]	on a Hypersheet $\mathfrak{H}^{2}$	Equation C

Theorem 17.9.1 In 2-Dimensional Space: Evaluation of Metric Determinate for Hypersheet $\mathfrak{H}^{2}$

1g	Given	Covariant Metric Determinate [a] for Hypersheet $\mathfrak{H}^{2}$	
Step	Hypothesis	$a = \begin{vmatrix} a_{11} & a_{12} \\ a_{21} & a_{22} \end{vmatrix}$ EQ A	$a = a_{11}\, a_{22} - a_{12}\, a_{21}$ EQ B
1	From 1g, D17.9.1B Gaussian Curvature or Total Curvature	$a = \begin{vmatrix} a_{11} & a_{12} \\ a_{21} & a_{22} \end{vmatrix}$	
2	From 1 and T6.10.2 Determinate of a 2x2 Matrix; 2x2 evaluated	$a = a_{11}\, a_{22} - a_{12}\, a_{21}$	
∴	From 1 and 2	$a = \begin{vmatrix} a_{11} & a_{12} \\ a_{21} & a_{22} \end{vmatrix}$ EQ A	$a = a_{11}\, a_{22} - a_{12}\, a_{21}$ EQ B

Theorem 17.9.2 In 2-Dimensional Space: The Contravariant Inverse Metric for Hypersheet $\mathfrak{H}^{2}$

1g	Given	Covariant Metric Determinate [a] for Hypersheet $\mathfrak{H}^{2}$	
Step	Hypothesis	$a^{11} = +a^{-1}\, a_{22}$ EQ A $a^{21} = -a^{-1}\, a_{12}$ EQ B	$a^{12} = -a^{-1}\, a_{21}$ EQ C $a^{22} = +a^{-1}\, a_{11}$ EQ D
1	From 1g and T6.10.3 Inverse of a 2x2 Matrix	$A_{\mathfrak{H}}^{-1} = \dfrac{1}{(a_{11} a_{22} - a_{12} a_{21})} \begin{pmatrix} +a_{22} & -a_{21} \\ -a_{12} & +a_{11} \end{pmatrix}$ for $(a_{11}a_{22} - a_{12}a_{21}) \neq 0$	
2	From 1, T17.9.1B In 2-Dimensional Space: Evaluation of Metric Determinate for Hypersheet $\mathfrak{H}^{2}$ and A4.2.3 Substitution	$A_{\mathfrak{H}}^{-1} = \dfrac{1}{a} \begin{pmatrix} +a_{22} & -a_{21} \\ -a_{12} & +a_{11} \end{pmatrix}$	for $a \neq 0$
3	From 2 and D17.1.3C Contravariant Metric Hypersheet Dyadic	$A_{\mathfrak{H}}^{-1} = \begin{pmatrix} a^{11} & a^{12} \\ a^{21} & a^{22} \end{pmatrix}$	
∴	From 2, 3 and equating elements	$a^{11} = +a^{-1}\, a_{22}$ EQ A $a^{21} = -a^{-1}\, a_{12}$ EQ B	$a^{12} = -a^{-1}\, a_{21}$ EQ C $a^{22} = +a^{-1}\, a_{11}$ EQ D

Theorem 17.9.3 In 2-Dimensional Space: Nonvanishing Components of the Riemann Tensor

1g	Given	Nonvanishing component of the Riemann tensor R_{1212} for Hypersheet $\mathfrak{H}^2$	
Step	Hypothesis	$R_{\alpha\alpha\beta\gamma} = R_{\alpha\beta\gamma\gamma} = \mathbf{0}$ EQ A	$R_{1212} = R_{2121} = -R_{2112} = -R_{1221}$ EQ B
1	T17.5.66 Riemann-Christoffel Tensor Second Kind; Duplicate last two indices and T17.5.7 Riemann-Christoffel Tensor First Kind to Second Kind	$R_{\alpha\alpha\beta\gamma} = R_{\alpha\beta\gamma\gamma} = \mathbf{0}$	
2	From T17.5.68 Riemann-Christoffel Tensor First Kind; Transpose First Two Indices	$R_{\alpha g\beta\gamma} = -R_{g\alpha\beta\gamma}$	
3	From 1g and 2; $(\alpha g\beta\gamma) \rightarrow (1212)$	$R_{1212} = -R_{2112}$	
4	From T17.5.69 Riemann-Christoffel Tensor First Kind; Transpose Last Two Indices	$R_{g\alpha\gamma\beta} = -R_{g\alpha\beta\gamma}$	
5	From 1g and 4; $(g\alpha\gamma\beta) \rightarrow (2121)$	$R_{2121} = -R_{2112}$	
6	From 1g and 4; $(g\alpha\gamma\beta) \rightarrow (1212)$	$R_{1212} = -R_{1221}$	
7	From 3, 5 and A4.2.3 Substitution	$R_{1212} = R_{2121}$	
∴	From 1, 5, 6, 7 and A4.2.2 Equality (A, B)	$R_{\alpha\alpha\beta\gamma} = R_{\alpha\beta\gamma\gamma} = \mathbf{0}$ EQ A	$R_{1212} = R_{2121} = -R_{2112} = -R_{1221}$ EQ B

Observation 17.9.1 R1212 is a Constant on a 2-Dimensional Hypersurface

From theorem T17.9.2 "In 2-Dimensional Space: Nonvanishing Components of the Riemann Tensor" R_{1212} is a nonvanishing component everywhere and the Riemann tensor equal to itself or its negative. It is from this observation that the Gaussian Curvature definition is derived. Since only metric coefficients $[a_{\alpha\beta}]$ are involved in this definition, the properties described by [K] are intrinsic properties of the hypersheet $\mathfrak{H}^2$

From theorem T17.9.3 (A, B) "In 2-Dimensional Space: Nonvanishing Components of the Riemann Tensor" in two dimensions there are a total of $2^4 = 16$ components out of which there is only one distinct nonvanishing component R_{1212}. It is seen that this tensor characterizes an extremely important property of all higher dimensional surfaces $\mathfrak{H}^{(n-1)}$. This is the reason that the Gaussian curvature is defined as exclusively a R_{1212} component.

Theorem 17.9.4 In 2-Dimensional Space: Gauss's Curvature Ratio Second-First Fundamental Forms

1g	Given	Nonvanishing component of the Riemann tensor R_{1212} for Hypersheet $\mathfrak{H}^2$	
Step	Hypothesis	$K = (c_{11}\, c_{22} - c_{12}\, c_{21}) / (a_{11}\, a_{22} - a_{12}\, a_{21})$ EQ A	$K = c / a$ EQ B
1	From D17.9.1C Covariant Matrix and Gaussian Curvature, T17.8.11A Codazzi-Gauss Equations: Equations of Gauss and A4.2.2A Equality	$a\, K = R_{\sigma\alpha\beta\gamma}$	
2	From 1g, 1, T4.4.6B Equalities: Left Cancellation by Multiplication and match up indices $(\sigma\alpha\beta\gamma) \rightarrow (1212)$	$K = R_{1212} / a$	
3	From 2, T17.8.11A Codazzi-Gauss Equations: Equations of Gauss, T17.9.1B In 2-Dimensional Space: Evaluation of Metric Determinate for Hypersheet $\mathfrak{H}^2$ and A4.2.3 Substitution	$K = (c_{22}\, c_{11} - c_{21}\, c_{12}) / (a_{11}\, a_{22} - a_{12}\, a_{21})$	
4	From 3 and A4.2.10 Commutative Multp	$K = (c_{11}\, c_{22} - c_{12}\, c_{21}) / (a_{11}\, a_{22} - a_{12}\, a_{21})$	
5	From 4, D17.8.2C Covariant Determinate of Gauss, T17.9.1B In 2-Dimensional Space: Evaluation of Metric Determinate for Hypersheet $\mathfrak{H}^2$ and A4.2.3 Substitution	$K = c / a$	
∴	From 4 and 5	$K = (c_{11}\, c_{22} - c_{12}\, c_{21}) / (a_{11}\, a_{22} - a_{12}\, a_{21})$ EQ A	$K = c / a$ EQ B

[17.9.1]Note: [K] the 2-dimensional Gaussian curvature is defined here to express as a ratio of two constants in the second-first fundamental quadratic forms [SOK67, pg 167], so it fits into a sequence of steps. It's contrived and in the next section it is more generally developed and its definition changes.

Section 17.10 Hypersphere Approximating Curvature to Hypersurface $\mathfrak{H}^{n-1}$

Just as the need arose to be able to measure the deviation from a curve in section 12.13 "Flat Plane Curvature", in this section a similar problem arises, measuring deviation of curvature to a hyperplane. In a like manner the problem can be solved. This can be done similarly by using a hypersphere in n-dimensions in place of its lesser 2-dimensional form the circle.

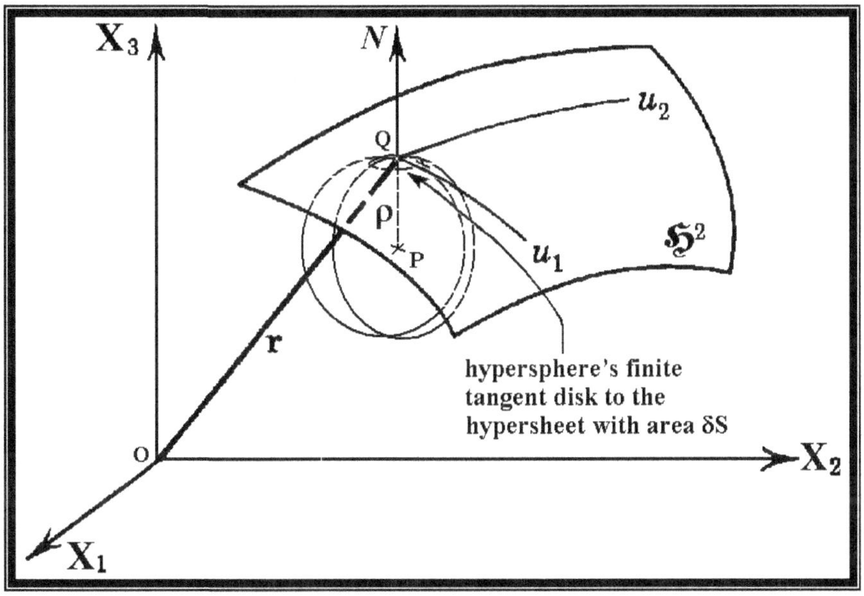

Figure 17.10.1 A Hypersphere Tangent at a Point-Q to a Hypersheet

Definition 17.10.1 Radius of Hypersurface Curvature

The radius of hypersurface curvature [ρ] arises from an infinitesimal segment along the curve from principal hypersphere's great circles that match the curvature of the segment from coordinate curve-u_i at the point-Q along the hypersurface's path. The radius of the circle lays parallel along the unit normal vector and center of the circle terminates at the opposite end of the radius at point-P.

Definition 17.10.2 Finite Tangent Disk to the Hypersurface $\mathfrak{H}^{(n-1)}$

A small disk capping the point-Q, called the ***tangent disk***, with area δS can be circularly inscribed having a span of an isosceles angle [δθ] about that point with a small, but finite radius [δr] lying tangent to the hypersurface. This area can be approximated from the product of small ***principal grid*** segments [$\cdot\delta u_i\cdot$] along the ***principal curvature coordinate***-u_i or variable optimally separated paths. Small, but finite arcs [δs_i] are parts of principal great circles and as a product they correspond to the following elements.

$$\delta S \approx \prod_i^{(n-1)} \delta s_i \qquad \text{Equation A}$$

$$\prod_i^{(n-1)} \delta s_i \approx \prod_i^{(n-1)} \delta u_i \qquad \text{Equation B}$$

Definition 17.10.3 Principal Curvatures to Variable Optimally Separated Paths

The measure of ***principal curvature*** [κ_i] is the reciprocal of the ***principal radii*** [r_i] from the planes of the great circles cutting through the hypersphere along the curves-u_i.

$\kappa_i \equiv 1 / r_i$	Measure of Principal Curvatures	Equation A
$\boldsymbol{\kappa}_i \equiv \kappa_i \mathbf{N}$	Curvature Vector	Equation B
$R_i \equiv \mid r_i \mid$	Radii of Principal Curvatures	Equation C

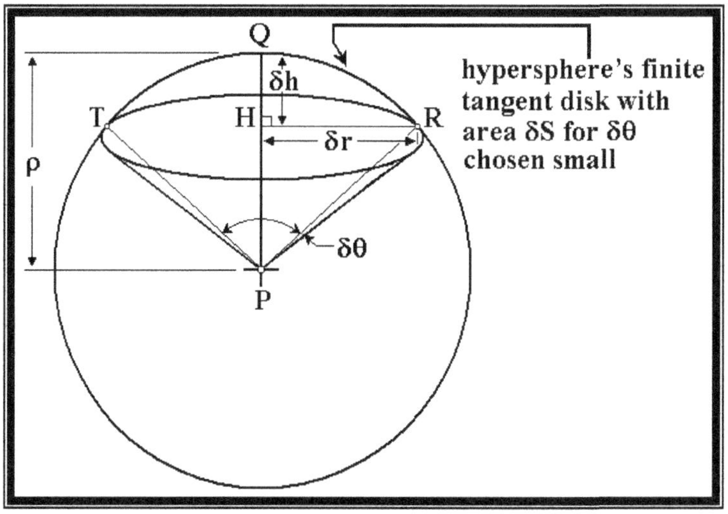

Figure 17.10.2 A Hypersphere Tangent Disk at a Point-Q

Theorem 17.10.1 Tangent Disk: Best Approximation to Hypersurface $\mathfrak{H}^2$

1g	Given	see F17.10.2 A Hypersphere Tangent Disk at a Point-Q
2g		$QP \equiv RP \equiv \rho$
3g		$\angle TPR \equiv \delta\theta$
4g		$HQ \equiv \delta h$
5g		$\delta S \equiv S_+ - \Delta S$
6g	δθ chosen such that; see OK.2.1(A, B) Best Approximate Limit of Sine and Cosine for a Small Angle	$(½\delta\theta) << 0.15$ radians or relatively smaller than 8.5°
Step	Hypothesis	$\delta S \sim \pi\rho^2 \sin(½\delta\theta)$
1	D17.10.2A Finite Tangent Disk to the Hypersurface $\mathfrak{H}^{(n-1)}$ and LxE.4.7.3 Spherical Figures: Zero and Segment of One Base	$\delta S = 2\pi\ QP\ HQ$
2	From 1g and A5.2.8 GII Order of Direction of Points on a $^{£}SL$	$QP = HQ + PH$
3	Form 2g, 4g, 2 and A5.2.15 GIV Transitivity of Segments	$\rho = \delta h + PH$
4	From 3 and T4.3.5B Equalities: Left Cancellation by Addition	$\delta h = \rho - PH$
5	From 1g and DxE.1.1.2 cosine of A	$PH = RP \cos \angle HPR$
6	From 1g, 2g and D5.1.16 Triangle; Isosceles	ΔTPR is isosceles
7	From 5 and T5.9.6 Isosceles Triangle: Right Triangles Bisect Vertex Angle	$\angle HPR = ½ \angle TPR$
8	From 3g, 7 and T5.2.18 GIV Transitivity of Angles	$\angle HPR = ½\ \delta\theta$
9	From 2g, 5, 8 and A4.2.3 Substitution	$PH = \rho \cos (½\ \delta\theta)$
10	From 4, 9 and A4.2.3 Substitution	$\delta h = \rho - \rho \cos (½\ \delta\theta)$
11	From 2g, 4g, 1 and A4.2.3 Substitution	$\delta S = 2\pi\ \rho\ \delta h$
12	From 10, 11 and A4.2.3 Substitution	$\delta S = 2\pi\ \rho\ [\rho - \rho \cos (½\ \delta\theta)]$
13	From 12, A4.2.12 Identity Multp, D4.1.20 Negative Coefficient and A4.2.14 Distribution	$\delta S = 2\pi\ \rho\rho\ [1 - \cos (½\ \delta\theta)]$
14	From 13 and D4.2.17 Exponential Notation	$\delta S = 2\pi\rho^2\ [1 - \cos (½\ \delta\theta)]$
15	From 14, DxK.3.1.2 Definition of a Derivative as a Limit	$(S_+ - \Delta S)\ \Delta\theta = 2\pi\rho^2\ [1 - \cos ½\ (\delta\theta + \Delta\theta)]$ $- 2\pi\rho^2\ [1 - \cos (½\delta\theta)]$

16	From 15, A4.2.14 Distribution, D4.1.20 Negative Coefficient and T4.8.2 Integer Exponents: Negative One Squared	$(S_+ - \Delta S)\ \Delta\theta = 2\pi\rho^2\ [1 - \cos(½\delta\theta + ½\Delta\theta) - 1 + \cos(½\delta\theta)]$
17	From 16, A4.2.5 Commutative Add, A4.2.8 Inverse Add and A4.2.7 Identity Add	$(S_+ - \Delta S)\ \Delta\theta = 2\pi\rho^2\ [\cos(½\delta\theta) - \cos(½\delta\theta + ½\ \Delta\theta)]$
18	From 17, LxE.3.1.24 Angle-Sum and Angle-Difference Relations, A4.2.14 Distribution, D4.1.20 Negative Coefficient and T4.8.2 Integer Exponents: Negative One Squared	$(S_+ - \Delta S)\ \Delta\theta = 2\pi\rho^2\ [\cos(½\delta\theta) - \cos(½\delta\theta)\cos(½\Delta\theta) + \sin(½\delta\theta)\sin(½\Delta\theta)]$
19	From 18, A4.2.12 Identity Multp, D4.1.20 Negative Coefficient and A4.2.14 Distribution	$(S_+ - \Delta S)\ \Delta\theta = 2\pi\rho^2\ \{\cos(½\delta\theta)\ [\ 1 - \cos(½\Delta\theta)\ ] + \sin(½\delta\theta)\sin(½\Delta\theta)\}$
20	From 6g, 19 and A4.2.3 Substitution	$(S_+ - \Delta S)\ \Delta\theta \approx 2\pi\rho^2\ \{\cos(½\delta\theta)\ [\ 1 - 1\ ] + \sin(½\delta\theta)\ ½\Delta\theta\}$
21	From 20, A4.2.8 Inverse Add and A4.2.10 Commutative Multp	$(S_+ - \Delta S)\ \Delta\theta \approx 2\pi\rho^2\ \{\cos(½\delta\theta)\ 0 + ½\sin(½\delta\theta)\ \Delta\theta\}$
22	From 21, T4.4.1 Equalities: Any Quantity Multiplied by Zero is Zero, A4.2.7 Identity Add and A4.2.10 Commutative Multp	$(S_+ - \Delta S)\ \Delta\theta \approx ½\ 2\pi\rho^2\ \sin(½\delta\theta)\ \Delta\theta$
23	From 22 and T4.4.18B Equalities: Product by Division, Factorization by 2	$(S_+ - \Delta S)\ \Delta\theta \approx \pi\rho^2\ \sin(½\delta\theta)\ \Delta\theta$
24	From 5g, 23 and A4.2.3 Substitution	$\delta S\ \Delta\theta \approx \pi\rho^2\ \sin(½\delta\theta)\ \Delta\theta$
∴	From 24 and T4.4.3 Equalities: Cancellation by Multiplication	$\delta S \sim \pi\rho^2\ \sin(½\delta\theta)$

Theorem 17.10.2 Tangent Disk: Best Approximation to Hypersurface $\mathfrak{H}^2$

1g	Given	see F17.10.2 A Hypersphere Tangent Disk at a Point-Q
2g		$C = 2\pi\rho$ Circumference of a great circle
3g		$\delta s = \rho\delta\theta$ Small fraction of a great circle arc
4g	$\delta\theta$ chosen such that; see OK.2.1A Best Approximate Limit of Sine and Cosine for a Small Angle	$(½\delta\theta) << 0.15$ radians or relatively smaller than 8.5°
Step	Hypothesis	$\delta S \sim ¼\ C\ \delta s$
1	From T17.10.1 Tangent Disk: Best Approximation to Hypersurface $\mathfrak{H}^2$	$\delta S \sim \pi\rho^2 \sin(½\delta\theta)$
2	From 4g, 1 and A4.2.3 Substitution	$\delta S \sim \pi\rho^2\ ½\delta\theta$
3	From 2 and A4.2.10 Commutative Multp	$\delta S \sim ½\pi\rho^2\ \delta\theta$
4	From 3 and D4.2.17 Exponential Notation	$\delta S \sim ½\pi\rho\ \rho\delta\theta$
5	From 4, T4.4.18B Equalities: Product by Division, Factorization by 2 and D4.1.19 Primitive Definition for Rational Arithmetic	$\delta S \sim ¼\ 2\pi\rho\ \rho\delta\theta$
∴	From 2g, 3g, 5 and A4.2.3 Substitution	$\delta S \sim ¼\ C\ \delta s$

Theorem 17.10.3 Tangent Disk: Cannot be Reduced to a None Singular Value

1g	Given	$\delta s \equiv \rho\ \delta\theta$		
2g		$\zeta \equiv ¼\ C\rho$		
3g		$a = \lvert a_{\alpha\beta}\rvert \neq 0$ by assumption; the covariant determinate of a hypersheet $\mathfrak{H}^{n-1}$		
Step	Hypothesis	$0 \neq \prod_i^2 \delta s_i$	$0 \neq \prod_i^2 r_i\ \delta\varphi_I$	$0 \neq \prod_i^2 \delta u_i$
1		A	B	C
2	From D17.10.2(A, B) Finite Tangent Disk to the Hypersurface $\mathfrak{H}^2$	$\delta S \approx \prod_i^2 \delta s_i$	$\prod_i^2 \delta s_i \approx \prod_i^2 r_i\ \delta\varphi_i$	$\prod_i^2 r_i\ \delta\varphi_i \approx \prod_i^2 \delta u_i$
3	From 2A; 2A, 2B; 2A, 2B, 2C; and A4.2.3 Substitution	$\delta S \approx \prod_i^2 \delta s_i$	$\delta S \approx \prod_i^2 r_i\ \delta\varphi_i$	$\delta S \approx \prod_i^2 \delta u_i$
4	From 3, T17.10.2 Tangent Disk: Best Approximation to Hypersurface $\mathfrak{H}^2$ and A4.2.3 Substitution	$¼\ C\ \delta s \approx \prod_i^2 \delta s_i$	$¼\ C\ \delta s \approx \prod_i^2 r_i\delta\varphi_i$	$¼\ C\ \delta s \approx \prod_i^2 \delta u_i$
5	From 1g, 4, A4.2.3 Substitution and A4.2.11 Associative Multp	$¼\ C\rho\ \delta\theta \approx \prod_i^2 \delta s_i$	$¼C\rho\ \delta\theta \approx \prod_i^2 r_i\delta\varphi_i$	$¼C\rho\ \delta\theta \approx \prod_i^2 \delta u_i$
6	From 2g, 5 and A4.2.3 Substitution	$\zeta\ \delta\theta \approx \prod_i^2 \delta s_i$	$\zeta\ \delta\theta \approx \prod_i^2 r_i\ \delta\varphi_i$	$\zeta\ \delta\theta \approx \prod_i^2 \delta u_i$

7	From 6 and as a smaller circularly inscribed circular tangential disk is taken such that $\delta\theta \rightarrow 0$	$\zeta\ 0 \approx \prod_i{}^2\ \delta s_i$	$\zeta\ 0 \approx \prod_i{}^2\ r_i\ \delta\varphi_i$	$\zeta\ 0 \approx \prod_i{}^2\ \delta u_i$
8	From 7 and T4.4.1 Equalities: Any Quantity Multiplied by Zero is Zero	$0 \approx \prod_i{}^2\ \delta s_i$	$0 \approx \prod_i{}^2\ r_i\ \delta\varphi_i$	$0 \approx \prod_i{}^2\ \delta u_i$
9	From 3g, 8, A17.1.10 Covariant Determinate Defines the Existence of a Hypersheet $\mathfrak{H}^{n-1}$, hence the inscribed principal arc with its corresponding angle and principal grid cannot be reduced to zero	$0 < \lvert\delta s_i\rvert$ as $\delta\theta \rightarrow 0$	$0 < \lvert\delta\varphi_i\rvert$ as $\delta\theta \rightarrow 0$	$0 < \lvert\delta u_i\rvert$ as $\delta\theta \rightarrow 0$
∴	From 9 and so by identity; it follows that the elements of the hypersurface cannot be reduced to zero.	$0 \neq \prod_i{}^2\ \delta s_i$ $0 < \lvert\delta s_i\rvert$ as $\delta\theta \rightarrow 0$	$0 \neq \prod_i{}^2\ r_i\ \delta\varphi_I$ $0 < \lvert\delta\varphi_i\rvert$ as $\delta\theta \rightarrow 0$	$0 \neq \prod_i{}^2\ \delta u_i$ $0 < \lvert\delta u_i\rvert$ as $\delta\theta \rightarrow 0$

Observation 17.10.1 Improper Modeling of a Tangent Disk

In section 17.9 "Gaussian Total Curvatures of a Surface in 2-Dimensions" the definition D17.9.1C "Covariant Matrix and Gaussian Curvature" is contrived and when applied to Einstein's constant curvature, T14.4.7 "Riemann-Christoffel Tensor Rank-0: Contraction of Einstein's Covariant Tensor" is not proportional to it. From theorem T17.10.3 "Tangent Disk: Cannot be Reduced to a None Singular Value" clearly our notion of a surface curvature and how to properly model it is incorrect. Not only that but also this model does not naturally extend itself to higher dimensions. These types of constructions are common models promoted by a number of authors. Yet intuitively the idea of measuring curvature relative to a tangential hypersphere is a very practical idea and a natural extension of curvature from a 2-dimensional curve to a tangential circle. With the hypersphere's n-dimensional representation one would think it to be a natural way to measure the deviation of curvature on a hypersurface $\mathfrak{H}^{n-1}$, indeed in the next section it is seen that a more rigorous development yields a proper definition of a Gaussian curvature. The development gives a correct definition based on its principal curvatures that are derived by eigenvalue methods, which in tern allows an appropriate development of curvature on a hypersurface $\mathfrak{H}^{n-1}$.

Observation 17.10.2 The Use of the Zero and Segment of One Base as a Model

Why did LxE.4.7.3 "Spherical Figures: Zero and Segment of One Base" fail when extending it to higher then three dimensions?

1) It is designed in only 3-dimensions; as such it cannot be extended into higher spaces.
2) There is no chance of a dimensional residue to be left behind so the surface vanishes by identity as smaller and smaller inscribed tangent disks are constructed.
3) Since LxE.4.7.3 vanishes identically on the left hand side of the equation the right hand side's discreet elements are forced to vanish, a contradiction since by definition the hypersurface exists.

Section 17.11 The Tangential Hypersphere to Gaussian Curvature

Definition 17.11.1 Hypersphere

A hypersphere is the locus of all points at a given distance from a fixed point-Q(• x_i •). The fixed point is called the ***center of the hypersphere*** and the fixed distance the ***radius*** [ρ]. Let the center point be represented by the point-P(• a_i •) in a Riemannian $\mathfrak{R}^n$ Space.

Definition 17.11.2 Total Gaussian Curvature

The accumulation of all minor principal curvatures for a hypersurface is the ***total Gaussian curvature***.

$K \equiv 1 / \rho^2$ Equation A

$[L^{-2}]$ Dimensionally Equation B

Theorem 17.11.1 Hypersphere: Auxiliary-1 Normalized Vector Dot Product

1g	Given	$\mathbf{r} = x^i\, \mathbf{b}_i$	positional head vector
2g		$\mathbf{r}_0 = a^i\, \mathbf{b}_i$	positional tail vector
3g		$\rho = \mid \mathbf{r} - \mathbf{r}_0 \mid$ a fixed radial distance from all locus of points on the hypersurface.	
4g		$\mathbf{b}_i \bullet \mathbf{b}_i = g_{ii}$	Riemannian $\mathfrak{R}^n$ manifold metric
5g		$\eta_i = (x_i - a_i) / \rho$	normalized contravariant vector
6g		$\eta^i = (x^i - a^i) / \rho$	normalized covariant vector

Step	Hypothesis	$1 = \eta_i\, \eta^i$ EQ A	$1 = \eta^i\, \eta_i$ EQ B
1	From 1g, 2g, 3g and A7.9.2 Substitution of Vectors	$\rho = \mid x^i\, \mathbf{b}_i - a^i\, \mathbf{b}_i \mid$	
2	From 1 and T7.10.9 Distribution of Vector over Addition of Scalars	$\rho = \mid (x^i - a^i)\, \mathbf{b}_i \mid$	
3	From 2 and T8.3.4 Dot Product: Magnitude	$\rho^2 = [(x^i - a^i)\, \mathbf{b}_i] \bullet [(x^j - a^j)\, \mathbf{b}_j]$	
4	From 3, T12.3.16A Magnitude of Covariant Vector, T10.3.7 Product of Tensors: Commutative by Multiplication of Rank p + q → q + p, T10.3.8 Product of Tensors: Association by Multiplication of Rank for p + q + r and A10.2.14 Correspondence of Tensors and Tensor Coefficients	$\rho^2 = (x^i - a^i)\, [(x^j - a^j)\, g_{ij}]$	
5	From 4, T10.3.20 Addition of Tensors: Distribution of a Tensor over Addition of Tensors and A10.2.14 Correspondence of Tensors and Tensor Coefficients	$\rho^2 = (x^i - a^i)\, (x^j\, g_{ij} - a^j\, g_{ij})$	
6	From 5 and O12.3.5A Rules for Raising or Lowering Tensor Indices	$\rho^2 = (x^i - a^i)\, (x_i - a_i)$	

7	From 6, A4.2.11 Associative Multp and T4.4.6B Equalities: Left Cancellation by Multiplication	$1 = [(x^i - a^i)\ (x_i - a_i)] / \rho^2$	
8	From 7, T7.10.3 Distribution of a Scalar over Addition of Vectors, D4.1.17 Exponential Notation, A4.2.10 Commutative Multp, A4.2.11 Associative Multp and D4.1.4(A, B) Rational Numbers	$1 = [(x^i - a^i) / \rho][(x_i - a_i) / \rho]$	
9	From 5g, 6g, 8 and A7.9.2 Substitution	$1 = \eta^i\ \eta_i$	
∴	From 9 and T8.3.3 Dot Product: Commutative Operation	$1 = \eta_i\ \eta^i$ EQ A	$1 = \eta^i\ \eta_i$ EQ B

Theorem 17.11.2 Hypersphere: Auxiliary-2 Normalized Differential Vector

1g	Given	$\eta_i = (x_i - a_i) / \rho$ normalized contravariant vector	
2g		$\eta^i = (x^i - a^i) / \rho$ normalized covariant vector	
Step	Hypothesis	$\eta_{i;\alpha} = (x_{i;\alpha}) / \rho$ EQ A	$\eta^i{}_\alpha = (x^i{}_\alpha) / \rho$ EQ B
1	From 1g and 2g	$\eta_i = (x_i - a_i) / \rho$	$\eta^i = (x^i - a^i) / \rho$
2	From 1 and T12.7.1B Tensor Uniqueness: Tensor Differentiation	$(\eta_i)_{;\alpha} = [(x_i - a_i) / \rho]_{;\alpha}$	$(\eta^i)_{;\alpha} = [(x^i - a^i) / \rho]_{;\alpha}$
3	From 2 and LxK.3.3.3D Differentiation of a polynomial	$\eta_{i;\alpha} = (x_i - a_i)_{;\alpha} / \rho$	$\eta^i{}_{;\alpha} = (x^i - a^i)_{;\alpha} / \rho$
4	From 3 and LxK.3.2.2B Differential as a Linear Operator	$\eta_{i;\alpha} = [(x_i)_{;\alpha} - (a_i)_{;\alpha}] / \rho$	$\eta^i{}_{;\alpha} = [(x^i)_{;\alpha} - (a^i)_{;\alpha}] / \rho$
5	From 4 and LxK.3.3.1B Differential of a Constant	$\eta_{i;\alpha} = [(x_i)_{;\alpha} - 0] / \rho$	$\eta^i{}_{;\alpha} = [(x^i)_{;\alpha} - 0] / \rho$
6	From 5, A7.8.6 Parallelogram Law: Multiplying a Negative Number Times a Vector, T7.10.1A Uniqueness of Scalar Multiplication to Vectors and T7.10.4 Grassmann's Zero Vector	$\eta_{i;\alpha} = (x_{i;\alpha} + 0) / \rho$	$\eta^i{}_{;\alpha} = (x^i{}_{;\alpha} + 0) / \rho$
∴	From 6, A7.9.11 Identity Vector Object Add and D17.1.9 Riemannian Manifold $\mathfrak{R}^n$ Mixed Tensor of Rank-2 on a Hypersheet $\mathfrak{H}^{n-1}$	$\eta_{i;\alpha} = (x_{i;\alpha}) / \rho$ EQ A	$\eta^i{}_\alpha = (x^i{}_\alpha) / \rho$ EQ B

Observation 17.11.1 Proper Modeling of a Tangent Disk

From theorem T17.11.2 "Hypersphere: Auxiliary-2 Normalized Differential Vector" brings back the idea of a tangent disk making an infinitesimal contact at the point-Q from the hypersphere tangent vector $[x^i{}_\alpha]$ to the hypersurface along the normal vector $[\eta^i{}_\alpha]$ on the hypersheet. While at the same time provides a radius $[\rho]$ showing a measured curvature at point-Q. Clearly a much more comprehensive and general approach then the contrived ideas in section 17.10 "Hypersphere Approximating Curvature to Hypersurface $\mathfrak{H}^{n-1}$".

Theorem 17.11.3 Hypersphere: Auxiliary-3 Unit Normal Form Exchange of Indices

Step	Hypothesis	$c_{\alpha\beta} = -x^i_{\ \alpha}\,\eta_{i;\beta}$
1	From T17.3.7C Surface Normal Vector: $\mathfrak{R}^n$ Transformation and Normal as Orthogonal System	$0 = x^i_{\ \alpha}\,\eta_i$
2	From 1 and T12.7.1B Tensor Uniqueness: Tensor Differentiation	$(0)_{;\beta} = (x^i_{\ \alpha}\,\eta_i)_{;\beta}$
3	From 2, LxK.3.3.1B Differential of a Constant, T12.7.8 Tensor Product: Chain Rule for Mixed Binary Product and A10.2.14 Correspondence of Tensors and Tensor Coefficients	$0 = (x^i_{\ \alpha})_{;\beta}\,\eta_i + x^i_{\ \alpha}\,(\eta_i)_{;\beta}$
4	From 3 and D17.1.9 Riemannian Manifold $\mathfrak{R}^n$ Mixed Tensor of Rank-2 on a Hypersheet $\mathfrak{H}^{n-1}$	$0 = x^i_{\ \alpha;\beta}\,\eta_i + x^i_{\ \alpha}\,\eta_{i;\beta}$
5	From 4, T17.7.8 Second Fundamental Covariant Metric: in Unit Normal Form, T10.2.5 Tool: Tensor Substitution and T10.2.15 Correspondence of Tensors and Tensor Coefficients	$0 = c_{\alpha\beta} + x^i_{\ \alpha}\,\eta_{i;\beta}$
∴	From 5, T10.3.20B Addition of Tensors: Right Cancellation by Addition, T10.3.18 Addition of Tensors: Identity by Addition and T10.2.15 Correspondence of Tensors and Tensor Coefficients	$c_{\alpha\beta} = -x^i_{\ \alpha}\,\eta_{i;\beta}$

Theorem 17.11.4 Hypersphere: Auxiliary-4 Metric Tangent to a Hypersurface $\mathfrak{H}^{n-1}$

Step	Hypothesis	$c_{\alpha\beta} = -a_{\alpha\beta} / \rho$ hypersphere tangent to the hypersurface
1	From T17.11.3 Hypersphere: Auxiliary-3 Unit Normal Form Exchange of Indices	$c_{\alpha\beta} = -x^i_{\ \alpha}\, \eta_{i;\beta}$
2	From 1, T17.11.2A Hypersphere: Auxiliary-2 Normalized Differential Vector, T10.2.5 Tool: Tensor Substitution and T10.2.15 Correspondence of Tensors and Tensor Coefficients	$c_{\alpha\beta} = -x^i_{\ \alpha}\, (x_{i;\beta}) / \rho$
3	From 2 and O12.3.5A Rules for Raising or Lowering Tensor Indices	$c_{\alpha\beta} = -x^i_{\ \alpha}\, (g_{ij}\, x^j_{\ ;\beta}) / \rho$
4	From 3, T10.3.8 Product of Tensors: Association by Multiplication of Rank for p + q + r, T10.3.7 Product of Tensors: Commutative by Multiplication of Rank p + q → q + p and A10.2.14 Correspondence of Tensors and Tensor Coefficients	$c_{\alpha\beta} = -(x^i_{\ \alpha}\, x^j_{\ \beta}\, g_{ij}) / \rho$
∴	From 4, T17.1.9B Manifold-Hypersheet Transformation: h-g Metrics With Change in Direction, T10.2.5 Tool: Tensor Substitution and T10.2.15 Correspondence of Tensors and Tensor Coefficients	$c_{\alpha\beta} = -a_{\alpha\beta} / \rho$ hypersphere tangent to the hypersurface

Theorem 17.11.5 Hypersphere: Relationship of Gauss's Equation to Total Curvature

Step	Hypothesis	$R_{\sigma\alpha\beta\gamma} = K\,(a_{\alpha\gamma}\, a_{\sigma\beta} - a_{\alpha\beta}\, a_{\sigma\gamma})$
1	From T17.8.11A Codazzi-Gauss Equations: Equation of Gauss	$R_{\sigma\alpha\beta\gamma} = c_{\alpha\gamma}\, c_{\sigma\beta} - c_{\alpha\beta}\, c_{\sigma\gamma}$
2	From 1, T17.11.4 Hypersphere: Auxiliary-4 Metric Tangent to a Hypersurface $\mathfrak{H}^{n-1}$, T10.2.5 Tool: Tensor Substitution and A10.2.14 Correspondence of Tensors and Tensor Coefficients	$R_{\sigma\alpha\beta\gamma} = (-a_{\alpha\gamma} / \rho)(-a_{\sigma\beta} / \rho) - (-a_{\alpha\beta} / \rho)(-a_{\sigma\gamma} / \rho)$
3	From 2, D4.1.20A Negative Coefficient, D4.1.4 (A, B) Rational Numbers, T10.4.8 Scalars and Tensors: Scalar Commutative with Tensors and A10.2.14 Correspondence of Tensors and Tensor Coefficients	$R_{\sigma\alpha\beta\gamma} = (-1)\,(-1)\,(a_{\alpha\gamma}\, a_{\sigma\beta}) / \rho\,\rho - (-1)\,(-1)\,(a_{\alpha\beta}\, a_{\sigma\gamma}) / \rho\,\rho$
4	From 3, T4.8.2B Integer Exponents: Negative One Squared, A4.2.12 Identity Multp, D4.1.17 Exponential Notation, T10.4.6 Scalars and Tensors: Distribution of a Scalar over Addition of Tensors and T10.2.15 Correspondence of Tensors and Tensor Coefficients	$R_{\sigma\alpha\beta\gamma} = (a_{\alpha\gamma}\, a_{\sigma\beta} - a_{\alpha\beta}\, a_{\sigma\gamma}) / \rho^2$
5	From 4, T10.4.8 Scalars and Tensors: Scalar Commutative with Tensors and T10.2.15 Correspondence of Tensors and Tensor Coefficients	$R_{\sigma\alpha\beta\gamma} = (1 / \rho^2)\,(a_{\alpha\gamma}\, a_{\sigma\beta} - a_{\alpha\beta}\, a_{\sigma\gamma})$
∴	From 5, D17.11.2 Total Gaussian Curvature and A4.2.3 Substitution	$R_{\sigma\alpha\beta\gamma} = K\,(a_{\alpha\gamma}\, a_{\sigma\beta} - a_{\alpha\beta}\, a_{\sigma\gamma})$

Theorem 17.11.6 Hypersphere: Relationship of Riemann's to Gauss's Total Curvature

Step	Hypothesis	$R = n(1-n)\ K$
1	From T17.11.5 Hypersphere: Relationship of Gauss's Equation to Total Curvature	$R_{\sigma\alpha\beta\gamma} = K\ (a_{\alpha\gamma}\ a_{\sigma\beta} - a_{\alpha\beta}\ a_{\sigma\gamma})$
2	From T17.5.71 Riemann-Christoffel Tensor First Kind: Contracted to Spatial Curvature	$R = a^{\alpha\beta}\ a^{\sigma\gamma}\ R_{\sigma\alpha\beta\gamma}$
3	From 1, 2, T10.2.5 Tool: Tensor Substitution and A10.2.14 Correspondence of Tensors and Tensor Coefficients	$R = a^{\alpha\beta}\ a^{\sigma\gamma}\ K\ (a_{\alpha\gamma}\ a_{\sigma\beta} - a_{\alpha\beta}\ a_{\sigma\gamma})$
4	From 3, T10.4.8 Scalars and Tensors: Scalar Commutative with Tensors and A10.2.14 Correspondence of Tensors and Tensor Coefficients	$R = a^{\alpha\beta}\ a^{\sigma\gamma}\ (a_{\alpha\gamma}\ a_{\sigma\beta} - a_{\alpha\beta}\ a_{\sigma\gamma})\ K$
5	From 4, T10.3.7 Product of Tensors: Commutative by Multiplication of Rank p + q → q + p, T10.3.20 Addition of Tensors: Distribution of a Tensor over Addition of Tensors and A10.2.14 Correspondence of Tensors and Tensor Coefficients	$R = a^{\sigma\gamma}\ (a_{\alpha\gamma}\ a^{\alpha\beta}\ a_{\sigma\beta} - a^{\alpha\beta}\ a_{\alpha\beta}\ a_{\sigma\gamma})\ K$
6	From 5 and T17.4.11 Hypersheet: Covariant Metric Inner Product is Symmetrical	$R = a^{\sigma\gamma}\ (a_{\alpha\gamma}\ a^{\alpha\beta}\ a_{\beta\sigma} - a^{\alpha\beta}\ a_{\beta\alpha}\ a_{\sigma\gamma})\ K$
7	From 6 and T17.4.6B Hypersheet: Covariant and Contravariant Metric Inverse	$R = a^{\sigma\gamma}\ (a_{\gamma\alpha}\ \delta^{\alpha}{}_{\sigma} - \delta^{\alpha}{}_{\alpha}\ a_{\sigma\gamma})\ K$
8	From 7, T17.4.14 Hypersheet: Covariant-Contravariant Kronecker Delta Summation of like Indices, T17.5.37B Contracted Covariant Bases Vector with Contravariant Vector, T10.2.5 Tool: Tensor Substitution and A10.2.14 Correspondence of Tensors and Tensor Coefficients	$R = a^{\sigma\gamma}\ (a_{\gamma\sigma} - n\ a_{\sigma\gamma})\ K$
9	From 9 and T17.4.11 Hypersheet: Covariant Metric Inner Product is Symmetrical	$R = a^{\sigma\gamma}\ (a_{\gamma\sigma} - n\ a_{\gamma\sigma})\ K$
10	From 10, T10.2.10 Tool: Identity Multiplication of a Tensor and T10.4.7 Scalars and Tensors: Distribution of a Tensor over Addition of Scalars	$R = a^{\sigma\gamma}\ a_{\gamma\sigma}\ (1 - n)\ K$

11	From 11 and T17.4.6B Hypersheet: Covariant and Contravariant Metric Inverse	$R = \delta^{\sigma}{}_{\sigma} (1 - n) K$
∴	From 12, T17.4.14 Hypersheet: Covariant-Contravariant Kronecker Delta Summation of like Indices, T10.2.5 Tool: Tensor Substitution and A10.2.14 Correspondence of Tensors and Tensor Coefficients	$R = n(1-n) K$

Theorem 17.11.7 Hypersphere: Relationship of Einstein's to Gauss's Total Curvature

Step	Hypothesis	$T = ½\, n\, (n - 1)\, (n - 2)\, K$
1	From T17.11.6 Hypersphere: Relationship of Riemann's to Gauss's Total Curvature	$R = n(1 - n) K$
2	From 1 and T17.5.63 Riemann-Christoffel Tensor Rank-0: Contraction of Einstein's Covariant Tensor	$T\, \xi = -(½\, n - 1)\, R$
3	From 1, 2 and A4.2.3 Substitution	$T = -(½\, n - 1)\, n(1 - n)\, K$
4	From 3, A4.2.13 Inverse Multp, D4.1.4B Rational Numbers, A4.2.14 Distribution and A4.2.10 Commutative Multp	$T = -½\, n\, (n - 2)\, (1 - n)\, K$
∴	From 4, D4.1.20 Negative Coefficient, A4.2.10 Commutative Multp, A4.2.14 Distribution, T4.8.2B Integer Exponents: Negative One Squared and A4.2.5 Commutative Add	$T = ½\, n\, (n - 1)\, (n - 2)\, K$

Theorem 17.11.8 Hypersphere: Riemannian Total Curvature Einstein and Gauss's Relationship

Step	Hypothesis	$T\ \xi = -\frac{1}{2}\ (n - 2)\ R$ EQ A	$K = [1\ /\ n(1-n)]\ R$ EQ B
1	From T17.11.6 Hypersphere: Relationship of Riemann's to Gauss's Total Curvature and T4.4.4B Equalities: Right Cancellation by Multiplication	$K = [1\ /\ n(1-n)]\ R$	
2	From 1 and T14.11.7 Hypersphere: Relationship of Einstein's to Gauss's Total Curvature	$T = \frac{1}{2}\ n\ (n - 1)\ (n - 2)\ K$	
3	From 1, 2 and A4.2.3 Substitution	$T\ \xi = \frac{1}{2}\ n\ (n - 1)\ (n - 2)\ [1\ /\ n(1-n)]\ R$	
4	From 3, D4.1.20 Negative Coefficient, A4.2.10 Commutative Multp, A4.2.14 Distribution, T4.8.2B Integer Exponents: Negative One Squared and A4.2.5 Commutative Add	$T\ \xi = -\frac{1}{2}\ n\ (n - 1)\ (n - 2)\ [1\ /\ n\ (n - 1)]\ R$	
∴	From 1, 4, A4.2.10 Commutative Multp, A4.2.13 Inverse Multp and A4.2.12 Identity Multp	$T\ \xi = -\frac{1}{2}\ (n - 2)\ R$ EQ A	$K = [1\ /\ n(1-n)]\ R$ EQ B

Theorem 17.11.9 Hypersphere: Riemannian Total Curvature in Determinate Form

Step	Hypothesis	$R = a^{\alpha\beta} [(\nabla_\beta(\ln \lvert\sqrt{a}\rvert)_{,\alpha} - \nabla_\gamma \Gamma_{\alpha\beta}{}^{\gamma}) + \sum_b(\Gamma_{b\beta}{}^{\gamma} \Gamma_{\alpha\gamma}{}^{b} - (\ln \lvert\sqrt{a}\rvert)_{,b}\Gamma_{\alpha\beta}{}^{b})]$ EQ A $R = a^{\alpha\beta} \left(\begin{vmatrix} \nabla_\beta & \nabla_\gamma \\ \Gamma_{\alpha\beta}{}^{\gamma} & (\ln \lvert\sqrt{a}\rvert)_{,\alpha} \end{vmatrix} + \sum_b \begin{vmatrix} \Gamma_{b\beta}{}^{\gamma} & (\ln \lvert\sqrt{a}\rvert)_{,b} \\ \Gamma_{\alpha\beta}{}^{b} & \Gamma_{\alpha\gamma}{}^{b} \end{vmatrix} \right)$ EQ B
1	T17.5.71 Riemann-Christoffel Tensor First Kind: Contracted to Spatial Curvature	$R = a^{\alpha\beta} a^{g\gamma} R_{g\alpha\beta\gamma}$
2	From 2, T14.2.2 Riemann-Christoffel Tensor First Kind and A17.5.1 Theorem Construction on Manifolds of Different Dimension	$R_{g\alpha\beta\gamma} = (\nabla_\beta \Gamma_{\alpha\gamma,g} - \nabla_\gamma \Gamma_{\alpha\beta,g}) + \sum_b (\Gamma_{g\gamma,b} \Gamma_{\alpha\beta}{}^{b} - \Gamma_{g\beta,b} \Gamma_{\alpha\gamma}{}^{b})$
3	From 1, 2, T10.2.5 Tool: Tensor Substitution, T10.3.20 Addition of Tensors: Distribution of a Tensor over Addition of Tensors and A10.2.14 Correspondence of Tensors and Tensor Coefficients	$R = a^{\alpha\beta} [(a^{g\gamma} \nabla_\beta \Gamma_{\alpha\gamma,g} - a^{g\gamma} \nabla_\gamma \Gamma_{\alpha\beta,g})$ $+ \sum_b (\Gamma_{g\gamma,b}\, a^{g\gamma} \Gamma_{\alpha\beta}{}^{b} - \Gamma_{g\beta,b}\, a^{g\gamma} \Gamma_{\alpha\gamma}{}^{b}]$
4	From 3 and T17.5.50 Differentiation of Contravariant Metric and Covariant Coefficient	$R = a^{\alpha\beta} [(\nabla_\beta \Gamma_{\alpha\gamma,g}\, a^{g\gamma} - \nabla_\gamma \Gamma_{\alpha\beta,g}\, a^{g\gamma})$ $+ \sum_b (\Gamma_{g\gamma,b}\, a^{g\gamma} \Gamma_{\alpha\beta}{}^{b} - \Gamma_{g\beta,b}\, a^{g\gamma} \Gamma_{\alpha\gamma}{}^{b}]$
5	From 4, D17.5.2 Hypersurface Christoffel Symbol of the Second Kind, T10.2.5 Tool: Tensor Substitution and A10.2.14 Correspondence of Tensors and Tensor Coefficients	$R = a^{\alpha\beta} [(\nabla_\beta \Gamma_{\alpha\gamma}{}^{\gamma} - \nabla_\gamma \Gamma_{\alpha\beta}{}^{\gamma})$ $+ \sum_b (\Gamma_{g\gamma,b}\, a^{g\gamma} \Gamma_{\alpha\beta}{}^{b} - \Gamma_{g\beta,b}\, a^{g\gamma} \Gamma_{\alpha\gamma}{}^{b}]$
6	From 5 and T17.5.18A Christoffel Symbol: First Kind Asymmetry of Last Two Indices	$\Gamma_{g\gamma,b} = (a_{gb})_{,\gamma} - \Gamma_{b\gamma,g}$
7	From 5 and T17.5.18A Christoffel Symbol: First Kind Asymmetry of Last Two Indices	$\Gamma_{g\beta,b} = (a_{gb})_{,\beta} - \Gamma_{b\beta,g}$
8	From 5, 6, 7, T10.2.5 Tool: Tensor Substitution and A10.2.14 Correspondence of Tensors and Tensor Coefficients	$R = a^{\alpha\beta} [(\nabla_\beta \Gamma_{\alpha\gamma}{}^{\gamma} - \nabla_\gamma \Gamma_{\alpha\beta}{}^{\gamma})$ $+ \sum_b ([(a_{gb})_{,\gamma} - \Gamma_{b\gamma,g}]\, a^{g\gamma} \Gamma_{\alpha\beta}{}^{b}$ $- [(a_{gb})_{,\beta} - \Gamma_{b\beta,g}]\, a^{g\gamma} \Gamma_{\alpha\gamma}{}^{b})]$

9	From 8, T10.3.20 Addition of Tensors: Distribution of a Tensor over Addition of Tensors, T17.5.50 Differentiation of Contravariant Metric and Covariant Coefficient and A10.2.14 Correspondence of Tensors and Tensor Coefficients	$R = a^{\alpha\beta} [(\nabla_\beta \Gamma_{\alpha\gamma}{}^{\gamma} - \nabla_\gamma \Gamma_{\alpha\beta}{}^{\gamma}) + \Sigma_b ([(a_{gb} a^{g\gamma}),_\gamma - \Gamma_{b\gamma,g} a^{g\gamma}] \Gamma_{\alpha\beta}{}^{b} - [(a_{gb} a^{g\gamma}),_\beta - \Gamma_{b\beta,g} a^{g\gamma}] \Gamma_{\alpha\gamma}{}^{b})]$
10	From 9, D17.5.2 Hypersurface Christoffel Symbol of the Second Kind, T17.4.11 Hypersheet: Covariant Metric Inner Product is Symmetrical and T17.4.6A Hypersheet: Covariant and Contravariant Metric Inverse	$R = a^{\alpha\beta} [(\nabla_\beta \Gamma_{\alpha\gamma}{}^{\gamma} - \nabla_\gamma \Gamma_{\alpha\beta}{}^{\gamma}) + \Sigma_b ([(\delta_b{}^{\gamma}),_\gamma - \Gamma_{b\gamma}{}^{\gamma}] \Gamma_{\alpha\beta}{}^{b} - [(\delta_b{}^{\gamma}),_\beta - \Gamma_{b\beta}{}^{\gamma}] \Gamma_{\alpha\gamma}{}^{b})]$
11	From 10 and T17.5.44B Differentiated Kronecker Delta and Fundamental Tensors	$R = a^{\alpha\beta} [(\nabla_\beta \Gamma_{\alpha\gamma}{}^{\gamma} - \nabla_\gamma \Gamma_{\alpha\beta}{}^{\gamma}) + \Sigma_b ([0 - \Gamma_{b\gamma}{}^{\gamma}] \Gamma_{\alpha\beta}{}^{b} - [0 - \Gamma_{b\beta}{}^{\gamma}] \Gamma_{\alpha\gamma}{}^{b})]$
12	From 11, T10.3.18 Addition of Tensors: Identity by Addition, and A10.2.14 Correspondence of Tensors and Tensor Coefficients, D4.1.20 Negative Coefficient and T4.8.2B Integer Exponents: Negative One Squared	$R = a^{\alpha\beta} [(\nabla_\beta \Gamma_{\alpha\gamma}{}^{\gamma} - \nabla_\gamma \Gamma_{\alpha\beta}{}^{\gamma}) + \Sigma_b (\Gamma_{b\beta}{}^{\gamma} \Gamma_{\alpha\gamma}{}^{b} - \Gamma_{b\gamma}{}^{\gamma} \Gamma_{\alpha\beta}{}^{b})]$
13	From 12, T17.5.36B Derivative of the Natural Log of the Covariant Determinate, T10.2.5 Tool: Tensor Substitution and A10.2.14 Correspondence of Tensors and Tensor Coefficients	$R = a^{\alpha\beta} [(\nabla_\beta(\ln \lvert\sqrt{a}\rvert),_\alpha - \nabla_\gamma \Gamma_{\alpha\beta}{}^{\gamma}) + \Sigma_b(\Gamma_{b\beta}{}^{\gamma} \Gamma_{\alpha\gamma}{}^{b} - (\ln \lvert\sqrt{a}\rvert),_b \Gamma_{\alpha\beta}{}^{b})]$
∴	From 13 and T6.10.2 Determinate of a 2x2 Matrix	$R = a^{\alpha\beta} [(\nabla_\beta(\ln \lvert\sqrt{a}\rvert),_\alpha - \nabla_\gamma \Gamma_{\alpha\beta}{}^{\gamma}) + \Sigma_b(\Gamma_{b\beta}{}^{\gamma} \Gamma_{\alpha\gamma}{}^{b} - (\ln \lvert\sqrt{a}\rvert),_b \Gamma_{\alpha\beta}{}^{b})]$ EQ A
		$R = a^{\alpha\beta} \left(\begin{vmatrix} \nabla_\beta & \nabla_\gamma \\ \Gamma_{\alpha\beta}{}^{\gamma} & (\ln \lvert\sqrt{a}\rvert),_\alpha \end{vmatrix} + \Sigma_b \begin{vmatrix} \Gamma_{b\beta}{}^{\gamma} & (\ln \lvert\sqrt{a}\rvert),_b \\ \Gamma_{\alpha\beta}{}^{b} & \Gamma_{\alpha\gamma}{}^{b} \end{vmatrix} \right)$ EQ B

Observation 17.11.2 Calculating Einstein and Gauss from a Riemannian Total Curvature

It is seen from theorems T17.11.8 “Hypersphere: Riemannian Total Curvature Einstein and Gauss’s Relationship” equations A and B is a simple matter by substituting T17.11.9 “Hypersphere: Riemannian Total Curvature in Determinate Form” A or B into them. This gives away to exclusively calculate them in terms of their spatial geometry.

Section 17.12 Minor Principal Curvatures on a Hypersurface $\mathfrak{H}^{n-1}$

From figure F17.10.1 "A Hypersphere Tangent at a Point-Q to a Hypersheet" the 3-dimensional sphere shows two great circles tangent to the coordinate curves on the embedded 2-dimensional sheet at the point-Q. When talking about 2-dimensions it is **not** implied that the embedded sheet is a flat, 2-space, Euclidian plane, but that there is a one-to-one correspondence between the great circles of the tangent sphere and the coordinate curves. Hence there is only as many great circles as coordinate curves, for this space only two are embedded in a 3-dimensional Riemannian space. While a great circle at point-Q can be rotated around that point generating an infinite number of great circles, yet only two principal great circles will match up with the coordinate curves to measure the curvatures along the coordinate paths. Also the coordinate paths are optimally spaced apart; it ensues that likewise the principal great circles will be optimal as well. In an orthogonal coordinate system the optimal spacing would be ninety degrees apart introducing the idea of degrees of freedom. In other words in this special case the optimal space of ninety degrees would only allow two coordinates to exist on the sheet or two degrees of freedom to travel about on the sheet. Pulling all of this together the principal coordinates have associated with them ***minor principal curvatures*** $[\kappa_i]$, hence the origins of the names, with the same number of degrees of freedom. This can be restated in a set of surmised propositions about sheets. These rules can be stated in a general way so they can be extrapolated to higher dimensions or ***hypersheets*** $\mathfrak{H}^{n-1}$.

Axiom 17.12.1 Propositions of Hypersheets or Hypersurfaces $\mathfrak{H}^{n-1}$

P1) Hypersheets coordinate nets are optimally spaced apart so that there can only have n–1 degrees of freedom.

P2) Hypersheets embedded in an n-dimensional space have (n–1) coordinate dimensions.

P3) Hypersheets have as many corresponding number of principal curvatures $[\kappa_i]$ as coordinate paths $[u_i]$.

P4) Hypersheets have unique independent coordinate paths; implying principal curvatures are also unique and different.

Theorem 17.12.1 Minor Principal Curvatures: Second Fundamental Quadratic Normal Curvature

1g	Given	$\mathbf{dr} \rightarrow [L]$	
2g		$ds \rightarrow [L]$	
3g		$x^i \rightarrow [L]$	
4g		$u^\alpha \rightarrow [L]$	
5g		r_n	some normal principal radius
6g		$\kappa_n \equiv 1 / r_n$	some normal principal curvature
Step	Hypothesis	$\kappa_n = -(\mathbf{dr} / ds) \bullet (\mathbf{dN} / ds)$ EQ A	$(\mathbf{dr} / ds) \bullet (\mathbf{dN} / ds) \rightarrow [L^{-1}]$ EQ B
1	From D17.7.2 Arc Length of a Second Fundamental Quadratic Covariant Metric	$d\beta^2 \equiv -\mathbf{dr} \bullet \mathbf{dN}$	
2	From 1, LxK.3.1.7B Uniqueness Theorem of Differentiation and DxK.3.2.1H Second Parametric Differentiation with fraction	$d\beta^2 / ds^2 \equiv -(\mathbf{dr} \bullet \mathbf{dN}) / ds^2$	
3	From 2, T4.8.7 Integer Exponents: Distribution Across a Rational Number, D4.1.17 Exponential Notation, A4.2.10 Commutative Multp and T4.4.15A Equalities: Product by Division	$(d\beta / ds)^2 \equiv -(\mathbf{dr} / ds) \bullet (\mathbf{dN} / ds)$	
4	From 1g, 2g, 3 and AA.3.14C Relationship of an Algebraic Quantity or Operation to its Dimension	$\mathbf{dr} / ds \rightarrow [L] / [L]$	
5	From 4 and TA.3.3A Unit Operator: Inverse Dimensional Identity	$\mathbf{dr} / ds \rightarrow [\varnothing]$	
6	From 3, D17.3.1 The Minor of a Normal Vector; covariant parameter infinitesimal linear transformation at a Point for a Riemannian manifold and T12.6.2B Differentiation of Tensor: Rank-1 Covariant	$x^{i;\alpha} \rightarrow \partial x^i / \partial u^\alpha$	
7	From 3g, 4g and AA.3.14C Relationship of an Algebraic Quantity or Operation to its Dimension	$x^{i;\alpha} \rightarrow [L] / [L]$	
8	From 3 and TA.3.3A Unit Operator: Inverse Dimensional Identity	$x^{i;\alpha} \rightarrow [\varnothing]$	
9	From 2g, 3, 8 and AA.3.14C Relationship of an Algebraic Quantity or Operation to its Dimension	$d\, x^{i;\alpha} / ds \rightarrow [\varnothing] / [L]$	
10	From 9, TA.3.2 Unit Operator: Exponential Reciprocal and AA.3.8 Identity Dim	$d\, x^{i;\alpha} / ds \rightarrow [L^{-1}]$	

11	From 3 and D17.3.1 The Minor of a Normal Vector	$d\mathbf{N} / ds \rightarrow d\, x^{i;\alpha} / ds$	
12	From 10, 11 and AA.3.15A Relationship between Quantities and Their Dimensional Equivalence	$d\mathbf{N} / ds \rightarrow [L^{-1}]$	
13	From 3, 5, 12 and AA.3.14B Relationship of an Algebraic Quantity or Operation to its Dimension	$(d\mathbf{r} / ds) \bullet (d\mathbf{N} / ds) \rightarrow [\varnothing][L^{-1}]$	
14	From 13 and A.3.8 Identity Dim	$(d\mathbf{r} / ds) \bullet (d\mathbf{N} / ds) \rightarrow [L^{-1}]$	
15	From 3, 14 and AA.3.15A Relationship between Quantities and Their Dimensional Equivalence	$(d\beta / ds)^2 \rightarrow [L^{-1}]$	
16	From 5g and some normal principal radius along the normal	$(1 / r_n) \equiv -(d\mathbf{r} / ds) \bullet (d\mathbf{N} / ds)$	
∴	From 6g, 14, 16 and A4.2.3 Substitution	$\kappa_n = -(d\mathbf{r} / ds) \bullet (d\mathbf{N} / ds)$ EQ A	$(d\mathbf{r} / ds) \bullet (d\mathbf{N} / ds) \rightarrow [L^{-1}]$ EQ B

Theorem 17.12.2 Minor Principal Curvatures: Auxiliary-1 Pseudo Vector Differential Equations

1g	Given	$\lambda^\alpha \equiv du^\alpha / ds$ tangent vector to the coordinate curve- u^α	
Step	Hypothesis	$1 = a_{\alpha\beta}\, \lambda^\alpha \lambda^\beta$ EQ A	$\kappa_n = c_{\alpha\beta}\, \lambda^\alpha \lambda^\beta$ EQ B
1	From T17.4.20B First Fundamental Quadratic Form of the Hypersurface $\mathfrak{H}^{n-1}$	$ds^2 = a_{\alpha\beta}\, du^\alpha\, du^\beta$	
2	From 1, A4.2.12 Identity Multp, T4.4.6B Equalities: Left Cancellation by Multiplication, A4.2.14 Distribution, D4.1.17 Exponential Notation T4.4.15A Equalities: Product by Division	$1 = a_{\alpha\beta}\, (du^\alpha / ds)(du^\beta / ds)$	
3	From 1g, 2 and A4.2.3 Substitution	$1 = a_{\alpha\beta}\, \lambda^\alpha \lambda^\beta$	
4	From T17.7.2A Second Fundamental Quadratic Covariant Metric: Measured Distance	$-(d\mathbf{r} / ds) \bullet (d\mathbf{N} / ds) = c_{\alpha\beta}\, (du^\alpha / ds)(du^\beta / ds)$	
5	From 4, T14.12.1 Minor Principal Curvatures: Normal Curvature for Second Fundamental Quadratic and A4.2.3 Substitution	$\kappa_n = c_{\alpha\beta}\, (du^\alpha / ds)(du^\beta / ds)$	
6	From 1g, 5 and A4.2.3 Substitution	$\kappa_n = c_{\alpha\beta}\, \lambda^\alpha \lambda^\beta$	
∴	From 3 and 6	$1 = a_{\alpha\beta}\, \lambda^\alpha \lambda^\beta$ EQ A	$\kappa_n = c_{\alpha\beta}\, \lambda^\alpha \lambda^\beta$ EQ B

Theorem 17.12.3 Minor Principal Curvatures: Auxiliary-2 Min Length and Normal Curvature

1g	Given	$f(\cdot\lambda^{\alpha}\cdot, \in) = \kappa_n + \in (a_{\alpha\beta}\, \lambda^{\alpha} \lambda^{\beta} - 1)$ Lagrange function of maximum and minimum for $\in$ the Lagrange multiplier	
2g		$\in \neq e(\lambda^{\gamma})$	independent of $[\lambda^{\gamma}]$
3g		$\kappa_n \neq g(\in)$	independent of $[\in]$
4g		$(a_{\alpha\beta}\, \lambda^{\alpha} \lambda^{\beta} - 1) \neq h(\in)$	independent of $[\in]$
Step	Hypothesis	$\partial f(\cdot\lambda^{\alpha}\cdot, \in) / \partial\lambda^{\gamma} =$ $2\,(c_{\gamma\alpha} + \in a_{\gamma\alpha})\lambda^{\alpha}$ EQ A	$\partial f(\cdot\lambda^{\alpha}\cdot, \in) / \partial\in =$ $(a_{\alpha\beta}\, \lambda^{\alpha} \lambda^{\beta} - 1)$ EQ B
1	From 1g and T17.12.2(A, B) Minor Principal Curvatures: Auxiliary-1 Pseudo Vector Differential Equations	$f(\cdot\lambda^{\alpha}\cdot, \in) = \kappa_n + \in (a_{\alpha\beta}\, \lambda^{\alpha} \lambda^{\beta} - 1)$	
2	From 1, T12.7.1B Tensor Uniqueness: Tensor Differentiation and TK.5.1 Derivative Distribution Across Vector Components	$\partial f(\cdot\lambda^{\alpha}\cdot, \in) / \partial\lambda^{\gamma} = \partial\,(c_{\alpha\beta}\, \lambda^{\alpha} \lambda^{\beta}) / \partial\lambda^{\gamma} +$ $\in \partial\,(a_{\alpha\beta}\, \lambda^{\alpha} \lambda^{\beta}) / \partial\lambda^{\gamma} - \in \partial\,(1) / \partial\lambda^{\gamma} +$ $(a_{\alpha\beta}\, \lambda^{\alpha} \lambda^{\beta} - 1)\, \partial \in / \partial\lambda^{\gamma}$	
3	From 2g, 2 and LxK.3.3.1B Differential of a Constant	$\partial f(\cdot\lambda^{\alpha}\cdot, \in) / \partial\lambda^{\gamma} = \partial\,(c_{\alpha\beta}\, \lambda^{\alpha} \lambda^{\beta}) / \partial\lambda^{\gamma} +$ $\in \partial\,(a_{\alpha\beta}\, \lambda^{\alpha} \lambda^{\beta}) / \partial\lambda^{\gamma} - \in 0 +$ $(a_{\alpha\beta}\, \lambda^{\alpha} \lambda^{\beta} - 1)\, 0$	
4	From 3, T4.4.1 Equalities: Any Quantity Multiplied by Zero is Zero and A4.2.7 Identity Add	$\partial f(\cdot\lambda^{\alpha}\cdot, \in) / \partial\lambda^{\gamma} = \partial\,(c_{\alpha\beta}\, \lambda^{\alpha} \lambda^{\beta}) / \partial\lambda^{\gamma} + \in \partial\,(a_{\alpha\beta}\, \lambda^{\alpha} \lambda^{\beta}) / \partial\lambda^{\gamma}$	
5	From 4, T12.7.8 Tensor Product: Chain Rule for Mixed Binary Product and A10.2.14 Correspondence of Tensors and Tensor Coefficients	$\partial f(\cdot\lambda^{\alpha}\cdot, \in) / \partial\lambda^{\gamma} =$ $\partial\,(c_{\alpha\beta}) / \partial\lambda^{\gamma}\, \lambda^{\alpha} \lambda^{\beta} +$ $c_{\alpha\beta}\, (\lambda^{\alpha}) / \partial\lambda^{\gamma}\, \lambda^{\beta} +$ $c_{\alpha\beta}\, \lambda^{\alpha}\, (\lambda^{\beta}) / \partial\lambda^{\gamma} +$ $\in \partial\,(a_{\alpha\beta}) / \partial\lambda^{\gamma}\, \lambda^{\alpha} \lambda^{\beta} +$ $\in a_{\alpha\beta}\, (\lambda^{\alpha}) / \partial\lambda^{\gamma}\, \lambda^{\beta} +$ $\in a_{\alpha\beta}\, \lambda^{\alpha}\, (\lambda^{\beta}) / \partial\lambda^{\gamma}$	

6	From 7, T17.5.32B Differentiation of the Covariant Metric Tensor (Ricci's Theorem-A), D6.1.6B Identity Matrix and differential Kronecker Delta	$\partial f(\bullet\lambda^\alpha\bullet, \in) / \partial\lambda^\gamma =$ $0 +$ $c_{\alpha\beta}\, \delta^\alpha{}_\gamma\, \lambda^\beta +$ $c_{\alpha\beta}\, \lambda^\alpha\, \delta^\beta{}_\gamma +$ $0 +$ $\in a_{\alpha\beta}\, \delta^\alpha{}_\gamma\, \lambda^\beta +$ $\in a_{\alpha\beta}\, \lambda^\alpha\, \delta^\beta{}_\gamma$	
7	From 6, O12.3.5H Rules for Raising or Lowering Tensor Indices, T17.4.11 Hypersheet: Covariant Metric Inner Product is Symmetrical and A4.2.7 Identity Add; Transposing dummy indices β → α	$\partial f(\bullet\lambda^\alpha\bullet, \in) / \partial\lambda^\gamma = c_{\gamma\alpha}\, \lambda^\alpha + c_{\gamma\alpha}\, \lambda^\alpha + \in a_{\gamma\alpha}\, \lambda^\alpha + \in a_{\gamma\alpha}\, \lambda^\alpha$	
8	From 7 and T4.3.10 Equalities: Summation of Repeated Terms by 2	$\partial f(\bullet\lambda^\alpha\bullet, \in) / \partial\lambda^\gamma = 2\, c_{\gamma\alpha}\, \lambda^\alpha + \in 2\, a_{\gamma\alpha}\, \lambda^\alpha$	
9	From 8 and A4.2.14 Distribution	$\partial f(\bullet\lambda^\alpha\bullet, \in) / \partial\lambda^\gamma = 2\, (c_{\gamma\alpha} + \in a_{\gamma\alpha})\lambda^\alpha$	
10	From 1, T12.7.1B Tensor Uniqueness: Tensor Differentiation and TK.5.1 Derivative Distribution Across Vector Components	$\partial f(\bullet\lambda^\alpha\bullet, \in) / \partial\in =$ $\partial (\kappa_n) / \partial\in +$ $\partial (\in) / \partial\in\; (a_{\alpha\beta}\, \lambda^\alpha\, \lambda^\beta - 1) +$ $\in \partial (a_{\alpha\beta}\, \lambda^\alpha\, \lambda^\beta - 1) / \partial\in$	
11	From 3g, 4g, 10 and LxK.3.3.1B Differential of a Constant	$\partial f(\bullet\lambda^\alpha\bullet, \in) / \partial\in =$ $0 +$ $\partial (\in) / \partial\in\; (a_{\alpha\beta}\, \lambda^\alpha\, \lambda^\beta - 1) +$ 0	
12	From 11 and A4.2.7 Identity Add	$\partial f(\bullet\lambda^\alpha\bullet, \in) / \partial\in = (a_{\alpha\beta}\, \lambda^\alpha\, \lambda^\beta - 1)$	
∴	From 9 and 12	$\partial f(\bullet\lambda^\alpha\bullet, \in) / \partial\lambda^\gamma =$ $2\, (c_{\gamma\alpha} + \in a_{\gamma\alpha})\lambda^\alpha$ EQ A	$\partial f(\bullet\lambda^\alpha\bullet, \in) / \partial\in =$ $(a_{\alpha\beta}\, \lambda^\alpha\, \lambda^\beta - 1)$ EQ B

Theorem 17.12.4 Minor Principal Curvatures: Minor Principal Curvature Matrix

1g	Given	$\in\ \equiv -\kappa$	scalar principal curvature
Step	Hypothesis	$\mathbf{0} = (c_{\gamma\alpha} - \kappa\ a_{\gamma\alpha})$ EQ A	$0 = \vert(c_{\gamma\alpha}) - \kappa\ (a_{\gamma\alpha})\vert$ EQ B
1	From T17.12.3B Minor Principal Curvatures: Auxiliary-2 Min Length and Normal Curvature	$\partial f(\bullet\lambda^{\alpha}\bullet, \in\ ) / \partial\in\ = a_{\alpha\beta}\ \lambda^{\alpha}\ \lambda^{\beta} - 1$	
2	From 1 and finding maximum and minimal critical points	$0 = a_{\alpha\beta}\ \lambda^{\alpha}\ \lambda^{\beta} - 1$	
3	From 2 and T4.3.4A Equalities: Reversal of Right Cancellation by Addition	$1 = a_{\alpha\beta}\ \lambda^{\alpha}\ \lambda^{\beta}$ retrieved the original side condition	
4	From T17.12.3A Minor Principal Curvatures: Auxiliary-2 Min Length and Normal Curvature	$\partial f(\bullet\lambda^{\alpha}\bullet, \in\ ) / \partial\lambda^{\gamma} = 2\ (c_{\gamma\alpha} + \in\ a_{\gamma\alpha})\lambda^{\alpha}$	
5	From 4 and finding maximum and minimal critical points	$0 = 2\ (c_{\gamma\alpha} + \in\ a_{\gamma\alpha})\lambda^{\alpha}$	
6	From 5 and the none trivial solution	$0 = c_{\gamma\alpha} + \in\ a_{\gamma\alpha}$	
7	From 6 and placing the quantity in matrix form of (n–1)x(n–1)	$\mathbf{0} = (c_{\gamma\alpha} + \in\ a_{\gamma\alpha})$ Principal Curvature Matrix	
8	From 7 and T6.9.1 Uniqueness of a Determinate; [$\in$]'s are (n–1) optimally spaced Minor Principal Curvatures [κ]	$\vert\mathbf{0}\vert = \vert(c_{\gamma\alpha}) - \kappa\ (a_{\gamma\alpha})\vert$	
9	From 8 and T6.9.6 The determinant of a matrix having a row or column of zeros is zero	$0 = \vert(c_{\gamma\alpha}) - \kappa\ (a_{\gamma\alpha})\vert$	
∴	From 1g, 6, 9 and A4.2.3 Substitution	$\mathbf{0} = (c_{\gamma\alpha} - \kappa\ a_{\gamma\alpha})$ EQ A	$0 = \vert(c_{\gamma\alpha}) - \kappa\ (a_{\gamma\alpha})\vert$ EQ B

Theorem 17.12.5 Minor Principal Curvatures: Calculating the Curvatures

Step	Hypothesis	$\kappa_\alpha = c_{\alpha\gamma} A_{\gamma\alpha} / a$ EQ A for $a \neq 0$; $(n-1)$ optimally spaced curvatures κ_α	$\kappa\delta_{\alpha\beta} = c_{\alpha\gamma} A_{\gamma\beta} / a$ EQ B
1	From T17.12.4B Minor Principal Curvatures: Minor Principal Curvature Matrix	$0 = \lvert(c_{\gamma\alpha}) - \kappa\,(a_{\gamma\alpha})\rvert$	
2	From 1 and T6.3.13 Matrix Identity left of Multiplication	$0 = \lvert I\,[(c_{\gamma\alpha}) - \kappa\,(a_{\gamma\alpha})]\rvert$	
3	From 2 and T6.9.19 Closure with respect to the Identity Matrix for Multiplication	$0 = \lvert(a_{\gamma\alpha})(a_{\gamma\alpha})^{-1}\,[(c_{\gamma\alpha}) - \kappa\,(a_{\gamma\alpha})]\rvert$	
4	From 3 and T6.3.17 Left Distribution for Matrices	$0 = \lvert(a_{\gamma\alpha})\,[(c_{\gamma\alpha})(a_{\gamma\alpha})^{-1} - \kappa\,(a_{\gamma\alpha})(a_{\gamma\alpha})^{-1}]\rvert$	
5	From 4 and T6.9.19 Closure with respect to the Identity Matrix for Multiplication	$0 = \lvert(a_{\gamma\alpha})\,[(c_{\gamma\alpha})(a_{\gamma\alpha})^{-1} - \kappa\,I]\rvert$	
6	From 5 and T6.9.27 The product of two matrices can be factored as individual determinants	$0 = \lvert a_{\gamma\alpha}\rvert\,\lvert\,[(c_{\gamma\alpha})(a_{\gamma\alpha})^{-1} - \kappa\,I]\,\rvert$	
7	From 6 and D17.1.5 Covariant Metric Determinate for Hypersheet $\mathfrak{H}^{n-1}$	$0 = a\,\lvert\,[(c_{\gamma\alpha})(a_{\gamma\alpha})^{-1} - \kappa\,I]\,\rvert$	
8	From 7, T6.9.18 Inverse Square Matrix for Multiplication and A17.1.10B Covariant Determinate Defines the Existence of a Hypersheet $\mathfrak{H}^{n-1}$; hence $a \neq 0$ resulting in the nontrivial solution	$0 = \lvert\,[(c_{\gamma\alpha})(A_{\alpha\gamma} / a) - \kappa\,I]\,\rvert$	
9	From 8 and A6.2.10C Matrix Multiplication	$0 = \lvert\,[(c_{\alpha\gamma}\,A_{\gamma\beta} / a) - \kappa\,I]\,\rvert$	
10	From 9 and D6.1.6A Identity Matrix	$0 = \lvert\,[(c_{\alpha\gamma}\,A_{\gamma\beta} / a) - \kappa\,(\delta_{\alpha\beta})]\,\rvert$	
11	From 10 and A6.2.9 Matrix Multiplication by a Scalar	$0 = \lvert\,[(c_{\alpha\gamma}\,A_{\gamma\beta} / a) - (\kappa\delta_{\alpha\beta})]\,\rvert$	
12	From 11 and A6.2.8C Matrix Addition	$0 = \lvert\,(c_{\alpha\gamma}\,A_{\gamma\beta} / a - \kappa\delta_{\alpha\beta})\,\rvert$	
13	From 12, T6.9.6 The determinant of a matrix having a row or column of zeros is zero and T6.9.1 Uniqueness of a Determinate; for every matrix element	$0 = c_{\alpha\gamma}\,A_{\gamma\beta} / a - \kappa\delta_{\alpha\beta}$	

14	From 13, T10.3.22A Addition of Tensors: Reversal of Right Cancellation by Addition, T10.3.18 Addition of Tensors: Identity by Addition and A10.2.14 Correspondence of Tensors and Tensor Coefficients	$\kappa\delta_{\alpha\beta} = c_{\alpha\gamma} A_{\gamma\beta} / a$	
∴	From 14 and D6.1.6A Identity Matrix; hence the principal curvature κ_α are the non-zero Kronecker Delta's diagonal elements	$\kappa_\alpha = c_{\alpha\gamma} A_{\gamma\alpha} / a$ EQ A for $a \neq 0$; (n–1) optimally spaced curvatures κ_α	$\kappa\delta_{\alpha\beta} = c_{\alpha\gamma} A_{\gamma\beta} / a$ EQ B

Theorem 17.12.6 Minor Principal Curvatures: Product of Eigen-Characteristic Roots

Step	Hypothesis	$0 = \prod_{\alpha=1}^{n-1} (\kappa - \kappa_\alpha)$
1	From 17.12.5B Minor Principal Curvatures: Calculating the Principal Curvatures	$\kappa\delta_{\alpha\beta} = c_{\alpha\gamma} A_{\gamma\beta} / a$
2	From 1, 17.12.4A Minor Principal Curvatures: Calculating the Principal Curvatures and A4.2.3 Substitution; of corresponding elements	$\kappa\delta_{\alpha\beta} = \kappa_\alpha\delta_{\alpha\beta}$
3	From 2, T10.3.18 Addition of Tensors: Identity by Addition, T10.3.23B Addition of Tensors: Left Cancellation by Addition and A10.2.14 Correspondence of Tensors and Tensor Coefficients	$\mathbf{0} = \kappa\delta_{\alpha\beta} - \kappa_\alpha\delta_{\alpha\beta}$
4	From 3, T10.4.7 Scalars and Tensors: Distribution of a Tensor over Addition of Scalars and A10.2.14 Correspondence of Tensors and Tensor Coefficients	$\mathbf{0} = (\kappa - \kappa_\alpha)\, \delta_{\alpha\beta}$
5	From 4 and T6.9.1 Uniqueness of a Determinate	$\lvert\mathbf{0}\rvert = \lvert(\kappa - \kappa_\alpha)\, \delta_{\alpha\beta}\rvert$
6	From 5 and T6.9.6 The determinant of a matrix having a row or column of zeros is zero	$0 = \lvert(\kappa - \kappa_\alpha)\, \delta_{\alpha\beta}\rvert$
∴	From 6 and C6.9.20.1 Determinate of a Diagonal Matrix	$0 = \prod_{\alpha=1}^{n-1} (\kappa - \kappa_\alpha)$

Theorem 17.12.7 Minor Principal Curvatures: Eigen-Characteristic Roots Expanded Product

Step	Hypothesis		
Step	Hypothesis	$\prod_{\alpha=1}^{n-1} (\kappa - \kappa_\alpha) = \sum_{\beta=1}^{n} (-1)^{\beta-1} \chi_{\beta-1} \kappa^{n-\beta}$	EQ A
		$\chi_{\beta-1} = \sum_{\rho(k) \in Lr} \prod_{\nu=1}^{\beta} \kappa_{\sigma[\nu, k]}$	EQ B
		$\rho(\beta, k) = \{\sigma[\nu, k] \mid \text{for } \sigma[\nu, k] < \sigma[\nu+1, k]$ for all $\nu = 1, 2, \ldots, \beta\}$	EQ C
		$L_r = \{\rho(\beta, 1), \ldots, \rho(\beta, k), \ldots, \rho(\beta, r)\}$	EQ D
1	From 17.12.6 Minor Principal Curvatures: As a Polynomial Factored Product	$0 = \prod_{\alpha=1}^{n-1} (\kappa - \kappa_\alpha)$	
∴	From 1 and expanding samples n = 2, 3, 4, and 5 and applying A3.11.1 Induction Property of Integers and D1.6.3 Set Summation Notation over Long Indices; applying the nomenclature of long indices.	$\prod_{\alpha=1}^{n-1} (\kappa - \kappa_\alpha) = \sum_{\beta=1}^{n} (-1)^{\beta-1} \chi_{\beta-1} \kappa^{n-\beta}$	EQ A
		$\chi_{\beta-1} = \sum_{\rho(k) \in Lr} \prod_{\nu=1}^{\beta} \kappa_{\sigma[\nu, k]}$.	EQ B
		$\rho(\beta, k) = \{\sigma[\nu, k] \mid \text{for } \sigma[\nu, k] < \sigma[\nu+1, k]$ for all $\nu = 1, 2, \ldots, \beta\}$	EQ C
		$L_r = \{\rho(\beta, 1), \ldots, \rho(\beta, k), \ldots, \rho(\beta, r), r = {}_nC_{n-\beta}\}$	EQ D

Theorem 17.12.8 Minor Principal Curvatures: Relation to Gaussian Principal Curvatures

1g	Given	$\in \equiv -\kappa$ scalar principal curvature	
Step	Hypothesis	$\frac{1}{2} n(n-1) K = -\sum_{\alpha<\beta} \kappa_\alpha \kappa_\beta$ EQ A	$K = -[2/n(n-1)] \sum_{\alpha<\beta} \kappa_\alpha \kappa_\beta$ EQ B
1	From T17.12.3A Minor Principal Curvatures: Auxiliary-2 Min Length and Normal Curvature and A4.4.4A Equalities: Right Cancellation by Multiplication	$\frac{1}{2} \partial f(\bullet\lambda^\alpha\bullet, \in) / \partial\lambda^\gamma = (c_{\alpha\beta} + \in a_{\alpha\beta})\lambda^\beta$	
2	From 1g, 1 and A4.2.3 Substitution; finding maximum and minimal critical points and $[\in]$'s are (n–1) optimally spaced curvatures $[\kappa_\alpha]$	$0 = (c_{\alpha\beta} - \kappa_\alpha a_{\alpha\beta})\lambda^\beta$	
3	From 2 and T17.5.3B Dot Product of Mixed Vectors (Tensor Contraction)	$0 = [(c_{\alpha\beta} - \kappa_\alpha a_{\alpha\beta})\mathbf{b}^\beta] \bullet (\lambda^\gamma \mathbf{b}_\gamma)$	
4	From 3 and yields a new vector characterizing curvature dependent only on the first and second fundamental metrics of a curved surface.	$\mathbf{C}_\alpha \equiv (c_{\alpha\beta} - \kappa_\alpha a_{\alpha\beta})\, \mathbf{b}^\beta$	

5	From 4 and taking the conjugate magnitude of the Minor Principal Curvatures vector. With the optimized eigenvalues having the angel between them maximized at ninety degrees, creates an orthogonal set of vectors. This reinforces the idea the Minor Principal Curvatures are optimally spaced apart.	$0 \equiv \mathbf{C}_\alpha \bullet \mathbf{C}_\beta^*$
6	From 4, 5 and A7.9.2 Substitution of Vectors	$0 = (c_{\alpha\tau} - \kappa_\alpha\, a_{\alpha\tau})b^\tau \bullet (c_{\beta\eta} + \kappa_\beta\, a_{\beta\eta})b^\eta$
7	From 6 and T17.5.2A Dot Product of Contravariant Vectors	$0 = (c_{\alpha\tau}\, c_{\beta\eta} - \kappa_\alpha\, a_{\alpha\tau}\, c_{\beta\eta} + \kappa_\beta\, a_{\beta\eta}\, c_{\alpha\tau} - \kappa_\alpha\kappa_\beta\, a_{\alpha\tau}\, a_{\beta\eta})\, a^{\tau\eta}$
8	From 7 and permuting the last indices of metric pairs	$0 = (c_{\alpha\eta}\, c_{\beta\tau} - \kappa_\alpha\, a_{\alpha\eta}\, c_{\beta\tau} + \kappa_\beta\, a_{\alpha\tau}\, c_{\beta\eta} - \kappa_\alpha\kappa_\beta\, a_{\alpha\eta}\, a_{\beta\tau})\, a^{\tau\eta}$
9	From 7 and 8; subtracting term-by-term	$0 = (c_{\alpha\tau}\, c_{\beta\eta} - c_{\alpha\eta}\, c_{\beta\tau} - \kappa_\alpha\, a_{\alpha\tau}\, c_{\beta\eta} + \kappa_\alpha\, a_{\alpha\eta}\, c_{\beta\tau} +$ $+ \kappa_\beta\, a_{\beta\eta}\, c_{\alpha\tau} - \kappa_\beta\, a_{\alpha\tau}\, c_{\beta\eta} +$ $-\kappa_\alpha\kappa_\beta\, a_{\alpha\tau}\, a_{\beta\eta} + \kappa_\alpha\kappa_\beta\, a_{\alpha\eta}\, a_{\beta\tau})\, a^{\tau\eta}$
10	From 9, T10.4.6 Scalars and Tensors: Distribution of a Scalar over Addition of Tensors, T17.4.12 Hypersheet: Contravariant Metric Inner Product is Symmetrical and A10.2.14 Correspondence of Tensors and Tensor Coefficients	$0 = [c_{\alpha\tau}\, c_{\beta\eta} - c_{\alpha\eta}\, c_{\beta\tau} + \kappa_\alpha\, (a_{\alpha\eta}\, c_{\beta\tau} - a_{\alpha\tau}\, c_{\beta\eta}) +$ $\kappa_\beta\, (a_{\alpha\tau}\, c_{\beta\eta} - a_{\beta\eta}\, c_{\alpha\tau}) +$ $\kappa_\alpha\kappa_\beta\, (a_{\alpha\eta}\, a_{\beta\tau} - a_{\alpha\tau}\, a_{\beta\eta})]\, a^{\eta\tau}$
11	From 10, A4.2.12 Identity Multp, D4.1.20A Negative Coefficient, T4.8.2B Integer Exponents: Negative One Squared, A4.2.14 Distribution and transposing dummy indices β ←→ α	$0 = [c_{\alpha\tau}\, c_{\beta\eta} - c_{\alpha\eta}\, c_{\beta\tau} + \kappa_\alpha\, (a_{\alpha\eta}\, c_{\beta\tau} - a_{\alpha\tau}\, c_{\beta\eta}) +$ $-\, \kappa_\alpha\, (a_{\alpha\eta}\, c_{\beta\tau} - a_{\alpha\tau}\, c_{\beta\eta}) +$ $-\kappa_\alpha\kappa_\beta\, (a_{\alpha\tau}\, a_{\beta\eta} - a_{\alpha\eta}\, a_{\beta\tau})]\, a^{\eta\tau}$
12	From 11, T10.3.19 Addition of Tensors: Inverse by Addition, T10.3.18 Addition of Tensors: Identity by Addition and A10.2.14 Correspondence of Tensors and Tensor Coefficients	$0 = [c_{\alpha\tau}\, c_{\beta\eta} - c_{\alpha\eta}\, c_{\beta\tau} - \kappa_\alpha\kappa_\beta\, (a_{\alpha\tau}\, a_{\beta\eta} - a_{\alpha\eta}\, a_{\beta\tau})]\, a^{\eta\tau}$
13	From 12, T4.4.17A Equalities: Product by Division, Factorization and A4.2.10 Commutative Multp	$0 = [c_{\alpha\tau}\, c_{\beta\eta} - c_{\alpha\eta}\, c_{\beta\tau} - (\kappa_\alpha\kappa_\beta / K)\, K(a_{\alpha\tau}\, a_{\beta\eta} - a_{\alpha\eta}\, a_{\beta\tau})]\, a^{\eta\tau}$
14	From 13, T17.8.11A Codazzi-Gauss Equations: Equation of Gauss, T17.11.5 Hypersphere: Relationship of Gauss's Equation to Total Curvature, T10.2.5 Tool: Tensor Substitution and A10.2.14 Correspondence of Tensors and Tensor Coefficients	$0 = [R_{\beta\alpha\eta\tau} - (\kappa_\beta\, \kappa_\alpha / K)\, R_{\beta\alpha\eta\tau}]\, a^{\eta\tau}$

15	From 14, T10.2.10 Tool: Identity Multiplication of a Tensor, T10.4.7 Scalars and Tensors: Distribution of a Tensor over Addition of Scalars and A10.2.14 Correspondence of Tensors and Tensor Coefficients	$0 = [1 - (\kappa_\alpha \kappa_\beta / K)]\, R_{\beta\alpha\eta\tau}\, a^{\eta\tau}$	
16	From 15 and taking the none trivial solution for every term	$0 = 1 - (\kappa_\alpha \kappa_\beta / K)$	
17	From 16, T4.3.5B Equalities: Left Cancellation by Addition and A4.2.7 Identity Add	$1 = (\kappa_\alpha \kappa_\beta / K)$	
18	From 17 and T4.4.5A Equalities: Reversal of Right Cancellation by Multiplication	$K = \kappa_\alpha \kappa_\beta$	
19	From 18 and the equality holds for all independent eigenvalues, hence summing over all values	$\sum_{i<j} K = \sum_{i<j} \kappa_\alpha \kappa_\beta$	
20	From 19, A4.2.12 Identity Multp, A4.2.14 Distribution and TE.3.2A Series: Summing Number of Triangular Elements	$½\, n(n-1)\, K = \sum_{i<j} \kappa_\alpha \kappa_\beta$	
21	From 20, T4.4.7B Equalities: Reversal of Left Cancellation by Multiplication and T4.4.6B Equalities: Left Cancellation by Multiplication	$K = [2/n(n-1)] \sum_{i<j} \kappa_\alpha \kappa_\beta$	
∴	From 20 and 21	$½\, n(n-1)\, K = \sum_{\alpha<\beta} \kappa_\alpha \kappa_\beta$ EQ A	$K = [2/n(n-1)] \sum_{\alpha<\beta} \kappa_\alpha \kappa_\beta$ EQ B

Theorem 17.12.9 Minor Principal Curvatures: Gaussian to Series Coefficient Characteristic Roots

1g	Given	$\beta = 2$	
2g		$\sigma[1, k] = \alpha$	
3g		$\sigma[2, k] = \beta$	
Step	Hypothesis	$K = [2/n(n-1)]\, \chi_1$ EQ A	$\chi_1 = ½\, n(n-1)\, K$ EQ B
1	From T17.12.7B Minor Principal Curvatures: Eigen-Characteristic Roots Expanded Product	$\chi_{\beta-1} = \sum_{\rho(k) \in Lr} \prod_{v=1}{}^{\beta} \kappa_{\sigma[v, k]}$	
2	From 1g, 1 and A4.2.3 Substitution	$\chi_1 = \sum_{\rho(k) \in Lr} \prod_{v=1}{}^{2} \kappa_{\sigma[v, k]}$	
3	From 2 and expand product	$\chi_1 = \sum_{\rho(k) \in Lr} \kappa_{\sigma[1, k]}\, \kappa_{\sigma[2, k]}$	
4	From 3 and T17.12.7C Minor Principal Curvatures: Eigen-Characteristic Roots Expanded Product	$\chi_1 = \sum_{\sigma[1, k]<\sigma[2, k]} \kappa_{\sigma[1, k]}\, \kappa_{\sigma[2, k]}$	
5	From 2g, 3g, 4 and A4.2.3 Substitution	$\chi_1 = \sum_{\alpha<\beta} \kappa_\alpha \kappa_\beta$	
6	From 5, T17.12.8B Minor Principal Curvatures: Relation to Gaussian Principal Curvatures and A4.2.3 Substitution	$K = [2/n(n-1)]\, \chi_1$	
7	From 5, T17.12.8A Minor Principal Curvatures: Relation to Gaussian Principal Curvatures, A4.2.3 Substitution, T4.4.2 Equalities: Uniqueness of Multiplication and T4.8.2B Integer Exponents: Negative One Squared	$\chi_1 = ½\, n(n-1)\, K$	
∴	From 6 and 7	$K = [2/n(n-1)]\, \chi_1$ EQ A	$\chi_1 = ½\, n(n-1)\, K$ EQ B

Observation 17.12.1 Solvability of Gaussian Principal Curvature

Theorem T17.12.7B "Minor Principal Curvatures: Eigen-Characteristic Roots Expanded Product" is an important equation, because if the principal curvatures cannot be found the Gaussian curvature surely can be computed. No further work in factoring the eigen-characteristic equation for roots is required.

Section 17.13 Major Principal Curvatures on a Hypersurface $\mathfrak{H}^{n-1}$

Definition 17.13.1 Major Principal Curvatures

Another way of looking at the ***minor of the Principal curvatures*** is through a common binary set of principal curvatures on a hypersurface. These special curvatures are in a matrix-determinate equation, which in turn yields the ***major principal curvatures***.

Definition 17.13.2 Mean Curvature

The ***arithmetic mean curvature*** is the normalized average mean Weingarten tensor of the major principle curvature.

$$H \equiv w_{\alpha}{}^{\alpha} / n(n-1)$$ Equation A

Definition 17.13.3 Surface of Zero Curvature

Any hypersurface having a Gaussian Total Curvature of [0 = K] is called a ***surface of zero curvature*** **(0)**, or a ***flat plane*** or ***Euclidian plane***.

Definition 17.13.4 Surface of Positive Curvature

Any hypersurface having a Gaussian Total Curvature of [0 < K] is called a ***surface of positive curvature*** (+).

Definition 17.13.5 Surface of Negative Curvature

Any hypersurface having a Gaussian Total Curvature of [0 > K] is called a ***surface of negative curvature*** (–).

Definition 17.13.6 Surface of Constant Curvature

Any hypersurface just as the need arose to be able to measure the deviation from a curve in section 12.13 "Flat Plane Curvature", in this section a similar problem arises, measuring deviation of curvature to a hyperplane. In a like manner the problem can be solved. This can be done similarly by using a hypersphere in n-dimensions in place of its lesser 2-dimensional form the circle.

Definition 17.13.7 Radius of Hypersurface Curvature

The radius of hypersurface curvature [ρ] arises from an infinitesimal segment along the curve from principal hypersphere's great circles that match the curvature of the segment from coordinate curve-u_i at the point-Q along the hypersurface's path. The radius of the circle lays parallel along the unit normal vector and center of the circle terminates at the opposite end of the radius at point-P.

Definition 17.13.8 Finite Tangent Disk to the Hypersurface $\mathfrak{H}^{(n-1)}$

A small disk capping the point-Q, called the ***tangent disk***, with area δS can be circularly inscribed having a span of an isosceles angle [δθ] about that point with a small, but finite radius [δr] lying tangent to the hypersurface. This area can be approximated from the product of small ***principal grid*** segments [**.**δu_i**.**] along the ***principal curvature coordinate***-u_i or variable optimally separated paths. Small, but finite arcs [δs_i] are parts of principal great circles and as a product they correspond to the following elements.

$$\delta S \approx \prod_i^{(n-1)} \delta s_i$$ Equation A

$$\prod_i^{(n-1)} \delta s_i \approx \prod_i^{(n-1)} \delta u_i$$ Equation B

Definition 17.13.9 Principal Curvatures to Variable Optimally Separated Paths

The measure of ***principal curvature*** [κ_i] is the reciprocal of the ***principal radii*** [r_i] from the planes of the great circles cutting through the hypersphere along the curves-u_i.

$\kappa_i \equiv 1 / r_i$	Measure of Principal Curvatures	Equation A
$\boldsymbol{\kappa}_i \equiv \kappa_i \mathbf{N}$	Curvature Vector	Equation B
$R_i \equiv \mid r_i \mid$	Radii of Principal Curvatures	Equation C

Observation 17.13.1 Improper Modeling of a Tangent Disk

In section 17.9 "Gaussian Total Curvatures of a Surface in 2-Dimensions" the definition D17.9.1C "Covariant Matrix and Gaussian Curvature" is contrived and when applied to Einstein's constant curvature, T14.4.7 "Riemann-Christoffel Tensor Rank-0: Contraction of Einstein's Covariant Tensor" is not proportional to it. From theorem T17.10.3 "Tangent Disk: Cannot be Reduced to a None Singular Value" clearly our notion of a surface curvature and how to properly model it is incorrect. Not only that but also this model does not naturally extend itself to higher dimensions. These types of constructions are common models promoted by a number of authors. Yet intuitively the idea of measuring curvature relative to a tangential hypersphere is a very practical idea and a natural extension of curvature from a 2-dimensional curve to a tangential circle. With the hypersphere's n-dimensional representation one would think it to be a natural way to measure the deviation of curvature on a hypersurface $\mathfrak{H}^{n-1}$, indeed in the next section it is seen that a more rigorous development yields a proper definition of a Gaussian curvature. The development gives a correct definition based on its principal curvatures that are derived by eigenvalue methods, which in tern allows an appropriate development of curvature on a hypersurface $\mathfrak{H}^{n-1}$.

Observation 17.13.2 The Use of the Zero and Segment of One Base as a Model

Why did LxE.4.7.3 "Spherical Figures: Zero and Segment of One Base" fail when extending it to higher then three dimensions?

1) It is designed in only 3-dimensions; as such it cannot be extended into higher spaces.
2) There is no chance of a dimensional residue to be left behind so the surface vanishes by identity as smaller and smaller inscribed tangent disks are constructed.
3) Since LxE.4.7.3 vanishes identically on the left hand side of the equation the right hand side's discreet elements are forced to vanish, a contradiction since by definition the hypersurface exists.

Definition 17.13.10 Hypersphere

A hypersphere is the locus of all points at a given distance from a fixed point-Q($\bullet$ x_i $\bullet$). The fixed point is called the ***center of the hypersphere*** and the fixed distance the ***radius*** [ρ]. Let the center point be represented by the point-P($\bullet$ a_i $\bullet$) in a Riemannian $\mathfrak{R}^n$ Space.

Observation 17.13.3 Proper Modeling of a Tangent Disk

From theorem T17.11.2 "Hypersphere: Auxiliary-2 Normalized Differential Vector" brings back the idea of a tangent disk making an infinitesimal contact at the point-Q from the hypersphere tangent vector $[x^i_\alpha]$ to the hypersurface along the normal vector $[\eta^i_\alpha]$ on the hypersheet. While at the same time provides a radius $[\rho]$ showing a measured curvature at point-Q. Clearly a much more comprehensive and general approach then the contrived ideas in section 17.10 "Hypersphere Approximating Curvature to Hypersurface $\mathfrak{H}^{n-1}$".

Observation 17.13.4 Calculating Einstein and Gauss from a Riemannian Total Curvature

It is seen from theorems T17.11.8 "Hypersphere: Riemannian Total Curvature Einstein and Gauss's Relationship" equations A and B is a simple matter by substituting T17.11.9 "Hypersphere: Riemannian Total Curvature in Determinate Form" A or B into them. This gives away to exclusively calculate them in terms of their spatial geometry.

From figure F17.10.1 "A Hypersphere Tangent at a Point-Q to a Hypersheet" the 3-dimensional sphere shows two great circles tangent to the coordinate curves on the embedded 2-dimensional sheet at the point-Q. When talking about 2-dimensions it is **not** implied that the embedded sheet is a flat, 2-space, Euclidian plane, but that there is a one-to-one correspondence between the great circles of the tangent sphere and the coordinate curves. Hence there is only as many great circles as coordinate curves, for this space only two are embedded in a 3-dimensional Riemannian space. While a great circle at point-Q can be rotated around that point generating an infinite number of great circles only two principal great circles will match up with the coordinate curves to measure the curvatures along the coordinate paths. Also the coordinate paths are optimally spaced apart; it ensues that likewise the principal great circles will be optimal as well. In an orthogonal coordinate system the optimal spacing would be ninety degrees apart introducing the idea of degrees of freedom. In other words in this special case the optimal space of ninety degrees would only allow two coordinates to exist on the sheet or two degrees of freedom to travel about on the sheet. Pulling all of this together the principal coordinates have associated with them ***minor principal curvatures*** $[\kappa_i]$, hence the origins of the names, with the same number of degrees of freedom. This can be restated in a set of surmised propositions about sheets. These rules can be stated in a general way so they can be extrapolated to higher dimensions or ***hypersheets*** $\mathfrak{H}^{n-1}$.

Observation 17.13.5 Solvability of Gaussian Principal Curvature

Theorem T17.12.7B "Minor Principal Curvatures: Eigen-Characteristic Roots Expanded Product" is an important equation, because if the principal curvatures cannot be found the Gaussian curvature surely can be computed. No further work in factoring the eigen-characteristic equation for roots is required.

Theorem 17.13.1 Major Principal Curvature: General Quadratic Equation on a Hypersurface $\mathfrak{H}^{n-1}$

1g	Given	$C = \begin{pmatrix} c_{\alpha\tau} & c_{\alpha\eta} \\ c_{\beta\tau} & c_{\beta\eta} \end{pmatrix}$
2g		$A = \begin{pmatrix} a_{\alpha\tau} & a_{\alpha\eta} \\ a_{\beta\tau} & a_{\beta\eta} \end{pmatrix}$
Step	Hypothesis	$0 = \kappa^2 - 2H\,\kappa + K$
1	From T17.8.11A Codazzi-Gauss Equations: Equation of Gauss	$R_{\beta\alpha\eta\tau} = \begin{vmatrix} c_{\alpha\tau} & c_{\alpha\eta} \\ c_{\beta\tau} & c_{\beta\eta} \end{vmatrix}$
2	From T17.11.5 Hypersphere: Relationship of Gauss's Equation to Total Curvature	$R_{\beta\alpha\eta\tau} = K\begin{vmatrix} a_{\alpha\tau} & a_{\alpha\eta} \\ a_{\beta\tau} & a_{\beta\eta} \end{vmatrix}$
3	From T17.12.4B Minor Principal Curvatures: Minor Principal Curvature Matrix	$0 = \lvert(c_{\alpha\beta}) - \kappa\,(a_{\alpha\beta})\rvert$
4	From 3 and placing in formal matrix form	$0 = \lvert C - \kappa A \rvert\, a^{\tau\eta}$
5	From 1g, 2g, 4, A6.2.9A Matrix Multiplication by a Scalar and A6.2.8C Matrix Addition	$0 = \begin{vmatrix} c_{\alpha\tau} - \kappa a_{\alpha\tau} & c_{\alpha\eta} - \kappa a_{\alpha\eta} \\ c_{\beta\tau} - \kappa a_{\beta\tau} & c_{\beta\eta} - \kappa a_{\beta\eta} \end{vmatrix}$
6	From 5 and T6.10.2A Determinate of a 2x2 Matrix	$0 = (c_{\alpha\tau} - \kappa a_{\alpha\tau})(c_{\beta\eta} - \kappa a_{\beta\eta}) - (c_{\alpha\eta} - \kappa a_{\alpha\eta})(c_{\beta\tau} - \kappa a_{\beta\tau})$
7	From 6, T4.12.4 Polynomial Quadratic: Product of N-Quantities, D4.1.20A Negative Coefficient and A4.2.14 Distribution	$0 = c_{\alpha\tau}\, c_{\beta\eta} + \kappa c_{\alpha\tau}\, a_{\beta\eta} - \kappa a_{\alpha\tau}\, c_{\beta\eta} + \kappa a_{\alpha\tau}\kappa a_{\beta\eta}$ $- c_{\alpha\eta}\, c_{\beta\tau} + \kappa c_{\alpha\eta}\, a_{\beta\tau} + \kappa a_{\alpha\eta}\, c_{\beta\tau} - \kappa a_{\alpha\eta}\kappa a_{\beta\tau}$
8	From 7, T10.3.16 Addition of Tensors: Commutative by Addition, T10.3.17 Addition of Tensors: Associative by Addition; by metric product types and A10.2.14 Correspondence of Tensors and Tensor Coefficients	$0 = (c_{\alpha\tau}\, c_{\beta\eta} - c_{\alpha\eta}\, c_{\beta\tau}) - \kappa c_{\alpha\tau}\, a_{\beta\eta} + \kappa\, c_{\alpha\eta}\, a_{\beta\tau}$ $+ \kappa a_{\alpha\tau}\, c_{\beta\eta} + \kappa a_{\alpha\eta}\, c_{\beta\tau} + (\kappa a_{\alpha\tau}\kappa a_{\beta\eta} - \kappa a_{\alpha\eta}\kappa a_{\beta\tau})$
9	From 8 and T10.4.6 Scalars and Tensors: Distribution of a Scalar over Addition of Tensors	$0 = (c_{\alpha\tau}\, c_{\beta\eta} - c_{\alpha\eta}\, c_{\beta\tau}) + (-c_{\alpha\tau}\, a_{\beta\eta} + c_{\alpha\eta}\, a_{\beta\tau})\kappa$ $+ (a_{\alpha\tau}\, c_{\beta\eta} + a_{\alpha\eta}\, c_{\beta\tau})\kappa + (a_{\alpha\tau}a_{\beta\eta} - a_{\alpha\eta}a_{\beta\tau})\kappa\kappa$
10	From 9, D4.1.17 Exponential Notation, T10.3.7 Product of Tensors: Commutative by Multiplication of Rank p + q → q + p and A10.2.14 Correspondence of Tensors and Tensor Coefficients	$0 = (c_{\alpha\tau}\, c_{\beta\eta} - c_{\alpha\eta}\, c_{\beta\tau}) + (-c_{\alpha\tau}\, a_{\beta\eta} + c_{\alpha\eta}\, a_{\beta\tau})\kappa$ $+ (c_{\beta\eta}\, a_{\alpha\tau} + c_{\beta\tau}\, a_{\alpha\eta})\kappa + (a_{\alpha\tau}a_{\beta\eta} - a_{\alpha\eta}a_{\beta\tau})\kappa^2$

11	From 10, T10.3.16 Addition of Tensors: Commutative by Addition and A10.2.14 Correspondence of Tensors and Tensor Coefficients	$0 = (c_{\alpha\tau}\, c_{\beta\eta} - c_{\alpha\eta}\, c_{\beta\tau}) + (-c_{\alpha\tau}\, a_{\beta\eta} + c_{\alpha\eta}\, a_{\beta\tau})\kappa + (c_{\beta\tau}\, a_{\alpha\eta} + c_{\beta\eta}\, a_{\alpha\tau})\kappa + (a_{\alpha\tau}a_{\beta\eta} - a_{\alpha\eta}a_{\beta\tau})\kappa^2$
12	From 11 and Transposing dummy indices $\beta \leftarrow \alpha$ and $\alpha \rightarrow \beta$	$0 = (c_{\alpha\tau}\, c_{\beta\eta} - c_{\alpha\eta}\, c_{\beta\tau}) + [(-c_{\alpha\tau}\, a_{\beta\eta} + c_{\alpha\eta}\, a_{\beta\tau}) + (c_{\alpha\tau}\, a_{\beta\eta} + c_{\alpha\eta}\, a_{\beta\tau})]\kappa + (a_{\alpha\tau}a_{\beta\eta} - a_{\alpha\eta}a_{\beta\tau})\kappa^2$
13	From 12, T10.3.16 Addition of Tensors: Commutative by Addition and A10.2.14 Correspondence of Tensors and Tensor Coefficients; pairing by common metrics	$0 = (c_{\alpha\tau}\, c_{\beta\eta} - c_{\alpha\eta}\, c_{\beta\tau}) + (-c_{\alpha\tau}\, a_{\beta\eta} + c_{\alpha\tau}\, a_{\beta\eta} + c_{\alpha\eta}\, a_{\beta\tau} + c_{\alpha\eta}\, a_{\beta\tau})\,\kappa + (a_{\alpha\tau}a_{\beta\eta} - a_{\alpha\eta}a_{\beta\tau})\,\kappa^2$
14	From 13, T10.3.19 Addition of Tensors: Inverse by Addition, T10.3.18 Addition of Tensors: Identity by Addition, T10.2.10A Tool: Identity Multiplication of a Tensor, T10.4.7 Scalars and Tensors: Distribution of a Tensor over Addition of Scalars, A10.2.14 Correspondence of Tensors and Tensor Coefficients and T4.3.10 Equalities: Summation of Repeated Terms by 2	$0 = (c_{\alpha\tau}\, c_{\beta\eta} - c_{\alpha\eta}\, c_{\beta\tau}) + 2\, c_{\alpha\eta}\, a_{\beta\tau}\, \kappa + (a_{\alpha\tau}a_{\beta\eta} - a_{\alpha\eta}a_{\beta\tau})\, \kappa^2$
15	From 1, 2, 14, A4.2.12 Identity Multp, A4.2.13 Inverse Multp, T10.2.5 Tool: Tensor Substitution and A10.2.14 Correspondence of Tensors and Tensor Coefficients	$0 = R_{\beta\alpha\eta\tau} + 2\, c_{\alpha\eta}\, a_{\beta\tau}\, \kappa + (1/K)\, R_{\beta\alpha\eta\tau}\, \kappa^2$
16	From 15, T17.5.71 Riemann-Christoffel Tensor First Kind: Contracted to Spatial Curvature, T10.3.6 Product of Tensors: Uniqueness by Multiplication of Rank for p + q	$0 = [R_{\beta\alpha\eta\tau} + 2\, c_{\alpha\eta}\, a_{\beta\tau}\, \kappa + (1/K)\, R_{\beta\alpha\eta\tau}\, \kappa^2]\, a^{\alpha\beta}\, a^{\eta\tau}$
17	From 16, T10.3.20 Addition of Tensors: Distribution of a Tensor over Addition of Tensors and A10.2.14 Correspondence of Tensors and Tensor Coefficients	$0 = R_{\beta\alpha\eta\tau}\, a^{\alpha\beta}\, a^{\eta\tau} + 2\, c_{\alpha\eta}\, a_{\beta\tau}\, a^{\alpha\beta}\, a^{\eta\tau}\kappa + (1/K)\, R_{\beta\alpha\eta\tau}\, a^{\alpha\beta}\, a^{\eta\tau}\kappa^2$
18	From 17, T17.5.71 Riemann-Christoffel Tensor First Kind: Contracted to Spatial Curvature, A4.2.3 Substitution, T17.4.12 Hypersheet: Contravariant Metric Inner Product is Symmetrical and T17.4.6A Hypersheet: Covariant and Contravariant Metric Inverse	$0 = R + 2\, c_{\alpha\eta}\, \delta_{\beta}^{\ \eta}\, a^{\beta\alpha}\, \kappa + (1/K)\, R\kappa^2$

19	From 18 and O12.3.5D Rules for Raising or Lowering Tensor Indices	$0 = R + 2\, c_{\alpha\beta}\, a^{\alpha\beta} \kappa + (1/K)\, R\, \kappa^2$
20	From 19, T17.11.6 Hypersphere: Relationship of Riemann's to Gauss's Total Curvature and A4.2.3 Substitution	$0 = n(1-n)\, K + 2\, c_{\alpha\beta}\, a^{\alpha\beta} \kappa + (1/K)\, n(1-n)\, K\, \kappa^2$
21	From 20, A4.2.10 Commutative Multp, T4.4.16B Equalities: Product by Division, Common Factor and A4.2.12 Identity Multp	$0 = n(1-n)\, K + 2\, c_{\alpha\beta}\, a^{\alpha\beta} \kappa + n(1-n)\, \kappa^2$
22	From 21, T4.4.2 Equalities: Uniqueness of Multiplication, T4.4.7B Equalities: Product by Division, Factorization, T4.4.16B Equalities: Product by Division, Common Factor, A4.2.12 Identity Multp and T4.4.15B Equalities: Product by Division	$0 = K + 2\, [c_{\alpha\beta}\, a^{\alpha\beta} / n(1-n)\,]\, \kappa + \kappa^2$
23	From 22, A4.2.5 Commutative Add, D4.1.20 Negative Coefficient, A4.2.12 Identity Multp, T4.8.2B Integer Exponents: Negative One Squared and A4.2.14 Distribution	$0 = \kappa^2 - 2\, [c_{\alpha\beta}\, a^{\alpha\beta} / n(n-1)\,]\, \kappa + K$
24	From 23 and T17.4.12 Hypersheet: Contravariant Metric Inner Product is Symmetrical	$0 = \kappa^2 - 2\, [c_{\alpha\beta}\, a^{\beta\alpha} / n(n-1)\,]\, \kappa + K$
25	From 24 and D17.8.1 Weingarten Tensor on a Hypersheet $\mathfrak{H}^{n-1}$	$0 = \kappa^2 - 2\, [w_{\alpha}{}^{\alpha} / n(n-1)\,]\, \kappa + K$
∴	From 23, D17.13.2A Mean Curvature and A4.2.3 Substitution	$0 = \kappa^2 - 2H\, \kappa + K$

Theorem 17.13.2 Major Principal Curvatures: Characteristic Roots of General Quadratic Equation

Step	Hypothesis	$\kappa_{M1} = H + \sqrt{H^2 - K}$ EQ A	$\kappa_{M2} = H - \sqrt{H^2 - K}$ EQ B
1	From T17.13.1 Major Principal Curvature: General Quadratic Equation on a Hypersurface $\mathfrak{H}^{n-1}$	$0 = \kappa^2 - 2H\, \kappa + K$	
2	From 1 and T4.12.6 Polynomial Quadratic: Quadratic Formula (Factoring a Binomial)	$\kappa = H \pm \sqrt{H^2 - K}$	
∴	From 2 and parsing the plus, and minus into two characteristic roots κ_{M1} and κ_{M2}	$\kappa_{M1} = H + \sqrt{H^2 - K}$ EQ A	$\kappa_{M2} = H - \sqrt{H^2 - K}$ EQ B

Observation 17.13.6 Major Principal Curvatures Comes Full Circle

With minor principal curvatures having (n–1) curvatures on a hypersurface is cumbersome, but by rendering them to just two ***major*** principal curvatures it becomes easier to manipulate and visualize them just as it was in 3-space.

Theorem 17.13.3 Major Principal Curvature: Mean Curvature on a Hypersurface $\mathfrak{H}^{n-1}$

Step	Hypothesis	$H = ½ (\kappa_{M1} + \kappa_{M2})$ EQ A	$\kappa_{M1} + \kappa_{M2} = 2H$ EQ B
1	From A4.2.2 Equality	$\kappa_{M1} + \kappa_{M2} = \kappa_{M1} + \kappa_{M2}$	
2	From 1, T17.13.2(A, B) Minor Principal Curvatures: Characteristic Roots of General Quadratic Equation and A4.2.3 Substitution	$\kappa_{M1} + \kappa_{M2} = H + \sqrt{H^2 - K} + H - \sqrt{H^2 - K}$	
3	From 2 and A4.2.5 Commutative Add	$\kappa_{M1} + \kappa_{M2} = H + H + \sqrt{H^2 - K} - \sqrt{H^2 - K}$	
4	From 3, T4.3.10 Equalities: Summation of Repeated Terms by 2, A4.2.8 Inverse Add and A4.2.7 Identity Add	$\kappa_{M1} + \kappa_{M2} = 2H$	
5	From 4, T4.4.23 Equalities: Multiplication of a Constant and T4.4.18B Equalities: Product by Division, Factorization by 2	$H = ½ (\kappa_{M1} + \kappa_{M2})$	
∴	From 4 and 5	$H = ½ (\kappa_{M1} + \kappa_{M2})$ EQ A	$\kappa_{M1} + \kappa_{M2} = 2H$ EQ B

Theorem 17.13.4 Major Principal Curvature: Gaussian Curvature on a Hypersurface $\mathfrak{H}^{n-1}$

Step	Hypothesis	$K = \kappa_{M1}\,\kappa_{M2}$
1	From A4.2.2 Equality	$\kappa_{M1}\,\kappa_{M2} = \kappa_{M1}\,\kappa_{M2}$
2	From 1, T17.13.2(A, B) Minor Principal Curvatures: Characteristic Roots of General Quadratic Equation and A4.2.3 Substitution	$\kappa_{M1}\,\kappa_{M2} = (H + \sqrt{H^2 - K})\,(H - \sqrt{H^2 - K})$
3	From 2 and T4.12.3 Polynomial Quadratic: Difference of Two Squares	$\kappa_{M1}\,\kappa_{M2} = H^2 - (\sqrt{H^2 - K})^2$
4	From 3 and T4.10.4 Radicals: Identity Radical Raised to a Power	$\kappa_{M1}\,\kappa_{M2} = H^2 - (H^2 - K)$
5	From 4, D4.1.20 Negative Coefficient and T4.8.2B Integer Exponents: Negative One Squared	$\kappa_{M1}\,\kappa_{M2} = H^2 - H^2 + K$
∴	From 5, A4.2.8 Inverse Add, A4.2.7 Identity Add and A4.2.2B Equality	$K = \kappa_{M1}\,\kappa_{M2}$

Table 17.13.1 Hypersurface Major Principle Curvature Geometry[SOK67I, pg 193]

κ_{M1}	κ_{M2}	H	K	Hypersurface Type
$0 = \kappa_{M1}$	$0 = \kappa_{M2}$	$0 = H$	$0 = K$	Flat Euclidian Plane
$0 < \kappa_{M1}$	$0 = \kappa_{M2}$	$0 < H = ½\,\kappa_{M1}$	$0 = K$	Positive Parabolic Cylinder -κ_{M1} opens up
$0 = \kappa_{M1}$	$0 < \kappa_{M2}$	$0 < H = ½\,\kappa_{M2}$	$0 = K$	Positive Parabolic Cylinder -κ_{M2} opens up
$0 < \kappa_{M1}$	$0 = \kappa_{M2}$	$0 > H = ½\,\kappa_{M1}$	$0 = K$	Negative Parabolic Cylinder-κ_{M1} opens down
$0 = \kappa_{M1}$	$0 < \kappa_{M2}$	$0 > H = ½\,\kappa_{M2}$	$0 = K$	Negative Parabolic Cylinder-κ_{M2} opens down
$0 < \kappa_{M1}$	$0 < \kappa_{M2}$	$0 < H$ $0 < ½\,(\kappa_{M1} + \kappa_{M2})$	$0 < K$ $0 < \kappa_{M1}\,\kappa_{M2}$	Elliptic
$0 > \kappa_{M1}$	$0 > \kappa_{M2}$	$0 > H$ $0 > ½\,(\kappa_{M1} + \kappa_{M2})$	$0 < K$ $0 < \kappa_{M1}\,\kappa_{M2}$	Hyperbolic
$\kappa_{M1} = \kappa_{M2} = \kappa_M$	$\kappa_{M2} = \kappa_{M1} = \kappa_M$	$0 < \kappa_M$	$0 < K$ $0 < \kappa_M^2$	Spherical
$0 < \kappa_{M1}$	$0 > \kappa_{M2}$	$0 \le H$ $0 \le ½\,\lvert\kappa_{M1} + \kappa_{M2}\rvert$	$0 > K$ $0 > \kappa_{M1}\,\kappa_{M2}$	Positive Hyperbolic Saddle κ_{M1}[KRE59I, pg 131] opens up and κ_{M2} opens down
$0 > \kappa_{M1}$	$0 < \kappa_{M2}$	$0 \le H$ $0 \le ½\,\lvert\kappa_{M1} + \kappa_{M2}\rvert$	$0 > K$ $0 > \kappa_{M1}\,\kappa_{M2}$	Negative Hyperbolic Saddle κ_{M1}[KRE59I, pg 131] opens down and κ_{M2} opens up

Observation 17.13.7 Minor and Major Hypersurface Geometry at a Point

From theorem T17.13.3 "Major Principal Curvature: Mean Curvature on a Hypersurface $\mathfrak{H}^{n-1}$" and T17.13.4 "Major Principal Curvature: Gaussian Curvature on a Hypersurface $\mathfrak{H}^{n-1}$" the major principle curvatures deal with the generalized differential geometry at a point that has been rendered down to just two major curvatures. So the manifold hypersurface does not actually look like these descriptions except in 3-space, remember the point is really comprised of (n–1) minor principle curvatures, which determines the actual geometry and anything above that diverges. Something us mere mortals cannot really visualize in (n–1)-dimensions. Visualizing geometry in 2 and 3-dimensions is well within our abilities and allows manipulation of such structures, hence the artifice of this math and table.

Word of warning most books on this topic do not make a clear distinction between minor and major principle curvatures in dimensions higher than three and freely interchange them leading to confusing and ironies conclusions.

Section 17.14 Meusnier Theorems: Other Curves on a Hypersurface

pg 187 [a][b]

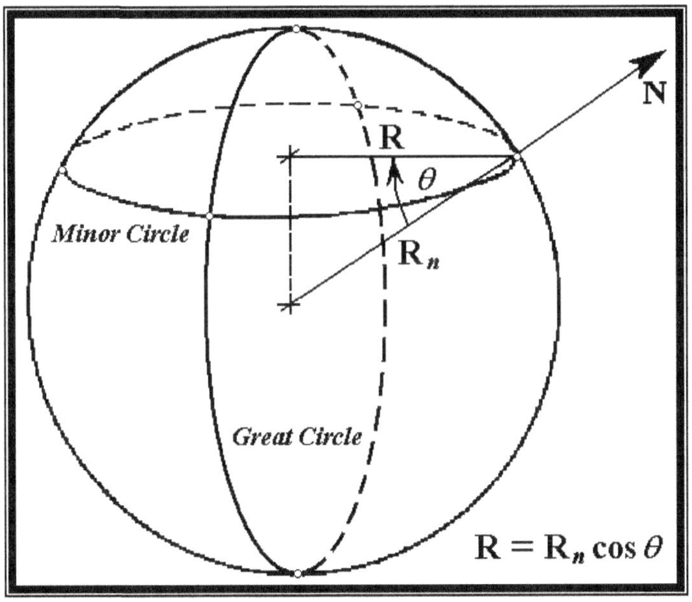

Figure 17.14.1 Meusnier Theorems: Curvature of the Minor Circle

Being able to measure curvature of a hypersurface or curvature of arcs on a hypersurface plays an important role in studying the geometry of manifolds, so it only follows that many different kinds of curvatures have been discovered and exploited. In this section one other curvature is developed and investigated, it will be shown there is a unique interrelationship to a different aspect of the geometry.

The infinitesimal hypersphere can measure conventional curvature, but there is another unique property of a hypersphere. A hypersphere is comprised of great circles, but another type of circle can be exploited the minor circle. The great circle can be thought of as longitudinal circle on a globe and the minor circle a perpendicular circle less in size, than the great circle, akin to latitude this provides a new and different radius or curvature.

Theorem 17.14.1 Meusnier Theorems: Pythagorean Relationship of Minor Curvatures

1g	Given	$\delta^i_{\ i} = (\mu^i \bullet \mu_i) = (\nu^i \bullet \nu_i) = (\eta^i \bullet \eta_i)$	
2g		$0 = (\nu^i \bullet \eta_i) = (\eta^i \bullet \nu_i)$	
3g		$(\nu^i \perp \eta^j)$ and $(\mu^i \perp \eta_j)$ Pseudo vector differential equations are an orthogonal system	
Step	Hypothesis	$\kappa^2 = \kappa_\alpha^{\ 2} + \kappa_n^{\ 2}$ EQ A	$1 = (\kappa_\alpha / \kappa)^2 + (\kappa_n / \kappa)^2$ EQ B
1	From T17.12.2B Minor Principal Curvatures: Auxiliary-1 Pseudo Vector Differential Equations	$\kappa_n = c_{\alpha\beta}\, \lambda^\alpha\, \lambda^\beta$	
2	From T7.7.12 Second Fundamental Covariant Metric: in Inverse Unit Normal Form	$x^i_{\ \alpha;\beta} = c_{\alpha\beta}\, \eta^i$	
3	T12.12.6 On Space Curve-C: First Order Derivative of Tangent Vector	$d\mathbf{T} / ds = \kappa\, \mathbf{N}$	
4	T17.1.1A Manifold-Hypersheet Transformation: Exact Differential	$dx_i = x_{i,\alpha}\, du_\alpha$	
5	From 4, LxK.3.1.7B Uniqueness Theorem of Differentiation, LxK.3.2.2A Differential as a Linear Operator and D16.1.5 Intrinsic Covariant Differential Notation	$x^{i:s} = x^{i;\alpha}\, u^{\alpha:s}$	
6	From 5, D16.1.7C Intrinsic Differential Pseudo Vectors and D17.1.9 Riemannian Manifold $\mathfrak{R}^n$ Mixed Tensor of Rank-2 on a Hypersheet $\mathfrak{H}^{n-1}$	$\lambda^i = x^i_{\ \alpha}\, \lambda^\alpha$	
7	From 3 and D16.1.7C Intrinsic Differential Pseudo Vectors; Riemannian Manifold $\mathfrak{R}^n$	$\delta\lambda^i / \delta s = \kappa\, \mu^i$	
8	From 3 and D16.1.7C Intrinsic Differential Pseudo Vectors; Hypersheet $\mathfrak{H}^{n-1}$	$\delta\lambda^\alpha / \delta s = \kappa_\alpha\, \nu^\alpha$	
9	From 7 and LxK.3.1.7B Uniqueness Theorem of Differentiation	$\delta\, (\lambda^i) / \delta s = \delta\, (\lambda^i) / \delta s$	
10	From 6, 9 and A7.9.2 Substitution of Vectors	$\delta\, (\lambda^i) / \delta s = \delta\, (x^i_{\ \alpha}\, \lambda^\alpha) / \delta s$	
11	From 10 and TK.3.9 Exact Differential, Parametric, Chain Rule	$\delta\, (\lambda^i) / \delta s = x^i_{\ \alpha}\, \delta\, (\lambda^\alpha) / \delta s + \lambda^\alpha\, \delta\, (x^i_{\ \alpha}) / \delta s$	

12	From 11, TK.3.5 Single Variable Chain Rule Alternate and T17.5.26A Differentiation of Tensor: Rank-1 Covariant	$\delta(\lambda^i) / \delta s = x^i_\alpha\, \delta(\lambda^\alpha) / \delta s + \lambda^\alpha (x^i_{\alpha;\beta})\, \delta u_\beta / \delta s$
13	From 2, 12, D16.1.7C Intrinsic Differential Pseudo Vectors, T10.2.5 Tool: Tensor Substitution, T10.3.7 Product of Tensors: Commutative by Multiplication of Rank p + q → q + p and A10.2.14 Correspondence of Tensors and Tensor Coefficients	$\delta(\lambda^i) / \delta s = \kappa_\alpha\, x^i_\alpha\, \nu^\alpha + c_{\alpha\beta}\, \lambda^\alpha \lambda^\beta \eta^i$
14	From 1, 7, 13 and A7.9.2 Substitution of Vectors	$\kappa\, \mu^i = \kappa_\alpha\, x^i_\alpha\, \nu^\alpha + \kappa_n\, \eta^i$
15	From 6, 14 and A7.9.2 Substitution of Vectors	$\kappa\, \mu^i = \kappa_\alpha\, \nu^i + \kappa_n\, \eta^i$
16	From 15 and O12.3.4E Rules of Tensor Contraction	$\kappa\, \mu_j = \kappa_\alpha\, \nu_j + \kappa_n\, \eta_j$
17	From 15, 16 and T8.3.1 Dot Product: Uniqueness Operation	$\kappa\, \mu^i \bullet \kappa_j\, \mu_j = (\kappa_\alpha\, \nu^i + \kappa_n\, \eta^i) \bullet (\kappa_\alpha\, \nu_j + \kappa_n\, \eta_j)$
18	From 17, T8.3.7 Dot Product: Distribution of Dot Product Across Addition of Vectors and T8.3.8 Dot Product: Distribution of Dot Product Across Another Vector	$\kappa\, \kappa\, (\mu^i \bullet \mu_j) = \kappa_\alpha\, \kappa_\alpha\, (\nu^i \bullet \nu_j) + \kappa_\alpha\, \kappa_n\, (\nu^i \bullet \eta_j) +$ $\kappa_n\, \kappa_\alpha\, (\eta^i \bullet \nu_j) + \kappa_n\, \kappa_n\, (\eta^i \bullet \eta_j)$
19	From 1g, 2g, 3g, 18 and A4.2.3 Substitution	$\kappa\, \kappa\, \delta^i_j = \kappa_\alpha\, \kappa_\alpha\, \delta^i_j + \kappa_\alpha\, \kappa_n\, 0 + \kappa_n\, \kappa_\alpha\, 0 + \kappa_n\, \kappa_n\, \delta^i_j$
20	From 19, T4.4.1 Equalities: Any Quantity Multiplied by Zero is Zero, A4.2.7 Identity Add, D6.1.6B Identity Matrix, LxE.3.1.37 Sums of Powers of 0 for the First n-Integers and A4.2.14 Distribution	$\kappa^2\, n = (\kappa_\alpha^{\,2} + \kappa_n^{\,2})\, n$
21	From 20 and T4.4.3 Equalities: Cancellation by Multiplication	$\kappa^2 = \kappa_\alpha^{\,2} + \kappa_n^{\,2}$
22	From 21, A4.2.12 Identity Multp, T4.4.4B Equalities: Right Cancellation by Multiplication, A4.3.14 Distribution and T4.8.7 Integer Exponents: Distribution Across a Rational Number	$1 = (\kappa_\alpha / \kappa)^2 + (\kappa_n / \kappa)^2$
∴	From 21 and 22	$\kappa^2 = \kappa_\alpha^{\,2} + \kappa_n^{\,2}$ EQ A | $1 = (\kappa_\alpha / \kappa)^2 + (\kappa_n / \kappa)^2$ EQ B

Theorem 17.14.2 Meusnier Theorems: Curvature of the Minor Circle (Theorem of Meusnier)

1g 2g	Given	Figure F17.14.1 Theorem of Meusnier Curvature of the Minor Circle $\xi \equiv |\mathbf{c}_\alpha|\lambda^\alpha \, |\mathbf{c}_\beta|\lambda^\beta$	
Step	Hypothesis	$R = \pm R_n \cos\theta$ EQ A	$\kappa \cos\theta = c_{\alpha\beta}\, \lambda^\alpha \lambda^\beta$ EQ B
1	From T17.12.2B Minor Principal Curvatures: Auxiliary-1 Pseudo Vector Differential Equations	$\kappa_n = c_{\alpha\beta}\, \lambda^\alpha \lambda^\beta$	
2	From 1g, 1, D17.7.1B Second Fundamental Quadratic Dyadic of Covariant Metric Form, D8.3.1 The Inner or Dot Product Operator and	$c_{\alpha\beta} = |\mathbf{c}_\alpha| \, |\mathbf{c}_\beta| \cos\theta$	
3	From 1, 2, T10.2.5 Tool: Tensor Substitution, T10.3.7 Product of Tensors: Commutative by Multiplication of Rank p + q → q + p, A10.2.14 Correspondence of Tensors and Tensor Coefficients, A4.2.14 Distribution and A4.2.11 Associative Multp	$(|\mathbf{c}_\alpha|\lambda^\alpha)(|\mathbf{c}_\beta|\lambda^\beta) \cos\theta = \kappa_n$	
4	From 2g, 3 and A4.2.3 Substitution	$\xi \cos\theta = \kappa_n$	
5	From 4 and T4.4.6B Equalities: Left Cancellation by Multiplication	$\cos\theta = \kappa_n / \xi$	
6	From 6 and T17.14.1B Meusnier Theorems: Pythagorean Relationship of Minor Curvatures	$1 = (\kappa_\alpha / \kappa)^2 + (\kappa_n / \kappa)^2$	
7	From LxE.3.1.19A Pythagorean Relations	$1 = (\sin\theta)^2 + (\cos\theta)^2$	
8	From 6, 7 and equating arguments of terms	$\sin\theta = \kappa_\alpha / \kappa$	
9	From 6, 7 and equating arguments of terms	$\cos\theta = \kappa_n / \kappa$	
10	From 5, 9 and A4.2.3 Substitution	$\kappa_n / \xi = \kappa_n / \kappa$	
11	From T4.4.8 Equalities: Cross Product of Proportions	$\kappa_n\, \kappa = \kappa_n\, \xi$	
12	From 11 and T4.4.3 Equalities: Cancellation by Multiplication	$\xi = \kappa$	
13	From 8, 12 and T4.4.5B Equalities: Reversal of Right Cancellation by Multiplication	$\kappa_\alpha = \kappa \sin\theta$	
14	From 9, 12 and T4.4.5B Equalities: Reversal of Right Cancellation by Multiplication	$\kappa_n = \kappa \cos\theta$	

15	From 14, A4.2.3 Substitution and D12.12.7A Curvature to a Spatial Curve-C	$1 / R_n = (1 / R) \cos \theta$	
16	From 15 and T4.4.8 Equalities: Cross Product of Proportions; plus or minus for either side of the minor circle	$R = \pm R_n \cos \theta$	
17	From 1, 14 and A4.2.3 Substitution	$\kappa \cos \theta = c_{\alpha\beta} \lambda^{\alpha} \lambda^{\beta}$	
∴	From 16 and 17	$R = \pm R_n \cos \theta$ EQ A	$\kappa \cos \theta = c_{\alpha\beta} \lambda^{\alpha} \lambda^{\beta}$ EQ B

Section 17.15 Parallel Hypersurfaces $\mathfrak{H}^{n-1}$

pg 195 [a][b]

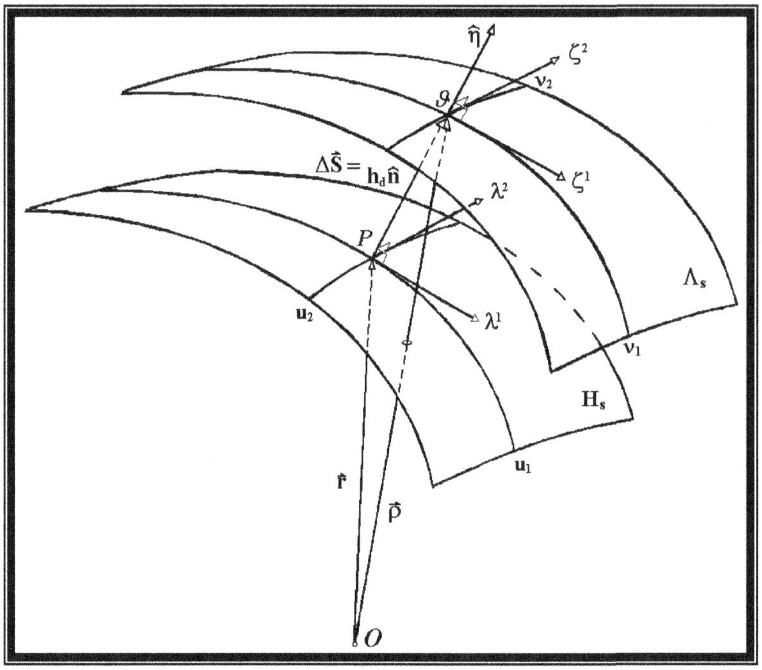

Figure 17.15.1 Parallel Hypersurfaces $\mathfrak{H}^{n-1}$ Spaced |Pϑ| Apart

Definition 17.15.1 Parallel Hypersurface's $\mathfrak{H}^{n-1}$ Positional Vector ΔS Spanning the Gap

The ***positional spanning distance*** between two Hypersurfaces $\mathfrak{H}^{n-1}$ is traced out by the positional vectors

$\Delta \mathbf{S} = \boldsymbol{\rho} - \mathbf{r}$ Equation A

Definition 17.15.2 Gap in the Strict Sense for Riemannian Curvature

When the gap distance between parallel hypersurfaces is small enough for the Riemannian Curvature spanning it exactly with little or no error, then it is said, "to be in the ***strict sense***."

$(h_d\, n^i)_{;\alpha;\beta;\gamma} \equiv x^i{}_{\alpha;\beta;\gamma} - x^i{}_{\alpha;\gamma;\beta}$ Equation A

Definition 17.15.3 Gap in the Wide Sense for Riemannian Curvature

When the gap distance between parallel hypersurfaces is just wide enough for the Riemannian Curvature to span it approximately with acceptable error, then it is said, "to be in the ***wide sense***."

$(h_d\, n^i)_{;\alpha;\beta;\gamma} \approx x^i{}_{\alpha;\beta;\gamma} - x^i{}_{\alpha;\gamma;\beta}$ Equation A

Axiom 17.15.1 Properties of Parallel Planes

P1) Normal unit vectors **n** and $\boldsymbol{\eta}$ to planes H_s to Λ_s respectively are collinear.

P2) Planes H_s to Λ_s are distanced apart along the normal between the planes by a length

$h_d \equiv |P\vartheta|$ Equation A

or

$h_d \equiv |\Delta \mathbf{S}|$ Equation B.

P3) Pseudo vectors $[\lambda^i, \zeta^i]$ and tangent vectors $[x_\rho{}^i{}_\kappa, x^i{}_\kappa]$ are collinear on planes H_s and Λ_s form parallel vector fields for all i:

$x_\rho{}^i{}_\kappa \parallel x^i{}_\kappa$ Equation A

and

$|x_\rho{}^i{}_\kappa| \equiv |x^i{}_\kappa|$ Equation B

P4) Normal vectors are perpendicular to pseudo vectors in either plane H_s to Λ_s.

P5) Sets of basis vectors for one plane is the same set for the other, the two planes are parts of the same whole system. It follows than that vectors constructed in the H_s can be constructed in the Λ_s from the same basis vector set.

P6) Iff minor curvatures at a point, on one parallel plane exist then it implies the same curvatures exists on the opposite surface at the corresponding point.

P7) Gap curvature is none commutative, noted by transpose of indices, between parallel hypersurfaces.

Axiom 17.15.2 Constant Curvature between Parallel Hypersurfaces $\mathfrak{H}^{n-1}$

Iff [K] is constant everywhere over a hypersurface, then the opposite hypersurface has the same constant curvature as well and the distance between the gap will not vary or vanish over the hypersurfaces:

$0 = (h_d)_{,\alpha,\beta,\gamma}$ Equation A

$0 = (h_d)_{,\alpha,\beta}$ Equation B

$0 = (h_d)_{,\gamma}$ Equation C

and

$0 \neq h_d$ Equation D

Theorem 17.15.1 Parallel Hypersurfaces $\mathfrak{H}^{n-1}$: Initial Model Setup between Hypersurfaces

1g	Given	Figure F17.15.1 Parallel Hypersurfaces $\mathfrak{H}^{n-1}$ Spaced $\lvert P\vartheta \rvert$ Apart	
Step	Hypothesis	$\boldsymbol{\rho} \equiv \mathbf{r} + \Delta\mathbf{S}$ EQ A	$\Delta\mathbf{S} = \boldsymbol{\rho} - \mathbf{r}$ EQ B
1	From 1g and $\mathbf{O\vartheta} \equiv \mathbf{OP} + \mathbf{P\vartheta}$	$\boldsymbol{\rho} \equiv \mathbf{r} + \Delta\mathbf{S}$	
2	From 1, A7.9.3 Commutative by Vector Addition and T7.11.4B Vector Addition: Right Cancellation by Addition	$\Delta\mathbf{S} = \boldsymbol{\rho} - \mathbf{r}$	
∴	From 1 and 2	$\boldsymbol{\rho} \equiv \mathbf{r} + \Delta\mathbf{S}$ EQ A	$\Delta\mathbf{S} = \boldsymbol{\rho} - \mathbf{r}$ EQ B

Observation 17.15.1 The Action Takes Place between the Parallel Hypersurfaces $\mathfrak{H}^{n-1}$

In the discussion of parallel hypersurfaces most authors I.S. Sokolnikoff [SOK67], Erwin Kreyszig [KRE59] and others define the connection between planes based on T17.15.1A "Parallel Hypersurfaces $\mathfrak{H}^{n-1}$: Initial Model Setup Between Hypersurfaces $\mathfrak{H}^{n-1}$", but any change to hyperplanes H_s and Λ_s is quit independent of each other. The reality is that all the action takes place between the planes. So as any given plane is changed, then that change is reflected relative to the other, therefore in this development equation T17.15.1B is used.

Theorem 17.15.2 Parallel Hypersurfaces $\mathfrak{H}^{n-1}$: Spanning the Gap in the Wide Sense

1g	Given	From figure F17.15.1 Parallel Hypersurfaces $\mathfrak{H}^{n-1}$ Spaced $\lvert P\vartheta \rvert$ Apart
Step	Hypothesis	$(h_d\, n^i)_{;\alpha;\beta;\gamma} \approx x^i{}_{\alpha;\beta;\gamma} - x^i{}_{\alpha;\gamma;\beta}$
1	From 17.15.1B Parallel Hypersurfaces $\mathfrak{H}^{n-1}$: Initial Model Setup Between Hypersurfaces $\mathfrak{H}^{n-1}$	$\Delta\mathbf{S} = \boldsymbol{\rho} - \mathbf{r}$
2	From 1 and T17.5.28A Differentiation of Tensor: Rank-1 Uniqueness of Differentiation	$(\Delta\mathbf{S})_{;\alpha;\beta;\gamma} = (\boldsymbol{\rho} - \mathbf{r})_{;\alpha;\beta;\gamma}$
3	From 2 and LxK.3.2.2B Differential as a Linear Operator	$(\Delta\mathbf{S})_{;\alpha;\beta;\gamma} = \boldsymbol{\rho}_{;\alpha;\beta;\gamma} - \mathbf{r}_{;\alpha;\beta;\gamma}$
4	From 1g, 3, A17.15.1P2A Properties of Parallel Planes, A17.1.8 Fundamental Property of Continuum for a Hypersheet and A7.9.2 Substitution of Vectors	$(h_d \sum_i n^i\, \mathbf{b}_i)_{;\alpha;\beta;\gamma} = (\sum_i x_{\rho i}\, \mathbf{b}_i)_{;\alpha;\beta;\gamma} - (\sum_i x_i\, \mathbf{b}_i)_{;\alpha;\beta;\gamma}$
5	From 4, LxK.3.2.3B Distribute Across Constant Product and T17.5.26A Differentiation of Tensor: Rank-1 Covariant	$(h_d \sum_i n^i{}_{;\alpha;\beta;\gamma}\, \mathbf{b}_i) = (\sum_i x_\rho{}^{i;\alpha}{}_{;\beta;\gamma}\, \mathbf{b}_i) - (\sum_i x^{i;\alpha}{}_{;\beta;\gamma}\, \mathbf{b}_i)$

6	From 5, D4.1.20A Negative Coefficient, T10.4.6 Scalars and Tensors: Distribution of a Scalar over Addition of Tensors, T10.3.20 Addition of Tensors: Distribution of a Tensor over Addition of Tensors and A10.2.14 Correspondence of Tensors and Tensor Coefficients; equating basis vectors-to-basis vectors	$(h_d \sum_i n^i)_{;\alpha;\beta;\gamma} = \sum_i (x_\rho{}^i{}_{\alpha;\beta;\gamma} - x^i{}_{\alpha;\beta;\gamma})$
∴	From 6, A17.15.1P7 Properties of Parallel Planes and T14.1.7 Changing Direction Changes Curvature; Thus as ρ → r in the ***wide sense***, then the difference can be set equal, because there will exist a unique value of gap distance $[h_d]$ to satisfy the equality and the transpose of indices allows the separation symbol-ρ to be dropped.	$(h_d\, n^i)_{;\alpha;\beta;\gamma} \approx x^i{}_{\alpha;\beta;\gamma} - x^i{}_{\alpha;\gamma;\beta}$

Observation 17.15.2 Equating Gap Distance to Differential Span; Possible Point of Error

The approximate equality comes about in order to error towards uncertainty. Unlike T17.8.9 "Codazzi-Gauss Equations: Riemann-Christoffel Tensor on a Hypersheet" which can be set exactly equal by construction, this case relies exclusively on being able to find the right gap distance $[h_d]$. This is the place and the reason if the equation should fail it will be here, because it is empirically set.

Theorem 17.15.3 Parallel Hypersurfaces $\mathfrak{H}^{n-1}$: Wide Sense Riemannian Curvature between Gap

Step	Hypothesis	$(h_d\, n^i)_{;\alpha;\beta;\gamma} \approx \sum_\nu x^i{}_\nu R^\nu{}_{\alpha\beta\gamma}$
1	From T17.15.2 Parallel Hypersurfaces $\mathfrak{H}^{n-1}$: Spanning the Gap in the Wide Sense	$(h_d\, n^i)_{;\alpha;\beta;\gamma} \approx x^i{}_{\alpha;\beta;\gamma} - x^i{}_{\alpha;\gamma;\beta}$
∴	From 2, T17.8.7 Riemann-Christoffel Tensor on a Hypersheet: Vector Relation, T10.2.5 Tool: Tensor Substitution and A10.2.14 Correspondence of Tensors and Tensor Coefficients	$(h_d\, n^i)_{;\alpha;\beta;\gamma} \approx \sum_\nu x^i{}_\nu R^\nu{}_{\alpha\beta\gamma}$

Theorem 17.15.4 Parallel Hypersurfaces $\mathfrak{H}^{n-1}$: Wide Sense Gaussian Curvature between Gap

Step	Hypothesis	$(n - 1)\ K\ x^i_\beta \approx \Sigma_\alpha \Sigma_\gamma a^{\alpha\gamma} (h_d\ n^i)_{;\alpha;\beta;\gamma}$
1	From 17.15.3 Parallel Hypersurfaces $\mathfrak{H}^{n-1}$: Wide Sense Riemannian Curvature between Gap	$(h_d\ n^i)_{;\alpha;\beta;\gamma} \approx \Sigma_\nu\ x^i_\nu\ R^\nu{}_{\alpha\beta\gamma}$
2	From 2, T17.5.67 Riemann-Christoffel Tensor First Kind to Second Kind, T10.2.5 Tool: Tensor Substitution and A10.2.14 Correspondence of Tensors and Tensor Coefficients	$(h_d\ n^i)_{;\alpha;\beta;\gamma} \approx \Sigma_\nu \Sigma_\eta\ x^i_\nu\ a^{\nu\eta}\ R_{\eta\alpha\beta\gamma}$
3	From 2, T17.11.5 Hypersphere: Relationship of Gauss's Equation to Total Curvature, T10.2.5 Tool: Tensor Substitution and A10.2.14 Correspondence of Tensors and Tensor Coefficients	$(h_d\ n^i)_{;\alpha;\beta;\gamma} \approx \Sigma_\nu \Sigma_\eta\ x^i_\nu\ a^{\nu\eta}\ K\ (a_{\alpha\gamma}\ a_{\eta\beta} - a_{\alpha\beta}\ a_{\eta\gamma})$
4	From 3, T10.4.6 Scalars and Tensors: Distribution of a Scalar over Addition of Tensors and A10.2.14 Correspondence of Tensors and Tensor Coefficients; exchange summation ν → η	$(h_d\ n^i)_{;\alpha;\beta;\gamma} \approx K\ \Sigma_\eta \Sigma_\nu\ x^i_\nu\ a^{\nu\eta}\ (a_{\alpha\gamma}\ a_{\eta\beta} - a_{\alpha\beta}\ a_{\eta\gamma})$
5	From 4, T10.3.6 Product of Tensors: Uniqueness by Multiplication of Rank for p + q and A10.2.14 Correspondence of Tensors and Tensor Coefficients; summing over all indices [α, γ]	$\Sigma_\alpha \Sigma_\gamma a^{\alpha\gamma} (h_d\ n^i)_{;\alpha;\beta;\gamma} \approx K\ \Sigma_\alpha \Sigma_\gamma a^{\alpha\gamma}$ $[\Sigma_\eta \Sigma_\nu\ x^i_\nu\ a^{\nu\eta}\ (a_{\gamma\alpha}\ a_{\eta\beta} - a_{\alpha\beta}\ a_{\gamma\eta})]$
6	From 5 and T10.3.20 Addition of Tensors: Distribution of a Tensor over Addition of Tensors and A10.2.14 Correspondence of Tensors and Tensor Coefficients; exchange summation [α, γ] → [η, ν]	$\Sigma_\alpha \Sigma_\gamma a^{\alpha\gamma} (h_d\ n^i)_{;\alpha;\beta;\gamma} \approx K\ \Sigma_\eta \Sigma_\nu\ x^i_\nu\ a^{\nu\eta}$ $(\Sigma_\alpha \Sigma_\gamma a^{\alpha\gamma}\ a_{\gamma\alpha}\ a_{\eta\beta} - a_{\alpha\beta}\ \Sigma_\alpha \Sigma_\gamma a^{\alpha\gamma}\ a_{\gamma\eta})$
7	From 6, T17.4.6B Hypersheet: Covariant and Contravariant Metric Inverse, T10.2.5 Tool: Tensor Substitution and A10.2.14 Correspondence of Tensors and Tensor Coefficients	$\Sigma_\alpha \Sigma_\gamma a^{\alpha\gamma} (h_d\ n^i)_{;\alpha;\beta;\gamma} \approx K\ \Sigma_\eta \Sigma_\nu\ x^i_\nu\ a^{\nu\eta}$ $(\Sigma_\alpha\ \delta^\alpha{}_\alpha\ a_{\eta\beta} - \Sigma_\alpha\ a_{\alpha\beta}\ \delta^\alpha{}_\eta)$

8	From 7, T17.4.14 Hypersheet: Covariant-Contravariant Kronecker Delta Summation of like Indices, T17.5.37B Contracted Covariant Bases Vector with Contravariant Vector, T10.2.5 Tool: Tensor Substitution and A10.2.14 Correspondence of Tensors and Tensor Coefficients	$\Sigma_\alpha \Sigma_\gamma a^{\alpha\gamma} (h_d n^i)_{;\alpha;\beta;\gamma} \approx K \Sigma_\eta \Sigma_\nu x^i_\nu a^{\nu\eta} (n a_{\eta\beta} - a_{\eta\beta})$
9	From 8, T10.2.10 Tool: Identity Multiplication of a Tensor, T10.4.7 Scalars and Tensors: Distribution of a Tensor over Addition of Scalars and A10.2.14 Correspondence of Tensors and Tensor Coefficients	$\Sigma_\alpha \Sigma_\gamma a^{\alpha\gamma} (h_d n^i)_{;\alpha;\beta;\gamma} \approx K \Sigma_\eta \Sigma_\nu x^i_\nu a^{\nu\eta} (n - 1) a_{\eta\beta}$
10	From 9, T10.4.6 Scalars and Tensors: Distribution of a Scalar over Addition of Tensors and A10.2.14 Correspondence of Tensors and Tensor Coefficients; exchange summation $\eta \rightarrow \nu$	$\Sigma_\alpha \Sigma_\gamma a^{\alpha\gamma} (h_d n^i)_{;\alpha;\beta;\gamma} \approx (n - 1) K \Sigma_\nu x^i_\nu \Sigma_\eta a^{\nu\eta} a_{\eta\beta}$
11	From 10, T17.4.6B Hypersheet: Covariant and Contravariant Metric Inverse, T10.2.5 Tool: Tensor Substitution and A10.2.14 Correspondence of Tensors and Tensor Coefficients	$\Sigma_\alpha \Sigma_\gamma a^{\alpha\gamma} (h_d n^i)_{;\alpha;\beta;\gamma} \approx (n - 1) K \Sigma_\nu x^i_\nu \delta^\nu_\beta$
12	From 11, T17.5.37B Contracted Covariant Bases Vector with Contravariant Vector, T10.2.5 Tool: Tensor Substitution and A10.2.14 Correspondence of Tensors and Tensor Coefficients	$\Sigma_\alpha \Sigma_\gamma a^{\alpha\gamma} (h_d n^i)_{;\alpha;\beta;\gamma} \approx (n - 1) K x^i_\beta$
∴	From 12, A10.2.6B Equality of Tensors and A10.2.14 Correspondence of Tensors and Tensor Coefficients	$(n - 1) K x^i_\beta \approx \Sigma_\alpha \Sigma_\gamma a^{\alpha\gamma} (h_d n^i)_{;\alpha;\beta;\gamma}$

Observation 17.15.3 Deformation of Gaussian Curvature between Parallel Hypersurfaces

Notice that theorem T17.15.4 "Parallel Hypersurfaces $\mathfrak{H}^{n-1}$: Wide Sense Gaussian Curvature between Gap" has the Gaussian curvature [K] as a common constant of proportionality between them and lies along the tangent coordinate net of the hypersurface. This implies that as the parallel hyperplane deform relative to one another they do so together, and the curvature of one hypersheet remains the same for its opposing hypersheet, hence remaining constant under deformation.

Theorem 17.15.5 Parallel Hypersurfaces $\mathfrak{H}^{n-1}$: Differential of the Gap

1g	Given	From figure F17.15.1 Parallel Hypersurfaces $\mathfrak{H}^{n-1}$ Spaced $\vert P\vartheta\vert$ Apart
2g		$\mathbf{n} = \Sigma_i\, n_i\, \mathbf{b}^i$
Step	Hypothesis	$(\Delta\mathbf{S})_{;\alpha;\beta;\gamma} = (h_d\, \Sigma_i\, n^i)_{;\alpha;\beta;\gamma}\, \mathbf{b}_i$
1	From 1g and A17.15.1P2A Properties of Parallel Planes	$\Delta\mathbf{S} \equiv h_d\, \mathbf{n}$
2	From 2g, 1, T17.3.7A Surface Normal Vector: $\mathfrak{R}^n$ Transformation and Normal as Orthogonal System and A7.9.2 Substitution of Vectors	$\Delta\mathbf{S} = (h_d\, \Sigma_i\, n^i)\, \mathbf{b}_i$
3	From 2 and T17.5.28A Differentiation of Tensor: Rank-1 Uniqueness of Differentiation	$(\Delta\mathbf{S})_{;\alpha} = (h_d\, \Sigma_i\, n^i)_{;\alpha}\, \mathbf{b}_i$
4	From 3 and T17.5.28A Differentiation of Tensor: Rank-1 Uniqueness of Differentiation	$(\Delta\mathbf{S})_{;\alpha;\beta} = (h_d\, \Sigma_i\, n^i)_{;\alpha;\beta}\, \mathbf{b}_i$
∴	From 4 and T17.5.28A Differentiation of Tensor: Rank-1 Uniqueness of Differentiation	$(\Delta\mathbf{S})_{;\alpha;\beta;\gamma} = (h_d\, \Sigma_i\, n^i)_{;\alpha;\beta;\gamma}\, \mathbf{b}_i$

Theorem 17.15.6 Parallel Hypersurfaces $\mathfrak{H}^{n-1}$: Spanning the Gap with a Given Length-h_d

1g	Given	From figure F17.15.1 Parallel Hypersurfaces $\mathfrak{H}^{n-1}$ Spaced $\vert P\vartheta\vert$ Apart
Step	Hypothesis	$(\Delta\mathbf{S})_{;\alpha;\beta;\gamma} =$ $[(h_d)_{,\alpha,\beta,\gamma}\, \mathbf{n}]$ $+ [3(h_d)_{,\alpha,\beta}\, \mathbf{n}_{;\gamma}]$ $+ [3(h_d)_{,\gamma}\, \mathbf{n}_{;\alpha;\beta}]$ $+ [h_d\, (\mathbf{n})_{;\alpha;\beta;\gamma}]$
1	From 1g and A17.15.1P2A Properties of Parallel Planes	$\Delta\mathbf{S} \equiv h_d\, \mathbf{n}$
2	From 1 and T17.5.28A Differentiation of Tensor: Rank-1 Uniqueness of Differentiation,	$(\Delta\mathbf{S})_{;\alpha} = (h_d\, \mathbf{n})_{;\alpha}$
3	From 2, T12.7.8 Tensor Product: Chain Rule for Mixed Binary Product and D12.6.1B Differental Tensor Notation	$(\Delta\mathbf{S})_{;\alpha} = (h_d)_{,\alpha}\, \mathbf{n} + (h_d)\, (\mathbf{n})_{;\alpha}$
4	From 3 and T17.5.28A Differentiation of Tensor: Rank-1 Uniqueness of Differentiation	$[(\Delta\mathbf{S})_{;\alpha}]_{;\beta} = [(h_d)_{,\alpha}\, \mathbf{n} + (h_d)\, (\mathbf{n})_{;\alpha}]_{;\beta}$
5	From 4 and LxK.3.2.2B Differential as a Linear Operator	$(\Delta\mathbf{S})_{;\alpha;\beta} = [(h_d)_{,\alpha}\, \mathbf{n}]_{;\beta} + [(h_d)\, (\mathbf{n})_{;\alpha}]_{;\beta}$

6	From 5, T12.7.8 Tensor Product: Chain Rule for Mixed Binary Product and D12.6.1B Differental Tensor Notation	$(\mathbf{\Delta S})_{;\alpha;\beta} = (h_d)_{,\alpha,\beta}\, \mathbf{n} + (h_d)_{,\alpha}\, \mathbf{n}_{;\beta} + (h_{d,\beta})\, (\mathbf{n})_{;\alpha} + (h_d)\, (\mathbf{n})_{;\alpha;\beta}$
7	From 6, exchange dummy indices β → α, T7.10.5 Identity with Scalar Multiplication to Vectors, T7.10.9 Distribution of Vector over Addition of Scalars and D4.1.19 Primitive Definition for Rational Arithmetic	$(\mathbf{\Delta S})_{;\alpha;\beta} = (h_d)_{,\alpha,\beta}\, \mathbf{n} + 2(h_d)_{,\beta}\, \mathbf{n}_{;\alpha} + (h_d)\, (\mathbf{n})_{;\alpha;\beta}$
8	From 7 and T17.5.28A Differentiation of Tensor: Rank-1 Uniqueness of Differentiation	$[(\mathbf{\Delta S})_{;\alpha;\beta}]_{;\gamma} = [(h_d)_{,\alpha,\beta}\, \mathbf{n} + 2(h_d)_{,\beta}\, \mathbf{n}_{;\alpha} + (h_d)\, (\mathbf{n})_{;\alpha;\beta}]_{;\gamma}$
9	From 8 and LxK.3.2.2B Differential as a Linear Operator	$(\mathbf{\Delta S})_{;\alpha;\beta;\gamma} = [(h_d)_{;\alpha;\beta}\, \mathbf{n}]_{;\gamma} + [2(h_d)_{,\beta}\, \mathbf{n}_{;\alpha}]_{;\gamma} + [(h_d)\, (\mathbf{n})_{;\alpha;\beta}]_{;\gamma}$
10	From 9, T12.7.8 Tensor Product: Chain Rule for Mixed Binary Product and D12.6.1B Differental Tensor Notation	$(\mathbf{\Delta S})_{;\alpha;\beta;\gamma} = [(h_d)_{,\alpha,\beta,\gamma}\, \mathbf{n}]$ $+ [(h_d)_{,\alpha,\beta}\, \mathbf{n}_{;\gamma}]$ $+ [2(h_d)_{,\beta,\gamma}\, \mathbf{n}_{;\alpha}] + [2(h_d)_{,\beta}\, \mathbf{n}_{;\alpha;\gamma}]$ $+ [(h_d)_{;\gamma}\, (\mathbf{n})_{;\alpha;\beta}] + [h_d\, (\mathbf{n})_{;\alpha;\beta;\gamma}]$
11	From 10 and exchanging dummy indices γ → α and β → γ	$(\mathbf{\Delta S})_{;\alpha;\beta;\gamma} = [(h_d)_{,\alpha,\beta,\gamma}\, \mathbf{n}]$ $+ [(h_d)_{,\alpha,\beta}\, \mathbf{n}_{;\gamma}]$ $+ [2(h_d)_{,\beta,\alpha}\, \mathbf{n}_{;\gamma}] + [2(h_d)_{,\gamma}\, \mathbf{n}_{;\alpha;\beta}]$ $+ [(h_d)_{;\gamma}\, (\mathbf{n})_{;\alpha;\beta}] + [h_d\, (\mathbf{n})_{;\alpha;\beta;\gamma}]$
12	From 11 and DxK.3.2.3 Leibniz's Commutative Differential Denominators	$(\mathbf{\Delta S})_{;\alpha;\beta;\gamma} = [(h_d)_{,\alpha,\beta,\gamma}\, \mathbf{n}]$ $+ [(h_d)_{,\alpha,\beta}\, \mathbf{n}_{;\gamma}]$ $+ [2(h_d)_{,\alpha,\beta}\, \mathbf{n}_{;\gamma}] + [2(h_d)_{,\gamma}\, \mathbf{n}_{;\alpha;\beta}]$ $+ [(h_d)_{;\gamma}\, (\mathbf{n})_{;\alpha;\beta}] + [h_d\, (\mathbf{n})_{;\alpha;\beta;\gamma}]$
∴	From 12, T7.10.5 Identity with Scalar Multiplication to Vectors, T7.10.9 Distribution of Vector over Addition of Scalars and D4.1.19 Primitive Definition for Rational Arithmetic	$(\mathbf{\Delta S})_{;\alpha;\beta;\gamma} =$ $[(h_d)_{,\alpha,\beta,\gamma}\, \mathbf{n}]$ $+ [3(h_d)_{,\alpha,\beta}\, \mathbf{n}_{;\gamma}]$ $+ [3(h_d)_{,\gamma}\, \mathbf{n}_{;\alpha;\beta}]$ $+ [h_d\, (\mathbf{n})_{;\alpha;\beta;\gamma}]$

Theorem 17.15.7 Parallel Hypersurfaces $\mathfrak{H}^{n-1}$: First Differential of Normal Hypersurface Vector

Step	Hypothesis	$\mathbf{n}_{;\gamma} = -(\Sigma_i \Sigma_\nu w_\gamma^{\ \nu} x^i_{\ \nu})\ \mathbf{b}_i$
1	From T17.3.7A Surface Normal Vector: $\mathfrak{R}^n$ Transformation and Normal as Orthogonal System and T17.5.28A Differentiation of Tensor: Rank-1 Uniqueness of Differentiation	$\mathbf{n}_{;\gamma} = \Sigma_i n^{i;\gamma}\ \mathbf{b}_i$
2	From 1 and D17.1.9 Riemannian Manifold $\mathfrak{R}^n$ Mixed Tensor of Rank-2 on a Hypersheet $\mathfrak{H}^{n-1}$	$\mathbf{n}_{;\gamma} = \Sigma_i n^i_{\ \gamma}\ \mathbf{b}_i$
∴	From 2, T17.8.5 Weingarten Formulas: Unit Normal Differential Solution (Weingarten Formula), T10.2.5 Tool: Tensor Substitution, T10.3.7 Product of Tensors: Commutative by Multiplication of Rank p + q → q + p and A10.2.14 Correspondence of Tensors and Tensor Coefficients, D4.1.20A Negative Coefficient and T7.10.3 Distribution of a Scalar over Addition of Vectors	$\mathbf{n}_{;\gamma} = -(\Sigma_i \Sigma_\nu w_\gamma^{\ \nu} x^i_{\ \nu})\ \mathbf{b}_i$

Theorem 17.15.8 Parallel Hypersurfaces $\mathfrak{H}^{n-1}$: Second Differential of Normal Hypersurface Vector

Step	Hypothesis	$\mathbf{n}_{;\alpha;\beta} = -\Sigma_i \Sigma_\nu (w_\alpha^{\ \nu}{}_{;\beta}\ x^i_{\ \nu} + w_\alpha^{\ \nu}\ c_{\nu\beta}\ n^i)\ \mathbf{b}_i$
1	From T17.15.7 Parallel Hypersurfaces $\mathfrak{H}^{n-1}$: First Differential of Normal Hypersurface Vector and T17.5.28A Differentiation of Tensor: Rank-1 Uniqueness of Differentiation	$[\mathbf{n}_{;\alpha}]_{;\beta} = [-(\Sigma_i \Sigma_\nu w_\alpha^{\ \nu} x^i_{\ \nu})\ \mathbf{b}_i]_{;\beta}$
2	From 1, T12.7.8 Tensor Product: Chain Rule for Mixed Binary Product and D12.6.1B Differental Tensor Notation	$\mathbf{n}_{;\alpha;\beta} = [-(\Sigma_i \Sigma_\nu w_\alpha^{\ \nu}{}_{;\beta}\ x^i_{\ \nu})\ \mathbf{b}_i] + [-(\Sigma_i \Sigma_\nu w_\alpha^{\ \nu}\ x^i_{\ \nu;\beta})\ \mathbf{b}_i]$

3	From 2, D4.1.20 Negative Coefficient, T10.4.6 Scalars and Tensors: Distribution of a Scalar over Addition of Tensors, T10.3.20 Addition of Tensors: Distribution of a Tensor over Addition of Tensors and A10.2.14 Correspondence of Tensors and Tensor Coefficients	$\mathbf{n}_{;\alpha;\beta} = -\sum_i \sum_\nu (w_\alpha{}^\nu{}_{;\beta}\, x^i{}_\nu + w_\alpha{}^\nu\, x^i{}_{\nu;\beta})\, \mathbf{b}_i$
∴	T17.7.12 Second Fundamental Covariant Metric: in Inverse Unit Normal Form, T10.2.5 Tool: Tensor Substitution and A10.2.14 Correspondence of Tensors and Tensor Coefficients	$\mathbf{n}_{;\alpha;\beta} = -\sum_i \sum_\nu (w_\alpha{}^\nu{}_{;\beta}\, x^i{}_\nu + w_\alpha{}^\nu\, c_{\nu\beta}\, n^i)\, \mathbf{b}_i$

Theorem 17.15.9 Parallel Hypersurfaces $\mathfrak{H}^{n-1}$: Third Differential of Normal Hypersurface Vector

Step	Hypothesis	$\mathbf{n}_{;\alpha;\beta;\gamma} = -\sum_i \sum_\nu \{$ $[(w_\alpha{}^\nu{}_{;\beta;\gamma} - (\sum_\eta c_{\alpha\eta}\, w_\beta{}^\eta)\, w_\gamma{}^\nu)]\, x^i{}_\nu$ $+ [(w_\alpha{}^\nu{}_{;\beta}\, c_{\nu\gamma}) + (w_\alpha{}^\nu{}_{;\gamma}\, c_{\nu\beta}) + (w_\alpha{}^\nu\, c_{\nu\beta;\gamma})]\, n^i \}\, \mathbf{b}_i$
1	From T17.15.8 Parallel Hypersurfaces $\mathfrak{H}^{n-1}$: Second Differential of Normal Hypersurface Vector and T17.5.28A Differentiation of Tensor: Rank-1 Uniqueness of Differentiation	$[\mathbf{n}_{;\alpha;\beta}]_{;\gamma} = [\, -\sum_i \sum_\nu (w_\alpha{}^\nu{}_{;\beta}\, x^i{}_\nu + w_\alpha{}^\nu\, c_{\nu\beta}\, n^i)\, \mathbf{b}_i\,]_{;\gamma}$
2	From 1 and LxK.3.2.2B Differential as a Linear Operator	$\mathbf{n}_{;\alpha;\beta;\gamma} = -\sum_i \sum_\nu [\, (w_\alpha{}^\nu{}_{;\beta}\, x^i{}_\nu)_{;\gamma} + (w_\alpha{}^\nu\, c_{\nu\beta}\, n^i)_{;\gamma}\,]\, \mathbf{b}_i$
3	From 2, T12.7.8 Tensor Product: Chain Rule for Mixed Binary Product, D12.6.1B Differental Tensor Notation and D17.1.9 Riemannian Manifold $\mathfrak{R}^n$ Mixed Tensor of Rank-2 on a Hypersheet $\mathfrak{H}^{n-1}$	$\mathbf{n}_{;\alpha;\beta;\gamma} = -\sum_i \sum_\nu [$ $(w_\alpha{}^\nu{}_{;\beta;\gamma}\, x^i{}_\nu) + (w_\alpha{}^\nu{}_{;\beta}\, x^i{}_{\nu;\gamma})$ $+ (w_\alpha{}^\nu{}_{;\gamma}\, c_{\nu\beta}\, n^i) + (w_\alpha{}^\nu\, c_{\nu\beta;\gamma}\, n^i) + (w_\alpha{}^\nu\, c_{\nu\beta}\, n^i{}_{;\gamma})]\, \mathbf{b}_i$

4	D4.1.20A Negative Coefficient and T7.10.3 Distribution of a Scalar over Addition of Vectors, T17.7.12 Second Fundamental Covariant Metric: in Inverse Unit Normal Form, T17.8.5 Weingarten Formulas: Unit Normal Differential Solution (Weingarten Formula), T10.2.5 Tool: Tensor Substitution, T10.3.7 Product of Tensors: Commutative by Multiplication of Rank p + q → q + p and A10.2.14 Correspondence of Tensors and Tensor Coefficients	$\mathbf{n}_{;\alpha;\beta;\gamma} = -\sum_i \sum_\nu [$ $(w_\alpha{}^\nu{}_{;\beta;\gamma}\, x^i{}_\nu) + (w_\alpha{}^\nu{}_{;\beta}\, c_{\nu\gamma}\, n^i)$ $+ (w_\alpha{}^\nu{}_{;\gamma}\, c_{\nu\beta}\, n^i) + (w_\alpha{}^\nu\, c_{\nu\beta;\gamma}\, n^i) - (w_\alpha{}^\nu\, c_{\nu\beta} \sum_\kappa w_\gamma{}^\kappa\, x^i{}_\kappa)]\, \mathbf{b}_i$
5	From 4, T10.3.16 Addition of Tensors: Commutative by Addition, T10.3.20 Addition of Tensors: Distribution of a Tensor over Addition of Tensors and A10.2.14 Correspondence of Tensors and Tensor Coefficients	$\mathbf{n}_{;\alpha;\beta;\gamma} = -\sum_i \{$ $\sum_\nu (w_\alpha{}^\nu{}_{;\beta;\gamma}\, x^i{}_\nu) - (\sum_\nu w_\alpha{}^\nu\, c_{\nu\beta} \sum_\kappa w_\gamma{}^\kappa\, x^i{}_\kappa)$ $+ \sum_\nu (w_\alpha{}^\nu{}_{;\beta}\, c_{\nu\gamma} + w_\alpha{}^\nu{}_{;\gamma}\, c_{\nu\beta} + w_\alpha{}^\nu\, c_{\nu\beta;\gamma})\, n^i \}\, \mathbf{b}_i$
6	From 5, D17.8.1 Weingarten Tensor on a Hypersheet $\mathfrak{H}^{n-1}$, T10.2.5 Tool: Tensor Substitution and and A10.2.14 Correspondence of Tensors and Tensor Coefficients; exchange summations ν → η	$\mathbf{n}_{;\alpha;\beta;\gamma} = -\sum_i \{$ $\sum_\nu (w_\alpha{}^\nu{}_{;\beta;\gamma}\, x^i{}_\nu) - (\sum_\eta c_{\alpha\eta} (\sum_\nu a^{\eta\nu}\, c_{\nu\beta}) \sum_\kappa w_\gamma{}^\kappa\, x^i{}_\kappa)$ $+ \sum_\nu (w_\alpha{}^\nu{}_{;\beta}\, c_{\nu\gamma} + w_\alpha{}^\nu{}_{;\gamma}\, c_{\nu\beta} + w_\alpha{}^\nu\, c_{\nu\beta;\gamma})\, n^i \}\, \mathbf{b}_i$
7	From 6, T10.3.7 Product of Tensors: Commutative by Multiplication of Rank p + q → q + p and A10.2.14 Correspondence of Tensors and Tensor Coefficients	$\mathbf{n}_{;\alpha;\beta;\gamma} = -\sum_i \{$ $\sum_\nu (w_\alpha{}^\nu{}_{;\beta;\gamma}\, x^i{}_\nu) - (\sum_\eta c_{\alpha\eta} (\sum_\nu c_{\beta\nu}\, a^{\nu\eta}) \sum_\kappa w_\gamma{}^\kappa\, x^i{}_\kappa)$ $+ \sum_\nu (w_\alpha{}^\nu{}_{;\beta}\, c_{\nu\gamma} + w_\alpha{}^\nu{}_{;\gamma}\, c_{\nu\beta} + w_\alpha{}^\nu\, c_{\nu\beta;\gamma})\, n^i \}\, \mathbf{b}_i$
8	From 7, T17.8.5 Weingarten Formulas: Unit Normal Differential Solution (Weingarten Formula), T10.2.5 Tool: Tensor Substitution and A10.2.14 Correspondence of Tensors and Tensor Coefficients	$\mathbf{n}_{;\alpha;\beta;\gamma} = -\sum_i \{$ $\sum_\nu (w_\alpha{}^\nu{}_{;\beta;\gamma}\, x^i{}_\nu) - (\sum_\eta c_{\alpha\eta}\, w_\beta{}^\eta) \sum_\kappa w_\gamma{}^\kappa\, x^i{}_\kappa)$ $+ \sum_\nu (w_\alpha{}^\nu{}_{;\beta}\, c_{\nu\gamma} + w_\alpha{}^\nu{}_{;\gamma}\, c_{\nu\beta} + w_\alpha{}^\nu\, c_{\nu\beta;\gamma})\, n^i \}\, \mathbf{b}_i$
9	From 8, T10.3.8 Product of Tensors: Association by Multiplication of Rank for p + q + r, A10.2.14 Correspondence of Tensors and Tensor Coefficients and changing dummy indices κ → ν; exchange summations η → ν	$\mathbf{n}_{;\alpha;\beta;\gamma} = -\sum_i \{$ $\sum_\nu (w_\alpha{}^\nu{}_{;\beta;\gamma}\, x^i{}_\nu) - \sum_\nu (\sum_\eta c_{\alpha\eta}\, w_\beta{}^\eta\, w_\gamma{}^\nu)\, x^i{}_\nu)$ $+ \sum_\nu (w_\alpha{}^\nu{}_{;\beta}\, c_{\nu\gamma} + w_\alpha{}^\nu{}_{;\gamma}\, c_{\nu\beta} + w_\alpha{}^\nu\, c_{\nu\beta;\gamma})\, n^i \}\, \mathbf{b}_i$

10	From 9; distribute summation [ν] out term-by-term, T10.3.8 Product of Tensors: Association by Multiplication of Rank for p + q + r and A10.2.14 Correspondence of Tensors and Tensor Coefficients	$\mathbf{n}_{;\alpha;\beta;\gamma} = -\sum_i \sum_\nu \{$ $(w_\alpha{}^\nu{}_{;\beta;\gamma}\, x^i{}_\nu) - (\sum_\eta c_{\alpha\eta}\, w_\beta{}^\eta)\, w_\gamma{}^\nu\, x^i{}_\nu)$ $+ (w_\alpha{}^\nu{}_{;\beta}\, c_{\nu\gamma} + w_\alpha{}^\nu{}_{;\gamma}\, c_{\nu\beta} + w_\alpha{}^\nu\, c_{\nu\beta;\gamma})\, n^i \}\, \mathbf{b}_i$
∴	From 10, T10.3.20 Addition of Tensors: Distribution of a Tensor over Addition of Tensors and A10.2.14 Correspondence of Tensors and Tensor Coefficients	$\mathbf{n}_{;\alpha;\beta;\gamma} = -\sum_i \sum_\nu \{$ $[(w_\alpha{}^\nu{}_{;\beta;\gamma} - (\sum_\eta c_{\alpha\eta}\, w_\beta{}^\eta)\, w_\gamma{}^\nu)]\, x^i{}_\nu$ $+ (w_\alpha{}^\nu{}_{;\beta}\, c_{\nu\gamma} + w_\alpha{}^\nu{}_{;\gamma}\, c_{\nu\beta} + w_\alpha{}^\nu\, c_{\nu\beta;\gamma})\, n^i \}\, \mathbf{b}_i$

Theorem 17.15.10 Parallel Hypersurfaces $\mathfrak{H}^{n-1}$: Gap Parsed into Orthogonal Components

1g	Given	$\mathbf{n} = \sum_i n_i\, \mathbf{b}^i$ Normalized vector
Step	Hypothesis	$(h_d \sum_i n^i)_{;\alpha;\beta;\gamma} =$ $\sum_\nu$ $\{h_d [(\sum_\eta c_{\alpha\eta}\, w_\beta{}^\eta)\, w_\gamma{}^\nu - w_\alpha{}^\nu{}_{;\beta;\gamma}] - 3(h_d)_{,\alpha,\beta}\, w_\gamma{}^\nu - 3(h_d)_{,\gamma}\, w_\alpha{}^\nu{}_{;\beta}$ $\}\, x^i{}_\nu$ $+$ $\{(h_d)_{,\alpha,\beta,\gamma} - \sum_\nu$ $3(h_d)_{,\gamma}\, w_\alpha{}^\nu\, c_{\nu\beta} + h_d\, (w_\alpha{}^\nu{}_{;\beta}\, c_{\nu\gamma} + w_\alpha{}^\nu{}_{;\gamma}\, c_{\nu\beta} + w_\alpha{}^\nu\, c_{\nu\beta;\gamma})$ $\}\, n^i$
1	From 17.15.5 Parallel Hypersurfaces $\mathfrak{H}^{n-1}$: Differential of the Gap	$(\Delta\mathbf{S})_{;\alpha;\beta;\gamma} = \sum_i (h_d \sum_i n^i)_{;\alpha;\beta;\gamma}\, \mathbf{b}_i$
2	From 1, T17.15.6 Parallel Hypersurfaces $\mathfrak{H}^{n-1}$: Spanning the Gap with a Given Length-h_d and T7.9.2 Substitution of Vectors	$\sum_i (h_d \sum_i n^i)_{;\alpha;\beta;\gamma}\, \mathbf{b}_i =$ $[(h_d)_{,\alpha,\beta,\gamma}\, \mathbf{n}]$ $+ [3(h_d)_{,\alpha,\beta}\, \mathbf{n}_{;\gamma}]$ $+ [3(h_d)_{,\gamma}\, \mathbf{n}_{;\alpha;\beta}]$ $+ [h_d\, (\mathbf{n})_{;\alpha;\beta;\gamma}]$
3	From 2, T17.3.7A Surface Normal Vector: $\mathfrak{R}^n$ Transformation and Normal as Orthogonal System and T7.9.2 Substitution of Vectors, T17.15.8 Parallel Hypersurfaces $\mathfrak{H}^{n-1}$: Second Differential of Normal Hypersurface Vector, T17.15.9 Parallel Hypersurfaces $\mathfrak{H}^{n-1}$: Third Differential of Normal Hypersurface Vector and T7.9.2 Substitution of Vectors	$\sum_i (h_d \sum_i n^i)_{;\alpha;\beta;\gamma}\, \mathbf{b}_i =$ $[(h_d)_{,\alpha,\beta,\gamma} \sum_i n^i\, \mathbf{b}_i]$ $- [3(h_d)_{,\alpha,\beta} \sum_i \sum_\nu (w_\gamma{}^\nu\, x^i{}_\nu)\, \mathbf{b}_i]$ $- [3(h_d)_{,\gamma} \sum_i \sum_\nu (w_\alpha{}^\nu{}_{;\beta}\, x^i{}_\nu)\, \mathbf{b}_i]$ $- [3(h_d)_{,\gamma} \sum_i \sum_\nu (w_\alpha{}^\nu\, c_{\nu\beta}\, n^i)\, \mathbf{b}_i]$ $- h_d \sum_i \sum_\nu \{$ $[(w_\alpha{}^\nu{}_{;\beta;\gamma} - (\sum_\eta c_{\alpha\eta}\, w_\beta{}^\eta)\, w_\gamma{}^\nu)]\, x^i{}_\nu\, \mathbf{b}_i$ $+ (w_\alpha{}^\nu{}_{;\beta}\, c_{\nu\gamma} + w_\alpha{}^\nu{}_{;\gamma}\, c_{\nu\beta} + w_\alpha{}^\nu\, c_{\nu\beta;\gamma})\, n^i \}\, \mathbf{b}_i$

4	From 3, T10.3.20 Addition of Tensors: Distribution of a Tensor over Addition of Tensors and A10.2.14 Correspondence of Tensors and Tensor Coefficients	$\Sigma_i (h_d \Sigma_i n^i)_{;\alpha;\beta;\gamma} \mathbf{b}_i = \Sigma_i \{$ $[(h_d)_{,\alpha,\beta,\gamma} n^i]$ $- [3(h_d)_{,\alpha,\beta} \Sigma_v (w_\gamma{}^v x^i{}_v)]$ $- [3(h_d)_{,\gamma} \Sigma_v (w_\alpha{}^v c_{v\beta} n^i)]$ $- [3(h_d)_{,\gamma} \Sigma_v (w_\alpha{}^v{}_{;\beta} x^i{}_v)]$ $- h_d \Sigma_v$ $[(w_\alpha{}^v{}_{;\beta;\gamma} - (\Sigma_\eta c_{\alpha\eta} w_\beta{}^\eta) w_\gamma{}^v)] x^i{}_v$ $+ (w_\alpha{}^v{}_{;\beta} c_{v\gamma} + w_\alpha{}^v{}_{;\gamma} c_{v\beta} + w_\alpha{}^v c_{v\beta;\gamma}) n^i \} \mathbf{b}_i$
5	From 4 and equating basis vectors term-by-term	$(h_d \Sigma_i n^i)_{;\alpha;\beta;\gamma} =$ $[(h_d)_{,\alpha,\beta,\gamma} n^i]$ $- [3(h_d)_{,\alpha,\beta} \Sigma_v (w_\gamma{}^v x^i{}_v)]$ $- [3(h_d)_{,\gamma} \Sigma_v (w_\alpha{}^v{}_{;\beta} x^i{}_v)]$ $- [3(h_d)_{,\gamma} \Sigma_v w_\alpha{}^v c_{v\beta} n^i)]$ $- h_d \Sigma_v [(w_\alpha{}^v{}_{;\beta;\gamma} - (\Sigma_\eta c_{\alpha\eta} w_\beta{}^\eta) w_\gamma{}^v)] x^i{}_v$ $- h_d \Sigma_v (w_\alpha{}^v{}_{;\beta} c_{v\gamma} + w_\alpha{}^v{}_{;\gamma} c_{v\beta} + w_\alpha{}^v c_{v\beta;\gamma}) n^i$
6	From 5, T10.3.16 Addition of Tensors: Commutative by Addition and A10.2.14 Correspondence of Tensors and Tensor Coefficients; by matching up orthogonal hypersurface components	$(h_d \Sigma_i n^i)_{;\alpha;\beta;\gamma} =$ $- h_d \Sigma_v [(w_\alpha{}^v{}_{;\beta;\gamma} - (\Sigma_\eta c_{\alpha\eta} w_\beta{}^\eta) w_\gamma{}^v)] x^i{}_v$ $- [3(h_d)_{,\alpha,\beta} \Sigma_v (w_\gamma{}^v x^i{}_v)]$ $- [3(h_d)_{,\gamma} \Sigma_v (w_\alpha{}^v{}_{;\beta} x^i{}_v)]$ $+$ $[(h_d)_{,\alpha,\beta,\gamma}] n^i$ $- h_d \Sigma_v (w_\alpha{}^v{}_{;\beta} c_{v\gamma} + w_\alpha{}^v{}_{;\gamma} c_{v\beta} + w_\alpha{}^v c_{v\beta;\gamma}) n^i$ $- 3(h_d)_{,\gamma} [\Sigma_v w_\alpha{}^v c_{v\beta}] n^i$
7	From 6, T10.3.20 Addition of Tensors: Distribution of a Tensor over Addition of Tensors and A10.2.14 Correspondence of Tensors and Tensor Coefficients	$(h_d \Sigma_i n^i)_{;\alpha;\beta;\gamma} =$ $-\Sigma_v h_d [w_\alpha{}^v{}_{;\beta;\gamma} - (\Sigma_\eta c_{\alpha\eta} w_\beta{}^\eta) w_\gamma{}^v] x^i{}_v$ $- \Sigma_v [3(h_d)_{,\alpha,\beta} w_\gamma{}^v] x^i{}_v$ $- \Sigma_v [3(h_d)_{,\gamma} w_\alpha{}^v{}_{;\beta}] x^i{}_v$ $+$ $[(h_d)_{,\alpha,\beta,\gamma}] n^i$ $- h_d \Sigma_v (w_\alpha{}^v{}_{;\beta} c_{v\gamma} + w_\alpha{}^v{}_{;\gamma} c_{v\beta} + w_\alpha{}^v c_{v\beta;\gamma}) n^i$ $-3(h_d)_{,\gamma} \Sigma_v [w_\alpha{}^v c_{v\beta}] n^i$
∴	From 7, D4.1.20A Negative Coefficient, T10.4.6 Scalars and Tensors: Distribution of a Scalar over Addition of Tensors, T10.3.20 Addition of Tensors: Distribution of a Tensor over Addition of Tensors and A10.2.14 Correspondence of Tensors and Tensor Coefficients; factoring summation by comparing orthogonal component-by-component	$(h_d \Sigma_i n^i)_{;\alpha;\beta;\gamma} =$ Σ_v $\{h_d [(\Sigma_\eta c_{\alpha\eta} w_\beta{}^\eta) w_\gamma{}^v - w_\alpha{}^v{}_{;\beta;\gamma}] - 3(h_d)_{,\alpha,\beta} w_\gamma{}^v - 3(h_d)_{,\gamma} w_\alpha{}^v{}_{;\beta}$ $\} x^i{}_v$ $+$ $\{(h_d)_{,\alpha,\beta,\gamma} - \Sigma_v$ $h_d (w_\alpha{}^v{}_{;\beta} c_{v\gamma} + w_\alpha{}^v{}_{;\gamma} c_{v\beta} + w_\alpha{}^v c_{v\beta;\gamma}) + 3(h_d)_{,\gamma} w_\alpha{}^v c_{v\beta}$ $\} n^i$

Theorem 17.15.11 Parallel Hypersurfaces $\mathfrak{H}^{n-1}$: Parsed into Orthogonal Gaussian Gap

Step	Hypothesis	$(n-1)\, K\, x^i_{\beta} \approx$ $\Sigma_\alpha \Sigma_\gamma a^{\alpha\gamma} \Sigma_\nu$ $\{h_d [(\Sigma_\eta c_{\alpha\eta} w_\beta{}^\eta)\, w_\gamma{}^\nu - w_\alpha{}^\nu{}_{;\beta;\gamma}] - 3(h_d)_{,\alpha,\beta}\, w_\gamma{}^\nu - 3(h_d)_{,\gamma}\, w_\alpha{}^\nu{}_{;\beta}$ $\}\, x^i_\nu$ $+$ $\Sigma_\alpha \Sigma_\gamma a^{\alpha\gamma} \{(h_d)_{,\alpha,\beta,\gamma} - \Sigma_\nu$ $h_d\, (w_\alpha{}^\nu{}_{;\beta}\, c_{\nu\gamma} + w_\alpha{}^\nu{}_{;\gamma}\, c_{\nu\beta} + w_\alpha{}^\nu\, c_{\nu\beta;\gamma}) + 3(h_d)_{,\gamma}\, w_\alpha{}^\nu\, c_{\nu\beta}$ $\}\, n^i$
1	From T17.15.10 Parallel Hypersurfaces $\mathfrak{H}^{n-1}$: Gap Parsed into Orthogonal Components	$(h_d\, \Sigma_i\, n^i)_{;\alpha;\beta;\gamma} =$ Σ_ν $\{h_d [(\Sigma_\eta c_{\alpha\eta} w_\beta{}^\eta)\, w_\gamma{}^\nu - w_\alpha{}^\nu{}_{;\beta;\gamma}] - 3(h_d)_{,\alpha,\beta}\, w_\gamma{}^\nu - 3(h_d)_{,\gamma}\, w_\alpha{}^\nu{}_{;\beta}$ $\}\, x^i_\nu$ $+$ $\{(h_d)_{,\alpha,\beta,\gamma} - \Sigma_\nu$ $h_d\, (w_\alpha{}^\nu{}_{;\beta}\, c_{\nu\gamma} + w_\alpha{}^\nu{}_{;\gamma}\, c_{\nu\beta} + w_\alpha{}^\nu\, c_{\nu\beta;\gamma}) + 3(h_d)_{,\gamma}\, w_\alpha{}^\nu\, c_{\nu\beta}$ $\}\, n^i$
2	From 2, T10.3.6 Product of Tensors: Uniqueness by Multiplication of Rank for p + q, D4.1.20A Negative Coefficient, T10.4.6 Scalars and Tensors: Distribution of a Scalar over Addition of Tensors, A10.2.14 Correspondence of Tensors and Tensor Coefficients and T4.8.2B Integer Exponents: Negative One Squared	$\Sigma_\alpha \Sigma_\gamma a^{\alpha\gamma} (h_d\, \Sigma_i\, n^i)_{;\alpha;\beta;\gamma} =$ $\Sigma_\alpha \Sigma_\gamma a^{\alpha\gamma} \Sigma_\nu$ $\{h_d [(\Sigma_\eta c_{\alpha\eta} w_\beta{}^\eta)\, w_\gamma{}^\nu - w_\alpha{}^\nu{}_{;\beta;\gamma}] - 3(h_d)_{,\alpha,\beta}\, w_\gamma{}^\nu - 3(h_d)_{,\gamma}\, w_\alpha{}^\nu{}_{;\beta}$ $\}\, x^i_\nu$ $+$ $\Sigma_\alpha \Sigma_\gamma a^{\alpha\gamma} \{(h_d)_{,\alpha,\beta,\gamma} - \Sigma_\nu$ $h_d\, (w_\alpha{}^\nu{}_{;\beta}\, c_{\nu\gamma} + w_\alpha{}^\nu{}_{;\gamma}\, c_{\nu\beta} + w_\alpha{}^\nu\, c_{\nu\beta;\gamma}) + 3(h_d)_{,\gamma}\, w_\alpha{}^\nu\, c_{\nu\beta}$ $\}\, n^i$
∴	From 2, T17.15.4 Parallel Hypersurfaces $\mathfrak{H}^{n-1}$: Wide Sense Gaussian Curvature between Gap, T10.2.5 Tool: Tensor Substitution and A10.2.14 Correspondence of Tensors and Tensor Coefficients	$(n-1)\, K\, x^i_{\beta} \approx$ $\Sigma_\alpha \Sigma_\gamma a^{\alpha\gamma} \Sigma_\nu$ $\{h_d [(\Sigma_\eta c_{\alpha\eta} w_\beta{}^\eta)\, w_\gamma{}^\nu - w_\alpha{}^\nu{}_{;\beta;\gamma}] - 3(h_d)_{,\gamma}\, w_\alpha{}^\nu{}_{;\beta} - 3(h_d)_{,\alpha,\beta}\, w_\gamma{}^\nu$ $\}\, x^i_\nu$ $+$ $\Sigma_\alpha \Sigma_\gamma a^{\alpha\gamma} \{(h_d)_{,\alpha,\beta,\gamma} - \Sigma_\nu$ $h_d\, (w_\alpha{}^\nu{}_{;\beta}\, c_{\nu\gamma} + w_\alpha{}^\nu{}_{;\gamma}\, c_{\nu\beta} + w_\alpha{}^\nu\, c_{\nu\beta;\gamma}) + 3(h_d)_{,\gamma}\, w_\alpha{}^\nu\, c_{\nu\beta}$ $\}\, n^i$

Theorem 17.15.12 Parallel Hypersurfaces $\mathfrak{H}^{n-1}$: Parsed into Normal Varying Gap

1g	Given	$\mathbf{n} = \sum_i n_i \mathbf{b}^i$ Normalized vector
Step	Hypothesis	$(h_d)_{,\alpha,\beta,\gamma} = \sum_v [h_d (w_\alpha{}^v{}_{;\beta} c_{v\gamma} + w_\alpha{}^v{}_{;\gamma} c_{v\beta} + w_\alpha{}^v c_{v\beta;\gamma}) + 3(h_d)_{,\gamma} w_\alpha{}^v c_{v\beta}]$
1	From 17.15.11 Parallel Hypersurfaces $\mathfrak{H}^{n-1}$: Parsed into Orthogonal Gaussian Gap	$(n-1) K x^i{}_\beta \approx \sum_\alpha \sum_\gamma a^{\alpha\gamma} \sum_v \{h_d [(\sum_\eta c_{\alpha\eta} w_\beta{}^\eta) w_\gamma{}^v - w_\alpha{}^v{}_{;\beta;\gamma}] - 3(h_d)_{,\alpha,\beta} w_\gamma{}^v - 3(h_d)_{,\gamma} w_\alpha{}^v{}_{;\beta} \} x^i{}_v + \sum_\alpha \sum_\gamma a^{\alpha\gamma} \{(h_d)_{,\alpha,\beta,\gamma} - \sum_v h_d (w_\alpha{}^v{}_{;\beta} c_{v\gamma} + w_\alpha{}^v{}_{;\gamma} c_{v\beta} + w_\alpha{}^v c_{v\beta;\gamma}) + 3(h_d)_{,\gamma} w_\alpha{}^v c_{v\beta} \} n^i$
2	From 1 and T17.3.15 Dual Covariant Tensors: Identity Addition	$(n-1) K x^i{}_\beta + 0\, n^i \approx \sum_\alpha \sum_\gamma a^{\alpha\gamma} \sum_v \{h_d [(\sum_\eta c_{\alpha\eta} w_\beta{}^\eta) w_\gamma{}^v - w_\alpha{}^v{}_{;\beta;\gamma}] - 3(h_d)_{,\alpha,\beta} w_\gamma{}^v - 3(h_d)_{,\gamma} w_\alpha{}^v{}_{;\beta} \} x^i{}_v + \sum_\alpha \sum_\gamma a^{\alpha\gamma} \{(h_d)_{,\alpha,\beta,\gamma} - \sum_v h_d (w_\alpha{}^v{}_{;\beta} c_{v\gamma} + w_\alpha{}^v{}_{;\gamma} c_{v\beta} + w_\alpha{}^v c_{v\beta;\gamma}) + 3(h_d)_{,\gamma} w_\alpha{}^v c_{v\beta} \} n^i$
3	From 1g, 2 and T17.3.9 Dual Covariant Tensors: Mutually Exclusive Normalized Tensor Components	$0 \approx \sum_\alpha \sum_\gamma a^{\alpha\gamma} \{(h_d)_{,\alpha,\beta,\gamma} - \sum_v h_d (w_\alpha{}^v{}_{;\beta} c_{v\gamma} + w_\alpha{}^v{}_{;\gamma} c_{v\beta} + w_\alpha{}^v c_{v\beta;\gamma}) + 3(h_d)_{,\gamma} w_\alpha{}^v c_{v\beta} \}$
4	From 3; for non-trivial case $a^{\alpha\gamma} \neq 0$, then all tensor coefficients must be zero	$0 = (h_d)_{,\alpha,\beta,\gamma} - \sum_v h_d (w_\alpha{}^v{}_{;\beta} c_{v\gamma} + w_\alpha{}^v{}_{;\gamma} c_{v\beta} + w_\alpha{}^v c_{v\beta;\gamma}) + 3(h_d)_{,\gamma} w_\alpha{}^v c_{v\beta}$
∴	From 4, T10.3.22 Addition of Tensors: Reversal of Right Cancellation by Addition and A10.2.14 Correspondence of Tensors and Tensor Coefficients	$(h_d)_{,\alpha,\beta,\gamma} = \sum_v [h_d (w_\alpha{}^v{}_{;\beta} c_{v\gamma} + w_\alpha{}^v{}_{;\gamma} c_{v\beta} + w_\alpha{}^v c_{v\beta;\gamma}) + 3(h_d)_{,\gamma} w_\alpha{}^v c_{v\beta}]$

Theorem 17.15.13 Parallel Hypersurfaces $\mathfrak{H}^{n-1}$: Gaussian Curvature in Dynamic Tangential Gap

1g	Given	$\mathbf{a}_\alpha = \sum_i x^i{}_\alpha \mathbf{b}_i$ Standard hypersurface vector
Step	Hypothesis	$(n-1)\,K\,x^i{}_\beta \approx \sum_\nu \{h_d\,[(\sum_\gamma \sum_\eta w_\gamma{}^\nu\, w_\eta{}^\gamma\, w_\beta{}^\eta) - \sum_\alpha \sum_\gamma a^{\alpha\gamma}\, w_\alpha{}^\nu{}_{;\beta;\gamma}] - 3\sum_\alpha \sum_\gamma (h_d)_{,\alpha,\beta}\, a^{\alpha\gamma}\, w_\gamma{}^\nu - 3\sum_\alpha \sum_\gamma (h_d)_{,\gamma}\, a^{\alpha\gamma}\, w_\alpha{}^\nu{}_{;\beta}\}\, x^i{}_\nu$
1	From 17.15.11 Parallel Hypersurfaces $\mathfrak{H}^{n-1}$: Gaussian Gap Parsed into Orthogonal Components	$(n-1)\,K\,x^i{}_\beta \approx \sum_\alpha \sum_\gamma a^{\alpha\gamma} \sum_\nu \{h_d\,[(\sum_\eta c_{\alpha\eta}\, w_\beta{}^\eta)\, w_\gamma{}^\nu - w_\alpha{}^\nu{}_{;\beta;\gamma}] - 3(h_d)_{,\alpha,\beta}\, w_\gamma{}^\nu - 3(h_d)_{,\gamma}\, w_\alpha{}^\nu{}_{;\beta}\}\, x^i{}_\nu + \sum_\alpha \sum_\gamma a^{\alpha\gamma}\, \{(h_d)_{,\alpha,\beta,\gamma} - \sum_\nu h_d\,(w_\alpha{}^\nu{}_{;\beta}\, c_{\nu\gamma} + w_\alpha{}^\nu{}_{;\gamma}\, c_{\nu\beta} + w_\alpha{}^\nu\, c_{\nu\beta;\gamma}) + 3(h_d)_{,\gamma}\, w_\alpha{}^\nu\, c_{\nu\beta}\}\, n^i$
2	From 1 and T17.3.15 Dual Covariant Tensors: Identity Addition	$(n-1)\,K\,x^i{}_\beta + 0\,n^i \approx \sum_\alpha \sum_\gamma a^{\alpha\gamma} \sum_\nu \{h_d\,[(\sum_\eta c_{\alpha\eta}\, w_\beta{}^\eta)\, w_\gamma{}^\nu - w_\alpha{}^\nu{}_{;\beta;\gamma}] - 3(h_d)_{,\alpha,\beta}\, w_\gamma{}^\nu - 3(h_d)_{,\gamma}\, w_\alpha{}^\nu{}_{;\beta}\}\, x^i{}_\nu + \sum_\alpha \sum_\gamma a^{\alpha\gamma}\, \{(h_d)_{,\alpha,\beta,\gamma} - \sum_\nu h_d\,(w_\alpha{}^\nu{}_{;\beta}\, c_{\nu\gamma} + w_\alpha{}^\nu{}_{;\gamma}\, c_{\nu\beta} + w_\alpha{}^\nu\, c_{\nu\beta;\gamma}) + 3(h_d)_{,\gamma}\, w_\alpha{}^\nu\, c_{\nu\beta}\}\, n^i$
3	From 1g, 2, T17.3.9 Dual Covariant Tensors: Mutually Exclusive Tangential Tensor Components and T17.3.10 Dual Covariant Tensors: Mutually Exclusive Normalized Tensor Components; by comparing tangential tensor components term-by-term	$(n-1)\,K\,x^i{}_\beta \approx \sum_\alpha \sum_\nu \sum_\gamma a^{\alpha\gamma}\, x^i{}_\nu\, \{h_d\,[(\sum_\eta c_{\alpha\eta}\, w_\beta{}^\eta)\, w_\gamma{}^\nu - w_\alpha{}^\nu{}_{;\beta;\gamma}] - 3(h_d)_{,\alpha,\beta}\, w_\gamma{}^\nu - 3(h_d)_{,\gamma}\, w_\alpha{}^\nu{}_{;\beta}\}$
4	From 3, T17.7.7 Second Fundamental Covariant Metric: Transpose of Indices, T10.3.20 Addition of Tensors: Distribution of a Tensor over Addition of Tensors, T10.3.7 Product of Tensors: Commutative by Multiplication of Rank p + q → q + p, T10.4.6 Scalars and Tensors: Distribution of a Scalar over Addition of Tensors and A10.2.14 Correspondence of Tensors and Tensor Coefficients	$(n-1)\,K\,x^i{}_\beta \approx \sum_\nu \{\, h_d\,[(\sum_\eta \sum_\gamma (\sum_\alpha c_{\eta\alpha}\, a^{\alpha\gamma})\, w_\beta{}^\eta)\, w_\gamma{}^\nu - \sum_\alpha \sum_\gamma a^{\alpha\gamma}\, w_\alpha{}^\nu{}_{;\beta;\gamma}] - 3\sum_\alpha \sum_\gamma (h_d)_{,\alpha,\beta}\, a^{\alpha\gamma}\, w_\gamma{}^\nu - 3\sum_\alpha \sum_\gamma (h_d)_{,\gamma}\, a^{\alpha\gamma}\, w_\alpha{}^\nu{}_{;\beta}\}\, x^i{}_\nu$
∴	From 4, D17.8.1 Weingarten Tensor on a Hypersheet $\mathfrak{H}^{n-1}$, T10.2.5 Tool: Tensor Substitution and A10.2.14 Correspondence of Tensors and Tensor Coefficients	$(n-1)\,K\,x^i{}_\beta \approx \sum_\nu \{h_d\,[(\sum_\gamma \sum_\eta w_\gamma{}^\nu\, w_\eta{}^\gamma\, w_\beta{}^\eta) - \sum_\alpha \sum_\gamma a^{\alpha\gamma}\, w_\alpha{}^\nu{}_{;\beta;\gamma}] - 3\sum_\alpha \sum_\gamma (h_d)_{,\alpha,\beta}\, a^{\alpha\gamma}\, w_\gamma{}^\nu - 3\sum_\alpha \sum_\gamma (h_d)_{,\gamma}\, a^{\alpha\gamma}\, w_\alpha{}^\nu{}_{;\beta}\}\, x^i{}_\nu$

Theorem 17.15.14 Parallel Hypersurfaces $\mathfrak{H}^{n-1}$: Gaussian Curvature with a Dynamic Gap

1g	Given	$W^{\nu}{}_{\beta} = w_{\gamma}{}^{\nu}\, w_{\eta}{}^{\gamma}\, w_{\beta}{}^{\eta}$ Weingarten mixed curvature tensor
Step	Hypothesis	$(n-1)\,K \approx h_d\, a^{\beta\kappa}\, [W^{\nu}{}_{\beta} - a^{\alpha\gamma}\, w_{\alpha}{}^{\nu}{}_{;\beta;\gamma}]\, a_{\kappa\nu}$ $-\,3\,(h_d)_{,\gamma}\, a_{\kappa\nu} w_{\alpha}{}^{\nu}{}_{;\beta}\, a^{\beta\kappa}\, a^{\alpha\gamma}$ $-\,3\,(h_d)_{,\alpha,\beta}\, a^{\beta\kappa}\, w_{\gamma}{}^{\nu}\, a_{\kappa\nu}\, a^{\alpha\gamma}$
1	From T17.15.13 Parallel Hypersurfaces $\mathfrak{H}^{n-1}$: Gaussian Curvature in Dynamic Tangential Gap	$(n-1)\,K\, x^{i}{}_{\beta} \approx$ $\Sigma_{\nu}\, \{h_d\, [(\Sigma_{\gamma}\, \Sigma_{\eta}\, w_{\gamma}{}^{\nu}\, w_{\eta}{}^{\gamma}\, w_{\beta}{}^{\eta}) - \Sigma_{\alpha}\, \Sigma_{\gamma}\, a^{\alpha\gamma}\, w_{\alpha}{}^{\nu}{}_{;\beta;\gamma}]$ $-\,3\Sigma_{\alpha}\, \Sigma_{\gamma}\, (h_d)_{,\alpha,\beta}\, a^{\alpha\gamma}\, w_{\gamma}{}^{\nu}$ $-\,3\Sigma_{\alpha}\, \Sigma_{\gamma}\, (h_d)_{,\gamma}\, a^{\alpha\gamma}\, w_{\alpha}{}^{\nu}{}_{;\beta}\}\, x^{i}{}_{\nu}$
2	From 1g, 1, T10.2.5 Tool: Tensor Substitution and using Einsteinian Notation	$(n-1)\,K\, x^{i}{}_{\beta} \approx \{h_d\, [W^{\nu}{}_{\beta} - a^{\alpha\gamma}\, w_{\alpha}{}^{\nu}{}_{;\beta;\gamma}]$ $-\,3\,(h_d)_{,\gamma}\, a^{\alpha\gamma}\, w_{\alpha}{}^{\nu}{}_{;\beta}$ $-\,3\,(h_d)_{,\alpha,\beta}\, a^{\alpha\gamma}\, w_{\gamma}{}^{\nu}\ \}\, x^{i}{}_{\nu}$
3	From 2, T10.3.6 Product of Tensors: Uniqueness by Multiplication of Rank for p + q and T10.3.20 Addition of Tensors: Distribution of a Tensor over Addition of Tensors	$(n-1)\,K\, x^{i}{}_{\beta}\, a^{\beta}{}_{\nu} \approx \{h_d\, [W^{\nu}{}_{\beta} - a^{\alpha\gamma}\, w_{\alpha}{}^{\nu}{}_{;\beta;\gamma}]\, a^{\beta}{}_{\nu}$ $-\,3\,(h_d)_{,\gamma}\, a^{\alpha\gamma}\, w_{\alpha}{}^{\nu}{}_{;\beta}\, a^{\beta}{}_{\nu}$ $-\,3\,(h_d)_{,\alpha,\beta}\, a^{\alpha\gamma}\, w_{\gamma}{}^{\nu}\, a^{\beta}{}_{\nu}\ \}\, x^{i}{}_{\nu}$
4	From 3 and T12.3.3B Contracted Covariant Bases Vector with Contravariant Vector	$(n-1)\,K\, x^{i}{}_{\nu} \approx \{h_d\, [W^{\nu}{}_{\beta} - a^{\alpha\gamma}\, w_{\alpha}{}^{\nu}{}_{;\beta;\gamma}]\, a^{\beta}{}_{\nu}$ $-\,3\,(h_d)_{,\gamma}\, a^{\alpha\gamma}\, w_{\alpha}{}^{\nu}{}_{;\beta}\, a^{\beta}{}_{\nu}$ $-\,3\,(h_d)_{,\alpha,\beta}\, a^{\alpha\gamma}\, w_{\gamma}{}^{\nu}\, a^{\beta}{}_{\nu}\ \}\, x^{i}{}_{\nu}$
5	From 4 and equating term-by-term	$(n-1)\,K \approx h_d\, [W^{\nu}{}_{\beta} - a^{\alpha\gamma}\, w_{\alpha}{}^{\nu}{}_{;\beta;\gamma}]\, a^{\beta}{}_{\nu}$ $-\,3\,(h_d)_{,\gamma}\, a^{\alpha\gamma}\, w_{\alpha}{}^{\nu}{}_{;\beta}\, a^{\beta}{}_{\nu}$ $-\,3\,(h_d)_{,\alpha,\beta}\, a^{\alpha\gamma}\, w_{\gamma}{}^{\nu}\, a^{\beta}{}_{\nu}$
6	From 5 and T17.4.6B Hypersheet: Covariant and Contravariant Metric Inverse	$(n-1)\,K \approx h_d\, [W^{\nu}{}_{\beta} - a^{\alpha\gamma}\, w_{\alpha}{}^{\nu}{}_{;\beta;\gamma}]\, a^{\beta\kappa}\, a_{\kappa\nu}$ $-\,3\,(h_d)_{,\gamma}\, w_{\alpha}{}^{\nu}{}_{;\beta}\, a^{\beta\kappa}\, a_{\kappa\nu}\, a^{\alpha\gamma}$ $-\,3\,(h_d)_{,\alpha,\beta}\, w_{\gamma}{}^{\nu}\, a^{\beta\kappa}\, a_{\kappa\nu}\, a^{\alpha\gamma}$
∴	From 6 and T10.3.7 Product of Tensors: Commutative by Multiplication of Rank p + q → q + p	$(n-1)\,K \approx h_d\, a^{\beta\kappa}\, [W^{\nu}{}_{\beta} - a^{\alpha\gamma}\, w_{\alpha}{}^{\nu}{}_{;\beta;\gamma}]\, a_{\kappa\nu}$ $-\,3\,(h_d)_{,\gamma}\, a_{\kappa\nu} w_{\alpha}{}^{\nu}{}_{;\beta}\, a^{\beta\kappa}\, a^{\alpha\gamma}$ $-\,3\,(h_d)_{,\alpha,\beta}\, a^{\beta\kappa}\, w_{\gamma}{}^{\nu}\, a_{\kappa\nu}\, a^{\alpha\gamma}$

Theorem 17.15.15 Parallel Hypersurfaces $\mathfrak{H}^{n-1}$: Static Normal Gap

1g	Given	[K] constant curvature
Step	Hypothesis	$0 = \sum_\nu (w_\alpha{}^\nu{}_{;\beta}\, c_{\nu\gamma} + w_\alpha{}^\nu{}_{;\gamma}\, c_{\nu\beta} + w_\alpha{}^\nu\, c_{\nu\beta;\gamma})$
1	From 17.15.12 Parallel Hypersurfaces $\mathfrak{H}^{n-1}$: Equations of Varying Gap	$(h_d)_{,\alpha,\beta,\gamma} =$ $\sum_\nu [h_d (w_\alpha{}^\nu{}_{;\beta}\, c_{\nu\gamma} + w_\alpha{}^\nu{}_{;\gamma}\, c_{\nu\beta} + w_\alpha{}^\nu\, c_{\nu\beta;\gamma}) + 3(h_d)_{,\gamma}\, w_\alpha{}^\nu\, c_{\nu\beta}]$
2	From 1g, 1, A17.15.2(A,B,C) Constant Curvature Between Parallel Hypersurfaces $\mathfrak{H}^{n-1}$ and A4.2.3 Substitution	$0 =$ $\sum_\nu [h_d (w_\alpha{}^\nu{}_{;\beta}\, c_{\nu\gamma} + w_\alpha{}^\nu{}_{;\gamma}\, c_{\nu\beta} + w_\alpha{}^\nu\, c_{\nu\beta;\gamma}) + 3\ 0\ w_\alpha{}^\nu\, c_{\nu\beta}]$
3	From 2, T10.2.7 Tool: Zero Times a Tensor is a Zero Tensor, T10.3.18 Addition of Tensors: Identity by Addition and A10.2.14 Correspondence of Tensors and Tensor Coefficients	$0 = h_d \sum_\nu (w_\alpha{}^\nu{}_{;\beta}\, c_{\nu\gamma} + w_\alpha{}^\nu{}_{;\gamma}\, c_{\nu\beta} + w_\alpha{}^\nu\, c_{\nu\beta;\gamma})$
∴	From 3, A17.15.2D Constant Curvature Between Parallel Hypersurfaces $\mathfrak{H}^{n-1}$; hence the sum being none trivial must be zero for constant curvature.	$0 = \sum_\nu (w_\alpha{}^\nu{}_{;\beta}\, c_{\nu\gamma} + w_\alpha{}^\nu{}_{;\gamma}\, c_{\nu\beta} + w_\alpha{}^\nu\, c_{\nu\beta;\gamma})$

Theorem 17.15.16 Parallel Hypersurfaces $\mathfrak{H}^{n-1}$: Gaussian Curvature in Static Tangential Gap

1g	Given	[K] constant curvature
Step	Hypothesis	$(n-1)\, K\, x^i{}_\beta \approx$ $h_d \sum_\nu (\sum_\gamma \sum_\eta w_\gamma{}^\nu\, w_\eta{}^\gamma\, w_\beta{}^\eta - \sum_\gamma \sum_\alpha a^{\alpha\gamma}\, w_\alpha{}^\nu{}_{;\beta;\gamma})\, x^i{}_\nu$
1	From 17.15.13 Parallel Hypersurfaces $\mathfrak{H}^{n-1}$: Gaussian Curvature in Varying Tangential Gap	$(n-1)\, K\, x^i{}_\beta \approx$ $\sum_\nu \{h_d [(\sum_\gamma \sum_\eta w_\gamma{}^\nu\, w_\eta{}^\gamma\, w_\beta{}^\eta) - \sum_\alpha \sum_\gamma a^{\alpha\gamma}\, w_\alpha{}^\nu{}_{;\beta;\gamma}] -$ $3\sum_\alpha \sum_\gamma (h_d)_{,\alpha,\beta}\, a^{\alpha\gamma}\, w_\gamma{}^\nu - 3\sum_\alpha \sum_\gamma (h_d)_{,\gamma}\, a^{\alpha\gamma}\, w_\alpha{}^\nu{}_{;\beta}\}\, x^i{}_\nu$
2	From 1g, 1, A17.15.2(A,B,C) Constant Curvature Between Parallel Hypersurfaces $\mathfrak{H}^{n-1}$ and A4.2.3 Substitution	$(n-1)\, K\, x^i{}_\beta \approx$ $\sum_\nu \{h_d [(\sum_\gamma \sum_\eta w_\gamma{}^\nu\, w_\eta{}^\gamma\, w_\beta{}^\eta) - \sum_\alpha \sum_\gamma a^{\alpha\gamma}\, w_\alpha{}^\nu{}_{;\beta;\gamma}] -$ $3\sum_\alpha \sum_\gamma 0\, a^{\alpha\gamma}\, w_\gamma{}^\nu - 3\sum_\alpha \sum_\gamma 0\, a^{\alpha\gamma}\, w_\alpha{}^\nu{}_{;\beta}\}\, x^i{}_\nu$
3	From 2, T10.2.7 Tool: Zero Times a Tensor is a Zero Tensor, T10.3.18 Addition of Tensors: Identity by Addition and A10.2.14 Correspondence of Tensors and Tensor Coefficients	$(n-1)\, K\, x^i{}_\beta \approx$ $\sum_\nu h_d [(\sum_\gamma \sum_\eta w_\gamma{}^\nu\, w_\eta{}^\gamma\, w_\beta{}^\eta) - \sum_\alpha \sum_\gamma a^{\alpha\gamma}\, w_\alpha{}^\nu{}_{;\beta;\gamma}]\, x^i{}_\nu$
∴	From 5, T10.4.6 Scalars and Tensors: Distribution of a Scalar over Addition of Tensors, T10.3.7 Product of Tensors: Commutative by Multiplication of Rank p + q → q + p and A10.2.14 Correspondence of Tensors and Tensor Coefficients	$(n-1)\, K\, x^i{}_\beta \approx$ $h_d \sum_\nu (\sum_\gamma \sum_\eta w_\gamma{}^\nu\, w_\eta{}^\gamma\, w_\beta{}^\eta - \sum_\alpha \sum_\gamma a^{\alpha\gamma}\, w_\alpha{}^\nu{}_{;\beta;\gamma})\, x^i{}_\nu$

Theorem 17.15.17 Intrinsic Hypersurfaces $\mathfrak{H}^{n-1}$: Right First Order Differential of Gap

1g	Given	From figure F17.15.1 Parallel Hypersurfaces $\mathfrak{H}^{n-1}$ Spaced $\lvert P\vartheta \rvert$ Apart
2g		$v \equiv ds / dt$
Step	Hypothesis	$d\,\Delta\mathbf{S} / ds = \sum_i \sum_\kappa [h_{d,\kappa}\, n^i - h_d (\sum_\gamma w_\kappa{}^\gamma x^i{}_\gamma)]\, \lambda^\kappa\, \mathbf{b}_i$
1	From 1g and A17.15.1P2A Properties of Parallel Planes	$\Delta\mathbf{S} = h_d\, \mathbf{n}$
2	From 1, LxK.3.1.7B Uniqueness Theorem of Differentiation and DxK.3.2.1B First Parametric Differentiation with Newton's dot mark notation, and Leibniz fraction notation	$d\,\Delta\mathbf{S} / dt = d\,(h_d\, \mathbf{n}\,) / dt$
3	From 2 and TK.3.4B Single Variable Chain Rule	$[d\,\Delta\mathbf{S} / ds]\,[ds / dt] = [d\,(h_d\, \mathbf{n}\,) / ds]\,[ds / dt]$
4	From 3 and TK.3.9 Exact Differential, Parametric, Chain Rule	$[d\,\Delta\mathbf{S} / ds]\,[ds / dt] = \sum_\kappa [\partial\,(h_d\, \mathbf{n}\,) / \partial u_\kappa\,]\,[\,du_\kappa / ds]\,[ds / dt]$
5	From 2g, 4, A4.2.3 Substitution and O12.6.1R2 Rules for Covariant and Contravariant Differential Notation; or T17.5.26A Differentiation of Tensor: Rank-1 Covariant	$[d\,\Delta\mathbf{S} / ds]\, v = \sum_\kappa [h_{d,\kappa}\, \mathbf{n} + h_d\, \mathbf{n}^{;\kappa}]\, \lambda^\kappa\, v$
6	From 5, A7.8.3B Scaling Collinear Vectors, T17.3.7A Hypersurface Normal Vector: Orthogonal System $\mathfrak{R}^n$ Transformation and Normal and A7.9.2 Substitution of Vectors	$d\,\Delta\mathbf{S} / ds = \sum_\kappa [h_{d,\kappa} \sum_i n^i\, \mathbf{b}_i + h_d (\sum_i n^i\, \mathbf{b}_i)^{;\kappa}]\, \lambda^\kappa$
7	From 6 and T7.10.3 Distribution of a Scalar over Addition of Vectors; exchanging summation [κ → i]	$d\,\Delta\mathbf{S} / ds = \sum_i \sum_\kappa [h_{d,\kappa}\, n^i + h_d\, (n^i)^{;\kappa}]\, \lambda^\kappa\, \mathbf{b}_i$
8	From 7, T17.5.26A Differentiation of Tensor: Rank-1 Covariant and D17.1.9 Riemannian Manifold $\mathfrak{R}^n$ Mixed Tensor of Rank-2 on a Hypersheet $\mathfrak{H}^{n-1}$	$d\,\Delta\mathbf{S} / ds = \sum_i \sum_\kappa (h_{d,\kappa}\, n^i + h_d\, n^i{}_\kappa)\, \lambda^\kappa\, \mathbf{b}_i$

9	From 8, T17.8.5 Weingarten Formulas: Unit Normal Differential Solution (Weingarten Formula), T10.2.5 Tool: Tensor Substitution and A10.2.14 Correspondence of Tensors and Tensor Coefficients	$d\,\Delta\mathbf{S} / ds = \Sigma_i \Sigma_\kappa [h_{d,\kappa} n^i + h_d (-\Sigma_\gamma w_\kappa{}^\gamma x^i{}_\gamma)] \lambda^\kappa \mathbf{b}_i$
∴	From 9, D4.1.20A Negative Coefficient, T10.4.6 Scalars and Tensors: Distribution of a Scalar over Addition of Tensors and A10.2.14 Correspondence of Tensors and Tensor Coefficients	$d\,\Delta\mathbf{S} / ds = \Sigma_i \Sigma_\kappa [h_{d,\kappa} n^i - h_d (\Sigma_\gamma w_\kappa{}^\gamma x^i{}_\gamma)] \lambda^\kappa \mathbf{b}_i$

Theorem 17.15.18 Intrinsic Hypersurfaces $\mathfrak{H}^{n-1}$: Left First Order Differential of Gap

1g	Given	Figure F17.15.1 Parallel Hypersurfaces $\mathfrak{H}^{n-1}$ Spaced $\lvert P\vartheta\rvert$ Apart	
Step	Hypothesis	$d\,\Delta\mathbf{S} / ds = \mathbf{0}$ EQ A	$d\,\Delta\mathbf{S} / ds = \Sigma_i \Sigma_\kappa (x_\rho{}^i{}_\kappa - x^i{}_\kappa) \lambda^\kappa \mathbf{b}_i$ EQ B
1	From 1, LxK.3.1.7B Uniqueness Theorem of Differentiation and DxK.3.2.1B First Parametric Differentiation with Newton's dot mark notation, and Leibniz fraction notation	$d\,\Delta\mathbf{S} / ds = d\,(\boldsymbol{\rho} - \mathbf{r}) / ds$	
2	From 1g, 1, A17.1.8A Fundamental Property of Continuum for a Hypersheet and A7.9.2 Substitution of Vectors	$d\,\Delta\mathbf{S} / ds = d\,(\Sigma_i x_{\rho i} \mathbf{b}_i - \Sigma_i x_i \mathbf{b}_i) / ds$	
3	From 2 and LxK.3.2.2A Differential as a Linear Operator	$d\,\Delta\mathbf{S} / ds = \Sigma_i [d\,(x_{\rho i} \mathbf{b}_i) / ds - d\,(x_i \mathbf{b}_i) / ds]$	
4	From 3 and TK.3.9 Exact Differential, Parametric, Chain Rule	$d\,\Delta\mathbf{S} / ds = \Sigma_i \{\Sigma_\kappa \partial (x_{\rho i} \mathbf{b}_i) / \partial u_\kappa [du_\kappa / ds] - \Sigma_\kappa \partial (x_i \mathbf{b}_i) / \partial u_\kappa [du_\kappa / ds]\}$	
5	From 4 and D16.1.7A Intrinsic Differential Pseudo Vectors	$d\,\Delta\mathbf{S} / ds = \Sigma_i \{\Sigma_\kappa \partial (x_{\rho i} \mathbf{b}_i) / \partial u_\kappa \lambda^\kappa - \Sigma_\kappa \partial (x_i \mathbf{b}_i) / \partial u_\kappa \lambda^\kappa\}$	
6	From 5 and T7.10.3 Distribution of a Scalar over Addition of Vectors	$d\,\Delta\mathbf{S} / ds = \Sigma_\kappa \Sigma_i \{\partial (x_{\rho i} \mathbf{b}_i) / \partial u_\kappa - \partial (x_i \mathbf{b}_i) / \partial u_\kappa\} \lambda^\kappa$	
7	From 6, T17.5.26A Differentiation and T7.10.9 Distribution of Vector over Addition of Scalars	$d\,\Delta\mathbf{S} / ds = \Sigma_i \Sigma_\kappa (x_\rho{}^{i;\kappa} - x^{i;\kappa}) \lambda^\kappa \mathbf{b}_i$	
8	From 7 and D17.1.9 Riemannian Manifold $\mathfrak{R}^n$ Mixed Tensor of Rank-2 on a Hypersheet $\mathfrak{H}^{n-1}$	$d\,\Delta\mathbf{S} / ds = \Sigma_i \Sigma_\kappa (x_\rho{}^i{}_\kappa - x^i{}_\kappa) \lambda^\kappa \mathbf{b}_i$	

9	From 8, A17.15.1P3B Properties of Parallel Planes, T10.3.19 Addition of Tensors: Inverse by Addition and A10.2.14 Correspondence of Tensors and Tensor Coefficients	$d\,\Delta\mathbf{S} / ds = \sum_i \sum_\kappa 0\, \lambda^\kappa\, \mathbf{b}_i$	
10	From 9, T4.4.1 Equalities: Any Quantity Multiplied by Zero is Zero and LxE.3.1.36 TE.17: Sums of Zeros	$d\,\Delta\mathbf{S} / ds = \sum_i 0\, \mathbf{b}_i$	
∴	From 8, 10 and D7.6.8C Zero Vector	$d\,\Delta\mathbf{S} / ds = \mathbf{0}$ EQ A	$d\,\Delta\mathbf{S} / ds = \sum_i \sum_\kappa (x_{\rho}{}^{i}{}_{\kappa} - x^{i}{}_{\kappa})\, \lambda^\kappa\, \mathbf{b}_i$ EQ B

Theorem 17.15.19 Intrinsic Hypersurfaces $\mathfrak{H}^{n-1}$: Differential of Non-Collinear Pseudo Vector

1g	Given	$\lambda^{\kappa,\xi} \equiv \partial\, \lambda^\kappa / \partial\, u_\xi$
2g		$\kappa \neq \xi$ then $\lambda^\kappa \perp u_\xi$ hypersurface curve
3g		$\lambda^\kappa \neq f(\, u_\xi\,)$ hence independent and constant with respect the parameter $[u_\xi]$
Step	Hypothesis	$\lambda^{\kappa,\xi} = \mathbf{0}$ for $\kappa \neq \xi$ orthogonal
1	From 1g	$\lambda^{\kappa,\xi} = \partial\, \lambda^\kappa / \partial\, u_\xi$
2	From 2g, 3g and LxK.3.3.1B Differential of a Constant	$\partial\, \lambda^\kappa / \partial\, u_\xi = \mathbf{0}$
∴	From 1 and 2	$\lambda^{\kappa,\xi} = \mathbf{0}$ for $\kappa \neq \xi$ orthogonal

Theorem 17.15.20 Intrinsic Hypersurfaces $\mathfrak{H}^{n-1}$: Left Second Order Differential of Gap

1g	Given	$v \equiv ds / dt$	
Step	Hypothesis	$d^2 \Delta\mathbf{S} / ds^2 = \mathbf{0}$ EQ A	$d^2 \Delta\mathbf{S} / ds^2 = \sum_i \sum_\xi \sum_\kappa (0) \lambda^\kappa \lambda^\xi \mathbf{b}_i$ EQ B
1	From T17.15.17B Intrinsic Hypersurfaces $\mathfrak{H}^{n-1}$: Left First Order Differential of Gap, LxK.3.1.7B Uniqueness Theorem of Differentiation and DxK.3.2.1C Second Parametric Differentiation with mark, fraction; with dot notation and Leibniz fraction notation	$d (d \Delta\mathbf{S} / ds) / dt = d \{ \sum_i \sum_\kappa ({x_\rho}^i{}_\kappa - x^i{}_\kappa) \lambda^\kappa \mathbf{b}_i \} / dt$	
2	From 1 and TK.3.4B Single Variable Chain Rule	$d (d \Delta\mathbf{S} / ds\, ds) [ds / dt] = d \{ \sum_i \sum_\kappa ({x_\rho}^i{}_\kappa - x^i{}_\kappa) \lambda^\kappa \mathbf{b}_i \} / ds\ [ds / dt]$	
3	From 1g, 2 and A4.2.3 Substitution	$d (d \Delta\mathbf{S} / ds\, ds)\, v = d \{ \sum_i \sum_\kappa ({x_\rho}^i{}_\kappa - x^i{}_\kappa) \lambda^\kappa \mathbf{b}_i \} / ds\ v$	
4	From 3, TK.3.9 Exact Differential, Parametric, Chain Rule and A7.8.3B Scaling Collinear Vectors	$d^2 \Delta\mathbf{S} / ds^2 = \sum_\xi \partial\{ \sum_i \sum_\kappa ({x_\rho}^i{}_\kappa - x^i{}_\kappa) \lambda^\kappa \mathbf{b}_i \} / \partial u_\xi\ [du_\xi / ds]$	
5	From 4 and D16.1.7A Intrinsic Differential Pseudo Vectors	$d^2 \Delta\mathbf{S} / ds^2 = \sum_\xi \partial\{ \sum_i \sum_\kappa ({x_\rho}^i{}_\kappa - x^i{}_\kappa) \lambda^\kappa \mathbf{b}_i \} / \partial u_\xi\ \lambda^\xi$	
6	From 10, and T17.5.72 Tensor Product: Differential of Covariant Tensor Sets	$d^2 \Delta\mathbf{S} / ds^2 = \sum_\xi \sum_i \sum_\kappa \{ ({x_\rho}^i{}_\kappa - x^i{}_\kappa)^{;\xi} \lambda^\kappa + ({x_\rho}^i{}_\kappa - x^i{}_\kappa) \lambda^{\kappa;\xi} \} \mathbf{b}_i\ \lambda^\xi$	
7	From 6, T17.15.18 Intrinsic Hypersurfaces $\mathfrak{H}^{n-1}$: Differential of Non-Collinear Pseudo Vector, T10.3.7 Product of Tensors: Commutative by Multiplication of Rank p + q → q + p, T10.2.5 Tool: Tensor Substitution and A10.2.14 Correspondence of Tensors and Tensor Coefficients	$d^2 \Delta\mathbf{S} / ds^2 = \sum_i \sum_\xi \sum_\kappa \{ ({x_\rho}^i{}_\kappa - x^i{}_\kappa)^{;\xi} \lambda^\kappa + ({x_\rho}^i{}_\kappa - x^i{}_\kappa)\, 0 \} \lambda^\xi \mathbf{b}_i$	
8	From 7, T10.2.7 Tool: Zero Times a Tensor is a Zero Tensor, T10.3.18 Addition of Tensors: Identity by Addition, T10.2.5 Tool: Tensor Substitution and A10.2.14 Correspondence of Tensors and Tensor Coefficients	$d^2 \Delta\mathbf{S} / ds^2 = \sum_i \sum_\xi \sum_\kappa \{ ({x_\rho}^i{}_\kappa - x^i{}_\kappa)^{;\xi} \lambda^\kappa \} \lambda^\xi \mathbf{b}_i$	
9	From 8, T10.3.20 Addition of Tensors: Distribution of a Tensor over Addition of Tensors and A10.2.14 Correspondence of Tensors and Tensor Coefficients	$d^2 \Delta\mathbf{S} / ds^2 = \sum_i \sum_\xi \sum_\kappa \{ ({x_\rho}^i{}_\kappa - x^i{}_\kappa)^{;\xi} \} \lambda^\kappa \lambda^\xi \mathbf{b}_i$	

10	From 9, A17.15.1P3B Properties of Parallel Planes, T10.3.19 Addition of Tensors: Inverse by Addition and A10.2.14 Correspondence of Tensors and Tensor Coefficients	$d^2 \Delta\mathbf{S} / ds^2 = \Sigma_i \Sigma_\xi \Sigma_\kappa \{ (0)^{;\xi} \} \lambda^\kappa \lambda^\xi \mathbf{b}_i$	
11	From 10 and LxK.3.3.1B Differential of a Constant	$d^2 \Delta\mathbf{S} / ds^2 = \Sigma_i \Sigma_\xi \Sigma_\kappa (0) \lambda^\kappa \lambda^\xi \mathbf{b}_i$	
12	From 9, T4.4.1 Equalities: Any Quantity Multiplied by Zero is Zero and LxE.3.1.36 TE.17: Sums of Zeros	$d^2 \Delta\mathbf{S} / ds^2 = \Sigma_i 0\, \mathbf{b}_i$	
∴	From 11, 12 and D7.6.8C Zero Vector	$d^2 \Delta\mathbf{S} / ds^2 = \mathbf{0}$ EQ A	$d^2 \Delta\mathbf{S} / ds^2 = \Sigma_i \Sigma_\xi \Sigma_\kappa (0) \lambda^\kappa \lambda^\xi \mathbf{b}_i$ EQ B

Theorem 17.15.21 Intrinsic Hypersurfaces $\mathfrak{H}^{n-1}$: Right Second Order Differential of Gap

1g	Given	$v \equiv ds / dt$
Step	Hypothesis	$d^2 \Delta\mathbf{S} / ds^2 = \Sigma_i \Sigma_\kappa \Sigma_\xi$ $[- \Sigma_\gamma (h_{d,\kappa} w_\xi{}^\gamma + h_{d,\xi} w_\kappa{}^\gamma + h_d\ w_\kappa{}^\gamma{}_{;\xi})\, x^i{}_\gamma +$ $(h_{d,\kappa,\xi} - h_d \Sigma_\gamma w_\kappa{}^\gamma c_{\gamma\xi})\, n^i]\, \lambda^\kappa \lambda^\xi\ \mathbf{b}_i$
1	From T17.15.16 Intrinsic Hypersurfaces $\mathfrak{H}^{n-1}$: Right First Order Differential of Gap	$d\, \Delta\mathbf{S} / ds = \Sigma_i \Sigma_\kappa [h_{d,\kappa}\, n^i - h_d (\Sigma_\gamma w_\kappa{}^\gamma x^i{}_\gamma)]\, \lambda^\kappa \mathbf{b}_i$
2	From 1, LxK.3.1.7B Uniqueness Theorem of Differentiation and DxK.3.2.1C Second Parametric Differentiation with mark, fraction; with dot notation and Leibniz fraction notation	$d\, (d\, \Delta\mathbf{S} / ds) / dt = d \{ \Sigma_i \Sigma_\kappa [h_{d,\kappa}\, n^i - h_d (\Sigma_\gamma w_\kappa{}^\gamma x^i{}_\gamma)]\, \lambda^\kappa \mathbf{b}_i \} / dt$
3	From 2 and TK.3.4B Single Variable Chain Rule	$d\, (d\, \Delta\mathbf{S} / ds\, ds)\, [ds / dt] = d \{ \Sigma_i \Sigma_\kappa [h_{d,\kappa}\, n^i - h_d (\Sigma_\gamma w_\kappa{}^\gamma x^i{}_\gamma)]\, \lambda^\kappa \mathbf{b}_i \} / ds\, [ds / dt]$
4	From 1g, 3 and A4.2.3 Substitution	$d^2 \Delta\mathbf{S} / ds^2\, v = d \{ \Sigma_i \Sigma_\kappa [h_{d,\kappa}\, n^i - h_d (\Sigma_\gamma w_\kappa{}^\gamma x^i{}_\gamma)]\, \lambda^\kappa \mathbf{b}_i \} / ds\ v$
5	From 4, TK.3.9 Exact Differential, Parametric, Chain Rule and A7.8.3B Scaling Collinear Vectors	$d^2 \Delta\mathbf{S} / ds^2 = \Sigma_\xi [du_\xi / ds]\, \partial\{ \Sigma_i \Sigma_\kappa [h_{d,\kappa}\, n^i - h_d (\Sigma_\gamma w_\kappa{}^\gamma x^i{}_\gamma)]\, \lambda^\kappa \mathbf{b}_i \} / \partial u_\xi$
6	From 5 and D16.1.7A Intrinsic Differential Pseudo Vectors	$d^2 \Delta\mathbf{S} / ds^2 = \Sigma_\xi \lambda^\xi\, \partial\{ \Sigma_i \Sigma_\kappa [h_{d,\kappa}\, n^i - h_d (\Sigma_\gamma w_\kappa{}^\gamma x^i{}_\gamma)]\, \lambda^\kappa \mathbf{b}_i \} / \partial u_\xi$
7	From 6, LxK.3.2.2A Differential as a Linear Operator, T17.5.26A Differentiation of Tensor: Rank-1 Covariant and T7.10.9 Distribution of Vector over Addition of Scalars	$d^2 \Delta\mathbf{S} / ds^2 = \Sigma_i \Sigma_\kappa \Sigma_\xi \{ [h_{d,\kappa}\, n^i \lambda^\kappa]^{;\xi} - [h_d (\Sigma_\gamma w_\kappa{}^\gamma x^i{}_\gamma)\, \lambda^\kappa]^{;\xi} \}\, \lambda^\xi\ \mathbf{b}_i$
8	From 7 and T17.5.72 Tensor Product: Differential of Covariant Tensor Sets	$d^2 \Delta\mathbf{S} / ds^2 = \Sigma_i \Sigma_\kappa \Sigma_\xi \{ [h_{d,\kappa,\xi}\, n^i \lambda^\kappa + h_{d,\kappa}\, n^{i;\xi} \lambda^\kappa + h_{d,\kappa}\, n^i \lambda^{\kappa,\xi}] - [h_{d,\xi} (\Sigma_\gamma w_\kappa{}^\gamma x^i{}_\gamma)\, \lambda^\kappa + h_d (\Sigma_\gamma w_\kappa{}^\gamma x^i{}_\gamma)^{;\xi} \lambda^\kappa + h_d (\Sigma_\gamma w_\kappa{}^\gamma x^i{}_\gamma)\, \lambda^{\kappa,\xi}] \}\, \lambda^\xi\ \mathbf{b}_i$

9	From 8, T17.15.18 Intrinsic Hypersurfaces $\mathfrak{H}^{n-1}$: Differential of Non-Collinear Pseudo Vector, T10.2.5 Tool: Tensor Substitution and A10.2.14 Correspondence of Tensors and Tensor Coefficients	$d^2\, \Delta\mathbf{S} / ds^2 = \sum_i \sum_\kappa \sum_\xi \{ [h_{d,\kappa,\xi}\, n^i \lambda^\kappa + h_{d,\kappa}\, n^{i;\xi} \lambda^\kappa + h_{d,\kappa}\, n^i\, \mathbf{0}] - [h_{d,\xi} (\sum_\gamma w_\kappa{}^\gamma x^i{}_\gamma) \lambda^\kappa + h_d (\sum_\gamma w_\kappa{}^\gamma{}_{;\xi}\, x^i{}_\gamma + w_\kappa{}^\gamma x^i{}_{\gamma;\xi}) \lambda^\kappa + h_d (\sum_\gamma w_\kappa{}^\gamma x^i{}_\gamma)\, \mathbf{0}] \}\, \lambda^\xi\, \mathbf{b}_i$
10	From 9, T10.2.7 Tool: Zero Times a Tensor is a Zero Tensor, T10.3.18 Addition of Tensors: Identity by Addition, T10.2.5 Tool: Tensor Substitution and A10.2.14 Correspondence of Tensors and Tensor Coefficients	$d^2\, \Delta\mathbf{S} / ds^2 = \sum_i \sum_\kappa \sum_\xi \{ [h_{d,\kappa,\xi}\, n^i \lambda^\kappa + h_{d,\kappa}\, n^{i;\xi} \lambda^\kappa] - [h_{d,\xi} (\sum_\gamma w_\kappa{}^\gamma x^i{}_\gamma) \lambda^\kappa + h_d (\sum_\gamma w_\kappa{}^\gamma{}_{;\xi}\, x^i{}_\gamma + w_\kappa{}^\gamma x^i{}_{\gamma;\xi}) \lambda^\kappa] \}\, \lambda^\xi\, \mathbf{b}_i$
11	From 10, T10.3.20 Addition of Tensors: Distribution of a Tensor over Addition of Tensors and A10.2.14 Correspondence of Tensors and Tensor Coefficients	$d^2\, \Delta\mathbf{S} / ds^2 = \sum_i \sum_\kappa \sum_\xi \{ [h_{d,\kappa,\xi}\, n^i + h_{d,\kappa}\, n^{i;\xi}] - [h_{d,\xi} (\sum_\gamma w_\kappa{}^\gamma x^i{}_\gamma) + h_d (\sum_\gamma w_\kappa{}^\gamma{}_{;\xi}\, x^i{}_\gamma + w_\kappa{}^\gamma x^i{}_{\gamma;\xi})] \}\, \lambda^\kappa \lambda^\xi\, \mathbf{b}_i$
12	From 11 and D17.1.9 Riemannian Manifold $\mathfrak{R}^n$ Mixed Tensor of Rank-2 on a Hypersheet $\mathfrak{H}^{n-1}$	$d^2\, \Delta\mathbf{S} / ds^2 = \sum_i \sum_\kappa \sum_\xi \{ [h_{d,\kappa,\xi}\, n^i + h_{d,\kappa}\, n^i{}_\xi] - [h_{d,\xi} (\sum_\gamma w_\kappa{}^\gamma x^i{}_\gamma) + h_d (\sum_\gamma w_\kappa{}^\gamma{}_{;\xi}\, x^i{}_\gamma + w_\kappa{}^\gamma x^i{}_{\gamma;\xi})] \}\, \lambda^\kappa \lambda^\xi\, \mathbf{b}_i$
13	From 12, T17.8.5 Weingarten Formulas: Unit Normal Differential Solution (Weingarten Formula), T17.7.12 Second Fundamental Covariant Metric: in Inverse Unit Normal Form, T10.2.5 Tool: Tensor Substitution and A10.2.14 Correspondence of Tensors and Tensor Coefficients	$d^2\, \Delta\mathbf{S} / ds^2 = \sum_i \sum_\kappa \sum_\xi \{ [h_{d,\kappa,\xi}\, n^i + h_{d,\kappa} (-\sum_\gamma w_\xi{}^\gamma x^i{}_\gamma)] - [h_{d,\xi} (\sum_\gamma w_\kappa{}^\gamma x^i{}_\gamma) + h_d (\sum_\gamma w_\kappa{}^\gamma{}_{;\xi}\, x^i{}_\gamma + w_\kappa{}^\gamma c_{\gamma\xi}\, n^i)] \}\, \lambda^\kappa \lambda^\xi\, \mathbf{b}_i$
14	From 13, D4.1.20A Negative Coefficient, A4.2.10 Commutative Multp, T10.4.6 Scalars and Tensors: Distribution of a Scalar over Addition of Tensors and A10.2.14 Correspondence of Tensors and Tensor Coefficients	$d^2\, \Delta\mathbf{S} / ds^2 = \sum_i \sum_\kappa \sum_\xi \{ h_{d,\kappa,\xi}\, n^i - h_{d,\kappa} (\sum_\gamma w_\xi{}^\gamma x^i{}_\gamma) - h_{d,\xi} (\sum_\gamma w_\kappa{}^\gamma x^i{}_\gamma) - h_d (\sum_\gamma w_\kappa{}^\gamma{}_{;\xi}\, x^i{}_\gamma + w_\kappa{}^\gamma c_{\gamma\xi}\, n^i) \}\, \lambda^\kappa \lambda^\xi\, \mathbf{b}_i$
15	From 14, T10.4.6 Scalars and Tensors: Distribution of a Scalar over Addition of Tensors and A10.2.14 Correspondence of Tensors and Tensor Coefficients	$d^2\, \Delta\mathbf{S} / ds^2 = \sum_i \sum_\kappa \sum_\xi \{ h_{d,\kappa,\xi}\, n^i - h_{d,\kappa} (\sum_\gamma w_\xi{}^\gamma x^i{}_\gamma) - h_{d,\xi} \sum_\gamma w_\kappa{}^\gamma x^i{}_\gamma - h_d \sum_\gamma w_\kappa{}^\gamma{}_{;\xi}\, x^i{}_\gamma - h_d \sum_\gamma w_\kappa{}^\gamma c_{\gamma\xi}\, n^i \}\, \lambda^\kappa \lambda^\xi\, \mathbf{b}_i$

16	From 15, T10.3.16 Addition of Tensors: Commutative by Addition and A10.2.14 Correspondence of Tensors and Tensor Coefficients; order summations by normal and tangent components	$d^2 \Delta\mathbf{S} / ds^2 = \Sigma_i \Sigma_\kappa \Sigma_\xi [- h_{d,\kappa} \Sigma_\gamma w_\xi{}^\gamma x^i{}_\gamma - h_{d,\xi} \Sigma_\gamma w_\kappa{}^\gamma x^i{}_\gamma - h_d \Sigma_\gamma w_\kappa{}^\gamma{}_{;\xi} x^i{}_\gamma + h_{d,\kappa,\xi} n^i - h_d \Sigma_\gamma w_\kappa{}^\gamma c_{\gamma\xi} n^i] \lambda^\kappa \lambda^\xi \mathbf{b}_i$
∴	From 16, D4.1.20A Negative Coefficient, T10.4.6 Scalars and Tensors: Distribution of a Scalar over Addition of Tensors, T10.3.20 Addition of Tensors: Distribution of a Tensor over Addition of Tensors and A10.2.14 Correspondence of Tensors and Tensor Coefficients	$d^2 \Delta\mathbf{S} / ds^2 = \Sigma_i \Sigma_\kappa \Sigma_\xi$ $[- \Sigma_\gamma (h_{d,\kappa} w_\xi{}^\gamma + h_{d,\xi} w_\kappa{}^\gamma + h_d \ w_\kappa{}^\gamma{}_{;\xi}) x^i{}_\gamma +$ $(h_{d,\kappa,\xi} - h_d \Sigma_\gamma w_\kappa{}^\gamma c_{\gamma\xi}) n^i] \lambda^\kappa \lambda^\xi \mathbf{b}_i$

Theorem 17.15.22 Intrinsic Hypersurfaces $\mathfrak{H}^{n-1}$: Left Second Order Differential of Gap

Step	Hypothesis	$d^2 \Delta\mathbf{S} / ds^2 = \Sigma_i \Sigma_\xi \Sigma_\kappa (\mathbf{0}) \lambda^\kappa \lambda^\xi x^i{}_\alpha \mathbf{b}_i + \Sigma_i \Sigma_\xi \Sigma_\kappa (\mathbf{0}) \lambda^\kappa \lambda^\xi n^i \mathbf{b}_i$
1	From T17.15.20 Intrinsic Hypersurfaces $\mathfrak{H}^{n-1}$: Right Second Order Differential of Gap	$d^2 \Delta\mathbf{S} / ds^2 = \Sigma_\xi \Sigma_i \Sigma_\kappa$ $[- \Sigma_\gamma (h_{d,\kappa} w_\xi{}^\gamma + h_{d,\xi} w_\kappa{}^\gamma + h_d \ w_\kappa{}^\gamma{}_{;\xi}) x^i{}_\gamma +$ $(h_{d,\kappa,\xi} - h_d \Sigma_\gamma w_\kappa{}^\gamma c_{\gamma\xi}) n^i] \lambda^\kappa \lambda^\xi \mathbf{b}_i$
2	From 1, T10.3.20 Addition of Tensors: Distribution of a Tensor over Addition of Tensors and A10.2.14 Correspondence of Tensors and Tensor Coefficients	$d^2 \Delta\mathbf{S} / ds^2 =$ $\Sigma_i \Sigma_\xi \Sigma_\kappa [- \Sigma_\gamma (h_{d,\kappa} w_\xi{}^\gamma + h_{d,\xi} w_\kappa{}^\gamma + h_d \ w_\kappa{}^\gamma{}_{;\xi}) \lambda^\kappa \lambda^\xi] x^i{}_\gamma \mathbf{b}_i +$ $\Sigma_i \Sigma_\xi \Sigma_\kappa [(h_{d,\kappa,\xi} - h_d \Sigma_\gamma w_\kappa{}^\gamma c_{\gamma\xi}) \lambda^\kappa \lambda^\xi] n^i \mathbf{b}_i$
3	From 2 and T17.15.19B Intrinsic Hypersurfaces $\mathfrak{H}^{n-1}$: Left Second Order Differential of Gap	$d^2 \Delta\mathbf{S} / ds^2 = \Sigma_i \Sigma_\xi \Sigma_\kappa (0) \lambda^\kappa \lambda^\xi \mathbf{b}_i$
4	From 3 and T17.3.18 Dual Covariant Tensors: Zero Tensor Rank-2; step 2 clearly states that the second order gap equation is a dual covariant tensor, since the left hand side kernel is zero then it must be concluded the core is a dual zero tensor of rank-2.	$d^2 \Delta\mathbf{S} / ds^2 \equiv \Sigma_i \Sigma_\xi \Sigma_\kappa (\overset{*}{Z}_2) \lambda^\kappa \lambda^\xi \mathbf{b}_i$
5	From 4, T17.3.18 Dual Covariant Tensors: Zero Tensor Rank-2, T10.2.5 Tool: Tensor Substitution and A10.2.14 Correspondence of Tensors and Tensor Coefficients	$d^2 \Delta\mathbf{S} / ds^2 = \Sigma_i \Sigma_\xi \Sigma_\kappa (\mathbf{0}\ x^i{}_\alpha + \mathbf{0}\ n^i) \lambda^\kappa \lambda^\xi \mathbf{b}_i$
∴	From 5, T10.3.20 Addition of Tensors: Distribution of a Tensor over Addition of Tensors and A10.2.14 Correspondence of Tensors and Tensor Coefficients	$d^2 \Delta\mathbf{S} / ds^2 = \Sigma_i \Sigma_\xi \Sigma_\kappa (\mathbf{0}) \lambda^\kappa \lambda^\xi x^i{}_\alpha \mathbf{b}_i + \Sigma_i \Sigma_\xi \Sigma_\kappa (\mathbf{0}) \lambda^\kappa \lambda^\xi n^i \mathbf{b}_i$

Theorem 17.15.23 Intrinsic Hypersurfaces $\mathfrak{H}^{n-1}$: Tangent and Normal Differential Gap

Step	Hypothesis	$\mathbf{0} = h_{d,\kappa}\, w_{\xi}{}^{\gamma} + h_{d,\xi}\, w_{\kappa}{}^{\gamma} + h_d\, w_{\kappa}{}^{\gamma}{}_{;\xi}$ EQ A	$h_{d,\kappa,\xi} = h_d \sum_{\gamma} w_{\kappa}{}^{\gamma}\, c_{\gamma\xi}$ EQ B
1	From T17.15.21 Intrinsic Hypersurfaces $\mathfrak{H}^{n-1}$: Left Second Order Differential of Gap, T17.15.20 Intrinsic Hypersurfaces $\mathfrak{H}^{n-1}$: Right Second Order Differential of Gap, T10.3.20 Addition of Tensors: Distribution of a Tensor over Addition of Tensors and A10.2.14 Correspondence of Tensors and Tensor Coefficients	$\sum_i \sum_{\xi} \sum_{\kappa} (\mathbf{0})\, \lambda^{\kappa}\lambda^{\xi}\, x^{i}{}_{\alpha}\, \mathbf{b}_i + \sum_i \sum_{\xi} \sum_{\kappa} (\mathbf{0})\, \lambda^{\kappa}\lambda^{\xi}\, n^{i}\, \mathbf{b}_i =$ $\sum_i \sum_{\xi} \sum_{\kappa} [- \sum_{\gamma} (h_{d,\kappa}\, w_{\xi}{}^{\gamma} + h_{d,\xi}\, w_{\kappa}{}^{\gamma} + h_d\, w_{\kappa}{}^{\gamma}{}_{;\xi})\, \lambda^{\kappa}\lambda^{\xi}\, x^{i}{}_{\gamma}\, \mathbf{b}_i] +$ $[(h_{d,\kappa,\xi} - h_d \sum_{\gamma} w_{\kappa}{}^{\gamma}\, c_{\gamma\xi})\, \lambda^{\kappa}\lambda^{\xi}\, n^{i}\, \mathbf{b}_i]$	
2	From 1 and T17.3.9 Dual Covariant Tensors: Mutually Exclusive Tangential Tensor Components; equating basis and pseudo vectors	$\mathbf{0} = -\sum_{\gamma} h_{d,\kappa}\, w_{\xi}{}^{\gamma} + h_{d,\xi}\, w_{\kappa}{}^{\gamma} + h_d\, w_{\kappa}{}^{\gamma}{}_{;\xi}$	
3	From 2, T10.4.1 Scalars and Tensors: Uniqueness of Scalar Multiplication to a Tensors, T4.4.1 Equalities: Any Quantity Multiplied by Zero is Zero and T4.8.2B Integer Exponents: Negative One Squared	$\mathbf{0} = \sum_{\gamma} h_{d,\kappa}\, w_{\xi}{}^{\gamma} + h_{d,\xi}\, w_{\kappa}{}^{\gamma} + h_d\, w_{\kappa}{}^{\gamma}{}_{;\xi}$	
4	From 1 and T17.3.10 Dual Covariant Tensors: Mutually Exclusive Normalized Tensor Components; equating basis and pseudo vectors	$0 = h_{d,\kappa,\xi} - h_d \sum_{\gamma} w_{\kappa}{}^{\gamma}\, c_{\gamma\xi}$	
5	From 4 and T4.3.6A Equalities: Reversal of Left Cancellation by Addition	$h_{d,\kappa,\xi} = h_d \sum_{\gamma} w_{\kappa}{}^{\gamma}\, c_{\gamma\xi}$	
∴	From 3 and 5	$\mathbf{0} = h_{d,\kappa}\, w_{\xi}{}^{\gamma} + h_{d,\xi}\, w_{\kappa}{}^{\gamma} + h_d\, w_{\kappa}{}^{\gamma}{}_{;\xi}$ EQ A	$h_{d,\kappa,\xi} = h_d \sum_{\gamma} w_{\kappa}{}^{\gamma}\, c_{\gamma\xi}$ EQ B

Theorem 17.15.24 Intrinsic Hypersurfaces $\mathfrak{H}^{n-1}$: Covariant Tangent Constant Curvature

1g	Given	[K] constant curvature
Step	Hypothesis	$\mathbf{0} = \sum_\gamma \sum_\xi \sum_\kappa w_\kappa{}^\gamma{}_{;\xi}\, \lambda^\kappa \lambda^\xi\, x^i{}_\gamma$
1	From T17.15.22S1 Intrinsic Hypersurfaces $\mathfrak{H}^{n-1}$: Tangent and Normal Differential Gap and T17.3.9 Dual Covariant Tensors: Mutually Exclusive Tangential Tensor Components	$\mathbf{0} = \sum_\gamma \sum_\xi \sum_\kappa (h_{d,\kappa}\, w_\xi{}^\gamma + h_{d,\xi}\, w_\kappa{}^\gamma + h_d\, w_\kappa{}^\gamma{}_{;\xi})\, \lambda^\kappa \lambda^\xi\, x^i{}_\gamma$
2	From 1g, 1, A17.15.2C Constant Curvature Between Parallel Hypersurfaces $\mathfrak{H}^{n-1}$, T10.2.7A Tool: Zero Times a Tensor is a Zero Tensor and T10.3.18 Addition of Tensors: Identity by Addition	$\mathbf{0} = \sum_\gamma \sum_\xi \sum_\kappa (h_d\, w_\kappa{}^\gamma{}_{;\xi})\, \lambda^\kappa \lambda^\xi\, x^i{}_\gamma$
3	From 2 and T10.4.6 Scalars and Tensors: Distribution of a Scalar over Addition of Tensors	$\mathbf{0} = h_d \sum_\gamma \sum_\xi \sum_\kappa (w_\kappa{}^\gamma{}_{;\xi})\, \lambda^\kappa \lambda^\xi\, x^i{}_\gamma$
∴	From 3, T10.2.8 Tool: Zero Tensor Times a Tensor is a Zero Tensor; for q = 0 and $\overset{*}{A}_0 = h_d$; and T10.4.1A Scalars and Tensors: Uniqueness of Scalar Multiplication to a Tensors	$\mathbf{0} = \sum_\gamma \sum_\xi \sum_\kappa w_\kappa{}^\gamma{}_{;\xi}\, \lambda^\kappa \lambda^\xi\, x^i{}_\gamma$

Theorem 17.15.25 Intrinsic Hypersurfaces $\mathfrak{H}^{n-1}$: Covariant Normal Constant Curvature

1g	Given	[K] constant curvature
Step	Hypothesis	$\mathbf{0} = \sum_\gamma w_\kappa{}^\gamma\, c_{\gamma\xi}$
1	From 17.15.22B Intrinsic Hypersurfaces $\mathfrak{H}^{n-1}$: Tangent and Normal Differential Gap	$h_{d,\kappa,\xi} = h_d \sum_\gamma w_\kappa{}^\gamma\, c_{\gamma\xi}$
2	From 1g and 1 and A17.15.2B Constant Curvature Between Parallel Hypersurfaces $\mathfrak{H}^{n-1}$	$\mathbf{0} = h_d \sum_\gamma w_\kappa{}^\gamma\, c_{\gamma\xi}$
∴	From 2, T10.2.8 Tool: Zero Tensor Times a Tensor is a Zero Tensor; for q = 0 and $\overset{*}{A}_0 = h_d$; and T10.4.1A Scalars and Tensors: Uniqueness of Scalar Multiplication to a Tensors	$\mathbf{0} = \sum_\gamma w_\kappa{}^\gamma\, c_{\gamma\xi}$

Observation 17.15.4 Second Fundamental Covariant Metric vs Weingarten Dyadic

The first and second order differential of gap relations T17.15.16 "Intrinsic Hypersurfaces $\mathfrak{H}^{n-1}$: Right First Order Differential of Gap", T17.15.20 "Intrinsic Hypersurfaces $\mathfrak{H}^{n-1}$: Right Second Order Differential of Gap" for parallel hypersurfaces are reminiscent of the Equation of Gauss T17.8.11 "Codazzi-Gauss Equations: Equation of Gauss" on a hypersheet. The Equation of Gauss on a hypersheet is independent of all other tensor relationships of the hyperplane, because it is exclusively defined in terms of the second fundamental covariant metric dyadic likewise the Weingarten mixed metric dyadic. This makes them general and independent.

Observation 17.15.5 Why Constant Curvature?

Why examine constant curvature? We live in a world of constant curvature so many of our physical models deal with constant curvature. Without spaces of constant curvature our physical universe as we know it would be torn apart right down to the very atoms themselves. Such spaces exist near or in massive stars and black holes; since such spaces subsist in the extremes of gravity wells this analysis is useful as an exploratory tool, for the study of constant curvature in the physical world.

Tensor Calculus & Physics: A General Treatise

Table of Contents

List of Tables

List of Figures

List of Observations

Chapter 18 Advance Tensors and Properties of Hypersurfaces $\mathfrak{H}^{n-1}$

References Volume I

[ABR72] Handbook of Mathematical Functions with Formulas, Graphs and Mathematical Tables, tenth edition by Milton Abramowitz and Irene A. Stegun, published by Dover Publications, Inc., New York, 1972

[ACZ96] Fermat's Last Teorem by Amir D. Aczel, published by Four Walls Eight Windows, 1996

[ADL66] A New Look at Geometry by Irving Adler, published by The John Day Company, 1966

[AYR54] Schaum's Outline Series Theory and problems of Plane and Spherical Trigonometry, by Frank Ayres, Jr., published by McGraw-Hill, Inc., 1954

[BEY88] CRC Standard Mathematical Tables 28th Edition, by William H. Beyer, published by CRC Press Inc., 1988

[BRI60] Encyclopædia Britannica, published by Encyclopædia Britannica, 1960

[BRO73] The Ascent of Man by J. Bronowski, from the PBS TV series and published by Little, Brown and Company, 1973

[CAJ71] Sir Isaac Newton Principia Vol I and II, The Motion of Bodies Motte's Translation by Cajori Rev. published by University of California Press, 1971

[CHE70] Introduction to Linear System Theory by Chi-Tsong Chen, published by Holt, Rinehart and Winston, Inc., 1970

[CLA71] Einstein The Life and Times by Ronald W. Clark, published by World Publishing Times Mirror, 1971 pg 95

[COP68] Introduction to Logic by Irving M. Copi, published by The Macmillan Company, 1968

[DIR75] Directions in Physics, by P.A.M. Dirac, F.R.S., O.M., published John Wiley & Sons, 1975

[DOY30] The Complete Sherlock Holmes, by Sir Arthur Conan Doyle, published by Doulbleday & Company, Inc., Fourth Edition, 1930

[EIS70] Applied Matrix and Tensor Analysis, by John A. Eisele and Robert M. Mason, published by Wiley-Interscience a Division of John Wiley & Sons, Inc., 1970

[EVE76] An Introduction to the History of Mathematics, by Howard Eves, published by Holt, Rienhart and Winston, 1976

[GER62] Lectures on Tensor Calculus and Differential Geometry, by Johan C. H. Gerretsen, published by P. Noordhoff N.V. Groningen, 1962

[GIE71] Thermophysics, by Warren H. Giedt, University of California, Davis, published by Van Nostrand Rienhold Company, 1971

[GRA65] Table of Integrals, Series, and Products, I.S. Gradshteyn and I.M. Ryzhik; Fourth Edition Prepared by Yu. V. Geronimus/ M. Yu. Tseytlin; Translated from the Russioan by Scripta Technica, Inc. Translatlion edited by Alan Jeffrey, published by Academic Press, 1965

[HAI83] "Doctor Who: A Celebration Two Decades Through Time and Space" by Peter Haining, published by W.H. Allen London, 1983

[HEA56] "EUCLID; The thirteen Books of THE ELEMENTS", by Sir Thomas L. Heath, published by Dover Publications, Inc. New York, 1956 Second Edition Revised with Additions VOL I, II and III.

[HUA67] "Engineering Mechanics Dynamics", by T.C. Huang, published by Addison-Wesley Publishing Company, Reading/ Massachusetts/ Palo Alto/ London/ Don Mills/ Ontario, 1969 Second Edition VOL II

[ITT75] Reference Data for Radio Engineers, by International Telephone and Telegraph Corporation, published by Howard W. Sams & Co., Inc. Indianapolis/Kansas City/New York, 1977

[JAIβ04] Schaum's Outline Abstract Algebra, by Lloyd R. Jaisingh and Frank Ayres, Jr., published by McGraw-Hill, Inc., 2004

[JOH69] College Algebra by R.E. Johnson, L.L. Lendsey, W.E. Slesnick & G.E. Bates, published by Cummings Publishing Company, 1969

[KAVβ04] Surveying with Construction Applications, by Barry F. Kavanagh, published by Pearson/Prentice Hall, fifth edition 2004

[KRE59] Differential Geometry, by Erwin Kreyszig, published by Dover Publications, Inc. New York, 1991, republication 1963, first publication 1959
[LAW62] Tensor Calculus and Relativity, by Derek F. Lawden, published by Spottiswoode Ballantyne & Co. Ltd. at London and Cochester Great Britain, 1962
[LEF97]: "GEOMETRY; The Easy Way", by Lawrence S. Leff, published by Barron's Educational Series, Inc., 1997
[LIP64] Schaum's Outline Series Theory and problems of Set Theory and Related Topics by Seymour Lipschutz; published by McGraw-Hill, Inc., 1964
[LIP68] Schaum's Outline Series Theory and problems of Linear Algebra by Seymour Lipschutz; published by McGraw-Hill, Inc., 1968
[MER71] Statics Second Edition by J.L. Meriam, published by John wiley & Sons, Inc., 1971
[MAX98] A Treatise on Electricity and Magnetism by Oxford classics series, published by Oxford University Press, 1998 Vol I and II
[MAR64] The new mathematics dictionary and handbook, by Robert W. Marks, published by Bantam Books, Inc. 1964
[NAY73] Perturbation Methods, by Ali Hasan Nayfeh, published by John Wiley & Sons, A Wiley-Interscience Publication, 1973
[NISβ04] "What is the origin of hours, minutes and seconds?"; published by National Institute of standards and technology, Physics Laboratory Time and Frequency Division, Boulder, CO 80305-3328, 2004
http://www.boulder.nist.gov/timefreq/general/history.htm#Anchor-16126
[NOL98] Schaum's Outline Series Logic, by John Nolt, Dennis Rohatyn and Achille Varzi, published by McGraw-Hill Publishing Company, 1998
[OREβ03] Long Division Algorithms Collected in the European Union, by Daniel Clark Orey, published http://www.csus.edu/indiv/o/oreyd/ACP.htm_files/longdiv.eu.2003.htm and
http://www.csus.edu/indiv/o/oreyd/ACP.htm_files/EducationalStudiesinMathematics.htm, 2003
[OXF71] The Oxford English Dictionary, published by Oxford University Press, 1971
Company Boston/Toronto, 1973
[PAR60] Modern Probability Theory and Its Applications, by Emanuel Parzen, published by John Wiley & Sons, Inc., 1960
[PLA86] Schaum's Outline Series Theory and Problems of Computer Graphics, by Roy A. Plastock and Gordon Kalley, published by McGraw-Hill Publishing Company, 1986
[PLU81] The Torah: A Modern Commentary, by Editor W. Gunther Plaut, published by The Union of American Hebrew Congregations – New York, 1981
[PRO70] College Calculus with Analytic Geometry, second edition, by Murray H. Potter and Charles B. Morrey Jr., published by Addison Wesley Publishing Company, 1970
[SHVβ01] How2 Lecture Series: C++ Programming Language, by Randall H. Shev, published by UCSC Extension, 2001
[SIN97] Fermat's Enigma "The epic quest to solve the world's greatest mathematical problem", by Simon Singh, published by Anchor Books a division of Random House, Inc. New York, 1997
[SOK67] Tensor Analysis Theory and Applications to Geometry and Mechanics of Continua, by I.S. Socolnikoff, published by John Wiley & Sons, Inc, 1967
[SPI59] Schaum's Outline Series Theory and Problems of Vector Analysis by Murray R. Spiegel, published by Schaum Publishing Company, 1959
[SPI63] Schaum's Outline Series Theory and Problems of Advanced Calculus by Murray R. Spiegel, published by Schaum Publishing Company, 1963
[SPI64] Schaum's Outline Series Complex Variables, by Murray R. Spiegel, published by Schaum Publishing Company, 1964
[SPI67] Applied Differential Equations, by Murray R. Spiegel, published by Prentice-Hall, INC. second edition, 1967
[SYK93] "The Best Bind Since Einstein", produced and directed by Christopher Sykes, a NOVA production for BBC-TV in association with WEBH/Boston, 1993 WGBH Education Foundation.

[WORβ04] "Where did Degrees, Minutes and Seconds come from?", by Sam Wormley, published by GPS Information http://gpsinformation.net/articles/degreesminutesseconds.htm

This quote is from Genesis Rabbah, which is from Midrash Rabbah, a collection of Midrashim that the rabbis wrote on the Jewish bible. The Midrash is from Gensis Rabbah Chapter 1, verse 10. Just to give you the traditional understanding of the Midrash: "...just as the letter *bet* is enclosed on three sides but open to the front, we are not to speculate on the origins of God or what may have existed before creation. The purpose of such a comment is not to limit scientific inquiry into the origins of the universe but to discourage efforts to prove the un-provable." Quoted from Etz Chayim (a new commentary from Conservative movement).

Other Author(s): Odeberg, Hugo, 1898-

Title: Midrash rabbah. Genesis.
The Aramaic portions of Bereshit rabba with grammar of Galilæan Aramaic / by Hugo Odeberg.

Subject(s): Aramaic language--Grammar.
Publisher: Lund : C.W.K. Gleerup ; Leipzig : O. Harrassowitz, 1939.

Description: 2 v.; 25 cm.
Series: Lunds universitets årsskrift ; n.f., bd. 36, nr. 3-4

Notes: Bibliography: v. 2, p. [166]-171.
Table of Contents: v. 1. Text with transcription v. 2. Short grammar of Galilæan Aramaic.

Call Number: BM517 .M63 1939

The Modern Elements
Volume II: Joining Gravitational and Electrical Fields

Tensor Calculus & Physics: A General Treatise

Table of Contents

List of Figures

List of Tables

List of Observations

List of Givens

List of Equations

Chapter1 *Uniting Attractive and Repulsive Forces at a Distance*

Section 1.1 And So It Begins

TIME: 9:39am (Discovery)
DATE: 11 March 2006
ENGR: R.H. Shev

SUBJECT Monocentric Particle Model with Unification of Fundamental Divergent Forces

MONOCENTRIC particle systems have beauty in that they are simple in logic and do not lead to digressively, degenerate set of arguments for a theory on particles. In a polycentric particle system one could be lead to a process of explanation that is circular or worse recursive, a set of sub-particles make up a particle, but what is that particle comprised of, and it in turn, and so on. Such degenerate and possible recursive arguments have arisen in other disciplines such as reproduction. It used to be thought humans came from the male sperm cell and in them there resided complete homunculi, and in the homunculi's gonads and sperm cells of another complete homunculi ad-infinitum always a miniature complete human without end. Now we know it starts and ends with the information coded on a DNA molecule.

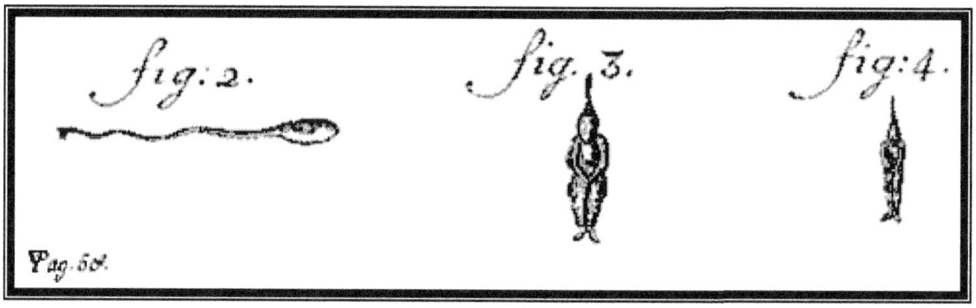

Figure 1.1.1 Antoni Van Leeuwenhoek's Notion of Reproduction [ANT1719]

The theory that I am proposing here has the beauty of being a monocentric particle system, based on a single photon in spin having a finite fixed Einsteinian quantum of energy, hν.

Let's consider what a photon in spin might be by doing an Einstein Gedunkanexpermentum (Thought-experiment):

A photon in spin can be thought of as taking a linear moving photon propagating across space and having it deflected in a new direction by a mirror having a special surface made out of Dirac idealized point nuclear particles. Idealized meaning conservation of energy is perfectly conserved the quantum energy going in being the same coming out. Another mirror is setup to deflect it again and another and another until with enough mirrors, it comes back to the original mirror and so the photon traces out a discrete circular path. Now let us assume it's possible to increase the number of mirrors and reduce their path length between by making the distance and the size of the mirrors smaller and smaller till the photon moves about in an ever tighter near continuous circle. Finally, in the limit with a radius of zero we pull all the mirrors away leaving the photon having a pure spin like a top at its point.

Such a physical enigma gives rise to a series of questions. If such a photon state could be made to exist what would its physical properties be like? Would it still be a photon or something else? What would it take to maintain it in such a state? In other words, what would keep it from flying apart and moving off in some other linear and random direction?

Section 1.2 Definitions of a Monocentric Particle System

Definition 1.2.1 Monocentric particle system

Monocentric particle system is a model of a nuclear particle that builds a system out of a single particle and explains all physical aspects that are attributed to that constituent. An example in modern physical theory is the concept of a string whose attributes are determined by harmonic resonance setup in the closed loop (though this work predates that hypothesis).

Definition 1.2.2 Polycentric particle system

Polycentric particle system is a model of a nuclear particle that builds the particle out of two or more sub-particles and explains all physical aspects that are attributed to that element. An example in modern physical theory is quarks, which intern is built out of other particles (eventually reduced to a monocentric particle system with the introduction of strings to avoid this circular digressive logical trap, strings becoming the fundamental building blocks of quarks).

Definition 1.2.3 Photon (Einsteinian definition) [HAL66, pg 1182]

A ***photon*** is a bundle or packet of energy that travels through space as a wave with velocity-c, but impacts on material media with a specific energy and momentum of a particle. Its mechanism for propagating through physical space is done through an electric and magnetic field component, as one-field collapses it induces the other, an adjacent distance away, thereby relaying its quantum of energy and moving forward from point-to-point in space. This time varying, field propagation can be modeled by Maxwell's Field Equations in free space, see Appendix A Table A.3.4 "Maxwell's Differential Field Equations":

Table 1.2.1 Physical Properties of a Photon

Algebraic Description	Description	Equation
$c = \lambda\nu$ [1.2.5]	photon celerity [1.2.1]	Equation A
• $\vec{s}_m = \varepsilon\vec{E} \times \vec{B}$ [1.2.2]	Poynting Field Momentum Density [1.2.4]	Equation B
$u_m = ½(\varepsilon\vec{E} \bullet \vec{E} + \mu^{-1}\vec{B} \bullet \vec{B})$ [1.2.3]	Poynting Field Energy Density [1.2.4]	Equation C
• $E = h\nu = \int_V u_m \, dV$	photon energy packet [1.2.1] [1.2..2]	Equation D
• $\vec{P} = (h/\lambda)\,\hat{s}_m = \int_V \vec{s}_m \, dV$	photon momentum packet [1.2.1] [1.2..2]	Equation E
• $q_p = 0$	no monopole exists for a photon[1.2.6]	Equation F
• $m_p = 0$		Equation G

Definition 1.2.4 Transmitter

A ***transmitter*** is any primary nuclear particle of origin for an emitted photon.

Definition 1.2.5 Receiver

A ***receiver*** is any primary nuclear particle that receives a photon and absorbs it. Also, a receiver is within the observer's frame of reference with respect to the origin of the photon.

Definition 1.2.6 Physical Constant

A ***physical*** or ***universal constant*** is any measurable number in the physical world that remains unaltered by position (homogenous space), or time (isotropic period), or relative motion, a truly invariant constant within any physical system. Examples are:

Sym	Name	Physical Constants	PC
c	Speed of Light	CXPPC.A.2.3.5	A
γ_0	Reciprocal Newtonian constant	CXPPC.A.2.3.2	B
ε_0	Electrical Permittivity	CXPPC.A.2.3.8	C
μ_0	Magnetic Permeability	CXPPC.A.2.3.9	D
q_0	Quantum Charge	CXPPC.A.2.3.7	E
α^{-1}	Fine Structure	CXPPC.A.2.3.6	F
h	Plank's Constant	CXPPC.A.2.3.3	G

Definition 1.2.7 Quantum (noun) [HAL66, pg 593]

Any system or quantity having a non-continuous, discrete, finite level of energy is called a ***quantum***. Here, as Einstein had shown, through his equation ($E_m = mc^2$), energy and mass are interchangeable; hence if energy is a quantum so is the counterpart mass, such relationships are called ***equivalent quantum*** or simply ***counterparts***. In system theory, such a level is called a ***discrete state*** as verse to the antonym ***continuous state***.

Definition 1.2.8 Quantized (verb) [HAL66, pg 593]

Quantized is the act of placing equal multiples of partitioned energy into discrete finite levels.

Definition 1.2.9 Quantum Constant

A ***quantum constant*** is a physical constant in a quantized system.

Definition 1.2.10 Event-one

Event-one is the origin, center-point and exact moment at which the universe was created and started its expansion. This event is the very first event, in the physical universe.

Definition 1.2.11 Field of Force

In a continuum of points of force any given point within that region of space, one can find a measurable finite amount of energy, which intern can be called upon to work on moving a test particle, in a specific direction and through a finite distance, such a collection of points of force are called a ***field of force*** or simply ***force field***.

Definition 1.2.12 Monopole of Force

A ***monopole of force*** or simply ***monopole*** is a single particle source, giving rise to a field of force. Such a field can be represented as a singularity, a dimensionless point of origin, with forces diverging all around, out of, or into, a solid angle.

Definition 1.2.13 Physical Field of Force

A ***physical field of force*** has a monopole as its point of origin.

Definition 1.2.14 Apparent Field of Force

An ***apparent field of force*** or simply ***apparent field*** has no monopole as its anchor and as such the field is not divergent. Its origins arise from outside of a system of particles from the perspective of an observer in a non-stationary frame of reference, hence a field of force is apparently seen and thereby created free to circulate or curl.

Definition 1.2.15 Curled Field of Force

A ***curled field of force*** is an apparent field of force.

Law 1.2.1 Photon Point Spin: Tangential Velocity

The tangential spin velocity moves at a constant speed of light-c. Also the particle is an Einsteinium quantum photon it follows that it must be moving at the velocity of light relative to the center of its spin, see D1.2.3A.

$$c = r\,\omega \qquad \text{Equation A}$$

Law 1.2.2 Photon Point Spin: Radius

Do to, Law 1.2.1 "Tangential Velocity", it follows than the radii must be defined as

$$r = c\,/\,\omega \qquad \text{Equation B}$$

Law 1.2.3 Photon Point Spin: Angular Rate

Do to, Law 1.2.2 "Photon Point Spin: Radius", it follows than the angular rate must be defined as

$$\omega = c\,/\,r \qquad \text{Equation C}$$

[1.2.1]Note: Einstein's equation of photon quantum energy packet that he developed in his paper on the Photoelectric Effect. [EINα05]

[1.2..2]Note: From LxMXE.9.4.5, this is how the momentum of a photon at a point is imparted throughout the finite region of the packet giving rise to its total momentum as a particle.

[1.2.3]Note: From LxMXE.9.4.6, this is how energy at a point is distributed throughout the finite region of the photon packet. It is the integration throughout this limited volume that gives the photon a finite quantum of energy D1.2.3C. Within that volume as the electric and magnetic fields alternately exchange their energy, that the packet is driven trough space as a traveling wave, hence giving rise to its wave nature.

[1.2.4]Note: Equations D1.2.3D and D1.2.3E give the properties of duality to a photon being a particle and a wave.

[1.2.5]Note: Let λ have a dimensional unit [L] its wavelength and ν $[T^{-1}]$ its frequency of propagation with the velocity-c the speed of light $[LT^{-1}]$.

[1.2.6]Note: No charge monopole then no mass monopole. Here charge and mass are only different sides of the same coin, hence inseparable with similar characteristics.

Section 1.3 PHYSICAL AND PLANCK CONSTANTS

Some years prior to my proposed hypothesis while I was at UC Davis I had been reading selected works by George Gamow's and in particularly "Thirty Years that Shook Physics" [GAM66, pg 155] in this book he speculates the use of dimensional analysis based on naturally occurring physical units of the universe. These constants are formalized in the book *Gravitation* [MIS73, pg 1215] by Misner, Thorne and Wheeler's and their reflections on the use of three physical constants c-speed of light, h-Planck's constant and G_0-Newton's Universal Gravitational constant to construct three fundamental, dimensional units. I realized that they could be arranged into other units to derive other dimensions and later I found that Max Planck, some hundred years prior, had the same thoughts in his 1899 paper [PLA1899] in terms of mass, length, time and entropy in [°K] Kelvin. At the time he presented the units, quantum mechanics had not, been discovered. He had not yet published his theory of blackbody radiation (first complete theory published December 1901), based on that paper they apparently came about as an afterthought as he groped for an understanding of the radiation phenomena he was studying and stated them at the very end of the paper, later on, these dimensions would become more concisely quantified into the following units:

$M_{plk} = \sqrt{(\hbar\, c/G_0)}$ mass [M]
$L_{plk} = \sqrt{(\hbar\, G_0/c^3)}$ length [L]
$T_{plk} = \sqrt{(\hbar\, G_0/c^5)}$ time [T]

Planck's dimensions can be caste into the following form by substituting CxPPC.A.2.3.2 into the above set:

Table 1.3.1 Planck's Unit Dimensions

M_{plk}	=	$\sqrt{(4\pi\gamma_0\hbar c)}$	=	$2.176629623\text{x}10^{-19}$ kg	[M]	Equation 1.3.1
L_{plk}	=	$\sqrt{(\hbar /4\pi\gamma_0 c^3)}$	=	$1.616140686\text{x}10^{-35}$ m	[L]	Equation 1.3.2
T_{plk}	=	$\sqrt{(\hbar /4\pi\gamma_0 c^5)}$	=	$5.390782033\text{x}10^{-44}$ s	[T]	Equation 1.3.3

When I first laid Planck's units out it bothered me that charge was not represented. Upon looking about, I found the electrical permittivity constant, contained the dimension for charge. Integrating it with the two main constants [h and c], charge could be represented and calculated.

Theorem 1.3.1 Summerfield: Quantum Electrical Charge

Step	Hypothesis	$2x = 2hc\varepsilon_0 / q^2 = 137.0360615$
1	From C3 Planck constant and C7 Permittivity constant for free space; approximates the charge squared per unit volt	$[Q^2][V^{-1}] \sim h\varepsilon_0$
2	From 1, multiplying by C4 [c] gives the correct dimensions of just charge squared.	$[Q^2] \sim hc\varepsilon_0$
3	From 2 and evaluate	$hc\varepsilon_0 = 1.75887035\text{x}10^{-36}\ C^2$ to match the dimension the square root is taken,
4	From 2, 3 and T4.10.1 Radicals: Uniqueness of Radicals	$[Q] \sim \sqrt{hc\varepsilon_0}$ evaluating the product
5	From 4 and evaluate	$\sqrt{hc\varepsilon_0} = 1.326224095\text{x}10^{-18}$ C
	Sometimes known as Plank's charge. Which is not quit the Coulomb quantum charge. Clearly, a constant is required to make up the difference. The reciprocal is taken in order to obtain a large decimal number:	
6	From 5 and equating to the reciprocal of [x]	$q^2 = hc\varepsilon_0 / x$ solving for x,

7	From 6 and T4.4.14 Equalities: Formal Cross Product	$x = hc\varepsilon_0 / q^2 = 68.51803077$ now multiplying by 2 gives
∴	From 7 and T4.4.2 Equalities: Uniqueness of Multiplication	$2x = 2hc\varepsilon_0 / q^2 = 137.0360615$ which is the fine structure constant as seen from the above table.

Theorem 1.3.2 Summerfield: Evaluation of Physical Charge Constants

1g 2g	Given	$\varepsilon_{rq} \equiv 1 / \alpha = 1 / 2x$ fine-structure constant $\varepsilon_q \equiv \varepsilon_{rq}\varepsilon_0$		
Step	Hypothesis	$q = \sqrt{4\pi\hbar c\varepsilon_q}$ $q = 1.6021917\text{x}10^{-19}$ $\varepsilon_{rq} = 0.007297351$ $\varepsilon_q = 6.461209(78)\text{x}10^{-14}$ $\varepsilon_q = \varepsilon_{rq}\varepsilon_0$	Coulomb [Q] C [Q] Null [∅] $C^2/nt\text{-}m^2$ $[M^{-1}][L^{-3}][T^2][Q^2]$ $C^2/nt\text{-}m^2$ $[M^{-1}][L^{-3}][T^2][Q^2]$	EQ A EQ B EQ C EQ D EQ E
1	From T1.3.1 The Summerfield Quantum of Electrical Charge	$2x = 2hc\varepsilon_0 / q^2$		
2	From 2 and T4.4.14 Equalities: Formal Cross Product	$q^2 = 2hc\varepsilon_0 / 2x$		
3	From 1g, 2, A4.2.3 Substitution and A4.2.10 Commutative Multp	$q^2 = 2hc\ \varepsilon_{rq}\varepsilon_0$		
4	From 3 and T4.10.1 Radicals: Uniqueness of Radicals	$q = \sqrt{2hc\varepsilon_{rq}\varepsilon_0}$		
5	From 4, A4.2.12 Identity Multp, A4.2.13 Inverse Multp and A4.2.10 Commutative Multp	$q = \sqrt{4\pi (h / 2\pi) c\varepsilon_{rq}\varepsilon_0}$		
6	From 5, CxPPC.A.2.3.4 Normalized Planck constant and A4.2.3 Substitution	$q = \sqrt{4\pi\hbar c\varepsilon_{rq}\varepsilon_0}$		
7	From 2g, 6 and A4.2.3 Substitution	$q = \sqrt{4\pi\hbar c\varepsilon_q}$		
8	From 7 and evaluate	$q = 1.6021917\text{x}10^{-19}$ C [Q]		
9	From 1g	$\varepsilon_{rq} = 1 / \alpha$		
10	From 9 and evaluate	$\varepsilon_{rq} = 0.007297351$ [∅]		
11	From 2g	$\varepsilon_q = \varepsilon_{rq}\varepsilon_0$		
12	From 11 and evaluate	$\varepsilon_q = 6.461209(78)\text{x}10^{-14}$ $[M^{-1}][L^{-3}][T^2][Q^2]$		
∴	From 7, 8, 10, 11 and 12	$q = \sqrt{4\pi\hbar c\varepsilon_q}$ $q = 1.6021917\text{x}10^{-19}$ $\varepsilon_{rq} = 0.007297351$ $\varepsilon_q = 6.461209(78)\text{x}10^{-14}$ $\varepsilon_q = \varepsilon_{rq}\varepsilon_0$	C [Q] C [Q] Null [∅] $C^2/nt\text{-}m^2$ $[M^{-1}][L^{-3}][T^2][Q^2]$ $C^2/nt\text{-}m^2$ $[M^{-1}][L^{-3}][T^2][Q^2]$	EQ A EQ B EQ C EQ D EQ E

The fine-structure constant or known as the Summerfield fine-structure constant, usually denoted by [α], is a fundamental physical constant characterizing the strength of the electromagnetic interaction. Arnold Summerfield originally introduced it into physics in 1916, as a measure of the quantum deviations in atomic spectral lines from the predictions made by the Bohr model. The fine-structure constant is a dimensionless quantity, and its numerical value is independent of the system of units used.

The relative quantum permittivity coefficient [ε_{rq}] is associated with light passing through a material media reducing the photon to move at sub-light speeds:

Observation 1.3.1 On the absence of the universal gravitational constant

- Unlike other Plank's constants why is the universal gravitational constant, which represents gravity not embodied in T1.3.2A "Summerfield: Evaluation of Physical Charge Constants"?
- Could there be a physical model that would explain this and be folded back into the original hypothesis?

Now sub-light and hyper-light speeds in a material media:

Maxwell's equations predicted the unification of electric and magnetic fields with the speed of light in a vacuum:

Theorem 1.3.3 Propagation of Light: Index of Refraction

1g 2g	Given	$c = 1/\sqrt{\varepsilon_0\mu_0}$ m/s [L][T^{-1}] $v = [\,1 / \sqrt{(\varepsilon_r\varepsilon_0\,\mu_r\mu_0)}\,]$ m/s [L][T^{-1}] light propagating at a speed [$v \le c$] through a material media[HEC74, pg 38]
Step	Hypothesis	$\frac{c}{v} = \sqrt{\varepsilon_r\mu_r}$ EQ A $\frac{c}{v} \equiv n_x$ EQ B
1	From 1g, 2g as a dimensionless ratio	$\frac{c}{v} = \frac{\sqrt{(\varepsilon_r\varepsilon_0\,\mu_r\mu_0)}}{\sqrt{\varepsilon_0\mu_0}}$
2	From 1, A4.2.21 Distribution Exp, and T4.4.9 Equalities: Any Quantity Divided by One is that Quantity	$\frac{c}{v} = \frac{\sqrt{\varepsilon_r\mu_r}\,\sqrt{\varepsilon_0\mu_0}}{1\;\;\sqrt{\varepsilon_0\mu_0}}$
3	From 2, A4.2.11 Associative Multp, A4.2.13 Inverse Multp and A4.2.12 Identity Multp	$\frac{c}{v} = \sqrt{\varepsilon_r\mu_r}$
4	By Definition	This introduces a new dimensionless parameter called the ***index of refraction*** **[n_x]**
5	From 3 and 4	$\frac{c}{v} \equiv n_x$
∴	From 3 and 5	$\frac{c}{v} = \sqrt{\varepsilon_r\mu_r}$ EQ A $\frac{c}{v} \equiv n_x$ EQ B

Observation 1.3.2 Conditions on Index of Refraction

1) if $v = c$ then $(c/v = n_0) = 1$ in free space.

In a material media the velocity slows down below light speed, hence

2) If $v < c$ then $(c/v = n_x) > 1$.

It would follow than if a faster than light media should exist

3) If $v > c$ then $(c/v = n_t) < 1$.

For $\mu_0 = 4\pi \times 10^{-7}$ Henry m^{-1} this constant has been carefully chosen to simplify calculations so that the relative charge permeability $\mu_{rq} = 1$ for a stationary quantum charge, which does not generate nor has any external magnetic field acting on it, now applying the above parameters to theorem T1.3.3A "Propagation of Light: Index of Refraction":

$$\frac{c}{v} = n_t \equiv \sqrt{\varepsilon_{rq}} = 0.085424524 \qquad \text{Equation 1.3.4}$$

It is now seen that inside a nuclear particle light speeds up traveling about 11.7 times faster than the speed of light. Let us impose an external magnetic field on our charge creating a typical relative charge permeability of

$$\mu_{rq} \approx 400 \text{ [HAL66, pg 940]} \qquad \text{Equation 1.3.5}$$

than from equation E1.3.3 and E1.3.4 the index of refraction has the magnitude approximating the material media of barium dense flint glass [HEC74 pg 189]:

$$n_t = 1.708490474 \qquad \text{Equation 1.3.6}$$

It is now seen that inside a nuclear particle light may be able to travel 11.7 times faster and with magnetic field damping it down to 0.09 times slower than the speed of light. If this explanation is true than this would go a long way to clarify why very fine splitting of the spectral lines may be occurring. The application of a magnetic moment a non-homogeneous space would be created causing the photon to try and travel at different speeds forcing it to split its energy into two new photons with dissimilar frequencies. This model is not ignoring quantum modeling because quantum effects can be shown to generate internal magnetic fields between atoms with a similar order of magnitude. This is simply an alternative model to the more traditional quantum representation explaining the fine splitting of the spectral lines.

This would be analogous to the ***black-ice*** phenomenon. This phenomenon occurs on wet, cold days when patches of ice form on black asphalt roads. A car drives onto this invisible region of road, friction between the tire-or-tires disappears and the motion of the car is realized with its full momentum, it speeds up and starts moving in a chaotic direction that the driver had not intended to go. The results are usually devastating for the car and its occupants. Here the photon that is absorbed within the particle hits the spatially non-homogeneous frictionless region of space, but with the addition of an external magnetic field, friction increases within the space forcing it to travel at different velocities, thereby parsing its energy and splitting the photon into two separate particles.

All of this I had deduced independently only to find the fine structure constant and its equation already existed in Sommerfeld's research. Now this equation can be added to Planck's dimensions, likewise one could find other dimensions using other physical constants such as Boltzmann constant for temperature as Planck had shown in his 1899 paper that Heat (Entropy) is temperature dependent DxUNT.A.2.1.8:

$$H_{plk} = \sqrt{(4\pi\gamma_0\hbar\, c^5 / k_0{}^2)} = 1.416910125\text{x}10^{21}\ ^\circ\text{K} \quad [H] \qquad \text{Equation 1.3.7}$$

Unlike the other dimensional equations the charge formula does not have Newton's gravitational constant not only that, but it suggests with the relative permittivity coefficient that there is a region of space within a nuclear particle that has a different speed limit to propagate light.

Finally, by playing with Plank's numbers a constant of force can be constructed finalizing the base set of his physical relations:

$$F_{plk} = M_{plk}\, L_{plk}\, T_{plk}{}^{-2} \qquad \text{Equation 1.3.8}$$

Applying physical constants

$$F_{plk} = 4\pi\gamma_0 c^4 = 1.210450132\text{x}10^{44}\ \text{N} \qquad [M][L][T^{-2}] \qquad \text{Equation 1.3.9}$$

Section 1.4 Unification of Divergent Photon Spin System

From hypothesis in paper "Photon0788.doc", it is speculated whether an independent physical phenomenon can be found that might support the idea that nuclear particles are independent of distances? The answer must be a resounding YES! This can be proven if the ratio of electrical to gravitational force is taken with respect to each other. Let the electrical and gravitational forces acting between any two relatively placed nuclear particles embedded in a space that is isotropic, homogenous, and stationary within the neighborhood of particles attributes $[q_{n1}, q_{n2}]$ and $[m_{n1}, m_{n2}]$. Also let the forces act along a Euclidean straight line (line of sight) at a distance [d], representing the path of travel for particle exchange, between them then the following is true:

If $\vec{F}_e$ parallel to $\vec{F}_g$ then the magnitudes of $|\vec{F}_e| = F_e$ and $|\vec{F}_g| = F_g$ so can be analyzed without consideration to direction, except along the line of sight between the two particles with an appropriate sign for selected charge.

Theorem 1.4.1 Particle-Particle Force Independent of Distance

1g	Given	$\lvert\pm q_{n1}\rvert = \lvert\pm q_{n2}\rvert = q$ single charge particle system
Step	Hypothesis	$\dfrac{F_e}{F_g} = \dfrac{-M_{plk}^2}{m_{n1} * m_{n2}}\, \varepsilon_{rq}$
1	From LA.3.2 Coulomb's Static Electrical Field of Force and LA.3.1 Newton's Static Gravitation Field of Force; as a ratio	$\dfrac{F_e}{F_g} = \dfrac{(-1)\,(\pm q_{n1}) * (\pm q_{n2}) / 4\pi\varepsilon_0\, d^2}{(+1)\, m_{n1} * m_{n2} / 4\pi\gamma_0\, d^2}$
2	From 1 and T4.5.3 Equalities: Compound Reciprocal Products	$\dfrac{F_e}{F_g} = \dfrac{4\pi\gamma_0\, d^2\, (-1)\,(\pm q_{n1}) * (\pm q_{n2})}{4\pi\varepsilon_0\, d^2\, (+1)\, m_{n1} * m_{n2}}$
3	From 2, A4.2.11 Associative Multp, A4.2.13 Inverse Multp and A4.2.12 Identity Multp	$\dfrac{F_e}{F_g} = \dfrac{\gamma_0\, (-1)\,(\pm q) * (\pm q)}{\varepsilon_0\, (+1)\, m_{n1} * m_{n2}}$
4	From 3, D4.1.4G Rational Numbers, A4.2.12 Identity Multp, A4.2.10 Commutative Multp, D4.1.20 Negative Coefficient and D4.1.17 Exponential Notation	$\dfrac{F_e}{F_g} = \dfrac{(-1)\,(\pm 1)^2 \gamma_0\, q^2}{\varepsilon_0\, m_{n1} * m_{n2}}$
5	From 4, T4.8.2A Integer Exponents: Negative One Squared, A4.2.12 Identity Multp and D4.1.20A Negative Coefficient	$\dfrac{F_e}{F_g} = \dfrac{-1}{m_{n1} * m_{n2}}\, \dfrac{\gamma_0\, q^2}{\varepsilon_0}$
6	From 5, T1.3.2A Summerfield: Evaluation of Physical Charge Constants and A4.2.3 Substitution	$\dfrac{F_e}{F_g} = \dfrac{-1}{m_{n1} * m_{n2}}\, \dfrac{\gamma_0\, 4\pi\hbar c \varepsilon_{rq} \varepsilon_0}{\varepsilon_0}$
7	From 6, A4.2.10 Commutative Multp, A4.2.11 Associative Multp, A4.2.13 Inverse Multp and A4.2.12 Identity Multp	$\dfrac{F_e}{F_g} = \dfrac{-1}{m_{n1} * m_{n2}}\, 4\pi\gamma_0\hbar c \varepsilon_{rq}$

8	From 7, E1.3.1 Planck's Unit of Mass and A4.2.3 Substitution	$\frac{F_e}{F_g} = \frac{-1}{m_{n1} * m_{n2}} M_{plk}{}^2\, \varepsilon_{rq}$
∴	From 8, A4.2.10 Commutative Multp and D4.1.20A Negative Coefficient	$\frac{F_e}{F_g} = \frac{-M_{plk}{}^2}{m_{n1} * m_{n2}} \varepsilon_{rq}$

Theorem 1.4.2 Particle-Antiparticle Force Independent of Distance

1g	Given	$\lvert \pm q_{n1} \rvert = \lvert \mp q_{n2} \rvert = q$ single charge anti-charge particle system
Step	Hypothesis	$\frac{F_e}{F_g} = \frac{M_{plk}{}^2}{m_{n1} * m_{n2}} \varepsilon_{rq}$
1	From DB.4.1 Antiparticle, LA.3.2 Coulomb's Static Electrical Field of Force and LA.3.1 Newton's Static Gravitation Field of Force; as a ratio	$\frac{F_e}{F_g} = \frac{(-1)\, (\pm q_{n1})*(\mp q_{n2}) / 4\pi\varepsilon_0\, d^2}{(+1)\, m_{n1}*m_{n2} / 4\pi\gamma_0\, d^2}$
2	From 1 and T4.5.3 Equalities: Compound Reciprocal Products	$\frac{F_e}{F_g} = \frac{4\pi\gamma_0\, d^2\, (-1)\, (\pm q_{n1})*(\mp q_{n2})}{4\pi\varepsilon_0\, d^2\, (+1)\, m_{n1}*m_{n2}}$
3	From 2, A4.2.11 Associative Multp, A4.2.13 Inverse Multp and A4.2.12 Identity Multp	$\frac{F_e}{F_g} = \frac{\gamma_0\, (-1)\, (\pm q)*(\mp q)}{\varepsilon_0\, (+1)\, m_{n1} * m_{n2}}$
4	From 3, D4.1.4G Rational Numbers, A4.2.12 Identity Multp, A4.2.10 Commutative Multp and D4.1.20 Negative Coefficient	$\frac{F_e}{F_g} = \frac{(-1)\, (\pm 1)(\mp 1) \gamma_0\, q^2}{\varepsilon_0\, m_{n1} * m_{n2}}$
5	From 4, A4.2.12 Identity Multp and D4.1.17 Exponential Notation	$\frac{F_e}{F_g} = \frac{(-1)^2}{m_{n1} * m_{n2}} \frac{\gamma_0\, q^2}{\varepsilon_0}$
6	From 5, T4.8.2A Integer Exponents: Negative One Squared, T1.3.2A Summerfield: Evaluation of Physical Charge Constants and A4.2.3 Substitution	$\frac{F_e}{F_g} = \frac{1}{m_{n1} * m_{n2}} \frac{\gamma_0\, 4\pi\hbar c \varepsilon_{rq} \varepsilon_0}{\varepsilon_0}$
7	From 6, A4.2.10 Commutative Multp, A4.2.11 Associative Multp, A4.2.13 Inverse Multp and A4.2.12 Identity Multp	$\frac{F_e}{F_g} = \frac{1}{m_{n1} * m_{n2}} 4\pi\gamma_0 \hbar c \varepsilon_{rq}$
8	From 7, E1.3.1 Planck's Unit of Mass and A4.2.3 Substitution	$\frac{F_e}{F_g} = \frac{1}{m_{n1} * m_{n2}} M_{plk}{}^2\, \varepsilon_{rq}$
∴	From 8, A4.2.10 Commutative Multp and D4.1.20A Negative Coefficient	$\frac{F_e}{F_g} = \frac{M_{plk}{}^2}{m_{n1} * m_{n2}} \varepsilon_{rq}$

Observation 1.4.1 G-E Fields in Opposition
From theorem T1.4.1 "Particle-Particle Force Independence of Distance" all paired particles of the same sine Electrical and Gravitational forces are in opposition to one another. While theorem T1.4.2 "Particle-Antiparticle Force Independence of Distance" all antiparticles of the differing sine Electrical and Gravitational forces ally to one another.

Observation 1.4.2 G-E Fields Dependence on Mass
While G-E Fields by T1.4.1 "Particle-Particle Force Independence of Distance" and T1.4.2 "Particle-Antiparticle Force Independence of Distance" show independence of distance and charge, but they are directly dependent on mass.

Observation 1.4.3 G-E Field Invariance of Charge
Clearly the forces are independent of distance, but in opposition to one another by the negative sign of the electrical field and any variation of the sign of charge. Combining all resulting signs, for matter or antimatter and taking the absolute magnitude of charge no further contribution is made to direction:

$$\frac{F_e}{F_g} = \frac{(\mp 1)\,\gamma_0\,|q_{n1}|*|q_{n2}|}{\varepsilon_0\,m_{n1}*m_{n2}} \qquad \text{Equation 1.4.4}$$

Millikan and Fletcher showed 1911 in their famous oil drop experiment that quantum charge is independent of position in any stationary frame of reference, that is no matter where the experiment is preformed in our universe it will always be measured having quantum value of [q]. [MILα10, Vol. 19 pg 209] In the 1903 experiment by Trouton and Noble they tried to show that a magnetic field arises on a capacitor plate in relation to the earth's relative motion about the sun. Within experiential limits none was observed. [TROα03, Vol. 72 pg 132] Einstein would later go on to show mathematically that in any moving or non-stationary frame of reference the charge is invariant, see his work on relativistic moving plasma charges. Physical evidence is continuously collected every day with working X-ray systems that use linear accelerators or klystrons, which add or remove RF energy to electron plasma beams. The amount of charge entered by the electron gun is the same as found at the target end. Charge is conserved[1.4.1] while the RF-energy in the guide only modulates the density of the plasma. The actual measure is still the number of individual quantum charged particles in and out [n x q_0]. Unlike mass, which gains weight, do to relativistic effects; energy goes into altering mass only, not the charge. Just as the value of light is, a physical constant so too is the quantum charge, likewise the same for other constants that are considered in this paper. Also, Lorrain and Corson [LOR62 pg 228] give an alternate example with charge mass ratio under relativistic effects showing charge as invariant.

Now for only dual, charged, nuclear particle, systems they can be combined into the following equality:

$$|q_{n1}| = |q_{n2}| = q \qquad \text{Equation 1.4.5}$$

So, equation E1.4.5 is independent of distance. Without distance existing between the two nuclear particles, every point in the universe appears to map onto a single point and communicate with a Newtonian sense of instantaneous time and only at the very moment the photon completes the special circuit, the Law B.5.V "Concurrency". Yet, the signs of inertia and charge only have an effect on directions of attracting or repelling.

[1.4.1]Note: This is not quite true, phantom electrons are added to the beam; causing dark radiation, due to the resonant RF field striping away free electrons off the cavity wall and pressuring them onward.

Let the paired particles be two electrons, positrons or electron-positron duet than
$m_e = m_{n1} = m_{n2} = 9.109558(54)x10^{-31}$ kg the electron rest mass

$$M_{plk}^2\ \varepsilon_{rq} = 4\pi\gamma_0\hbar c\varepsilon_{rq} = 3.457277562x10^{-40}\ kg^2 \qquad \text{Equation 1.4.10}$$

$$\frac{M_{plk}^2\ \varepsilon_{rq}}{m_e^2} = 4.1661949243x10^{20} \qquad \text{Equation 1.4.11}$$

A very large number magnifying the gravitational force, which explains why the electrical field is so much stronger compared to gravity.

Now the Higgs particle / electron mass ~ 112,000 MeV / 0.511 MeV = 219,000 is the biggest observed particle.

$$\frac{M_{plk}^2\ \varepsilon_{rq}}{m_{hg}^2} = 8.6866x10^{14} \qquad \text{Equation 1.4.12}$$

Clearly as the nuclear mass particle increases, the electrical force decreases or weakens and the gravitational field strengthens while the electrical effects on a nuclear level contribute only as a constant $[\varepsilon_{rq}]$ in the background.

In summary as most physical problems handle inertial and electrical effects independently; clearly, they are interrelated suggesting the possibility of a unifying model.

Section 1.5 *The Monocentric Particle Model*

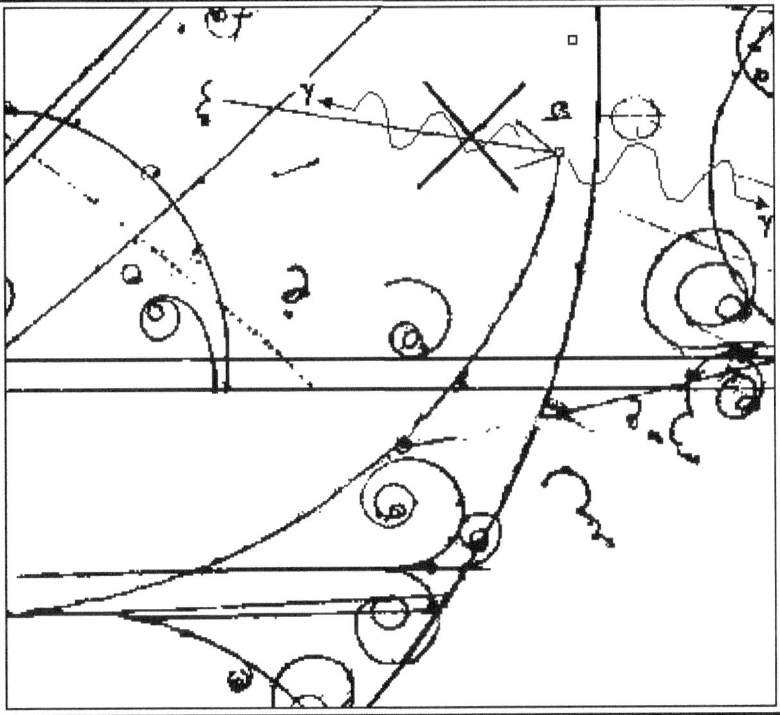

Photograph from CERN: Above opposite process electron-positron annihilation pointed to by the arrow and the two resulting gamma rays drawn in because photons will not leave a track in a bubble chamber.

$$e^{-} + e^{+} \rightarrow \gamma + \gamma$$

Fig. A Annihilation

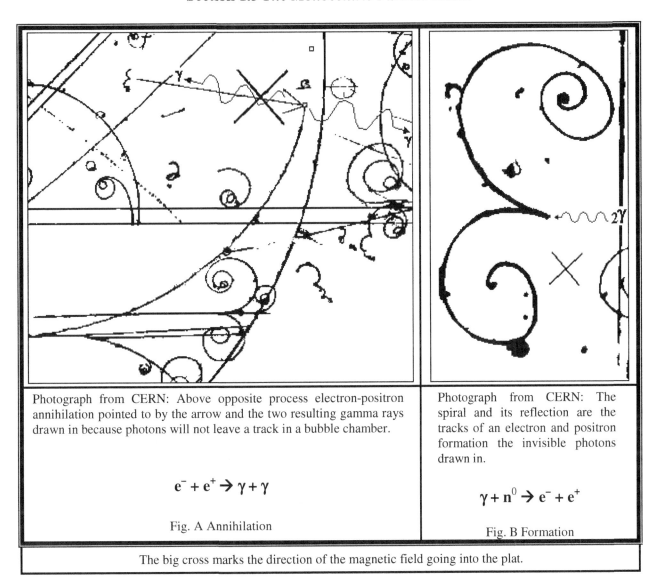

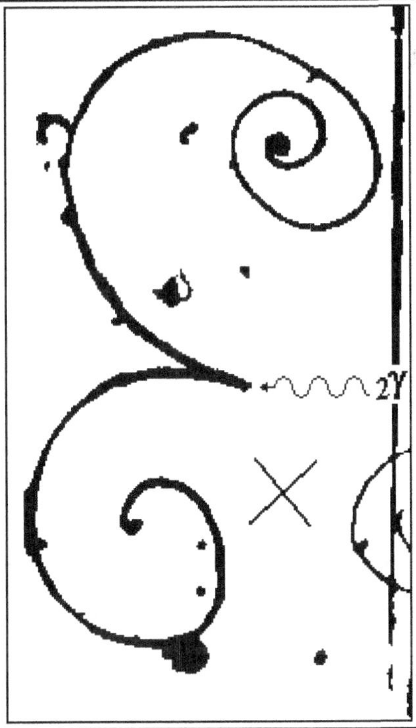

Photograph from CERN: The spiral and its reflection are the tracks of an electron and positron formation the invisible photons drawn in.

$$\gamma + n^{0} \rightarrow e^{-} + e^{+}$$

Fig. B Formation

The big cross marks the direction of the magnetic field going into the plat.

Figure 1.5.1 Electron-Positron particle Annihilation and Formation

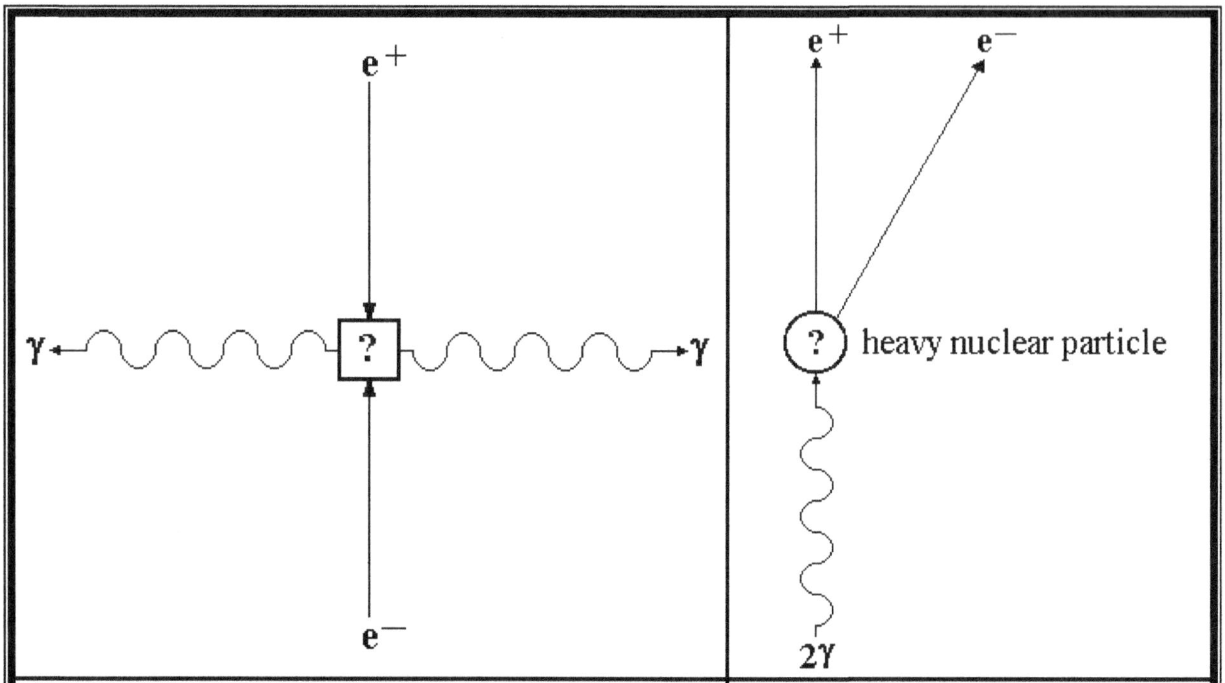

Idealized representation of (e^-, e^+) annihilation without the magnetic field.
Fig. A Annihilation

Idealized representation of (e^-, e^+) formation without the magnetic field.
Fig. B Formation

Figure 1.5.2 Electron-positron Particle Annihilation and Formation Diagrams

From the particle experiment within the stationary region of transformation energy for the electron and positron are brought into the system as $2mc^2$ [1.4.1] and leave with $2h\nu$ for pair annihilation, definition-3A. While the reverse is true, $2h\nu$ enters and $2mc^2$ leaves the system. Another physical phenomena is also seen, the particle momentum 2mc enters and leaves with the photon momentum $2h/\lambda$ for particle annihilation, definition-3B. The reverse follows photon momentum $2h/\lambda$ enters and $2mc^2$ leaves the system marked with the bound [**?**] question mark. Here are the first elementary nuclear particle principles seen for a process of creation and annihilation they are called ***conservation of energy and momentum for pair formation***. They are stated here in two physical laws:

Law 1.5.1 Pair Formation Conservation Energy
As particles are created or annihilated, energy is conserved $mc^2 = h\nu$.

Law 1.5.2 Pair Formation Conservation of Momentum
As particles are created or annihilated, momentum is conserved $mc = h/\lambda$.

Clearly, the transformation process for annihilation and formation are symmetrical. Evidently, within the region of transformation [**?**], there must be a common mechanism. The machine converts electrons and positrons to high-energy gamma rays and vice versa. What is happening within that black box–[**?**] ? Whatever that mechanism is, it would have to be able to assemble or disassemble itself. Looking at equation E1.3.5 it is comprised of four physical constants all having to do with some aspect of a photon. There are no inertial or mass constants. Whatever creates charge is totally derived from a photon. This leads to considering a model based on a monocentric particle system using a photon to construct it from. The problem with this idea is that mass particles can either be at rest in a stationary frame of reference or only travel at sub-light speeds. Yet a light photon must travel at the velocity-c (see Appendix B Law 2.II "Invariance of the speed of light") no matter what frame it is in so if a particle is comprised of a photon how can it exist at sub-light speed or even in a stationary frame?

Theorem 1.5.1 Monocentric Particle: Properties of a Photon in Spin

1g 2g	Given	$\lambda = 2\pi r_m$ $c = c$		
Step	Hypothesis	$m_0c = \hbar/r_m$ EQ A photon spin momentum	$m_0c^2 = \hbar c / r_m$ EQ B photon spin energy	$r_m = \hbar/cm_0$ EQ C spin radius
1	From 1g, L1.5.2 Pair Formation Conservation of Momentum, T4.4.9 Equalities: Any Quantity Divided by One is that Quantity, T4.4.15B Equalities: Product by Division and CxPPC.A.2.3.4 Planck's Normalized Constant	$m_0c \;=\; \hbar/r_m$ photon spin momentum		
2	From 1, CxPPC.A.2.3.5 Speed of Light, T4.4.2 Equalities: Uniqueness of Multiplication and D4.1.17 Exponential Notation	$m_0c^2 \;=\; \hbar c / r_m$ photon spin energy		
3	From 1, 2g, T4.4.13 Equalities: Cancellation by Division, A4.2.13 Inverse Multp and A4.2.12 Identity Multp	$m_0 \;=\; \hbar/cr_m$ mass in terms of spin radius		
4	From 3, T4.4.5A Equalities: Reversal of Right Cancellation by Multiplication and T4.4.6B Equalities: Left Cancellation by Multiplication	$r_m \;=\; \hbar/cm_0$ spin radius		
∴	From 1, 2 and 4	$m_0c = \hbar/r_m$ EQ A photon spin momentum	$m_0c^2 = \hbar c / r_m$ EQ B photon spin energy	$r_m = \hbar/cm_0$ EQ C spin radius

Substituting actual constants CxPPC.A.2.3.4 "Normalized Planck's constant", CxPPC.A.2.3.5 "Speed of Light" and CxPPC.A.2.3.11 "Electron Rest Mass" into equation T1.5.1C "Monocentric Particle: Properties of a Photon in Spin" an electron or positron, the spin radius would have a radius of

$$r_m = \frac{1.0545919(80)\text{x}10^{-34} \text{ J*sec}}{2.9979250(10)\text{x}10^{8} \text{ m sec}^{-1} * 9.109558(54)\text{x}10^{-31} \text{ kg}}$$

$$r_m = 3.861591839\text{x}10^{-13} \text{ m} \qquad \text{Equation 1.5.1}$$

For a Higgs particle 219,000 bigger than an electron

$$r_m = 1.763283945\text{x}10^{-18} \text{ m} \qquad \text{Equation 1.5.2}$$

From their radius of spin the greater the mass the smaller the spin radius as defined by equation 1.5.1 and 1.5.2.

Section 1.6 *Inside-Outside the Spinning Particle*

So far, this model has dealt with what is happing on the inside, but what of external forces acting on the particle? If the model is true than a transmitter particle within its neighborhood, let's say a distance [d] away, would cause it to generate a predetermine ability to move in the appropriate direction[1.6.1]. Since the only energy available to it is the energy stored within than by conservation of energy, the energies must be equivalent. So, the electrical and gravitational energy can be equated to an inertial potential energy of the particle.

Theorem 1.6.1 Monocentric Particle Outside: Newtonian

1g 2g	Given	$M \equiv m_g$ dual nuclear particle system $m_p\, c^2 \equiv E_g$ conservation of rest energy to the energy within the field
Step	Hypothesis	$dc^2 = \dfrac{m_g}{4\pi\gamma_0}$
1	From 2g; conservation of rest energy, to the gravitational field	$E_p = E_g$
2	From 1, TA.3.1 Gravitational Potential Energy and A4.2.3 Substitution	$m_p c^2 = \dfrac{1}{4\pi\gamma_0} \dfrac{M\, m_g}{d}$
3	From 1g, 2, A4.2.3 Substitution and D4.1.17 Exponential Notation	$m_g c^2 = \dfrac{1}{4\pi\gamma_0} \dfrac{m_g^{\,2}}{d}$
4	From 3, D4.1.17 Exponential Notation and T4.4.3 Equalities: Cancellation by Multiplication	$c^2 = \dfrac{1}{4\pi\gamma_0} \dfrac{m_g}{d}$
∴	From 4, A4.2.12 Identity Multp and T4.4.5A Equalities: Reversal of Right Cancellation by Multiplication	$dc^2 = \dfrac{m_g}{4\pi\gamma_0}$

Theorem 1.6.2 Monocentric Particle Outside: Coulomb

1g	Given	$Q \equiv q$ dual nuclear particle system
2g		$q_p \equiv q$ the only energy is the energy within the particle itself
3g		$E_p \equiv m_p c^2$ the energy within the particle is its own rest energy
Step	Hypothesis	$m_e = \frac{1}{4\pi\varepsilon_0} \frac{2hc\varepsilon_q}{dc^2}$
1	From the energy of the static electrical field is its own rest energy	$E_p = E_Q$
2	From 3g, 1, TA.3.2 Electrical Potential Energy and A4.2.3 Substitution	$m_p c^2 = \frac{1}{4\pi\varepsilon_0} \frac{Q\,q_p}{d}$
3	From 1g, 2g, 2, A4.2.3 Substitution and D4.1.17 Exponential Notation	$m_e c^2 = \frac{1}{4\pi\varepsilon_0} \frac{q^2}{d}$
4	From 3, T1.3.2A Summerfield: Evaluation of Physical Charge Constants, T4.10.4 Radicals: Identity Radical Raised to a Power and A4.2.3 Substitution	$m_e c^2 = \frac{1}{4\pi\varepsilon_0} \frac{4\pi\hbar c\varepsilon_q}{d}$
∴	From 4, CxPPC.A.2.3.4 Normalized Planck's constant, A4.2.13 Inverse Multp, A4.2.12 Identity Multp, D4.1.19 Primitive Definition for Rational Arithmetic and T4.4.5A Equalities: Reversal of Right Cancellation by Multiplication	$m_e = \frac{1}{4\pi\varepsilon_0} \frac{2hc\varepsilon_q}{d\,c^2}$

Observation 1.6.1 Self Attraction of Rest Mass to Itself

Now m_e is the electrical mass not to be confused with the spin-rest mass m_0 and a relative parameter. Since the transmitter and receiver particle are mass particles, they will attract one another with the same electrical and gravitational energy.

Newton is very clear about his universal gravitational law the transmitter mass attracts the receiver's platform, but the observer, also attracts the receiver particle so the twin pair formation nuclear particle having the same mass can be thought of as attracting itself, in this special case, with the same amount of energy that it posses.

Observation 1.6.2 Receiver Mass is its Own Rest Mass

The gravitational energy of the particle must be only the energy, found in the particle itself, an individual twin particle would appear as if being generated in mirrored pair formation. The transmitter particle can now be set to the particle mass itself, ($M \equiv m_o$).

Theorem 1.6.3 Monocentric Particle Outside: G-E Field as Paired Plank Mass

Step	Hypothesis	$m_e\, m_g = M_{plk}{}^2 \varepsilon_{rq}$ EQ A	$m_e m_g = 4\pi\gamma_0 \hbar c\ \varepsilon_{rq}$ EQ B
1	From T1.6.2 Monocentric Particle Outside: Coulomb	$m_e = \dfrac{1}{4\pi\varepsilon_0} \dfrac{2hc\varepsilon_q}{dc^2}$	
2	From 1, T1.6.1 Monocentric Particle Outside: Newtonian and A4.2.3 Substitution	$m_e = \dfrac{1}{4\pi\varepsilon_0} \dfrac{2hc\varepsilon_q}{m_g / (4\pi\gamma_0)}$	
3	From 2, T4.5.3 Equalities: Compound Reciprocal Products and A4.2.10 Commutative Multp	$m_e = \dfrac{4\pi\gamma_0}{4\pi\varepsilon_0} \dfrac{2hc\varepsilon_q}{m_g}$	
4	From 3, A4.2.13 Inverse Multp, A4.2.12 Identity Multp, T1.3.2E Summerfield: Evaluation of Physical Charge Constants, A4.2.3 Substitution and T4.4.5 Equalities: Reversal of Right Cancellation by Multiplication	$m_e m_g = \dfrac{\gamma_0}{\varepsilon_0}\ 2hc\varepsilon_{rq}\varepsilon_0$	
5	From 4, A4.2.10 Commutative Multp, A4.2.13 Inverse Multp and A4.2.12 Identity Multp	$m_e m_g = 2\gamma_0 hc\ \varepsilon_{rq}$	
6	From 5, A4.2.12 Identity Multp, A4.2.13 Inverse Multp, D4.1.19 Primitive Definition for Rational Arithmetic, A4.2.10 Commutative Multp, A4.2.11 Associative Multp, CxPPC.A.2.3.4, A4.2.3 Substitution	$m_e m_g = 4\pi\gamma_0 \hbar c\ \varepsilon_{rq}$	
∴	From 5, E1.3.1 Plank's Constant of Mass, T4.10.4 Radicals: Identity Radical Raised to a Power and A4.2.3 Substitution	$m_e m_g = M_{plk}{}^2 \varepsilon_{rq}$ EQ A	$m_e m_g = 4\pi\gamma_0 \hbar c\ \varepsilon_{rq}$ EQ B

[1.6.1]Note: At this point in the theory's development I do not say there is motion, because of force, but only the potential to move in a particular direction. In order to have motion an exchange of a particle must take place, which does result in what we perceive as a force of action.

Theorem 1.6.4 Monocentric Particle Outside: Gaussian Curvature of Dual Spins

Step	Hypothesis	$(1/r_e)\,(1/r_g) = K_{eg}$ EQ A $\quad \kappa_e\,\kappa_g = K_{eg}$ EQ B $\quad K_{eg} = \varepsilon_{rq} / L_{plk}^2$ EQ C		
1	From T1.6.3A Monocentric Particle: G-E Field as Paired Plank Mass	$m_e\,m_g$	=	$M_{plk}^2\varepsilon_{rq}$
2	From 1, T4.4.23 Equalities: Multiplication of a Constant and D4.1.17 Exponential Notation	$(m_e\,c)\,(m_g\,c)$	=	$c^2M_{plk}^2\varepsilon_{rq}$
3	From 2, T1.5.1A Monocentric Particle: Properties of a Photon in Spin and A4.2.3 Substitution	$(\hbar/r_e)\,(\hbar/r_g)$	=	$c^2M_{plk}^2\varepsilon_{rq}$
4	From 3, A4.2.10 Commutative Multp, A4.2.11 Associative Multp, D4.1.17 Exponential Notation and T4.4.6B Equalities: Left Cancellation by Multiplication	$(1/r_e)\,(1/r_g)$	=	$c^2\,(M_{plk}^2\varepsilon_{rq}) / \hbar^2$
5	From 4, E1.6.3B Monocentric Particle Outside: G-E Field as Paired Plank Mass and A4.2.3 Substitution	$(1/r_e)\,(1/r_g)$	=	$c^2(4\pi\gamma_0\hbar c)\,\varepsilon_{rq} / \hbar^2$
6	From 5, D4.1.17 Exponential Notation, A4.2.10 Commutative Multp, A4.2.11 Associative Multp, A4.2.13 Inverse Multp, A4.2.12 Identity Multp	$(1/r_e)\,(1/r_g)$	=	$(4\pi\gamma_0\,c^3)\,\varepsilon_{rq} / \hbar$
7	From 6, A4.2.10 Commutative Multp, A4.2.11 Associative Multp and D4.1.17 Exponential Notation	$(1/r_e)\,(1/r_g)$	=	$\varepsilon_{rq}\,4\pi\gamma_0\,c^3 / \hbar$
8	From 7, E1.3.2 Plank's Length, T4.10.4 Radicals: Identity Radical Raised to a Power, T4.5.9 Equalities: Ratio and Reciprocal of a Ratio and A4.2.3 Substitution	$(1/r_e)\,(1/r_g)$	=	$\varepsilon_{rq} / L_{plk}^2$ The principle curvatures of spin
9	From 8, D12.12.7A Curvature to a Spatial Curve-C and A4.2.3 Substitution	$\kappa_e\,\kappa_g$	=	$\varepsilon_{rq} / L_{plk}^2$
10	From 9, T17.12.8B Minor Principal Curvatures: Relation to Gaussian Principal Curvatures and A4.2.3 Substitution; for n=2 minor curvature system	K_{eg}	=	$\varepsilon_{rq} / L_{plk}^2$
∴	From 8, 9 and 10	$(1/r_e)\,(1/r_g) = K_{eg}$ EQ A $\quad \kappa_e\,\kappa_g = K_{eg}$ EQ B $\quad K_{eg} = \varepsilon_{rq} / L_{plk}^2$ EQ C		

Observation 1.6.3 G-E Field Unification with Particle Gaussian Curvature

Here is the unification of electrical and gravitational fields with a Gaussian type curvature for the particle.

Theorem 1.6.5 Monocentric Particle Outside: Ordering G-E Spin Radii

1g	Given E1.3.2 Plank's Length	L_{plk}	=	$1.616140686\text{x}10^{-35}$ m			
2g	Relative Permittivity	ε_{rq}	=	$7.297351\ \text{x}10^{-3}$			
3g	E1.5.1 Inertial mass radius	r_g	=	$3.861591839\text{x}10^{-13}$ m			
Step	Hypothesis	r_e	<	r_g EQ A	m_g	<	m_e EQ B
1	From 1g, T4.8.6 Integer Exponents: Uniqueness of Exponents, D4.1.17 Exponential Notation and D4.1.19 Primitive Definition for Rational Arithmetic	L_{plk}^2	=	$2.611910717\text{x}10^{-70}\ \text{m}^2$			
2	From T1.6.4(A,C) Monocentric Particle: Gaussian Curvature of Dual Spins and T4.4.7B Equalities: Reversal of Left Cancellation by Multiplication	r_e	=	$L_{plk}^2/(\varepsilon_{rq}\ r_g)$			
3	From 1g, 2g, 3g, 2 and A4.2.3 Substitution	r_e	=	$(2.611910717\text{x}10^{-70}\ \text{m}^2)$ / $[(7.297351\ \text{x}10^{-3})\ (3.861591839\text{x}10^{-13}\ \text{m})]$			
4	From 3 and D4.1.19 Primitive Definition for Rational Arithmetic	r_e	=	$9.268868553\ \text{x}10^{-56}$ m			
5	From 3g, 4 and D4.1.7 Greater Than Inequality [>]	r_e	<	r_g			
6	From 5 and T4.7.33 Inequalities: Cross Multiplication with Positive Reciprocal Product	$1/r_g$	<	$1/r_e$			
7	From 6 and T4.7.7 Inequalities: Multiplication by Positive Number	$\hbar\ /cr_g$	<	$\hbar\ /c\ r_e$			
8	From 7, T1.5.1;3 Monocentric Particle: Properties of a Photon in Spin and A4.2.3 Substitution	m_g	<	m_e			
∴	From 5 and 8	r_e	<	r_g EQ A	m_e	>	m_g EQ B

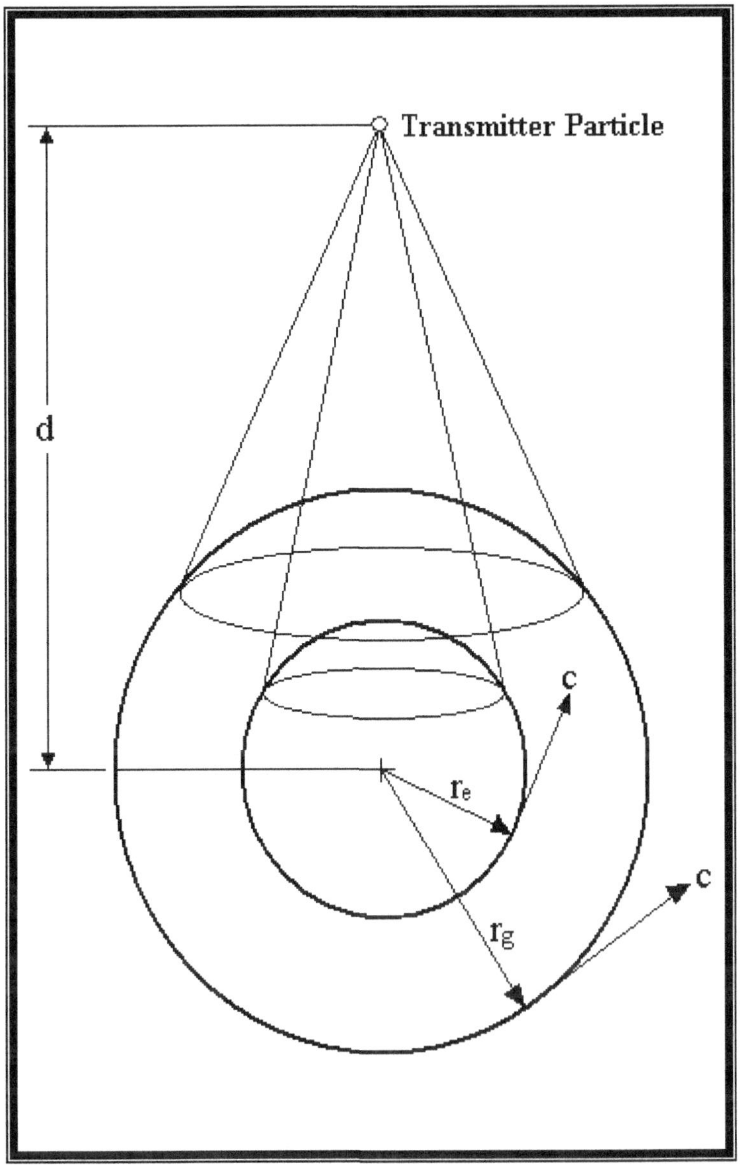

This mechanism is a 2-dimensional Gaussian Curvature for the internal space of the spinning particle. Two different fields of force act externally causing the spinning photon to do something very sophisticated and split into two concentric spinning balls. These two radii are reminiscent of the event horizon for a black hole with a Schwarzschild radius, but unlike a black hole it has two radii. They are sort of two-event horizons for a micro black hole.

In 1974 [HAW74], Stephen Hawking made certain calculations and conjectured that during the formation of the universe micro black holes were created having just barely enough mass to form, but do to emission of subatomic particles at their event horizon they would loss mass-energy, bleeding away and eventually pop like a gamma ray soap bubble. Certainly the matter we are made of came from event-one maybe Hawking came closer than he could have imagined to the answer although as far as I know no one has ever considered a black hole with two event horizons.

Figure 1.6.1 Receiver-Particle with Two Event Horizons

This figure shows that the transmitter particle acts over the entire surface of the concentric spheres, gravitational and electrical fields of the photon as it spins with corresponding equivalent energy. It also shows that at any point on their surface the velocity of spin is tangential and measured at the velocity-c.

Physicist **Dr. Brian Greene** (born February 9, 1963) is a string theorist, author of *The Elegant Universe*. As of 2003, a professor at Columbia University has suggested that the electron may be a micro black hole. Small black holes would look like an elementary particle, because it would be completely defined by their mass, charge and spin. The problem with this conjecture is that since the electron has not been observed to evaporate for such a small black hole, an additional explanation for the electron's stability is needed.

Section 1.7 *Flux Across the Event Surface*

Flux Inside the Particle Event Surface

The net flux passing through the ball of spin would be;

Theorem 1.7.1 Monocentric Particle Inside: Quantizing Photon Spin

1g	Given	$d \equiv r$ potential of energy acting on itself in spin
2g		$F_s \equiv -E_s / r$ internal centripetal spin, of force
3g		$E_s \equiv \hbar c / r$ Plank's photon spin energy
4g		$dS \equiv r^2 \sin\phi\, d\theta d\phi$ Area element on the surface of a sphere
Step	Hypothesis	$\Phi_s = 2hc$ $[M][L^3][T^{-2}]$
1	From LxVIE.9.7.1.2 Divergence Theorem of Gauss	$\Phi_s = \oint_s \mathbf{F}_s \bullet d\mathbf{S}$
2	From 1, 1g, 2g, 3g and A4.2.3 Substitution	$F_s = -(\hbar c / r) / r$
3	From 2, T4.5.1 Equalities: Reciprocal Products and D4.1.17 Exponential Notation	$F_s = -\hbar c / r^2$
4	From 4g, 1, 3 and A4.2.3 Substitution; integrated over the entire spin surface of the particle to get the net flux	$\Phi_s = \int_0^{2\pi}\int_0^{\pi} (-\hbar c / r^2)\, r^2 \sin\varphi\, d\theta d\phi$
5	From 4, D4.1.20A Negative Coefficient, A4.2.10 Commutative Multp, Lx9.4.2.1 Basic integration of a constant, Lx9.4.3.1 Distribution of Constant Out Of Integrand and A4.2.11 Associative Multp	$\Phi_s = -2\pi\hbar c \int_0^{\pi} \sin\phi\, d\phi$
6	From 5 and Lx9.4.2.7 Basic integration of the sine	$\Phi_s = -2\pi\hbar c\, (-\cos\phi \Big\vert_0^{\pi}$
7	From 6, Dx9.4.2.4 The Limit of an Integral, DxE.1.6.13 cosine of PI and DxE.1.6.1 cosine of 0	$\Phi_s = -2\pi\hbar c\, (-1 - 1)$
8	From 7, D4.1.19 Primitive Definition for Rational Arithmetic, D4.1.20A Negative Coefficient, A4.2.10 Commutative Multp, T4.8.2B Integer Exponents: Negative One Squared, A4.2.12 Identity Multp	$\Phi_s = 4\pi\hbar c$
∴	From 8, CxPPC.A.2.3.4 Normalized Planck's constant and D4.1.19 Primitive Definition for Rational Arithmetic	$\Phi_s = 2hc$ $[M][L^3][T^{-2}]$

Observation 1.7.1 Monocentric Particle Inside: Quantized Flux

The inside flux reveals a quantity that is quantized and contains the first two physical constants that are found in the Planck's dimensional charge.

Flux Outside the Particle Event Surface

Observe that integration is different than in classical Newtonian analysis. Newtonian field analysis and for that matter Einsteinian tensor analysis are based on space being made up of a very large number of infinitesimal points in a continuum so that calculus limits can approximate the model fairly well and develop models of this complex phenomenon. Here the premise is the use of individual nuclear particles so the mathematical space is quantized. In the world of Newton and Einstein, the universe is a continuum for the macro and/or cosmic having a large number of small particles that approximate a continuum. It follows than that, Einstein's Theory of General Relativity breaks down and does not work on a micro scale, but since Special Relativity was developed based on relative singular particle motion it still applies.

Also the phrase "acting on it" implies that any DB.5.2 "Relative Binary Reference Frames" have the added condition, that the binary particles are common, that is coming from the same pair production they would have identical mass and spin radius so it would be as if the particle were acting on itself, center-to-center.

Theorem 1.7.2 Monocentric Particle Outside: Self Induced Gravitational Flux

1g	Given: Law A.3.1 Newton's Static Gravitation Field	$F_g \equiv -m_g M / 4\pi\gamma_0 d^2$	Induced force by centripetal spin
2g		$d \equiv r_g$	Self induced potential spin energy
3g		$M \equiv m_g$	Dual nuclear particle system
4g		$dS \equiv r^2 \sin\phi \, d\phi \, d\theta$	Area element on surface sphere
Step	Hypothesis	$\Phi_g = m_g^2 / 4\pi\gamma_0$	$[M][L^3][T^{-2}]$
1	From LxVIE.9.7.1.2 Divergence Theorem of Gauss	$\Phi_g = \oint_S F_g \bullet dS$	
2	From 1g, 2g, 3g, A4.2.3 Substitution and D4.1.17 Exponential Notation	$F_g = \dfrac{-1}{4\pi\gamma_0} \dfrac{m_g^2}{r_g^2}$	
3	From 4g, 1, 2 and A4.2.3 Substitution; integrated over the entire spin surface of the particle to get the net flux	$\Phi_g = \int_0^{2\pi}\int_0^{\pi} (-m_g^2 / 4\pi\gamma_0 r_g^2) r_g^2 \sin\phi \, d\phi \, d\theta$	
4	From 3, D4.1.20A Negative Coefficient, A4.2.10 Commutative Multp, Lx9.4.2.1 Basic integration of a constant over variable-θ, Lx9.4.3.1 Distribution of Constant Out Of Integrand and A4.2.11 Associative Multp	$\Phi_g = -2\pi(m_g^2 / 4\pi\gamma_0) \int_0^{\pi} \sin\phi \, d\phi$	
5	From 4 and Lx9.4.2.7 Basic integration of the sine	$\Phi_g = -2\pi(m_g^2 / 4\pi\gamma_0) (-\cos\phi \Big|_0^{\pi}$	
6	From 5, Dx9.4.2.4 The Limit of an Integral, DxE.1.6.13 cosine of PI and DxE.1.6.1 cosine of 0	$\Phi_g = -2\pi(m_g^2 / 4\pi\gamma_0) (-1 -1)$	

7	From 6, D4.1.19 Primitive Definition for Rational Arithmetic, D4.1.20A Negative Coefficient, A4.2.10 Commutative Multp, T4.8.2B Integer Exponents: Negative One Squared, A4.2.12 Identity Multp	$\Phi_g = 4\pi(m_g^2 / 4\pi\gamma_0)$
∴	From 7 and D4.1.19 Primitive Definition for Rational Arithmetic	$\Phi_g = m_g^2 / \gamma_0$ $[M][L^3][T^{-2}]$

Observation 1.7.2 Quantum of Spinning Flux
T1.7.2 "Monocentric Particle Inside: Electrical Photon Spin" can be interpreted as a quantum of flux inside the spinning particle or directly proportional to the square of the inertial mass.

Theorem 1.7.3 Monocentric Particle Outside: Self Induced Electrical Flux

1g	Given: Law A.3.2 Coulomb's Static Electrical Field	$F_e \equiv -Q\, q_e / 4\pi\varepsilon_0\, d^2$ Induced force by centripetal spin
2g		$d \equiv r_e$ Self induced potential spin energy
3g		$Q \equiv q_e$ Dual nuclear particle system
4g		$dS \equiv r^2 \sin\phi\, d\phi\, d\theta$ Area element on surface sphere
Step	Hypothesis	$\Phi_e = \pm 4\pi\hbar c\varepsilon_{rg}$ $[M][L^3][T^{-2}]$
1	From LxVIE.9.7.1.2 Divergence Theorem of Gauss	$\Phi_e = \oint_s \mathbf{F}_e \bullet d\mathbf{S}$
2	From 1g, 2g, 3g, A4.2.3 Substitution and D4.1.17 Exponential Notation	$F_e = \dfrac{\mp 1}{4\pi\varepsilon_0} \dfrac{Q\, q_e}{r^2}$
3	From 4g, 1, 2 and A4.2.3 Substitution; integrated over the entire spin surface of the particle to get the net flux	$\Phi_e = \int_0^{2\pi}\int_0^{\pi} (\mp q_e^2 / 4\pi\varepsilon_0\, r_e^2)\, r_e^2 \sin\phi\, d\phi\, d\theta$
4	From 3, D4.1.20A Negative Coefficient, A4.2.10 Commutative Multp, Lx9.4.2.1 Basic integration of a constant over variable-θ, Lx9.4.3.1 Distribution of Constant Out Of Integrand and A4.2.11 Associative Multp	$\Phi_e = \mp 2\pi(q_e^2 / 4\pi\varepsilon_0) \int_0^{\pi} \sin\phi\, d\phi$
5	From 4 and Lx9.4.2.7 Basic integration of the sine	$\Phi_e = \mp 2\pi(q_e^2 / 4\pi\varepsilon_0) (-\cos\phi \Big|_0^{\pi}$
6	From 5, Dx9.4.2.4 The Limit of an Integral, DxE.1.6.13 cosine of PI and DxE.1.6.1 cosine of 0	$\Phi_e = \mp 2\pi(q_e^2 / 4\pi\varepsilon_0) (-1 -1)$

7	From 6, D4.1.19 Primitive Definition for Rational Arithmetic, D4.1.20A Negative Coefficient, A4.2.10 Commutative Multp, T4.8.2B Integer Exponents: Negative One Squared and A4.2.12 Identity Multp	$\Phi_e = \pm 4\pi(q_e^2 / 4\pi\varepsilon_0)$
8	From 7, and D4.1.19 Primitive Definition for Rational Arithmetic	$\Phi_e = \pm q_e^2 / \varepsilon_0$
9	From 8, T1.3.2A Summerfield: Evaluation of Physical Charge Constants, T4.8.6 Integer Exponents: Uniqueness of Exponents and T4.10.4 Radicals: Identity Radical Raised to a Power	$\Phi_e = \pm 4\pi\hbar c\varepsilon_q / \varepsilon_0$
10	From 9, T1.3.2;2g Summerfield: Evaluation of Physical Charge Constants and A4.2.3 Substitution	$\Phi_e = \pm 4\pi\hbar c\varepsilon_{rq}\varepsilon_0 / \varepsilon_0$
∴	From 10, A4.2.11 Associative Multp, A4.2.13 Inverse Multp and A4.2.12 Identity Multp	$\Phi_e = \pm 4\pi\hbar c\varepsilon_{rq}$ $[M][L^3][T^{-2}]$

Observation 1.7.3 Finial Definition of Sign for Electrical Charge
Finally it can be seen that the sign of charge comes from Coulomb's equation of static charge and has a direct meaning to the electric flux of a photon and its rotation of spin with the relative direction of travel resulting between the two charges.

Theorem 1.7.4 Monocentric Particle Outside: G-E Flux Ratio

Step	Hypothesis	$\dfrac{\Phi_e}{\Phi_g} = \dfrac{\pm M_{plk}^2\,\varepsilon_{rq}}{m_g^2}$
1	From T1.7.3 Monocentric Particle Outside: Self Induced Electrical Flux, and T1.7.2 Monocentric Particle Outside: Self Induced Gravitational Flux	$\dfrac{\Phi_e}{\Phi_g} = \dfrac{\pm 4\pi\hbar c\varepsilon_{rq}}{m_g^2 / \gamma_0}$
2	From 1 and T4.5.9 Equalities: Ratio and Reciprocal of a Ratio	$\dfrac{\Phi_e}{\Phi_g} = \dfrac{\pm 4\pi\gamma_0\hbar c\varepsilon_{rq}}{m_g^2}$
∴	From 2, A4.2.11 Associative Multp, E1.3.1 Planck's Mass, T4.8.6 Integer Exponents: Uniqueness of Exponents and T4.10.4 Radicals: Identity Radical Raised to a Power	$\dfrac{\Phi_e}{\Phi_g} = \dfrac{\pm M_{plk}^2\,\varepsilon_{rq}}{m_g^2}$

Section 1.8 Rotation Inertial Density & Mass

Definition 1.8.1 Geometric Mean for Rotation Inertial Density

Let the product be equivalent to the square of the geometric mean for rotation inertial density

$\bar{\eta}^2 \quad \equiv \quad \eta_e \eta_g$ geometric mean for rotation inertial density Equation A

Theorem 1.8.1 Monocentric Particle: Rotational Density and Gradient of Mass

Step	Hypothesis	$\eta_r = \dfrac{\hbar}{c\, r^2}$ EQ A	$\eta_r = \dfrac{-\partial m}{\partial r}$ EQ B
1	From LxVIE.9.7.1.1 Divergence Theorem of Gauss Scalar Multiplication	$\oint_S m\, d\vec{S} = \int_V \nabla m\, dV$	
2	From 1 and the dimensionality of LxRFXA.1.1.7 Rotational inertia a vector can be defined	$\vec{\eta} \equiv -\nabla m$	$[M][R^{-1}]$
3	From T1.5.1A Monocentric Particle: Properties of a Photon in Spin and T4.4.6B Equalities: Left Cancellation by Multiplication	$m = \hbar / r_m c$	
4	Dx9.6.1.1B Covariant Differential Operator, DEL	$\nabla \equiv \sum_i \vec{b}^i\, \nabla_i$	
5	From 4 and the gradient in spherical coordinates	$\nabla \equiv \mathbf{r}\, \nabla_r + (\boldsymbol{\theta}\ 1/r)\nabla_\theta + (\boldsymbol{\phi}\ 1/ r \sin\theta)\nabla_\phi$	
6	From 3, 5 there is no change in direction for $(\boldsymbol{\theta}, \boldsymbol{\phi})$	$\nabla_\theta m = \nabla_\phi m = 0$	
7	From 5, 6 and T7.11.2 Vector Addition: Identity	$\nabla m = \mathbf{r}\, \nabla_r m$	
8	From 2 and 7	$\eta_r = -\nabla_r m$	
9	From 3, 8 and A4.2.3 Substitution	$\eta_r = -\dfrac{\partial (\hbar/rc)}{\partial r}$	
∴	From 9, Lx9.3.2.3B Distribute Constant in/out of Product and Lx9.3.3.3F Differential of a reciprocal, T4.8.2B Integer Exponents: Negative One Squared, A4.2.12 Identity Multp and 3, 9 and A4.2.3 Substitution	$\eta_r = \dfrac{\hbar}{c\, r^2}$ EQ A	$\eta_r = \dfrac{-\partial m}{\partial r}$ EQ B

Theorem 1.8.2 Monocentric Particle: Gaussian Curvature of G-E Rotational Density

Step	Hypothesis		
		$\eta_e\eta_g = \dfrac{M_{plk}{}^2\varepsilon_{rq}}{L_{plk}{}^2}$ EQ A	$\eta_e\eta_g = 16\pi^2\gamma_0{}^2\, c^4\varepsilon_{rq}{}^2$ EQ B
1	From T1.8.1 Monocentric Particle: Rotational Density and Gradient of Mass; as a dual product	$\eta_e\eta_g = \dfrac{\hbar}{cr_e{}^2}\ \dfrac{\hbar}{cr_g{}^2}$	
2	From 1, T4.5.4 Equalities: Product of Two Rational Fractions, A4.2.10 Commutative Multp and D4.1.17 Exponential Notation	$\eta_e\eta_g = \dfrac{\hbar^2}{c^2}\ \dfrac{1}{r_e{}^2\, r_g{}^2}$	
3	From 2, T1.6.4(A,C) Monocentric Particle Outside: Gaussian Curvature of Dual Spins and A4.2.3 Substitution	$\eta_e\eta_g = \dfrac{\hbar^2}{c^2}\ K_{eg}{}^4$	
4	From 3, T1.6.4C Monocentric Particle Outside: Gaussian Curvature of Dual Spins and A4.2.3 Substitution	$\eta_e\eta_g = \dfrac{\hbar^2}{c^2}\ \dfrac{\varepsilon_{rq}{}^2}{L_{plk}{}^4}$	
5	From 4, E1.3.2 Planck's Length, T4.10.4 Radicals: Identity Radical Raised to a Power, D4.1.19 Primitive Definition for Rational Arithmetic and A4.2.3 Substitution	$\eta_e\eta_g = \dfrac{\hbar^2}{c^2}\ \dfrac{\varepsilon_{rq}{}^2}{\hbar^2/16\pi^2\gamma_0{}^2c^6}$	
6	From 5, T4.5.7 Equalities: Product and Reciprocal of a Product, T4.5.4 Equalities: Product of Two Rational Fractions, A4.2.10 Commutative Multp and A4.2.13 Inverse Multp	$\eta_e\eta_g = \dfrac{1}{c^2}\ \dfrac{16\pi^2\gamma_0{}^2c^6\varepsilon_{rq}{}^2}{1}$	
7	From 6, A4.2.12 Identity Multp and A4.2.19 Difference Exp	$\eta_e\eta_g = 16\pi^2\gamma_0{}^2c^4\varepsilon_{rq}{}^2$	
8	From 7, E1.3.1 Plank's Mass, E1.3.2 Plank's Length, T4.2.13 Inverse Multp and A4.2.12 Identity Multp	$M_{plk}{}^2 / L_{plk}{}^2 = \dfrac{4\pi\gamma_0\hbar c}{\hbar /4\pi\gamma_0 c^3}$	
9	From 8, T4.5.7 Equalities: Product and Reciprocal of a Product, A4.2.10 Commutative Multp and A4.2.18 Summation Exp	$M_{plk}{}^2 / L_{plk}{}^2 = 16\pi^2\gamma_0{}^2c^4$	
∴	From 7, 9 and A4.2.3 Substitution	$\eta_e\eta_g = \dfrac{M_{plk}{}^2\varepsilon_{rq}{}^2}{L_{plk}{}^2}$ EQ A	$\eta_e\eta_g = 16\pi^2\gamma_0{}^2\, c^4\varepsilon_{rq}{}^2$ EQ B

Theorem 1.8.3 Monocentric Particle: Geometric Mean for Rotational Inertial Density

Step	Hypothesis	
		$\bar{\eta} = \pm 4\pi\gamma_0\, c^2\varepsilon_{rq}$
1	From D1.8.1 Geometric Mean for Rotation Inertial Density	$\bar{\eta}^2 = \eta_e\eta_g$
2	From 1, T1.8.2B Monocentric Particle: Gaussian Curvature of G-E Rotational Density and A4.2.3 Substitution	$\bar{\eta}^2 = 16\pi^2\gamma_0^{\ 2}\, c^4\varepsilon_{rq}^{\ 2}$
3	From 2, T4.10.7 Radicals: Reciprocal Exponent by Positive Square Root	$\bar{\eta} = \pm\sqrt{16\pi^2\gamma_0^{\ 2}\, c^4\varepsilon_{rq}^{\ 2}}$
∴	From 3, T4.10.5 Radicals: Distribution Across a Product and T4.10.3 Radicals: Identity Power Raised to a Radical	$\bar{\eta} = \pm 4\pi\gamma_0\, c^2\varepsilon_{rq}$

Observation 1.8.1 About the Geometric Mean for Rotational Density
So, the geometric mean for rotation inertial density is quantized and made from the gravitational and light constants.

This is the core of the theory, because the physical interpretation of the plus and minus sign is the direction of the photon spin. Here at the very structural crux of the particle the rotation inertial density conveys whether the electrical force will attract (be directed toward) or repelled (be directed away) from another charged particle.

Now all the pieces of the jigsaw puzzle are present. Two spinning photons have relative directed signs of spin. That information is transferred by the exchange of a photon within the ***unique binary frame of reference***, which completes the spatial circuit; by Minkowski's spatial continuum law of ***concurrency*** it makes the particle system distance independent. It follows that no modulation of the photon with information is required invalidating Tesla's assumption of photon modulation being required. The model is analogous to spinning a top on the surface of a mirror and the direction of the mirror image is opposite to the actual top, but distance independent. This is a bit odd because it makes no difference whether the particles are just touching each other or a trillion light years apart. The moment that connection is made by the intermediary photon, charge and direction of motion for the particles are known.

This leads to another physical law the "Law of Equivalency of Linear and Spin Momentum".

Law 1.8.2 Law of Equivalency of Linear and Spin Momentum

Linear Momentum [P] is implemented from the spin momentum [P_s] from two nuclear particles in a unique binary frame of reference.

$$P \equiv P_s \qquad \text{Equation A}$$

Observation 1.8.2 Unique Binary Frame of Reference is a Relative frame of Reference

Also observe that the moment the spatial circuit is complete that the receiver is now within the frame of reference for the origin of the photon. Since the process of linking the transmitter to the receiver is random the origin of the photon cannot be an absolute frame of reference, thereby upholding the fundamental principle of being *relatively* positioned one to the other.

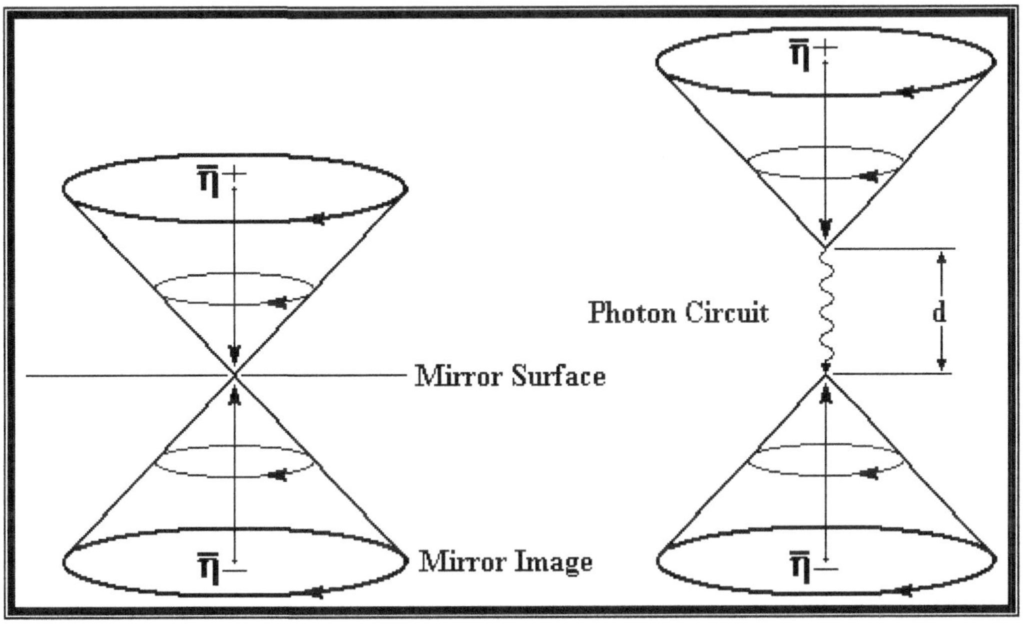

Figure 1.8.1 Model for Attraction of Unlike Signed Particles

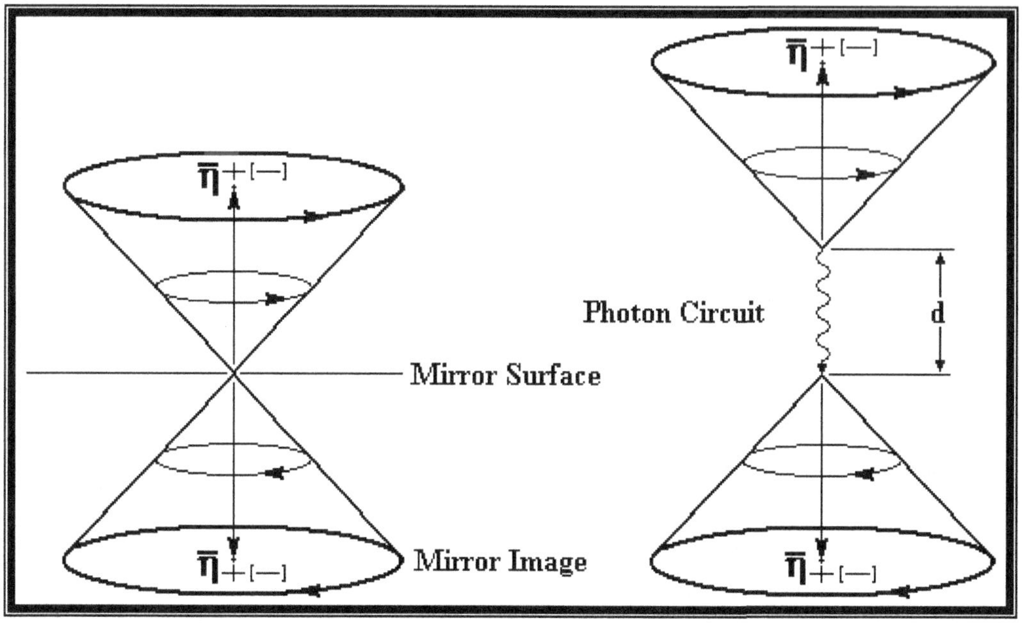

Figure 1.8.2 Model for Repulsion of Like Signed Particles

Now the geometric mean of the rotation inertial density predicts that in particle pair formation that charge is conserved. As particles are formed, they will always come into being by twos with a plus and minus charge. The name for this process is called ***conservation of charge for pair formation***. It is stated in the two following physical laws:

Law 1.8.3 Charge Pair Formation
As particles are created or annihilated, they will always come into being or out by twos with a plus and minus charge.

Law 1.8.4 Zero Net Pair Formation
The net charge in particle pair formation is zero, or the sum of the charge must add to zero.
$q^+ - q^- \equiv 0$ for $m^+ - m^- = 0$

Note working backwards the origins for mass comes from T1.8.1B "Monocentric Particle: Rotational Density and Gradient of Mass"

Theorem 1.8.4 Monocentric Particle: Mass from Rotational Inertial Density

Step	Hypothesis	$m = -\int_r \eta_r \, \partial r$
1	From T1.8.1B Monocentric Particle: Rotational Density and Gradient of Mass	$\eta_r = \frac{-\partial m}{\partial r}$
∴	From 1, T4.4.5B Equalities: Reversal of Right Cancellation by Multiplication, T9.4.2 Integrals: Uniqueness of Integration, Lx9.4.2.1 Integration of power function	$m = -\int_r \eta_r \, \partial r$

Section 1.9 *Net Inward Centripetal Force*

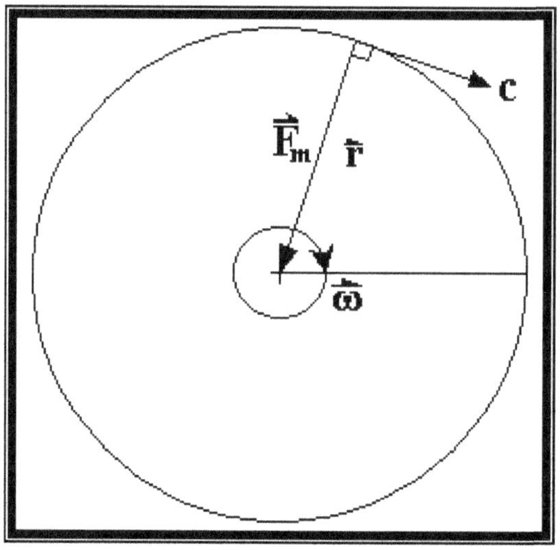

Figure 1.9.1 Net Inward Centripetal Force

It would require a tremendous amount of inward centripetal force to keep a photon in spin and not fly apart. This would preclude any curl in the energy field, hence

Theorem 1.9.1 Monocentric Particle: Inward Centripetal Force of a Mass Particle

1g	Given:	$\mathbf{0}$	$\equiv$	$\nabla\times\vec{F}_g$ for a photon in spin	
2g				For a Higgs particle 219,000 bigger than an electron	
Step	Hypothesis	F_g	$=$	$2.12017882\text{x}10^{-1}$ N	electron
		F_g	$=$	$4.64319162\text{x}10^{4}$ N	higgs particle
1	From 1g and Lx9.6.1.11 Curl of Gradient Operator	$\vec{F}_g$	$=$	$-\nabla E_g$	
2	From 1 and the gradient in spherical coordinates	∇	$\equiv$	$\mathbf{r}\,\nabla_r + (\boldsymbol{\theta}\ 1/r)\nabla_\theta + (\boldsymbol{\phi}\ 1/\ r\sin\theta)\nabla_\phi$	
3	From 1, 2 and T1.5.1B Monocentric Particle: Properties of a Photon in Spin	F_g	$=$	$-\partial E_g / \partial r$	
4	From 3; resulting in no change in direction for $(\boldsymbol{\theta}, \boldsymbol{\phi})$	$\nabla_\theta E_g$	$=$	$\nabla_\phi E_g = 0$	
5	From 1, 3, A4.2.3 Substitution and T7.11.2 Vector Addition: Identity	F_g	$=$	$-\partial\ (\hbar c/r_g) / \partial r$	
6	From 5, Lx9.3.2.3B Distribute Constant in/out of Product, Lx9.3.3.3F Differential of a reciprocal, T4.8.2B Integer Exponents: Negative One Squared and A4.2.12 Identity Multp	F_g	$=$	$\hbar c/r_g^{\,2}$	

7	From 6, CxPPC.A.2.3.4 Normalized Plank's Constant, CxPPC.A.2.3.4 Speed of Light, and E1.5.1 Electron Radius and A4.2.3 Substitution	F_g	=	$\dfrac{1.0545919(80)\text{x}10^{-34}\text{ J sec}*2.9979250(10)\text{x}10^{8}\text{ m sec}^{-1}}{(3.861591839\text{x}10^{-13}\text{ m})^2}$	
∴	From 7; Evaluate	F_g	=	$2.12017882\text{x}10^{-1}$ N	electron
	From 2g; Evaluate	F_g	=	$4.64319162\text{x}10^{4}$ N	higgs particle

Clearly, a large force is required to keep the particle together even with a greater mass the force is small. If both electrical and gravitational centripetal force contributes their net force as a unified field than there would be sufficient force to maintain spin.

As with the rotation inertial density let the product be equivalent to the geometric mean of the centripetal forces.

Theorem 1.9.2 Monocentric Particle: Inward Dual Force of an G-E Particle

1g	Given:	$\bar{F}^2$	≡	$F_e\,F_g$ As with the rotation inertial density let the product be equivalent to the geometric mean of the centripetal forces.
Step	Hypothesis	$\bar{F}$	=	$-4\pi\gamma_0 c^4 \varepsilon_{rq}$
1	From T1.9.1;6 Monocentric Particle: Inward Centripetal Force of a Mass Particle: for and electron particle	F_e	=	$\hbar c/r_e^{\,2}$ electrical
2	From T1.9.1;6 Monocentric Particle: Inward Centripetal Force of a Mass Particle: for and inertial particle	F_g	=	$\hbar c/r_g^{\,2}$ inertial
3	From 1, 2 and the product	$F_e\,F_g$	=	$(\hbar c/r_e^{\,2})\,(\hbar c/r_g^{\,2})$
4	From 3, T4.5.4 Equalities: Product of Two Rational Fractions, A4.2.10 Commutative Multp, D4.1.17 Exponential Notation and A4.2.22 Distribution Exp	$F_e\,F_g$	=	$\dfrac{(\hbar c)^2}{(r_e\,r_g)^2}$
5	From 4, T1.6.4(A, C) Monocentric Particle Outside: Gaussian Curvature of Dual Spins, A4.2.3 Substitution and A4.2.11 Associative Multp and T4.4.9 Equalities: Any Quantity Divided by One is that Quantity	$F_e\,F_g$	=	$\dfrac{(\hbar c)^2}{1}\;\dfrac{\varepsilon_{rq}^{\,2}}{L_{plk}^{\,4}}$
6	From 5, E1.3.2 Plank's Length and T4.10.4 Radicals: Identity Radical Raised to a Power	$F_e\,F_g$	=	$\dfrac{(\hbar c)^2}{1}\;\dfrac{\varepsilon_{rq}^{\,2}}{\hbar^2/16\pi^2\gamma_0^{\,2}c^6}$

7	From 6, T4.5.9 Equalities: Ratio and Reciprocal of a Ratio and A4.2.22 Distribution Exp	$F_e\,F_g = \dfrac{\hbar^2 c^2}{1} \quad \dfrac{16\pi^2\gamma_0{}^2 c^6 \varepsilon_{rq}{}^2}{\hbar^2}$
8	From 7, A4.2.10 Commutative Multp, A4.2.13 Inverse Multp, A4.2.12 Identity Multp and A4.2.18 Summation Exp	$F_e\,F_g = 16\pi^2\gamma_0{}^2 c^8 \varepsilon_{rq}{}^2$
9	From 1g and T4.10.7 Reciprocal Exp	$\bar{F} = \pm\sqrt{F_e\,F_g}$
10	From 9, 8, A4.2.3 Substitution; negative sign is selected for inward centripetal force	$\bar{F} = -\sqrt{16\pi^2\gamma_0{}^2 c^8 \varepsilon_{rq}{}^2}$
∴	From 10, T4.10.5 Radicals: Distribution Across a Product and T4.10.3 Radicals: Identity Power Raised to a Radical	$\bar{F} = -4\pi\gamma_0 c^4 \varepsilon_{rq}$

But, note the interesting relationships that rotation inertial density and Planck's force, play in defining the role that the mean centripetal force plays:

The above theorem gives a physical interpretation to Plank's dimension of force as being a centripetal force whose role is similar to the function that gravity plays.

Conjecture: Might this inward attractive force be the origin for the positive directed action exerted by gravity and why there has never been found a counter force (anti-gravity)?

Without the sign being negative the particle could not hold together, if it had been positive it would fly apart and nothing physical would remain in existence. Given the validity of the theory and the fact that we are here that cannot be the case. So we are forced to the conclusion that it's negative.

Theorem 1.9.3 Monocentric Particle: Inward Dual Force of an G-E Particle

Step	Hypothesis	$\bar{F} = -8.833375188\text{x}10^{41}$ N
1	From T1.9.2 Monocentric Particle: Inward Dual Force of an G-E Particle, CxPPC.A.2.3.2 Reciprocal Newtonian constant, CxPPC.A.2.3.5,	$\bar{F} = -4\pi\,(1.19248789\text{x}10^9)\,(2.997925010\text{x}10^8)^4\,(7.297351\text{x}10^{-3})$
2	From 1 and evaluate	$\bar{F} = -4\pi(7.029376627\text{x}10^{40})$
∴	From 2, D4.1.19 Primitive Definition for Rational Arithmetic	$\bar{F} = -8.833375188\text{x}10^{41}$ N

Observation 1.9.1 The latch on the spinning photon

Clearly, mean centripetal force in theorem T1.9.3 "Monocentric Particle: Inward Dual Force of an G-E Particle" provides sufficient force ~10^{41} N to maintain a stable particle.

The above theorem shows that every point within the spinning photon acts inward toward the center, thereby giving rise to a net mean force, which intern keeps the photon from flying apart and moving in a rectilinear direction. This force is the latch on the door locking the photon into its eternal spin.

Theorem 1.9.4 Monocentric Particle: Mean Inward Dual Force to Rotation Inertial Density

Step	Hypothesis	$\bar{F} = -\bar{\eta}\, c^2$
1	From T1.9.2 Monocentric Particle: Inward Dual Force of an G-E Particle	$\bar{F} = -4\pi\gamma_0 c^4 \varepsilon_{rq}$
2	From 1, D4.1.19 Primitive Definition for Rational Arithmetic, A4.2.18 Summation Exp and A4.2.10 Commutative Multp	$\bar{F} = -4\pi\gamma_0 c^2 \varepsilon_{rq}\, c^2$
∴	From 2, T1.8.3 Monocentric Particle: Geometric Mean for Rotational Inertial Density and A4.2.3 Substitution	$\bar{F} = -\bar{\eta}\, c^2$

Theorem 1.9.5 Monocentric Particle: Mean Inward Dual Force to Plank's Force

Step	Hypothesis	$\bar{F} = -F_{plk}\, \varepsilon_{rq}$
1	From T1.9.2 Monocentric Particle: Inward Dual Force of an G-E Particle	$\bar{F} = -4\pi\gamma_0 c^4 \varepsilon_{rq}$
2	From 1 and A4.2.11 Associative Multp	$\bar{F} = (-4\pi\gamma_0 c^4)\, \varepsilon_{rq}$
∴	From 2, E1.3.9 Plank's Force and A4.2.3 Substitution; pg 1330	$\bar{F} = -F_{plk}\, \varepsilon_{rq}$

Theorem 1.9.6 Monocentric Particle: Dimensionality of Mean Inward Dual Force

Step	Hypothesis	$\bar{F} = [M][L][T^{-2}]$
1	From T1.9.2 Monocentric Particle: Inward Dual Force of an G-E Particle	$\bar{F} = -4\pi\gamma_0 c^4 \varepsilon_{rq}$
2	From 2, CxPPC.A.2.3.2 Reciprocal Newtonian constant and CxPPC.A.2.3.5 Velocity of Light for free space	$\bar{F} = ([M][L^{-3}][T^{2}])\, [L^{4}][T^{-4}]$
∴	From 2, AA.4.2.5 Commutative Dim and AA.4.2.11 Exponent Summation Dim	$\bar{F} = [M][L][T^{-2}]$

This also gives a possible reason why no such thing as antigravity has ever been found, which would require a positive sign. The meaning of the negative and positive sign can be succinctly stated in the following ***physical laws of existence for a particle of matter***.

Law 1.9.2 Particle Existence
A photon attaining spin can only exist as a particle of matter and hold together with a Plank's mean negative inward centripetal force.

Law 1.9.3 Particle Nonexistence
Particle Nonexistence: A photon not being able to attain spin is nonexistent as a particle of matter and will fly outward apart with an equivalent initial positive centripetal force.

Since the centripetal force sets up an exclusive attraction between particles, they behave like electrical charged components. Of course this means repulsion would be excluded and only attraction permitted, analogous to the Pauli's exclusion principle where like charges can only coexist about the atomic nucleus if their orbit is in opposite directions nullifying like charge repulsion. Here the internal centripetal force only permits attraction.

Using the idea of centripetal force one can conclude the existence of a phenomenon of something akin to charge, but for gravity:

Rules of gravitational charge:

- ***Unlike gravitational charges are precluded due to the direction of internal centripetal force.***
- ***Like gravitational charges are equivalent, because they yield the same directed momentum.***

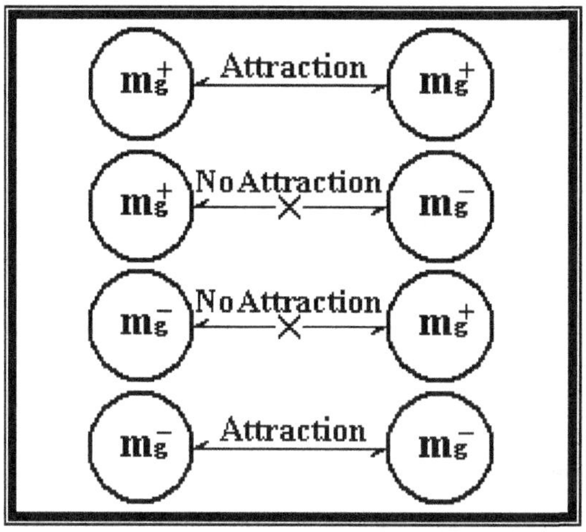

Figure 1.9.2 Relative Paired Gravitational Charge Attract

Law 1.9.4 Relatively Paired Gravitational Charge Attraction
The principle of centripetal force only allows inertial particles to attract regardless of the sign for spin between any two gravitational charges.

Section 1.10 Force at a Distance; Paradox of Attractive Momentum

With the law on concurrency there is an issue that is not being addressed. The state of two nuclear particles some distance apart and the direction of relative momentum between them. Einstein's photoelectric effect clearly states that such particles can and do emit or absorb energy as a quantum [hv] [HAL66 pg 1184]. Compton with his experiments in observing momentum imparted by a photon to a nuclear particle defines the relative direction of motion for the two particles [HAL66 pg 1184]. Compton's model however is only one sided. For a true exchange of a photon between two nuclear particles it says nothing about relative direction of their momentum. Since such a model of binary frames has not been considered before some extraordinary effects happen that have not been observed. This will lead us to the mechanics for a discrete model of all forces that act at a distance.

An analysis in terms of momentum reveals that attractive and repulsive forces at a distance do not conform to classical Newtonian physics as historically was thought. This reveals a new way of looking at how the mechanics of positive and negative repulsion work forcing particles to move toward and away from one another.

For force at a distance to work, a photon exchange must take place. How the mechanism for photons being absorbed and their direction of reemission comes about at this time is unknown, but when done the particle determines the direction of motion. This phenomenon is known by Newton's law of momentum:

Law 1.10.1 Newton's Law of Momentum III [CAJ71, Vol. I pg 13]

To every action there is always opposed an equal reaction: or, the mutual actions of two bodies upon each other are always equal, and directed to contrary parts.

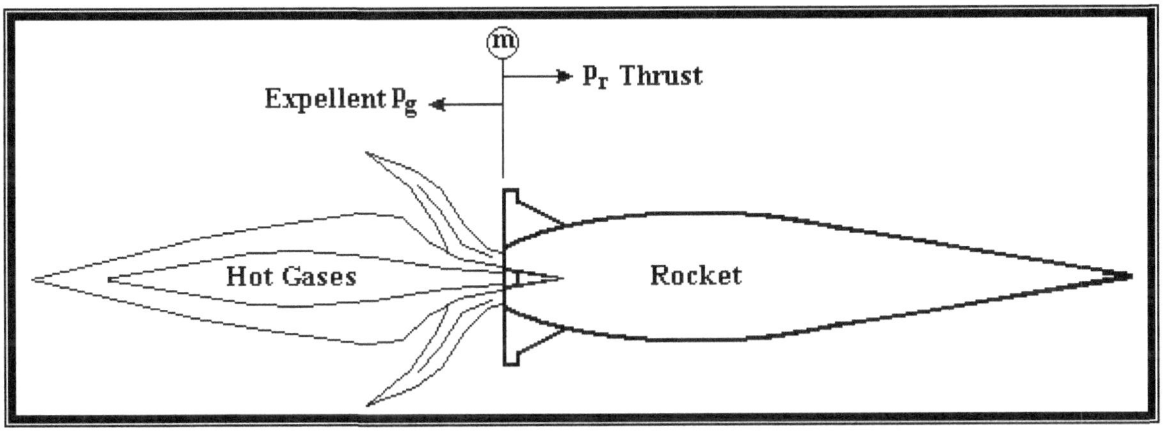

Figure 1.10.1 Moment Diagram for a Rocket

Based on Newton's third law the expellent is opposite to the momentum resulting in thrust as seen in the above diagram. This is classical Newtonian mechanics. Now let's consider the mechanical moment of electrical and gravitational forces at a distance. Here nuclear particles can be thought of as little rockets absorbing photons for fuel and expelling photons in place of hot rocket gases.

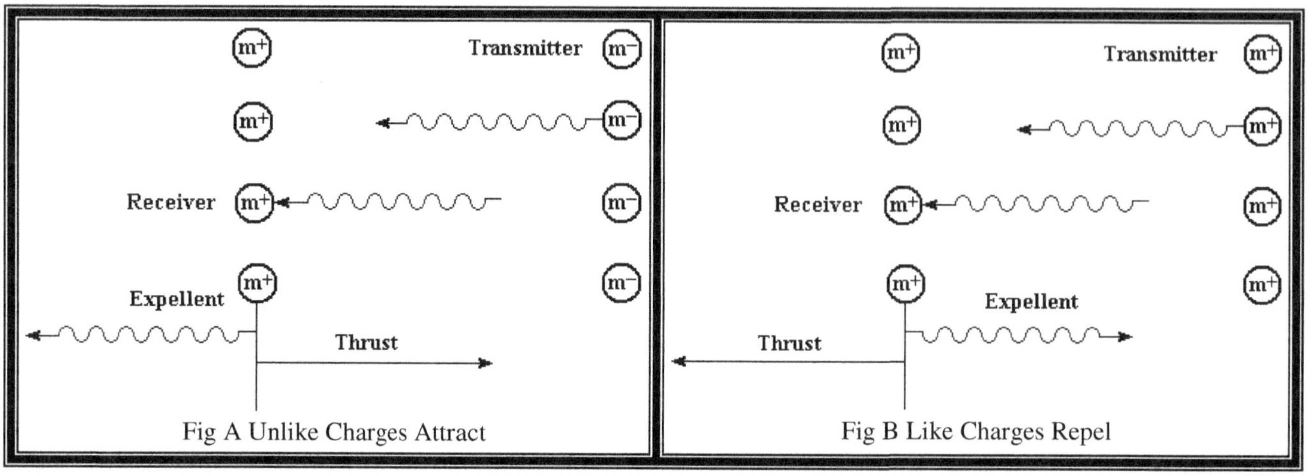

Fig A Unlike Charges Attract

Fig B Like Charges Repel

Figure 1.10.2 Moment Diagram for Particle Photon Exchange

As seen from Figure 1.10.2A particles naturally move away from each other for unlike charges and Newtonian mechanics holds. Coulomb's Law of like charges, E1.4.1, guarantees that a volley of continuous photon exchange is maintained between particles implementing a repulsive force; however this is not true for attractive forces. In order to maintain the proper order of thrust momentum to expellent only one photon can be exchanged and force at a distance for unlike charges falls apart. This means that all unlike charges for electrical interactions would fail yet experience shows otherwise as demonstrated in 1785 by Coulomb's experiments on charge [HAL66, pg 650]. Gravity being an attractive force as well would also be repealed, something that no one so far has observed, see the Cavendish experiment based on Newton's Law [HAL66, pg 389] which was done in 1798.

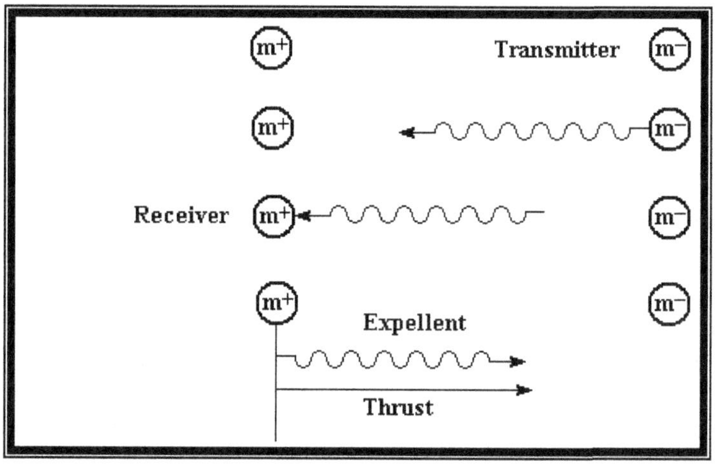

Figure 1.10.3 Moment Diagram for Unlike Charge Photon Exchange

The only way to make this work is to realign the direction of expellent and thrust momentums, thereby violating Newton's Law of momentum. So, a new ***law of momentum for attractive forces*** must be stated for nuclear particles having electrical and gravitational fields of force at a distance. As so many things do not work in a quantum world, as common sense says it should work, here a new property is reviled with a contrary physical result in the micro world.

Law 1.10.2 G-E Momentum Attraction at a Distance

To every action there is always a similarly an equal reaction: or the mutual actions of two bodies upon each other are always equal, and directed toward comparable parts.

Law 1.10.3 G-E Alternate Momentum Attraction at a Distance

Alternatively, expellent and thrust momentums align themselves in the same direction to attract one another.

Summary: G-E Transmitter Directed Action Relative to its Place of Origin

	m_e (+)	m_e (−)	m_g (+)	m_g (−)
m_e (+)	Rotation in opposition anti mirror image; repulsion	Rotation in cohesion mirror image; attraction		
m_e (−)	Rotation in cohesion mirror image; attraction	Rotation in opposition anti mirror image; repulsion		
m_g (+)			Inward centripetal rotation; attraction	Inward centripetal rotation; attraction
m_g (−)			Inward centripetal rotation; attraction	Inward centripetal rotation; attraction

The following rules are constrained by Law 1.10.3 "G-E Alternate Momentum Attraction at a Distance".

With respect to the transmitter particle there is a directed action from its relative place of origin given by the following rules:

Rule of Repulsion: *Thrust is always negative or out of phase by 180°.*
Rule of Attraction: *Thrust is always positive or in phase by 0°.*
Rule of Expellent: *Expellent is always positive* for *repulsion and attraction.*

Summary: G-E Transmitter Directed Phase Action Relative to its Place of Origin

	m_e (+)	m_e (−)	m_g (+)	m_g (−)
m_e (+)	180° out of phase; repulsion	0° in phase; attraction		
m_e (−)	0° in phase; attraction	180° out of phase; repulsion		
m_g (+)			0° in phase; attraction	0° in phase; attraction
m_g (−)			0° in phase; attraction	0° in phase; attraction

The previous momentum-diagrams are an idealized model for photon-particle interaction. They do not permit for the photon's scattering and deflection angles with relativistic velocity for the particle mass. Allowing for the directed action from the transmitter photon's place of origin and using a modified "Compton Effect model" [HAL66, pg 1184] a more general system can be constructed.

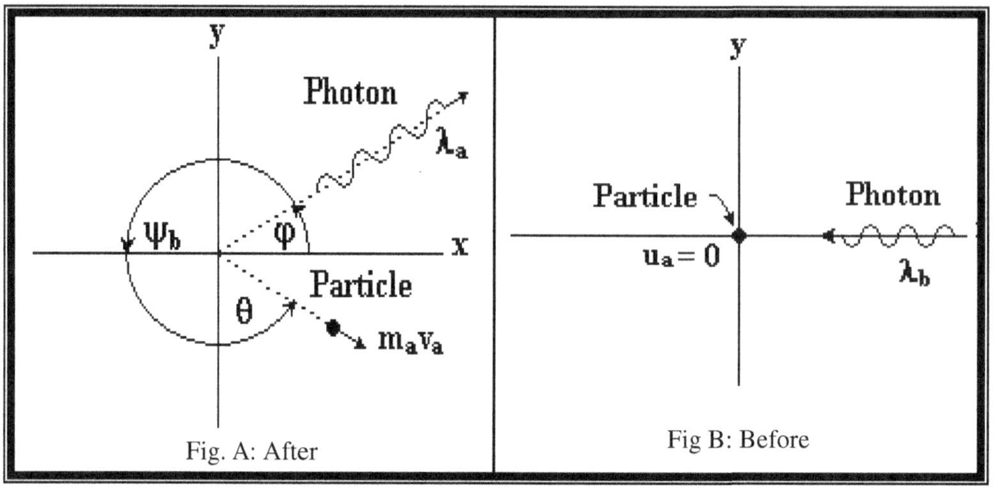

Figure 1.10.4 Compton Effect with Scatter-φ and Deflection-θ Angles

Given LxEMT.B.4.1b in a (1–2) frame of reference such that $u_a = 0$ than the momentum for the particle on the y-axis is, see figure 1.10.4A

$$0 = (h/\lambda_a)\sin(\varphi) - \gamma(m_a v_a)\sin(\theta) \quad \text{for } 0 \le \varphi \qquad \text{Equation 1.10.1}$$

Given LxEMT.B.4.1b in a (1–2) frame of reference such that u = 0 than the momentum for the particle on the x-axis is, see figure 1.10.4A and B

$$(h/\lambda_b) = (h/\lambda_a)\cos(\varphi) + \gamma(m_a v_a)\cos(\theta) \quad \text{for } 0 \le \theta \le 2\pi \qquad \text{Equation 1.10.2}$$

for the photon's directed phase action relative to its place of origin for an electrical field

$$\psi_e = \begin{cases} 2n\pi \text{ attraction} \\ (2n-1)\pi \text{ repulsion} \end{cases} \quad n = 0, \pm1, \pm2, \ldots \qquad \text{Equation 1.10.3}$$

for the photon's directed phase action relative to its place of origin for a gravitational field

$$\psi_g = 2m\pi \quad \text{attraction} \quad m = 0, \pm1, \pm2, \ldots \qquad \text{Equation 1.10.4}$$

In order to model the idealize photon exchange let $\varphi = 0$ and $\theta = \psi_b$

Section 1.11 *G-E Momentum and Mechanism of Action at a Distance*

Definition 1.11.1 Photon Spin Momentum: Geometric Mean of Singular Momentum

Just as the geometric mean centripetal force is the product of the electrical and gravitational spin force theorem T1.9.2 "Monocentric Particle: Inward Dual Force of an G-E Particle" so to the geometric mean moment can be written as a product.

$$\bar{P}_s^2 \equiv P_e P_g \qquad \text{Equation A}$$

Definition 1.11.2 Photon Spin Momentum: Geometric Mean of Rotational Inertia

The geometric mean moment can be written as a product:

$$I_s^2 \equiv I_e I_g \qquad \text{Equation A}$$

or as a root magnitude

$$I_s \equiv |\sqrt{I_e I_g}| \qquad \text{Equation B}$$

From conservation of momentum (Law 1.5.2 "Pair Formation Conservation of Momentum") and LB.5.V "Concurrency" what happens to the momentum of the expellent photon must be true for both electrical and gravitational momentum; hence momentum is set up for both components as follows:

Law 1.11.1 Photon Spin Momentum: Electrical and Gravitational

Particle momentum is limited to just the particle's momentum so momentum must be parsed between the electrical and gravitational inertial as follows:

$$P_e = -I_e \omega_e^2 / c \qquad \text{Equation A}$$

and

$$P_g = -I_g \omega_g^2 / c \qquad \text{Equation B}$$

Law 1.11.2 Alternate Law of Relatively Paired Gravitational Charge Attraction

Gravitational and Electrical angular spin are equivalent; (see law L1.9.3 "Paired Gravitational Charge Attraction") where gravity contributes the same spin, hence the same net momentum to attract.

$$g \equiv e \qquad \text{Equation A}$$

Theorem 1.11.1 Photon Spin Momentum: Singular Particle

1g	Given:	$\mathbf{L}$	$\equiv$	$\mathbf{r} \times \mathbf{P}$ Angular momentum
2g		$\hbar$	$\equiv$	$\lvert\mathbf{L}\rvert$ Plank's constant a quantum of angular momentum
3g		$\mathbf{L}$	$\perp$	$\boldsymbol{\omega}$ see F12.16.2 perpendicular idealized kinematic spin system
Step	Hypothesis	P_s	$=$	$-I\ \omega^2 / c$
1	From 1g and LxRFXA.1.1.8 Angular momentum; rotation about a fixed axis	$\mathbf{L}$	$\equiv$	$I\ \boldsymbol{\omega}$
2	From 1, T8.3.1 Dot Product: Uniqueness Operation	$\mathbf{L} \bullet \boldsymbol{\omega}$	$=$	$I\ \boldsymbol{\omega} \bullet \boldsymbol{\omega}$
3	From 3g, 2, D3.8.1 The Inner or Dot Product Operator, T8.3.4 Dot Product: Magnitude, A4.2.3 Substitution and A4.2.2B Equality	$I\ \omega^2$	$=$	$\hbar\ \omega \cos(\pi/2)$
4	From 3, DxE.1.6.1 Cosine, A4.2.12 Identity Multp, D4.1.17 Exponential Notation and T4.4.3 Equalities: Cancellation by Multiplication	$\hbar$	$=$	$I\ \omega$
5	From 1g, 4 and T8.3.2 Dot Product: Equality	$\mathbf{L} \bullet \boldsymbol{\omega}$	$\equiv$	$(\mathbf{r} \times \mathbf{P}_s) \bullet \boldsymbol{\omega}$
6	From 5 and Lx9.5.1.6 Volume of Parallelepiped	$\hbar\ \omega$	$=$	$-(\mathbf{r} \times \boldsymbol{\omega}) \bullet \mathbf{P}_s$
7	From 6 and T12.16.3 Kinematics: Rotation of Positional Vector about a Point; v-tangential spin velocity	$\hbar\ \omega$	$=$	$-\mathbf{v} \bullet \mathbf{P}_s$
8	From 3g, 7 and Law 1.2.1 Photon Point Spin: Tangential Velocity; is at the velocity of light-c	$\hbar\ \omega$	$=$	$-c\ P_s$
9	From 8, D4.1.20 Negative Coefficient and T4.4.4A Equalities: Right Cancellation by Multiplication	P_s	$=$	$-\hbar\ \omega / c$
10	From 9, Law 1.2.2 Photon Point Spin: Radius, T4.4.9 Equalities: Any Quantity Divided by One is that Quantity and T4.4.14 Equalities: Formal Cross Product	P_s	$=$	$-\hbar / r$

11	From 10, D4.1.20 Negative Coefficient, T4.4.5A Equalities: Reversal of Right Cancellation by Multiplication	$\hbar$	=	$-P_s r$
12	From 5, 11 and A4.2.3 Substitution	$I \omega$	=	$-P_s r$
13	From 12, T4.4.23 Equalities: Multiplication of a Constant and D4.1.17 Exponential Notation	$I \omega^2$	=	$-P_s r \omega$
14	From 13 and Law 1.2.1 Photon Point Spin: Tangential Velocity	$I \omega^2$	=	$-P_s c$
∴	From 14 and T4.4.4A Equalities: Right Cancellation by Multiplication	P_s	=	$-I \omega^2 / c$

Theorem 1.11.2 Photon Spin Momentum: Geometric Mean of Singular Momentum

1g	Given:	(e, g)	=	{ 1 for spin CCW; 0 for spin CW plus and minus sign of a square root are interrupted as direction of spin
2g		sgn(e, g)	≡	$(-1)^{e+g}$
Step	Hypothesis	P	=	$\text{sgn}(e+g)\ I_s\ (\omega_e\ \omega_g) / c$
1	From D1.11.1 Photon Spin Momentum: Geometric Mean of Singular Momentum	$\bar{P}_s^{\,2}$	≡	$P_e\ P_g$
2	From 1, L1.11.1(A,B) Photon Spin Momentum: Electrical and Gravitational and A4.2.3 Substitution	$\bar{P}_s^{\,2}$	=	$(-I_e\ \omega_e^2 / c\ )(-I_g\ \omega_g^2 / c)$
3	From 2, D4.1.20 Negative Coefficient, A4.2.10 Commutative Multp, T4.8.2B Integer Exponents: Negative One Squared and D4.1.17 Exponential Notation	$\bar{P}_s^{\,2}$	=	$+I_e\ I_g\ \omega_e^2\ \omega_g^2 / c^2$
4	From 1g, 3, T4.10.7 Radicals: Reciprocal Exponent by Positive Square Root, T4.10.5 Radicals: Distribution Across a Product, T4.10.3 Radicals: Identity Power Raised to a Radical	$\bar{P}_s$	=	$(-1)^{e+g}\ \lvert\sqrt{(I_e\ I_g)}\rvert\ (\omega_e\ \omega_g) / c$

5	From 2g, 4, D1.11.2B Photon Spin Momentum: Geometric Mean of Rotational Inertia L1.8.1A Law of Equivalency of Linear and Spin Momentum and A4.2.3 Substitution	$\bar{P}_s = \text{sgn}(e+g)\ I_s\ (\omega_e\ \omega_g)\ /\ c$
∴	From 5, L1.8.1A Law of Equivalency of Linear and Spin Momentum and A4.2.3 Substitution	$P = \text{sgn}(e+g)\ I_s\ (\omega_e\ \omega_g)\ /\ c$

Theorem 1.11.3 Photon Spin Momentum: Finding Rotational Inertia

Step	Hypothesis	$I = \int_V \eta\ dV_R$
1	From LxRFXA.1.1.13 Rotation inertial density	$\eta = dI/dV_R$
2	From 1 and T4.4.5B Equalities: Reversal of Right Cancellation by Multiplication	$dI = \eta\ dV_R$
∴	From 2, T9.4.2 Integrals: Uniqueness of Integration and Lx9.4.2.1 Integration of power functions	$I = \int_V \eta\ dV_R$

Table 1.11.1 Photon Spin Momentum: Rotation Inertial

		Description
Lx11.1.1	$I_r = \int_V \eta_r\ dV$	rotation inertial density for spin
Lx11.1.2	$I_g = \int_V -\eta_g\ dV$	rotation inertial density conveys direction of rotation than applying equation T1.11.2 "Photon Spin Momentum: Geometric Mean of Singular Momentum" for gravity.
Lx11.1.3	$I_g = \int_V \eta_g\ dV$	rotation inertial density for electrical field of spin
Lx11.1.4	$I_e = \pm \int_V \eta_e\ dV$	rotation inertial density for the electrical field of spin

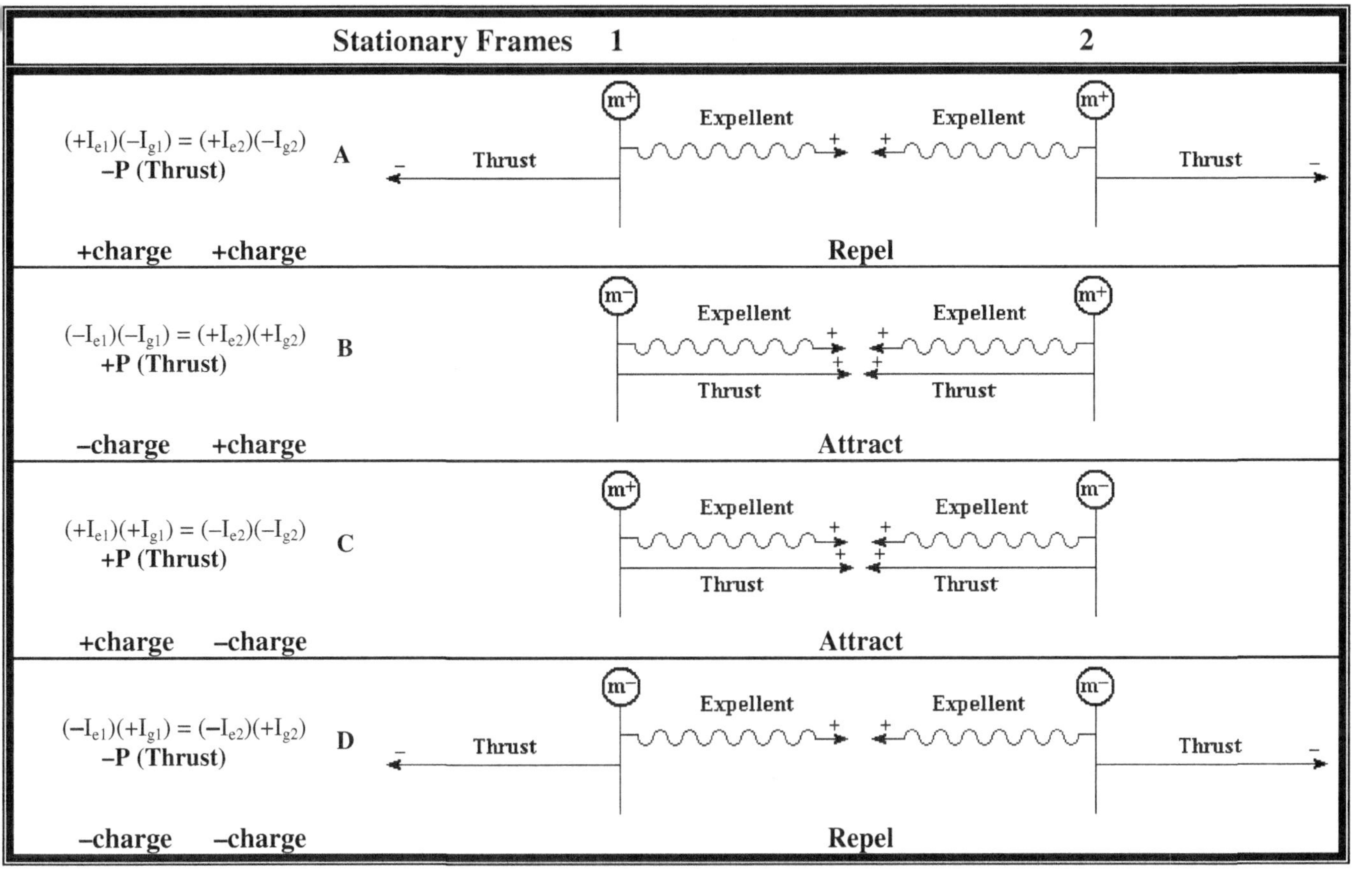

Figure 1.11.1 Charged Particle Attraction-Repulsion by Rotational Inertia

Observation 1.11.1 Gravity a Dominate Force Over a Charged Continuum

$F_e > F_g$ for nuclear particles, however given a large aggregate of charged particles they pair up into molecules with their positive and negative charges neutralized, which results in $F_e << F_g$ over shadowing the effect of free electrical charges. So only the centripetal force particles dominate, hence large aggregates of particles can only attract contributing to the domination of the gravitational system.

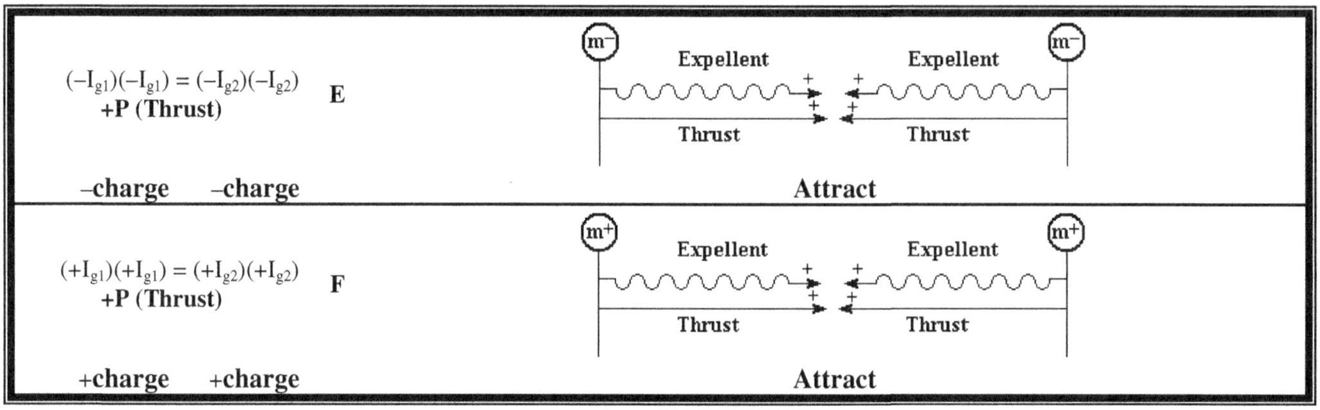

Figure 1.11.2 Mass Particle Attraction by Spin Inertia

Observation 1.11.2 Nontraditional Gravitational Attraction

Notice that gravity plays a key role in attraction of charged particles and is not independent as in traditional analysis.

$$E = \pm c \sqrt{(P_o^2 + P^2)} - E_o \qquad \text{Equation 1.12.1}$$

From EB.4.2 Dirac-Einstein's Energy Equation

Theorem 1.11.4 Particle Energy-Momentum: Mathematical Model of Thrust

1g	Given	$P \equiv E / c$ thrust-momentum and energy imparted to particle
2g		$E \equiv h\nu$ excellent-energy supplied to particle by photon
3g		$P_0 \equiv E_0 / c$ stationary momentum and energy of particle
4g		$\nu / c = 1 / \lambda$
Step	Hypothesis	$P = \pm\sqrt{(h / \lambda) [(h / \lambda) + 2P_0]}$
1	From E1.12.1 and T4.3.4A Equalities: Reversal of Right Cancellation by Addition	$E_0 + E = \pm c \sqrt{(P_o^2 + P^2)}$
2	From 1, T4.8.6 Integer Exponents: Uniqueness of Exponents, A4.2.21 Distribution Exp and T4.8.2A Integer Exponents: Negative One Squared	$(E_0 + E)^2 = c^2 (P_o^2 + P^2)$
3	From 2, T4.4.4A Equalities: Right Cancellation by Multiplication and T4.8.7 Integer Exponents: Distribution Across a Rational Number	$((E_0 + E) / c)^2 = P_o^2 + P^2$

4	From 3 and A4.2.12 Distribution	$(E_0 / c + E / c)^2 = P_o^2 + P^2$
5	From 2g, 3g, 4 and A4.2.3 Substitution	$(P_0 + h\nu / c)^2 = P_o^2 + P^2$
6	From 4g, 5 and A4.2.3 Substitution	$(P_0 + h / \lambda)^2 = P_o^2 + P^2$
7	From 6 and T4.12.1 Polynomial Quadratic: The Perfect Square	$P_0^2 + (h / \lambda)^2 + 2(P_0h / \lambda) = P_o^2 + P^2$
8	From 7 and T4.3.2 Equalities: Cancellation by Addition	$(h / \lambda)^2 + 2(P_0h / \lambda) = P^2$
9	From 8 and A4.2.2B Equality	$P^2 = (h / \lambda)^2 + 2(P_0h / \lambda)$
10	From 9, D4.1.17 Exponential Notation, A4.2.11 Associative Multp and A4.2.12 Distribution	$P^2 = (h / \lambda) [(h / \lambda) + 2P_0]$
∴	From 10 and T4.10.7 Radicals: Reciprocal Exponent by Positive Square Root	$P = \pm\sqrt{} (h / \lambda) [(h / \lambda) + 2P_0]$

Theorem 1.11.5 Particle Energy-Momentum: Expellent Negative Momentum Exchange

1g	Given	$P = 0$ expellent contributes no net momentum between particles
Step	Hypothesis	$P_{ex} = -h / (2\lambda)$
1	From T1.12.1 Particle Energy-Momentum: Mathematical Model; positive energy exchange	$P = \pm\sqrt{} (h / \lambda) [(h / \lambda) + 2P_{ex}]$
2	From 1g, 1, A4.2.3 Substitution, T4.8.6 Integer Exponents: Uniqueness of Exponents and T4.8.5A Integer Exponents: Zero Raised to the Positive Power-n	$0 = \{ \pm\sqrt{} (h / \lambda) [(h / \lambda) + 2P_{ex}] \}^2$
3	From 2, T4.10.4 Radicals: Identity Radical Raised to a Power and A4.2.21 Distribution Exp	$0 = (h / \lambda)^2 [(h / \lambda) + 2P_{ex}]^2$
4	From 3, T4.8.5A Integer Exponents: Zero Raised to the Positive Power-n and; equating factors	$0 = (h / \lambda) + 2P_{ex}$
∴	From 4, T4.3.3B Equalities: Right Cancellation by Addition and A4.2.7 Identity Add	$P_{ex} = -h / (2\lambda)$

Theorem 1.11.6 Particle Energy-Momentum: Thrust of G-E Field Momentum

1g	Given	F1.11.1 Charged Particle Attraction-Repulsion by Rotational Inertia
Step	Hypothesis	$P = (-1)^{e+g} \sqrt{(h / \lambda)\, [\, (h / \lambda) + 2[(\gamma_0/\varepsilon_0)\, (q^2 / m_o)\, c]}$
1	From E1.6.5.8 Momentum of an Electrical Stationary Mass Relative to Rest Mass; dimensional proportional coefficient of mass to charge $[M^2][Q^{-2}]$	$P_e = (\gamma_0/\varepsilon_0)\, (q^2 / m_o)\, c$
2	From T1.12.1 Particle Energy-Momentum: Mathematical Model	$P = \pm\sqrt{(h / \lambda)\, [\, (h / \lambda) + 2P_e\,]}$
∴	From 1g, 1, 2, E1.11.17 G-E Directed Angular Spin and A4.2.3 Substitution	$P = (-1)^{e+g} \sqrt{(h / \lambda)\, [\, (h / \lambda) + 2[(\gamma_0/\varepsilon_0)\, (q^2 / m_o)\, c]}$

Theorem 1.11.7 Particle Energy-Momentum: Thrust of Gravity Field Momentum

1g	Given	F1.12.2 Mass Particle Attraction by Spin Inertia
Step	Hypothesis	$P = +\sqrt{(h / \lambda)\, [\, (h / \lambda) + 2\, m_o c]}$
1	From E1.6.5.8 Momentum of an Electrical Stationary Mass Relative to Rest Mass	$P_g = m_o c$
2	From T1.12.1 Particle Energy-Momentum: Mathematical Model	$P = \pm\sqrt{(h / \lambda)\, [\, (h / \lambda) + 2P_g\,]}$
3	From 1g, 1, 2, E1.11.17 G-E Directed Angular Spin, L1.11.1A Alternate Law of Relatively Paired Gravitational Charge Attraction and A4.2.3 Substitution	$P = (-1)^{g+g} \sqrt{(h / \lambda)\, [\, (h / \lambda) + 2\, m_o c]}$
4	From 3 and T4.3.10 Equalities: Summation of Repeated Terms by 2	$P = (-1)^{2g} \sqrt{(h / \lambda)\, [\, (h / \lambda) + 2\, m_o c]}$
5	From 4 and A4.2.22 Product Exp	$P = [(-1)^2]^g \sqrt{(h / \lambda)\, [\, (h / \lambda) + 2\, m_o c]}$
∴	From 5, T4.8.2A Integer Exponents: Negative One Squared, T4.8.1 Integer Exponents: Unity Raised to any Integer Value and A4.2.12 Identity Multp	$P = +\sqrt{(h / \lambda)\, [\, (h / \lambda) + 2\, m_o c]}$

Observation 1.11.3 Origin of Positive Gravitational Field

The electrical charge is the same as the gravitational charge (see Figure 1.12.2 "Mass Particle Attraction by Rotational Inertia"), hence one and the same, which leads to a positive momentum so all gravitational fields attract as proven in Theorem 1.12.4 "Particle Energy-Momentum: Thrust of Gravity Field Momentum".

Section 1.12 Mass Density Tensor a Discrete Manifold

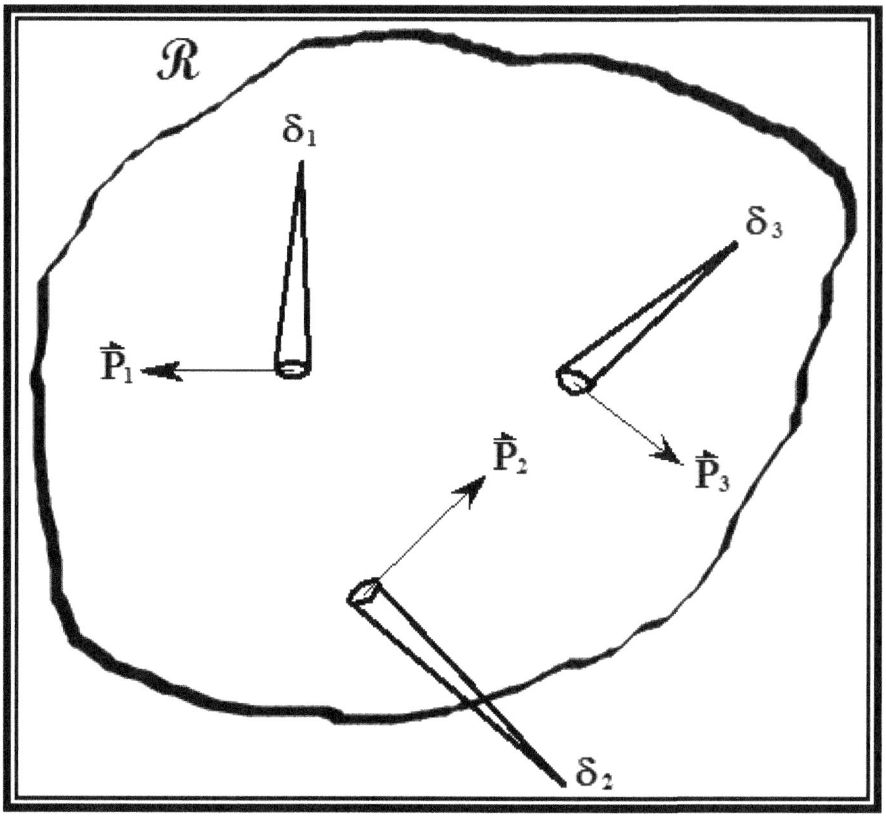

Figure 1.12.1 Particle Momentum Entering and Leaving a Region of Space

Force at a-distance weather an electrical or gravitational field plays its actions out within a system or region of space defining a field, which acts as the source. The particles leave or enter the region with a certain momentum by crossing through a hyperspace boundary. The particles within the space act like actors upon a stage and play out their actions as they interact with each other.

These particles within this given region of space carry out their actions having momentum, charge and inertial mass. These quantities can best be modeled as a mathematical tensor mass density within the region.

Theorem 1.12.1 Mass Density Tensor with Position, Velocity and Momentum

1g	Given	$E_{i\alpha}{}^{\beta}$	≡	$(\partial M_{i\alpha}{}^{\beta} / \partial V)\, c^2$ Einsteinian energy
2g	The position of a particle is marked within that region (see figure F1.12.1) with an impulse function [GEO67II pg 44] and momentum density as four-dimensional vector.	$\mathbf{P}_{vi}{}^{\alpha}(t)$	≡	$p_{vxi}{}^{\alpha}\, \mathbf{i} + p_{vyi}{}^{\alpha}\, \mathbf{j} + p_{vzi}{}^{\alpha}\, \mathbf{k} + [E_{vti\alpha}{}^{\beta} / c]\, \mathbf{w}$
3g	Dimensionality of momentum density			$[M][L^{-2}][T^{-1}]$
Step	Hypothesis	$T_{\alpha}{}^{\beta}$	=	$\Sigma_i\, (\mathbf{v}_{i\alpha} / c) \bullet (\mathbf{P}_{vi}{}^{\beta} / c)\, \delta(.x - x_i(t).)$
1	Let the density for a small finite region of space be	$T_{\alpha}{}^{\beta}$	≈	$\delta\rho_{\alpha}{}^{\beta}$
2	Let the sum of all nuclear particle masses be over that finite volume at a given position-$\delta(x_i)$	$\delta\rho_{\alpha}{}^{\beta}$	≈	$[(\Sigma_i\, \Delta M_{i\alpha}{}^{\beta}) / \Delta V]\, \delta(.x - x_i(t).)$ [1.12.1]
3	For a large number of particles, on the order of Avogadro's number (see, CxPPC.A.2.3.14) can approximate a continuum. A continuum with fine enough divisibility that it can be mathematically modeled as partial differential limit for the mass density and marking its position with a Dirac delta function; for $n > N_a \sim 10^{23}$	$\rho_{\alpha}{}^{\beta}$	≡	$\Sigma_i\, (\partial M_{i\alpha}{}^{\beta} / \partial V)\, \delta(.x - x_i(t).)$
4	From 1, 3 and A4.2.3 Substitution	$T_{\alpha}{}^{\beta}$	=	$\Sigma_i\, (\partial M_{i\alpha}{}^{\beta} / \partial V)\, \delta(.x - x_i(t).)$
5	From 1g	$E_{i\alpha}{}^{\beta}$	=	$(\partial M_{i\alpha}{}^{\beta} / \partial V)\, c^2$
6	From 5 and can best be represented by the inner product of the particle's velocity and momentum	$(\partial M_{i\alpha}{}^{\beta} / \partial V)\, c^2$	≡	$\mathbf{v}_{i\alpha} \bullet \mathbf{P}_{vi}{}^{\beta}$
7	From 6 and T4.4.5B Equalities: Reversal of Right Cancellation by Multiplication	$(\partial M_{i\alpha}{}^{\beta} / \partial V)$	=	$\mathbf{v}_{i\alpha} \bullet \mathbf{P}_{vi}{}^{\beta} / c^2$
8	From 7, D4.1.18C Negative Exponential, A4.2.10 Commutative Multp and A4.2.11 Associative Multp	$(\partial M_{i\alpha}{}^{\beta} / \partial V)$	=	$(\mathbf{v}_{i\alpha} / c) \bullet (\mathbf{P}_{vi}{}^{\beta} / c)$
∴	From 4, 8 and A4.2.3 Substitution	$T_{\alpha}{}^{\beta}$	=	$\Sigma_i\, (\mathbf{v}_{i\alpha} / c) \bullet (\mathbf{P}_{vi}{}^{\beta} / c)\, \delta(.x - x_i(t).)$

From step-1 it is dimensionally represented simply as density, were it gets its third name ***Mass Density Tensor***.

$$T_{\alpha}{}^{\beta} \rightarrow [M]\,[L^{-3}] \qquad \text{Equation A}$$

[1.12.1]Note: Development of the Dirac delta function. [CHE70 pg 76]

Section 1.13 G-E Continuum Field Equations

In the history of the modern field equation the models are of a physical space that acts continuously on a particle source, as shown by Maxwell and Einstein with a set of continuum field equations. As shown in table Tb1.13.1 "Source-Field Equations" start on the left hand side with a model of the monopole source. This arrangement of the monopole, as point source, provides away of driving the field on the right hand side. This simple form is what gave Einstein the idea for a gravity system and kept him on track when developing his famous dynamic law of gravity.

Table 1.13.1 Source-Field Equations

Monopole Sources		Field Equation	Lemma & Theorem	Name of Law	Developer
ρ / ε	=	$\nabla \bullet \vec{E}$	LxMXE.A.3.4.1	Coulomb's Law	Maxwell
0	=	$\nabla \bullet \vec{B}$	LxMXE.A.3.4.2	Gauss's Law	Maxwell
$\mu(\varepsilon\, \partial\vec{E} / \partial t + \vec{J})$	=	$\nabla \times \vec{B}$	LxMXE.A.3.4.3	Ampère's Circuital Law	Maxwell
$-\partial\vec{B} / \partial t$	=	$\nabla \times \vec{E}$	LxMXE.A.3.4.4	Faraday's Law	Maxwell
$-\rho_e / \varepsilon_0$	=	$\Box^2 \Phi$	TA.3.6	Coulomb's Quaternion Law	Maxwell
$-\rho_g / \gamma_0$	=	$\nabla^2 u$	TA.3.5	Newton's Poisson Law	Newton
$(8\pi G_0 / c^2)\, T_j^{\ i}$	=	$R_j^{\ i} - ½\, \delta_j^{\ i} R$	TB.6.5	Einstein's Gravity Field Law	Einstein
$(1 / \gamma_0)\, T_j^{\ i}$	=	$R_j^{\ i} - ½\, \delta_j^{\ i} R$	T1.13.4	Einstein's G-E Field Law	Shevtzov
$(1 / \gamma_0)\, T$	=	$-(½n - 1)\, R$	T1.15.3	Anti G-E Joined Fields	Shevtzov
K_{EG}	=	$h_d\, a^{\beta\kappa}\, [a^{\alpha\gamma}\, w_{\alpha}{}^{\nu}{}_{;\beta;\gamma} - W^{\nu}{}_{\beta}]\, a_{\kappa\nu}$	T17.15.15	Shevtzov's G-E Particle Law	Shevtzov

From the above table there are two theorems TB.6.5 "Einstein's Gravity Field Law" and T1.13.4 "Einstein's G-E Field Law", on the source side they have a common mixed tensor $T_j^{\ i}$ described by Einstein as an Energy-Tensor [PER52II pg 148] yet Wienberg [WEI72II pg 43] calls it out as an Energy-Momentum Tenor so, which one is it? Actually both give a correct description in some way, but equation E1.13.9 is an exact description ***Mass Density Tensor***, which is preferred.

Observation 1.13.1 Codazzi vs. Weingarten Metrics

Einstein built his law on gravity based on Codazzi's metrics so he could describe a dynamic gravity field. Likewise Weingarten's metrics can be used to model a single monopole source with two common properties charge and mass. Just as a single nuclear particle internalizes charge and mass together so to the fields must be represented as one using the Weingarten Curvature Tensor.

Observation 1.13.2 Continuum vs. Discrete Field Equations

Historical physical space acts on its particle source(s) continuously without any desecrate disruptions. This is why all the development of equations represents a set of formula in a continuum. If they had used discrete disruptions on the motion of particles it would have been possible to take the fields towards a quantum direction, something Einstein missed out on, which explains why his theory was not able to be unified with quantum mechanics.

Observation 1.13.3 Have I actually done it?

The need for a whole new field description beyond Maxwell's field equations is required in order to combine electric and gravitational fields this might be possible by using T1.14.2 "G-E SNP: Shevtzov's Mixed Curvature Tensor with Constant Gap"? It's one drawback that it models a set of contracted event horizons as a scalar Gaussian curvature, rather than like a monopole on the left hand side of the equation. This model should be able to add and subtract energy into the system, thereby observing the production or annihilation of particles.

Or does it?

Mightn't there be a tensor field equation that can neatly bind together the two G-E fields? Certainly Einstein wasn't able to do it with his model of a singular parametric field law? Theorems TA.3.5 "Newton's Poisson Law" and TA.3.6 "Coulomb's Quaternion Law" still is a strong beckon to me as it was for Einstein, they are so strongly similar, differing only by one term in time. Yet the two fields appear to be completely independent from one another; hence they might be best modeled by separating them by ninety degrees by the use of a pure imaginary number imbedded in the tensor field?

Law 1.13.1 G-E Ninety Degrees Out Of Phase
Gravitational and electrical fields are separated by [$j = \sqrt{-1}$] a pure imaginary number placing the two fields with their monopole sources independently ninety degrees out of phase to one another, gravity being the real part.

Observation 1.13.4 Retarded Potential of a Time Varying Gravitational Field

Just as Maxwell found when trying to create a set of symmetrical field equations he had to extend his magnetic field equation to include induction by electrical currents [HAL66II pg 966]. Similarly this idea will be followed in order to combine the two fields. There has been a great deal of investigation into time varying gravitational fields, with many ingénues experiments developed to try and prove them, at last with not much success, due to the incredible weakness of gravity. Gravitational fields really only show themselves on a cosmic scale were relatively huge masses are involved, while the relatively weak strength of electric fields reveal themselves on micro or macro-scale. In order to properly see the effects of gravity it requires enormous masses to detect a small change in the resulting large forces, therefore requiring the most sensitive instruments to detect these variations, see vibrating bar detector [MIS73 pg 1004]. It is due to this that not much success has been achieved investigating time varying gravitational fields.

Yet a more general proof can be made using a thought experiment to support a time varying gravitational field and gravity waves. When a star collapses and goes supernova it loses a large amount of mass and energy in a very short time. One moment the massive star has a huge amount of mass the next its mass is expelled and drastically reduced. So observing this activity from earth the expelled ionized gases will have a lot of energy pumped into them and will give off radiation. The indirectly glowing ions would allow us to measure the illumination dropping off; consequently we would see a huge ***gravitational collapse*** as a time varying step function. The star is some light years away this would come to us with a retarded potential [$t - (r/c)$], or time delay traveling at the speed of light, which can only be true if it is a traveling wave of gravity. This traveling wave phenomenon implies that gravity must have a time varying component. So a time varying term for gravity must be added into the gravitational field dynamics. This would allow a linear distribution of gravity and electricity, thereby giving away to join them.

A reverse argument could also be made for ram pressures of intergalactic clouds crashing into each other in a nebula causing a seeding process with a gravitational wave traveling in the opposite direction, ***gravitational compression***, thereby creating a mechanism to start forming a star or stars.

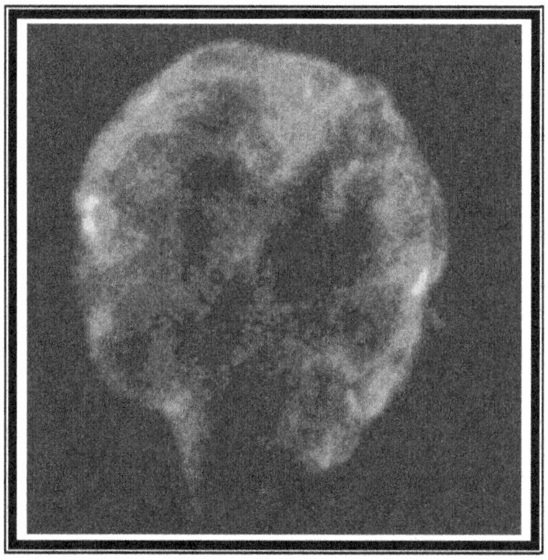

Figure 1.13.1 Collapsing Cygnus Loop Generates X-rays
[http://imagine.gsfc.nasa.gov/Images/basic/xray/rosat06_small.gif]

All that remains of the original Cygnus Loop star is a small, super-dense core composed almost entirely of neutrons — a neutron star. Or, if the original star is very massive indeed (say 15 or more times the mass of our Sun), even the neutrons cannot survive the core collapse...and a ***black hole*** forms.

Figure 1.13.2 NGC6745 Bird's eye Nebula Collision Results in Compression
[http://www.cstarsoas.org/NGC_6745.jpg]

High relative velocities cause ram pressures at the surface of contact between the interacting interstellar clouds. This pressure, in turn, produces material densities sufficiently extreme as to trigger star formation threw gravitational compression.

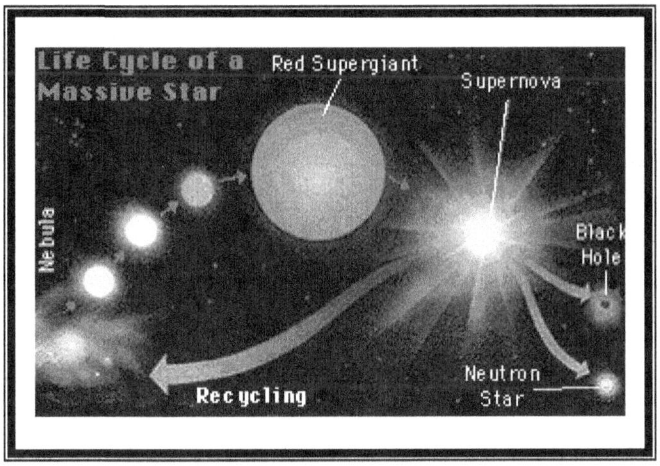

Figure 1.13.3 Supernovae from Single, Massive Stars
[http://imagine.gsfc.nasa.gov/Images/basic/xray/supernova_cycle.gif]

Law 1.13.2 Retarded Potential of a Gravitational Field with Time Term
Gravity travels with a retarded potential time delay [t – (r/c)] at the speed of light, resulting in a covariant time term between the D'Alembert Quaternion, and the Covariant Laplacian Operators. Thus extending the potential energy in a Riemannian coordinate system to four-dimensions, thereby allowing the combining of operators Dx9.6.2.1A and Dx9.6.10.1A to include the missing time varying term, and by inspection;

$$\Box^2 u \sim \nabla^2 u + g_{44}\, u_{,4,4} \qquad \text{Equation A}$$

For a space-time continuum the system can exactly be extended to a specific four term expression, by letting n = 4, in its general tensor form of the T15.3.1 "Laplacian Operator: Definition of Laplace", which gives;

$$\Box^2 u \equiv (\sqrt{g}\, \Sigma_i \Sigma_j\, g^{ij}\, u_{,j})_{,i} / \sqrt{g} \qquad \text{Equation B}$$

from the diagonal, such that

$$x_4 \equiv jct \qquad \text{time varying space term} \qquad \text{Equation C}$$

Section 1.14 *Newtonian Acceleration as an Intrinsic Limit*

In Newtonian space a particle moves freely in a gravitational field along a trajectory path with respect to time. It can best be represented by the equation of an affine spatial curve, see theorem T16.3.3 "Affine Field: Affine Spatial Curve" and T16.3.4 "Affine Field: Affine Temporal Time-Line" with no external acceleration $a^i = 0$.

It is immediately seen that on the left hand side the intrinsic double derivative with respect to time is a Newtonian acceleration term as a limit.

For no external acceleration the velocity cross terms are zero only differentials with respect to themselves on the diagonal remain.

Observation 1.14.1 Approximating Slow Velocities

With respect to T16.3.4 "Affine Field: Affine Temporal Time-Line"; let the gravitational field be quasistatic field ($0 = g_{i\alpha,j}$), by confining ourselves to the case where the motion of matter generating the gravitational field is slow moving (in comparison with the speed of light), in such a case a small difference then becomes a small differentiation, hence negligible and if appropriate may be neglected with respect to space and time.

Section 1.15 G-E Fields: Combined with a Proportionality Ratio

Theorem 1.15.1 G-E Fields: Proportionality Ratio

Step	Hypothesis	$\sqrt{(\varepsilon_0 / \gamma_0)} = 8.616825824 \text{ x}10^{-11}$ C / kg EQ A	[Q][M^{-1}] EQ B
1	From CxPPC.A.2.3.8 Permittivity constant for free space	$\varepsilon_0 = 8.854185(29) \text{ x}10^{-12}$ F m^{-1}	
2	From 1; CxPPC.A.2.3.2 Reciprocal Newtonian constant	$\gamma_0 = 1.1924(87)89 \text{ x } 10^{+9}$ N^{-1}*m^{-2} kg^2	
3	From 1 and 2 divide	$\varepsilon_0 / \gamma_0 = 7.424968729 \text{ x}10^{-21}$ C^{+2} / kg^{+2}	
∴	From 3, T4.10.1 Radicals: Uniqueness of Radicals, DxUNT.A.2.1.4 Charge and DxUNT.A.2.1.1 Mass	$\sqrt{(\varepsilon_0 / \gamma_0)} = 8.616825824 \text{ x}10^{-11}$ C / kg EQ A	[Q][M^{-1}] EQ B

Observation 1.15.1 An Order of Magnitude

Note that for Einstein's Gravity Field Law and the G-E Field Law are related by a constant of proportionality $\sqrt{(\varepsilon_0 / \gamma_0)}$ within the Riemannian curvature field. The question arises with two different constants, what effect would the difference in magnitude have on all the calculations that have been done in the past hundred years of their use in Einstein's gravity field law? In using these proportional constants the magnitude to one another comes out very close; $(1/4\pi\gamma_0) \sim 6.7 \text{ x } 10^{-11}$ and $\sqrt{(\varepsilon_0 / \gamma_0)} \sim 8.6 \text{ x } 10^{-11}$, in short they are perceptually small in comparison. As the old equation is exchanged for the new one then the units of change give the same approximate order of magnitude yielding similar results

Theorem 1.15.2 G-E Fields: Joining Gravitational and Electrical Fields

Step	Hypothesis	$(-1 / \gamma_0) [\rho_g + j \sqrt{(\gamma_0/\varepsilon_0)} \rho_e] = \Box^2 [u + j \sqrt{(\varepsilon_0 / \gamma_0)} \Phi]$	
1	From TA.3.6 NLDA: Coulomb's Law for Covariant D'Alembert Quaternion Operator	$(-\rho_e / \varepsilon_0) = \Box^2 \Phi$ Coulomb's Quaternion Law	[M][Q^{-1}][T^{-2}]
2	From TA.3.5 NLPE: Newton's Law for Poisson's Equation and L1.13.2(A, B) Gravitational Field with Time Term	$(-\rho_g / \gamma_0) = \Box^2 u$ Newton's Poisson Law	[T^{-2}]
3	From 1, T1.15.1 G-E Fields: Proportionality Ratio and T4.4.2 Equalities: Uniqueness of Multiplication: dimensionally correct [T^{-2}]	$\sqrt{(\varepsilon_0 / \gamma_0)} (-\rho_e / \varepsilon_0) = \sqrt{(\varepsilon_0 / \gamma_0)} \Box^2 \Phi$	
4	From 1, 2 and L1.13.1 G-E Ninety Degrees Out Of Phase	$(-\rho_g / \gamma_0) + j \sqrt{(\varepsilon_0 / \gamma_0)} (-\rho_e / \varepsilon_0) = \Box^2 u + j \sqrt{(\varepsilon_0 / \gamma_0)} \Box^2 \Phi$	
5	From 4, A4.2.12 Identity Multp, A4.2.13 Inverse Multp, D4.1.17 Exponential Notation and T4.10.3 Radicals: Identity Power Raised to a Radical	$(-\rho_g / \gamma_0) + j \sqrt{(\gamma_0\varepsilon_0)} / \gamma_0 (-\rho_e / \varepsilon_0) = \Box^2 u + j \sqrt{(\varepsilon_0 / \gamma_0)} \Box^2 \Phi$	

6	From 5, D4.1.20 Negative Coefficient, A4.2.14 Distribution, A4.2.12 Identity Multp, A4.2.13 Inverse Multp, D4.1.17 Exponential Notation, and T4.10.3 Radicals: Identity Power Raised to a Radical,	$(-1/\gamma_0)\,[(\rho_g) + j\,\varepsilon_0\sqrt{(\gamma_0/\varepsilon_0)}\,(\rho_e/\varepsilon_0)] = \Box^2 u + j\sqrt{(\varepsilon_0/\gamma_0)}\,\Box^2 \Phi$
∴	A4.2.10 Commutative Multp, A4.2.13 Inverse Multp, A4.2.12 Identity Multp and Lx9.3.2.2B Differential as a Linear Operator	$(-1/\gamma_0)\,[\rho_g + j\sqrt{(\gamma_0/\varepsilon_0)}\,\rho_e] = \Box^2\,[u + j\sqrt{(\varepsilon_0/\gamma_0)}\,\Phi]$ Gravitational-Electrical anti-point source generating a complex potential energy field

Observation 1.15.2 It is done!

A special tensor field (see observation 1.13.3) is not required instead only a simple complex number does the trick and the gravitational and electrical fields are joined. Einstein had been working too hard, all the way to his death bed. Too bad he didn't notice the missing time term in the Poisson gravity equation. [EINα19, page 191] [WEYLα18, page 201] [ISAβ07, page 543]

Theorem 1.15.3 G-E Fields: Joined Fields in Newtonian Geodesic space $\mathfrak{R}^n$

1g	Given	$N \equiv \rho_g + j\sqrt{(\gamma_0/\varepsilon_0)}\,\rho_e$ $[T^{-2}]$ Normal Point Source
2g		$U \equiv u + j\sqrt{(\varepsilon_0/\gamma_0)}\,\Phi$ $[L^2]\,[T^{-2}]$ G-E Potential Energy Point Source
3g		$T \equiv ½\,(½\,n - 1)\,N$ Newton's energy-momentum tensor
4g		$\sqrt{g} \equiv 1$ Euclidian (Newtonian) Space
5g		$g^{ij} \equiv \delta^{ij}$
Step	Hypothesis	$(1/\gamma_0)\,T = -(½n - 1)\,R$ Anti-Energy field Newtonian Complex Subspace $\mathfrak{R}^n$
1	From T1.15.2 G-E Fields: Combining Gravitational and Electrical Fields	$(-1/\gamma_0)\,[\rho_g + j\sqrt{(\gamma_0/\varepsilon_0)}\,\rho_e] = \Box^2\,[u + j\sqrt{(\varepsilon_0/\gamma_0)}\,\Phi]$
2	From 1g, 2g, 4 and A4.2.3 Substitution	$(-1/\gamma_0)\,N = \Box^2 U$
3	From 4g, 5g, 2 and T15.3.1 Laplacian Operator: Definition of Laplacian and A4.2.3 Substitution in an approximated Euclidian Space $\mathfrak{E}^n$	$(-1/\gamma_0)\,N \approx (1\,\Sigma_i\,\Sigma_j\,\delta^{ij}\,U_{,j})_{,i}/1$
4	From 3, D4.1.20A Negative Coefficient and T4.8.2B Integer Exponents Negative One Squared; Contravariant D'Alembert Quaternion Operator	$(-1)(1/\gamma_0)\,N \approx (-1)(-1)\,\Sigma_i\,\Sigma_j\,(\delta^{ij}\,U_{j,i})$

5	From 4, T4.4.3 Equalities: Cancellation by Multiplication, T12.2.21A Covariant Metric Determinate in Euclidian Manifold $\mathfrak{E}^n$ and D4.1.20A Negative Coefficient; for an orthogonal Newtonian subspace	$(1 / \gamma_0)\, N \approx (-1) \sum_i \sum_j (\delta^{ij} U_{j,i})$
6	From 5, T10.2.10A Tool: Identity Multiplication of a Tensor, C12.8.5.4 Differentiation of Contravariant Metric and Covariant Coefficient	$(1 / \gamma_0)\, N \approx (-1) (\sum_i U_{i,i})$
7	From 6, T14.5.3 Geodesic Space $\mathfrak{G}^n$: Laplacian Contravariant Metric and T10.2.5 Tool: Tensor Substitution	$(1 / \gamma_0)\, ½\, N \equiv (-1)R$
8	From 7 and T10.4.1B Scalars and Tensors: Uniqueness of Scalar Multiplication to a Tensors	$(1 / \gamma_0)\, ½\, (½n - 1)\, N = -(½n - 1)R$
∴	From 3g, 8 and T10.2.5 Tool: Tensor Substitution	$(1 / \gamma_0)\, T = -(½n - 1)\, R$ Iff true in Newtonian Subspace $\mathfrak{S}^n$ and Newtonian Geodesic space $\mathfrak{G}^n$ then it is true in all Riemannian spaces $\mathfrak{R}^n$ in general.

Observation 1.15.3 G-E Fields: Opposition of Potential Energy to Gravitational Contraction

The above theorem demonstrates the total net potential energy in opposition to repulsion of matter by the combination of G-E fields (see step 6). This anti-potential or repulsive potential could be in opposition to the contracting force of gravity and thereby shape the cosmos by changing the rate of gravitational collapse.

Theorem 1.15.4 Einstein's Dual Parametric General G-E Field Equation

Steps	Hypothesis	$(1 / \gamma_0)\, T_i^{\ j} = R_i^{\ j} - ½\, \delta_i^{\ j} R$
1	From T14.4.7 "Riemann-Christoffel Tensor Rank-0: Contraction of Einstein's Covariant Tensor", T1.13.3 G-E Fields: Combined Fields in Newtonian Geodesic space $\mathfrak{R}^n$ and T10.2.5 Tool: Tensor Substitution	$T\, \xi = T\, (1 / \gamma_0)$
2	From 1 and equating constants	$\xi = (1 / \gamma_0)$

∴	From 2, T14.4.5 Riemann-Christoffel Tensor Rank-2: Einstein's Mixed Tensor, T12.2.7B Transpose of Mixed Kronecker Delta and A4.2.3 Substitution; substitution of dummy indices k → j and a → i	$(1 / \gamma_0)\, T_i^{\ j}$	=	$R_i^{\ j} - \frac{1}{2}\, \delta_i^{\ j} R$

Observation 1.15.4 Einstein's Dual Parametric General G-E Field Equation in Completed Form

Einstein's general gravity field equation can now be stated in its complete form from Newton's gravity equation: "Newton's Law for Poisson's Equation" and Coulomb's electrical field equation "Coulomb's Law for Covariant D'Alembert Quaternion Operator", finally it can be seen how they complement one another. It is not surprising that since they are linearly related that they have the same tensor form as Einstein's gravity field equation.

Theorem 1.15.5 Einstein Gravity Field Equation of Zero Rank

Steps	Hypothesis	$(8\pi G_0 / c^2)\, T$	=	$(\frac{1}{2}n - 1)\, R$
1	From T17.11.6 Hypersphere: Relationship of Riemann's to Gauss's Total Curvature	R	=	$n(1-n)\, K$
2	From 1, T17.11.8A Hypersphere: Riemannian Total Curvature Einstein and Gauss's Relationship, TB.6.4 Newton-Einstein: Einstein's Gravitational Equality Factor and A4.2.3 Substitution	$(8\pi G_0 / c^2)\, T$	=	$(\frac{1}{2}n - 1)\, n(n-1)\, K$
∴	From 1, 2 and A4.2.3 Substitution	$(8\pi G_0 / c^2)\, T$	=	$(\frac{1}{2}n - 1)\, R$

Section 1.16 Shevtzov's G-E Structure for Nuclear Particles

Field systems can now be considered that are completely independent of Maxwell's electromagnetic equations, yet describe the interrelation between gravity and electrical fields modeling a nuclear particle.

Theorem 1.16.1 G-E SNP: Shevtzov's General G-E Particle Field Equation

1g	Given	Physical model of spinning photon with 2 event horizons, for electrical and gravitational fields; n = 2 field space.
2g		$K_{GE} \equiv \kappa_g \kappa_e$ Model of G-E Gaussian curvature to principle curvatures
3g		$\kappa_1 \equiv \kappa_e$ Electrical principle curvature
4g		$\kappa_2 \equiv \kappa_g$ Gravitational principle curvature
Step	Hypothesis	$K_{GE} \approx h_d\, a^{\beta\kappa}\, [a^{\alpha\gamma}\, w_\alpha{}^\nu{}_{;\beta;\gamma} - W^\nu{}_\beta]\, a_{\kappa\nu}$ $+ 3\, (h_d)_{,\gamma}\, a_{\kappa\nu} w_\alpha{}^\nu{}_{;\beta}\, a^{\beta\kappa}\, a^{\alpha\gamma}$ $+ 3\, (h_d)_{,\alpha,\beta}\, a^{\beta\kappa}\, w_\gamma{}^\nu\, a_{\kappa\nu}\, a^{\alpha\gamma}$
1	From T17.15.14 Parallel Hypersurfaces $\mathfrak{H}^{n-1}$: Gaussian Curvature in a Dynamic Gap	$(n-1)\, K \approx h_d\, a^{\beta\kappa}\, [W^\nu{}_\beta - a^{\alpha\gamma}\, w_\alpha{}^\nu{}_{;\beta;\gamma}]\, a_{\kappa\nu}$ $- 3\, (h_d)_{,\gamma}\, a_{\kappa\nu} w_\alpha{}^\nu{}_{;\beta}\, a^{\beta\kappa}\, a^{\alpha\gamma}$ $- 3\, (h_d)_{,\alpha,\beta}\, a^{\beta\kappa}\, w_\gamma{}^\nu\, a_{\kappa\nu}\, a^{\alpha\gamma}$
2	From T17.12.8B Minor Principal Curvatures: Relation to Gaussian Principal Curvatures	$K = -[2/n(n-1)]\, \Sigma_{\alpha<\beta} \kappa_\alpha \kappa_\beta$
3	From 1g, 2 and A4.2.3 Substitution; evaluating the summation of the inequality	$K = -[2/2(2-1)]\, \kappa_1 \kappa_2$
4	From 3g, 4g, 3, D4.1.19 Primitive Definition for Rational Arithmetic and A4.2.12 Identity Multp and A4.2.3 Substitution	$K = -\kappa_g \kappa_e$
5	From 2g, 4, 9 and A4.2.3 Substitution	$K = -K_{GE}$
6	From 4, 5 and A4.2.3 Substitution	$K_{GE} = \kappa_g \kappa_e$
7	From 6, D17.10.3A Principal Curvatures to Variable Optimally Separated Paths and A4.2.3 Substitution	$\kappa_g \kappa_e = 1 / r_g r_e$
8	From 7, T1.6.4(A,B) Monocentric Particle Outside: Gaussian Curvature of Dual Spins	$1 / r_g r_e = \varepsilon_{rq} / L_{plk}{}^2$
9	From 8, Equation 1.3.2 Plank's Length, T4.10.4 Radicals: Identity Radical Raised to a Power and A4.2.3 Substitution	$\varepsilon_{rq} / L_{plk}{}^2 = 4\pi\gamma_0 c^3 \varepsilon_{rq} / \hbar$
10	From 6, 7, 8, 9 and A4.2.3 Substitution	$K_{GE} = 4\pi\gamma_0 c^3 \varepsilon_{rq} / \hbar$

11	From 1g, 1, 5 and A4.2.3 Substitution	$(2 - 1)(-K_{GE}) \approx h_d\, a^{\beta\kappa}\, [W^{\nu}{}_{\beta} - a^{\alpha\gamma}\, w_{\alpha}{}^{\nu}{}_{;\beta;\gamma}]\, a_{\kappa\nu}$ $- 3\, (h_d)_{,\gamma}\, a_{\kappa\nu} w_{\alpha}{}^{\nu}{}_{;\beta}\, a^{\beta\kappa}\, a^{\alpha\gamma}$ $- 3\, (h_d)_{,\alpha,\beta}\, a^{\beta\kappa}\, w_{\gamma}{}^{\nu}\, a_{\kappa\nu}\, a^{\alpha\gamma}$
12	From 11, D4.1.19 Primitive Definition for Rational Arithmetic, D4.1.20 Negative Coefficient, A4.2.14 Distribution T4.8.2B Integer Exponents: Negative One Squared and T4.4.3 Equalities: Cancellation by Multiplication	$K_{GE} \approx h_d\, a^{\beta\kappa}\, [-W^{\nu}{}_{\beta} + a^{\alpha\gamma}\, w_{\alpha}{}^{\nu}{}_{;\beta;\gamma}]\, a_{\kappa\nu}$ $+ 3\, (h_d)_{,\gamma}\, a_{\kappa\nu} w_{\alpha}{}^{\nu}{}_{;\beta}\, a^{\beta\kappa}\, a^{\alpha\gamma}$ $+ 3\, (h_d)_{,\alpha,\beta}\, a^{\beta\kappa}\, w_{\gamma}{}^{\nu}\, a_{\kappa\nu}\, a^{\alpha\gamma}$
∴	From 12 and T10.3.16 Addition of Tensors: Commutative by Addition	$K_{GE} \approx h_d\, a^{\beta\kappa}\, [a^{\alpha\gamma}\, w_{\alpha}{}^{\nu}{}_{;\beta;\gamma} - W^{\nu}{}_{\beta}]\, a_{\kappa\nu}$ $+ 3\, (h_d)_{,\gamma}\, a_{\kappa\nu} w_{\alpha}{}^{\nu}{}_{;\beta}\, a^{\beta\kappa}\, a^{\alpha\gamma}$ $+ 3\, (h_d)_{,\alpha,\beta}\, a^{\beta\kappa}\, w_{\gamma}{}^{\nu}\, a_{\kappa\nu}\, a^{\alpha\gamma}$

[Date completed April 27 2010; 4:10pm]

Theorem 1.16.2 G-E SNP: Shevtzov's G-E Particle Field Equation with Constant Gap

1g	Given	$h_d = \text{constant}$
Step	Hypothesis	$K_{GE} \approx h_d\, a^{\beta\kappa}\, [a^{\alpha\gamma}\, w_{\alpha}{}^{\nu}{}_{;\beta;\gamma} - W^{\nu}{}_{\beta}]\, a_{\kappa\nu}$
1	From T1.14.1 G-E SNP: Shevtzov's General G-E Particle Field Equation	$K_{GE} \approx h_d\, a^{\beta\kappa}\, [a^{\alpha\gamma}\, w_{\alpha}{}^{\nu}{}_{;\beta;\gamma} - W^{\nu}{}_{\beta}]\, a_{\kappa\nu}$ $+ 3\, (h_d)_{,\gamma}\, a_{\kappa\nu} w_{\alpha}{}^{\nu}{}_{;\beta}\, a^{\beta\kappa}\, a^{\alpha\gamma}$ $+ 3\, (h_d)_{,\alpha,\beta}\, a^{\beta\kappa}\, w_{\gamma}{}^{\nu}\, a_{\kappa\nu}\, a^{\alpha\gamma}$
∴	From 1g, 1, Lx9.3.3.1B Differential of a Constant, T10.2.7A Tool: Zero Times a Tensor is a Zero Tensor, T10.2.9 Tool: Zero Tensor Plus a Zero Tensor is a Zero Tensor and T10.3.18 Addition of Tensors: Identity by Addition	$K_{GE} \approx h_d\, a^{\beta\kappa}\, [a^{\alpha\gamma}\, w_{\alpha}{}^{\nu}{}_{;\beta;\gamma} - W^{\nu}{}_{\beta}]\, a_{\kappa\nu}$

Observation 1.16.1 Proof of Theory by Experimentation

Einstein at the end of every paper he wrote proposed experiment to verify his theory. The written theory was self-evident having been derived from first principles. Yet a sense of confidence needed to be established for the theory so he made a prediction that could be verified by experimentation. This theory was designed to predict particle formation and the gravitational and electric fields they generate.

Section 1.17 Spatial and Time Varying Gap between Concentric Surfaces

Looking at the "Logarithmic Charge Particle Electron Mass Levels" there appears to be a predictable deterministic change in particle formation in discrete levels. This would imply a type of quantum development that could only be described by a wave of formation between the gap of the GE field. Clearly the wave would have spatial position between the gaps and in order to go from level-to-level would have to change with time.

Let's consider two parallel surfaces for a GE field that has a varying gap distance h_d with concentric spherical surfaces in space and time.

Definition 1.17.1 Spatial and Time Varying GE Wave Gap

$$h_d(\alpha, \beta, \gamma, t) \equiv r_e - r_g \qquad \text{Equation A}$$

$$h_d(\alpha, \beta, \gamma, t) > 0 \qquad \text{Equation B}$$

$$h_d(\alpha, \beta, \gamma, t) \equiv h_0\, w_p(\kappa_p \bullet r_p - j\omega_p t) \qquad \text{Equation C}$$

Section 1.18 *The Fossil Record at the Instant of Event One*

Universal Physical Quantum Constants (UPQC) it was speculated by Plank that the physical constants might vary with time in order to account for their extreme reciprocal magnitudes. This has lead to a large number of creative suppositions, among them Gorge Gamov proposing what the consequences would be if this premises were true. This thesis is based on the interrelation between these constants and their independence of time, which have clear physical representations that direct how the working of the universe's machinery operates. As an example holding all UPQC's constant, but letting the quantum charge change for electrostatic force very would shape another universe that would be quite different as observed today. Based on this idea one could conclude that these constants must have been time invariant both at the instant of event-one and as measured presently. It would follow then that we have a window looking at the forces at that moment, barrowing the nomenclature from Paleontology, the UPQC's can be thought of as a physical fossil record giving us measurements and the physics at the time of conception for event-one.

One might also ask the question what if these constants vary with other event-one's giving rise to other completely different universes in physical space? As an example in some universes it might be possible to measure the speed of light at another velocity? If true it can be concluded that the constants are conditional, initializing creations of that particular universe.

On the other hand if they turn out to be the same for all universes than these constants are truly universal and no matter where we are in space they are invariant and will always be measured the same regardless of the frame of reference an observer would gauge them in, so at this time Plank and Einstein remain unchallenged.

Section 1.19 Nuclear & Neutral Particle Formation

This section reviews the fundamental nuclear particles and their hierarchy of formation hopefully to revile correspondences with possible links and connections that might lead to a mechanism to predict particle and neutral formation.

Observation 1.19.1 GE-Field and a Weak Nuclear Force

Clearly there are other intermediate particle exchanges inside the atomic nucleus giving rise to other non-GE forces. If a spinning photon in the immediate vicinity, or local neighborhood of two composite particles such as alpha-particles were to unwind and then rewind into a spinning particle again it could generate a local moment of force for attraction under the law of concurrency. This force would only exist just in the local vicinity and be observed as a weak force of attraction.

Table 1.19.1 Particle Structure

	Particle Family	Decay Formula	m_e	Mass MeV/c^2			
1	Lepton Family	$e^- = e^-$	1.0	0.511			
2	[1.19.1]	$e^+ = e^+$	1.0	0.511			
3	[1.19.2]	$\mu^- \rightarrow e^- + `\nu_e + \nu_\mu$	206.8				
4		$\pi^+ \rightarrow (\mu^- + \nu_\mu) + e^+ + e^+$	273.2				
5		$\pi^- \rightarrow (\mu^- + \nu_\mu) + e^- + e^-$	273.2				
6		$\pi^0 \rightarrow 2\gamma$	264.2				
7	Mesons Family	$K^+ \rightarrow \mu^+ + \nu_\mu$	966.2				
8		$K^+ \rightarrow \pi^+ + \pi^0$	966.2				
9		$K^- \rightarrow \pi^- + \pi^0$	966.2				
10		$K^- \rightarrow \mu^- + `\nu_\mu$	966.2				
11		$K^- \rightarrow \pi^0 + \mu^- + `\nu_\mu$	966.2				
12		$K^0 \rightarrow \pi^+ + (\pi^- + K^+)$	974.0				
13		$K^0 \rightarrow 2\pi^0$	974.0				
14		$K^0 \rightarrow \pi^+ + e^- + `\nu_e$	974.0				
15		$\eta^+ \rightarrow 3\mu^+$ or 2γ	1074.0				
16		$\rho^+ \rightarrow \pi^+ + \pi^0$	1506.8				
17		$\rho^- \rightarrow \pi^- + \pi^0$	1506.8				
18		$\rho^0 \rightarrow \pi^- + \pi^+$	1506.8				
19		$\omega^0 \rightarrow (\pi^- + \pi^+) + \pi^0$	1530.3				

Table 1.19.1 (Continue)

20	Baryon Family [1.19.1]	$p^+ \rightarrow p^+$	1836.2				
21		$p^- \rightarrow p^-$	1836.2				
22		$n^0 \rightarrow p^+ + e^- +$ `ν	1838.7				
23		$\eta^0 \rightarrow \pi^+ + (\pi^- + \eta^+)$	1874.8				
24		$\varphi+ \rightarrow K^- + K^+$	1996.1				
25		$\varphi+ \rightarrow K^0 +$ `K^0	1996.1				
26		$\Lambda^0 \rightarrow p^+ + \pi^-$	2183.7				
27		$\Lambda^0 \rightarrow n^0 + \pi^0$	2183.7				
28		$\Sigma^+ \rightarrow p^+ + \pi^0$	2327.6				
29		$\Sigma^+ \rightarrow n^0 + \pi^+$	2327.6				
30		$\Sigma^0 \rightarrow \Lambda^0 + \gamma$	2333.7				
31		$\Sigma^- \rightarrow n^0 + \pi^-$	2343.1				
32		$\Delta^{++} \rightarrow p^+ + \pi^+$	2411.0				
33		$\Delta^+ \rightarrow p^+ + \pi^0$	2411.0				
34		$\Delta^0 \rightarrow n^0 + \pi^0$	2411.0				
35		$\Delta^- \rightarrow n^0 + \pi^-$	2411.0				
36		$\Xi^0 \rightarrow \Lambda^0 + \pi^0$	2573.4				
37		$\Xi^- \rightarrow \Lambda^0 + \pi^-$	2585.1				
38		$\Omega^- \rightarrow \Xi^0 + \pi^-$	3272.0				
39		$\Omega^- \rightarrow \Lambda^0 + K^-$	3272.0				
40		`$D^0 \rightarrow K^+ + \pi^-$	3648.9				
41		$D^+ \rightarrow K^- + \pi^+ + \pi^+$	3853.2				

Table 1.19.2 Particle Families

Particle Family	Particles by Family	Grouped by Electron Mass
Bosons	γ, ν_e, `$\nu_e$, $\nu_\mu$, `ν_μ, ν_τ, `ν_τ,	$0\ m_e = M$[1.19.2]
Leptons	e^-, e^+, μ^-, μ^+, π^-, π^+, π^0	$1m_e \leq M < 207m_e$
Mesons	K^+, K^-, K^0, η^+, ρ^-, ρ^+, ρ^0, ω^0	$273m_e < M < 1531m_e$
Baryons	p^+, p^-, n^0, η^0, $\varphi+$, Λ^0, Σ^+, Σ^0, Σ^-, Δ^{++}, Δ^+, Δ^0, Δ^-, Ξ^0, Ξ^+, Ξ^-, Ω^-, `D^0, D^+	$1836m_e < M$

[1.19.2]Note: For this study particles are only categorized by mass. Formally, however other properties are used to group them, such as spin and type of interaction. [GAU99 pg 226]

Table 1.19.3 Charged Particles Only

	Particle Family	Decay Formula	m_e	Mass MeV/c^2			
1	Lepton Family	$e^- = e^-$	1.0				
2		$e^+ = e^+$	1.0				
3		$\mu^- \rightarrow e^- +$ `$\nu_e + \nu_\mu$	206.8				
4		$\pi^+ \rightarrow (\mu^- + \nu_\mu) + e^+ + e^+$	273.2				
5		$\pi^- \rightarrow (\mu^- + \nu_\mu) + e^- + e^-$	273.2				
7	Meson Family	$K^+ \rightarrow \mu^+ + \nu_\mu$	966.2				
8		$K^+ \rightarrow \pi^+ + \pi^0$	966.2				
9		$K^- \rightarrow \pi^- + \pi^0$	966.2				
10		$K^- \rightarrow \mu^- +$ `ν_μ	966.2				
11		$K^- \rightarrow \pi^0 + \mu^- +$ `ν_μ	966.2				
15		$\eta^+ \rightarrow 3\mu^+$ or 2γ	1074.0				
16		$\rho^+ \rightarrow \pi^+ + \pi^0$	1506.8				
17		$\rho^- \rightarrow \pi^- + \pi^0$	1506.8				
20	Baryon Family	$p^+ \rightarrow p^+$	1836.2				
21		$p^- \rightarrow p^-$	1836.2				
24		$\varphi+ \rightarrow K^- + K^+$	1996.1				
25		$\varphi+ \rightarrow K^0 +$ `K^0	1996.1				
28		$\Sigma^+ \rightarrow p^+ + \pi^0$	2327.6				
29		$\Sigma^+ \rightarrow n^0 + \pi^+$	2327.6				
31		$\Sigma^- \rightarrow n^0 + \pi^-$	2343.1				
32		$\Delta^{++} \rightarrow p^+ + \pi^+$	2411.0				
33		$\Delta^+ \rightarrow p^+ + \pi^0$	2411.0				
35		$\Delta^- \rightarrow n^0 + \pi^-$	2411.0				
37		$\Xi^- \rightarrow \Lambda^0 + \pi^-$	2585.1				
38		$\Omega^- \rightarrow \Xi^0 + \pi^-$	3272.0				
39		$\Omega^- \rightarrow \Lambda^0 + K^-$	3272.0				
41		$D^+ \rightarrow K^- + \pi^+ + \pi^+$	3853.2				
42	Hadron Family	$H_d \rightarrow \pi + \rho + \eta + \eta' + \varphi + \omega + J/\psi + Y + \theta + K + B + D + T$					

Observation 1.19.2 Hadron Family Partials

Hadron family is the largest partials speculated and may lead to the largest particle that can be created. It is clear from the "Logarithm Charge Particle Electron Mass Levels" chart the particles are eigen (discreet sized) values, but appear to level out as they become larger possible ending with the largest Hadron particle, but there is no clear proof of this. There may be a maximum upper limit to how much energy can be placed into a particle restricted by the impedance of the particle to take on a maximum energy, which would make the curve end asymptotically.

[1.19.1]Note: Particle is fundamental with no other constituent particles comprising it.

[1.19.2]Note: [`] symbolizes antiparticle also if charge, as an example is [+] than the opposite sign [-] implies an antiparticle, see [NAVβ06] and [GAU99 pg 225]

Table 1.19.4 Neutral Particles Only

	Particle Family	Decay Formula	m_e	Mass MeV/c^2			
6	Lepton Family	$\pi^0 \rightarrow 2\gamma$	264.2				
12	Mesons Family	$K^0 \rightarrow \pi^+ + (\pi^- + K^+)$	974.0				
13		$K^0 \rightarrow \pi^0 + \pi^0$	974.0				
14		$K^0 \rightarrow \pi^+ + e^- + \grave{}\nu_e$	974.0				
19		$\rho^0 \rightarrow \pi^- + \pi^+$	1506.8				
20		$\omega^0 \rightarrow (\pi^- + \pi^+) + \pi^0$	1530.3				
22	Baryon Family	$n^0 \rightarrow p^+ + e^- + \grave{}\nu$	1838.7				
23		$\eta^0 \rightarrow \pi^+ + (\pi^- + \eta^+)$	1874.8				
26		$\Lambda^0 \rightarrow p^+ + \pi^-$	2183.7				
27		$\Lambda^0 \rightarrow n^0 + \pi^0$	2183.7				
30		$\Sigma^0 \rightarrow \Lambda^0 + \gamma$	2333.7				
34		$\Delta^0 \rightarrow n^0 + \pi^0$	2411.0				
36		$\Xi^0 \rightarrow \Lambda^0 + \pi^0$	2573.4				
40		$D^0 \rightarrow K^+ + \pi^-$	3648.9				

Observation 1.19.3 Why Neutral Particles Only?

For K^0 and η^0 charge is not conserved leading to the conclusion that they are incorrect do to a type-o, a poor experimental interpretation or a bad result.

The above theory gives a mechanism to build positive and negative particles but what of neutral particles? The sums of charges add to zero creating a neutral particle following the (L1.8.4) zero net pair formation. Since neutral particles can always be decomposed into individual charged components, their origins are self-evident; this is why the focus is only on charged particle pair formation where I believe the greater mystery resides.

Particle formation is well founded for dual particle creation. Based on the above table to create neutral particles it is comprised of a paired set of opposite charged particles canceling their charge out and one exchange particle to pass energy between the particles providing weak attractive force z^0 to bind them together.

$$\pi^0 \rightarrow 2\gamma \qquad \text{Law of Existence EQ A}$$
$$X^0 \rightarrow x^0 + z^0 \qquad \text{Law of Equality EQ B}$$
$$X^0 \rightarrow (x^+ + x^-) + z^0 \qquad \text{Law of Identity EQ C}$$

Table 1.19.5 Laws for Existence of a Neutral Particle

Laws	Laws of Nuclear Sets	Description
Law 1.19.1 Closure in Existence	$\prod^0 \leftarrow\rightarrow \pi^0 = 2\gamma$	Closure with respect to Existence
Law 1.19.2 Equality of Existence	$X^0 \leftarrow\rightarrow x^0$	Equality with respect to Existence
Law 1.19.3 Commutative in Existence	$x^0 + z^0 = z^0 + x^0$	Commutative in Existence
Law 1.19.4 Associative in Existence	$x^0 + (y^0 + z^0) = (x^0 + y^0) + z^0$	Associative in Existence
Law 1.19.5 Identity in Existence	$z^0 + \pi^0 = z^0$	Identity in Existence
Law 1.19.6 Inverse in Existence	$z^+ + z^- = \pi^0$	Inverse in Existence

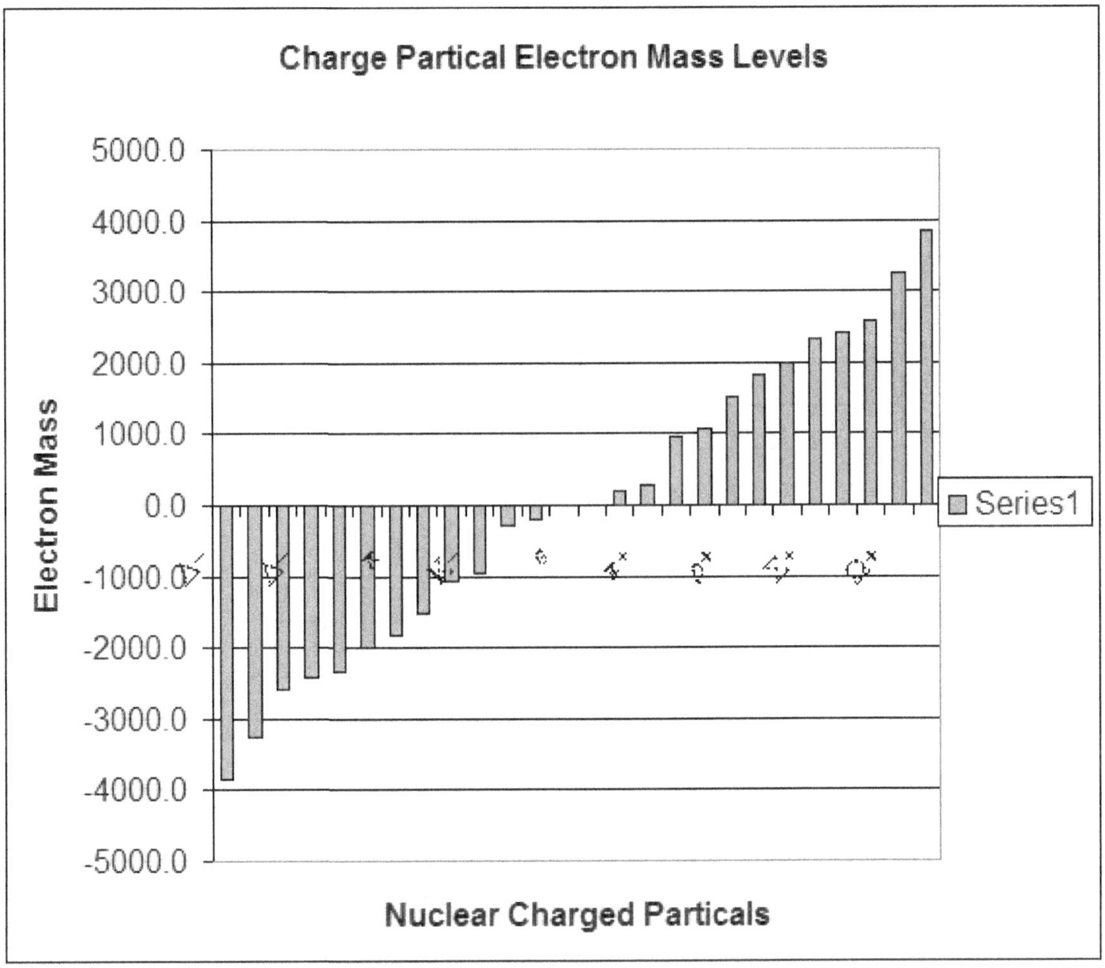

Figure 1.19.1 Electron Mass vs. Nuclear Charged Particles

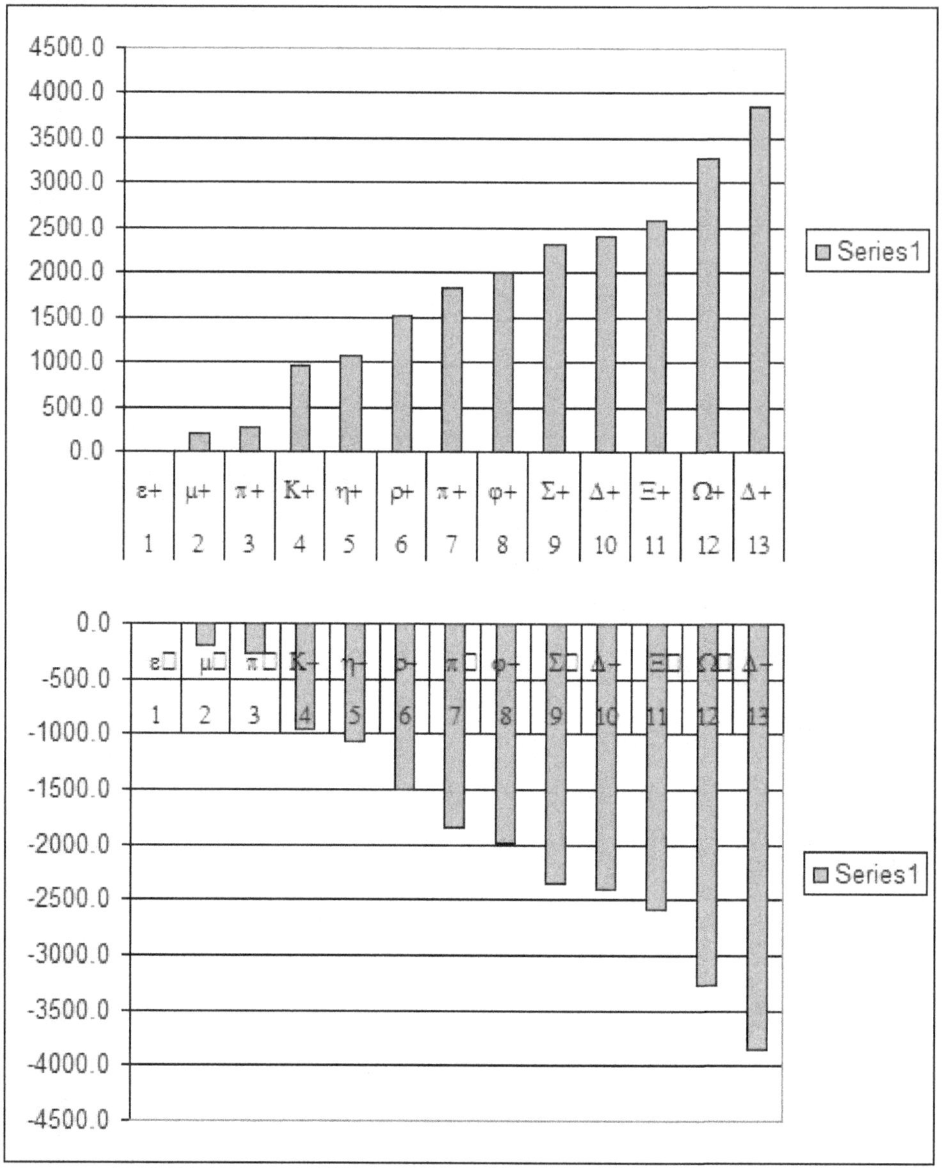

Figure 1.19.2 Nuclear Charged Particle-Antiparticle

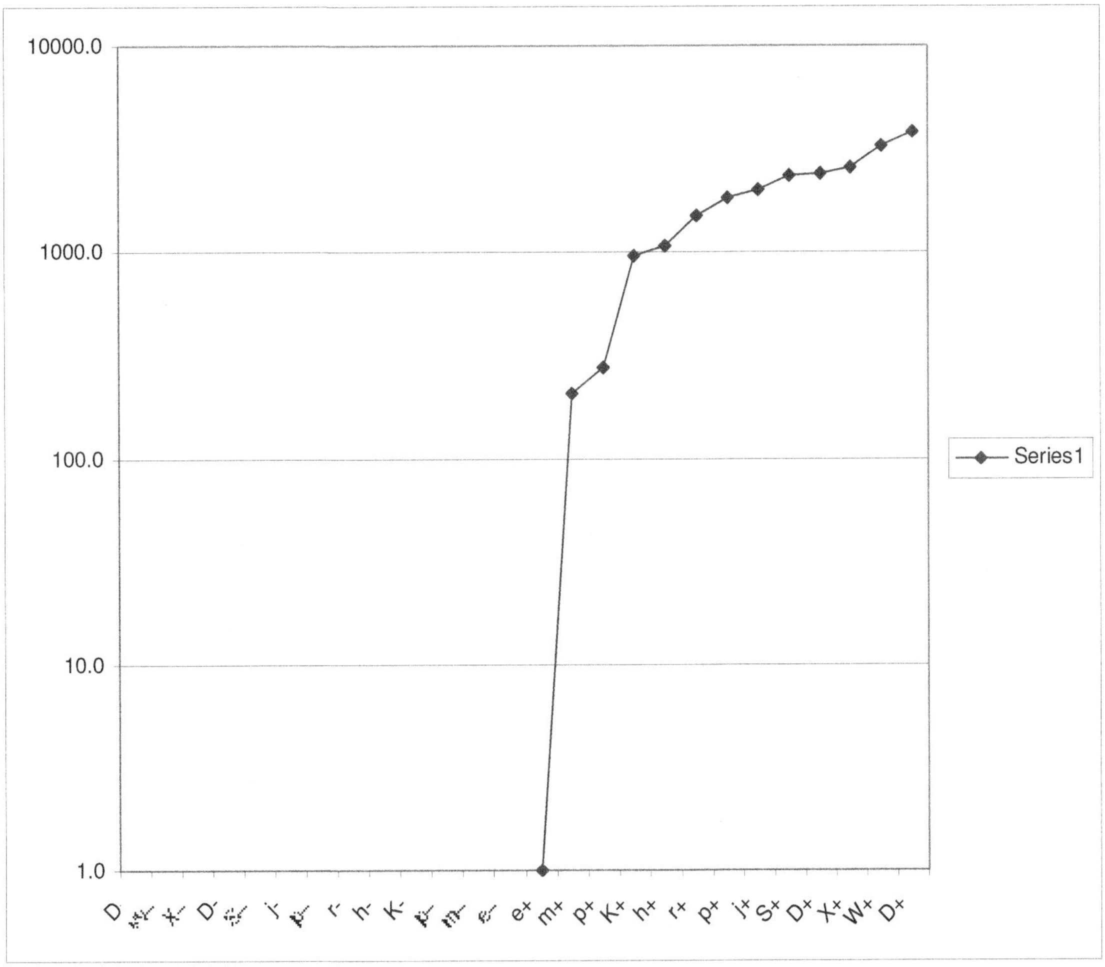

Figure 1.19.3 Logarithm Charge Particle Electron Mass Levels

Table 1.19.6 Fermions (half-integer spin)

Generation	Name/Flavor		Electric charge	(e) Mass (MeV)	Antiquark	
1	up	(u)	+2/3	1.5 to 4.0	antiup quark	($\underline{u}$)
	down	(d)	−1/3	4 to 8	antidown quark	($\underline{d}$)
2	Strange	(s)	−1/3	80 to 130	antistrange	($\underline{s}$)
	Charm	(c)	+2/3	1,150 to 1,350	anticharm quark	($\underline{c}$)
3	Bottom	(b)	−1/3	4,100 to 4,400	antibottom quark	($\underline{b}$)
	Top	(t)	+2/3	178,000 ± 4,300	antitop quark	($\underline{t}$)

Table 1.19.7 Charged lepton / antiparticle

Name	Symbol	Electric Charge (e)	Mass (MeV)
Electron / Positron	e^-/e^+	−1 / +1	0.511
Muon	μ^-/μ^+	−1 / +1	105.7
Tau lepton	τ^-/τ^+	−1 / +1	1,777

Table 1.19.8 Bosons (integer spin)

Name	Charge (e)	Spin	Mass (GeV)	Force mediated
Photon	0	1	0	EG-Field
W	+1 / −1	1	83.4	Weak nuclear
Z^0	0	1	91.2	Weak nuclear
Gluon	0	1	0	Strong nuclear
Higgs	0	0	>112	See below

The Higgs boson (spin-0) is predicted by electro-weak theory, and is the only Standard Model particle not yet observer. In the Higgs mechanism of the Standard Model, the massive Higgs boson is created by spontaneous symmetry breaking of the Higgs field. The intrinsic masses of the elementary particles (particularly the massive W and Z bosons) would be explained by their interactions with this field. Many physicists expect the Higgs to be discovered at the Large Hardron Collider (LHC) particle accelerated now under construction at CEDN

Definition 1.19.2 State of a system

at a given instant in time is the amount of information at that instant that together with the initial conditions, input parameters, determines uniquely the behavior of the system for all time thereafter, Chen[8].

Definition 1.19.3 State of equilibrium zeroed input response

is when the state of a system behaves as if it has a zeroed input Response, Chen[8].

Definition 1.19.4 State stability

in the sense of Lyapunov is when the response due to any initial state that is sufficiently near to equilibrium state will not move far away From the equilibrium state, Chen[8]

Definition 1.19.5 State of an object

is its status, phase, situation, or activity, Embley[5]

Definition 1.19.6 Trigged

is a set of events and conditions that activate a state transition, Embley[5].

Definition 1.19.7 Transition

is a process of changing the state of an object, Embley[5].

Definition 1.19.8 Action

may cause events, create or destroy objects and relationships, observe objects and relationships, and send or receive messages. It can also direct to the next state, Embley[5].

References Volume II

[ANT1719] “Epistolae ad societatem regiam anglicam” by Leeuwenhoek, Antoni van (1632-1723) Lugduni Batavorum: apud Joh. Arnold. Langerak, 1719

[EINα05] "On a Heuristic Viewpoint Concerning the Production and Transformation of Light" by Einstein, Albert published by *Annalen der Physik* **17**: 132–148. This annus mirabilis paper on the photoelectric effect was received by Annalen der Physik March 18, 1905

[EINα19] “Do Gravitational Fields Play An Essenntial Part In The Structure Of The Ellementary Particles Of Matter?” by A. Einstein, translated from Sitzungsberichte der Preussichen Akad d. Wissenschasten, 1919

[FEY63] “The Feynman Lectures on Physics by Feynman, Leighton and Sands”, published by California Institute of Technology, 1963

[GAM66] “Thirty Years that Shook Physics” by George Gamow, published by Educational Services Incorporated (The Science Study Series edition), 1966

[GAU99] “Schaum's Outline Theory and Problems of Modern Physics”, by Ronald Gautreau Ph.D. and William Savin Ph.D., published by McGraw-Hill, Inc., second edition, 1999

[GEO67] “Methods of Signal and System Analysis”, by George R. Cooper and Clare D. McGillem, published by Holt, Rinehart and Winston, Inc., School of Electrical Engineering Purdue University, 1967

[GERβ06] “Who was Nicola Tesla and what where his greatest inventions” by Frank Germano published by Global Energy Technologies, (Inc at International Turbine and Power, LLC Bladeless Disk Technologies, former cooperate entity), 931 Rumsey Ave, P.O. Box 550 Cody Wyoming 82414, 2006

[HAL66] “Physics” by Halliday and Resnick, published by John Wiley & Sons, Inc., 1966

[HAW74] “Black Hole Explosions” by Steven W. Hawking, published in Nature Vol. 248 pg 30–31, 1974

[HAW80] “Superspace and supergravity” : proceedings of the Nuffield workshop, Cambridge, June 16 - July 12, 1980, by S W Hawking; M Roček; Nuffield Foundation, published by Cambridge ; New York : Cambridge University Press, 1981 < ISBN: 0521239087 9780521239080, OCLC: 7530038>

[HEC74] “Optics” by Eugene Hecht and Alfred Zahac, published by Addison – Wesley Publishing Company, 1974

[ISAβ07] “Einstein His Life and Universe” by Walter Isaacson, 2007

[ITT77] “Reference Data for Radio Engineers”, sixth edition, by International Telephone and Telegraph, published by Howard W. Sams & Co., Inc., Indianapolis, 1977

[LOR62] “Electromagnetic Fields and Waves”, second edition, by Paul Lorrain University of Montreal and Dale R. Corson Cornell University, published by W. H. Freeman and Company San Francisco, 1962

[MAL67] “Modern Physics,” by Isaac Maleh, published by Charles E. Merrill Books, Inc., Columbus, Ohio, 1967

[MILα10] “A new modification of the cloud method of determining the elementary electrical charge and the most probable value of that charge”, by Robert Millikan and Harvey Fletcher, published by Phys. Mag. XIX, p. 209, 1910

[MIS73] “Gravitation”, third edition, by Charles W. Misner U. Maryland, Kip S. Thorne California Institute of Technology and John Archibald Wheeler U. Princeton, published by W. H. Freeman and Company San Francisco, 1973

[MOO60] “Traveling-Wave Engineering” by Richard K. Moore, published by McGraw-Hill Book Company, 1960

[NAVβ06] “Hyper Physics” by Carl R. Nave published by Department of Physics and Astronomy George State University, 2006 see http://hyperphysics.phy-astr.gsu.edu/hbase/particles/

[PAN62] “Classical Electricity and Magnetism” by Wolfgang K. H. Panofsky Stanford University and Melba Phillips Washington University, published by Addison-Wesley Publishing Company, INC. second edition 1962

[PLA1899] (Paper presented to the Prussian Academy of Sciences) Max Planck: “Über irreversible Strahlungsvorgänge'. *Sitzungsberichte der Preußischen Akademie der Wissenschaften”*, vol. 5, p. 479, 1899

[PER52] “The principle of Relativity, A collection of original memoirs on the Special and General Theory of Relativity” by W. Perrett, G. B. Jeffery and A. Sommerfeld, published by Dover Publications, Inc., 1952
[POY1884] “On the Transfer of Energy in the Electromagnetic Field” by John Henry Poyning, published by *Philosophical Transactions of the Royal Society of London*, Vol. 175, 1884
[SES71] “Fundamentals of Transmission Lines and Electromagnetic Fields”, by S.R. Seshadri, published by Addison-Wesley Publishing Company, 1971
[SLE74] “A mathematical theory of communication” by C.E. Shannon (Bell Systems Technology Journal, July 1948) Selected paper by editor David Slepian published by IEEE Press, 1974
[SPI59] “Schaum's Outline Series Vector Analysis” by Murray R. Spiegel, published by McGraw-Hill Book Company, 1959
[SPL77] “Digital Communications by Satellite” by James J. Spilker, Jr., published by Prentice-Hall, Inc., 1977
[TROα03] “The forces acting on a charged condenser moving through space” by F. T. Trouton and H. R. Noble, published by Proceedings of the Royal Society, Vol. 72, p. 132, 1903
[WEI72] “Gravitation and Cosmology Principles and Applications of the General Theory of Relativity by Steven Weinberg”, published by John Wiley & Sons, Inc., 1972
[WEYLα18] “Gravitation and Electricity” by H. Weyl translated from Sitzungsberichte der Preussichen Akad d. Wissenschasten, 1918

Working References:

Stephen W. Hawking primordial black holes

[13] Hawking, SW (1974). "Black Hole Explosions". *Nature* **248** (1): 30-31. Retrieved on 2007-03-23. The Quantum Mechanics of Black Holes," by S. W. Hawking; *Scientific American*, January 1977].

Black-hole evaporation
It is often said that nothing can escape from a black hole. But in 1974, Stephen Hawking realized that, owing to quantum effects, black holes should emit particles with a thermal distribution of energies — as if the black hole had a temperature inversely proportional to its mass. In addition to putting black-hole thermodynamics on a firmer footing, this discovery led Hawking to postulate 'black hole explosions', as primordial black holes end their lives in an accelerating release of energy.
Nature **248,** 30–31 (1974)

Stephen W. Hawking *Superspace and Supergravity* 1981
The Very Early Universe (1983)

References Book I Superspace and supergravity : proceedings of the Nuffield workshop, Cambridge, June 16 - July 12, 1980
by S W Hawking; M Roček; Nuffield Foundation.

- **Type:** English : Book
- **Publisher:** Cambridge ; New York : Cambridge University Press, 1981.
- **ISBN:** 0521239087 9780521239080
- **OCLC:** 7530038

Section I.1 References

Papers and Letters By Albert Einstein

- Einstein, Albert (1901), "Folgerungen aus den Capillaritätserscheinungen (Conclusions Drawn from the Phenomena of Capillarity)", *Annalen der Physik* **4**: 513
- Einstein, Albert (1905a), "On a Heuristic Viewpoint Concerning the Production and Transformation of Light", *Annalen der Physik* **17**: 132–148. This annus mirabilis paper on the photoelectric effect was received by Annalen der Physik March 18.
- Einstein, Albert (1905b), *A new determination of molecular dimensions*. This PhD thesis was completed April 30 and submitted July 20.
- Einstein, Albert (1905c), "On the Motion—Required by the Molecular Kinetic Theory of Heat—of Small Particles Suspended in a Stationary Liquid", *Annalen der Physik* **17**: 549–560. This annus mirabilis paper on Brownian motion was received May 11.
- Einstein, Albert (1905d), "On the Electrodynamics of Moving Bodies", *Annalen der Physik* **17**: 891–921. This annus mirabilis paper on special relativity received June 30.

- Einstein, Albert (1905e), "Does the Inertia of a Body Depend Upon Its Energy Content?", *Annalen der Physik* **18**: 639–641. This annus mirabilis paper on mass-energy equivalence was received September 27.
- Einstein, Albert (1915), "Die Feldgleichungen der Gravitation (The Field Equations of Gravitation)", *Koniglich Preussische Akademie der Wissenschaften*: 844–847
- Einstein, Albert (1917a), "Kosmologische Betrachtungen zur allgemeinen Relativitätstheorie (Cosmological Considerations in the General Theory of Relativity)", *Koniglich Preussische Akademie der Wissenschaften*
- Einstein, Albert (1917b), "Zur Quantentheorie der Strahlung (On the Quantum Mechanics of Radiation)", *Physikalische Zeitschrift* **18**: 121–128
- Einstein, Albert (July 11, 1923), "Fundamental Ideas and Problems of the Theory of Relativity", *Nobel Lectures, Physics 1901–1921*, Amsterdam: Elsevier Publishing Company. Retrieved on 2007-03-25
- Einstein, Albert (1924), "Quantentheorie des einatomigen idealen Gases (Quantum theory of monatomic ideal gases)", *Sitzungsberichte der Preussichen Akademie der Wissenschaften Physikalisch—Mathematische Klasse*: 261–267. First of a series of papers on this topic.
- Einstein, Albert (1926), "Die Ursache der Mäanderbildung der Flussläufe und des sogenannten Baerschen Gesetzes", *Die Naturwissenschaften*: 223-224. On Baer's law and meanders in the courses of rivers.
- Einstein, Albert; Boris Podolsky & Nathan Rosen (May 15, 1935), "Can Quantum-Mechanical Description of Physical Reality Be Considered Complete?", *Physical Review* **47** (10): 777–780
- Einstein, Albert (1940), "On Science and Religion", *Nature* **146**
- Einstein, Albert, *et al.* (December 4, 1948), "To the editors", *New York Times*
- Einstein, Albert (May 1949), "Why Socialism?", *Monthly Review*. Retrieved on 2006-01-16
- Einstein, Albert (1950), "On the Generalized Theory of Gravitation", *Scientific American* **CLXXXII** (4): 13–17
- Einstein, Albert (1954), *Ideas and Opinions*, New York: Random House, ISBN 0-517-00393-7
- Einstein, Albert (1969), *Albert Einstein, Hedwig und Max Born: Briefwechsel 1916–1955*, Munich: Nymphenburger Verlagshandlung
- Einstein, Albert (1979), *Autobiographical Notes* (Centennial ed.), Chicago: Open Court, ISBN 0-875-48352-6. The chasing a light beam thought experiment is described on pages 48–51.
- Collected Papers: Stachel, John, Martin J. Klein, a. J. Kox, Michel Janssen, R. Schulmann, Diana Komos Buchwald and others (Eds.) (1987–2006). *The Collected Papers of Albert Einstein, Vol 1–10.* Princeton University Press. Further information about the volumes published so far can be found on the webpages of the Einstein Papers Project.

Volume II Appendix II

TIME: 10:00am (Discovery)
DATE: 01 JULY 1988
ENGR: R.H. Shev

SUBJECT: Origin of Unification for G-E Field Theory At A Distance

No, matter what model was chosen for photon exchange and setting up a direction for momentum all lead to a paradox. Upon rethinking the problem, I went back to fundamentals of special relativity. In classical theory, coulomb's law gives the relationship of charge to force;

$$F = \frac{-1\ (+/-q)*(+/-q)}{4\pi\epsilon_0 r^2}$$

In this equation for a continuum, charge determines direction for force or momentum, light travels at a finite velocity, and all reference frames are absolute. In my theory, it follows relativity theory where frames of reference are independent of one another and can only relate to each other by exchanging information relative to one another through the media of a photon. In classical theory the sign of the charge some how determines direction of a force acting at a distance in a manner that is continues. In my theory, this appears to brake down because the model leads to contradictions, charge appearing always to be independent of one another. Yet, the direction of momentum is somehow related to the charge knowing of each other's existence. This can only happen if information of each other's being is somehow communicated.

Possible models that occur to me are:

In the presence of one another space may be deformed either positively for positive momentum to bring charge together (+/–) or (–/+), or repulse each other if space is deformed negatively (+/+) or (–/–). However, this still requires an exchange of information and that can only be done through the postulation of the photon being the only way of making the bridge. One of the basic premise of this theory is that space is a void and Euclidean in nature, hence refutes the idea that it is warped in some manner.

UPDATE: 20 SEPT 1988

The other model that comes to mind is that some how the information for the existence of charge with its sign somehow is carried by the only thing traveling between each frame the photon. The number of ways to modulate a photon to allow it to carry information is seven. Modulation can ether be carried out by amplitude, phase of the polarization of the EM fields, frequency, or combinations of the three. If this where the case in some physical experiment somewhere someone would have observed the ability to change the direction of motion of a charged particle by simply modulating the light. That has not been the case. Classical theory or other wise has not predicted such an effect. So for one reason or another, these models can be eliminated.

What then can I propose as a mechanism for charge by which it can know of the existence of another charge particle in a relative frame of reference? From special relativity, one reference frame to another light remains invariant. That is no matter what reference frame we choose we will always measure light at the same velocity. To do this light forms a Minkowski's light cone, which if you where traveling at the speed of light in the photons frame of reference, all points in the universe exist simultaneously, therefore light moves instantly from one point to another with zero length in a frozen moment of time existing in a perpetual present! This of course is not true for the sub-light world and all the famous space and time distortions occur consequently as everything tries to adjust itself accordingly. It occurred to me that in my theory I have other quantities that remain invariant in relative reference frames, such as charge, the sign of charge, rest mass, and the sign of the mass. This comes about because the nature of matter is light in spin. I can also show the existence of a charge-mass cone embedded in the fabric of the space-time continuum. Therefore a charged particle has a half of a cone in its reference frame and when the photon bridges to another frame it passes through all points in the universe simultaneously, thereby uniting the two half cones and creating a single conic cone in the space-time continuum. This intern causes the direction of the momentum to be selected for the appropriate orientation and the reference frames move away or towards each other accordingly. So it is the invariant of the sign for the charge being embedded in the vary fabric of space itself that allows the information to pass and allows force at a distance to exist. These thoughts can be extended to gravitational fields as well if we remember we are dealing with a charge-mass light cone and in this theory the sign of the mass is invariant as well, hence mass being positive can only move in one direction towards each other. Hence the mechanism of gravity.

In causing our 4-dimensional space to intersect with the mass-charge, light cone the normal vector to the plane was defined in spatial terms rather than temporal, this allowed the 4-dimensional spin of the photon to be projected into 3-space, as we perceive it. However, if we setup another reference frame and look at our particle from a distance by allowing a photon to traverse the space, a new light or spatial cone can be defined by the resulting geometry. This cone is one of inertial force at a distance. The inertial-force light cone differs from the Einsteinian light cone in that it terminates into a particle rather then free space. It can be thought that the termination point of a photon is a like a node which ends the path of the light particle, hence the terminology arises to distinguish the different termination points, a FREE SPACE NODE verses a PARTICLE NODE. Clearly light will continue on the same path forever as long as there is no physical obstruction such as a particle of matter. So a free space node is not real but a virtual node having only mathematical significance in modeling any specific position that the photon has traveled to at that particular instant in time.

Tensor Calculus & Physics: A General Treatise

Table of Contents

List of Tables

List of Figures

List of Observations

List of Givens

Appendix A Newton and Maxwell Field Systems

Section A.1 *Rectilinear vs. Rotational Physical Systems*

Table A.1 Correspondence between Rectilinear vs. Rotational Physical Equations

Rectilinear Motion			
Law & Definitions	Name	Parameter	Dimensions
LxRCMA.1.1.1	Displacement in time	t-time taken along a line of travel	[T]
LxRCMA.1.1.2	Displacement in space	x, y, z-distance along three axes	[L]
LxRCMA.1.1.3	Directed displacement or vector	$\vec{r} = x\hat{i} + y\hat{j} + z\hat{k} = (x, y, z)$	[L]
LxRCMA.1.1.4	Speed or celerity	$\dot{x} = dx/dt$, $\dot{y} = dy/dt$, $\dot{z} = dz/dt$	$[L][T^{-1}]$
LxRCMA.1.1.5	Velocity	$\vec{v} = d\vec{r}/dt$	$[L][T^{-1}]$
LxRCMA.1.1.6	Acceleration	$\vec{a} = d\vec{r}/dt^2 = d\vec{v}/dt$	$[L][T^{-2}]$
LxRCMA.1.1.7	Mass	m	[M]
LxRCMA.1.1.8	Linear momentum	$\vec{P} = \int \vec{F}\, dt = m\,\vec{v}$	$[M][L][T^{-1}]$
LxRCMA.1.1.9	Force	$\vec{F} = d\vec{P}/dt$	$[M][L][T^{-2}]$
LxRCMA.1.1.10	Work	$W = \int \vec{F} \bullet d\vec{r}$	$[M][L^2][T^{-2}]$
LxRCMA.1.1.11	Kinetic energy	$KE = ½mv^2$	$[M][L^2][T^{-2}]$
LxRCMA.1.1.12	Power	$P_w = \vec{F} \bullet \vec{v}$	$[M][L^2][T^{-3}]$
LxRCMA.1.1.13	Inertial density	$\rho = d\,m / d\,V$	$[M][L^{-3}]$

Rotation about a Fixed Axis [HAL66 pg 278]			
Law & Definitions	Name	Parameter	Dimensions
LxRFXA.1.1.1	Period of one rotation	T time per revolution or cycle	[T]
LxRFXA.1.1.2	Frequency of rotations	$\nu = 1/T$ Period per revolution	$[T^{-1}]$
LxRFXA.1.1.3	Angular displacement	θ	[A]
LxRFXA.1.1.4	Arclength displacement	$s = \theta r$	$[S] \equiv [A][R]$
LxRFXA.1.1.5	Angular velocity	$\vec{\omega} = d\vec{\theta}/dt = 2\pi\nu\,\hat{\omega}$	$[A][T^{-1}]$
LxRFXA.1.1.6	Angular acceleration	$\vec{\alpha} = d\vec{\omega}/dt$	$[A][T^{-2}]$
LxRFXA.1.1.7	Rotational inertia	$I = mr^2$	$[M][R^2]$
LxRFXA.1.1.8	Angular momentum	$\vec{L} = \int \vec{\tau}\, dt = I\,\vec{\omega} = \vec{r} \times \vec{P}$	$[M][S][T^{-1}]$ [R]
LxRFXA.1.1.9	Torque	$\vec{\tau} = d\vec{L}/dt = \vec{r} \times \vec{F}$	$[M][S][T^{-2}]$ [R]
LxRFXA.1.1.10	Work in spin	$W = \int \vec{\tau} \bullet d\vec{\theta}$	$[M][S^2][T^{-2}]$
LxRFXA.1.1.11	Kinetic energy in spin	$KE = ½I\omega^2$	$[M][S^2][T^{-2}]$
LxRFXA.1.1.12	Power in spin	$P_w = \vec{\tau} \bullet \vec{\omega}$	$[M][S^2][T^{-3}]$
LxRFXA.1.1.13	Rotation inertial density	$\eta = dI/dV_R$	$[M][R^{-1}]$

Section A.2 *Physical and Numerical Dimensional Analysis*

Table A.2 Definitions Fundamental Physical Dimensions [HAL66 Apx F pg 53], [MOO60]

Definition Dimensions	Dimension	Units (MKS) System	Abbreviated Symbol
DxUNT.A.2.1.1 Mass	[M]	gram or kilograms	gm or kg
DxUNT.A.2.1.2 Length	[L]	meter	m
DxUNT.A.2.1.3 Time	[T]	second	s
DxUNT.A.2.1.4 Charge	[Q]	coulomb	coul
DxUNT.A.2.1.5 Angle measured in	[A]	degree or radian	[°] or [r]
DxUNT.A.2.1.6 Arclength	[S]	meter	m
DxUNT.A.2.1.7 Radial Length	[R]	meter	m
DxUNT.A.2.1.8 Heat (Entropy)	[H]	temp (Temperature)	°T
DxUNT.A.2.1.9 Nothing Measured	[∅]	NULL measured unit	1

Fundamental physical quantities are atomized; they cannot be broken down into any other product other than itself as a unit of measure.

Table A.3 Symbols, Dimensions and Units for Physical Quantities [HAL66 Apx F pg 53], [MOO60]

Definitions	Quantity	Symbol	Dimensions	Units
DxELC.A.2.2.1	Energy, total	E	$[M][L^2][T^{-2}]$	joule
DxELC.A.2.2.2	kinetic	K	$[M][L^2][T^{-2}]$	joule
DxELC.A.2.2.3	potential	U	$[M][L^2][T^{-2}]$	joule
DxELC.A.2.2.4	work	W	$[M][L^2][T^{-2}]$	joule
DxELC.A.2.2.5	Force	$\vec{F}$	$[M][L][T^{-2}]$	newton or nt
DxELC.A.2.2.6	Power	P	$[M][L^2][T^{-3}]$	watt
DxELC.A.2.2.7	Resistance	R	$[M][L^2][T^{-1}][Q^{-2}]$	ohm or Ω
DxELC.A.2.2.8	Impedance	Z	$[M][L^2][T^{-1}][Q^{-2}]$	ohm (complex)
DxELC.A.2.2.9	Conductance	$G \equiv 1/R$	$[M^{-1}][L^{-2}][T][Q^2]$	mho or Ω^{-1}
DxELC.A.2.2.10	Admittance	$Y \equiv 1/Z$	$[M^{-1}][L^{-2}][T][Q^2]$	mho (complex)
DxELC.A.2.2.11	Resistivity	ρ_r	$[M][L^3][T^{-1}][Q^{-2}]$	ohm-meters
DxELC.A.2.2.12	Conductivity	σ	$[M^{-1}][L^{-3}][T][Q^2]$	mhos/meter
DxELC.A.2.2.13	Inductance	L	$[M][L^2][Q^{-2}]$	henry
DxELC.A.2.2.14	Mutual inductance	M	$[M][L^2][Q^{-2}]$	henry
DxELC.A.2.2.15	Capacitance	C	$[M^{-1}][L^{-2}][T^2][Q^2]$	farad
DxELC.A.2.2.16	Charge	q	$[Q]$	coulomb or coul
DxELC.A.2.2.17	Current	i	$[T^{-1}][Q]$	ampere or amp
DxELC.A.2.2.18	Charge density	ρ_q	$[L^{-3}][Q]$	coul/meter^3
DxELC.A.2.2.19	Current density	$\vec{J}$	$[L^{-2}][T^{-1}][Q]$	amps/meter^2
DxELC.A.2.2.20	Electric potential	V	$[M][L^2][T^{-2}][Q^{-1}]$	voltage or volt
DxELC.A.2.2.21	Electromotive force	$\mathcal{E}$	$[M][L^2][T^{-2}][Q^{-1}]$	volts
DxELC.A.2.2.22	Electric Field Strength	$\vec{E}$	$[M][L][T^{-2}][Q^{-1}]$	volts/meter
DxELC.A.2.2.23	Electric displacement	$\vec{D}$	$[M^{-2}][Q]$	coul-meters^2
DxELC.A.2.2.24	Electric polarization	$\vec{P}$	$[M^{-2}][Q]$	coul-meters^2
DxELC.A.2.2.25	Electric flux	Φ_E	$[M][L^3][T^{-2}][Q^{-1}]$	volts-meters
DxELC.A.2.2.26	Magnetic induction or Magnetic flux density	$\vec{B}$	$[M][T^{-1}][Q^{-1}]$	tesla ≡ webers/meter^2
DxELC.A.2.2.27	Magnetization,	$\vec{M}$	$[L^{-1}][T^{-1}][Q]$	amps/meter
DxELC.A.2.2.28	Magnetic Field	$\vec{H}$	$[L^{-1}][T^{-1}][Q]$	amps/meter
DxELC.A.2.2.29	Magnetic flux	Φ_B	$[M][L^2][T^{-1}][Q^{-1}]$	weber ≡ volt-sec
DxELC.A.2.2.30	Permittivity constant	ε	$[M^{-1}][L^{-3}][T^2][Q^2]$	farads/meter[A.2.2.1]
DxELC.A.2.2.31	Permeability constant	μ	$[M][L][Q^{-2}]$	henrys/meter[A.2.2.1]

[A.2.2.1]Note: $\varepsilon\mu \rightarrow [M^{-1}][L^{-3}][T^2][Q^2] * [M][L][Q^{-2}] \equiv [L^{-2}][T^2] \rightarrow v^{-2}$

the reciprocal square of velocity for a photon passing though in a material media. [$\mathcal{E}$ $\mathcal{E}$ ε]

Table A.4 Planck's and Other Physical Constants

Using the following values for those constants [HAL66], [ITT77 pg 3–11] and [WIK06]

	Physical Constants	Value MKS System of Measure	Error (ppm)	
CxPPC.A.2.3.1	G_0	$6.6732(31)\text{x}10^{-11}$ N*m^{2}kg^{-2}	460	Newtonian Gravitational constant
CxPPC.A.2.3.2	$\gamma_0 = 1/4\pi\, G_0^{[A.2.1]}$	1.1924(87)89 x 10^{+9} N^{-1}*m^{-2} kg^2	460	Reciprocal Newtonian constant
CxPPC.A.2.3.3	h	$6.6261968(34)\text{x}10^{-34}$ J sec	7.6	Planck's constant
CxPPC.A.2.3.4	$\hbar \equiv h/(2\pi)$	$1.0545919(80)\text{x}10^{-34}$ J sec	7.6	Normalized Planck's constant
CxPPC.A.2.3.5	c	$2.9979250(10)\text{x}10^8$ m sec^{-1}	0.33	Velocity of Light for free space
CxPPC.A.2.3.6	α^{-1}	137.03602(21)	1.5	Fine structure constant
CxPPC.A.2.3.7	q	$1.6021917(70)\text{x}10^{-19}$ C	4.4	Charge in Coulombs
CxPPC.A.2.3.8	ε_0	$8.854185(29)\text{x}10^{-12}$ F m^{-1}	1.9	Permittivity constant for free space
CxPPC.A.2.3.9	μ_0	4π x10^{-7} Henry m^{-1}	0.0	Permeability constant for free space
CxPPC.A.2.3.10	k	$1.3806505(24)\text{x}10^{-23}$ J °K^{-1}	0.01	Boltzmann's constant
CxPPC.A.2.3.11	m_e	$9.109558(54)\text{x}10^{-31}$ kg	6.0	Electron rest mass
CxPPC.A.2.3.12	m_p	$1.672614(11)\text{x}10^{-27}$ kg	6.6	Proton rest mass
CxPPC.A.2.3.13	m_n	$1.674920(11)\text{x}10^{-27}$ kg	6.6	Neutron rest mass
CxPPC.A.2.3.14	N_a	6.022169(40) x10^{23} particles moles^{-1}	6.6	Avogadro's Number
CxPPC.A.2.3.15	m_ec^2/e	$5.119940991\text{x}10^5$ J C^{-1}	2.3	Volts; rest energy of an electron

	Physical Constants	Dimensional Units	Name
CxPPC.A.2.3.1	G_0	$[M^{-1}][L^3][T^{-2}]$	Newtonian Gravitational constant
CxPPC.A.2.3.2	$\gamma_0 = 1/4\pi\, G_0^{[A.2.1]}$	$[M][L^{-3}][T^2]$	Reciprocal Newtonian constant
CxPPC.A.2.3.3	h	$[M][L^2][T^{-1}]$	Planck's constant
CxPPC.A.2.3.4	$\hbar \equiv h/(2\pi)$	$[M][L^2][T^{-1}]$	Normalized Planck's constant
CxPPC.A.2.3.5	c	$[L][T^{-1}]$	Velocity of Light for free space
CxPPC.A.2.3.6	α^{-1}	$[\varnothing]$	Fine structure constant
CxPPC.A.2.3.7	q	$[Q]$	Charge in coulombs
CxPPC.A.2.3.8	ε_0	$[M^{-1}][L^{-3}][T^2][Q^2]$	Permittivity constant for free space
CxPPC.A.2.3.9	μ_0	$[M][L][Q^{-2}]$	Permeability constant for free space
CxPPC.A.2.3.10	k	$[M][L^2][T^{-2}][H^{-1}]$	Boltzmann's constant
CxPPC.A.2.3.11	m_e	$[M]$	Electron rest mass
CxPPC.A.2.3.12	m_p	$[M]$	Proton rest mass
CxPPC.A.2.3.13	m_n	$[M]$	Neutron rest mass
CxPPC.A.2.3.14	N_a	$[N][moles^{-1}]$	Avogadro's Number
CxPPC.A.2.3.15	m_ec^2/e	$[M][L^2][T^{-2}][Q^{-1}]$	Volts; rest energy of an electron

∅ – Null, normalized or no dimensional units
M – mass in [kg] kilograms
L – length in [m] meters
T – time in [s] seconds
Q – charge in [C] Coulombs
H – heat or temperature in [°K] Kelvin and energy in [J] Joules. If Boltzmann's constant is in Centigrade [°C] or Kelvin (an absolute temperature), Fahrenheit [°F] or Rankine [°R] (also an absolute temperature), then the temperature transformations are
[T in °C] = (°C/°K) [T in °K] – 273.15°C,
[T in °F] = (°F/°R) [T in °R] – 459.67°F and
[T in °F] = 1.8(°F/°C) [T in °C] + 32.00°F.
Ideally, Boltzmann's constant should be in an absolute temperature to avoid dealing with negative temperatures. [GIE71, pg 40] While index [°], is spoken as "degrees of unit temperature".
N – number count of any quantity

[A.2.1]Note: Equation CxPPC.A.2.3.1 is cast into this form to setup a direct correspondence in kind between the two stationary field equations Coulomb and Newton's equations. This is done with a generality that alludes to a possible relationship between the two classical fields. This generalization is exploited for the development of this treatise in order to merge these fields.

The following are axioms of operations performed on dimensions.

Table A.5 Axioms for Unit Operator Field Dimensions

Axiom	Axioms by Unit Operator	Description
Axiom A.2.4.1 Existence Dim	$\exists!$ a field U $\ni$, quantities ([$\varnothing$], [M], [L], [T], [Q], [A], [S], [R]) $\in$ U	Existence of a dimensional field
Axiom A.2.4.2 Equality Dim	$[U_x] = [U_y]$ Equation A $[U_y] = [U_x]$ Equation B	Equality of unit operator
Axiom A.2.4.3 Substitution Dim	$[U_y] = [U_x]$ and $[U_z] = [U_y]$ $\therefore [U_z] = [U_x]$	Substitution of unit operator

Axiom	Axioms by Unit Operator	Description
Axiom A.2.4.4 Closure Dim	$([U_x]\,[U_y] \wedge [U_y]\,[U_x]) \in U$	Closure with respect to unit operator
Axiom A.2.4.5 Commutative Dim	$[U_x]\,[U_y] = [U_y]\,[U_x]$	Commutative by unit operator
Axiom A.2.4.6 Associative Dim	$[U_x]\,([U_y]\,[U_z]) = ([U_x]\,[U_y])\,[U_z]$	Associative by unit operator
Axiom A.2.4.7 Identity Dim	$[U_x]\,[\varnothing] = [U_x]$	Identity by unit operator
Axiom A.2.4.8 Exponent Inverse Dim	$[U_x^0] = [\varnothing]$	Exponent Null result for unit operator
Axiom A.2.4.9 Exponent Identity Dim	$[U_x] = [U_x^1]$	Exponent identity by unit operator
Axiom A.2.4.10 Exponent Product Dim	$[U_x]^n = [U_x^n]$	Exponent product under unit operator
Axiom A.2.4.11 Exponent Summation Dim	$[U_x^n]\,[U_x^m] = [U_x^{n+m}]$	Exponent addition of unit operator
Axiom A.2.4.12 Dimensional Unit as an Algebraic Quantity	The dimension $[U_x]$ is an algebraic quantity and as such all-algebraic rules for definitions D4.n.m and axioms A4.n.m theorems T4.n.m for all n and m, apply in defining and manipulating the dimensions.	Dimensional unit as an algebraic quantity
Axiom A.2.4.13 Algebraic Quantity Related to Dimensional Operator	An algebraic quantity [x] is related to its dimension $[U_x]$ in the following way: $x \leftrightarrow [U_x]$ $x \lozenge y \leftrightarrow [U_x]\,[U_y]$ $x / y \leftrightarrow [U_x] / [U_y]$	Algebraic Quantity related to dimensional Operation Equation A Equation B Equation C
Axiom A.2.4.14 Implication of Quantities and Their Dimensional	if $(y \rightarrow [U_y])$ and $(x \rightarrow y)$ then $(x \rightarrow [U_y])$	

Theorem A.2.1 Unit Operator: Uniqueness of Dimensions

1g	Given	$([U_f], [U_x], [U_y]\ [U_z], [U_w]) \in U$
2g		$[U_x] = [U_y]$
3g		$[U_z] = [U_w]$
Step	Hypothesis	$[U_x]\ [U_z] = [U_y]\ [U_w]$
1	From 1g and AA.2.4.5 Closure Dim	$[U_x]\ [U_z] = [U_f]$
2	From 2g, 3g, 1 and AA.2.4.4 Substitution Dim	$[U_y]\ [U_w] = [U_f]$
∴	From 2 and AA.2.4.3A Equality Dim	$[U_x]\ [U_z] = [U_y]\ [U_w]$

Theorem A.2.2 Unit Operator: Exponential Reciprocal

1g	Given	$[U_x^{-1}] = 1 / [U_x]$	by assumption
Step	Hypothesis	$[U_x^{-1}] = 1 / [U_x]$	
1	From 1g	$[U_x^{-1}] = 1 / [U_x]$	
2	From 1 and D4.1.18 Negative Exponential	$[U_x^{-1}] = [U_x]^{-1}$	
3	From 2 and AA.2.4.21 Exponent Product Dim	$[U_x^{-1}] = [U_x^{-1}]$	
∴	From 1g, 3 and by identity	$[U_x^{-1}] = 1 / [U_x]$	

Theorem A.2.3 Unit Operator: Inverse Dimensional Identity

1g	Given	$[U_x^0] = [U_x] / [U_x]$	by assumption		
Step	Hypothesis	$[U_x] / [U_x] = [\varnothing]$	EQ A	$[U_x]\ [U_x^{-1}] = [\varnothing]$	EQ B
1	From 1g	$[U_x^0] = [U_x] / [U_x]$			
2	From 1 and D4.1.18 Negative Exponential	$[U_x^0] = [U_x]\ [U_x]^{-1}$			
3	From 2 and AA.2.4.11 Exponent Product Dim	$[U_x^0] = [U_x]\ [U_x^{-1}]$			
4	From 3 and AA.2.4.10 Exponent Identity Dim	$[U_x^0] = [U_x^1]\ [U_x^{-1}]$			
5	From 4 and AA.2.4.12 Exponent Summation Dim	$[U_x^0] = [U_x^{1-1}]$			
6	From 5 and D4.1.19 Primitive Definition for Rational Arithmetic	$[U_x^0] = [U_x^0]$			
∴	From 1g, 3, 6, AA3.9 Exponent Inverse Dim, AA.2.4.4 Substitution Dim and by identity	$[U_x] / [U_x] = [\varnothing]$	EQ A	$[U_x]\ [U_x^{-1}] = [\varnothing]$	EQ B

Section A.3 *Newtonian and Maxwellian Field Physics*

Table A.6 Maxwell and Poynting Field Parameters [HAL66 pg 957, 987]

Lemma	Name	Equation	Dimensions	MKS Units
LxMXP.A.3.1.1	Coulomb Charge Density	$\rho_q = \frac{q}{V_{olume}}$	$[Q][L^{-3}]$	coulombs /meter^3
LxMXP.A.3.1.2	Current Density	$\vec{J} = \rho_q \vec{V}$	$[L^{-2}][T^{-1}][Q]$	amps/meter^2
LxMXP.A.3.1.3	Coulomb Force Density	$\vec{f}_q = \rho_q \vec{E}$	$[M][L^{-2}][T^{-2}]$	nts/meter^3
LxMXE.A.3.1.4	Lorentz Force Density	$\vec{f}_L = \rho_q(\vec{E} + \vec{V} \times \vec{B})$	$[M][L^{-2}][T^{-2}]$	nts/meter^3
LxMXE.A.3.1.5	Poynting Field Power Density	$p_P = \vec{E} \bullet \vec{J}$	$[M][L^{-1}][T^{-3}]$	watts/meter^3
LxMXE.A.3.1.6	Electrical displacement	$\vec{D} = \varepsilon\vec{E} + \vec{P}$	$[M^{-2}][Q]$	coul-meters^2
LxMXP.A.3.1.7	Magnetic induction	$\vec{B} = \mu(\vec{H} + \vec{M})$	$[M][T^{-1}][Q^{-1}]$	webers/meter^2
LxMXE.A.3.1.8	Poynting Field Vector	$\vec{s}_p = \vec{E} \times \vec{B} = \mu\vec{E} \times \vec{H}$	$[M][L^{-2}][T^{-1}]$	watts / meter^2 [A.3.1.1]
LxMXE.A.3.1.9	Poynting Field Energy Density	$u_p = ½(\mu\vec{H}\bullet\vec{H} + \varepsilon\vec{E}\bullet\vec{E})$	$[M][L^{-1}][T^{-2}]$	joules/meter^3 [A.3.1.2]
LxMXE.A.3.1.10	Poynting Field Force Density-1	$\vec{f}_{1p} = \mu(\vec{H}\bullet\nabla)\vec{H} + \varepsilon(\vec{E}\bullet\nabla)\vec{E} + \vec{f}_q$	$[M][L^{-2}][T^{-2}]$	nts/meter^3
LxMXE.A.3.1.11	Poynting Power Density-2	$\vec{f}_{2p} = [(\vec{H}\bullet\nabla)\vec{E} - (\vec{E}\bullet\nabla)\vec{H}]$	$[M][L^{-1}][T^{-3}]$	watt/meter^3
LxMXE.A.3.1.12	Poynting Power Density-3	$\vec{f}_{3p} = (\vec{B}\bullet\nabla)\vec{B} + v^{-2}(\vec{E}\bullet\nabla)\vec{E}$		

[A.3.1.1]Note: Poynting's vector $\vec{E} \times \vec{H}$ from the paper "On the Transfer of Energy in the Electromagnetic Field" [POY1884, pg 343-361] formally is the rate of energy transfer through a unit of area. In this paper a modified Poynting Momentum Vector $\varepsilon\vec{E} \times \vec{B}$ is used to model the momentum at every infinitesimal point throughout the volume of the energy packet. None polarizing electrical–magnetic field components are used to simplify manipulation.

[A.3.1.2]Note: In equation LxMXE.A.3.1.9 and LxMXE.A.3.1.13 the photon energy empties from one collapsing field into the other inducing the next field and then back again giving rise to a perpetual mechanism allowing the photon to pull itself through space like a slinky bouncing downstairs as it tows its spring body behind it. Unlike a mechanical system, Maxwell's equations, in free space show no friction or resistance within the photon. A photon is truly a perpetual motion machine.

Table A.3.1 Maxwell's and Poynting's Field Parameters (Continued)

LxMXE.A.3.1.13	Poynting Momentum Field Vector	$\vec{s}_m = \varepsilon\vec{E}\times\vec{B} = v^{-2}\vec{E}\times\vec{H}$	$[M][L^{-2}][T^{-1}]$	(kg meters/second) / meter^3 [A.3.5.1]
LxMXE.A.3.1.14	Poynting Momentum Field Energy Density	$u_m = ½\,(\mu^{-1}\vec{B}\bullet\vec{B}+\varepsilon\vec{E}\bullet\vec{E})$	$[M][L^{-1}][T^{-2}]$	joules/meter^3 [A.3.5.2]
LxMXE.A.3.1.15	Poynting Momentum Field Force Density-1	$\vec{f}_{1m} = \mu^{-1}(\vec{B}\bullet\nabla)\vec{B} + \varepsilon(\vec{E}\bullet\nabla)\vec{E} + \vec{f}_q$	$[M][L^{-2}][T^{-2}]$	nts/meter^3
LxMXE.A.3.1.16	Poynting Momentum Power Density-2	$\vec{f}_{2m} = \varepsilon\,[(\vec{B}\bullet\nabla)\vec{E} - (\vec{E}\bullet\nabla)\vec{B}]$	$[M][L^{-3}][T^{-1}]$	

Table A.7 Maxwell's Integral Field Equations [HAL66 pg 964]

Lemma	Name	Equation	Crucial Experiment
LxMXE.A.3.2.1	Gauss's Law for a closed electrical surface	$\varepsilon \oint_s \vec{E}\bullet d\vec{S} = q$	• Coulomb's Law • Charge moves to the outer surface
LxMXE.A.3.2.2	Gauss's Law for a closed magnetic surface	$\oint_s \vec{B}\bullet d\vec{S} = 0$	Gauss's Law for no magnetic monopole
LxMXE.A.3.2.3	Ampère's Circuital Law as extended by Maxwell	$\oint_c \vec{B}\bullet d\vec{L} = v^{-2}\dfrac{d\Phi_E}{dt} + \mu\, i$	• Speed of light in a material media • Moving charge raises a magnetic field
LxMXE.A.3.2.4	Faraday's Law	$\oint_c \vec{E}\bullet d\vec{L} = -\dfrac{d\Phi_B}{dt}$	Current flows as Magnetic flux cuts a wire

For dielectric and magnetic materials are present, hence $\varepsilon = \varepsilon_r\varepsilon_0$ and $\mu = \mu_r\mu_0$

Table A.8 Maxwell's Differential Field Equations [HAL66 Supplementary Topic V pg 20]

Lemma	Name	Equation	Crucial Experiment
LxMXE.A.3.3.1	Gauss's Law for a closed electrical surface	$\nabla \bullet \vec{E} = \rho / \varepsilon$	• Coulomb's Law • Charge moves to the outer surface
LxMXE.A.3.3.2	Gauss's Law for a closed magnetic surface	$\nabla \bullet \vec{B} = 0$	Gauss's Law for no magnetic monopole
LxMXE.A.3.3.3	Ampère's Circuital Law as extended by Maxwell	$\nabla \times \vec{B} = v^{-2} \dfrac{\partial \vec{E}}{\partial t} + \mu \vec{J}$	• Speed of light in a material media • Moving charge raises a magnetic field
LxMXE.A.3.3.4	Faraday's Law	$\nabla \times \vec{E} = -\dfrac{\partial \vec{B}}{\partial t}$	Current flows as Magnetic flux cuts a wire
LxMXE.A.3.3.5	Material Media Velocity	$v^2 = \dfrac{1}{\mu\varepsilon}$	Material media velocity constant
LxMXE.A.3.3.6	Free Space Velocity	$c^2 = \dfrac{1}{\mu_0\varepsilon_0}$	Free space velocity constant; at light speed
LxMXE.A.3.3.7	Index of Refraction	$n_x = \dfrac{c}{v} = \sqrt{\mu_r\varepsilon_r}$	Relative media index

How is a field of force detected? Answer; by dropping a test particle within a field generated by a source [M or Q] and watching the particles direction of motion. So for gravity let us drop a mass particle [m], likewise for an electric field a charge particle [q]. This gives rise to the path of the particles revealing the monopole field of force.

Table A.9 Definitions of Physical Vector Fields

Definition A.3.7 **Physical Vector Field Strength**		
	Field Vector	**Field Strength**
DxPVFS.A.3.7.1	$\mathbf{A} \equiv \lim_{m \to 0} \dfrac{\mathbf{F}}{m}$	Inertial or accelerated field strength
DxPVFS.A.3.7.2	$\mathbf{E} \equiv \lim_{q \to 0} \dfrac{\mathbf{F}}{q}$	Electrical field strength

Observation A.3.1 Continuous verses Quantized Physical Limits

A non-atom model for a quantity of mass can be continually subdivided without limit till it reaches zero. While acceleration becomes infinitely large following Lorenz's rule, hence exactly modeling the rule of a limit. However charge is another matter. Charge is quantized so for a large number of quantum charges. Let's say on the order of Avogadro's number approximately 10^{23} particles, the concept of a limit would still apply, but when the numbers reduced near to a single quantum charge used as a one-sided limit to approaching this finite number.

$$q_0 \equiv 1.6021917(70) \times 10^{-19} \text{ C (CxPPC.A.2.3.7 "Charge in Coulombs") EQ A}$$

The concept of a limit must be carefully applied in approaching a singular quantum charge. So an electrical field can never tend to infinity, but some relatively large finite number determined by an Einsteinian force.

The following table looks at the potential of finite velocity of charge propagating for electric, magnetic and inertial fields. For example, if the charge density changes in one region of space, the integrals imply that Φ changes simultaneously throughout all space. However of course this would be inadmissible the potentials Φ and Θ cannot vary instantaneously with time [t] so cannot correspond instantaneously to a change with charge, current or inertial distributions at the particular moment. The analogy with astronomy is obvious: one cannot see a star as it is now, but only as it was thousands or millions of years ago, because the speed of light is finite. So the brackets indicate that the distributions are those at a previous time [t – (r/c)]. These are the ***retarded potentials***; see [SES71 pg 427].

From now on, subscripted square brackets $[\bullet]_{rt}$ will be used exclusively to identify retarded values of charge density $[\rho_e]$, current density $[\mathbf{J}]$, mass density $[\rho_m]$, position, velocity, etc. These integrals are used in a continuous, linear, homogeneous, or isotropic medium with $(\varepsilon, \mu, \gamma)$ rather than $(\varepsilon_0, \mu_0, \gamma_0)$ in free space.

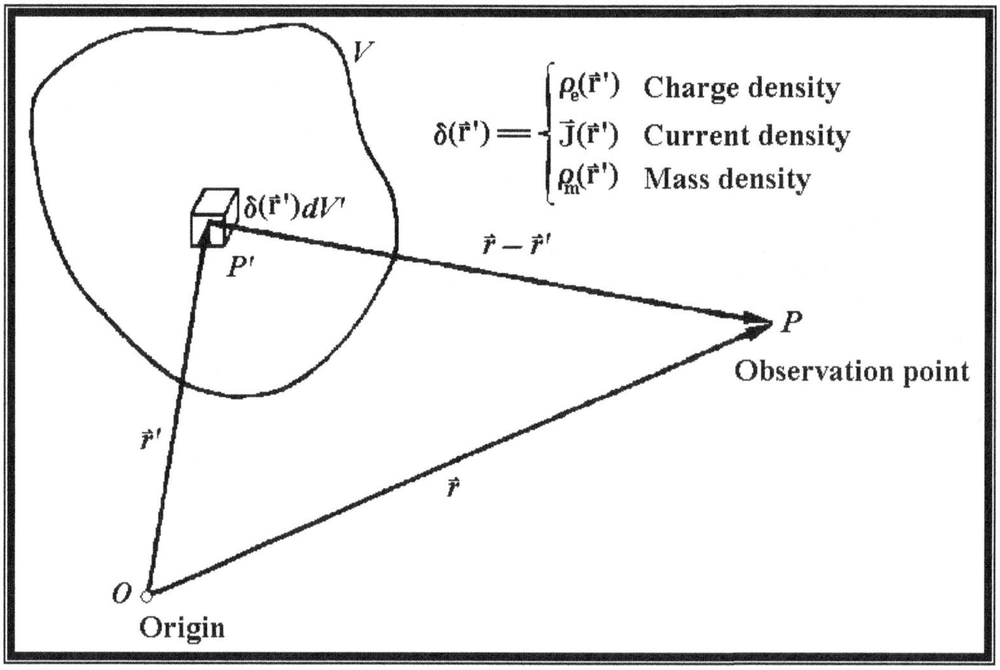

Figure A.3.1 Potential Units Due to Density Distribution

see [SES71 pg 171 and 192]

Table A.10 Maxwell and Newton's Retarded Potential Energy

Lemma	Name	Equation	Crucial Comments
LxMXE.A.3.5.1	Potential of Electric Field	$\Phi = \dfrac{-1}{4\pi\varepsilon} \int_v \dfrac{[\rho_e(\mathbf{r}')]_{rt}}{\lvert\mathbf{r} - \mathbf{r}'\rvert} dv'$	Integrated over the volume of charge per unit of charge [SES71 pg 427]
LxMXE.A.3.5.2	Potential of Changing Magnetic Field	$\boldsymbol{\Theta} = \dfrac{-\mu}{4\pi} \int_v \dfrac{[\mathbf{J}(\mathbf{r}')]_{rt}}{\lvert\mathbf{r} - \mathbf{r}'\rvert} dv'$	Integrated over the volume of current density per unit of current [SES71 pg 195]
LxMXE.A.3.5.3	Potential of Gravitational Field	$u = \dfrac{-1}{4\pi\gamma_0} \int_v \dfrac{[\rho_m(\mathbf{r}')]_{rt}}{\lvert\mathbf{r} - \mathbf{r}'\rvert} dv'$	Integrated over the volume of mass per unit of mass [WEI72 pg 78]
LxMXE.A.3.5.4	Lorentz Condition	$\dfrac{\partial \Phi}{\partial t} = -c^2\, \nabla \bullet \boldsymbol{\Theta}$	General relation linking both potentials [LOR62 pg 271]

Table A.11 Maxwell and Newton's Vector Fields in Potential Form

Lemma	Name	Equation	Crucial Comments
LxMXE.A.3.6.1	Potential of Electric Field	$\mathbf{E} = -\nabla\Phi - \dfrac{\partial \boldsymbol{\Theta}}{\partial t}$	Electric field in potential form [LOR62 pg 426]
LxMXE.A.3.6.2	Potential of Changing Magnetic Field	$\mathbf{B} = \nabla \times \boldsymbol{\Theta}$	Magnetic flux density field in potential form [LOR62 pg 426]
LxMXE.A.3.6.3	Potential of Gravitational Field	$\mathbf{A} = -\nabla u$	Gravitational force in potential form, for $U \equiv M\, u$ [WEI72 pg 78]

Table A.12 Newton and Coulomb Laws of Static Force Fields

Law of Static Force Fields		**Equation**	**Description**
Law A.3.1	Newton's Static Gravitation Field of Force	$\mathbf{F} = \dfrac{-1}{4\pi\gamma_0} \dfrac{M\, m_g}{r^2}\, \mathbf{r}$	The force between any two particles having masses M and m_g separated by a distance [r] has attraction acting along a line of sight joining the particles with a magnitude of force.
Law A.3.2	Coulomb's Static Electrical Field of Force	$\mathbf{F} = \dfrac{\pm 1}{4\pi\varepsilon_0} \dfrac{Q\, q_e}{r^2}\, \mathbf{r}$	The force between any two particles having a charge Q and q_e separated by a distance [r] has attraction or repulsion acting along a line of sight joining the particles with a magnitude of force.

Theorem A.3.1 Gravitational Potential Energy

Steps	Hypothesis	$U = \dfrac{-1}{4\pi\gamma_0} \dfrac{M\, m_g}{r}$
1	From the potential work done in a gravitational system	$U = -W^r_\infty$ The work it takes to move a particle from infinity to [r]
2	From 1 and the definition of work force through distance	$U = -\oint_S \mathbf{F} \bullet d\,\mathbf{S}$
3	From 2 and T8.4.2 Dot Product: Analytic Relationship	$U = -\int_\infty^r F(r)\, d\,r$
4	From 3 and LA.1.1 Universal Gravitation Field of Force	$U = -\int_\infty^r \dfrac{1}{4\pi\gamma_0} \dfrac{M\, m_g}{r^2}\, d\,r$
5	From 4 and Dx9.3.2.4 Definition of a Definite Integral evaluation of [f] bound a(x) to b(x).	$U = \dfrac{-1}{4\pi\gamma_0} \dfrac{M\, m_g}{r} \Big\vert_\infty^r$
∴	From 5 and T9.2.2 Reciprocal Numbers Tend to Zero as Denominator Increases without Limit	$U = \dfrac{-1}{4\pi\gamma_0} \dfrac{M\, m_g}{r}$

Theorem A.3.2 Electrical Potential Energy

Steps	Hypothesis	$\Phi = \dfrac{-1}{4\pi\varepsilon_0} \dfrac{Q\, q_e}{r}$
1	From the potential work done in a electrical system	$\Phi = -W^r_\infty$ The work it takes to move a particle from infinity to [r]
2	From 1 and the definition of work force through distance	$\Phi = -\oint_S \mathbf{F} \bullet d\,\mathbf{s}$
3	From 2 and T8.4.2 Dot Product: Analytic Relationship	$\Phi = -\int_\infty^r F(r)\, d\,r$
4	From 3 and LA.1.2 Coulomb's Electrical Field of Force	$\Phi = -\int_\infty^r \dfrac{1}{4\pi\varepsilon_0} \dfrac{Q\, q_e}{r^2}\, d\,r$
5	From 4 and Dx9.3.2.4 Definition of a Definite Integral evaluation of [f] bound a(x) to b(x).	$\Phi = \dfrac{-1}{4\pi\varepsilon_0} \dfrac{Q\, q_e}{r} \Big\vert_\infty^r$
∴	From 5 and T9.2.2 Reciprocal Numbers Tend to Zero as Denominator Increases without Limit	$\Phi = \dfrac{-1}{4\pi\varepsilon_0} \dfrac{Q\, q_e}{r}$

Observation A.3.2 Direction in a Gravitational and Electrical Potential Energy

Where [–] compensates for the direction of the coordinate system pointing down to the center of the earth's gravity system and [+] points up away from the system center. [HAL66 pg 407]

Theorem A.3.3 NLGS: Newton's Law for a Closed Gravitational Surface

1g	Given	$\mathbf{F} = \dfrac{1}{4\pi\gamma_0} \dfrac{M\, m_g}{r^2}\, \mathbf{r}$
2g		$\mathbf{A} \equiv \dfrac{\mathbf{F}}{M}$
3g		$m_g \equiv \int_V \rho_g\, dv$
4g		$d\,\mathbf{S} \equiv dA_{\theta\varphi}\, \mathbf{r}$
5g		$dA_{\theta\varphi} \equiv r^2 \sin\theta\, d\theta d\varphi$ divergence outwardly through a spherical envelope
Steps	Hypothesis	$\nabla \bullet \mathbf{A} = \dfrac{\rho_g}{\gamma_0}$
1	From A4.2.2 Equality	$\oint_S \mathbf{A} \bullet d\,\mathbf{S} = \oint_S \mathbf{A} \bullet d\,\mathbf{S}$
2	From 1g, 2g, 5g, 1 and A7.9.2 Substitution of Vectors	$\oint_S \mathbf{A} \bullet d\,\mathbf{S} = \oint_S \dfrac{1}{4\pi\gamma_0} \dfrac{m_g}{r^2}\, \mathbf{r} \bullet dA_{\theta\varphi}\, \mathbf{r}$
3	From 2, Lx9.3.1.3 Integrable on the integrand with a constant and T8.3.9B Dot Product: Left Scalar-Vector Association	$\oint_S \mathbf{A} \bullet d\,\mathbf{S} = m_g \dfrac{1}{4\pi\gamma_0} \oint_S \dfrac{1}{r^2}\, r^2 \sin\theta\, d\theta d\varphi\, \mathbf{r} \bullet \mathbf{r}$
4	From 3, T6.4.5 Dot Product: Unit Vector, A4.2.13 Inverse Multp and A4.2.12 Identity Multp	$\oint_S \mathbf{A} \bullet d\,\mathbf{S} = m_g \dfrac{1}{4\pi\gamma_0} \int_0^{2\pi}\!\int_0^{\pi} \sin\theta\, d\theta d\varphi$
5	From 4 and Dx9.3.1.4 Definition of a Definite Integral evaluation of [f] bound [a to b]	$\oint_S \mathbf{A} \bullet d\,\mathbf{S} = m_g \dfrac{1}{4\pi\gamma_0} \int_0^{2\pi} -\cos\theta\, \vert_0^{\pi}\, d\varphi$
6	From 5 and evaluated	$\oint_S \mathbf{A} \bullet d\,\mathbf{S} = m_g \dfrac{1}{4\pi\gamma_0} \int_0^{2\pi} (-\cos\pi + \cos 0)\, d\varphi$
7	From 6, DxE.1.6.13, DxE.1.6.13 Trigonometric Functions of Some Special Angles and D4.1.19 Primitive Definition for Rational Arithmetic	$\oint_S \mathbf{A} \bullet d\,\mathbf{S} = m_g \dfrac{2}{4\pi\gamma_0}\, 2\pi - 0$
8	From 7, A4.1.7 Identity Add D4.1.19 Primitive Definition for Rational Arithmetic	$\oint_S \mathbf{A} \bullet d\,\mathbf{S} = m_g \dfrac{4\pi}{4\pi\gamma_0}$
9	From 8, A4.1.13 Inverse Multp and A4.1.12 Identity Multp	$\oint_S \mathbf{A} \bullet d\,\mathbf{S} = \dfrac{m_g}{\gamma_0}$

10	From 3g, 9 and A4.2.3 Substitution	$\oint_S \mathbf{A} \bullet d\,\mathbf{S} = \frac{1}{\gamma_0} \int_V \rho_g\, dv$
	From 10, Lx9.3.1.3 Integrable on the integrand with a constant, LxVIE.6.1	
11	Divergence Theorem of Gauss Scalar Multiplication and A7.9.2 Substitution of Vectors	$\int_V \nabla \bullet \mathbf{A}\, dv = \int_V \frac{\rho_g}{\gamma_0}\, dv$
∴	From 11 and equivalence of the integral kernel	$\nabla \bullet \mathbf{A} = \frac{\rho_g}{\gamma_0}$

Theorem A.3.4 NLGS: Coulomb's Law for a Closed Electrical Surface

1g	Given	$\mathbf{F} = \frac{1}{4\pi\varepsilon_0} \frac{Q\, q_e}{r^2}\, \mathbf{r}$
2g		$\mathbf{E} \equiv \frac{\mathbf{F}}{Q}$
3g		$q_e \equiv \int_V \rho_e\, dv$
4g		$d\,\mathbf{S} \equiv dA_{\theta\varphi}\, \mathbf{r}$
5g		$dA_{\theta\varphi} \equiv r^2 \sin\theta\, d\theta d\varphi$ divergence outwardly through a spherical envelope
Steps	Hypothesis	$\nabla \bullet \mathbf{E} = \frac{\rho_e}{\varepsilon_0}$
1	From A4.2.2 Equality	$\oint_S \mathbf{E} \bullet d\,\mathbf{S} = \oint_S \mathbf{E} \bullet d\,\mathbf{S}$
2	From 1g, 2g, 5g, 1 and A7.9.2 Substitution of Vectors	$\oint_S \mathbf{E} \bullet d\,\mathbf{S} = \oint_S \frac{1}{4\pi\varepsilon_0} \frac{q_e}{r^2}\, \mathbf{r} \bullet dA_{\theta\varphi}\, \mathbf{r}$
3	From 2, Lx9.3.1.3Integrable on the integrand with a constant and T8.3.9B Dot Product: Left Scalar-Vector Association	$\oint_S \mathbf{E} \bullet d\,\mathbf{S} = q_e \frac{1}{4\pi\varepsilon_0} \oint_S \frac{1}{r^2}\, r^2 \sin\theta\, d\theta d\varphi\, \mathbf{r} \bullet \mathbf{r}$
4	From 3, T6.4.5 Dot Product: Unit Vector, A4.2.13 Inverse Multp and A4.2.12 Identity Multp	$\oint_S \mathbf{E} \bullet d\,\mathbf{S} = q_e \frac{1}{4\pi\varepsilon_0} \int_0^{2\pi}\int_0^{\pi} \sin\theta\, d\theta d\varphi$
5	From 4 and Dx9.3.1.4 Definition of a Definite Integral evaluation of [f] bound [a to b]	$\oint_S \mathbf{E} \bullet d\,\mathbf{S} = q_e \frac{1}{4\pi\varepsilon_0} \int_0^{2\pi} -\cos\theta\, \vert_0^{\pi}\, d\varphi$
6	From 5 and evaluated	$\oint_S \mathbf{E} \bullet d\,\mathbf{S} = q_e \frac{1}{4\pi\varepsilon_0} \int_0^{2\pi} (-\cos\pi + \cos 0)\, d\varphi$
7	From 6, DxE.1.6.13, DxE.1.6.13 Trigonometric Functions of Some Special Angles and D4.1.19 Primitive Definition for Rational Arithmetic	$\oint_S \mathbf{E} \bullet d\,\mathbf{S} = q_e \frac{2}{4\pi\varepsilon_0}\, 2\pi - 0$
8	From 7, A4.1.7 Identity Add D4.1.19 Primitive Definition for Rational Arithmetic	$\oint_S \mathbf{E} \bullet d\,\mathbf{S} = q_e \frac{4\pi}{4\pi\varepsilon_0}$

9	From 8, A4.1.13 Inverse Multp and A4.1.12 Identity Multp	$\oint_s \mathbf{E} \bullet d\,\mathbf{S} = \frac{q_e}{\varepsilon_0}$
10	From 3g, 9 and A4.2.3 Substitution	$\oint_s \mathbf{E} \bullet d\,\mathbf{S} = \frac{1}{\varepsilon_0} \int_v \rho_e\, dv$
11	From 10, Lx9.3.1.3 Integrable on the integrand with a constant, LxVIE.6.1 Divergence Theorem of Gauss Scalar Multiplication and A7.9.2 Substitution of Vectors	$\int_v \nabla \bullet \mathbf{E}\, dv = \int_v \frac{\rho_e}{\varepsilon_0}\, dv$
∴	From 11 and equivalence of the integral kernel	$\nabla \bullet \mathbf{E} = \frac{\rho_e}{\varepsilon_0}$

Theorem A.3.5 NLPE: Newton-Gravitational Formula Cast as a Poisson's Equation

Steps	Hypothesis	$\nabla^2 u = \dfrac{-\rho_g}{\gamma_0} \rightarrow [T^{-2}]$
1	From TA.3.3 NLGS: Newton's Law for a closed gravitational surface	$\nabla \bullet \mathbf{A} = \dfrac{\rho_g}{\gamma_0}$
2	From 1, LxMXE.A.3.6.3 Gravitational force in potential form and A7.9.2 Substitution of Vectors	$\nabla \bullet (-\nabla u) = \dfrac{\rho_g}{\gamma_0}$
3	From 2, D4.1.20A Negative Coefficient, T8.3.10A Dot Product: Right Scalar-Vector Association, T4.4.23 Equalities: Multiplication of a Constant, T4.8.2B Integer Exponents: Negative One Squared	$\nabla \bullet (\nabla u) = \dfrac{-\rho_g}{\gamma_0}$
∴	From 3, Lx9.6.1.10 Covariant Laplacian Operator, and AA.3.14C and TA.3.2 Unit Operator: Exponential Reciprocal	$\nabla^2 u = \dfrac{-\rho_g}{\gamma_0} \rightarrow [T^{-2}]$

Theorem A.3.6 NLDA: Coulomb's Law for Covariant D'Alembert Quaternion Operator

1g	Given	$w \equiv j\,ct$
Steps	Hypothesis	$\Box^2 \Phi = \dfrac{-\rho_e}{\varepsilon}$ $[M][Q^{-1}][T^{-2}]$
1	From LxMXE.A.3.4.1 Gauss's Law for a closed electrical surface	$\dfrac{\rho_e}{\varepsilon} = \nabla \bullet \vec{E}$
2	From 1, LxMXE.A.3.6.1 Potential of Electric Field and A7.9.2 Substitution of Vectors	$\dfrac{\rho_e}{\varepsilon} = \nabla \bullet (-\nabla\Phi - \dfrac{\partial \Theta}{\partial t})$
3	From 2, D4.1.20A Negative Coefficient, A4.2.14 Distribution, T8.3.10A Dot Product: Right Scalar-Vector Association, T4.4.23 Equalities: Multiplication of a Constant, T4.8.2B Integer Exponents: Negative One Squared	$\dfrac{\rho_e}{\varepsilon} = -\nabla \bullet \nabla\Phi + \dfrac{\partial (-\nabla \bullet \Theta)}{\partial t}$

4	From LxMXE.A.3.5.4 Lorentz Condition, Covariant Laplacian Operator, T4.4.4A Equalities: Right Cancellation by Multiplication and A4.2.3 Substitution	$\frac{\rho_e}{\varepsilon} = -\nabla^2 \Phi + \frac{\partial^2 \Phi}{c^2 \partial t^2}$
5	From 4 and T4.8.2B Integer Exponents: Negative One Squared	$\frac{\rho_e}{\varepsilon} = -\nabla^2 \Phi - \frac{\partial^2 \Phi}{(-c^2) \partial t^2}$
6	From 5 and T4.9.5B Rational Exponent: Square Root of a Negative One	$\frac{\rho_e}{\varepsilon} = -\nabla^2 \Phi - \frac{\partial^2 \Phi}{(j^2 c^2) \partial t^2}$
7	From 6 and A4.2.21 Distribution Exp	$\frac{\rho_e}{\varepsilon} = -\nabla^2 \Phi - \frac{\partial^2 \Phi}{\partial (jct)^2}$
8	From 1g, 7 and A4.2.3 Substitution	$\frac{\rho_e}{\varepsilon} = -\nabla^2 \Phi - \frac{\partial^2 \Phi}{\partial w^2}$
∴	From 8, Dx9.6.10.1A Covariant D'Alembert Quaternion Operator, Operator and AA.3.14C and TA.3.2 Unit Operator: Exponential Reciprocal	$\Box^2 \Phi = \frac{-\rho_e}{\varepsilon} \rightarrow [M][Q^{-1}][T^{-2}]$

Observation A.3.3 Einstein's Conclusion: Comparison between Gravitational and Electrical Fields

From theorems TA.3.4 "NLPE: Newton's Law for Poisson's Equation" and TA.3.5 "NLDA: Coulomb's Law for Covariant D'Alembert Quaternion Operator" it is seen that the two equations are almost identical, differing only by a single time-term by adding a fourth dimension to the Poisson equation to attain an electrical field. This clear and simple difference between gravitational and electrical fields is a clue that there must be an interrelation between the two physical fields. It is this incredible similarity between gravity and electricity that made Einstein believed there surely must be a unifying connection between the two fields of force, which encouraged him in a lifetime quest to find that unification.

Theorem A.3.7 NG2PE: Alternate Newton-Gravitational Formula Cast as Poisson's Equation

1g	Given	$A_m \equiv \dfrac{1}{4\pi\gamma_0} \dfrac{m_g}{r^2}$
Steps	Hypothesis	$\nabla^2 u = \dfrac{-\rho_g}{\gamma_0}$
1	From DxPVFS.A.3.7.1 Inertial or accelerated field strength and LA.3.1 Newton's Static Gravitation Field of Force	$\mathbf{A} = A_m \hat{\mathbf{r}} + 0\,\hat{\boldsymbol{\vartheta}} + 0\,\hat{\boldsymbol{\varphi}}$
2	From 1g, 1 and T9.6.3 GRV: Reciprocal Squared Functions Have No Curl	$\nabla\times \mathbf{A} = \mathbf{0}$
3	From 2 and T9.6.2B GRV: Curl of Divergence for Scalar Function	$\mathbf{A} = \pm\nabla u$
4	From 3 and Lx9.6.1.1B Uniqueness of Divergence Operator	$\nabla\bullet \mathbf{A} = \pm\nabla\bullet\nabla u$
5	From 4 and Lx9.6.1.10 Covariant Laplacian Operator; OA.3.2 selecting negative sign for a gravitational system	$\nabla\bullet \mathbf{A} = -\nabla^2 u$
∴	From 5, TA.3.3 NLGS: Newton's Law for a Closed Gravitational Surface and A4.2.3 Substitution	$\nabla^2 u = \dfrac{-\rho_g}{\gamma_0}$

Section A.3 *Newtonian and Maxwellian Field Physics (continued)*

Theorem A.2.4 PVVT: Electric Field Wave Equation Nonsteady-Nonhomogenous Fields

Steps	Hypothesis	$\nabla^2\vec{E} = \frac{1}{v^2}\frac{\partial^2\vec{E}}{\partial t^2} + \frac{\mu\partial\vec{J}}{\partial t} + \nabla(\rho/\varepsilon)$
1	From Lx9.6.1.13 Curl of Curl Operator; of an Electric Field Vector	$\nabla\times\nabla\times\vec{E} = \nabla(\nabla\bullet\vec{E}) - \nabla^2\vec{E}$
2	From 1, LxMXE.A.3.4.1 Gauss's Law for a closed electrical surface and A4.2.3 Substitution	$\nabla\times\nabla\times\vec{E} = \nabla(\rho/\varepsilon) - \nabla^2\vec{E}$
3	From A7.9.1 Equivalence of Vector and Lx9.6.1.13 Curl of Curl Operator	$\nabla\times\nabla\times\vec{E} = \nabla\times(\nabla\times\vec{E})$
4	From 3, LxMXE.A.3.4.4 Faraday's Law and A7.9.2 Substitution of Vectors	$\nabla\times\nabla\times\vec{E} = \nabla\times(-\frac{\partial\vec{B}}{\partial t})$
5	From 4, D4.1.20A Negative Coefficient, T6.9.11 Multiply a row or column of a determinant [A] by a scalar k, D9.3.1.3 Definition of a Derivative as a Limit, Lx9.2.1.11 Dual Function Commutativety of Limits Coefficient	$\nabla\times\nabla\times\vec{E} = -\frac{\partial\nabla\times\vec{B}}{\partial t}$
6	From 5, D4.1.20A Negative Coefficient, T7.10.3 Distribution of a Scalar over Addition of Vectors, LxMXE.A.3.4.3 Ampère's Circuital Law, A7.9.2 Substitution of Vectors and Lx9.3.2.2B Differential as a Linear Operator	$\nabla\times\nabla\times\vec{E} = \frac{\partial}{\partial t}\frac{\partial\vec{E}}{v^2\partial t} - \frac{\partial\mu\vec{J}}{\partial t}$
7	From 6, Distribute Constant in/out of Product and Dx9.3.2.2E Multiple Partial Differential Operator	$\nabla\times\nabla\times\vec{E} = -\frac{1}{v^2}\frac{\partial^2\vec{E}}{\partial t^2} - \frac{\partial\mu\vec{J}}{\partial t}$
8	From 2, 7 and A7.9.2 Substitution of Vectors	$\nabla(\rho/\varepsilon) - \nabla^2\vec{E} = -\frac{1}{v^2}\frac{\partial^2\vec{E}}{\partial t^2} - \frac{\partial\mu\vec{J}}{\partial t}$
∴	From 8, T7.11.7A Vector Addition: Reversal of Left Cancellation by Addition and T7.11.5B Vector Addition: Reversal of Right Cancellation by Addition	$\nabla^2\vec{E} = \frac{1}{v^2}\frac{\partial^2\vec{E}}{\partial t^2} + \frac{\mu\partial\vec{J}}{\partial t} + \nabla(\rho/\varepsilon)$

Theorem A.2.5 PVVT: Magnetic Field Wave Equation Nonsteady-Nonhomogenous Fields

Steps	Hypothesis	$\nabla^2 \vec{B} = \frac{1}{v^2} \frac{\partial^2 \vec{B}}{\partial t^2} - \mu \nabla \times \vec{J}$
1	From Lx9.6.1.13 Curl of Curl Operator; of an Magnetic Field Vector	$\nabla \times \nabla \times \vec{B} = \nabla (\nabla \bullet \vec{B}) - \nabla^2 \vec{B}$
2	From 1, LxMXE.A.3.4.2 Gauss's Law for a closed magnetic surface and A4.2.3 Substitution	$\nabla \times \nabla \times \vec{B} = \nabla (0) - \nabla^2 \vec{B}$
3	From 2, Dx9.6.3.1A Covariant Gradient Operator, Lx9.3.3.1B Differential of a Constant, D7.6.5C Zero Vector and T7.11.2 Vector Addition: Identity	$\nabla \times \nabla \times \vec{B} = -\nabla^2 \vec{B}$
4	From A7.9.1 Equivalence of Vector and Lx9.6.1.13 Curl of Curl Operator	$\nabla \times \nabla \times \vec{B} = \nabla \times (\nabla \times \vec{B})$
5	From 4, LxMXE.A.3.4.3 Ampère's Circuital Law as extended by Maxwell and A7.9.2 Substitution of Vectors	$\nabla \times \nabla \times \vec{B} = \nabla \times \left(\frac{\partial \vec{E}}{v^2 \partial t} + \mu \vec{J} \right)$
6	From 5, T6.9.11 Multiply a row or column of a determinant [A] by a scalar k, D9.3.1.3 Definition of a Derivative as a Limit, Lx9.2.1.11 Dual Function Commutativety of Limits Coefficient	$\nabla \times \nabla \times \vec{B} = \frac{1}{v^2} \frac{\partial \nabla \times \vec{E}}{\partial t} + \nabla \times \mu \vec{J}$
7	From 6, D4.1.20A Negative Coefficient, T7.10.3 Distribution of a Scalar over Addition of Vectors, LxMXE.A.3.4.4 Faraday's Law, A7.9.2 Substitution of Vectors and Lx9.3.2.2B Differential as a Linear Operator	$\nabla \times \nabla \times \vec{B} = - \frac{1\, \partial}{v^2 \partial} \frac{\partial \vec{B}}{\partial t} + \nabla \times \mu \vec{J}$

8	From 3, 7, A7.9.2 Substitution of Vectors Distribute Constant in/out of Product, Dx9.3.2.2E Multiple Partial Differential Operator and T6.9.11 Multiply a row or column of a determinant [A] by a scalar k,	$-\nabla^2\vec{B} = -\frac{1}{v^2}\frac{\partial^2\vec{B}}{\partial t^2} + \mu\nabla\times\vec{J}$
∴	From 8, D4.1.20A Negative Coefficient, T7.10.3 Distribution of a Scalar over Addition of Vectors, T4.8.3 Integer Exponents: Negative One Raised to an Even Number and T4.8.4 Integer Exponents: Negative One Raised to an Odd Number	$\nabla^2\vec{B} = \frac{1}{v^2}\frac{\partial^2\vec{B}}{\partial t^2} - \mu\nabla\times\vec{J}$

Now theorems Theorem A.10.1 PVVT and Theorem A.10.2 PVVT give current density results in eddy currents with curls, but what happens when it diverges?

Theorem A.2.6 PVVT: Divergent Current Density Nonsteady-Nonhomogenous Fields

Steps	Hypothesis	$\nabla\bullet\vec{J} = -\frac{\partial\rho}{\partial t}$
1	From LxMXE.A.3.4.3 Ampère's Circuital Law as extended by Maxwell and T8.3.1 Dot Product: Uniqueness Operation	$\nabla\bullet(\nabla\times\vec{B}) = \nabla\bullet(\frac{1}{v^2}\frac{\partial\vec{E}}{\partial t} + \mu\vec{J})$
2	From 1 and Lx9.6.1.12 Divergence of Curl Operator	$0 = \frac{1}{v^2}\frac{\partial\nabla\bullet\vec{E}}{\partial t} + \nabla\bullet\mu\vec{J}$
3	From 2, LxMXE.A.3.4.1 Gauss's Law for a closed electrical surface, A4.2.3 Substitution, Lx9.3.2.3B Distribute Constant in/out of Product and A4.2.14 Distribution	$0 = \frac{1}{v^2}\frac{\partial(\rho/\varepsilon)}{\partial t} + \mu\nabla\bullet\vec{J}$
4	From 3, T4.2.2 Equalities: Uniqueness of Multiplication, A4.2.13 Inverse Multp and A4.2.12 Identity Multp	$0 = \frac{1}{v^2\varepsilon\mu}\frac{\partial\rho}{\partial t} + \nabla\bullet\vec{J}$
5	From 4, LxMXE.A.3.4.5 Material Media Velocity and A4.2.3 Substitution	$0 = \frac{v^2}{v^2}\frac{\partial\rho}{\partial t} + \nabla\bullet\vec{J}$
∴	From 5, A4.2.13 Inverse Multp and A4.2.12 Identity Multp	$\nabla\bullet\vec{J} = -\frac{\partial\rho}{\partial t}$

Theorem A.3.9 PVVT also goes by the name of conservation of charge do to the net charge being zero.

Table A.13 Current Density field equations

	Hypothesis	Equation	P-Variables	Parameters
LxCDF.A.3.5.1	Theorem A.3.3 PVVT: Divergent Current Density For Nonsteady, Nonhomogenous Fields	$\nabla \bullet \vec{J} = -\frac{\partial \rho}{\partial t}$	$\vec{J}$ ρ	μ v ε
LxCDF.A.3.5.2	Theorem A.3.1 PVVT: Electric Field Wave Equation For Nonsteady, Nonhomogenous Fields	$\frac{\mu \partial \vec{J}}{\partial t} + \nabla (\rho/\varepsilon) = \Box^2 \vec{E}$	$\vec{J}$ ρ $\vec{E}$	μ ε
LxCDF.A.3.5.3	Theorem A.3.5 PVVT: Magnetic Field Wave Equation Nonsteady-Nonhomogenous Fields	$-\mu \nabla \times \vec{J} = \Box^2 \vec{B}$	$\vec{J}$ $\vec{B}$	μ

The D'Alembert and Poincare field operator can be separated from the right side of the equation for a photon when it equates to zero or the field is steady state being constant and independent of the parametric variable(s) on the left side of the equality.

Note that in Theorem A.3.1 "PVVT: Electric Field Wave Equation For Nonsteady, Nonhomogenous Fields" currents impart an electric field wave or visa versa electric field waves generate time varying currents. Likewise from Theorem A.3.2 "PVVT: Magnetic Field Wave Equation For Nonsteady, Nonhomogenous Fields" the curl of the current density or eddy currents generate a magnetic field wave and symmetrically magnetic field waves generate eddy currents.

A photon moving at light speed has no charge [$\rho_q = 0$] or mass to slow it down so it travels at full velocity [$v = c$].

Table A.14: Maxwell's and Poynting's Base Parameters in Free Space

Lemma	Name	Equation	Dimensions	MKS Units
LxMXP.A.3.5.1	Velocity of Light	$v = c$	$[L][T^{-1}]$	meters/second
LxMXP.A.3.5.2	Coulomb Charge Density	$\rho_q = \dfrac{q_p}{V_{olume}} = 0$	$[Q][L^{-3}]$	coul/meter^3
LxMXP.A.3.5.3	Current Density	$\vec{J}_p = \rho_p \vec{v} = \mathbf{0}$	$[L^{-2}][T^{-1}][Q]$	amps/meter^2
LxMXE.A.3.5.4	Coulomb Force Density	$\vec{f}_q = \rho_p \vec{E} = \mathbf{0}$	$[M][L^{-2}][T^{-2}]$	nts/meter^3
LxMXE.A.3.5.5	Lorentz Force Density	$\vec{f}_L = \rho_q(\vec{E} + \vec{v} \times \vec{B}) = \mathbf{0}$	$[M][L^{-2}][T^{-2}]$	nts/meter^3
LxMXE.A.3.5.6	Poynting Field Power Density	$p_p = \vec{E} \bullet \vec{J}_p = 0$	$[M][L^{-1}][T^{-3}]$	watts/meter^3
LxMXE.A.3.5.7	Poynting Momentum Field Force Density-1	$\vec{f}_{1m} = \mu^{-1}(\vec{B} \bullet \nabla)\vec{B} + \varepsilon(\vec{E} \bullet \nabla)\vec{E} + \vec{f}_q = \mu_0^{-1}(\vec{B} \bullet \nabla)\vec{B} + \varepsilon_0(\vec{E} \bullet \nabla)\vec{E}$	$[M][L^{-2}][T^{-2}]$	nts/meter^3

By substituting Table A.3.6 into Table A.3.4 "Maxwell's Differential Field Equations", eliminates the base parameters for the Maxwell's field equations for a photon in free space.

Table A.15: Maxwell's Differential Field Equations in Free Space

Lemma	Name	Equation	Crucial Experiment
LxMXE.A.3.6.1	Gauss's Law for a closed electrical surface	$\nabla \bullet \vec{E} = 0$	• Coulomb's Law for no charge monopole in free space
LxMXE.A.3.6.2	Gauss's Law for a closed magnetic surface	$\nabla \bullet \vec{B} = 0$	• Gauss's Law for no magnetic monopole
LxMXE.A.3.6.3	Ampère's Circuital Law as extended by Maxwell	$\nabla \times \vec{B} = \dfrac{1}{c^2} \dfrac{\partial \vec{E}}{\partial t}$	• Speed of light in free space
LxMXE.A.3.6.4	Faraday's Law	$\nabla \times \vec{E} = - \dfrac{\partial \vec{B}}{\partial t}$	

So it follows that space is a conduit having an admittance limiting the speed at which light travels. This is due to the rate at which the internal energy is transferred from one collapsing field component to building the next induced field LxMXE.A.3.2.13 "Poynting Momentum Field Energy Density", as shown in table "Maxwell's and Poynting's Field Parameters" [TA.3.2]. For description of wave motion for the field components see theorem A.3.4 "PVVT: Electric Field Wave Equation in Free Space" and A.3.5 "PVVT: Magnetic Field Wave Equation in Free Space".

Given A.3.1 Non-material media

1g	Given	$\mu = 0$
2g		$\rho = 0$

Theorem A.2.7 PVVT: Electric Field Wave Equation in Free Space

Steps	Hypothesis	$\nabla^2\vec{E} = \frac{1}{c^2}\frac{\partial^2\vec{E}}{\partial t^2}$
1	From TA.3.4 PVVT: Electric Field Wave Equation Nonsteady-Nonhomogenous Fields	$\nabla^2\vec{E} = \frac{1}{v^2}\frac{\partial^2\vec{E}}{\partial t^2} + \frac{\mu\partial\vec{J}}{\partial t} + \nabla(\rho/\varepsilon)$
∴	From 1g, 2g, Dx9.6.3.1A Covariant Gradient Operator, Lx9.3.3.1B Differential of a Constant, D7.6.5C Zero Vector and T7.11.2 Vector Addition: Identity	$\nabla^2\vec{E} = \frac{1}{c^2}\frac{\partial^2\vec{E}}{\partial t^2}$

Theorem A.2.8 PVVT: Magnetic Field Wave Equation in Free Space

Steps	Hypothesis	$\nabla^2\vec{B} = \frac{1}{c^2}\frac{\partial^2\vec{B}}{\partial t^2}$
1	From TA.3.5 PVVT: Magnetic Field Wave Equation Nonsteady-Nonhomogenous Fields	$\nabla^2\vec{B} = \frac{1}{v^2}\frac{\partial^2\vec{B}}{\partial t^2} - \mu\nabla\times\vec{J}$
∴	From 1g, 2g, Dx9.6.3.1A Covariant Gradient Operator, Lx9.3.3.1B Differential of a Constant, D7.6.5C Zero Vector and T7.11.2 Vector Addition: Identity	$\nabla^2\vec{B} = \frac{1}{c^2}\frac{\partial^2\vec{B}}{\partial t^2}$

Section A.4 *Poynting Vector Field Equations*

Theorem A.4.1 PVVT: Poynting Vector Differentiated With Respect to Time

Steps	Hypothesis	
	Hypothesis	$\dfrac{\partial \vec{s}_p}{\partial t} = v^2(-2\nabla u_p + \vec{f}_{1p} - \vec{f}_L)$
1	From LxMXE.A.3.5.8 Poynting Field Vector and T9.3.2 Uniqueness of Differentials	$\dfrac{\partial \vec{s}_p}{\partial t} = \dfrac{\partial(\vec{E}\times\vec{H})}{\partial t}$
2	From 1 and Lx9.2.2.3 Differential of Cross Product	$\dfrac{\partial \vec{s}_p}{\partial t} = \dfrac{\partial \vec{E}}{\partial t} \times \vec{H} + \vec{E} \times \dfrac{\partial \vec{H}}{\partial t}$
3	From LxMXE.A.3.4.3 Ampère's Circuital Law; as a Magnetic Field	$\nabla\times\vec{H} = \varepsilon \dfrac{\partial \vec{E}}{\partial t} + \vec{J}$
4	From LxMXE.A.3.4.4 Faraday's Law; as a Magnetic Field	$\nabla\times\vec{E} = -\mu \dfrac{\partial \vec{H}}{\partial t}$
5	From 2, 3, 4, T7.11.4A Vector Addition: Right Cancellation by Addition, T4.4.4A Equalities: Right Cancellation by Multiplication and A7.9.2 Substitution of Vectors	$\dfrac{\partial \vec{s}_p}{\partial t} = \dfrac{1}{\varepsilon}(\nabla\times\vec{H} - \vec{J}) \times \vec{H} - \dfrac{1}{\mu}\vec{E}\times(\nabla\times\vec{E})$
6	From 5, A4.2.12 Identity Multp, A4.2.13 Inverse Multp, T7.10.3 Distribution of a Scalar over Addition of Vectors and Lx9.5.1.1 Asymmetric Property of Cross Product	$\dfrac{\partial \vec{s}_p}{\partial t} = \dfrac{1}{\varepsilon\mu}[\mu(-\vec{H}\times\nabla\times\vec{H} - \vec{J}\times\vec{H}) - \varepsilon(\vec{E}\times\nabla\times\vec{E})]$
7	From 6, LxMXE.A.3.4.5 Material Media Velocity, D4.1.20A Negative Coefficient and T7.10.3 Distribution of a Scalar over Addition of Vectors	$\dfrac{\partial \vec{s}_p}{\partial t} = -v^2(\mu\vec{H}\times\nabla\times\vec{H} + \varepsilon\vec{E}\times\nabla\times\vec{E} + \mu\vec{J}\times\vec{H})$

8	From 7, T4.4.4A Equalities: Right Cancellation by Multiplication, Lx9.6.1.14 Triple Product of Curl Operator, T7.10.3 Distribution of a Scalar over Addition of Vectors, A7.9.3 Commutative by Vector Addition, A7.9.4 Associative by Vector Addition	$-v^{-2} \dfrac{\partial \vec{s}_p}{\partial t} = \nabla [(\mu\vec{H}\bullet\vec{H} + \varepsilon\vec{E} \bullet\vec{E})] - \mu (\vec{H} \bullet\nabla) \vec{H} - \varepsilon (\vec{E} \bullet\nabla) \vec{E} + \mu\vec{J} \times \vec{H}$
9	From 8, T4.4.18A Equalities: Product by Division-Factorization by 2, Lx9.3.2.3B Distribute Constant in/out of Product, A7.9.12 Inverse Vector Object Add, T6.9.11 Multiply a row or column of a determinant [A] by a scalar k; The static coulomb force balances itself out so it needs to be reintroduced at this point	$-v^{-2} \dfrac{\partial \vec{s}_p}{\partial t} = 2 \nabla [\tfrac{1}{2}(\mu\vec{H}\bullet\vec{H} + \varepsilon\vec{E} \bullet\vec{E})] - \mu (\vec{H} \bullet\nabla) \vec{H} - \varepsilon (\vec{E} \bullet\nabla) \vec{E} - \rho_q \vec{E} + \rho_q \vec{E} + \vec{J} \times \mu\vec{H}$
10	From 9, D4.1.20 Negative Coefficient, T7.10.3 Distribution of a Scalar over Addition of Vectors, LxMXP.A.3.5.7 Magnetic induction, LxMXE.A.3.5.13 Poynting Momentum Field Energy Density as a Magnetic Field, LxMXE.A.3.5.14 Poynting Momentum Field Force Density-1 as a Magnetic Field, A4.2.3 Substitution and A7.9.2 Substitution of Vectors	$-v^{-2} \dfrac{\partial \vec{s}_p}{\partial t} = 2 \nabla u_p - \vec{f}_{1p} + \rho_q (\vec{E} + \vec{v} \times \vec{B})$
∴	From 10 and T4.4.7B Equalities: Reversal of Left Cancellation by Multiplication	$\dfrac{\partial \vec{s}_p}{\partial t} = v^2 (-2\nabla u_p + \vec{f}_{1p} - \vec{f}_L)$

Theorem A.2.9 PVVT: Divergence of the Poynting Vector

Steps	Hypothesis	
		$\nabla\bullet\vec{s}_p = -\dfrac{2\partial\, u_p}{\partial t} - p_p$
1	From LxMXE.A.3.5.8 Poynting Field Vector and Lx9.6.1.1B Uniqueness of Divergence Operator	$\nabla\bullet\vec{s}_p = \nabla\bullet(\vec{E}\times\vec{H})$
2	From 1 and Lx9.6.1.7 Divergence of Cross Product	$\nabla\bullet\vec{s}_p = (\vec{H}\bullet\nabla\times\vec{E}) - (\vec{E}\bullet\nabla\times\vec{H})$
3	From 2, LxMXE.A.3.4.4 Faraday's Law; as a Magnetic Field, LxMXE.A.3.4.3 Ampère's Circuital Law; as a Magnetic Field and A4.2.3 Substitution	$\nabla\bullet\vec{s}_p = -\mu\,\vec{H}\bullet\dfrac{\partial\vec{H}}{\partial t} - \varepsilon\,\vec{E}\bullet\dfrac{\partial\vec{E}}{\partial t} - \vec{E}\bullet\vec{J}$
4	From 3, Lx9.2.2.7 Differential of Identical Dot Product and T4.4.4A Equalities: Right Cancellation by Multiplication	$\nabla\bullet\vec{s}_p = -\dfrac{\partial(\mu\,\vec{H}\bullet\vec{H})}{\partial t} - \dfrac{\partial(\varepsilon\,\vec{E}\bullet\vec{E})}{\partial t} - \vec{E}\bullet\vec{J}$
5	From 4 and Lx9.3.2.2B Differential as a Linear Operator, Lx9.3.2.3B Distribute Constant in/out of Product and T4.4.18B Equalities: Product by Division-Factorization by 2	$\nabla\bullet\vec{s}_p = -\dfrac{2\,\partial(\frac{1}{2}[\mu\,\vec{H}\bullet\vec{H} + \varepsilon\,\vec{E}\bullet\vec{E}])}{\partial t} - \vec{E}\bullet\vec{J}$
∴	From 5, LxMXE.A.3.5.9 Poynting Field Energy Density, LxMXE.A.3.5.5 Poynting Field Power Density and A4.2.3 Substitution	$\nabla\bullet\vec{s}_p = -\dfrac{2\partial\, u_p}{\partial t} - p_p$

Theorem A.2.10 PVVT: Curl of the Poynting Vector

Steps	Hypothesis	$\nabla\times\vec{s}_p = -(\rho/\varepsilon)\,\vec{H} + \vec{f}_{2p}$
1	From LxMXE.A.3.5.8 Poynting Field Vector and Lx9.6.1.1C Uniqueness of Curl Operator	$\nabla\times\vec{s}_p = \nabla\times(\vec{E}\times\vec{H})$
2	From 1 and Lx9.6.1.8 Curl of Cross Product	$\nabla\times\vec{s}_p = (\vec{H}\bullet\nabla)\vec{E} - (\vec{E}\bullet\nabla)\vec{H} - \vec{H}\,(\nabla\bullet\vec{E}) + \vec{E}\,(\nabla\bullet\vec{H})$
3	From 2, LxMXE.A.3.4.1 Gauss's Law for a closed electrical surface, LxMXE.A.3.4.2 Gauss's Law for a closed magnetic surface; as a Magnetic Field	$\nabla\times\vec{s}_p = (\vec{H}\bullet\nabla)\vec{E} - (\vec{E}\bullet\nabla)\vec{H} - \vec{H}\,(\rho/\varepsilon) + \vec{E}\,(0)$
4	From 3, A7.9.3 Commutative by Vector Addition, T7.10.4 Grassmann's Zero Vector and T7.11.2 Vector Addition: Identity	$\nabla\times\vec{s}_p = -(\rho/\varepsilon)\,\vec{H} + (\vec{H}\bullet\nabla)\vec{E} - (\vec{E}\bullet\nabla)\vec{H}$
∴	From 4, LxMXE.A.3.5.11 Poynting Power Density-2 and A7.9.2 Substitution of Vectors	$\nabla\times\vec{s}_p = -(\rho/\varepsilon)\,\vec{H} + \vec{f}_{2p}$

Theorem A.2.11 PVVT: Divergence Poynting Field Energy Density

Steps	Hypothesis	$\nabla u_p = -\,v^{-2}\mu^{-1}\,\dfrac{\partial\,\vec{s}_p}{\partial t} + \mu^{-1}\vec{f}_{3p} + \vec{E}\times\vec{J}$
1	From LxMXE.A.3.5.9 Poynting Field Energy Density	$u_p = ½\,(\,\mu\vec{H}\bullet\vec{H}+\varepsilon\vec{E}\bullet\vec{E}\,)$
2	From 1 and Lx9.6.1.1A Uniqueness of Gradient Operator	$\nabla u_p = \nabla\,½\,(\,\mu\vec{H}\bullet\vec{H}+\varepsilon\vec{E}\bullet\vec{E}\,)$
3	From 2, Lx9.3.2.3B Distribute Constant in/out of Product, Lx9.6.1.2A Distribution of Gradient and T7.10.3 Distribution of a Scalar over Addition of Vectors	$\nabla u_p = ½\,\mu\nabla(\vec{H}\bullet\vec{H}) + ½\,\varepsilon\nabla(\vec{E}\bullet\vec{E}\,)$
4	From 3 and Lx9.6.1.9 Gradient of Dot Product	$\nabla u_p = ½\,\mu[\,(\vec{H}\bullet\nabla)\vec{H} + (\vec{H}\bullet\nabla)\vec{H} + \vec{H}\times(\nabla\times\vec{H}) + \vec{H}\times(\nabla\times\vec{H})] + ½\,\varepsilon[\,(\vec{E}\bullet\nabla)\vec{E} + (\vec{E}\bullet\nabla)\vec{E} + \vec{E}\times(\nabla\times\vec{E}) + \vec{E}\times(\nabla\times\vec{E})]$

5	From 4, A4.2.12 Identity Multp, A4.2.13 Inverse Multp, T7.10.3 Distribution of a Scalar over Addition of Vectors, D4.1.19 Primitive Definition for Rational Arithmetic and T4.4.18 Equalities: Product by Division-Factorization by 2	$\nabla u_p = \mu[\,(\vec{H}\bullet\nabla)\vec{H} + \vec{H}\times(\nabla\times\vec{H})] + \varepsilon[\,(\vec{E}\bullet\nabla)\vec{E} + \vec{E}\times(\nabla\times\vec{E})]$
6	From 5, T7.10.3 Distribution of a Scalar over Addition of Vectors and A7.9.3 Commutative by Vector Addition	$\nabla u_p = (\mu\vec{H}\bullet\nabla)\vec{H} + \varepsilon\,(\vec{E}\bullet\nabla)\vec{E} + \mu\,\vec{H}\times(\nabla\times\vec{H}) + \varepsilon\,\vec{E}\times(\nabla\times\vec{E})$
7	From 6, LxMXP.A.3.5.7 Magnetic induction, A4.2.12 Identity Multp, A4.2.13 Inverse Multp and	$\nabla u_p = \mu^{-1}\,[(\vec{B}\bullet\nabla)\vec{B} + v^{-2}\,(\vec{E}\bullet\nabla)\vec{E} + \vec{B}\times(\nabla\times\vec{B}) + v^{-2}\,\vec{E}\times(\nabla\times\vec{E})]$
8	From 3, LxMXE.A.3.4.3 Ampère's Circuital Law as extended by Maxwell, LxMXE.A.3.4.4 Faraday's Law, A7.9.2 Substitution of Vectors and T7.10.3 Distribution of a Scalar over Addition of Vectors	$\nabla u_p = \mu^{-1}\,[(\vec{B}\bullet\nabla)\vec{B} + v^{-2}\,(\vec{E}\bullet\nabla)\vec{E} + \vec{B}\times(v^{-2}\,\frac{\partial\vec{E}}{\partial t} + \mu\vec{J}) - v^{-2}\,\vec{E}\times\frac{\partial\vec{B}}{\partial t}\,]$
9	From 8, A7.9.3 Commutative by Vector Addition, T6.9.11 Multiply a row or column of a determinant [A] by a scalar k and T7.10.3 Distribution of a Scalar over Addition of Vectors	$\nabla u_p = \mu^{-1}\,[(\vec{B}\bullet\nabla)\vec{B} + v^{-2}\,(\vec{E}\bullet\nabla)\vec{E} + \mu\vec{B}\times\vec{J} + v^{-2}\,(\vec{B}\times\frac{\partial\vec{E}}{\partial t} - \vec{E}\times\frac{\partial\vec{B}}{\partial t})]$
10	From LxMXE.A.3.5.8 Poynting Field Vector and Lx9.2.2.3 Differential of Cross Product	$\frac{\partial\,\vec{E}\times\vec{B}}{\partial t} = \vec{E}\times\frac{\partial\,\vec{B}}{\partial t} - \vec{B}\times\frac{\partial\vec{E}}{\partial t}$
11	From 9, 10, T7.11.4 Vector Addition: Right Cancellation by Addition, T7.11.5 Vector Addition: Left Cancellation by Addition, A7.9.2 Substitution of Vectors and T7.10.3 Distribution of a Scalar over Addition of Vectors	$\nabla u_p = \mu^{-1}\,[(\vec{B}\bullet\nabla)\vec{B} + v^{-2}\,(\vec{E}\bullet\nabla)\vec{E} + \mu\vec{B}\times\vec{J}\; v^{-2}\,(-\frac{\partial\,\vec{E}\times\vec{B}}{\partial t} + \vec{E}\times\frac{\partial\,\vec{B}}{\partial t} - \vec{E}\times\frac{\partial\vec{B}}{\partial t})]$

12	From 11, T7.11.3 Vector Addition: Inverse, T7.11.2 Vector Addition: Identity and D4.1.20A Negative Coefficient	∇u_p	=	$\mu^{-1} [(\vec{B}\bullet\nabla)\vec{B} + v^{-2} (\vec{E}\bullet\nabla)\vec{E} + \mu\vec{B}\times\vec{J} - v^{-2} \frac{\partial \vec{E}\times\vec{B}}{\partial t}]$
13	From 1, LxMXE.A.3.5.8 Poynting Field Vector and A7.9.2 Substitution of Vectors	∇u_p	=	$\mu^{-1}[(\vec{B}\bullet\nabla)\vec{B} + v^{-2} (\vec{E}\bullet\nabla)\vec{E} + \mu\vec{B}\times\vec{J} - v^{-2} \frac{\partial \vec{s}_p}{\partial t}]$
∴	From 4, LxMXE.A.3.5.11 Poynting Power Density-2 and A7.9.2 Substitution of Vectors and A7.9.3 Commutative by Vector Addition, T7.10.3 Distribution of a Scalar over Addition of Vectors, A4.2.13 Inverse Multp and T7.10.5 Identity with Scalar Multiplication to Vectors	∇u_p	=	$- v^{-2}\mu^{-1} \frac{\partial \vec{s}_p}{\partial t} + \mu^{-1}\vec{f}_{3p} + \vec{B}\times\vec{J}$

Table A.16 Poynting Field Equations

	Hypothesis	Equation	P-Variables		Parameters	
LxPFE.A.4.1.1	TA.10.4 PVVT: Divergence of the Poynting Vector	$\nabla\bullet\vec{s}_p = - \frac{\partial u_p}{\partial t} - p_p$	$\vec{s}_p$	u_p	v	∅
LxPFE.A.4.1.2	TA.10.3 PVVT: Lorentz Force Density from Maxwell's Equations	$\nabla u_p = - \frac{1}{v^2} \frac{\partial \vec{s}_p}{\partial t} + \vec{f}_{1p} - \vec{f}_L$				

Poynting Momentum Vector Field Equations

Theorem A.2.12 PVVT: Poynting Momentum Vector Differentiated With Respect to Time

Steps	Hypothesis	$\dfrac{\partial \vec{s}_m}{\partial t} = -2\nabla u_m + \vec{f}_{Im} - \vec{f}_L$
1	From LxMXE.A.3.5.12 Poynting Momentum Field Vector and Lx9.3.1.7B Uniqueness Theorem of Differentiation	$\dfrac{\partial \vec{s}_m}{\partial t} = \dfrac{\partial (\varepsilon\vec{E}\times\vec{B})}{\partial t}$
2	From 1 and Lx9.2.2.3 Differential of Cross Product	$\dfrac{\partial \vec{s}_m}{\partial t} = \dfrac{\varepsilon\partial\vec{E}}{\partial t} \times \vec{B} + \varepsilon\vec{E} \times \dfrac{\partial \vec{B}}{\partial t}$
3	From LxMXE.A.3.4.3 Ampère's Circuital Law	$\nabla\times\vec{B} = v^{-2} \dfrac{\partial\vec{E}}{\partial t} + \mu\vec{J}$
4	From LxMXE.A.3.4.4 Faraday's Law	$\nabla\times\vec{E} = -\dfrac{\partial\vec{B}}{\partial t}$
5	From 2, 3, 4, T7.11.4A Vector Addition: Right Cancellation by Addition, T4.4.4A Equalities: Right Cancellation by Multiplication and A7.9.2 Substitution of Vectors	$\dfrac{\partial \vec{s}_m}{\partial t} = \varepsilon\, v^2(\nabla\times\vec{B} - \mu\vec{J}) \times \vec{B} - \varepsilon\vec{E} \times\nabla\times\vec{E}$
6	From 5 and T7.10.3 Distribution of a Scalar over Addition of Vectors	$\dfrac{\partial \vec{s}_m}{\partial t} = -\varepsilon\, v^2\, \vec{B} \times\nabla\times\vec{B} - \mu\varepsilon\, v^2\vec{J} \times \vec{B} - \varepsilon(\vec{E} \times\nabla\times\vec{E})$
7	From 6, A7.9.3 Commutative by Vector Addition and LxMXE.A.3.4.5 Material Media Velocity	$-\dfrac{\partial \vec{s}_m}{\partial t} = (\varepsilon/\varepsilon\mu)\vec{B} \times\nabla\times\vec{B} + \varepsilon(\vec{E} \times\nabla\times\vec{E}) + (v^2 / v^2)\, \vec{J}\times \vec{B}$
8	From 7, A4.2.12 Identity Multp and A4.2.13 Inverse Multp	$-\dfrac{\partial \vec{s}_m}{\partial t} = (1/\mu)(\vec{B} \times\nabla\times\vec{B}) + \varepsilon(\vec{E} \times\nabla\times\vec{E}) + \vec{J}\times \vec{B}$
9	From 8, Lx9.6.1.14 Triple Product of Curl Operator, T7.10.3 Distribution of a Scalar over Addition of Vectors, A7.9.3 Commutative by Vector Addition, A7.9.4 Associative by Vector Addition.	$-\dfrac{\partial \vec{s}_m}{\partial t} = (1/\mu)(\nabla(\vec{B}\bullet\vec{B}) - (\vec{B}\bullet\nabla)\vec{B}) + \varepsilon(\nabla(\vec{E}\bullet\vec{E}) - (\vec{E}\bullet\nabla)\vec{E}) + \vec{J}\times \vec{B}$

10	From 9, Lx9.3.2.3B Distribute Constant in/out of Product and T7.10.3 Distribution of a Scalar over Addition of Vectors	$-\dfrac{\partial \vec{s}_m}{\partial t} = [\nabla(1/\mu)(\vec{B}\bullet\vec{B}) - (1/\mu)(\vec{B}\bullet\nabla)\vec{B}] + [\nabla\varepsilon(\vec{E}\bullet\vec{E}) - \varepsilon(\vec{E}\bullet\nabla)\vec{E}] + \vec{J}\times\vec{B}$
11	From 10, A7.9.4 Associative by Vector Addition, A7.9.3 Commutative by Vector Addition and Lx9.6.1.2A Distribution of Gradient	$-\dfrac{\partial \vec{s}_m}{\partial t} = \nabla[(1/\mu)(\vec{B}\bullet\vec{B}) + \varepsilon(\vec{E}\bullet\vec{E})] - (1/\mu)(\vec{B}\bullet\nabla)\vec{B} - \varepsilon(\vec{E}\bullet\nabla)\vec{E} + \vec{J}\times\vec{B}$
12	From 11 and T4.4.18A Equalities: Product by Division-Factorization by 2 and Lx9.3.2.3B Distribute Constant in/out of Product	$-\dfrac{\partial \vec{s}_m}{\partial t} = 2\nabla\tfrac{1}{2}[(1/\mu)(\vec{B}\bullet\vec{B}) + \varepsilon(\vec{E}\bullet\vec{E})] - (1/\mu)(\vec{B}\bullet\nabla)\vec{B} - \varepsilon(\vec{E}\bullet\nabla)\vec{E} + \vec{J}\times\vec{B}$
13	From 12, D4.1.20A Negative Coefficient and T7.10.3 Distribution of a Scalar over Addition of Vectors, A7.9.12 Inverse Vector Object Add; The static coulomb force balances itself out so it needs to be reintroduced at this point	$-\dfrac{\partial \vec{s}_m}{\partial t} = 2\nabla\tfrac{1}{2}[(1/\mu)(\vec{B}\bullet\vec{B}) + \varepsilon(\vec{E}\bullet\vec{E})] - [(1/\mu)(\vec{B}\bullet\nabla)\vec{B}) + \varepsilon(\vec{E}\bullet\nabla)\vec{E}) + \rho_q\vec{E}] + \rho_q\vec{E} + \vec{J}\times\vec{B}$
14	From 13, LxMXP.A.3.5.2 Current Density, A7.9.2 Substitution of Vectors and T7.10.3 Distribution of a Scalar over Addition of Vectors	$-\dfrac{\partial \vec{s}_m}{\partial t} = 2\nabla\tfrac{1}{2}[(1/\mu)(\vec{B}\bullet\vec{B}) + \varepsilon(\vec{E}\bullet\vec{E})] - [(1/\mu)(\vec{B}\bullet\nabla)\vec{B}) + \varepsilon(\vec{E}\bullet\nabla)\vec{E}) + \rho_q\vec{E}] + \rho_q(\vec{E} + \vec{V}\times\vec{B})$

∴	From 14, T7.10.1A Uniqueness of Scalar Multiplication to Vectors, T7.10.3 Distribution of a Scalar over Addition of Vectors, T4.8.2B Integer Exponents: Negative One Squared, T4.8.4 Integer Exponents: Negative One Raised to an Odd Number, LxMXE.A.3.5.13 Poynting Momentum Field Energy Density, LxMXE.A.3.5.14 Poynting Momentum Field Force Density-1, LxMXE.A.3.5.4 Lorentz Force Density, A4.2.3 Substitution and A7.9.2 Substitution of Vectors	$\frac{\partial \vec{s}_m}{\partial t} = -2\nabla u_m + \vec{f}_{1m} - \vec{f}_L$

Theorem A.2.13 PVVT: Divergence of the Poynting Momentum Vector

Steps	Hypothesis	$\nabla\bullet\vec{s}_m = -v^{-2}(\frac{2\,\partial u_m}{\partial t} + p_p)$
1	From LxMXE.A.3.5.8 Poynting Field Vector and Lx9.6.1.1B Uniqueness of Divergence Operator	$\nabla\bullet\vec{s}_m = \nabla\bullet(\varepsilon\vec{E}\times\vec{B})$
2	From 1 and Lx9.6.1.7 Divergence of Cross Product	$\nabla\bullet\vec{s}_m = \varepsilon(\vec{B}\bullet\nabla\times\vec{E} - \vec{E}\bullet\nabla\times\vec{B})$
3	From 2, LxMXE.A.3.4.4 Faraday's Law; as a Magnetic Field, LxMXE.A.3.4.3 Ampère's Circuital Law and A4.2.3 Substitution	$\nabla\bullet\vec{s}_m = -\varepsilon\vec{B}\bullet\frac{\partial\vec{B}}{\partial t} - \mu\varepsilon^2\vec{E}\bullet\frac{\partial\vec{E}}{\partial t} - \mu\varepsilon\vec{E}\bullet\vec{J}$
4	From 3, Lx9.2.2.7 Differential of Identical Dot Product, A4.2.12 Identity Multp, A4.2.13 Inverse Multp, T4.4.18B Equalities: Product by Division-Factorization by 2 and Lx9.3.2.3B Distribute Constant in/out of Product	$\nabla\bullet\vec{s}_m = -\mu\varepsilon[\frac{2\mu^{-1}\partial(\frac{1}{2}\vec{B}\bullet\vec{B})}{\partial t} + \frac{2\partial(\frac{1}{2}\varepsilon\vec{E}\bullet\vec{E})}{\partial t} + \vec{E}\bullet\vec{J}]$

5	From 4, LxMXE.A.3.4.5 Material Media Velocity, and Lx9.3.2.2B Differential as a Linear Operator, Lx9.3.2.3B Distribute Constant in/out of Product and T4.4.14 Equalities: Formal Cross Product	$\nabla\bullet\vec{s}_m = -v^{-2}\dfrac{2\,\partial\,[\,½\,(\mu^{-1}\,\vec{B}\bullet\vec{B} + \varepsilon\vec{E}\bullet\vec{E})]}{\partial t} - v^{-2}\,\vec{E}\bullet\vec{J}$
∴	From 5, LxMXE.A.3.5.13 Poynting Momentum Field Energy Density, LxMXE.A.3.5.5 Poynting Field Power Density and A4.2.3 Substitution	$\nabla\bullet\vec{s}_m = -v^{-2}(\dfrac{2\,\partial\,u_m}{\partial t} + p_p)$

Theorem A.2.14 PVVT: Curl of the Poynting Momentum Vector

Steps	Hypothesis	$\nabla\times\vec{s}_m = -\rho\vec{B} + \vec{f}_{2m}$
1	From LxMXE.A.3.5.12 Poynting Momentum Field Vector and Lx9.6.1.1C Uniqueness of Curl Operator	$\nabla\times\vec{s}_m = \nabla\times(\varepsilon\vec{E}\times\vec{B})$
2	From 1, Lx9.6.1.8 Curl of Cross Product and Lx9.3.2.3B Distribute Constant in/out of Product	$\nabla\times\vec{s}_m = \varepsilon\,(\vec{B}\bullet\nabla)\vec{E} - (\varepsilon\vec{E}\bullet\nabla)\vec{B} - \varepsilon\vec{B}\,(\nabla\bullet\vec{E}) + \varepsilon\vec{E}\,(\nabla\bullet\vec{B})$
3	From 2, LxMXE.A.3.4.1 Gauss's Law for a closed electrical surface, LxMXE.A.3.4.2 Gauss's Law for a closed magnetic surface; as a Magnetic Field	$\nabla\times\vec{s}_m = \varepsilon\,(\vec{B}\bullet\nabla)\vec{E} - (\varepsilon\vec{E}\bullet\nabla)\vec{B} - \varepsilon\vec{B}\,(\rho/\varepsilon) + \varepsilon\vec{E}\,(0)$
4	From 3, A4.2.13 Inverse Multp, A4.2.12 Identity Multp, A7.9.3 Commutative by Vector Addition, T7.10.4 Grassmann's Zero Vector and T7.11.2 Vector Addition: Identity	$\nabla\times\vec{s}_m = -\rho\vec{B} + \varepsilon[(\vec{B}\bullet\nabla)\vec{E} - (\vec{E}\bullet\nabla)\vec{B}]$
∴	From 4, LxMXE.A.3.5.15 Poynting Momentum Power Density-2 and A7.9.2 Substitution of Vectors	$\nabla\times\vec{s}_m = -\rho\vec{B} + \vec{f}_{2m}$

Theorem A.2.15 PVVT: Integrated Lorentz Force Density

Steps	Hypothesis	Equation	
		$\vec{F}_L = Q\,[\,\vec{E} + \vec{V} \times \vec{B}\,]$	The Lorentz force
1	From LxMXE.A.3.5.4 Lorentz Force Density	$\vec{F}_L = \int_V \rho_q(\,\vec{E} + \vec{V} \times \vec{B}\,)\,dV$	
2	From 1 and integrated over the region of charge	$\vec{F}_L = [\,\vec{E} + \vec{V} \times \vec{B}\,] \int_V \rho_q\,dV$	
∴	From 2; the net charge in motion	$\vec{F}_L = [\,\vec{E} + \vec{V} \times \vec{B}\,]\,Q$	The Lorentz force

Theorem A.2.16 PVVT: Integrated Current Force Density with a Constant

Steps	Hypothesis	Equation	
		$\vec{F}_L = Q\,(\,\vec{E} + \vec{V} \times \vec{B}\,)$	The Lorentz force
1	From LxMXE.A.3.5.8 Poynting Field Vector and D9.4.6 Indefinite Integral	$\vec{F}_L = \int_V \rho_q\,(\,\vec{V} \times \vec{B}\,)\,dV + \vec{C}$	with constant of integration
2	From 1 and integrated over the region of charge	$\vec{F}_L = (\vec{V} \times \vec{B}) \int_V \rho_q\,dV + \vec{C}$	
3	From 2; the net charge in motion	$\vec{F}_L = (\vec{V} \times \vec{B})\,Q + \vec{C}$	
4	From 3; singular stationary charge remains its static coulomb field still exerts itself	$\vec{F}_L = \vec{C}$	for $\vec{V} = \mathbf{0}$
5	From 4; singular stationary charge generates a static electric field	$Q\vec{E} = \vec{C}$	for $\vec{V} = \mathbf{0}$
6	From 3, 5 and A7.9.2 Substitution of Vectors	$\vec{F}_L = Q\,(\vec{V} \times \vec{B}) + Q\,\vec{E}$	
∴	From 6 and T7.10.3 Distribution of a Scalar over Addition of Vectors	$\vec{F}_L = Q\,(\,\vec{E} + \vec{V} \times \vec{B}\,)$	The Lorentz force

This proof clearly shows the origins of Lorentz force that it comes directly from Maxwell's equations and provides away to interpret terms on the right hand side of the Poynting equations. It also demonstrates that when modeling electromagnetic phenomena it's important to use field density in order to assign a physical property to every point within the region. The Poynting field equations can be summarized as follows:

Table A.17 Poynting Momentum Field Equations

	Hypothesis	Equation	P-Variables		Parameters	
LxPME.A.5.1.1	TA.5.2 PVVT: Divergence of the Poynting Vector	$\nabla \bullet \vec{s}_m = -\,v^{-2} \dfrac{\partial\,u_m}{\partial t} - v^{-2} p_P$	$\vec{s}_m$	u_m	v	∅
LxPME.A.5.1.2	TA.5.4 PVVT: Lorentz Force Density from Maxwell's Equations	$\nabla\,u_m = -\dfrac{\partial\,\vec{s}_m}{\partial\,t} + \vec{f}_{lm} - \vec{f}_L$				

Theorem A.2.17 PVVT: Divergence of Poynting Field Power Density

Steps	Hypothesis			
		∇p_p	=	$(\vec{J}\bullet\nabla)\vec{E} + (\vec{E}\bullet\nabla)\vec{J} + \vec{E}\times(\nabla\times\vec{J}) - \vec{J}\times\dfrac{\partial\vec{B}}{\partial t}$
1	From LxMXE.A.3.2.5 Poynting Field Power Density	p_p	=	$\vec{E}\bullet\vec{J}$
2	From 1 and Lx9.6.1.1A Uniqueness of Gradient Operator	∇p_p	=	$\nabla(\vec{E}\bullet\vec{J})$
3	From 2 and Lx9.6.1.9 Gradient of Dot Product	∇p_p	=	$(\vec{J}\bullet\nabla)\vec{E} + (\vec{E}\bullet\nabla)\vec{J} + \vec{E}\times(\nabla\times\vec{J}) + \vec{J}\times(\nabla\times\vec{E})$
∴	From 3, LxMXE.A.3.3.4 Faraday's Law, A7.9.2 Substitution of Vectors, A7.8.6 Parallelogram Law: Multiplying a Negative Number Times a Vector and T6.9.11 Multiply a row or column of a determinant [A] by a scalar k	∇p_p	=	$(\vec{J}\bullet\nabla)\vec{E} + (\vec{E}\bullet\nabla)\vec{J} + \vec{E}\times(\nabla\times\vec{J}) - \vec{J}\times\dfrac{\partial\vec{B}}{\partial t}$

Theorem A.2.18 PVVT: Divergence of Coulomb Force Density

Steps	Hypothesis			
		$\varepsilon\nabla\bullet\vec{f}_q$	=	$\varepsilon\vec{E}\bullet\nabla\rho_q + \rho_q^{\ 2}$
1	From LxMXP.A.3.2.3 Coulomb Force Density	$\vec{f}_q$	=	$\rho_q\,\vec{E}$
2	From 1 and Lx9.6.1.1B Uniqueness of Divergence Operator	$\nabla\bullet\vec{f}_q$	=	$\nabla\bullet(\rho_q\,\vec{E})$
3	From 2 and Lx9.6.1.5 Divergence of Scalar Vector Product	$\nabla\bullet\vec{f}_q$	=	$\vec{E}\bullet\nabla\rho_q + \rho_q\nabla\bullet\vec{E}$
4	From 3, LxMXE.A.3.3.1 Gauss's Law for a closed electrical surface and A4.2.18 Summation Exp	$\nabla\bullet\vec{f}_q$	=	$\vec{E}\bullet\nabla\rho_q + \rho_q^{\ 2}/\varepsilon$
∴	From 4, T4.4.2 Equalities: Uniqueness of Multiplication, A4.2.14 Distribution, A4.2.13 Inverse Multp and A4.2.12 Identity Multp	$\varepsilon\nabla\bullet\vec{f}_q$	=	$\varepsilon\vec{E}\bullet\nabla\rho_q + \rho_q^{\ 2}$

Theorem A.2.19 PVVT: Divergence of Lorentz Force Density

Steps	Hypothesis	
		$\nabla\bullet\vec{f}_L = \rho_q^2/\varepsilon + \vec{E}\bullet\nabla\rho_q + \vec{B}\bullet(\nabla\times\vec{J}) - \mu(\vec{J}\bullet\vec{J}) - \vec{J}\bullet\dfrac{\partial\vec{E}}{v^2\partial t}$
1	From LxMXE.A.3.2.4 Lorentz Force Density	$\vec{f}_L = \rho_q(\vec{E} + \vec{V}\times\vec{B})$
2	From 1 and T7.10.3 Distribution of a Scalar over Addition of Vectors	$\vec{f}_L = \rho_q\vec{E} + \rho_q\vec{V}\times\vec{B}$
3	From 2, LxMXP.A.3.5.3 Current Density and A7.9.2 Substitution of Vectors	$\vec{f}_L = \rho_q\vec{E} + \vec{J}\times\vec{B}$
4	From 3 and Lx9.6.1.1B Uniqueness of Divergence Operator	$\nabla\bullet\vec{f}_L = \nabla\bullet(\rho_q\vec{E} + \vec{J}\times\vec{B})$
5	From 4 and Lx9.6.1.3A Distribution of Divergence	$\nabla\bullet\vec{f}_L \quad \nabla\bullet(\rho_q\vec{E}) + \nabla\bullet(\vec{J}\times\vec{B})$
6	From 5 and Lx9.6.1.5 Divergence of Scalar Vector Product	$\nabla\bullet\vec{f}_L = \vec{E}\bullet\nabla\rho_q + \rho_q\nabla\bullet\vec{E} + \nabla\bullet(\vec{J}\times\vec{B})$
7	From 6 and Lx9.6.1.7 Divergence of Cross Product	$\nabla\bullet\vec{f}_L = \vec{E}\bullet\nabla\rho_q + \rho_q(\rho_q/\varepsilon) + \vec{B}\bullet(\nabla\times\vec{J}) - \vec{J}\bullet(\nabla\times\vec{B})$
8	From 7, LxMXE.A.3.3.3 Ampère's Circuital Law as extended by Maxwell, A7.9.2 Substitution of Vectors and T8.3.7 Dot Product: Distribution of Dot Product Across Addition of Vectors	$\nabla\bullet\vec{f}_L = \vec{E}\bullet\nabla\rho_q + \rho_q^2/\varepsilon + \vec{B}\bullet(\nabla\times\vec{J}) - \vec{J}\bullet(\mu\vec{J}) - \vec{J}\bullet\dfrac{\partial\vec{E}}{v^2\partial t}$
∴	From 8 and T8.3.10A Dot Product: Right Scalar-Vector Association	$\nabla\bullet\vec{f}_L = \rho_q^2/\varepsilon + \vec{E}\bullet\nabla\rho_q + \vec{B}\bullet(\nabla\times\vec{J}) - \mu(\vec{J}\bullet\vec{J}) - \vec{J}\bullet\dfrac{\partial\vec{E}}{v^2\partial t}$

Theorem A.2.20 PVVT: Divergence of Poynting Momentum Field Density-1

Steps	Hypothesis			
		$\nabla\bullet\vec{f}_{1m}$	=	$\nabla[\mu^{-1}](\vec{B}\bullet\nabla)\vec{B}+\mu^{-1}\nabla\bullet[(\vec{B}\bullet\nabla)\vec{B}]+$ $\nabla[\varepsilon](\vec{E}\bullet\nabla)\vec{E}]+\varepsilon\nabla\bullet[(\vec{E}\bullet\nabla)\vec{E}]+\vec{E}\bullet\nabla\rho_q+\rho_q^{\,2}/\varepsilon$
1	From LxMXE.A.3.2.14 Poynting Momentum Field Force Density-1	$\vec{f}_{1m}$	=	$\mu^{-1}(\vec{B}\bullet\nabla)\vec{B}+\varepsilon(\vec{E}\bullet\nabla)\vec{E}+\vec{f}_q$
2	From 1 and Lx9.6.1.1B Uniqueness of Divergence Operator	$\nabla\bullet\vec{f}_{1m}$	=	$\nabla\bullet[\mu^{-1}(\vec{B}\bullet\nabla)\vec{B}+\varepsilon(\vec{E}\bullet\nabla)\vec{E}+\vec{f}_q]$
3	From 2 and Lx9.6.1.3A Distribution of Divergence	$\nabla\bullet\vec{f}_{1m}$	=	$\nabla\bullet[\mu^{-1}(\vec{B}\bullet\nabla)\vec{B}]+\nabla\bullet[\varepsilon(\vec{E}\bullet\nabla)\vec{E}]+\nabla\bullet\vec{f}_q$
∴	From 3, Lx9.6.1.5 Divergence of Scalar Vector Product, TA2.18S5 PVVT: Divergence of Coulomb Force Density and A4.2.3 Substitution	$\nabla\bullet\vec{f}_{1m}$	=	$\nabla[\mu^{-1}](\vec{B}\bullet\nabla)\vec{B}+\mu^{-1}\nabla\bullet[(\vec{B}\bullet\nabla)\vec{B}]+$ $\nabla[\varepsilon](\vec{E}\bullet\nabla)\vec{E}]+\varepsilon\nabla\bullet[(\vec{E}\bullet\nabla)\vec{E}]+\vec{E}\bullet\nabla\rho_q+\rho_q^{\,2}/\varepsilon$

Theorem A.2.21 PVVT: Divergence of Poynting Momentum Field Density-2

Steps	Hypothesis			
		$\nabla\bullet\vec{f}_{2m}$	=	$+\nabla(\varepsilon v)(\vec{B}\bullet\nabla)\vec{E}+(\varepsilon v)\nabla\bullet[(\vec{B}\bullet\nabla)\vec{E}]$ $\nabla(\varepsilon v)(\vec{E}\bullet\nabla)\vec{B}-(\varepsilon v)\nabla\bullet[(\vec{E}\bullet\nabla)\vec{B}]$
1	From LxMXE.A.3.2.15 Poynting Momentum Field Force Density-2	$\vec{f}_{2m}$	=	$\varepsilon[(\vec{B}\bullet\nabla)\vec{E}-(\vec{E}\bullet\nabla)\vec{B}]v$
2	From 1, Lx9.6.1.1B Uniqueness of Divergence Operator and T7.10.1 Uniqueness of Scalar Multiplication to Vectors	$\nabla\bullet\vec{f}_{2m}$	=	$\nabla\bullet(\varepsilon\, v\,[(\vec{B}\bullet\nabla)\vec{E}-(\vec{E}\bullet\nabla)\vec{B}])$
∴	From 2, T7.10.3 Distribution of a Scalar over Addition of Vectors, Lx9.6.1.3A Distribution of Divergence and Lx9.6.1.5 Divergence of Scalar Vector Product	$\nabla\bullet\vec{f}_{2m}$	=	$+\nabla(\varepsilon v)(\vec{B}\bullet\nabla)\vec{E}+(\varepsilon v)\nabla\bullet[(\vec{B}\bullet\nabla)\vec{E}]$ $\nabla(\varepsilon v)(\vec{E}\bullet\nabla)\vec{B}-(\varepsilon v)\nabla\bullet[(\vec{E}\bullet\nabla)\vec{B}]$

Theorem A.2.22 PVVT: Curl of Poynting Momentum Field Density-2

Steps	Hypothesis		
		$\nabla\times\vec{f}_{2m}$ =	$+\nabla(\varepsilon v)[(\vec{B}\bullet\nabla)\vec{E} + \varepsilon v\,\nabla\times[(\vec{B}\bullet\nabla)\vec{E}]$ $-\nabla(\varepsilon v)(\vec{E}\bullet\nabla)\vec{B} - \varepsilon v\,\nabla\times[(\vec{E}\bullet\nabla)\vec{B}]$
1	From LxMXE.10.2.15	$\vec{f}_{2m}$ =	$\varepsilon[(\vec{B}\bullet\nabla)\vec{E} - (\vec{E}\bullet\nabla)\vec{B}]v$
2	From 1 and Lx9.6.1.1C Uniqueness of Curl Operator	$\nabla\times\vec{f}_{2m}$ =	$\nabla\times(\varepsilon[(\vec{B}\bullet\nabla)\vec{E} - (\vec{E}\bullet\nabla)\vec{B}]v)$
3	From 2 and T7.10.1 Uniqueness of Scalar Multiplication to Vectors	$\nabla\times\vec{f}_{2m}$ =	$\nabla\times(\varepsilon v[(\vec{B}\bullet\nabla)\vec{E} - (\vec{E}\bullet\nabla)\vec{B}])$
∴	From 3, T7.10.3 Distribution of a Scalar over Addition of Vectors, Lx9.6.1.4A Distribution of Curl and Lx9.6.1.6 Curl of Scalar Vector Product	$\nabla\times\vec{f}_{2m}$ =	$+\nabla(\varepsilon v)[(\vec{B}\bullet\nabla)\vec{E} + \varepsilon v\,\nabla\times[(\vec{B}\bullet\nabla)\vec{E}]$ $-\nabla(\varepsilon v)(\vec{E}\bullet\nabla)\vec{B} - \varepsilon v\,\nabla\times[(\vec{E}\bullet\nabla)\vec{B}]$

Theorem A.2.23 PVVT: Divergence Poynting Momentum Field Energy Density

Steps	Hypothesis			
		∇u_m	=	$-\dfrac{\partial \vec{s}_m}{\partial t} + \mu^{-1}\vec{f}_{3p} + \vec{B}\times\vec{J}$
1	From LxMXE.A.3.5.14 Poynting Momentum Field Energy Density	u_m	=	$½ (\mu^{-1}\vec{B}\bullet\vec{B}+\varepsilon\vec{E}\bullet\vec{E})$
2	From 1 and Lx9.6.1.1A Uniqueness of Gradient Operator	∇u_m	=	$\nabla ½ (\mu^{-1}\vec{B}\bullet\vec{B}+\varepsilon\vec{E}\bullet\vec{E})$
3	From 2, Lx9.3.2.3B Distribute Constant in/out of Product, Lx9.6.1.2A Distribution of Gradient and T7.10.3 Distribution of a Scalar over Addition of Vectors	∇u_m	=	$½ \mu^{-1}\nabla(\vec{B}\bullet\vec{B}) + ½ \varepsilon\nabla(\vec{E}\bullet\vec{E})$
4	From 3 and Lx9.6.1.9 Gradient of Dot Product	∇u_m	=	$½ \mu^{-1} [(\vec{B}\bullet\nabla)\vec{B} + (\vec{B}\bullet\nabla)\vec{B} + \vec{B}\times(\nabla\times\vec{B}) + \vec{B}\times(\nabla\times\vec{B})] +$ $½ \varepsilon[(\vec{E}\bullet\nabla)\vec{E} + (\vec{E}\bullet\nabla)\vec{E} + \vec{E}\times(\nabla\times\vec{E}) + \vec{E}\times(\nabla\times\vec{E})]$
5	From 4, A4.2.12 Identity Multp, A4.2.13 Inverse Multp, T7.10.3 Distribution of a Scalar over Addition of Vectors, D4.1.19 Primitive Definition for Rational Arithmetic and T4.4.18 Equalities: Product by Division-Factorization by 2	∇u_m	=	$\mu^{-1} [(\vec{B}\bullet\nabla)\vec{B} + \vec{B}\times(\nabla\times\vec{B})] + \varepsilon[(\vec{E}\bullet\nabla)\vec{E}+ \vec{E}\times(\nabla\times\vec{E})]$
6	From 5, T7.10.3 Distribution of a Scalar over Addition of Vectors and A7.9.3 Commutative by Vector Addition	∇u_m	=	$(\mu^{-1}\vec{B}\bullet\nabla)\vec{B} + \varepsilon (\vec{E}\bullet\nabla)\vec{E} + \mu^{-1} \vec{B}\times(\nabla\times\vec{B}) + \varepsilon \vec{E}\times(\nabla\times\vec{E})$
7	From 6, LxMXE.A.3.4.5 Material Media Velocity, A4.2.12 Identity Multp and A4.2.13 Inverse Multp	∇u_m	=	$\mu^{-1} [(\vec{B}\bullet\nabla)\vec{B} + v^{-2} (\vec{E}\bullet\nabla)\vec{E} +$ $\vec{B}\times(\nabla\times\vec{B}) + v^{-2} \vec{E}\times(\nabla\times\vec{E})]$
8	From 3, LxMXE.A.3.4.3 Ampère's Circuital Law as extended by Maxwell, LxMXE.A.3.4.4 Faraday's Law, A7.9.2 Substitution of Vectors and T7.10.3 Distribution of a Scalar over Addition of Vectors	∇u_m	=	$\mu^{-1} [(\vec{B}\bullet\nabla)\vec{B} + v^{-2} (\vec{E}\bullet\nabla)\vec{E} +$ $\vec{B}\times(v^{-2} \dfrac{\partial\vec{E}}{\partial t} + \mu\vec{J}) - v^{-2}\vec{E}\times \dfrac{\partial\vec{B}}{\partial t}]$

9	From 8, A7.9.3 Commutative by Vector Addition, T6.9.11 Multiply a row or column of a determinant [A] by a scalar k and T7.10.3 Distribution of a Scalar over Addition of Vectors	∇u_m	=	$\mu^{-1}[(\vec{B}\bullet\nabla)\vec{B} + v^{-2}(\vec{E}\bullet\nabla)\vec{E} + \mu\vec{B}\times\vec{J} + v^{-2}(\vec{B}\times\frac{\partial\vec{E}}{\partial t} - \vec{E}\times\frac{\partial\vec{B}}{\partial t})]$
10	From LxMXE.A.3.5.8 Poynting Field Vector and Lx9.2.2.3 Differential of Cross Product	$\frac{\partial\vec{E}\times\vec{B}}{\partial t}$	=	$\vec{E}\times\frac{\partial\vec{B}}{\partial t} - \vec{B}\times\frac{\partial\vec{E}}{\partial t}$
11	From 9, 10, T7.11.4 Vector Addition: Right Cancellation by Addition, T7.11.5 Vector Addition: Left Cancellation by Addition, A7.9.2 Substitution of Vectors and T7.10.3 Distribution of a Scalar over Addition of Vectors	∇u_m	=	$\mu^{-1}[(\vec{B}\bullet\nabla)\vec{B} + v^{-2}(\vec{E}\bullet\nabla)\vec{E} + \mu\vec{B}\times\vec{J}$ $v^{-2}(-\frac{\partial\vec{E}\times\vec{B}}{\partial t} + \vec{E}\times\frac{\partial\vec{B}}{\partial t} - \vec{E}\times\frac{\partial\vec{B}}{\partial t})]$
12	From 11, T7.11.3 Vector Addition: Inverse, T7.11.2 Vector Addition: Identity and D4.1.20A Negative Coefficient	∇u_m	=	$\mu^{-1}[(\vec{B}\bullet\nabla)\vec{B} + v^{-2}(\vec{E}\bullet\nabla)\vec{E} + \mu\vec{B}\times\vec{J} - \mu\varepsilon\frac{\partial\vec{E}\times\vec{B}}{\partial t}]$
13	From 1, LxMXE.A.3.5.13 Poynting Momentum Field Vector and A7.9.2 Substitution of Vectors	∇u_m	=	$\mu^{-1}[(\vec{B}\bullet\nabla)\vec{B} + v^{-2}(\vec{E}\bullet\nabla)\vec{E} + \mu\vec{B}\times\vec{J} - \mu\frac{\partial\vec{s}_m}{\partial t}]$
∴	From 4, LxMXE.A.3.5.11 Poynting Power Density-2 and A7.9.2 Substitution of Vectors and A7.9.3 Commutative by Vector Addition, T7.10.3 Distribution of a Scalar over Addition of Vectors, A4.2.12 Identity Multp, A4.2.13 Inverse Multp and T7.10.5 Identity with Scalar Multiplication to Vectors	∇u_m	=	$-\frac{\partial\vec{s}_m}{\partial t} + \mu^{-1}\vec{f}_{3p} + \vec{B}\times\vec{J}$

Theorem A.2.24 PVVT: Poynting Energy Wave Equation

Steps	Hypothesis	
		$\nabla^2 u_m = -v^{-2}\left(\dfrac{\partial^2 u_m}{\partial t^2} + \dfrac{\partial p_p}{\partial t}\right) + \nabla\bullet\mathbf{f}_{1m} - \nabla\bullet\mathbf{f}_L$
1	From LxPME.A.5.1.2 Divergent Energy Density, A7.9.2 Substitution of Vectors and Dx9.6.4.1 Divergent Operating on a Covariant Vector	$\nabla\bullet\nabla u_m = \nabla\bullet\left(-\dfrac{\partial \mathbf{s}_m}{\partial t} + \mathbf{f}_{1m} - \mathbf{f}_L\right)$
2	From 1, Lx9.6.1.3A Distribution of Divergence and Lx9.2.1.11 Dual Function Commutativety of Limits	$\nabla^2 u_m = -\dfrac{\partial \nabla\bullet\mathbf{s}_m}{\partial t} + \nabla\bullet\mathbf{f}_{1m} - \nabla\bullet\mathbf{f}_L$
∴	From 2, TA.5.5 PVVT: Divergence of the Poynting Momentum Vector and A4.2.3 Substitution	$\nabla^2 u_m = -v^{-2}\left(\dfrac{\partial^2 u_m}{\partial t^2} + \dfrac{\partial p_p}{\partial t}\right) + \nabla\bullet\mathbf{f}_{1m} - \nabla\bullet\mathbf{f}_L$

Theorem A.2.25 PVVT: Poynting Momentum Wave Equation

Steps	Hypothesis	
	Hypothesis	$\nabla^2 \vec{s}_m = -2v^{-2} \frac{\partial^2 \vec{s}_m}{\partial t^2} + 2\mu^{-1}v^{-2}\vec{f}_{3p,t} - 2v^{-2}(\vec{E}\times\vec{J})_{,t} + 2v^{-2}\nabla p_p - \rho v^{-2}\vec{E}_{,t} + v^{-1}(\vec{f}_{1m} + \vec{f}_q) + (\vec{E}\bullet\nabla)\vec{J} - \rho\,\mu\vec{J}$
1	From TA.5.5 PVVT: Divergence of the Poynting Momentum Vector and Lx9.6.1.1A Uniqueness of Gradient Operator	$\nabla(\nabla\bullet\vec{s}_m) = -v^{-2}\nabla\left(\frac{2\partial u_m}{\partial t} + v^{-2}p_p\right)$
2	From 1, Lx9.6.1.2A Distribution of Gradient, Lx9.3.2.3B Distribute Constant in/out of Product, Lx9.2.1.11 Dual Function Commutativety of Limits	$\nabla(\nabla\bullet\vec{s}_m) = -v^{-2}\frac{2\partial\nabla u_m}{\partial t} - v^{-2}\nabla p_p$
3	From 2, TA.5.1.5 PVVT: Divergence Poynting Momentum Field Energy Density, T7.10.3 Distribution of a Scalar over Addition of Vectors and A7.9.2 Substitution of Vectors	$\nabla(\nabla\bullet\vec{s}_m) = -2v^{-2}\left(\frac{\partial^2\vec{s}_m}{\partial t^2} + \mu^{-1}\frac{\partial\vec{f}_{3p}}{\partial t} - \frac{\partial\vec{E}\times\vec{J}}{\partial t} + \nabla p_p\right)$
4	From Lx9.6.1.13 Curl of Curl Operator, T7.11.5A Vector Addition: Reversal of Right Cancellation by Addition and T7.11.6B Vector Addition: Left Cancellation by Addition	$\nabla^2\vec{s}_m = \nabla(\nabla\bullet\vec{s}_m) - \nabla\times(\nabla\times\vec{s}_m)$
5	From 4, TA.5.6 PVVT: Curl of the Poynting Momentum Vector	$\nabla^2\vec{s}_m = \nabla(\nabla\bullet\vec{s}_m) - \nabla\times(-\rho\vec{B} + \vec{f}_{2m})$
6	From 5 and Lx9.6.1.4A Distribution of Curl	$\nabla^2\vec{s}_m = \nabla(\nabla\bullet\vec{s}_m) - \rho\nabla\times\vec{B} - \nabla\times\vec{f}_{2m}$
7	From 6, TA.5.514 PVVT: Curl of Poynting Momentum Filed Density-2 and A7.9.2 Substitution of Vectors	$\nabla^2\vec{s}_m = \nabla(\nabla\bullet\vec{s}_m) - \rho\,(v^{-2}\vec{E}_{,t} + \mu\vec{J}) + v^{-1}(\vec{f}_{1m} + \vec{f}_q) + (\vec{E}\bullet\nabla)\vec{J}$
8	From 7 and T7.10.3 Distribution of a Scalar over Addition of Vectors	$\nabla^2\vec{s}_m = \nabla(\nabla\bullet\vec{s}_m) - \rho v^{-2}\vec{E}_{,t} - \rho\,\mu\vec{J} + v^{-1}(\vec{f}_{1m} + \vec{f}_q) + (\vec{E}\bullet\nabla)\vec{J}$

9	From 8 and A7.9.3 Commutative by Vector Addition	$\nabla^2 \vec{s}_m \;=\; \nabla(\nabla \bullet \vec{s}_m) - \rho v^{-2}\,\vec{E}_{,t} + v^{-1}(\vec{f}_{1m} + \vec{f}_q) + (\vec{E} \bullet \nabla)\vec{J} - \rho\,\mu\vec{J}$
∴	From 3, 9 and A7.9.2 Substitution of Vectors	$\nabla^2 \vec{s}_m \;=\; -2v^{-2}\dfrac{\partial^2 \vec{s}_m}{\partial t^2} + 2\mu^{-1}v^{-2}\vec{f}_{3p,t} - 2v^{-2}(\vec{E}\times\vec{J})_{,t} + 2v^{-2}\nabla p_p - \rho v^{-2}\,\vec{E}_{,t} + v^{-1}(\vec{f}_{1m} + \vec{f}_q) + (\vec{E} \bullet \nabla)\vec{J} - \rho\,\mu\vec{J}$

Theorems TA.5.13 "PVVT: Poynting Energy Wave Equation" and TA.5.14 "PVVT: Poynting Momentum Wave Equation" clearly prove the duality property of charged particles. Most importantly they prove that the particle is comprised of a wave-packet of finite energy and momentum, derived directly from Maxwell's equations and independent of any either media in space, while reconfirming De Broglie's notion of a charge particle having the property of a wave.

Theorem A.2.26 PVVT: Poynting Energy Wave Equation in Free Space

1g	Given	$\vec{f}_L = \mathbf{0}$	From LxMXE.A.3.5.5 Lorentz Force Density
2g		$p_p = 0$	From LxMXE.A.3.5.6 Poynting Field Power
Steps	Hypothesis	$\nabla^2 u_m = -\dfrac{1}{c^2}\dfrac{\partial^2 u_m}{\partial t^2}$	
1	From Theorem A.10.19 PVVT: Poynting Energy Wave Equation	$\nabla^2 u_m = -v^{-2}\left(\dfrac{\partial^2 u_m}{\partial t^2} + \dfrac{\partial p_p}{\partial t}\right) + \nabla \bullet \vec{f}_{1m} - \nabla \bullet \vec{f}_L$	
2	From 1g, 2g, 1, A4.2.3 Substitution, A7.9.2 Substitution of Vectors and Dx9.6.4.1 Divergent Operating on a Covariant Vector	$\nabla^2 u_m = -c^{-2}\left(\dfrac{\partial^2 u_m}{\partial t^2} + \dfrac{\partial 0}{\partial t}\right) + \nabla \bullet \vec{f}_{1m} - 0$	
∴	From 2 and A4.2.7 Identity Add	$\nabla^2 u_m = -\dfrac{1}{c^2}\dfrac{\partial^2 u_m}{\partial t^2} + \nabla \bullet \vec{f}_{1m}$	

Theorem A.2.27 PVVT: Poynting Momentum Wave Equation in Free Space

1g	Given	$v = c$	From LxMXP.A.3.5.1 Velocity of light
2g		$\rho = 0$	From LxMXP.A.3.5.2 Coulomb Charge Density
3g		$\vec{J} = \mathbf{0}$	From LxMXP.A.3.5.3 Current Density
4g		$\vec{f}_L = \mathbf{0}$	From LxMXE.A.3.5.5 Lorentz Force Density
5g		$p_p = 0$	From LxMXE.A.3.5.6 Poynting Field Power Density
Steps	Hypothesis	$\nabla^2 \vec{s}_m = -c^{-2}\left(\frac{\partial^2 \vec{s}_m}{\partial t^2} + \frac{\partial \vec{f}_{1m}}{\partial t}\right) - c^{-1}\nabla\times\vec{f}_{2m}$	
1	From Theorem A.10.20 PVVT: Poynting Momentum Wave Equation	$\nabla^2 \vec{s}_m = -v^{-2}\left(\frac{\partial^2 \vec{s}_m}{\partial t^2} + \frac{\partial \vec{f}_{1m}}{\partial t} - \frac{\partial \vec{f}_L}{\partial t} + \nabla p_p\right) - v^{-1}\nabla\times\vec{f}_{2m} + (\nabla\rho)\times\vec{B} + \frac{\rho\partial\vec{E}}{v^2\partial t} + \rho\vec{J}$	
2	From 1g, 2g, 3g, 4g, 5g, 1, A4.2.3 Substitution, A7.9.2 Substitution of Vectors and Dx9.6.3.1A Covariant Gradient Operator	$\nabla^2 \vec{s}_m = -c^{-2}\left(\frac{\partial^2 \vec{s}_m}{\partial t^2} + \frac{\partial \vec{f}_{1m}}{\partial t} - \frac{\partial \mathbf{0}}{\partial t} + \nabla 0\right) - c^{-1}\nabla\times\vec{f}_{2m} + (\nabla 0)\times\vec{B} + \frac{0\partial\vec{E}}{c^2\partial t} + \mathbf{0}$	
∴	From 2, A4.2.7 Identity Add and T7.11.2 Vector Addition: Identity	$\nabla^2 \vec{s}_m = -c^{-2}\left(\frac{\partial^2 \vec{s}_m}{\partial t^2} + \frac{\partial \vec{f}_{1m}}{\partial t}\right) - c^{-1}\nabla\times\vec{f}_{2m}$	

Theorems "A.5.15 PVVT: Poynting Energy Wave Equation in Free Space" and "A.5.16 PVVT: Poynting Momentum Wave Equation in Free Space" clearly prove the duality property of a photon in free space. Most importantly they prove that the photon is a wave-packet of finite energy and momentum, derived directly from Maxwell's equations and independent of any either media in space, while reconfirming Einstein's notion of a photon as a quantum particle.

Tensor Calculus & Physics: A General Treatise

Table of Contents

List of Tables

List of Figures

List of Observations

List of Givens

List of Equations

Appendix B Special Relativity Review

Section B.1 Introduction to Special Relativity

A more intuitive explanation of the Lorentz force is demonstrated by Relative observers in different frames of reference watching moving charged particles near the velocity of light. Here a new force arises with different dimensions, not from a monopole as with charged particles, but simply a new place to stand that is perpendicular to the direction of travel, called a magnetic force. Lorrain and Corson elaborate on this theory in their book "Electromagnetic Fields and Waves" [LOR62 pg 248]. This confirms that electrical and gravitational fields are fundamental; consequently, all else arising from them are ***apparent fields of force***, and having no singular monopole as a source (point of origin) as a requirement for the fields' existence. This is why in Maxwell's set of field equations for Gauss's Law is zero.

In order to develop these proofs it requires the reader to have some background in the derivation of Relativistic Physics, Lorrain and Corson, have an excellent tutorial [LOR62 pg 193], therefore only a basic review summarizing this physics is developed in the following proofs and tables. Let's start with following idea found in relatively, moving frames of reference:

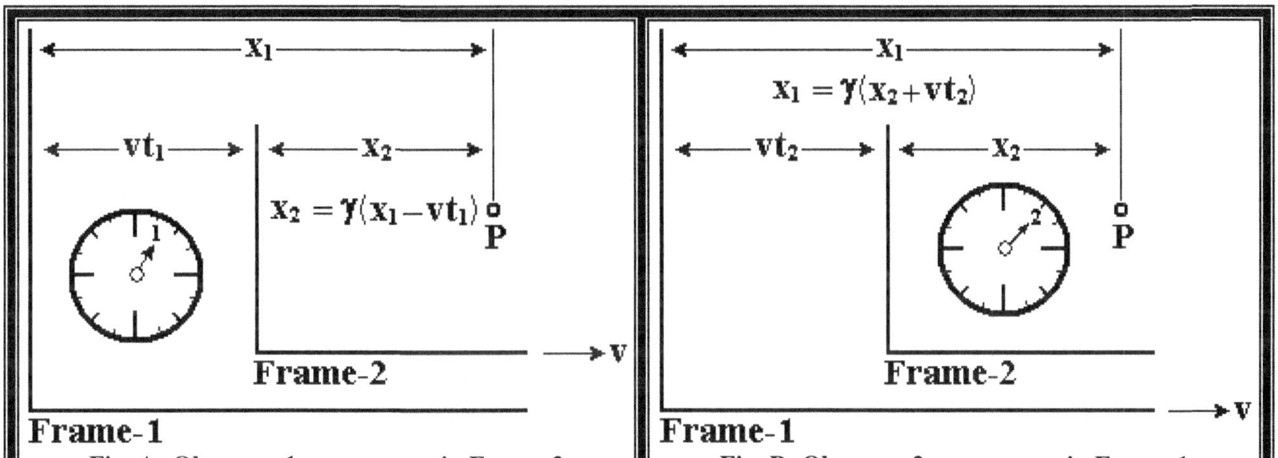

Fig. A: Observer-1 measures x_2 in Frame-2

Fig. B: Observer-2 measures x_1 in Frame-1

Figure B.1.1 Frames of Reference in Relative Motion

Einstein showed that in the direction of motion, length contacts, time slows down and mass gets fatter, outside a relative frame of reference moving at a constant velocity–v near the speed of light. Also as seen From Figure B.1.1A and B.1.1B that they are proportional to [γ] a dilation of distance, and symmetrical with a Galilean translation between observer–1 and –2:

Section B.2 Definitions, Laws and Lemmas Set Down by Einstein

Definition B.2.1 Stationary Frame of Reference

A frame of reference, where the equations of Newtonian mechanics hold (to a fist order approximation, v << c) is called a ***stationary system or stationary frame of reference***. When Relative velocity between two frames of reference being exactly zero (v = 0), then either frame of reference is said to be at ***rest*** with respect to one another. In such a system, any material point can be defined relatively thereto by the employment of rigid standards of measurement and the methods of Euclidean geometry, and can be expressed in Cartesian Coordinates.
Modified from [PER52 pg 38].

Definition B.2.1 **Simultaneous Event**

A ***simultaneous event*** (action, such as a train arriving at a train station at a specific time) takes place in a stationary frame of reference at the exact moment measured by a clock or watch within that frame. Modified From [PER52 pg 39].

Definition B.2.2 **Synchronization of Time**

Synchronization of time in any two-reference frames, between two clocks A and B in stationary reference frames A and B is a process that uses light as an intermediary to set the time for the occurrence of an event for watch-B a distance AB apart. Modified From [PER52 pg 40].

Count	Reference Frame-A	Reference Frame -B
1	t_{a1} event occurs at [A]	t_{b1} light beam received at [B]
2	t_{a2} light beam received back at [A]	t_{b2} light beam received again back at [B]
By 1,2	$t_{a2} - t_{a1} = 2\ AB / c = 2\tau$	$t_{b2} - t_{b1} = 2\ AB / c = 2\tau$

a) $t_{a2} - t_{a1} = t_{b2} - t_{b1} = 2\tau$ time delay calculated so clocks can be set and synchronized
b) $t_{a1} = t_{b1} - \tau$ clock-B can now be synchronized to clock-A.
and substituting (a) into (b) gives
c) $t_{a1} = t_{b2} - 2\tau - \tau = t_{b2} - 3\tau$

Definition B.2.3 **Velocity**

velocity ≡ light path / time interval

where time interval is to be taken in the sense of the definition B.2.3.

Law B.2.I Symmetry for the Synchronization of Clocks[3.14.3]

If the clock at [B] synchronizes with the clock at [A], then clock at [A] can synchronize with the clock at [B]. Modified From [PER52 pg 40].

Law B.2.2 Transitivity (Syllogism) for Synchronization of Clocks

If the clock at [A] synchronizes with the clock at [B] and [C] locks onto [B], then clocks [A] and [C] are also, synchronized with each other. (Assuming perfect synchronization, no time delays take place.) Modified From [PER52 pg 40].

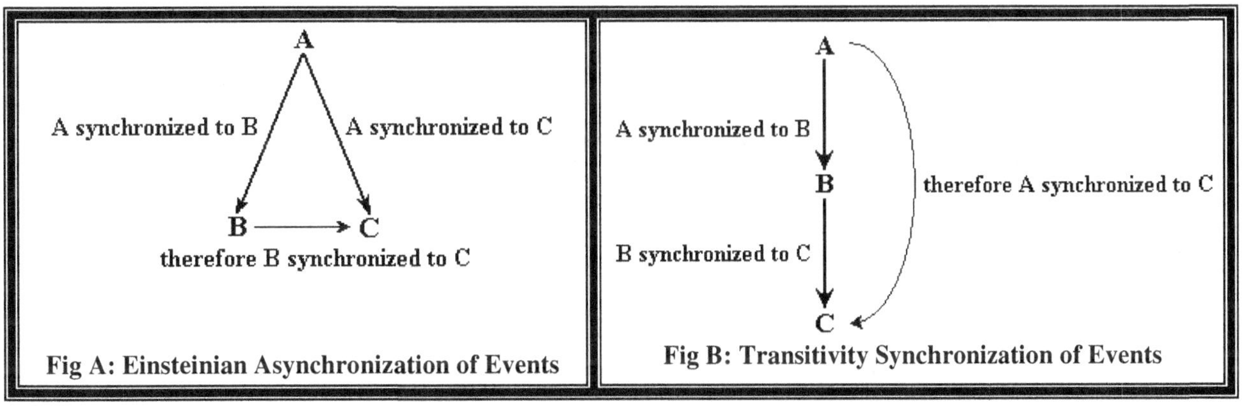

Figure B.2.1 Synchronization of Events

Notice that [(A → B)$_a$ ≡ (A → B)$_b$], [(A → C)$_a$ ≡ (B → C)$_b$] and [(B → C)$_a$ ≡ (A → C)$_b$] topographically are have a one-to-one correspondence to one another, hence are the same system of events and therefore Figure B.2.1A can be interchanged with Figure B.2.1B. It follows as long as event A does not precede B or C (violating the law of causality) than the secondary events are relatively symmetrical and can be ordered in anyway one so chooses.

Einstein chose to let his synchronization of events to be asynchronous as shown in Figure B.2.1A but in order to allow his theory to follow a systematic axiom development Law B.2.II is stated in syllogistic format. It makes no diffidence, which is used because in the end all events are still synchronized back to the original event, hence independent of whatever path might be taken, using his method of light, clock and metric rod.

Law B.2.3 Invariance of State for Physical Systems Undergoing Change

The laws by which the states of physical systems undergo change are not affected, whether these changes of state be referred to the one or the other of two systems for coordinates in uniform translated motion. [PER52 pg 41]

Law B.2.4 Invariance of the Speed of Light in any Frame of Reference

Any ray of light moves in the "stationary" system of coordinates with the determined velocity c,

$c \equiv 2.9979250(10)\text{x}10^8 \text{ m sec}^{-1}$ (CxPPC.A.2.3.5 "Velocity of Light for free space") EQ A

whether the ray be emitted by a stationary or by a moving body. It follows by definition of "Velocity" [DB.2.4] where the time interval is to be taken in the sense of definition of "Synchronization of Time" [DB.2.3]. [PER52 pg 41]

Table B.2.1 Basic Fitzgerald–Lorentz Formula [PER52 pg 14]

Auxiliary Support	Definition			
DxLZF.B.2.1.1	$\beta \equiv v/c$ and $v < c$	[∅]	NA	relative, light speed, velocity ratio between any two frames of Reference.
DxLZF.B.2.1.2	$\beta_{wn} \equiv v_{wn}/c$ and $v_{wn} < c$	[∅]	NA	relative, light speed, velocity ratio between axial Reference frame-w and frame-n
DxLZF.B.2.1.3	$\eta \equiv \sqrt{1 - \beta^2}$	[∅]	NA	Fitzgerald contraction factor
DxLZF.B.2.1.4	$\gamma \equiv 1/\eta$	[∅]	NA	Lorentz dilation (expansion) factor
Lemma	Transformation	Dimensions	Derived Units	Explanation
LxLZF.B.2.1.1	$\Delta l_1 = \eta\, \Delta l_2$ [B.2.1]	[L]	meters	Contraction of a rigid, straight bar
LxLZF.B.2.1.2	$\Delta t_1 = \gamma\, \Delta t_2$	[T]	seconds	Time dilation, time slowing down
LxLZF.B.2.1.3	$\Delta m_1 = \gamma\, \Delta m_2$	[M]	kilograms	Getting fatter

Table B.2.2 Einstein's Coordinate Transformations in the x-Direction [LOR62 pg204]

Lemma	Transformation 2-1		Lemma	Transformation 1-2
LxECT.B.2.2.1a	$x_1 = \gamma\, [x_2 + vt_2]$		LxECT.B.2.2.1b	$x_2 = \gamma\, [x_1 - vt_1]$
LxECT.B.2.2.2a	$y_1 = y_2$		LxECT.B.2.2.2b	$y_2 = y_1$
LxECT.B.2.2.3a	$z_1 = z_2$		LxECT.B.2.2.3b	$z_2 = z_1$
LxECT.B.2.2.4a	$t_1 = \gamma\, [t_2 + vx_2 / c^2]$		LxECT.B.2.2.4b	$t_2 = \gamma\, [t_1 - vx_1 / c^2]$

Table B.2.3 Einstein's Velocity Transformations in the x-Direction [LOR62 pg216]

Lemma	Transformation 2-1		Lemma	Transformation 1-2
LxEVT.B.2.3.1a	$v_{x1} = \dfrac{v_{x2} + v}{1 + (v_{x2}v/c^2)}$		LxEVT.B.2.3.1b	$v_{x2} = \dfrac{v_{x1} - v}{1 - (v_{x1}v/c^2)}$
LxEVT.B.2.3.2a	$v_{y1} = \dfrac{v_{y2}}{\gamma[1 + (v_{x2}v/c^2)]}$		LxEVT.B.23.2b	$v_{y2} = \dfrac{v_{y1}}{\gamma[1 - (v_{x1}v/c^2)]}$
LxEVT.B.2.3.3a	$v_{z1} = \dfrac{v_{z2}}{\gamma[1 + (v_{x2}v/c^2)]}$		LxEVT.B.2.3.3b	$v_{z2} = \dfrac{v_{z1}}{\gamma[1 - (v_{x1}v/c^2)]}$

Table B.2.4 Einstein's Momentum Transformations in the x-Direction [LOR62 pg223]

Lemma	Transformation 2-1		Lemma	Transformation 1-2
LxEMT.B.2.4.1a	$p_{x1} = \gamma\, [p_{x2} + v(\mathcal{E}_2/c^2)]$		LxEMT.B.2.4.1b	$p_{x2} = \gamma\, [p_{x1} - v(\mathcal{E}_1/c^2)]$
LxEMT.B.2.4.2a	$p_{y1} = p_{y2}$		LxEMT.B.2.4.2b	$p_{y2} = p_{y1}$
LxEMT.B.2.4.3a	$p_{z1} = p_{z2}$		LxEMT.B.2.4.3b	$p_{z2} = p_{z1}$
LxEMT.B.2.4.4a	$\mathcal{E}_1/c = \gamma\, [\mathcal{E}_2/c + (v/c)\, p_{x2}]$		LxEMT.B.2.4.4b	$\mathcal{E}_2/c = \gamma\, [\mathcal{E}_1/c - (v/c)\, p_{x1}]$

Table B.2.5a Einstein's Acceleration Transformations in the x-Direction [LOR62 pg216]

Lemma	Transformation 2-1
LxEAT.B.2.5.1a	$a_{x1} = \dfrac{a_{x2}}{\gamma^3 (1 + \beta_{x2}\beta)^3}$
LxEAT.B.2.5.2a	$a_{y1} = \dfrac{1}{\gamma^2 (1 + \beta_{x2}\beta)^2} \left[a_{z2} - \left(\dfrac{\beta_{y2}\beta}{1 + \beta_{x2}\beta}\right) a_{x2}\right]$
LxEAT.B.2.5.3a	$a_{z1} = \dfrac{1}{\gamma^2 (1 + \beta_{x2}\beta)^2} \left[a_{z2} - \left(\dfrac{\beta_{z2}\beta}{1 + \beta_{x2}\beta}\right) a_{x2}\right]$

Table B.2..5b Einstein's Acceleration Transformations in the x-Direction (continued)

Lemma	Transformation 1-2
LxEAT.B.2.5.1b	$a_{x2} = \dfrac{a_{x1}}{\gamma^3 (1 - \beta_{x1}\beta)^3}$
LxEAT.B.2.5.2b	$a_{y2} = \dfrac{1}{\gamma^2 (1 - \beta_{x1}\beta)^2} \left[a_{y1} + \left(\dfrac{\beta_{y1}\beta}{1 - \beta_{x1}\beta}\right) a_{x1}\right]$
LxEAT.B.2.5.3b	$a_{z2} = \dfrac{1}{\gamma^2 (1 - \beta_{x1}\beta)^2} \left[a_{z1} + \left(\dfrac{\beta_{z1}\beta}{1 - \beta_{x1}\beta}\right) a_{x1}\right]$

Table B.2.6a Einstein's Force Transformations in the x-Direction [LOR62 pg226]

Lemma	Transformation 2-1
LxEFT.B.2.6.1a	$F_{x1} = F_{x2} + \dfrac{\beta(\beta_{y2} F_{y2} + \beta_{z2} F_{z2})}{(1 + \beta_{x2}\beta)}$
LxEFT.B.2.6.2a	$F_{y1} = \dfrac{F_{y2}}{\gamma (1 + \beta_{x2}\beta)}$
LxEFT.B.2.6.3a	$F_{z1} = \dfrac{F_{z2}}{\gamma (1 + \beta_{x2}\beta)}$
LxEFT.B.2.6.4a	$\dot{\mathcal{E}}_1/c = \dfrac{\dot{\mathcal{E}}_2/c + \beta F_{x2}}{(1 + \beta_{x2}\beta)}$

Table B.2.6b Einstein's Force Transformations in the x-Direction (continued)

Lemma	Transformation 1-2
LxEFT.B.2.6.1b	$F_{x2} = F_{x1} - \dfrac{\beta(\beta_{y1} F_{y1} + \beta_{z1} F_{z1})}{(1 - \beta_{x1}\beta)}$
LxEFT.B.2.6.2b	$F_{y2} = \dfrac{F_{y1}}{\gamma (1 - \beta_{x1}\beta)}$
LxEFT.B.2.6.3b	$F_{z2} = \dfrac{F_{z1}}{\gamma (1 - \beta_{x1}\beta)}$
LxEFT.B.2.6.4b	$\dot{\mathcal{E}}_2/c = \dfrac{\dot{\mathcal{E}}_1/c - \beta F_{x1}}{(1 - \beta_{x1}\beta)}$

Table B.2.7 Einstein's Volume Transformations in the x-Direction [LOR62 pg228]

Lemma	Transformation 2-1		Lemma	Transformation 1-2
LxEVT.B.2.7.1a	$\upsilon_{x1} = \dfrac{\upsilon_{x2}}{\gamma (1 + \beta_{x2}\beta)}$		LxEVT.B.2.7.1b	$\upsilon_{x2} = \dfrac{\upsilon_{x1}}{\gamma (1 - \beta_{x1}\beta)}$

Table B.2.8 Einstein's Time Transformations in the x-Direction [LOR62 pg210, 211, 212]

Lemma	Transformation 2-1		Lemma	Transformation 1-2	
LxETT.B.2.8.1a	$\Delta t_1 = \gamma \Delta t_2$		LxETT.B.2.8.1b	$\Delta t_2 = \gamma \Delta t_1$	Period
LxETT.B.2.8.2a	$t_1 = t_2 \sqrt{\dfrac{1-\beta}{1+\beta}}$		LxETT.B.2.8.2b	$t_2 = t_1 \sqrt{\dfrac{1-\beta}{1+\beta}}$	Clock
LxETT.B.2.8.3a	$\nu_d = \nu_s \gamma [1 - \cos(\phi)\beta]$		LxETT.B.2.8.3b	$\nu_d = \nu_s \gamma$	Doppler
Source moving at an angle ϕ toward the detectoD			SouDce moving away From detectoD $\phi = ½\pi$		
LxETT.B.2.8.4a	$\nu_d = \nu_s \sqrt{\dfrac{1+\beta}{1-\beta}}$		LxETT.B.2.8.4b	$\nu_d = \nu_s \sqrt{\dfrac{1-\beta}{1+\beta}}$	Doppler
Source moving away From detector $\phi = \pi$			Source moving toward the detector $\phi = 0$		

These Transformations lead to a natural development of a set of four-dimensional quaternion type vectors, hence

Table B.2.9 Four Dimensional Quaternion Vectors

Definition	4-Dimensional Quaternion Vector	Composed From
Dx4DQV.B.2.9.1	(x, y, z, ict)	Table B.2.2: Einstein's Coordinate Transformations in the x-Direction
Dx4DQV.B.2.9.2	$(p_{x1}, p_{x1}, p_{x1}, i\mathcal{E}/c)$	Table B.2.4: Einstein's Momentum Transformations in the x-Direction
Dx4DQV.B.2.9.3	$(F_x, F_y, F_z, i\dot{\mathcal{E}}/c)$	Table B.2.6: Einstein's Force Transformations in the x-Direction
Dx4DQV.B.2.9.4	$(J_x, J_y, J_z, iv\rho)$	Theorem A.10.3 PVVT: Divergent Current Density For Nonsteady, Nonhomogenous Fields

Table B.2.10 Galilean Coordinate Transformations in the x-Direction [LOR62 pg204]

Lemma	Transformation 2-1		Lemma	Transformation 1-2
LxECT.B.2.10.1a	$x_1 = x_2 + vt_2$		LxECT.B.2.10.1b	$x_2 = x_1 - vt_1$
LxECT.B.2.10.2a	$y_1 = y_2$		LxECT.B.2.10.2b	$y_2 = y_1$
LxECT.B.2.10.3a	$z_1 = z_2$		LxECT.B.2.10.3b	$z_2 = z_1$
LxECT.B.2.10.4a	$t_1 = t_2$ absolute time		LxECT.B.2.10.4b	$t_2 = t_1$ absolute time

Einstein developed his transformations of coordinates by deducing their form based on the Galilean coordinate Transformations and the constraints of Lorentz and Fitzgerald length and time for ridged bodies. Since the Galilean coordinate transformations are invariant it would have to follow that his Transformations must also be invariant. He was able to test these ideas by use Theorem B.3.1 "STC: Space-Time Continuum Invariance of Relativistic Magnitude" to properly deduce the forms.

[B.2.1]Note: Subscripts refer to frames of reference where observer–1 stands in frame–1 with rod and clock and observed distance and time for observer-2 whose frame moves in a straight line at a Relative velocity–v. Likewise, this idea is symmetrical and can be reversed for observer–2 observing, observer–1. Discern that everything in the observer's frame is normal, but looking into the other frame distorts distance and time.

Einstein's Transversal Light Beam Clock

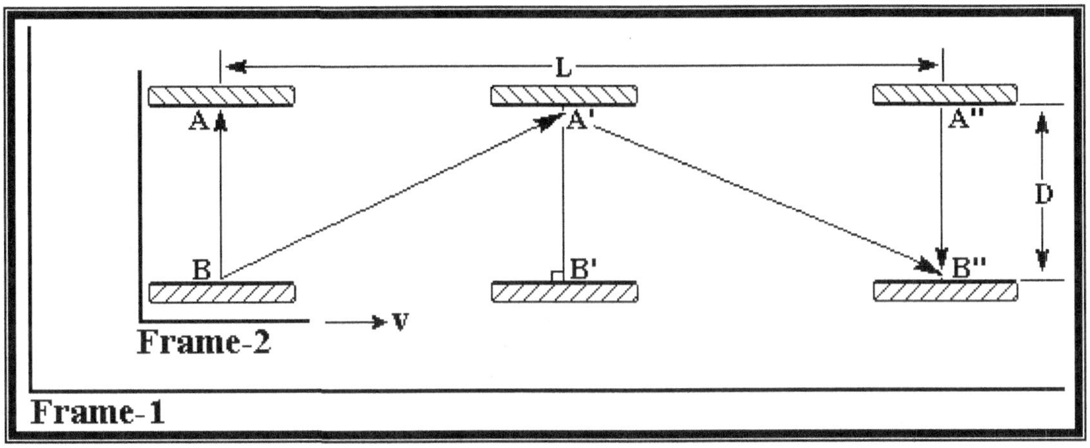

Figure B.2.2 Einstein's Transversal Light Beam Clock

Let's bounce a narrow beam of light in a transversal direction between two perfectly mirrored plates A and B a constant distance apart D, see Figure B.2.2 above.

The frame of reference-2 embedding the mirrored plates is stationary Relative to reference frame-1 and points A, A', A'', B, B' and B'' are coincident between the two frames. Our clock keeps good time with a period ΔT_2.

Now allow Reference Frame-2 to move at a constant velocity [v], less than the speed of light [c]. This changes things for points A, A', A'', B, B' and B'', which are no longer coincident, but spaced over a distance L for a single count of the clock.

From the point of view of observer-1 the time it takes for the beam of light to trace out its path BA' to A'B'' is:

Theorem B.2.1 FTD: Fitzgerald Time Dilation

1g	Given		Unless otherwise specified References are with Respect to Frame-1
2g	From geometry Figure B.2.2	L	$\equiv$ v ΔT_1 longitudinal distance: one clock cycle in frame-1
3g	From geometry Figure B.2.2	D	$\equiv$ ½c ΔT_2 transversal distance: one clock cycle in frame-2
4g	From geometry Figure B.2.2	BB"	$\equiv$ c ΔT_1 duration light traveled over path BA'B"
5g	From geometry Figure B.2.2	A'B'	$\equiv$ D
6g	From geometry Figure B.2.2	BB'	$\equiv$ ½ L
Steps	Hypothesis	ΔT_1	$= \gamma \Delta T_2$ Fitzgerald Time Dilation
1	From T5.8.3 Pythagorean Theorem A and geometry Figure B.2.2	BB"	$= 2\sqrt{A'B'^2 + BB'^2}$ light path length traveled over BA'B"
2	From 4g, 5g, 6g, 1 and A4.2.3 Substitution	$c\Delta T_1$	$= 2[\sqrt{D^2 + (½L)^2}]$
3	From 2g, 2 and A4.2.3 Substitution	$c\Delta T_1$	$= 2[\sqrt{D^2 + (½ v \Delta T_1)^2}]$
4	From 3, T4.4.3A Equalities: right Cancellation by Multiplication and A4.2.11 Associative Multp	$(½ c\Delta T_1)$	$= \sqrt{D^2 + (½ v \Delta T_1)^2}$
5	From 4 and A4.2.25 Reciprocal Exp	$(½ c\Delta T_1)^2$	$= D^2 + (½ v \Delta T_1)^2$
6	From 5, T4.3.5A Equalities: Right Cancellation by Addition and A4.2.2B Equality	D^2	$= (½ c\Delta T_1)^2 - (½ v \Delta T_1)^2$
7	From 6, A4.2.10 Commutative Multp, A4.2.21 Distribution Exp, and A4.2.14 Distribution	D^2	$= (c^2 - v^2)(½\Delta T_1)^2$
8	From 7, A4.2.13 Inverse Multp, A4.2.10 Commutative Multp and A4.2.12 Identity Multp	D^2	$= [c^2\, 1 - c^2(v^2/c^2)](½\Delta T_1)^2$
9	From 8 and A4.2.14 Distribution	D^2	$= [1 - (v^2/c^2)](½\Delta T_1)^2 c^2$
10	From 9, A4.2.21 Distribution Exp and A4.2.10 Commutative Multp	D^2	$= [1 - (v/c)^2](½c\Delta T_1)^2$

11	From 10, DxLZF.B.2.1.1, A4.2.3 Substitution, A4.2.21 Distribution Exp and T4.10.3 Radicals: Identity	D	$= (\sqrt{1 - \beta^2})\ (½c\Delta T_1)$
12	From 11, DxLZF.B.2.1.3 and A4.2.3 Substitution	D	$= \eta\ (½c\Delta T_1)$
13	From 12, A4.2.11 Associative Multp and A4.2.10 Commutative Multp	D	$= (½c)\ (\eta\ \Delta T_1)$
14	From 3g, 1 and A4.2.3 Substitution	$(½c)\ \Delta T_2$	$= (½c)\ (\eta\ \Delta T_1)$
15	From 14 and T4.4.2 Equalities: Cancellation by Multiplication	ΔT_2	$= \eta\ \Delta T_1$
∴	From 15, T4.4.3A Equalities: Right Cancellation by Multiplication, DxLZF.B.2.1.4, A4.2.3 Substitution and A4.2.2B Equality	ΔT_1	$= \gamma\ \Delta T_2$ Fitzgerald Time Dilation

In the moving frame time expands, called the Fitzgerald Time Dilation, slowing down relative to the outside observer-1

If the relative frames were a satellite and its ground station than the signal transmission from the orbiting satellite would expands time dilating it on the order of 4.6 microseconds, see Spilker [SPL77 pg 497].

Einstein's Longitudinal Light Beam Measuring Rode

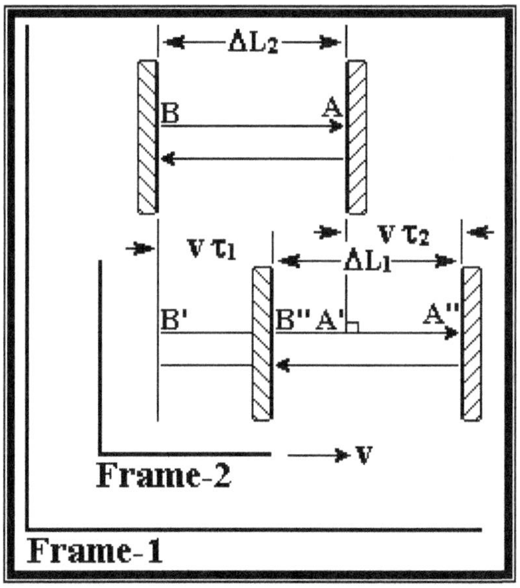

Figure B.2.3 Einstein's Longitudinal Light Beam Measuring Rode

Lets bounce a narrow beam of light in a longitudinal direction between two perfectly mirrored plates A and B mounted on a ridged rod of length ΔL_2, see Figure B.2.3 above.

Notice that Einstein does not make the assumption that ΔL_1 and ΔL_2 are the same in length as a result he discovers that length shortens in the direction of motion as it approaches the speed of light.

Now allow Reference Frame-2 to move at a constant velocity [v], less than the speed of light [c] and points A', A", B' and B", are no longer coincident, but spaced over a comparative distance. From the point of view of observer-1 the time it takes for the beam of light to trace out its path B'A" to A"B'' is:

Theorem B.2.1 OMSLC: Outside Mirror System Lorentz Contraction

1g	Given	Unless otherwise specified references are with respect to frame-1		
2g	From geometry Figure B.2.3	ΔT_2	$\equiv$	$2\Delta L_2 / c$ one clock cycle in frame-1
3g	From geometry Figure B.2.3	B"A"	$\equiv$	ΔL_1
4g	From geometry Figure B.2.3	A'B"	$\equiv$	$v\tau_1$ distance of travel at the velocity-v relative to initial position of starting mirror in frame-2. [B.1.2]
5g	From geometry Figure B.2.3	τ_1	$\equiv$	B'A" / c
6g	From geometry Figure B.2.3	B'A"	$\equiv$	B'B" + B"A" Path outside for single count of the clock.
Steps	Hypothesis	B'A"	=	$\Delta L_1 / (1 - \beta)$
1	From 4g, 5g, A4.2.3 Substitution, A4.2.10 Commutative Multp and A4.2.11 Associative Multp	B'B"	=	(v / c) B'A"
2	From 3g, 6g, 1 and A4.2.3 Substitution	B'A"	=	(v / c) B'A" + ΔL_1
3	From 2, DxLZF.B.2.1.1 and A4.2.3 Substitution	B'A"	=	β B'A" + ΔL_1
4	From 3, T4.3.5A Equalities: Right Cancellation by Addition and A4.2.2B Equality	ΔL_1	=	B'A" – β B'A"
5	From 4, A4.2.12 Identity Multp and A4.2.14 Distribution	ΔL_1	=	B'A" $(1 - \beta)$
∴	From 5 and T4.4.3B Equalities: Right Cancellation by Multiplication	B'A"	=	$\Delta L_1 / (1 - \beta)$

Theorem B.2.2 IMSLC: Inside Mirror System Lorentz Contraction

1g	Given	Unless otherwise specified references are with respect to Frame-1		
2g	From geometry Figure B.2.3	B"A"	$\equiv$	ΔL_1
3g	From geometry Figure B.2.3	A'A"	$\equiv$	$v\tau_2$ distance of travel at the velocity-v relative to initial position of starting mirror in frame-2. [B.2.2]
4g	From geometry Figure B.2.3	τ_2	$\equiv$	B"A' / c
5g	From geometry Figure B.2.3	B"A'	$\equiv$	B"A" – A'A" Path inside for single count of the clock.
Steps	Hypothesis	B"A'	=	$\Delta L_1 / (1 + \beta)$
1	From 3g, 4g, A4.2.3 Substitution, A4.2.10 Commutative Multp and A4.2.11 Associative Multp	A'A"	=	(v / c) B"A'
2	From 4g, 5g, 1 and A4.2.3 Substitution	B"A'	=	ΔL_1 – (v / c) B"A'
3	From 2, DxLZF.B.2.1.1 and A4.2.3 Substitution	B"A'	=	$\Delta L_1 - \beta$ B"A'
4	From 3, T4.3.6B Equalities: Reversal of Right Cancellation by Addition and A4.2.2B Equality	ΔL_1	=	B"A' + β B"A'
5	From 4, A4.2.12 Identity Multp and A4.2.14 Distribution	ΔL_1	=	B"A' $(1 + \beta)$
∴	From 5 and T4.4.3B Equalities: Right Cancellation by Multiplication	B"A'	=	$\Delta L_1 / (1 + \beta)$

[B.2.2]Note: Because of the difference for the physical distances |B'A"| / c > |B"A'| / c, see Figure B.2.3, it follows that the internal times $\tau_1 > \tau_2$ cannot be equal, but are uniquely different times.

Theorem B.2.3 LTD: Lorentz Length Contraction

1g	Given	Unless otherwise specified References are with respect to Frame-1		
2g	From geometry Figure B.2.3	B"A"	≡	$c\ \Delta T_1$ the period of time light transverses length ΔL_1.
3g	From geometry Figure B.2.3	B"A"/c	≡	(B"A' + B'A") / c the time it takes light to travel over the path length start to finish.
4g	From geometry Figure B.2.3	$c\Delta T_2$	≡	$2\Delta L_2$ ridged body length in frame-2
Steps	Hypothesis	ΔL_1	=	$\eta \Delta L_2$ Lorentz Length Contraction
1	From 3g, T4.1.2 OMSFC: Outside Mirror System Fitzgerald Contraction, T4.1.3 IMSFC: Inside Mirror System Fitzgerald Contraction and A4.2.3 Substitution	B"A"/c	=	$[\Delta L_1 / (1 + \beta) + \Delta L_1 / (1 - \beta)] / c$
2	From 1, T4.4.2 Equalities: Cancellation by Multiplication and T4.5.5 Equalities: Addition of Two Rational Fractions	B"A"	=	$[(1 - \beta) + (1 + \beta)]\ \Delta L_1 / (1 + \beta)(1 - \beta)$
3	From 2, A4.2.6 Associative Add, A4.2.5 Commutative Add, D4.1.14 Primitive Definition for Rational Arithmetic and A4.2.8 Inverse Add	B"A"	=	$2\ \Delta L_1 / (1 + \beta)(1 - \beta)$
4	From 3 and T4.12.3 Polynomial Quadratic: Difference of Two Squares	B"A"	=	$2\ \Delta L_1 / (1 - \beta^2)$
5	From 2g, 4, DxLZF.B.2.1.3, A4.2.25 Reciprocal Exp and A4.2.3 Substitution	$c\ \Delta T_1$	=	$2\ \Delta L_1 / \eta^2$
6	From 5, DxLZF.B.2.1.4 and TB.2.1 LTD: Lorentz Time Dilation	$c\ \gamma\ \Delta T_2$	=	$2\ \Delta L_1\ \gamma^2$
7	From 4g, 6, A4.2.10 Commutative Multp, A4.2.11 Associative Multp, A4.2.18 Summation Exp and A4.2.3 Substitution	$2\gamma\ \Delta L_2$	=	$2\gamma\ \gamma \Delta L_1$

8	From 7, T4.4.2 Equalities: Cancellation by Multiplication	ΔL_2	$= \gamma\Delta L_1$	
∴	From 8, T4.3.3B Equalities: Right Cancellation by Multiplication, DxLZF.B.2.1.4 and A4.2.3 Substitution	ΔL_1	$= \eta\Delta L_2$	Lorentz Length Contraction

In the moving frame length shrinks, called the Lorentz Length Contraction, shortening relative to the outside observer-1

In a relative set of frames were two nuclear charged particles, of the same charge, with a distance ΔL_2 apart, and accelerated under an external electrical and magnetic fields. For a bunch of such particles, ignoring internal repulsive forces, the group would contract in their forward direction of motion, thereby increasing the charge density. Conservation of charge guarantees, charges can neither be added nor reduced from the group. We know this to be true, because this physical property can be seen in the design of Klystrons and Linear Accelerators. Were the design of the resonant cavities are increasingly lengthen, thereby increasing electrical field time duration to effectively couple into the ever foreshortening or fattening of the accelerated electron group.

In their book [PAN62 pg 290-293], Panofsky and Phillips, develop these proofs for dilation and contraction.

Theorem B.2.4 ILFC: Invariance of Length, Fitzgerald Contraction

1g 2g	Given	$\Delta l_2 = x_{2a} - x_{2b}$ Magnitude of a rigid stationary rod in frame–2 Frame–2 moves at a velocity v in an interval of time $t_1 = t_{1a} = t_{1b}$.	$\Delta l_1 = x_{1a} - x_{1b}$ Magnitude of a rigid stationary rod in frame–1 Frame–1 moves at a velocity v in an interval of time $t_2 = t_{2a} = t_{2b}$.
Steps	Hypothesis	$\Delta l_1 = \eta\ \Delta l_2$ Ridged bar in frame-2 is seen to contract From frame-1	$\Delta l_2 = \eta\ \Delta l_1$ Ridged bar in frame-1 is seen to contract From frame-2
1	From 1g	$\Delta l_2 = x_{2a} - x_{2b}$	$\Delta l_1 = x_{1a} - x_{1b}$
2	From 1, LxECT.B.2.2.1a, LxECT.B.2.2.1b and A4.2.3 Substitution	$\Delta l_2 = \gamma[x_{1a} - vt_{1a}] - \gamma[x_{1b} - vt_{1b}]$	$\Delta l_1 = \gamma[x_{2a} + vt_{2a}] - \gamma[x_{2b} + vt_{2b}]$
3	From 2, A4.2.14 Distribution and D4.1.19 Primitive Definition for Rational Arithmetic	$\Delta l_2 = \gamma\,[x_{1a} - vt_{1a} - x_{1b} + vt_{1b}]$	$\Delta l_1 = \gamma\,[x_{2a} + vt_{2a} - x_{2b} - vt_{2b}]$
4	From 2g, 3, A4.2.5 Commutative Add and A4.2.3 Substitution	$\Delta l_2 = \gamma\,[x_{1a} - x_{1b} - vt_1 + vt_1]$	$\Delta l_1 = \gamma\,[x_{2a} - x_{2b} + vt_2 - vt_2]$
5	From 4, A4.2.8 Inverse Add and A4.2.7 Identity Add	$\Delta l_2 = \gamma\,[x_{1a} - x_{1b}]$	$\Delta l_1 = \gamma\,[x_{2a} - x_{2b}]$
6	From 5, 1g, DxLZF.B.2.4 and A4.2.3 Substitution	$\Delta l_2 = \Delta l_1 / \eta$ but observer-1 is looking at Rigid bar in frame-2	$\Delta l_1 = \Delta l_2 / \eta$ but observer-2 is looking at rigid bar in frame-1
∴	From 6 and T4.4.4B Equalities: reversal of right Cancellation by Multiplication	$\Delta l_1 = \eta\ \Delta l_2$ ridged bar in frame-2 is seen to contract From frame-1	$\Delta l_2 = \eta\ \Delta l_1$ ridged bar in frame-1 is seen to contract From frame-2

Theorem B.2.5 ITLD: Invariance of Time, Lorentz Dilation

1g	Given	$\Delta t_2 = t_{2a} - t_{2b}$ Period of time in Frame–2	$\Delta t_1 = t_{1a} - t_{1b}$ Period of time in Frame–1
2g		Frame–2 moves at a velocity v in an interval of time $t_1 = t_{1a} = t_{1b}$.	Frame–1 moves at a velocity v in an interval of time $t_2 = t_{2a} = t_{2b}$.
3g		$\Delta l_2 = x_{2a} - x_{2b}$ Difference between two stationary points in frame–2	$\Delta l_1 = x_{1a} - x_{1b}$ Difference between two stationary points in frame–1
4g		$\Delta t_1 \equiv \Delta l_1 \, v/c^2$ Two simultaneous events at different points in frame–1	$\Delta t_2 \equiv \Delta l_2 \, v/c^2$ Two simultaneous events at different points in frame–2
Steps	Hypothesis	$\Delta t_2 = \gamma \, \Delta t_1$	$\Delta t_1 = \gamma \, \Delta t_2$
1	From 1g	$\Delta t_2 = t_{2a} - t_{2b}$	$\Delta t_1 = t_{1a} - t_{1b}$
2	From 1, LxECT.B.2.2.4a, LxECT.B.2.2.4b and A4.2.3 Substitution	$\Delta t_2 = \gamma \, [t_{1a} - vx_{1a} / c^2] - \gamma[t_{1b} - vx_{1b} / c^2]$	$\Delta t_1 = \gamma \, [t_{2a} + vx_{2a} / c^2] - \gamma[t_{2b} + vx_{2b} / c^2]$
3	From 2, A4.2.14 Distribution, D4.1.20 Negative Coefficient and D4.1.19 Primitive Definition for Rational Arithmetic	$\Delta t_2 = \gamma \, [t_{1a} - vx_{1a} / c^2 - t_{1b} + vx_{1b} / c^2]$	$\Delta t_1 = \gamma \, [t_{2a} + vx_{2a} / c^2 - t_{2b} - vx_{2b} / c^2]$
4	From 2g, 3, A4.2.5 Commutative Add and A4.2.3 Substitution	$\Delta t_2 = \gamma \, [t_1 - t_1 - vx_1/ c^2 + vx_1/ c^2]$	$\Delta t_1 = \gamma \, [t_2 - t_2 + vx_2/ c^2 - vx_2/ c^2]$
5	From 4, A4.2.8 Inverse Add and A4.2.7 Identity Add	$\Delta t_2 = \gamma \, [x_{1a} - x_{1b}] \, v/c^2$	$\Delta t_1 = \gamma \, [x_{2a} - x_{2b}] \, v/c^2$
6	From 3g, 5 and A4.2.3 Substitution	$\Delta t_2 = \gamma \, \Delta l_1 \, v/c^2$	$\Delta t_1 = \gamma \, \Delta l_2 \, v/c^2$
∴	From 5, 4g and A4.2.3 Substitution	$\Delta t_2 = \gamma \, \Delta t_1$	$\Delta t_1 = \gamma \, \Delta t_2$

Section B.3 Relativistic Work, Energy and Momentum

Theorem B.3.1 EWFN: Einstein's Work Function For a Nuclear Charged Particle

1g	Given	$\dfrac{d^2x}{dt^2} = \dfrac{\varepsilon X}{m_0 \gamma^3}$	Equation of particle motion [PER52 pg 62, EQ A]
2g		$W_x = \int F_x\, dx$	LxRCMA.1.1.10, Work in moving a nuclear particle without irradiating along the x-axis.
3g		$F_x = \varepsilon X$	Force exerted by the electric field along the x-axis.
4g		$v = \dfrac{dx}{dt}$	Differential velocity
Steps	Hypothesis	$W_x = mc^2 - m_0c^2$	
1	From 1g, T4.4.4 Equalities: Reversal of Right Cancellation by Multiplication, 3g and A4.2.3 Substitution	$F_x = m_0 \gamma^3 \dfrac{d^2x}{dt^2}$	
2	From 1 and Dx9.3.2.1F Definitions on Differential Notation	$F_x = m_0 \gamma^3 \dfrac{d\,(dx/dt)}{dt}$	
3	From 4g, 2 and A4.2.3 Substitution	$F_x = m_0 \gamma^3 \dfrac{dv}{dt}$	
4	From 3, A4.2.2 Equality and A4.2.10 Commutative Multp	$F_x\, dx = m_0 \gamma^3 \dfrac{dx}{dt}\, dv$	
5	From 4,4g and A4.2.3 Substitution	$F_x\, dx = m_0 \gamma^3\, v dv$	
6	From 5, 2g and A4.2.3 Substitution	$W_x = \int_0^v m_0 \gamma^3\, v dv$	
7	From 6 and A4.2.14 Distribution	$W_x = m_0 \int_0^v \gamma^3\, v dv$	
8	From DxLZF.B.2.1.1 and T4.4.4B Equalities: Reversal of Right Cancellation by Multiplication	$v = \beta c$	
9	From 7, 8, A4.2.3 Substitution and A4.2.14 Distribution	$W_x = m_0c^2 \int_0^\beta \gamma^3\, \beta d\beta$	
10	From 9 and Lx9.4.4.9	$W_x = m_0c^2 \left. \dfrac{1}{\sqrt{1 - \beta^2}} \right\|_0^\beta$	
11	From 10 and evaluating the bounding values	$W_x = m_0c^2 \left(\dfrac{1}{\sqrt{1 - \beta^2}} - 1 \right)$	
12	From 11, DxLZF.B.2.1.4 and A4.2.3 Substitution	$W_x = (m_0\gamma)c^2 - m_0c^2$	
∴	From 12, LxLZF.B.2.1.3 and A4.2.3 Substitution	$W_x = mc^2 - m_0c^2$	

Theorem B.3.2 AE2N: Approximating Einstein's Work Function to Newton's Kinetic Energy

1g	Given	$KE \approx \frac{1}{2} m_0 v^2$ Newtonian kinetic energy
Steps	Hypothesis	$KE = mc^2 - m_0c^2$
1	From TB.3 LDT: Einstein's Work Function For a Nuclear Charged Particle	$W_x = mc^2 - m_0c^2$
2	From 1, LxLZF.B.2.1.3 and A4.2.3 Substitution	$W_x = (m_0\gamma)c^2 - m_0c^2$
3	From 2, DxLZF.B.2.1.4, DxLZF.B.2.3 and A4.2.3 Substitution	$W_x = m_0c^2 \left(\frac{1}{\sqrt{1 - \beta^2}} - 1 \right)$
4	From 3, LxE.3.11, A4.2.3 Substitution and expanding LOT.	$W_x = m_0c^2 (1 + \frac{1}{2}\beta^2 + (3/8) \beta^4 + (7/16) \beta^6 + HOT - 1)$
5	From 4, A4.2.5 Commutative Add, D4.1.19 Primitive Definition for Rational Arithmetic and conditions of HOT are to small to make any significant contributions.	$W_x \approx \frac{1}{2}\beta^2 m_0c^2$ for $0 \le \beta << 1$
6	From 5, A4.2.10 Commutative Multp, DxLZF.B.2.1.1 and A4.2.3 Substitution	$W_x \approx \frac{1}{2} m_0 (v/c)^2 c^2$
7	From 6, T4.8.7 Integer Exponents: Distribution Across a rational Number, D4.1.4B rational Numbers, A4.2.10 Commutative Multp, A4.2.13 Inverse Multp and A4.2.12 Identity Multp	$W_x \approx \frac{1}{2} m_0v^2$ Newtonian kinetic energy
8	From 7, 1g and its implied that the work in moving an electrical particle in and electric field is the kinetic energy of the particle system.	$W_x = KE$ kinetic energy
∴	From 1, 8 and A4.2.3 Substitution	$KE = mc^2 - m_0c^2$

Hence true kinetic energy (KE) is measured relative to its rest energy (RE).

Theorem B.3.3 TNKE: Total Energy to Net Kinetic Energy

Steps	Hypothesis			
Steps	Hypothesis	TE	=	$KE + m_0c^2$
1	From TB.2.2 AE2N: Approximating Einstein's Work Function to Newton's Kinetic Energy	KE	=	$mc^2 - m_0c^2$
2	From 1 and T4.3.6B Equalities: Reversal of right Cancellation by Addition	mc^2	=	$KE + m_0c^2 \equiv$ TE total energy of the kinetic system
∴	From 2	TE	=	$KE + m_0c^2$

Just as velocity in Einsteinian physics is relative to the velocity of light so to the total energy TE for the system can be measured relative RE or its rest mass (RM) of a nuclear particle. The total energy is simply the sum of the rest mass and the kinetic energy. Classically in order to conserve energy the total energy of a system had to be the sum of the kinetic and potential energy, held constant. This allowed an-interplay to exist between the two energy types while holding the net system steady or stable. A potential energy can still exist to allow for such interaction. If the TE is conserved and the RE being constant for a particular particle than the kinetic energy is conserved as well and can be defined as follows:

Theorem B.3.4 CTKE: Conservation of the Total Kinetic Energy

1g	Given	KE	=	$KE_n + PE_n$
2g		DE	=	m_0c^2
Steps	Hypothesis	TE	=	$KE_n + PE_n + RE$
1	From TB.2.3 TNKE: Total Energy to Net Kinetic Energy	TE	=	$KE + m_0c^2$
∴	From 1, 1g, 2g, A4.2.3 Substitution and the TE conserved.	TE	=	$KE_n + PE_n + RE$

Theorem B.3.5 CTKE: Conservation of Total Kinetic Energy for Multiple Particle System

1g	Given	$TE_s = \sum TE_{ni}$ for m-particle system
2g		$KE_s = \sum KE_{ni}$ for m-particle system
3g		$PE_s = \sum PE_{ni}$ for m-particle system
4g		$RE_s = \sum RE_{ni}$ for m-particle system
Steps	Hypothesis	$TE = KE + PE + RE$ Total relativistic energy for a conserved system.
1	From TB.2.4 CTKE: Conservation of the Total Kinetic Energy for any individual particle-i	$TE_i = KE_{ni} + PE_{ni} + RE_{ni}$ for i = 1, 2, … m particle system
2	From 1 and summed over all possible m-particle energies	$\sum TE_i = \sum KE_{ni} + \sum PE_{ni} + \sum RE_{ni}$ for m-particle system
3	From 2, 1g, 2g, 3g, 4g and A4.2.3 Substitution	$TE_s = KE_s + PE_s + RE_s$
∴	From 3, as an aggregate of particles there individual identities are obscured hence the subscript-s can be dropped.	$TE = KE + PE + RE$ Total relativistic energy for a conserved system.

Theorem B.3.6 REAS: Rest Energy for an Aggregate Particle System

1g	Given	$m_0 = \sum m_{0i}$ for m-particle system
Steps	Hypothesis	$RE = m_0c^2$ for aggregate particle system
1	From TB.2.5 CTKE: Conservation of Total Kinetic Energy for Multiple Particle System, Step 4g	$RE_s = \sum RE_{ni}$ for m-particle system
2	From 1, TB.2.4 CTKE: Conservation of the Total Kinetic Energy, Step 2g and A4.2.3 Substitution	$RE_s = \sum m_{0i}c^2$ for m-particle system
3	From 2 and A4.2.14 Distribution	$RE_s = (\sum m_{0i})\, c^2$ for m-particle system
∴	From 3, 1g and A4.2.3 Substitution	$RE = m_0c^2$ for aggregate particle system

It is from Theorem B.3.6 "REAS: Rest Energy for an Aggregate Particle System" that Einstein's famous equation of energy comes from, where given any amount of matter it has the potential to convert its entire rest mass into energy and being symmetrical by ***conservation of energy and momentum for pair production*** Law 4.1, radiant energy can be converted back into mass, $E = mc^2$ or $m = E / c^2$.

Theorem B.3.7 IRMT: Invariance of Relativistic Momentum under Lorenz Transformation

1g	Given	Q_{m2}	$\equiv$	$(\vec{p}_2, i\mathcal{E}_2/c)$
2g		$\vec{p}_2$	=	(p_{x2}, p_{x2}, p_{x2})
Steps	Hypothesis	$Q_{m2} \bullet Q_{m2}$	=	$Q_{m1} \bullet Q_{m1}$
1	From 1g, 2g, Dx4DQV.B.2.9.2 and D7.5.2A	$Q_{m2} \bullet Q_{m2}$	=	$p_{x2}^2 + p_{x2}^2 + p_{x2}^2 - (\mathcal{E}_2/c)^2$
2	From 1, LxEMT.B.2.4.1b, LxEMT.B.2.4.2b, LxEMT.B.2.4.3b, LxEMT.B.2.4.4b and A4.2.3 Substitution	$Q_{m2} \bullet Q_{m2}$	=	$\gamma^2 [p_{x1} - v(\mathcal{E}_1/c^2)]^2 + p_{x1}^2 + p_{x1}^2 - \gamma^2 [\mathcal{E}_1/c - (v/c)\, p_{x1}]^2$
3	From 2, A4.2.5 Commutative Add, and A4.2.14 Distribution	$Q_{m2} \bullet Q_{m2}$	=	$\gamma^2 \{[p_{x1} - v(\mathcal{E}_1/c^2)]^2 + [-\mathcal{E}_1/c + (v/c)\, p_{x1}]^2\} + p_{x1}^2 + p_{x1}^2$
4	From 3, T4.12.2 Polynomial Quadratic: The Perfect Square by Difference and A4.2.21 Distribution Exp	$Q_{m2} \bullet Q_{m2}$	=	$\gamma^2 \{[p_{x1}^2 - 2v(\mathcal{E}_1/c^2)\, p_{x1} + v^2(\mathcal{E}_1/c^2)^2] + [-(\mathcal{E}_1/c)^2 + 2(v/c)\, p_{x1}(\mathcal{E}_1/c) - (v/c)^2\, p_{x1}^2]\} + p_{x1}^2 + p_{x1}^2$
5	From 4, T4.8.7 Integer Exponents: Distribution Across a Rational Number, A4.2.22 Product Exp	$Q_{m2} \bullet Q_{m2}$	=	$\gamma^2 \{[p_{x1}^2 - 2v(\mathcal{E}_1/c^2)\, p_{x1} + v^2(\mathcal{E}_1^2/c^4)] + [-(\mathcal{E}_1/c)^2 + 2(v/c)\, p_{x1}(\mathcal{E}_1/c) - (v/c)^2\, p_{x1}^2]\} + p_{x1}^2 + p_{x1}^2$
6	From 5 and D4.1.19 Primitive Definition for Rational Arithmetic	$Q_{m2} \bullet Q_{m2}$	=	$\gamma^2 \{[p_{x1}^2 - 2v(\mathcal{E}_1/c^{1+1})\, p_{x1} + v^2(\mathcal{E}_1^2/c^{2+2})] + [-(\mathcal{E}_1/c)^2 + 2(v/c)\, p_{x1}(\mathcal{E}_1/c) - (v/c)^2\, p_{x1}^2]\} + p_{x1}^2 + p_{x1}^2$
7	From 6, A4.2.18 Summation Exp, T4.5.1 Equalities: Reciprocal Products, A4.2.10 Commutative Multp and T4.8.7 Integer Exponents: Distribution Across a Rational Number	$Q_{m2} \bullet Q_{m2}$	=	$\gamma^2 \{[p_{x1}^2 - 2(v/c)(\mathcal{E}_1/c)\, p_{x1} + (v/c)^2(\mathcal{E}_1/c)^2] + [-(\mathcal{E}_1/c)^2 + 2(v/c)\, p_{x1}(\mathcal{E}_1/c) - (v/c)^2\, p_{x1}^2]\} + p_{x1}^2 + p_{x1}^2$
8	From 7, DxLZF.B.2.1.1 and A4.2.3 Substitution	$Q_{m2} \bullet Q_{m2}$	=	$\gamma^2 \{[p_{x1}^2 - 2\beta(\mathcal{E}_1/c)\, p_{x1} + \beta^2(\mathcal{E}_1/c)^2] + [-(\mathcal{E}_1/c)^2 + 2\beta\, p_{x1}(\mathcal{E}_1/c) - \beta^2\, p_{x1}^2]\} + p_{x1}^2 + p_{x1}^2$
9	From 8, A4.2.10 Commutative Multp and A4.2.5 Commutative Add	$Q_{m2} \bullet Q_{m2}$	=	$\gamma^2 \{[p_{x1}^2 - \beta^2\, p_{x1}^2 - 2\beta(\mathcal{E}_1/c)\, p_{x1} + 2\beta(\mathcal{E}_1/c)\, p_{x1} - (\mathcal{E}_1/c)^2 + \beta^2(\mathcal{E}_1/c)^2]\} + p_{x1}^2 + p_{x1}^2$

10	From 9, A4.2.8 Inverse Add, A4.2.7 Identity Add D4.1.20 Negative Coefficient and A4.2.14 Distribution	$Q_{m2} \bullet Q_{m2}$	=	$\gamma^2 \{[(1 - \beta^2) p_{x1}{}^2 - (1 - \beta^2)(\mathcal{E}_1/c)^2]\} + p_{x1}{}^2 + p_{x1}{}^2$
11	From 10 and A4.2.14 Distribution	$Q_{m2} \bullet Q_{m2}$	=	$\gamma^2 (1 - \beta^2)[p_{x1}{}^2 - (\mathcal{E}_1/c)^2] + p_{x1}{}^2 + p_{x1}{}^2$
12	From 11, DxLZF.B.2.1.3, DxLZF.B.2.1.4, A4.2.3 Substitution and A4.2.10 Commutative Multp	$Q_{m2} \bullet Q_{m2}$	=	$\eta^2 (1/\eta^2)[p_{x1}{}^2 - (\mathcal{E}_1/c)^2] + p_{x1}{}^2 + p_{x1}{}^2$
13	From 12, T4.4.13 Equalities: Product by Division, Common Factor	$Q_{m2} \bullet Q_{m2}$	=	$1\,[p_{x1}{}^2 - (\mathcal{E}_1/c)^2] + p_{x1}{}^2 + p_{x1}{}^2$
14	From 13 A4.2.12 Identity Multp	$Q_{m2} \bullet Q_{m2}$	=	$p_{x1}{}^2 - (\mathcal{E}_1/c)^2 + p_{x1}{}^2 + p_{x1}{}^2$
15	From 14 and A4.2.5 Commutative Add	$Q_{m2} \bullet Q_{m2}$	=	$p_{x1}{}^2 + p_{x1}{}^2 + p_{x1}{}^2 - (\mathcal{E}_1/c)^2$
∴	From 15 and D6A	$Q_{m2} \bullet Q_{m2}$	=	$Q_{m1} \bullet Q_{m1}$

Now the Relativistic total energy must be the same amount of energy of the particle itself for Einstein's 4-dimensional momentum.

Theorem B.3.8 EDEE: Einstein's Relativistic Energy Equation

1g	Given	Q_m	≡	$(\vec{P}, ip_0)$
2g		$\vec{p}$	=	$(p_x, p_x, p_x) = m\,(v_x, v_x, v_x) = m\,\vec{v}$
3g		PE	=	0 for an non-conservative TE
4g		p_0	=	m_0c Rest momentum
5g		$\vec{p} \bullet \vec{p}$	=	$p^2 = m^2(\vec{v} \bullet \vec{v}) = m^2v^2$
Steps	Hypothesis	KE	=	$[c\sqrt{(p^2 + p_0^2)}] - RE$
1	From 1g, 2g, Dx4DQV.B.2.9.2, D7.5.2C, 5g and A4.2.3 Substitution	$Q_m \bullet Q_m^*$	=	$p^2 + p_0^2$
2	From 3g, TB.2.5 CTKE: Conservation of the Total Kinetic Energy for a Multiple Particle System	TE	=	KE + RE
3	From 2, T4.4.1 Equalities: Uniqueness of Multiplication and T4.8.6 Integer Exponents: Uniqueness of Exponents	$(TE/c)^2$	=	$[(KE + RE)/c]^2$
4	From 1, 3 and the total energy to move a particle is also the momentum of the particle.	$(TE/c)^2$	≡	$Q_m \bullet Q_m^*$
5	From 1, 3, 4 and A4.2.3 Substitution	$[(KE + RE)/c]^2$	=	$p^2 + p_0^2$
6	From 5 and A4.2.25 Reciprocal Exp	$(KE + RE)/c$	=	$\sqrt{p^2 + p_0^2}$
7	From 6, T4.4.6B Equalities: Reversal of Left Cancellation by Multiplication and A4.2.10 Commutative Multp	KE + RE	=	$c\sqrt{p^2 + p_0^2}$
∴	From 7 and T4.3.5B Equalities: right Cancellation by Addition	KE	=	$[c\sqrt{(p^2 + p_0^2)}] - RE$

With Theorem B.3.8 "EREE: Einstein's Relativistic Energy Equation" Einstein the great unifier was able to unify energy and momentum.

Section B.4 Rebirth of Quaternions

As theorized by Paul M. Dirac and proven in experiment by Carl D. Anderson there are sets of fundamental particles that are antiparticles.

Definition B.4.1 **Antiparticle**

For any given fundamental nuclear particle they have an antiparticle, identical in mass, but opposite in charge.

In Book I chapter 1 in footnote [1.4.1] I say Hamilton lost the race of three-dimensional vectors to Grassman, however, Yogi Barra of the New York Yankees, once said, "It ain't over 'till it's over." Well he was right! In 1927 Paul A.M. Dirac, Professor of Physics, was at the Sovay Conference and was about to revitalize quaternions in an unforeseen way, at that time, in a little known field of physics that would become known as ***Quantum Mechanics***, thereby unifying them both [DIR75, pg 2] (also see Book II, TB.3.8 "EDEE: Einstein's Relativistic Energy Equation"). Einstein had reveled in his special theory of relativity how energy is related relative to its rest energy and reduces to Newton's kinetic energy equation at slow velocities below the speed of light.

$E = ½m_o v^2$ Newton EQA	$E = (c \sqrt{p_o^2 + p^2}) – E_o$ Einstein EQB

Equation B.4.1: Newtonian-Einsteinian Energy Equation

Where $[m_o]$ is the rest mass, $[p_o]$ the potential momentum at rest and $[E_o]$ the rest energy of the particle in a stationary frame of reference. Also where [c] is the velocity of light and [p] the momentum of the particle carrying the energy. Einstein's equation is the general form of Newton's which deals with low velocity particles having speeds, such that v << c. This can be seen in the following proof:

Theorem B.4.1 Newton's Limit to Einstein's Energy Equation at Low Velocity

1g	Given	$E_o \equiv m_o c^2$ Rest Energy
2g		$p_o \equiv m_o c$ Rest Momentum
3g		$p \equiv m_o v$
4g		$x \equiv v / c$
Steps	Hypothesis	$E \approx ½m_o v^2$ for $v << c$ Newtonian kinetic energy
1	From TB.3.8 EDEE: Einstein's Relativistic Energy Equation, A4.2.12 Identity Multp, A4.2.13 Inverse Multp, A4.2.10 Commutative Multp and A4.2.11 Associative Multp	$E = (c \sqrt{p_o^2 + p_o^2 (p^2 / p_o^2)}) – E_o$
2	From 1, A4.2.21 Distribution Exp and A4.2.14 Distribution	$E = (c \sqrt{p_o^2 (1 + (p / p_o)^2}) – E_o$
3	From 2, T4.10.3 Radicals: Identity, 2g, 3g and A4.2.3 Substitution	$E = (c\, p_o \sqrt{(1 + (m_o v / m_o c)^2}) – E_o$
4	From 3, 2g, D4.1.19 Primitive Definition for Rational Arithmetic, A4.2.10 Commutative Multp and A4.2.11 Associative Multp	$E = (m_o c^2 (\sqrt{1 + ((m_o / m_o) (v / c))^2}) – E_o$

5	From 4, 1g, A4.2.3 Substitution, A4.2.13 Inverse Multp and A4.2.12 Identity Multp	$E = E_o ((\sqrt{1 + (v / c)^2}) - 1)$
6	From LxE.3.1.10 Binomial Series Square Root	$(1 + x^2)^{1/2} = 1 + \frac{1}{2}x^2 + \frac{1}{8}x^4 + \text{HOT}$
7	From 5, 6, 4g and A4.2.4 Closure Add first two terms	$E \approx E_o (1 + \frac{1}{2} (v / c)^2 - 1)$
8	From 7, A4.2.5 Commutative Add and A4.2.8 Inverse Add	$E \approx E_o (\frac{1}{2} (v / c)^2)$
9	From 8, 1g, A4.2.10 Commutative Multp and A4.2.21 Distribution Exp	$E \approx \frac{1}{2} m_o c^2 (v^2 / c^2)$
10	From 9 and A4.2.10 Commutative Multp	$E \approx \frac{1}{2} m_o v^2 (c^2 / c^2)$
∴	From 10, A4.2.13 Inverse Multp and A4.2.12 Identity Multp	$E \approx \frac{1}{2} m_o v^2$ for $v << c$ Newtonian kinetic energy

What Dirac observed was that something was missing in Einstein's new energy equation. He noticed the equation had a radical and any time one takes the square root of a quantity you have to assign a plus and minus sign to the radical. If he did, Einstein's equation would have to be altered in the following way:

$$E = \pm c \sqrt{(p_o^2 + p^2)} - E_o$$

Equation B.4.2: Dirac-Einstein's Energy Equation

If true than the plus and minus, as indicated by the equation, refers to positive as well as negative mass-energy. This implication would be that negative particles could be considered having the same mass but opposite charge as measured from the rest mass. As he thought about it, he realized there would be nothing in the physical world that would preclude the existences of such a particle. If such a particle existed than it would have to be mirrored as an antiparticle to conventional known matter.

Now in Newtonian and Einsteinian physics negative energies are not necessary, hence not dealt with, however in Quantum Mechanics the mechanism exists to be able to jump from one discrete energy state to another and in fact Dirac and Weyl showed it to be necessary to have negative energies for nuclear particles to properly move up and down the energy rungs (levels) in the orbits about an atomic nucleus. So, this new idea in Dirac's mind seemed to naturally work.

Now one other problem with Einstein's equation is it does not say how a nuclear, charged, particle finding itself immersed in quantized electric and magnetic, time varying field might move. However, Dirac knew that the general Schrödinger wave equation provided just such a vehicle, [GIE71, pg 154].

$H \equiv E - (\hbar^2 / 2m_o)\nabla^2$ Hamilton's Energy Operator EQ A	$iH\Psi + \hbar\, \psi_{,t} \equiv 0$ Schrödinger Wave Equation EQ B
where $\nabla^2 \equiv \partial^2_{,x,x} + \partial^2_{,y,y} + \partial^2_{,z,z}$ Laplacian or Del Squared Operator EQ C from Green's Theorem on Integral Volume-Surface Areas	
and $\hbar \equiv h / 2\pi$ for $h \equiv 6.62619650 \times 10^{-34}$ joule-sec, Planck's constant [ITT75, pg 3-11]	

Equation B.4.3: General Schrödinger Wave Equation

If one, where to a write a wave equation by inspection, as Dirac did, and it embodies the energy of Einstein's equation as a function of space and time, than the following equation derived would have to have the form of Maxwell's wave equation:

$p^2\Psi + \hbar^2 [(1 / c^2) \partial^2_{,t,t} - \nabla^2] \Psi \equiv 0$ where $p \equiv mc$ and [m] is the relativistic mass.

Equation B.4.4: Dirac-Maxwell Wave Equation for Energy

Also [Ψ] is a special multivariable function called the ***wave function*** and taking the magnitude $\Psi\Psi^*$ as defined by C7.4.13.1 "Magnitude: Squared" gives the ***probability wave function*** as to where the most likely place is to find the particle as it moves through space and time. The actual probability is given by

$$P(x, y, z, t) \equiv \int_{-\infty}^{x} \int_{-\infty}^{y} \int_{-\infty}^{z} \int_{-\infty}^{t} \psi \, \psi^* \, dxdydzdt$$

Equation B.4.5: Calculating Probability of Particle Position in Space and Time

So far so good, however Dirac was now bothered by an inconsistency between Schrödinger and his wave equation. The Schrödinger equation has a first partial of the wave function in time while his wave equation requires a second partial differentiation with respect to time. It seemed to him that his wave equation should be a linear first order differential rather than a quadratic of partials, if so, than it would be more compatible with Schrödinger's equation. What if there was a way to simply factor the quadratic differential and render it into a first order differential, but doing that to differential equations is fraught with danger in violation of differential limits and such. Well throwing caution to the wind, and not considering that second-order wave equations normally have parametric first order differential cross terms, as an example, the diffusion equations for diffusing particles through a material media. He went ahead anyway and created a linear first order differential operator by inspection.

Theorem B.4.2 Taking the Magnitude of the Dirac Wave Equation

1g	Given	$Q^2 \equiv (1 / c^2) \partial^2_{,t,t} - \nabla^2$
2g		$S \equiv p_o + i\hbar Q$
Steps	Hypothesis	$\Psi\lvert S\rvert^2 = 0$
1	From 1g, EB.4.4 Dirac-Maxwell Wave Equation for Energy and A4.2.3 Substitution	$p_o^2\Psi + \hbar^2 Q^2\Psi = 0$
2	From C7.4.13.1 Magnitude: Squared	$\lvert S\rvert^2 = S\, S^*$
3	From 2g, 2 and A4.2.3 Substitution	$\lvert S\rvert^2 = (p_o + i\hbar Q)\,(p_o - i\hbar Q)$
4	From 3 and T7.4.11 Complex Difference of Two Squares	$\lvert S\rvert^2 = p_o^2 + \hbar^2 Q^2$
5	From A4.2.2A Equality	$\Psi = \Psi$
6	From 4, 5 and T4.4.1 Equalities: Uniqueness of Multiplication	$\Psi\lvert S\rvert^2 = (p_o^2 + \hbar^2 Q^2)\,\Psi$
7	From 6 and A4.2.14 Distribution	$\Psi\lvert S\rvert^2 = p_o^2\Psi + \hbar^2 Q^2\Psi$
∴	From 1, 7 and A4.2.3 Substitution	$\Psi\lvert S\rvert^2 = 0$

Which shows as a complex number EB.4.4 "Dirac-Maxwell Wave Equation for Energy" is factorable this provided proof that lead Dirac to postulate four new number types [γ_t, γ_x, γ_y, γ_z] called ***spin numbers***, such that when combined in the product of Q^2 the Dirac quaternion operator gives a distributed outcome.

$Q \equiv (\gamma_t / c)\partial_{,t} + \gamma_x\partial_{,x} + \gamma_y\partial_{,y} + \gamma_z\partial_{,z}$	$p_o\Psi + i\hbar\, Q\Psi \equiv 0$
Dirac Quaternion Operator EQA	Dirac Energy Wave Equation EQB

Equation B.4.6: Dirac First Order Wave Equation

Now he had created a new type of wave equation that mirrored Schrödinger's wave equation EB.4.3, working backwards by factoring EB4.4 "Dirac-Maxwell Wave Equation for Energy" yielding EB4.6(A,B) "Dirac First Order Wave Equation". He would then only have to determine the properties of the coefficients his new equation would need to have:

Theorem B.4.3 Solving for Dirac Spin Numbers

1g	Given	$Q^2 \equiv \gamma_I [(1 / c^2)\, \partial^2_{,t,t} - \nabla^2]$			
Steps	Hypothesis	Solving for Dirac spin numbers $[\gamma_t, \gamma_x, \gamma_y, \gamma_z]$			
1	From 1g	$\gamma_I [(1 / c^2)\, \partial^2_{,t,t} - \nabla^2] = Q^2$			
2	From 1, EB4.6A Dirac First Order Wave Equation and A4.2.3 Substitution	$\gamma_I [(1 / c^2)\, \partial^2_{,t,t} - \nabla^2] =$ $((\gamma_I / c)\partial_{,t} + \gamma_x\partial_{,x} + \gamma_y\partial_{,y} + \gamma_z\partial_{,z})$ $((\gamma_I / c)\partial_{,t} + \gamma_x\partial_{,x} + \gamma_y\partial_{,y} + \gamma_z\partial_{,z})$			
3	From 2 and A4.2.14 Distribution, for the right side of equality	$(\gamma_t\gamma_t / c^2)$ $\partial^2_{,t,t}$	$(\gamma_t\gamma_x / c)$ $\partial^2_{,t,x}$	$(\gamma_t\gamma_y / c)$ $\partial^2_{,t,y}$	$(\gamma_t\gamma_z / c)$ $\partial^2_{,t,z}$
		$(\gamma_x\gamma_t / c)$ $\partial^2_{,x,t}$	$(\gamma_x\gamma_x)$ $\partial^2_{,x,x}$	$(\gamma_x\gamma_y)$ $\partial^2_{,x,y}$	$(\gamma_x\gamma_z)$ $\partial^2_{,x,z}$
		$(\gamma_y\gamma_t / c)$ $\partial^2_{,y,t}$	$(\gamma_y\gamma_x)$ $\partial^2_{,y,x}$	$(\gamma_y\gamma_y)$ $\partial^2_{,y,y}$	$(\gamma_y\gamma_z)$ $\partial^2_{,y,z}$
		$(\gamma_z\gamma_t / c)$ $\partial^2_{,z,t}$	$(\gamma_z\gamma_x)$ $\partial^2_{,z,x}$	$(\gamma_z\gamma_y)$ $\partial^2_{,z,y}$	$(\gamma_z\gamma_z)$ $\partial^2_{,z,z}$
4	From 3 and left side of the equality	(γ_I / c^2) $\partial^2_{,t,t}$	0	0	0
		0	(γ_I) $\partial^2_{,x,x}$	0	0
		0	0	(γ_I) $\partial^2_{,y,y}$	0
		0	0	0	(γ_I) $\partial^2_{,z,z}$

The only way step 4 can be valid is by using Hamilton's non-commutative law for skew asymmetric, complex, quaternion numbers and extending the properties to Dirac's numbers:

5	From 3 and applying skew asymmetry.	(γ_I / c^2) $\partial^2_{,t,t}$	$(\gamma_t\gamma_x / c)$ $\partial^2_{,t,x}$	$(\gamma_t\gamma_y / c)$ $\partial^2_{,t,y}$	$(\gamma_t\gamma_z / c)$ $\partial^2_{,t,z}$
	From 4 and equating terms across the equality	$(-\gamma_t\gamma_x / c)$ $\partial^2_{,t,x}$	(γ_I) $\partial^2_{,x,x}$	$(\gamma_x\gamma_y)$ $\partial^2_{,x,y}$	$(\gamma_x\gamma_z)$ $\partial^2_{,x,z}$
		$(-\gamma_t\gamma_y / c)$ $\partial^2_{,t,y}$	$(-\gamma_x\gamma_y)$ $\partial^2_{,x,y}$	(γ_I) $\partial^2_{,y,y}$	$(\gamma_y\gamma_z)$ $\partial^2_{,y,z}$
		$(-\gamma_t\gamma_z / c)$ $\partial^2_{,t,z}$	$(-\gamma_x\gamma_z)$ $\partial^2_{,x,z}$	$(-\gamma_y\gamma_z)$ $\partial^2_{,y,z}$	(γ_I) $\partial^2_{,z,z}$
6	From 4, 5 and equating terms	$\gamma_I = \gamma_t\gamma_t$	$\gamma_t\gamma_x$	$\gamma_t\gamma_y$	$\gamma_t\gamma_z$
		$\gamma_x\gamma_t = -\gamma_t\gamma_x$	$\gamma_I = \gamma_x\gamma_x$	$\gamma_x\gamma_y$	$\gamma_x\gamma_z$
		$\gamma_y\gamma_t = -\gamma_t\gamma_y$	$\gamma_y\gamma_x = -\gamma_x\gamma_y$	$\gamma_I = \gamma_y\gamma_y$	$\gamma_y\gamma_z$
		$\gamma_z\gamma_t = -\gamma_t\gamma_z$	$\gamma_z\gamma_x = -\gamma_x\gamma_z$	$\gamma_z\gamma_y = -\gamma_y\gamma_z$	$\gamma_I = \gamma_z\gamma_z$

Up until now, Dirac has taken a logical path gaining his insight through natural mathematical properties and pushing them to their logical conclusion, but at this point, he takes quite literally a quantum leap in logic. Why he did this is not elaborated in his autobiographic description of ***The Development of Quantum Mechanics*** [DIR75]. It may have been because of his keen awareness of Heisenberg's metrification of Bohr's model for an atom to explain the quantumnization of states providing a mechanism to move from one energy level to the next. Whatever his reason he then proceeded to choose a set of four-by-four complex matrices to represent his quaternion numbers; this eliminated the need to create any new number artifices. Or it may have been simply that the use of matrices and complex numbers provided him a rich medium to be creative in? However, he than builds his arrays out of Pauli Spin Matrices, which also maybe a clue as to where he might have acquired the idea. Whatever his thoughts the summation on the identity's diagonal elements result in the correct format for the Dirac-Maxwell Wave Equation for Energy.

∴	From 6 and applying results from Eisele and Mason [EIS70 pg 323]	$\gamma_I = \begin{pmatrix} I & 0 \\ 0 & I \end{pmatrix} = I$
		$\gamma_x = \begin{pmatrix} 0 & -\sigma_x \\ -\sigma_x & 0 \end{pmatrix}$
		$\gamma_y = \begin{pmatrix} 0 & -\sigma_y \\ -\sigma_y & 0 \end{pmatrix}$
		$\gamma_z = \begin{pmatrix} 0 & -\sigma_z \\ -\sigma_z & 0 \end{pmatrix}$
		$\gamma_t = \begin{pmatrix} I & 0 \\ 0 & -I \end{pmatrix}$

Where Pauli Spin Matrices are:

σ_I	=	$\begin{pmatrix} 1 & 0 \\ 0 & 1 \end{pmatrix}$	EQA
σ_x	=	$\begin{pmatrix} 0 & 1 \\ 1 & 0 \end{pmatrix}$	EQB
σ_y	=	$\begin{pmatrix} 0 & -i \\ i & 0 \end{pmatrix}$	EQC
σ_z	=	$\begin{pmatrix} 1 & 0 \\ 0 & -1 \end{pmatrix}$	EQD
$\sigma_i\sigma_i$	=	I	EQE
$\sigma_i\sigma_i$	=	$-\sigma_i\sigma_i$ for $i \neq j$	EQF
$\sigma_i\sigma_i$	=	$i\,\sigma_k$ cycling i, j, k	EQG

Equation B.4.7: Pauli Spin Matrices and Properties

Dirac's prediction of matter having an anti-counterpart with ½-spin was verified in experiments soon after confirming the remarkable prediction of antimatter in his theory. In 1931, Carl D. Anderson preformed a series of critical experiments by photographing positron paths in cloud chambers as cosmic rays hit a lead plate. The direction of the curvature of the path, caused by a magnetic field, indicated that the particles where positively charged, but with the same mass as an electron.

Dirac's prediction applied not only to the electron, but was proven to apply to all the fundamental constituents of matter (nuclear particles). Each type of particle must have a corresponding antiparticle type. The mass of any antiparticle is identical to that of the particle. All the rest of its properties are also closely related, but with the signs of all charges reversed. For example, a proton has a positive electric charge, but an antiproton has a negative electric charge. The existence of antimatter partners for all matter particles is now a well-verified phenomenon, for hundreds of particle types having been observed with pairings.

<table>
<tr><td></td><td>The Nobel Prize in Physics 1933 "for writing a series of papers leading to his relativistic theory of the electron (1928) and the theory of holes (1930), which predicted particles having the same mass and opposite charge of an electron."</td><td></td><td>The Nobel Prize in Physics 1936 "for his discovery of the positron"</td></tr>
<tr><td colspan="2">Paul Adrien Maurice Dirac Cambridge University, Cambridge, Great Britain 1902 - 1984</td><td colspan="2">Carl David Anderson USA California Institute of Technology Pasadena, CA, USA 1905 – 1991</td></tr>
</table>

Figure B.4.1 Paul Dirac and Carl Anderson

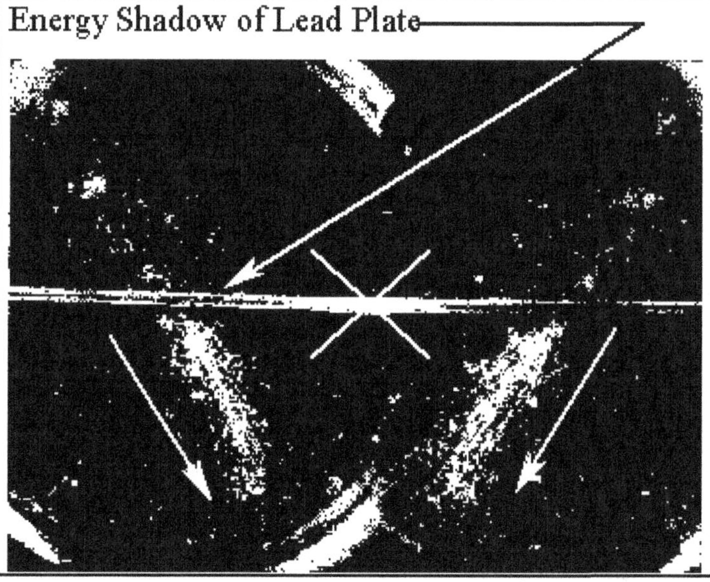

A cosmic-ray shower of more than one hundred positive and negative electrons (Magnetic field 7,900 gauss, 4,300 meters above sea level.) Photographic plate from "CARL D. ANDERSON The production and properties of positrons *Nobel Lecture, December 12, 1936*"

The white cross in the center of the plate is the direction of magnetic field going into the photographic plate and white arrows follow the magnetically segregated electron and positron shower created by the high-energy cosmic rays striking the lead plate.

Figure B.4.2 One of Carl Anderson's Original Plates Showing a Positron Shower

Section B.5 Minkowski's Space and Time Manifolds

The definitions of temporal periods in step-4g in Theorem B.2.6 "ITLD: Invariance of Time, Lorentz Dilation" Results from an idiosyncrasy or a hidden property of Relativity. This property is remarkably fundamental, and as such not obvious, which also lead Einstein to miss it. If he had found it, it might have allowed him to find his long sought after unification of forces at a distance.

Observation B.1 Temporal Period

The $\Delta t_n = \Delta l_n\ v/c^2$ in TB.2.6, 4g "ITLD: Invariance of Time, Lorentz Dilation" is entirely embedded in the fabric of the structure for the reference frame-n, yet it contains the velocity-v the motion of Reference Frame-m moving relative to [n^{th}] frame. How does reference frame-n know of the existence of frame-m in order to incorporate that velocity?

In a simplistic way if the two embedded points that make up the ends of the interval Δl_n were to exchange a photon the time or distance traveled between would have to be given by Einstein's definition of velocity DB.2.4, $\Delta t_n = \Delta l_n /c$ or $\Delta l_n = c\ \Delta t_n$ instead there is an embedded relativistic, light speed, velocity ratio, $\Delta l_n\ v/c^2$ or $\Delta l_n\ \beta/c$. Where does the relative, light speed, velocity ratio come from?

The answer is that when Einstein setup his relativistic transformation equations he modeled it by exchanging or passing a photon between any two given, frames of reference. By doing so an exchange of a quantum of energy [hv] was carried by the photon, which allows information to be passed, and establishing the frames existence. This binds the two reference frames together in a unique way. Now Reference Frame-n can now know the presence of Reference Frame-m, and at the moment of contact by the photon the relative velocity between them would also be known. It's as if there is no distance between the two reference frames and they know instantly of each other, of course this is not true, because they are limited by the photon's finite rate of travel, but it is enough to introduce a binding transformation factor β. So with respect to one another when these reference frames are adjoined they become a unique pair. This is the idea that Einstein missed, that reference frames could be uniquely attached to one another; this does not mean they are absolute, because any arbitrary pair of reference frames can be selected to match up and comprise these relative binary frames. Such unique pairs of frames work together and constitute a binary system called a ***relative binary frame system***.

Not only can information about relative velocity be passed between reference frames, but with gravitational or electrical forces the direction of action can also be passed. This than is how our universe hangs together by Binary Relative Reference Frames.

Definition B.5.2 Relative Binary Reference Frames

Relative Binary Reference Frames are any paired relative frames that have exchanged a photon and as a result forms a bridge to one another; resulting in information from the transmitting frame being successfully processed by the receiver. This ***process*** can be defined here as meaning some sort of operation transforming space, time, energy, momentum or force into an action that works on a nuclear particle.

From Theorem B.4.2 STC: "Space-Time Continuum Invariance of Relativistic Magnitude" is not necessarily zero, but equated to the transformation of magnitude in another frame of reference. If so two cases can be considered, let $r^2 \equiv x^2 + y^2 + z^2$ be ***spatial and temporal magnitude*** then

Definition B.5.3 Minkowski's Spatial Constant D_σ

$D_\sigma^2 \equiv r^2 - (ct)^2$ Equation A

A 4-space, elliptic hyperboloid made from One-Sheet of Rotation about the t-axis for a Minkowski's Spatial Constant D_σ, with dimensional units [L].

Definition B.5.4 Minkowski's Temporal Constant D_τ

$D_\tau^2 \equiv (ct)^2 - r^2$ Equation A

A 4-space, elliptic hyperboloid made from Two-Sheets of Rotation about the t-axis having a Minkowski's Temporal Constant D_τ, with dimensional units [L].

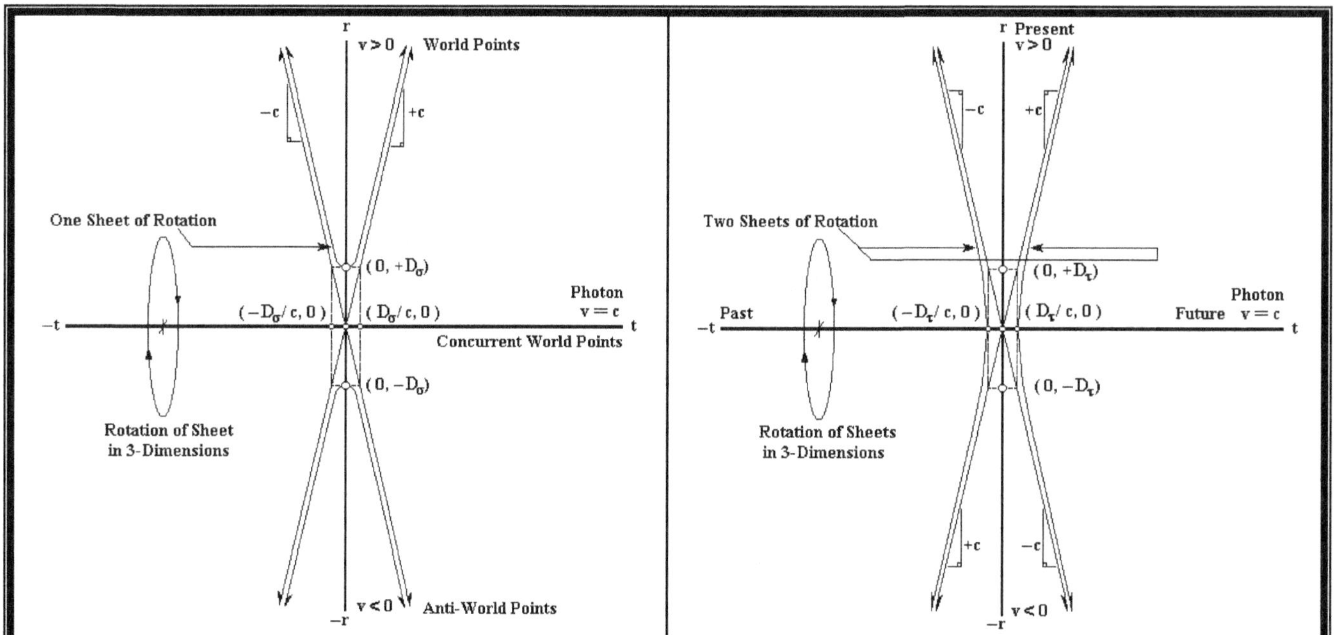

Fig. A Spatial Continuum **Fig. B Temporal Continuum**

Figure B.5.1 Minkowskian Space-Time Manifolds

These types of space-times were first explored by H. Minkowski in his paper on "Space and Time" [PER52 pg 84] and are known by his name as Minkowski's spatial manifold [LAW62I pg. 19] in Figure B.5.1A and the alternative case Minkowski's temporal manifold in Figure B.5.1B.

His terminology is used here with the four-dimensional coordinate positions in his manifold, which he calls ***world points*** and the hyperbolic branch(s) called ***world sheet***(s). As Minkowski uses the definition for ***world line*** as the curve path that the particle takes while it moves through space and time. Einstein in his theory on General Relativity calls the optimal path a particle takes through a field of force (gravitational or electrical) a ***geodesic path*** in actuality they are both one and the same, since the traveling particle can exist inside or outside of the field. The word optimal can be taken in the Einsteinian sense to mean that the particle is neutrally balanced in the field of force generating no additional aberrational forces or deviations in its natural trek.

The use of the word "world" is the stage, were the particle plays its role upon. The descriptive word "world" is really not appropriate a more accurate description would be the set of points comprising the entire universe for which the particle is found in. One might use the term universal point, universal sheet or universal line. World is an archaic word having a limited description, however Minkowski defined it as such and the scientific community has excepted it for many years now, so it will continue to be used as such in this paper.

From the spatial continuum diagram, Figure B.5.1.A, that section of space is divided up into three main Regions by the asymptotic envelope, Anti-World Points, Concurrent World Points and World Points. A light cone, when $D_\sigma = 0$, is concurrent with all other, world points throughout the universe. When $D_\sigma^2 > 0$ a pair of mass particles are embedded in a single sheet of the elliptic hyperboloid; hence have a measurable distance between them. For space being asymptotic to the speed of light than a mass particle can only travel at sub-light speeds, while the photon has the panoramic view of the whole universe.

Law B.5.V Concurrency

All world points in the universe are concurrent from the perspective of the photon, distance does not exist, and time of travel between any two points in the universe has slowed down to an eternal present; see Minkowski's spatial continuum Figure B.5.1B. If Relative frames of reference are associated at every world point in the universe than any two represent a single pair of frames and the world points are ***concurrent***, hence creating a ***unique binary frame of reference*** with respect to each other.

Observation B.2 On Asynchronous Events

In Figure B.2.1A "Synchronization of Events" the asynchronous event alludes to the idea that events are independent of origin preventing the possibility of a cause preceding an event. This is in volition of another of Einstein's Law, the Law of Causality. In Einstein's original 1905 paper, he had not considered this and upon later examination of time dilation for elementary pair production and annihilation, he was forced to reflect on this possibility leading him to introduce a new law to patch the hole in his theory.

In the Minkowskiain Regions of space and time strange things happen to nuclear particles as well as photons. Let's go back to the original questions that Einstein pondered as a fifteen year old, "What would it be like to travel at the speed of light? What would the universe look like from the vantage point of traveling with a photon?" With the concepts of Minkowskiain space and time these questions can be answered. From the temporal continuum diagram, Figure B.5.1.B, that section of space is divided up into three main regions by the asymptotic envelopes of past, present and future. It is a light cone when $D_\tau = 0$ existing only in the present it has no past, nor future. When $D_\tau^2 > 0$ a mass particle having had an event occurring in the past cannot have another event occurring in the future as a consequence of its history. The past may not precede the future, because those events are physically embedded within their own respective, non-connecting sheet, of the elliptic hyperboloid's space-time continuum, this then provides the validity for the Law of Causality.

Law B.5.VI Causality

Causality is where two events occurring at two different locations are related, that is, if one causes the other, then all observers will record the same temporal sequence. Although the exact time difference will vary from observer to observer, by time dilation, event–1 will always occur before event–2. Causality is preserved because neither signals nor observers can travel faster than the speed of light. [MAL67 pg 112] A photon exists in the perpetual state of the present see Minkowski's temporal continuum Fig B.5.1B. Since it was first stated as a law most people have known it as the "Law of Causality" per se being part of Einstein's theory. This is why it is stated here as such, but actually it can be proven from first principles, hence it may be really a theorem of Relativity TB.5.1 "TCCE: Theorem of Causality and Conservation of Energy", also see Lorrain and Corson for a formal proof of this physical base proposition [LOR62 pg 214].

Theorem B.5.1 TCCE: Theorem of Causality and Conservation of Energy

1g	Given	(cause) t_{A1} Assume (cause) t_{A1}	$<$ $>$	t_{B1} (effect) on the time line. Event B (effect) comes before Event A (cause) t_{B1} (effect) for a single nuclear particle of mass m_0
2g		0	$<$	v
3g		0	$<$	v_{2A}
4g		0	$<$	v_{2B}
5g		0	$<$	$1/c^2$
6g		0	$<$	γ_A
7g		0	$<$	γ_B
8g	From A4.2.16 The Trichotomy Law of Ordered Numbers	0	$<$	1
Steps	Hypothesis	t_{A2}	$<$	t_{B2} Event A cause must come before Event B effect to conserve energy
1	From DB.2.4 Velocity	x_{A2}	$\equiv$	$v_A t_{A2}$
2	From DB.3.4 Velocity	x_{B2}	$\equiv$	$v_B t_{B2}$
3	From 1g, LxECT.B.2.2.4b and A4.2.3 Substitution	$\gamma_A\,[t_{A2} - v_A x_{A2} / c^2]$	$>$	$\gamma_B\,[t_{B2} - v_B x_{B2} / c^2]$
4	From 1, 2, 3 and A4.2.3 Substitution	$\gamma_A\,[t_{A2} - v_A(v_A t_{A2}) / c^2]$	$>$	$\gamma_B\,[t_{B2} - v_B(v_B t_{B2}) / c^2]$
5	From 4, T4.7.17 Inequalities: Cancellation by Multiplication for a Positive Number and A4.2.11 Associative Multp	$\gamma_A\,[t_{A2} - v_A^2 t_{A2} / c^2]$	$>$	$\gamma_B\,[t_{B2} - v_B^2 t_{B2} / c^2]$
6	From 5 and A4.2.14 Distribution	$\gamma_A t_{A2}\,[1 - v_A^2 / c^2]$	$>$	$\gamma_B t_{B2}\,[1 - v_B^2 / c^2]$
7	From 6, DxLZF.B.2.1.2 and A4.2.3 Substitution	$\gamma_A t_{A2}\,(1 - \beta_A^2)$	$>$	$\gamma_B t_{B2}\,(1 - \beta_B^2)$
8	From 7, DxLZF.B.2.1.3, DxLZF.B.2.1.4 and A4.2.3 Substitution	$\eta_A^{-1} t_{A2}\, \eta_A$	$>$	$\eta_B^{-1} t_{B2}\, \eta_B$
9	From 8, A4.2.10 Commutative Multp, A4.2.13 Inverse Multp and A4.2.12 Identity Multp	t_{A2}	$>$	t_{B2}

7	From 2g, 3g, 5g and T4.7.19 Inequalities: Multiplication for Positive Numbers	$0 < v_A^2 / c^2$
8	From 2g, 4g, 5g and T4.7.19 Inequalities: Multiplication for Positive Numbers	$0 < v_B^2 / c^2$
9	From 6g, 8g, 7, T4.7.7 Inequalities: Addition of (a < b) + (c < d) and T4.7.5 Inequalities: Multiplication by Positive Number	$0 < \gamma_A (1 + v_A^2 / c^2)$
10	From 7g, 8g, 8, T4.7.7 Inequalities: Addition of (a < b) + (c < d) and T4.7.5 Inequalities: Multiplication by Positive Number	$0 < \gamma_B (1 + v_B^2 / c^2)$
11	From 9, 10 and assume Case I	$\gamma_A (1 + v_A^2 / c^2) > \gamma_B (1 + v_B^2 / c^2)$
12	From 1g, 11 and T4.7.19 Inequalities: Multiplication for Positive Numbers	$\gamma_A t_{A2} [1 + v_A^2 / c^2] > \gamma_B t_{B2} [1 + v_B^2 / c^2]$
13	From 12 validates 6	
14	From 11 and T4.7.16 Inequalities: Cancellation by Addition	$v_A^2 / c^2 > v_B^2 / c^2$
15	From 14 and T4.7.6 Inequalities: Multiplication by Negative Number Reverses Order	$-v_A^2 / c^2 < -v_B^2 / c^2$
16	From 15 and T4.7.2 Inequalities: Uniqueness of Addition by a Positive Number	$1 - v_A^2 / c^2 < 1 - v_B^2 / c^2$
17	From 16 and DxLZF.B.2.1.3	$\eta_A < \eta_B$
18	From 17 and T4.7.31 Inequalities: Reciprocal Cross Multiplication	$1/\eta_B < 1/\eta_A$
19	From 18 and DxLZF.B.2.1.4	$\gamma_B < \gamma_A$
20	From 1g, 19, A4.2.2 Equality and T4.7.5 Inequalities: Multiplication by Positive Number	$m_0\gamma_B < m_0\gamma_A$
21	From 20 and LxLZF.B.2.1.3	$m_B < m_A$

22	From 21, A4.2.2 Equality and T4.7.5 Inequalities: Multiplication by Positive Number	m_Bc^2	<	m_Ac^2
23	From 1g and 22	(cause) m_Ac^2	<	m_Bc^2 (effect)
∴	From 23 it is implied more energy results at the end, creating the effect, which cannot be. Conservation of energy guarantees energy cannot be created out of nothing.	t_{A2}	<	t_{B2} hence it must be concluded Event A (cause) comes before Event B (effect) the original premise

The Theorem B.5.1 "TCCE: Theorem of Causality and Conservation of Energy" is not a very rigorous proof, but it does show that the Law of Causality may not be a first principle and can be derived from other primary propositions, hence not a law.

Table B.5.1 LTM Asymptotic Relationships to the Velocity of Light

Length	Time	Mass		Velocity	At extremes
Δl_0	Δt_0	Δm_0		0	At Rest
0	∞	∞	0	c	At velocity-c
Theorem B.2.4 LTD: Lorentz Length Contraction	Theorem B.2.1 FTD: Fitzgerald Time Dilation	LxLZF.B.2.1.3 EMD: Einstein Mass Dilation	D1.1.3G Definition of a photon		

These phenomena, concurrency and causality, are symmetrical laws of the space-time continuum. Under the Lorenz transformations, as used by Einstein, time and distance are distorted in order to fit within the asymptotic boundaries of the speed of light. This than is the corner stone upon which the theory of force at a distance is based and what was identified so many years ago in the 1988 paper "Unification of Field Theory at a Distance".

Now let us consider the transformation of relativistic magnitude between relative reference frames:

Theorem B.5.2 STC: Space-Time Continuum Invariance of Relativistic Magnitude

1g	Given	w	=	ict for $i = \sqrt{-1}$
2g		β	=	v/c
3g		γ^2	=	$1 / (1 - \beta^2)$
Steps	Hypothesis	$x_2^2 + y_2^2 + z_2^2 + w_2^2$	=	$x_1^2 + y_1^2 + z_1^2 + w_1^2$
1	From LxECT.B.2.2.1b and T4.8.6 Integer Exponents: Uniqueness of Exponents	x_2^2	=	$(\gamma [x_1 - vt_1])^2$ square both sides
2	From 1 and A4.2.8 Summation Exp	x_2^2	=	$\gamma^2 (x_1 - vt_1)^2$
3	From 2 and T4.12.2 Polynomial Quadratic: The Perfect Square by Difference	x_2^2	=	$\gamma^2 [x_1^2 - 2 x_1(vt_1) + (vt_1)^2]$
4	From LxECT.B.2.2.4b and T4.8.6 Integer Exponents: Uniqueness of Exponents	t_2^2	=	$(\gamma[t_1 - vx_1 / c^2])^2$
5	From 4 and A4.2.8 Summation Exp	t_2^2	=	$\gamma^2(t_1 - vx_1 / c^2)^2$
6	From 5 and T4.12.2 Polynomial Quadratic: The Perfect Square by Difference	t_2^2	=	$\gamma^2 [t_1 - 2 x_1(v t_1) / c^2 + x_1^2v^2 / c^4]$
7	From 6, A4.2.2 Equality, T4.4.1 Equalities: Uniqueness of Multiplication, A4.2.14 Distribution, A4.2.10 Commutative Multp, A4.2.13 Inverse Multp and A4.2.21 Distribution Exp	$(ct_2)^2$	=	$\gamma^2 [(ct_1)^2 - 2 x_1(v t_1) + x_1^2v^2 / c^2]$
8	From 3, 7, T4.3.9 Equalities: Uniqueness of Subtraction, A4.2.14 Distribution, D4.1.20 Negative Coefficient and D4.1.19 Primitive Definition for Rational Arithmetic	$x_2^2 - (ct_2)^2$	=	$\gamma^2 [x_1^2 - 2 x_1(vt_1) + (vt_1)^2$ $- x_1^2v^2 / c^2 + 2 x_1(v t_1) - (ct_1)^2]$

9	From 8, A4.2.8 Inverse Add, A4.2.7 Identity Add, T4.8.7 Integer Exponents: Distribution Across a Rational Number and A4.2.10 Commutative Multp	$x_2^2 - (ct_2)^2$	=	$\gamma^2 [x_1^2 - (v/c)^2 x_1^2 + (vt_1)^2 - (ct_1)^2]$
10	From 2g, 9, A4.2.21 Distribution Exp, A4.2.13 Inverse Multp, A4.2.14 Distribution and A4.2.3 Substitution	$x_2^2 - (ct_2)^2$	=	$\gamma^2 (x_1^2 - \beta^2 x_1^2 + \beta^2 c^2 t_1^2 - c^2 t_1^2)$
11	From 10, A4.2.12 Identity Multp, T4.8.2 Integer Exponents: Negative One Squared, D4.1.20 Negative Coefficient, A4.2.5 Commutative Add and A4.2.14 Distribution	$x_2^2 - (ct_2)^2$	=	$\gamma^2 (1 - \beta^2)\, x_1^2 - \gamma^2 (1 - \beta^2)\, c^2 t_1^2$
12	From 3g, 11, A4.2.3 Substitution, A4.2.13 Inverse Multp A4.2.12 Identity Multp and A4.2.21 Distribution Exp	$x_2^2 - (ct_2)^2$	=	$x_1^2 - (ct_1)^2$
13	From 12, LxECT.B.2.2.2b, LxECT.B.2.2.3b, T4.8.6 Integer Exponents: Uniqueness of Exponents and T4.3.1 Equalities: Uniqueness of Addition	$x_2^2 + y_2^2 + z_2^2 - (ct_2)^2$	=	$x_1^2 + y_1^2 + z_1^2 - (ct_1)^2$
14	From 13 and T4.9.4 Rational Exponent: Square Root of a Positive and Negative One	$x_2^2 + y_2^2 + z_2^2 + i^2(ct_2)^2$	=	$x_1^2 + y_1^2 + z_1^2 + i^2 (ct_1)^2$
15	From 14 and A4.2.21 Distribution Exp	$x_2^2 + y_2^2 + z_2^2 + (ict_2)^2$	=	$x_1^2 + y_1^2 + z_1^2 + (ict_1)^2$
∴	From 1g, 15 and A4.2.3 Substitution	$x_2^2 + y_2^2 + z_2^2 + w_2^2$	=	$x_1^2 + y_1^2 + z_1^2 + w_1^2$

Here is a four–dimensional Pythagorean Theorem based on the unification of space and time in a dimensional continuum. Time being a parametric parameter completes the transformation from S_1 to S_2, like the Galilean coordinates the Einsteinian system remains invariant under transformation for magnitude in space-time.

Iff $0 = x^2 + y^2 + z^2 - (ct)^2$ then $(ct)^2 = x^2 + y^2 + z^2$ is a four-dimensionally geometric elliptic cone of light, known as a ***light-cone*** with a time like ***spherical radius*** = ct. As a light wave moves radially outward from a relativistic 4-point source (x, y, z, ict) it traces out a spherical region of space for every instant of time-t during its travel. Since it radially expands outward everywhere its intensity dissipates with the inverse radial square of the distance. As nuclear particles pass energy by exchanging photons this gives rise to Coulomb and Newton's famous inverse square laws for force at a distance.

Theorem B.5.3 MSTCR: Minkowski's Spatial-Temporal Constant Relationship

Steps	Hypothesis	$D_\sigma = \pm iD_\tau$ EQ A	$D_\sigma^2 = -D_\tau^2$ EQ B
1	From Case I	$D_\sigma^2 = r^2 - (ct)^2$	
2	From 1, T4.8.2 Integer Exponents: Negative One Squared, D4.1.17 Exponential Notation, and D4.1.20 Negative Coefficient	$D_\sigma^2 = (-1)(-1)r^2 + (-1)(ct)^2$	
3	From 2 and A4.2.14 Distribution	$D_\sigma^2 = (-1)[(-1)r^2 + (ct)^2]$	
4	From 3 and D4.1.20 Negative Coefficient	$D_\sigma^2 = -[-r^2 + (ct)^2]$	
5	From 4 and A4.2.5 Commutative Add	$D_\sigma^2 = -[(ct)^2 - r^2]$	
6	From 5, Case II and A4.2.3 Substitution	$D_\sigma^2 = -D_\tau^2$	
7	From 6 and T4.10.10 Radicals: Squaring the Inverse by Negative Square Root	$D_\sigma = \pm i\sqrt{D_\tau^2}$	
∴	From 6, 7, T4.10.4 Radicals: Identity Radical Raised to a Power	$D_\sigma = \pm iD_\tau$ EQ A	$D_\sigma^2 = -D_\tau^2$ EQ B

The two manifolds are dimensionally ninety degrees to one another hence independent forming space and time; see Figure B.3.1 A and B.

Theorem B.5.4 MSAS: Minkowski's Spatial Asymptotic Slopes

1g	Given	$m_s \equiv \lim_{t \to \infty} \dfrac{dr}{dt}$ Minkowski's spatial asymptotic slope
Steps	Hypothesis	$m_s = \pm c$
1	From Case I	$D_\sigma^2 = r^2 - (ct)^2$
2	From 1, Differentiating both sides of the equality with respect to time.	$\dfrac{d D_\sigma^2}{dt} = \dfrac{d(r^2 - (ct)^2)}{dt}$
3	From 2, Lx9.1.3.1B Differential of a Constant and Lx9.3.2.2B Differential as a Linear Operator	$0 = \dfrac{d(r^2)}{dt} - \dfrac{d[(ct)^2]}{dt}$
4	From 3, Lx9.3.2.3B Distribute Constant in/out of Product, Lx9.3.3.3B Differential of a power, Lx9.3.3.2B Differential Identity and A4.2.12 Identity Multp	$0 = 2r \dfrac{dr}{dt} - 2c^2t$
5	From 4, T4.3.6A Equalities: Reversal of Right Cancellation by Addition and T4.4.2 Equalities: Cancellation by Multiplication	$r \dfrac{dr}{dt} = c^2t$
6	From 5 and T4.4.3B Equalities: Right Cancellation by Multiplication	$\dfrac{dr}{dt} = \dfrac{c^2t}{r}$
7	From 1 and T4.3.6B Equalities: reversal of Right Cancellation by Addition	$r^2 = D_\sigma^2 + (ct)^2$
8	From 7 and A4.2.21 Distribution Exp	$r^2 = D_\sigma^2 + c^2t^2$
9	From 8, A4.2.12 Identity Multp, A4.2.13 Inverse Multp and A4.2.10 Commutative Multp	$r^2 = (D_\sigma^2 / t^2)\, t^2 + c^2t^2$
10	From 9 and A4.2.14 Distribution	$r^2 = t^2 [(D_\sigma^2 / t^2) + c^2]$
11	From 10, T4.10.7 Radicals: Reciprocal Exponent by Positive Square Root	$r = \pm\sqrt{t^2 [(D_\sigma^2 / t^2) + c^2]}$

12	From 11, T4.10.5 Radicals: Distribution Across a Product and T4.10.3 Radicals: Identity Power Raised to a Radical	$r = \pm t \sqrt{[(D_\sigma^2 / t^2) + c^2]}$
13	From 6, 12 and A4.2.3 Substitution	$\frac{dr}{dt} = \frac{\pm t\, c^2}{t \sqrt{[(D_\sigma^2 / t^2) + c^2]}}$
14	From 13, A4.2.13 Inverse Multp and A4.2.12 Identity Multp	$\frac{dr}{dt} = \frac{\pm c^2}{\sqrt{[(D_\sigma^2 / t^2) + c^2]}}$
15	From 14 and Lx9.2.1.1 Uniqueness of Limits	$\lim_{t \to \infty} \frac{dr}{dt} = \lim_{t \to \infty} \frac{\pm c^2}{\sqrt{[(D_\sigma^2 / t^2) + c^2]}}$
16	From 1g, 15, Lx9.2.1.2 Limit of a Constant, A4.2.3 Substitution Lx9.2.1.7 Limit of a Composite Function, Lx9.2.1.5 Limit of the Products and T9.2.2 Reciprocal Numbers Tend to Zero as Denominator Increases without Limit	$m_s = \frac{\pm c^2}{\sqrt{[0 + c^2]}}$
17	From 16, A4.2.7 Identity Add and T4.10.3 Radicals: Identity	$m_s = \pm c^2 / c$
∴	From 17, D4.1.17 Exponential Notation, A4.2.13 Inverse Multp and A4.2.12 Identity Multp	$m_s = \pm c$ slope at infinity to the asymptote

Theorem B.5.5 MTAS: Minkowski's Temporal Asymptotic Slopes

1g	Given	$m_t \equiv \lim_{t \to \infty} \dfrac{dr}{dt}$ Minkowski's temporal asymptotic slope
Steps	Hypothesis	$m_t = \pm c$
1	From Case I	$D_\tau^2 = (ct)^2 - r^2$
2	From 1, Differentiating both sides of the equality with Respect to time.	$\dfrac{d D_\tau^2}{dt} = \dfrac{d((ct)^2 - r^2)}{dt}$
3	From 2, Lx9.1.3.1B Differential of a Constant and Lx9.3.2.2B Differential as a Linear Operator	$0 = \dfrac{d[(ct)^2]}{dt} - \dfrac{d(r^2)}{dt}$
4	From 3, Lx9.3.2.3B Distribute Constant in/out of Product, Lx9.3.3.3B Differential of a power, Lx9.3.3.2B Differential Identity and A4.2.12 Identity Multp	$0 = 2c^2t - 2r \dfrac{dr}{dt}$
5	From 4, T4.3.6A Equalities: reversal of Right Cancellation by Addition and T4.4.2 Equalities: Cancellation by Multiplication	$r \dfrac{dr}{dt} = c^2t$
6	From 5 and T4.4.3B Equalities: Right Cancellation by Multiplication	$\dfrac{dr}{dt} = \dfrac{c^2t}{r}$
7	From 1 and T4.3.6B Equalities: reversal of Right Cancellation by Addition	$r^2 = (ct)^2 + D_\tau^2$
8	From 7 and A4.2.21 Distribution Exp	$r^2 = c^2t^2 + D_\tau^2$
9	From 8, A4.2.12 Identity Multp, A4.2.13 Inverse Multp and A4.2.10 Commutative Multp	$r^2 = c^2t^2 + (D_\tau^2/ t^2)\, t^2$
10	From 9 and A4.2.14 Distribution	$r^2 = t^2\, [c^2 + (D_\tau^2/ t^2)]$
11	From 10, T4.10.7 Radicals: Reciprocal Exponent by Positive Square Root	$r^2 = \pm\sqrt{}\, t^2\, [c^2 + (D_\tau^2/ t^2)]$

12	From 11, T4.10.5 Radicals: Distribution Across a Product and T4.10.3 Radicals: Identity Power Raised to a Radical	$r = \pm t \sqrt{[c^2 + (D_\tau^2 / t^2)]}$
13	From 6, 12 and A4.2.3 Substitution	$\frac{dr}{dt} = \frac{\pm t\, c^2}{t \sqrt{[c^2 + (D_\tau^2 / t^2)]}}$
14	From 13, A4.2.13 Inverse Multp and A4.2.12 Identity Multp	$\frac{dr}{dt} = \frac{\pm c^2}{\sqrt{[c^2 + (D_\tau^2 / t^2)]}}$
15	From 14 and Lx9.2.1.1 Uniqueness of Limits	$\lim_{t \to \infty} \frac{dr}{dt} = \lim_{t \to \infty} \frac{\pm c^2}{\sqrt{[c^2 + (D_\tau^2 / t^2)]}}$
16	From 1g, 15, Lx9.2.1.2 Limit of a Constant, A4.2.3 Substitution Lx9.2.1.7 Limit of a Composite Function, Lx9.2.1.5 Limit of the Products and T9.2.2 Reciprocal Numbers Tend to Zero as Denominator Increases without Limit	$m_s = \frac{\pm c^2}{\sqrt{[c^2 + 0]}}$
17	From 16, A4.2.7 Identity Add and T4.10.3 Radicals: Identity	$m_s = \pm c^2 / c$
∴	From 17, D4.1.17 Exponential Notation, A4.2.13 Inverse Multp and A4.2.12 Identity Multp	$m_s = \pm c$ slope at infinity to the asymptote

Theorem B.5.6 MVQ: Minkowski's Velocity Quaternions

1g	Given	$\vec{s}_4$	$\equiv$	$(\vec{r}, \text{i tc})$ displacement quaternion
2g		$\vec{v}_4$	$\equiv$	$\dot{\vec{s}}$ velocity quaternion
Steps	Hypothesis	$\vec{v}_4$	$=$	$(\dot{\vec{r}}, \text{ic})$ velocity quaternion
1	From 1g Minkowski's Displacement Quaternion	$\vec{s}_4$	$=$	$(\vec{r}, \text{i tc})$ displacement quaternion
2	From 1 and Lx9.3.1.7B Uniqueness Theorem of Differentiation	$\dfrac{d\,\vec{s}_4}{dt}$	$=$	$\dfrac{d\,(\vec{r}, \text{i tc})}{dt}$
3	From 2, Lx9.3.2.2B Differential as a Linear Operator, Lx9.3.3.2A Differential Identity and A4.2.12 Identity Multp	$\dot{\vec{s}}_4$	$=$	$(\dot{\vec{r}}, \text{i c})$
∴	From 2g, 3 and A7.2.9 Substitution of Vectors	$\vec{v}_4$	$=$	$(\dot{\vec{r}}, \text{ic})$ velocity quaternion

Theorem B.5.7 MAQ: Minkowski's Acceleration Quaternions

1g	Given	$\vec{a}_4$	$\equiv$	$\dot{\vec{v}}_4$ acceleration quaternion
Steps	Hypothesis	$\vec{a}_4$	$=$	$(\ddot{\vec{r}}, \text{i } 0)$ acceleration quaternion
1	From TB.3.6 MVQ: Minkowski's Velocity Quaternions	$\vec{v}_4$	$=$	$(\dot{\vec{r}}, \text{i c})$ velocity quaternion
2	From 1 and Lx9.3.1.7B Uniqueness Theorem of Differentiation	$\dfrac{d\,\vec{v}_4}{dt}$	$=$	$\dfrac{d\,(\dot{\vec{r}}, \text{i c})}{dt}$
3	From 2, Lx9.3.2.2B Differential as a Linear Operator and Lx9.3.3.1B Differential of a Constant	$\dot{\vec{v}}_4$	$=$	$(\ddot{\vec{r}}, \text{i } 0)$
∴	From 1g, 3 and A7.2.9 Substitution of Vectors	$\vec{a}_4$	$=$	$(\ddot{\vec{r}}, \text{i } 0)$ acceleration quaternion

Theorem B.5.8 MOVDQ: Minkowski's Orthogonal Velocity-Displacement Quaternions

1g	Given	$\vec{v}_4 \equiv (\dot{\vec{r}}, i\ c)$ velocity quaternion
2g		$\vec{s}_4 \equiv (\vec{r}, i\ tc)$ displacement quaternion
Steps	Hypothesis	$0 = \vec{s}_4 \bullet \vec{v}_4$
1	From DB.3.3 Minkowski's Spatial Constant D_s	$D_s^2 = r^2 - (ct)^2$
2	From 1 and TK3.3 Uniqueness of Differentials	$\frac{d}{dt}(D_s^2) = \frac{d}{dt}[r^2 - (ct)^2]$
3	From 2, Lx9.3.3.1A Differential of a Constant, Lx9.3.2.2A Differential as a Linear Operator, Lx9.3.3.3A Differential of a powder, Lx9.3.3.2A Differential Identity and Lx9.3.2.3A Distribute Constant in/out of Product	$0 = 2\vec{r} \bullet \dot{\vec{r}} - 2\ c\ t\ c$
4	From 3, A4.2.14 Distribution, A4.2.10 Commutative Multp and T4.3.2 Equalities: Any Quantity Multiplied by Zero is Zero	$2 * 0 = 2(\vec{r} \bullet \dot{\vec{r}} - t\ c\ c)$
5	From 4, T4.4.2 Equalities: Cancellation by Multiplication and A4.2.11 Associative Multp	$0 = \vec{r} \bullet \dot{\vec{r}} - (t\ c)\ c$
6	From 5 and D7.5.2B Inner Product Of Four-Dimensional Quaternion	$0 = (\vec{r}, i\ tc) \bullet (\dot{\vec{r}}, ic)$
∴	From 1g, 2g, 6 and A4.2.3 Substitution	$0 = \vec{s}_4 \bullet \vec{v}_4$

Theorem B.5.9 MIADQ: Minkowski's Inner Product Acceleration-Displacement Quaternions

1g	Given	$\overset{*}{a}_4$	$\equiv$	$(\ddot{\mathbf{r}}, i\ 0)$ acceleration quaternion
2g		$\overset{*}{v}_4$	$\equiv$	$(\dot{\mathbf{r}}, i\ c)$ velocity quaternion
3g		$\overset{*}{s}_4$	$\equiv$	$(\mathbf{r}, i\ tc)$ displacement quaternion
4g		v^2	$\equiv$	$\dot{\mathbf{r}} \bullet \dot{\mathbf{r}}$
Steps	Hypothesis	$c^2\ \eta^2$	$=$	$\overset{*}{s}_4 \bullet \overset{*}{a}_4$ asymptotic to [c]
1	From TB.3.8 MOVDQ: Minkowski's Orthogonal Velocity-Displacement Quaternions and D7.5.2B Inner Product Of Four-Dimensional Quaternions	0	$=$	$\mathbf{r} \bullet \dot{\mathbf{r}} - (t\ c)\ c$
2	From 1 and T9.3.2 Uniqueness of Differential	$\frac{d}{dt}\ (0)$	$=$	$\frac{d}{dt}\ [\mathbf{r} \bullet \dot{\mathbf{r}} - (t\ c)\ c]$
3	From 2, Lx9.3.3.1A Differential of a Constant, Lx9.3.2.2A Differential as a Linear Operator, Lx9.3.3.3A Differential of a powder, Lx9.3.3.2A Differential Identity and Lx9.3.2.3A Distribute Constant in/out of Product	0	$=$	$\mathbf{r} \bullet \ddot{\mathbf{r}} + \dot{\mathbf{r}} \bullet \dot{\mathbf{r}} - c\ c$
4	From 4g, 3, D4.1.17 Exponential Notation and A4.2.3 Substitution	0	$=$	$\mathbf{r} \bullet \ddot{\mathbf{r}} + v^2 - c^2$
5	From 4, T4.3.6A Equalities: Reversal of Right Cancellation by Addition and T4.3.5A Equalities: Right Cancellation by Addition	$c^2 - v^2$	$=$	$\mathbf{r} \bullet \ddot{\mathbf{r}}$
6	From 5, A4.2.12 Identity Multp, A4.2.13 Inverse Multp and A4.2.14 Distribution	$c^2\ [\ 1 - (v/c)^2\ ]$	$=$	$\mathbf{r} \bullet \ddot{\mathbf{r}}$
7	From 6, DxLZF.B.2.1.1 velocity ratio, DxLZF.B.2.1.3 Fitzgerald contraction factor and A4.2.3 Substitution	$c^2\ \eta^2$	$=$	$\mathbf{r} \bullet \ddot{\mathbf{r}}$
8	From 1g, 3g, D7.5.2B Inner Product Of Four-Dimensional Quaternions	$\mathbf{r} \bullet \ddot{\mathbf{r}}$	$=$	$\overset{*}{s}_4 \bullet \overset{*}{a}_4$
∴	From 7, 8 and A4.2.3 Substitution	$c^2\ \eta^2$	$=$	$\overset{*}{s}_4 \bullet \overset{*}{a}_4$ asymptotic to [c]

Theorem B.5.10 MOADQ: Minkowski's Orthogonal Acceleration-Displacement Quaternions

1g	Given	$v = c$
Steps	Hypothesis	$0 = \vec{s}_4 \bullet \vec{a}_4$ orthogonal
1	From 1g, DxLZF.B.2.1.1 velocity ratio, A4.2.3 Substitution and A4.2.13 Inverse Multp	$\beta = 1$
2	From 1, DxLZF.B.2.1.3 Fitzgerald contraction factor, A4.2.3 Substitution and A.2.8 Inverse Add	$\eta = 0$
∴	From 2, TB.3.9 MIADQ: Minkowski's Inner Product Acceleration-Displacement Quaternions, A4.2.3 Substitution and T4.3.2 Equalities: Any Quantity Multiplied by Zero is Zero	$0 = \vec{s}_4 \bullet \vec{a}_4$ at [c] displacement is orthogonal to acceleration

If one goes further into Minkowski's paper [PER52 pg 84] he does something that herald's General Relativity. By using the curvature of the path he binds his temporal manifold of time, to every point along the trajectory of the particle. By using Gaussian curvature Einstein achieved the same effect by using a more encompassing method, the use of optimized path in a tensor manifold, thereby associating special Relativity Transformations at every world point within the universe.

Observation B.3 Velocity Quaternion at Every Point on the World Line

What Minkowski noticed in his differentiation in theorem TB.3.8 "MOVDQ: Minkowski's Orthogonal Velocity-Displacement Quaternions" the resulting quaternion vector expression is analogous to the inner product of two orthogonal vectors. Throwing away the displacement quaternion he was left with a velocity quaternion that varied from point to point along a spatial world line, which will be called Σ.

$$\vec{v}_4 = \dot{\vec{\sigma}}_4 = (\dot{\vec{r}}, ic) \qquad \text{Equation A}$$

Observation B.4 Minkowski's Spatial and Temporal Variables σ-τ

In order to model a spatial and temporal manifold at every point along the world line-Σ–T lets consider DB.3.2 Minkowski's Spatial and Temporal Constants D_σ and D_τ and Observation B.3.3A, subsequently the existence of variable σ and τ forces a conclusion that D_σ and D_τ can no longer be held constant. So on world line-Σ–T a new model must be adapted from equation DB.3.2A

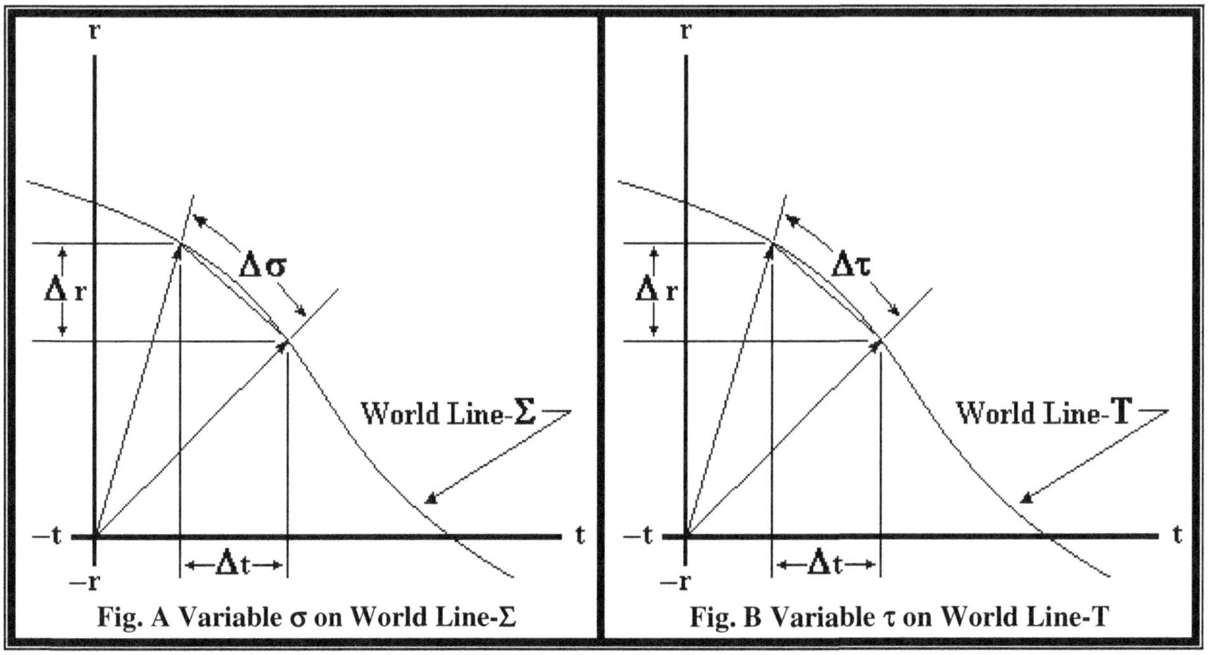

Figure B.5.2 Variable on World Line

$$\Delta\sigma^2 = \Delta r^2 - (c\Delta t)^2 \qquad \text{Equation A}$$

If that difference were made minute, but finite we would have a ***virtual spatial displacement*** in space-along the world line.

$$\delta\sigma^2 = \delta r^2 - (c\delta t)^2 \qquad \text{Equation B}$$

and its counterpart for a temporal space:

$$\Delta\tau^2 = (c\Delta t)^2 - \Delta r^2 \qquad \text{Equation C}$$

If that difference were made minute, but finite we would have a ***virtual temporal displacement*** in time along the world line.

$$\delta\tau^2 = (c\delta t)^2 - \delta r^2 \qquad \text{Equation D}$$

Theorem B.5.11 ISWL: Infinitesimal Space Along the World Line

Steps	Hypothesis	$(d\sigma / dt)^2$	=	$(dr / dt)^2 - c^2$
1	From OB.3.4B	$\delta\sigma^2$	=	$\delta r^2 - (c\delta t)^2$
2	From 1 and A4.2.21 Distribution Exp	$\delta\sigma^2$	=	$\delta r^2 - c^2\delta t^2$
3	From 2, A4.2.12 Identity Multp, A4.2.13 Inverse Multp, A4.2.10 Commutative Multp and A4.2.11 Associative Multp	$\delta t^2 (\delta\sigma^2 / \delta t^2)$	=	$\delta t^2 (\delta r^2 / \delta t^2) - \delta t^2 c^2$
4	From 3, A4.2.14 Distribution and T4.4.2 Equalities: Cancellation by Multiplication	$(\delta\sigma^2 / \delta t^2)$	=	$(\delta r^2 / \delta t^2) - c^2$
5	From 4, T4.8.7 Integer Exponents: Distribution Across a Rational Number	$(\delta\sigma / \delta t)^2$	=	$(\delta r / \delta t)^2 - c^2$
6	From 5, Lx9.2.1.1 Uniqueness of Limits and Lx9.2.1.4 Limits of Sums	$\lim_{\delta t \to 0} (\delta\sigma / \delta t)^2$	=	$\lim_{\delta t \to 0} (\delta r / \delta t)^2 - \lim_{\delta t \to 0} c^2$
∴	From 5, Lx9.2.1.2 Limit of Constant and Dx9.3.1.3 Definition of a Derivative Alternate Form	$(d\sigma / dt)^2$	=	$(dr / dt)^2 - c^2$

Theorem B.5.12 IAL: Infinitesimal Arc Length

Steps	Hypothesis	$d\sigma / d\theta$	=	ρ_σ Any curve can have a radius at every infinitesimal point along the curve so to can the world line-Σ.
1	From D5.1.36B Subtended Arc length of a Cord	σ	=	$\theta\, \rho_\sigma$ arc length for a circle
2	From 1 and a minute, but finite displacement of a circle	$\delta\sigma$	=	$\delta\theta\, \rho_\sigma$ minute arc length holding radius constant
3	From 2 and T4.4.3A Equalities: Right Cancellation by Multiplication	$\delta\sigma / \delta\theta$	=	ρ_σ
∴	From 3, Dx9.3.1.3 Definition of a Derivative Alternate Form and Dx9.3.2.1G First Parametric Differentiation with Lagrange's prime mark notation	$d\sigma / d\theta$	=	ρ_σ

Theorem B.5.13 RCWL: Radius of Curvature for a World Line-Σ

Steps	Hypothesis			
	Hypothesis	ρ_σ^2	=	$(dr / d\theta)^2 - c^2 (dt / d\theta)^2$
1	From TB.3.11 ISWL: Infinitesimal Space Along the World Line	$(d\sigma / dt)^2$	=	$(dr / dt)^2 - c^2$
2	From 1, Lx9.3.2.1A On Product Functions, Chain Rule, A4.2.12 Identity Multp, A4.2.13 Inverse Multp, A4.2.10 Commutative Multp and A4.2.11 Associative Multp	$[(d\sigma / d\theta)(d\theta / dt)]^2$	=	$[(dr / d\theta)(d\theta / dt)]^2 - c^2 [(d\theta / dt)^2 / (d\theta / dt)^2]$
3	From 2, A4.2.21 Distribution Exp, and T4.4.2 Equalities: Cancellation by Multiplication	$(d\sigma / d\theta)^2$	=	$(dr / d\theta)^2 - c^2 / (d\theta / dt)^2$
4	From 3, TB.3.12 IAL: Infinitesimal Arc Length and A4.2.3 Substitution	ρ_σ^2	=	$(dr / d\theta)^2 - c^2 / (d\theta / dt)^2$
∴	From 4 and T4.5.7 Equalities: Product and Reciprocal of a Product	ρ_σ^2	=	$(dr / d\theta)^2 - c^2 (dt / d\theta)^2$

Theorem B.5.14 RCWL: Radius of Curvature for a World Line-T

Steps	Hypothesis			
	Hypothesis	ρ_τ^2	=	$c^2 (dt / d\theta)^2 - (dr / d\theta)^2$
1	From TB.3.3B MSTCR: Minkowski's Spatial-Temporal Constant Relationship	$(d\tau / dt)^2$	=	$-(d\sigma / dt)^2$
2	From 1 and Lx9.3.2.1A On Product Functions, Chain Rule	$[(d\tau / d\theta)(d\theta / dt)]^2$	=	$-[(d\sigma / d\theta)(d\theta / dt)]^2$
3	From 2, A4.2.21 Distribution Exp and T4.4.2 Equalities: Cancellation by Multiplication	$(d\tau / d\theta)^2$	=	$-(d\sigma / d\theta)^2$
4	From 3, TB.3.12 IAL: Infinitesimal Arc Length and A4.2.3 Substitution	ρ_τ^2	=	$-\rho_\sigma^2$
∴	From 4, TB.3.13 RCWL: Radius of Curvature for a World Line-Σ, A4.2.14 Distribution and D4.1.19 Primitive Definition for Rational Arithmetic	ρ_τ^2	=	$c^2 (dt / d\theta)^2 - (dr / d\theta)^2$

Section B.6 Relativistic Magnetic Fields

Given two charge particles in a frame of Reference–2 and the velocity between it and frame of Reference–1 is zero or the two frames are stationary relative to each other. It follows than that the force exerted by each particle on them is given by the coulomb equation of force [SES71 pg 165].

Theorem B.6.1 EFFD: Electric Field of Force in the x-Direction

1g	Given [SES71 pg 165]	$\vec{F} = \dfrac{Q\,q}{4\pi\varepsilon_0} \dfrac{\hat{r}}{D^2}$ for v = 0 between frames 1 and 2
2g		$\hat{r} = \dfrac{\vec{r}}{D}$
3g		$\hat{r} = \dfrac{(\Delta x, \Delta y, \Delta z)}{\sqrt{\Delta x^2 + \Delta y^2 + \Delta z^2}}$
Steps	Hypothesis	$\vec{F} = \dfrac{Q\,q}{4\pi\varepsilon_0} \dfrac{(\Delta x, \Delta y, \Delta z)}{(\Delta x^2 + \Delta y^2 + \Delta z^2)^{3/2}}$
1	From 1g, 2g, T4.4.12A Equalities: Product by Division and A4.2.3 Substitution	$\vec{F} = \dfrac{Q\,q}{4\pi\varepsilon_0} \dfrac{1}{D^2} \dfrac{\vec{r}}{D}$
2	From 3g, 1 and A4.2.3 Substitution	$\vec{F} = \dfrac{Q\,q}{4\pi\varepsilon_0} \dfrac{1}{(\Delta x^2 + \Delta y^2 + \Delta z^2)} \dfrac{(\Delta x, \Delta y, \Delta z)}{\sqrt{\Delta x^2 + \Delta y^2 + \Delta z^2}}$
∴	From 2 and T4.4.12A Equalities: Product by Division	$\vec{F} = \dfrac{Q\,q}{4\pi\varepsilon_0} \dfrac{(\Delta x, \Delta y, \Delta z)}{(\Delta x^2 + \Delta y^2 + \Delta z^2)^{3/2}}$

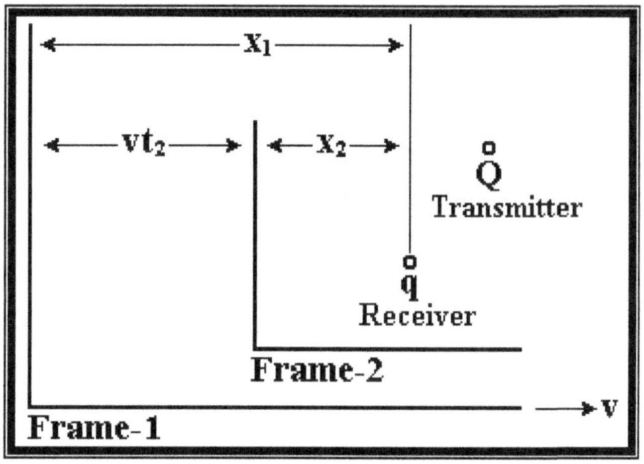

Figure B.6.1 Transformation 1-2: Relative moving charges in the x-y plane

Given B.6.1 Basic Propositions of Relative Two-Charge Particle System

1g	Given	Frame-2 will now move at a constant velocity-v Relative to frame-1
2g		$\Delta z_2 = 0$ Now the two particles lie within the x-y plane of frame-2. It follows that the observer stands there looking back into frame–1 so, only Einstein's Transformations 1-2 are Referenced:
3g		$t_1 = 0$ at the moment the two frames start to move apart or the argument can be made at the moment they pass each other.
4g		$t_2 = 0$ in frame–2 no travel time passes

Lema B.6.1 PRPS: Initial Propositions of Relative Two charge Particle System

Steps	Hypothesis	
1 ∴	From 1g, 3g and LxECT.B.2.2.1b	$\Delta x_2 = \gamma\, \Delta x_1$
2 ∴	From 1g and LxECT.B.2.2.2b	$\Delta y_2 = \Delta y_1$
3 ∴	From 2g and LxECT.B.2.2.3b	$\Delta z_1 = 0$
4 ∴	From Fig. B.4.1 distances constant	$v_{x2} = v_{y2} = v_{z2} = 0$
5 ∴	From 4 and LxEVT.B.2.3.1b	$v_{x1} = v$
6 ∴	From 4, LxEVT.B.3.2b and LxEVT.B.3.3b	$v_{y1} = v_{z1} = 0$
7 ∴	From 5, dividing both sides by c and applying DxLZF.B.2.1.2	$\beta_{x1} = \beta$
8 ∴	From 6, dividing both sides by c and applying DxLZF.B.2.1.2	$\beta_{y1} = \beta_{z1} = 0$
9 ∴	From 8 and by LxEFT.B.2.6.1b	$F_{x2} = F_{x1}$

10	From 7, LxEFT.B.2.6.2b, DxLZF.B.2.1.1, T4.5.7 Equalities: Product and Reciprocal of a Product, A4.2.3 Substitution and D4.1.17 Exponential Notation	$F_{y2} = \dfrac{\eta F_{y1}}{1-\beta^2}$
11 ∴	From 10, DxLZF.B.2.1.3, A4.2.25 Reciprocal Exp, D4.1.17 Exponential Notation, A4.2.13 Inverse Multp, A4.2.12 Identity Multp, DxLZF.B.2.1.4 and A4.2.3 Substitution	$F_{y2} = \gamma F_{y1}$
12 ∴	From 2g since the two charge particles lie in the x-y plane no force is acting in the direction of the z-axis.	$F_{z2} = 0$
13 ∴	From 2g, EFF: Electric Field of Force in the x-Direction and A4.2.3 Substitution	$\vec{F}_2 = \dfrac{Q\,q}{4\pi\varepsilon_0}\dfrac{(\Delta x_2, \Delta y_2, 0)}{(\Delta x_2^{\,2} + \Delta y_2^{\,2} + 0)^{3/2}}$

Theorem B.6.2 EFFX: Electric Field of Force Transformation 2-1

Steps	Hypothesis	$\vec{F}_1 = \dfrac{Q\,q}{4\pi\varepsilon_0}\dfrac{(\gamma\Delta x_1, \gamma\,\Delta y_1, 0)}{(\gamma^2\Delta x_1^{\,2} + \Delta y_1^{\,2})^{3/2}} + \dfrac{Q\,q}{4\pi}\dfrac{(0, -\gamma\,\mu_0\,v^2\Delta y_1, 0)}{(\gamma^2\Delta x_1^{\,2} + \Delta y_1^{\,2})^{3/2}}$
1	From LB.4.1.13 PDP, D7.6.5 Euclidian Vector Space E^n of n-Dimensions, LB.4.1.9 PDP, LB.1.11 PDP, LB.1.12 PDP, LB.4.1.1 PDP, LB.4.1.2 PDP, A4.2.21 Distribution Exp and A4.2.3 Substitution	$(F_{x1}, \gamma F_{y1}, 0)_2 = \dfrac{Q\,q}{4\pi\varepsilon_0}\dfrac{(\gamma\,\Delta x_1, \Delta y_1, 0)}{(\gamma^2\Delta x_1^{\,2} + \Delta y_1^{\,2})^{3/2}}$
2	From 1 and T4.4.5 Equalities: Left Cancellation by Multiplication on y-vector component	$\vec{F}_1 = \dfrac{Q\,q}{4\pi\varepsilon_0}\dfrac{(\gamma\Delta x_1, \Delta y_1 / \gamma, 0)}{(\gamma^2\Delta x_1^{\,2} + \Delta y_1^{\,2})^{3/2}}$
3	From 2 and T4.4.14A Equalities: Product by Division, Factorization	$\vec{F}_1 = \dfrac{Q\,q}{4\pi\varepsilon_0}\dfrac{(\gamma\Delta x_1, \gamma\,\Delta y_1 / \gamma^2, 0)}{(\gamma^2\Delta x_1^{\,2} + \Delta y_1^{\,2})^{3/2}}$
4	From 3 and DxLZF.B.2.1.4	$\vec{F}_1 = \dfrac{Q\,q}{4\pi\varepsilon_0}\dfrac{(\gamma\Delta x_1, \gamma\,\Delta y_1\,\eta^2, 0)}{(\gamma^2\Delta x_1^{\,2} + \Delta y_1^{\,2})^{3/2}}$
5	From 4 and DxLZF.B.2.1.3	$\vec{F}_1 = \dfrac{Q\,q}{4\pi\varepsilon_0}\dfrac{(\gamma\Delta x_1, \gamma\,\Delta y_1\,(1-\beta^2), 0)}{(\gamma^2\Delta x_1^{\,2} + \Delta y_1^{\,2})^{3/2}}$
6	From 5 and A4.2.14 Distribution	$\vec{F}_1 = \dfrac{Q\,q}{4\pi\varepsilon_0}\dfrac{(\gamma\Delta x_1, \gamma\,\Delta y_1, 0)}{(\gamma^2\Delta x_1^{\,2} + \Delta y_1^{\,2})^{3/2}} + \dfrac{Q\,q}{4\pi\varepsilon_0}\dfrac{(0, -\gamma\,\beta^2\Delta y_1, 0)}{(\gamma^2\Delta x_1^{\,2} + \Delta y_1^{\,2})^{3/2}}$

7	From 6, DxLZF.B.2.1.1, T4.4.12B Equalities: Product by Division	$\vec{F}_1 = \frac{Q\,q}{4\pi\varepsilon_0} \frac{(\gamma\Delta x_1, \gamma\,\Delta y_1, 0)}{(\gamma^2\Delta x_1^{\,2} + \Delta y_1^{\,2})^{3/2}} + \frac{Q\,q}{4\pi\varepsilon_0 c^2} \frac{(0, -\gamma\, v^2\Delta y_1, 0)}{(\gamma^2\Delta x_1^{\,2} + \Delta y_1^{\,2})^{3/2}}$
∴	From 7 and Equation 2.7	$\vec{F}_1 = \frac{Q\,q}{4\pi\varepsilon_0} \frac{(\gamma\Delta x_1, \gamma\,\Delta y_1, 0)}{(\gamma^2\Delta x_1^{\,2} + \Delta y_1^{\,2})^{3/2}} + \frac{Q\,q}{4\pi} \frac{(0, -\gamma\,\mu_0\, v^2\Delta y_1, 0)}{(\gamma^2\Delta x_1^{\,2} + \Delta y_1^{\,2})^{3/2}}$

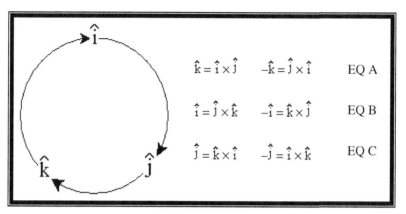

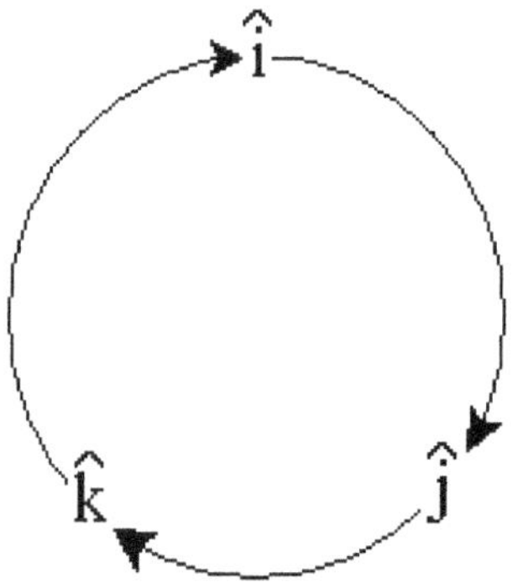

$\hat{k} = \hat{i} \times \hat{j}$ $\quad -\hat{k} = \hat{j} \times \hat{i}$ EQ A

$\hat{i} = \hat{j} \times \hat{k}$ $\quad -\hat{i} = \hat{k} \times \hat{j}$ EQ B

$\hat{j} = \hat{k} \times \hat{i}$ $\quad -\hat{j} = \hat{i} \times \hat{k}$ EQ C

Figure B.6.2 Hamilton's Product of Permuted Unit Vectors

Theorem B.6.3 EFFX: Permutation of Vector Y-Component

1g	Given	$-\hat{j} = \hat{i} \times \hat{k}$ Figure B.2, EQ C
2g		$D_{b1}^{\;3} \equiv (\gamma^2\Delta x_1^{\,2} + \Delta y_1^{\,2})^{3/2}$
3g		$q\,\mu_0 v\Delta y_1 / D_{b1}^{\;3} \rightarrow$ $[Q][M][L][Q^{-2}]\,[L][T^{-1}]\,[L]\,[L^{-3}] = [M][T^{-1}][Q^{-1}] \rightarrow B_1$
4g		$q\,\gamma\Delta x_1 / (4\pi\varepsilon_0\, D_{b1}^{\;3}) = E_{x1}$
5g		$q\,\gamma\Delta y_1 / (4\pi\varepsilon_0\, D_{b1}^{\;3}) = E_{y1}$
Steps	Hypothesis	$(0, -\,q\,\gamma\,\mu_0\, v^2\Delta y_1 / 4\pi D_b^{\;3}, 0) = (v, 0, 0) \times (0, 0, B_1)$
1	From 1g and TB.4.4 EFFX: Electric Field of Force Transformation 2-1, second term and A4.2.3 Substitution	$\gamma\,\mu_0\, v^2\Delta y_1\,(-\hat{j}\,) = \gamma\,\mu_0\, v^2\Delta y_1\,(\,\hat{i} \times \hat{k}\,)$
2	From 1, D4.1.17 Exponential Notation, A4.2.10 Commutative Multp and A4.2.11 Associative Multp	$-\gamma\,\mu_0\, v^2\Delta y_1\,\hat{j} = (v\,\hat{i}) \times (\gamma\mu_0 v\Delta y_1\,\hat{k})$
3	From 2, TB.4.4 EFFX: Electric Field of Force Transformation 2-1, 2g, substituting back into parenthesis vector notation	$(0, -\,q\,\gamma\,\mu_0\, v^2\Delta y_1 / 4\pi D_{b1}^{\;3}, 0) = (v, 0, 0) \times (0, 0, q\,\gamma\mu_0 v\Delta y_1 / 4\pi D_{b1}^{\;3})$
∴	From 3g, 4g, 5g, 3 and A4.2.3 Substitution	$(0, -\,q\,\gamma\,\mu_0\, v^2\Delta y_1 / 4\pi D_b^{\;3}, 0) = (v, 0, 0) \times (0, 0, B_1)$

Theorem B.6.4 EFFX: Lorentz Force Vector

1g	Given	$\dfrac{q}{4\pi\varepsilon_0} \dfrac{\gamma\,\Delta x_1}{(\gamma^2\Delta x_1{}^2 + \Delta y_1{}^2)^{3/2}} \equiv E_{x1}$
2g		$\dfrac{q}{4\pi\varepsilon_0} \dfrac{\gamma\,\Delta y_1}{(\gamma^2\Delta x_1{}^2 + \Delta y_1{}^2)^{3/2}} \equiv E_{y1}$
3g		$(E_{x1}, E_{x1}, 0) \equiv \vec{E}_1$
4g		$(v, 0, 0) \times (0, 0, B_1) \equiv \vec{V} \times \vec{B}_1$
Steps	Hypothesis	$\vec{F}_1 = Q\,(\vec{E}_1 + \vec{V} \times \vec{B}_1)$ the Lorentz Force Vector
1	From 1g, 2g, TB.6.2 EFFX: Electric Field of Force Transformation 2-1, TB.6.3 EFFX: Permutation of Vector Y-Component and A4.2.3 Substitution	$\vec{F}_1 = Q\,(E_{x1}, E_{x1}, 0) + Q\,(v, 0, 0) \times (0, 0, B_1)$
2	From 3g, 4g, 1 and A4.2.3 Substitution	$\vec{F}_1 = Q\,\vec{E}_1 + Q\,(\vec{V} \times \vec{B}_1)$
∴	From 2 and A4.2.14 Distribution	$\vec{F}_1 = Q\,(\vec{E}_1 + \vec{V} \times \vec{B}_1)$ the Lorentz Force Vector

So from Lemma B.6.1 "PDP: Initial Propositions of Relative Two charge Particle System" the observer-2 sees a normal coulomb electrical force $\vec{F}_2$, but observer-1 see a modified force $\vec{F}_1$ with an additional electrical field component $\vec{V} \times \vec{B}_1$. The new ostensible electric field incorporates a velocity vector $\vec{V}$ and a special perpendicular field, a magnetic flux density $\vec{B}_1$. Just as frame of reference can be found that neutralizes the effects of gravity along the geodesic path in free fall, so too one can find a frame were the magnetic field vanishes. These forces are apparent having no singular monopole to generate them, but arise as simply a matter of where your frame of Reference is.

Deriving the Lorentz Force From first principles is a good exercise in the application of Relativistic physics. It demonstrates the methods that Einstein used in implementing his physics see his original 1905 paper. [PER52 pg 47 and 50].

Observation B.5 The Third Term

This analysis only considers a moving relative charge particle at constant relativistic velocities, but what happens if we take into account an accelerated particle frame? As shown by Feynman [FEY63, Vol. I pg 2.3 and Vol. II pg 21.1] a third term would arise, but no new apparent field emerges. The third term represents emission of radiant energy. Accelerated charges irradiate. Since our particle has limited finite capacity to store energy the pumping in of more energy to accelerate or decelerated, has no place to go and must be expelled to balance the internal energy of the particle. Like a leaking boat, the occupants must bail to stay afloat, and analogous to a rocket motor nozzle expelling hot gases, causes the particle to accelerate or decelerate opposite in direction the occupant is bailing. The constant tangential spin velocity of light limits energy going in and out of the system bring about the ***leaky boat phenomenon***.

Feynman's three term electric field equation

$$\mathbf{E} = \frac{q}{4\pi\epsilon_0}\frac{\mathbf{e}_r}{r^2} + \frac{\mathbf{r}}{c}\frac{d}{dt}\left(\frac{\mathbf{e}_r}{r^2}\right) + \frac{1}{c^2}\frac{d^2}{dt^2}\mathbf{e}_r \qquad \text{Equation A}$$

results in generating a ***synchrotron magnetic field***

$$c\mathbf{B} = \mathbf{e}_r \times \mathbf{E} \qquad \text{Equation B}$$

and an acceleration term.

$$\frac{1}{c^2}\frac{d^2}{dt^2}\mathbf{e}_r \qquad \text{Equation C}$$

Section B.7 Newtonian Connection to General Relativity

What is a Newtonian space?

The Principia denotes the Newtonian manifold as a Euclidian flat region in every direction, it's orthogonal, and time is absolute. It's a geodesic space implying that a vector is constant, so as a vector travels through space and changes direction, it contributes no distortion to the vector [$\Gamma_{ij}^{k} = 0$]. These ideas are quite obvious in light of the modern topic and notation of this paper, but in Newton's day working with Euclidian constructed diagrams and not denoting variables as we use them in modern algebra, it was not. They lay buried in the obscure and primitive geometric symbolism of his time. Einstein would have to somehow show that his theory in a tensor geodesic space could be simplified to a Newtonian space as an approximation, thereby showing his space to be a natural extension of Newton's. So to do all of this Einstein had to satisfy the following four conditions:

1) In the special case of Einstein's it must revert to Newton's static, weak gravitational field and thereby describe Newton's familiar laws of gravity and motion.
2) It must preserve the laws of classical physics, the conservation of energy and momentum.
3) It should satisfy the principle of equivalence, which holds that observations made by an observer who is uniformly accelerating would be equivalent to those made by an observer standing in a comparable gravitational field.
4) It must follow naturally the suggested generally covariant quantities supplied by the mathematics of Ricci and Levi-Civita. [PER52II pg 123]

With all of these constraining conditions and the added complexity of tensor mathematics, these ideas are deeply buried; it was quite a formidable task.

However in reconstructing Einstein's solution it became clear that these proofs did not naturally follow a logical deductive mathematical progression. Einstein with his unique way of internalizing physics used that sense to guide his way through the problem-using math to support various parts of the ongoing arguments. Was he wrong? Time and experimentation has shown he was not; however it was not a clean proof, which is a progressive series of deductive logical steps. What I did was to use a fundamental idea that Einstein originally identified, which encouraged him to go with the use of tensors in the first place. The use of tensors in a geodesic space can be transformed into a simpler more compatible space, the problem proved and then mapped back to a more general space, thereby making it a universal proof in all spaces. Also there were critical pivot ideas that I picked out and used to keep the proof consistent forcing it to follow a parallel development to Einstein's.

Theorem B.7.1 Riemann-Christoffel Tensor as a Linear Approximation

1g	Given	$\sqrt{g}$	=	$\lvert \pm\rho_v \rvert$ orthogonal and spatially constant
2g		$\sqrt{g}$	=	± 1 Euclidian submanifold
3g		u	≡	$\sum_b g^{bb}\, g_{bb}$ $[L^2][T^{-2}]$ contraction to a potential energy scalar in a Newtonian space, which implies the contravariant metric has the reciprocal potential energy of the space $g^{ij} \rightarrow [L^{-2}][T^2]$ so the covariant Riemann-Christoffel Tensor Rank-2 contracts to a Gaussian Curvature
Steps	Hypothesis	R	=	$\frac{1}{2} \sum_i \sum_j g^{ij}\, u_{j,i}$ for orthogonal constant space
1	From D14.4.2 Riemannian or Einstein Spatial Curvature	R	=	$g^{ij}\, R_{ij}$
2	From 1, T14.4.1A Riemann-Christoffel Tensor Rank-2: Ricci Tensor and T10.2.5 Tool: Tensor Substitution	R	=	$g^{ij}\, (\nabla_j [\nabla_i (\sqrt{g}) / \sqrt{g}\,] - \sum_\alpha \nabla_\alpha \Gamma_{ij}{}^{\alpha}) + \sum_\beta [\sum_\alpha \Gamma_{\beta j}{}^{\alpha} \Gamma_{i\alpha}{}^{\beta} - \Gamma_{ij}{}^{\beta} \nabla_\beta \ln (\sqrt{g})]$
3	From 1g, 2 and A4.2.3 Substitution	R	=	$g^{ij}\, (\nabla_j [\nabla_i (\rho_v) / \rho_v\,] - \sum_\alpha \nabla_\alpha \Gamma_{ij}{}^{\alpha}) + \sum_\beta [\sum_\alpha \Gamma_{\beta i}{}^{\alpha} \Gamma_{i\alpha}{}^{\beta} - \Gamma_{ij}{}^{\beta} \nabla_\beta \ln (\rho_v)]$
4	From 3, Lx9.3.3.1B Differential of a Constant, A4.2.7 Identity Add and T12.2.21 Covariant Metric Determinate in Euclidian Manifold $\mathfrak{E}^n$	R	=	$g^{ij}\, [(-\sum_\alpha \nabla_\alpha \Gamma_{ij}{}^{\alpha}) + \sum_\alpha \sum_\beta \Gamma_{\beta j}{}^{\alpha} \Gamma_{i\alpha}{}^{\beta}]$
5	From 4, D4.1.20A Negative Coefficient and T4.8.2B Integer Exponents: Negative One Squared	R	=	$g^{ij}\, [(-1)\sum_\alpha \nabla_\alpha \Gamma_{ij}{}^{\alpha} + (-1)(-1)\sum_\alpha \sum_\beta \Gamma_{\beta j}{}^{\alpha} \Gamma_{i\alpha}{}^{\beta}]$
6	From 5, A4.2.14 Distribution and D4.1.20A Negative Coefficient	R	=	$-g^{ij}\, [\sum_\alpha (\nabla_\alpha \Gamma_{ij}{}^{\alpha}) - \sum_\alpha (\sum_\beta \Gamma_{\beta j}{}^{\alpha} \Gamma_{i\alpha}{}^{\beta})]$
7	From 6 and equating term-by-term moves the summation to the outside of equation	R	=	$-g^{ij} \sum_\alpha (\nabla_\alpha \Gamma_{ij}{}^{\alpha} - \sum_\beta \Gamma_{\beta j}{}^{\alpha}\, \Gamma_{i\alpha}{}^{\beta})$
8	From 7 and T12.6.3A Differentiation of Tensor: Rank-1 Contravariant	R	=	$-g^{ij} \sum_\alpha (\Gamma_{ij}{}^{\alpha})_{;\alpha}$
9	From 1g, 8 and T12.6.3A Differentiation of Tensor: Rank-1 Contravariant	R	=	$-\sum_b g^{bb} \sum_\alpha (\Gamma_{bb}{}^{\alpha})_{;\alpha}$ orthogonal

10	From 9, T12.9.4 Christoffel Symbol: Second Kind Diagonal Indices and T10.2.5 Tool: Tensor Substitution; orthogonal coordinates	$R = -\Sigma_b g^{bb} \Sigma_\alpha(-½ \Sigma_\beta g_{bb,\beta} g^{\beta\alpha})_{;\alpha}$
11	From 10, D4.1.20A Negative Coefficient, A4.2.14 Distribution and T4.8.2B Integer Exponents: Negative One Squared	$R = ½ \Sigma_\alpha \Sigma_\beta (\Sigma_b g^{bb} g_{bb})_{,\beta,\alpha} g^{\beta\alpha}$
12	From 3g, 11 and T10.2.5 Tool: Tensor Substitution; orthogonal coordinates	$R = ½ \Sigma_\alpha \Sigma_\beta u_{,\beta,\alpha} g^{\beta\alpha}$
∴	From 12 and T10.3.7 Product of Tensors: Commutative by Multiplication of Rank p + q → q + p; for orthogonal constant in a Newtonian space	$R = ½ \Sigma_\alpha \Sigma_\beta g^{\alpha\beta} u_{,\beta,\alpha}$

Theorem B.7.2 Approximating Einstein & Newton's Energy-Momentum Tensor

1g	Given	$N \equiv 4\pi \rho_g v^2$ Newton's energy density
2g		$\sqrt{g} = \lvert\pm\rho_v\rvert$ orthogonal and spatially constant
3g		$T \equiv ½ (½ n - 1) N$ Newton's mass-density
Steps	Hypothesis	$T \xi = -(½n - 1) R$ Newtonian Subspace $\mathfrak{R}^n$
1	From TA.3.5 NLPE: Newton's Law for Poisson's Equation	$\frac{-\rho_g}{\gamma_0} = \nabla^2 u$
2	From 1, T15.3.1 Laplacian Operator: Definition of Laplacian and T10.2.5 Tool: Tensor Substitution	$\frac{-\rho_g}{\gamma_0} = \nabla_i (\sqrt{g}\, g^{ij} \nabla_j u) / \sqrt{g}$
3	From 2, CxPPC.A.2.3.2 Reciprocal Newtonian constant, T4.5.7 Equalities: Product and Reciprocal of a Product, D4.1.20A Negative Coefficient and T4.8.2B Integer Exponents: Negative One Squared	$(-1)\, 4\pi G_0 \rho_g = (-1)(-1)\nabla_i (\sqrt{g}\, g^{ij} \nabla_j u) / \sqrt{g}$

4	From 2g, 3, T4.4.3 Equalities: Cancellation by Multiplication and D4.1.20A Negative Coefficient; substitution of dummy indices i → α and j → β for the orthogonal Newtonian subspace	$4\pi\ \rho_g\ G_0$	≈	$-\Sigma_\alpha\ \Sigma_\beta\ \nabla_\alpha\ g^{\alpha\beta}\ (\nabla_\beta\ u)$ orthogonal and spatially constant
5	From 4, C12.8.5.4 Differentiation of Contravariant Metric with Covariant Coefficient and T10.2.5 Tool: Tensor Substitution; missing velocity squared dimension	$4\pi\ \rho_g\ G_0\ [L^2][T^{-2}]$	≈	$-\Sigma_\alpha\ \Sigma_\beta\ g^{\alpha\beta}\ u_{,\alpha,\beta}$
6	From 5, A4.2.12 Identity Multp, A4.2.13 Inverse Multp, A4.2.10 Commutative Multp and A4.2.11 Associative Multp; compensate for missing velocity squared dimension	$(4\pi\ \rho_g\ v^2)\ G_0\ /\ v^2$	≈	$-\Sigma_\alpha\ \Sigma_\beta\ g^{\alpha\beta}\ u_{,\alpha,\beta}$
7	From 1g, 6 and T10.2.5 Tool: Tensor Substitution	$N\ G_0\ /\ v^2$	≈	$-\Sigma_\alpha\ \Sigma_\beta\ g^{\alpha\beta}\ u_{,\alpha,\beta}$
8	From 7 and T10.4.1B Scalars and Tensors: Uniqueness of Scalar Multiplication to a Tensors	$½\ N\ G_0\ /\ v^2$	≈	$-½\ \Sigma_\alpha\ \Sigma_\beta\ g^{\alpha\beta}\ u_{,\alpha,\beta}$
9	From 8, D4.1.20A Negative Coefficient and A4.2.11 Associative Multp	$½\ N\ G_0\ /\ v^2$	≈	$-\ (\ ½\ \Sigma_\alpha\ \Sigma_\beta\ g^{\alpha\beta}\ u_{,\alpha,\beta})$
10	From 9, TB.5.1 Riemann-Christoffel Tensor of Curvature Linear Approximation and T10.2.5 Tool: Tensor Substitution	$½\ N\ G_0\ /\ v^2$	≈	$-R$
11	From 10 and T10.4.1B Scalars and Tensors: Uniqueness of Scalar Multiplication to a Tensors	$½\ (\ ½n - 1\ )\ N\ G_0\ /\ v^2$	≈	$-(\ ½n - 1\ )R$
∴	From 3g, 11, T14.4.7 Riemann-Christoffel Tensor Rank-0: Contraction of Einstein's Covariant Tensor and T10.2.5 Tool: Tensor Substitution;	$T\ \xi$	=	$-(\ ½n - 1\ )\ R$

What is true in Newtonian subspace $\mathfrak{R}_s^n$ and Newtonian Geodesic space $\mathfrak{R}_g^n$, and transforming to a general Riemannian spaces $\mathfrak{R}^n$, implies it's true for all spaces.

Observation B.6 Newtonian Subspace as a Geodesic Space $\mathfrak{R}^n$

Theorem B.5.2 "Approximating Einstein & Newton's Energy-Momentum Tensor" shows that Einstein's theorem T17.5.61 "Riemann-Christoffel Tensor Rank-2: Einstein's Mixed Tensor" holds under its contracted form in Newtonian space, hence holds for all metric spaces. Also the constant of proportionality [ξ] establishes equivalency converting the approximation to an equality and is comprised of Newton's gravitational constant of proportionality divided by some velocity squared. Now [ξ] having been developed by Einstein is called ***Einstein's gravitational equality factor***.

Observation B.7 Einstein's Velocity Constant v^2 for Gravity

In theorem TB.5.3 "Einstein's General Gravity Field Equation" the constant of proportionality for velocity is unknown, however fortunately Einstein came up with a way of evaluating it to a first approximation as the velocity of light. This is not surprising since from the law of LB.5.V "Concurrency" all points are congruent iff the particle exchanged between any two mass-electrical particles is an Einsteinian photon with a quantum amount of energy. It follows than that electrical and gravity field effects are limited to the bridge speed limit that of light.

Theorem B.7.3 Riemannian Spatial Curvature: Newtonian Time Potential Energy as a Limit

1g	Given	$\sqrt{g}$	~	1 approximates a Euclidean Space $\mathfrak{E}^4$
2g	For Riemannian space tends to flatten out approximating a Euclidean region $\mathfrak{R}^n \rightarrow \mathfrak{E}^n$ spatial contravariant metrics vanish everywhere in space except the diagonal	$g^{\beta\alpha}$	~	$\delta^{\beta\alpha}$ approximates a Euclidean Space $\mathfrak{E}^4$
3g		$\sum_\beta \Gamma_{\beta j}{}^{\alpha} \Gamma_{i\alpha}{}^{\beta}$	=	0 approximates a Euclidean Space $\mathfrak{E}^4$
4g	Same argument as 2g	$g^{i\alpha}$	~	$\delta^{\beta\alpha}$ approximates a Euclidean Space $\mathfrak{E}^4$
5g	For the last index of covariant does not vanish.	g_{44}	≠	0
6g		T	~	ρ approximates a Euclidean Space $\mathfrak{E}^4$
7g		$x^{k:t}$	~	1 approximates a Euclidean Space $\mathfrak{E}^4$
Steps	Hypothesis	$-2\rho\,\xi$	≈	$\nabla^2 g_{44}$
1	From D14.4.2 Riemannian or Einstein Spatial Curvature; n = 4	R	≡	$g^{ij} R_{ij}$
2	From 1g, 1, T14.4.1A Riemann-Christoffel Tensor Rank-2: Ricci Tensor, A4.2.3 Substitution and T10.2.5 Tool: Tensor Substitution	R	=	$g^{ij} \{(\nabla_j [\nabla_i (1) / 1] - \nabla_\alpha \Gamma_{ij}{}^{\alpha}) + \sum_\beta [\Gamma_{\beta j}{}^{\alpha} \Gamma_{i\alpha}{}^{\beta} - \Gamma_{ij}{}^{\beta} \nabla_\beta \ln (1)]\}$
3	From 2, Lx9.3.3.1 Differential of a Constant and T4.11.2 Real Exponents: The Logarithm of Unity	R	=	$g^{ij} [(\nabla_j [0 / 1] - \nabla_\alpha \Gamma_{ij}{}^{\alpha}) + \sum_\beta (\Gamma_{\beta j}{}^{\alpha} \Gamma_{i\alpha}{}^{\beta} - \Gamma_{ij}{}^{\beta} \nabla_\beta 0)]$
4	From 3, T4.4.10 Equalities: Zero Divided by a Non-Zero Number and Lx9.3.3.1 Differential of a Constant	R	=	$g^{ij} [(0 - \nabla_\alpha \Gamma_{ij}{}^{\alpha}) + \sum_\beta (\Gamma_{\beta j}{}^{\alpha} \Gamma_{i\alpha}{}^{\beta} - \Gamma_{ij}{}^{\beta}\, 0)]$
5	From 4, T10.3.18 Addition of Tensors: Identity by Addition and T10.2.7B Tool: Zero Times a Tensor is a Zero Tensor	R	=	$g^{ij} (-\nabla_\alpha \Gamma_{ij}{}^{\alpha} + \sum_\beta \Gamma_{\beta j}{}^{\alpha} \Gamma_{i\alpha}{}^{\beta})$
6	From 5, D12.5.2 Christoffel Symbol of the Second Kind, T12.5.3B Christoffel Symbol of the First Kind and T10.2.5 Tool: Tensor Substitution	$\Gamma_{ij}{}^{\alpha}$	=	$\sum_\beta \frac{1}{2} (g_{i\beta,j} + g_{j\beta,i} - g_{ij,\beta})\, g^{\beta\alpha}$
7	From 2g, 6 and T10.2.5 Tool: Tensor Substitution	$\Gamma_{ij}{}^{\alpha}$	≈	$\sum_\beta \frac{1}{2} (g_{i\beta,j} + g_{j\beta,i} - g_{ij,\beta})\, \delta^{\beta\alpha}$
8	From 7 and Kronecker Delta evaluated; β → α	$\Gamma_{ij}{}^{\alpha}$	≈	$\frac{1}{2} (g_{i\alpha,j} + g_{j\alpha,i} - g_{ij,\alpha})$

9	From 8, Lx9.3.1.7B Uniqueness Theorem of Differentiation and Lx9.3.2.3 Distribute Across Constant Product	$\nabla_\alpha \Gamma_{ij}{}^\alpha$	$\approx$	$½\, \nabla_\alpha\, (g_{i\alpha,j} + g_{j\alpha,i} - g_{ij,\alpha})$
10	From 9, Lx9.3.2.2B Differential as a Linear Operator and substituting dummy index j → i	$\nabla_\alpha \Gamma_{ij}{}^\alpha$	$\approx$	$½\, (g_{i\alpha,j,\alpha} + g_{j\alpha,i,\alpha} - g_{ij,\alpha,\alpha})$
11	From 10, T10.2.10 Tool: Identity Multiplication of a Tensor, T10.4.7 Scalars and Tensors: Distribution of a Tensor over Addition of Scalars and D4.1.19 Primitive Definition for Rational Arithmetic	$\nabla_\alpha \Gamma_{ij}{}^\alpha$	$\approx$	$½\, (2\, g_{i\alpha,j,\alpha} - g_{ij,\alpha,\alpha})$
12	From 3g, Lx9.3.3.1 Differential of a Constant, T10.3.18 Addition of Tensors: Identity by Addition and Dx9.6.2.3A Covariant Laplacian Operator	$\nabla_\alpha \Gamma_{ij}{}^\alpha$	$\approx$	$-½\, \nabla^2 g_{ij}$
13	From 3g, 5, 12, T10.3.18 Addition of Tensors: Identity by Addition and T10.2.5 Tool: Tensor Substitution	R	$\approx$	$g^{ij}\, [-(-½\, \nabla^2 g_{ij})]$
14	From 13, D4.1.20A Negative Coefficient and T4.8.2B Integer Exponents: Negative One Squared	R	$\approx$	$½ g^{ij}\, \nabla^2 g_{ij}$
15	From 4g, 14 and T10.2.5 Tool: Tensor Substitution	R	$\approx$	$½ \delta^{ij}\, \nabla^2 g_{ij}$
16	From 15 and Kronecker Delta evaluated; j → i	R	$\approx$	$½ \nabla^2 g_{ii}$
17	From 5g, 16 and T10.2.5 Tool: Tensor Substitution	R	$\approx$	$½ \nabla^2 g_{44}$
18	From 17, TB.7.2 Approximating Einstein & Newton's Energy-Momentum Tensor and T10.2.5 Tool: Tensor Substitution; in the space-time continuum n=4	$T\, \xi$	$\approx$	$-(\, ½4 - 1\,)\, (½\, \nabla^2 g_{44})$
19	From 18 and D4.1.19 Primitive Definition for Rational Arithmetic	$\rho\, \xi$	$\approx$	$-½\, \nabla^2 g_{44}$
19	From 6g, 19 and T10.2.5 Tool: Tensor Substitution	$\rho\, \xi$	$\approx$	$-½\, \nabla^2 g_{44}$
∴	From 19, T4.4.5A Equalities: Reversal of Right Cancellation by Multiplication, D4.1.20A Negative Coefficient and D4.1.19 Primitive Definition for Rational Arithmetic	$-2\rho\, \xi$	$\approx$	$\nabla^2 g_{44}$

Observation B.8 What Einstein Sought?

What Einstein was seeking was Gravitational manifold reduced to a near Euclidian flat region in every direction, orthogonal, and time absolute to represent a Newtonian approximation of a space. By comparing equation of TB.7.3 "Riemannian Spatial Curvature: Newtonian Time Potential Energy as a Limit"

$[-\rho\ 2\xi \approx \nabla^2 g_{44}]$

to equation TA.3.5 "NLPE: Newton-Gravitational Formula Cast as a Poisson's Equation"

$[-\rho_g / \gamma_0 = \nabla^2 u]$

Einstein did exactly that providing a direct one-to-one correspondence as an estimate.

Theorem B.7.4 Newton-Einstein: Solving for Einstein's Gravitational Equality Factor

1g	Given: Energy imparted by the test particle m_g in the source field	$\delta U \equiv -m_g c^2 (\xi / 8\pi) \dfrac{\delta M}{r}$
2g	Mass density	$\delta M \equiv \rho\, dv$
Steps	Hypothesis	$\xi = \dfrac{8\pi G_0}{c^2}$
1	From TA.3.1 Gravitational Potential Energy; finite difference in the potential energy of the gravitational field	$\delta U = \dfrac{-1}{4\pi\gamma_0} \dfrac{m_g\, \delta M}{r}$
2	From 1g, 1, T4.2.3 Substitution and TB.7.3 Riemannian Spatial Curvature: Newtonian Time Potential Energy as a Limit; the potential energy is balanced against the test particle of the system	$-m_g c^2 (\xi / 8\pi) \dfrac{\delta M}{r} = \dfrac{-1}{4\pi\gamma_0} \dfrac{m_g\, \delta M}{r}$
3	From 2, T4.4.6B Equalities: Left Cancellation by Multiplication, A4.2.13 Inverse Multp and A4.2.12 Identity Multp	$\dfrac{-\xi}{8\pi} \dfrac{\delta M}{r} = \dfrac{-1}{4\pi\gamma_0 c^2} \dfrac{\delta M}{r}$
4	From 2g, 3 and A4.2.3 Substitution; then integrating over the region of the source field	$\dfrac{-\xi}{8\pi} \int_V \dfrac{\rho}{r}\, dv = \dfrac{-1}{4\pi\gamma_0 c^2} \int_V \dfrac{\rho}{r}\, dv$
∴	From 4, T4.4.3 Equalities: Cancellation by Multiplication, CxPPC.A.2.3.2 Reciprocal Newtonian constant, T4.2.3 Substitution and T4.4.7B Equalities: Reversal of Left Cancellation by Multiplication	$\xi = \dfrac{8\pi G_0}{c^2}$

Theorem B.7.5 Einstein's General Gravity Field Equation

Steps	Hypothesis	$T_i^{\ j} (8\pi G_0 / c^2) \quad = \quad R_i^{\ j} - ½ \delta^j_{\ i} R$
1	From T14.4.5 Riemann-Christoffel Tensor Rank-2: Einstein's Mixed Tensor and T12.2.7B Transpose of Mixed Kronecker Delta	$T_i^{\ j} \xi \quad = \quad R_i^{\ j} - ½ \delta_i^{\ j} R$
∴	From 1, TB.6.4 Newton-Einstein: Einstein's Gravitational Equality Factor and T4.2.3 Substitution	$T_i^{\ j} (8\pi G_0 / c^2) \quad = \quad R_i^{\ j} - ½ \delta_i^{\ j} R$

Observation B.9 Einstein's General Gravity Field Equation in Completed Form

Einstein's general gravity field equation can now be stated in its complete form from Newton's gravity field equation TA.3.5 "NLPE: Newton's Law for Poisson's Equation".

Tensor Calculus & Physics: A General Treatise

Table of Contents

List of Tables

List of Figures

List of Observations

List of Givens

Appendix C Entropy and Force at a Distance

In Maxwellian physics, his equations require a photon with energy [hν] to pass between charged particles for them to interact. The speed that this interaction took place was so fast it was always assumed that like charges repel and unlike attract immediately, but he showed light particles did the work, however photons have a finite speed so an immediate response would have a retarded potential [t – (r/c)].

Let's consider an Einsteinian Gedunkanexpermentum (Thought-experiment):
Two charged particles are some distance apart and relatively stationary with respect to one another. No other particles are present within their neighborhood of space to influence them. What action takes place between the two particles? The answer is nothing happens they neither proceed toward one another nor recede. Without an intermediary particle to shuttle energy between them nothing happens. This awareness of each other remains an unknown.

In fact what would happen if only one photon where emitted by one nuclear particle and missed another? Would they be aware of each other? Again the answer would have to be no. Positive energy transfer would have to take place to denote existence.

Section C.1 Duality of Nuclear Particle Matter

These fundamental minute bits of matter are not just charge particles, but have mass, hence exert gravitational force as well as electrical. If charge particles cannot move relative to each other without exchange of an intermediate photon it follows, that this must be true for gravity as well. So, the intermediate particle for gravity is also the photon[C.1.1].

Law C.1.1 Force at a Distance Acts at the Speed of Light
Electrical and Gravitational fields exert a force at a distance, upon any pair of nuclear particles, by exchanging an intermediate photon, a quantum of energy [hν] is passed at the speed of light; hence action is carried out by the intermediate particle's photon celerity.

[C.1.1]Note: Another argument could be made that there exists a shadow particle representing the intermediary particle for gravitational force, for lack of a better name a graviton. If such a particle did exist than it would have to work in tandem along with the photon, otherwise there would be a time delay between electrical and gravitational actions. In the billions of experiments, either consciously or subconsciously, that have been performed in particle physics no such time delay has ever been observed, it would follow that the intermediary particle of exchange must be the same particle, the photon.

Section C.2 The Continuum

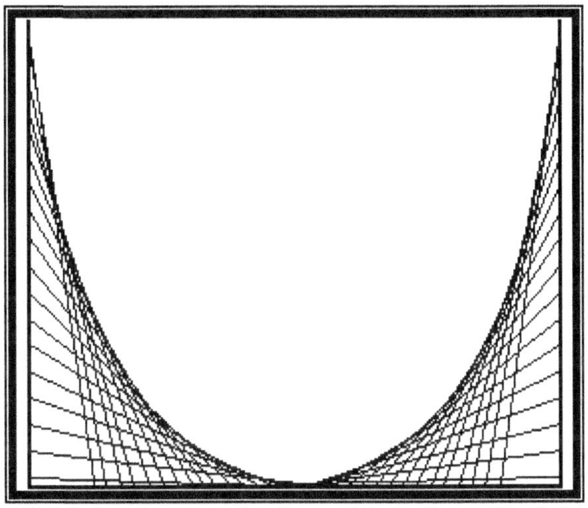

Figure C.2.1 Continuum of a Curve Made From Straight Lines

The other reason that particles appear to interact with one another continuously and instantaneously is the concept of a continuum (see Figure C.2.1 "the continuum of a curve line") here the curve is not real. It is an optical illusion created by a set of straight lines tracing out an envelope of a curve and the mind's eye averages them smoothly together giving the illusion of a continuous curve. We live in a region of space, which has vast quantities of physical particles that are constantly interacting by bumping into one other and by so doing exchanging mechanical and photon energy; thereby energy appears to continuously exchange in an arbitrary way. These chaotic systems of particles have average patterns of behavior; hence appear not to be purely random. So, we assign laws to those average patterns that are predictable, hence with a large number of particles on the order of Avogadro's number [CxPPC.A.2.3.14], gases behave under certain idealized gas laws of the kinetic theory of molecules [HAL66 pg 571]. Yet treating these particles individually none of those principles seem to exist. Interaction is not simultaneous, but at the average rate of travel through distance for the particle impacting into one another. Scale also plays an important role in the physical world as particle numbers decrease and likewise increase with volume. We have the nuclear size less than ($< 10^{-8}$ meters) (***micro-scale***) man size or height of the atmosphere less than ($< 10^{5}$ meters) (***macro-scale***) and planets, suns, solar systems, star clusters, galaxies, galaxy clusters and the universe itself greater than ($> 10^{6}$ meters) (***cosmological-scale***). Einstein's world of General Relativity works on a cosmological scale, and since gravity is a relatively weak force, a large number of particles create an apparent continuum of space as if it were warped (deformed) in the presence of matter.

With vast quantities of photons whizzing about, and interacting, the strong electrical forces would seem as if they should be continuous, instantaneous and dominate. Not true. For local electrical fields they will be strong; but with a large number of neutrally pared charges, the large mass systems over shadow electrical fields and mass rules on a cosmological-scale.

Section C.3 Piercing the Continuum's Vale

As can be seen from Figure C.2.1 "the continuum of a curve line" sometimes the underling structure can show through revealing its core-assembly when this happens we say that the ***continuum's vale has been pierced***. As an example, individually two relatively stationary nuclear particles demonstrate and establish an electric field of force, yet a central conglomerate of charged particles could demonstrate so on a nuclear and macro scale the phenomena would show through the vale of mass. Another example is the Ideal Gas Law, which reveals nothing of molecular kinetics from which it is derived from. Here through the continuum of large numbers create an averaging effect, which shrouds the mechanics of individual particles, thereby veiling them.

Section C.4 Probability of Entropy for Binary Particle Systems

In order for two particles to move at a distance relative to one another, they must make contact by exchanging energy between them threw an intermediate particle. If contact is made, it can be represent by a positive one $[1]_1$ a hit while a miss would be zero $[0]_2$. If contact is made, than information is passed to the receptor particle it now knows of the existence of the transmitter particle at that moment. So, the number of occurrences of hit events would be measured as {1} and while it's counterpart for misses would be {2}. It would follow that the probability of a hit to a miss would be 1:2.

Information for the direction of motion is determined by the Law on Concurrency acting on the relative spin of particles which can be represented by the binary states of a hit or miss intern described by Shannon's formula for entropy of information [SLE74 pg 9 Theorem 2]. When direction has been established information has been passed and can be measured by a binary number system state by a [0] and [1] the simplest form information can be rendered in. For this reason the units are called ***bits of information***. Likewise, in this particle system, a binary exchange is the lowest numerical state and why log is used in base-2.

Section C.5 (IPE) Intermediate Particle Exchange

Law C.5.1 Minimum Number of Particles for Existence of Entropy
Temperature can only exist with a minimum of two particles to exchange energy; hence one particle by itself has no entropy, it follows that the expanse between two particles makes them distance dependent.

Theorem C.5.1 IPE Entropy: Mean Possibilities of a Hit or Miss

1g	Given:	T	=	distance between particles / c the speed of light. Also the period of the event to occur.
2g		μ	≡	$T\,\nu$ m–possibilities of a hit or a miss
Step	Hypothesis	μ	=	$(E_p\, d) / (h\, c)$
1	From D1.2.3D Photon (Einsteinian definition)	E_p	=	$h\nu$
2	From 1 and T4.4.4B Equalities: Right Cancellation by Multiplication; Frequency of energy exchange	ν	=	E_p / h
3	From 1g, 2g, 1 and A4.2.3 Substitution	μ	=	$(d / c)\,(E_p / h)$
∴	From 3, A4.2.10 Commutative Multp and A4.2.11 Associative Multp	μ	=	$(E_p\, d) / (h\, c)$

Theorem C.5.2 IPE Entropy: Shannon's Formula for Entropy

1g	Given: "A mathematical theory of communication" [SLE74 pg 9 Thm 2] referred to as bits per spatial circuit connection made by the intermediate particle carrying information of direction for the electrical or gravitational field	H	=	$\mu\,(\,-p \log_2 p - q \log_2 q\,)$
2g	Probability for a particle hitting another particle if equally likely	p_1	=	½ Hit
3g	Probability for a particle missing another particle if equally likely	q_0	=	½ Miss
Step	Hypothesis	H	=	$(E_p\, d) / (h\, c)$
1	From 1g, 2g, 3g, TC.4.1 IPE Entropy: Mean Possibilities of a Hit or Miss and A4.2.3 Substitution	H	=	$(E_p\, d) / (h\, c)\,(\,-½ \log_2 ½ - ½ \log_2 ½\,)$
2	From 1, T4.11.6 Real Exponents: Difference of Logarithms, T4.11.2 Real Exponents: The Logarithm of Unity, A2.7 Identity Add, T4.1.20 Negative Coefficient and D4.1.19 Primitive Definition for Rational Arithmetic	H	=	$(E_p\, d) / (h\, c)\,(\,½ \log_2 2 + ½ \log_2 2\,)$

3	From 2 and T4.11.3 Real Exponents: The Logarithm of the Base	$H = (E_p\ d) / (h\ c) (\ ½\ 1 + ½\ 1\ )$
∴	From 3 and D4.1.19 Primitive Definition for Rational Arithmetic; information capacity or entropy	$H = (E_p\ d) / (h\ c)$

Observation C.5.1 Temperature Associated With Photon Exchange
The energy imparted by the information must be equivalent to the electrical and gravitational field upon contact by the intermediate photon. To match that energy Boltzmann's constant must be used to transform the information into energy with the use of temperature. Hence, photon exchange provides a quantum amount of energy which intern raises the temperature of the system as information is passed. It follows that intermediate particle exchange; temperature and information are intricately related.

Theorem C.5.3 IPE Entropy: Gravitational Inertia and Entropy

1g	Given: Boltzmann's energy for particle exchange in the direction of attraction	$E_h \equiv -(k\ T)\ H$
2g	Conservation of rest energy to the energy within the field	$E_g \equiv m_g\ c^2$
Step	Hypothesis	$E_g = \pm d\ c^2 \sqrt{\dfrac{(4\pi\gamma_0 k\ T\ E_p)}{(h\ c)}}$
1	From 1g, TC.4.2 IPE Entropy: Shannon's Formula for Entropy and A4.2.3 Substitution	$E_h = (k\ T)\ (E_p\ d) / (h\ c)$
2	From 1, TA.3.1 Gravitational Potential Energy, A4.2.3 Substitution, T4.1.20 Negative Coefficient and T4.8.2 Integer Exponents: Negative One Squared	$\dfrac{m_g^{\ 2}}{4\pi\gamma_0 d} = \dfrac{(k\ T)\ (E_p\ d)}{(h\ c)}$
3	From 2, T4.4.7B Equalities: Reversal of Left Cancellation by Multiplication, A4.2.10 Commutative Multp, A4.2.11 Associative Multp and D4.1.17 Exponential Notation	$m_g = \pm\sqrt{\dfrac{(k\ T)\ (E_p\ 4\pi\gamma_0 d^2)}{(h\ c)}}$
4	From 3, T4.10.5 Radicals: Distribution Across a Product and T4.10.3 Radicals: Identity Power Raised to a Radical	$m_g = \pm d \sqrt{\dfrac{(4\pi\gamma_0 k\ T\ E_p)}{(h\ c)}}$
5	From 4, T4.4.2 Equalities: Uniqueness of Multiplication and a constant	$m_g\ c^2 = \pm d\ c^2 \sqrt{\dfrac{(4\pi\gamma_0 k\ T\ E_p)}{(h\ c)}}$
∴	From 2g, 3 and A4.2.3 Substitution	$E_g = \pm d\ c^2 \sqrt{\dfrac{(4\pi\gamma_0 k\ T\ E_p)}{(h\ c)}}$

Observation C.5.2 Entropy not a function of inverse distance

Entropy is not a function of the inverse distance between particles as a parametric parameter, but directly, therefore cannot be independent and removed. Entropy does not follow Minkowski's notion of concurrency with all points in the universe being same at the speed of light.

Section C.6 Total Radiant Energy Emitted By a Blackbody

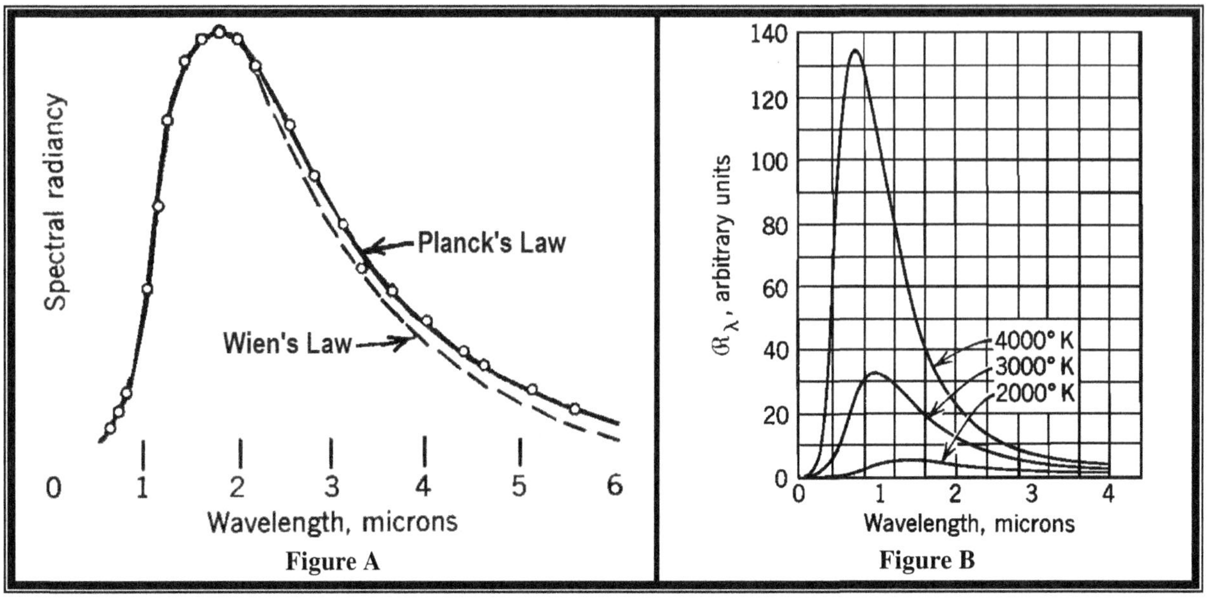

Figure C.6.1 Blackbody Radiant Energy Curves at Different Peak Temperatures

A ***cavity radiator*** emits radiation in an idealized way over a spectrum of wavelengths for any heated solid. Its light-emitting properties prove to be independent of any particular material and to vary in a simple way with temperature. Wilhelm Wien was the first to empirically derive a law that described this effect, but Planck generalized it in a formula that exactly matched the phenomena. This was the beginning of quantum mechanics, because in generalizing Wien's law (see FC.6.1A "Blackbody Radiant Energy Curves at Different Peak Temperatures"), Planck was forced to use a bed spring model of atoms embedded in the walls of the cavity. The atoms vibrating by the uniform temperature of the cavity emitted photons at a discrete frequency of energy. With the Newtonian physics of the day this puzzled Plank, because physical objects were not discrete but continuous? It took Einstein to throw out Newton, and explain that the photons were self contained packets of energy (hν) that traveled through space at Maxwell's light velocity-c and having a unique single frequency that was directly proportional to its energy given by Plank's constant-h. Plank's constant acts as a constant of proportionality, which Einstein deduced from his own experiment the "Photo Electric Effect", see Tb 1.2.1D "Physical Properties of a Photon".

If one were to heat up an iron crowbar in a furnace till it glows a uniform red color this can only happen, because the majority of atoms are at that single temperature and vibrating near the same wavelength. So why is the radiant energy curve not a flat line radiating at a monochromatic value? In fact in reality the cavity is not truly uniformly heated, in actually the narrow pin hole of the cavity filters out low and long wavelengths and the eye's narrow bandwidth prevents us from seeing the total curve. The atoms are actually vibrating at many different frequencies bounded by the problematic distribution curve of the radiant energy. Where the curve peaks at an average temperature characterizing the curve (see FC.6.1B "Blackbody Radiant Energy Curves at Different Peak Temperatures"). This curve is identical to the Boltzmann-Maxwell velocity curve for the kinetics of gas molecules [HAL66 pg 603], which also peaks at typical average temperatures, because the probabilities are the same even though the physics are totally independent of one another. What is common in both theories is that they deal with large number of molecules in motion and bounded at high and low ends as energy tapers off so maybe it should not have been surprising that the distribution curves would be shaped in the same way. Finding other similar phenomena is a good way of deducing that the experimenter is on the correct track.

Wilhelm Carl Werner Otto Fritz Franz Wien
Figure A

Karl Ernst Ludwig Marx Planck
Figure B

Figure C.6.2 Wilhelm Wien and Max Planck Founders of the Blackbody Radiant Curve

Theorem C.6.1 Blackbody Radiant Energy: Total Integration for Frequency

1g	Given: From Plank's first derived radiation constant for wavelength	$c_1 \equiv 2\pi c^2 h$	in watts-m^2
2g	From Plank's second derived radiation constant for wavelength	$c_2 \equiv hc/k$	in m-°K
3g	From Plank's first derived radiation constant for frequency	$k_1 = 2\pi h/c^2$	in watts-sec^4-m^2
4g	From Plank's second derived radiation constant for frequency	$k_2 = h/k$	in sec-°K
Step	Hypothesis	$\Re \equiv \int_o^\infty \Re_\nu \, d\nu$	EQ A
	for	$\Re_\nu \equiv \nu^3 k_1 \dfrac{1}{e^{(k2\ \nu/T)} - 1}$	EQ B
1	From Plank's cavity spectral radiance formula [HAL66 pg 1177] in terms of wavelength	$\Re_\lambda \equiv \dfrac{c_1}{\lambda^5} \dfrac{1}{e^{c2/\lambda T} - 1}$	in watts / m^3
2	From Plank's blackbody total radiance formula [HAL66 pg 1173]	$\Re \equiv \int_o^\infty \Re_\lambda \, d\lambda$	in watts / m^2
3	From Tb1.2.1A Photon Celerity; converting wavelength into frequency, for light speed c = λν	$\lambda = \dfrac{c}{\nu}$	
4	From 5, Lx9.3.2.6 Differential of a Binary Ratio and Lx9.3.3.1B Differential of a Constant	$d\lambda = -\dfrac{c\, d\nu}{\nu^2}$	in m
5	From 4, 6 and A4.2.3 Substitution	$\Re \equiv -\int_\infty^o \dfrac{c\, \Re'_\nu}{\nu^2} d\nu$	
6	From 1, 5, 7, A4.2.3 Substitution, Lx9.4.1.10 and Integrability on a reverse interval	$\Re = \int_o^\infty (c \dfrac{\nu^5 c_1}{c^5 \nu^2}) \dfrac{1}{e^{(c2/c)(\nu/T)} - 1} d\nu$	
7	From 8, A4.1.17 Exponential Notation Multp, A4.2.10 Commutative Multp, A4.2.11 Associative and A4.2.13 Inverse Multp	$\Re = \int_o^\infty \dfrac{\nu^3 c_1}{c^4} \dfrac{1}{e^{(c2/c)(\nu/T)} - 1} d\nu$	
8	From 3g, 4g, 7 and A4.2.3 Substitution	$\Re = \int_o^\infty \nu^3 k_1 \dfrac{1}{e^{(k2\ \nu/T)} - 1} d\nu$	
9	From 8, Plank's cavity spectral radiance formula [HAL66 pg 1177] in terms of frequency	$\Re_\nu \equiv \nu^3 k_1 \dfrac{1}{e^{(k2\ \nu/T)} - 1}$	
∴	From 8; for frequency	$\Re \equiv \int_o^\infty \Re_\nu \, d\nu$	EQ A
	From 9	$\Re_\nu \equiv \nu^3 k_1 \dfrac{1}{e^{(k2\ \nu/T)} - 1}$	EQ B

Theorem C.6.2 Blackbody Radiant Energy: Total Energy under Blackbody Curve

1g	Given:	$x = (k_2/T)\ \nu$
2g		$\nu = (T/k_2)\ x$
Step		$E_v = \frac{3}{4}\ \pi\sigma\ d^3 T^4$
1	From C.7.2 Blackbody Radiant Energy: Total Integration for Wavelength or Frequency	$\Re \equiv \int_0^\infty \Re_\nu\ d\nu$ in watts / m^2
2	From 1 and integrating over the cavity volume gives the total radiant energy	$E_v \equiv (4/c) \int_V \int_0^\infty \Re_\nu\ d\nu\ dV$ in joules
3	From 2 and Lx9.4.1.3 Integral on the integrand with a constant; The volume is independent of the frequencies (in a more rigorous development this might not be true) [GAU99 pg 289 problem 26.1]	$E_v = (4V/c) \int_0^\infty \Re_\nu\ d\nu$
4	From 3, TC.7.1 Blackbody Radiant Energy: Total Integration for Wavelength or Frequency and A4.2.3 Substitution	$E_v = \dfrac{4V\ k_1}{c} \int_0^\infty \dfrac{\nu^3}{e^{(k2\ \nu/T)} - 1}\ d\nu$
5	From 2g and Lx9.3.1.7A Uniqueness Theorem of Differentiation	$d\nu = (T/k_2)\ dx$
6	From 2g, 5, A4.2.3 Substitution and Lx9.4.1.3 Integral on the integrand with a constant	$E_v = \dfrac{4V\ k_1\ T^4}{c\ k_2^4} \int_0^\infty \dfrac{x^3}{e^x - 1}\ dx$
7	From 6 and Lx9.4.5.5 Combinations of rational functions of powers and exponentials	$E_v = \dfrac{4V\ k_1\ T^4}{c\ k_2^4}\ \Gamma(3 + 1)\ \zeta(3 + 1)$
8	From 7 and D4.1.19 Primitive Definition for Rational Arithmetic	$E_v = \dfrac{4V\ k_1\ T^4}{c\ k_2^4}\ \Gamma(4)\ \zeta(4)$
9	From 8 and Lx9.4.5.5 Combinations of rational functions of powers and exponentials, Lx9.4.5.2 Factorial, Lx9.4.5.3 Riemann Zeta Function, Lx9.4.5.4 Bernoulli Numbers, D4.1.19 Primitive Definition for Rational Arithmetic and A4.2.3 Substitution	$E_v = \dfrac{4V\ k_1\ T^4}{c\ k_2^4}\ 3!\ (\pi^4 / 90)$

10	From 9, A4.2.10 Commutative Multp, A4.211 Associative Multp and D4.1.19 Primitive Definition for Rational Arithmetic	E_v	=	$\dfrac{\pi^4 6\ 4V\ k_1\ T^4}{90c\ k_2^4}$
11	From 10, TC.6.1;3g, 4g, A4.2.3 Substitution, A4.2.21 Distribution Exp,	E_v	=	$\dfrac{2\pi h\ \pi^4 6\ 4V\ (kT)^4}{h^4\ 90\ c\ c^2}$
12	From 11, A4.2.10 Commutative Multp, A4.211 Associative Multp, A4.2.18 Summation Exp, A4.2.13 Inverse Multp, A4.2.12 Identity Multp, A4.1.17 Exponential Notation and D4.1.19 Primitive Definition for Rational Arithmetic	E_v	=	$\dfrac{8\pi^5\ V}{15\ (hc)^3}\ (kT)^4$
13	From 12, T4.5.7 Equalities: Product and Reciprocal of a Product, CxPPC.A.2.3.4 Normalized Planck's constant and A4.4.21 Distribution Exp	E_v	=	$\dfrac{2\pi^2\ V}{15\ (\hbar c)^3}\ (kT)^4$
14	From 13, A4.4.21 Distribution Exp, A4.2.10 Commutative Multp and A4.211 Associative Multp; by definition	σ	≡	$\dfrac{2\pi^2\ k^4}{15(\hbar c)^3}$ Stefan-Boltzmann constant [HAL66 pg1175]
15	From 14, CxPPC.A.2.3.4.(4, 5, 10), A4.2.3 Substitution and D4.1.19 Primitive Definition for Rational Arithmetic; evaluate	σ	=	$1.513065893 \mathrm{x} 10^{-15}\ J°/m^3K^4$
16	From 13, 14 and A4.2.3 Substitution	E_v	=	$\sigma\ V\ T^4$
17	From LxE.4.7.2 Volume of a sphere; Let the irradiating body be a planetoid or sun, and shaped as some near spherical object then the volume would be a sphere:	V	≡	$\frac{3}{4}\ \pi\ d^3$
∴	From 16, 17, A4.2.3 Substitution, A4.2.10 Commutative Multp and A4.211 Associative Multp	E_v	=	$\frac{3}{4}\ \pi\sigma\ d^3T^4$

Section C.7 Finding the Peak of a Blackbody Curve

Differentiating the radiance curve TC.6.1B "Blackbody Radiant Energy: Total Integration for Wavelength or Frequency" and setting it to zero gives the peak temperature of the curve:

Theorem C.7.1 Blackbody Radiant Energy: Peak of Blackbody Curve

1g	Given:	$\Re_\nu$ Is a convex curve having a well defined peak at point (T_{mx}, ν_{mx})
2g		$\gamma_\tau \equiv \dfrac{k_2\nu_{mx}}{T_{mx}}$ Gamma-tau constant
Step		$T_{mx} = \dfrac{h\nu_{mx}}{k\gamma_\tau}$ Peak temperature
1	From TC.7.1B Blackbody Radiant Energy: Total Integration for Frequency	$\Re_\nu \equiv \nu^3 k_1 \dfrac{1}{e^{(k2\ \nu/T)} - 1}$
2	From 1, T9.3.2 Uniqueness of Differentials, Lx9.3.2.4B Chain Rule Across Products, Lx9.3.2.6 Differential of a Binary Ratio, Lx9.3.2.2B Differential as a Linear Operator, Lx9.3.3.1B Differential of a Constant and Lx9.3.3.6 Differential of a Natural base [e]	$\dfrac{d\Re_\nu}{d\nu} = 3\nu^2 k_1 \dfrac{1}{e^{(k2\ \nu/T)} - 1} - \nu^3 k_1 \dfrac{(k_2/T)e^{(k2\ \nu/T)}}{(e^{(k2\ \nu/T)} - 1)^2}$
3	From 2 and A4.2.14 Distribution	$\dfrac{d\Re_\nu}{d\nu} = \dfrac{3\nu^2 k_1 - \nu^3 k_1 (k_2/T)\, e^{(k2\ \nu/T)}}{e^{(k2\ \nu/T)} - 1}$
4	From 1g, 3, D9.3.10 Maximum, Lx9.3.1.1 Rolle's Theorem, A4.2.3 Substitution and Tb4.11.1 Definition of Logarithm Types	$0 = \dfrac{3\nu_{mx}^2 k_1 - \nu_{mx}^3 k_1 (k_2/T_{mx}) \exp(k_2\, \nu_{mx}/T_{mx})}{e^{(k2\ \nu mx/Tmx)} - 1}$
5	From 4, T4.4.4B Equalities: Right Cancellation by Multiplication and T4.4.1 Equalities: Any Quantity Multiplied by Zero is Zero	$0 = 3 - (k_2\nu_{mx}/T_{mx}) \exp(k_2\, \nu_{mx}/T_{mx})$
6	From 2g, 5 and A4.2.3 Substitution	$3 = \gamma_\tau \exp(\gamma_\tau)$
7	From 6 and numerical algorithm by half integers	By the use the numerical algorithm by half integers a computer can be used to find the limiting root to the accuracy of its finite state $\gamma_\tau = 2.821439845^+$ with an accuracy of $\pm 6\text{x}10^{-9}$ or 6 ppb
∴	From 7, T4.4.8 Equalities: Cross Product of Proportions and T4.4.6B Equalities: Cross Product of Proportions	$T_{mx} = \dfrac{h\nu_{mx}}{k\gamma_\tau}$ Hence any peak frequency will have a max temperature for a radiating body.

Section C.8 Quotes and Events From Nicola Tesla's Life [Gerβ06]

§1: Despite his accomplishments, by 1915, at age 60, Tesla was living on credit and drifting from one cheap hotel to another, a victim of his own poor business decisions, underdeveloped ideas and inability to create another innovation as profound as the AC paradigm. **In 1931, at the age of 75, Tesla received birthday greetings from Lee de Forest and Albert Einstein**. In his later years he spent most of his time at the New York Public Library or feeding pigeons that he called- "my sincere friends".

§2: Concerning Albert Einstein's relativity theory, Tesla stated that '...the relativity theory, by the way, is much older than its present proponents. It was advanced over 200 years ago by my illustrious countryman Boskovic, the great philosopher, who, not withstanding other and multifold obligations, wrote a thousand volumes of excellent literature on a vast variety of subjects. Boskovic dealt with relativity, including the so-called time-space continuum...', (1936 unpublished interview, quoted in Anderson, L, ed. Nikola Tesla: Lecture Before the New York Academy of Sciences: The Streams of Lenard and Roentgen and Novel Apparatus for Their Production, April 6, 1897, reconstructed 1994).

§3: When he was eighty-one, **Tesla stated he had completed a dynamic theory of gravity**. He stated that it was "***worked out in all details***" and hoped to give to the world the theory soon. The theory was, unfortunately, never published. At the time of his announcement, it was considered by the scientific establishment to exceed the bounds of reason. While Tesla had "worked out a dynamic theory of gravity" that he soon hoped to give to the world, he died before he publicized any details. Few details were revealed by Tesla about his theory in the announcement. Tesla's critique in the announcement was the opening clash between him and modern experimental physics[1]. Tesla may have viewed his principles in such a manner as to not be in conflict with other modern theories (besides Einstein's).

§4: The bulk of the theory was developed between 1892 and 1894, during the period that he was conducting experiments with high frequency and high potential electromagnetism and patenting devices for their utilization. It was completed, according to Tesla, by the end of the 1930s. **Tesla's theory explained gravity using electrodynamics consisting of transverse waves (to a lesser extent) and longitudinal waves (for the majority)**. Tesla stated in 1925 that,

§5: "There is no thing endowed with life - from man, who is enslaving the elements, to the nimblest creature - in all this world that does not sway in it's turn. Whenever action is born from force, though it be infinitesimal, the cosmic balance is upset and the universal motion results."

§6: Tesla was critical of Einstein's (theory of) relativity work,
[a] magnificent mathematical garb which fascinates, dazzles and makes people blind to the underlying errors. The theory is like a beggar clothed in purple whom ignorant people take for a king...., its exponents are brilliant men but they are metaphysicists rather than scientists... (New York Times, July 11, 1935, p23, D.8).

§7: Tesla also stated that:
"I hold that space cannot be curved, for the simple reason that it can have no properties. It might as well be said that God has properties. He has not, but only attributes and these are of our own making. Of properties we can only speak when dealing with matter filling the space. To say that in the presence of large bodies space becomes curved is equivalent to stating that something can act upon nothing. I, for one, refuse to subscribe to such a view." (New York Hearald Tribune, September 11, 1932)

[1]Note: This phrase is essentially incorrect Tesla was an excellent modern experimental physicists it would have been better phrased as a conflict between experimental physics and theoretical physics, which even today is an ongoing debate in the physics community especially when you need to liven up a cocktail party of physicists.

Section C.9 My Point Of View on Tesla

Tesla was raised as a Serbian Catholic and sad to say also in all their beliefs and prejudices of that century. As is true about that part of the world their distrust, misunderstandings and religious dogma prevailed and are truths to these people. So when this upstart Jew Einstein comes along with a theory and mathematics that was beyond his own understanding and mathematical skills, which for him were very good for that era and in his own field of study, applied electromagnetism, this was too much so he went on the defensive. Notice he never directly attacked Einstein, but the foundations of his theory, as he perceived it $\S_6\S_7$. While Einstein admired and respected him $\S_1$, sadly the feeling was not mutual nor with a lifetime of entrenched miss-beliefs could it have been otherwise. Interesting though as shown by the above quote $\S_7$ he was, I believe, right, but for the wrong reasons, clearly his feelings were clouding his better judgment. This actually shows that his keen practical nature and experience were intuitively guiding him. From what I know of his life's history, he was never overtly a vicious or vindictive man, but always a gentleman as only his upbringing could have allowed him to be. Was he prejudice against Jews in general? Nothing in his adult life history while in America would indicate this; certainly as he rose up through American society, he would have had to work with Jews and in some cases they would have financed his work. I think in his highly focused world of experimentation there would have been no room for such feelings. What I do believe is that when Einstein inadvertently invaded his world all of these latent feelings bubbled to the surface. I personally will not let such a very small smudge on his soul tarnish my admiration and respect for this creative and fascinating genius.

As for his belief neither that transverse nor longitudinal electromagnetic waves can carry information to move particles in an appropriate direction, in my own experience as an Electrical and Communications Engineer I have never observed gravitational effects by any kind of modulation or polarization of electromagnetic waves, nor as far as I know has anybody else. This is why these ideas were the first to be considered and the first to be dismissed in my own theory.

Also, if anyone could have created an experiment to demonstrate such a thing Tesla would have been the one. His creative genius, in detail and experimentation with electromagnetism were very complete and in considerable depth of conceptual understanding. He certainly was a peer to Faraday if not more so in his experimental work.

RHS

Printed in the United States
By Bookmasters